HANDBOOK OF ELECTRICAL POWER SYSTEM DYNAMICS

HANDBOOK OF ELECTRICAL POWER SYSTEM DYNAMICS

Modeling, Stability, and Control

Edited by

Mircea Eremia
Electrical Power Systems Department
University "Politehnica" of Bucharest

Mohammad Shahidehpour
Illinois Institute of Technology

IEEE
IEEE PRESS

IEEE PRESS SERIES ON POWER ENGINEERING
Mohamed E. El-Hawary, *Series Editor*

WILEY

For general information on our other products and services or for technical support, please contact our Customer Care Department within the United States at (800) 762-2974, outside the United States at (317) 572-3993 or fax (317) 572-4002.

Wiley also publishes its books in a variety of electronic formats. Some content that appears in print may not be available in electronic formats. For more information about Wiley products, visit our web site at www.wiley.com.

Library of Congress Cataloging-in-Publication Data:

Handbook of electrical power system dynamics : modeling, stability, and control / edited by Mircea Eremia, Mohammad Shahidehpour.
pages cm
Includes bibliographical references.
ISBN 978-1-118-49717-3 (cloth)
1. Electric power system stability–Mathematical models–Handbooks, manuals, etc. 2. Electric power systems–Control–Handbooks, manuals, etc. 3. Electric machinery–Dynamics–Handbooks, manuals, etc. I. Eremia, Mircea. II. Shahidehpour, M., 1955-
TK1001.H35 2012
621.31—dc23

2012032673

10 9 8 7 6 5 4 3 2 1

CONTENTS

FOREWORD

Electricity is the driving factor of the modern world. Humanity is demanding more and more energy as the demand for better life quality, and industry development is increasing. The history of modern civilization begun more than one century ago when electricity generators and infrastructure for electricity transmission were invented. As the demand for electrical power has increased, the electrical power systems have been expanded over large distances and become more complex. There has been, thus, a continuous need for innovation to create more efficient and reliable components.

Recently, the electrical power systems have gone through a deregulation process, and electricity market has been created aiming to stimulate competition, achieve fair electrical energy price, encourage the investments for modernization and commissioning new power plants, etc. However, the immediate effects of the electricity market were additional problems in power system operation.

The limited conventional energy resources and the need for environment protection, on one hand, and the advantages of actual robust simulation hardware and software tools, on the other hand, encouraged the humanity to successfully exploit the Aeolian, solar, and other nonconventional resources. The share of electricity generation from renewable energy sources has significantly increased in the last years, and the targets are very ambitious for the future. Large wind farms are developed onshore and offshore, resulting in significant change in the generation pattern and thus changes in the power flow. Moreover, under the increasing share of generation from renewables, changes in power flows may sometimes occur quite often during one hour. This problem, in effect, requires strengthening the transmission grid.

The power system operators are, thus, facing bigger challenges than that in the past, such as limitations in scheduling and handling generation resources due to the electricity market, operation of the transmission networks close to their technical limits due to difficulties in constructing new transmission facilities, and generation uncertainties due to the intermittency and less inaccurate forecasts of the renewable energy sources, or even due to natural forces like earthquakes and storms.

The major grid blackouts experienced in the last years prove that investments and innovation are always required in the power system infrastructure, management, and education. The operational manual of the ENTSO-E network has been updated in order to prevent major incidents that occurred in the past due to permissive rules. In a strongly interconnected continental power system, as it is the ENTSO-E network, collaboration between power system operators based on clear rules is critical.

As a reaction to the technical issues of power systems, new concepts are under development. It is expected that the new ideas for more intelligent electrical networks (Smart Grids) and creation of continental supergrids may improve the power system security while satisfying the customers' needs as regards the quantity and quality. This may be seen as a new era of electricity.

This book is a successful collection of theories and applications, from modeling for dynamic analysis, methods for stability assessment and control strategies that finally help the reader to understand the causes and effects of power system blackouts and, on one hand, to understand why some preventive actions are required in order to ensure appropriate security levels and avoid the blackouts. The authors of this book, both from academia and industry, are active specialists in CIGRE and IEEE-PES activities.

Education has been a critical ingredient for creating a sustainable electricity industry. Investment in education is the minimum condition to create professionals.

André Merlin
President of CIGRE

ACKNOWLEDGMENTS

The authors wish to take this opportunity to acknowledge all persons and institutions that contributed directly or indirectly in carrying out this book, either by technical or by editorial support.

Special acknowledgments are addressed to Prof. Prabha Kundur, president of Kundur Power System Solutions Inc., for his support and inspiring recommendations. His book on power system dynamics, a reference work in the field, has been also inspiring for several theories presented in this book.

For some chapters the authors benefited by the kindness of some institutions or companies, which permitted reprinting or adapting figures, equations, or excerpts. Special thanks are thus addressed to Institute of Electrical and Electronics Engineers (IEEE), International Electrotechnical Committee (IEC), Conseil International des Grands Réseaux Électriques – International Council on Large Electric Systems (CIGRE), as well as John Wiley & Sons, Inc. for their reprinting permission and support. Acknowledgments are also addressed to Schneider Electric for the implicit permission to reprint some excerpts.

The authors express their gratitude to Prof. Ronald Harley (GeorgiaTech) for his permission to reprint some excerpts in Chapter 2. A stamp on the theory presented in Chapter 2 is also due to Prof. Eugeniu Potolea (University "Politehnica" of Bucharest) to whom the authors would like to address special thanks. Acknowledgments are extended to Prof. Mihaela Morega for recommendations at Chapter 2 and Prof. Florin Alexe for the recommendations at Chapter 3 (both are from University "Politehnica" of Bucharest).

Acknowledgments are addressed also to Prof. Daniel Roye and Prof. Seddik Bacha (INP Grenoble) for granting reprinting permission of some figures in Chapter 4 and for providing valuable recommendations.

In carrying out Chapter 5, Prof. Nicolae Golovanov (University "Politehnica" of Bucharest) has offered his expertise for which the authors would like to address thanks. Valuable ideas have been used in Chapter 5 following the collaboration with Dr. Jay C. Das, a consultant of AMEC Inc.

The authors would like to express their gratitude to Prof. Wilson Xu (University of Alberta) for granting the permission to reprint some excerpts in Chapter 11. Acknowledgments are extended to Dr. Mrinal K. Pal (an independent consultant of MKPalConsulting) for granting the use of some ideas in Chapter 11. Some theories presented in Chapter 11 are the result also following the collaboration with Prof. Thierry Van Cutsem (Université de Liège) and emeritus professor Jacques Trecat (Faculté Polytechnique de Mons).

The authors of Chapter 13 would like to address thanks to Dmitry N. Efimov (Energy Systems Institute of the Russian Academy of Science, Irkutsk) and Dr. Lu Wei, a PhD of INP Grenoble, for their valuable contribution.

Writing a book is a complex work. The authors would like to extend their gratitude to Dr. Mircea Scutariu (Mott McDonald, Glasgow), Dr. Constantin Surdu (EdF France),

Dr. Valentin Ilea, Dr. Ioana Pisică, Dr. Petre Răzuşi, Dr. Florin Cătălin Ionescu and PhD students Cristian Virgil Cristea, Alexandru Mandiş, and Valeriu Iulian Presadă (University "Politehnica" of Bucharest) for their help in drawing figures and editing text and equations.

The authors gratefully acknowledge the excellent collaboration with the IEEE Press and John Wiley & Sons, and address many thanks to Taisuke Soda, Mary Hatcher, Sanchari Sil, and Danielle LaCourciere for their patience and professionalism in carrying out the printed book.

Mircea Eremia
Mohammad Shahidehpour

CONTRIBUTORS

Alberto Berizzi, Dipartimento di Elettrotecnica, Politecnico di Milano, Piazza Leonardo da Vinci, 32, 20133 Milano, Italy

Yvon Besanger, INP Grenoble, LEG, 961 rue de la Houille Blanche, 38402 Saint Martin d'Heres, Cedex, France

Alberto Borghetti, Department of Electrical Engineering, University of Bologna, Viale Risorgimento 2, 40136 Bologna, Italy

Klaus-Peter Brand, Power Systems, ABB Switzerland Ltd., 72 Bruggerstrasse, CH-5400 Baden, Switzerland

Constantin Bulac, Department of Electrical Power Systems, University "Politehnica" of Bucharest, 313, Spl. Independenţei, 060042 Bucharest, Romania

Sandro Corsi, CESI, Via Rubattino 54, 20134 Milano, Italy

Mircea Eremia, Department of Electrical Power Systems, University "Politehnica" of Bucharest, 313, Spl. Independenţei, 060042 Bucharest, Romania

Feng Gao, Technische Universität Berlin, Einsteinnufer 11 (EMH-1), D-10587 Berlin, Germany

Nouredine Hadjsaid, INP Grenoble, LEG, 961 rue de la Houille Blanche, 38402 Saint Martin d'Heres, Cedex, France

Roberto Marconato, Dipartimento di Elettrotecnica, Politecnico di Milano, Piazza Leonardo da Vinci, 32, 20133 Milano, Italy

Ivan De Mesmaeker, Power Systems, ABB Switzerland Ltd., 72 Bruggerstrasse, CH-5400 Baden, Switzerland

Carlo Alberto Nucci, Department of Electrical Engineering, University of Bologna, Viale Risorgimento 2, 40136 Bologna, Italy

Mario Paolone, Department of Electrical Engineering, University of Bologna, Viale Risorgimento 2, 40136 Bologna, Italy

Les Pereira, Northern California Power Agency, 180 Cirby Way, Roseville, CA 95678, USA

Mohammad Shahidehpour, Electrical and Computer Engineering Department, Illinois Institute of Technology, 3301 South Dearborn Street, Chicago, IL 60616-3793, USA

Kai Strunz, Technische Universität Berlin, Einsteinnufer 11 (EMH-1), D-10587 Berlin, Germany

Lucian Toma, Department of Electrical Power Systems, University "Politehnica" of Bucharest, 313, Spl. Independenţei, 060042 Bucharest, Romania

Ion Triştiu, Department of Electrical Power Systems, University "Politehnica" of Bucharest, 313, Spl. Independenţei, 060042 Bucharest, Romania

S.S. (Mani) Venkata, Alstom Grid, 10865 Willows Road, NE, Redmond, WA 98052-2502, USA

Nikolai Voropai, Siberian Branch of the Russian Academy of Sciences, Energy Systems Institute, 130 Lermontov Street, Irkutsk 664033, Russia

1

INTRODUCTION

Mircea Eremia and Mohammad Shahidehpour

Motto: The "millennial" generation have also been called the Net generation because they are the first generation that has not known life without the Internet.

Power systems are the most complex systems ever devised by man. This is primarily for several reasons. The electrical transmission networks can consist of hundred or even thousands of substations to which power plants, lines, transformers, and/or consumers are connected. These substations are typically distributed across wide geographical areas, which can be separated by tens, hundreds, or even thousands of kilometers. Finally, generation sources that exist within these networks can at times be located at great distances from each other, yet must operate synchronously to continuously balance the load.

In order to achieve operation security and economical benefits, the electrical power systems have been expanded at all levels and interconnections have been developed linking neighbor electrical networks from local to national level and from national to continental and even to planetary dimension.

Expanding the power system interconnections is critical to allow transfer of the electrical energy over long distances from locations where the primary energy resources are available to the large consumers that include cities and industries. The actual trend is to exploit the clean energy sources, for example, wind and solar, which causes changing of the usual generation pattern. The strong wind and high-rated power wind turbines suitable for the offshore locations in North Sea, the U.S. coast Mid-Atlantic, and so

Handbook of Electrical Power System Dynamics: Modeling, Stability, and Control. Edited by Mircea Eremia and Mohammad Shahidehpour.

on are attractive "ingredients" for designing large generation projects. Similar attention is paid to the sunny land in North Africa or Australia to construct large photovoltaic power plants.

However, changing the generation pattern may require strengthening parts of the electrical network by increasing the transmission capacity or even by developing new transmission corridors when the new clean resources-based power plants are located very far from the consumption areas. Engineers think thereby to design supergrids that allow transferring large amount of electrical energy over large distances. In most cases, superhighways consisting of high-voltage direct current (HVDC) links are required. The power electronic-based technologies become vital for the power system operation. In the AC transmission systems, the FACTS devices may significantly contribute to controlling the state parameters and deal with transient processes.

Furthermore, advanced digital technologies have been implemented for protection and control functions, able to anticipate, detect, and correct the changes occurring in the power system parameters.

Most important is to realize that, although there are significant benefits provided by all upgrades in the power systems, the amount of information has increased significantly. This is in fact the biggest challenge for the power system operators that are responsible to safely operate the power system. Furthermore, in the planning and operation activity, the system operators need accurate modeling of the power system components and phenomena as well as robust tools for both steady-state and dynamic analyses.

Although there are a large number of benefits of the power system interconnections, adverse effects are also recorded, which include increased risk of system instability, especially related to voltage, generators' angles, and low-frequency interarea oscillations.

On the other hand, the transition toward open access electrical networks and development of the electricity market lead to increased stressing conditions for the power system operation and current congestions. In interconnected power systems, the physical paths do not match the commercial contracts and "parasite" flow may create additional problems. These may also explain the major grid blackouts that have occurred recently even in highly developed regions from United States, Japan, and European countries, and so on. For these reasons, the power system stability problem is one of the most important concerns of the power systems engineers.

However, the electrical energy is one of the humankind's development "pylons." Stable and economical operation of power systems, subjected to environment protection, is the main concern for power system engineers, who are continuously looking for solutions to improve the power quality to the final consumer. This may include the upgrade toward more "intelligent/smart" networks following the advancements in telecommunications, hardware and software computation technologies, power electronics, sensors, metering systems, wide area coordination systems, new electrotechnical materials, new generation technologies, and so on. *Smart grids* might be our wish to create opportunities for better technologies able to coexist in harmony with the nature.

Sir Isaac Newton once said about himself "If I have seen further than others it is because I have stood on the shoulders of giants." It is therefore our duty to acknowledge prior work of great specialists, as mentioned also in the large number of references. In the field of power system stability, it is worth mentioning some of the most important books that were published over the years:

Crary, S.B. *Power system stability*, 1947.	Zhdanov, P.S. – *Power system stability* (in Russian), 1948.
Kimbark, E.W. *Power system stability*, 1948, 1950, 1956. (reprinted in 1995)	Venikov, V.A. *Transient phenomena in electric power systems*, 1965.
Anderson, P.M. and Fouad, A.A. *Power system control and stability*, 1977 (2nd edition in 2003).	Kundur, P. *Power system stability and control*, 1994.
Taylor, C. *Power system voltage stability*, 1994.	Pavella, M., Murthy, P.G. *Transient stability of power systems: Theory and practice*, 1994.
Padyar, K.R. *Power system dynamics. Stability and control*, 1996.	Machowschi, J., Bialek, J., Bumby, J. *Power system dynamics and stability*, 1997 (2nd edition in 2008).
Sauer, P., Pai, M.A. *Power system dynamics and stability*, 1998.	Ilic, M., Zaborsky, J. *Dynamic and control in large electric power systems*, 2000.
Marconato, R. *Electric power systems. Vol. III. Dynamic behaviour, stability and emergency controls*, 2008.	Van Cutsem, T., Vournas, C.D. *Voltage stability of electric power systems*, 1998.

The idea of this book was born 7–8 years ago, encouraged by Professor Prabha Kundur after the visit to Romania as Distinguished Lecturer of IEEE. This is also the result of a strong relationship between coauthors after collaboration in a large number of activities organized under IEEE-PES and CIGRE, projects developed at European level or various educational programs. Several schools from seven countries have therefore joined to share their knowledge with other engineers. Each chapter reflects mainly the tradition and experience in topic of power system dynamics, stability, and control of the contributors' schools.

Part I of the book, comprising Chapters 2–7, is devoted to "Power system modeling and control."

The theory and modeling of the synchronous generator in steady state, under dynamic behavior as well as when subjected to a short circuit, are presented in the first part of Chapter 2. Modeling of various excitation systems and power system stabilizers is also included in the first part, as they are vital components of the synchronous generator for its stable operation. The second section is devoted to the induction motor, which is the most important dynamic component of the load. Starting from the known form of the electric circuits equations, the fundamental equations in the *d-q* reference frame are developed for steady-state and transient operation studies. Other theories of load modeling are provided in Chapter 11.

Chapter 3 presents the modeling of the main components of the classical power plants, which include steam, gas, or hydraulic turbines. The first part provides the steam systems configurations, the corresponding mathematical models, and the governing systems. The next part includes the basic configurations and the model block diagrams of combined cycle power plants. The last part of the chapter includes the modeling of hydraulic prime movers and the main auxiliary control systems used in power system studies.

The wind turbine generating systems have gained their place in the power systems because the share of electrical energy produced by wind sources reaches high percentages in some countries and is continuously growing. From the power system operation point of view special attention must be dedicated to wind turbine systems due to their intermittency in power generation. Chapter 4 describes the theory of energy extraction from wind and transformation into electrical energy. The classification of wind turbine systems and types of electrical generators are then provided. The second part presents the modeling and the vector control of the doubly fed induction generator and the permanent magnet synchronous generator-based wind turbine systems. The fault ride-through capability of these generation systems are described in the final.

Appropriate design of power system components and their protection systems is very important in order to avoid the effects of undesired but inevitable disturbances, for example, short circuits. Chapter 5 focuses on "Short-circuit currents calculation" methods in meshed and nonmeshed networks, based on the newest recommendations of the IEC Standards 60900-1–4. Three applications are provided in the final where calculation of the single-phase and the symmetrical three-phase short-circuit currents is performed for three different network configurations. These help the reader to understand what steps are needed to perform such calculations.

The main electrical quantities of the power system that define the electrical energy are the frequency and the voltage. As the load is affected by variation of either quantity, appropriate control actions are of great importance. Keeping the frequency and the voltage in predefined limits is also important for the stability of the power system.

The basic theory of "Active power and frequency control" is shown in Chapter 6. After a short presentation on the situation in Europe and United States, the chapter continues with the fundamental characteristics of frequency control, that is, system dynamics, inertia, droop, regulation, and dynamic frequency response. In order to understand the power system frequency performances, it is important to adequately model the power plants governors. Special attention is paid here to the thermal governor and hydraulic governor, respectively. In the second part of the chapter, development of a new model of thermal governor is presented. Then, the theory of area generation control in isolated systems as well as in multiarea systems is shown.

A more exhaustive approach of the "Voltage and reactive power control" is observed in Chapter 7. The principles and technologies used for controlling the voltage either by directly focusing on voltage or by acting on the reactive power are discussed in this chapter. Equations for a simple radial network are first written; then more complex models are developed for voltage control at regional and national levels, where centralized and coordinated actions are usually performed. This latter approach, which assumes a hierarchization of the control actions, similar to frequency control, is developed in detail. Greater attention is paid to the secondary voltage regulation system based on the Italian experience. This chapter also includes examples of hierarchical grid voltage control systems already in operation in some power systems (Italy and France) and implementation studies in other systems.

The second part of the book, titled "Power system stability and protection," comprising Chapters 8–12, is devoted to evaluation of different forms of power system stability as well as analysis of critical situation experienced in power systems together with possible preventive or remedial solutions.

A "Background of power system stability" is provided in Chapter 8, as a preamble for the next chapters. Classification of the types of stability in a power system and a short description of each type are provided, based on reference publications of IEEE and CIGRE working groups. Before beginning to evaluate the stability of an electrical generator or the power system as a whole, it is very important to understand the nature of the problem that may cause instability and what quantity is most affected.

Small disturbances can occur at any moment in power systems as both the loads and the power plants are characterized by a dynamic behavior. Safe operation of the interconnected power systems assumes also that all generators are coherent to each other, that is, they do not oscillate against each other. The theory of "Small-disturbance angle stability and electromechanical oscillations damping" in power systems is presented in Chapter 9, where all dynamic models are developed from an automatic system point of view, appropriate for dynamic simulations. The causes of electromechanical oscillations and the major factors affecting the damping of oscillations are first identified, and then solutions for improvement of oscillations damping are proposed. One of these solutions also includes the design of the power system stabilizer and tuning its parameters. To better understand the theory, various applications are presented in the chapter.

Chapter 10, titled "Transient stability" covers the angle stability problem in the case of large disturbances. The power system behavior, in general, and the synchronous generator, in particular, are evaluated from simple to complex. The transient stability analysis begins from the classical theory of the equal area criterion and integration methods and moves to the more complex Lyapunov-based methods, emphasizing the Russian school in mathematics and applications of Lyapunov theory in power systems. Dynamic equivalents are also included in this chapter as a tool in transient stability assessment of large interconnected power systems. An educational exercise with equal area criterion and integration of swing equation using both the trapeze method and the Runge–Kutta method on a simple power system is presented at the end of chapter.

Small-disturbance and large-disturbance angular stabilities are related to the synchronous generator. However, stability problems in power systems, as regards keeping the voltage within secure operating limits, may appear due to the load. Chapter 11 relates "Voltage stability" issues and explains the interaction between network, loads, and control systems and the mechanisms leading to voltage instability. Voltage stability is assessed here by methods and indices such as sensitivity analysis, modal analysis, the smallest singular value, participation factors, or bifurcation theory. When a power system is subjected to under-voltage problems, the means employed to bring the voltage within security limits, for example, on-load tap changers or generators excitation systems, can sometimes have opposite results worsening the voltage level. Based on such issues, some countermeasures are discussed in the final.

Beyond any evaluation of the power system dynamic behavior, either seen as a whole or decomposed in individual components, timely and correct actions are necessary through dedicated protection systems in critical moments. Chapter 12, titled "Power system protection," presents a vision on the protection systems in the modern power systems, which are based on digital devices, advanced telecommunication infrastructure, and wide area coordination. Reliability, selectivity, speed of performance, recording for statistic analysis, or adaptation to the power system conditions are the advantages of the new protection systems, which are briefly discussed in this chapter.

The third part of the book approaches the problems of "Grid blackouts and restoration processes." The power systems are continuously subjected to critical events or stressed operating conditions, which may push the system beyond the safety limit. The technology has continuously advanced, but the major grid blackouts that happened around the world in the last two decades shows that the power systems are still vulnerable. Maybe the reason is that new causes of system security challenges have been added, such as: still aged transmission infrastructure, increased number of natural disasters, higher number of uncertainty factors caused by the electricity market, rapid changes in network operation due to intermittent renewable energy sources, insufficient training of human operators, or even military attack. However, many of these causes may reside in financial aspects.

On this line, Chapter 13, titled "Major grid blackouts: analysis, classification, and prevention," discusses the most important partial or total blackouts that occurred in power systems over the years. The chain of power system conditions and events that have led to blackout, also including the resynchronization, are first explained. Learning from history is very important for the human in order to design better machinery or to better train the engineers assisting the machinery, since the blackouts have major economical and social implications.

On the other hand, Chapter 14, titled "Restoration processes after blackouts," presents the way in which a power system can be reenergized after a blackout by contribution of individual power plants, through their black-start capability. The analysis is focusing on modeling the gas turbine and combined cycle power plants in the dynamic process of system restoration.

A deeper evaluation of the transient phenomena is presented in Chapter 15, based on "Computer simulation of scale-bridging transients in power systems." The chapter shows how various scale-bridging phenomena ranging from electromagnetic to electromechanical transients are modeled efficiently. Simulation of a blackout based on this theory is also provided.

The book is written in a didactic fashion and addressed to a broad category of power system engineers, from master and Ph.D. students to practicing engineers involved in power system operation and design, equipment manufacturers, telecommunications, and so on.

PART I

POWER SYSTEM MODELING AND CONTROL

Courtesy of Alstom

Handbook of Electrical Power System Dynamics: Modeling, Stability, and Control. Edited by Mircea Eremia and Mohammad Shahidehpour.

2

SYNCHRONOUS GENERATOR AND INDUCTION MOTOR

Mircea Eremia and Constantin Bulac

2.1 THEORY AND MODELING OF SYNCHRONOUS GENERATOR[1]

Synchronous generators are the principal supplying source in power systems, converting the mechanical power, provided by a hydraulic, steam, gas, or wind turbine, into electrical energy.

2.1.1 Design and Operation Principles

The main components of a synchronous generator are (Figure 2.1):

- The *stator* is the stationary part, made up in cylindrical form, provided on the inner surface with equidistant slots recessed by the windings of the three phases (*a*, *b*, and *c*), displaced in space from each other by 120 electric degrees.
- The *rotor* is the mobile part, rotating inside the stator, which contains also a field (excitation) winding (supplied with a direct current) and damper or amortisseur windings consisting of bars connected to end rings that form the so-called squirrel cage (in steady state passed by no current). The direct current in the field winding produces a magnetic field, which is stationary relative to the rotor, but rotating relative to the stator with a speed equal to the rotor speed. In this way, alternating

[1] This chapter reflects, to some extent, theories from Dr. Prabha Kundur's book *Power System Stability and Control* [1].

Handbook of Electrical Power System Dynamics: Modeling, Stability, and Control. Edited by Mircea Eremia and Mohammad Shahidehpour.

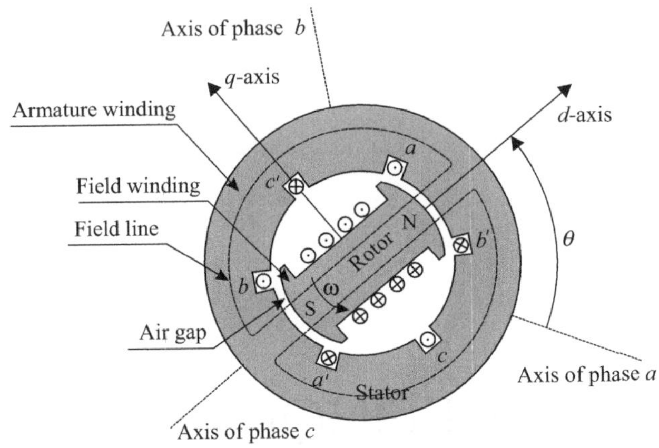

Figure 2.1. Schematic diagram of a three-phase synchronous machine.

voltages displaced by 120° in time are induced in the stator windings, having the frequency given by the rotor speed and the number of magnetic pairs of poles.

The magnetic circuit of the machine consists of five basic sections: the air gap, the stator teeth, the back of the stator, the pole core with shoes, and the rotor yoke (Figure 2.7). By summing the total current in all five sections, we find the total current (ampere-turns) of pole excitation necessary to produce a given flux in the air gap of the machine.

In any machine, the currents in all the windings combine to produce the resultant flux. The action of the machine depends on the facts, first that the flux induces voltages in the windings and second that the flux interacts with the currents to produce torque.

Figure 2.1 shows a cross section of a synchronous generator with one pair of poles [1].

The flux spreads throughout the whole machine, but its effect depends primarily on the distribution of flux density round the air gap; the attention is therefore focused on this region. At any instant the curve of flux density around the air gap circumference may be of any form and is not necessarily sinusoidal. The *main flux* of an AC machine is defined to be that determined by the fundamental component of the curve of air gap flux density, and the radial line where the fundamental density is a maximum is called the *axis of the flux*. The main flux is then completely defined by a magnitude and a direction [2].

In order to calculate the flux due to a given system of currents it is first necessary to determine the *magnetomotive force* (mmf) due to the currents. Figure 2.2 is a developed diagram for a two-pole machine extending between angular position 0 and 2π. The currents in the conductors of a coil are distributed in slots as indicated and form two bands, symmetrically distributed about the points A and B.

The currents in two bands flow in opposite directions. Since the current distribution is known, the mmf round any closed path can be found; in particular, the mmf round a path crossing the air gap, such as ACDFGH. Because of the high permeability of the iron, all the mmf round the closed path can be assumed to appear across the air gap at the point F, if A is chosen as a point of zero air gap flux density, which, for a symmetrical distribution of current, must occur at the center of each band. Hence, a curve of mmf distribution round the air gap can be drawn for any value of current flowing in the coil. Thus, although magnetomotive force is fundamentally a line integral round a closed path, a value can be associated with each point along the air gap circumference, giving *the space distributed*

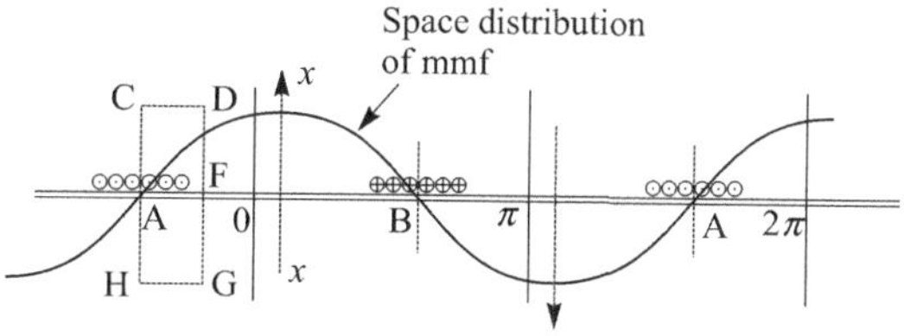

Figure 2.2. The developed diagram of the air gap circumference along a straight line and the fundamental sine wave of the magnetic flux density [2].

mmf curve of the machine. If the conductors are assumed to be located at points around the air gap circumference, the mmf curve is a stepped curve, but the fundamental component, which, by symmetry, is zero at the points A and B, can be drawn as in Figure 2.2. The radial line at the point of maximum mmf (*x*–*x* in Figure 2.2) is called the axis of the mmf, and since this depends only on the conductor distribution, it is also the axis of the coil. The curve shows the instantaneous magnitude of the mmf, which depends on the instantaneous value of the current.

The mmf curve determines the flux density curve:

- If the machine has a uniform air gap, and if saturation is neglected, the flux density is everywhere proportional to the mmf.
- In a salient pole machine, because the flux density is not proportional with mmf, in order to calculate the flux, it becomes necessary to resolve the mmf wave into component wave along the direct and quadrature axes.

On either axis a sinusoidal mmf wave produces a flux density distribution, which can be determined [2]. The flux density curve is not sinusoidal, but because of the symmetry of the pole about its axis, its fundamental component has the same axis as the mmf producing it. Hence, if the harmonics are ignored, an mmf wave on either the direct or the quadrature axes produces a proportional sinusoidal flux density wave on the same axis, the factors of proportionality being different for the two axes. By this means the flux components on the two axes, due to any current, can be found and, if there is no saturation, the resultant flux, and hence the resultant flux density wave, is obtained by combining them.

In practice, the armature windings and round rotor machine field windings are distributed in many slots so that the resulting mmf and flux waveforms have nearly sinusoidal space distribution.

To obtain a sinusoidal emf in the conductors of the stator winding requires a sinusoidal distribution of induction around the periphery of the rotor. In the case of salient poles, it is advisable for this purpose to use the shape of the pole pieces: it is necessary to construct a salient pole machine with a nonuniform air gap gradually increasing from the center of the pole to the edges.

When the generator is connected to the network, the stator windings carry alternating currents with the pulsation equal to that of the network. These currents generate in turn a rotating magnetic field interacting with the field produced by the excitation current, resulting in an electromagnetic torque (C_e) that, under steady-state conditions, is equal to the mechanical torque applied to the rotor (C_m). To obtain a constant torque, both the stator and the rotor magnetic fields must rotate with the same speed (to be synchronous).

If C_m is not equal to C_e, the rotor speed and so the rotor electromagnetic field speed is different from the stator electromagnetic field speed. Therefore, eddy currents occur in the damper bars, so is in the solid steel of the rotor (the solid steel of the rotor can be considered as an equivalent damper winding), with a frequency equal to the slip frequency. These induced currents produce an electromagnetic field, which, according to the Lenz's law, oppose the change that has produced them. Therefore, if the rotor speed is oversynchronous, the induced currents cause an increase in the electromagnetic torque C_e, and the rotor decelerates. Otherwise, if the rotor speed is subsynchronous, the induced currents cause the decrease in electromagnetic torque and the rotor accelerates.

The number of the magnetic poles (P) determinates the mechanical speed ω_m of the rotor and the frequency f of stator currents. Denoting by $p = P/2$ the number of pairs of poles, and n the rotor speed (rot/min), the frequency (Hz) of the induced emf is given by

$$f = \frac{\omega(\mathrm{rad/s})}{2\pi} = \frac{p \cdot \omega_m(\mathrm{rad/s})}{2\pi} = \frac{(p \cdot 2\pi \cdot n)/60}{2\pi} = p \cdot \frac{n}{60} = p \cdot f_m \quad (2.1)$$

which in steady state is equal to the power system frequency.

In terms of the design and rotation speed [2] of the rotor, the synchronous generators are classified in two categories:

- *Round (cylindrical) Rotor Generators or Turbogenerators.*
  - have cylindrical shaped rotor;
  - are driven by high-speed steam or gas turbines (they have one or maximum two pairs of poles);
  - the conductors wounded on the field winding are placed into longitudinal slots, providing a constant air gap; because of the symmetry these machines are called isotropic.
- *Salient Pole Generators or Hydrogenerators.*
  - the rotor is made up of several pairs of salient poles, distributed on the periphery of a polar wheel;
  - are driven by low-speed hydraulic turbines, imposing a large number of pairs of poles ($p \geq 3$).

Generally, the angles are measured in electrical radians or electrical degrees. The angle covered by a pair of poles is 2π rad or 360 electric degrees (Figure 2.3).

In order to establish the relationship between electrical degrees, used in voltage and current wave expressions, and geometrical degrees defining the stator position, equation (2.1) is multiplied by $2\pi t$, obtaining

$$2\pi f t = p 2\pi f_m t$$

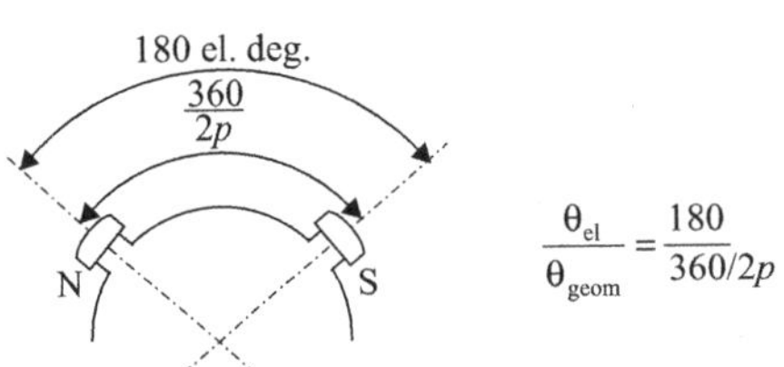

Figure 2.3. The relationship between mechanical and electrical angles.

[2] The synchronous mechanical speed depends on the frequency and the number of pairs of poles. At the frequency of 50 Hz, the synchronous speed is 3000 rot/min for one pair of poles, 1500 rot/min for two pairs of poles, 1000 rot/min for three pairs of poles, and so on.

or

$$\omega t = p\omega_{\mathrm{m}} t$$

It results

$$\theta_{\mathrm{el}} = p\theta_{\mathrm{geom}}$$

On the other hand, the relationship between the angle expressed in radians and the angle expressed in electrical degrees is given by the equality:

$$\frac{\delta(\mathrm{rad})}{\delta(\mathrm{el.\,deg.})} = \frac{2\pi f}{360 f}$$

or

$$\delta(\mathrm{rad}) = \frac{314}{360 f}\delta(\mathrm{el.\,deg.}) = \frac{\delta(\mathrm{el.\,deg.})}{57.3}$$

2.1.2 Electromechanical Model of Synchronous Generator: Swing Equation

The *swing equation* relates the differential equation that describes the acceleration (deceleration) of the synchronous generator and turbine (prime mover) due to any imbalance between mechanical torque and electromagnetic torque:

$$J\frac{\mathrm{d}\omega_{\mathrm{m}}}{\mathrm{d}t} = C_{\mathrm{a}}^{\mathrm{a}} = C_{\mathrm{m}}^{\mathrm{a}} - C_{\mathrm{e}}^{\mathrm{a}} \tag{2.2}$$

where J is the combined moment of inertia of the rotating mass, in kg $\times$ m^2; ω_{m} is the angular velocity of the rotor, in mechanical radians per second (mech. rad/s); $C_{\mathrm{a}}^{\mathrm{a}}$ is the accelerating torque, in N $\times$ m; $C_{\mathrm{m}}^{\mathrm{a}}$ is the mechanical torque, in N $\times$ m; $C_{\mathrm{e}}^{\mathrm{a}}$ is the electromagnetic torque, in N $\times$ m; and t is the time, in seconds.

The differential equation (2.2) can be normalized and expressed in per unit (p.u.) defining the inertia constant H as the rotor kinetic energy stored in rotating mass (in Joule = Watt $\times$ second) referred to the generator base rating (S_{b} in VA):

$$H = \frac{1}{2}\cdot\frac{J\omega_{0\mathrm{m}}^2}{S_{\mathrm{b}}} \quad (\mathrm{s}) \tag{2.3}$$

where $\omega_{0\mathrm{m}}$ is the rated angular velocity of the rotor, in mech. rad/s.

Expressing the moment of inertia J from (2.3) as

$$J = \frac{2HS_{\mathrm{b}}}{\omega_{0\mathrm{m}}^2}$$

and substituting in equation (2.2) obtain

$$\frac{2H}{\omega_{0\mathrm{m}}^2}S_{\mathrm{b}}\frac{\mathrm{d}\omega_{\mathrm{m}}}{\mathrm{d}t} = C_{\mathrm{m}}^{\mathrm{a}} - C_{\mathrm{e}}^{\mathrm{a}}$$

or

$$2H\frac{\mathrm{d}}{\mathrm{d}t}\left(\frac{\omega_{\mathrm{m}}}{\omega_{0\mathrm{m}}}\right)=\frac{C_{\mathrm{m}}^{\mathrm{a}}-C_{\mathrm{e}}^{\mathrm{a}}}{S_{\mathrm{b}}/\omega_{0\mathrm{m}}} \tag{2.4}$$

For the left-hand side of equation (2.4), we take into account that

$$\frac{\omega_{\mathrm{m}}}{\omega_{0\mathrm{m}}}=\frac{\omega_{\mathrm{r}}/p}{\omega_0/p}=\frac{\omega_{\mathrm{r}}}{\omega_0}$$

where ω_{r} is the angular velocity of the rotor, in el. rad/s; ω_0 is the rated angular velocity of the rotor, in el. rad/s; and p is the number of pairs of poles of the synchronous generator.

The right-hand side of equation (2.4), representing the difference between the mechanical and electromagnetic torques, expressed in p.u., is approximately equal to the difference between the mechanical input power and electrical output power:

$$\frac{C_{\mathrm{m}}^{\mathrm{a}}-C_{\mathrm{e}}^{\mathrm{a}}}{S_{\mathrm{b}}/\omega_{0\mathrm{m}}}=C_{\mathrm{a}}=C_{\mathrm{m}}-C_{\mathrm{e}}\approx P_{\mathrm{m}}-P_{\mathrm{e}} \tag{2.4$'$}$$

where $C_{\mathrm{m}}, C_{\mathrm{e}}$ are the mechanical and electromagnetic torques, in p.u.; $P_{\mathrm{m}}, P_{\mathrm{e}}$ are the mechanical power and electrical power, in p.u., referred to S_{b}; and C_{a} is the accelerating torque, in p.u.

From (2.4) it results

$$2H\frac{\mathrm{d}}{\mathrm{d}t}\left(\frac{\omega_{\mathrm{r}}}{\omega_0}\right)=C_{\mathrm{a}}=C_{\mathrm{m}}-C_{\mathrm{e}}\approx P_{\mathrm{m}}-P_{\mathrm{e}} \tag{2.4$''$}$$

In addition, denoting by

$$\omega=\frac{\omega_{\mathrm{r}}-\omega_0}{\omega_0}\left(=\frac{\omega_{\mathrm{r}}}{\omega_0}-1\right) \tag{2.5}$$

and taking into account that

$$\frac{\mathrm{d}\omega}{\mathrm{d}t}=\frac{\mathrm{d}}{\mathrm{d}t}\left(\frac{\omega_{\mathrm{r}}}{\omega_0}\right) \tag{2.5$'$}$$

from equation (2.4″) it results the equation of motion in per unit

$$2H\frac{\mathrm{d}\omega}{\mathrm{d}t}=C_{\mathrm{m}}-C_{\mathrm{e}}\approx P_{\mathrm{m}}-P_{\mathrm{e}} \tag{2.6}$$

Denoting by δ the angle (in electric radians) giving the position of rotor, at an instant t, with respect to a synchronously rotating reference system (with ω_0), and by δ_0 its value at the instant $t=0$ (see Figure 2.4a and b), then

$$\delta=\omega_{\mathrm{r}}t+\delta_0-\omega_0 t=(\omega_{\mathrm{r}}-\omega_0)t+\delta_0$$

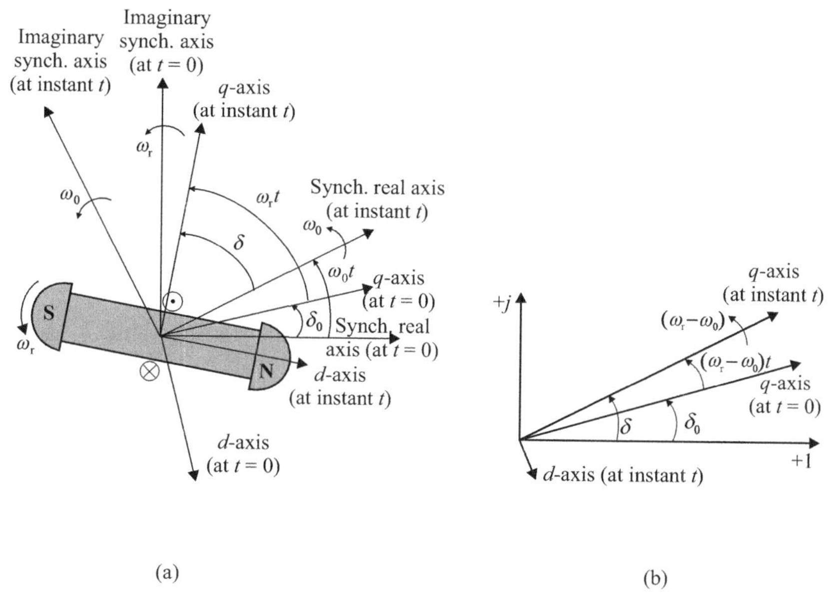

Figure 2.4. The definition of angles.

resulting that

$$\frac{d\delta}{dt} = \omega_r - \omega_0 = \omega_0\left(\frac{\omega_r - \omega_0}{\omega_0}\right) = \omega_0\omega \tag{2.7}$$

and

$$\frac{d^2\delta}{dt^2} = \omega_0\frac{d\omega}{dt} = \frac{\omega_0}{2H}(C_m - C_e)$$

and thus we obtain another form of the equation of motion

$$\frac{2H}{\omega_0}\frac{d^2\delta}{dt^2} = C_m - C_e \approx P_m - P_e \tag{2.8}$$

Usually, the differential equation of motion contains also a damping torque component, obtained by adding a term proportional to the speed deviation ω, in equations (2.6) and (2.8), respectively,

$$2H\frac{d\omega}{dt} + D\omega = C_m - C_e \approx P_m - P_e \tag{2.9}$$

$$\frac{2H}{\omega_0}\frac{d^2\delta}{dt^2} + \frac{D}{\omega_0}\frac{d\delta}{dt} = C_m - C_e \approx P_m - P_e \tag{2.9$'$}$$

where D, the damping coefficient, may be determined either from design data or by test; D is in p.u. torque/p.u. speed deviation.

The time interval in which the synchronous generator, starting from standstill, reaches the synchronous speed ω_0, when an acceleration torque equal to $C_a = 1$ p.u. is applied, is defined as the *mechanical starting time of the generator*, denoted by M.[3]

[3] In the technical literature, the mechanical starting time is denoted by also T_a. This notation will be used in a further chapter.

From equation (2.4″), we achieve

$$\frac{d}{dt}\left(\frac{\omega_r}{\omega_0}\right) = \frac{C_a}{2H}$$

Integrating the last equation in time domain, it results

$$\frac{\omega_r}{\omega_0} = \frac{1}{2H}\int_0^t C_a\,dt$$

Therefore, for $\omega_r/\omega_0 = 1$ and $C_a = 1$, as well as for the starting value $\omega_r/\omega_0 = 0$ (the machine is in standstill), we obtain

$$1 = \frac{1}{2H}\int_0^M dt = \frac{M}{2H}$$

Thus, the mechanical starting time of the generator, in seconds, is

$$M = 2H$$

The *electromechanical model of the synchronous generator*, in p.u., is given by the following differential equations, also called swing equations:

$$\left\| \begin{aligned} & M\frac{d\omega}{dt} + D\omega = C_m - C_e \approx P_m - P_e \\ & \frac{d\delta}{dt} = \omega_0\omega \end{aligned} \right. \tag{2.10}$$

or

$$\left\| \frac{M}{\omega_0}\frac{d^2\delta}{dt^2} + \frac{D}{\omega_0}\frac{d\delta}{dt} = C_m - C_e \cong P_m - P_e \right. \tag{2.10'}$$

At the equilibrium point, the rotor speed ω reaches a stable value, that is, $\omega_r = \omega_0$ and $\omega = 0$, respectively, and therefore, from (2.10) and (2.10′) it results

$$\left.\frac{d\delta}{dt}\right|_{\delta=\delta_0} = 0; \quad \left.\frac{d^2\delta}{dt^2}\right|_{\delta=\delta_0} = 0$$

where δ_0 is the rotor angle at the equilibrium point.

Considering the last two equations, from equation (2.10′) it results that at the equilibrium point

$$P_m = P_e(\delta)$$

that is, the mechanical power is equal to the electrical power.

Replacing d/dt by s, the Laplace operator, the following block diagram representing the swing equation is obtained (see Figure 2.5).

The swing equation can also be expressed in absolute values. Thus, considering expressions (2.4′), (2.5), and (2.5′), equation (2.9) becomes

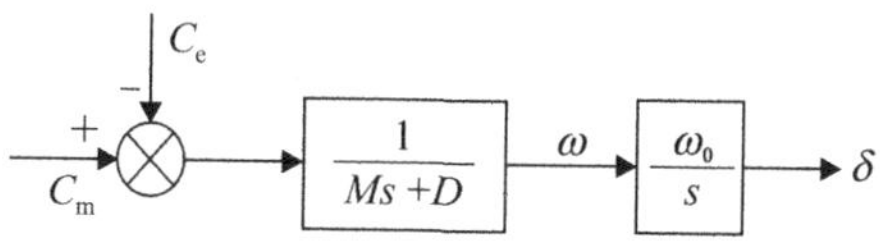

Figure 2.5. Block diagram of the swing equation.

$$\frac{M}{\omega_0}\frac{d\omega_r}{dt} + D\frac{\omega_r - \omega_0}{\omega_0} = \frac{C_m^a - C_e^a}{S_b/\omega_0} \tag{2.11}$$

It results that

$$\frac{d\omega_r}{dt} + \frac{D}{M}(\omega_r - \omega_0) = \frac{\omega_0}{MS_b}\left(\omega_0 C_m^a - \omega_0 C_e^a\right) \tag{2.11$'$}$$

or

$$\frac{d\omega_r}{dt} + \frac{D}{M}(\omega_r - \omega_0) = \frac{\omega_0}{MS_b}\left(P_m^a - P_e^a\right) \tag{2.11$''$}$$

where P_m^a, P_e^a are the mechanical and electrical powers, in MW.

Equation (2.11″) together with equation (2.7) forms *the electromechanical model of the synchronous generator expressed in absolute values*:

$$\left\| \begin{aligned} &\frac{d\omega_r}{dt} + \frac{D}{M}(\omega_r - \omega_0) = \frac{\omega_0}{MS_b}\left(P_m^a - P_e^a\right) \\ &\frac{d\delta}{dt} = \omega_r - \omega_0 \end{aligned} \right. \tag{2.12}$$

This model is also known as the *second-order model* since it contains two first-order differential equations.

2.1.3 Electromagnetic Model of Synchronous Generator

2.1.3.1 Basic Equations. In developing the equations of a synchronous machine, the following assumptions are made [1]:

(i) The stator windings are sinusoidally distributed along the air gap as far as the mutual effects with the rotor are concerned.

(ii) The stator slots cause no appreciable variation of the rotor inductances with rotor position.

(iii) Magnetic hysteresis is negligible.

(iv) Magnetic saturation effects are negligible; the machine equations will be developed first by assuming linear flux–current relationships.

In order to study the power system operating conditions, the synchronous generators are represented as a number of equivalent windings, magnetic coupled and rotating.

From Figure 2.1, the following circuits are identified (Figure 2.6):

- *Stator Circuits.* The three stator windings *a–a*′, *b–b*′, and *c–c*′, distributed 120° apart in trigonometric rotation direction. The voltages at the stator winding terminals

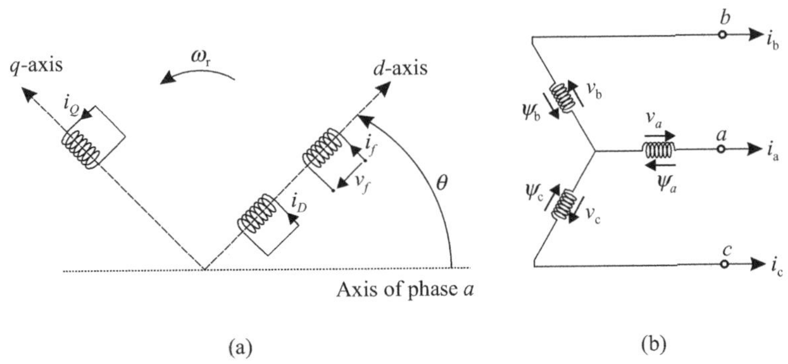

Figure 2.6. The circuits of the synchronous generator: (a) rotor circuit; (b) stator circuit.

are v_a, v_b, and v_c, and the currents are i_a, i_b, and i_c. Using the generator convention, the stator current is considered positive when it is out of the machine.

- *Rotor Circuits.* The field (excitation) winding *f–f′* is taken as the direct axis, which is simply called *d*-axis, and the quadrature axis, which is 90° ahead of the direct axis in the rotational direction, is called *q*-axis. In this case, it is usual to say that the *q*-axis is leading the *d*-axis. Choosing the positive direction of the *q*-axis as reference is arbitrarily. Moreover, consider that the damper windings are disposed on the two rotor axes. The number of these windings characterizes the accuracy of the generator modeling. Common models use one damper winding in the *d*-axis, called *D*, and one in the *q*-axis, called *Q*. These windings are permanently short-circuited.

The rotor position is given by the electrical angle θ (electric radians) measured between the *d*-axis and the axis of phase "*a*" ($\theta = \omega_r t$).

In order to establish the equations describing the operation of synchronous generator, the state quantities at the circuits terminals—voltages and currents—are associated according to the generator convention for stator circuits *a*, *b*, *c*, while for the field circuit *f* they are associated according to the load convention. Arbitrarily directions can be chosen for the damper circuit currents since they are in short circuit.

By applying the Faraday's law of induction, the following instantaneous voltage equations and magnetic flux equations, expressed in p.u., are obtained.

Voltage Equations.

- *For the Stator Windings*

$$v_a(t) = -R_a \cdot i_a(t) + \frac{d\psi_a(t)}{dt} \tag{2.13a}$$

$$v_b(t) = -R_b \cdot i_b(t) + \frac{d\psi_b(t)}{dt} \tag{2.13b}$$

$$v_c(t) = -R_c \cdot i_c(t) + \frac{d\psi_c(t)}{dt} \tag{2.13c}$$

where R_a, R_b, and R_c are the resistances of phases *a*, *b*, and *c*, respectively; ψ is the flux linkage of stator windings *a*, *b*, and *c*.

These equations can be expressed in matrix form:

$$[v_S] = -[R_S][i_S] + \frac{d}{dt}[\psi_S] \tag{2.14}$$

where

$$[R_S] = \begin{bmatrix} R_a & & \\ & R_b & \\ & & R_c \end{bmatrix}$$

- *For the Rotor Windings*

$$v_f(t) = R_f \cdot i_f(t) + \frac{d\psi_f(t)}{dt} \tag{2.15a}$$

$$0 = R_D \cdot i_D(t) + \frac{d\psi_D(t)}{dt} \tag{2.15b}$$

$$0 = R_Q \cdot i_Q(t) + \frac{d\psi_Q(t)}{dt} \tag{2.15c}$$

or in matrix form

$$[v_R] = [R_R][i_R] + \frac{d}{dt}[\psi_R] \tag{2.16}$$

where

$$[R_R] = \begin{bmatrix} R_f & & \\ & R_D & \\ & & R_Q \end{bmatrix}$$

MAGNETIC FLUX EQUATIONS. The flux linkage in the phase a winding at any instant is given as follows:

$$\psi_a(t) = -l_{aa}i_a - l_{ab}i_b - l_{ac}i_c + l_{af}i_f + l_{aD}i_D + l_{aQ}i_Q$$

Similar expressions can be written for the flux linkages of windings b and c, as well as for the others windings of the synchronous generator:

$$\begin{bmatrix} \psi_a(t) \\ \psi_b(t) \\ \psi_c(t) \\ \hdashline \psi_f(t) \\ \psi_D(t) \\ \psi_Q(t) \end{bmatrix} = \overset{\text{Stator} \qquad\qquad \text{Rotor}}{\left[\begin{array}{ccc:ccc} l_{aa} & l_{ab} & l_{ac} & l_{af} & l_{aD} & l_{aQ} \\ l_{ba} & l_{bb} & l_{bc} & l_{bf} & l_{bD} & l_{bQ} \\ l_{ca} & l_{cb} & l_{cc} & l_{cf} & l_{cD} & l_{cQ} \\ \hdashline l_{fa} & l_{fb} & l_{fc} & l_{ff} & l_{fD} & l_{fQ} \\ l_{Da} & l_{Db} & l_{Dc} & l_{Df} & l_{DD} & l_{DQ} \\ l_{Qa} & l_{Qb} & l_{Qc} & l_{Qf} & l_{QD} & l_{QQ} \end{array}\right]} \cdot \begin{bmatrix} -i_a \\ -i_b \\ -i_c \\ \hdashline i_f \\ i_D \\ i_Q \end{bmatrix} = \left[\begin{array}{c:c} L_{SS} & L_{SR} \\ \hdashline L_{RS} = L_{SR}^T & L_{RR} \end{array}\right] \begin{bmatrix} -i_a \\ -i_b \\ -i_c \\ \hdashline i_f \\ i_D \\ i_Q \end{bmatrix} \tag{2.17}$$

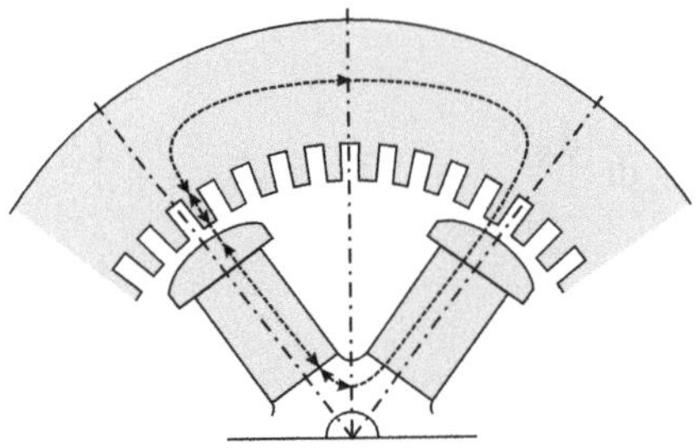

Figure 2.7. Detail of the magnetic circuit.

Note that the negative sign associated with the stator winding currents is due to the adopted direction (the receivers rule but with negative sign).

Before calculating self- and mutual inductances from the magnetic flux equations, some additional explanations are necessary. The self- and mutual inductances of the stator circuits vary with the rotor position and magnetic circuit permeance (**P**).

The magnetic flux produced by a stator winding follows a path through the stator iron, air gap, rotor iron, and back to the stator through the air gap (Figure 2.7).

Since the periphery of rotor is not uniform, there are differences along the air gap created between the rotor and stator that are more pronounced for salient pole generators. As the rotor changes its position in time, changes in the permeance of the magnetic path occur. Decomposition in Fourier series of the real variation curves results in one constant term and variable harmonic terms. As the north and south poles of the phase windings have symmetrical influences, the permeances follow a second-order harmonic variation. In practical calculations only the first term is retained, higher order even harmonics being small enough to be neglected. The permeance can therefore be written as [1]

$$\mathbf{P} = \mathbf{P}_0 + \mathbf{P}_2 \cos 2\alpha$$

where α stands for the angular position of a certain point located on the periphery of rotor measured from the d-axis (Figure 2.8).

(i) *Self-Inductance of the Stator Windings.* The self-inductance of the stator phase a winding l_{aa} is equal to the ratio of the flux linking the phase a winding to the current i_a, while the influence of the other currents is neglected. The self-inductance reaches its maximum when the phase axis overlaps on the d-axis, that is, when the angular position is $\theta = 0°$, then reaches its minimum for $\theta = 90°$ and again a maximum for $\theta = 180°$, and so on (Figure 2.9). The inductance is directly proportional to the permeance, which has a second-order harmonic variation.

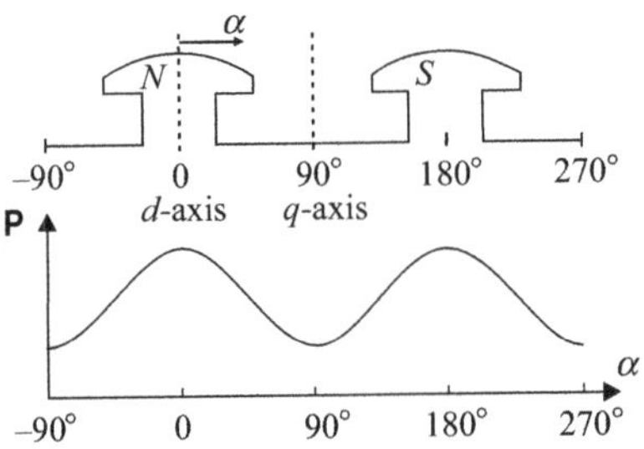

Figure 2.8. Variation of permeance with rotor position [1].

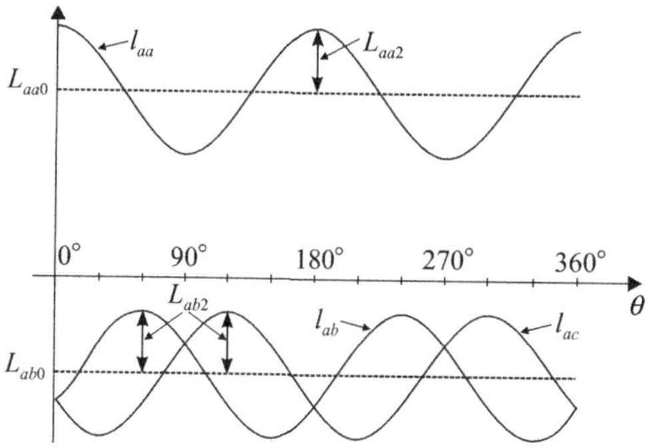

Figure 2.9. Variation of self-inductance l_{aa} and mutual inductances l_{ab} and l_{ac} of the stator winding a in terms of the angular position θ.

The self-inductance l_{aa} of the phase a, due to the air gap magnetic flux, is given by [1]

$$l_{gaa} = L_{g0} + L_{aa2} \cos 2\theta$$

In order to find the self-inductance l_{aa}, we should add the stator leakage inductance L_{al}, due to the leakage flux lines (not crossing the air gap), to l_{gaa}, which gives

$$l_{aa} = L_{al} + l_{gaa} = L_{al} + L_{g0} + L_{aa2} \cos 2\theta = L_{aa0} + L_{aa2} \cos 2\theta \qquad (2.18a)$$

Similarly, the expressions for the other stator self-inductances are obtained:

$$l_{bb} = L_{bb0} + L_{bb2} \cos 2\left(\theta - \frac{2\pi}{3}\right)$$

$$l_{cc} = L_{cc0} + L_{cc2} \cos 2\left(\theta + \frac{2\pi}{3}\right)$$

Since the windings of the phases b and c are designed identical to that of phase a, but displaced from it by 120° and 240°, respectively, we next consider that $L_{aa0} = L_{bb0} = L_{cc0}$ and $L_{aa2} = L_{bb2} = L_{cc2}$, resulting

$$l_{bb} = L_{aa0} + L_{aa2} \cos 2\left(\theta - \frac{2\pi}{3}\right) \qquad (2.18b)$$

$$l_{cc} = L_{aa0} + L_{aa2} \cos 2\left(\theta + \frac{2\pi}{3}\right) \qquad (2.18c)$$

In equations (2.18a), (2.18b), and (2.18c), the stator self-inductances comprises a second harmonic term superimposed on a constant term. Higher order harmonic terms have been neglected.

(ii) *Stator Mutual Inductances.* The mutual inductance between any two stator windings is always negative and exhibits a second harmonic variation because of the rotor shape. It has a maximum value when the direct axis coincides with the bisectrix of the angle formed by their axes. Therefore, the inductance l_{ab} has a maximum absolute value when $\theta = -30°\ (-\pi/6)$ or $\theta = 150°(5\pi/6)$. The variation of the mutual inductance between phases a and b, and between phases a and c in terms of angle θ is shown in Figure 2.9.

Thus, the variations of stator mutual inductances can be represented as follows:

$$l_{ab} = l_{ba} = -L_{ab0} + L_{ab2}\cos\left(2\theta - \frac{2\pi}{3}\right) = -L_{ab0} - L_{ab2}\cos\left(2\theta + \frac{\pi}{3}\right) \tag{2.19a}$$

$$l_{bc} = l_{cb} = -L_{ab0} - L_{ab2}\cos(2\theta - \pi) \tag{2.19b}$$

$$l_{ca} = l_{ac} = -L_{ab0} - L_{ab2}\cos\left(2\theta - \frac{\pi}{3}\right) \tag{2.19c}$$

Note that $L_{ab2} = L_{aa2}$ and $L_{ab0} \cong L_{aa2}/2$.

(iii) *Mutual Inductance Between Stator and Rotor Windings.* With the air gap variation due to stator slots neglected, the rotor circuit permeance may be considered constant (the permeance is not fluctuating). However, the mutual inductance variations are due to the relative motion between the windings themselves. A coupling between the stator and rotor windings does not exist when they are perpendicular on each other, thus the mutual inductance is zero; the coupling has a maximum when the windings are lined up and also the mutual inductance is maximum. Given the sinusoidal distribution for mmf and flux waves, it results

$$l_{af} = l_{fa} = L_{md}\cos\theta \tag{2.20a}$$

$$l_{aD} = l_{Da} = L_{mD}\cos\theta \tag{2.20b}$$

$$l_{aQ} = l_{Qa} = L_{mQ}\cos\left(\theta + \frac{\pi}{2}\right) = -L_{mQ}\sin\theta \tag{2.20c}$$

The mutual inductances between phase b winding and the rotor circuits are obtained by replacing θ with $(\theta - 2\pi/3)$, and by $(\theta + 2\pi/3)$ for phase c winding.

(iv) *Rotor Inductances.* All the rotor self-inductances are constant since the effects of stator slots and saturation are neglected. They are represented with single subscript notation:

$l_{ff} = L_{ff} = L_f$ denotes self-inductance of the rotor winding;

$l_{DD} = L_{DD} = L_D$ denotes self-inductance of the damper winding in the D-axis;

$l_{QQ} = L_{QQ} = L_Q$ denotes self-inductance of the damper winding in the Q-axis.

Moreover, because the d-axis and q-axis are displaced by $\pi/2$ from each other, the mutual inductances between the windings placed on different axes are zero. Thus,

$l_{fD} = l_{Df} = L_{fD}$ denotes mutual inductances of the windings in the d-axis (damper D and field f);

$l_{fQ} = l_{Qf} = 0$ denotes that the mutual inductances between windings located in different axes are zero;

$l_{DQ} = l_{QD} = 0$.

Substituting the expressions of these inductances into equation (2.17), it results

$$\psi_a = -i_a[L_{aa0} + L_{aa2}\cos 2\theta] + i_b\left[L_{ab0} + L_{aa2}\cos\left(2\theta + \frac{\pi}{3}\right)\right]$$
$$+ i_c\left[L_{ab0} + L_{aa2}\cos\left(2\theta - \frac{\pi}{3}\right)\right] + i_f L_{md}\cos\theta + i_D L_{mD}\cos\theta - i_Q L_{mQ}\sin\theta \quad (2.21a)$$

$$\psi_b = i_a\left[L_{ab0} + L_{aa2}\cos\left(2\theta + \frac{\pi}{3}\right)\right] - i_b\left[L_{aa0} + L_{aa2}\cos 2\left(\theta - \frac{2\pi}{3}\right)\right]$$
$$+ i_c[L_{ab0} + L_{aa2}\cos(2\theta - \pi)] + i_f L_{md}\cos\left(\theta - \frac{2\pi}{3}\right) \quad (2.21b)$$
$$+ i_D L_{mD}\cos\left(\theta - \frac{2\pi}{3}\right) - i_Q L_{mQ}\sin\left(\theta - \frac{2\pi}{3}\right)$$

$$\psi_c = i_a\left[L_{ab0} + L_{aa2}\cos\left(2\theta - \frac{\pi}{3}\right)\right] + i_b[L_{ab0} + L_{aa2}\cos(2\theta - \pi)]$$
$$- i_c\left[L_{aa0} + L_{aa2}\cos 2\left(\theta + \frac{2\pi}{3}\right)\right] + i_f L_{md}\cos\left(\theta + \frac{2\pi}{3}\right) \quad (2.21c)$$
$$+ i_D L_{mD}\cos\left(\theta + \frac{2\pi}{3}\right) - i_Q L_{mQ}\sin\left(\theta + \frac{2\pi}{3}\right)$$

Similarly, the rotor circuit flux linkages equations become

$$\psi_f = L_f i_f + L_{fD} i_D - L_{md}\left[i_a\cos\theta + i_b\cos\left(\theta - \frac{2\pi}{3}\right) + i_c\cos\left(\theta + \frac{2\pi}{3}\right)\right] \quad (2.22a)$$

$$\psi_D = L_{Df} i_f + L_{DD} i_D - L_{mD}\left[i_a\cos\theta + i_b\cos\left(\theta - \frac{2\pi}{3}\right) + i_c\cos\left(\theta + \frac{2\pi}{3}\right)\right] \quad (2.22b)$$

$$\psi_Q = L_{QQ} i_Q + L_{mQ}\left[i_a\sin\theta + i_b\sin\left(\theta - \frac{2\pi}{3}\right) + i_c\sin\left(\theta + \frac{2\pi}{3}\right)\right] \quad (2.22c)$$

If the magnetic flux equations (2.21) and (2.22) are written in matrix form, it results

$$\begin{bmatrix} \psi_a \\ \psi_b \\ \psi_c \\ \hline \psi_f \\ \psi_D \\ \psi_Q \end{bmatrix} = \overset{\text{Stator} \qquad\qquad\qquad \text{Rotor}}{\left[\begin{array}{c:c} [L_{SS}(\theta)] & [L_{SR}(\theta)] \\ \hdashline [L_{RS}(\theta)] = [L_{SR}(\theta)]^T & [L_{RR}] \end{array}\right]} \cdot \begin{bmatrix} -i_a \\ -i_b \\ -i_c \\ \hline i_f \\ i_D \\ i_Q \end{bmatrix} \quad (2.23a)$$

or

$$\begin{bmatrix} \psi_S(\theta) \\ \psi_R(\theta) \end{bmatrix} = \begin{bmatrix} [L_{SS}(\theta)] & [L_{SR}(\theta)] \\ [L_{SR}(\theta)]^T & [L_{RR}] \end{bmatrix} \cdot \begin{bmatrix} [i_S] \\ [i_R] \end{bmatrix} \tag{2.23b}$$

where θ is the rotor angle position, defined by convention as the angle between rotor d-axis and axis of phase a.

Concluding, it results that the inductances matrices from equations (2.23) are

$$[L_{SS}] = \begin{bmatrix} L_{aa0} + L_{aa2}\cos(2\theta) & -L_{ab0} - L_{ab2}\cos\left(2\theta + \frac{2\pi}{3}\right) & -L_{ab0} - L_{ab2}\cos\left(2\theta - \frac{\pi}{3}\right) \\ -L_{ab0} - L_{ab2}\cos\left(2\theta + \frac{2\pi}{3}\right) & L_{aa0} + L_{aa2}\cos\left(2\theta - \frac{2\pi}{3}\right) & -L_{ab0} - L_{ab2}\cos(2\theta - \pi) \\ -L_{ab0} - L_{ab2}\cos\left(2\theta - \frac{\pi}{3}\right) & -L_{ab0} - L_{ab2}\cos(2\theta - \pi) & L_{aa0} + L_{aa2}\cos\left(2\theta + \frac{2\pi}{3}\right) \end{bmatrix}$$

$$[L_{SR}] = [L_{SR}]^T = \begin{bmatrix} L_{md}\cos\theta & L_{mD}\cos\theta & -L_{mQ}\sin\theta \\ L_{md}\cos\left(\theta - \frac{2\pi}{3}\right) & L_{mD}\cos\left(\theta - \frac{2\pi}{3}\right) & -L_{mQ}\sin\left(\theta - \frac{2\pi}{3}\right) \\ L_{md}\cos\left(\theta + \frac{2\pi}{3}\right) & L_{mD}\cos\left(\theta + \frac{2\pi}{3}\right) & -L_{mQ}\sin\left(2\theta + \frac{2\pi}{3}\right) \end{bmatrix}$$

$$[L_{RR}] = \begin{bmatrix} L_{ff} & L_{fD} & 0 \\ L_{fd} & L_{DD} & 0 \\ 0 & 0 & L_{QQ} \end{bmatrix}$$

In establishing the terms of inductances matrices equations, the following equalities were taken into account: $L_{aa0} = L_{bb0} = L_{cc0}$; $L_{ab2} = L_{aa2}$, and $L_{ab0} \cong L_{aa0}/2$.

Note that the inductance matrices $L_{SS}(\theta)$ and $L_{SR}(\theta)$ depend on the rotor position, that is θ, whereas the terms of matrix L_{RR} are constant. Denoting by ω_r, the rotor speed then $\theta = \omega_r t$ resulting that $L_{SS}(\theta)$ and $L_{SR}(\theta)$ are periodically time variable functions.

Using the matrix equation (2.23b) for studying the synchronous generator operation leads to a complex calculus. This justifies the approach of new variables more adequate comparative to the stator phase quantities: i_a, i_b, i_c, v_a, v_b, v_c, and so on. This indispensable transformation is given by Park transformation.

2.1.3.2 Park Transformation. The transformation proposed by Park is based on the two-reaction theory originally developed by Blondel [3] and the further exposition of the concept by Doherty and Nickle [4].

The Park transformation, proposed initially in 1929 [5], consists in replacing the three stator windings a, b, and c by three fictitious rotor windings called d, q, 0 (rotor coordinates system) (Figure 2.10):

- The windings disposed in the d-axis and q-axis are rotating together with the rotor, and there are no mutual couplings between them.
- The 0-axis is independent of the d-axis and q-axis since zero-sequence currents flow through it only in unbalanced conditions.

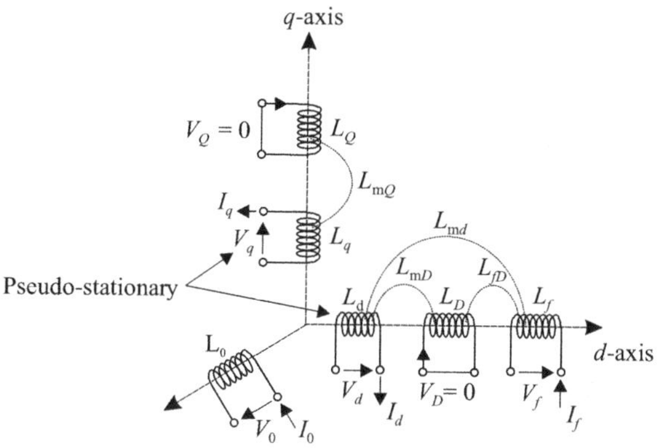

Figure 2.10. Representation of the synchronous generator windings after Park transformation.

The main advantage of the Park transformation is that all windings from Figure 2.10 are stationary to each other, thus self- and mutual inductances are constant in time. This observation helps considering more simple equations using d, q, 0 components instead of phase quantities a, b, c.

Considering the flux equations (2.21) of the rotor circuit windings

$$\psi_f = L_f i_f + L_{fD} i_D - L_{md}\left[i_a \cos\theta + i_b \cos\left(\theta - \frac{2\pi}{3}\right) + i_c \cos\left(\theta + \frac{2\pi}{3}\right)\right]$$

$$\psi_D = L_{Df} i_f + L_{DD} i_D - L_{mD}\left[i_a \cos\theta + i_b \cos\left(\theta - \frac{2\pi}{3}\right) + i_c \cos\left(\theta + \frac{2\pi}{3}\right)\right]$$

$$\psi_Q = L_{QQ} i_Q + L_{mQ}\left[i_a \sin\theta + i_b \sin\left(\theta - \frac{2\pi}{3}\right) + i_c \sin\left(\theta + \frac{2\pi}{3}\right)\right]$$

it can be seen that the stator currents are similarly combined in each axis. This suggests the transformation of the stator phase currents into new state variables. Therefore, the following relations for the currents in d, q, 0 axes are proposed [1]:

$$I_d = k_d\left[i_a \cos\theta + i_b \cos\left(\theta - \frac{2\pi}{3}\right) + i_c \cos\left(\theta + \frac{2\pi}{3}\right)\right] \tag{2.24a}$$

$$I_q = -k_q\left[i_a \sin\theta + i_b \sin\left(\theta - \frac{2\pi}{3}\right) + i_c \sin\left(\theta + \frac{2\pi}{3}\right)\right] \tag{2.24b}$$

The terms k_d and k_q are chosen so as to simplify the expressions of related equations. Usually, they are taken equal to 2/3, although $\sqrt{2/3}$ can also lead to a simple form.

In balanced conditions, the expressions of the stator currents are

$$\begin{aligned} i_a &= I_m \sin(\omega_s t) \\ i_b &= I_m \sin\left(\omega_s t - \frac{2\pi}{3}\right) \\ i_c &= I_m \sin\left(\omega_s t + \frac{2\pi}{3}\right) \end{aligned} \tag{2.25}$$

where $\omega_s = 2\pi f$ is the angular velocity for stator currents, in electrical radians/second (el. rad/s) and I_m is the peak value of the stator current.

Substituting equations (2.25) in (2.24a) gives

$$I_d = k_d\left[I_m \sin(\omega_s t)\cos\theta + I_m \sin\left(\omega_s t - \frac{2\pi}{3}\right)\cos\left(\theta - \frac{2\pi}{3}\right) + I_m \sin\left(\omega_s t + \frac{2\pi}{3}\right)\cos\left(\theta + \frac{2\pi}{3}\right)\right]$$
$$= k_d I_m\left\{\sin(\omega_s t)\cos\theta + \left[\sin(\omega_s t)\cos\frac{2\pi}{3} - \sin\frac{2\pi}{3}\cos(\omega_s t)\right]\cdot\left[\cos\theta\cos\frac{2\pi}{3} + \sin\theta\sin\frac{2\pi}{3}\right]\right.$$
$$\left. + \left[\sin(\omega_s t)\cos\frac{2\pi}{3} + \cos(\omega_s t)\sin\frac{2\pi}{3}\right]\cdot\left[\cos\theta\cos\frac{2\pi}{3} - \sin\theta\sin\frac{2\pi}{3}\right]\right\}$$

After appropriate calculations and reduction of equation terms, it results

$$I_d = k_d I_m\left\{\sin(\omega_s t)\cos\theta\left[1 + 2\cos^2\left(\frac{2\pi}{3}\right)\right] - 2\cos(\omega_s t)\sin\theta\sin^2\left(\frac{2\pi}{3}\right)\right\}$$
$$= \frac{3}{2}k_d I_m[\sin(\omega_s t)\cos\theta - \cos(\omega_s t)\sin\theta]$$

or

$$I_d = \frac{3}{2}k_d I_m \sin(\omega_s t - \theta) \tag{2.26a}$$

Choosing k_d equal to 2/3, it results equal peak values for the currents I_d and I_m. From equation (2.24b), under balanced load conditions, it results

$$I_q = -k_q\frac{3}{2}I_m\cos(\omega_s t - \theta) \tag{2.26b}$$

The best choice is again $k_q = 2/3$ for which the peak value of the q-axis current I_q is equal to the peak value of the stator current I_m.

The two currents, I_d and I_q, together produce a field identical to that produced by the original phase currents, i_a, i_b, and i_c. Therefore, the third current I_0 should have no contribution to the mmf produced in the air gap. The current I_0 is the zero-sequence current adopted from the theory of symmetrical components:

$$I_0 = \frac{1}{3}(i_a + i_b + i_c)$$

Under balanced conditions we have $i_a + i_b + i_c = 0$, which gives

$$I_0 = 0 \tag{2.26c}$$

Note that I_0 is the instantaneous value of current, which may vary with the time in any manner.

Park transformation from the a, b, c system to the d, q, 0 system gives

$$\begin{bmatrix} I_d \\ I_q \\ I_0 \end{bmatrix} = \frac{2}{3}\begin{bmatrix} \cos\theta & \cos\left(\theta - \frac{2\pi}{3}\right) & \cos\left(\theta + \frac{2\pi}{3}\right) \\ -\sin\theta & -\sin\left(\theta - \frac{2\pi}{3}\right) & -\sin\left(\theta + \frac{2\pi}{3}\right) \\ \frac{1}{2} & \frac{1}{2} & \frac{1}{2} \end{bmatrix}\cdot\begin{bmatrix} i_a \\ i_b \\ i_c \end{bmatrix} = [P]\begin{bmatrix} i_a \\ i_b \\ i_c \end{bmatrix} \tag{2.27}$$

The inverse transformation from d, q, 0 to a, b, c variables is given by

$$\begin{bmatrix} i_a \\ i_b \\ i_c \end{bmatrix} = \begin{bmatrix} \cos\theta & -\sin\theta & 1 \\ \cos\left(\theta - \frac{2\pi}{3}\right) & -\sin\left(\theta - \frac{2\pi}{3}\right) & 1 \\ \cos\left(\theta + \frac{2\pi}{3}\right) & -\sin\left(\theta + \frac{2\pi}{3}\right) & 1 \end{bmatrix} \cdot \begin{bmatrix} I_d \\ I_q \\ I_0 \end{bmatrix} = [P]^{-1} \begin{bmatrix} I_d \\ I_q \\ I_0 \end{bmatrix} \tag{2.27'}$$

Park has chosen the coefficients equal to 2/3 and 1/2 (for zero-sequence component) in order to simplify the expressions. Note that when using these values, the transformation matrix $[P]$ is not orthogonal, that is, $[P]^{-1} \neq [P]^{\mathrm{T}}$.

***2.1.3.3 Park Equations of Synchronous Generator*[4].** STATOR FLUX LINKAGES IN d, q, 0 COMPONENTS. Stator and rotor flux linkage equations can be written under matrix form:

$$\begin{bmatrix} [\psi_{\mathrm{S}}] \\ [\psi_{\mathrm{R}}] \end{bmatrix} = \begin{bmatrix} [L_{\mathrm{SS}}] & [L_{\mathrm{SR}}] \\ [L_{\mathrm{SR}}] & [L_{\mathrm{RR}}] \end{bmatrix} \cdot \begin{bmatrix} [i_{\mathrm{S}}] \\ [i_{\mathrm{R}}] \end{bmatrix} \tag{2.23b}$$

Applying the inverse transformation only for the stator currents in the right member of equation (2.23b), we obtain

$$\begin{bmatrix} [\psi_{\mathrm{S}}] \\ [\psi_{\mathrm{R}}] \end{bmatrix} = \begin{bmatrix} [L_{\mathrm{SS}}] & [L_{\mathrm{SR}}] \\ [L_{\mathrm{RS}}] & [L_{\mathrm{RR}}] \end{bmatrix} \cdot \begin{bmatrix} [P]^{-1} & 0 \\ 0 & [1] \end{bmatrix} \cdot \begin{bmatrix} [I_{\mathrm{P}}] \\ [i_{\mathrm{R}}] \end{bmatrix} \tag{2.28}$$

where

$$[I_{\mathrm{P}}]^{\mathrm{T}} = [I_d \ I_q \ I_0]$$

Appling the Park transformation for stator flux linkages, it results

$$\begin{bmatrix} [\psi_{\mathrm{P}}] \\ [\psi_{\mathrm{R}}] \end{bmatrix} = \begin{bmatrix} [P] & 0 \\ 0 & [1] \end{bmatrix} \cdot \begin{bmatrix} [\psi_{\mathrm{S}}] \\ [\psi_{\mathrm{R}}] \end{bmatrix} \tag{2.29}$$

where

$$[\psi_{\mathrm{P}}]^{\mathrm{T}} = [\psi_d \ \psi_q \ \psi_0]$$

From equations (2.29) and (2.28), it results

$$\begin{bmatrix} [\psi_{\mathrm{P}}] \\ [\psi_{\mathrm{R}}] \end{bmatrix} = \begin{bmatrix} [P] & 0 \\ 0 & [1] \end{bmatrix} \cdot \begin{bmatrix} [\psi_{\mathrm{S}}] \\ [\psi_{\mathrm{R}}] \end{bmatrix} = \begin{bmatrix} [P] & 0 \\ 0 & [1] \end{bmatrix} \cdot \begin{bmatrix} [L_{\mathrm{SS}}] & [L_{\mathrm{SR}}] \\ [L_{\mathrm{RS}}] & [L_{\mathrm{RR}}] \end{bmatrix} \cdot \begin{bmatrix} [P]^{-1} & 0 \\ 0 & [1] \end{bmatrix} \cdot \begin{bmatrix} [I_{\mathrm{P}}] \\ [i_{\mathrm{R}}] \end{bmatrix}$$

$$= \begin{bmatrix} [P] \cdot [L_{\mathrm{SS}}] & [P] \cdot [L_{\mathrm{SR}}] \\ [L_{\mathrm{RS}}] & [L_{\mathrm{RR}}] \end{bmatrix} \cdot \begin{bmatrix} [P]^{-1} & 0 \\ 0 & [1] \end{bmatrix} \cdot \begin{bmatrix} [I_{\mathrm{P}}] \\ [i_{\mathrm{R}}] \end{bmatrix}$$

[4] Note that although lower cases are used to designate instantaneous values, upper cases are used in the following in order to represent quantities obtained by Park transformation.

then
$$\begin{bmatrix}[\psi_P]\\[\psi_R]\end{bmatrix}=\begin{bmatrix}[P]\cdot[L_{SS}]\cdot[P]^{-1} & [P]\cdot[L_{SR}]\\ [L_{RS}]\cdot[P]^{-1} & [L_{RR}]\end{bmatrix}\cdot\begin{bmatrix}[I_P]\\[i_R]\end{bmatrix}=\begin{bmatrix}[L'_{SS}] & [L'_{SR}]\\ [L'_{RS}] & [L_{RR}]\end{bmatrix}\cdot\begin{bmatrix}[I_P]\\[i_R]\end{bmatrix} \tag{2.30}$$

After appropriate substitutions and calculations (including trigonometric reductions), it results

$$\begin{aligned}\psi_d&=-\left(L_{aa0}+L_{ab0}+\frac{3}{2}L_{aa2}\right)I_d+L_{md}I_f+L_{mD}I_D\\ \psi_q&=-\left(L_{aa0}+L_{ab0}-\frac{3}{2}L_{aa2}\right)I_q+L_{mQ}I_Q\\ \psi_0&=-(L_{aa0}-2L_{ab0})I_0\end{aligned} \tag{2.30$'$}$$

Note that although the rotor currents do not suffer any modification, for simplification of notations upper cases I_f, I_D, I_Q have been used instead of i_f, i_D, i_Q.

Defining new inductances

$$\begin{aligned}L_d&=L_{aa0}+L_{ab0}+\frac{3}{2}L_{aa2}\\ L_q&=L_{aa0}+L_{ab0}-\frac{3}{2}L_{aa2}\\ L_0&=L_{aa0}-2L_{ab0}\end{aligned} \tag{2.31}$$

or

$$[L'_{SS}]=\begin{bmatrix}L_d & 0 & 0\\ 0 & L_q & 0\\ 0 & 0 & L_0\end{bmatrix} \tag{2.31$'$}$$

then flux linkages equations in d, q, 0 coordinates (2.30′) become

$$\psi_d=-L_dI_d+L_{md}I_f+L_{mD}I_D \tag{2.32a}$$

$$\psi_q=-L_qI_q+L_{mQ}I_Q \tag{2.32b}$$

$$\psi_0=-L_0I_0 \tag{2.32c}$$

It results also that

$$[L'_{SR}]=[P]\cdot[L_{SR}]=\begin{bmatrix}L_{md} & L_{mD} & 0\\ 0 & 0 & L_{mQ}\\ 0 & 0 & 0\end{bmatrix} \tag{2.33}$$

$$[L'_{RS}]=[L_{RS}]\cdot[P]^{-1}=\begin{bmatrix}\frac{3}{2}L_{md} & 0 & 0\\ \frac{3}{2}L_{mD} & 0 & 0\\ 0 & \frac{3}{2}L_{mQ} & 0\end{bmatrix} \tag{2.33$'$}$$

The d, q, 0 components of the stator flux linkages are seen to be related to the components of stator and rotor currents through constant inductances, which are the main benefit of the Park transformation.

ROTOR FLUX LINKAGES IN d, q, 0 COMPONENTS. Substituting the currents I_d and I_q from expressions (2.24) in equations (2.22) of rotor flux linkages, it results

$$\psi_f = L_f I_f + L_{fD} I_D - \frac{3}{2} L_{\mathrm{md}} I_d \tag{2.34a}$$

$$\psi_D = L_{Df} I_f + L_{DD} I_D - \frac{3}{2} L_{\mathrm{m}D} I_d \tag{2.34b}$$

$$\psi_Q = L_{QQ} I_Q - \frac{3}{2} L_{\mathrm{m}Q} I_q \tag{2.34c}$$

The Park transformation has resulted in constant inductances in the stator and rotor linkage equations, that is, they do not change with the rotor position. However, the mutual inductances between the stator and rotor quantities are not reciprocal. This can be seen in equations (2.34a) and (2.32a), where the mutual inductance associated with the flux linking the field winding due to current I_d flowing in the d-axis stator winding includes the term 2/3, whereas the mutual inductance associated with flux linking the d-axis stator winding due to field current I_f does not include the term 2/3. By appropriate choice of the per unit system for the rotor quantities this issue can be avoided (see Section 2.1.3.4).

Note that the saturation effects are neglected. Furthermore, as shown earlier, the component I_0 does not appear in the rotor flux linkage equations since it produces no mmf in the air gap.

STATOR VOLTAGE EQUATIONS IN d, q, 0 COMPONENTS. Consider the stator voltage equations (2.13a), (2.13b), and (2.13c) of the three phases, in matrix form:

$$[v_{\mathrm{S}}] = -R \cdot [i_{\mathrm{S}}] + \frac{\mathrm{d}}{\mathrm{d}t} [\psi_{\mathrm{S}}] \tag{2.14}$$

where $R_a = R_b = R_c = R$ represents the resistance of the stator winding.

Appling the inverse Park transformation, it results

$$[P]^{-1} [V_{\mathrm{P}}] = -R [P]^{-1} [I_{\mathrm{P}}] + \frac{\mathrm{d}}{\mathrm{d}t} \left([P]^{-1} [\psi_{\mathrm{P}}] \right) \tag{2.35}$$

Because $[P]^{-1}$ is a function of angle θ, the last term of the right-hand side can be expressed under the form

$$\frac{\mathrm{d}}{\mathrm{d}t} \left([P]^{-1} [\psi_{\mathrm{P}}] \right) = [P]^{-1} \frac{\mathrm{d}}{\mathrm{d}t} [\psi_{\mathrm{P}}] + \frac{\mathrm{d}\theta}{\mathrm{d}t} \left(\frac{\mathrm{d}}{\mathrm{d}\theta} [P]^{-1} \right) [\psi_{\mathrm{P}}] \tag{2.36}$$

where

$$\frac{\mathrm{d}[P]^{-1}}{\mathrm{d}\theta}=\frac{\mathrm{d}}{\mathrm{d}\theta}\begin{bmatrix}\cos\theta & -\sin\theta & 1\\ \cos\left(\theta-\frac{2\pi}{3}\right) & -\sin\left(\theta-\frac{2\pi}{3}\right) & 1\\ \cos\left(\theta+\frac{2\pi}{3}\right) & -\sin\left(\theta+\frac{2\pi}{3}\right) & 1\end{bmatrix}=\begin{bmatrix}-\sin\theta & -\cos\theta & 0\\ -\sin\left(\theta-\frac{2\pi}{3}\right) & -\cos\left(\theta-\frac{2\pi}{3}\right) & 0\\ -\sin\left(\theta+\frac{2\pi}{3}\right) & -\cos\left(\theta+\frac{2\pi}{3}\right) & 0\end{bmatrix}$$

$$=\begin{bmatrix}\cos\theta & -\sin\theta & 1\\ \cos\left(\theta-\frac{2\pi}{3}\right) & -\sin\left(\theta-\frac{2\pi}{3}\right) & 1\\ \cos\left(\theta+\frac{2\pi}{3}\right) & -\sin\left(\theta+\frac{2\pi}{3}\right) & 1\end{bmatrix}\cdot\underbrace{\begin{bmatrix}0 & -1 & 0\\ 1 & 0 & 0\\ 0 & 0 & 0\end{bmatrix}}_{[P_1]}=[P]^{-1}[P_1] \tag{2.37}$$

where $[P_1]$ is a 90° rotational operator in the d–q plan.

Taking into account equations (2.36) and (2.37), equation (2.35) can be written as

$$[P]^{-1}[V_\mathrm{P}]=-R[P]^{-1}[I_\mathrm{P}]+[P]^{-1}\frac{\mathrm{d}[\psi_\mathrm{P}]}{\mathrm{d}t}+\frac{\mathrm{d}\theta}{\mathrm{d}t}[P]^{-1}[P_1][\psi_\mathrm{P}]$$

Multiplying the last equation by $[P]$, we obtain

$$[V_\mathrm{P}]=-R[I_\mathrm{P}]+\frac{\mathrm{d}\theta}{\mathrm{d}t}[P_1][\psi_\mathrm{P}]+\frac{\mathrm{d}\psi_\mathrm{P}}{\mathrm{d}t} \tag{2.38}$$

where

$$[P_1][\psi_\mathrm{P}]=\begin{bmatrix}0 & -1 & 0\\ 1 & 0 & 0\\ 0 & 0 & 0\end{bmatrix}\begin{bmatrix}\psi_d\\ \psi_q\\ \psi_0\end{bmatrix}=\begin{bmatrix}-\psi_q\\ \psi_d\\ 0\end{bmatrix}$$

Developing equation (2.38), it results

$$V_d=-RI_d-\psi_q\frac{\mathrm{d}\theta}{\mathrm{d}t}+\frac{\mathrm{d}\psi_d}{\mathrm{d}t} \tag{2.39a}$$

$$V_q=-RI_q+\psi_d\frac{\mathrm{d}\theta}{\mathrm{d}t}+\frac{\mathrm{d}\psi_q}{\mathrm{d}t} \tag{2.39b}$$

$$V_0=-RI_0+\frac{\mathrm{d}\psi_0}{\mathrm{d}t} \tag{2.39c}$$

where V_d, V_q are the voltages induced in the d-axis and q-axis; I_d, I_q are the currents in the d-axis and q-axis; ψ_d, ψ_q, ψ_0 are the d, q, and 0 windings flux linkages; $\mathrm{d}\theta/\mathrm{d}t=\omega$ is the rotor angular velocity; for $f=50\,\mathrm{Hz}$, $p\theta=\omega=\omega_\mathrm{S}=314$ el. rad/s.

The following observations can be drawn:

- Equations (2.39) have a similar form to those of a static coil, except for the $\psi_q(\mathrm{d}\theta/\mathrm{d}t)$ and $\psi_d(\mathrm{d}\theta/\mathrm{d}t)$ terms. They are called *speed voltages* and are due to the flux change in space by the rotation of the field windings. The terms $\mathrm{d}\psi_d/\mathrm{d}t$ and $\mathrm{d}\psi_q/\mathrm{d}t$ are called *transformer voltages* and are due to the flux change in time.
- The transformer voltages can be, in general, neglected and thus the speed voltages are dominant terms in the stator voltage expressions; under these circumstances, it can be inferred that the voltage V_q in the q-axis is induced by the flux attached to the d-axis; this is because we have assumed that the q-axis leads the d-axis by 90°. Also, the voltage V_d in the d-axis is induced by a flux attached to an axis lagging the d-axis by 90°, that is, the negative q-axis [1].

Rotor Voltage Equations. The voltage equations of the rotor circuits, which remain unchanged by Park transformation, are

$$V_f = R_f I_f + \frac{\mathrm{d}\psi_f}{\mathrm{d}t} \tag{2.40a}$$

$$0 = R_D I_D + \frac{\mathrm{d}\psi_D}{\mathrm{d}t} \tag{2.40b}$$

$$0 = R_Q I_Q + \frac{\mathrm{d}\psi_Q}{\mathrm{d}t} \tag{2.40c}$$

where V_f is the field-winding voltage, R_f is the field-winding resistance, R_D, R_Q are the damper windings resistances.

Under balanced conditions the zero-sequence circuit can be neglected.

Regrouping, on one hand, the circuits d, f, and D, and, on the other hand, the circuits q and Q, the Park equations describing the synchronous machine operation can be written under the form

$$\begin{bmatrix} V_d \\ -V_f \\ 0 \end{bmatrix} = -\begin{bmatrix} R & 0 & 0 \\ 0 & R_f & 0 \\ 0 & 0 & R_D \end{bmatrix} \cdot \begin{bmatrix} I_d \\ I_f \\ I_D \end{bmatrix} - \begin{bmatrix} \psi_q \frac{\mathrm{d}\theta}{\mathrm{d}t} \\ 0 \\ 0 \end{bmatrix} + \begin{bmatrix} \frac{\mathrm{d}\psi_d}{\mathrm{d}t} \\ -\frac{\mathrm{d}\psi_f}{\mathrm{d}t} \\ \frac{\mathrm{d}\psi_D}{\mathrm{d}t} \end{bmatrix} \tag{2.41}$$

$$\begin{bmatrix} V_q \\ 0 \end{bmatrix} = \begin{bmatrix} R & 0 \\ 0 & R_Q \end{bmatrix} \cdot \begin{bmatrix} -I_q \\ I_Q \end{bmatrix} + \begin{bmatrix} \psi_d \frac{\mathrm{d}\theta}{\mathrm{d}t} \\ 0 \end{bmatrix} + \begin{bmatrix} \frac{\mathrm{d}\psi_q}{\mathrm{d}t} \\ \frac{\mathrm{d}\psi_Q}{\mathrm{d}t} \end{bmatrix} \tag{2.42}$$

with the relationships between flux linkages and currents:

$$\begin{bmatrix} \psi_d \\ \psi_f \\ \psi_D \end{bmatrix} = \begin{bmatrix} L_d & L_{\mathrm{md}} & L_{\mathrm{mD}} \\ \frac{3}{2} L_{\mathrm{md}} & L_f & L_{fD} \\ \frac{3}{2} L_{\mathrm{mD}} & L_{Df} & L_{DD} \end{bmatrix} \cdot \begin{bmatrix} -I_d \\ I_f \\ I_D \end{bmatrix} \tag{2.43}$$

$$\begin{bmatrix} \psi_q \\ \psi_Q \end{bmatrix} = \begin{bmatrix} L_q & L_{mQ} \\ \frac{3}{2}L_{mQ} & L_{QQ} \end{bmatrix} \cdot \begin{bmatrix} -I_q \\ I_Q \end{bmatrix} \tag{2.44}$$

ELECTRIC POWER AND TORQUE. The instantaneous three-phase power produced by the electrical machine through the stator is

$$P(t) = v_a i_a + v_b i_b + v_c i_c \tag{2.45}$$

Replacing the a, b, c components with d, q, 0 components gives

$$P(t) = \frac{3}{2}\left(V_d I_d + V_q I_q + 2V_0 I_0\right) \tag{2.45$'$}$$

Substitution of voltage components from equations (2.39) into equation (2.45′) gives

$$P(t) = \frac{3}{2}\left[\left(-RI_d - \omega_r\psi_q + \frac{d\psi_d}{dt}\right)I_d + \left(-RI_q + \omega_r\psi_d + \frac{d\psi_q}{dt}\right)I_q + 2\left(-RI_0 + \frac{d\psi_0}{dt}\right)I_0\right]$$

where the rotor speed is $\omega_r = d\theta/dt$.

Rearranging of terms, we obtain

$$P(t) = \frac{3}{2}\left[\underbrace{\left(I_d\frac{d\psi_d}{dt} + I_q\frac{d\psi_q}{dt} + 2I_0\frac{d\psi_0}{dt}\right)}_{\text{Rate of change of armature magnetic energy}} + \underbrace{(\psi_d I_q - \psi_q I_d)\omega_r}_{\text{Power transferred accross the air gap}} - \underbrace{\left(I_d^2 + I_q^2 + 2I_0^2\right)R}_{\text{Armature resistance loss}}\right] \tag{2.45$''$}$$

The air gap torque C_e is obtained by dividing the power transferred across the air gap by the rotor speed expressed in mechanical radians per second:

$$C_e = \frac{3}{2}\left(\psi_d I_q - \psi_q I_d\right)\frac{\omega_r}{\omega_m} = \frac{3}{2}\left(\psi_d I_q - \psi_q I_d\right)\frac{p_f}{2} \tag{2.46}$$

TRANSIENT PROCESSES OR STATES IN SYNCHRONOUS GENERATOR. Synchronous generators are subjected to the following transient processes or states:

(i) *Stator transient states* are associated with the transformer voltages. After a sudden change in the system, the transformer voltages disappear and the speed voltages are dominant in the system response. For example, due to a short circuit the transformer voltages generate DC components of the stator phase currents, which attenuates in a fraction of a second, since the resistance is present, a period relatively short comparative to the time frame taken in stability studies. Consequently, a possible simplification consists in neglecting the transformer voltages in the stator equations.

(ii) *Rotor transient states* are associated with the terms $d\psi_f/dt$, $d\psi_D/dt$, and so on, of the rotor windings equations. Two dynamic behaviors are identified:

- *subtransient dynamic response*, associated with the damper windings and eddy currents;

- *transient dynamic response*, associated with the field winding;

(iii) *Mechanical states*, associated with the shaft motion.

2.1.3.4 Representation of Synchronous Generator Equations in Per Unit.

PER UNIT SYSTEM DEFINITION. Expressing the electrical quantities in the per unit system (p.u.) is usually preferred in the power system analysis in order to normalize the variables or the system quantities and to simplify the calculations by eliminating the actual system of units. Expressing the electrical quantities in per units help, in some cases, improving the convergence of computation algorithms.

A certain quantity X expressed in the international system of units (SI) is transformed into the per unit system, usually represented with the subscript *, by referring it to a base value X_b given also in the international system of units, that is,

$$X_*\,(\text{p.u.}) = \frac{X\,(\text{SI})}{X_b\,(\text{SI})}$$

When transforming the electrical quantities in per units, only a part of the base values can be chosen independently, the other ones being determined using the existing relationship between the power system quantities. Usually, the base values are chosen so that the main power system quantities, usually the voltages, expressed in p.u., are equal to 1 for rated conditions.

In the case of synchronous generators, per unit system is used to eliminate some arbitrary constants and simplify the mathematical equations, so that equivalent circuits (ECs) be easier to obtained. In this respect, the base values are chosen in a way to satisfy the following [1]:

a. The per unit mutual inductances between different windings are to be reciprocal (e.g., $L_{af*} = L_{fa*}$). This hypothesis allows the synchronous generator model to be represented by equivalent circuits.
b. All the per unit inductances between rotor and stator circuits in each axis are to be equal (e.g., $L_{af*} = L_{aD*}$).

Per Unit System for the Stator Quantities. The base values of the stator quantities, indicated with the subscript sb, are defined as follows:

(i) The independent base quantities are

- base voltage (V_{sb})—peak phase-to-neutral rated voltage, in V;
- base current (I_{sb})—peak rated current, in A;
- base frequency (f_b)—rated frequency in Hz (it is unique in the system and thus the subscript s is omitted).

(ii) The remaining base values are determined as follows:

- Base three-phase power (S_{sb}) is taken equal to the rated three-phase apparent power of the generator, in VA:

$$S_{sb} = S_{ng} = 3V_{sn}I_{sn} = 3\frac{V_{sb}}{\sqrt{2}}\cdot\frac{I_{sb}}{\sqrt{2}} = \frac{3}{2}V_{sb}I_{sb}$$

where V_{sn} and I_{sn} are the rotor voltage and current of the generator.

- Base impedance (Z_{sb}):

$$Z_{sb} = \frac{V_{sb}}{I_{sb}}$$

- Base angular speed (ω_b):

$$\omega_b = 2\pi f_b (\text{in el. rad/s})$$

respectively, mechanical angular base speed (ω_{mb}):

$$\omega_{mb} = \frac{1}{p}\omega_b (\text{in mechanical rad/s})$$

(p being the number of pair of poles).

- Base inductance (L_{sb}):

$$L_{sb} = \frac{Z_{sb}}{\omega_{sb}} (\text{in H})$$

- Base flux (ψ_{sb}):

$$\psi_{sb} = L_{sb} I_{sb} = \frac{V_{sb}}{\omega_{sb}} (\text{in Wb})$$

- Base torque (C_{sb}):

$$C_{sb} = \frac{S_{sb}}{\omega_{sb}} = \frac{3}{2} \cdot \frac{V_{sb} I_{sb}}{\omega_{sb}} \cdot p = \frac{3}{2} \cdot p \cdot \psi_{sb} \cdot I_{sb} (\text{in N} \times \text{m})$$

- Base time (t_b) is defined as the time interval required for the rotor to perform a rotation of 1 el. rad at synchronous speed:

$$t_b = \frac{1}{\omega_b} = \frac{1}{2\pi f_b}$$

- Time derivation operator:

$$\frac{d(\cdot)}{dt_*} = \frac{1}{\omega_b} \cdot \frac{d(\cdot)}{dt}$$

Per Unit System for Rotor Quantities. In order to represent the rotor voltages and the fluxes crossing the rotor in the per unit system, the base quantities attached to the field winding (*f*) are indicated with the subscript *f*b, while the quantities attached to the damper windings (*D*) and (*Q*) are indicated with the subscripts *D*b and *Q*b.

VOLTAGE EQUATIONS IN PER UNIT. In order to express *the stator voltage equations in per units*, equations (2.39) are divided to V_{sb} and the angular speed $d\theta/dt$ is replaced by ω_r.

For example, dividing equation (2.39a) to V_{sb},

$$V_d = -RI_d - \omega_r \psi_q + \frac{d\psi_d}{dt} \tag{2.39a}$$

and taking into account that

$$V_{\mathrm{sb}} = I_{\mathrm{sb}} Z_{\mathrm{sb}} = \frac{\psi_{\mathrm{sb}}}{L_{\mathrm{sb}}} \omega_{\mathrm{b}} L_{\mathrm{sb}} = \omega_{\mathrm{b}} \psi_{\mathrm{sb}}$$

it results

$$\frac{V_{\mathrm{d}}}{V_{\mathrm{sb}}} = -\frac{R}{Z_{\mathrm{sb}}} \cdot \frac{I_{\mathrm{d}}}{I_{\mathrm{sb}}} - \frac{\omega_{\mathrm{r}}}{\omega_{\mathrm{b}}} \cdot \frac{\psi_{\mathrm{q}}}{\psi_{\mathrm{sb}}} + \frac{\mathrm{d}}{\mathrm{d}t}\left(\frac{1}{\omega_{\mathrm{b}}} \cdot \frac{\psi_{\mathrm{d}}}{\psi_{\mathrm{sb}}}\right)$$

or, using the per unit system notations

$$V_{d*} = -R_* I_{d*} - \omega_{\mathrm{r}*} \psi_{q*} + \frac{\mathrm{d}\psi_{d*}}{\mathrm{d}t_*} \quad (2.39'\mathrm{a})$$

Similarly

$$V_{q*} = -R_* I_{q*} + \omega_{\mathrm{s}*} \psi_{d*} + \frac{\mathrm{d}\psi_{q*}}{\mathrm{d}t_*} \quad (2.39'\mathrm{b})$$

$$V_{0*} = -R_* I_{0*} + \frac{\mathrm{d}\psi_{0*}}{\mathrm{d}t_*} \quad (2.39'\mathrm{c})$$

Note that if all quantities are expressed in p.u., the stator equations (2.39′a), (2.39′b), and (2.39′c) holds the original form.

In the same manner, *the rotor voltage equations* can also be expressed in p.u. In this respect, taking into account that $V_{f\mathrm{b}} = \omega_{\mathrm{b}} \psi_{f\mathrm{b}} = Z_{f\mathrm{b}} I_{f\mathrm{b}}$, by a similar calculation as the previously used one to deduce equation (2.39′a), from (2.40a) it results

$$V_{f*} = R_{f*} I_{f*} + \frac{\mathrm{d}\psi_{f*}}{\mathrm{d}t_*} \quad (2.40'\mathrm{a})$$

Expressing equations (2.40b) and (2.40c) in p.u., we obtain

$$0 = R_{D*} I_{D*} + \frac{\mathrm{d}\psi_{D*}}{\mathrm{d}t_*} \quad (2.40'\mathrm{b})$$

$$0 = R_{Q*} I_{Q*} + \frac{\mathrm{d}\psi_{Q*}}{\mathrm{d}t_*} \quad (2.40'\mathrm{c})$$

Flux Linkage Equations in Per Unit. The stator flux linkages are expressed in per units by dividing all terms of equation (2.32) to the base flux $\psi_{\mathrm{sb}} = L_{\mathrm{sb}} I_{\mathrm{sb}}$, which gives

$$\psi_{d*} = -L_{d*} I_{d*} + L_{\mathrm{md}*} I_{f*} + L_{\mathrm{m}D*} I_{D*} \quad (2.32'\mathrm{a})$$

$$\psi_{q*} = -L_{q*} I_{q*} + L_{\mathrm{m}Q*} I_{Q*} \quad (2.32'\mathrm{b})$$

$$\psi_{0*} = -L_{0*} I_{0*} \quad (2.32'\mathrm{c})$$

where

$$L_{\mathrm{md}*} = \frac{L_{\mathrm{md}} I_{f\mathrm{b}}}{L_{\mathrm{sb}} I_{\mathrm{sb}}}; \; L_{\mathrm{m}D*} = \frac{L_{\mathrm{m}D} I_{D\mathrm{b}}}{L_{\mathrm{sb}} I_{\mathrm{sb}}}; \; L_{\mathrm{m}Q*} = \frac{L_{\mathrm{m}Q} I_{Q\mathrm{b}}}{L_{\mathrm{sb}} I_{\mathrm{sb}}} \quad (2.47)$$

The rotor flux linkages from equations (2.34) can also be expressed in per units if all terms of each rotor flux are divided to appropriate base flux. Thus,

$$\psi_{f*} = L_{f*}I_{f*} + L_{fD*}I_{D*} - L'_{md*}I_{d*} \quad (2.34'a)$$

$$\psi_{D*} = L_{Df*}I_{f*} + L_{D*}I_{D*} - L'_{mD*}I_{d*} \quad (2.34'b)$$

$$\psi_{Q*} = L_{Q*}I_{Q*} - L'_{mQ}I_{q*} \quad (2.34'c)$$

where

$$L_{fD*} = \frac{L_{fD}I_{Db}}{L_{fb}I_{fb}}; \quad L'_{md*} = \frac{3}{2}\cdot\frac{L_{md}I_{sb}}{L_{fb}I_{fb}}; \quad L_{Df*} = \frac{L_{Df}I_{fb}}{L_{Db}I_{Db}}$$
$$L'_{mD*} = \frac{3}{2}\cdot\frac{L_{mD}I_{sb}}{L_{Db}I_{Db}}; \quad L'_{mQ*} = \frac{3}{2}\cdot\frac{L_{mQ}I_{sb}}{L_{Qb}I_{Qb}} \quad (2.48)$$

Let us now consider that the mutual inductances between two rotor windings are reciprocal. For instance for $L_{fD*} = L_{Df*}$ from (2.48), we achieve

$$\frac{L_{fD}I_{Db}}{L_{fb}I_{fb}} = \frac{L_{Df}I_{fb}}{L_{Ds}I_{Db}}$$

Using also the equality $L_{fD} = L_{Df}$, we obtain

$$L_{Db}I_{Db}^2 = L_{fb}I_{fb}^2$$

Multiplying both terms of the last expression by ω_b and taking into account that $\omega_b L_{Db} I_{Db} = V_{Db}$ and $\omega_b L_{fb} I_{fb} = V_{fb}$, it results

$$V_{Db}I_{Db} = V_{fb}I_{fb}$$

Therefore, the reciprocity of mutual inductances is valid only if the last equality holds.

Moreover, assuming that the stator and rotor mutual inductances on each axis are equal, that is, $L_{md*} = L'_{md*}$, we obtain

$$\frac{L_{md}I_{fb}}{L_{sb}I_{sb}} = \frac{3}{2}\frac{L_{md}I_{sb}}{L_{fb}I_{fb}} \quad \Rightarrow \quad L_{fb}I_{fb}^2 = \frac{3}{2}L_{sb}I_{sb}^2$$

Multiplying both members by ω_b and taking account of $\omega_b L_{sb} I_{sb} = V_{sb}$, it results

$$V_{fb}I_{fb} = \frac{3}{2}V_{sb}I_{sb} = S_{sb}$$

Therefore, the base apparent power for the rotor circuits must be equal to the base apparent power for the stator circuits and equal to the rated apparent power of the synchronous generator. Thus, the product of base voltage and base current attached to each rotor winding is uniquely defined by this condition.

Therefore, in order to define a complete per unit system it is necessary to know either the base voltage or the base current for the rotor circuits. The stator self-inductances L_{d*} and L_{q*}, associated with the fluxes produced by the currents I_d and I_q, respectively, are the leakage inductance (L_l) due to flux that does not link any rotor circuit and the mutual inductances due to flux that links the rotor circuits in the d-axis (L_{ad}), and q-axis (L_{aq}), that is,

$$L_{d*} = L_{l*} + L_{ad*}; \quad L_{q*} = L_{l*} + L_{aq*}$$

One way of defining the base current in the rotor circuits is to consider the equality between mutual inductances of the windings disposed on each axis, expressed in per unit [1].

Thus, as stated earlier $L_{ad*} = L_{md*} = L_{mD*}$, which gives

$$I_{fb} = \frac{L_{ad}}{L_{md}} I_{sb}; \quad I_{Db} = \frac{L_{ad}}{L_{mD}} I_{sb}$$

whereas considering $L_{aq*} = L_{mQ*}$, we obtain

$$L_{Qb} = \frac{L_{aq}}{L_{mQ}} I_{sb}$$

so that the per unit system is completely defined.

This per unit system is known in literature as L_{ad} or X_{ad} base reciprocal per unit system, because the base current in any rotor circuit is defined as that which induces in each phase a per unit voltage equal to per unit L_{ad*}, that is, the same voltage as balanced three-phase unit-peak armature currents [1].

Overview of Electrical Equations in p.u.: The dqfDQK Model[5]. Besides the d, q, and f circuits, the complete set of equations assumes two amortisseur circuits in the q-axis, denoted by Q and K, and one amortisseur circuit in the d-axis, denoted by D. Therefore, the per unit equations are as follows:

- *Stator Voltage Equations.*

$$V_d = -RI_d - \omega_r \psi_q + \frac{d\psi_d}{dt} \tag{2.39'a}$$

$$V_q = -RI_q + \omega_r \psi_d + \frac{d\psi_q}{dt} \tag{2.39'b}$$

$$V_0 = -RI_0 + \frac{d\psi_0}{dt} \tag{2.39'c}$$

- *Rotor Voltage Equations.*

$$V_f = R_f I_f + \frac{d\psi_f}{dt} \tag{2.40'a}$$

[5] Although all quantities are expressed in per unit, for simplicity, we next drop the notation $*$.

$$0 = R_D I_D + \frac{d\psi_D}{dt} \tag{2.40'b}$$

$$0 = R_Q I_Q + \frac{d\psi_Q}{dt} \tag{2.40'c}$$

$$0 = R_K I_K + \frac{d\psi_K}{dt} \tag{2.49}$$

- *Stator Flux Linkage Equations.*

$$\psi_d = -(L_{md} + L_l)I_d + L_{md}I_f + L_{md}I_D \tag{2.50}$$

$$\psi_q = -(L_{mq} + L_l)I_q + L_{mq}I_Q + L_{mq}I_K \tag{2.51}$$

$$\psi_0 = -L_0 I_0 \tag{2.52}$$

- *Rotor Flux Linkage Equations.*

$$\psi_f = L_f I_f + L_{fD}I_D - L'_{md}I_d \tag{2.53}$$

$$\psi_D = L_{Df}I_f + L_{1D}I_D - L'_{mD}I_d \tag{2.54}$$

$$\psi_Q = L_{1Q}I_Q + L_{mq}I_K - L_{mq}I_q \tag{2.55a}$$

$$\psi_K = L_{mq}I_Q + L_{2K}I_K - L_{mq}I_q \tag{2.55b}$$

Additional to the systems of equations presented earlier, we have also added here equations corresponding the winding K attached to the q-axis, that is, equation (2.49) to voltages, equation (2.55b) to flux linkages, as well as new terms to equations (2.51) and (2.55a).

Observations [1]

- Equations (2.55a) and (2.55b) hold true in the assumption that the per unit mutual inductances L_{QK} and L_{mq} are equal, which means that the q-axis stator and rotor circuits link the mutual flux L_{mq} only.
- For the sake of simplicity, expression in per unit both for the machine's quantities and for the electrical network' quantities is preferred in power system stability analysis. The exception is the time t, which is expressed in seconds, and the per unit term $d\psi/dt$ in equations (2.39′), (2.40′), and (2.49) is replaced by $(1/\omega_b) \cdot (d\psi/dt)$.
- If the base frequency is taken equal to the frequency of the stator quantities, the per unit reactance of a stator winding is equal to the per unit inductance since the angular velocity is equal to unity.

2.1.3.5 Equivalent Circuits for the *d*- and *q*-Axes.

Developing the equivalent circuits is a common practice for the synchronous generator representation although the generator performance can be directly evaluated using equations (2.39′), (2.40′), (2.49)–(2.55). Furthermore, the reciprocal per unit system is used.

The d-axis equivalent circuit can be drawn using the flux linkage equations for rotor and stator circuits in d-axis:

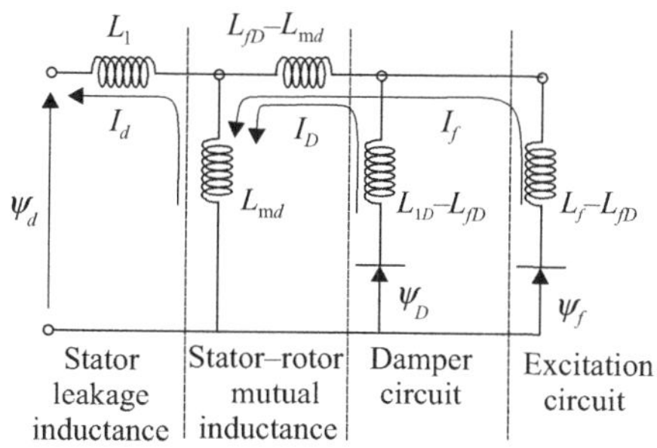

Figure 2.11. Equivalent circuit illustrating the relationship $\psi - I$ in the d-axis.

$$\psi_d = -(L_{md} + L_l)I_d + L_{md}I_f + L_{md}I_D \tag{2.50}$$

$$\psi_f = L_f I_f + L_{fD}I_D - L'_{md}I_d \tag{2.53}$$

$$\psi_D = L_{Df}I_f + L_{1D}I_D - L'_{md}I_d \tag{2.54}$$

where L_l and L_{md} are the leakage inductance and the mutual inductance, respectively, and $L_d = L_l + L_{md}$, knowing that in per unit $L_{md} = L_{mD} = L_{ad}$ and $L_{Df} = L_{fD}$. It may also be written that $L_q = L_l + L_{mq}$, where L_{mq} is the mutual inductance.

The equivalent circuit shown in Figure 2.11, developed using the above presented equations, illustrates the relationship between flux linkages and currents in the d-axis.

The relationships between flux linkages and currents in the q-axis are

$$\psi_Q = L_{1Q}I_Q + L_{mq}I_K - L_{mq}I_q \tag{2.55a}$$

$$\psi_K = L_{mq}I_Q + L_{2K}I_K - L_{mq}I_q \tag{2.55b}$$

which also helps in developing an equivalent circuit for the q-axis.

Introducing the rotor circuit per unit leakage inductances as given by

$$L_{fd} = L_f - L_{fD} \tag{2.56a}$$

$$L_D = L_{1D} - L_{fD} \tag{2.56b}$$

$$L_Q = L_{1Q} - L_{mq} \tag{2.56c}$$

$$L_K = L_{2K} - L_{mq} \tag{2.56d}$$

and the voltages and flux linkages, we obtain the circuits with complete characteristics as shown in Figure 2.12 [1].

The damper windings are placed in slots near the air gap and therefore the flux linking the damper circuit is almost the same with the flux linking the air gap. For this reason, the series inductance ($L_{fD} - L_{md}$) represented in the d-axis equivalent circuit is usually neglected (Figure 2.12b). However, this is not strictly valid for short-pitched damper circuits and solid rotor iron paths [1].

In the q-axis equivalent circuit, the field winding is not represented since the amortisseurs represent the overall effects in this axis. We may, thus, assume that the q-axis stator and rotor circuits link the mutual flux L_{mq} only (Figure 2.12b).

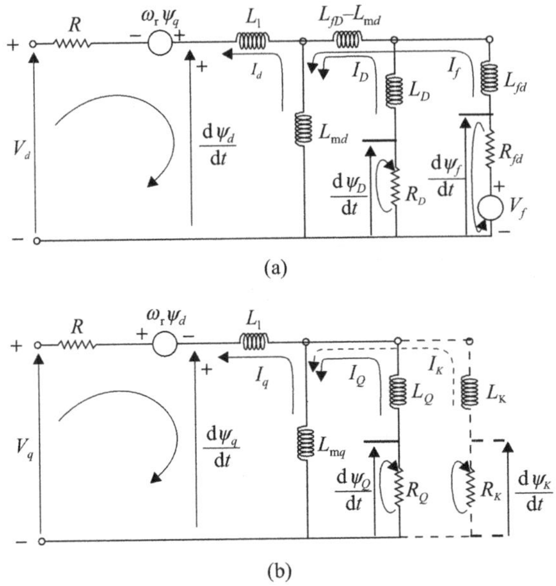

Figure 2.12. Complete *d*- and *q*-axis equivalent circuits: (a) *d*-axis equivalent circuit; (b) *q*-axis equivalent circuit.

The *d*- and *q*-axis equivalent circuits shown in Figure 2.13 do not include the stator resistance and the speed voltages.

Another version of the simplified equivalent circuits for the direct- and quadrature-axis model with a single damper winding from Figure 2.13a and b is given in IEEE Std. 1110-2002 [6] (Figure 2.14a–c).

The equivalent circuit from Figure 2.14a includes an ideal transformer representing the fact that there are different numbers of turns on the armature and field winding, just as in the case of the primary and secondary windings of a transformer. The variables V_f' and I_f' correspond to the actual values of field voltage and current that would be measured at the field-winding terminals. The variables V_f and I_f correspond to the values of field voltage and current reflected to the armature winding through the field in direct-axis armature winding turns ratio, N_{afd}.

It is common to represent synchronous machine using a per unit representation, rather than actual units, in which case an ideal transformer may or may not be required, depending on the choice of the base for the per unit system. Whether in actual units or in per unit, the ideal transformer is typically left out of the equivalent circuit, resulting in the equivalent circuit of Figure 2.14b, in which the field voltage and current are as reflected to the

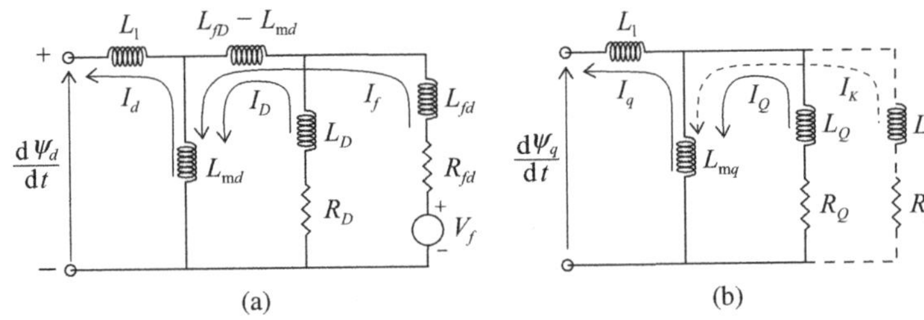

Figure 2.13. Simplified equivalent circuits: (a) *d*-axis equivalent circuit; (b) *q*-axis equivalent circuit [1].

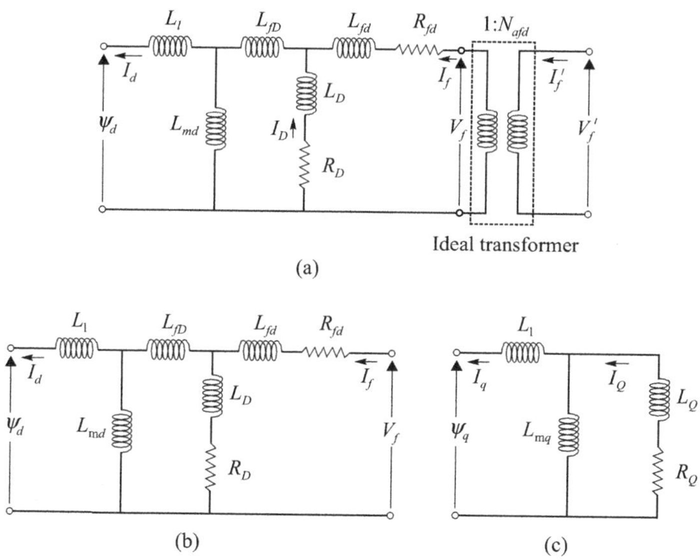

Figure 2.14. d-axis equivalent circuits with a single *d*-axis damper winding including an ideal transformer (a) and without ideal transformer (b); *q*-axis equivalent circuit with a single *q*-axis damper winding (c).

armature winding. The choice of the field-winding base values is equivalent to selecting a turns ratio in the equivalent circuit of Figure 2.14a.

2.1.3.6 Steady-State Operation of the Synchronous Generator. Application of Park transformation (d, q, 0) for stator currents under balanced conditions allows us to obtain direct-sequence currents; similar situation applies to stator voltages and flux linkages.

Under steady state, the rotor quantities are also constant, thus all time derivative terms are set to zero, whereas the zero-sequence components are absent and $\omega_r = \omega_s = 1$ p.u. From equations

$$0 = R_D I_D + \frac{d\psi_D}{dt} \tag{2.40$'$b}$$

$$0 = R_Q I_Q + \frac{d\psi_Q}{dt} \tag{2.40$'$c}$$

yield $R_D I_D = R_Q I_Q = 0$. Therefore, the damping currents I_D and I_Q are zero. This situation is to be expected since, under steady-state conditions, the rotating magnetic field due to the stator currents is stationary with respect to the rotor.

Equivalent Circuit and Phasor Diagram. Taking into account the previous considerations, the equations of stator and rotor voltages and flux linkages become

$$V_d = -RI_d - \omega_r \psi_q = -RI_d + X_q I_q \tag{2.57a}$$

$$V_q = -RI_q + \omega_r \psi_d = -RI_q - X_d I_d + X_{md} I_f = -RI_q - X_d I_d + E_f \tag{2.57b}$$

$$V_f = R_f I_f \tag{2.40''a}$$

$$\psi_d = -L_d I_d + L_{md} I_f; \quad \psi_q = -L_q I_q$$

$$\psi_f = L_{md} I_f - L_{md} I_d$$

$$\psi_D = L_{mD} I_f - L_{md} I_d; \quad \psi_Q = -L_{mQ} I_q$$

where $X_d = \omega_0 L_d$ and $X_q = \omega_0 L_q$ are called direct-axis and quadrature-axis synchronous reactance, respectively;

$$E_f = X_{md} I_f = \omega_0 L_{md} \frac{V_f}{R_f} \tag{2.58}$$

is the emf induced by the field current I_f under no-load conditions of the synchronous generator.

Indeed, under no-load conditions when $I_d = I_q = 0$, from (2.57a) and (2.57b) it results that $V_d = 0$ and $V_q = E_f$. Therefore, the phasor $\underline{E}_f$ lies along the q-axis.

The steady-state equations (2.57a and b) were written in the (d, q) reference frame attached to each generator. In order to use them in the algebraic-differential equations of the power system mathematical model, it is necessary to transform these equations to the (+1, +j) reference frame of the grid. For this, the connection between the two reference frames is therefore considered:

$$\underline{A} = \bar{A}\, e^{j(\delta - (\pi/2))}$$

The voltage and current vectors at the generator terminals are defined as

$$\bar{V} \triangleq V_d + jV_q$$

$$\bar{I} \triangleq I_d + jI_q$$

Substituting the expressions of V_d and V_q from the expressions (2.57a) and (2.57b), give

$$\bar{V} = -R(I_d + jI_q) + jE_f + X_q I_q - jX_d I_d$$

If the term $jX_q\bar{I} = jX_q(I_d + jI_q)$ is added and subtracted, then

$$\begin{aligned}\bar{V} &= -R\bar{I} + jE_f - jX_q\bar{I} + jX_q\bar{I} + X_q I_q - jX_d I_d \\ &= -(R + jX_q)\bar{I} + j[E_f - (X_d - X_q)I_d]\end{aligned} \tag{2.59}$$

Finally, the steady-state equation of the synchronous generator in the (d, q) reference frame can be written as

$$\bar{V} = \bar{E} - (R + jX_q)\bar{I} \tag{2.60}$$

where

$$\bar{E} = 0 + jE_q = j\left[E_f - (X_d - X_q)I_d\right]$$

and

$$E_q = E_f - (X_d - X_q)I_d \tag{2.61}$$

Considering that $X_d > X_q$ it results that $E_f > E_q$.

Taking into account the equation that links the two reference frames, the voltage phasor expression becomes

$$\underline{V} = \bar{V}\,e^{j(\delta - \pi/2)}$$

therefore, by multiplying the terms from equation (2.60) with $e^{j(\delta-\pi/2)}$, it results

$$\underline{V} = \underline{E} - (R + jX_q)\underline{I} \tag{2.60$'$}$$

The steady-state equivalent circuit of the synchronous generator can be drawn (Figure 2.15) using equation (2.60′).

If one assumes that $R < X_q$, the synchronous generator can be represented in steady state by an emf orientated along the q-axis:

$$\underline{V} = \underline{E} - jX_q\underline{I} \tag{2.60$''$}$$

In the case of isotropic synchronous generators (round rotor generators), $X_d = X_q = X_s$, from (2.61) and (2.58) achieve

$$E_q = E_f = X_{md}I_f \tag{2.61$'$}$$

Based on equation (2.60′) and the equivalent circuit from Figure 2.15, the phasor diagram of the synchronous machine under balanced steady state can be drawn (see Figure 2.16), where δ_i (internal rotor angle) is the phase shift between the emf E_q and the

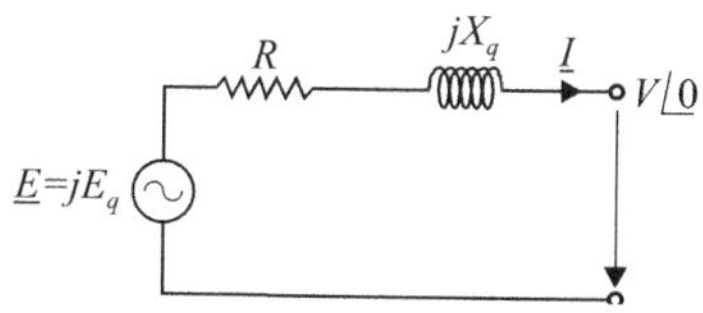

Figure 2.15. The equivalent circuit of the synchronous generator.

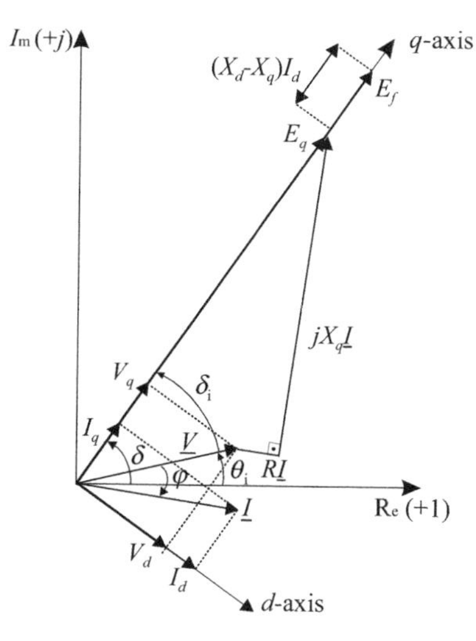

Figure 2.16. Phasor diagram of the synchronous machine under balanced steady state.

terminal voltage $\underline{V}$, while φ (power factor angle) is the phase shift between the terminal voltage $\underline{V}$ and the load current $\underline{I}$.

Computation of Steady-State Quantities.

Step a. Given the active power P, reactive power Q, and voltage V at the generator terminals, the current and power factor angle are computed as follows:

$$\underline{I} = \left(\frac{\underline{S}_g}{\underline{V}}\right)^* = \frac{P - jQ}{V\,e^{-j\theta_i}} = \frac{P - jQ}{V}\,e^{j\theta_i}$$

or

$$I = \frac{\sqrt{P^2 + Q^2}}{V}; \quad \varphi = \text{acos}\left(\frac{P}{VI}\right) \tag{2.62}$$

Step b. Computation of internal rotor angle δ_i.

From equation (2.60″), the emf $\underline{E}$ and, at the same time, the rotor position in the general reference frame (+1, +j) can be calculated:

$$\underline{E} = \underline{V} + jX_q\underline{I} = E_q\,e^{j\delta_i} \tag{2.60‴}$$

Next, from equation (2.61), E_f, the emf induced by the field current of the synchronous machine under no-load conditions, is computed:

$$E_f = E_q + (X_d - X_q)I_d \tag{2.61″}$$

Step c. Given the value of the rotor angle δ_i, the stator components of currents and voltages are computed as follows:

$$\begin{aligned} V_d &= V\sin(\delta - \theta_i) = V\sin\delta_i \quad ; \quad V_q = V\cos(\delta - \theta_i) = V\cos\delta_i \\ I_d &= I\sin(\delta_i + \varphi) \quad ; \quad I_q = I\cos(\delta_i + \varphi) \end{aligned} \tag{2.63}$$

where $\delta_i = \delta - \theta_i$ is the internal angle, that is, the phase shift between emf E_q and the voltage V at the machine terminals.

Step d. The other quantities are determined as follows:

For $\omega_r = 1$:

- from equation (2.57b)

$$V_q = -RI_q + \omega_r\psi_d \quad \Rightarrow \quad \psi_d = V_q + RI_q$$

- from equation (2.57a)

$$V_d = -RI_d - \omega_r\psi_q \quad \Rightarrow \quad \psi_q = -V_d - RI_d$$

- from equation (2.57b), it follows

$$
\begin{aligned}
V_q &= -RI_q - X_d I_d + X_{\mathrm{md}} I_f \quad \Rightarrow \quad I_f = \left(V_q + RI_q + X_d I_d\right)/X_{\mathrm{md}} \\
V_f &= R_f I_f \\
C_{\mathrm{e}} &= P + RI^2
\end{aligned}
$$

Power Equations and the Salient Pole Synchronous Generator Characteristic. Consider the expression of the complex apparent power:

$$
\begin{aligned}
\underline{S} &= \underline{V}\,\underline{I}^* = \bar{V} e^{j(\delta-\pi/2)} \cdot \bar{I}^* e^{-j(\delta-\pi/2)} = \bar{V}\bar{I}^* \\
&= \left(V_d + jV_q\right)\left(I_d - jI_q\right) = P + jQ
\end{aligned}
$$

Equating the real and the imaginary parts, it results

$$P = V_d I_d + V_q I_q \tag{2.64a}$$

$$Q = V_q I_d - V_d I_q \tag{2.64b}$$

Neglecting the armature resistance R in equations (2.57a) and (2.57b), gives

$$I_q = \frac{V_d}{X_q} \tag{2.65a}$$

$$I_d = \frac{E_f - V_q}{X_d} \tag{2.65b}$$

$$V_d = V \sin \delta_{\mathrm{i}} \tag{2.65c}$$

$$V_q = V \cos\delta_{\mathrm{i}} \tag{2.65d}$$

Substituting equations (2.65a), (2.65b), (2.65c), (2.65d) in (2.64a) and (2.64b), we obtain the expressions of active and reactive powers:

$$P = \frac{E_f V}{X_d} \sin \delta_{\mathrm{i}} + \frac{V^2}{2}\left(\frac{X_d - X_q}{X_d X_q}\right) \sin 2\delta_{\mathrm{i}} \tag{2.66}$$

$$Q = \frac{E_f V}{X_d} \cos \delta_{\mathrm{i}} - \left(\frac{\cos^2 \delta_{\mathrm{i}}}{X_d} + \frac{\sin^2 \delta_{\mathrm{i}}}{X_q}\right) V^2 \tag{2.67}$$

Considering the *case of a round rotor generator*, for which $X_d = X_q = X_s$, the expressions of the active and reactive powers from (2.66) and (2.67) reduce to

$$P = \frac{E_f V}{X_{\mathrm{s}}} \sin \delta_{\mathrm{i}} \tag{2.68a}$$

$$Q = \frac{E_f V}{X_{\mathrm{s}}} \cos \delta_{\mathrm{i}} - \frac{V^2}{X_{\mathrm{s}}} \tag{2.68b}$$

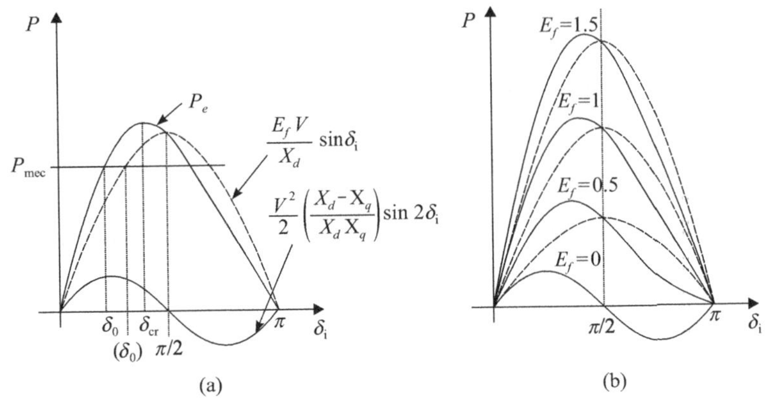

Figure 2.17. Power-angle characteristics for salient pole generators.

As shown in equation (2.66), the *active power characteristic of a salient pole generator* (Figure 2.17a) is composed of a main sinusoidal component, overlapped by a second component of double frequency sinusoidal form. The magnitude of the second component is proportional to the difference between the magnitudes of the d- and q-axis components of the synchronous reactance and does not depend on the emf E_f.

The second-order harmonic displaces the maximum of the nonsalient pole generator power characteristic; therefore, the critical angle δ_{cr} for which the maximum power is obtained is smaller than 90°. This aspect should not be seen as a stability aggravation since the initial angle δ_0 decreases simultaneously with it for the same mechanical power P_{mec}. On the contrary, the power characteristic magnitude P_e increases for the same values E_f and X_d compared to the salient pole machine characteristic.

Note that the increase of power P_e is important only for very small values of the emf E_f, when the magnitude of the fundamental power characteristic $(E_f V/X_d)$ has the same order as the magnitude of the second harmonic $((V^2/2)(X_d - X_q)/(X_d X_q))$. Usually, when the emf E_f has a considerable large value, the magnitude of the second harmonic is less than 10–15% of the fundamental magnitude and, consequently, its influence on the power characteristic magnitude is very small (Figure 2.17b).

Observation

Since the representation of salient pole generators (for which $X_d \neq X_q$) requires large computational effort, in static stability analysis the saliency is omitted, and their equivalent reactance is taken equal to the direct synchronous reactance X_d. This assumption allows us to neglect the second-order harmonic of the power characteristic.

A salient pole machine is represented as a round rotor machine even when, for accuracy, the effect of saliency needs to be taken into account. A fictitious emf is introduced in the round rotor generator model, which changes in terms of the generator operation, so that the active and reactive powers of the equivalent generator reflect with high accuracy the operation of the real salient pole generator. In order to obtain a correct value for the angle δ_i, the synchronous reactance of the equivalent round rotor generator, of variable emf, is taken equal to the q-axis reactance X_q of the salient pole generator.

2.1.3.7 Synchronous Generator Behavior on Terminal Short Circuit. SHORT CIRCUIT AT THE TERMINALS OF A SIMPLE R-L CIRCUIT. The electrical transient characteristics of a synchronous generator can be simply visualized by analyzing the behavior when a

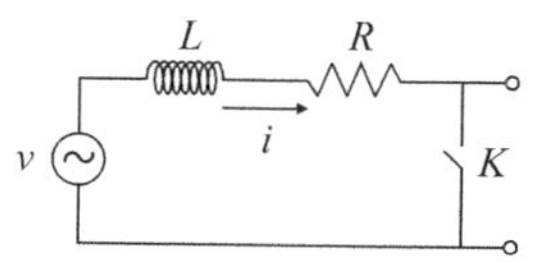

Figure 2.18. Basic R-L circuit supplied with an AC voltage.

three-phase short circuit occurs at the generator terminals. Such an analysis allows, on one hand, the identification of phenomena occurring in the synchronous generator under transient state and, on the other hand, the identification of some approximations necessary for its representation in large-scale stability studies.

Therefore, given the inductive character of the synchronous generator electric circuits, it is useful to analyze the transient response of a simple R-L circuit (Figure 2.18) supplied with an AC sinusoidal voltage, having the form

$$v(t) = V_m \sin(\omega t + \alpha)$$

where α is the angle on the voltage wave at which the fault occurs.

As known, the functioning equation of such a circuit is given by the Kirchhoff's second theorem written for instantaneous quantities, that is,

$$v = L\frac{di}{dt} + Ri \qquad (2.69)$$

from which, by integration, it results the expression of current

$$i = \frac{V_m}{Z}\sin(\omega t + \alpha - \varphi) + C\,e^{-(R/L)t} \qquad (2.70)$$

where $Z = \sqrt{R^2 + \omega^2 L^2}$ is the magnitude of the circuit impedance $\underline{Z} = R + j\omega L$ and $\varphi = \text{atan}(\omega L/R)$ is the impedance angle.

The integration constant C is determined from the initial conditions. Therefore, if at the instant $t_0 = 0$ a short circuit occurs at the circuit terminals by closing the switch K, then

$$C = i_0 - \frac{V_m}{Z}\sin(\alpha - \varphi)$$

However, since at $t_0 = 0$, $i_0 = i(0) = 0$ (for open circuit), the above expression can be written as

$$C = -\frac{V_m}{Z}\sin(\alpha - \varphi)$$

Then

$$i = I_m \sin(\omega t + \alpha - \varphi) - I_m \sin(\alpha - \varphi)e^{-(R/L)t} = i_{AC} + i_{DC} \qquad (2.71)$$

where $I_m = V_m/Z$ is the maximum steady-state current. For a high X/R ratio, $\varphi \cong 90°$.

There are two cases (Figure 2.19):

First Case (Figure 2.19*a*). If the short circuit occurs at an instant for which the instantaneous value of the voltage is zero, $\alpha = 0$ at $t = 0$, the instantaneous

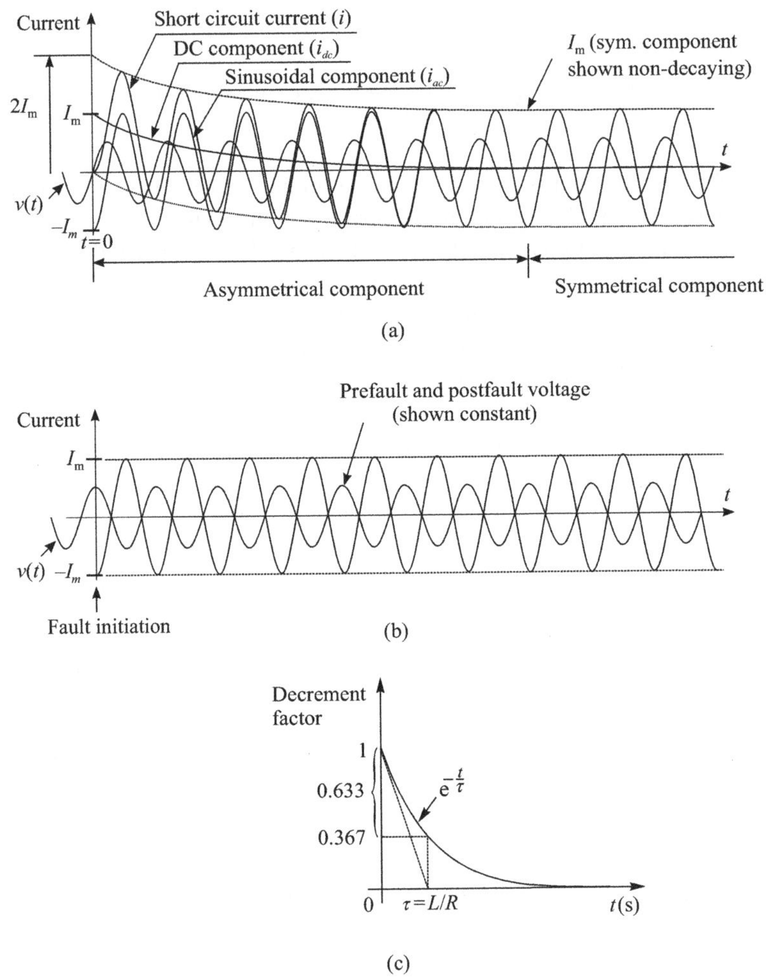

Figure 2.19. Transients of currents upon a short circuit occurrence: (a) current waveforms with maximum asymmetry; (b) symmetrical current waveform; (c) time constant of DC component decay [7].

short-circuit current is composed of two components: an AC short-circuit current component—the first term of equations (2.71)—and a transient current component—the second term of equation (2.71)—called DC component, with initial value equal but opposite to the AC component and that decays at an exponential rate.

For $\alpha = 0$, equation (2.71) becomes

$$i = I_m \sin(\omega t - \varphi) - I_{DC} \sin(-\varphi) e^{-(R/L)t} \qquad (2.71')$$

where the initial value of $I_{DC} = I_m$.

In this first case, at $t_0 = 0$ the maximum value of the superior envelop of the current from equation (2.71) is $2I_m$:

$$I_{env}^{max} = |I_m \sin(\omega t - \varphi)| + I_{DC}\, e^{-(R/L)t} = |I_m \sin(-90°)| + I_m e^0 = 2I_m$$

The decaying component decreases with an exponential decrement factor defined as its value any time after a short circuit. Assuming that the time constant is $\tau = L/R$, then the second term from (2.71) becomes $I_{DC}\,e^{-t/\tau}$. After an elapsed time equal to the time constant, the current decays to about 63.3% from its initial value, that is, the transitory component is reduced to 0.367 per unit (Figure 2.19c).

The presence of the DC component in the short-circuit current results in a current wave shape envelope asymmetrical about the zero line and axis of the wave (Figure 2.19b). If the DC transient component decays to zero in a short time, then the short-circuit current is a symmetrical wave.

Second Case (Figure 2.19*b*). If the short circuit occurs at an instant when the instantaneous voltage is maximum, $\alpha = \pi/2$ at $t = 0$, there is no transient component, and the short-circuit current is

$$i = I_{\mathrm{m}} \sin \omega t$$

which has the same shape as the steady-state curve with a peak value $I_{\mathrm{m}} = V_{\mathrm{m}}/Z$.

THREE-PHASE SHORT CIRCUIT AT THE TERMINALS OF A SYNCHRONOUS GENERATOR. In the analysis of the three-phase short circuit at the terminals of a synchronous generator, we take into account the following assumptions [7]:

- The impedances of the electrical network components (lines, transformers, reactors) do not change in time.
- The rotating machines (generators and motors) are the main sources of short-circuit currents. The flux linking each armature winding is trapped at the instant of short-circuit occurrence. The flux, and so the inductance L, will then decay in terms of machine time constants. Therefore, the decay of the AC component must also be considered. Note also that the induction machines exhibit a pattern of the decay different from the synchronous machines.
- The load currents have been neglected since, theoretically, they are much smaller than the short-circuit currents. The load current is generally only a percentage of the short-circuit current and determines the effective voltages of short-circuit sources prior to the fault.
- The control systems, for example, excitation system and turbine governor, affect the transient process for extended duration of the short circuit. The duration of a short-circuit current depends mainly on the neutral grounding solution and on the performances of the protective devices.
- In a three-phase electrical system, the currents and voltages are 120 electrical degrees apart from each other. When a symmetrical three-phase short circuit occurs, the (short circuit) instantaneous currents on the three phases of the synchronous generator are as shown in Figure 2.20, while their sum is zero at any instant, $i_a + i_b + i_c = 0$.

Envelope lines are drawn through the peaks of the alternating current waves. Also, a dotted line is drawn half-way between the envelope curves to represents the wave stabilization [2]. Thus, mathematically, the current can be split into two components: a *unidirectional* or *asymmetrical component* and an *alternating symmetrical component* of

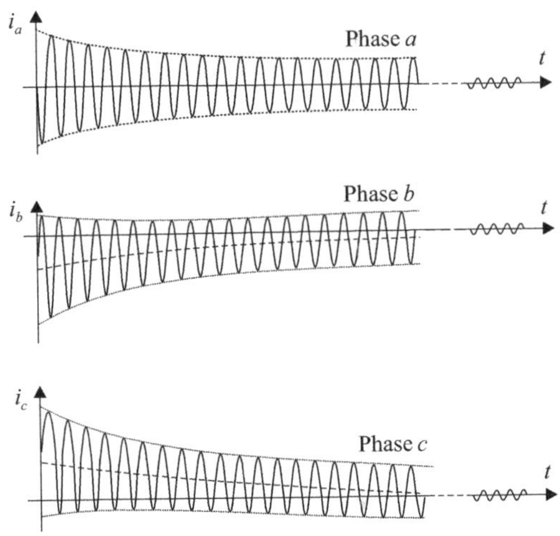

Figure 2.20. Instantaneous phase currents of a synchronous machine during a three-phase short circuit at its terminals.

frequency given by the power grid. The word "asymmetrical" describes the shape of the AC wave about the zero axis.

The Stator Current During a Three-Phase Short Circuit. In general, if a short circuit occurs at or very near the terminals of an alternator, the stator current on one of the three phases shown by an oscillogram would be of the form given in Figure 2.21.

The shape of the stator current wave, upon a short circuit occurring at the generator terminals, shows that it contains two components:

- An alternating component, which decays initially very rapidly (in a few cycles) and then relatively slow (in several seconds) until it reaches a steady-state value.
- A DC component, which decays exponentially in several cycles.

The variation with time of the *alternating component of the current*, for example, of the a phase, is shown in Figure 2.22. This component is symmetrical about the zero axis.

The processes following a fault occurrence can be split into three time intervals: the subtransient period, the transient period, and the steady state. The subtransient period lasts for 1–5 cycles (on a 60 Hz basis), while the transient period can last for 5–200 cycles. Normally, if the short circuit occurs near the generator terminals, the protection systems disconnect the generator from the network so that the steady-state conditions of a short circuit is not reached.

Analyzing the envelopes, it is obvious from Figure 2.22 that the reduction of the short-circuit current follows an exponential evolution. Short-circuit current damping is due to the mutual coupling between the stator and rotor windings as well as between these windings

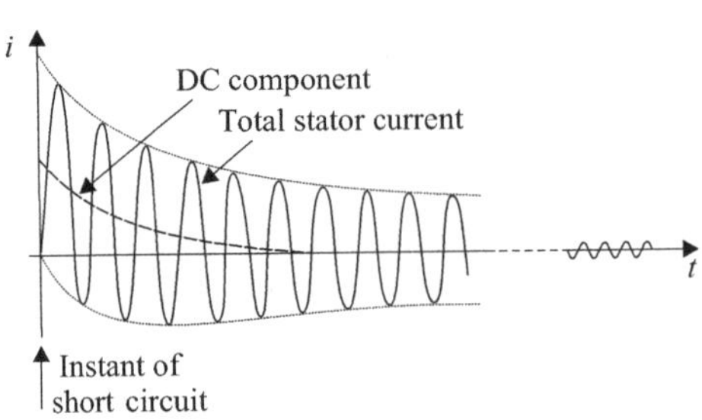

Figure 2.21. Stator current upon a short-circuit occurrence.

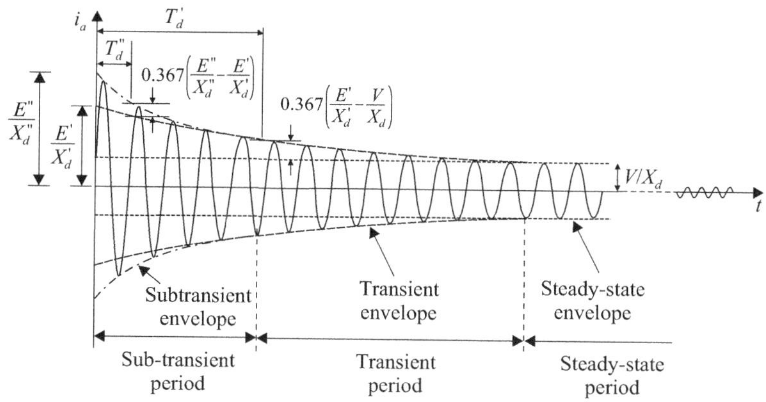

Figure 2.22. Alternating component of the short-circuit armature current.

and the damper windings and any path of the induced currents through the body and poles of the rotor [8].

The rms value of the armature current as function of time, t, under short-circuit conditions, may be written as

$$I_{AC} = \left(\frac{E''}{X_d''} - \frac{E'}{X_d'}\right) e^{-t/T_d''} + \left(\frac{E'}{X_d'} - \frac{V}{X_d}\right) e^{-t/T_d'} + \frac{V}{X_d} \tag{2.72}$$

where $((E''/X_d'') - (E'/X_d'))\, e^{-t/T_d''}$ is the decaying subtransient component; $((E'/X_d') - (V/X_d))\, e^{-t/T_d'}$ is the decaying transient component; V/X_d is the steady-state component; E'' and E' are the generator internal voltage behind subtransient (X_d'') and transient (X_d') reactances, respectively; T_d'' and T_d' are the short-circuit direct axis subtransient and transient time constants.

The current i_{AC}, given in Figure 2.22, can be divided into three components, corresponding to the three periods of the short circuit:

- The subtransient component given by the difference between the subtransient envelope and the transient envelope. This component has an initial value equal to $((E''/X_d'') - (E'/X_d'))$ and decays with the d-axis short-circuit subtransient time constant T_d'', representing the time in which the difference between the subtransient envelope and transient envelope is $1/e = 0.367$ of the initial difference.
- The transient component given by the difference between the transient envelope and the steady-state envelope. Its initial value is $((E'/X_d') - (V/X_d))$, and decays with the d-axis short-circuit transient time constant T_d', representing the time in which the difference between the transient envelope and steady-state envelope is $1/e = 0.367$ of the initial difference.
- The constant component, steady value of short-circuit current, with the value V/X_d.

The values of the d-axis transient and subtransient reactances and time constants can be deduced from the short-circuit oscillograms [2].

Except for the case when the short circuit occurs at the zero crossing of the current, the AC components of short-circuit currents on the three phases include a *DC component* as an offset of current waves with respect to the zero axis, which decays exponentially (Figure 2.21). The initial value of this DC component depends on the instant at which the short circuit occurs ($t = 0$); it reaches a maximum (which coincides with the peak value of the initial AC current component of the phase in question) when the voltage behind the subtransient reactance is zero. On the other hand, the DC current component is absent when the short circuit occurs at the instant at which the voltage behind the subtransient reactance is maximum. The DC current component decays with a time constant equal to the subtransient short-circuit time constant T''_d [8].

The power system analysis in the case of a short circuit considering both the DC (asymmetrical) component and the alternating component involves high computational effort. In order to simplify the computation, in many cases, *the DC offset is either neglected or treated separately* [1].

The DC offset can be eliminated from the armature current by considering that the fluxes are independent of time ($d\psi_d/dt = 0, d\psi_q/dt = 0$) in the stator voltage equations (2.39′a) and (2.39′b). The terms $d\psi/dt$, called transformer voltages, represent the stator transients. In steady state, these terms are zero whereas upon a fault occurrence they increase instantaneously.

The Field Current During a Three-Phase Short Circuit. When a short circuit occurs at the synchronous generator terminals, the rotor flux, so is the field current, is influenced. Assuming constant field voltage, the field current varies with time as shown in Figure 2.23.

Envelope lines can also be drawn along the extreme values of the oscillatory wave of the field current after the short circuit, and a line is drawn half-way between the envelope lines. The field current can thus be divided into

- an alternating component, which reaches a maximum in the brief instant after the short-circuit occurrence then it decreases in a damped oscillatory mod; it may be considered as decayed after a time approximately equal to $4T_a$ (T_a is the armature short-circuit time constant);
- a unidirectional component, which starts from an initial value equal with the steady value i_{f0}, rises suddenly upon the occurrence of the short circuit and follows the dotted curve until it finally returns to the steady value.

The synchronous machine behavior under short-circuit conditions presented above assumed constant field (excitation) voltage. In practice, however, the automatic voltage regulator (AVR) reacts immediately trying to restore the terminal voltage at the preset value. In extreme conditions, such as short circuits, terminal voltage restoration is done by forcing the field voltage to reach the maximum value called ceiling voltage in very short time with the help of a high initial response circuit. In turn, this will cause the short-circuit current to increase. An example of stator current pattern of a synchronous generator under no load upon the occurrence of a three-phase short circuit is shown in Figure 2.24. The effect of increased stator current of the AVR is obvious [8].

Solution of the Three-Phase Symmetrical Fault Currents Under Load Conditions. In real operation, it is most probable for a generator to experience a short circuit at its terminal

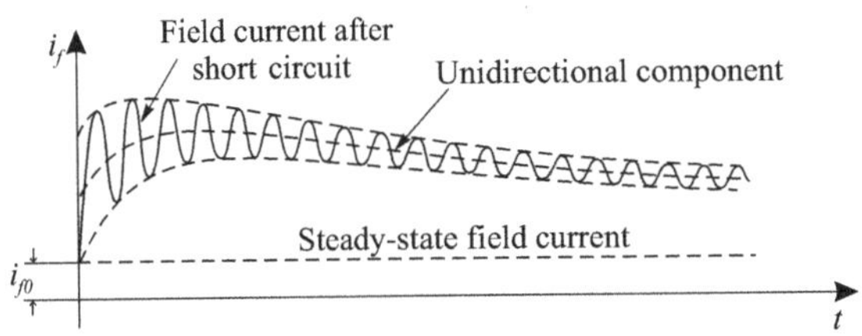

Figure 2.23. Field current of a synchronous machine during a three-phase short circuit at its terminals.

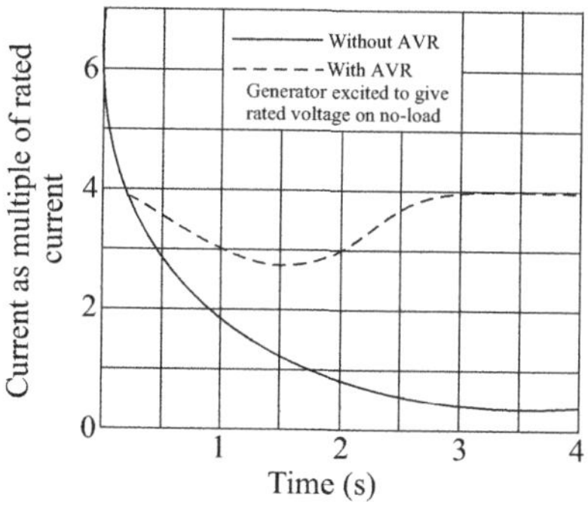

Figure 2.24. The effect of a typical regulator AVR for a terminal three-phase short circuit on an unload generator [8].

when it delivers power. Two methods might be used in the solution of the three-phase symmetrical fault currents [9]:

- *Represent the Generator by an Internal Voltage Behind a Reactance.* During each stage of the dynamic process following the short circuit (subtransient, transient, and steady state), the internal voltage can be calculated by one of the following equation:

$$\begin{aligned}\underline{E}'' &= \underline{V} + jX''_d \underline{I}_g \\ \underline{E}' &= \underline{V} + jX'_d \underline{I}_g \\ \underline{E} &= \underline{V} + jX_d \underline{I}_g\end{aligned}$$

  where $\underline{E}''$, $\underline{E}'$, and $\underline{E}$ are the subtransient, transient, and steady-state internal voltages, respectively; X''_d, X'_d, X_d are the subtransient, transient, and steady-state reactances, respectively; $\underline{V}$ is the terminal voltage; and $\underline{I}_g$ is the prefault load current delivered by the generator.
- *Applying the Thévenin's Theorem and Superposition.* The fault current is first determined in the absence of load as the ratio of the voltage at the fault location $\underline{V}_{\mathrm{Th}}$ to the equivalent reactance of the network seen from the fault location $\underline{Z}_{\mathrm{Th}}$, both calculated using the Thévenin's Theorem, that is, $\underline{I}_{\mathrm{fault}} = \underline{V}_{\mathrm{Th}}/\underline{Z}_{\mathrm{Th}}$. The total current is then determined by superposing the fault current $\underline{I}_{\mathrm{fault}}$ with the load current $\underline{I}_g$.

The mathematical models implemented in computer programs to perform simulation of the short-circuit currents assume some simplifying hypothesis:

- The generator is represented by an equivalent circuit consisting of a voltage source and an impedance of which value is chosen so as to represent the worst conditions after the fault.
- The effect of generator load and speed change are ignored and the excitation action is assumed constant.

Equivalent Circuits of Synchronous Generator During Fault. The d- and q-axis equivalent circuits used to represent a salient pole synchronous machine for short-circuit conditions help in deriving the short-circuit current equations and simulate the decaying AC component as shown in Figure 2.23. These circuits are shown in Figure 2.25 for the subtransient, transient, and steady-state periods.

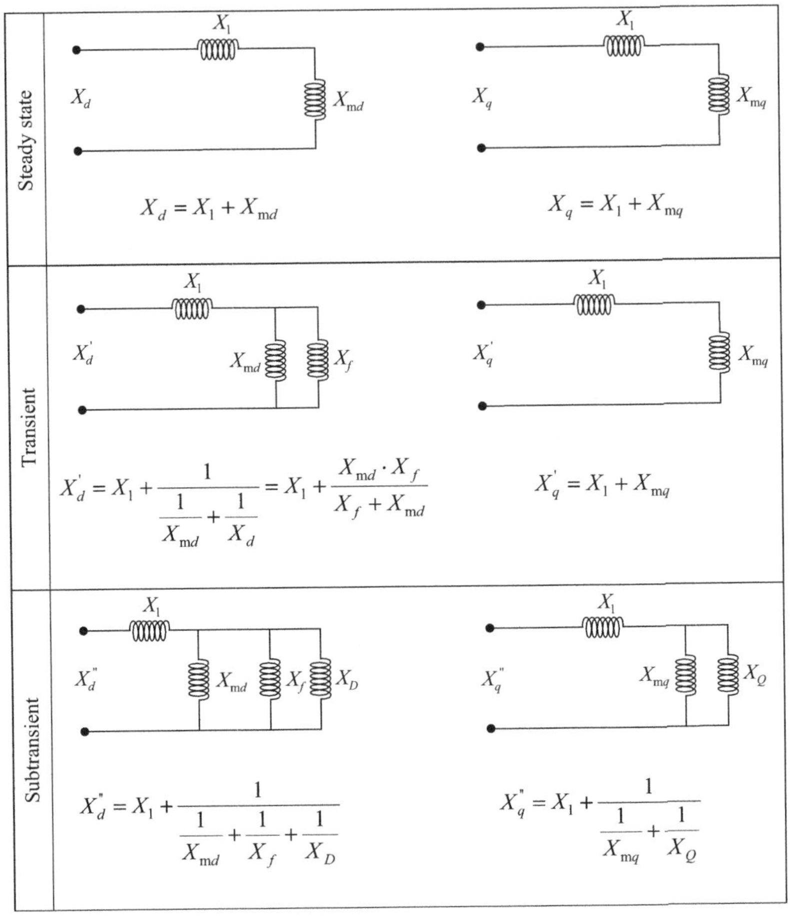

Figure 2.25. Equivalent circuits of a synchronous generator during subtransient, transient, and steady state, after a terminal fault.

The reactance of the synchronous machine changes during disturbances. The most important definitions are as follows:

- *Leakage Reactance* X_1. The armature leakage reactance is a parameter that accounts for that part of the magnetic flux produced by the armature that does not cross the air gap to link with the rotor winding. The flux leakage is formed by end-winding leakage, slot leakage, and tooth tip leakage fluxes.
- *Subtransient reactance* X''_d is associated with the subtransient period and is due to the flux leakage reactance and the armature flux crossing the air gap and penetrating the rotor circuits.
- *Transient reactance* X'_d is associated with the transient period. The damper winding current and the current in the rotor surface decay to zero after few cycles, associated with the subtransient period. Thus, the transient reactance is due to the flux leakage and the armature flux that penetrates the rotor to the field windings.
- *Synchronous reactance* X_d determines the steady-state current. If a short circuit is involved, the synchronous reactance is defined after all currents in the field winding

and dumper circuits have decayed. The synchronous reactance is, therefore, the sum of the leakage reactance and the fictitious mutual reactance X_{md}, which is much larger than the leakage reactance. It has different values depending on weather or not the magnetic core is saturated. The saturated synchronous reactance, determined by neglecting the armature resistance, is the ratio of the open-circuit test voltage to the armature short-circuit test current, for a given field excitation. The unsaturated synchronous reactance is given by the ratio of the open-circuit test voltage on air gap characteristic to the armature short-circuit test current (see Figure 2.32). The saturated value may be only 60–80% of the unsaturated value.

- *Quadrature-axis reactances (X''_q, X'_q, and X_q).* are defined mainly for salient pole generators, due to the constructional features of the rotor to accommodate the windings. In salient pole generators, the peak of density of the flux wave is not in phase with the peak of density of the mmf wave as it is the case of a round rotor generator.

2.1.4 Synchronous Generator Parameters

The inductances and resistances of the stator and rotor circuits, as used in Section 2.1.3 in the synchronous generator equations, are referred to as fundamental or basic parameters. These parameters are also used to identify the parameters of the d- and q-axis equivalent circuits shown in Figures 2.13a, b and 2.14a–c. In a compact form, these parameters may be handled in a vector form, as shown below:

$$\theta_{\text{EC}} = \left[R_{fd}, R_D, R_Q, L_{fd}, L_D, L_Q, L_1, L_{\text{md}}, L_{mq}, L_{fD}, N_{afd}\right] \tag{2.73}$$

Note that the last parameter of this vector is the armature-to-field turn-ratio N_{afd}, which relates the actual field-winding variables (voltage, current, resistance, inductance) to their values referred to the stator winding as in Figure 2.14. Because some of the parameters from equation (2.73), such as N_{afd} and the leakage inductance L_1, cannot be uniquely determined from measured responses of the synchronous generators, derived parameters related to the observed behavior of the generator viewed from the terminals under suitable test conditions are determined [10].

2.1.4.1 Operational Parameters. The operational parameters relating the stator and rotor terminal quantities are in general used to determine the electrical characteristics of the synchronous generator.

Use of the d–q transformation permits us to consider the generator as being composed of two separate electrical networks representing the d- and q-axis as seen from the generator rotor. The d-axis network represents those portions of the generator in which the flow of current creates flux that links the field winding. The q-axis network represents those portions of the generator in which the flow of current creates flux that does not link the field winding.

A schematic representation of those two networks is shown in Figure 2.26a and b [11].

The d-axis network has two ports (terminal pairs): the field winding with terminal variables V_f (field voltage) and I_f (field current), and the d-axis coil with terminal variables

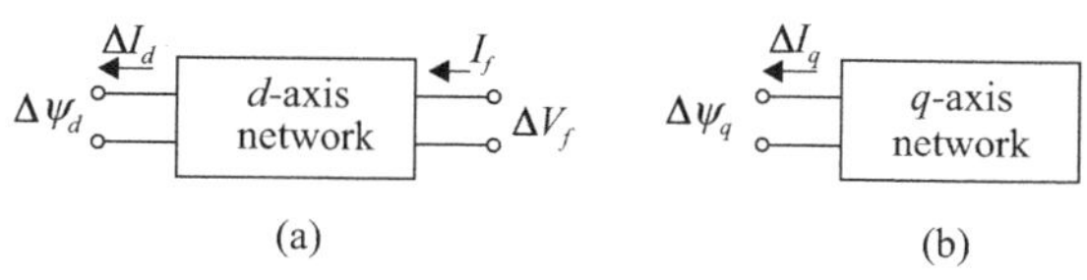

Figure 2.26. The d- and q-axis networks identifying terminal quantities.

ψ_d (d-axis flux linkage) and I_d (d-axis current). The q-axis has only one port, with terminal variables ψ_q (q-axis flux linkage) and I_q (q-axis current).

The process of transformation to the rotating (d–q) reference frame results in the following set of coupled equations that can be used to compute the actual fixed frame armature voltage and currents:

$$V_d = -RI_d - \omega_r \psi_q + \frac{d\psi_d}{dt} \tag{2.39'a}$$

$$V_q = -RI_q + \omega_r \psi_d + \frac{d\psi_q}{dt} \tag{2.39'b}$$

The coupling is due to the speed voltage terms $\omega_r \psi_d$ and $\omega_r \psi_q$. The d- and q-axis networks are defined in such a fashion that the armature resistance R is external to the networks.

It will be assumed that the d- and q-axis networks can be modeled as lumped element networks composed of linear inductors and resistors. Such models should certainly be valid for studies in which the generator can be considered to be operating in a perturbed condition around some operating point.

The equivalent circuits of the synchronous machine in the d- and q-axes, as shown in Figure 2.27, are widely used in stability studies. The stator is modeled through the leakage inductance, the rotor is represented by the field winding and a damper winding in the d-axis as well as one or two damper windings in the q-axis, whereas the mutual coupling between stator and rotor is represented by the mutual inductances L_{md} and L_{mq}.

In the representation from Figure 2.27, the mutual inductances L_{fD} and L_{md} (see Figure 2.11) are assumed to be equal and ($L_{fD} - L_{md}$) disappeared. In these conditions, equations (2.50), (2.53), and (2.54) for d-axis flux linkages in the operational form become [1]

$$\psi_d(s) = -L_d I_d(s) + L_{md} I_f(s) + L_{md} I_D(s) \tag{2.74}$$

$$\psi_f(s) = -L_{md} I_d(s) + L_f I_f(s) + L_{md} I_D(s) \tag{2.75}$$

$$\psi_D(s) = -L_{md} I_d(s) + L_{md} I_f(s) + L_{1D} I_D(s) \tag{2.76}$$

The fact that the d-axis network has two terminal pairs whereas the q-axis network has one only is extremely significant [11].

For the linear and passive two-terminal network of the q-axis equivalent circuit (Figure 2.27b), the following equation holds:

$$s\psi_q(s) = -Z_q(s) I_q \tag{2.77}$$

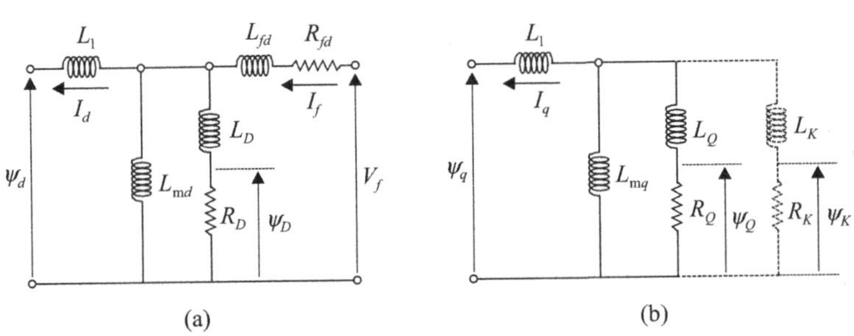

Figure 2.27. Structure of commonly used model [1,10,11].

where $Z_q(s)$ is the input impedance of the q-axis network:

$$Z_q(s) = sL_l + \frac{sL_{mq}(R_Q + sL_Q)}{R_Q + s(L_{mq} + L_Q)} \tag{2.78}$$

with only one damper winding Q in the q-axis.

Hence, we may deduce the usual expression

$$\psi_q(s) = -L_q(s)I_q \tag{2.79}$$

where $L_q(s)$ is the so-called operational inductance of the q-axis.

Let us consider the linear and passive parts of the d-axis comprised between the inputs V_f, I_f and the outputs $s\psi_d$, I_d of Figure 2.27a.

The equations of this four-terminal network, in terms of operational impedances, are of the form

$$\begin{bmatrix} s\psi_d(s) \\ V_f \end{bmatrix} = \begin{bmatrix} \underline{Z}_{11}(s) & \underline{Z}_{12}(s) \\ \underline{Z}_{21}(s) & \underline{Z}_{22}(s) \end{bmatrix} \begin{bmatrix} -I_d \\ I_f \end{bmatrix} \tag{2.80}$$

with

$$\underline{Z}_{12}(s) = \underline{Z}_{21}(s) \tag{2.81}$$

since the four-terminal network is reciprocal.

Note that the d-axis network requires three parameters to completely specify its properties, that is, $\underline{Z}_{11}(s)$, $\underline{Z}_{12}(s)$, and $\underline{Z}_{22}(s)$. Moreover, as it has a T-configuration (Figure 2.28), we get

$$\begin{aligned} \underline{Z}_{11}(s) &= \underline{Z}_1(s) + \underline{Z}_3(s) \\ \underline{Z}_{22}(s) &= \underline{Z}_2(s) + \underline{Z}_3(s) \\ \underline{Z}_{12}(s) &= \underline{Z}_3(s) \end{aligned} \tag{2.82}$$

where

$$\underline{Z}_1(s) = sL_l; \quad \underline{Z}_2(s) = R_{fd} + sL_{fd}; \quad \underline{Z}_3(s) = \frac{sL_{md}(R_D + sL_D)}{R_D + s(L_{md} + L_D)} \tag{2.83}$$

Instead of the form (2.80), we generally use the following hybrid form, whose inputs are the d-axis component of the stator current and the excitation voltage, and outputs are the d-axis of the stator flux and the excitation current [11]:

$$\begin{bmatrix} \psi_d(s) \\ I_f \end{bmatrix} = \begin{bmatrix} L_d(s) & \dfrac{A(s)}{s} \\ -A(s) & \dfrac{1}{\underline{Z}_{22}(s)} \end{bmatrix} \cdot \begin{bmatrix} -I_d \\ V_f \end{bmatrix} \tag{2.84}$$

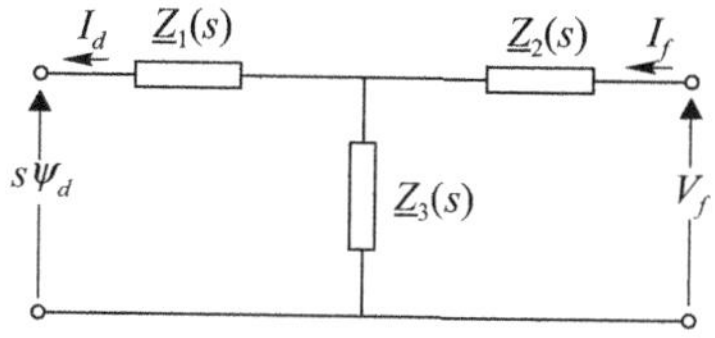

Figure 2.28. Linear passive and reciprocal four-terminal T-network along the d-axis.

where

$$L_d(s) \triangleq \frac{Z_{11}(s)Z_{22}(s) - Z_{12}^2(s)}{sZ_{22}(s)} \tag{2.85}$$

$$A(s) \triangleq \frac{Z_{12}(s)}{Z_{22}(s)} \tag{2.86}$$

The two parameters used in deriving the d-axis network presented here are $L_d(s)$ and $A(s)$. Note from equation (2.84) that $A(s)$ is the ratio of induced field current to d-axis armature current (with field-winding short-circuited) and $L_d(s)$ is the d-axis operational inductance.

Hence, the relationship between the values of the terminal quantities from Figure 2.26 are given by

$$\psi_d(s) = A(s)\,V_f(s) - L_d(s)\,I_d(s) \tag{2.87}$$

$$\psi_q(s) = -L_q(s)\,I_q(s) \tag{2.79}$$

where $A(s)$ is the stator to field operational transfer function, $L_d(s)$ is the d-axis operational inductance, and $L_q(s)$ is the q-axis operational inductance.

Taking into account equations (2.82) and (2.83) and carrying out the steps of equations (2.85) and (2.86), the following expressions for the two d-axis operational transfer functions $A(s)$ and $L_d(s)$, in terms of circuit elements of Figure 2.27a, are obtained [1]:

$$L_d(s) = L_d \frac{1 + (T_4 + T_5)s + T_4T_6s^2}{1 + (T_1 + T_2)s + T_1T_3s^2} \tag{2.88}$$

$$A(s) = A_0 \frac{1 + sT_{kd}}{1 + (T_1 + T_2)s + T_1T_3s^2} \tag{2.89}$$

where

$$A_0 = \frac{L_{md}}{R_{fd}} \tag{2.90a}$$

$$T_{kd} = \frac{L_D}{R_D} \tag{2.90b}$$

$$T_1 = \frac{L_{md} + L_{fd}}{R_{fd}} \tag{2.91a}$$

$$T_2 = \frac{L_{md} + L_D}{R_D} \tag{2.91b}$$

$$T_3 = \frac{1}{R_D}\left(L_D + \frac{L_{md}L_{fd}}{L_{md} + L_{fd}}\right) \tag{2.91c}$$

$$T_4 = \frac{1}{R_{fd}}\left(L_{fd} + \frac{L_{md}L_l}{L_{md} + L_l}\right) \tag{2.91d}$$

$$T_5 = \frac{1}{R_D}\left(L_D + \frac{L_{md}L_l}{L_{md} + L_l}\right) \tag{2.91e}$$

$$T_6 = \frac{1}{R_D}\left(L_D + \frac{L_{md}L_{fd}L_l}{L_{md}L_l + L_{md}L_{fd} + L_{fd}L_l}\right) \tag{2.91f}$$

It is worth noting that both the numerator and the denominator of $L_d(s)$ are second-degree polynomials in s. Thus, each of them admits two roots, which should be real, in that the equivalent circuits have ohmic-inductive branches, that is, these roots may be associated with time constants. We deduce that the transfer functions from equations (2.88) and (2.89) may be written in more usual (factored) form [1,12]:

$$L_d(s) = L_d\frac{(1+sT'_d)(1+sT''_d)}{(1+sT'_{d0})(1+sT''_{d0})} \tag{2.88'}$$

$$A(s) = \frac{L_{md}}{R_{fd}}\frac{1+sT_{kd}}{(1+sT'_{d0})(1+sT''_{d0})} \tag{2.89'}$$

Based on similarities between d-axis and q-axis equivalent circuits, we may write the expression for the q-axis operational inductance as

$$L_q(s) = L_q\frac{(1+sT'_q)(1+sT''_q)}{(1+sT'_{q0})(1+sT''_{q0})} \tag{2.92}$$

The models in Figure 2.27a and b, and equations (2.88′), (2.89′), and (2.92) share in common three inductances and one resistance that characterize the machine steady-state behavior [10]:

- L_l, *The Armature Leakage Inductance*. It is generally computed by the manufacturer at the design stage. There is no standard test that allows its determination in a practical manner (IEEE Std. 115-1995).
- $L_d = L_{md} + L_l$, *The Direct-Axis Synchronous Inductance*. It is found by performing an open-circuit test (IEEE Std. 115-1995, 10.3) and a sustained or steady-state short-circuit test.
- $L_q = L_{mq} + L_l$, *The Quadrature-Axis Synchronous Inductance*. It is found by performing a slip test (IEEE Std. 115-1995, 10.4.2).
- R_{fd} is the field resistance referred to the stator.

The information about the determination of the remaining elements of the parameters vectors θ_{EC}, also shown in equation (2.73), or θ_{OI} (for operational inductance of the equivalent circuits referred to the stator terminal) by dynamic test, can be found in the IEEE Std. 1110-2002 (Sections 7.2.2 and 7.3). Another analytical method was introduced by Canay [13] that yields an equivalent circuit fitting exactly the input operational inductance in a given axis, $L_d(s)$ or $L_q(s)$, for an arbitrary number of damper windings.

2.1.4.2 Standard Parameters. The standard parameters are used to represent the synchronous machine electrical characteristics in the form of effective inductances or reactances seen from the machine terminals. These characteristics are determined so as to reflect sustained transient and subtransient conditions. The corresponding time constants determine the rate of decay of currents and voltages. The standard parameters can be determined from the expressions of the operational parameters $L_d(s)$, $L_q(s)$, and $A(s)$ [1].

New Forms of the d-Axis Time Constants. The time constants of the synchronous machine T'_{d0}, T''_{d0}, T'_d, and T''_d can be expressed in terms of the fundamental parameters by equating the numerators and denominators of equations (2.88) and (2.88′), that is,

$$\left(1 + sT'_{d0}\right)\left(1 + sT''_{d0}\right) = 1 + s(T_1 + T_2) + s^2(T_1T_3) \tag{2.93}$$

$$\left(1 + sT'_{d}\right)\left(1 + sT''_{d}\right) = 1 + s(T_4 + T_5) + s^2(T_4T_6) \tag{2.94}$$

The new time constants are related to the terms (T_1, . . . , T_6) by appropriate identification of the left and right terms of equations (2.93) and (2.94), that is [1]

$$\begin{matrix} T'_{d0}T''_{d0} = T_1T_3 & & T'_dT''_d = T_4T_6 \\ & ; & \\ T'_{d0} + T''_{d0} = T_1 + T_2 & & T'_d + T''_d = T_4 + T_5 \end{matrix} \tag{2.95}$$

and may be obtained together with the appropriate typical parameters of a generator through appropriate experimental identification test [12]. Using the four expressions from (2.95), the expressions of the time constants that can result would be very complex.

In order to simplify the calculation of solutions of equations (2.93) and (2.94), we take into account that in practice $R_D >> R_{fd}$ and therefore based on equations (2.91a)–(2.91f) the following inequalities hold:

$$\begin{matrix} T_4 \gg T_5; \; T_1 \gg T_2 \\ T_4 \gg T_6; \; T_1 \gg T_3 \end{matrix} \tag{2.96}$$

Hence, the right-hand side of equations (2.93) and (2.94) may be approximated as

$$1 + s(T_1 + T_2) + s^2(T_1T_3) \approx (1 + sT_1)(1 + sT_3) \tag{2.97′}$$

$$1 + s(T_4 + T_5) + s^2(T_4T_6) \approx (1 + sT_4)(1 + sT_6) \tag{2.97″}$$

Comparing the right-hand side of the expressions (2.97′) and (2.97″) with the denominator and numerator of $L_d(s)$, we achieve the classical identities [2]:

$$T'_{d0} \approx T_1; \quad T''_{d0} \approx T_3; \quad T'_d \approx T_4; \quad T''_d \approx T_6 \tag{2.98}$$

The parameters T_1 to T_6 given in equations (2.91) are expressed in per unit (or radians). Conversion in seconds is performed by dividing them to $\omega_0 = 2\pi f$.

The meaning of the open-circuit time constants can be explained if assumes that no load is applied at the stator terminals ($\Delta I_d = 0$). Therefore, the change in the d-axis flux linkage from (2.87), taking into account (2.89′), is [1]

$$\Delta\psi_d(s) = \frac{L_{md}}{R_{fd}} \frac{1 + sT_{kd}}{(1 + sT'_{d0})(1 + sT''_{d0})} \Delta V_f \tag{2.99}$$

Equation (2.99) shows that, under open-circuit conditions, the time response of the d-axis stator flux to any change in the field voltage depends on the time constants T'_{d0} and T''_{d0}. Given that $R_D \gg R_{fd}$, it results that $T_1 \gg T_3$ and also that $T'_{d0} \gg T''_{d0}$.

The time constants of synchronous machine are as follows [7]:

(i) *The d-axis open-circuit subtransient time constant* T''_{d0} characterizes the initial decay of the transients in the d-axis variables of the synchronous machine with the stator winding open-circuited. This time constant is characteristic to the immediate period following a disturbance during which all effects in the damper windings are considered and the field-winding resistance is very small.

(ii) *The d-axis open-circuit transient time constant* T'_{d0} characterizes the decay of transients in the d-axis variables of the synchronous machine with the stator winding open-circuited, following the subtransient period, but prior to the steady state, during which any effect in the damper windings is negligible.

(iii) *The subtransient short-circuit time constant* T''_{d} defines the rate of decay of the subtransient component of the stator current following a three-phase short circuit that occurs at the machine terminals.

(iv) *The transient short-circuit time constant* T'_{d} defines the rate of decay of the transient component of the stator current when a three-phase short circuit that occurs at the machine terminals.

(v) *The armature time constant* T_{ar} defines the rate of decay of the DC component of the stator current under the same conditions.

TRANSIENT, SUBTRANSIENT, AND SYNCHRONOUS INDUCTANCES. In addition to the time constants defined so far (T'_{d0}, T''_{d0}, T'_{d}, and T''_{d}), there are other typical parameters that derive from the d- and q-axis operational inductances $L_d(s)$ and $L_q(s)$.

The inductance $L_d(s)$ under steady-state, transient, and subtransient conditions is determined assuming that during the subtransient period $R_{fd} = R_Q = 0$, and that during the transient period $R_D = R_K = \infty$ [1].

- In steady state, since $s = 0$, the d-axis operational inductance from equation (2.88′) becomes

$$L_d(0) = L_d \tag{2.100}$$

- Following a sudden change, as $s \rightarrow \infty$, $L_d(s)$ tends to

$$L(\infty) = L''_d = L_d \frac{T'_d T''_d}{T'_{d0} T''_{d0}} \tag{2.101}$$

which is called d-axis subtransient inductance.

- If the damper winding is neglected, at the limit, we obtain

$$L(\infty) = L'_d = L_d \frac{T'_d}{T'_{d0}} \tag{2.102}$$

which is called d-axis transient inductance.

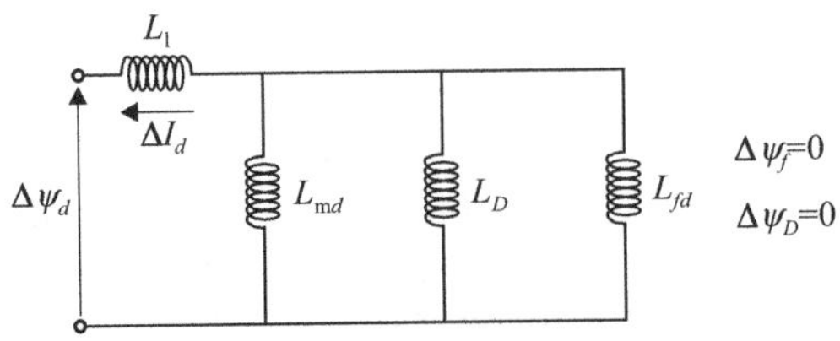

Figure 2.29. Equivalent circuit in the d-axis for incremental values, immediately following a disturbance.

Substituting the expressions of the time constants from equations (2.91) and (2.98) in equations (2.101) and (2.102), we obtain

$$L''_d = L_l + \frac{L_{md}L_{fd}L_D}{L_{md}L_{fd} + L_{md}L_D + L_{fd}L_D} \tag{2.101$'$}$$

$$L'_d = L_l + \frac{L_{md}L_{fd}}{L_{md} + L_{fd}} \tag{2.102$'$}$$

According to the law of constant flux linkages, we may assume that the rotor flux linkages do not change instantly following a disturbance, that is, $\Delta\psi_f = 0$ and also $\Delta\psi_D = 0$, and the d-axis equivalent circuit from Figure 2.11 reduces to the circuit illustrated in Figure 2.29 [1].

Calculating the equivalent inductance of the circuit from Figure 2.29, we get the effective inductance $\Delta\psi_d/\Delta I_d$ with the same expression as that in equation (2.101′) representing L''_d. In the absence of the damper winding, with $L_D = \infty$, the inductance corresponds to L'_d and takes the form given by equation (2.102′).

Based on the similarities between the d-axis and q-axis equivalent circuits, *the q-axis open-circuit transient and subtransient time constants* are achieved [1]:

$$T'_{q0} = \frac{L_{mq} + L_Q}{R_Q} \tag{2.103}$$

$$T''_{q0} = \frac{1}{R_K}\left(L_K + \frac{L_{mq}L_Q}{L_{mq} + L_Q}\right) \tag{2.104}$$

where $L_Q = L_{1Q} - L_{mq}$ is the q-axis inductance of the damper winding; $L_K = L_{2K} - L_{mq}$ is the q-axis inductance of the second damper winding; R_K is the resistance of the second damper winding.

The q-axis subtransient and transient inductances are therefore given by

$$L''_q = L_l + \frac{L_{mq}L_QL_K}{L_{mq}L_Q + L_{mq}L_K + L_QL_K} \tag{2.105}$$

$$L'_q = L_l + \frac{L_{mq}L_Q}{L_{mq} + L_Q} \tag{2.106}$$

In steady state, since $s = 0$, the q-axis operational inductance $L_q(s)$ is

$$L_q(0) = L_q \tag{2.107}$$

TABLE 2.1. Expressions for Standard Parameters of Synchronous Machine [1]

Parameter	T'_{d0}	T'_d	T''_{d0}	T''_d	L'_d	L''_d
Classical expression	T_1	T_4	T_3	T_6	$L_d\left(\frac{T_4}{T_1}\right)$	$L_d\left(\frac{T_4T_6}{T_1T_3}\right)$
Accurate expression	T_1+T_2	T_4+T_5	$T_3\left(\frac{T_1}{T_1+T_2}\right)$	$T_6\left(\frac{T_4}{T_4+T_5}\right)$	$L_d\left(\frac{T_4+T_5}{T_1+T_2}\right)$	$L_d\left(\frac{T_4T_6}{T_1T_3}\right)$

Note that the exact values of T'_{d0} and T''_{d0} are given by the poles of $L_d(s)$ and those of T'_d and T''_d by the zeros of $L_d(s)$. The exact and approximate (classical) expressions for the standard parameters are summarized in Table 2.1. These expressions apply to a synchronous machine model represented by the equivalent circuit shown in Figure 2.27 that considers two rotor circuits in each axis with equal mutual inductances [1].

The expressions of T_1 to T_6 from Table 2.1 are given in equations (2.91a)–(2.91f). Note that

(i) similar expressions apply to the q-axis parameters;

(ii) all parameters are in per unit and all mutual inductances in the d-axis are assumed equal;

(iii) time constants in seconds are obtained by dividing the per unit values given in Table 2.1 by $\omega_0 = 2\pi f$.

Considering Unequal Mutual Effects. In deriving the parameters obtained above, we have assumed that the mutual inductances in the d-axis are equal. Under real conditions, however, the mutual inductances between the field winding and the damper could have different values. Whereas calculation of the armature variables is performed with good accuracy, the calculation of the field current could lead to large errors [1].

Consideration of unequal values of the mutual inductances require addition of a series inductance L_{pl} in the d-axis equivalent circuit (Figure 2.30) [14]. This series inductance, given by the difference between the two mutual inductances, that is, the mutual inductance L_{fD} of the path linking the field winding and an equivalent rotor iron circuit or rotor damper bar circuit and the mutual inductance L_{md} of the path linking the field winding and the stator, $L_{pl} = L_{fD} - L_{md}$, is called differential leakage inductance, and corresponds to the peripheral leakage flux (ϕ_{pl}) that links only the field winding and damper [1].

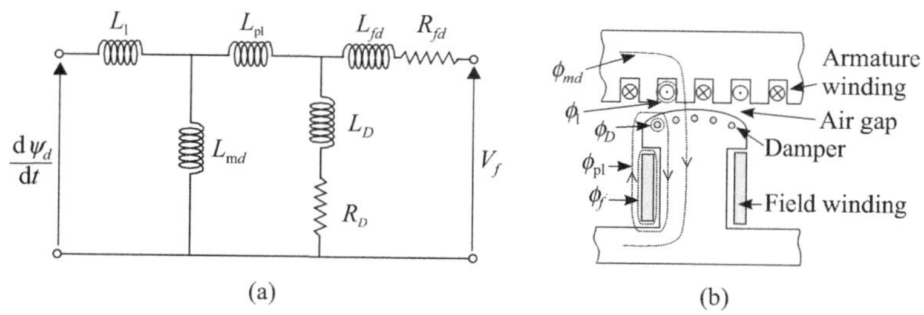

Figure 2.30. Unequal mutual effects in d-axis: (a) equivalent circuit; (b) flux paths [1,14].

Assuming again that during the subtransient period $R_{fd} = 0$ and that during the transient period $R_D = \infty$, the standard parameters become

$$\left\| \begin{array}{l} L_d = L_{md} + L_l \\ L'_d = L_l + \dfrac{1}{(1/L_{md}) + 1/(L_{fd} + L_{pl})} = L_l + \dfrac{L_{md}(L_{fd} + L_{pl})}{L_{md} + L_{fd} + L_{pl}} \\ L''_d = L_l + \dfrac{L_D L_{fd} L_{md} + L_D L_{pl} L_{md} + L_{md} L_{fd} L_{pl}}{L_{md} L_{fd} + L_{md} L_D + L_D L_{fd} + L_D L_{pl} + L_{fd} L_{pl}} \end{array} \right. \qquad (2.107)$$

$$\left\| \begin{array}{l} T'_{d0} = \dfrac{L_{md} + L_{fd} + L_{pl}}{R_{fd}} \\ T''_{d0} = \dfrac{1}{R_D}\left(L_D + \dfrac{L_{fd}(L_{md} + L_{pl})}{L_{pl} + L_{fd} + L_{md}}\right) \\ T'_d = \dfrac{1}{R_{fd}}\left(L_{fd} + L_{pl} + \dfrac{L_{md} L_l}{L_{md} + L_l}\right) \\ T''_d = \dfrac{1}{R_D}\left(L_D + \dfrac{L_{md} L_{pl} L_{fd} + L_l L_{fd} L_{md} + L_l L_{fd} L_{pl}}{L_{fd} L_{md} + L_{fd} L_l + L_{pl} L_{md} + L_{pl} L_l + L_{md} L_l}\right) \end{array} \right. \qquad (2.108)$$

Expressions (2.107) and (2.108) are based on the approximations associated with the classical definition of the parameters [1].

Salient Pole Machine Parameters. The expressions of the standard parameters derived earlier are applicable only for turbogenerators (round rotor machines) because two rotor circuits in each axis were considered. Hydrogenerators (salient pole machines) are usually modeled with only one damper circuit in the q-axis, that is, the Q circuit. The damper circuit is associated with the subtransient period only and thus the transient parameters L'_q (or X'_q) and T'_{q0} are not specific for these generators. Thereby, the expressions for the q-axis parameters of a salient pole machine are [1]

$$\left\| \begin{array}{l} L_q = L_l + L_{mq} \\ L''_q = L_l + \dfrac{L_{mq} L_Q}{L_{mq} + L_Q} \\ T''_{q0} = \dfrac{L_{mq} + L_Q}{R_Q} \end{array} \right. \qquad (2.109)$$

As for the d-axis, the same circuits (field and damper) are considered for both salient pole and round rotor generators and thus the expressions previously derived are also valid for the d-axis.

In per units, the reactances and their corresponding inductances are equal. Hence, in practice it is more usual to define the synchronous machine parameters as reactances. Table 2.2 gives the main parameters of the synchronous generators with salient pole and round rotor. The reactances are in per unit with stator base values equal to the corresponding machine rated values [15].

TABLE 2.2. The Typical Values of the Parameters of Synchronous Generators [15]

Parameter			Round Rotor Generators: Air Cooled	Round Rotor Generators: Hydrogen Cooled	Round Rotor Generators: Hydrogen/ Water Cooled	Salient Pole Generators: 4 Pole	Salient Pole Generators: Multipole
Synchronous reactance	X_d	p.u.	2.0–2.8	2.1–2.4	2.1–2.6	1.75–3.0	1.4–1.9
	X_q	p.u.	1.8–2.7	1.9–2.4	2.0–2.5	0.9–1.5	0.8–1.0
Transient reactance	X'_d	p.u.	0.2–0.3	0.27–0.33	0.3–0.36	0.26–0.35	0.24–0.4
	X'_q	p.u.	—	—	—	0.3–1.0	
Subtransient reactance	X''_d	p.u.	0.15–0.23	0.19–0.23	0.21–0.27	0.19–0.25	0.16–0.25
	X''_q	p.u.	0.16–0.25	0.19–0.23	0.21–0.28	0.19–0.35	0.18–0.24
Transient time constants	T'_d	s	0.6–1.3	0.7–1.0	0.75–1.0	0.4–1.1	0.25–1
Subtransient time constants	T''_d	s	0.013–0.022	0.017–0.025	0.022–0.03	0.02–0.04	0.02–0.06
	T''_q	s	0.013–0.022	0.018–0.027	0.02–0.03	0.025–0.04	0.025–0.08
Transient open-circuit time constants	T'_{d0}	s	6–12	6–10	6–9.5	3–9	1.7–4.0
	T'_{q0}	s	—	—	—	0.5–2.0	
Subtransient open-circuit time constants	T''_{d0}	s	0.018–0.03	0.023–0.032	0.025–0.035	0.035–0.06	0.03–0.1
	T''_{q0}	s	0.026–0.045	0.03–0.05	0.04–0.065	0.13–0.2	0.1–0.35
Stator leakage reactance	X_1	p.u.		0.1–0.2		0.1–0.2	
Stator resistance	R_a	p.u.		0.002–0.02		0.0015–0.005	

Note: All values are unsaturated.

Analyzing the expressions for machine parameters given in Table 2.1 and from expressions (2.90a) and (2.90b), and (2.91a)–(2.91f), we may infer that [1]

$$X_d \geq X_q > X'_q \geq X'_d > X''_q \geq X''_d \tag{2.110}$$

and

$$\begin{aligned} T'_{d0} > T'_d > T''_{d0} > T''_d > T_{kd} \\ T'_{q0} > T'_q > T''_{q0} > T''_q \end{aligned} \tag{2.111}$$

The armature time constant T_{ar} is associated with the subtransient conditions of the synchronous machine due to DC components of the armature current following a short circuit, and hence it involves the subtransient inductances L''_d and L''_q. In per unit, the armature time constant is given by

$$T_{\mathrm{ar}} = \frac{1}{R_a}\left(\frac{L''_d + L''_q}{2}\right) \tag{2.112}$$

whereas its usual values are between 0.03 and 0.35 s.

The conventional *method of determining synchronous machine parameters* is from short-circuit tests on unloaded machines. The test procedures are specified in IEEE Standard 115-1995 [10]. These tests provide X_d, X_q, X'_d, X''_d, T'_{d0}, T'_d, T''_{d0}, and T''_d. They do not, however, provide q-axis transient and subtransient constants. In addition, they do not include measurement of the field circuit during the short-circuit tests, and consequently the field circuit is not specifically identified.

Several alternative testing and analytical methods have been proposed and used to obtain better models: enhanced sudden short-circuit tests [16,17], stator decrement test [17–19], frequency-response tests [20–24]. For more information, the reader can refer to Refs [1,10].

2.1.5 Magnetic Saturation

The effects of stator and rotor iron saturation were neglected so far in the development of basic equations of the synchronous generator and the analysis of its characteristics. However, as known, in the case of a ferromagnetic circuit, the saturation occurs in the iron leading to considerable nonlinearity. Therefore, in the general case of magnetic circuits with air gap, as for the synchronous generators, where the closing path of the flux linkages contains the magnetic iron and the air gap, a dependency exists between the flux linkages ψ and the magnetomotive force v_m as indicated by the characteristics in Figure 2.31.

In the absence of saturation, this dependency is illustrated by the line OA called *air gap line* or *air gap characteristic*, indicating the field current (or mmf) required to overcome the reluctance of the air gap. In this case, the magnetic circuit reluctance is dominated by the air gap reluctance.

In reality, the more the magnetomotive force increases, the more the iron saturates and the characteristic ψ–v_m diverts from the air gap line following the saturation curve OB. Consequently, the mmf v_{mt} necessary to obtain a given value ψ_t of the total flux consists of the mmf corresponding to the air gap line v_{mi} and the mmf corresponding to the iron v_{mf}, that is, $v_{mt} = v_{mi} + v_{mf}$.

Under these considerations, the degree of saturation in the rotor and stator iron can be quantified by the *saturation coefficient* K_S, defined as the ratio of the mmf necessary to obtain the flux ψ_t in the absence of saturation to the mmf necessary to obtain the same flux considering the saturation effect. Therefore, observing the similarity of triangles (Figure 2.31), it results

$$K_S = \frac{v_{mi}}{v_{mi} + v_{mf}} = \frac{\psi_t}{\psi_t + \psi_S} = \frac{1}{1 + (\psi_S/\psi_t)} = \frac{1}{1 + S} \tag{2.113}$$

where $S = \psi_S/\psi_t$ is the saturation function of the magnetic iron.

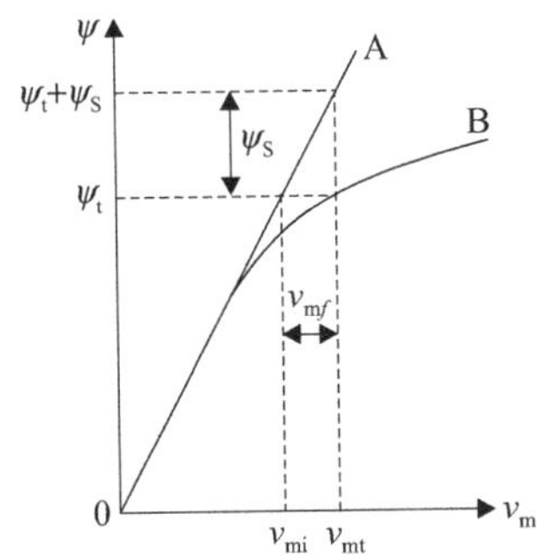

Figure 2.31. Flux–mmf characteristics for a magnetic iron circuit.

By neglecting the magnetic saturation, linear dependency relationships are obtained between the flux linkages, respectively between electromotive forces and the corresponding currents, which in the case of synchronous generator the computational effort is much simplified. Instead, by considering the iron saturation through considerable nonlinearities makes more complicated the analysis of power system transients. Under these conditions, from the dynamic analysis point of view, rigorously representation of synchronous generator saturation is inadequate, since, in general, appropriate information is not available for such approach. For this reason, approximate methods are used in practice to represent the magnetic saturation [1,25,26].

2.1.5.1 Open-Circuit and Short-Circuit Characteristics. For a synchronous generator, the data essential to the treatment of saturation and its influence on the ratio performance/cost are given by the open-circuit and short-circuit characteristics drawn based on measurements.

The open-circuit characteristic (OCC) gives the variation of terminal voltage v in terms of the field current I_f under no-load and rated speed conditions. Knowing that under no-load conditions the current is zero, that is $I_d = I_q = 0$, from the synchronous generator equations, it results

$$\begin{aligned} \psi_q &= 0; \qquad v_d = 0 \\ \psi_d &= L_{\mathrm{md}} I_f; \; v = v_q = E_f = X_{\mathrm{md}} I_f \end{aligned} \tag{2.114}$$

from which, taking into account that the angular speed in per unit is $\omega_{\mathrm{r}} = 1$, it results that the flux ψ_d and terminal voltage v, in per unit, are equal. Therefore, the open-circuit characteristic gives either the variation of the terminal voltage or the variation of the flux linkage in terms of the field current.

A typical open-circuit characteristic is shown in Figure 2.32. It can be seen that for values of voltage at the armature terminals lower than 0.8, in per unit, the saturation is practically inexistent and the open-circuit characteristic follows the air gap line indicating the field current required to overcome the air gap magnetic resistance (reluctance).

However, for values of the terminal voltage greater than 0.8 p.u., magnetic saturation occurs and, therefore, the open-circuit characteristic diverts from the air gap line. Taking into account equations (2.114) for open-circuit conditions, it results that the departure of this characteristic is an indication of the degree of saturation in the d-axis of the synchronous generator.

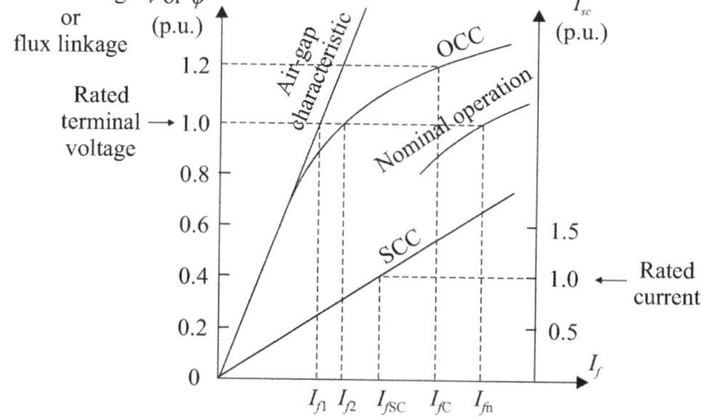

Figure 2.32. Typical open-circuit and short-circuit characteristics.

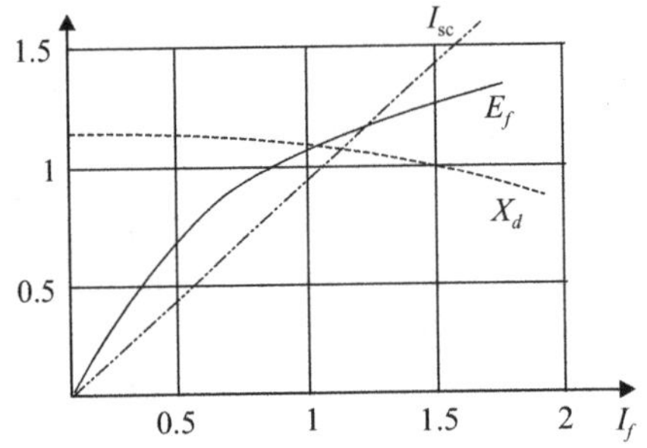

Figure 2.33. Variation of synchronous reactance.

The short-circuit characteristic (SCC) gives the variation of the short-circuit current in terms of the field current I_f, with the generator operating at rated frequency in the steady state and the armature terminals short-circuited (Figure 2.32). The SCC is linear since the saturation is low or absent due to the demagnetizing effect of the armature reaction [1].

Under short-circuit conditions, if the stator resistance (R) is neglected, the terminal voltage being equal to zero $v_d = v_q = 0$, from the steady-state equations of the synchronous generator, it results that the induced emf is

$$E_f = X_d I_{SC} \tag{2.115}$$

and, therefore, the unsaturated value of d-axis synchronous reactance, given by relation $X_d = E_f / I_{SC}$, is constant.

It is obvious that, in the presence of magnetic saturation, the ratio E_f/I_{SC} is not constant, decreasing with the increase of the field (Figure 2.33). Therefore, the d-axis synchronous reactance X_d is no longer an independent parameter of the synchronous generator, taking different values depending on the operating state.

If the saliency is neglected ($X_d = X_q = X_S$), and taking into account that the induced emf is directly proportional to the field current intensity, then equation (2.115) can be written as

$$KI_f = X_S I_{SC} \tag{2.116}$$

Since under rated short-circuit state $I_{SC} = 1$ p.u., and the corresponding field current is I_{fSC} (see Figure 2.32), from equation (2.116) it results

$$KI_{fSC} = 1.0 \cdot X_{S,\text{unsat}} \tag{2.117}$$

where $X_{S,\text{unsat}}$ is the unsaturated value of synchronous reactance.

On the other hand, considering that under no-load conditions the terminal voltage is 1 p.u., according to Figure 2.32, the value of the field current corresponding to this voltage, on the air gap characteristic, is I_{f1} and therefore $KI_{f1} = 1$. Taking into account this equality, from equation (2.117), the unsaturated value of the synchronous reactance is obtained:

$$X_{S,\text{unsat}} = \frac{I_{fSC}}{I_{f1}} \tag{2.118}$$

If the effect of saturation is considered, then, according to Figure 2.32, in order to obtain a terminal voltage equal to 1 p.u. (rated no-load voltage), the value of the necessary field current is $I_{f2} > I_{f1}$, and the saturated value of the synchronous reactance is

$$X_{S,sat} = \frac{I_{fSC}}{I_{f2}} < \frac{I_{fSC}}{I_{f1}} = X_{S,unsat} \tag{2.119}$$

In conclusion, in the case of a nonlinear characteristic of the no-load condition, the values of the synchronous reactance that should be taken into account in steady-state stability studies under normal operating condition of the generator, do not perfectly coincide with the values X_d, obtained from short-circuit measurements; however, the characteristic of X_d variation—decreasing for large field currents—is also maintained under these conditions.

Usually, information about synchronous reactance of the generators always contains unsaturated values of the reactance. In steady-state stability assessment, in the calculation of power characteristics, and so on, these values should be lowered by at least 20–30%.

For round rotor generators, with uniform air gap, the saturation has the same influence both on the d- and q-axis synchronous reactances. Instead, for salient pole generators, the higher value of the air gap along the q-axis removes the influence of the iron characteristic on the q-axis synchronous reactance X_q, practically being independent of the saturation.

The short-circuit ratio (SCR) is a measure of the relative strength of rotor (field) and stator ampere-turns. It is the ratio of the field current required to produce rated stator voltage at rated speed on the open-circuit curve to the field current required to produce rated stator current at short circuit. The unsaturated value is given by SCR $= I_{f1}/I_{fSC}$, and is the reciprocal of the unsaturated per unit synchronous impedance. The saturated short-circuit ratio is given by I_{f2}/I_{fSC}, which is slightly different than the reciprocal of the per unit synchronous reactance and may exhibit small fluctuations.

SCR is an indicator reflecting the degree of saturation having a practical significance with respect to both the performance of the generator and its cost. Therefore, the lower the degree of saturation, the lower is the SCR and has a minimum value, equal to the inverse of reactance $X_{S,unsat}$, in the absence of saturation [1].

2.1.5.2 Considering the Saturation in Stability Studies. In the representation of magnetic saturation for stability studies, a simple method is necessary to determine the saturation factor K_S, defined according to (2.113), for any operating condition and to introduce it in the calculation of synchronous generator parameters.

In developing a saturation model, the following assumptions are usually made [27]:

(i) Usually the only saturation data available for a machine is its open-circuit characteristic. It is commonly assumed that the d-axis saturation characteristic of the loaded generator is the same as the open-circuit characteristic. In using the OCC when the generator is loaded it is necessary to assume some flux level to identify the operating point. In some models the total air gap flux is used, while in others the voltage behind subtransient reactance is used.

(ii) It is often assumed that for round rotor machines the d- and q-axis saturation characteristics are the same. Several investigations have revealed that the q-axis of round rotor machines saturates significantly more than the d-axis, leading to errors in the calculation of the initial rotor angle and field excitation. Ideally, a generator saturation model should allow for different d- and q-axis saturation characteristics.

(iii) Saturation functions can be used to represent saturation characteristics, including two-piece linear, exponential, nonlinear, and quadratic. The choice has little effect on the accuracy of the model and will normally be determined by ease of implementation.

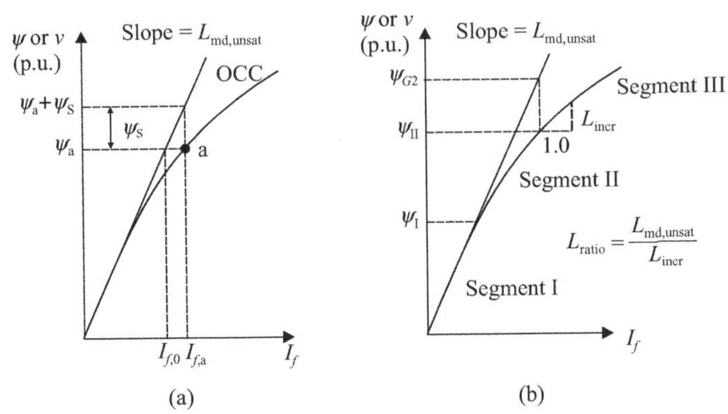

Figure 2.34. Open-circuit saturation characteristic of the synchronous generator used for evaluation of saturation effects under loaded conditions: (a) definition of the saturation factor in the d-axis; (b) definition of saturation segments [1].

(iv) In the coupled-circuit representation of saturation, it is usually assumed that the leakage reactances are independent of saturation and that the leakage fluxes have reduced contribution to the iron saturation. Hence, only the magnetizing reactances X_{md} and X_{mq} saturate and their saturation is determined by the air gap flux linkages.

The saturated values of the mutual inductances are determined by

$$\begin{aligned} L_{md} &= K_{Sd} L_{md,\text{unsat}} \\ L_{mq} &= K_{Sq} L_{mq,\text{unsat}} \end{aligned} \tag{2.120}$$

where K_{Sd} and K_{Sq} represent the degree of saturation in the d-axis and q-axis, respectively, and $L_{md,\text{unsat}}$ and $L_{mq,\text{unsat}}$ are the unsaturated values of L_{md} and L_{mq}.

The factor K_{Sd} can be determined from the OCC. Let us consider an operating point marked with "a" on the OCC to which the field current $I_{f,\text{a}}$ and flux ψ_a corresponds (Figure 2.34a).

Taking into account equation (2.113) of the saturation factor and the similarity of triangles, it results

$$K_{Sd} = \frac{\psi_\text{a}}{\psi_\text{a} + \psi_\text{S}} = \frac{I_{f,0}}{I_{f,\text{a}}} \tag{2.121}$$

Regarding the determination of the degree of saturation in the q-axis, the following issues are taken into account:

- For salient pole generators (hydrogenerators), since the closing path of the flux is mainly air, the iron saturation along the q-axis is insignificant and therefore it can be approximated that $K_{Sq} = 1$ for all loading conditions.
- For round rotor generators (turbogenerators), magnetic saturation exists also in the q-axis, and the factor K_{Sq} should be determined from the no-load saturation characteristic along the q-axis. Since data referring to this characteristic are not available (the magnetizing characteristic along the d-axis is obtained through open-circuit

measurement) it can be approximated, with good accuracy that $K_{Sq} = K_{Sd}$. This approximation is equivalent with the assumption that the air gap and the magnetic path reluctance, respectively, are nonuniform along the rotor circumference.

Taking into account the above-presented issues, it results that, in order to represent the effects of saturation, it is necessary to identify a suitable mathematical function, which can quantify the deviation of OCC from the air gap line. In this regard, the OCC is divided into three segments (Figure 2.34b) [1]:

- *Segment I*, corresponding to the interval in which saturation phenomenon does not occur, is characterized by values of the flux linkages lower than the threshold value ψ_{I}, which is usually 0.8 p.u. In this case, it results

$$\psi_{\mathrm{S}} = 0 \tag{2.122}$$

- *Segment II*, corresponding to a partially saturation of the magnetic iron, is characterized by values of the flux linkages greater than ψ_{I} and lower than ψ_{II}. The threshold value ψ_{II} over which full saturation of the magnetic iron occurs is usually 1.2 p.u. In this case, ψ_{S} can be expressed by an exponential function:

$$\psi_{\mathrm{S}} = A_{\mathrm{S}}\, \mathrm{e}^{B_{\mathrm{S}}(\psi_{\mathrm{a}} - \psi_{\mathrm{I}})} \tag{2.123}$$

where A_{S} and B_{S} are constants depending on the saturation characteristic in the segment II. When $\psi_{\mathrm{a}} = \psi_{\mathrm{I}}$, from equation (2.123) it results $\psi_{\mathrm{S}} = A_{\mathrm{S}}$ and therefore this representation results in a small discontinuity at the junction of segments I and II. However, A_{S} is normally very small and the discontinuity is inconsequential.

- *Segment III*, corresponding to full saturation of the iron, is characterized by values of the flux linkages larger than ψ_{II}. In this case, the deviation from the air gap characteristic can be evaluated by

$$\psi_{\mathrm{S}} = \psi_{\mathrm{G2}} + L_{\mathrm{ratio}}(\psi_{\mathrm{a}} - \psi_{\mathrm{II}}) - \psi_{\mathrm{a}} \tag{2.124}$$

where $L_{\mathrm{ratio}} = L_{\mathrm{m}d,\mathrm{unsat}}/L_{\mathrm{incr}}$ is the ratio of the slope of the air gap line, equal to the unsaturated value of the inductivity $L_{\mathrm{m}d,\mathrm{unsat}}$, to the incremental slope of segment III of the OCC.

Then, saturation characteristic for any given synchronous generator is completely defined by parameters ψ_{I}, ψ_{II}, ψ_{G2}, A_{S}, B_{S}, and L_{ratio}.

The value of $K_{\mathrm{S}d}$, for any given operating condition, is computed as a function of the corresponding air gap flux linkage given by

$$\psi_{\mathrm{a}} = \sqrt{\psi_{\mathrm{m}d}^2 + \psi_{\mathrm{m}q}^2} \tag{2.125}$$

where $\psi_{\mathrm{m}d}$ and $\psi_{\mathrm{m}q}$ are the d- and q-axis components of the air gap or mutual flux linkages. The values of the two components $\psi_{\mathrm{m}d}$ and $\psi_{\mathrm{m}q}$ are given by

$$\begin{aligned} \psi_{\mathrm{m}d} &= \psi_d + L_{\mathrm{l}} I_d \\ \psi_{\mathrm{m}q} &= \psi_q + L_{\mathrm{l}} I_q \end{aligned} \tag{2.126}$$

where L_{l} is the leakage inductance.

Multiplying equations (2.126) by ω_r and taking into account the stator equations

$$V_d = -RI_d - \omega_r\psi_q$$
$$V_q = -RI_q + \omega_r\psi_d$$

resulted by neglecting the transformer emf $d\psi_d/dt$ and $d\psi_q/dt$ in equations (2.39a) and (2.39b), obtain

$$\begin{aligned}\omega_r\psi_{md} &= \omega_r\psi_d + \omega_r L_l I_d = V_q + RI_q + \omega_r L_l I_d = V_{mq} \\ \omega_r\psi_{mq} &= \omega_r\psi_q + \omega_r L_l I_q = -V_d - RI_d + \omega_r L_l I_q = -V_{md}\end{aligned} \tag{2.127}$$

from which it results

$$V_{md} + jV_{mq} = V_d + jV_q + R(I_d + jI_q) + j\omega_r L_l(I_d + jI_q) \tag{2.128a}$$

and

$$\underline{V}_a = \underline{V} + (R + jX_l)\underline{I} \tag{2.128b}$$

In the per unit system, given that $\omega_r = 1$, from equations (2.127) and (2.128) it results that the flux linkage ψ_a is equal to the air gap voltage magnitude V_a.

Under these conditions, for any operating state characterized by values $\underline{V}$ and $\underline{I}$ of the terminal voltage and armature current, respectively, the computation of the saturation factors K_{Sd} and K_{Sq} is performed using the following algorithm:

1. Compute, using equation (2.128b), the emf $\underline{V}_a$ and its components V_{md} and V_{mq}.
2. Compute the air gap flux linkage $\psi_a = \sqrt{\psi_{md}^2 + \psi_{mq}^2} = \sqrt{V_{md}^2 + V_{mq}^2}$.
3. In terms of the value of ψ_a computed at step 2, the deviation ψ_S from the air gap line is determined using equations (2.122), (2.123), or (2.124).
4. Using equation (2.121), determine K_{Sd}.
5. If the generator has round rotor then set $K_{Sq} = K_{Sd}$, whereas if the generator has salient rotor set $K_{Sq} = 1$.

In detailed machine modeling, where the equations are expressed explicitly in terms of mutual and leakage reactances, saturation can be accounted for simply adjusting the values of X_{md} and X_{mq} during the step-by-step computation, according to

$$X_{md,S} = \frac{X_{md}}{1 + K_{Sd}}; \quad X_{mq,S} = \frac{X_{mq}}{1 + K_{Sq}}$$

Cross-Magnetizing Phenomenon. It has been recognized that the magnetic coupling between the direct and quadrature axes of saturated synchronous machine, known as cross-magnetizing phenomenon, which in effect results in the accurate representation of the saturation effect in the axis of the resultant ampere-turns, plays an important role in their analysis. This magnetic coupling causes changes in the flux linkages in their direct and quadrature axes. This could be represented in the phasor diagram in the form of two d- and

q-axis voltage drops, E_{dq} and E_{qd}, which are proportional to the change in the d- and q-axis flux linkages, respectively, and which can be called the cross-magnetizing voltages. The values of these cross-magnetizing voltages are functions of the ampere-turns of both the direct and quadrature axes [28]. In this case, the projections of the internal voltage behind the leakage and Potier reactances on the direct and quadrature axes are used to define the operating point on the corresponding saturation curves to determine K_d and K_q, respectively. More details can be found in the IEEE Std. 1110-2002 [[6], Section 6.2.3].

2.1.6 Modeling in Dynamic State

2.1.6.1 Simplified Electromagnetic Models. The power system stability studies are usually based on the following assumptions:

- Transformer emf due to variation in time of the flux linkages ($\mathrm{d}\psi_d/\mathrm{d}t$ and $\mathrm{d}\psi_q/\mathrm{d}t$) is neglected.
- The rotor speed $\omega_\mathrm{r} = \mathrm{d}\theta/\mathrm{d}t$ is assumed equal to ω_0, since the deviations of the angular speed are small.
- The stator winding resistance R, which has low value, is neglected.
- The magnetic saturation is neglected.

Synchronous Generator with Rotor Field Winding Only. In this particular case assumes that only the field winding is located on the rotor besides the two fictitious windings d and q, respectively, which are equivalent to the stator winding. Therefore, the Park equations of the synchronous generator, expressed in p.u., become

$$V_d \cong -\omega_0 \psi_q \tag{2.129a}$$

$$V_q \cong \omega_0 \psi_d \tag{2.129b}$$

$$\psi_d = -L_d I_d + L_{\mathrm{md}} I_f \tag{2.130a}$$

$$\psi_q = -L_q I_q \tag{2.130b}$$

$$\psi_f = -L_{\mathrm{md}} I_d + L_f I_f \tag{2.130c}$$

$$V_f = R_f I_f + \frac{\mathrm{d}\psi_f}{\mathrm{d}t} \quad \Rightarrow \quad \frac{\mathrm{d}\psi_f}{\mathrm{d}t} = V_f - R_f I_f \tag{2.131}$$

where L_d, L_q are the d- and q-axis synchronous inductances, L_f is the field-winding inductance, L_{md} is the mutual inductance between the field winding f and the fictitious d winding.

Substituting the flux equations from (2.130a) and (2.130b) in (2.129a) and (2.129b), it results

$$V_d \cong X_q I_q \tag{2.129$'$a}$$

$$V_q \cong -\omega_0 L_d I_d + \omega_0 L_{\mathrm{md}} I_f = -X_d I_d + E_{\mathrm{I}q} \tag{2.129$'$b}$$

$$E_{\mathrm{I}q} = \omega_0 L_{\mathrm{md}} I_f = X_{\mathrm{md}} I_f \tag{2.132}$$

where $X_d = \omega_0 L_d$ is the d-axis synchronous reactance, $X_q = \omega_0 L_q$ is the q-axis synchronous reactance, and $E_{\mathrm{I}q}$ is the emf proportional to the field current.

Under no-load conditions, when $I_d = I_q = 0$, from (2.129′b) it results that $V_q = E_{Iq}$. Therefore, E_{Iq} is called open-circuit electromotive force or open-circuit voltage.

The field current I_f is expressed from equation (2.130c) of the rotor circuit flux linkages and substituted in equation (2.130a) of the d-axis flux linkages, which gives

$$\psi_d = -\left(L_d - \frac{L_{md}^2}{L_f}\right)I_d + \frac{L_{md}}{L_f}\psi_f = -L'_d I_d + \frac{L_{md}}{L_f}\psi_f \tag{2.133}$$

Taking into account (2.133), from (2.129b) obtain

$$V_q = -X'_d I_d + E'_q \tag{2.134}$$

where X'_d is the d-axis transient reactance, with

$$X'_d = \omega_0 L'_d = \omega_0\left(L_d - \frac{L_{md}^2}{L_f}\right) \tag{2.135}$$

and E'_q is the emf proportional to the rotor circuit flux linkages or the emf behind the transient reactance:

$$E'_q = \omega_0 \frac{L_{md}}{L_f}\psi_f \tag{2.136}$$

It can be seen that, in the case of the simplified model, equation (2.129a) remains unchanged since there is no rotor winding in the q-axis.

The relationship between E_{Iq} and E'_q may be easily obtained by equating the two forms (2.129′b) and (2.134) of the voltage V_q:

$$-X_d I_d + E_{Iq} = -X'_d I_d + E'_q$$

that is

$$E_{Iq} = (X_d - X'_d)I_d + E'_q \tag{2.137}$$

In order to establish the variation in time of the voltage E'_q, its definition expression (2.136) is differentiated, taking also into account equation (2.131). It then results

$$\frac{dE'_q}{dt} = \omega_0 \frac{L_{md}}{L_f}\frac{d\psi_f}{dt} = \omega_0 \frac{L_{md}}{L_f} V_f - \omega_0 L_{md}\frac{R_f}{L_f} I_f \tag{2.138}$$

Taking into account that

$$E_f = \omega_0 \frac{L_{md}}{R_f} V_f \tag{2.58}$$

is the induced emf by the field current under no-load conditions (see Section "Equivalent Circuit and Phasor Diagram"), and

$$T'_{d0} = \frac{L_f}{R_f} \tag{2.139}$$

is the d-axis open-circuit time constant, that is, the time constant of the rotor winding when the stator winding is open-circuited, the differential equation (2.138) becomes

$$\frac{\mathrm{d}E'_q}{\mathrm{d}t} = \frac{\omega_0 L_{\mathrm{md}}}{R_f}\frac{R_f}{L_f}V_f - \omega_0 L_{\mathrm{md}} I_f \frac{R_f}{L_f} = \frac{(\omega_0 L_{\mathrm{md}} V_f)/R_f - \omega_0 L_{\mathrm{md}} I_f}{L_f/R_f}$$

or

$$\frac{\mathrm{d}E'_q}{\mathrm{d}t} = \frac{E_f - E_{\mathrm{I}q}}{T'_{d0}} \tag{2.140}$$

Also, from (2.140) the relationship between E_f and $E_{\mathrm{I}q}$ is obtained under the form

$$E_f = E_{\mathrm{I}q} + T'_{d0}\frac{\mathrm{d}E'_q}{\mathrm{d}t} \tag{2.141}$$

Substituting equation (2.137) of the emf $E_{\mathrm{I}q}$ in (2.140), it results

$$\frac{\mathrm{d}E'_q}{\mathrm{d}t} = \frac{1}{T'_{d0}}\left[-E'_q + E_f - (X_d - X'_d)I_d\right] \tag{2.140$'$}$$

Equations (2.140) and (2.140′), called damping equations of the field flux, expressing the variation of the magnetic flux in the field winding (represented by E'_q in (2.136)) under the influence of the excitation (E_f), as well as of the armature reaction (I_d).

Another form of the differential equation representing the variation in time of the emf E'_q is obtained by eliminating the variables $E_{\mathrm{I}q}$ and I_d from equations (2.129′b), (2.134), and (2.141). In this respect, from (2.129′b) the current I_d is expressed and substituted in (2.134), which gives

$$V_q = -\frac{X'_d}{X_d}(E_{\mathrm{I}q} - V_q) + E'_q \tag{2.142}$$

Expressing $E_{\mathrm{I}q}$ from (2.141) and substituting in (2.142) gives

$$V_q = -\frac{X'_d}{X_d}E_f + T'_{d0}\frac{X'_d}{X_d}\frac{\mathrm{d}E'_q}{\mathrm{d}t} + \frac{X'_d}{X_d}V_q + E'_q$$

or

$$T'_d\frac{\mathrm{d}E'_q}{\mathrm{d}t} + E'_q = \frac{X'_d}{X_d}E_f + \frac{X_d - X'_d}{X_d}V_q \tag{2.143}$$

where

$$T'_d = T'_{d0}\frac{X'_d}{X_d} \tag{2.144}$$

is the transient short-circuit time constant defined assuming that the stator windings are short-circuited.

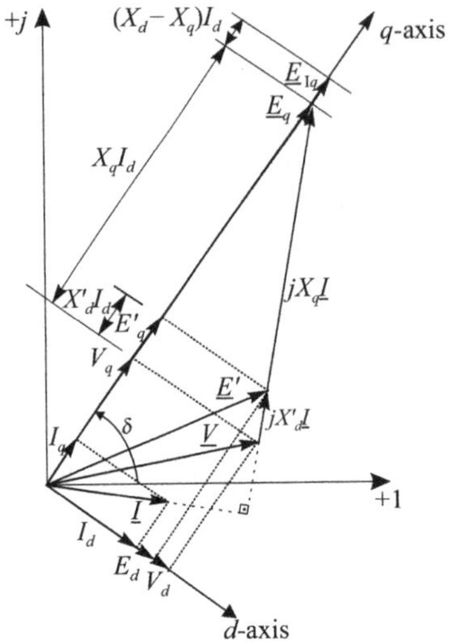

Figure 2.35. Synchronous generator phasor diagram with damper winding neglected, under transient conditions, in terms of E_q, E_{1q}, and E'_q.

From equation (2.143) it results, by numerical integration, the variation in time of the quantity E'_q under transient conditions.

In order to develop the *phasor diagram of the synchronous generator with damper winding neglected* (only with field winding) *under transient conditions* (Figure 2.35), it is necessary to express the emf E'_q, E_{1q}, and E_q in terms of the d- and q-axis components of the stator voltages and currents.

Therefore, if equation (2.129′b) is multiplied with j, gives

$$jE_{1q} = jV_q + jX_dI_d$$

or using phasor notations

$$\underline{E}_{1q} = \underline{V}_q + jX_d\underline{I}_d \tag{2.145}$$

Substituting E_{1q} from (2.129′b) in (2.137) gives

$$E'_q = E_{1q} - (X_d - X'_d)I_d = V_q + X'_dI_d$$

Multiplying with j and using phasor notations gives

$$\underline{E}'_q = \underline{V}_q + jX'_d\underline{I}_d \tag{2.146}$$

From equations (2.145) and (2.146), it can be seen that the emf phasors $\underline{E}_{1q}$ and $\underline{E}'_q$ are located along the q-axis; previously it was shown that the emf $\underline{E}_q$ is also located along the q-axis. The voltage $\underline{E}'_q$ is the projection on the q-axis of the emf $\underline{E}'$ (behind the reactance X'_d).

If in equation (2.60)

$$\underline{E}_q = j[X_{md}I_f - (X_d - X_q)I_d] \tag{2.60}$$

we substitute equation (2.132) $E_{1q} = X_{md} I_f$, gives

$$\underline{E}_{1q} = \underline{E}_q + j(X_d - X_q)\underline{I}_d \tag{2.147}$$

The simplified model—a *third-order model*—of the synchronous generator, which assumes only the existence of the field winding in the rotor (without damper circuits in the d- and q-axis), is used in the dynamic analysis of the small-signal stability and less in transient stability studies. The simplified model consists of the following equations:

a. The differential equation defining the component of transient emf E'_q:

$$\frac{\mathrm{d}E'_q}{\mathrm{d}t} = \frac{1}{T'_{d0}}\left[-E'_q + E_f - (X_d - X'_d)I_d\right] \tag{2.140$'$}$$

b. The swing equations (2.10) and (2.10′), defining the variables ω and δ:

$$\begin{aligned} M\frac{\mathrm{d}\omega}{\mathrm{d}t} + D\omega &\cong P_\mathrm{m} - P_\mathrm{e} \\ \frac{\mathrm{d}\delta}{\mathrm{d}t} &= \omega_0\omega \end{aligned} \tag{2.10}$$

or

$$\frac{M}{\omega_0}\frac{\mathrm{d}^2\delta}{\mathrm{d}t^2} + \frac{D}{\omega_0}\frac{\mathrm{d}\delta}{\mathrm{d}t} \approx P_\mathrm{m} - P_\mathrm{e} \tag{2.10$'$}$$

c. Algebraic equations:

$$I_d = \frac{E'_q - V_q}{X'_d} \tag{2.135$'$}$$

$$I_q \cong \frac{V_d}{X_q} \tag{2.129$'$a}$$

A particular form of this model can be obtained if consider $E'_q = \text{ct.}$ on the time interval of the analyzed dynamic condition.

SYNCHRONOUS GENERATOR WITH ONE DAMPER WINDING IN THE q-AXIS. In completing the set of equations developed for the previous case—synchronous generator without damper windings—the Park equations are written also for the damper winding Q in the q-axis.

Flux linkages equations (ψ_q) and (ψ_Q):

$$\psi_q = -L_q I_q + L_{mQ} I_Q \tag{2.32$'$b}$$

$$\psi_Q = -L'_{mQ} I_q + L_Q I_Q \tag{2.34$'$c}$$

where I_q is the component of armature current in the q-axis and I_Q is the current in the damper circuit.

Equation of the d-axis terminal voltage component:

$$V_d \cong -\omega_0 \psi_q \tag{2.129a}$$

Equation of the damper circuit Q in the q-axis:

$$0 = R_Q I_Q + \frac{d\psi_Q}{dt} \tag{2.40$'$c}$$

where R_Q is the damper circuit resistance.

Rearranging the terms from equations (2.32′b), (2.34′c), (2.129a), and (2.40′c) yields [1]

$$V_d = X_q I_q + E_{Id} \tag{2.148}$$

$$V_d = X'_q I_q + E'_d \tag{2.149}$$

$$0 = E_{Id} + T'_{q0} \frac{dE'_d}{dt} \tag{2.150}$$

where $X_q = \omega_0 L_q$ is the q-axis synchronous reactance;
and

$$X'_q = \omega_0 L'_q = \omega_0 (L_q - (L^2_{mq}/L_Q)) \tag{2.151}$$

is the q-axis transient reactance;

$$E_{Id} = -\omega_0 L_{mq} I_Q \tag{2.152}$$

is the voltage proportional to the damper circuit current (q-axis);

$$E'_d = -\omega_0 (L_{mq}/L_Q)\psi_Q \tag{2.153}$$

is the voltage proportional to the damper circuit flux linkages (q-axis); and

$$T'_{q0} = L_Q/R_Q \tag{2.154}$$

is the open-circuit time constant of the q-axis damper winding.

In order to eliminate the variables I_q and E_{Id}, the current I_q is expressed from (2.148) and substituted in (2.149), which gives

$$V_d = \frac{X'_q}{X_q}(V_d - E_{Id}) + E'_d \tag{2.149$'$}$$

Expressing E_{Id} from equation (2.150) and substituting in (2.149′), gives the differential equation:

$$T'_q \frac{dE'_d}{dt} + E'_d = \left(\frac{X_q - X'_q}{X_q}\right) V_d \tag{2.155}$$

where

$$T'_q = T'_{q0} \frac{X'_q}{X_q} \tag{2.156}$$

is the short-circuit time constant of the q-axis damper winding.

Another form of the differential equation (2.155) may be obtained by eliminating V_d. Therefore, taking into account equations (2.149) and (2.156), it results

$$\frac{\mathrm{d}E'_d}{\mathrm{d}t} = \frac{1}{T'_{q0}} \left[-E'_d + \left(X_q - X'_q \right) I_q \right] \tag{2.155'}$$

From equation (2.155) or (2.155′), by numerical integration, it results the variation in time of the quantity E'_d in transient state.

Equation (2.155′) together with equations (2.140′) and (2.10) forms the model of the synchronous generator with one damper winding in the q-axis, which is a *fourth-order model*:

$$\frac{\mathrm{d}E'_d}{\mathrm{d}t} = \frac{1}{T'_{q0}} (-E'_d + (X_q - X'_d) I_q) \tag{2.155'}$$

$$\frac{\mathrm{d}E'_q}{\mathrm{d}t} = \frac{1}{T'_{d0}} \left(-E'_q + E_f - (X_d - X'_d) I_d \right) \tag{2.140'}$$

$$\begin{aligned} M \frac{\mathrm{d}\omega}{\mathrm{d}t} + D\omega &\cong P_\mathrm{m} - P_\mathrm{e} \\ \frac{\mathrm{d}\delta}{\mathrm{d}t} &= \omega_0 \omega \end{aligned} \tag{2.10}$$

or

$$\frac{M}{\omega_0} \frac{\mathrm{d}^2\delta}{\mathrm{d}t^2} + \frac{D}{\omega_0} \frac{\mathrm{d}\delta}{\mathrm{d}t} \approx P_\mathrm{m} - P_\mathrm{e} \tag{2.10'}$$

$$I_d = \frac{E'_q - V_q}{X'_d} \tag{2.135'}$$

$$I_q = \frac{V_d - E'_d}{X'_q} \tag{2.149''}$$

The phasor representation of the quantities corresponding to the simplified model is shown in Figure 2.36.

The reference coordinates system (d, q), individual for each synchronous machine, was introduced, which rotates at the speed ω with respect to the general reference coordinates system ($+1$, $+j$).

The expression of the active power for a salient pole generator is

$$P_\mathrm{e} = \mathrm{Re}\{\underline{V} \cdot \underline{I}^*\} = V_d I_d + V_q I_q = \frac{E'_q V_d}{X'_d} - \frac{E'_d V_q}{X'_q} + V_d V_q \left(\frac{1}{X'_q} - \frac{1}{X'_d} \right) \tag{2.157}$$

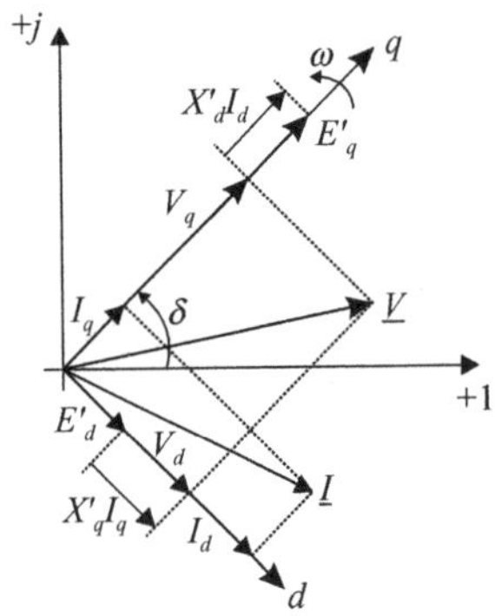

Figure 2.36. Phasor representation of the voltages and currents in transient state.

Synchronous Generator with One Damper Winding in the q-Axis, and the Transient Saliency Neglected ($X'_q = X'_d$). Replacing X'_q by X'_d in equations (2.149) and (2.134), we achieve the voltage phasor as

$$\underline{V} = V_d + jV_q = E'_d + X'_d I_q + j\left(E'_q - X'_d I_d\right) = \underline{E}' - jX'_d \underline{I} \tag{2.158}$$

where

$$\underline{E}' = E'_d + jE'_q \tag{2.159}$$

is the phasor of transient emf behind the d-axis transient reactance X'_d.

It can be seen that the phasor $\underline{E}'$ has the two components defined earlier:

$$E'_d = -\omega_0 \frac{L_{mq}}{L_Q} \psi_Q \tag{2.153}$$

$$E'_q = \omega_0 \frac{L_{md}}{L_f} \psi_f \tag{2.136}$$

Usually, transient stability computations are extended over a short time period (several seconds), period in which first swing power system stability or instability is decided. If, in this period, assumes that the rotor flux linkages ψ_f and ψ_Q are approximately constant and equal to those before the disturbance, it results other assumptions that lead to *the classical model of the synchronous generator for transient stability studies* (for short term). In this case, according to equations (2.136) and (2.153), the components E'_q and E'_d of the phasor $\underline{E}'$ are constant under transient conditions with respect to the coordinates reference system (d, q), individual for each synchronous generator. It results that the phasor $\underline{E}'$ has constant magnitude and constant direction in the reference system (d, q) that rotates with the speed ω.

This particular model offers considerable computational simplicity, required for simulation in time of the transient state, since it simplifies the interconnecting procedure of the synchronous generator to the electrical network.

Equation (2.158) written for each synchronous generator, together with the algebraic equations of nodal voltage that defines the operating state of the electrical network (assuming the loads modeled by constant admittances), allows direct computation of the transient state at the instants t_+.

Knowing the operating state of the system at a given instant, in order to obtain the dynamic response for the variables E'_q and E'_d, equations (2.143) and (2.155), where X'_q is replaced by X'_d, are integrated and gives

$$T'_d \frac{\mathrm{d}E'_q}{\mathrm{d}t} + E'_q = \frac{X'_d}{X_d} E_f + \frac{X_d - X'_d}{X_d} V_q \tag{2.143}$$

$$T'_q \frac{\mathrm{d}E'_d}{\mathrm{d}t} + E'_d = \frac{X_q - X'_d}{X_q} V_d \tag{2.155''}$$

$$\begin{aligned} M\frac{\mathrm{d}\omega}{\mathrm{d}t} + D\omega &\cong P_{\mathrm{m}} - P_{\mathrm{e}} \\ \frac{\mathrm{d}\delta}{\mathrm{d}t} &= \omega_0 \omega \end{aligned} \tag{2.10}$$

or

$$\frac{M}{\omega_0}\frac{\mathrm{d}^2\delta}{\mathrm{d}t^2} + \frac{D}{\omega_0}\frac{\mathrm{d}\delta}{\mathrm{d}t} \cong P_{\mathrm{m}} - P_{\mathrm{e}} \tag{2.10'}$$

After obtaining, for each synchronous generator, the values of E'_q and E'_d, at the end of an integration step, the computation is continued by determining a new operating point of the system, followed by a new integration step, and so on.

Under these considerations ($X'_q = X'_d$), the expression (2.157) of the active power of synchronous generator becomes

$$P_{\mathrm{e}} = \frac{E'_q V_d - E'_d V_q}{X'_q} \tag{2.160}$$

Figure 2.37 illustrates the phasor diagram of voltages and currents assuming $X'_q = X'_d$ and the transient model of the synchronous generator.

When the rotor speed changes, the angle δ' of the phasor $\underline{E}'$ relative to the synchronous reference system (+1, +j) may be used instead of angle δ to measure the change in rotor position.

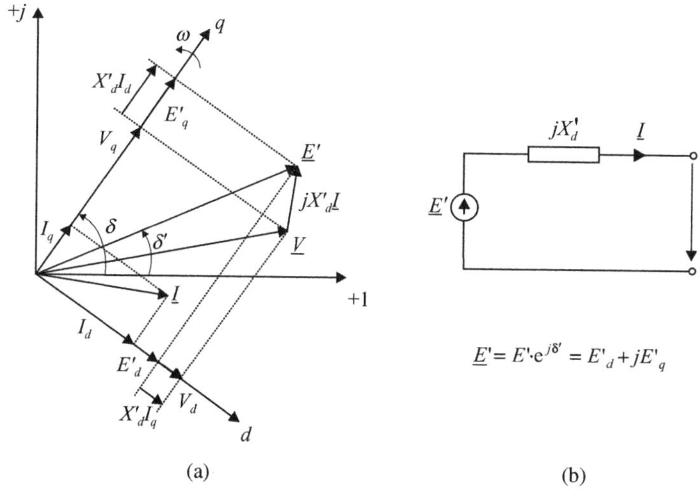

Figure 2.37. Phasor diagram (a) and transient model (b) of the synchronous generator assuming $X'_q = X'_d$.

Therefore, *the classical model* used for transient stability studies consists of two electromechanical differential equations and one phasor equation as follows:

$$\frac{M}{\omega_0}\frac{\mathrm{d}^2\delta'}{\mathrm{d}t^2}+\frac{D}{\omega_0}\frac{\mathrm{d}\delta'}{\mathrm{d}t}\cong P_\mathrm{m}-P_\mathrm{e} \tag{2.161}$$

$$\frac{\mathrm{d}\delta'}{\mathrm{d}t}=\omega_0\omega \tag{2.162}$$

$$\underline{E}'=E'\,\mathrm{e}^{j\delta'}=\underline{V}+jX'_d\underline{I} \tag{2.163}$$

where the magnitude of emf $\underline{E}'$ is constant during the transient period.

When the classical model is used, the active power of the synchronous generator has the expression

$$P_\mathrm{e}=\frac{E'V}{X'_d}\sin\delta' \tag{2.164}$$

2.1.6.2 Detailed Model in Dynamic State. Modeling the Electromagnetic Subtransient Phenomena. In order to obtain a high accuracy in transient stability computation, the synchronous generators are modeled by systems of nonlinear differential equations obtained based on Park transformation where the rotor is represented by four equivalent windings: the field winding, a damper circuit in the d-axis, and two damper circuits in the q-axis, respectively.

The mathematical model is based on the following main assumptions:

- In the rotor d-axis there exists the field winding (f) and a damper circuit (D); in the rotor q-axis, the magnetic influence of two damper circuits (Q and K) is considered.
- The magnetic saturation is neglected, whereas in the equations of the voltage components V_d and V_q, the transformer emf and the voltage drop across the armature resistance ($R_\mathrm{a}=0$) are neglected.

Under these assumptions, the operational equations with external parameters that give the synchronous generator performances in transient operation, expressed in the coordinates system (d, q) solidary with the rotor, are

- components of the terminal voltage:

$$V_d\cong-\omega\psi_q;\quad V_q\cong\omega\psi_d$$

- components of stator flux linkages:

$$\psi_d=A(s)V_f-L_d(s)I_d \tag{2.87}$$

$$\psi_q=-L_q(s)I_q \tag{2.79}$$

where s denotes the operator d/dt.

We observe from (2.87) that the d-axis component of the stator flux is dynamically related to the field voltage and to the d-axis component of the armature current. Also, from

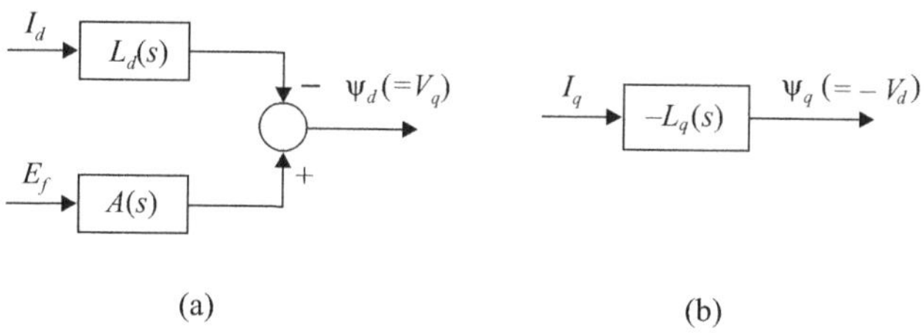

Figure 2.38. General dynamic model of the electromagnetic part of a synchronous generator for analysis of electromechanical phenomena.

(2.79) we see that the q-axis component of the stator flux is related only to the q-axis component of the armature current. Generic dynamic models (Figure 2.38) can be drawn based on equations (2.87) and (2.79) [12].

In the sixth-order dynamic model there are a field circuit and an additional rotor circuit that acts along the d-axis and two additional rotor circuits that act along the q-axis. The transfer functions $L_d(s)$, $L_q(s)$, and $A(s)$ are expressed under the form [1,2,12,29]

$$L_d(s) = L_d \frac{(1+sT'_d)(1+sT''_d)}{(1+sT'_{d0})(1+sT''_{d0})} \tag{2.88$'$}$$

$$L_q(s) = L_q \frac{\left(1+sT'_q\right)\left(1+sT''_q\right)}{\left(1+sT'_{q0}\right)\left(1+sT''_{q0}\right)} \tag{2.92}$$

$$A(s) = \frac{L_{md}}{R_f} \frac{(1+sT_{kd})}{(1+sT'_{d0})(1+sT''_{d0})} \tag{2.89$'$}$$

Substituting the expressions from (2.88′), (2.92), and (2.89′) in equations (2.87) and (2.79), the following operational equations used to define the components I_d and I_q of the stator current are obtained:

$$I_d = \frac{E_f}{X_d} \frac{(1+sT_{kd})}{(1+sT'_d)(1+sT''_d)} - \frac{V_q}{X_d} \frac{(1+sT'_{d0})(1+sT''_{d0})}{(1+sT'_d)(1+sT''_d)} \tag{2.165}$$

$$I_q = \frac{V_d}{X_q} \frac{\left(1+sT'_{q0}\right)\left(1+sT''_{q0}\right)}{\left(1+sT'_q\right)\left(1+sT''_q\right)} \tag{2.166}$$

where $E_f = \omega(L_{md}/R_f)V_f$, $X_d = \omega L_d$, and $X_q = \omega L_q$.

Further on, the mathematical model of the synchronous generator is presented, where the differential equations that defines the transient and subtransient emf in the reference coordinates system (d, q) are obtained by decomposing in simple fractions the operational equations of the stator current components. The result is a simple system of four first-order differential equations that model the electromagnetic transients of the synchronous generator.

In order to define the subtransient E''_d and transient E'_d emf in the d-axis, the operational equation (2.166) is decomposed as follows:

$$I_q = \frac{V_d}{X_q}\left[A_q + \frac{B_q}{1+sT''_q} + \frac{C_q}{\left(1+sT''_q\right)\left(1+sT'_q\right)}\right] \tag{2.167}$$

where the constants A_q, B_q, and C_q can be obtained by equating equations (2.166) and (2.167); the following conditions result:

$$\begin{aligned} &A_q + B_q + C_q = 1 \\ &T'_{q0} + T''_{q0} = A_q\left(T'_q + T''_q\right) + B_q T'_q \\ &T'_{q0}T''_{q0} = A_q T'_q T''_q \end{aligned} \tag{2.168}$$

then we achieve

$$\begin{aligned} A_q &= \frac{T'_{q0}T''_{q0}}{T'_qT''_q} \equiv \frac{X_q}{X''_q} \\ B_q &= \frac{1}{T'_q}\left[\left(T'_{q0} + T''_{q0}\right) - \frac{X_q}{X''_q}\left(T'_q + T''_q\right)\right] \\ C_q &= 1 + \frac{X_q}{X''_q}\frac{T''_q}{T'_q} - \frac{T'_{q0} + T''_{q0}}{T'_q} \end{aligned} \tag{2.169}$$

Substituting the constant A_q from (2.169) in (2.167) gives

$$I_q = \frac{V_d}{X''_q} + \frac{V_d}{X_q}\frac{1}{1+sT''_q}\left(B_q + \frac{C_q}{1+sT'_q}\right) \tag{2.170}$$

The d-axis subtransient emf E''_d is defined using the equation

$$E''_d = V_d - X''_q I_q \tag{2.171}$$

from which we obtain

$$I_q = \frac{V_d - E''_d}{X''_q}$$

Equating with equation (2.170), it results

$$E''_d = -\frac{X''_d}{X_q}\frac{V_d}{1+sT''_q}\left(B_q + \frac{C_q}{1+sT'_q}\right) \tag{2.171'}$$

The d-axis transient emf E'_d is also defined by

$$E'_d = -C_q\frac{V_d}{1+sT'_q} \tag{2.172}$$

The operational equation (2.171′) becomes

$$E''_d = -\frac{1}{1+sT''_q}\frac{X''_q}{X_q}\left(B_q V_d - E'_d\right) \tag{2.171''}$$

The following first-order differential system of equations corresponds to the operational equations (2.172) and (2.171″):

$$T'_q \frac{dE'_d}{dt} + E'_d = -C_q V_d \tag{2.173}$$

$$T''_q \frac{dE''_d}{dt} + E''_d = -\frac{X''_q}{X_q}\left(B_q V_d - E'_d\right) \tag{2.174}$$

If the q-axis transient reactance X'_q is additionally introduced, given by equation (2.156)

$$X'_q = X_q \frac{T'_q}{T'_{q0}} \tag{2.175}$$

and neglecting the subtransient time since they have small influence, the following approximate values are obtained for the constants B_q and C_q:

$$B_q \cong \frac{X_q}{X''_q}\frac{X''_q - X'_q}{X'_q} \tag{2.176a}$$

$$C_q \cong \frac{X'_q - X_q}{X'_q} \tag{2.176b}$$

It results, then, the following form of the differential equations (2.173) and (2.174):

$$T'_q \frac{dE'_d}{dt} + E'_d = \frac{X_q - X'_q}{X'_q} V_d \tag{2.177}$$

$$T''_q \frac{dE''_d}{dt} + E''_d = \frac{X''_q}{X_q} E'_d + \frac{X'_q - X''_q}{X'_q} V_d \tag{2.178}$$

Integration of equations (2.177) and (2.178) gives the time response of the quantities E'_d and E''_d, used to determine the q-axis armature current.

In order to define the subtransient E''_q and transient E'_q emf in the q-axis, the operational equation (2.165) is decomposed under the form

$$\begin{aligned} I_d = {} & \frac{E_f}{X_d}\left[\frac{B'_d}{1+sT''_d} + \frac{C'_d}{\left(1+sT''_d\right)\left(1+sT'_d\right)}\right] \\ & - \frac{V_q}{X_d}\left[A_d + \frac{B_d}{1+sT''_d} + \frac{C_d}{\left(1+sT''_d\right)\left(1+T'_d\right)}\right] \end{aligned} \tag{2.179}$$

Equating equations (2.165) and (2.179) and identifying the terms, give

$$B'_d = \frac{T_Q}{T'_d} = \alpha \quad ; \quad C'_d = (1-\alpha) \quad ; \quad A_d = \frac{T'_{d0}T''_{d0}}{T'_dT''_d} \equiv \frac{X_d}{X''_d}$$

$$B_d = \frac{1}{T'_d}\left[(T'_{d0} + T''_{d0}) - \frac{X_d}{X''_d}(T'_d + T''_d)\right] \tag{2.180}$$

$$C_d = 1 + \frac{X_d}{X''_d}\frac{T''_d}{T'_d} - \frac{T'_{d0} + T''_{d0}}{T'_d}$$

Substituting the expression of A_d in equation (2.179), it results

$$I_d = \frac{E_f}{X_d}\frac{1}{1+sT''_d}\left(B'_d + \frac{C'_d}{1+sT'_d}\right) - \frac{V_q}{X_d}\frac{1}{1+sT''_d}\left(B_d + \frac{C_d}{1+sT''_d}\right) - \frac{V_q}{X''_d} \tag{2.181}$$

The subtransient emf E''_q is defined by

$$E''_q = V_q + X''_d I_d \tag{2.182}$$

which gives

$$I_d = \frac{E''_q - V_q}{X''_d}$$

Substituting the expression of I_d in equation (2.181), it results

$$E''_q = \frac{X''_d}{X_d}\frac{1}{1+sT''_d}\left[(B'_dE_f - B_dV_q) + \frac{1}{1+sT'_d}(C'_dE_f - C_dV_q)\right]$$

Let us denote by

$$E'_q = \frac{1}{1+sT'_d}(C'_dE_f - C_dV_q) \tag{2.183}$$

therefore

$$E''_q = \frac{1}{1+sT''_d}\frac{X''_d}{X_d}(B'_dE_f - B_dV_q + E'_d) \tag{2.184}$$

The following differential equations correspond to the operational equations (2.183) and (2.184):

$$T'_d\frac{\mathrm{d}E'_q}{\mathrm{d}t} + E'_q = C'_dE_f - C_dV_q \tag{2.185}$$

$$T''_d\frac{\mathrm{d}E''_q}{\mathrm{d}t} + E''_q = \frac{X''_d}{X_d}\left(B'_dE_f - B_dV_q + E'_q\right) \tag{2.186}$$

Using the expression of the d-axis transient reactance from (2.144) and applying simplifying assumptions, obtain the following approximate expressions for the sought

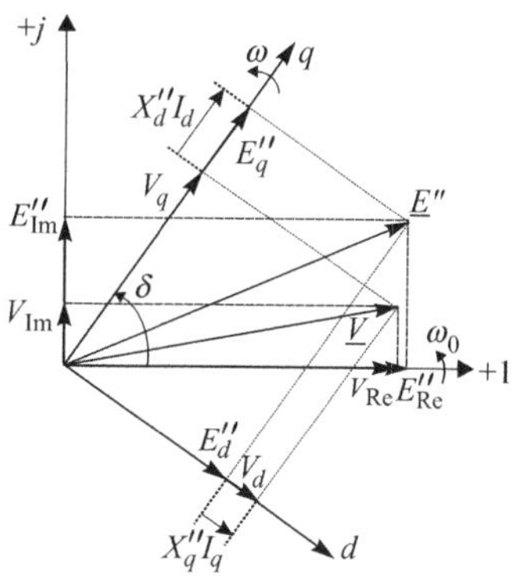

Figure 2.39. The phasor diagram of the voltages when considering the subtransient electromagnetic phenomena.

constants [29]:

$$B_d \cong \frac{X_d}{X''_d}\frac{X''_d - X'_d}{X'_d};\quad C_d \cong \frac{X'_d - X_d}{X'_d}$$
$$B'_d = \alpha;\qquad C'_d = (1-\alpha) \tag{2.187}$$

The following form of the differential system of equations (2.185) and (2.186) is obtained

$$T'_d\frac{dE'_q}{dt} + E'_q = (1-\alpha)E_f + \frac{X_d - X'_d}{X'_d}V_q \tag{2.188}$$

$$T''_d\frac{dE''_q}{dt} + E''_q = \frac{X''_d}{X_d}\left(\alpha E_f + E'_q\right) + \frac{X'_d - X''_d}{X'_d}V_q \tag{2.189}$$

Integration of differential equations (2.188) and (2.189) gives the time response of the variables E'_q and E''_q used to determine the d-axis armature current.

Figure 2.39 shows the phasor diagram of the voltages when considering the subtransient electromagnetic phenomena.

Interconnection of Synchronous Generator to the Electrical Grid. As previously shown, the Park equations of the synchronous generator are referred to a local reference system, individual for each machine (d- and q-axis), which rotates at the same speed ω_r as the rotor.

Interconnection of the synchronous generator to the electrical grid requires a common angular reference—the unique coordinates reference system $(+1, +j)$ or (Re, Im), which rotates at synchronous speed ω_0. Therefore, the rotational phasors of the electrical parameters and the d- and q-axis of the synchronous generator are represented on the same diagram (Figure 2.40). Let us denote by δ the angle by which the rotor q-axis leads the synchronous axis $+1$.

Under steady-state conditions, δ is the angle of the emf $\underline{E}_q$, whereas under dynamic conditions, δ varies in time with the rotor speed.

According to Figure 2.40, the coordinates system (d, q), associated to the synchronous generator, rotates counterclockwise with the angle $\delta - \pi/2$ with respect to the coordinates system $(+1, +j)$, associated to the electrical grid. Therefore, the voltage phasor $\underline{V}$ may be expressed in two ways [43].

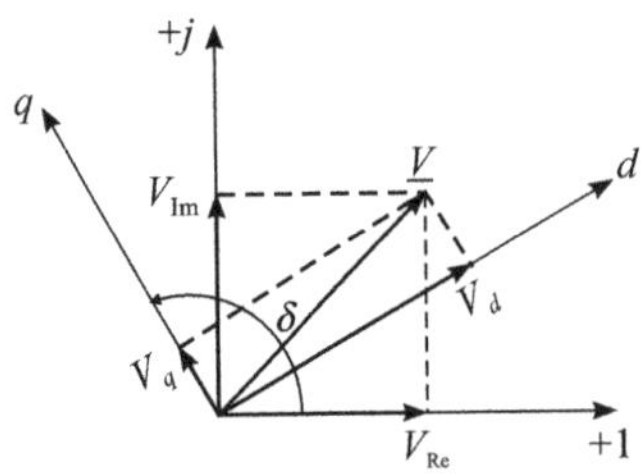

Figure 2.40. The dependence between the coordinates system of the power grid (+1, +j) and that of the synchronous generator (d, q).

(i) In terms of components V_d and V_q:

$$\underline{V} = (V_d + jV_q)e^{j(\delta-\pi/2)} \tag{2.190}$$

(ii) In terms of components V_{Re} and V_{Im}:

$$\underline{V} = V_{Re} + jV_{Im} = V\,e^{j\theta} \tag{2.191}$$

It, therefore, results

$$V_{Re} + jV_{Im} = (V_d + jV_q)e^{j(\delta-\pi/2)} = (V_d + jV_q)[\cos(\delta - \pi/2) + j\sin(\delta - \pi/2)]$$

Identifying the real and imaginary components, obtain the transformation equations from the coordinates system (d, q) to the coordinates system (+1, +j)

$$V_{Re} = V_d\cos(\delta - \pi/2) - V_q\sin(\delta - \pi/2) = V_d\sin\delta + V_q\cos\delta$$
$$V_{Im} = V_q\cos(\delta - \pi/2) + V_d\sin(\delta - \pi/2) = -V_d\cos\delta + V_q\sin\delta$$

or in matrix form

$$\begin{bmatrix} V_{Re} \\ V_{Im} \end{bmatrix} = \begin{bmatrix} \sin\delta & \cos\delta \\ -\cos\delta & \sin\delta \end{bmatrix} \begin{bmatrix} V_d \\ V_q \end{bmatrix}$$

The inverse transformation, that is, from the coordinates system (+1, +j) to the coordinates system (d, q), is given by the matrix relationship:

$$\begin{bmatrix} V_d \\ V_q \end{bmatrix} = \begin{bmatrix} \sin\delta & -\cos\delta \\ \cos\delta & \sin\delta \end{bmatrix} \begin{bmatrix} V_{Re} \\ V_{Im} \end{bmatrix}$$

For converting the components E''_{Re} and E''_{Im} of the phasor $\underline{E}''$, from the coordinates system (d, q) to the general coordinates system (+1,+j), the following equations are used:

$$\begin{aligned} E''_{Re} &= E''_d\sin\delta + E''_q\cos\delta \\ E''_{Im} &= E''_q\sin\delta - E''_d\cos\delta \end{aligned} \tag{2.192}$$

If the voltage equations in the coordinates system (d, q)

$$V_d = X''_q I_q + E''_d$$
$$V_q = -X''_d I_d + E''_q$$

are expressed in the unique reference system (+1, +j), using the expressions (2.190) and (2.192) give the following matrix equation:

$$\begin{bmatrix} V_{Re} \\ V_{Im} \end{bmatrix} = \begin{bmatrix} \left(X''_q - X''_d\right)\sin\delta\cos\delta & X''_q \sin^2\delta + X''_d \cos^2\delta \\ -\left(X''_q \cos^2\delta + X''_d \sin^2\delta\right) & -\left(X''_q - X''_d\right)\sin\delta\cos\delta \end{bmatrix} \cdot \begin{bmatrix} I_{Re} \\ I_{Im} \end{bmatrix} + \begin{bmatrix} E''_{Re} \\ E''_{Im} \end{bmatrix} \tag{2.193}$$

For a round rotor generators, with subtransient saliency neglected, the subtransient reactances are equal $X''_d = X''_q = X''$, and equation (2.193) becomes

$$\begin{bmatrix} V_{Re} \\ V_{Im} \end{bmatrix} = \begin{bmatrix} 0 & X'' \\ -X'' & 0 \end{bmatrix} \cdot \begin{bmatrix} I_{Re} \\ I_{Im} \end{bmatrix} + \begin{bmatrix} E''_{Re} \\ E''_{Im} \end{bmatrix} \tag{2.194}$$

or

$$V_{Re} = X'' I_{Im} + E''_{Re}$$
$$V_{Im} = -X'' I_{Re} + E''_{Im}$$

It results

$$\underline{V} = V_{Re} + jV_{Im} = \left(E''_{Re} + jE''_{Im}\right) - jX''(I_{Re} + jI_{Im}) = \underline{E}'' - jX''\underline{I} \tag{2.195}$$

where the phasor $\underline{E}''$ stands for the subtransient emf behind the reactance X''.

Equations (2.193), in the general case, or (2.195), in the case of $X''_d = X''_q$, are used for interconnecting each synchronous generator to the electrical grid. In this respect, from (2.193) obtain the phasor of currents injected by the synchronous generator into the electrical network, which can also be expressed in terms of the terminal voltage phasor and the subtransient emf phasor:

$$\begin{bmatrix} I_{Re} \\ I_{Im} \end{bmatrix} = \begin{bmatrix} \left(X''_q - X''_d\right)\sin\delta\cos\delta & X''_q \sin^2\delta + X''_d \cos^2\delta \\ -\left(X''_q \cos^2\delta + X''_d \sin^2\delta\right) & -\left(X''_q - X''_d\right)\sin\delta\cos\delta \end{bmatrix}^{-1} \cdot \left[\begin{bmatrix} V_{Re} \\ V_{Im} \end{bmatrix} - \begin{bmatrix} E''_{Re} \\ E''_{Im} \end{bmatrix}\right] \tag{2.196}$$

With subtransient saliency neglected, gives

$$I_{Re} = \frac{1}{X''}\left(E''_{Im} - V_{Im}\right)$$
$$I_{Im} = \frac{1}{X''}\left(V_{Re} - E''_{Re}\right)$$

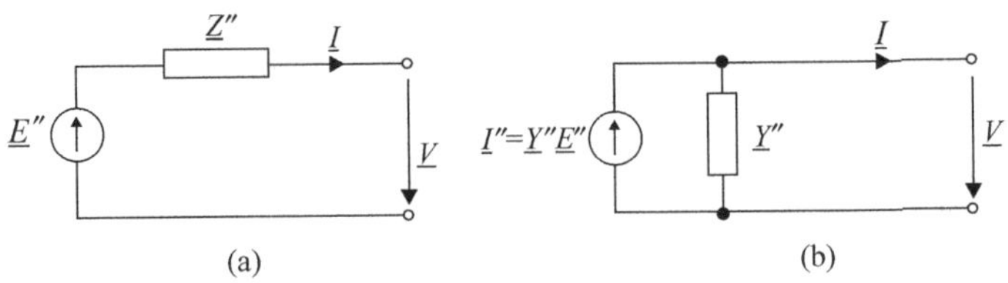

Figure 2.41. The equivalent circuits of the synchronous generator with magnetic saturation neglected: (a) Thévenin equivalent; (b) Norton equivalent.

or in phasor form

$$\underline{I} = \frac{\underline{E}''}{jX''} - \frac{1}{jX''}\underline{V} \tag{2.197}$$

that represents the Norton equivalent of the synchronous generator "seen" from its terminals.

Therefore, the synchronous generator can be represented by one of the equivalent circuits shown in Figure 2.41.

2.1.7 Reactive Capability Limits

2.1.7.1 Loading Capability Chart. The synchronous generator is the main equipment in a bulk power system used for voltage control. A generator maintains the voltage at its terminals to the predefined value by exchanging reactive power with the electrical network. The produced/absorbed amount of reactive power needed for the generator to control the voltage as required by the system operator may fall beyond the capability limits of the generator.

The reactive capabilities of the synchronous generator depend on the type of cooling employed and the environmental conditions such as pressure of cooling agent (e.g., hydrogen) or air temperature (which is variable with the season, e.g., summer or winter).

In order to develop the capability curves of the synchronous generator we start from the phasor diagram of the synchronous machine under balanced steady state from Figure 2.16, in which we neglect the resistance since it is much smaller then the reactance and considering the round rotor generator, with $X_S = X_d = X_q$. The saturation is also neglected as it involves negligible differences. In terms of powers, we obtain the so-called *loading capability chart* (Figure 2.42) [30].

In the following we define the three design limits of the generator that restrict provision of reactive power.

(i) *Armature Current Limit.* The current produced by the generator results in heating of the stator windings and significant damages if limiting actions are not taken. For this reason, the value of the current carried in the stator winding must be limited to a certain value. In the P–Q plane, this limit is a circle with center in the origin and radius equal to the rated apparent power (Figure 2.42). The complex output power is given by the expression

$$\underline{S} = P + jQ = \underline{V}_g \underline{I}_g^* = V_g I_g (\cos\varphi_g + j\sin\varphi_g)$$

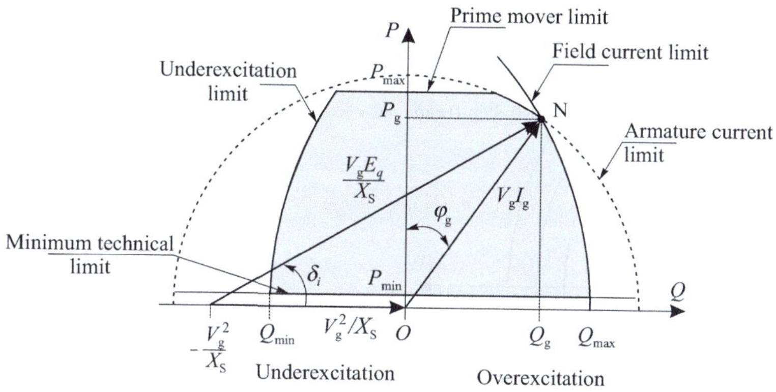

Figure 2.42. Loading capability chart of a synchronous generator [30].

Assuming nominal terminal voltage and armature current as well as nominal cooling conditions, the output apparent power is a combination of active and reactive powers. At the limit, more reactive power may require reduction of the active power and vice versa.

(ii) *Field Current Limit.* The excitation current carried in the rotor winding causes also heating and energy losses and therefore a second limit is defined. In the plane P–Q, this limit is given by the arc of the circle with origin in $(-V_{\mathrm{g}}^2/X_{\mathrm{S}})$ and radius $(V_{\mathrm{g}}E_q/X_{\mathrm{S}})$. The origin of this circle and, therefore, the length of the radius depends on the size of the generator reactance. This circle is obtained by multiplying the phasors of V_{g}, $E_q(=X_{\mathrm{md}}I_f)$, and $X_{\mathrm{S}}I_{\mathrm{g}}$ from Figure 2.16 with $V_{\mathrm{g}}/X_{\mathrm{S}}$. Therefore, the output powers can be expressed as

$$P = V_{\mathrm{g}}I_{\mathrm{g}}\cos\varphi_{\mathrm{g}} = \frac{E_qV_{\mathrm{g}}}{X_{\mathrm{S}}}\sin\delta_{\mathrm{i}} \tag{2.198}$$

$$Q = V_{\mathrm{g}}I_{\mathrm{g}}\sin\varphi_{\mathrm{g}} = \frac{E_qU_{\mathrm{g}}}{X_{\mathrm{S}}}\cos\delta_{\mathrm{i}} - \frac{V_{\mathrm{g}}^2}{X_{\mathrm{S}}} \tag{2.199}$$

As seen in Figure 2.42, the field current limit is more restrictive then the armature current limit and the two curves intersect at the point N. In some cases, the point N is located on the prime mover limit and the field current limit is the only restriction for the generated reactive power.

(iii) *End Region Heating Limit.* When operated in underexcitation the end-turn leakage flux, as illustrated in Figure 2.43, enters axially (perpendicular) to the

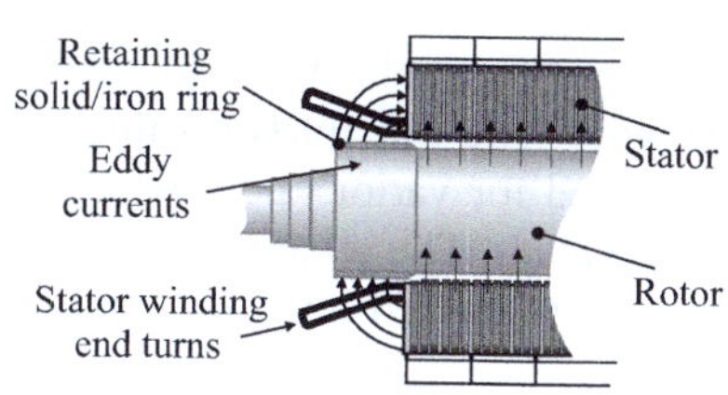

Figure 2.43. View of stator winding end region.

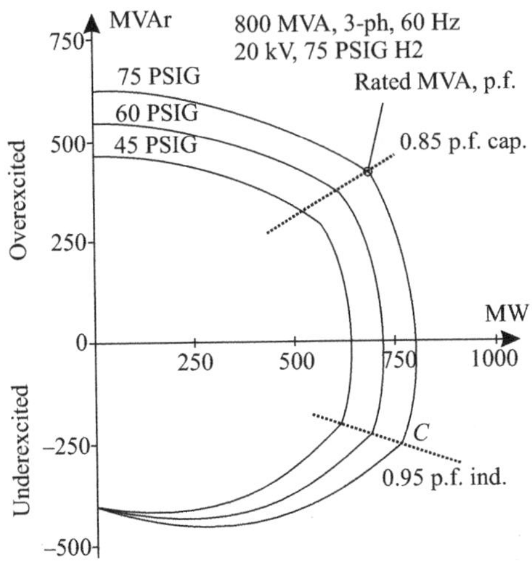

Figure 2.44. Reactive capability curve of a hydrogen-cooled 800 MVA rated generator [31].

stator laminations causing eddy currents, which results in localized heating in the end region of the armature [1]. The end region heating limits the amount of output reactive power in underexcitation operation. When overexcited, the end-turn leakage flux is small since the high field currents keep the retaining ring saturated.

The steam turbine generators are more severely affected by this limit, and, in general, operating in underexcitation is a highly stressing condition for the generator. For this reason, most of the thermal and nuclear power plants are not required by the system operator to provide leading reactive power. On the contrary, the hydrogenerators are capable to operate in underexcitation, and sometimes they are used to provide reactive power service only, similar to the synchronous condensers.

The reactive capability limits defined above are specific for the design characteristics of the machine. However, these limits change for different terminal voltages, the cooling conditions, and so on. Figure 2.44 shows the influence of the hydrogen pressure on the generator capabilities.

Note that is Figure 2.44, the term PSIG, identified also by psi, denotes pound-force per square inch; in metric system, 1 bar (10^5 Pa) $\approx$ 14.5 psi.

The reactive capability limits may also be subjected to stability issues. For this reason, in practical situations, more restrictions may be imposed by the use of excitation regulators. Operating the generator at the physical/stability limits may be risky and therefore the operational limits defined in the overexcitation and underexcitation regulators are more restrictive than the physical limits.

2.1.7.2 The V Curves. The relationship between the apparent output power or the armature current and the field current is sometimes represented graphically through the *V curves* (Figure 2.45) [1,32]. These curves are shown with solid line for two values of the output active power, in p.u. The *V* curves are expressed for various power factors while holding the terminal voltage and speed at the rated values. The relationship of the apparent power and the field current is also represented, with dashed line, for constant power factors (p.f.).

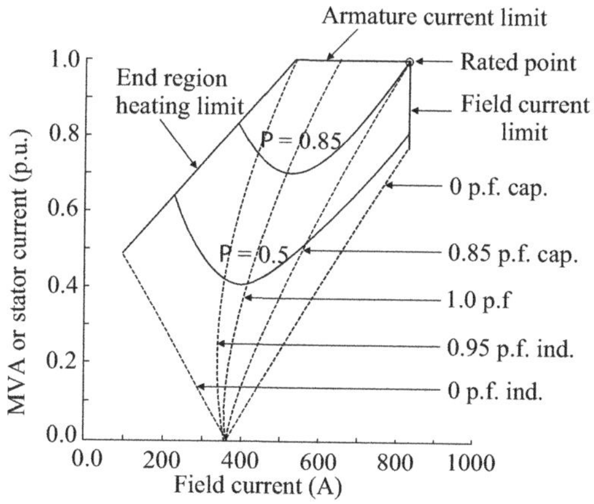

Figure 2.45. The V curves and compounding curves for a generator at rated armature voltage.

2.1.8 Description and Modeling of the Excitation Systems

2.1.8.1 Components and Performances of Excitation Control System. A general functional block diagram of the excitation control system presents the feedback control system that includes the synchronous machine and its excitation system (Figure 2.46). The excitation system is the equipment providing field current for the synchronous machine, including all power, regulating, control, and protective elements [33].

The main subsystems of the excitation system are as follows:

(i) *The exciter*, which is an auxiliary structure producing the power required by the synchronous generator field winding in form of DC voltage and current that can be quickly varied.

(ii) *The voltage regulator*, which processes and amplifies the input control signals to a level and form appropriate for control of the exciter. This includes both regulating and excitation system stabilizing functions. The voltage regulator is a synchronous machine regulator that functions to maintain the terminal voltage of the machine at a predetermined value or to vary it according to a predetermined plan. The *excitation system stabilizer* is an element or group of elements that

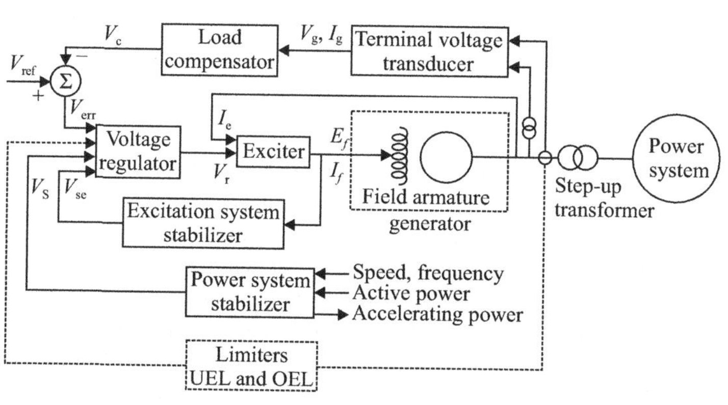

Figure 2.46. General functional block diagram of the excitation control system.

modify the forward signal by either series or feedback compensation to improve the dynamic performance of the excitation control system.

(iii) *Terminal Voltage Transducer and Load Compensator.* The voltage output from the transducer represents the main signal of the excitation control system. Load compensator is used to control a voltage, which represents the voltage at an internal or external point of the generator.

(iv) *Power system stabilizer (PSS)* is a compensation circuit aimed to provide additional damping torque through excitation control.

(v) *Limiters and protective circuits*, include a wide area of control and protective functions, which ensure that the capability limits of the exciter and synchronous generator are not exceeded.

The excitation voltage may range from few volts up to 700 V, while the DC field current may approach 8000 A on large turbogenerators.

If the excitation system if fed from the generator terminals, in case of a short circuit near the generator, the terminal voltage may drop very much and the field voltage is therefore affected. For this situation, and for others situations too, the excitation system must be capable of *field forcing*, which is a control action that rapidly drives the field voltage of a synchronous machine in the positive or in the negative direction for a limited time. The *positive ceiling voltage* is defined as the maximum direct voltage that the excitation system is designed to supply from its terminals under predefined conditions, which can be 2–2.5 times the rated DC field voltage [33]. Excitation forcing in the positive direction is required to maintain the stability of the synchronous machine when large network perturbations occur, such as short circuits, which cause sudden voltage drops at the machine terminals. In case of perturbations causing sudden increase of voltage at the machine terminals, the polarity of the DC excitation voltage is rapidly changed with the purpose of deexciting the machine. The *negative ceiling voltage* is the maximum opposite polarity voltage that the excitation system can apply for a limited time to the field circuit, which can be 1.5–2.5 times the rated DC field voltage with negative sign.

The excitation response time must be small so that the automatic voltage regulator be capable of controlling the terminal voltage when the generator is subjected to power system disturbances or transients. *The excitation system voltage response time* is the time in seconds for the excitation voltage to attain 95% of the difference between ceiling voltage and rated field voltage under specified conditions. A high initial response ensures a higher synchronizing torque required to maintain the generator stability. With the advent of power electronics and development of static excitation systems this time has decreased significantly.

2.1.8.2 Types and Modeling of Excitation Systems. The exciter generates the field current I_f or the field voltage $E_f(= \omega L_{md} V_f / R_f)$ that are applied to the field winding of the synchronous generator. A change in the field voltage causes a change in the same direction in the field current proportional to the rotor winding resistance. Excessive field current causes overheating of the rotor winding while a very low current weakens the coupling between the rotor and stator fields and the generator may loss the synchronism. In the early years of the power systems, the field voltage E_f was left constant with occasional manual adjustments. Currently, however, the field voltage is typically adjusted by a feedback control to regulate the terminal voltage of the generator. Types of excitation dynamics are determined by (a) the nature of the source of DC voltage E_f and (b) the nature of the feedback control arrangement [34].

Three distinctive types of excitation systems are identified by the IEEE Std. 421.1-2007 [35,36] on the basis of the excitation power sources:

- *Type DC excitation systems*, which utilize a direct current generator with a commutator typically at the end of the generator shaft as source of excitation system power.
- *Type AC excitation systems*, which utilize an alternator and either stationary or rotating rectifier to produce the direct current needed for the synchronous generator field.
- *Type ST excitation systems*, in which excitation power is supplied through transformers or auxiliary generator windings and thyristor rectifiers.

The types AC and ST excitation systems allow only positive current flow to the field of the machine; on the other hand, some systems allow negative voltage, facts that lead the current to zero. Special measures are made to allow the flow of negative field current when it is induced by the synchronous machine.

DC Exciters. Early excitation systems, as used in the years from 1920s to 1960s, were based on DC exciters, which directly supplied the main synchronous generator field through slip rings. Primitive excitation systems employed mechanical devices to control the exciter field by varying a control rheostat. However, this technology was slow responsive.

DC Generator–Commutator Exciters. The IEEE Std. 421.1-2007 defines four types of DC generator–commutator exciters [33]: with rotating amplifier, with static amplifier, with continuously acting regulator employing static amplifiers, and separately excited with noncontinuously acting rheostatic regulator.

One of the solutions was the use of a *rotating amplifier* (amplidyne) to supply the exciter field through a DC commutator and slip rings (Figure 2.47). Variation in the excitation current is obtained by an auxiliary series buck-boost exciter.

The auxiliary exciter (a permanent magnet generator, PMG) and the amplidyne are driven by a DC motor coupled to both machines. The exciter output supplies the rest of its

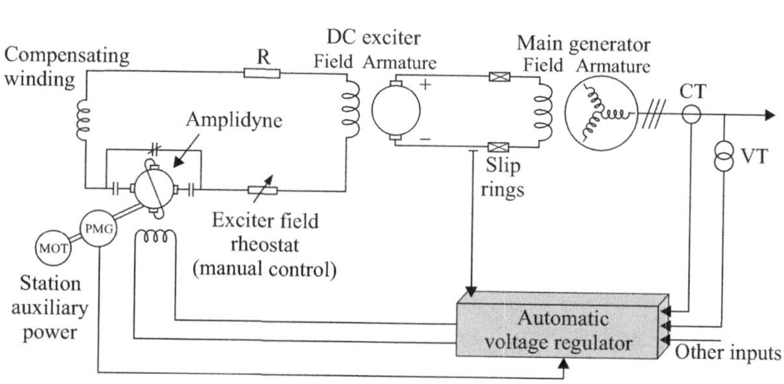

Figure 2.47. DC generator–commutator exciter with rotating amplifier. (Adapted from Ref. 33.)

own field by self-excitation. The amplidyne can force the exciter field in the direction indicated by the voltage regulator in order to maintain the generator terminal voltage at the predefined value. If the amplidyne regulator is not available, a rheostat will provide a manual control of the exciter field.

The *type DC1A exciter model* [36], shown in Figure 2.48, is used to represent *field-controlled DC commutator exciters* with continuously acting voltage regulators, especially with direct acting rheostat, rotating amplifier and magnetic amplifier types. Continuously acting regulator can initiate a corrective action for a sustained infinitesimal charge in the controlled variable.

The principal input to this model is the output V_c from the terminal voltage transducer and load compensator model (see Figures 2.46 and 2.61). At the summing junction, the terminal voltage transducer output V_c is subtracted from the set point reference, V_{ref}. To improve the performance of the excitation system, a soft negative feedback unit is used to provide a series of adjustment to field voltage of the generator; V_F is the output of the soft negative feedback unit of the excitation voltage [37]. The power system stabilizing signal V_S is added to produce an error voltage. In normal operating conditions these last two signals are zero. The resulting signal is amplified in the regulator.

These voltage regulators utilize power sources that are essentially unaffected by brief transients on the synchronous machine or auxiliaries buses. The time constants, T_B and T_C, may be used to model equivalent time constants inherent in the voltage regulator, but they are frequently small enough to be neglected, and provision should be made for zero input data.

The major time constant T_A and gain K_A associated with the voltage regulators are shown incorporating nonwindup limits typical of saturation or amplifier power supply limitations. The voltage regulator output V_R (see Figure 2.48) is used to control the exciter, which may be either separately excited or self-excited (Figure 2.49a and b) [35]. K_E is computed so that initially $V_R = 0$, and the load compensator is not used [1].

The exciter saturation S_E is a nonlinear function of the exciter output E_f and also should be interpreted as a multiplier of E_f, that is, $V_X = E_f S_E(E_f)$. A signal derived from the field

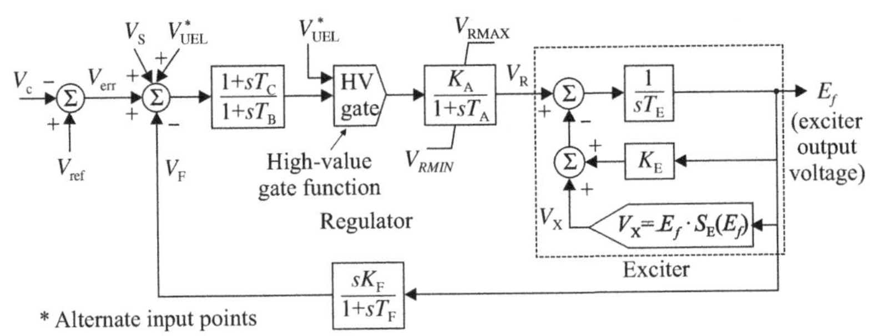

Figure 2.48. IEEE type DC1A—DC commutator exciter model. (Reprinted with permission from Ref. 36.)

Sample data: Self-excited DC exciter:

$$K_A = 46 \quad T_A = 0.06 \; T_B = 0; T_C = 0 \; T_E = 0.46$$

$$K_F = 0.1 \quad T_F = 1.0 \quad V_{RMAX} = 1.0 \quad V_{RMIN} = -0.9$$

$$S_E(E_{f1}) = 0.33 \; E_{f1} = 3.1 \; S_E(E_{f2}) = 0.1 \quad E_{f2} = 2.3$$

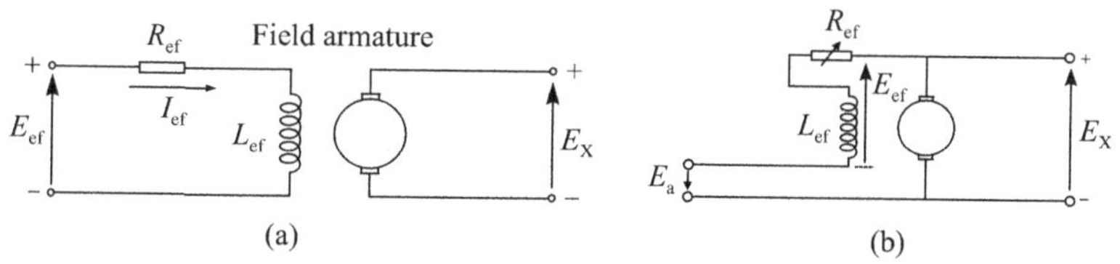

Figure 2.49. Separately excited DC exciter (a) and self-excited DC exciter model (b). (Reprinted with permission from Ref. 35.)

voltage is used to provide excitation system stabilization V_F, via the rate feedback with gain K_F and time constant, T_F.

Starting from the expression of the per unit voltage across the self-excited DC exciter field

$$E_{ef} = E_a + E_X \tag{2.200}$$

the block diagram of the DC exciter model (Figure 2.50a) is obtained (see [35]).

As shown in Ref. 35, taking account of the resistance R_{ef}, which comprises the resistance of the field winding including any external field circuit resistances, and the incremental inductance of the field circuit L_{ef}, the following expression results:

$$V_R = K_E E_f + S'_E E_f + T_E \frac{dE_f}{dt} \tag{2.201}$$

The block diagram from Figure 2.50a can be reduced to the form used in the models for DC1 and DC2 exciters (Figure 2.50b), where

$$K_E = \frac{R_{ef}}{R_{gb}} - 1; \quad T_E = \frac{L_{efu}}{R_{gb}}; \quad S'_E = \frac{R_{ef}}{R_{gb}} S_E; \quad V_R = E_{ef}; \quad E_f = E_X$$

$$L_{efu} = L_{ef} \left(\frac{dI_{ef}}{dE_X} \right) \Bigg|_{E_X = E_{X_0}}$$

with E_{X_0} being the value of E_X at the operating point, and I_{ef} is the exciter current.

For expression in per unit the base exciter resistance $R_{gb} = R_g$ was used.

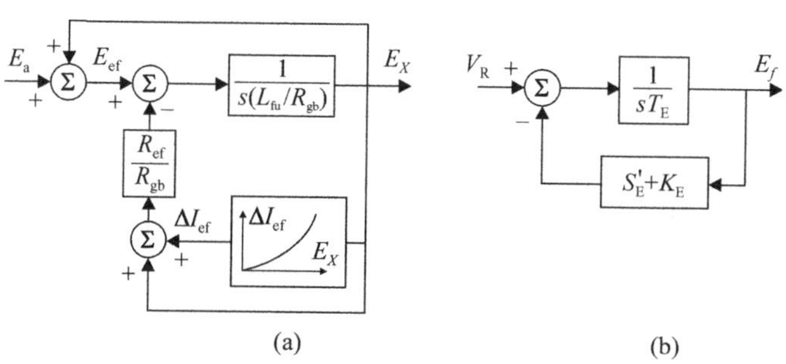

Figure 2.50. Block diagram of a self-excited DC exciter. (Reprinted with permission from Ref. 35.)

It can be seen that adjustment of R_{ef} affects the feedback path including the saturation function S'_E. The constant T_E in the forward loop remains unaffected by change in R_{ef}.

Others DC exciters models can be found in Ref. 36, that is, *type DC2A excitation system* used to represent field-controlled DC commutator exciters with continuously acting voltage regulators having supplies obtained from the generator or auxiliary bus, or *type DC3A excitation system* used to represent older systems, in particular those DC commutator exciters with noncontinuously acting regulators that were commonly employed before the development of the continuously acting regulators.

AC Alternator Supplied Rectifier Excitation. These excitation systems use an AC alternator and either stationary or rotating rectifiers to produce the direct current needed for the generator field. The early AC excitation systems used a combination of magnetic and rotating amplifier as regulators [38], but the new systems use electronic amplifier regulation.

Loading effects on such exciters are significant and the use of generator field current as an input to the models allows these effects to be represented accurately. These systems do not allow the supply of negative field current. Only the alternator supplied controlled rectifier excitation system (AC4A model) provides negative field voltage forcing [33].

(i) *Stationary Rectifier Systems* In order to improve the dynamic and transient stability of the synchronous generator, one required characteristic of an excitation system is the ability to produce high level of exciter ceiling voltage very rapidly following a change in the terminal voltage. With the advent of power electronics and rectifying systems, the performances of the excitation systems have been improving significantly.

Figure 2.51 shows the schematic diagram of an AC exciter (an alternator) supplying DC to the generator excitation through a stationary rectifier bridge. The exciter may be mounted on the same shaft as the main generator and driven by the prime mover. Since the rectifier bridge supplying the generator field is noncontrolled (it consists of diodes), the excitation system regulator controls the DC excitation current by controlling the exciter field through the exciter rectifier.

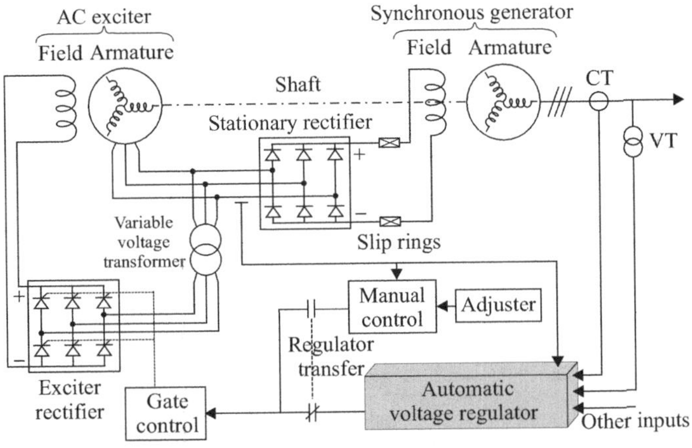

Figure 2.51. Alternator-rectifier exciter employing stationary noncontrolled rectifiers. (Adapted from Ref. 33.)

The exciter does not employ self-excitation, and the voltage regulator power is taken from a source that is not affected by external transients. The diode characteristic in the exciter output imposes a lower limit of zero on the exciter output voltage [36].

In case of an alternator-rectifier exciter employing controlled rectifiers, the diodes of the stationary rectifier are replaced with thyristors, and the firing angle is given by a separate gate circuitry. Also, the automatic voltage regulator controls the rectifier supplying the synchronous generator field, while the exciter field is controlled by a separate exciter regulator.

The excitation system of the synchronous machine is provided with two independent control systems:

- An automatic regulator, which automatically controls the DC excitation current based on current and voltage measurements (through a current transformer and a voltage transformer) from the generator terminals and other auxiliary control functions; the firing angle of the controlled rectifier is adjusted so that the voltage at the generator terminals is maintained at the desired value irrespective of the generator load.
- A manual regulator, which aims to maintain the generator field voltage at constant value, by manual settings, as a temporary action when the automatic regulator is not available for various reasons; the adjuster is a device or function by which the set point is determined.

The AC exciter is self-excited with an independent AC output taken from the exciter terminals through a power transformer and a power current transformer, which is then introduced in the exciter rectifier to provide a DC voltage to the exciter field.

Type AC1A excitation system model, shown in Figure 2.52, represents an alternator-rectifier excitation system with noncontrolled rectifiers and feedback from exciter field current.

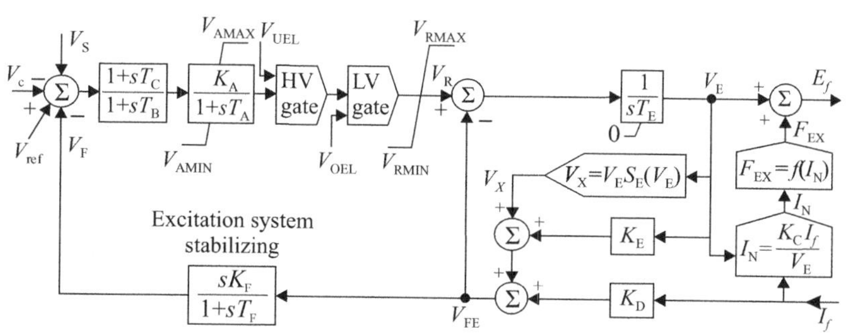

Figure 2.52. IEEE type AC1A—alternator-rectifier excitation system with noncontrolled rectifiers and feedback from exciter field current model. (Reprinted with permission from Ref. 36.)

Sample data:

$K_A = 400$	$T_A = 0.02$	$T_B = 0$	$T_C = 0$	$K_F = 0.03$
$T_F = 1.0$	$K_E = 1.0$	$T_E = 0.80$	$K_D = 0.38$	$K_C = 0.20$
$V_{AMAX} = 14.5$	$V_{AMIN} = -14.5$	$V_{RMAX} = 6.03$	$V_{RMIN} = -5.43$	
$S_E(V_{E1}) = 0.10$	$V_{E1} = 4.18$	$S_E(V_{E2}) = 0.03$	$V_{E2} = 3.14$	

The demagnetizing effect of the load excitation current I_f on the dynamics of the exciter alternator output voltage V_E is accounted for in the feedback path that includes the constant K_D. This constant is a function of the exciter alternator synchronous and transient reactances. The exciter output voltage drop due to rectifier regulation is simulated by inclusion of the constant K_C, which is a rectifier loading factor proportional to commutating reactance, and the rectifier regulation curve, $F_{EX} = f(I_N)$ [36].

In this model, a signal V_{FE} proportional to the exciter field current is derived from the summation of signals from exciter output voltage V_E multiplied by $K_E + S_E(V_E)$ and I_f multiplied by the demagnetization term, K_D. In some models, the signal proportional to the exciter field current signal V_{FE} is used as the input to the excitation system stabilizing block with output V_F.

(ii) *Rotating Rectifier Systems* A rotating rectifier exciter is one case of a brushless excitation system. The brushes and the slip rings are eliminated since the alternator-exciter and the diode rectifiers are rotating with the shaft. Elimination of the brushes and the slip rings eliminate the high-maintenance items.

The basic brushless excitation control system is illustrated in Figure 2.53. This system incorporates also an alternator-rectifier main exciter and a PMG pilot exciter, both driven from the synchronous machine shaft and rotating together with the diode rectifier and the generator excitation [24]. It is, however, impossible to meter any of the generator field quantities directly since all these components are rotating and no slip rings are used.

The voltage regulator senses generator terminal voltage derived from the potential transformers and compares this quantity with a constant reference quantity derived internally. The difference between the sensed value and the reference value determines the firing angle to the exciter rectifier thyristors.

The permanent magnet generator pilot exciter supplies DC voltage to the alternator-exciter field through a controlled exciter rectifier. In turn, the alternator provides excitation voltage to the main generator field through a diode rectifier.

In order to achieve high initial response performances, the excitation system provides a strong forcing action to increase the generator field excitation in response to reduced terminal voltage values. Forcing actions refers to the amount

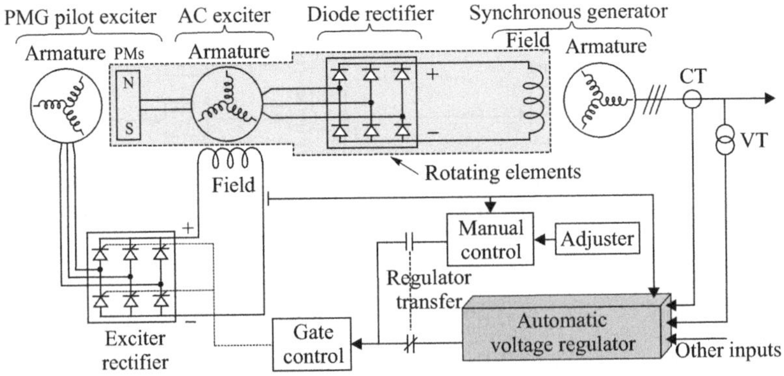

Figure 2.53. Alternator-rectifier exciter employing rotating noncontrolled rectifiers (brushless excitation control system). (Adapted from Refs [24,33,36].)

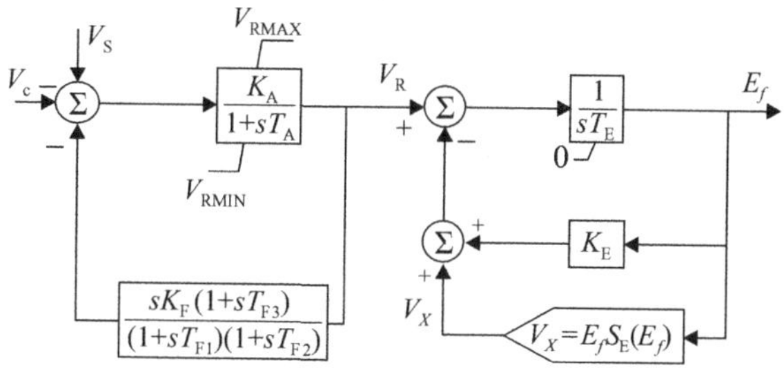

Figure 2.54. Simplified model of a rotating rectifier excitation system [36].

of voltage applied to the machine field in excess of the amount that will produce the desired machine output under given operating conditions.

The *AC5A excitation system model* (Figure 2.54) can be used to represent a brushless excitation system. The regulator is supplied from a source, such as a permanent magnet generator, which is not affected by system disturbances. Unlike other AC models, this model uses loaded rather than open-circuit exciter saturation data in the same way as it is used for the DC models [36].

The reader can find more details and also others types of excitation systems models in IEEE Std. 421.5-2005 [36].

Static Excitation Systems. A static excitation system has no moving part and consists basically of a transformer, a thyristor converter and a voltage regulator. A complete system also includes control, field flashing, and deexcitation circuits. The excitation power is generally derived from the generator terminals or auxiliary bus through the exciter transformer and the thyristor converter (rectifier). In some cases, the excitation power is derived from an auxiliary winding of the generator, eliminating therefore the transformer.

For many of the static systems, the exciter ceiling voltage is very high. For such systems, additional current field limiter circuits may be used to protect the exciter and the generator rotor.

(i) *Potential Source-Rectifier Exciter Employing Controlled Rectifiers* In such an excitation system, the excitation power is derived from the generator terminals through a power potential (exciter) transformer and a controlled rectifier (Figure 2.55). The voltage regulator controls the DC output of the rectifier thyristors via pulse-triggering unit controls.

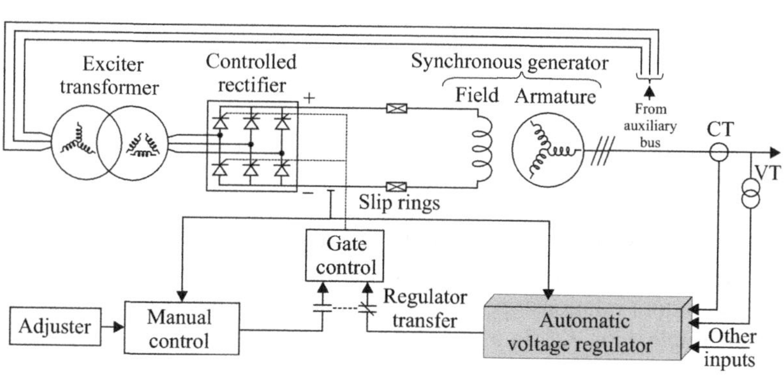

Figure 2.55. Potential source-rectifier exciter employing controlled rectifiers. (Adapted from Ref. 33.)

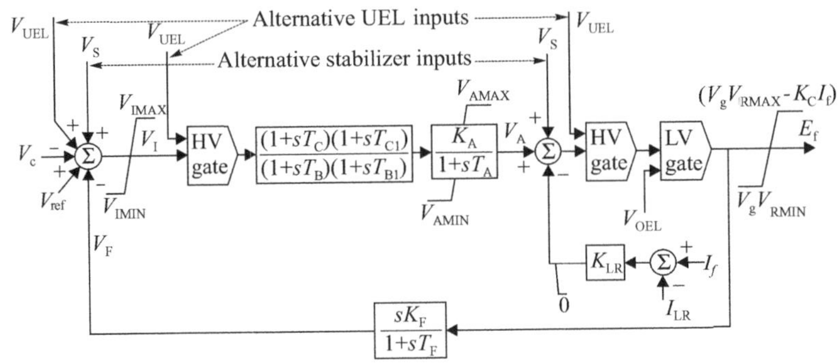

Figure 2.56. IEEE type ST1A potential source-controlled rectifier exciter model. (Reprinted with permission from Ref. 36.)

Sample data: Exciter and regulator:

$K_A = 210$	$T_A = 0$	$T_C = 1.0$	$T_B = 1.0$ $K_C = 0.038$
$T_{C1} = 0$	$T_{B1} = 0$ $K_F = 0$	$T_F = 0$ (not used)	$I_{LR} = 4.4$ $K_{LR} = 4.54$
$V_{RMAX} = 6.43$	$V_{RMIN} = -6.0$	V_{IMAX}, V_{IMIN} not represented	

V_{IMAX}, V_{IMIN} are the voltage regulator input limits and V_{RMAX}, V_{RMIN} are the maximum and minimum voltage regulator outputs.

Since the excitation power is derived from the generator terminals, the maximum exciter voltage available (ceiling voltage) is directly related to the generator terminal voltage. When a severe fault occurs near the generator, the terminal voltage is, therefore, affected and the ceiling voltage is reduced. This limitation is mainly offset by the excitation system initial response and high postfault field-forcing ability. The generators using such excitation systems perform satisfactory when connected to a strong power system.

In a simpler configuration, the excitation power is taken from an auxiliary winding of the generator, case in which the exciter transformer is eliminated.

The *type ST1A excitation system model*, shown in Figure 2.56, corresponds to the potential source-controlled rectifier exciter excitation system (Figure 2.55).

The inherent exciter time constants of this excitation system type are very small and exciter stabilization may not be required. Also, it may be desirable to reduce the transient gain of these systems for other reasons. This model is sufficiently versatile to represent transient gain reduction implemented either in the forward path via time constants, T_B and T_C (in which case K_F would normally set to zero), or in the feedback path by suitable choice of rate feedback parameters, K_F and T_F. Voltage regulator gain and any inherent excitation system time constant are represented by K_A and T_A, respectively. The time constants T_{C1} and T_{B1} allow for the possibility of representing transient gain increase, in which case T_{C1} would be greater than T_{B1}.

The way in which the firing angle for the bridge rectifiers is derived affects the input–output relationship, which is assumed to be linear in the model by choice of a simple gain, K_A. In a few systems, the bridge relationship is not linearized, leaving this nominally linear gain a sinusoidal function, the amplitude of which may be dependent on the supply voltage. As the gain is normally set very high, a linearization of this characteristic is normally satisfactory for modeling purpose. The representation of the ceiling is the same whether the characteristic is linear or sinusoidal.

In many cases, the internal limits on V_I can be neglected. The field voltage limits that are functions of both terminal voltage and synchronous machine field current should be modeled. A field current limiter is sometimes employed to protect the generator rotor and exciter, as a result of the very high forcing capability of these systems. The field current of generator I_f is constrained by K_{LR} (exciter output current limiter gain) and I_{LR} (exciter output current limiter reference) in the diagram. Proportional unit K_{LR} has a windup lower limit. To avoid this unit we can simply set K_{LR} to zero [36].

(ii) *Compound Source-Rectifier Exciter Employing Noncontrolled Rectifiers* Very rapid response and a largely self-regulating excitation action can be achieved in a static system by "compounding" the excitation inputs so that the excitation voltage is responsive both to generator output load current and to the main generator terminal voltage. This system is also a form of self-excitation system and the exciter input derives from the main generator terminals through a power potential transformer (PPT), as voltage source, and a saturable-current transformer (SCT), as current source. The regulator controls the excitation by controlling the saturation of the current transformer through a controlled rectifier (Figure 2.57). The voltage source supports operation during no-load conditions, while under loaded conditions, a portion of the field excitation is derived from the generator load current.

Self-excited units have the inherent disadvantage that the AC output voltage is low at the same time the exciter is attempting to correct the low voltage. However, combining the potential and current sources enable full excitation power to be supplied through system disturbances with severely depressed generator line voltage. This performance feature can be valuable in certain power system applications.

The compound source-rectifier exciter systems can be found in various configurations, designed by various companies. In Ref. 33, three other types of compound exciter systems are presented with various capabilities.

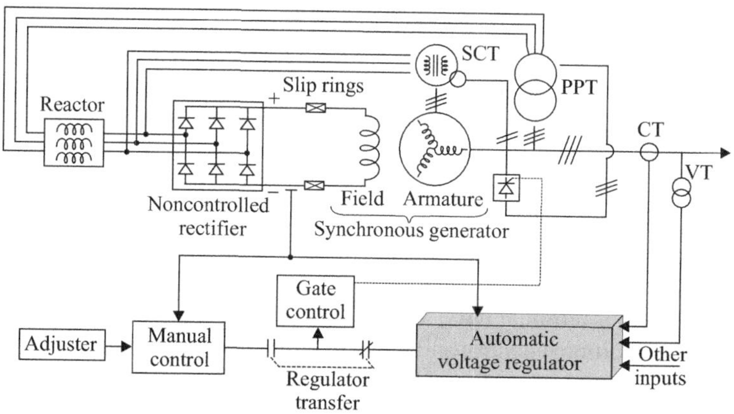

Figure 2.57. Compound source-rectifier exciter employing uncontrolled rectifiers. (Adapted from Ref. 33.)

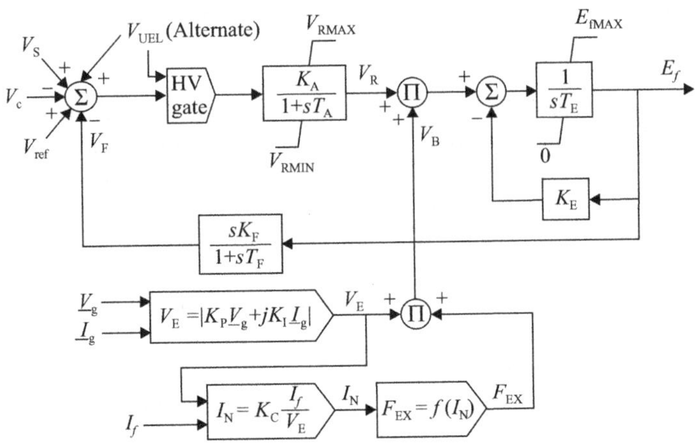

Figure 2.58. IEEE type ST2A—compound source-rectifier exciter. (Reprinted with permission from Ref. 36.)

Sample data:

$T_R = 0$ $T_E = 0.5$ $K_E = 1.0$ $K_A = 120$ $K_F = 0.05$ $V_{RMAX} = 1.0$

$T_A = 0.15$ $T_F = 1.0$ $K_P = 4.88$ $K_I = 8.0$ $K_C = 1.82$ $V_{RMIN} = 0$

$E_{fMAX} = 2.75$ times d-axis synchronous reactance of the synchronous machine in p.u

Type ST2A excitation system model (Figure 2.58) is used to represent the compound source-rectifier excitation system. It is necessary to form a model of the exciter power source utilizing a phasor combination of terminal voltage $\underline{V}_g$ and terminal current $\underline{I}_g$. T_E is a time constant associated with the inductance of the control winding. E_{fMAX} represents the limit on the exciter voltage due to saturation of the magnetic components.

The performances of power electronics and the advent of protection and control systems have contributed to designing various static excitation systems. Several other models of ST type excitation systems are presented in Ref. 36.

2.1.8.3 Control and Protective Functions. The excitation system of a synchronous generator includes a wide number of principal control and protective functions (Figure 2.59). In terms of structure of the excitation system, these functions may be customized or some of them can be eliminated.

Automatic and Manual Regulators. The excitation system includes an automatic voltage regulator and a manual voltage regulator. The aim of the automatic regulator is to maintain the generator terminal (AC) voltage to a predefined value. This action involves the use of various additional control and protective functions, as shown in Figure 2.59. The manual regulator is used when the automatic regulator is not available, including also testing and generator start-up. It acts in the way to maintain the generator field voltage at constant value. The manual regulator (DC) set point is adjusted by operator intervention only and therefore care must be taken to keep the generator in safe operation.

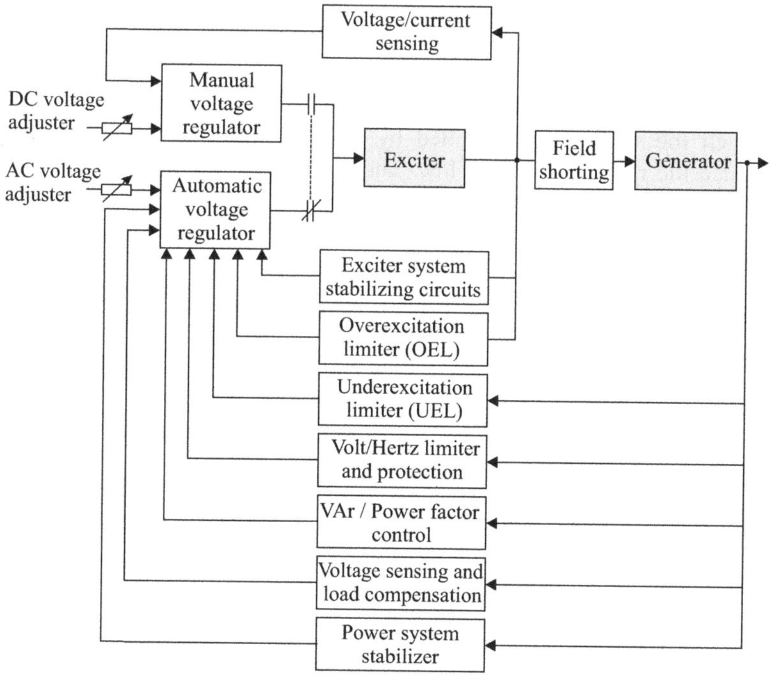

Figure 2.59. Excitation system control and protective circuits. (Adapted from Ref. 1.)

Excitation System Stabilizing Circuits. Some excitation systems are characterized by significantly large time delays and therefore poor performances under dynamic operating conditions. This is the case of all DC exciters and some of the AC exciters. When the generator operates under no load, since the terminal voltage varies insignificantly, a high regulator gain can leads to unstable excitation control. Therefore, excitation system stabilizing circuits are introduced either in series or as a feedback path to improve the dynamic performances of the excitation system. Figure 2.60 shows one simple example of feedback compensating reaction.

Modern excitation systems employ solid-state technologies and simple configurations and therefore are characterized by very fast response capabilities and very good dynamic performances.

Terminal Voltage Transducer and Load Compensation. The aim of the automatic voltage regulator is to maintain the voltage at the generator terminals at a desirable value. In some cases, a load compensation function is needed to perform this action. *Reactive-drop compensation* is one of the compensation methods, which provides regulation of a voltage that is not the generator terminal voltage, but other synthesized voltage using the terminal voltage, the terminal current, and a variable compensating impedance. The other method,

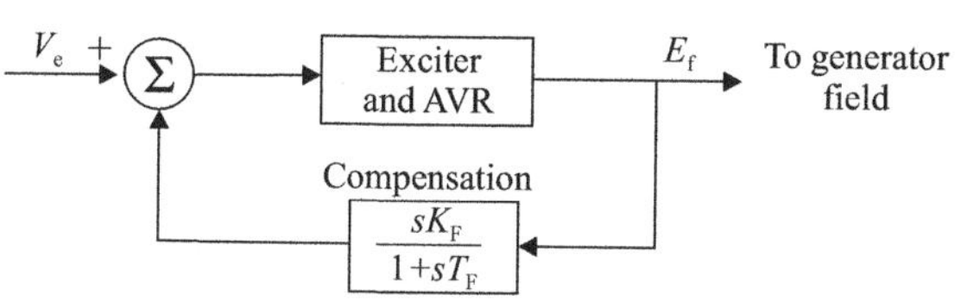

Figure 2.60. Derivative feedback excitation control system stabilization [1].

referred to as *line-drop compensation*, provides voltage regulation to a point beyond the machine terminals.

The compensation impedance, consisting of an adjustable resistance (R_c) and an adjustable inductive reactance (X_c), is located between the generator terminals and the point at which the voltage is controlled by this function. Normally the resistance is neglected since the ratio R/X is very low. The compensation impedance and the sensed current are used to generate a voltage drop that is added or subtracted from the sensed terminal voltage, and the result is a intermediary compensation voltage, given by

$$V_{c1} = \left| \underline{V}_g + (R_c + jX_c)\underline{I}_g \right| \tag{2.202}$$

Figure 2.61a shows the load compensation circuit, while Figure 2.61b shows the block diagram of the terminal voltage transducer and the load compensator.

For modeling purposes, the filtering associated with the voltage transducer may be simply reduced to a time constant T_R. For many excitation systems, this constant, associated to the whole process, from sensing to load compensation, is very small and therefore it may be set to zero. Taking T_R into account a compensation voltage is obtained, which is fed to the voltage regulator.

The compensation voltage V_c is compared to a reference voltage V_{ref} as shown in each of the excitation system models. The equivalent voltage regulator reference signal V_{ref} is calculated to satisfy the steady-state loop equations. If the load compensation is inactive or absent, the excitation system attempts to maintain the terminal voltage determined by the reference signal [36].

The reactive voltage compensation is visually used in one of the following two ways [36]:

(i) When more than one generator units are sharing the same step-up transformer, and there is no impedance interposed between them, they may attempt to control the voltage slightly different. Therefore, the load compensator is used to create artificial coupling impedance between the units so that they will share reactive power appropriately for the same purpose. In such cases, R_c and X_c would have

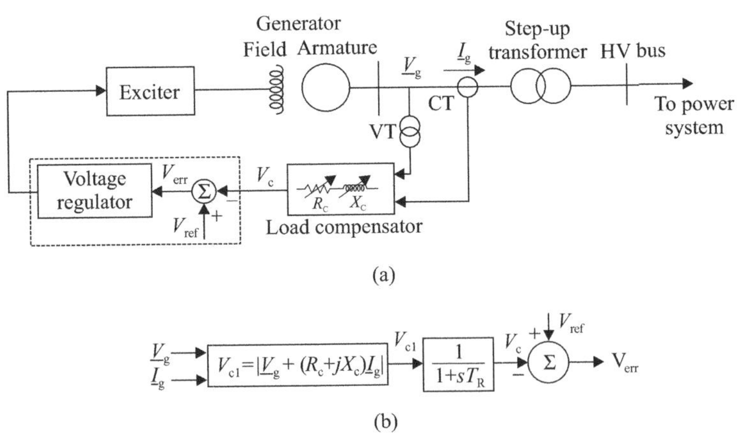

Figure 2.61. Load compensation circuit (a) and its block diagram (b) [36].

positive values. This situation corresponds to a regulating point within the synchronous machine.

(ii) When a single unit is connected to the power system through significant impedance, or when two or more generator units are connected through individual transformers, it may be desirable to set the regulation point of the voltage beyond the machine terminals. The compensation is normally about 50–80% of the transformer impedance, in such way to obtain balanced voltage drops across the transformers and allow paralleling for the generators. For these cases R_c and X_c would take on the appropriate negative values.

When load compensation is not employed ($R_c = X_c = 0$), the load compensation circuit reduces to a simple sensing circuit and comparator.

Excitation Limiters. Reactive power injection or absorption by the synchronous machine and voltage control is limited by stability or thermal limits. Therefore, excitation limiters are used to ensure proper operation of the generator (Figure 2.62a and b). They are part of the excitation system voltage regulator.

For stability reasons, most excitation systems installed on synchronous generators are designed with field-forcing capabilities, which allow field currents between the maximum continuous current and the maximum excitation current for a limited period of time. So, the aim of *the overexcitation limiter* (OEL) is to protect the generator against overheating, and the field-forcing function must go hand in hand with the limiter. Historically, OELs have not been modeled for power system simulations. Nowadays, the power systems are more stressed and the generators are often reaching this limit for longer periods, increasing the importance of appropriately modeling the OELs.

This limiter incorporates an integrator, with a lower limit of 0 at its output and a gain $1/T_o$, in order to comply with the rotor limit under steady-state conditions (Figure 2.62a). In this connection, a second limit of the excitation current, greater than $I_{f,\max}$, but which may be exceeded for a shorter time, should be considered. Thus, the only positive output

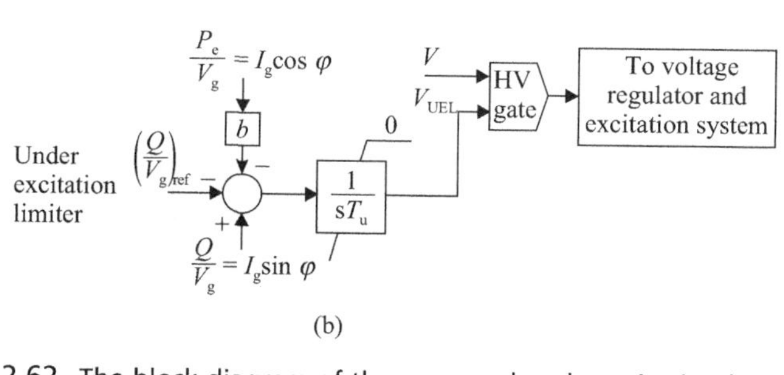

Figure 2.62. The block diagram of the over- and underexcitation limiters [12].

signal V_{OEL} from the integrator performs a function that is equivalent to the decrease of the reference V_{ref} of the voltage regulator. In other terms, the output signal lowers the excitation voltage and therefore the reactive power delivered, that is, it deexcites the generator in the attempt to bring back I_f to the limit value $I_{f,max}$ [12].

The overexcitation limiter output signal V_{OEL} generally enters the voltage regulator through a low-value gate. The output of the low-value gate function is the lowest value of the input values.

The underexcitation limiter (UEL) (Figure 2.62b) is designed to boost the generator excitation whenever it senses a condition in which excitation level is determined to be too low. The UEL is an auxiliary control circuit included in the excitation system that provides the limit to which the voltage regulator is allowed to demand underexcited reactive current (or reactive power) from the synchronous machine. UEL is typically applied to prevent operation that jeopardizes the stability of the machine or that would lead to overheating in the stator end region.

The UEL was originally implemented as a steady-state (slow) control, heaving little effect on the first swing transients. Subsequently, as the performances of the excitation have been improved, it was applied in faster control loops.

The UEL normally uses the terminal voltage and current, or a combination of the real and reactive powers as inputs to determine the limit start point and provide the necessary feedback. The limiter output signal V_{UEL} generally enters the voltage regulator through a high value (HV) gate or a summing junction as indicated in the block diagrams of the various excitation systems. The output of the high-value gate function is the highest value of the input values.

In the underexcitation limiter, an integrator with gain $1/T_u$, which is in the range of $0.1\ s^{-1}$, that is, $T_u \cong T_0 \cong 10\ s$, has an upper limit of 0. Therefore, if the underexcitation limit is exceeded, the input signal of the integrator is negative. The negative output V_{UEL}, which subtracts from V_{ref}, increases the excitation voltage; in other terms, it excites the generator in the attempt to go back to the underexcitation limit. Under steady-state condition, the integrator input is zero.

Thanks to the values of T_0 and T_u, the limiter loops are slow and provide control for slow phenomena, such as rotor and stator conductor heating. This means, for instance, that the rotor thermal limit may be exceeded for some seconds, if a transient due to a short circuit in the power system takes place. This function facilitates maintaining stability of the machine upon occurrence of a disturbance in the power system [12].

The *Volts-per-Hertz limiter and protection* is designed to protect the generator by limiting the ratio of terminal voltage to system frequency. A higher than nominal V/Hz ratio is an indication of excessive magnetic flux resulted from overvoltages and/or low frequencies. Overfluxing causes core overheating and may damage the generator and the step-up transformer. Typical maximum V/Hz ratios for the generator under steady-state operations are 105–110%. When the V/Hz limit is violated, an adjustment of the excitation level is activated so as to maintain the V/Hz value within the permissible range. However, since the synchronous machine can withstand a high level of magnetic flux for short periods of time, an inverse time delay is used before the limiter is activated. For lower than 95% V/Hz there is a derating of the machine capability and also lower stability margins. The V/Hz limiter is of greatest importance when the machine is operating at a non-synchronous speed, such as during a start-up or during an islanding condition. When the machine is synchronized to a system the V/Hz limiter functions essentially as an inverse time voltage limiter [39].

An example of a V/Hz limiter model is shown in Figure 2.63.

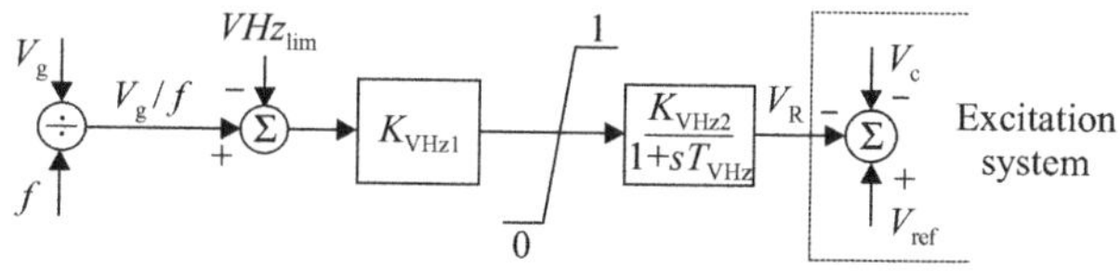

Figure 2.63. A model of the Volt/Hertz limiter.

When the per unit volt/hertz ratio V/Hz exceeds a predefined limit, VHz_{lim}, an additional negative signal is introduced in the excitation system forcing the excitation down. The V/Hz value is multiplied with the gain K_{VHz1}. Before entering in the summing point of the voltage regulator of the excitation system, a lag function is used to adjust the V/Hz time [40].

It is worth mentioning that [36]

(i) field current limiters are not normally represented in large system studies; they are becoming increasingly important in the representation of bus-fed static systems employing fast-acting limiters;

(ii) terminal voltage and V/Hz limiters are not normally represented in excitation system models. Some models, however, do provide a gate through which the output of a terminal voltage limiter V_{OEL} could enter the regulator loop. A terminal voltage limiter function is also included with one of the supplementary discontinuous excitation control models.

FIELD-SHORTING CIRCUITS[6]. For the AC and ST type exciters (static exciters), the current delivered by the exciter cannot be negative. Under some conditions—of pole slipping and system short circuits—a negative current may be induced in the field of the synchronous machine. If this current is not allowed to flow, a dangerously high voltage can result across the field circuit. In some cases, damper windings or solid iron rotor effects may limit the maximum voltage experienced by the field winding and rectifiers under such conditions. In other cases, special circuitry is usually provided to bypass the exciter to allow negative field current to flow. These take the form of either "crowbar" circuits (field shorting) or nonlinear resistors (varistors) (Figure 2.64a and b) [1,36].

In the case of the crowbar, a field discharge resistor (FDR) is inserted across the field of the synchronous machine by thyristors that are triggered on the overvoltage produced when the field current attempts to reverse and is blocked by the rectifiers on the output of the exciter.

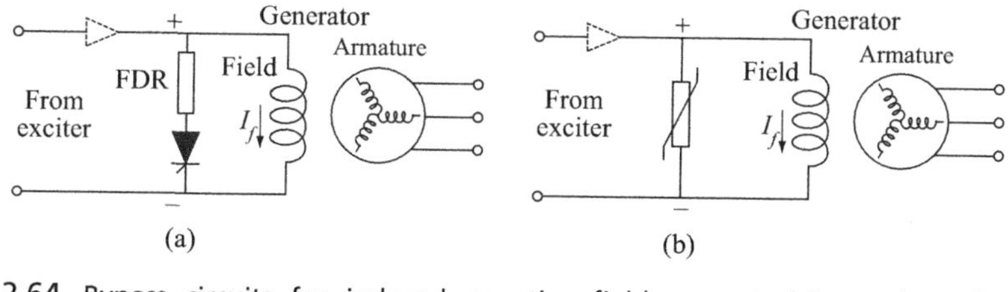

Figure 2.64. Bypass circuits for induced negative field current: (a) crowbar; (b) varistor. (Reprinted with permission from Ref. 36.)

[6] Reprinted with permission from Ref. 36.

Varistors are nonlinear resistors that are connected permanently across the field of the synchronous machine. During normal conditions, the resistance of these devices is very high, and little current flows through them. The varistor current increases very rapidly as the voltage across it is increased beyond a threshold level and thus limits the voltage seen by the field winding and the rectifiers on the output of the exciter.

For more details, the reader can refer to Refs [1,36].

POWER SYSTEM STABILIZER. Power system stabilizers are used to enhance damping of power system oscillation through excitation control. The expansion of electrical networks, for economical and security reasons, through interconnection lines can arise small-signal stability problems due to low-frequency *interarea oscillations* that may appear (sudden electromagnetic oscillations of 0.2–1 Hz). These oscillations correspond to a group of synchronous generators that swing coherently, but against one or more groups of generators in another area. There are also other types of oscillations within the power systems: *local oscillations*—when a synchronous machine oscillates against other generator in the same power plant or against the power system, and *subsynchronous oscillations*—when the mechanical modes of the turbine-generator shaft system are excited.

The interarea oscillations and the local oscillations are strongly related to the power system stability when subjected to small perturbations around the steady-state operation point (P_{nom}), and are visible in the generators speed response under steady-state conditions. The solution for the local and interarea oscillations problem is to add a control loop, sensitive to oscillations, called "power system stabilizer" that allows oscillations damping.

The AVR, contributing to power system stability enhancement under steady-state conditions, can be insufficient in the case of transient stability. In practice, on the other hand the torque added by the AVR to the synchronous generators is not sufficient to act against oscillations that may appear in the power system. Therefore, with the development of interconnections between power systems, high loading occurs on the interconnection lines (especially the very high voltage lines) that contribute to the instability phenomena.

In order to cope with the oscillations and instability problems, the PSS can be added as a reaction (correction) loop to the voltage regulator (AVR). This loop is designed to produce a torque acting against oscillatory modes that occur at the generators shafts. Commonly used inputs are shaft speed, terminal frequency, accelerating power, and terminal voltage [36]. Under steady-state operating conditions the PSS has a zero output. This is very effective method of enhancing small-signal stability performance. The transfer function of the PSS must compensate the existing phase lag between the excitation and the electric torque.

The stabilizer models provided in the following are generally consistent with the excitation models and may be applicable for investigation of control modes of instability that usually occur below 3 Hz [36]. Stabilizer parameters should be consistent with the type of input signal specified in the stabilizer model. Parameters for stabilizers with different input signals may look very different while providing similar damping characteristics.

For pumping-storage units, the stabilizer can be used with the synchronous machine operating in either the generating or pumping modes, but different parameters would usually be required for operation in two modes.

TYPE PSS1A—POWER SYSTEM STABILIZER. In Figure 2.65 is presented the generalized form of a power system stabilizer with a single input. Some common stabilizer input signals (V_{SI}) are speed, frequency, and power.

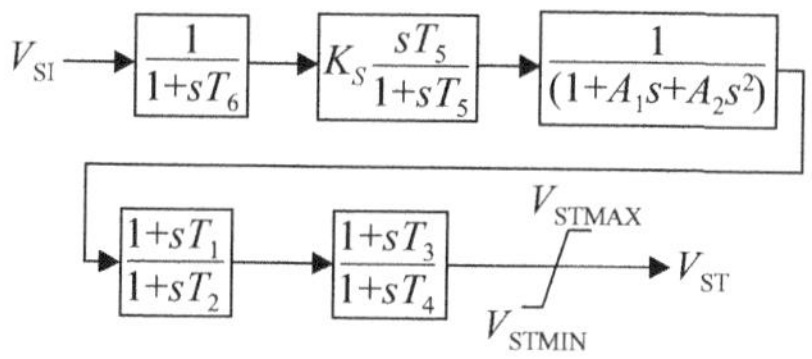

Figure 2.65. Type PSS1A—single-input power system stabilizer. (Reprinted with permission from Ref. 36.)

In Figure 2.65, T_6 may be used to represent a transducer time constant; stabilizer gain is set by the term K_S and signal washout is set by the time constant T_5. In the next block, A_1 and A_2 allow some of the low-frequency effects of high-frequency torsional filters to be accounted for. When not used for this purpose, the block can be used to assist in shaping the gain and phase characteristics of the stabilizer, if required. The next two blocks allow two stages of lead-lag compensation, as set by constants T_1 to T_4.

Stabilizer output can be limited in various ways. This model has only simple stabilizer output limits, V_{STMAX} and V_{STMIN}. In some systems, the stabilizer output is disconnected immediately from the regulator following a severe fault to avoid compromising the regulator action during the first swing. This is accomplished, as shown in Figure 2.66, by opening the output of the PSS for a predefined time T_{DR}, if the terminal voltage V_g drops below the minimum value V_{gMIN} [36].

In other systems, the stabilizer output is limited as a function of generator terminal voltage [35].

The stabilizer output V_{ST} is an input to the supplementary discontinuous control models. If the discontinuous control models are not used, $V_S = V_{ST}$.

Type PSS2B—Power System Stabilizer. A dual-input stabilizer model that usually uses combinations of power and speed or frequency to derive the stabilizing signal is shown in Figure 2.67.

In particular, this model can be used to represent two distinct types of dual-input stabilizer implementations as described below [36]:

a. Stabilizers that, in the frequency range of system oscillations, act as electrical power input stabilizers. These use the speed or frequency input for the generation of an equivalent mechanical power signal, to make the total signal insensitive to mechanical power change.
b. Stabilizers that use a combination of speed (or frequency) and electrical power. These systems usually use the speed directly, without phase-lead compensation, and add a signal proportional to electrical power to achieve the desired stabilizing signal shaping.

While the same model is used for two types of dual-input stabilizers mentioned above, the parameters used in the model for equivalent stabilizing action will be very different.

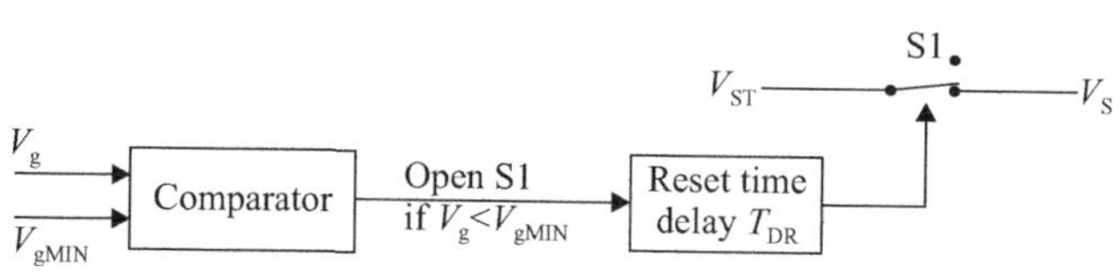

Figure 2.66. Type DEC3A—discontinuous excitation controller temporary interruption of stabilizing signal. (Reprinted with permission from Ref. 36.)

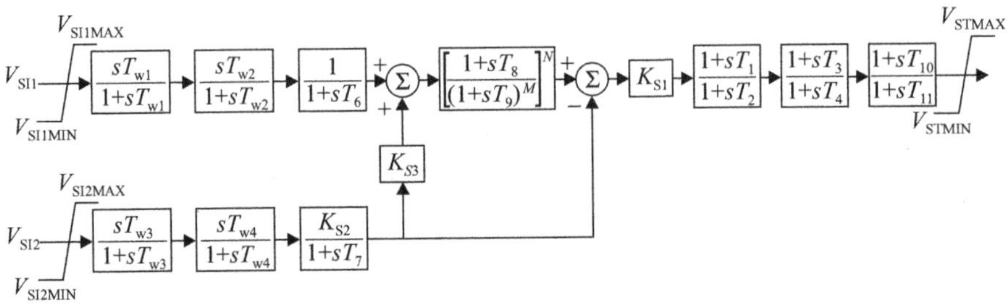

Figure 2.67. Type PSS2B—dual-input power system stabilizer. (Reprinted with permission from Ref. 36.)

In Figure 2.67, V_{SI1} would normally represent speed or frequency and V_{SI2} would be a power signal. For each input, two washouts can be represented—T_{w1} to T_{w4}—along with a transducer or integrator time constant (T_6, T_7). The washout circuit is a high-pass filter that prevents any steady change in frequency, speed, and power affecting the field voltage because of the PSS action. The PSS is expected to act only during transient changes in the stabilizing signal [41].

For the first type of dual-input stabilizer, K_{S3} would normally be 1 and K_{S2} would be equal to $T_7/2H$, where H is the inertia constant of the synchronous machine. These constants determine the amount of damping determined by the power system stabilizer.

The indices, N (an integer up to 4) and M (an integer up to 5), allow a "ramp-tracking" or simpler filter characteristic to be represented. Usual values are $N = 1$ and $M = 5$. The ramp-tracking filter produces a zero steady-state error to ramp changes in the input integral of electrical power signal.

Phase compensation is provided by the two lead-lag or lag-lead blocks—T_1 to T_4. The lead-lag and lag-lead functions are used to compensate the phase shift between the excitation voltage and the electrical torque of the machine. An additional block with lag time constant T_{11} and lead time constant T_{10} can be used to model stabilizers that incorporate a third lead-lag function.

Note: For many types of studies, the simpler single-input PSS1A model, with appropriate parameters, may be used instead of the two-input PSS2B model.

2.1.8.4 Example

To illustrate the deduction of the operating equations, we consider the simplified schemes for the exciter system, terminal voltage control block (VCB), and PSS. The IEEE ST1-Type exciter scheme with a PSS input is shown in Figure 2.68a.

The following notations are used in Figure 2.68:

V_C is the generator terminal voltage;

K_A is the voltage regulator gain;

T_A is the voltage regulator time constant;

V_R is the voltage regulator output;

T_R is the regulator input filter time constant;

K_F is the excitation control system stabilizer gain;

T_F is the excitation control system stabilizer time constant;

K_S is the stabilizer gain;

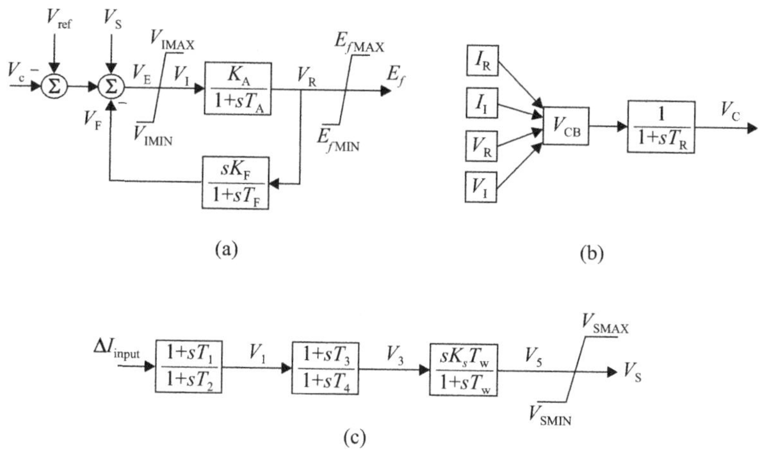

Figure 2.68. ST1-Type exciter with PSS input—V_S (a); terminal voltage control block (b); lead-lag power system stabilizer (c).

T_w is the PSS washout time constant;
T_i is the time constants;
V_i is the intermediary variables;
ΔI_{input} is the input signal.

The control equations are given by

$$
\left\|
\begin{array}{l}
V_E = V_{ref} - V_C + V_S - V_F \\
V_F = (K_F/T_F)V_R - V_{FL} \\
\dot{V}_{FL} = V_F \\
V_I = \begin{cases} V_{IMAX} & \text{if} \quad V_E \geq V_{IMAX} \\ V_E & \text{if} \quad V_{IMAX} > V_E > V_{IMIN} \\ V_{IMIN} & \text{if} \quad V_E < V_{IMIN} \end{cases} \\
\dot{V} = (K_A V_I - V_R)/T_A \\
\dot{E}_f = \begin{cases} E_{fMAX} & \text{if} \quad \dot{V}_R \geq E_{fMAX} \\ \dot{V}_R & \text{if} \quad E_{fMAX} > \dot{V}_R > E_{fMIN} \\ E_{fMIN} & \text{if} \quad \dot{V}_R < E_{fMIN} \end{cases}
\end{array}
\right. \tag{2.203}
$$

- *The terminal voltage* (V_C) (Figure 2.68b) is a function of the d- and q-axis voltages, which, in turn, depend on the real and imaginary portions of the voltage at the terminals (V_R, V_I), the current through the machine connection node (I_R, I_I) and the impedance seen from the synchronous machine terminals ($R_C + jX_C$). The complete set of equations for the terminal VCB is given by

$$\begin{aligned}\dot{V}_{\mathrm{C}} &= (V_{\mathrm{CB}} - V_{\mathrm{C}})\frac{1}{T_{\mathrm{R}}} \\ V_{\mathrm{CB}} &= \sqrt{(V_{\mathrm{R}} + I_{\mathrm{R}}X_{\mathrm{C}} + I_{\mathrm{I}}R_{\mathrm{C}})^2 + (V_{\mathrm{I}} + I_{\mathrm{R}}X_{\mathrm{C}} + I_{\mathrm{I}}R_{\mathrm{C}})^2}\end{aligned} \tag{2.204}$$

$$\begin{bmatrix} I_{\mathrm{R}} \\ I_{\mathrm{I}} \end{bmatrix} = \begin{bmatrix} \sin\theta & \cos\theta \\ -\cos\theta & \sin\theta \end{bmatrix} \cdot \begin{bmatrix} I_{\mathrm{R}} \\ I_q \end{bmatrix}$$

where I_d and I_q are the d-axis and q-axis currents, respectively.

- *The lead-lag PSS* structure is shown in Figure 2.68c. The control equations for this PSS are given by

$$\left\| \begin{aligned} &\Delta I_{\mathrm{input}} = \begin{cases} \Delta\omega = \omega_{\mathrm{n}} - \omega_0 \\ \text{or} \\ \Delta P_{\mathrm{a}} = P_{\mathrm{m}} - P_{\mathrm{e}} \end{cases} \\ &\dot{V}_1 = (T_1/T_2)\Delta\dot{I}_{\mathrm{input}} + (\Delta I_{\mathrm{input}} - V_1)/T_2 \\ &\dot{V}_3 = (T_3/T_4)\dot{V}_1 + (V_1 - V_3)/T_4 \\ &\dot{V}_5 = K_{\mathrm{S}}\dot{V}_3 - (V_3/T_{\mathrm{w}}) \\ &\dot{V}_{\mathrm{S}} = \begin{cases} V_{\mathrm{SMAX}} & \text{if} \quad \dot{V}_5 \geq V_{\mathrm{SMAX}} \\ \dot{V}_5 \quad \text{if} & V_{\mathrm{SMAX}} > \dot{V}_5 > V_{\mathrm{SMIN}} \\ V_{\mathrm{SMIN}} & \text{if} \quad \dot{V}_5 < V_{\mathrm{SMIN}} \end{cases} \end{aligned} \right. \tag{2.205}$$

2.2 THEORY AND MODELING OF THE INDUCTION MOTOR

The induction/asynchronous motors represent a major part of the loads supplied from the power system buses and are referred to as complex consumers (more than 60% of the electrical energy generated in a power system is consumed by induction motors). For this reason, the behavior and the dynamic characteristics of the load are, mainly, derived from the dynamic characteristics of induction motors.

2.2.1 Design and Operation Issues

Similar to the synchronous generator, the main constructive elements of the induction motor are the *stator* and the *rotor*, which are separated by the *air gap*. The stator is the inductor part and consists of a magnetic core filled by insulated coils (windings) spatially displaced by 120°. The rotor, like the stator, has a symmetrical structure and consists also of a magnetic core filled by rotor coils (windings) and represents the induced part. Unlike the synchronous generator, where the rotor winding (excitation winding) is connected to a direct current source, the rotor windings of the induction motor is three-phase wye connected and designed for the same pole pairs p as the stator windings. Their terminals are connected to a passive external circuit through slip rings, in case of wound rotor, or are short-circuited internally when the rotor has a simple or double squirrel cage type construction.

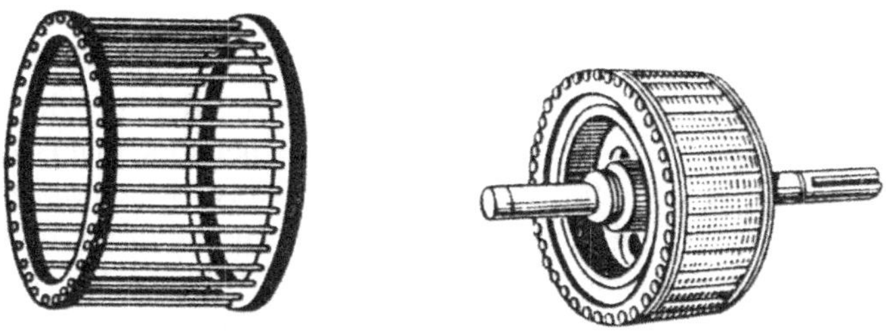

(a) (b)

Figure 2.69. Squirrel-cage rotor induction machine: (a) squirrel cage of rotor; (b) squirrel cage rotor [42].

Induction motor rotors are made in two forms: short-circuited (squirrel cage) and with slip rings. The first of these is simpler in construction and is more frequently used. The winding of such a rotor is in form of a cylindrical cage (the so-called "squirrel cage") of copper or aluminum bars, short-circuited around the ends by two rings. The bars of this winding are placed without insulation in the slots of the rotor (Figure 2.69a). Another method in use is to pour molten aluminum into the slots of the rotor.

A slip-ring rotor, also known as a phase-wound rotor, has a winding made of insulated wire (Figure 2.70a) and in most cases in three-phase wye connected. The free ends of this winding are brought out to the slip rings on the rotor shaft.

Brushes bear on the slip rings and connect the rotor winding to a three-phase rheostat (Figure 2.70b). This system permits the resistance of the rotor circuits to be varied. This is very important in starting the motor.

An asynchronous machine is similar to a transformer in the sense that the power is transferred from the stator (primary) to the rotor (secondary) winding only by mutual induction. For this reason an asynchronous machine is often called an induction machine.

The induction machine operation is due to the fact that the ratio of the rotor angular speed and the network frequency varies with the motor load and characteristics of supply. The induction machines can operate as motor, generator or asynchronous brake.

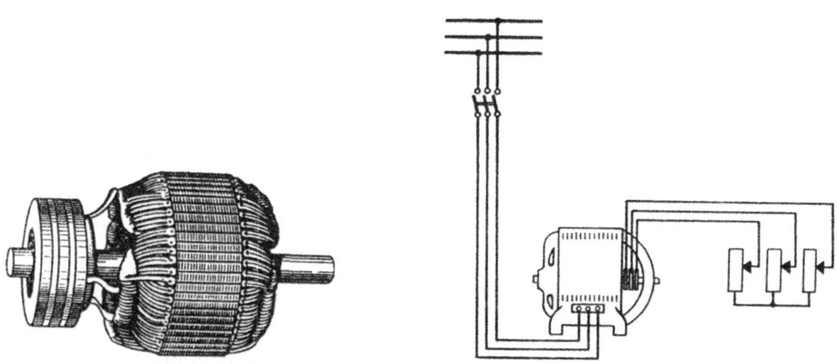

(a) (b)

Figure 2.70. Wound rotor induction machine: (a) rotor with slip rings; (b) slip-ring motor connected to starting rheostat [42].

When the induction motor is connected to the electrical grid (considered as a balanced direct-sequence three-phase voltage source, of frequency f_s) the currents carried in the stator windings produce a rotating magnetic field of which fundamental component has an angular frequency equal to that of the network, that is, $\omega_s = 2\pi f_s$, and an angular velocity $\omega_{ms} = \omega_s/p$ [mech. rad/s], respectively. If the rotor rotates with an angular velocity $\omega_{mr} = \omega_r/p$ [mech. rad/s] different from that of the rotating magnetic field, then voltages, of frequency f_r, directly proportional to the relative velocity $\omega_s - \omega_r = p(\omega_{ms} - \omega_{mr})$, are induced in the rotor windings, which produce induced currents, carried in the rotor windings.

The slip speed of the rotor in per unit of the synchronous speed is

$$s = \frac{\omega_{ms} - \omega_{mr}}{\omega_{ms}} = \frac{(\omega_s/p) - (\omega_r/p)}{\omega_s/p} = \frac{\omega_s - \omega_r}{\omega_s} = 1 - \frac{\omega_r}{\omega_s} \tag{2.206}$$

and the frequency of the induced rotor voltage and, therefore, of the rotor currents are equal to the slip frequency:

$$f_r = sf_s \tag{2.207}$$

The rotor currents, which are equal to the ratio of the induced voltages and the corresponding winding rotor impedance at the rotor frequency f_r, produce in turn a magnetic field, which reacts with the magnetic field produced by the stator creating therefore a torque. To obtain a positive torque (motor), the slip of the induction machine should be positive, that is, $\omega_s > \omega_r$.

At no load, the slip is positive, but very low. Instead, if an antagonist torque, due to a mechanical load, is applied to the rotor shaft, the rotor speed decreases and the slip increases leading to higher values of the induced voltages and currents and therefore achieve an increase in the produced motor torque. In this case the induction machine operates as a motor.

When the rotor is driven by a prime mover at a speed greater than that of the rotating magnetic field, that is, $\omega_s < \omega_r$, the slip is negative so the polarities of the induced voltages are reversed, and the induction machine operates as a generator.

2.2.2 General Equations of the Induction Motor

2.2.2.1 Electrical Circuit Equations. The mathematical model of the induction machine is developed in a similar way as that of the synchronous generator. First, the electrical circuits equations of the induction machine are written in terms of phase quantities appropriate for the stator (a, b, c) and rotor (A, B, C), then the Park transformation is applied to obtain the equations in rotor coordinates (d, q, 0). However, peculiar characteristics of the induction machine are to be considered [1,29]:

- The rotor speed varies with the load, aspect that should be considered when selecting the d–q reference frame.
- The rotor has a symmetrical structure and hence the equivalent circuits in the d- and q-axes are identical.
- The neutrals of the three-phase windings of the stator and rotor are isolated and hence the zero-sequence component is missing.

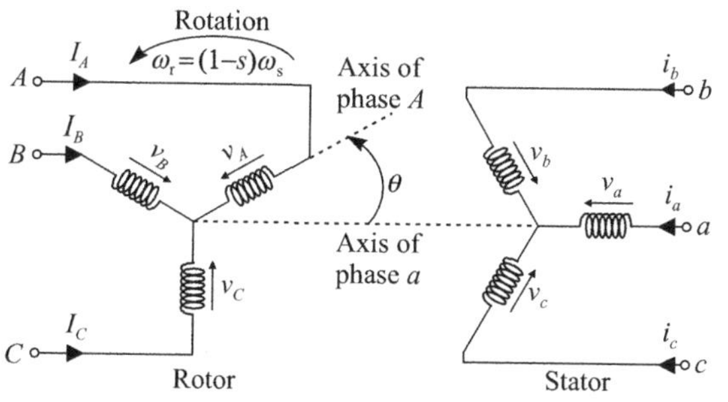

Figure 2.71. Stator and rotor circuits of the induction motor.

- Since there is no excitation source applied to the rotor windings, the dynamics of the rotor circuits are determined by the slip s, rather than by the excitation control system.

Furthermore, the loads rule is adopted to associate the direction of the currents in the rotor and stator circuits (Figure 2.71); commonly one could assume that

- the magnetic circuits saturation is negligible;
- the air gap magnetic flux has a sinusoidal distribution.

The voltage equations of the induction machine are

- for the stator circuits:

$$\begin{aligned} v_a &= R_s i_a + \frac{d\psi_a}{dt} \\ v_b &= R_s i_b + \frac{d\psi_b}{dt} \\ v_c &= R_s i_c + \frac{d\psi_c}{dt} \end{aligned} \tag{2.208}$$

- for the rotor circuits:

$$\begin{aligned} v_A &= R_r i_A + \frac{d\psi_A}{dt} \\ v_B &= R_r i_B + \frac{d\psi_B}{dt} \\ v_C &= R_r i_C + \frac{d\psi_C}{dt} \end{aligned} \tag{2.209}$$

Equations (2.208) and (2.209) can be written in matrix form as

$$[v_s] = R_s[i_s] + \left[\frac{d\psi_s}{dt}\right] \tag{2.208'}$$

$$[v_r] = R_r[i_r] + \left[\frac{d\psi_r}{dt}\right] \tag{2.209'}$$

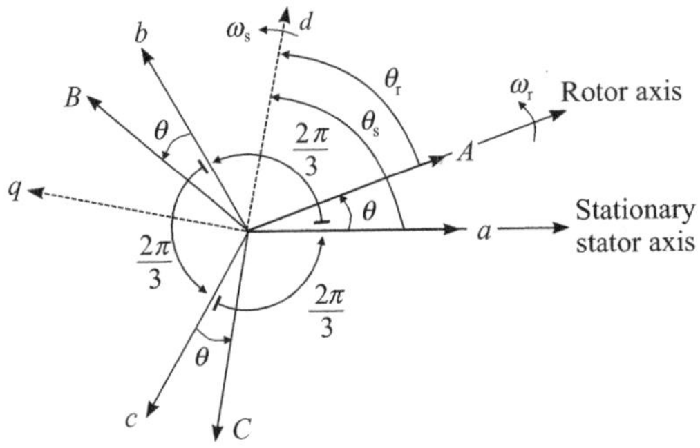

Figure 2.72. Definition of the coordinate systems and angles used in the induction machine theory.

where $[\psi_s] = [\psi_a, \psi_b, \psi_c]^t$ is the vector of the fluxes linking the stator windings, $[\psi_r] = [\psi_A, \psi_B, \psi_C]^t$ is the vector of the fluxes linking the rotor windings, $[i_s] = [i_a, i_b, i_c]^t$ is the vector of the stator windings currents, $[i_r] = [i_A, i_B, i_C]^t$ is the vector of the rotor windings currents, and R_s, R_r are the resistances of the stator and rotor windings, respectively.

Note that the positive directions of currents are into the windings.

The total fluxes linking the two windings are given by the following matrix form:

$$\begin{bmatrix} [\psi_s] \\ \hline [\psi_r] \end{bmatrix} = \left[\begin{array}{c|c} [l_{ss}] & [l_{sr}] \\ \hline [l_{sr}]^t & [l_{rr}] \end{array}\right] \begin{bmatrix} [i_s] \\ \hline [i_r] \end{bmatrix} \tag{2.210}$$

where $[l_{ss}]$ is the matrix of self- and mutual inductances of the stator windings, $[l_{sr}]$ is the matrix of mutual inductances between the stator and rotor windings, and $[l_{rr}]$ is the matrix of self- and mutual inductances of the rotor windings.

Considering the symmetrical structure of the rotor, it results that the self- and mutual inductances of the stator and rotor windings, respectively, are constant, while the mutual inductances between the two windings depend on the rotor position given by the angle θ (Figure 2.72):

$$\theta = \omega_r t + \theta_0 \tag{2.211}$$

where θ_0 is the value of θ at the instant $t = 0$.

Substituting ω_r from (2.206) in (2.211) achieve the relationship that describes the dependency of θ in terms of the slip s and the angular velocity of the rotor field ω_s, in el. rad/s:

$$\theta = (1 - s)\omega_s t + \theta_0 \tag{2.212}$$

Therefore, $l_{aa} = l_{bb} = l_{cc} = L_{aa}$ and $l_{ab} = l_{bc} = l_{ca} = L_{ab}$ for the stator windings, and $l_{AA} = l_{BB} = l_{CC} = L_{AA}$ and $l_{AB} = l_{BC} = l_{CA} = L_{AB}$ for the rotor windings, respectively. Hence,

$$[l_{ss}] = \left[\begin{array}{c|c|c} L_{aa} & L_{ab} & L_{ab} \\ \hline L_{ab} & L_{aa} & L_{ab} \\ \hline L_{ab} & L_{ab} & L_{aa} \end{array}\right] \tag{2.213a}$$

$$[l_{rr}] = \left[\begin{array}{c|c|c} L_{AA} & L_{AB} & L_{AB} \\ \hline L_{AB} & L_{AA} & L_{AB} \\ \hline L_{AB} & L_{AB} & L_{AA} \end{array}\right] \tag{2.213b}$$

Assuming a sinusoidal distribution of the magnetic flux in the air gap, the mutual inductances between the stator and rotor windings have a sinusoidal variation in terms of θ, which defines the rotor position. Therefore, the mutual inductance between a stator winding and a rotor winding is maximum when the axes of the two windings coincide, decreases as the angle θ increases and becomes zero when the axes are perpendicular, decreases again as the angle increases and reaches a maximum negative value when the axes are in opposition and so forth. Based on these aspects and considering the angles defined in Figure 2.72, it results

$$[l_{sr}] = \left[\begin{array}{c|c|c} L_{aA}\cos\theta & L_{aA}\cos\left(\theta+\frac{2\pi}{3}\right) & L_{aA}\cos\left(\theta-\frac{2\pi}{3}\right) \\ \hline L_{aA}\cos\left(\theta-\frac{2\pi}{3}\right) & L_{aA}\cos\theta & L_{aA}\cos\left(\theta+\frac{2\pi}{3}\right) \\ \hline L_{aA}\cos\left(\theta+\frac{2\pi}{3}\right) & L_{aA}\cos\left(\theta-\frac{2\pi}{3}\right) & L_{aA}\cos\theta \end{array}\right] \tag{2.213c}$$

$$[l_{rs}] = [l_{sr}]^t = \left[\begin{array}{c|c|c} L_{aA}\cos\theta & L_{aA}\cos\left(\theta-\frac{2\pi}{3}\right) & L_{aA}\cos\left(\theta+\frac{2\pi}{3}\right) \\ \hline L_{aA}\cos\left(\theta+\frac{2\pi}{3}\right) & L_{aA}\cos\theta & L_{aA}\cos\left(\theta-\frac{2\pi}{3}\right) \\ \hline L_{aA}\cos\left(\theta-\frac{2\pi}{3}\right) & L_{aA}\cos\left(\theta+\frac{2\pi}{3}\right) & L_{aA}\cos\theta \end{array}\right] \tag{2.213d}$$

where L_{aA} is the maximum value of the mutual inductances.

From equations (2.210), (2.213a), (2.213b), (2.213c), and (2.213d), it results the following expressions for the total flux linkage in the stator winding a and in the rotor winding A:

$$\begin{aligned} \psi_a &= L_{aa}i_a + L_{ab}(i_b+i_c) + L_{aA}\left[i_A\cos\theta + i_B\cos\left(\theta+\frac{2\pi}{3}\right) + i_C\cos\left(\theta-\frac{2\pi}{3}\right)\right] \\ \psi_A &= L_{AA}i_A + L_{AB}(i_B+i_C) + L_{aA}\left[i_a\cos\theta + i_b\cos\left(\theta-\frac{2\pi}{3}\right) + i_c\cos\left(\theta+\frac{2\pi}{3}\right)\right] \end{aligned} \tag{2.214}$$

With no neutral currents due to winding connections or balanced conditions, we have

$$\begin{cases} i_a + i_b + i_c = 0 \\ i_A + i_B + i_C = 0 \end{cases} \tag{2.215}$$

and equations (2.214) become

$$\begin{aligned}\psi_a &= L_{ss}i_a + L_{aA}\left[i_A\cos\theta + i_B\cos\left(\theta + \frac{2\pi}{3}\right) + i_C\cos\left(\theta - \frac{2\pi}{3}\right)\right]\\ \psi_A &= L_{rr}i_A + L_{aA}\left[i_a\cos\theta + i_b\cos\left(\theta - \frac{2\pi}{3}\right) + i_c\cos\left(\theta + \frac{2\pi}{3}\right)\right]\end{aligned} \tag{2.216}$$

where $L_{ss} = L_{aa} - L_{ab}$ and $L_{rr} = L_{AA} - L_{AB}$.

Similar expressions to (2.216) can be obtained also for the total flux linkage in the windings of the b and c phases and of the B and C phases, respectively.

2.2.2.2 The d–q Transformation. As in the case of the synchronous machine, the equations of the induction motor can be simplified by applying the change of variables. If the d–q frame attached to the rotor was chosen for the synchronous generator, for the induction motor it is more convenient to use a d–q frame rotating with the synchronous speed. The q-axis is assumed to be $\pi/2$ ahead of the d-axis in the direction of rotation [1] (Figure 2.72). Therefore, two Park transformation matrices are required: one to transform the quantities and equations attached to the stator windings and the other to transform the quantities and equations attached to the rotor windings.

Let θ_s be the angle between the d-axis and the fixed stator phase a-axis at the instant t, and θ_r be the angle between the d-axis and the rotational rotor phase A-axis at the same instant. Selecting as time reference the instant at which the three axes (d, a, and A) overlap then $\theta_0 = 0$ and, according to equation (2.211) and Figure 2.72, gives

$$\begin{cases}\theta = \omega_r t \quad ; \quad \theta_s = \omega_s t\\ \theta_r = \theta_s - \theta = (\omega_s - \omega_r)t = s\omega_s t\end{cases} \tag{2.217}$$

Therefore, the transformation matrices for the stator (a, b, c) and rotor (A, B, C) phase quantities in (d, q, 0) quantities, known as the Park transformation matrices for the induction motor, are

$$[P_s] = \frac{2}{3}\begin{bmatrix}\cos\theta_s & \cos\left(\theta_s - \frac{2\pi}{3}\right) & \cos\left(\theta_s + \frac{2\pi}{3}\right)\\ -\sin\theta_s & -\sin\left(\theta_s - \frac{2\pi}{3}\right) & -\sin\left(\theta_s + \frac{2\pi}{3}\right)\\ \frac{1}{2} & \frac{1}{2} & \frac{1}{2}\end{bmatrix} \tag{2.218}$$

and

$$[P_r] = \frac{2}{3}\begin{bmatrix}\cos\theta_r & \cos\left(\theta_r - \frac{2\pi}{3}\right) & \cos\left(\theta_r + \frac{2\pi}{3}\right)\\ -\sin\theta_r & -\sin\left(\theta_r - \frac{2\pi}{3}\right) & -\sin\left(\theta_r + \frac{2\pi}{3}\right)\\ \frac{1}{2} & \frac{1}{2} & \frac{1}{2}\end{bmatrix} \tag{2.218'}$$

The two matrices are similar, the only difference being given by the angles θ_s and θ_r, which define the relative position of the stator and rotor winding, respectively, with respect to the *d–q* reference frame.

By applying the Park transformation achieve the operating equations of the induction machine written in a unique reference frame rotating with the synchronous speed ω_s. The time dependence of the mutual inductances between stator and rotor windings is therefore eliminated, and the inductances are constant.

2.2.2.3 Basic Equations in the d–q Reference Frame. Under balanced operating conditions by applying the Park transformation to the systems of equations (2.208) and (2.209) we achieve the equations of the induction motor in *d–q* coordinates:

- *Stator Voltage Equations*

$$\begin{aligned} v_{ds} &= R_s i_{ds} + \frac{d\psi_{ds}}{dt} - \frac{d\theta_s}{dt}\psi_{qs} = R_s i_{ds} + \frac{d\psi_{ds}}{dt} - \omega_s \psi_{qs} \\ v_{qs} &= R_s i_{qs} + \frac{d\psi_{qs}}{dt} + \frac{d\theta_s}{dt}\psi_{ds} = R_s i_{qs} + \frac{d\psi_{qs}}{dt} + \omega_s \psi_{ds} \end{aligned} \tag{2.219}$$

- *Rotor Voltage Equations*

$$\begin{aligned} v_{dr} &= R_r i_{dr} + \frac{d\psi_{dr}}{dt} - \frac{d\theta_r}{dt}\psi_{qr} = R_r i_{dr} + \frac{d\psi_{dr}}{dt} - s\omega_s \psi_{qr} \\ v_{qr} &= R_r i_{qr} + \frac{d\psi_{qr}}{dt} + \frac{d\theta_r}{dt}\psi_{dr} = R_r i_{qr} + \frac{d\psi_{qr}}{dt} + s\omega_s \psi_{dr} \end{aligned} \tag{2.220}$$

These equations are similar to the stator equations of the synchronous generator. The expression of each voltage contains three terms: the resistive voltage drop Ri, the transient mmf $d\psi/dt$, due to the variation in time of the magnetic flux, and the speed voltage term. The first two terms are familiar terms associated with the voltage of any static coil, while the third term is peculiar to the specific situation at hand. Thus, the terms $\omega_s\psi_{qs}$ and $\omega_s\psi_{ds}$ in equations (2.219) represent the voltages induced in the stator windings by the synchronously rotating flux waves. Similarly, the terms $s\omega_s\psi_{qr}$ and $s\omega_s\psi_{dr}$ in equations (2.220) represent the voltages induced in the rotor windings that rotate at the slip speed $s\omega_s$ with respect to the synchronously rotating flux waves [1].

The *d–q* components of voltages, currents, and fluxes in equations (2.219) and (2.220) are determined using the relationships between the phase components and the (*d*, *q*, 0) components, defined by the Park transformation:

$$\begin{bmatrix} d_s \\ q_s \\ 0_s \end{bmatrix} = [P_s] \begin{bmatrix} a \\ b \\ c \end{bmatrix} \quad \text{or} \quad \begin{bmatrix} a \\ b \\ c \end{bmatrix} = [P_s]^{-1} \begin{bmatrix} d_s \\ q_s \\ 0_s \end{bmatrix} \tag{2.221}$$

for the stator quantities, and

$$\begin{bmatrix} d_r \\ q_r \\ 0_r \end{bmatrix} = [P_r] \begin{bmatrix} A \\ B \\ C \end{bmatrix} \quad \text{or} \quad \begin{bmatrix} A \\ B \\ C \end{bmatrix} = [P_r]^{-1} \begin{bmatrix} d_r \\ q_r \\ 0_r \end{bmatrix} \tag{2.222}$$

for the rotor quantities, respectively.

Therefore, applying the direct transformation (2.221) for the stator currents, gives

$$\begin{bmatrix} i_{ds} \\ i_{qs} \\ i_{0s} \end{bmatrix} = [P_s] \begin{bmatrix} i_a \\ i_b \\ i_c \end{bmatrix}$$

and considering equation (2.218), which defines the $[P_s]$ matrix, gives

$$\begin{aligned} i_{ds} &= \frac{2}{3}\left[i_a \cos\theta_s + i_b \cos\left(\theta_s - \frac{2\pi}{3}\right) + i_c \cos\left(\theta_s + \frac{2\pi}{3}\right)\right] \\ i_{qs} &= -\frac{2}{3}\left[i_a \sin\theta_s + i_b \sin\left(\theta_s - \frac{2\pi}{3}\right) + i_c \sin\left(\theta_s + \frac{2\pi}{3}\right)\right] \\ i_{0s} &= \frac{1}{3}(i_a + i_b + i_c) = 0 \end{aligned} \tag{2.223}$$

The inverse transformation is given by

$$\begin{aligned} i_a &= i_{ds} \cos\theta_s - i_{qs} \sin\theta_s \\ i_b &= i_{ds} \cos\left(\theta_s - \frac{2\pi}{3}\right) - i_{qs} \sin\left(\theta_s - \frac{2\pi}{3}\right) \\ i_c &= i_{ds} \cos\left(\theta_s + \frac{2\pi}{3}\right) - i_{qs} \sin\left(\theta_s + \frac{2\pi}{3}\right) \end{aligned} \tag{2.224}$$

Similar transformations apply to the expressions of the rotor currents and stator and rotor voltages, respectively.

To determine the $(d, q, 0)$ flux components, the matrix equation (2.210) is used taking also into account (2.221) and (2.222). The currents vectors expressed in $(d, q, 0)$ components, obtained by application of Park transformation, are

$$[i_{Ps}] = [P_s][i_s] = [i_{ds}, i_{qs}, i_{0s}]^t \quad \text{and} \quad [i_{Pr}] = [P_r][i_r] = [i_{dr}, i_{qr}, i_{0r}]^t$$

Therefore,

$$[i_s] = [P_s]^{-1}[i_{Ps}] \quad \text{and} \quad [i_r] = [P_r]^{-1}[i_{Pr}]$$

and equation (2.210) is then developed as

$$[\psi_s] = [l_{ss}][i_s] + [l_{sr}][i_r] = [l_{ss}][P_s]^{-1}[i_{Ps}] + [l_{sr}][P_r]^{-1}[i_{Pr}] \tag{2.225a}$$

$$[\psi_r] = [l_{rs}][i_s] + [l_{rr}][i_r] = [l_{rs}][P_s]^{-1}[i_{Ps}] + [l_{rr}][P_r]^{-1}[i_{Pr}] \tag{2.225b}$$

Equation (2.225a) is left-multiplied by $[P_s]$, and equation (2.225b) by $[P_r]$, which gives

$$[\psi_{Ps}] = [\psi_{ds}, \psi_{qs}, \psi_{0s}]^t = [P_s][l_{ss}][P_s]^{-1}[i_{Ps}] + [P_s][l_{sr}][P_r]^{-1}[i_{Pr}] \tag{2.226a}$$

$$[\psi_{\rm Pr}] = [\psi_{dr}, \psi_{qr}, \psi_{0r}]^t = [P_{\rm r}][l_{\rm rs}][P_{\rm s}]^{-1}[i_{\rm Ps}] + [P_{\rm r}][l_{\rm rr}][P_{\rm r}]^{-1}[i_{\rm Pr}] \tag{2.226b}$$

Substituting the inductances matrices and Park transformation matrix achieve the stator and rotor flux linkages in d–q components:

$$\begin{cases} \psi_{ds} = L_{\rm ss} i_{ds} + L_{\rm m} i_{dr} \\ \psi_{qs} = L_{\rm ss} i_{qs} + L_{\rm m} i_{qr} \\ \psi_{dr} = L_{\rm rr} i_{dr} + L_{\rm m} i_{ds} \\ \psi_{qr} = L_{\rm rr} i_{qr} + L_{\rm m} i_{qs} \end{cases} \tag{2.227}$$

with $L_{\rm m} = 3/2L_{aA}$, the magnetizing reactance.

In equation (2.227), we see that application of Park transformation for induction motor study corresponds, physically, to the replacement of the stator and rotor windings with two fictitious windings, ds and qs, and dr and qr, respectively, disposed on the d–q frame, which rotates with the synchronous speed $\omega_{\rm s}$ set by the power grid.

2.2.2.4 Electric Power and Torque. The instantaneous power input to the stator is

$$p_{\rm s} = v_a i_a + v_b i_b + v_c i_c \tag{2.228a}$$

or in terms of d–q components

$$p_{\rm s} = \frac{3}{2}\left(v_{ds} i_{ds} + v_{qs} i_{qs}\right) \tag{2.228b}$$

Similarly, the instantaneous power input to the rotor is

$$p_{\rm r} = \frac{3}{2}\left(v_{dr} i_{dr} + v_{qr} i_{qr}\right) \tag{2.229}$$

The electromagnetic torque developed is obtained as the power associated with the speed voltages (the instantaneous rotor power obtained by neglecting the winding losses and the transient mmf) divided by the shaft speed in mechanical radians per second. Substituting in equation (2.229) the voltages v_{dr} and v_{qr} with the corresponding speed voltages $-s\omega_{\rm s}\psi_{qr}$ and $s\omega_{\rm s}\psi_{dr}$, respectively, and taking into account that the rotor speed, in radians per second, with respect to the d–q frame, is $s\omega_{\rm s}/p$, achieve the expression of the electromagnetic torque:

$$C_{\rm e} = \frac{3}{2}\frac{\left(-s\omega_{\rm s}\psi_{qr} i_{dr} + s\omega_{\rm s}\psi_{dr} i_{qr}\right)}{s\omega_{\rm s}/p} = \frac{3}{2}p\left(\psi_{dr} i_{qr} - \psi_{qr} i_{dr}\right) \tag{2.230}$$

where p is the number of pole pairs.

2.2.3 Steady-State Operation of the Induction Motor

In steady-state operation, the derivatives in terms of time are zero. Hence, substituting the flux linkages, expressed in equations (2.227), in (2.219) and (2.220), respectively, gives

- stator voltage equations:

$$\begin{aligned} v_{ds} &= R_s i_{ds} - \omega_s \psi_{qs} = R_s i_{ds} - \omega_s L_{ss} i_{qs} - \omega_s L_m i_{qr} \\ v_{qs} &= R_s i_{qs} + \omega_s \psi_{ds} = R_s i_{qs} + \omega_s L_{ss} i_{ds} + \omega_s L_m i_{dr} \end{aligned} \tag{2.219$'$}$$

- rotor voltage equations:

$$\begin{aligned} v_{dr} &= R_r i_{dr} - s\omega_s \psi_{qr} = R_r i_{dr} - s\omega_s L_{rr} i_{qr} - s\omega_s L_m i_{qs} \\ v_{qr} &= R_r i_{qr} + s\omega_s \psi_{dr} = R_r i_{qr} + s\omega_s L_{rr} i_{dr} + s\omega_s L_m i_{ds} \end{aligned} \tag{2.220$'$}$$

On the other hand, under balanced steady-state conditions, the stator voltages may be written as

$$\begin{aligned} v_a &= V_m \cos(\omega_s t + \alpha) = \sqrt{2}V\cos(\omega_s t + \alpha) = \sqrt{2}\frac{U}{\sqrt{3}}\cos(\omega_s t + \alpha) \\ v_b &= V_m \cos\left(\omega_s t + \alpha - \frac{2\pi}{3}\right) = \sqrt{2}V\cos\left(\omega_s t + \alpha - \frac{2\pi}{3}\right) = \sqrt{2}\frac{U}{\sqrt{3}}\cos\left(\omega_s t + \alpha - \frac{2\pi}{3}\right) \\ v_c &= V_m \cos\left(\omega_s t + \alpha + \frac{2\pi}{3}\right) = \sqrt{2}V\cos\left(\omega_s t + \alpha + \frac{2\pi}{3}\right) = \sqrt{2}\frac{U}{\sqrt{3}}\cos\left(\omega_s t + \alpha + \frac{2\pi}{3}\right) \end{aligned} \tag{2.231}$$

where V_m and V are the phase-to-neutral peak and rms voltages, $U = \sqrt{3}V$ is the rms value of phase-to-phase voltage, and α is the phase angle of v_a with respect to the time origin.

Application of Park transformation, defined by (2.218) and (2.221), to equation (2.231), taking into account that, in accordance with (2.217), $\omega_s t = \theta_s$, leads to

$$\begin{aligned} v_{ds} = \frac{2}{3}\cdot V_m \bigg[&\cos(\theta_s + \alpha)\cos\theta_s + \cos\left(\theta_s + \alpha - \frac{2\pi}{3}\right)\cos\left(\theta_s - \frac{2\pi}{3}\right) \\ &+ \cos\left(\theta_s + \alpha + \frac{2\pi}{3}\right)\cos\left(\theta_s + \frac{2\pi}{3}\right)\bigg] \end{aligned} \tag{2.232a}$$

$$\begin{aligned} v_{qs} = \frac{2}{3}\cdot V_m \bigg[&\cos(\theta_s + \alpha)\sin\theta_s + \cos\left(\theta_s + \alpha - \frac{2\pi}{3}\right)\sin\left(\theta_s - \frac{2\pi}{3}\right) \\ &+ \cos\left(\theta_s + \alpha + \frac{2\pi}{3}\right)\sin\left(\theta_s + \frac{2\pi}{3}\right)\bigg] \end{aligned} \tag{2.232b}$$

$$v_{0s} = \frac{1}{3}\cdot V_m \left[\cos\theta_s + \cos\left(\theta_s - \frac{2\pi}{3}\right) + \cos\left(\theta_s + \frac{2\pi}{3}\right)\right] \tag{2.232c}$$

Taking into consideration the trigonometric identities

$$\cos x \cdot \cos y = \frac{1}{2}[\cos(x + y) + \cos(x - y)] \tag{2.233a}$$

$$\cos x \cdot \sin y = \frac{1}{2}[\sin(x + y) - \sin(x - y)] \tag{2.233b}$$

$$\cos x + \cos\left(x - \frac{2\pi}{3}\right) + \cos\left(x + \frac{2\pi}{3}\right) = 0 \tag{2.233c}$$

$$\sin x + \sin\left(x - \frac{2\pi}{3}\right) + \sin\left(x + \frac{2\pi}{3}\right) = 0 \tag{2.233d}$$

it can be written that

$$\cos(\theta_s + \alpha) \cdot \cos\theta_s = \frac{1}{2}[\cos(2\theta_s + \alpha) + \cos\alpha] \tag{2.234a}$$

$$\begin{aligned}\cos\left(\theta_s + \alpha - \frac{2\pi}{3}\right) \cdot \cos\left(\theta_s - \frac{2\pi}{3}\right) &= \frac{1}{2}\left[\cos\left(2\theta_s + \alpha - \frac{4\pi}{3}\right) + \cos\alpha\right]\\ &= \frac{1}{2}\left[\cos\left(2\theta_s + \alpha + \frac{2\pi}{3}\right) + \cos\alpha\right]\end{aligned} \tag{2.234b}$$

$$\begin{aligned}\cos\left(\theta_s + \alpha + \frac{2\pi}{3}\right) \cdot \cos\left(\theta_s + \frac{2\pi}{3}\right) &= \frac{1}{2}\left[\cos\left(2\theta_s + \alpha + \frac{4\pi}{3}\right) + \cos\alpha\right]\\ &= \frac{1}{2}\left[\cos\left(2\theta_s + \alpha - \frac{2\pi}{3}\right) + \cos\alpha\right]\end{aligned} \tag{2.234c}$$

Summing equations (2.234) and taking into account the trigonometric identity (2.233c), achieve

$$\begin{aligned}&\cos(\theta_s + \alpha) \cdot \cos\theta_s + \cos\left(\theta_s + \alpha - \frac{2\pi}{3}\right) \cdot \cos\left(\theta_s - \frac{2\pi}{3}\right)\\ &\quad + \cos\left(\theta_s + \alpha + \frac{2\pi}{3}\right) \cdot \cos\left(\theta_s + \frac{2\pi}{3}\right)\\ &= \frac{3}{2}\cos\alpha + \frac{1}{2}\left[\cos(2\theta_s + \alpha) + \cos\left(2\theta_s + \alpha - \frac{2\pi}{3}\right) + \cos\left(2\theta_s + \alpha + \frac{2\pi}{3}\right)\right] = \frac{3}{2}\cos\alpha\end{aligned}$$

Therefore, equation (2.232a), which defines the d-axis component of the stator voltage, becomes

$$v_{ds} = \frac{2}{3} \cdot V_m \cdot \frac{3}{2} \cdot \cos\alpha = V_m \cdot \cos\alpha \tag{2.235}$$

Similarly, the equation for the q-axis component of the stator voltage is obtained taking also into account the trigonometric identity (2.233b)

$$\cos(\theta_s + \alpha) \cdot \sin\theta_s = \frac{1}{2}[\sin(2\theta_s + \alpha) - \sin\alpha] \tag{2.236a}$$

$$\begin{aligned}\cos\left(\theta_s + \alpha - \frac{2\pi}{3}\right) \cdot \sin\left(\theta_s - \frac{2\pi}{3}\right) &= \frac{1}{2}\left[\sin\left(2\theta_s + \alpha - \frac{4\pi}{3}\right) - \sin\alpha\right]\\ &= \frac{1}{2}\left[\sin\left(2\theta_s + \alpha + \frac{2\pi}{3}\right) - \sin\alpha\right]\end{aligned} \tag{2.236b}$$

$$\cos\left(\theta_s+\alpha+\frac{2\pi}{3}\right)\cdot\sin\left(\theta_s+\frac{2\pi}{3}\right)=\frac{1}{2}\left[\sin\left(2\theta_s+\alpha+\frac{4\pi}{3}\right)-\sin\alpha\right]$$
$$=\frac{1}{2}\left[\sin\left(2\theta_s+\alpha-\frac{2\pi}{3}\right)-\sin\alpha\right] \quad (2.236c)$$

Summing equations (2.236) and taking into account the trigonometric identity (2.233d), achieve

$$\cos(\theta_s+\alpha)\cdot\sin\theta_s+\cos\left(\theta_s+\alpha-\frac{2\pi}{3}\right)\cdot\sin\left(\theta_s-\frac{2\pi}{3}\right)$$
$$+\cos\left(\theta_s+\alpha+\frac{2\pi}{3}\right)\cdot\sin\left(\theta_s+\frac{2\pi}{3}\right)$$
$$=-\frac{3}{2}\sin\alpha+\frac{1}{2}\left[\sin(2\theta_s+\alpha)+\sin\left(2\theta_s+\alpha-\frac{2\pi}{3}\right)+\sin\left(2\theta_s+\alpha+\frac{2\pi}{3}\right)\right]=-\frac{3}{2}\sin\alpha$$

Therefore, equation (2.232b) becomes

$$v_{qs}=-\frac{2}{3}\cdot V_m\cdot\left(-\frac{3}{2}\cdot\sin\alpha\right)=V_m\cdot\sin\alpha \quad (2.237)$$

Moreover, if the trigonometric identity (2.233c) is considered in (2.232c), the equation of v_{qs} becomes

$$v_{0s}=0 \quad (2.238)$$

The stator voltage components previously determined can be furthermore written as

$$v_{ds}=V_m\cos\alpha=\sqrt{2}V\cos\alpha=\sqrt{2}\frac{U}{\sqrt{3}}\cos\alpha$$
$$v_{qs}=V_m\sin\alpha=\sqrt{2}V\sin\alpha=\sqrt{2}\frac{U}{\sqrt{3}}\sin\alpha \quad (2.239)$$
$$v_{0s}=0$$

Similarly, applying the Park transformation to the instantaneous steady-state stator currents

$$i_a=I_m\cos(\omega_s t+\beta)=\sqrt{2}I\cos(\omega_s t+\beta)$$
$$i_b=I_m\cos\left(\omega_s t+\beta-\frac{2\pi}{3}\right)=\sqrt{2}I\cos\left(\omega_s t+\beta-\frac{2\pi}{3}\right) \quad (2.240)$$
$$i_c=I_m\cos\left(\omega_s t+\beta+\frac{2\pi}{3}\right)=\sqrt{2}I\cos\left(\omega_s t+\beta+\frac{2\pi}{3}\right)$$

and using the trigonometric identities (2.233), the following equations can be obtained:

$$i_{ds}=I_m\cos\beta=\sqrt{2}I\cos\beta$$
$$i_{qs}=I_m\sin\beta=\sqrt{2}I\sin\beta \quad (2.241)$$
$$i_{0s}=0$$

where I_m and I are the peak and rms currents and β is the phase angle of i_a with respect to the time origin.

Since we have considered balanced steady-state conditions, the computation will be further made with reference to a single phase.

Using the phasor representation for the phase quantities and taking into account equations (2.239) and (2.241), then

$$\begin{aligned} \underline{V}_{\mathrm{s}} &= V\,e^{j\alpha} = V\cos\alpha + jV\sin\alpha = \frac{v_{ds} + jv_{qs}}{\sqrt{2}} = V_{ds} + jV_{qs} \\ \underline{I}_{\mathrm{s}} &= I\,e^{j\beta} = I\cos\beta + jI\sin\beta = \frac{i_{ds} + ji_{qs}}{\sqrt{2}} = I_{ds} + jI_{qs} \end{aligned} \tag{2.242}$$

Similarly, it can be shown that

$$\begin{aligned} \underline{V}_{\mathrm{r}} &= \left(v_{d\mathrm{r}} + jv_{q\mathrm{r}}\right)/\sqrt{2} = V_{d\mathrm{r}} + jV_{q\mathrm{r}} \\ \underline{I}_{\mathrm{r}} &= \left(i_{d\mathrm{r}} + ji_{q\mathrm{r}}\right)/\sqrt{2} = I_{d\mathrm{r}} + jI_{q\mathrm{r}} \end{aligned} \tag{2.243}$$

From the steady-state stator equation (2.219′), it results

$$\begin{aligned} v_{ds} + jv_{qs} &= R_{\mathrm{s}}\left(i_{ds} + ji_{qs}\right) - \omega_{\mathrm{s}}L_{\mathrm{ss}}i_{qs} - \omega_{\mathrm{s}}L_{m}i_{q\mathrm{r}} + j(\omega_{\mathrm{s}}L_{\mathrm{ss}}i_{ds} + \omega_{\mathrm{s}}L_{m}i_{d\mathrm{r}}) \\ &= R_{\mathrm{s}}\left(i_{ds} + ji_{qs}\right) + j\omega_{\mathrm{s}}L_{\mathrm{ss}}\left(i_{ds} + ji_{qs}\right) + j\omega_{\mathrm{s}}L_{m}\left(i_{d\mathrm{r}} + ji_{q\mathrm{r}}\right) \end{aligned} \tag{2.244}$$

Dividing by $\sqrt{2}$ equation (2.244) and taking into account equations (2.242) and (2.243), it results

$$\begin{aligned} \underline{V}_{\mathrm{s}} &= R_{\mathrm{s}}\underline{I}_{\mathrm{s}} + j\omega_{\mathrm{s}}L_{\mathrm{ss}}\underline{I}_{\mathrm{s}} + j\omega_{\mathrm{s}}L_{m}\underline{I}_{\mathrm{r}} = R_{\mathrm{s}}\underline{I}_{\mathrm{s}} + j\omega_{\mathrm{s}}(L_{\mathrm{ss}} - L_{m})\underline{I}_{\mathrm{s}} + j\omega_{\mathrm{s}}L_{m}(\underline{I}_{\mathrm{s}} + \underline{I}_{\mathrm{r}}) \\ &= R_{\mathrm{s}}\underline{I}_{\mathrm{s}} + jX_{\mathrm{s}}\underline{I}_{\mathrm{s}} + jX_{m}(\underline{I}_{\mathrm{s}} + \underline{I}_{\mathrm{r}}) \end{aligned} \tag{2.244′}$$

where

$$X_{\mathrm{s}} = \omega_{\mathrm{s}}(L_{\mathrm{ss}} - L_{m}) \tag{2.245a}$$

is the stator leakage reactance.

Similarly, from the rotor equations (2.220′), it results

$$v_{d\mathrm{r}} + jv_{q\mathrm{r}} = R_{\mathrm{s}}\left(i_{d\mathrm{r}} + ji_{q\mathrm{r}}\right) + js\omega_{\mathrm{s}}L_{\mathrm{rr}}\left(i_{d\mathrm{r}} + ji_{q\mathrm{r}}\right) + js\omega_{\mathrm{s}}L_{m}\left(i_{ds} + ji_{qs}\right) \tag{2.246}$$

Taking into account equations (2.242) and (2.243), and dividing equation (2.246) by $\sqrt{2}$ obtain

$$\begin{aligned} \underline{V}_{\mathrm{r}} &= R_{\mathrm{r}}\underline{I}_{\mathrm{r}} + js\omega_{\mathrm{s}}L_{\mathrm{rr}}\underline{I}_{\mathrm{r}} + js\omega_{\mathrm{s}}L_{m}\underline{I}_{\mathrm{s}} = R_{\mathrm{r}}\underline{I}_{\mathrm{r}} + js\omega_{\mathrm{s}}(L_{\mathrm{rr}} - L_{m})\underline{I}_{\mathrm{r}} + js\omega_{\mathrm{s}}L_{m}(\underline{I}_{\mathrm{s}} + \underline{I}_{\mathrm{r}}) \\ &= R_{\mathrm{r}}\underline{I}_{\mathrm{r}} + jsX_{\mathrm{r}}\underline{I}_{\mathrm{r}} + jsX_{m}(\underline{I}_{\mathrm{s}} + \underline{I}_{\mathrm{r}}) \end{aligned} \tag{2.246′}$$

where

$$X_{\mathrm{r}} = \omega_{\mathrm{s}}(L_{\mathrm{rr}} - L_{\mathrm{m}}) \tag{2.245b}$$

is the rotor leakage reactance.

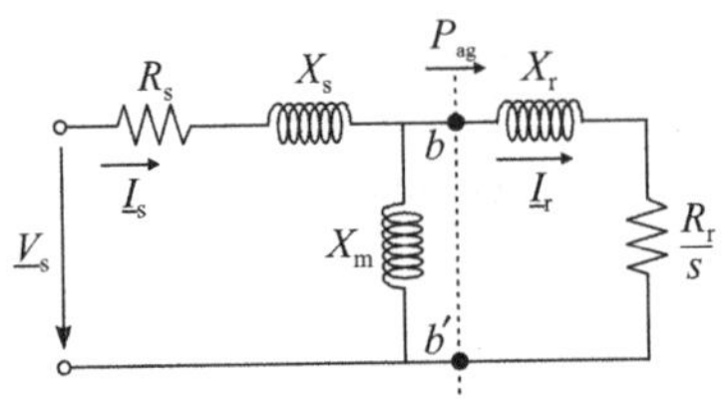

Figure 2.73. Equivalent circuit of an induction motor under steady-state conditions.

Since the rotor windings are not connected to a supply (they are either short-circuited or connected to a passive circuit), we have $\underline{V}_r = (v_{dr} + jv_{qr})/\sqrt{2} = 0$. Therefore, equation (2.246′) can be written as

$$\frac{R_r}{s}\underline{I}_r + jX_r\underline{I}_r + jX_m(\underline{I}_s + \underline{I}_r) = 0 \tag{2.247}$$

Equations (2.244′) and (2.247) can be represented by the equivalent one-phase circuit of the induction motor shown in Figure 2.73, where *all quantities have been referred to the stator.* Therefore, the power transferred across the air gap to the rotor is

$$P_{ag} = \frac{R_r}{s}I_r^2 \tag{2.248}$$

and the power losses in the rotor resistance are

$$\Delta P_r = R_r I_r^2 \tag{2.248′}$$

Therefore, the mechanical power transferred to the shaft is

$$P_{sh} = P_{ag} - \Delta P_r = \frac{R_r}{s}I_r^2 - R_r I_r^2 = \frac{1-s}{s}R_r I_r^2 \tag{2.249}$$

and the electromagnetic torque developed by the motor is

$$C_e = \frac{P_{sh}}{\omega_{mr}} = \frac{pP_{sh}}{\omega_r} = \frac{p}{(1-s)\omega_s}\frac{(1-s)}{s}R_r I_r^2 = \frac{p}{s\omega_s}R_r I_r^2 \tag{2.250}$$

Equation (2.250) was written taking into account that $\omega_{mr} = \omega_r/p$ and $\omega_r = (1-s)\omega_s$.

Since the above quantities P_{sh} and C_e are per phase, for a three-phase motor the electromagnetic torque developed at the shaft is

$$C_e = 3\frac{p}{s\omega_s}R_r I_r^2 \tag{2.250′}$$

As it can be seen in equations (2.250) and (2.250′), the electromagnetic torque depends on the slip s. For the analysis of the torque–slip relationship, the current I_r is to be eliminated from equation (2.250′). In this regard, the part to left of the nodes b–b' of the circuit of Figure 2.73 may be replaced by its Thévenin's equivalent. Figure 2.74 shows the resulting simplified equivalent circuit, where

$$\underline{V}_{Th} = \frac{jX_m\underline{V}_s}{R_s + j(X_m + X_s)}$$
$$R_{Th} + jX_{Th} = \frac{jX_m(R_s + jX_s)}{R_s + j(X_s + X_m)} \tag{2.251}$$

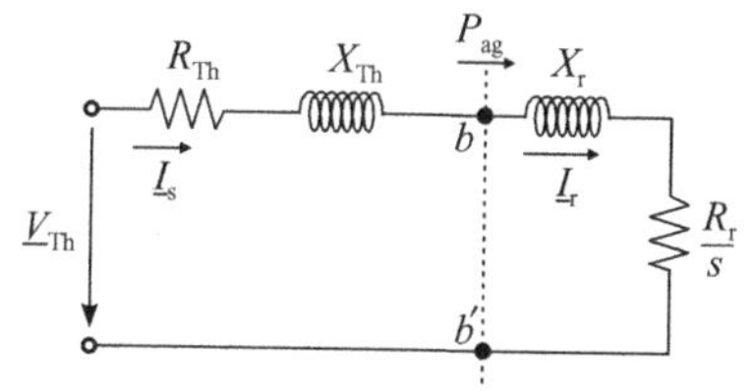

Figure 2.74. Equivalent circuit suitable for evaluating torque–slip relationship [1].

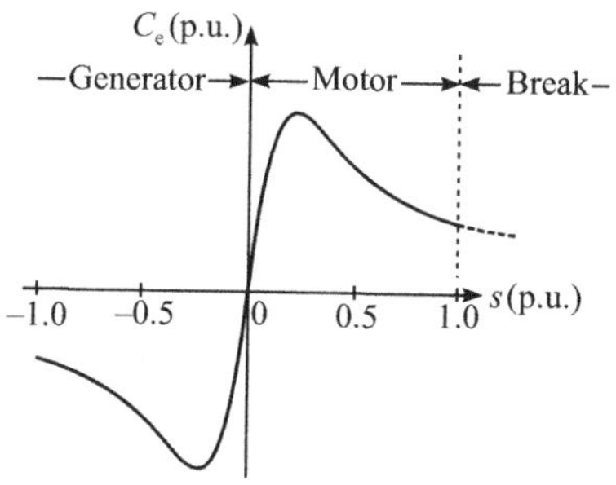

Figure 2.75. Torque–slip characteristic of an induction motor.

From Figure 2.74, the rotor current is

$$\underline{I}_r = \frac{\underline{V}_{Th}}{(R_{Th} + (R_r/s)) + j(X_{Th} + X_r)} \tag{2.252}$$

and the expression (2.250′) of the steady-state three-phase electromagnetic torque is

$$C_e = 3\frac{p}{s\omega_s}R_r I_r^2 = 3\frac{p}{s\omega_s}R_r \underline{I}_r \underline{I}_r^* = 3\frac{p}{s\omega_s}\frac{R_r V_{Th}^2}{(R_{Th} + (R_r/s))^2 + (X_{Th} + X_r)^2} \tag{2.253}$$

Replacing the quantities V_{Th}, R_{Th}, and X_{Th} in equation (2.253) with their expressions from (2.251) give the dependence $C_e = f(s)$, where the stator voltage V_s is a parameter. Figure 2.75 shows a typical relationship between electromagnetic torque and slip/speed.

2.2.4 Electromechanical Model of Induction Motor

The electromechanical equation or the *equation of motion* describes the motion of the induction motor rotor subjected to the electromagnetic torque C_e, which is a motor torque, and to the mechanical torque C_m of the driven mechanical load, which is a resistant torque. When the machine operates as a generator, the roles of the two torques reverses, that is, the mechanical torque C_m developed by the prime mover is a motor torque and the electromagnetic torque C_e at the shaft becomes a resistant torque.

The motor operation will be considered in the following and apply the fundamental law of mechanics. Therefore

$$J\frac{d\omega_{mr}}{dt} = C_e - C_m \tag{2.254}$$

where J is the combined moment of inertia of the rotating masses (rotor and driven load).

As in the case of the synchronous generator, the inertia constant H is defined as the rotor kinetic energy stored within rotating mass, at the synchronous speed $\omega_{0,\mathrm{ms}}$, divided by the base power S_b, which usually is taken as the apparent motor rating, that is,

$$H = \frac{(1/2)J\omega_{0,\mathrm{ms}}^2}{S_\mathrm{b}} \tag{2.255}$$

The term J is obtained from (2.255) then substituted in (2.254), which gives

$$\frac{2HS_\mathrm{b}}{\omega_{0,\mathrm{ms}}^2}\frac{\mathrm{d}\omega_\mathrm{mr}}{\mathrm{d}t} = C_\mathrm{e} - C_\mathrm{m} \tag{2.256}$$

or

$$2H\frac{\mathrm{d}(\omega_\mathrm{mr}/\omega_{0,\mathrm{ms}})}{\mathrm{d}t} = \frac{C_\mathrm{e} - C_\mathrm{m}}{S_\mathrm{b}/\omega_{0,\mathrm{ms}}} = C_{\mathrm{e}*} - C_{\mathrm{m}*} \tag{2.256$'$}$$

where $C_{\mathrm{e}*}$ and $C_{\mathrm{m}*}$ are the per unit torques, referred to the nominal torque:

$$C_\mathrm{n} = \frac{S_\mathrm{b}}{\omega_{0,\mathrm{ms}}} = \frac{S_\mathrm{n}}{\omega_{0,\mathrm{ms}}}$$

From equation (2.206) of the slip, it results that $\omega_\mathrm{mr}/\omega_\mathrm{ms} = 1 - s$, and the equation of motion (2.256$'$) becomes

$$2H\frac{\mathrm{d}(1-s)}{\mathrm{d}t} = C_{\mathrm{e}*} - C_{\mathrm{m}*} \tag{2.257}$$

and

$$2H\frac{\mathrm{d}s}{\mathrm{d}t} = -C_{\mathrm{e}*} + C_{\mathrm{m}*} \tag{2.257$'$}$$

Generally, the mechanical torque C_m is dependent on the speed and implicitly on the slip. To express this dependency, an exponential model is used [1]:

$$C_\mathrm{m} = C_\mathrm{m0}\left(\frac{\omega_\mathrm{mr}}{\omega_{0,\mathrm{ms}}}\right)^m \tag{2.258}$$

or a polynomial one

$$C_\mathrm{m} = C_\mathrm{m0}\left[A\left(\frac{\omega_\mathrm{mr}}{\omega_{0,\mathrm{ms}}}\right)^2 + B\frac{\omega_\mathrm{mr}}{\omega_{0,\mathrm{ms}}} + C\right] \tag{2.259}$$

where C_m0 stands for the resistant mechanical torque of the driven mechanism at the synchronous speed, and m, A, B, and C are the coefficients which depend on the driven mechanism type.

2.2.5 Electromagnetic Model of Induction Motor

In order to develop the mathematical model of the induction motor for stability studies, as for the synchronous generator, the stator transient phenomena are neglected, that is, in (2.219) the time derivatives are zero. Under these considerations, and taking into account that the rotor voltages are $v_{dr} = v_{qr} = 0$, the Park equations of the induction motor under transient conditions are

- stator voltage equations:

$$\begin{aligned} v_{ds} &= R_s i_{ds} - \omega_s \psi_{qs} \\ v_{qs} &= R_s i_{qs} + \omega_s \psi_{ds} \end{aligned} \qquad (2.219'')$$

- rotor voltage equations:

$$\begin{aligned} 0 &= R_r i_{dr} - s\omega_s \psi_{qr} + \frac{d\psi_{dr}}{dt} \\ 0 &= R_r i_{qr} + s\omega_s \psi_{dr} + \frac{d\psi_{qr}}{dt} \end{aligned} \qquad (2.260)$$

- stator flux linkage equations:

$$\begin{aligned} \psi_{ds} &= L_{ss} i_{ds} + L_m i_{dr} \\ \psi_{qs} &= L_{ss} i_{qs} + L_m i_{qr} \end{aligned} \qquad (2.227')$$

- rotor flux linkage equations:

$$\begin{aligned} \psi_{dr} &= L_{rr} i_{dr} + L_m i_{ds} \\ \psi_{qr} &= L_{rr} i_{qr} + L_m i_{qs} \end{aligned} \qquad (2.227'')$$

Next, the rotor currents i_{dr} and i_{qr} are eliminated and a relationship between the stator current $\underline{I}_s = (i_{ds} - j i_{qs})/\sqrt{2}$ and a voltage behind a transient reactance is sought. Thus, from equation (2.227″) the rotor currents are achieved:

$$i_{dr} = \frac{\psi_{dr} - L_m i_{ds}}{L_{rr}}; \quad i_{qr} = \frac{\psi_{qr} - L_m i_{qs}}{L_{rr}} \qquad (2.261)$$

which are substituted in (2.227′), resulting

$$\begin{aligned} \psi_{ds} &= \frac{L_m}{L_{rr}} \psi_{dr} + \left(L_{ss} - \frac{L_m^2}{L_{rr}} \right) i_{ds} \\ \psi_{qs} &= \frac{L_m}{L_{rr}} \psi_{qr} + \left(L_{ss} - \frac{L_m^2}{L_{rr}} \right) i_{qs} \end{aligned} \qquad (2.262)$$

Substituting equation (2.262) of the stator flux linkages in (2.219″), we obtain

$$\begin{aligned} v_{ds} &= R_s i_{ds} - \omega_s \frac{L_m}{L_{rr}} \psi_{qr} - \omega_s L'_{ss} i_{qs} = R_s i_{ds} - X'_s i_{qs} + e'_d \\ v_{qs} &= R_s i_{qs} + \omega_s \frac{L_m}{L_{rr}} \psi_{dr} + \omega_s L'_{ss} i_{ds} = R_s i_{qs} + X'_s i_{ds} + e'_q \end{aligned} \qquad (2.263)$$

where

$$X'_{\rm s} = \omega_{\rm s}\left(L_{\rm ss} - \frac{L_{\rm m}^2}{L_{\rm rr}}\right) \tag{2.264}$$

is the transient reactance of the induction motor,

$$e'_d = -\omega_{\rm s}\frac{L_{\rm m}}{L_{\rm rr}}\psi_{qr} \tag{2.265a}$$

is the d-axis transient mmf, and

$$e'_q = \omega_{\rm s}\frac{L_{\rm m}}{L_{\rm rr}}\psi_{dr} \tag{2.265b}$$

is the q-axis transient mmf.

Equations (2.263) can be written in a complex form:

$$v_{ds} + jv_{qs} = R_{\rm s}\left(i_{ds} + ji_{qs}\right) + jX'_{\rm s}\left(i_{ds} + ji_{qs}\right) + e'_d + je'_q$$

and dividing by $\sqrt{2}$, we obtain

$$\underline{V}_{\rm s} = \left(R_{\rm s} + jX'_{\rm s}\right)\underline{I}_{\rm s} + \underline{E}' \tag{2.266}$$

where $\underline{E}' = \frac{1}{\sqrt{2}}\left(e'_d + je'_q\right)$

Equation (2.266) is represented by the equivalent circuit of the induction motor for transient studies, given in Figure 2.76.

The time variation of the e'_d and e'_q components of the transient voltage $\underline{E}'$ is obtained using the rotor voltage equations (2.260), in which the expressions of the rotor currents from (2.261) are substituted. Therefore

$$\begin{aligned}\frac{{\rm d}\psi_{dr}}{{\rm d}t} &= -R_{\rm r}\left(\frac{\psi_{dr}}{L_{\rm rr}} - \frac{L_{\rm m}}{L_{\rm rr}}i_{ds}\right) + s\omega_{\rm s}\psi_{qr}\\ \frac{{\rm d}\psi_{qr}}{{\rm d}t} &= -R_{\rm r}\left(\frac{\psi_{qr}}{L_{\rm rr}} - \frac{L_{\rm m}}{L_{\rm rr}}i_{qs}\right) - s\omega_{\rm s}\psi_{dr}\end{aligned} \tag{2.267}$$

On the other hand, from the definition equations of voltages e'_d and e'_q, in (2.265a) and (2.265b), the rotor flux linkage equations are obtained:

$$\psi_{dr} = \frac{L_{\rm rr}}{\omega_{\rm s}L_{\rm m}}e'_q; \quad \psi_{qr} = -\frac{L_{\rm rr}}{\omega_{\rm s}L_{\rm m}}e'_d$$

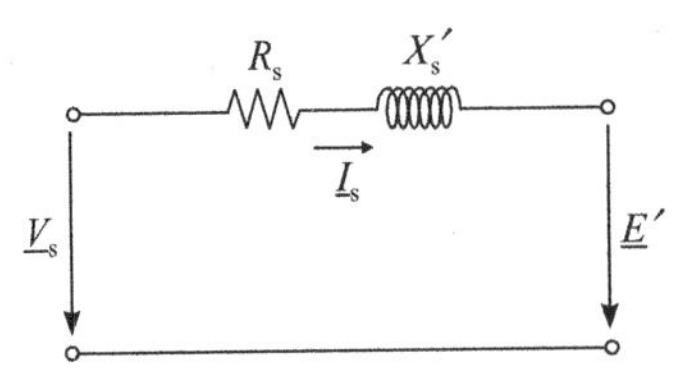

Figure 2.76. Equivalent circuit of the induction motor for transient studies.

which are substituted in equation (2.267). After computation achieve the system of differential equations:

$$\begin{aligned}\frac{\mathrm{d}e'_d}{\mathrm{d}t} &= -\frac{1}{T'_0}\left(e'_d + \omega_s \frac{L_m^2}{L_{rr}} i_{qs}\right) + s\omega_s e'_q \\ \frac{\mathrm{d}e'_q}{\mathrm{d}t} &= -\frac{1}{T'_0}\left(e'_q - \omega_s \frac{L_m^2}{L_{rr}} i_{ds}\right) - s\omega_s e'_d\end{aligned} \tag{2.268}$$

where

$$T'_0 = \frac{L_{rr}}{R_r} \tag{2.269}$$

is the transient open-circuit time constant (expressed in radians) of the induction motor. This equation characterizes the decay of the rotor transients when the stator is open-circuited [1].

By subtracting the reactances from equations (2.245a) and (2.264) gives

$$\omega_s \frac{L_m^2}{L_{rr}} = X_s + X_m - X'_s \tag{2.270}$$

Taking into account equation (2.270), the differential equations (2.268) become

$$\left\|\begin{aligned}\frac{\mathrm{d}e'_d}{\mathrm{d}t} &= -\frac{1}{T'_0}\left[e'_d + (X_s + X_m - X'_s)i_{qs}\right] + s\omega_s e'_q \\ \frac{\mathrm{d}e'_q}{\mathrm{d}t} &= -\frac{1}{T'_0}\left[e'_q - (X_s + X_m - X'_s)i_{ds}\right] - s\omega_s e'_d\end{aligned}\right. \tag{2.271}$$

Observations

(i) As in the case of the synchronous generator, the induction motor equations can be expressed in per unit. Since for the induction motor the rotor quantities are referred to the stator, the same base quantities are chosen for both rotor and stator. Therefore, the following independent base quantities are chosen:

- base voltage (V_b)—the peak value of the rated phase-to-earth voltage;
- base current (I_b)—the peak value of the rated current;
- base frequency (f_b)—nominal frequency.

The reminder of the base quantities is determined in a similar way as for the synchronous generator (see Section 2.1.3.4), using the relationship between them.

(ii) In many practical applications, especially for small-size induction motors, since the time constant T'_0 has low values, the transients from the rotor circuits can be neglected [1]. Therefore, for the induction motor representation, the steady-state equivalent circuit (Figure 2.73) is used together with equation (2.253) representing the *torque–slip* relationship and the rotor equation of motion.

The reader should note that the theory and performance of synchronous machines have also been covered in other remarkable books (please see Refs. 44, 45, 46)

REFERENCES

[1] Kundur, P. *Power system stability and control*, McGraw-Hill, Inc., New York, 1994.

[2] Adkins, B., Harley, R.G. *The general theory of alternating current machines: Application to practical problems*, Chapman and Hall, London, 1975.

[3] Blondel, A. The two-reaction method for study of oscillatory phenomena in coupled alternators, *Revue Generale de l'Electricité*, No. 13, pp. 235–251, Feb. 1923.

[4] Doherty, R.E., Nickle, C.A. Synchronous machines—Part I and II—An extension of Blondel's two reaction theory, *AIEE Trans.*, Vol. 45, pp. 912–942, 1926.

[5] Park, R.H. Two-reaction theory of synchronous machines. Generalized method of analysis. Part I, *AIEE Trans.*, Vol. 48, pp. 716–730, Jul. 1929.

[6] IEEE Standard 1110-2002, *IEEE Guide for synchronous generator modeling practices and applications in power system stability analysis.*

[7] Das, J.C. *Power system analysis. Short-circuit load flow and harmonics*, 2nd edition, CRC Press/Taylor & Francis Group, Boca Raton, London, New York, 2012.

[8] IET. *Power system protection. Volume 1: Principles and components.* Edited by The Electricity Training Association, IET London, Cambridge University Press, 1995.

[9] Saadat, H. *Power system analysis*, 3rd edition, PSA Publishing, 2010.

[10] ANSI/IEEE Standard 115-1995, *IEEE Guide: Test procedures for synchronous machines.*

[11] Umans, S.D., Mallick, J.A., Wilson, G.L. Modeling of solid rotor turbogenerators. Part I: Theory and techniques, *IEEE Trans. Power Apparatus Syst.*, Vol. PAS-97, No. 1, pp. 269–277, Jan./Feb. 1978.

[12] Marconato, R. *Electric power systems. Volume 2: Steady state behaviour, controls, shortcircuits and protection systems*, CEI—Italian Electrotechnical Committee, Milano, Italy, 2004.

[13] Canay, I.M. Determination of the model parameters of machines from the reactance operators $x_d(p)$, $x_q(p)$ (evaluation of standstill frequency response test), *IEEE Trans. Power Deliv.*, Vol. 8, No. 2, pp. 272–279, Jun. 1993.

[14] Salvatore, L., Savino, M. Exact relationship between parameters and test data for models of synchronous machines, *Electr. Mach. Power Syst.*, Vol. 8, pp. 169–184, 1983.

[15] Rush, P. (Coordinator), *Network Protection & Automation Guide*, Alstom, 2002.

[16] Takeda, Y., Adkins, B. Determination of synchronous machine parameters allowing for unequal mutual inductances, *Proc. IEE (London)*, Vol. 121, No. 12, pp. 1501–1504, Dec. 1974.

[17] Shackshaft, G., Poray, A.T. Implementation of new approach to determination of synchronous machine parameters from tests, *Proc. IEE (London)*, Vol. 124, No. 12, pp. 1170–1178, 1977.

[18] deMello, F.P., Ribeiro, J.R. Derivation of synchronous machine parameters from tests, *IEEE Trans.*, Vol. PAS-96, pp. 1211–1218, Jul./Aug. 1977.

[19] deMello, F.P., Hannett, L.N. Validation of synchronous machine models and determination of model parameters from tests, *IEEE Trans.*, Vol. PAS-100, pp. 662–672, Feb. 1981.

[20] EPRI Report EL-1424 *Determination of synchronous machine stability constants*, Vol. 2, prepared by Ontario Hydro, Dec. 1980.

[21] Coultes, M.E., Watson, W. Synchronous machine models by standstill frequency response tests, *IEEE Trans.*, Vol. PAS-100, pp. 1480–1489, Apr. 1981.

[22] Dandeno, P.L., Poray, A.T. Development of detailed turbogenerator equivalent circuits from standstill frequency response measurements, *IEEE Trans.*, Vol. PAS-100, pp. 1646–1653, Apr. 1981.

[23] IEEE Standard 115A-1984, *IEEE Trial use standard procedures for obtaining synchronous machine parameters by standstill frequency response testing.*

[24] Dillman, T.L., Keay, F.W., Raczkowski, C., Skooglund, J.W., South, W.H. Brushless excitation: today's high-initial-response brushless excitation systems offer improvements in turbine-generator performance and reliability, *IEEE Spectrum*, Vol. 9, No. 3, pp. 58–66, Mar. 1972.

[25] Anderson, P.M., Fouad, A.A. *Power system control and stability*, IEEE Press Power Engineering Series & Wiley-Interscience, 2003.

[26] Machowschi, J., Bialek, J., Bumby, J. *Power system dynamics and stability*, John Wiley & Sons, New York, 1997.

[27] Paserba, J. (convenor) *Analysis and control of power system oscillations,* CIGRE Study Committee 38, Task Force 07 of Advisory Group 01, Technical Brochure 111, Dec. 1996.

[28] El-Serafi, A.M., Demeter, E. Determination of the saturation curves in the intermediate axes of cylindrical-rotor synchronous machines, *Proceedings of the International Conference on Electrical Machines, Istanbul, Turkey*, Vol. 3, pp. 1060–1065, Sep. 2–4, 1998.

[29] Potolea, E. *Electric power system state calculations (in Romanian)*, Technical Publishing, Bucharest, 1977.

[30] Eremia, M. (editor), Song, Y.-H., Hatzyargyriou, N., et al. *Electric power systems. Volume I. Electric networks*, Publishing House of the Romanian Academy, Bucharest, 2006.

[31] Nilsson, N.E., Mercurio, J. Synchronous generator capability curve testing and evaluation, *IEEE Trans. Power Deliv.*, Vol. 9, No. 1, Jan. 1994.

[32] Klempner, G., Kerszenbaum, I. *Operation and maintenance of large turbo-generators*, IEEE & Wiley, New York, 2004.

[33] IEEE Standard 421.1-2007, *Standard definitions for excitation systems for synchronous machines.*

[34] Ilić, M., Zaborszky, J. *Dynamics and control of large electric power systems*, John Wiley & Sons, Inc., New York, 2000.

[35] Crenshaw, M.L. (Chairman) *Excitation system models for power system stability studies*, IEEE Committee Report prepared by the IEEE Working Group Computer Modeling of Excitation Systems, IEEE Trans., Vol. PAS-100, No. 2, pp. 494–509, Feb. 1981.

[36] IEEE Standard 421.5-2005, *IEEE Recommended practice for excitation system models for power system stability studies.*

[37] Wang, Xi-Fan, Song, Y.H., Irving, M. *Modern power systems analysis*, Springer Science, New York, 2008.

[38] Kimbark, E.W. *Power system stability. Volume III. Synchronous machines*, John Wiley & Sons, New York, 1956.

[39] Mummert, C.R. *Excitation system limiter models for use in system stability studies,* IEEE PES Winter Meeting, New York, pp. 187–192, Feb. 1999.

[40] Quiñonez-Varela, G., Cruden, A. *Experimental testing and model validation of a small-scale generator set for stability analysis*, Proceedings of 2003 IEEE Bologna PowerTech Conference, Bologna, Italy, 23–26 Jun. 2003.

[41] Pal, B., Chaudhuri, B. *Robust control in power systems*, Springer Science & Business Media, New York, 2005.

[42] Kasatkin, A., Perekalin, M. *Basic electrical engineering*, Mir Publishers, Moscow, 1970.

[43] Eremia, M., Trecat, J., Germond, A. Reseaux electriques. Aspects actuels, Technical Publishing, Bucharest, 2000.

[44] Concordia, C. *Synchronous machines. Theory and performance*, JohnWiley & Sons, New York, 1951.

[45] Adkins, B. *The general theory of electrical machines*, Chapman and Hall, London, 1957.

[46] Fitzgerald A.E., Kingsley, C. *Electric machinery*, 2nd Edition, McGraw-Hill, New York, 1961.

3

MODELING THE MAIN COMPONENTS OF THE CLASSICAL POWER PLANTS

Mohammad Shahidehpour, Mircea Eremia, and Lucian Toma

3.1 INTRODUCTION

The classical prime sources of electrical energy supplied by utilities are the kinetic energy of water and the thermal energy derived from fossil fuels and nuclear fission. The prime movers convert these sources of energy into mechanical energy that is, in turn, converted into electrical energy by synchronous generators. In the last two decades, however, big steps have been taken on the wind generation technology side. In general, wind turbine generators tend to be quite different in both mechanical and electrical construction from traditional large thermal, nuclear, and hydro power plants and they will be discussed in Chapter 4.

Figure 3.1 shows the functional relationship of the prime mover system in the context of the overall power system [1].

The electric system performance (voltages, power, frequency, etc.) is affected by the action of the generators, condition of the network, and behavior of loads. The prime mover system couples with the electrical system through mechanical power with its effects on generator rotor speed and angle.

The prime mover energy supply system responds to commands for generation changes from manual or automatic generation control (AGC) and from speed deviations. Internal plant variables that affect mechanical power are the turbine values and boiler main steam pressure.

The turbine control logic operating on control and intercept valves (IV) responds mainly to speed deviations. There may be override action from main steam pressure (initial pressure regulator) and electrical power (power load unbalance, fast valving, etc.). The automatic

Handbook of Electrical Power System Dynamics: Modeling, Stability, and Control. Edited by Mircea Eremia and Mohammad Shahidehpour.

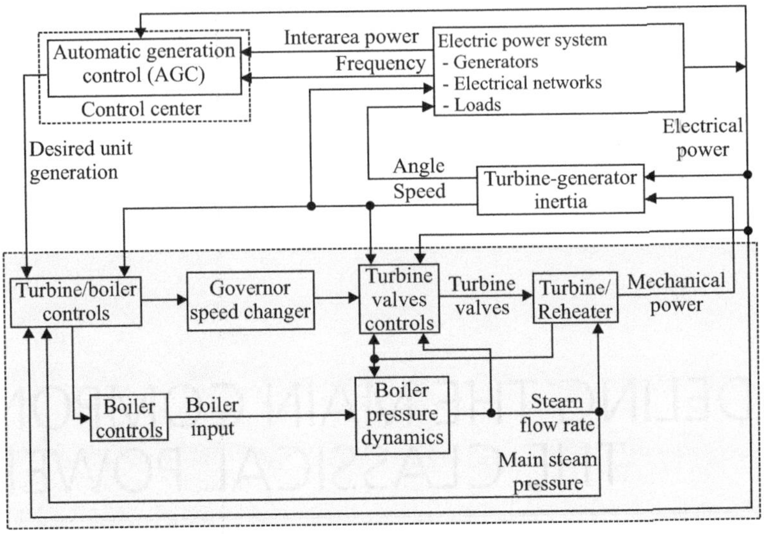

Figure 3.1. Functional block diagram showing relationship of prime mover systems control to complete system [1].

generation control determines the desired unit generation and imparts control action through the governor speed changer or through the turbine boiler coordinated controls [1].

3.2 TYPES OF TURBINES

3.2.1 Steam Turbines

A steam turbine is generally composed of two or more cylinders, each of them with several stages, that are driven by high-, medium- or low-pressure steam as described in Figure 3.2.

Between the high-pressure (HP) and intermediate-pressure (IP) cylinders there may or may not be a reheater, and, in case of large units, the low-pressure (LP) cylinder is normally split into two or more flows.

Depending on the way the cylinders are coupled, the next two turbine types are obtained:

- Turbines with shafts in tandem compound in which all turbine cylinders are coupled on a single shaft, on which the generator is also coupled. These turbines are rotating with 3000 rot/min (in the systems with $f = 50$ Hz) or with 3500 rot/min (in the systems with $f = 60$ Hz).

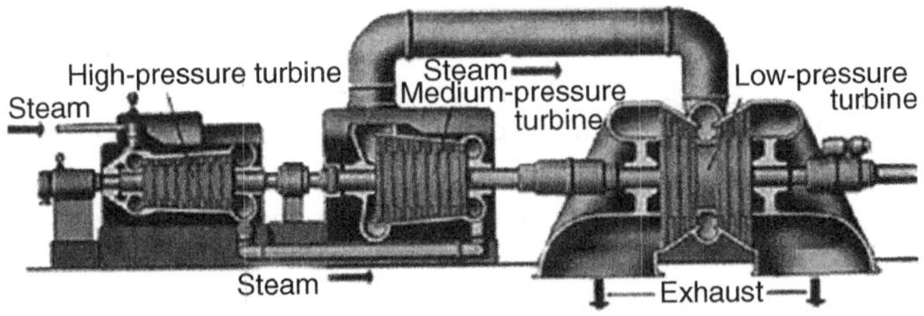

Figure 3.2. The structure of a steam turbine.

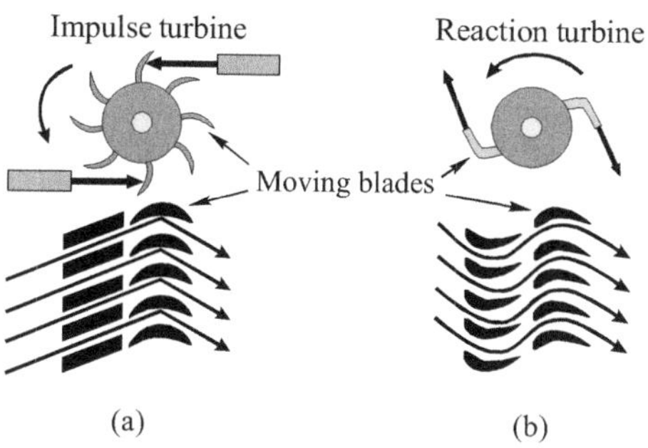

Figure 3.3. Two types of turbine blading: (a) impulse blading; (b) reaction blading.

- Steam turbine in "cross", which has two shafts, at every shaft being connected one generator. Usually, the second shaft rotates with a speed equal to half of the speed of the primary shaft.

Internal, the steam turbine consists of rows of blades designed to extract the heat and pressure energy of the steam and convert this energy into mechanical energy. To accomplish this goal, high-pressure steam is admitted through a set of control valves (CV) and allowed to expand as it passes through the turbine, to be exhausted, usually to a condenser, at relatively low pressure and temperature.

Thus, the type and arrangement of turbine blading are important in extracting all possible energy from the steam and converting this energy into mechanical work of spinning the turbine rotor and attached electric generator.

Two types of turbine blading are used: impulse blading and reaction blading (Figure 3.3) [2]:

- In *impulse blading*, the steam expands and its pressure drops as it passes through a nozzle, leaving the nozzle at high velocity. This kinetic energy is converted into mechanical energy as the steam strikes the moving turbine blades and pushes them forward.
- The *reaction blading* operates on a different principle. Here the "nozzle" through which the steam expands is moving with the shaft, giving the shaft a torque due to the unbalanced forces acting on the blade intake and exhaust surfaces.

3.2.2 Gas Turbines

The gas turbine installations are thermal machineries that convert the chemical energy, stored in the primary fuel, into mechanical energy. The working fluid used in this case is a gas (air, carbon dioxide, helium, etc.).

The modern gas turbines are based on a Brayton (or Joule) thermodynamic cycle that has a relatively low efficiency. The ideal Brayton cycle is made up of four completely irreversible processes (Figure 3.4): 1-2 isentropic compression, 2-3 constant pressure heat addition, 3-4 isentropic expansion, and 4-1 constant pressure heat rejection [3].

Figure 3.5 shows the typical internal combustion gas turbine power plant arrangement (in open cycle).

The air is aspirated through the filter that eliminates mechanical impurities to avoid erosion and gradually destroy of the compressor's blades. The air is then compressed

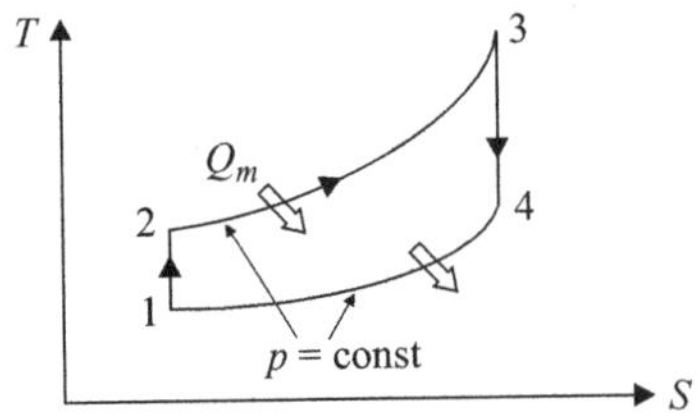

Figure 3.4. Ideal temperature/entropy diagram.

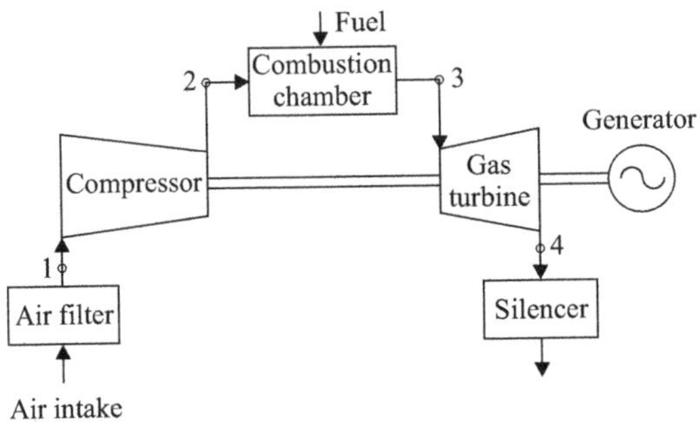

Figure 3.5. Typical gas turbine power plant arrangement.

through an adiabatic process with constant entropy within the compressor (process 1-2), which is usually axial. Pressure of 13–20 times that of atmospheric, appropriate for the operational value of the combustion chamber (burner), is achieved after the compression stage.

Fuel, either liquid or gas, is mixed with the compressed air and burnt in the combustor (process 2-3), after which the hot gases are expanded through the turbine (process 3-4) producing mechanical work. One part of the mechanical work is used to drive the blades of the turbine and consequently the shaft of the synchronous generator connected to it, and the remaining part is used to drive the compressor, which is mounted on the same shaft structure as the gas turbine. A silencer is used to reduce the noise produced by the gases exhausted from the gas turbine into the atmosphere.

3.2.3 Hydraulic Turbines

Hydraulic turbines are of two basic types: impulse turbines and reaction turbines [2]:

- The *impulse turbine* or the *Pelton wheel* is used in hydro power plants with heads from 15 to 1900 m, although it is most efficient for heads higher than 300 m. The installed power can reach up to 435 MW for a single unit. The impulse turbines are mounted on either a horizontal (Figure 3.6) or a vertical shaft. The turbine wheel is spun by directing forceful streams of water from nozzles against a series of spoon-shaped buckets mounted on the edge of the wheel and using the high momentum of the water to drive the wheel.

  The turbine speed is controlled by adjusting the flow of water through a needle valve that can increase or decrease the nozzle opening.

  The total drop in pressure of the water occurs in the nozzle and the resulting jet of water is directed tangentially at buckets on the wheel producing impulsive force on them at atmospheric pressure.

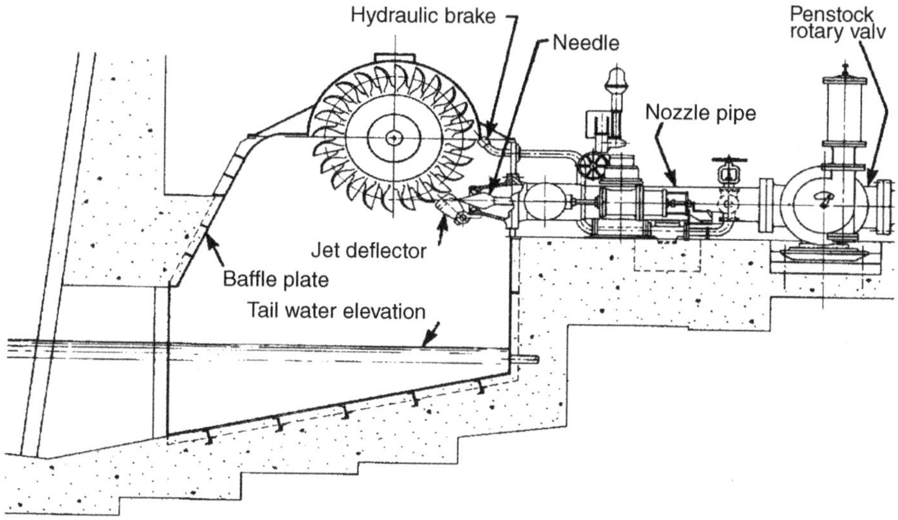

(a)

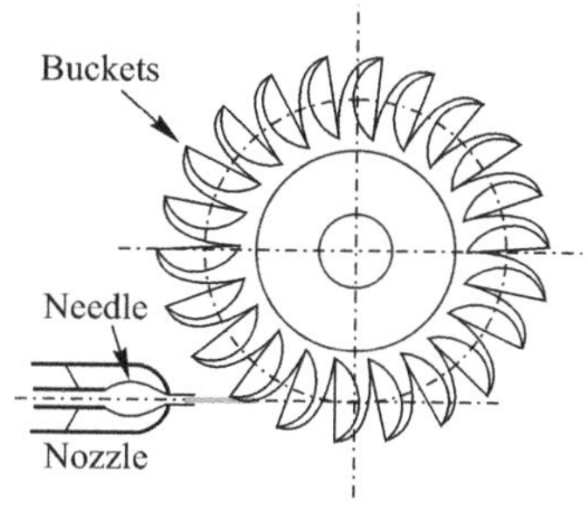

(b)

Figure 3.6. Pelton turbine.

- In *reaction turbines*, the water completely fills the cavity occupied by the runner and it changes the pressure as it moves through the turbine. As the name implies, the turbine is turned by reaction force caused by pressure or weight of fluid rather than by a direct impulse.

  The reaction turbines are used for heads ranging from 30 to 700 m and outputs up to 700 MW. However, satisfactorily operation may be obtained for heads lower than 30 m. The reaction turbines are classified as
- radial flow, where the water flows perpendicular to the shaft;
- axial flow, where the stationary vanes direct the water to flow parallel to the shaft;
- mixed flow, which is a combination of radial and axial flow.

There are two main subcategories of reaction turbine, that is, Francis and propeller. However, the *James Francis's turbine* is the most common water turbine in use today, which has a radial inward flow (Figure 3.7).

(a) (b)

Figure 3.7. A typical vertical shaft reaction turbine arrangement.

In the *Francis turbine* power plant, the water passes through the spiral case surrounding the blades and flows in an inward direction, then through the runner exerting pressure on the blades thus turning the runner. A *draft tube*, through which the water is then evacuated, is designed so that to allow the turbine runner be positioned above the tail water and also to reduce the kinetic energy losses at discharge.

The electric generator is usually directly connected to the runner shaft. The vertical shaft is the most common arrangement in the reaction turbine-based power plants.

The control of a reaction turbine is performed by means of movable guide vanes called *wicket gates* through which the water passes from the spiral case into the runner. The wicket gates are opened/closed simultaneously by means of a large "shift ring" to which the gates are attached. Rotating the gates requires a very large force, so servomotors are used. A second control device used in reaction turbines is a large *bypass valve*, which is actuated by the shifting ring [2].

Another widely used reaction turbine is the *propeller type turbine*. As the name implies, it uses propeller type wheels with adjustable blades (Figure 3.8). Viktor Kaplan combined the adjusted blades with adjusted wicket gates to create a more efficient turbine over a wide range of flow and water level. The *Kaplan turbine* employs water velocity to a great extent than the Francis turbine. The Kaplan turbines are preferred in locations with heads of 2–150 m. Compared to the Francis turbines, the Kaplan units operate at higher speeds for a given head and the water velocity through the turbine is greater, leaving the runner with a fast swirling motion. Thus, the draft tube design is important in Kaplan turbine applications [2].

Currently there are two trends in using the water turbines:

- The European trend is inclined to use Francis turbines for heads up to 500 m and has at base a performance higher with 1–2% in the case of Francis turbines; using this type of turbine at falls that great causes problems of mechanic nature.
- The American trend is inclined to use Pelton turbines for any falls greater than 300 m, considering their elasticity in exploitation, but with supplementary costs of the turbine and the generator (because of lower speed, at small falls, the diameter of the Pelton turbine and generator is increased).

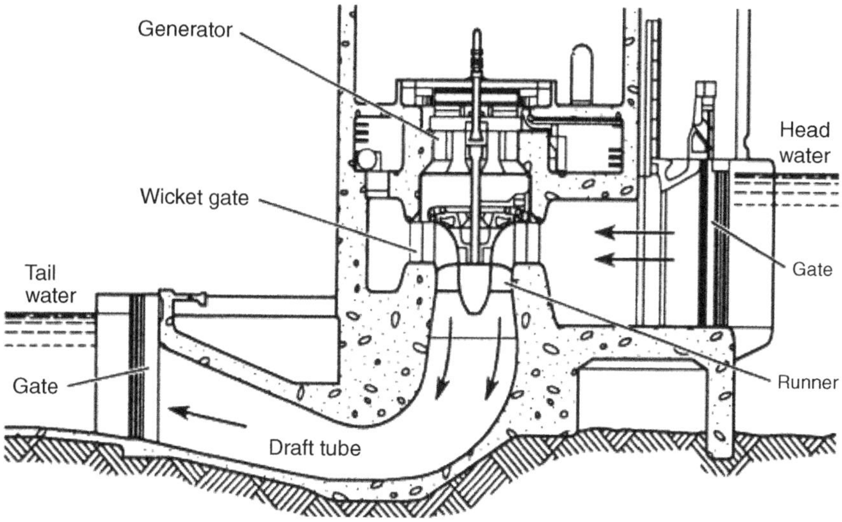

(a)

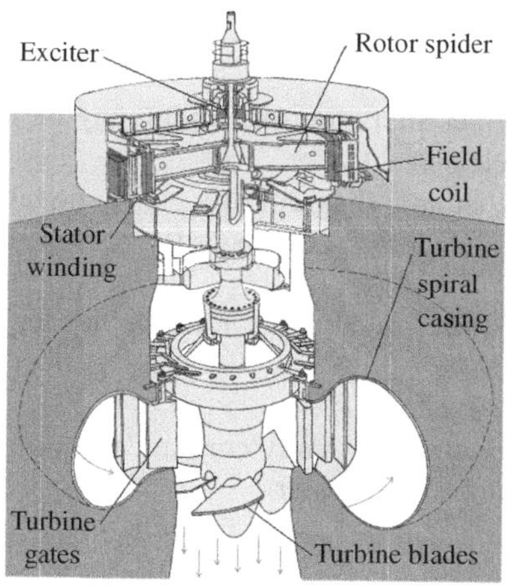

(b)

Figure 3.8. The Kaplan propeller turbine [4].

3.3 THERMAL POWER PLANTS

3.3.1 Generalities

Generation of mechanical energy by a steam turbine (prime mover) takes place via conversion of the heat transferred to the water and steam of the boiler into the kinetic energy of the steam that is fed to the blades of the turbine wheel. The water is compressed by extraction and feed pumps and moderately heated in low-pressure heaters and in the degasifier (PR_{LP}). The rotating energy is converted into electrical energy by the generator.

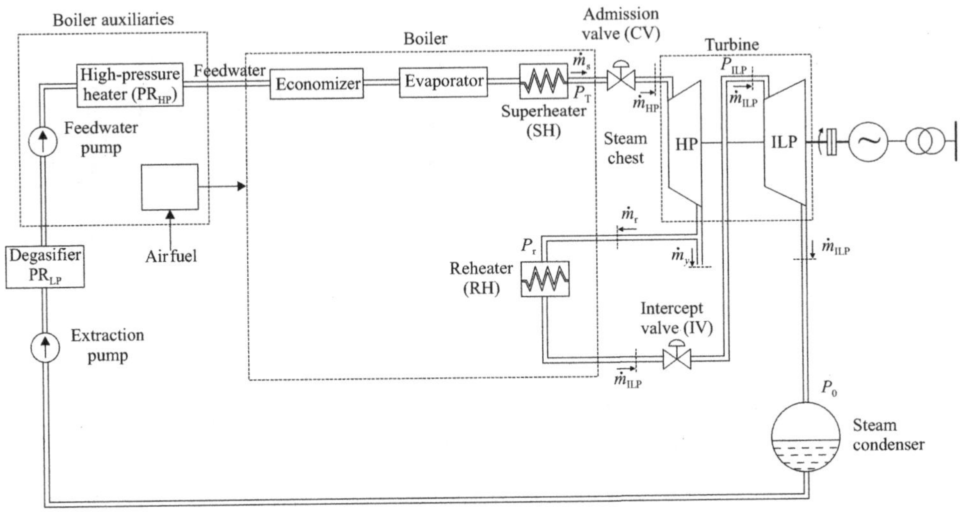

Figure 3.9. Schematic diagram of conventional steam-fired thermal power plant. (Adapted from Ref. 6.)

The heat source for the boiler supplying the steam may be furnace fired by fossil fuels (coal, oil, or gas) or a nuclear reactor.

Figure 3.9 shows a schematic diagram of a conventional steam-fired thermal power plant, whose IP and LP cylinders are supposed to be concentrated in single cylinder (ILP) [5].

The steam–water mixture, subject to Rankine cycle, consists of the following stages [5]:

(i) Heating of water at a practically constant pressure in the high-pressure heater (PR_{HP}), economizer, and boiler.

(ii) Vaporization of water at constant pressure and temperature in the boiler's evaporator.

(iii) Superheating of steam, practically isobaric at a pressure P_s, in the boiler.

(iv) Pressure drop, in part on the admission valve and in part in the HP cylinder of the turbine.

(v) Reheating of steam, practically isobaric at a pressure P_r, in the boiler.

(vi) Expansion of steam in the LP cylinder of the turbine.

(vii) Isothermal and isobaric conversion at a pressure P_0, of the exhausted steam into water, in the condenser.

Figure 3.10 shows the basic structure of the prime mover and energy supply system.

- The "*turbine/reheater*" block defines mechanical power as function of main steam pressure (P_{T}), control valve flow area (CV), and intercept valve flow area (IV).
- The "*boiler*" block models boiler main steam pressure (P_{T}) and steam flow rate ($\dot{m}_s$), as function of turbine control valve flow area (CV) and the fuel, air, and feedwater

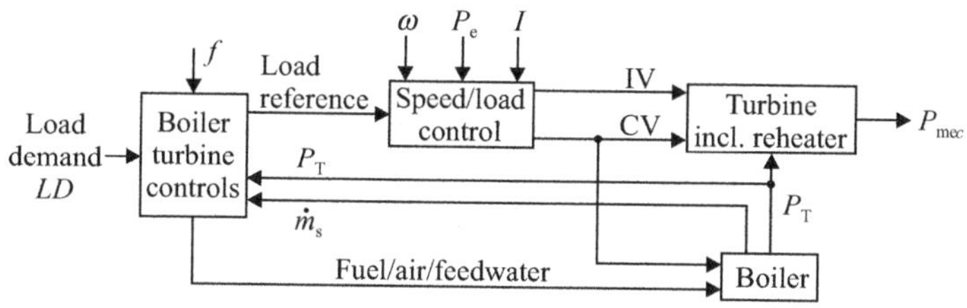

Figure 3.10. Elements of the prime mover and energy supply system [1].

that are essentially the energy inputs to the boiler. The pressure effect of the reheater is included in the turbine model.

- The "*speed/load control*" block details the turbine control logic in response to change in speed/load reference (LR), speed (ω), main steam pressure (P_T) through the initial pressure regulator, and possibly, in the case of fast valving applications, in response to changes in electrical power (P_e) and generator current (I).
- The "*boiler turbine control*" block develops the load reference, input to the speed/load controls in response to the load demand (LD) set either manually or by AGC. Other inputs to the control logic, depending on the type of coordinated controls being used, are plant frequency (f), main steam pressure (P_T), and steam flow rate ($\dot{m}_S$).

In its simplest form, the boiler and turbine controls are decoupled, with generation changes implemented directly through the load reference and the boiler controls responding to changes in the steam flow ($\dot{m}_S$) and pressure (P_T).

3.3.2 Boiler and Steam Chest Models

A simplified representation of the process physics in a boiler is illustrated in Figure 3.11, showing an equivalent lumped volume storing steam at an internal pressure-labeled drum pressure in series with superheater and steam leads with their friction drop effects [1].

The heat released in the furnace T is used to boil the water passing through the water walls and transform it into steam ($\dot{m}_W$). For easier simulation purposes, the transformation process of the thermal agent from water to steam is approximated by two lumped storage volumes connected through an equivalent orifice representing the pressure drop due to friction in the pipes (Figure 3.11a). The main storage of energy takes place in the water walls and drum. The mixture water–steam is separated in the drum, into saturated steam, which is sent to the superheater, and saturated liquid, which is sent for recirculation in the furnace circuits.

Figure 3.11b shows a low-order nonlinear model representing the steam flow rate, $\dot{m}_S$, as a function of the turbine valve position, CV, and energy input to the boiler. The nonlinearities in the boiler processes are due to the fact that the steam flow sent to the turbine is the product of the control valve flow area and throttle pressure P_T.

The steam flow rate $\dot{m}_S$ is directly related to the throttle pressure P_T, which in turn is proportional to the integral of flow in ($\dot{m}$) minus flow out ($\dot{m}_S$) of the superheater-related storage. The rate of change of the throttle pressure is given by the constant C_{SH}. The flow into the superheater-related storage and also out of the drum-related storage, $\dot{m}$, is

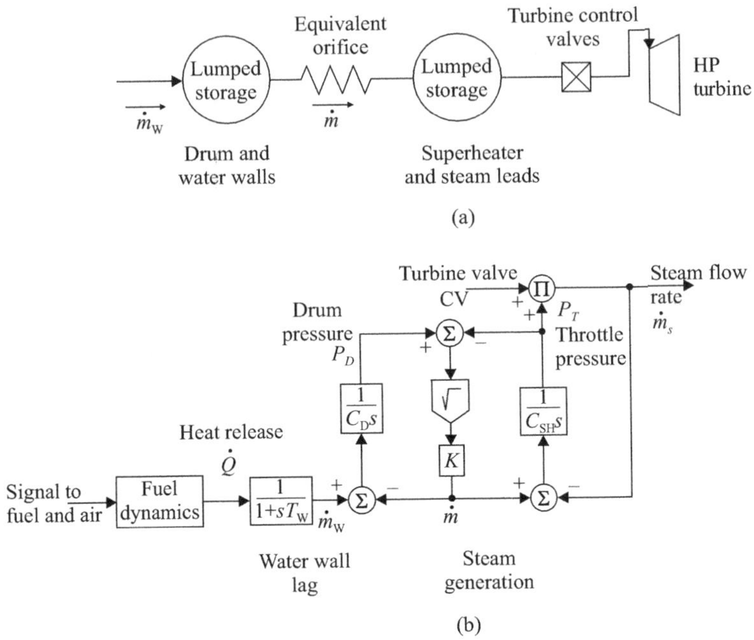

Figure 3.11. Model of boiler pressure effects. (Reprinted with permission from Ref. 1. © 1991 IEEE.)

Sample data:

$$T_W = 5 - 7\,\text{s}; \quad C_D = 90 - 300\,\text{s}; \quad C_{SH} = 5 - 15\,\text{s}; \quad K = 3.5$$
$$\text{CV} = 1\,\text{at full load, } 0\,\text{at no-load}$$

proportional to the square root of pressure difference between the two storage parts. The drum pressure P_D is proportional to the integral of the difference between the steam generation ($\dot{m}_w$) and steam flow ($\dot{m}$). The steam generation ($\dot{m}_w$) is given by the heat release in the furnace with a time lag T_w due to the delay in the heat transfer in the water walls. The process of heat generation and release in the furnace, either during start-up or regular loading or unloading of the generator unit, is subject to fuel system dynamics. For oil- or gas-fired power plants, this process is relative fast, while for coal-fired units, it can be quite slow. However, both of them are slow comparative to the hydro power plants.

The control of a *boiler–turbine–generator* assembly in a steam generator unit is usually performed as a single entity control. However, there are several *control modes* defined in the literature. The prime mover system and the control method for adjusting the output active power, as described previously, are representative for many units operating in the *turbine leading boiler mode* (Figure 3.12a).

In this control mode, any generation adjustment is initiated by turbine valves position adjustment and the boiler reacts with appropriate measures to changes in pressure and steam flow. This control mode has the disadvantage that it requires a good boiler–turbine coordination to avoid stability problems.

A more conventional control mode is the *boiler leading turbine* mode (Figure 3.12b), where turbine valve position is adjusted to regulate the boiler pressure and changes in

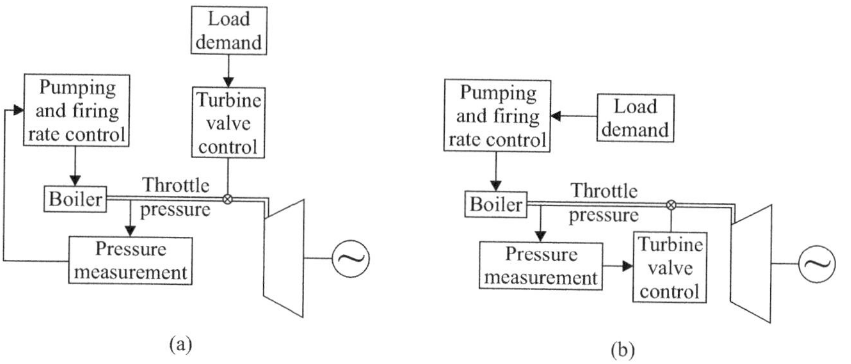

Figure 3.12. Power plant control modes: (a) turbine leading boiler mode; (b) boiler leading turbine mode [2].

generation are initiated by changing inputs to the boiler. To keep the pressure constant, the turbine valve control acts to change the steam flow by fast adjustment of the valve position. The power output to the turbine will, therefore, closely follow changes in the boiler steam generation as caused by changes in input to the boiler [1].

Steam Chest and High-Pressure Piping. The steam enters the HP turbine through the turbine control valves and the inlet piping. The valves are located in the turbine steam chest (Figure 3.13a). In large units the chest contains a series of steam valves. Changes in the HP turbine steam flow are characterized by a time delay, with a time constant T_{CH}, in the turbine valve position adjustment, as also shown in Figure 3.13b.

The following notations have been used in Figure 3.13 [6]:

P_T is the throttle pressure;

P_{SG} is an internal boiler pressure;

P_{GV} is the power at gate or valve outlet;

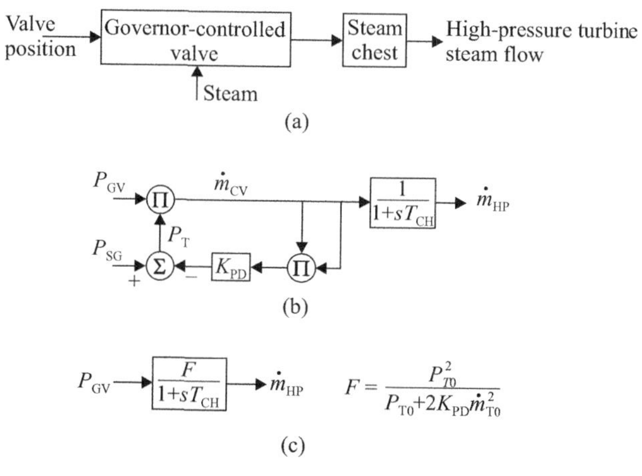

Figure 3.13. Steam chest and high-pressure piping representation: (a) functional blocks diagram; (b) approximate nonlinear model; (c) linear model. (Reprinted with permission from Ref. 6. © 1973 IEEE.)

P_{T0} is the initial (time 0^-) throttle pressure;
$\dot{m}_{T0}$ is the initial (time 0^-) throttle valve steam flow;
T_{CH} is steam chest time constant;
$\dot{m}_{HP}$ is the high-pressure turbine steam flow.

The block diagram shown in Figure 3.13c represents a linear model and takes into account the effect of the boiler tube drop. The pressure P_{SG} is assumed constant over the study interval while P_T is assumed a variable pressure. Denoting by K_{PD} the pipe-drop coefficient, the governor-controlled valve steam flow, $\dot{m}_{CV}$, is given by [6]

$$\dot{m}_{CV} = P_{GV}\left[P_{SG} - K_{PD} \cdot (\dot{m}_{CV})^2\right]$$

The effective gain of the governor controlled is reduced to the fraction *F*. If boiler tube drop is neglected, *F* is unity.

3.3.3 Steam System Configurations

Depending on the number of pressure cycles, a fossil-fueled units may consist of HP, IP, and LP turbine sections. In modern power plants, the steam exhausted from the HP turbine is rerouted to the power plant furnaces where it is reheated then go drive the next turbine section. Increasing the temperature of the steam before entering the next turbine section increases the efficiency of the Rankine cycle. More than one reheat stages are possible, but complications occur and the high capital cost may not justify the improvement in efficiency. The steam exhausted from the IP turbine enters the LP turbine through a crossover pipe. Figure 3.14 shows a tandem-compound steam turbine configuration of a fossil-fueled unit with a single reheat.

To better understand the operation of a basic steam system configuration, it is useful to know the position and the role of some components [6]:

- The *main steam stop valve* (MSV) is provided upstream of the steam chest and is one of the controlling means for admission of steam to the turbine during start-up and shutdown of the unit. It is also called emergency valve since it quickly stops the steam flow to prevent damages to the turbine when the unit overspeeds. The unit may be controlled automatically or manually, but usually is controlled automatically through a hydraulic control system. Because, for normal functioning of the control

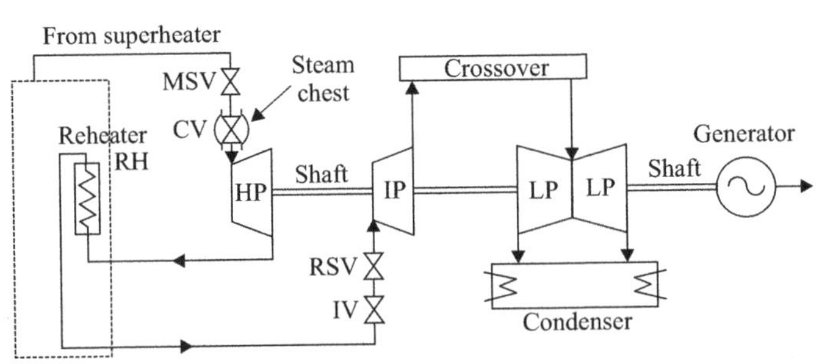

Figure 3.14. Configuration of tandem-compound steam turbine of fossil-fueled unit with a single reheat [7].

system it serves as a backup redundant means of control, the stop valve is not usually modeled.

- The *governor* or *admission valves*, also known as *control valves* (CV), are located in the turbine steam chest and they control the flow of steam to the high-pressure turbine. The number of governor valves depends on the unit size. The governor valves control the quantity of steam flowing to the turbine by changing the valve position. The mechanical power developed by the HP turbine depends on the quantity of steam flow admitted to the turbine through the valve.
- In thermal power plants designed with *reheater* (RH), located on the steam path between the HP turbine and the IP turbine, the pipe entering the IP turbine is provided with two valves, with functioning similar to the main steam stop valve and control valve:
  - The *reheat stop valve* (RSV), which offers backup protection for the IP turbine in the event the unit experiences shutdown, such as in an overspeed trip operation.
  - The *intercept valve* (IV), which throttle the steam flow to the IP turbine in case of loss of load in order to prevent overspeeding.
- The *crossover* is a large pipe through which the steam exhausted from the IP turbine is carried to the LP turbine(s) at low pressure and therefore in large volumes. Usually, the LP turbine is designed with double or triple flow.

Six common steam system configurations and corresponding mathematical models, shown in Figure 3.15, were proposed in Ref. 6. In the block diagrams, the time constants

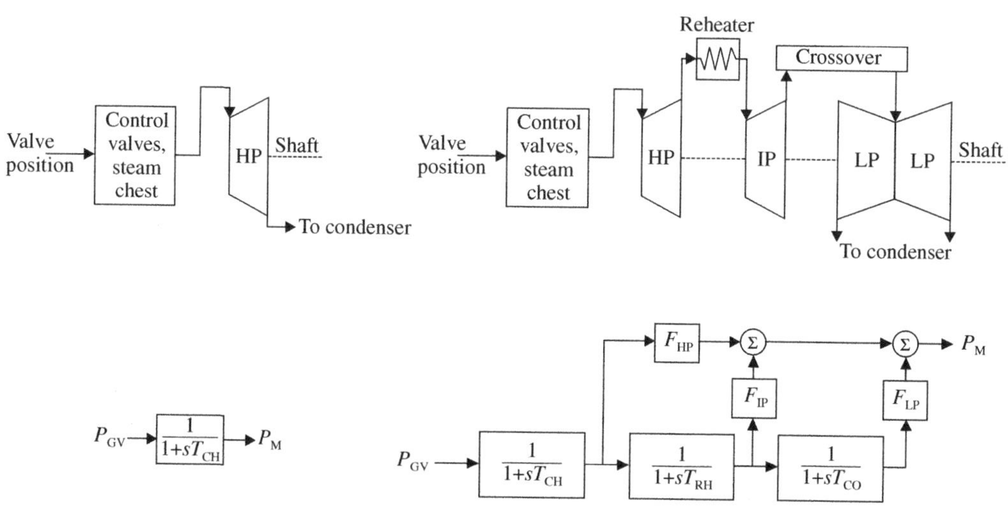

Sample data: $T_{CH} = 0.2\ldots0.5$ s

(a)

Sample data: $T_{CH} = 0.1\ldots0.4$ s; $T_{RH} = 4\ldots11$ s; $T_{CO} = 0.3\ldots0.5$ s; $F_{IP} = 0.4$; $F_{LP} = 0.3$

(b)

Figure 3.15. Steam system configurations and corresponding mathematical models. (a) Non-reheat; (b) tandem compound, single reheat, single shaft; (c) tandem compound, double reheat, single shaft; (d) cross compound, single reheat, double shaft, same speed; (e) cross compound, single reheat, double shaft, different speeds; (f) cross compound, double reheat, double shaft, same speed. (Reprinted with permission from Ref. 6. © 1973 IEEE.)

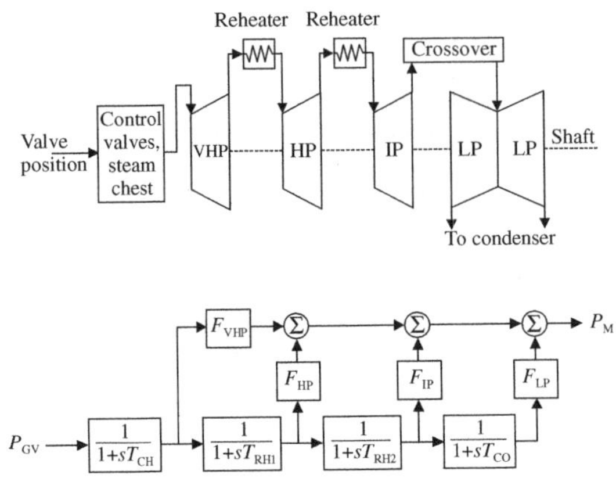

Sample data: $T_{CH} = 0.1\ldots0.4$ s; $T_{RH1} = 4\ldots11$ s; $T_{RH2} = 4\ldots11$ s; $T_{CO} = 0.3\ldots0.5$ s; $F_{VHP} = 0.22$ s; $F_{HP} = 0.22$ s; $F_{IP} = 0.30$ s; $F_{LP} = 0.26$ s

(c)

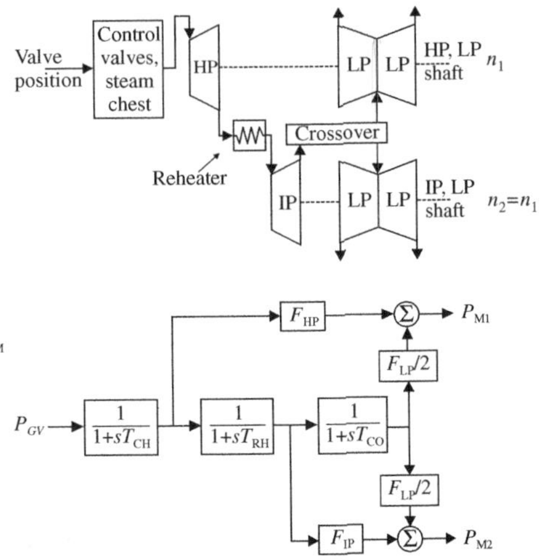

Sample data: $T_{CH} = 0.1\ldots0.4$ s; $T_{RH} = 4\ldots11$ s; $T_{CO} = 0.3\ldots0.5$ s; $F_{VHP} = 0.22$ s; $F_{HP} = 0.3$ s; $F_{IP} = 0.3$ s; $F_{LP} = 0.4$ s

(d)

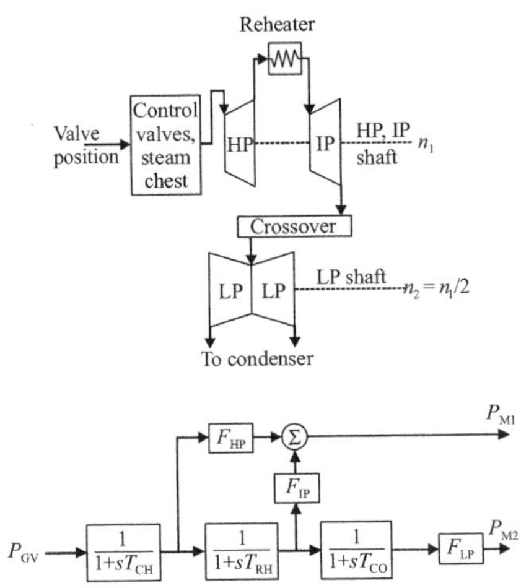

Sample data: $T_{CH} = 0.1\ldots0.4$ s; $T_{RH} = 4\ldots11$ s; $T_{CO} = 0.3\ldots0.5$ s; $F_{HP} = 0.25$ s; $F_{IP} = 0.25$ s; $F_{LP} = 0.5$ s

(e)

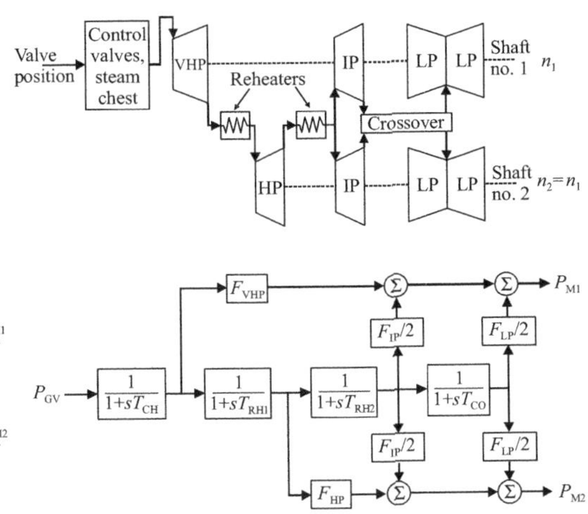

Sample data: $T_{CH} = 0.1\ldots0.4$ s; $T_{RH1} = 4\ldots11$ s; $T_{RH2} = 4\ldots11$ s; $T_{CO} = 0.3\ldots0.5$ s; $F_{VHP} = 0.22$ s; $F_{HP} = 0.22$ s; $F_{IP} = 0.28$ s; $F_{LP} = 0.28$ s

(f)

Figure 3.15. *(Continued)*

T_{CH}, T_{RH}, and T_{CO} represent delays due to the steam chest and inlet piping, reheaters, and crossover piping, respectively. The fractions F_{VHP}, F_{HP}, F_{IP}, and F_{LP} represent portions of the total turbine power developed in the various cylinders. Fractions from the flow of steam that enters the turbine have to satisfy the condition: $F_{VHP} + F_{HP} + F_{IP} + F_{LP} = 1$.

Each turbine model has a functional block at the entrance having the role of introducing a time delay between the adjustment instant in the valves position and the instant of steam flow change in the turbine. As explained before, the control of the steam admitted to the HP turbine, and also in this functional block, is performed by changes in the control valves position as a result of the speed governor action to change the mechanical power developed by the turbine at the shaft. Additionally, the intercept valve contributes to the generator unit stability by quickly routing the steam toward the condenser to unload the turbine in a very short time. This action is called "fast valving."

3.3.4 General Steam System Model

The steam driven turbine power plants are designed in a large number of configurations. But, provision of a common format model, with appropriate characteristics, is more convenient for implementation in computers programs.

A general model that would accommodate all types is shown in Figure 3.16 [1]. The coefficients K_1 to K_8 reflect the contributions from various turbine sections. They are functions of the efficiency and enthalpy drop across the stage. In turn, the enthalpy is function of the pressures across the stages. In the case of tandem-compound units, the two shaft mechanical powers P_{mechHP} and P_{mechLP} can be merged into a single mechanical power provided to the synchronous machine, whereas for cross-compounding units, the two individual shaft mechanical powers must be modeled explicitly. The turbine response time depends on the charging time of various volumes, that is, the high-pressure turbine bowl T_4, the reheater T_5, and the crossover T_6. In the case of double reheat units, the model also takes account of another time constant, T_7.

The model in Figure 3.16 includes the boiler pressure effects but does not consider the effect of the intercept valves. The effect of control valve can be modeled individually since it serves as input in the generic model. For a generator unit designed with reheat, Figure 3.17 shows an enhanced model that accounts for the effect of the intercept valve control action. A limit PR_{MAX} associated with the reheat pressure was introduced to account for the safety valve action.

In the models shown above, the control valve CV and the intercept valve IV are to be interpreted as flow areas. Any nonlinearity between control signal and valve flow areas is assumed to be included in the turbine control logic module.

The model described in Figure 3.17 works properly for normal dynamics. Under emergency conditions, such as fast valving, a more detailed representation of the turbine is required.

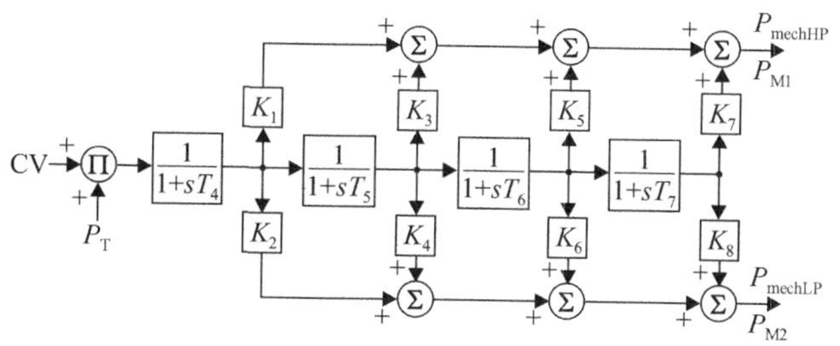

Figure 3.16. Generic turbine model. (Reprinted with permission from Ref. 1. © 1991 IEEE.)

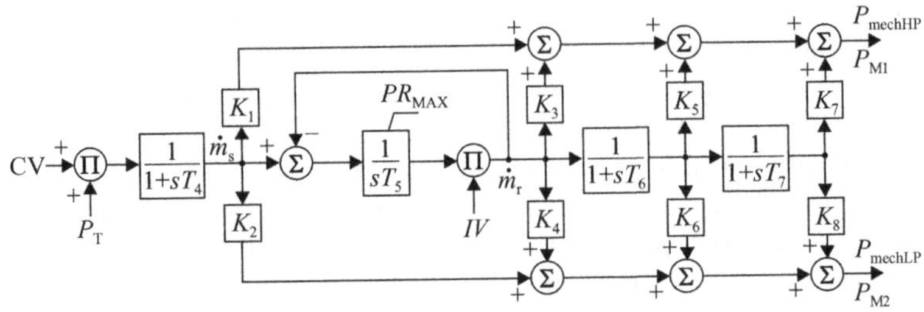

Figure 3.17. Generic turbine model including IV effects. (Reprinted with permission from Ref. 1. © 1991 IEEE.)

Table 3.1 presents the correspondence between the time constants and the fractions for each turbine type [6].

3.3.5 Governing Systems for Steam Turbines

The governing system is one of the most important parts of a steam turbine generator unit for the output power and/or frequency regulation. Adjustment of speed governor set point is currently the method for matching the generation to load, under various market forms, or simply for keeping the frequency at desired value.

A power system frequency is maintained constant and equal to a desired value if the total generation matches the total load. When a power imbalance occurs in the system, caused by either load or generation change, the system frequency deviates from the desired value. As the generator speed is directly related to the system frequency, any increase in frequency causes increase in the generator rotor speed and vice versa. In response to changes in the rotor speed, the speed governor varies automatically (increase or decrease as appropriate) the prime mover output (torque) through valve control. The amount of torque required to be developed by the turbine in order to keep the synchronous generator rotor speed at the desired value is given by the amount of current (power) absorbed by the system

TABLE 3.1 Interpretation of Parameters Used in General Model for Turbines [6]

	Time Constants				Fractions							
System Description	T_4	T_5	T_6	T_7	K_1	K_2	K_3	K_4	K_5	K_6	K_7	K_8
Nonreheat (Figure 3.15a)	T_{CH}	–	–	–	1	0	0	0	0	0	0	0
Tandem compound single reheat (Figure 3.15b)	T_{CH}	T_{RH}	T_{CO}	–	F_{HP}	0	F_{IP}	0	F_{LP}	0	0	0
Tandem compound double reheat (Figure 3.15c)	T_{CH}	T_{RH1}	T_{RH2}	T_{CO}	T_{VHP}	0	F_{HP}	0	F_{IP}	0	F_{LP}	0
Cross-compound single reheat (Figure 3.15d)	T_{CH}	T_{RH}	T_{CO}	–	F_{HP}	0	0	F_{IP}	$\frac{F_{LP}}{2}$	$\frac{F_{LP}}{2}$	0	0
Cross-compound single reheat (Figure 3.15e)	T_{CH}	T_{RH}	T_{CO}	–	F_{HP}	0	F_{IP}	0	0	F_{LP}	0	0
Cross-compound double reheat (Figure 3.15f)	T_{CH}	T_{RH1}	T_{RH2}	T_{CO}	T_{VHP}	0	0	F_{HP}	$\frac{F_{IP}}{2}$	$\frac{F_{IP}}{2}$	$\frac{F_{LP}}{2}$	$\frac{F_{LP}}{2}$
Hydro	0	$T_w/2$	–	–	−2	0	3	0	0	0	0	0

load from the generator. For small frequency variations, the speed governor controls the steam flow necessary to develop the required torque through a specific control logic. However, if large imbalances occur in the power system, the control is taken over by the automatic generation control system with appropriate time delay. The rate and magnitude of the governor response to a speed change are tuned for the characteristics of the generator unit and the power system. The inputs to the speed governor control logic are speed, acceleration, electrical power, generator current, and so on, whereas the outputs are the CV and IV flow areas.

The steam flow admitted to the turbine is controlled by the speed governor that sets the valve position (stroke) that is, therefore, proportional to speed. This stroke is mechanically compared to a preset reference position to give a position error proportional to the speed error. The force that controls this position error is small and must be amplified in both force and stroke. This is the purpose of the two amplifiers labeled speed relay and servomotor [2].

A simplified system block diagram of a steam turbine control is shown in Figure 3.18. Although there are various types of speed governors, that is, mechanical hydraulic, electrohydraulic, or digital electrohydraulic, they have similar steady-state speed-output characteristics and their application principle (for normal operation) is the same.

For normal load changes, the control valves are used with straightforward proportional control on speed error, whereas for more severe disturbances, the limitation of overspeed involves both IVs and CVs.

The turbine speed/load control model can usually be generic when limited to normal primary speed control and supplementary load control (AGC). For problems involving large accelerations enlisting the discrete and nonlinear actions provided in the particular design, the speed/load control block should be defined by the manufacturer.

Reference [6] presents approximate mathematical representations for typical mechanical hydraulic (MHC) and electrohydraulic (EHC) speed governing systems.

3.3.5.1 Mechanical Hydraulic Control (MHC).

The *speed governor* is basically a speed-sensitive mechanical device that senses the shaft speed and converts it into a valve position reference (Figure 3.19).

The mechanical governor operates similarly to a classical centrifugal device or a flyball governor that uses flying weights (flyballs) mounted on spring-loaded arms. The rotor speed signal (ω_r) is converted to linear displacement by means of the centrifugal forces opposed by the spring. When the shaft speed decreases, the flyballs spin more slowly

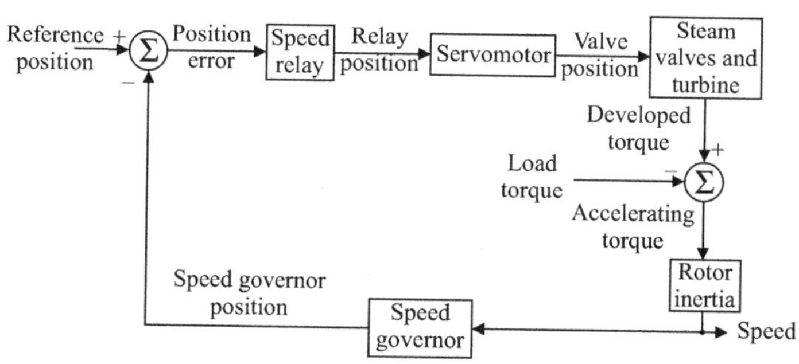

Figure 3.18. Block diagram of steam turbine control system [8].

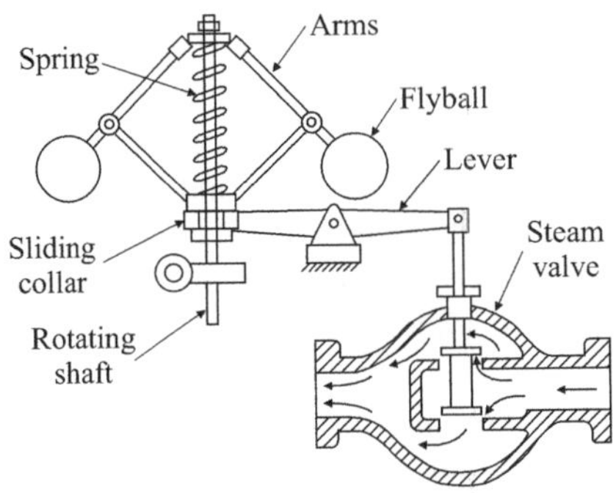

Figure 3.19. Mechanical speed governor [2].

and they move in causing the valve to move up and allow more steam to flow and vice versa.

Figure 3.20a shows the representation of a typical mechanical hydraulic speed governing system consisting of a speed governor, a speed relay, a hydraulic servomotor, and governor-controlled valves [6,7].

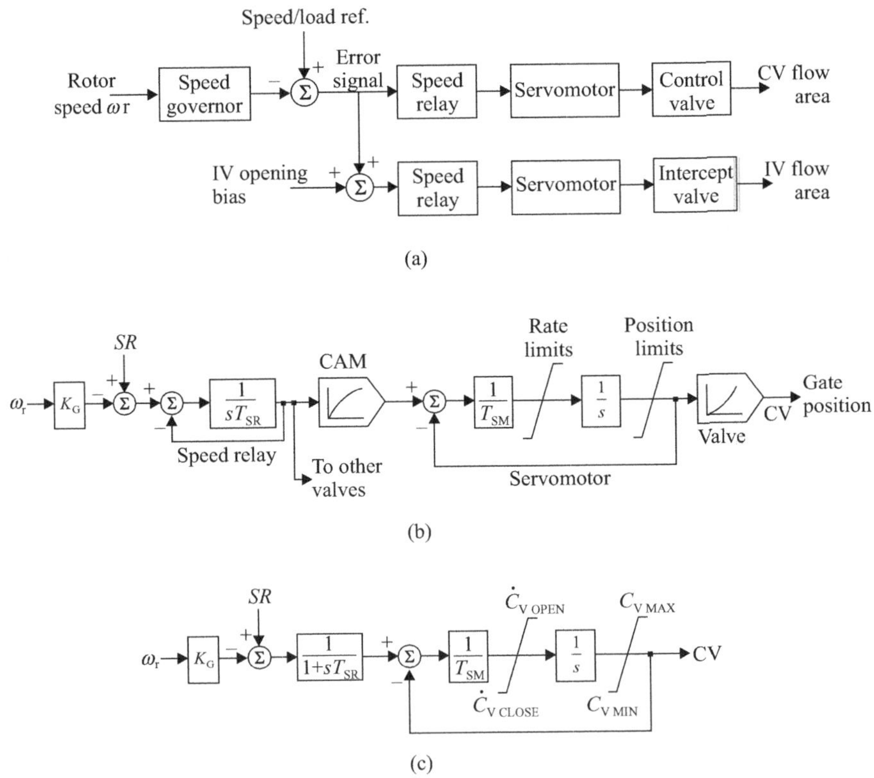

Figure 3.20. Mechanical hydraulic speed governor for steam turbines: (a) functional block diagram; (b) and (c) approximate mathematical representations [6,7].

Sample data for Figure 3.20:

$$K_G = 20.0 \quad T_{SR} = 0.1\,\text{s} \quad T_{SM} = 0.2 - 0.3\,\text{s}$$
$$\dot{C}_{VOPEN} = 0.1\,\text{p.u./s/valve}; \dot{C}_{VCLOSE} = 0.1\,\text{p.u./s/valve}.$$

As shown in Figure 3.20a, the speed governor output is compared with a speed/load reference signal resulting in an error signal that determines the position for the CVs and, when appropriate, for the IVs that is set by servomotors. However, under normal conditions the CVs are used for speed/load control whereas the IVs are held fully open by a bias (IV opening bias) signal. In case of overspeed, a large error signal occurs, the bias is overcome and a signal is sent to the servomotor to close the IVs rapidly. As soon as the shaft speed is brought back to the reference speed, the error signal is restored to a value less than the bias and the IVs are again fully opened.

An approximate nonlinear mathematical model for the chain acting on the control valve in response to changes in the shaft speed is presented in Figure 3.20b. The speed governor determines a valve position that is assumed linear to the shaft speed, with no time delay. The signal is amplified with a gain K_G, which is the reciprocal of regulation or droop (R), then compared to the signal SR, obtained from the governor speed changer (see Figure 3.1). The signal SR is determined by the AGC system.

A hydraulic servo, called *speed relay*, develops an output proportional to the load reference signal less any contribution from speed deviation ($\Delta\omega$) through the primary speed control gain ($1/R$). The speed relay is represented as an integrator with time constant T_{SR} and direct feedback. A nonlinear "cam" to compensate for valve nonlinearity is shown between the speed relay and the servomotor. On very large turbines, additional amplification to the energy levels necessary to move the steam valves is obtained using *hydraulic servomotors*. The servomotor is represented by an integrator with time constant T_{SM} and direct feedback. Rate limiting of the servomotor may occur for large, rapid speed deviations. These rate limits are shown at the input to the integrator representing the servomotor. Position limits are also indicated and may correspond to wide-open valves or the setting of a load limiter. In power system studies, nonlinearities in the speed control mechanism are usually neglected except for rate limits and the limits on value position (see Figure 3.20c) [6].

3.3.5.2 Electrohydraulic Control (EHC). An electrohydraulic speed control mechanism is more flexible than a mechanical hydraulic one through the use of electronic circuits.

Figure 3.21a shows the functional block diagram of a typical electrohydraulic control configuration. The steam flow (or first stage pressure) feedback and the servomotor feedback loop provide for improved linearity over the mechanical hydraulic system. The block diagram of Figure 3.21b shows approximate mathematical relationship for the speed governing function of the General Electric electrohydraulic control system with the steam flow feedback operative.

Comparative to the MHC systems (Section 3.3.5.1), the EHC governing systems are designed with another two speed control features [1]:

- A triggering system is activated and the IVs are fast closed whenever a load unbalance occurs causing an error signal to the IV servovalve greater that 0.1 p.u. This speed error represents a speed deviation of $\Delta\omega > 0.05(\text{LR}) + 0.002\,\text{p.u.}$ When the triggering system is activated, the IV servovalve control is blocked for 1 s, after which the IV servovalves are free to respond to speed control.
- Whenever a power/load unbalance occurs upon a load rejection, a power/load relay (PLU) is fast close the valves CV and IV.

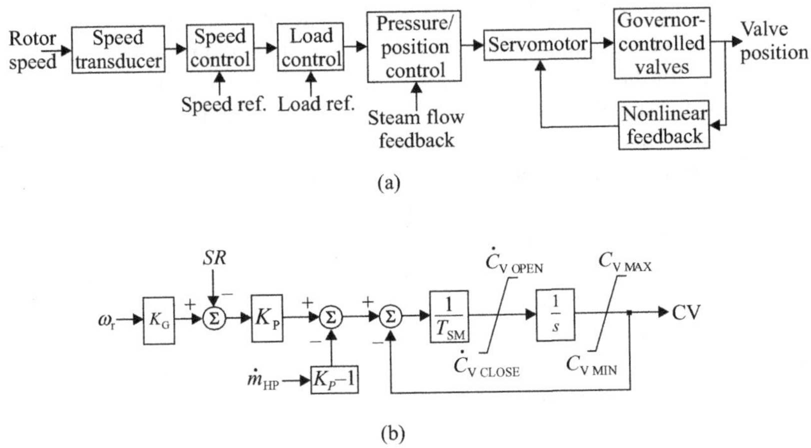

Figure 3.21. Electrohydraulic speed governor system for steam turbines: (a) functional block diagram; (b) approximate mathematical model for EHC system [1].

Sample values of the parameters for Figure 3.21:

$$K_G = 20.0; \quad K_P = 3.0 \text{ (with steam flow feedback)};$$
$$K_P = 1.0 \text{ (without steam flow feedback)}; \quad T_{SM} = 0.1 \text{ s}.$$

Rate limits:

$$\dot{C}_{VOPEN} = 0.1 \text{ p.u./s/valve};$$
$$\dot{C}_{VCLOSE} = 0.1 \text{ p.u./s/valve}.$$

A block diagram for an electrohydraulic control system designed by General Electric is shown in Figure 3.22 [1].

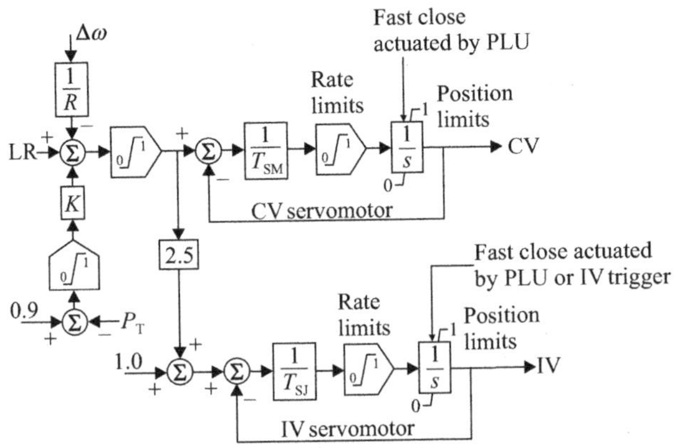

Figure 3.22. Block diagram for the electrohydraulic control of the General Electric [1].

Sample data for Figure 3.22:

$R = 0.05\,\text{p.u.};\quad T_{SM} = 0.1 - 0.2\,\text{s};\quad T_{SJ} = 0.1 - 0.2\,\text{s};$
Time to fast close $= 0.15\,\text{s};$
Rate limits CV $= -0.2;\ 0.1;$ rate limits IV $= -0.2;\ 0.1.$

In normal operation, the valve position set to the CV is determined by the error between the load reference signal and the speed deviation times the gain $1/R$. An additional action may take place in the closing direction of the CV if the main steam pressure falls below the initial pressure regulator set point (0.9). The IVs respond to overspeed in the same manner as in the MHC systems.

3.3.5.3 Digital Electrohydraulic Control (DEHC). The block diagram for a normal speed control function of a digital electrohydraulic control system designed by Westinghouse is shown Figure 3.23.

In addition to the normal speed control function, the governing system is provided with overspeed protection functions [1,7].

For partial loss of load, in order to improve the transient stability of the generator unit, a *close intercept valves* (CIV) function is sometimes provided, which is based on the mismatch between turbine mechanical power and electrical load. This function is generally referred to as fast valving.

For the complete loss of load, the digital electrohydraulic control system is designed with two overspeed protection components: the *load drop anticipator of overspeed* (LDAO) and *overspeed sensing*. The LDOA function closes the CV and IV when the turbine power is greater than 30% of rated, whereas the overspeed sensing function closes the CVs and LVs when speed exceeds 103% of rated.

3.3.5.4 General Model for Speed Governing Systems. A general speed governor model appropriate to represent either a mechanical hydraulic system or an electrohydraulic system for a steam turbine is presented in Figure 3.24a.

The following notations have been used in Figure 3.24:

P_0 is the initial (time $= 0^-$) mechanical power;

P_{GV} is the total power at gate or valve outlet;

$\dot{P}_{UP}$ and $\dot{P}_{DOWN}$ are the limits on rate of change of power imposed by control valve rate limits;

P_{MAX} and P_{MIN} are the power limits imposed by valve or gate excursion.

Rate limits are nominally 0.1 p.u./s except for mechanical hydraulic system where $\dot{P}_{DOWN}$ is 1.0 p.u./s. Nominally, $K = 100$ (% steady-state speed regulation).

Table 3.2 contains a list of typical parameters for the block diagram of Figure 3.24 [6].

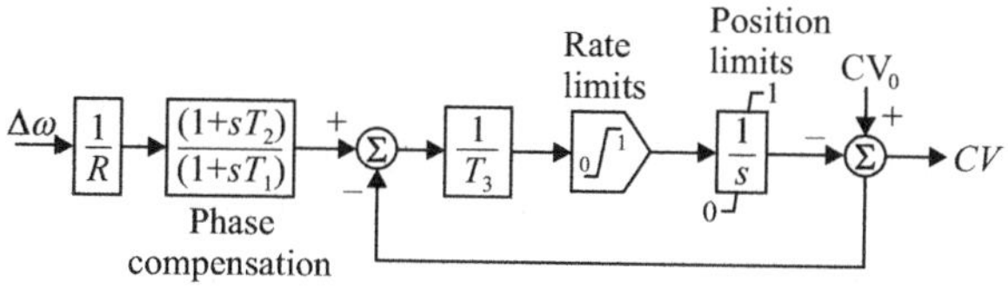

Figure 3.23. Digital electrohydraulic controls (Westinghouse) [1].

Sample data for Figure 3.23:

$$R = 0.05\,\text{p.u.};\quad T_1 = 7.5\,\text{s};\quad T_2 = 2.8\,\text{s};\quad T_3 = 0.1\,\text{s};$$
$$\text{Rate limits} = +0.5(\text{closing}); -0.4(\text{opening}).$$

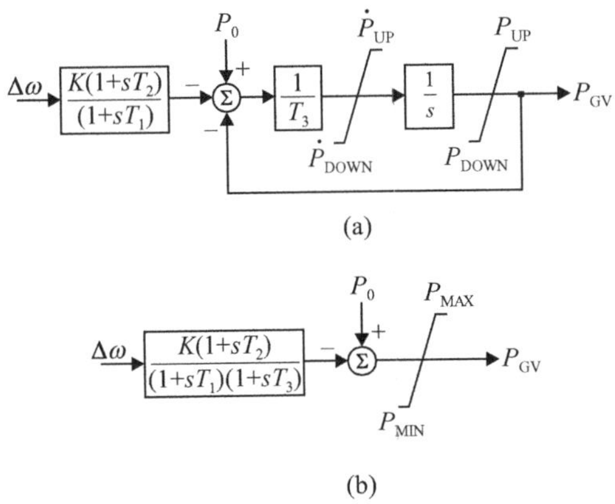

Figure 3.24. General models for speed governing systems: (a) steam turbine systems; (b) hydro systems [6].

TABLE 3.2 Speed Governing System Parameters for Use with Figure 3.24a [6]

	Time Constants (s)		
System	T_1	T_2	T_3
Mechanical hydraulic	0.2–0.3	0	0.1
	General Electric EH		
With steam feedback[a]	0	0	0.025
Without steam feedback	0	0	0.1
	Westinghouse EH		
With steam feedback[a]	2.8[b]	1.0[b]	0.15
Without steam feedback	0	0	0.1

[a]Steam flow feedback includes the steam chest time constant T_{CH} that must be modified when Figure 3.24a is used.
[b]These values may vary considerably from one unit to other.

3.4 COMBINED-CYCLE POWER PLANTS

3.4.1 Generalities

The share of power installed in combined-cycle power plants in the total generation capacity have increased in the last two decades in a number of power systems.

The minimum temperature of a gas turbine power plant, based on a Brayton cycle, is the same range as the maximum temperature of a conventional steam turbine power plant, based on a Rankine cycle. It is therefore feasible to create a gas-steam combined-cycle

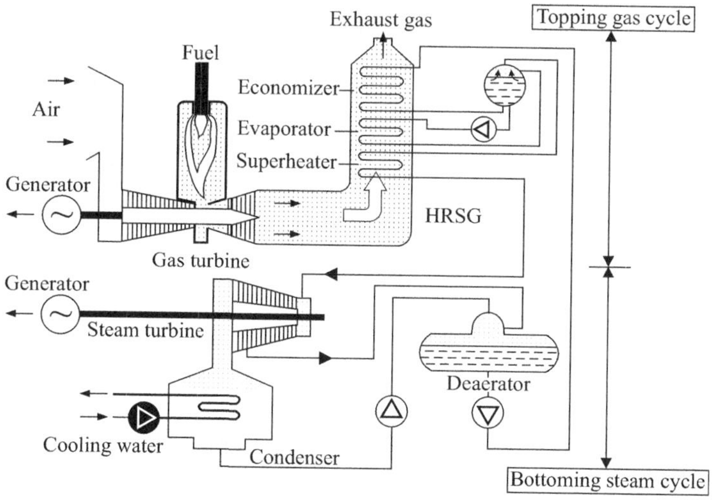

Figure 3.25. Typical gas-steam combined-cycle power plant. (*Source*: Alstom Power AG.)

thermodynamic cascade, where the higher (upper) temperature plant is a gas turbine circuit and the lower temperature plant is a steam turbine circuit. A significant amount of heat stored in the gas exhausted from the gas turbine is transferred to the working fluid of the steam turbine by means of a heat recovery steam generator (HRSG). Combining the two cycles thus results in significantly higher overall efficiency compared to a conventional fossil fuel plant. The higher efficiency is due to the greater utilization of the total enthalpy produced by combustion in the gas turbine. A typical simple cycle conventional fossil fuel plant has an efficiency of 35–43%, while a combined-cycle power plant can have efficiencies exceeding 58% [9]. However, because of the high capital costs, the combined-cycle power plants are feasible for large installed capacity.

A typical combined-cycle turbine arrangement is shown in Figure 3.25.

In the combined-cycle power plant, the prime mover duty is divided between gas/combustion turbine and the heat recovery steam turbine, with each turbine powering its own generator. The amount of electrical energy obtained in the steam cycle is strictly dependent on the rating of the gas turbine part.

3.4.2 Configurations of Combined-Cycle Power Plants

The combined-cycle power plants can be designed in a number of arrangements in terms of the installed power. They can be categorized into two main groups [9]:

- Single-shaft power plants, where the gas turbine, steam turbine, and electrical generator are all connected in tandem on a single rotating mechanical shaft (Figure 3.26a).
- Multishaft power plants, where one or more gas turbines, each designed with its own HRSG, feeding steam to a single steam turbine, all on separate shafts with separate generators (Figure 3.26b).

The upper temperature plant consists of an axial compressor, a combustion chamber, and a turbine. The air is aspired into the compressor through the air intake system. The

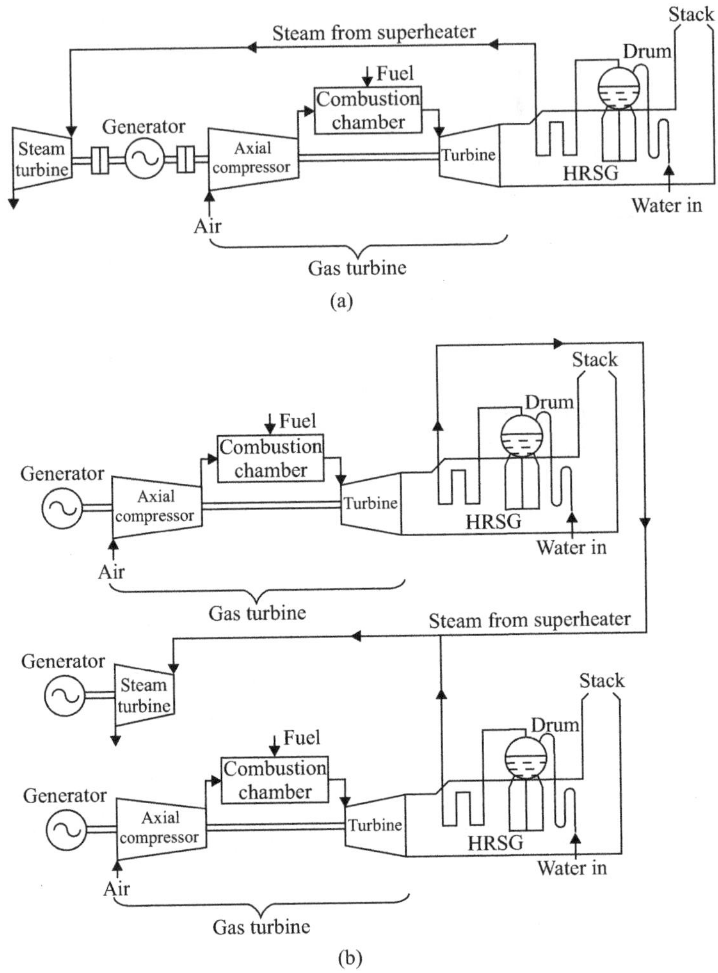

Figure 3.26. Combined-cycle power plant configuration: (a) single-shaft combined-cycle power plant; (b) multishaft combined-cycle power plant with two gas turbines [9,10].

compressor increases the air pressure to the operational value of the combustion chamber (burner). The compressed air is mixed with fuel in the combustion chamber where a burning process takes place. The hot gases exhausted from the combustion chamber are then expanded in the gas turbine producing mechanical work. The mechanical work is used to drive the compressor and the synchronous generator. An exhaust system provides the path for exhausting the gases from the turbine. The power plant is also designed with auxiliaries and control systems.

3.4.3 Model Block Diagrams of Combined-Cycle Power Plant

Figure 3.27 shows the simplified functional block diagram and the coupling between submodels of the combined-cycle power plant. The main blocks are the speed/load control, fuel and flow controls, gas turbine, and heat recovery steam generator with steam turbine. This model may be used with good accuracy in dynamic studies of power systems [11].

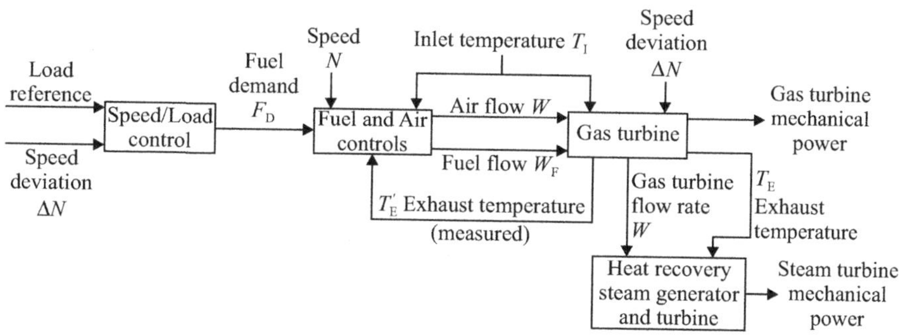

Figure 3.27. Subsystems of the combined-cycle power plant. (Reprinted with permission from Ref. 11. © 1994 IEEE.)

Some characteristics are to be emphasized for representing the connection between parts of the combined-cycle power plant:

- The mechanical power developed by the steam turbine is a function of the flow rate and temperature of the gas exhausted from the gas turbine.
- The mechanical power developed by the gas turbine is a function of the fuel flow and the airflow; the firing temperature is controlled and kept below a designed limit by adjusting the fuel flow and airflow, based on measurements of the exhaust temperature and compressor pressure ratio.
- The fuel demand at the gas turbine is set in terms of a load reference signal and the speed deviation ΔN.
- At reduced load of the gas turbine, a high exhaust temperature is obtained by reducing the airflow; the airflow is a function of the opening position of the inlet guide vanes (IGV), the atmospheric temperature and pressure, and shaft speed.

There are two most widely used models of the gas turbine, Rowen and IEEE, suitable for small and large disturbance stability studies. The main difference between the two models is the control action necessary to maintain a high firing temperature (turbine inlet temperature) that determines a low NO_X gases emission level.

Both models use the *speed governor* model shown in Figure 3.28. The inputs to the speed governor are the load demand V_L and the speed deviation ΔN. The speed governor determines the demanded fuel flow F_D in terms of the input variables and some functional and design characteristics. In Figure 3.28, W stands for the gain (1/droop) value, X is the governor lead time constant, Y is the governor lag time constant, and Z is the governor operation mode ($1 = \text{droop}$, $0 = \text{isocronous}$).

The *Rowen's model* [12], shown in Figure 3.29, includes the most important part of a heavy-duty gas turbine in a single cycle with no heat recovery, assuming that it operates at ambient temperature (15°C) and normal pressure (1.01 bar), and the rotor speed is maintained fairly constant within 95–107% of the rated speed.

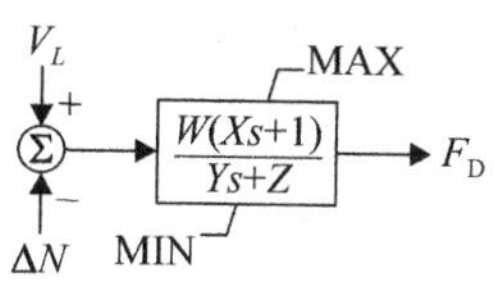

Figure 3.28. Combined-cycle speed-load controls [11].

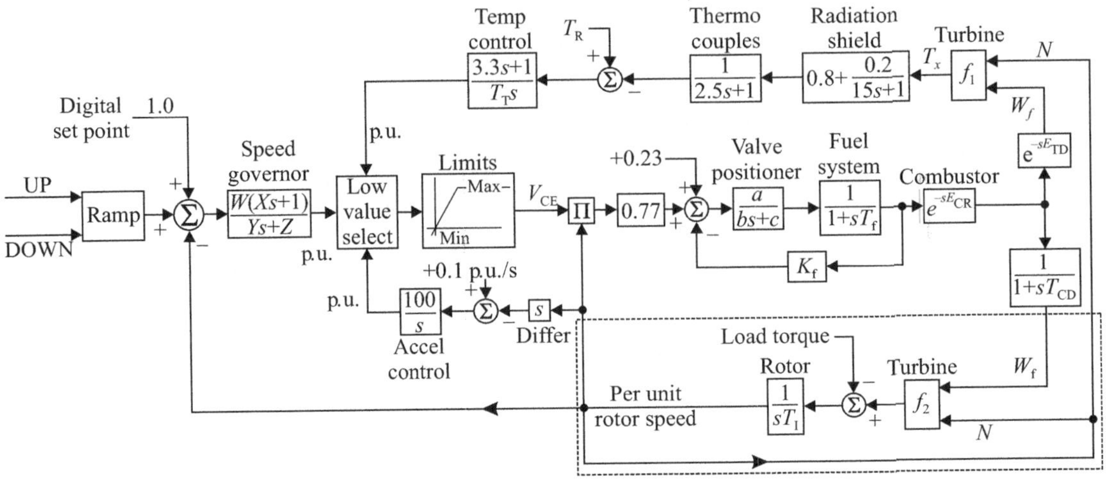

Figure 3.29. Rowen's model. (Reproduced from Ref. 12.)

Sample data, in English units [13]:

$$W = 16.7;\ X = 0.6\,\text{s};\ Y = 1.0\,\text{s};\ Z = 1;\ \text{MAX} = 1.5\,\text{p.u.};\ \text{MIN} = -0.1\,\text{p.u.};$$
$$a = 1;\ b = 0.05;\ c = 1;\ T_f = 0.4\,\text{s};\ K_f = 0;\ E_{CR} = 0.01\,\text{s};$$
$$E_{TD} = 0.04\,\text{s};\ T_R = 950\,(^\circ\text{F});\ T_T = 450\,(^\circ\text{F});\ T_{CD} = 0.2\,\text{s};\ T_I = 15.64;$$
$$f_1 = T_X = T_R - 700(1 - W_f) + 550(1 - N);\ f_2 = 1.3(W_f - 0.23) + 0.5(1 - N).$$

The model includes three major control loops: speed/load control loop, acceleration control loop, and temperature control loop. The outputs of the three loops enter in a low value select block aiming to set the least amount of fuel.

- The speed/load loop is associated with the speed governor.
- The acceleration control loop is designed to limit the fuel flow when the rotor begins to accelerate beyond a threshold value, in case of sudden loss of load.
- The temperature control loop is designed to limit the exhaust temperature of the gas turbine to the maximum limit against any action of the speed governor that may push the temperature beyond the limit.

Parameters a, b, and c are chosen to appropriately set the valve position. The fuel system introduces the fuel into the combustor with a fuel time control constant T_f. The combustion reaction is then characterized by a time delay E_{CR}.

The 0.23 offset takes into account the minimum fuel limit at no load, self-sustaining conditions, essential to maintain the compressor in operation.

The outputs of the gas turbine in the Rowen's model are determined by two function blocks, f_1 and f_2. Function f_1 determines the output temperature of the turbine in terms of the rotor speed and fuel flow. The fuel flow W_f is associated with the turbine and exhaust delay E_{TD}. The output temperature is also associated with the turbine-rated exhaust temperature T_R (in °F) and the temperature controller integration rate T_T (in °F). Function f_2 calculates the output turbine torque again in terms of the rotor speed and fuel flow, but with specific delay given by the compressor discharge volume time constant T_{CD}. The rotor has an inertia $T_I = 2H$.

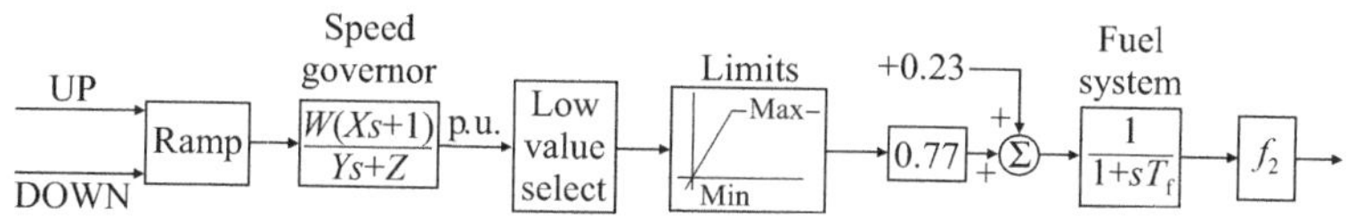

Figure 3.30. Simplified Rowen's model [14].

A more detailed model of the gas turbine representation may include the dynamics of the IGV, which change the position to adjust the airflow in the compressor. However, if the frequency variations are lower than 1% of the reference value, and the control of the gas turbine does not push the output temperature beyond its limits, the model from Figure 3.29 may be reduced to the model shown in Figure 3.30 [14,15].

The *IEEE model* is split into two parts: one part representing the control functions and the other part modeling the thermodynamic processes in the gas turbine. The first part includes the airflow control loop, fuel flow control loop, and temperature control loop.

(i) The *Gas Turbine Fuel and Air Controls* are shown in Figure 3.31.

The turbine exhaust temperature can be controlled by adjusting the airflow, over a limited range. At reduced load of the gas turbine, maintaining a high exhaust temperature results in improved overall efficiency. In such situations, the fuel flow and the airflow are controlled to maintain constant the gas turbine inlet temperature, for example to a reference exhaust temperature T_R.

The reference exhaust temperature in per unit of the absolute firing temperature at rated conditions is calculated with the expression [2,11]

$$T_R = T_f\left[1 - \left(1 - \frac{1}{X}\right)\eta_T\right] \tag{3.1}$$

where T_R is the reference exhaust temperature in p.u. of the absolute firing temperature at rated conditions, T_f is the turbine inlet temperature in p.u. of

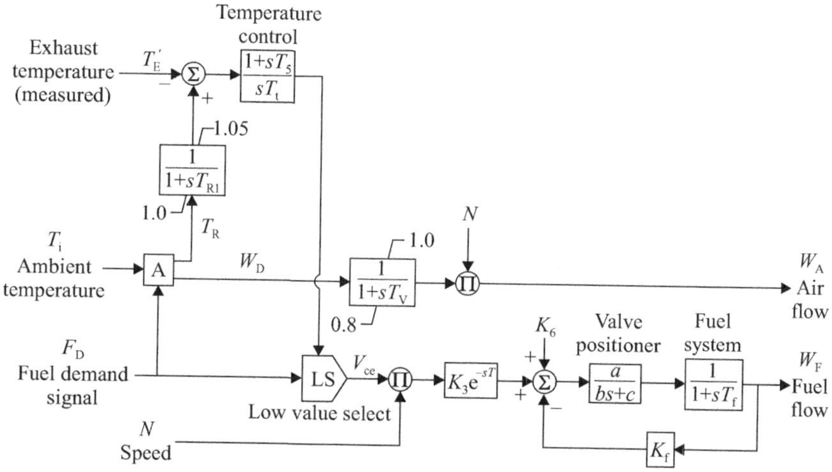

Figure 3.31. Gas turbine fuel and airflow controls. (Reprinted with permission from Ref. 11. © 1994 IEEE.)

Sample data for Figure 3.31:

$K_3 = 0.7$; $K_6 = 0.3$; $K_f = 1$; $T_f = 0.01$; $T_5 = 3.3$; $T_t = 0.45$; $T_V = 10$;
$a = 10$; $b = 1$; $c = 0$.

design absolute firing temperature, η_T is the turbine efficiency, and X is a cycle isentropic pressure ratio parameter and is given by

$$X = (P_R)^{(\gamma-1)/\gamma} \tag{3.2}$$

where $P_R = P_{R0} \cdot W$ is the isentropic cycle pressure ratio, P_{R0} is the design cycle pressure ratio, $\gamma = c_p/c_v$ is the ratio of specific heats, and W is the airflow in p.u. of design airflow (W_0).

The per unit airflow required to produce a specified power generation at the given gas turbine inlet temperature T_f is given by the turbine power balance equation [11]:

$$W = \frac{P_G \cdot K_0}{T_f(1-(1/X))\eta_T - T_i((X-1)/\eta_C)} \text{ p.u.} \tag{3.3}$$

where P_G is the design power output in p.u. of rated;

$$K_0 = \frac{kW_0 \cdot 3413}{W_0 \cdot T_{f0} \cdot C_P} \tag{3.4}$$

where kW_0 is the base net output p.u., W_0 is the airflow p.u., T_{f0} is the turbine inlet temperature in p.u. of the design absolute finite temperature, C_P is the average specific heat, T_i is the compressor inlet (ambient) temperature in p.u. of design absolute firing temperature, and η_C is the compressor efficiency.

For the sake of simplicity, influences of the combustor pressure drop, specific heat change, and treatment of cooling flow have been included in the compressor and turbine efficiency values, η_C and η_T.

The desired airflow W_D and the desired exhaust temperature reference T_R are calculated in the block A (Figure 3.31) in terms of the desired fuel F_D and ambient temperature T_i over the design range of airflow.

The measured exhaust temperature T'_E is compared to the exhaust temperature reference T_R and the resulting error acts on the temperature controller, which sets the fuel flow demand V_{ce} through the "low value select" block. The exhaust temperature reference is fed into the summator with a time delay T_{R1}. The actual fuel flow W_F results in terms of the valve positioner and fuel flow control.

The vane control of the airflow has a dynamics given by the time constant T_V with nonwindup limits. As shown, the actual airflow W_A is the product of the desired airflow and the shaft speed.

(ii) *The Gas Turbine Power Generation*. The computation of gas turbine mechanical power P_{MG} and exhaust temperature T_E is shown in Figure 3.32.

The net output power of the gas turbine is the difference between the turbine power and the compressor power and is determined from equation (3.3), and is a function of the turbine inlet temperature T_f and the airflow W. The turbine inlet temperature T_f can be determined from the combustor heat balance [11]:

$$T_f = T_{CD} + \frac{W_f K_2}{W} = T_i\left[1 + \frac{X-1}{\eta_C}\right] + \frac{W_f K_2}{W} \tag{3.5}$$

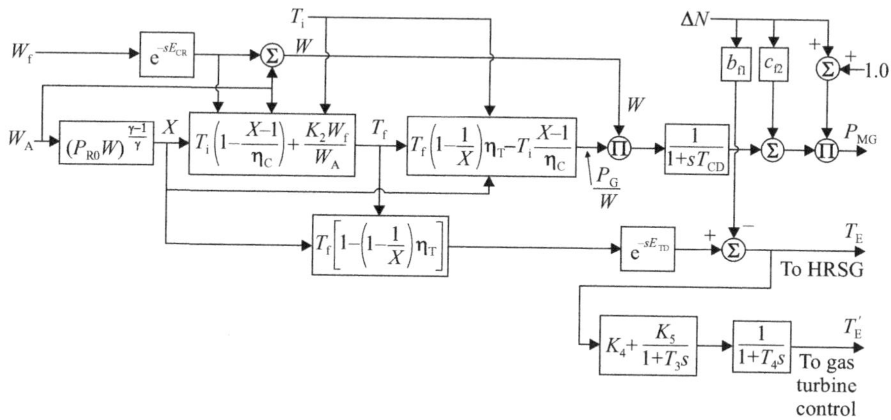

Figure 3.32. Gas turbine mechanical power and exhaust temperature model. (Reprinted with permission from Ref. 11. © 1994 IEEE.)

Sample data for Figure 3.32:

$$E_{CR} = 0.1;\ E_{TD} = 0.04;\ T_{CD} = 0.1;\ T_3 = 15;\ T_4 = 3;\ K_4 = 0.8;\ K_5 = 0.2$$

where $K_2 = \Delta T_0/T_{f_0}$ is the design combustor temperature rise in p.u. of absolute firing temperature, T_{CD} is the compressor discharge temperature in p.u. of design absolute firing temperature, W_f is the fuel flow in p.u. of design airflow (W_0).

The gas turbine exhaust temperature can be determined substituting T_E for T_R in equation (3.1). The mechanical power P_{MG} is a function of the ratio P_G/W as obtained from equation (3.3).

The block diagram also includes the combustor time delay E_{CR}, the compressor discharge volume time constant T_{CD}, and the turbine exhaust system transport delay E_{TD} [11].

(iii) *The Steam Turbine Power Generation.* The heat transferred from the gas circuit to the steam circuit in the heat recovery steam generator system depends on exhaust flow W and the exhaust temperature T_E from the gas turbine. The heat is delivered to the high- and low-pressure steam generation sections.

The thermodynamic processes in the compressor (compression) and turbine (expansion) are theoretically isentropic with no exchange of heat. However, in reality there are inefficiencies in the two adiabatic processes [3,16].

In the ideal compressor cycle, the temperature change is $T_{02} - T_{01}$ (cycles 1–2 in Figure 3.4), while in the real cycle, the temperature change would be $T'_{02} - T_{01}$. The compressor isentropic efficiency is therefore defined as

$$\eta_C = \frac{C_p\,\Delta T'_0}{C_p\,\Delta T_0} = \frac{T'_{02} - T_{01}}{T_{02} - T_{01}} \tag{3.6}$$

where C_p stands for the mean heat capacity of the gas, T'_0 is the real temperature, and T_0 is the temperature in the ideal process.

Equation (3.6) can be developed into

$$T_{02} - T_{01} = \frac{1}{\eta_C}\left(T'_{02} - T_{01}\right) = \frac{T_{01}}{\eta_C}\left(\frac{T'_{02}}{T_{01}} - 1\right) \tag{3.6$'$}$$

The change in pressure ratio of the cycle is proportional to the change in the temperature. Therefore, equation (3.6′) becomes

$$T_{02} - T_{01} = \frac{T_{01}}{\eta_C}\left[\left(\frac{P_{02}}{P_{01}}\right)^{(\gamma-1)/\gamma} - 1\right] \tag{3.6$''$}$$

Assuming that the heat addition in the combustor takes place at constant pressure (neglecting the pressure loss in combustor), with P_{02} equal to P_{03} (see Figure 3.5), the compressor pressure ratio P_{02}/P_{01} is equivalent to the cycle pressure ratio X, that is, the turbine inlet pressure divided by ambient pressure. Assuming that T_{01} is the ambient temperature, T_{02} (the compressor discharge temperature, T_{CD}), can be written as

$$T_{02} = T_{01} + \frac{T_{01}}{\eta_C}(X - 1) \tag{3.6$''$}$$

Using turbine isentropic efficiency instead, (3.1) can be similarly worked out as shown above [16]. A detailed derivation of both equations is shown in Ref. 3.

The power in kilowatts delivered by the high- and low-pressure steam turbine sections may be computed as [2]

$$kWg = \frac{m_{HP}E_{HP} + m_{LP}E_{LP}}{3413} \tag{3.7}$$

where E_{HP} and E_{LP} are the steam actual available energies in the HP and LP turbine sections [11].

The steam flows, m_{HP} and m_{LP}, are computed by [2]

$$m_{HP} = K_T \cdot P_{HP}$$
$$m_{HP} + m_{LP} = K' \cdot P_{LP}$$

where K_T is the throttle value flow coefficient, K' is the admission point flow coefficient, P_{HP} is the HP steam turbine section pressure, and P_{LP} is the LP steam turbine section pressure.

In the absence of information required by the models shown in Figures 3.31 and 3.32, the simplified steam power model shown in Figure 3.33 can be used with good accuracy for many types of studies [11]. The values of the time constants are $T_M = 5$ s; $T_B = 20$ s.

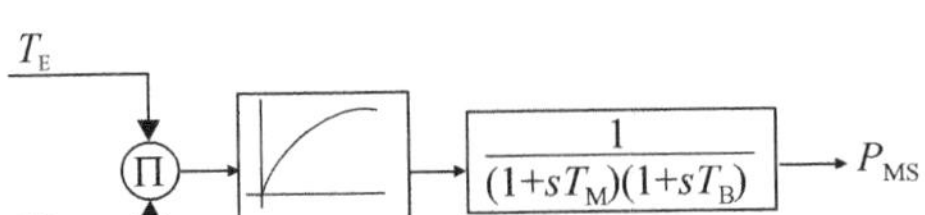

Figure 3.33. A simplified steam turbine power response model [11].

3.5 NUCLEAR POWER PLANTS

An electrical nuclear power plant (NPP) consists of two major parts (Figure 3.34):

- The nuclear island (the reactor), which converts the nuclear energy into thermal energy through fission process;
- The classical part of the power plant, which converts the thermal energy into mechanical energy.

The main characteristic elements of a nuclear reactor are as follows:

- The *nuclear fuel*: natural uranium, enriched uranium, or plutonium.
- The *cooling agent*, which carries away the heat generated through fission in the active zone of the reactor and transfers it to the classical part of the power plant. Gases (He, CO_2), light water, heavy water, and liquid metals (sodium, potassium) can be used as cooling agents.
- The *moderator*, which slows down the neutrons resulting from the nuclear fission to a level that is favorable for new fission reactions. In many cases, the *moderator* acts as cooling agent. Light water (H_2O), heavy water (D_2O), and graphite can be used as moderator.

Nuclear power plants can also be classified in terms of the number of circuits:

- NPP with one circuit (Figure 3.35a), in which case the cooling agent of the reactor is also used as working agent on the classical part of the power plant. Most of the reactors can be included in this category and use light water as moderator and cooling agent, in combination with enriched uranium as nuclear fuel (boiling water reactor, BWR).
- NPP with two circuits (Figure 3.35b), where the primary circuit is designed for reactor cooling and the secondary circuit carries the working agent in the classical part of the plant. This category includes the pressurized water reactor (PWR) and the pressurized heavy water reactor (PHWR).

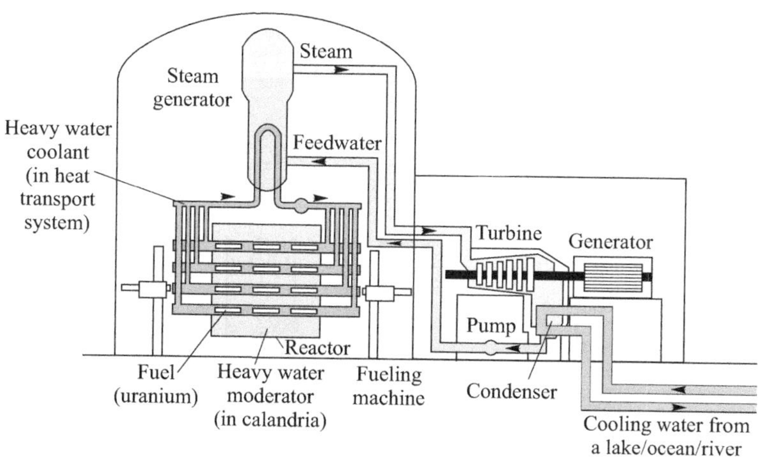

Figure 3.34. General structure of a nuclear power plant.

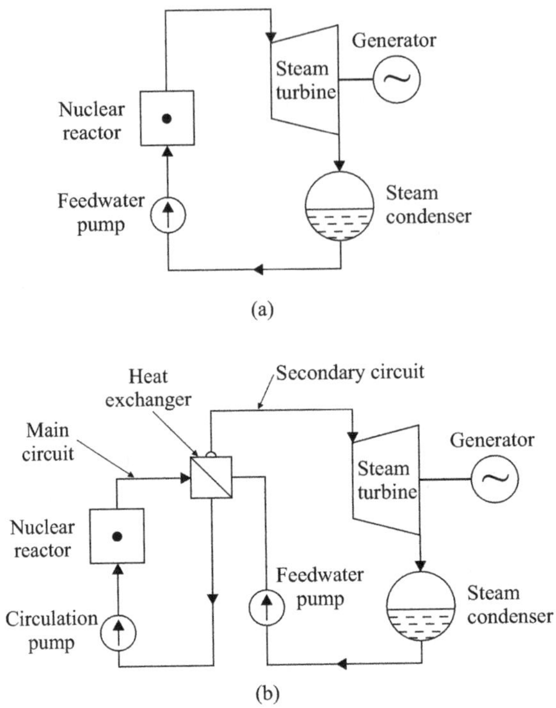

Figure 3.35. Schematic diagram of nuclear power plants: (a) with one circuit (BWR); (b) with two circuits (PWR, PHWR).

Nuclear units usually have tandem-compound turbines and generators with two pairs of poles running at 1500 rot/min (at 50 Hz) or 1800 rot/min (at 60 Hz). A double flux turbine configuration, specific for a CANDU type nuclear power plant, is shown in Figure 3.36. It consists of one HP section and three LP sections. A moisture-separator-reheater (MSR) is used to dry the exhaust steam leaving the HP turbine and to reheat the dry steam before it enters the LP turbines. The steam is reheated using high-pressure steam.

As shown in Figure 3.36, the nuclear power plant steam turbines may be equipped with four sets of valves: main inlet stop valve (MSV), control valve, intermediate stop valves (ISVs), and intercept valve. The MSV and ISV are used only in emergency conditions. The CV is regulated during normal operating conditions to respond to normal load changes. The CV and IV react to overspeed caused by sudden loss of electrical load.

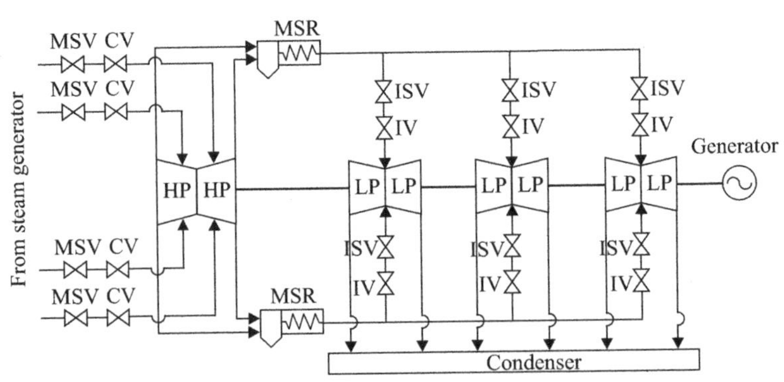

Figure 3.36. An example of nuclear unit turbine configuration.

3.6 HYDRAULIC POWER PLANTS

3.6.1 Generalities

The hydraulic power plant consists of four basic elements that are necessary to generate power from water: a means of creating head, a conduit to convey water, a hydraulic turbine, and an electric generator [17].

The *dam* creates the operating head necessary to move the turbines, establishes the amount of water storage available power production, and impounds the water supply necessary for daily or seasonal stream flow release pattern.

Water is the medium through which energy is delivered to a hydroelectric turbine for the purpose of generating electric power. There are constrains placed on water levels or pressures that require certain elements of the water-handling system to be included within the boundaries of the "controlled process" for the turbine governing [17].

Figure 3.37 is a schematic diagram of the system that withdraws water from the intake reservoir (or from the river, in the case of run-on-river hydraulic plants), and supplies it to the power plant.

A brief description of the main elements of the water system for a hydroelectric generating unit is presented below [17]:

- The *head pond* is the water impoundment used as an energy source for the hydroelectric unit. The size of the head pond storage (very large or no storage

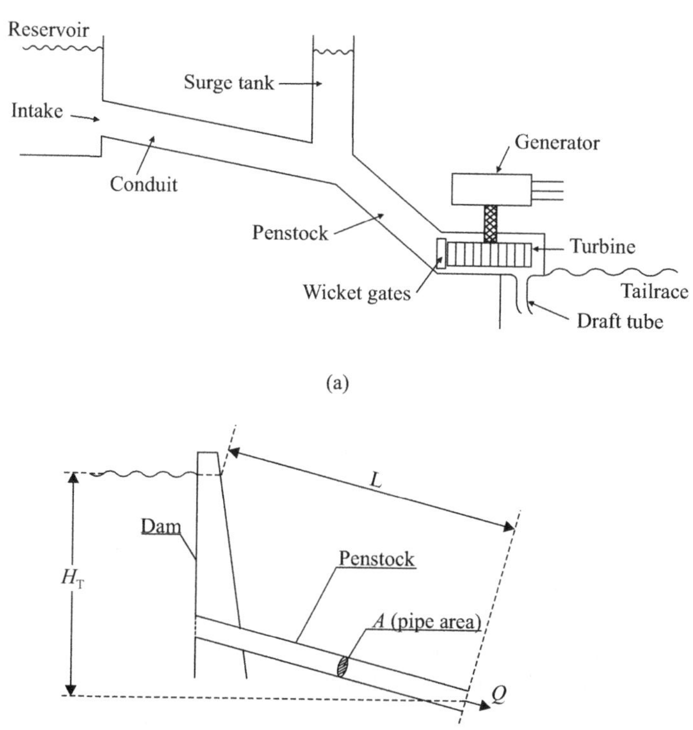

Figure 3.37. Schematic diagram of the water supply system of a hydraulic plant (a) and penstock (b).

capacity) affects the rate at which the hydroelectric generating unit affects the head pond water level. The head pond may also be called the reservoir, forebay, headrace, or head water. From the reservoir, water is drawn from an area called the forebay into a large pipe, and flows to the turbine. In some cases, a relatively level section of pipe, called the *conduit*, is necessary to move the water to a point where it begins a steep descent through the penstock to the turbine.

- The *water column* comprises all of the structures used to convey water from the head pond to the turbine. The water column may include an intake structure, a penstock, one or more surge tanks, and a spiral case. The composite water column inertias and elasticity of these structures contribute to the water hammer effect that impacts the performance of the turbine governing system.
- The *draft tube* conveys the water from the discharge side of the turbine to the tailrace. It is normally a part of the powerhouse structure and it is designed to minimize exit losses. The inertia of water in the draft tube also contributes to the total water inertia that impacts the performance of the turbine governing system. In some applications this effect is significant (tailrace and lower reservoir).
- The *tail pond* can be an open stream, the reservoir of a downstream project, a canal, or a tunnel exiting from an underground powerhouse. The level of the tail pond exerts back pressure on the turbine. This affects the power output of the turbine. The size of the tail pond affects the rate at which the hydroelectric generating unit can affect the water level. Impulse turbines usually rotate in open air, so tail pond level does not affect the output of the generating unit. The tail pond may also be called tailrace or lower reservoir.
- The *turbine* converts the potential energy of water into mechanical energy, which in turn drives the generator. Under pressure water enters the turbine through the wicket gates and is discharged through the draft tube after its energy is extracted. The amount of power the turbine is able to produce depends on the head on the turbine, the flow rate of water passing through the unit, and the efficiency of the turbine. Modern turbines can develop power from almost any combination of head and flow. Although there are many types of turbines, they fit into two basic categories: impulse turbines and reaction turbines.

The performance of a hydraulic turbine is influenced by the characteristics of the water column feeding the turbine; these include the effects of water inertia, water compressibility, and pipe wall elasticity in the penstock. The effect of water inertia is to cause changes in the turbine flow to lag behind changes in turbine gate opening. The effect of elasticity is to cause traveling waves of pressure and flow in the pipe; this phenomenon is commonly referred to as water hammer [7].

Water hammer is defined as the change in pressure, above or below normal value, caused by sudden changes in the rate of water flow. This causes a pressure wave to travel along the penstock, possibly subjecting the pipe walls to great stresses. A device often used to relieve the problems of both positive and negative water hammer is the surge tank, a large tank usually located between the conduit and penstock [2].

The mechanical power depends on the hydraulic power entering the turbine and can be varied by acting on the *water admission valve* (wicket gates) of the turbine. The admission valve is placed in the terminal part of the penstock. In effect, the water flow into the turbine can be controlled by adjusting the opening of the valve.

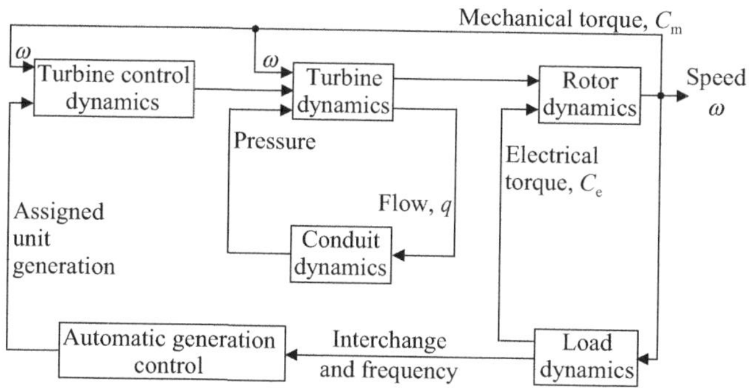

Figure 3.38. Functional block diagram showing relationship of hydro prime mover and controls to complete system. (Reprinted with permission from Ref. 18. © 1992 IEEE.)

3.6.2 Modeling of Hydro Prime Mover Systems and Controls

3.6.2.1 General Block Diagram. The dynamic performance of a hydro turbine is determined by the dynamics of water flow in the conduit and penstock. The mechanical power developed by the turbine is described by a nondynamic relationship in terms of the flow rate and head. In some dynamic performance simulations, the effect of the traveling wave phenomena on the water pressure and flow in the penstock needs to be considered. However, consideration of traveling wave phenomena for stability studies is not necessary [6]. A general block diagram representing the relationship of the hydro prime mover and the plant controls is shown in Figure 3.38 [18].

3.6.2.2 Modeling of Turbine Conduit Dynamics[1]. *Nonlinear Model Assuming a Nonelastic Water Column.* The block diagram for dynamic simulations of a hydraulic turbine with penstock, assuming unrestricted head and tailrace, is shown in Figure 3.39. The model allows also considering a surge tank of any dimension. The penstock is modeled assuming an incompressible fluid and a rigid conduit of length L and cross-section A (Figure 3.37b) [18].

The dynamics of the mechanical power developed by the turbine is given by the rate of change of water flow in the conduit, expressed as [18]

$$\frac{\mathrm{d}q}{\mathrm{d}t} = (h_0 - h - h_l)g\frac{A}{L} \tag{3.8}$$

where q is the turbine flow rate, in m^3/s; A is the penstock cross-section area, in m^2; L is the penstock length, in m; g is the gravitational acceleration, in m/s^2; h_0 is the static head of water column above the turbine, in m; h is the head of turbine admission, in m; and h_l is the head loss due to friction in the conduit, in m.

Penstock head losses h_l are proportional to the flow squared, and f_P is a head loss coefficient that is usually ignored.

Taking h_0 as the base head h_{base} and defining by q_{base} the turbine flow rate with gates fully open (gate position $G = 1$), expression of equation (3.8) in per unit gives

[1] Portions of the development here are from Ref. 18.

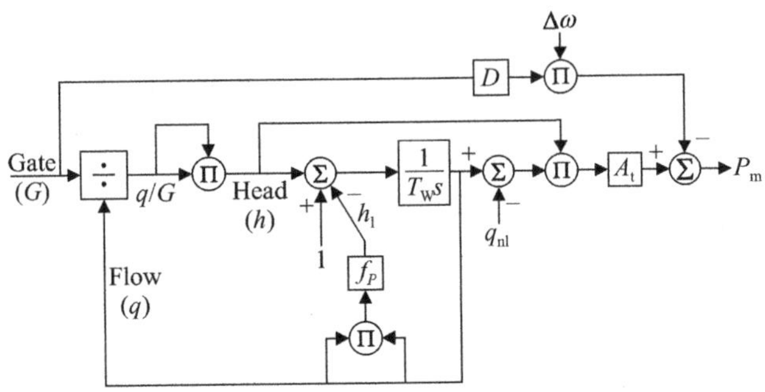

Figure 3.39. Nonlinear model of turbine and nonelastic water column. (Reprinted with permission from Ref. 18. © 1992 IEEE.)

$$\frac{dq_*}{dt} = \frac{1 - h_* - h_{l*}}{T_W} \tag{3.9}$$

where h_* and h_{l*} are the per unit values of the head at the turbine and head loss, respectively. The term T_W is a water time constant in the penstock, and is given by

$$T_W = \left(\frac{L}{A}\right)\frac{q_{base}}{h_{base}\cdot g}\ (s) \tag{3.10}$$

The base flow is a function of the base head and base gate position, $q = f(\text{gate}, \text{head})$. In per unit, the flow rate through the turbine is

$$q_* = G\sqrt{h_*} \tag{3.11}$$

The mechanical power developed by an ideal turbine is equal to the product of the flow rate and the head, multiplied by appropriate conversion factors. The real turbines are not 100% efficient; therefore, the no-load flow rate, accounting for turbine fixed power losses, is subtracted from the actual flow rate. The speed deviation damping effect, which is a function of gate opening, affects also the mechanical power.

Thus, the per unit turbine power P_m, on generator MVA base, is expressed as

$$P_m = A_t h_*(q_* - q_{nl*}) - G\cdot D\cdot \Delta\omega \tag{3.12}$$

where q_{nl*} is the per unit no-load flow and A_t is a proportionality factor.

The factor A_t is assumed constant and is calculated using turbine MW rating and generator MVA rating:

$$A_t = \frac{\text{Turbine MW rating}}{(\text{Generator MVA rating})h_{r*}(q_{r*} - q_{nl*})} \tag{3.13}$$

where h_{r*} is the per unit head at rated flow and q_{r*} is the per unit flow at rated load.

In some stability programs, the parameter A_t, also called turbine gain, is used to convert the actual gate position to the effective gate position, that is, $A_t = 1/(g_{FL} - g_{NL})$ as described in Ref. 7.

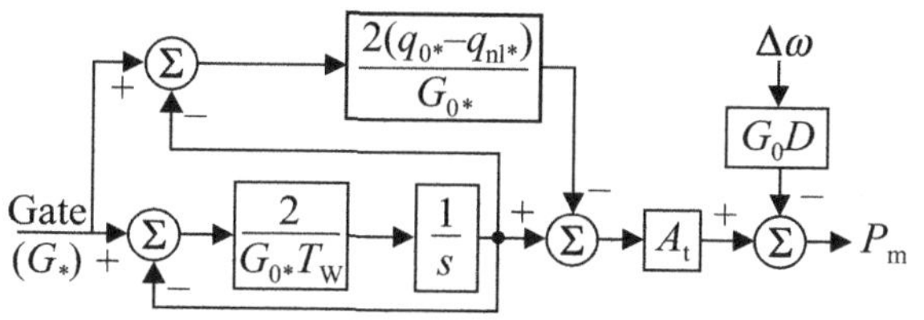

Figure 3.40. Linearized model of turbine with nonelastic water column. (Reprinted with permission from Ref. 18. © 1992 IEEE.)

LINEAR MODELS. Neglecting friction losses in the penstock, the simplified version of the block diagram shown in Figure 3.39 is presented in Figure 3.40 [18].

As can be inferred from Figure 3.40, the change in mechanical power output in terms of the gate position, speed deviation, and constructive parameters, is

$$\Delta P_{\rm m} = \frac{A_{\rm t}[1 - (q_{0*} - q_{\rm nl*})T_{\rm W}s]\Delta G}{1 + ((G_{0*}T_{\rm W})/2)s} - G_{0*}D\,\Delta\omega \tag{3.14}$$

where G_{0*} is the per unit gate opening and q_{0*} is the per unit steady-state flow rate, at operating point. Note that $G_{0*} = q_{0*}$.

Neglecting the damping $D = 0$, an expression similar to the classical penstock/turbine linear transfer function is obtained:

$$\frac{\Delta P_{\rm m*}}{\Delta G_*} = \frac{1 - G_{0*}T_{\rm W}s}{1 + ((G_{0*}T_{\rm W})/2)s}A_{\rm t} \tag{3.15}$$

where the term $G_{0*}T_{\rm W}$ is an approximation to the effective water time constant for small perturbation around the operating point.

The linear models, shown in Figure 3.40, are useful for control system tuning studies using linear analysis tools (frequency response, eigenvalues analysis, etc.).

OTHER MODELS. The two models described above are not always applicable. Therefore, other models may be used [18]:

(i) *Traveling Wave Models.* The model assuming inelastic water columns is adequate for *short to medium length penstocks*. However, for *long penstocks* the travel time of pressure and flow waves can be significant, and therefore the dynamics is different, due to the *elasticity of the steel in the penstock* and of the *compressibility of water.*

(ii) *Nonlinear Model Including Surge Tank Effects and Nonelastic Columns.* In hydro power plants with long supply conduits, the water hammer effect produces violent pressure oscillations causing damages to the turbine. In order to avoid this problem it is common practice to use a surge tank, located between the conduit and the penstock, but as closed as possible to the turbine. In many cases, the surge tank includes an orifice that dissipates the energy of hydraulic oscillations and produces damping. Dynamic performance simulations over seconds to minutes require consideration of the surge tank effects.

(iii) *Nonlinear Model Including Surge Tank Effects and Elastic Water Column in Penstock.* In the cases where the traveling waves have significant influence on the penstock, the upper penstock or tunnel is considered inelastic because the dynamic effects contributed by that system and surge chamber involve low frequency effects, while the high-frequency response components are contributed by the lower penstock that is subject to abrupt gate or flow area change.

3.6.3 Hydro Turbine Governor Control Systems

Since the driving agent in hydro power plants is the water, changes in the output mechanical/electrical power are done by changing the dynamics/parameters of water through various devices the hydro plant consists of, such as gates, blades, needles, or deflectors. The governor aims to control the rotor speed and therefore the output power based on feedback signal (actual speed) from the prime mover and the set point/reference value. Optional, other feedbacks from other parameters are introduced in the governor. Figure 3.41 illustrates a basic governor control system [17].

The governor controller compares the actual speed (the process output) with the reference speed (set point input). In terms of the resulted speed error, the governor sends the command to the actuator (Section 3.6.3.2), which in turn acts on the control device, for example change the gate opening position.

3.6.3.1 Set Point Controller. The reference point or set point is set according to the desired operating conditions of the hydroelectric generating unit. There are two strategies used to respond to changes in the power system frequency: speed droop or *permanent speed droop* and speed regulation or *power droop*.

In the permanent speed droop strategy, the wicket gate position of the output active power can be used as feedback. If the output power signal is used to develop the permanent speed droop characteristic, the permanent speed droop term is usually called *speed regulation* or *power droop*.

PERMANENT SPEED DROOP. The permanent speed droop characteristic determines the amount of change in the gate position a unit is set to produce in response to a unity change in the unit speed. Therefore, the permanent speed droop is defined as the ratio of the change in unit speed (in % rated speed) to the change in governor output (% gate position). The permanent droop is usually expressed in percents, and thus the ratio is multiplied by 100. Figure 3.42 shows a permanent droop speed loop attached to a typical governor controller.

In the presence of the permanent speed droop feedback loop, the governor acts in accordance with the operating characteristic shown in Figure 3.43, with a slope given by the constant b_p. A typical value for the permanent speed droop is 5%, which means that, when the generator is connected to an isolated load, a 20% change in the load, which needs a 20% change in wicket gate position, results in 1% change in the rotor speed (or 0.5 Hz on a 50 Hz system).

The governor commands the actuator to control the turbine operation based on the set point. By adjusting the set point, the generator can operate at the system frequency for any output power.

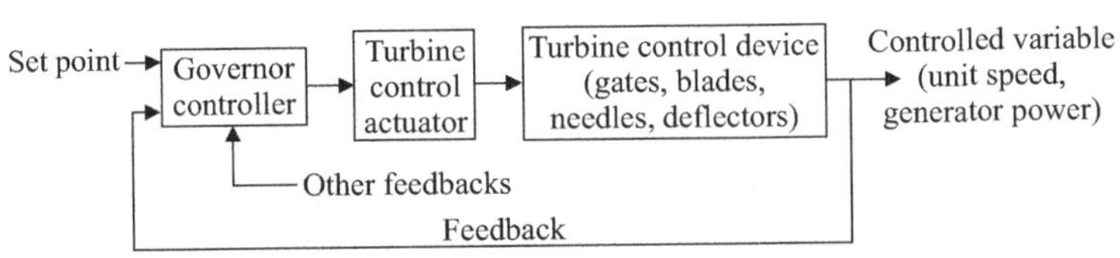

Figure 3.41. Basic governor control system. (Reprinted with permission from Ref. 17. © 2004 IEEE.)

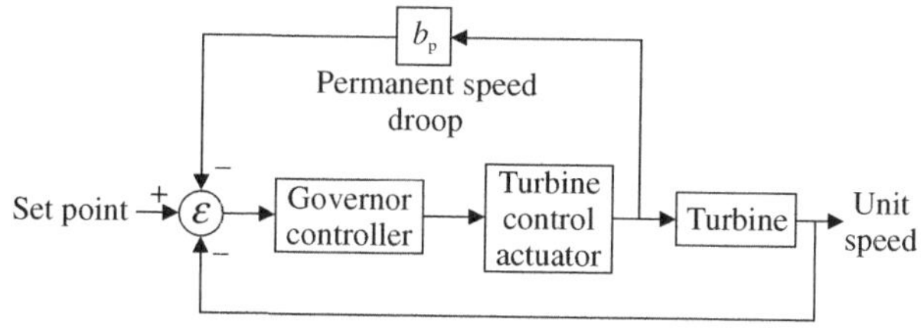

Figure 3.42. Typical governing system with permanent speed droop [17].

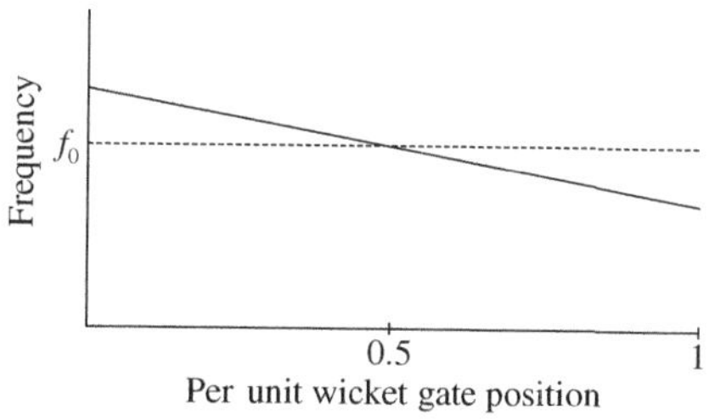

Figure 3.43. Permanent speed droop characteristic.

Speed Regulation. Comparative to the permanent speed droop governing system that uses the gate/actuator position as feedback, the *speed regulation*, also known as *power droop*, governing system uses the output power as an intermediate feedback to perform speed regulation. The error between the output power and the desired generation or set point is multiplied with the constant R_S then is added to the governor controller set point. The block diagram of a typical speed regulation governing system is shown in Figure 3.44.

The speed regulation characteristic is similar to the permanent speed droop characteristic, except that the "per-unit output" is used instead of the "wicket gate position" in the horizontal axis. The slope of the power droop characteristic is determined by the constant R_S.

The speed regulation governing systems are usually employed in large interconnected power systems where the influence of a single unit on the power system frequency is very low. However, a speed regulation-based unit is inherently less stable than a permanent speed droop governor because additional dynamic influences from the water column are included in the primary feedback path (generated power) of the turbine governing system [17].

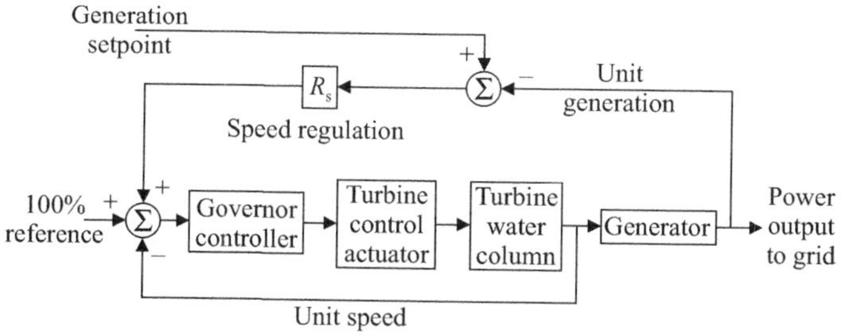

Figure 3.44. Governing system with speed regulation. (Reprinted with permission from Ref. 17.

3.6.3.2 The Actuator. An actuator is a mechanical device used for controlling/ moving a system/ mechanism as a response to the command from a controller (Figure 3.41). In the case of a hydraulic generating unit, the mechanism to be moved would be, for instance, the wicket gates system, and the variable to be controlled is gates opening.

For Kaplan or bulb turbines, which are provided with adjustable blades, separate actuator servomotors are generally used to independently control the wicket gates and the runner blades. In the case of Pelton turbines, one actuator is used for each injector needle to control the water flow to the wheel. Furthermore, a separate actuator servomotor is used to control the deflector mechanism, which acts to deflect the water stream away from the turbine runner to reduce the developed torque of the turbine more rapidly than can be done with the slower moving needle servomotors [17]. A sensor detects the position of each actuator and sends the information to the governor.

There are several types of actuators used to control the turbine operating parameters, which perform the same basic function [17].

(i) *Mechanical Actuators.* A mechanical actuator typically converts the rotary motion, for example of the turbine shaft, into a linear motion, for example the mechanical position. The mechanical actuators are primitive generations of actuators. With the advent of more efficient actuators, the mechanical actuators are now used only as backup in small generation units.

(ii) *Mechanical Hydraulic Actuators (*Figure 3.45*).* The mechanical hydraulic actuator is characterized by a mechanical set point position input that is amplified by a hydraulic amplifier driving the output servomotor [17].

The actuator servomotor sets the position of the turbine gate linkage. The flow of hydraulic oil to the gate servomotor is controlled by the distributing valve and its servomotor rate limiters. The pilot valve and distributing valve provide the necessary amplification of the position error (difference) between the set point input and the servomotor position output.

(iii) *Electromechanical Actuators (*Figure 3.46*).* An electromechanical actuator uses mechanical power developed by an electric motor as its source of energy. It uses a ballscrew or other gear reduction to transfer the mechanical power of the motor to actuate the controlled variable.

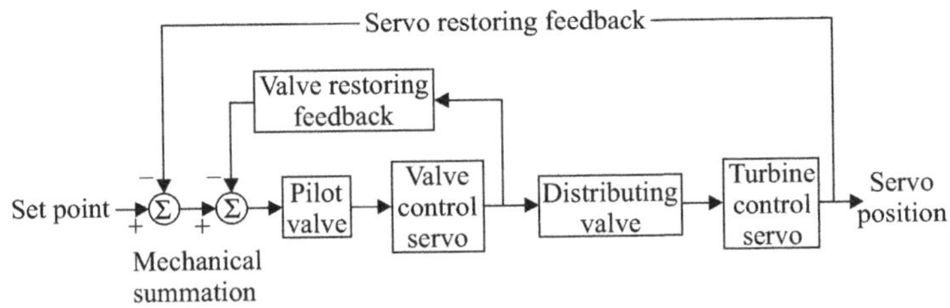

Figure 3.45. Typical mechanical hydraulic actuator. (Reprinted with permission from Ref. 17. © 2004 IEEE.)

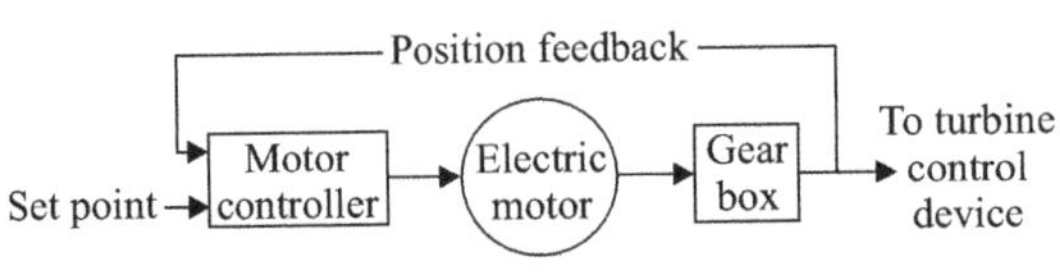

Figure 3.46. Block diagram of an electromechanical actuator [17].

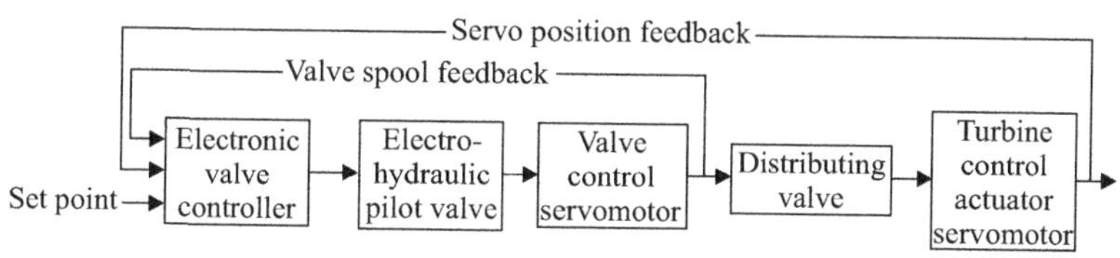

Figure 3.47. Typical electrohydraulic actuator. (Reprinted with permission from Ref. 17. © 2004 IEEE.)

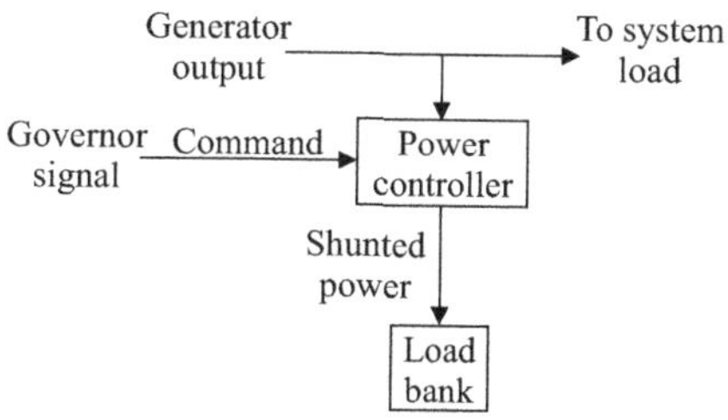

Figure 3.48. Electronic load actuator [17].

(iv) *Electrohydraulic actuators* (Figure 3.47). An electrohydraulic actuator uses an electronic set point signal as its input, and it amplifies this set point with an electrohydraulic amplifier to determine the position of the output hydraulic servomotor.

An electrohydraulic actuator system is typically employed on large turbines where high oil flow rates are required for the gate servomotor. Although Figure 3.47 does not show any hydraulic shutdown valves, such valves are often used to override the main valve output for the purpose of protective shutdown of the unit [17].

A *load "actuator"* is designed to control the electrical load on the generator so the generator rotor is maintained at the desired speed. Figure 3.48 is a block diagram of an electronic load controller.

Typically this type of actuator is used only on small hydroelectric units. The load bank is used to impose an electrical load on the generator, which is reflected as a mechanical load on the turbine. This mechanical loading is used to control the unit's speed. The load bank may be either air-cooled or water-cooled.

REFERENCES

[1] deMello, F.P. (Chairman) Dynamic models for fossil fueled steam units in power system studies, IEEE Working Group on Prime Mover and Energy Supply Models for System Dynamic Performance Studies, *IEEE Trans. Power Syst.*, Vol. 6, No. 2, pp. 753–761, May 1991.

[2] Anderson, P.M., Fouad, A.A. *Power system control and stability*, 2nd edition, IEEE Press Power Engineering Series, John Wiley & Sons, Piscataway, NJ, 2003.

[3] Cohen, H., Rogers, G., Saravanamuttoo, H. *Gas turbine theory*, 4th edition, Addison Wesley Longman, Reading, MA, 1996.

[4] Henry, P. *Turbomachines hydrauliques*, Presses Polytechniques et Universitaire Romandes, Lausanne, 1992.

[5] Marconato, R. *Electric power systems. Vol. 2. Steady-state behavior, controls, short-circuits and protection systems*, CEI-Italian Electrotechnical Committee, 2004

[6] Byerly, R.T. (Chairman) Dynamic models for steam and hydro turbines in power system studies, EEE Committee Report, Task Force on Overall Plant Response, *IEEE Trans.*, Vol. PAS-92, pp. 1904–1915, Nov.–Dec. 1973.

[7] Kundur, P. *Power system stability and control*, McGraw-Hill, Inc., New York, 1994.

[8] Eggenberger, M.A. A simplified analysis of the no-load stability of mechanical-hydraulic speed control systems for steam turbines, AMSE Paper 60-WA-34, Dec. 1960.

[9] Pourbeik, P. Modelling of combined-cycle power plants for power system studies, *IEEE Power Engineering Society General Meeting*, Vol. 3, July 13–17, 2003.

[10] Pourbeik, P. (Convenor) *Modeling of gas turbines and steam turbines in combined-cycle power plants*, CIGRE Task Force 38.02.25, 2003.

[11] deMello, F.P. (Chairman) Dynamic models for combined cycle plants in power system studies. IEEE Working Group on Prime mover and Energy supply models for system dynamic performance studies, *IEEE Trans. Power Syst.*, Vol. 9, No. 3, pp. 1698–1708, Aug. 1994.

[12] Rowen, W.I. Simplified mathematical representation of heavy-duty gas turbines, *Trans. ASME*, Vol. 105, No. 1, pp. 865–869, 1983.

[13] Hajagos, L.M., Bérubé, G.R. Utility experience with gas turbine testing and modeling, *IEEE PES Winter Meeting, Columbus, USA, 21 Jan.–1 Feb.* 2001.

[14] Rowen, W.I. Simplified mathematical representation of a single shaft gas turbines in mechanical drive service, International Gas Turbine and Aeroengine Congress and Exposition, Cologne, Germany, 1992.

[15] Yee, S.K., Milanovič, J.V., Hughes, F.M. Overview and comparative analysis of gas turbine models for power system stability studies, *IEEE Trans. Power Systems*, Vol. 23, No. 1, Feb. 2008.

[16] Cengel, Y.A., Boles, M.A. *Thermodynamics: An engineering approach*, McGraw-Hill, New York, 1994.

[17] Kornegay, D. (Chairman) *IEEE Guide for the application of turbine governing systems for hydroelectric generating units*, IEEE Standard 1207–2004, 2 Nov. 2004.

[18] deMello, F.P. (Chairman) Hydraulic turbine and turbine control models for system dynamic studies, Working Group on Prime Mover and Energy Supply Models for System Dynamic Performance Studies, *IEEE Trans. Power Sys.*, Vol. 7, No. 1, pp. 167–179, Feb. 1992.

[19] Ilić, M., Zaborszky, J. *Dynamics and control of large electric power systems*, John Wiley & Sons, Inc., New York, 2000.

[20] Machowschi, J., Bialek, J., Bumby, J. *Power system dynamics and stability*, John Wiley & Sons, New York, 1997.

[21] Bailie, R.C. *Energy conversion engineering*, Addison-Wesley, Reading, MA, 1978.

4

WIND POWER GENERATION

Mohammad Shahidehpour and Mircea Eremia

4.1 INTRODUCTION

The share of power generation from wind has significantly increased in the last decade, especially in the transmission network. For this reason, accurate modeling of the wind turbine systems is of great importance for power system performance analysis.

The wind energy is converted into electrical energy in two phases (Figure 4.1):

- at the turbine level, which extracts a part of kinetic energy of the wind into mechanical energy;
- at the generator level, where the mechanical energy is converted into electrical energy, then transmitted to the electrical network.

The main components of a modern wind turbine system are the turbine, the nacelle, and the tower. For horizontal axis turbines, a pitch control system and a yaw system are also used for steering in response to change in wind direction.

The *nacelle* accommodates the generator and cooling system, the gearbox, and other electronic control equipments that allow controlling different orientation mechanisms and the overall operation of the wind turbine.

The height of tower is very important since the wind speed increases with the height above earth.

The *gearbox* consists of a slow rotating shaft, which is placed in the hub, and a fast rotating shaft (1000–2000 rpm), which is coupled to the generator rotor shaft. It is

Handbook of Electrical Power System Dynamics: Modeling, Stability, and Control. Edited by Mircea Eremia and Mohammad Shahidehpour.

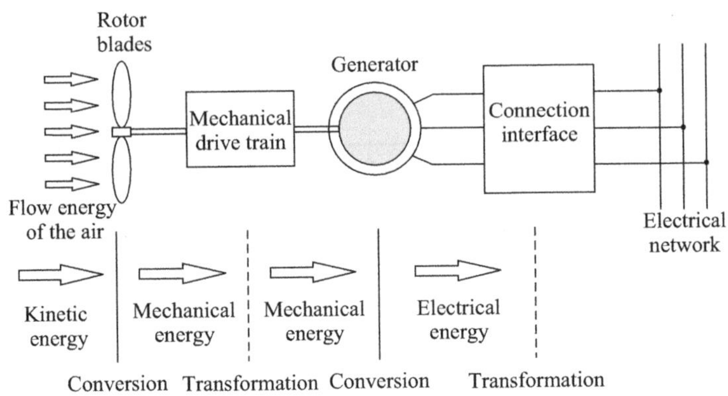

Figure 4.1. Functional chain and conversion stages of a wind energy converter.

also equipped with a mechanical disc brake, which is coupled to the generator, and a cooling system for oil. The gearbox introduces, besides a large weight (several of tons), the most maintenance issues. In general, the wind turbines are designed with three blades that better capture the wind energy and allow its transfer to the slower shaft.

A *pitch control* (*electromechanical*) *system* acts in certain conditions to change the position of the blades with respect to their longitudinal axis and thus to control the mechanical couple and to limit its power. Also, the blades can achieve an aerodynamic brake by "cut-in speed" (perpendicular to wind direction) or only by turning their extremities. The pitch control can be passive or active, as the blades are bolted onto the hub at fixed angle or the blades can be turned out or into the wind to achieve the highest efficiency. A *stall control system* is also used to control the power of the wind turbine by changing the aerodynamic design of the rotor blades. At low wind speeds, the stall control acts similarly to the pitch control, whereas at high speeds, the blades are pitched slightly opposite to that achieved by a pitch control.

The *yaw control system*, which uses electric actuators, allows orientation of the nacelle facing the wind. An anemometer and a wind vane located on the nacelle roof are used to provide the data necessary for the guidance control system to trigger or stop the wind turbine according to wind speed.

Some manufacturers have tried to suppress the gearbox by introducing the "direct attack" system, or to reduce it. This requires a special electrical generator capable of running at the same speed with turbine rotor speed, which means that the generator must be designed with a large number of poles pairs [1].

Figure 4.2 shows the arrangement of the components in the nacelle for two types of wind turbines. Figure 4.2a shows the conventional drive train design in the form of a geared transmission with a high-speed generator. Figure 4.2b, by contrast, shows the gearless variant with the generator being driven directly from the turbine.

A wind turbine can be equipped with an induction generator or a synchronous generator. In terms of the rotational speed, in general, wind turbine systems can be classified into two types: fixed speed and variable speed. The largest machines tend to operate at variable speed, whereas smaller and simpler turbines are of fixed speed.

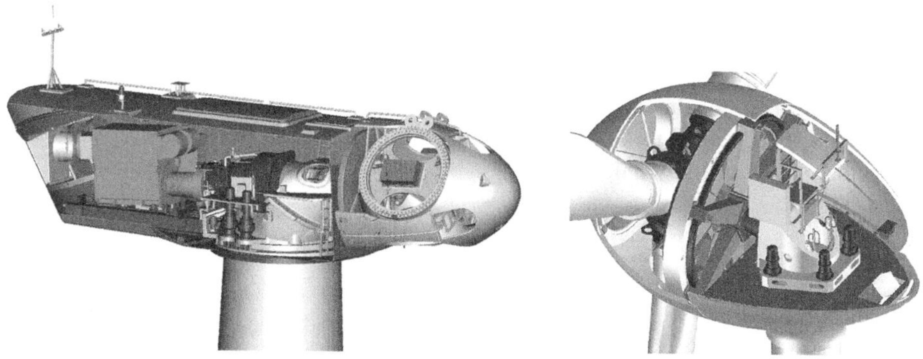

(a) (b)

Figure 4.2. Nacelle of a wind turbine with (a) gearbox and high speed (Courtesy of General Electric) [2] and (b) a gearless wind turbine (Courtesy of Enercon).

4.2 SOME CHARACTERISTICS OF WIND POWER GENERATION

The Aerodynamic Profile of Wind Turbine's Blades. Figure 4.3 shows the general aspect of a horizontal axis wind turbine as well as the blades aerodynamic profile.

Figure 4.3b and c shows the cross section of a rotor blade and the forces that act on the blade. In terms of the wind direction, the wind turbine rotor is turned round propelled by drag or lift forces. The lift force (L) is perpendicular to the direction of the relative wind velocity (V_{rel}), while the drag force (D) acts in the direction of it. The relative wind velocity is composed of the blade motion (V_{Blade}) and the wind velocity (V_{W}).

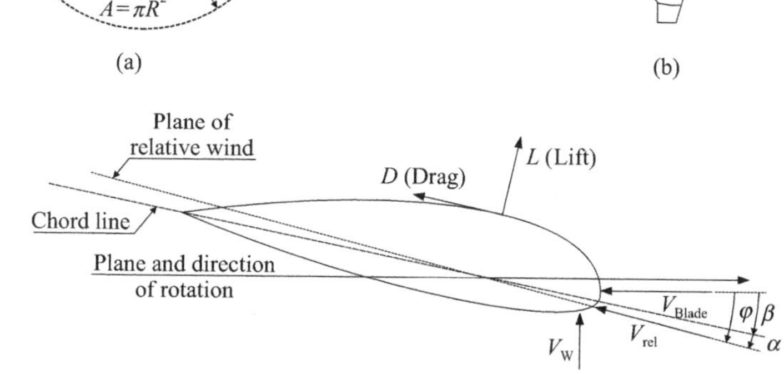

Figure 4.3. General view and cross section of a wind turbine blade with relative wind comprising the rotating wind (V_{Blade}) and ambient wind (V_{W}).

The wind forces acting on a blade section depend also on the angle of incidence φ between the plane of rotation of the rotor blades and the plane of relative wind (Figure 4.3c). The angle of incidence φ is determined by the ambient wind speed V_W and the speed of the blade $V_{Blade}(= -\omega_t R)$.

If the tip–speed ratio λ is defined as the ratio of the rotational speed of the tip of a blade to the actual speed of the wind, that is

$$\lambda = \frac{\omega_t R}{V_W} \tag{4.1}$$

the angle of incidence can be calculated as

$$\varphi = \text{atan}\left(\frac{V_W}{\omega_t R}\right) = \text{atan}\left(\frac{1}{\lambda}\right)$$

where ω_t is the mechanical angular speed of the turbine rotor and R is the wind turbine rotor radius.

The modern wind turbines are provided with control possibilities of the pitch angle β through a servomechanism. By turning the blade, the angle of attack α between the chord line of the blade and the plane of relative wind (V_{rel}) will change accordingly [3]. Older and simpler wind turbines have constant blade angle β, which is called stall (or passive stall) control.

MECHANICAL POWER OF WIND TURBINE. The power of an air mass that flows perpendicularly on an area A at the speed V_W is given by the expression [3]:

$$P_W = \frac{1}{2}\rho A V_W^3 \quad \text{(W)} \tag{4.2}$$

where ρ is the air density (kg/m^3), $A = \pi R^2$ is the area swept by the rotor blades (m^2), V_W is the average wind speed passing through the A surface (m/s), and R is the wind turbine rotor radius (m).

Integrated in time, the total available wind energy is obtained. The power in wind energy is converted into mechanical–rotational energy by means of the wind turbine rotor.

However, the wind energy cannot be entirely extracted by the wind turbine. The power extracted from wind by the turbine is determined by multiplying the available wind power P_W with an efficiency coefficient C_p that is,

$$P_T = C_p(\lambda, \beta)P_W = C_p(\lambda, \beta)\frac{1}{2}\rho A V_W^3 \tag{4.3}$$

where C_p is the performance coefficient, which depends on the aerodynamic properties of the blades, λ is the tip–speed ratio, and β is the blade pitch angle.

According to Betz's law [4], the theoretical maximum power that can be extracted from wind is only 59.3% of the available power, that is,

$$P_{Betz} = C_{pBetz}P_W = 0.59 \times \frac{1}{2}\rho A V_W^3 \tag{4.4}$$

In practical designs, modern turbines can achieve efficiencies below 40%. The efficiency of the turbine is a function of the wind speed, and therefore efficiency may vary from case to case.

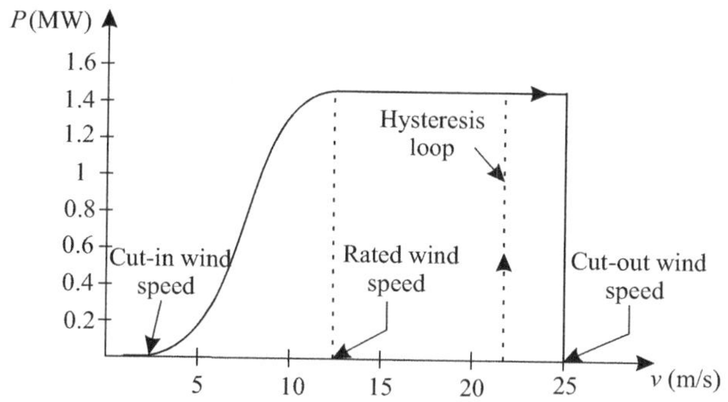

Figure 4.4. Power curve of a 1.5 MW ($d = 82$ m) wind turbine with active pitch control [3,5].

If the rotor of the wind turbine rotates too slowly, the energy captured by the turbine is too low, whereas in case of a very high velocity, the surface formed by the blades will appear as a wall that may be a danger for the wind turbine.

Assuming a wind turbine with the blades diameter $d = 82$ m, the maximum power that can be extracted by the wind turbine at the wind speed $V_W = 12$ m/s and a performance coefficient $C_p = 0.4$ is

$$P_T = C_p \times \frac{1}{2}\rho A V_W^3 = 0.4 \cdot \frac{1}{2} \cdot 1.225 \cdot \pi \cdot 41^2 \cdot 12^3 = 2.236\,\text{MW}$$

Power Curve. The power curve relates the power developed by the wind turbine and is represented for wind speed values between the cut-in speed and the cut-out speed (Figure 4.4). At wind speeds lower than the cut-in speed value and higher than the cut-out speed value, the turbine does not operate.

The rated capacity is reached at wind speed values between 12 and 16 m/s, depending on the individual wind turbine design. The power produced by the wind turbine above the rated wind speed is limited to the rated capacity, and therefore only one part of the available wind power is used. This can be achieved by pitch control or by stall control.

The power curve is sensitive to the air pressure (which vary with the height above sea), changes in the aerodynamics of the rotor blades (which can be affected by dirt or ice), shadowing or wake effect, and so on.

The wind turbine shuts down if the wind speed exceeds the cut-out speed, that is, 20–25 m/s. However, the wind speed may drop below then may exceed again the cut-out speed. For this reason there is a delay, referred to as *the hysteresis loop* (Figure 4.4), after which the wind turbine starts operating again. Usually, the wind turbine restarts if there is a drop in wind speed of 3–4 m/s below the cut-out speed.

For the power system, shutting down a large amount of power generation may create significant problems. To avoid this, the wind turbine is provided with step-by-step power reduction at increased wind speeds instead of sudden cut-out.

Performance Coefficient. The performance coefficient, denoted by C_p, shows the efficiency of power extraction by the wind turbine from the wind. It varies with the wind speed, the rotational speed of the wind turbine, and turbine blade parameters such as angle of attack and pitch angle. Usually, it is represented as a function $C_p(\lambda, \beta)$ in terms of the tip–speed ratio λ and blade pitch angle β. Figure 4.5 shows the performance coefficient

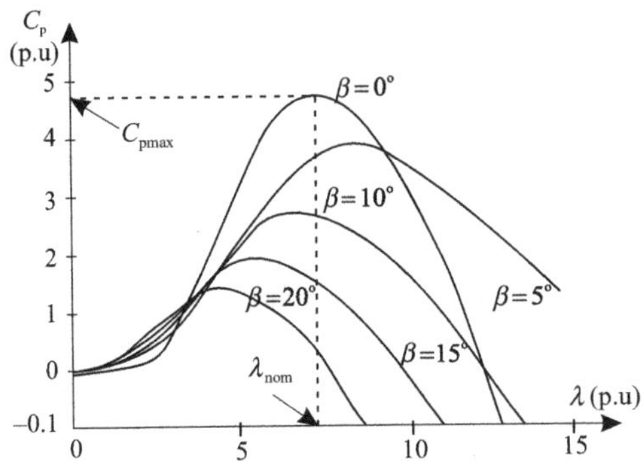

Figure 4.5. $C_p(\lambda, \beta)$ characteristics with control of the blade pitch angle [1].

characteristics for various blade pitch angles and tip–speed ratios. It has a peak at a given value of λ and drops off to zero at higher tip–speed ratios.

The wind turbines are designed/chosen in terms of the average wind speeds. A good wind site may have average wind speeds ranging around 7–10 m/s, and therefore the wind turbines should be designed to extract the maximum amount of wind energy possible at wind speeds between 10 and 15 m/s. Higher speeds have a low frequency of occurrence, and therefore using oversized turbines might be economically inefficient. Furthermore, higher speeds add significant stress on the turbine. As shown in Figure 4.4, the wind turbine operates at constant power if the wind speed exceeds the rated wind speed and keeps operating until the cut-out speed is reached. Therefore, the turbine needs to be provided with a control mechanism to regulate the generated power when the wind speed increases above the rated wind speed. This control can be achieved in two ways [3]:

- Fixed-speed designs that typically use stall.
- Variable-speed designs that typically use dynamic blade pitch control.

Capacity Factor. Directly related to the wind profile at the site where the turbines are located, one way to evaluate the economic efficiency of an individual wind turbine or of a wind power plant is by calculating the capacity factor:

$$\text{Capacity factor} = \frac{\text{Actual annual energy produced}}{\text{Annual energy that could be produced at full capacity}}$$

Investment in wind generation is typically reasonable economic for capacity factors above 0.25, while values above 0.3 are considered of high economic efficiency. On the other hand, offshore sites have higher economic efficiency compared to onshore sites, having capacity factors ranging from 0.35 to 0.45.

4.3 STATE OF THE ART TECHNOLOGIES

4.3.1 Overview of Generator Concepts

There are several types of three-phase generators that can be used in wind turbines, as shown in the classification presented below [3].

Induction (asynchronous) generator			Synchronous generator	
Squirrel cage induction generator (SCIG)	Wound rotor induction generator (WRIG)		Permanent magnet synchronous generator (PMSG)	Wound rotor synchronous generator (WRSG)
	Dynamic slip-controlled induction generator (DSIG)	Doubly fed induction generator (DFIG)		

The wind turbine generators (WTGs) can also be classified into two categories in terms of the rotational speed, that is, fixed-speed (FWTGs) and variable-speed (VWTGs). Fixed-speed wind turbine generators have simple design and are directly connected to the power grid, whereas the variable-speed wind turbine generators incorporate power electronic-based converters as interface for connection to the power grid.

4.3.1.1 General Description. *Induction Generator.* The induction generator is the most common type of electrical machine used in wind turbine systems because it is simple, reliable, lightweight, and cheaper. The main disadvantage is that it requires a reactive magnetizing current to produce the rotating magnetic field, which can be established once the generator is connected to the main power grid. Once connected, the generator can be self-excited. It is preferred that the reactive power be provided by an external source, such as a capacitor bank, and not from the main power grid.

The amount of active power produced by the induction generator is proportional to the slip, that is, the difference between rotor angular speed and the stator angular speed of the generator.

The rotor speed depends on the torque developed by the turbine. When the wind speed is very low, below the cut-in speed, the turbine cannot develop enough torque and the generator behaves as a motor absorbing current from the grid. To avoid this situation, the wind turbine is disconnected from the grid.

The rotor of an induction machine may be one of two types: the squirrel cage and the wound rotor.

Squirrel Cage Rotor. The rotor of a squirrel cage machine is designed with a series of bars placed in slots near the rotor surface, which are short-circuited by end rings at each end of the rotor (see Figure 2.69b).

The SCIG wind turbine is directly connected to the power grid and the influence between active power, reactive power, terminal voltage, and rotor speed follows a strict relationship. The generator is supposed to operate at fixed speed since the maximum variation in the rotor speed is around 2%. Furthermore, as the wind turbine produces more active power, the generator absorbs more reactive power from the external source. Because the wind speed varies continuously, the reactive power compensation must be performed dynamically.

Cage generators are remarkable for their extremely simple layout that permits robust construction and operational reliability even in the event of rough handling.

Wound Rotor. The wound rotor machine consists of a rotor core designed with three-phase windings instead of bars, but with the same number of poles as the stator. The advantage of using windings instead of bars is that the wires can be brought out and connected externally through slip rings and brushes or by means of power electronic converter (which may or may not require slip rings and brushes), so that the current through the windings can be controlled.

By using a power electronic converter, the power can be extracted from or injected to the rotor circuit and the induction generator can be magnetized from either the stator circuit or the rotor circuit. It is also possible to recover slip energy from the rotor circuit and feed it into the output of the stator. The most common configurations of wound rotor generators are the DSIG and the DFIG [3].

- The *dynamic slip-controlled induction generator* is a special configuration of a wound rotor induction generator in which a variable external resistance is attached to the rotor winding. This configuration allows the generator to have a variable slip in order to reduce the fluctuations in the torque and power output, especially during gusts. The slip is simply varied by changing the total rotor resistance using a power electronic converter. One type of dynamic slip-controlled induction generator is the OptiSlip design, a Vestas registered mark, where the external resistor and the switches are mounted on the rotor shaft and the control signal is transmitted via optical fiber.
- *The doubly fed induction generator* is another configuration of a wound rotor induction generator in which the rotor is separately controlled. The stator windings are directly connected to the power grid, and the rotor windings are separately connected to the grid through a bidirectional back-to-back IGBT-based voltage source converter (VSC).

Reactive Power Compensation. Either configuration of an induction generator needs a reactive magnetizing current to build up the magnetic field, although reactive power does not contribute to direct energy conversion. The higher the reactive current content in the overall current, the lower is the power factor. An induction machine is not "excited" like a synchronous machine, and therefore it takes the reactive power from the grid. Figure 4.6a shows the variation of the reactive power absorbed by the induction generator from

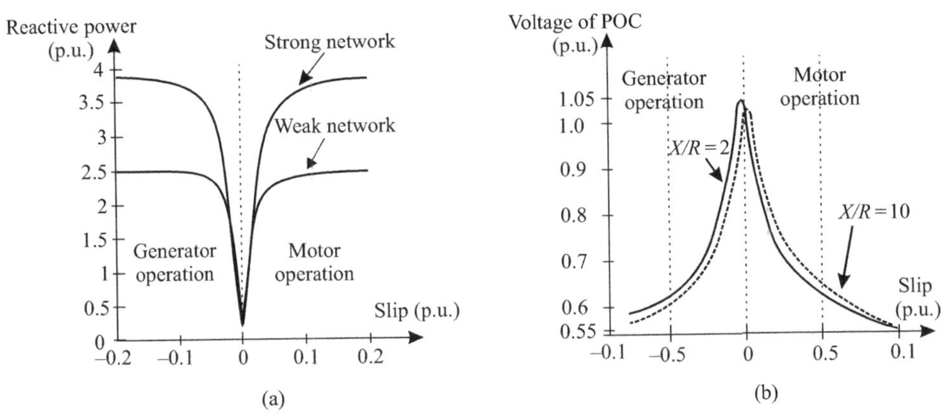

Figure 4.6. Reactive power drawn by an inductive generator and voltage at the point of connection: (a) reactive power variation with slip; (b) voltage at the point of connection. (Reprinted with permission from Ref. 5.)

the grid, and Figure 4.6b shows the voltage at the point of connection (POC) to the main grid [5].

Two cases are considered in Figure 4.6: a strong network (short-circuit power of 3600 MVA) and a weak network (short-circuit power of 360 MVA) through a transmission line having a X/R ratio of 10. It can be seen that as the slip or the active power generation increases, the amount of reactive power absorbed by the generator also increases. A large amount of reactive power absorbed from the grid results in significant voltage drop on the transmission line. The voltage at the point of connection with the network decreases as the slip increases (Figure 4.6b).

Synchronous Generator. There are two types of synchronous generators commonly used by manufacturers in the modern wind turbines [3]:

- The wound rotor synchronous generator.
- The permanent magnet synchronous generator.

The synchronous generator does not need a reactive magnetizing current since the magnetic field can be created by permanent magnets or a conventional field winding. Furthermore, if the generator is designed with a greater number of poles, the gearbox can be removed, case in which the generator operates at the same rotational speed as the turbine. In order to achieve a full control, the synchronous machine-based wind turbines are connected to the grid through a power electronic converter.

Today, the synchronous generators used for wind turbines lie in the range between a few kW to 7 MW. Conventionally constructed synchronous machines of this size use salient-pole rotors. The rotor comprises the pole shoes, the poles lying beneath, and the exciter windings. The stator consists of the stator core and AC windings.

The simplified configuration of a brushless generator is shown in Figure 4.7 [6]. The rotor is supplied by the exciter, which consists of exciter poles in the stator, rotary field windings, and a rectifier bridge at the end of the shaft. The exciter or the exciter coils in the stator are supplied from the pilot exciter. Rotating an outer permanent magnet generates current in the pilot exciter coils, which is fed via a voltage controller to the main exciter

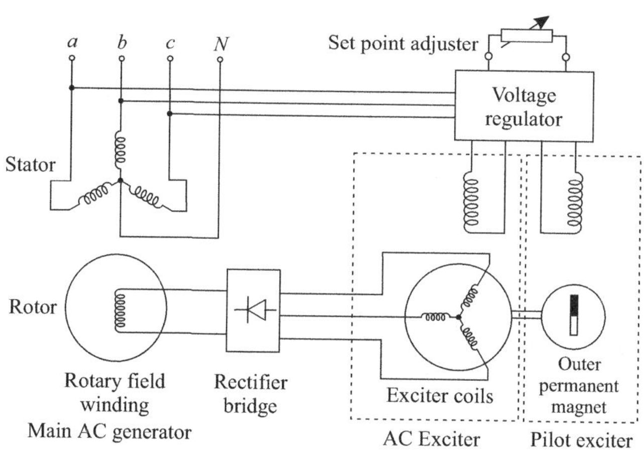

Figure 4.7. Synchronous generator structure of a self-regulating brushless model (Courtesy of AvK).

coils. In this way, the necessary magnetic field is set up in the poles of the exciter and crosses the air gap to the AC windings of the exciter [6].

This type of construction is preferred for grid-independent power supplies. If a grid is available to power the exciter windings, the pilot exciter becomes superfluous.

In brushless machines, the exciter current is fed from the stator windings, across the air gap of the exciter, and via the rotating rectifier bridge to the rotor. Its rotating magnetic field—rotary field—is transferred to the stator.

The much simpler construction of the machine and the gain in dynamic characteristics must, however, be set against the severe disadvantage of feeding electricity through brushes and slip rings. Higher frictional losses, brush and slip-ring erosion, and higher maintenance costs are the consequences.

4.3.1.2 Squirrel Cage Induction Generator. The simplest electrical topology of a wind turbine system incorporates a fixed-speed wind turbine with a squirrel cage induction generator. The basic configuration of the wind turbine system is shown in Figure 4.8 [7].

The main components of a SCIG wind turbine system are turbine aerodynamics, blade control system, mechanical drive train, induction generator, reactive power compensation device, coupling transformer, protection (especially under voltage protection).

This type of wind turbine is directly connected to the electrical network. For this reason it is simple and cheap. Furthermore, no synchronization device is required. However, the wind turbine has to operate at constant speed as the frequency of the grid determines the speed of the generator rotor. This requires a special mechanical structure capable to absorb the high mechanical stress caused by wind gusts.

Another disadvantage of the induction generators is the high starting current due to sudden magnetization in the instant of coupling to the network and the need for reactive power. Connection of the induction generator to the grid generates an in-rush current, which can be up to five to eight times the rated current, causing severe voltage disturbance on the power system. This current is limited by use of a soft-starter, which consists of two thyristors, connected antiparallel, as commutation devices in each phase. Connection of the generator to the electrical network is assisted by adjusting the firing angle of the thyristors in a predefined number of grid periods. Once the wind turbine is cut-in, the thyristors are short-circuited by switches. The reactive power required by the generator for maintaining the magnetization is provided by capacitor banks.

The main characteristics of a SCIG wind turbine are shown in Table 4.1 [7].

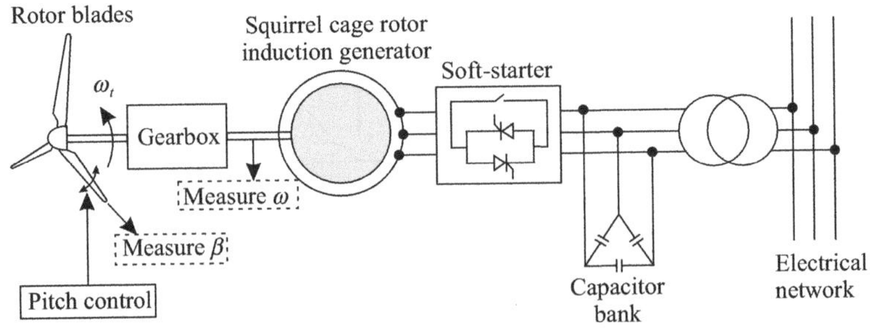

Figure 4.8. The basic configuration of a SCIG wind turbine [7].

TABLE 4.1. Main advantages and disadvantages of SCIG

Advantages	Disadvantages
• Robust standard generator • Relatively low cost • No power electronics	• The power produced is not optimized • Requires maintenance of gearbox • It cannot control the reactive power • The magnetization of the generator is made from the network

4.3.1.3 Dynamic Slip-Controlled Wound Rotor Induction Generator. In order to improve the torque-slip controllability of the induction generator, a variable resistor R_C is introduced in series with the rotor circuit (Figure 4.9) [7]. The external resistor is connected to the rotor windings through a power electronic converter, and the currents flow between the resistor and the rotor via slip rings. This solution is known as dynamic slip control and ensures a speed variation in the range of 1–10%. The power converter is designed for low voltage and high currents. Alternatively, the resistors and electronics can be mounted on the rotor, eliminating the slip rings (the Weir design). Use of power electronic allows the rapid control of the rotor currents so that the output power can be kept constant even during gusting conditions, and can help the machine to perform better during grid perturbations.

When operating below rated wind speed and power, the WRIG wind turbine acts similar to a fixed-speed wind turbine. However, above rated values, by controlling the resistance it is possible to control the air-gap torque and thus the slip. Besides the sensitive growth of the energy that can be captured from the wind, a reduction of the influence of wind gusts on the active power produced by the generator is obtained. The speed range is typically at most 10% as it depends upon the size of the resistance and the slip power is dissipated in the resistor [3,7].

The main characteristics of induction generator with control of the rotational speed by an external rotor resistance are presented in Table 4.2 [7].

This solution preserves the same drawbacks as the previous squirrel cage induction generator.

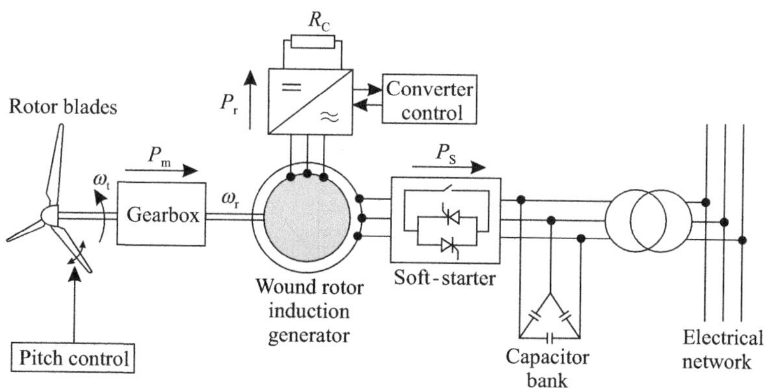

Figure 4.9. Basic configuration of a variable-speed wind turbine with external rotor resistor.

TABLE 4.2. Main advantages and disadvantages of WRIG

Advantages	Disadvantages
• Variable speed (0 to $+10\%\omega_{synchronism}$) • Robust design • Low power converter with power electronics	• The power produced is not optimized • Requires maintenance of gearbox • It cannot control the reactive power • The magnetization of the generator is made from the network

4.3.1.4 Doubly Fed Induction Generator. A large number of modern wind turbines are equipped with a DFIG that has the stator connected directly to the electrical network, meaning that it operates synchronously at the network frequency, and the three-phase wound rotor connected via a back-to-back voltage source converter and a transformer, as shown in Figure 4.10 [7].

The main advantage of the DFIG wind turbines is their ability to supply active power at a constant voltage and frequency while the rotor speed may be varied. As the magnetization of the DFIG can also be provided from the rotor circuit, the reactive power can be controlled independently of the active power. The rotor circuit can provide reactive power to the stator via the grid-side converter.

The back-to-back converter allows the control of the electromechanical torque and the rotor excitation. The size of the converter is a fraction of the generator rating, normally in the range between 15% and 30%. Since the power converter operates in a bidirectional way, the DFIG can be operated either in subsynchronous or in supersynchronous operational mode, with a variation of ±30% with respect to the synchronous speed.

On the other hand, the DFIG wind turbine may cause some problems during a grid fault when its rotor circuit, together with the back-to-back converter, is exposed to a high overcurrent induced by a high transient stator current. In this case, the wind turbine should be disconnected from the network in order to avoid some damage to the electrical or mechanical part. However, this solution is not acceptable in the case of short-term grid disturbances due to passive grid stability problems and thus some active protection systems need to be applied to keep the turbine connected to the network but also protected against any overcurrent.

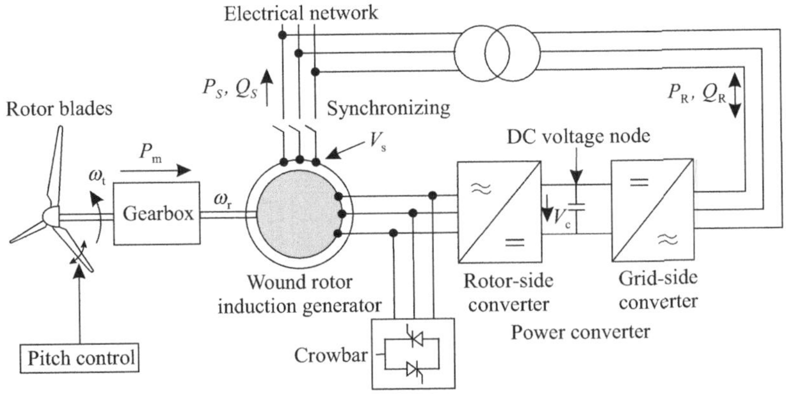

Figure 4.10. Basic configuration of a doubly fed induction generator. Variable speed drive [7].

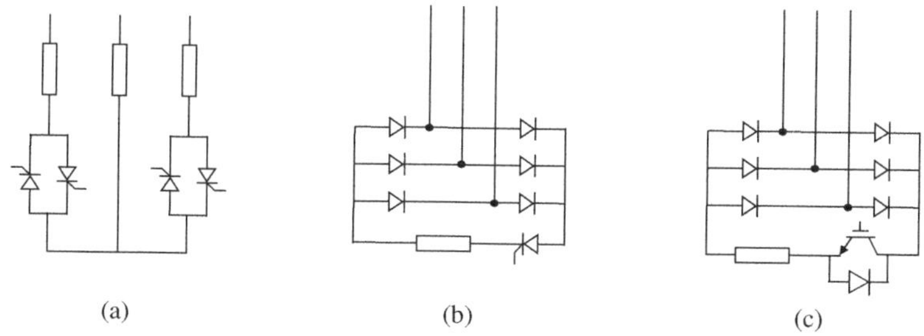

Figure 4.11. Crowbar arrangement [8,9].

TABLE 4.3. Main advantages and disadvantages of DFIG

Advantages	Disadvantages
• Possible speed regulation for optimal utilization of energy (±30% ω_{synch}) • The power electronics is designed at 30% P_n • Reactive power for magnetization of the machine is provided by the power converter • Standard asynchronous machine • Grid connection easy to administrate	• Requires maintenance of gearbox • The cost of the power electronic equipments • Control–command complex systems of the entire unit • Slip rings and brushes wear and tear, maintenance

One solution is to connect a crowbar (Figure 4.10) to the rotor terminals, designed to bypass the rotor-side converter, that is, to short circuit the rotor, in order to avoid overcurrent on the rotor-side converter as well as overvoltage on the DC-link capacitor.

Figure 4.11a shows *the architecture of a crowbar* consisting of antiparallel connected thyristors with external resistors. The crowbar can also be constructed by using diode bridge and a single thyristor, as shown in Figure 4.11b [8].

The crowbar is connected in case of either a DC-link overvoltage or a rotor-side converter overcurrent. Because of technical limitations of the two arrangements, the rotor current is not interrupted immediately. Furthermore, the crowbar remains connected until the stator is disconnected from the power grid. This is not acceptable under the present grid codes requirements. In order to remove the crowbar fast enough, an active crowbar is used, in which the thyristor is replaced with a GTO-thyristor or an IGBT (Figure 4.11c).

The main advantages and disadvantages of the doubly fed induction generator are presented in Table 4.3 [7,10].

4.3.1.5 Wound Rotor Synchronous Generator. The stator windings of a wound rotor synchronous generator are connected to the electrical network through a bidirectional power converter consisting of two back-to-back connected pulse width modulation (PWM) voltage source converters. The generator-side converter regulates the electromagnetic torque, whereas the network-side converter regulates the active and reactive powers produced.

The DC excitation current supplied to the rotor winding is provided by a rotating rectifier through slip rings and brushes; brushless exciter is also used. Therefore, no external source of reactive power is required. The frequency (in Hz) of the currents produced by the stator is determined by the mechanical speed (in rpm) of the rotor and the number of pole pairs.

The main advantages of the wound rotor synchronous generator are as follows:

- The efficiency of the machine is usually high since the whole stator current is used for the electromagnetic torque production.
- The WRSG with salient poles allows the direct control of the power factor, minimizing the reactive component of the stator current in any operating circumstances.
- Is it possible to design the generator with smaller pole pitch and therefore multiple pole pairs. This could be a very important characteristic in order to obtain low-speed multiplier machine, eliminating the gearbox.

However, the existence of a winding circuit in the rotor may be a drawback as compared to permanent magnet synchronous generator. In addition, in order to allow the generator to regulate both the active and reactive powers, the frequency converter is typically sized 1.2 times the rated active power.

Some variable-speed wind turbine manufacturers, for example Enercon and Lagerwey, use a low-speed multipole WRSG. The advantage is that the gearbox is removed, but the price is a large and heavy generator and a full-scale frequency converter sized to handle the full power of the wind turbine system. Another solution is the four-pole (high speed) WRSG used by Made, which requires a gearbox.

4.3.1.6 Permanent Magnet Synchronous Generator. The excitation of a permanent magnet synchronous generator is provided by the permanent magnets, and therefore no rotor winding is needed. Compared to the generators with excitation winding on the rotor, the PM machine has the advantage of lower rotor losses, smaller dimensions of the rotor, simpler cooling circuit (the rotor does not require cooling), and reduced failures. However, the costs for manufacturing the permanent magnets are very high, and an appropriate cooling system is required since the permanent magnets are sensitive to high temperatures.

The permanent magnet generator requires the use of a full-scale power converter for connection to the power grid in order to adjust the voltage and frequency at the generator terminal to those of the power system. Furthermore, during external short circuits and wind gusts the permanent magnet machine may cause very stiff performance. Although the converter is an added cost, it allows the generator to operate at any speed so as to fit the current conditions [3].

The rotor is provided with permanent magnet poles and, in terms of the rotational speed, it may have salient poles or may be round. Salient poles are more usual in low-speed machines and may be more suitable for wind turbines.

The most common types of PM machines are the radial flux machines, the axial flux machines (Figure 4.12), and the transverse flux machines [6].

The synchronous nature of the PMSG may cause problems during start-up, synchronization, and voltage regulation. It does not readily provide a constant voltage [3].

Architecture of Permanent Magnet Generator-Based Wind Turbine. Figure 4.13 shows the arrangement of a permanent magnet generator connected to the power grid through a frequency converter, consisting of a three-phase diode rectifier (passive), a PWM

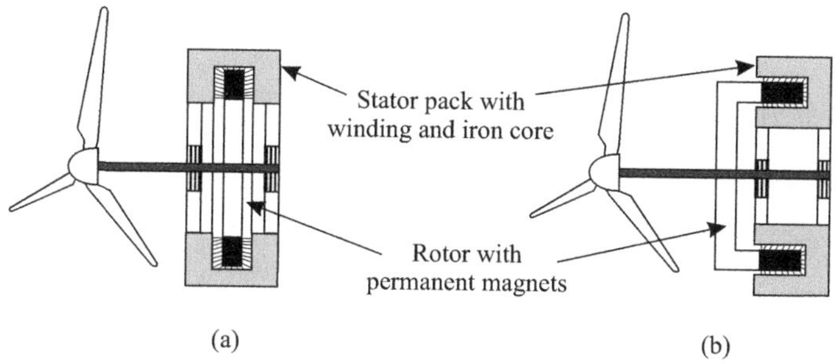

Figure 4.12. Schematic representation of double air gap designs for machines with (a) axial; (b) radial air gap. (Reproduced from Ref. 6.)

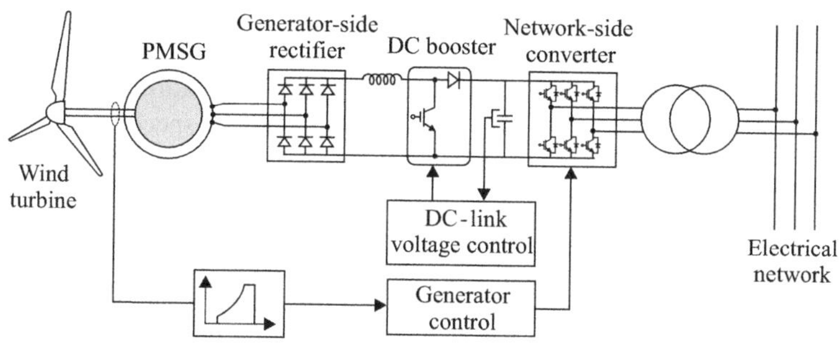

Figure 4.13. Permanent magnet synchronous generator with a boost chopper [5].

voltage source inverter, and a DC booster. The DC booster is used to regulate the DC-link voltage, whereas the network-side converter controls the operation of the generator. The reference power to the network-side converter is optimized according to the maximum power-speed characteristic shown in Figure 4.33. This configuration can include or not a gearbox.

The disadvantage of this arrangement is that the diode rectifier increases the level of harmonic distortion and the current amplitude of the PMSG. For this reason, this configuration has been considered for small-sized wind turbine systems ($\leq$50 kW). This issue was overcome using a PWM rectifier instead of the diode rectifier, shown in Figure 4.14.

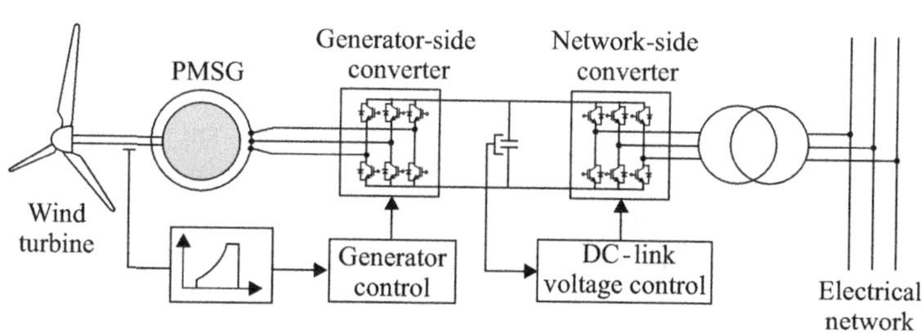

Figure 4.14. Permanent magnet synchronous generator with PWM converter [5].

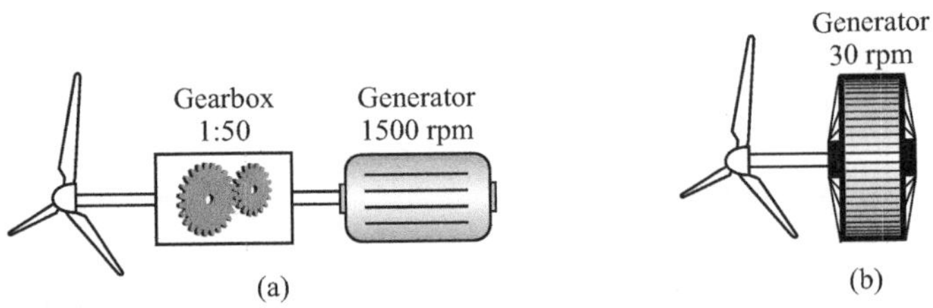

Figure 4.15. Drive trains of (a) conventional wind turbine; (b) direct-drive generator.

The controllable IGBTs allow the generator-side converter to control the generator operation, whereas the network-side converter controls the DC-link voltage and therefore the active and reactive powers exported to the network.

From High-Speed to Low-Speed Generators. Large wind turbines were classically designed with generators with small number of poles and high speeds, up to 1500 rpm at 50 Hz and 1800 rpm at 60 Hz. As the turbine speed is much lower than the generator speed, typically between 20 and 60 rpm, a gearbox was required between the turbine and generator to adapt the speed. In order to eliminate the problems introduced by the gearbox (e.g., high losses and high noise), nowadays there is a trend to use low-speed generators, which can be directly connected to the turbine shaft. Figure 4.15 shows comparatively the drive trains of a conventional wind turbine and one with a direct-drive generator [11].

Direct connection of the generator to the turbine involves a very high torque developed at the rotor shaft. For instance, a 500 kW direct-drive generator, with 30 rpm, has the same rated torque as a 50 MW steam turbine generator, but with 3000 rpm [5]. It is important to know that the size and the losses of a low-speed generator depend on the rated torque rather than on the rated power. High-rated torque direct-drive generators are usually heavier and less efficient than the conventional ones. In order to increase the efficiency and reduce the weight of the active parts of the wind turbine, these generators are usually designed with large diameter of the rotor and small pole pitch (which means a large number of poles).

Permanent Magnet Versus Wound Rotor. A synchronous generator is self-excited from the rotor excitation, consisting of either permanent magnets or current-carrying winding.

The excitation current of a wound rotor synchronous generator is adjustable and, consequently, the output voltage can be controlled independently of the load current [5]. This is why in the classical power plants, where the generators are connected directly to the power grid, the rotors are provided with wound rotors rather than permanent magnets. The generators of modern wind turbines are connected to the grid through a flexible power electronic-based interface; therefore, the advantage of controllable no-load voltage is of less importance. Instead, compared to the permanent magnet rotor, the wound rotor is heavier and has higher losses.

Permanent magnet excitation avoids the field current supply or reactive power compensation facilities needed by wound rotor synchronous generators and induction generator and it also removes the need for slip rings [12].

The main advantages and disadvantages of the permanent magnet synchronous generator configuration are presented in Table 4.4 [7]. As the diameter of the turbine rotor increases, the permanent magnet generators seem to be the first choice for the wind turbines manufacturers.

TABLE 4.4. Main advantages and disadvantages of PMSG

Advantages	Disadvantages
• Variable-speed operation (0–100% of the nominal speed) • Optimization of extracted wind power when the wind speed is low • The gearbox can be removed • Very low rotor losses • Simple rotor with no parts likely to wear and tear	• High diameter of the machine • The cost of the power electronic equipments (dimensioned at 100% P_n) and of the machine (which is not standard) • Power losses in power converter • High costs of permanent magnets • Possibility of demagnetization • Insufficient experience in construction and installation

4.3.2 Overview of Wind Turbines Concepts

4.3.2.1 Fixed-Speed Wind Turbines. The early large-sized wind turbine technologies were based on generators operating at fixed speed, that is, the rotor speed is constant regardless of the wind speed, and is determined by the power grid frequency, the generator characteristics, and the gear ratio. The induction generator (squirrel cage or wound rotor) is more suited to operate at fixed speeds and its stator can be connected directly to the power grid through a soft-starter (see Figure 4.8). In case of a squirrel cage generator, the slip and hence the rotor speed can be varied, but the variation is very small so that the wind turbine is referred to as a fixed-speed system.

Induction generators operate at higher speeds, and therefore a gearbox is used to transfer the mechanical energy from the low-speed shaft of the aerodynamic rotor to the high-speed shaft driving the generator. For size and cost reasons, the induction generator operates at a standard nominal speed of 1500 rpm. In terms of the rated power, the gearbox must provide a gear ratio of about 30–100. A solution to increase the power production was to design a squirrel cage induction generator running at two different, but constant, speeds. This is possible by changing the number of stator pole pairs from eight poles for low wind speeds to four to six poles for high wind speeds.

4.3.2.2 Variable-Speed Wind Turbines. As the technology became more mature and advancements have been done especially regarding the generator and power electronic-based interface with the electrical network, the wind turbines have been designed with variable speeds. The aim has been to capture as much as possible energy from the wind. Thus, the variable-speed wind turbines are designed to maximize the aerodynamic efficiency over a wide range of wind speeds.

The two variable-speed wind turbine topologies are the partial-scale converter wind turbine based on doubly fed induction generator (Figure 4.16) and the full-scale converter wind turbine (FCWT) based on synchronous or induction generator (Figure 4.17) [13].

Variable-speed operation of the wind turbine allows continuous change—increase or decrease—of the rotational speed, in terms of the wind speed V_W, so that the tip–speed ratio λ can be maintained constant at the optimal value, which corresponds to the maximum power coefficient. Contrary to a fixed-speed system, a variable-speed system maintains the generator torque fairly constant and the variations in wind are absorbed by changes in the generator rotor speed. To allow the wind turbine to operate at variable speed, the mechanical rotor speed and the electrical frequency of the grid must be decoupled.

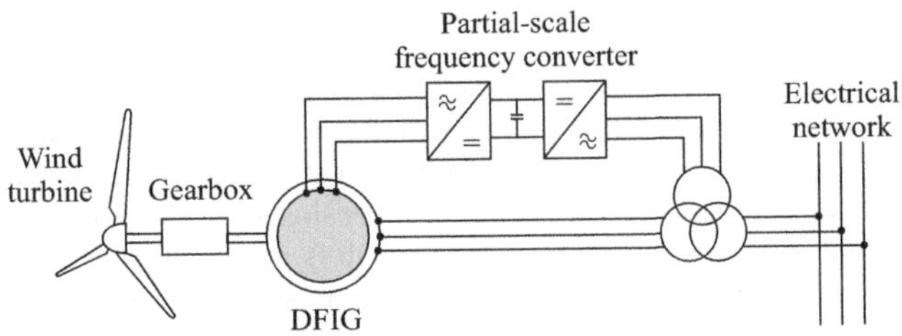

Figure 4.16. Partial variable-speed wind turbine system using doubly fed induction generator with wound rotor [5,13].

This means that the electrical frequency of the generator may vary as the wind speed changes, while the grid frequency remains unchanged [5]. For this reason, the variable-speed wind turbines are typically connected to the grid through a power converter, which controls the generator operation.

The rotor of a doubly fed induction generator (Figure 4.16) is connected to the electrical network through a *partial-scale frequency converter*, sized for just a fraction of the wind turbine rated power. In this way, the mechanical and electrical rotor frequencies are decoupled, and the electrical stator and the rotor frequencies can be matched independently of the mechanical rotor speed.

A typical configuration of a *full-scale frequency converter* wind turbine is shown in Figure 4.17. The generator is connected to the power grid through a back-to-back converter sized for the full rated power of the turbine, decoupling the generator frequency by the grid frequency. In this type of wind turbine, a wide range of electrical machine types may be employed: induction generator, wound field or permanent magnet synchronous generator. Depending on the type of generator used, the wind turbine may include or not a gearbox.

In the direct-drive (gearless) configuration, the turbine and the generator rotors are mounted on the same shaft, which means that the generator is designed with large-diameter

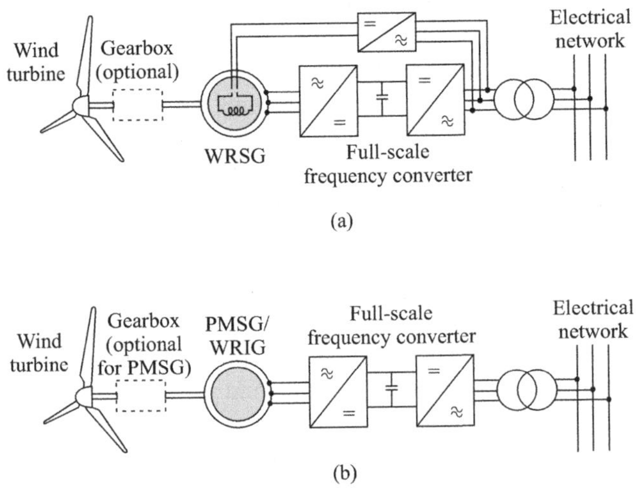

Figure 4.17. Full variable-speed wind turbine systems: (a) wound rotor synchronous generator; (b) wound rotor induction generator or permanent magnet synchronous generator [5,13].

and multiple poles, capable to run at low speeds. On the other hand, a smaller generator with a smaller number of poles requires a gearbox [14].

As the grid codes have become more restrictive regarding the immunity of the wind turbines to grid faults, variable speed and full-rated electronic interface are currently the only choice of the manufacturers for large wind turbines.

Compared to other alternatives, the full-scale wind turbine offers several advantages, such as [8]

- decoupling between the grid and the generator minimizes the effects of a grid fault from propagating to the generator, and therefore provides better fault response;
- the full-size generator-side converter facilitates a large operating speed range for the turbine, which improves the wind turbine power performance;
- the full-size grid-side converter gives more space for the wind turbine to provide reactive power into the grid, particularly during a grid fault.

The full-scale wind turbine concept poses, however, some disadvantages:

- Higher converter losses, than in a DFIG wind turbine, since the entire output power passes the converter.
- The high investment cost of the full converter, although this cost trend is decreasing.

In addition to these main wind turbine systems, there are some other configurations to be mentioned.

The *semivariable-speed turbine* is typically equipped with an induction generator of which total rotor resistance can be changed by addition of an external resistor controlled with power electronics (Figure 4.9). Any change in the rotor resistance results in shifting the torque/speed characteristic of the generator. In this way, the rotor speed can be varied (decreased) by about 10% of the rotor speed. Therefore, a semivariable capability is achieved at relatively low cost.

The main advantage of variable-speed wind turbines is that more energy can be generated for a certain wind speed regime. Although the electrical efficiency decreases due to the losses in the power electronic converters that are essential for variable-speed operation, the aerodynamic efficiency increases. The aerodynamic efficiency gain can exceed the electrical efficiency loss, resulting in a higher overall efficiency. In addition, the mechanical stress is less because the rotor acts as a flywheel (storing energy temporally as a buffer), thus reducing the drive-train torque variations.

The main drawback of variable-speed generating systems is that they are more expensive. However, using a variable-speed generating system can also give major savings in other subsystems of the turbine such as lighter foundations in offshore applications, limiting the overall cost increase [13].

4.3.3 Overview of Power Control Concepts

The wind turbine systems are sometimes subjected to high wind speeds causing high aerodynamic forces that act on the turbine blades and also that lead to increased rotor speed. In order to avoid any damages, the wind turbine is provided with various systems to control aerodynamic forces on the turbine rotor [3]. On the generator side, the command system can be split into two essentials functional levels (Figure 4.18):

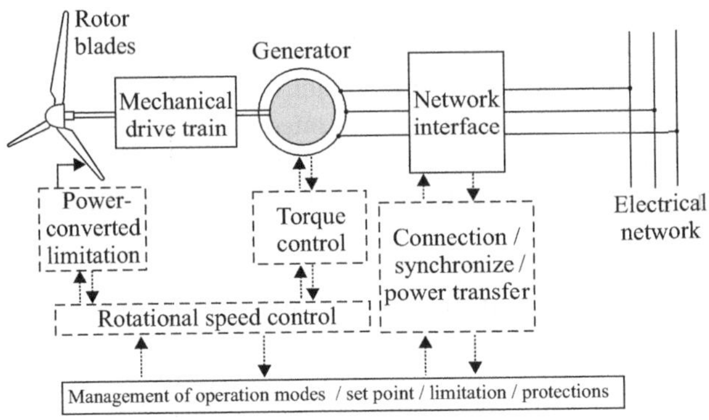

Figure 4.18. General control structure of a wind turbine [7].

- regulating systems, monitoring systems, and associated protections;
- management system of operation modes and protections management.

More general can be considered a third level corresponding to the overall management in a wind power plant designed in accordance with the grid codes requirements. The control of the power supplied to the main electrical network can be done at each of the two levels of conversion and in different ways according to operating conditions:

- At the turbine level, especially for limiting the power generated during high wind speeds.
- At the generator level, particularly for variable-speed wind turbine systems, where the energy captured is optimized at low or medium wind speeds. This requires the control of parameters that affect the generator operation (current, speed) or restrictions of the operating system (DC-link voltage, currents through the interface with the electrical network).

Limiting the mechanical power developed by the turbine so that the rated power is not exceeded is an essential aspect for a wind turbine. This can be done using different alternatives such as a pitch control, a passive stall control or an active stall control. The last two alternatives are commonly used in a fixed-speed wind turbine, while the first alternative, the pitch control, is used in variable-speed wind turbine.

The three control systems of the aerodynamic forces are [3]

(i) *The stall control* (passive control), is the simplest, most robust and cheapest control method, where the blades are bolted onto the hub at a fixed angle. The aerodynamic profile of the rotor blades causes the rotor to stall (loss the power surplus) when the wind speed exceeds a certain level [15]. The geometry of the blades limits the lift forces when the wind speed becomes too high and causes a progressive aerodynamic discharge of the blades that will reduce the power captured. This concept has the advantage of not requiring any mechanical or electrical auxiliary system. The disadvantage is that the power captured by the wind turbine is a function only of the wind speed and the rotation speed; there is

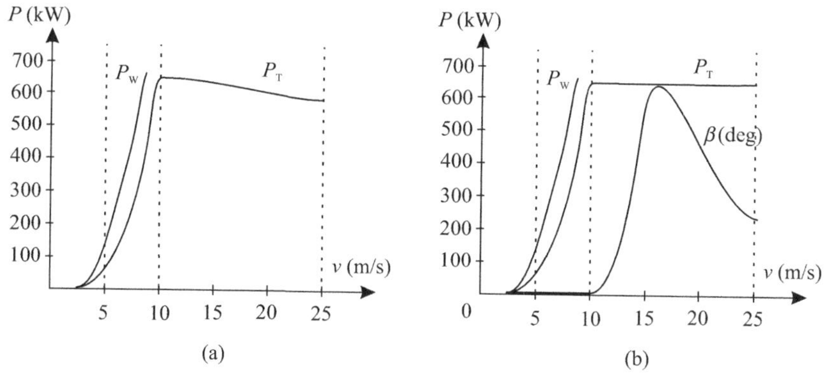

Figure 4.19. Typical power characteristics: (a) stall control; (b) pitch angle control [16].

no any possibility to adjust the pitch angle. The stall control causes less power fluctuations than a fast-pitch power regulation. However, this method is less efficient at low wind speeds, it has no assisted start-up, and the wind turbine system experiences variations in the maximum steady-state power due to variations in air density and grid frequencies.

In case of a stall control wind turbine system, in the event of a fault in the electrical grid, if the energy captured by the wind turbine cannot be transmitted to the grid, there must be a mechanical brake on the turbine shaft (high braking torque), dimensioned "to tire" the kinetic energy of the turbine. The brake can be fitted behind the gearbox where the torque is lower, and it is used only as a "parking" brake. An emergency stop can also be provided by the generator under the condition that this one has a resistive electrical circuit for energy recovery, which connects, in case of emergency, to the rheostatic brake.

The variation curve of available wind power P_W in terms of the wind speed is shown in Figure 4.19a [16].

(ii) *The pitch control* (active control) assumes that the blades can be turned out of the wind direction as the wind speed becomes too high or turned into the wind when the wind speed decreases below the rated value. This type of control provides a good power control and can contribute at start-up and during emergency stop. Compared to the stall control, where the output power decreases as the wind speed increases above the rated value, the pitch control ensures a constant output power, close to the rated power, for wind speeds above the rated value (Figure 4.19b).

The blades are rotated by using electrical or hydraulic devices and allow the pitch angle β to be varied between 0° and 25° (up to 30°) [16]. Turning the blades out of the wind, the aerodynamic forces acting on the blades can be reduced to acceptable limits. Because the axial pressing force is reduced, the efforts in the tower are also reduced. This advantage is amplified on variable speed because the extra energy during the gusts of wind, on which the variations are too brutal for blade orientation mechanism to compensate the effects, can be stored by the rotor inertia through its speed variation (if the generator can accept this energy), while the transmitted power remains constant. Mechanical brake is then a parking

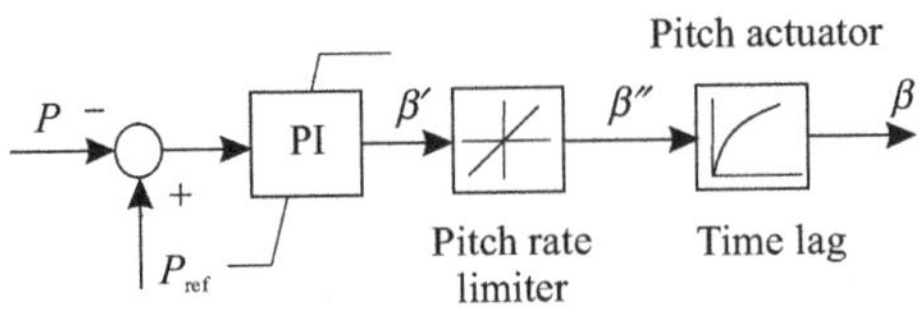

Figure 4.20. Pitch control for active stall wind turbine [8].

brake. When the wind speed drops below the cut-in speed, the power can be cancelled by turning the blades at $\beta = 90°$ [16].

The mechanism used to rotate the blades involves higher costs and maintenance. Because of the limited speed of the mechanism, the output electrical power exhibits more fluctuations than in the case of the stall control.

(iii) *The active stall control*, referred also to as "assisted stall" or "combi stall," is a combination of the stall control and pitch control. At low wind speeds, the blades are pitched similar to a pitch-controlled wind turbine in order to achieve maximum efficiency [3]. At high wind speeds, the blades are pitched toward stall, that is, in opposite direction than that employed by an active pitch control system. For this reason, the active stall control is sometimes called negative pitch control.

Compared to the pitch control system, aerodynamic braking requires pitch angles of only about $-20°$, so the travel of the pitch mechanism is very much reduced. Furthermore, in order to keep the output power constant at the rated value only small change of the blade pitch is required [18].

The active stall wind turbine achieves limited power fluctuations and also is able to compensate the variations in air density. Other advantage is that it can assist the wind turbine system in case of emergency stops and during start-up.

An example of active stall controller block is depicted in Figure 4.20 [8].

The measured output power is compared to the reference value, which is equal either to the optimum power or to the rated power. The resulted error is passed through a PI controller to provide the required pitch angle value β'. Because of the physical limitations, the real value must be limited by a pitch rate limiter, which provides the value β''. This value is then fed into a pitch actuator. The pitch actuator can be modeled as a first-order time lag system in order to model the hydraulic system of the actuator.

In order to avoid unnecessary continuous changes in pitch angle, which may wear the pitch mechanism, the mechanism is allowed to pitch the blades only at specified period of sample times and only if the difference between the new and the old set point exceeds a certain minimum value. For power system stability studies, however, this mechanism can be excluded from the model.

4.4 MODELING THE WIND TURBINE GENERATORS

4.4.1 Model of a Constant-Speed Wind Turbine

Figure 4.21 shows the simplified structure of a constant-speed wind turbine model. The most important parts of this wind turbine to be modeled are the turbine rotor, the drive train, and the generator. The input to the wind turbine model is the wind speed value V_W, and the link with the electrical network model is given by the active and reactive powers produced, P and Q, and by the actual voltage and frequency, U and f, of the network taken as reference for the wind turbine model.

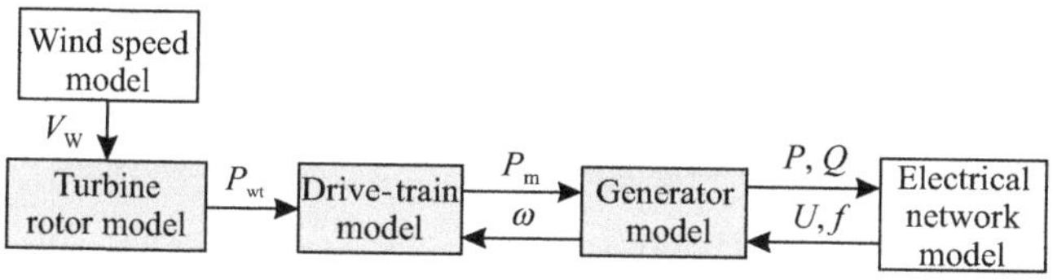

Figure 4.21. Simplified structure of a constant-speed wind turbine model [3,19].

WIND SPEED MODEL. One way to model the wind speed is to use historical data of the wind speed measured at the site where the wind power plant is/will be located. The wind speed may vary very much and may have very different characteristics from one moment to another. Therefore, simulations based only on "real" values are limited to those historical data available [20]. Simulation for a complete range of characteristics of the wind, such as wind gusts or turbulence, may require additional measured data.

On the other hand, analytical expressions can be established to model the wind based on characteristics chosen by the user, which may also include measured data. A general expression of the actual wind speed, which provides reasonable flexibility in power systems simulations, is made up by the sum of the following components [3,21]:

$$V_W(t) = V_{wa}(t) + V_{wr}(t) + V_{wg}(t) + V_{wt}(t) \tag{4.5}$$

where $V_W(t)$ is the wind speed at time t, $V_{wa}(t)$ is the average value of the wind speed, $V_{wr}(t)$ is the ramp component, $V_{wg}(t)$ is the gust component, and $V_{wt}(t)$ is the turbulence component.

The wind speed components are all in meter/second, and the time t is in seconds.

WIND TURBINE ROTOR MODEL. For electrical simulation, the wind turbine can be represented by the well-known algebraic equation (4.3) of the power extracted from the wind, which gives the relationship between the total available wind power and the performance coefficient. In off-line simulations, both the air density ρ and the area swept by the turbine blades are considered constant. The wind speed value is obtained from the wind speed model, and the performance coefficient is a controlled term since it is a function of the pitch angle β and the tip–speed ratio λ.

REPRESENTATION OF THE WIND TURBINE DRIVE TRAIN. A wind turbine drive train can be represented by a multimass system modeled by inertia, angular positions, and angular velocities. The masses are coupled between them through spring constants and damping coefficients.

The *three-mass model* of the constant-speed wind turbine drive train is shown in Figure 4.22. It consists of the following three inertias [8]:

- The inertia J_b, representing the flexible parts of the blades, which can be flexibly rotated by pitch control.
- The inertia J_h, representing the rigid parts of the blades bolted into the hub, the turbine hub, the low-speed shaft, and rotating parts of the gearbox that are stiffly connected to the low-speed shaft.
- The inertia J_g, representing the generator rotor, the high-speed shaft including also a disk brake, and the rotating parts of the gearbox that are stiffly connected to the high-speed shaft.

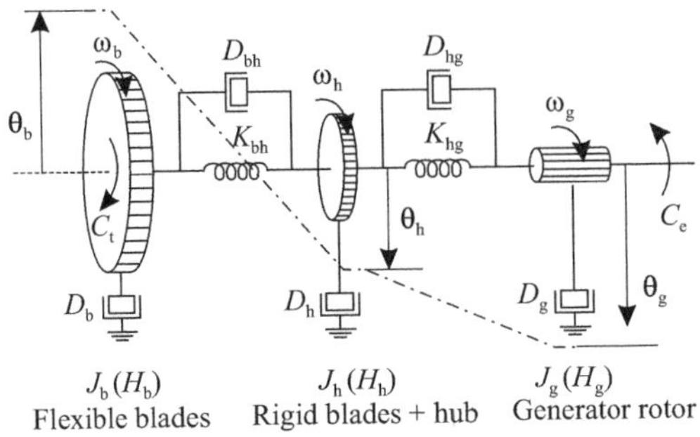

Figure 4.22. The three-mass model of drive train (Adapted from [8]).

In Figure 4.22, θ_b, θ_h, and θ_g represent angular positions of the blades, hub, and generator, ω_b, ω_h, and ω_g correspond to the angular velocities of the blades, hub, and generator, and H_b, H_h, and H_g are the inertia constants of the blades, hub, and generator, respectively.

The springs of blades and the low-speed shaft are considered flexible. The elasticity between adjacent masses is expressed by the spring constants: K_{bh} is the effective blade stiffness and K_{hg} represents the shaft stiffness, resultant of both the low- and high-speed shafts. The damping between masses is represented by damping coefficients: D_{bh} is the damping coefficient between the blades and the hub, and D_{hg} is the damping coefficient between the hub and the generator rotor [5,22].

The equations of motion of *the three-mass model*, expressed in per unit, can be written as [8,22]

$$\left\|\begin{aligned}
&2H_b\frac{d\omega_b}{dt} = C_t - K_{bh}(\theta_b-\theta_h) - D_{bh}(\omega_b-\omega_h)\\
&2H_h\frac{d\omega_h}{dt} = K_{bh}(\theta_b-\theta_h) - K_{hg}(\theta_h-\theta_g) - D_{bh}(\omega_b-\omega_h) - D_{hg}(\omega_h-\omega_g)\\
&2H_g\frac{d\omega_g}{dt} = -C_e + K_{hg}(\theta_h-\theta_g) - D_{hg}(\omega_h-\omega_g)\\
&\frac{d\theta_b}{dt} = \omega_b;\quad \frac{d\theta_h}{dt} = \omega_h;\quad \frac{d\theta_g}{dt} = \omega_g
\end{aligned}\right. \tag{4.6}$$

In equations (4.6), C_t is the aerodynamic torque determined as a function of the power extracted from wind by the turbine $P_T$and the blade speed ω_b, that is, $C_t = P_T/\omega_b$, whereas C_e stands for the electromagnetic torque of the generator. The terms corresponding to the damping coefficients of the blades D_b, the hub D_h, and the generator D_g have been neglected.

The three-mass model involves very large computation time in power system dynamic simulations. For this reason, a simpler model, with two masses, is preferred.

In the *two-mass model*, the drive train of a wind turbine is modeled as two inertias J_t and J_g that are connected to each other through a spring (Figure 4.23). The spring represents the low stiffness of the drive-train shaft. The first mass accounts for the blades, the hub, and the low-speed shaft, whereas the second mass accounts for the high-speed shaft, the gearbox, and the generator rotor. It is supposed that the gearbox is one of main sources of shaft flexibility [8,22,23].

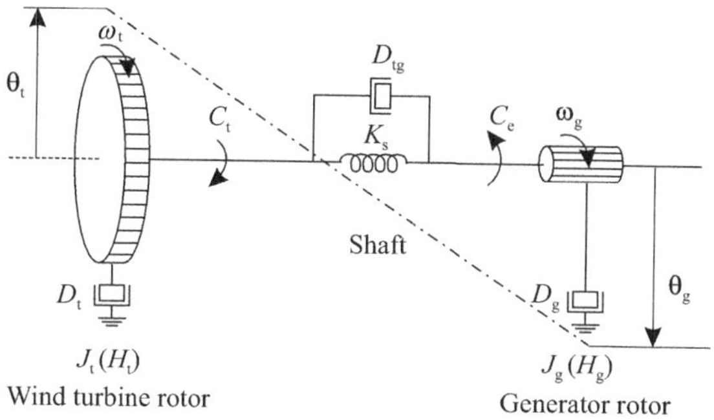

Figure 4.23. Two-mass model of a drive train (Adapted from [8]).

The equations of motion of *the two-mass model*, expressed in per unit, can be written as [8]

$$\begin{aligned}
2H_t\frac{d\omega_t}{dt} &= C_t - K_s(\theta_t - \theta_g) - D_{tg}(\omega_t - \omega_g)\\
2H_g\frac{d\omega_g}{dt} &= -C_e + K_s(\theta_t - \theta_g) + D_{tg}(\omega_t - \omega_g)\\
\frac{d\theta_t}{dt} &= \omega_t; \quad \frac{d\theta_g}{dt} = \omega_g
\end{aligned} \tag{4.7}$$

where H_t and H_g are the wind turbine and the generator rotor inertia constant, respectively; ω_t and ω_g are the turbine and the generator rotor angular velocities, respectively, in revolutions per minute; θ_t and θ_g are the turbine and the generator rotor angular positions, respectively, in mechanical degrees; D_{tg} is the mutual damping, caused by differences in speeds of the rotor and the turbine shaft; and K_s is the shaft stiffness.

In this model, the terms corresponding to the turbine self-damping D_t representing the aerodynamic resistance in the turbine blades, and the generator self-damping D_g representing the damping due to mechanical friction and windage, have been neglected.

The drive-train model can be simplified under the form of a *one-mass model* by removing shaft stiffness and mutual damping between masses. The equivalent inertia is the sum of the generator rotor and the turbine inertias. Therefore, assuming that the turbine angular velocity and the generator rotor angular velocity are equal, the equation of motion of one-mass drive-train model reduces to

$$2(H_t + H_g)\frac{d\omega_m}{dt} = C_t - C_e \tag{4.8}$$

Squirrel Cage Induction Generator Model. In order to write the stator and rotor voltage equations of an induction generator in the *d–q* reference frame, we make use of equations (2.219) and (2.220) of the induction motor, but assuming the generator convention, that is, the current leaving the machine is positive and the current entering the machine is negative. Thus, the voltage equations are [24]

$$\left\| \begin{aligned} u_{ds} &= -R_s i_{ds} - \omega_s \psi_{qs} + \frac{d\psi_{ds}}{dt} \\ u_{qs} &= -R_s i_{qs} + \omega_s \psi_{ds} + \frac{d\psi_{qs}}{dt} \\ u_{dr} &= -R_r i_{dr} - s\omega_s \psi_{qr} + \frac{d\psi_{dr}}{dt} \\ u_{qr} &= -R_r i_{qr} + s\omega_s \psi_{dr} + \frac{d\psi_{qr}}{dt} \end{aligned} \right. \quad (4.9)$$

The subscripts d and q stand for direct and quadrature axis component, respectively, and the subscripts r and s for rotor and stator, respectively.

The slip s is defined as

$$s = 1 - \frac{p}{2}\frac{\omega_m}{\omega_s} \quad (4.10)$$

where p is the number of poles and ω_m is the rotor mechanical angular velocity.

The impedances are expressed in per unit in order to make them independent of the voltage level and generator rating. Using again the generator convention, the stator and rotor flux linkages from equation (4.9) can be calculated as

$$\left\| \begin{aligned} \psi_{ds} &= -L_{ss} i_{ds} - L_m i_{dr} \\ \psi_{qs} &= -L_{ss} i_{qs} - L_m i_{qr} \\ \psi_{dr} &= -L_{rr} i_{dr} - L_m i_{ds} \\ \psi_{qr} &= -L_{rr} i_{qr} - L_m i_{qs} \end{aligned} \right. \quad (4.11)$$

where ω_s and ω_r are stator and rotor electrical angular velocities, in rad/s; L_{ss} and L_{rr} are stator and rotor inductances; and L_m is the mutual inductance between stator and rotor.

Substituting equation (4.11) in equation (4.9), and neglecting the stator transients, the voltage–current relationships become

$$\left\| \begin{aligned} u_{ds} &= -R_s i_{ds} + \omega_s (L_{ss} i_{qs} + L_m i_{qr}) \\ u_{qs} &= -R_s i_{qs} - \omega_s (L_{ss} i_{ds} + L_m i_{dr}) \\ u_{dr} &= 0 = -R_r i_{dr} + s\omega_s (L_{rr} i_{qr} + L_m i_{qs}) - \frac{d}{dt}(L_{rr} i_{dr} + L_m i_{ds}) \\ u_{qr} &= 0 = -R_r i_{qr} - s\omega_s (L_{rr} i_{dr} + L_m i_{ds}) - \frac{d}{dt}(L_{rr} i_{qr} + L_m i_{qs}) \end{aligned} \right. \quad (4.12)$$

The rotor voltages of a squirrel cage induction generator are equal to zero because the squirrel cage is short-circuited at the ends.

The electrical torque C_e is given by

$$C_e = \psi_{qr} i_{dr} - \psi_{dr} i_{qr} \quad (4.13)$$

and the equation of the motion of the generator is

$$\frac{d\omega_m}{dt} = \frac{1}{2H_m}(C_m - C_e) \tag{4.14}$$

The active power generated P and the reactive power consumed Q are

$$\begin{aligned} P_s &= u_{ds}i_{ds} + u_{qs}i_{qs} \\ Q_s &= u_{qs}i_{ds} - u_{ds}i_{qs} \end{aligned} \tag{4.15}$$

The generator exchanges active and reactive powers with the grid only through the stator terminals. Because the rotor is not connected to the grid *it does not need to be taken into account* [3].

4.4.2 Modeling the Doubly Fed Induction Generator Wind Turbine System

4.4.2.1 DFIG Model. A doubly fed induction generator is a special design of an induction machine, in which both the stator and the rotor are connected to the power grid. The stator, consisting of three-phase windings with two up to eight poles, is connected to the grid through a transformer. The rotor is also designed with three-phase windings, which are connected to an external stationary circuit via slip rings and brushes. This arrangement allows the rotor exchange power in both directions with the power grid. The general control block diagram of the DFIG wind turbine is illustrated in Figure 4.24 [8,25,26].

The main parts of the DFIG model are the models of the turbine rotor, the drive train, the generator, the rotor-side converter, the grid-side converter with the DC-link, as well as the interface and power grid models. Additional models represent different controls such as

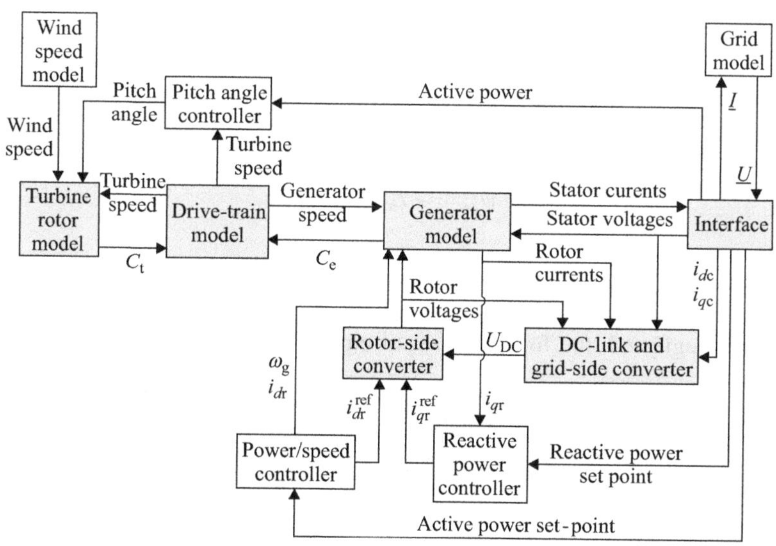

Figure 4.24. Control block diagram of a DFIG wind turbine (Adapted from [8]).

pitch angle controller, power/speed controller, reactive power controller, and a wind speed model.

Similar to variable-speed drive applications (induction motors), the control of the doubly fed induction machine is performed by decoupling the internal controls. The rotating *d–q* frame is aligned with any synchronously rotating variable, such as the stator flux. In this way a decoupled control between the electrical torque and the rotor excitation current is obtained. In contrast with the squirrel cage induction machine, the excitation to the machine rotor is provided by the rotor-side converter [27].

With appropriate PWM control, it is possible to control the torque and hence the speed of the generator rotor. The frequency of excitation varies with the wind speed and thus with the mechanical speed of the generator rotor.

Doubly Fed Induction Generator Model. For easier modeling of the doubly fed induction generator, the *d–q* reference frame is chosen. Furthermore, the generator convention is considered, which means that the currents are outputs instead of inputs, and active power and reactive power have a positive sign when they are fed into the grid.

Using the generator convention we achieve again the following set of stator and rotor equations [24]:

$$\left\| \begin{aligned} u_{ds} &= -R_{\mathrm{s}} i_{ds} - \omega_{\mathrm{s}} \psi_{qs} + \frac{\mathrm{d}\psi_{ds}}{\mathrm{d}t} \\ u_{qs} &= -R_{\mathrm{s}} i_{qs} + \omega_{\mathrm{s}} \psi_{ds} + \frac{\mathrm{d}\psi_{qs}}{\mathrm{d}t} \\ u_{dr} &= -R_{\mathrm{r}} i_{dr} - s\omega_{\mathrm{s}} \psi_{qr} + \frac{\mathrm{d}\psi_{dr}}{\mathrm{d}t} \\ u_{qr} &= -R_{\mathrm{r}} i_{qr} + s\omega_{\mathrm{s}} \psi_{dr} + \frac{\mathrm{d}\psi_{qr}}{\mathrm{d}t} \end{aligned} \right. \tag{4.16}$$

and the flux linkages:

$$\left\| \begin{aligned} \psi_{ds} &= -L_{\mathrm{ss}} i_{ds} - L_{\mathrm{m}} i_{dr} \\ \psi_{qs} &= -L_{\mathrm{ss}} i_{qs} - L_{\mathrm{m}} i_{qr} \\ \psi_{dr} &= -L_{\mathrm{rr}} i_{dr} - L_{\mathrm{m}} i_{ds} \\ \psi_{qr} &= -L_{\mathrm{rr}} i_{qr} - L_{\mathrm{m}} i_{qs} \end{aligned} \right. \tag{4.17}$$

The transients in the rotor fluxes are sometimes neglected because of the complexity in modeling the converter and due to computational speed required during simulations. With the transients neglected, the following set of equations results:

$$\left\| \begin{aligned} u_{ds} &= -R_{\mathrm{s}} i_{ds} + \omega_{\mathrm{s}} (L_{\mathrm{ss}} i_{qs} + L_{\mathrm{m}} i_{qr}) \\ u_{qs} &= -R_{\mathrm{s}} i_{qs} - \omega_{\mathrm{s}} (L_{\mathrm{ss}} i_{ds} + L_{\mathrm{m}} i_{dr}) \\ u_{dr} &= -R_{\mathrm{r}} i_{dr} + s\omega_{\mathrm{s}} (L_{\mathrm{rr}} i_{qr} + L_{\mathrm{m}} i_{qs}) \\ u_{qr} &= -R_{\mathrm{r}} i_{qr} - s\omega_{\mathrm{s}} (L_{\mathrm{rr}} i_{dr} + L_{\mathrm{m}} i_{ds}) \end{aligned} \right. \tag{4.18}$$

The electrical angular velocity of the rotor ω_r is

$$\omega_r = \frac{p}{2}\omega_m \tag{4.19}$$

where p is the number of poles and ω_m is the mechanical angular velocity (rad/s).

The electrical torque of the generator is given by

$$C_e = \psi_{ds} i_{qr} - \psi_{qs} i_{dr} \tag{4.20}$$

The total active and reactive powers exchanged by the DFIG with the electrical network are the sum of the stator and rotor powers:

$$P = P_s + P_r \tag{4.21a}$$

$$Q = Q_s + Q_r \tag{4.21b}$$

The stator active and reactive powers are

$$P_s = u_{ds} i_{ds} + u_{qs} i_{qs} \tag{4.22a}$$

$$Q_s = u_{qs} i_{ds} - u_{ds} i_{qs} \tag{4.22b}$$

and the rotor active and reactive powers are

$$P_r = u_{dr} i_{dr} + u_{qr} i_{qr} \tag{4.23a}$$

$$Q_r = u_{qr} i_{dr} - u_{dr} i_{qr} \tag{4.23b}$$

The reactive power Q expressed in equation (4.23b) is not necessarily equal to the reactive power exchanged by the machine with the electrical network. It depends on the control strategy for the grid-side converter that feeds the rotor winding. This is not valid for the active power, since the converter efficiency is incorporated in the rotor power [3].

4.4.2.2 Drive Train of DFIG. For a fixed-speed wind turbine, detailed drive-train dynamics might need to be considered especially in transient analysis. For a variable-speed wind turbine the drive-train characteristics have almost no effect on the network operation due to the decoupling effect of the power electronic converter.

The drive-train system of a DFIG wind turbine can be modeled using a two-mass model [28]. Despite the minor influence of shaft natural damping, it is important to note that a wind turbine with a power converter, such as the one with a DFIG, is often supplied with shaft torsional active damping [8].

The equations representing the shaft torsional mode of oscillation can be split into two equations:

- The *mechanical equations of the induction generator*:

$$\left\| \begin{aligned} & 2H_g \frac{d\omega_g}{dt} = \omega_e (C_{shaft} - C_e) \\ & \frac{d\theta_g}{dt} = \omega_g \end{aligned} \right. \tag{4.24}$$

- The *mechanical equations of the wind turbine*:

$$\left\| \begin{aligned} & 2H_t \frac{d\omega_t}{dt} = \omega_e (C_t - C_{shaft}) \\ & \frac{d\theta_t}{dt} = \omega_t \end{aligned} \right. \tag{4.25}$$

The incoming torque from the shaft to the induction generator, C_{shaft}, in both equations, consists of a term, $C_{torsion}$, representing elasticity of the shaft, and a term, $C_{damping}$, representing the damping torque of the shaft:

$$C_{shaft} = C_{torsion} + C_{damping} = K_s(\theta_t - \theta_g) + D_{tg}(\omega_t - \omega_g) \tag{4.26}$$

The mechanical torque produced by the wind turbine C_t is transferred to the generator through the multiple shafts to generate the electromagnetic torque C_e.

The equivalent shaft is characterized by the effective shaft stiffness K_s, expressed in p.u. torque/rad, and the damping torque D_{tg}, expressed in p.u. torque/(rad/s).

In equations (4.24) and (4.25), the speed of the induction machine ω_g and the speed of the wind turbine ω_t are measured in radians per second, whereas the angular position of the machine θ_g and the angular position of the wind turbine/hub θ_t are measured in radians. Also, the base angular speed is $\omega_e = 2\pi f$, where f is the power frequency, for example, 50 Hz.

Based on equations (4.24) and (4.25) we can draw the torsional model of the DFIG drive train with two masses (Figure 4.25).

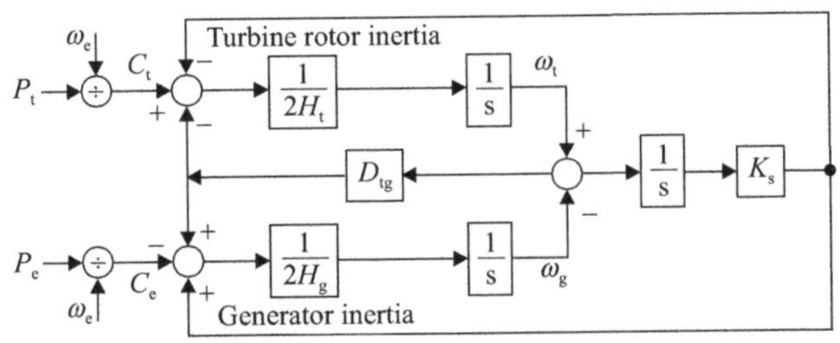

Figure 4.25. Two-mass torsional model [1].

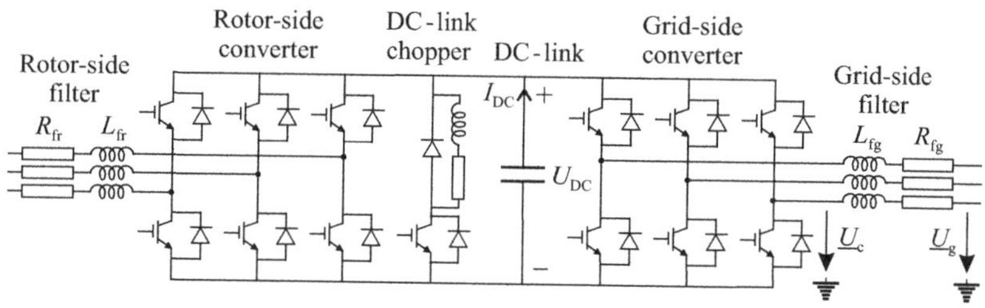

Figure 4.26. Back-to-back converter (Adapted from [8]).

4.4.2.3 Power Converter. The power electronic has gained its place in power system applications because of its flexibility in operation. A back-to-back converter is used to asynchronously interconnect the rotor of the DFIG system with the electrical network, allowing power transfer in both directions. The back-to-back converter consists of two VSCs and a DC link (Figure 4.26).

The back-to-back four-quadrant PWM–VSC, presented in Figure 4.26, is widely used in wind power systems today. The use of IGBT and PWM is a modern solution in power electronic-based power systems applications to minimize the harmonic content of the current. As a result, it reduces the torque pulsation on the generator and improves the quality of output power. The flexibility that may be achieved in controlling various parameters on either side of the converter with PWM techniques is also of great importance. By AC–DC–AC transformation, the generator rotor frequency is decoupled from the power grid frequency. For this reason, the converter is sometimes called frequency converter.

A DC link separates the two voltage source converters so that they can be controlled independently of each other.

On each side of the converter, R-L filters are used to remove the switching harmonics. The DC-link chopper is a protection circuit against DC-link overvoltages during grid faults. The arrangement consists of a resistor and an electronic switch, usually an IGBT.

The generator rotor-side converter controls the rotor currents of the machine, and thus controls the active and reactive powers of the machine as given by (4.23). The rotor-side converter is designed for the maximum slip power and reactive power control capability.

The grid-side converter exchanges power between the rotor-side converter and the power grid. It normally controls the DC-link voltage only. However, the converter might be employed to provide reactive power support during a grid fault, but this capability requires a larger converter rating.

The power rating of the grid-side converter is mainly given by maximum slip power since it usually operates at a unity power factor.

In stability studies, it is accepted to disregard the switching dynamics of the converter. In addition, converters are assumed to be able to follow the demanded values of the converter current [8].

4.4.2.4 Control Strategy for the DFIG. As the stator of the DFIG wind turbine is directly connected to the electrical network and the rotor is connected through power electronic-based interface, the only way to control the DFIG is through the control on the rotor operation. As the DFIG operates either as motor or as generator, the control system has to be designed so as to allow both operating states [29].

The most employed control strategy of the DFIG is based on the *rotor current vector control* in the *d–q* frame. Depending on the quantity required to be controlled, the *d*-axis is aligned either to the stator flux linkage vector [30] or to the stator voltage vector [31]. This procedure is chosen to allow decoupled control between active and reactive powers.

Flux magnitude and angle control (FMAC) is another control strategy, which aims to control the generator torque and terminal voltage by adjustment of the rotor flux magnitude and phase [32,33]. FMAC can also add auxiliary control loops to provide power system stabilizer, voltage support, and short-term frequency support capabilities.

However, conventional vector control and FMAC involve relatively complex transforms between the rotor and synchronous reference frame. In order to get the rotor speed and position information, an accurate position encoder has to be included or a sensorless algorithm should be employed. These methods increase the system complexity.

A *direct power control* (DPC) strategy is proposed in Ref. [34]. With appropriate rotor voltage vectors, DPC can also achieve decoupled active and reactive power control. However, the switching frequency is not constant with the variation of operating conditions. This makes the design of the harmonic filter of the rotor-side power converter difficult.

More recently, an *equivalent synchronous machine model* and a corresponding control scheme have been proposed in Ref. [35]. This control strategy relies on adjusting the magnitude and frequency of the rotor voltage to control the stator voltage and the active power. Coordinate transformation, rotor current, rotor speed, and position information are not required in the control strategy. Thus, the control system design is simplified.

Figure 4.27 shows the overall control system of a DFIG wind turbine, consisting of a generator control level and a wind turbine control level. This system follows mainly a *rotor current vector control* strategy.

The DFIG controller receives measured values of the currents and voltages in (*a*, *b*, *c*) coordinates from both sides of the power converter, as shown in Figure 4.27. As the DFIG

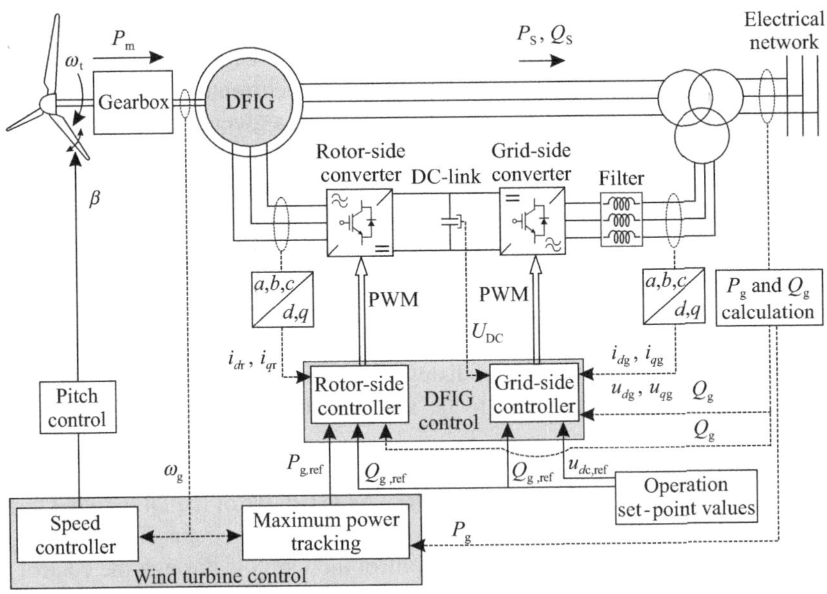

Figure 4.27. Overall control system of a DFIG wind turbine (Adapted from [10]).

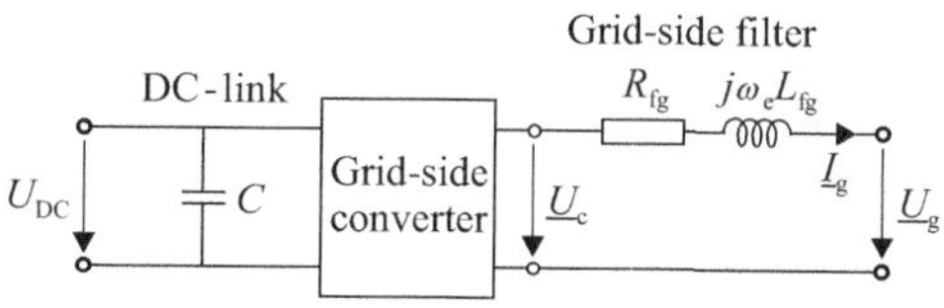

Figure 4.28. Schematic diagram of the grid-side converter.

control will be performed in the $(d–q)$ frame, a transformation $(a, b, c) \rightarrow (\alpha–\beta) \rightarrow (d–q)$ is required [29].

Grid-Side Converter Control. The general model of the grid-side converter is presented in Figure 4.28, which includes the current-controlled voltage source converter, the grid-side filter, and the DC-link capacitor [8]. The harmonic filter consists of a resistance R_{fg} and an inductance L_{fg}.

In the $d–q$ components, the voltage balance across the grid-side filter can be written as [8]

$$\begin{aligned} u_{dg} - u_{dc} &= R_{fg} i_{dg} + L_{fg} \frac{di_{dg}}{dt} - \omega_e L_{fg} i_{qg} \\ u_{qg} - u_{qc} &= R_{fg} i_{qg} + L_{fg} \frac{di_{qg}}{dt} + \omega_e L_{fg} i_{dg} \end{aligned} \tag{4.27}$$

where u_{dg} and u_{qg} denote the grid voltage vector components, which are obtained from the voltage $\underline{U}_c$, whereas u_{dc} and u_{qc} are the converter-side voltage vector components, which are obtained from the voltage $\underline{U}_g$. The grid pulsation is $\omega_e = 2\pi f$, where f is the grid frequency.

The last term in both equations can be seen as a disturbance on the controller, causing also a coupling of the two equations, which makes difficult to control the two currents, i_{dg} and i_{qg}, respectively.

In order to achieve decoupled control on the grid-side converter, we assume that the d-axis of the reference frame is aligned along the stator voltage vector position. Under these conditions, $u_{qg} = 0$ and $u_{dg} = u_s$. In per units, the grid-side voltages u_g are equal to the stator voltages u_s.

With $u_{qg} = 0$, we achieve

$$\begin{aligned} u_{dc} &= -u'_{dg} + \omega_e L_{fg} i_{qg} + u_{dg} \\ u_{qc} &= -u'_{qg} - \omega_e L_{fg} i_{dg} \end{aligned} \tag{4.28}$$

where

$$\begin{aligned} u'_{dg} &= R_{fg} i_{dg} + L_{fg} \frac{di_{dg}}{dt} \\ u'_{qg} &= R_{fg} i_{qg} + L_{fg} \frac{di_{qg}}{dt} \end{aligned} \tag{4.29}$$

Based on equation (4.28), we can draw the grid-side converter control scheme, which consists of two cascade control loops, providing the reference values to the converter $u_{dc,ref}$ and $u_{qc,ref}$ (Figure 4.29). The control of these voltages is performed through control of the

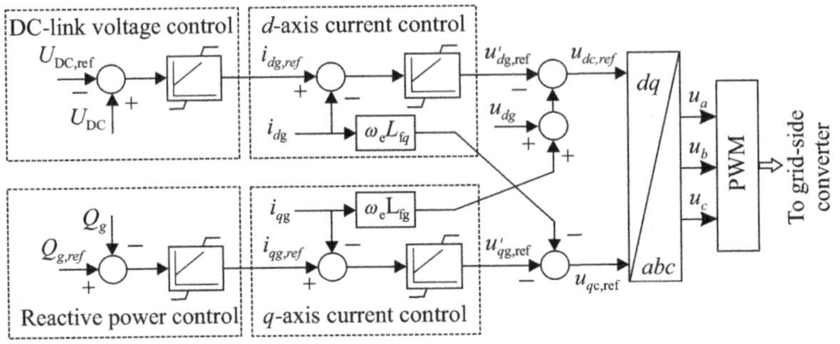

Figure 4.29. Grid-side converter control scheme (Adapted from [10]).

grid-side currents. The d-axis reference current is provided by a DC-link voltage control loop, whereas the q-axis reference current is provided by the reactive power control loop. Both loops consist of two control levels: a slower control level (the DC-link voltage control on the d-axis loop and the reactive power control on the q-axis loop) and a faster control level (for current control on either loop).

In the d–q frame, the active and reactive powers delivered to the grid through the rotor are

$$P_g = u_{dg} i_{dg} + u_{qg} i_{qg} \tag{4.30a}$$

$$Q_g = u_{qg} i_{dg} - u_{dg} i_{qg} \tag{4.30b}$$

Therefore, the active and reactive powers are proportional to i_{dg} and i_{qg}, respectively.

In general, the reactive power exchanged by the machine with the power grid through the grid-side converter is set to zero; thus, $i_{qg} = 0$. However, according to equation (4.30b), positive or negative values are feasible.

The main purpose of the grid-side converter is to keep constant the voltage across the DC link. Assuming that we have lossless converter and lossless DC link, the energy stored in the DC link is

$$E_C = \int P \, dt = \frac{1}{2} C \cdot U_{DC}^2 \tag{4.31}$$

where $P(= P_r - P_g)$ is the net active power flow into the capacitor, P_r is the rotor power inflow and P_g is the grid power outflow; C is the DC-link capacitance; and U_{DC} is the capacitor voltage.

The converter dynamics can therefore be described by

$$C \frac{dU_{DC}}{dt} = \frac{P_r - P_g}{U_{DC}}$$

Under the considered control strategy, since $u_{qg} = 0$, the active power delivered to the grid through the rotor is $P_g = u_{dg} i_{dg}$. Therefore, the DC-link voltage can be controlled by the control of the d-axis current component i_{dg} [10].

ROTOR-SIDE CONVERTER CONTROL. In the vector control of the rotor-side converter, we assume that the d-axis is oriented along the stator flux vector position. From power grid operation point of view, the stator flux is maintained almost constant because the stator voltages are almost constant in amplitude, frequency, and phase.

Due to the chosen reference frame, $\psi_{qs} = 0$ and $\mathrm{d}\psi_{qs}/\mathrm{d}t = 0$. Therefore, from (4.17) we achieve

$$i_{qs} = -\frac{L_{\mathrm{m}}}{L_{\mathrm{ss}}} i_{qr} \tag{4.32a}$$

The d-axis stator current becomes

$$i_{ds} = -\frac{1}{L_{\mathrm{ss}}}(\psi_{ds} + L_{\mathrm{m}} i_{dr}) \tag{4.32b}$$

As the stator flux ψ_{ds} does not vary much, the stator transients may also be neglected, that is, $\mathrm{d}\psi_{ds}/\mathrm{d}t = 0$. With $\psi_{qs} = 0$ and neglecting the stator resistance $R_{\mathrm{s}} = 0$, the stator voltages given in equations (4.16) become

$$\begin{aligned} u_{\mathrm{ds}} &= 0 \\ u_{qs} &= \omega_{\mathrm{s}} \psi_{ds} \end{aligned} \tag{4.33}$$

Replacing the d- and q-axis stator currents from equations (4.32a) and (4.32b) in equations (4.22), the expressions of the active and reactive powers delivered by the stator become

$$P_{\mathrm{s}} = u_{qs} i_{qs} = -u_{qs} \frac{L_{\mathrm{m}}}{L_{\mathrm{ss}}} i_{qr} \tag{4.34a}$$

$$Q_{\mathrm{s}} = u_{qs} i_{ds} = -\frac{u_{qs}}{L_{\mathrm{ss}}}(\psi_{ds} + L_{\mathrm{m}} i_{dr}) = -\frac{u_{qs}^2}{\omega_{\mathrm{s}} L_{\mathrm{ss}}} - \frac{L_{\mathrm{m}} u_{qs}}{L_{\mathrm{ss}}} i_{dr} \tag{4.34b}$$

The negative sign of the stator powers is in accordance with the generator convention.

From the above equations it is obvious that under flux-oriented control, the active power delivered by the stator can be controlled through the control of the q-axis rotor current and the reactive power (for constant d-axis fluxes) may be controlled through the control of the d-axis rotor current.

The machine control by PWM technique is generally performed by controlling the rotor voltages. Therefore, decoupling between the rotor voltages is required.

Replacing the expression of the d-axis rotor flux from (4.17) into the expression of the q-axis rotor voltage from (4.18), and taking into account expressions (4.32b) and (4.33), gives

$$u_{qr} = -R_{\mathrm{r}} i_{qr} + s\omega_{\mathrm{s}} \psi_{dr} = -R_{\mathrm{r}} i_{qr} - s\omega_{\mathrm{s}} \left[\left(L_{\mathrm{rr}} - \frac{L_{\mathrm{m}}^2}{L_{\mathrm{ss}}} \right) i_{dr} - \frac{L_{\mathrm{m}}}{\omega_{\mathrm{s}} L_{\mathrm{ss}}} u_{qs} \right] \tag{4.35}$$

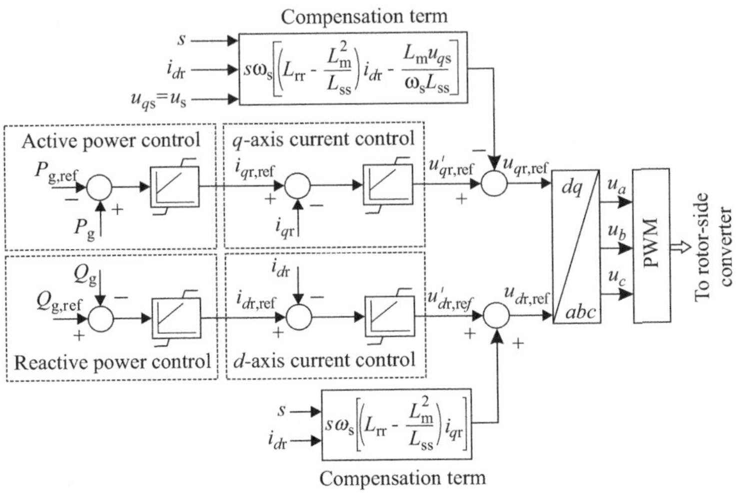

Figure 4.30. Rotor-side converter control scheme (Adapted from [5,8]).

Similarly, replacing the expression of the q-axis rotor flux from (4.17) into the expression of the d-axis rotor voltage from (4.18), and taking into account the expression (4.32a), gives:

$$u_{dr} = -R_r i_{dr} + s\omega_s\left(L_{rr} - \frac{L_m^2}{L_{ss}}\right) i_{qr} \tag{4.36}$$

Equations (4.35) and (4.36) are the necessary conditions for rotor voltage decoupling. While the rotor resistance may be neglected, the remaining terms can be added in the control loops as compensation terms.

Figure 4.30 illustrates the two cascading control loops of the rotor-side converter, which provides the reference values of the d- and q-axis rotor voltages.

The pulse width modulation procedure on the rotor-side converter is performed after appropriate transformation of the rotor voltages from the d–q frame to *abc* coordinates.

The input to the q-axis control loop is a reference active power obtained from a maximum power tracking (MPT) controller, whereas the input to the second control loop is a reference reactive power required to control the terminal voltage and the power factor.

(i) *Active Power Control*

In normal operation, the active power reference value for the active power control loop is provided by a maximum power tracking controller based on predefined P-ω lookup table. This table provides the maximum (optimal) power that can be extracted from wind for a certain wind speed. For high wind speeds there is a cross-coupling with the speed controller, which determines the optimal pitch angle. Figure 4.31 illustrates the simplified block diagram of the active power control scheme. In some cases the torque is controlled instead of active power. If frequency regulation loop would be included in the wind turbine control scheme, an additional signal adds to the optimal value extracted from the lookup table.

(ii) *Reactive Power Control*

The reactive power control loop is typically employed to control the terminal voltage or the power factor through the control of the d-axis rotor current using the rotor-side

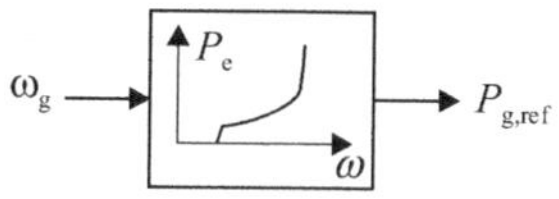

Figure 4.31. DFIG power control scheme [5].

converter. Reactive power can also be produced by the network-side converter, but the rotor-side converter is likely to be preferred in the case of a DFIG wind turbine [5].

4.4.2.5 Aerodynamic Model and Pitch Angle Controller. The wind turbine converts the kinetic energy of wind into mechanical energy in the form of mechanical torque through the turbine shaft that drives the electrical generator. The mechanical torque developed by the turbine is given by

$$C_t = \frac{P_t}{\omega_t} = \frac{1}{2\omega_t}\rho V_w^3 A C_p(\lambda, \beta) \tag{4.37}$$

If the wind speed increases too much, the electromagnetic torque is no longer sufficient to control the rotor speed from becoming too high and consequently, the generator can be damaged. Turbine blades can also be damaged because of high forces developed at high wind speeds. The solution is to reduce the power extracted from the wind.

One way of limiting the forces acting on the turbine blades at wind speeds greater than the rated value is by changing the pitch angle β, thus reducing the performance coefficient $C_p(\lambda, \beta)$ [17]. Rotation of blades around their longitudinal axis is performed by either hydraulic or electrical drives. Therefore, a pitch angle controller model (Figure 4.32) has to be integrated in the wind turbine system model [8].

As the wind speed cannot be measured precisely, the input to the controller can be the active power and the rotor speed. The lower part of the pitch controller shown in Figure 4.32 is the rotor speed regulator while the upper part is an aerodynamic power limiter.

Control of Wind Power Extraction of a DFIG. The rotor-side converter is responsible for controlling the mechanical torque as well as the stator terminal voltage or the power factor. The variable-speed operation of the DFIG is possible because the power converter decouples the mechanical rotor speed and the power system electrical frequency. The rotor speed control capability facilitates also the optimization of the power extracted from the wind [26,36]. Figure 4.33 shows an example of mechanical power characteristics

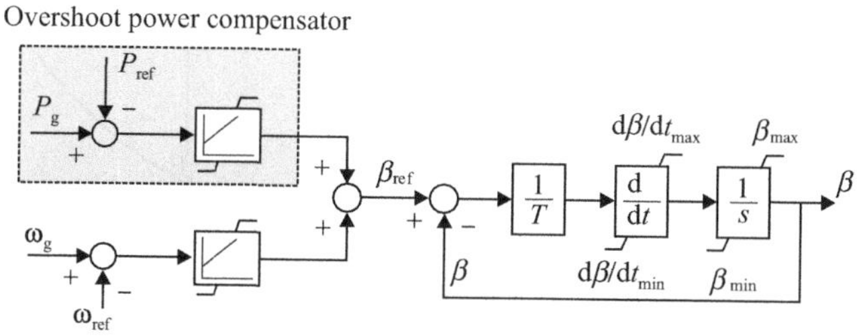

Figure 4.32. Pitch angle controller model [8].

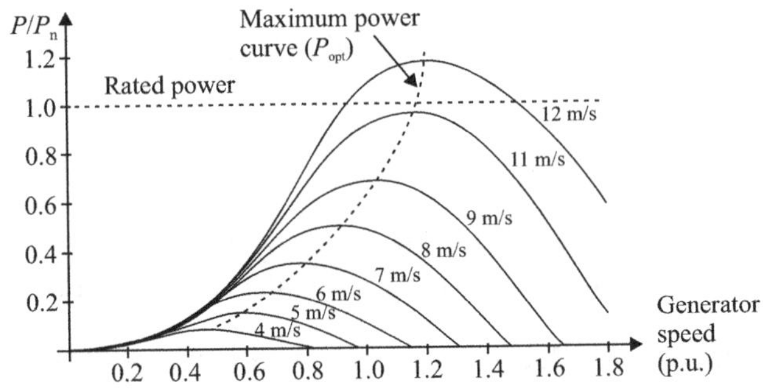

Figure 4.33. Wind turbine characteristic for maximum power extraction [6].

for various wind speeds. These characteristics have a maximum that corresponds to a certain generator rotor speed.

The control strategy of the DFIG aims to extract the maximum possible power from the wind. To achieve this, the active power is set so as to follow the curve of maximum power, P_{opt} (Figure 4.33).

The control strategy can be based either on the mechanical power developed versus wind speed characteristic (Figure 4.34a) or on the electrical power versus generator rotor speed characteristic (Figure 4.34b), both determined based on measured aerodynamic data of the turbine.

Figure 4.34a shows the optimal characteristic of the mechanical power that can be developed by the turbine in terms of the wind speed. Figure 4.34b shows the electrical power developed by the generator versus rotor speed for various wind speeds. The generator speed range is restricted to a limit between the minimum speed ω_{min} and the maximum speed ω_{max}.

The curve of optimal power P_{opt} extracted from the wind is defined by the relationship [30]:

$$P_{opt} = K_{opt} \cdot \omega_r^3 \tag{4.38}$$

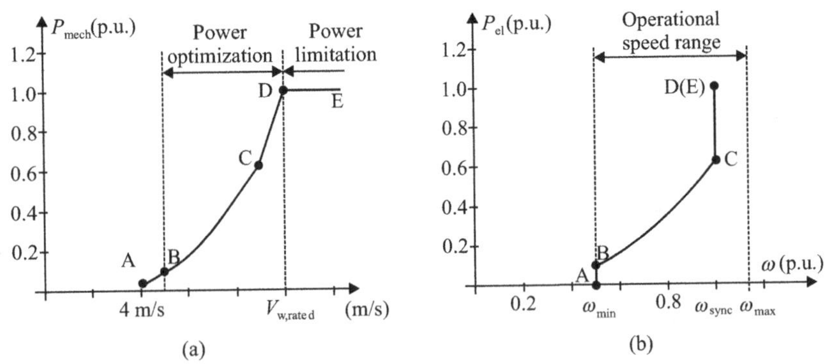

Figure 4.34. Wind turbine system characteristics [10]: (a) mechanical power versus wind speed; (b) electrical power versus rotor speed.

and the optimal torque is given by

$$C_{\text{opt}} = K_{\text{opt}} \cdot \omega_r^2 \tag{4.38'}$$

where K_{opt} is a parameter obtained from the aerodynamic performance of the wind turbine.

As shown in Figure 4.34, the optimal power curve is divided into several regions [8]:

- *Minimum Speed Operating Region (region A–B).* Due to high thermal stress on the IGBT power converter at low rotor frequencies, it is not practical to maintain the wind turbine in operation at low wind speeds. Therefore, a cut-in wind speed is introduced in the control strategy, as a limitation on the left side of the power characteristic. In this region, the wind turbine operates at almost constant rotor speed, which is usually around 50% below the synchronous speed.
- *Optimal Speed Operating Region (region B–C).* Within this operating range, characterized by low to medium wind speeds, the wind turbine operates at variable rotor speed. The control strategy is adapted so as to extract the maximum power from the wind, which is given by equation (4.38). However, during turbulent wind conditions, the operating point can be shifted from the optimal characteristic. In this region, the rated power of the wind turbine is not reached, and the control of the generator speed is provided by the rotor-side converter. This strategy is also called "wind-driven mode."
- *Maximum Speed—Partial Load Operating Region (region C–D).* By design, the generator reaches its nominal angular speed (synchronous speed) at wind speeds below the rated wind speed. In this region, the generator is maintained around the rated speed, although it operates at powers lower than the rated value. During turbulent wind conditions, the rotor speed may increase above the synchronous speed, but, due to power converter constraints, it cannot exceed 15–20% above the synchronous speed. Still the control is provided by the rotor-side converter only.
- *Maximum Speed—Full Load Operating Region (region D–E).* If the wind speed increases above the rating value, the rotor-side converter is no longer able to control the torque and this task is taken over by the pitch angle regulator. By pitching the blade angle, the regulator limits the aerodynamic input power so that the torque and the rotor speed can be maintained at constant value. For very high wind speeds, the pitch control will regulate the input power until the wind speed shutdown limit is reached.

Due to some operational restrictions, for example, size and efficiency of the generator or the acceptable noise emission, under normal operating conditions the wind turbine rotational speed must be limited to values between the minimum speed $\omega_{\min}$ and the nominal speed ω_{sync}. However, when wind gusts occur, the rotor speed may increase above the nominal speed. This is acceptable for a short transient period but must be limited to a maximum value $\omega_{\max}$ higher than the nominal speed. Therefore, there can be two control strategies of the DFIG: the power optimization strategy and the power limitation strategy [10].

4.4.2.6 Operating Modes. The operating modes of the DFIG can be differentiated in terms of the relative speed of the rotor over the synchronous speed and the power flow direction passing the power converter, namely supersynchronous, synchronous, and subsynchronous operations (Figure 4.35).

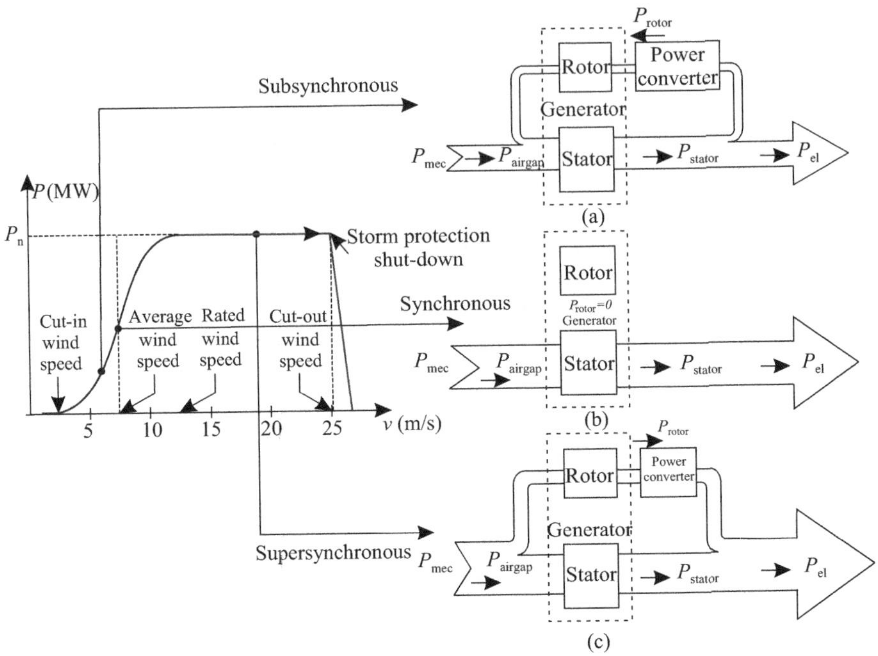

Figure 4.35. Operation modes of DFIG.

At low wind speeds, the DFIG operates at a rotor speed lower than the synchronous speed, that is, the slip is positive and the rotor absorbs power from the power grid through the power converter (Figure 4.35a). In the region C–D, when the generator operates at the synchronous speed, the generator slip is zero and there is no power flowing through the power converter (Figure 4.35b). During wind gusts, the rotor speed may exceed the synchronous speed and the slip become negative. In this case, the mechanical power from the shaft is split into two parts: the largest part of the power goes to the stator and a fraction of the power passes through the rotor (Figure 4.35c).

4.4.3 Full-Scale Converter Wind Turbine

4.4.3.1 General Model. The general structure of a variable-speed wind turbine with direct-drive synchronous generator is illustrated in Figure 4.36. As the generator is connected to the grid through a power electronic-based full-scale converter (frequency converter), the generator frequency, which is variable, is different from the power grid frequency (50 or 60 Hz). Comparative to a DFIG (a partial-scale converter wind turbine), the reactive power, the voltage, and the frequency are all determined by the converter, while the active power depends on the machine operation.

Full-scale converter wind turbines can include either wound rotor synchronous generator or permanent magnet synchronous generator, with or without a gearbox. The principle of controlling the frequency converter is similar in all cases.

Some observations are necessary [3]:

- The wind speed model is identical to that in the case of a constant-speed wind turbine model.

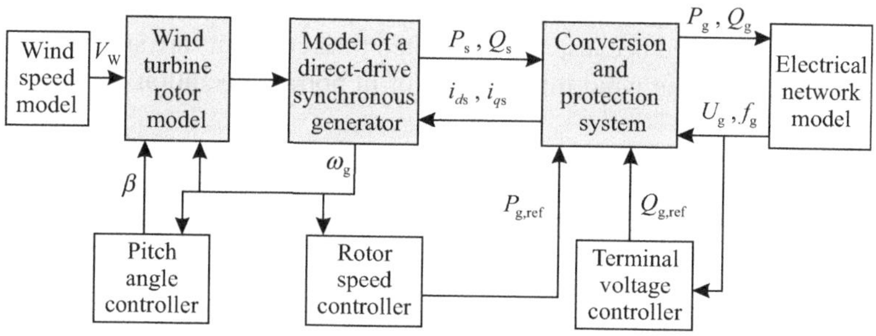

Figure 4.36. General structure of a variable-speed wind turbine with a direct-drive synchronous generator [3].

- The rotor model as well as the rotor speed and pitch angle controllers are identical to those used in the doubly fed induction generator.
- The converter and the protection system of a wind turbine with a DFIG are different from that of a direct-drive synchronous generator.

Taking into account the previous observations, only the generator model and the control systems are discussed below.

4.4.3.2 Model of a Direct-Drive Wind Turbine with Synchronous Generator.

Wound Rotor Synchronous Generator Model. The model of a wound rotor synchronous generator was demonstrated in Chapter 2. According to equations (2.39), in the *d–q* reference frame, the stator voltage equations are

$$\left\| \begin{aligned} u_{ds} &= -R_s i_{ds} - \omega_e \psi_{qs} + \frac{d\psi_{ds}}{dt} \\ u_{qs} &= -R_s i_{qs} + \omega_e \psi_{ds} + \frac{d\psi_{qs}}{dt} \end{aligned} \right. \tag{4.39}$$

and the rotor voltage equations are

$$\left\| \begin{aligned} u_f &= R_f i_f + \frac{d\psi_f}{dt} \\ 0 &= R_D i_D + \frac{d\psi_D}{dt} \\ 0 &= R_Q i_Q + \frac{d\psi_Q}{dt} \end{aligned} \right. \tag{4.40}$$

In order to complete the model, the flux linkage equations are added:

$$\left\| \begin{aligned} \psi_{ds} &= -L_{ds} i_{ds} + L_{md} i_f \\ \psi_{qs} &= -L_{qs} i_{qs} \\ \psi_f &= L_f I_f \end{aligned} \right. \tag{4.41}$$

Notice that all quantities in equations (4.39)–(4.41) are in per units.

The transformer voltages in the stator voltage equations, $d\psi/dt$, due to the flux change in space, may be neglected because the associated time constants are small. Therefore, the voltages of interest for the wind turbine control in normal operation become

$$\left\| \begin{aligned} u_{ds} &= -R_s i_{ds} + \omega_e L_{qs} i_{qs} \\ u_{qs} &= -R_s i_{qs} - \omega_e L_{ds} i_{ds} \\ u_f &= R_f i_f + \frac{d\psi_f}{dt} \end{aligned} \right. \tag{4.42}$$

The electromechanical torque is defined as

$$C_e = \psi_{ds} i_{qs} - \psi_{qs} i_{ds} \tag{4.43}$$

Some applications use the electromechanical torque in the rotor speed controller to provide the set point for the stator. However, more usual is the use of active power determined from maximum power point lookup table.

The active and reactive powers generated by the synchronous machine are given by

$$P_s = u_{ds} i_{ds} + u_{qs} i_{qs} \tag{4.44a}$$

$$Q_s = u_{qs} i_{ds} - u_{ds} i_{qs} \tag{4.44b}$$

Since the generator is fully decoupled from the grid by the frequency converter, the generator operation does not affect the power factor and the reactive power delivered/absorbed to/from the power grid. Therefore, the reactive power delivered/absorbed by the stator Q_s, as given in equation (4.44b), is of interest when sizing the converter and also to avoid exceeding the capability limits of the generator [3].

Permanent Magnet Synchronous Generator Model. In the PMSG model, the flux is assumed to be sinusoidally distributed along the air gap. In analogy to (4.39), under balanced steady-state conditions, the stator voltage equations are

$$\left\| \begin{aligned} u_{ds} &= -R_s i_{ds} - \omega_s \psi_{qs} + \frac{d\psi_{ds}}{dt} \\ u_{qs} &= -R_s i_{qs} + \omega_s \psi_{ds} + \frac{d\psi_{qs}}{dt} \end{aligned} \right. \tag{4.45}$$

In the case of a PMSG, the excitation winding is replaced by the permanent magnets and damping winding is removed. Therefore, the terms involving the field winding and the damping winding in equations (4.40) and (4.41) disappear. Furthermore, the flux ψ_{pm} of the permanent magnet mounted on the rotor that is coupled to the stator winding should be added in the d-axis stator flux expression. The flux linkages can be calculated using the following set of equations:

$$\left\| \begin{aligned} \psi_{ds} &= -L_{ds} i_{ds} + \psi_{pm} \\ \psi_{qs} &= -L_{qs} i_{qs} \end{aligned} \right. \tag{4.46}$$

where ψ_{pm} is the permanent magnet flux linkage, and L_{ds} and L_{qs} are the stator leakage inductances.

The electrical torque C_e of the permanent magnet synchronous machine is given by [16,37]

$$C_e = p\left[\left(L_{qs} - L_{ds}\right)i_{ds} + \psi_{pm}\right]i_{qs} \tag{4.47}$$

where p is the number of poles pairs.

For a *nonsalient-pole machine*, the stator inductances L_{ds} and L_{qs} are approximately equal. Consequently, the d-axis stator current i_{ds} does not influence the electromagnetic torque, and equation (4.51) becomes

$$C_e = p\psi_{pm}i_{qs} \tag{4.47$'$}$$

The turbine generator system can be described by the equation of motion:

$$\frac{d\omega_m}{dt} = \frac{1}{2H}\left(C_m - C_e\right) \tag{4.48}$$

where H is the inertia constant of the rotor, and C_m and C_e are the mechanical and electrical torque, respectively.

Note that the stator electrical angular velocity is given by

$$\omega_s = p\omega_m$$

where ω_m is the mechanical angular velocity (rad/s).

4.4.3.3 Control of Full-Scale Converter Wind Turbine. The control of a full-scale converter wind turbine depends on the configuration of the frequency converter and the machine control strategy, resulting in a large number of control possibilities. One control strategy may reside in the fact that the grid-side converter is responsible for controlling the voltage of the wind turbine system at the interface with the electrical network, whereas the generator-side converter operates at the stator voltage.

In the following, we assume that the wind turbine is equipped with a permanent magnet synchronous generator and that both the generator-side converter and the grid-side converter consist of PWM-controlled IGBTs (Figure 4.37).

The control strategy presented here is quite similar to the control strategy presented for the DFIG wind turbine. The difference resides in the fact that, under normal conditions, the power grid operation does not affect the generator operation.

GRID-SIDE CONVERTER CONTROL. The control of the grid-side converter is based on the assumption that the d-axis is oriented along the grid voltage vector position, that is, $u_{qg} = 0$, and thus the grid voltage has a d-axis component only u_{dg}. The voltage u_g is metered on the grid side of the wind turbine system.

The expressions of the active and reactive powers on the grid side are

$$P_g = u_{dg}i_{dg} + u_{qg}i_{qg} \tag{4.49a}$$

$$Q_g = u_{qg}i_{dg} - u_{dg}i_{qg} \tag{4.49b}$$

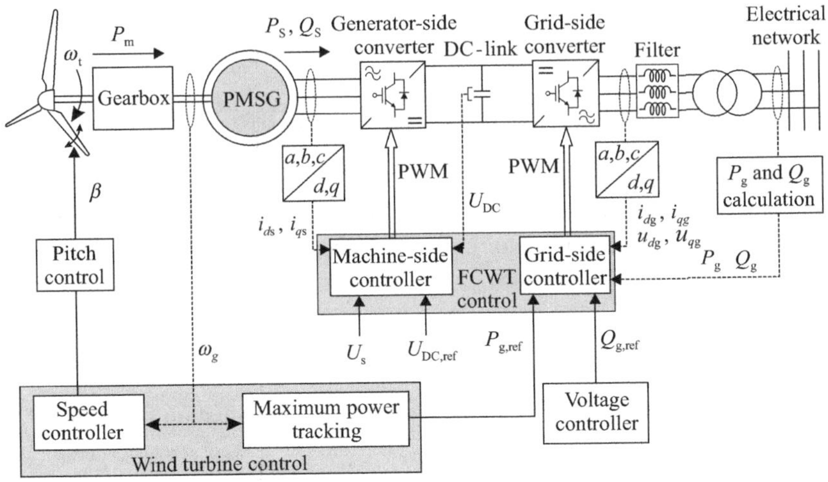

Figure 4.37. Overall control system of a FCWT system.

Under the above assumption, equations (4.49) become

$$P_g = u_{dg} i_{dg} \tag{4.49$'$a}$$

$$Q_g = -u_{dg} i_{qg} \tag{4.49$'$b}$$

showing that the active power is determined by the d-axis grid current, whereas the reactive power is determined by the q-axis grid current, metered on the grid side of the power converter.

Figure 4.38 shows the grid-side converter controller, which consists of two cascading loops, with the reference active power being the input to the d-axis control loop and the reactive power being the input to the q-axis control loop.

The reference to the active power control is obtained from the MPT characteristic. The aim is to optimize the generator operation so that to achieve the highest efficiency and deliver as much active power as possible.

Under normal conditions, because the generator is decoupled from the power grid through the frequency converter, the generator is not required to deliver/absorb reactive power. Therefore, the reference reactive power can be set to zero, that is, $Q_{g,ref} = 0$. However,

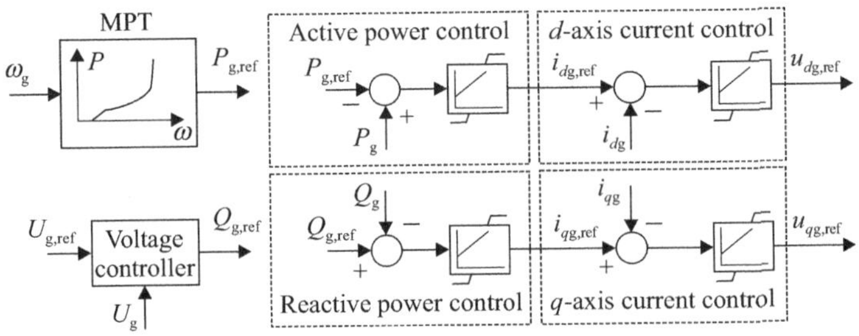

Figure 4.38. Grid-side converter controller [8,10].

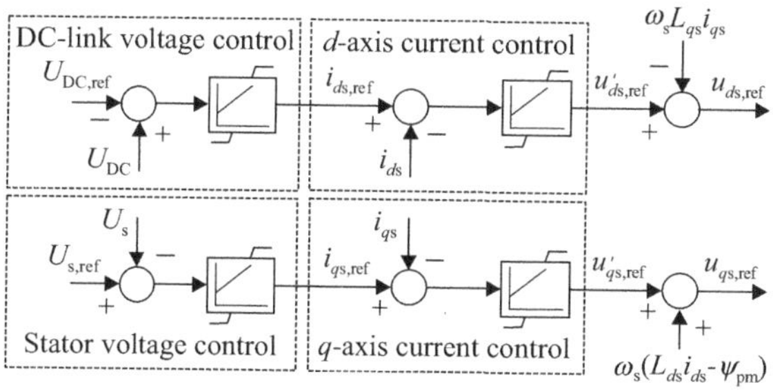

Figure 4.39. Generator-side converter control [10].

when a voltage support is required from the wind turbine system in normal operation or in case of faults occurring in the electrical network, reactive power exchange between the wind turbine system and the electrical network is required, that is, $Q_{g,ref} \neq 0$. A voltage regulator can be designed to provide voltage control at the point of connection through appropriate support of reactive power, which is the duty of the grid-side converter. As in the case of any electrical machine, the amount of reactive power exchanged by the wind turbine system with the power grid must be limited to the frequency converter rating [10].

Generator-Side Converter Control. From the voltage equations expressed in (4.45) and stator linkages equations in (4.46), it can be seen that there is a cross relationship between the two axes. Under steady-state conditions, the derivative terms in equations (4.45) reduce to zero. Under this assumption, replacing the stator flux linkages from equation (4.46) in (4.45), achieve

$$\begin{aligned} u_{ds} &= -R_s i_{ds} + \omega_s L_{qs} i_{qs} \\ u_{qs} &= -R_s i_{qs} - \omega_s \left(L_{ds} i_{ds} - \psi_{pm}\right) \end{aligned} \tag{4.50}$$

Neglecting also the stator resistance, we can see that the d-axis stator voltage depends on the q-axis stator current, and the q-axis stator voltage depends on the d-axis stator current. In order to obtain independent controllers for the two coordinates, the influence of the q-axis on the d-axis components and vice versa must be eliminated. The two components can be decoupled by considering the two terms on the right-hand side of equation (4.50) as compensation terms (Figure 4.39).

The aim of the generator-side converter is to keep constant the DC-link voltage and to control the generator stator voltage to its rated value, that is, $U_{s,ref} = U_{s,rated}$. The advantage of controlling the stator voltage at its rated value is that the power converter always operates at the rated voltage for which it was designed and optimized [10].

The reference value to the DC-link voltage control loop can be provided by a damping controller, which aims to generate a torque that dampens the speed oscillations.

4.5 FAULT RIDE-THROUGH CAPABILITY

4.5.1 Generalities

The power systems have sometimes undergone unpredictable disturbances, for example short circuits. A short circuit occurring close to a wind turbine system results in a voltage dip at the generator terminals and hence an active power drop [23,38]. In the past, wind turbines were

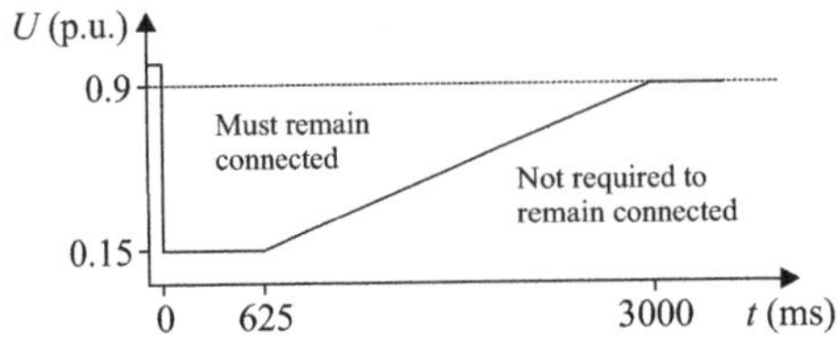

Figure 4.40. Border limit of a voltage dip occurring at the wind turbine system terminal.

allowed to be disconnected from the power grid during such a disturbance. As the share of wind turbine systems has increased in the power system generation, the grid codes stipulates now more restrictive conditions related to the case when the wind turbine systems are allowed to be disconnected. Figure 4.40 shows an example of the limit voltage dip, particularly in Ireland and Romania, beyond which the wind turbine generator may be disconnected.

According to Figure 4.40, for voltage values situated anywhere above the curve the wind turbine system must remain connected to the power grid, whereas in case of more severe faults with voltage dips below the curve the limit wind turbine system is allowed to disconnect.

A voltage dip experienced at the wind turbine generator terminals leads to a high transient rotor current, which may harm the rotor converter and also may cause overvoltage on the DC-link capacitor.

If the protective actions are less efficient, the fault must be cleared very quickly. Also, the generator may accelerate considerably beyond its rated speed leading to instability.

Different alternatives for fault ride-through schemes are described in the literature. Some of these schemes are as follows [8,39]:

(i) *Active Crowbar*
This solution is typically employed in doubly fed induction generator-based wind turbine systems. The crowbar resistance is sized for values between 1 and 10 times the rotor resistance [9]. When necessary to quickly damp the rotor transient current, the crowbar may be sized at higher resistance values, with the risk that a too high resistance value may lead to overvoltage on the converter. Therefore, the size of the crowbar resistance is a compromise between these two factors.

Occurrence of a fault affecting the wind turbine system is sensed by overvoltage protection attached to the DC-link capacitor and overcurrent protection attached on the rotor side. After protection triggering the power converter control is blocked and the crowbar is activated. Once the transients are damped, the power converter is deblocked and the crowbar is deactivated. The next step is resynchronization and return to normal operation.

(ii) *Switched Stator Resistance*
Disconnection of a faulty line causes a voltage phase-angle jump at the wind turbine generator terminal and hence a high transient current. One solution to this inconvenience can be the insertion of a switched resistance in series with the stator circuit. This resistance causes the stator and rotor transient current to decay more rapidly [40].

(iii) *DC-Link Chopper*
The chopper consists of an active switch (IGBT) and series-connected resistor placed in parallel with the DC-link capacitor (Figure 4.26). The DC-link chopper is a preferred fault ride-through device under the actual grid codes specifications as it helps the wind turbine system to remain connected to the power grid during grid faults.

The aim of the DC-link chopper is to provide smoothing of the DC-link voltages during power fluctuations in the intermediate link of the converter caused by grid faults. When the DC-link voltage increases above a threshold, the chopper is switched "on" and the surplus power is dissipated in the chopper resistor. When the DC-link voltage decreases below a lower uncritical threshold, the chopper is switched "off" by means of a switching device.

In the case of a DFIG-based wind turbine system, the DC-link chopper is used in combination with the crowbar, while in case of full-scale converter wind turbine systems, the chopper is a part of the fault ride-through strategy together with the pitch controller and damping controller.

4.5.2 Blade Pitch Angle Control for Fault Ride-Through

From the power system point of view, the dynamic influence of wind turbine systems depends very much on the technology used, that is, fixed-speed wind turbine systems, DFIG-based wind turbine systems, and full-scale converter variable-speed wind turbine systems.

The fixed-speed wind turbine systems using squirrel cage induction generators provide natural damping of power system oscillations as the stator is directly connected to the power grid, whereas in the case of variable-speed wind turbine systems, the power electronic interface contributes to enhancement of transient stability and voltage stability of the power system.

The fault ride-through capability of two types of wind turbine systems are presented in the following:

(i) *The Case of the Active Stall-Controlled Wind Turbine* [23]
An active stall control is usually applied to wind turbine systems equipped with squirrel cage induction generators, which are connected directly to the power grid. Therefore, in case of a transient fault, during which the generator experiences a voltage dip at its terminals, the rotor speed increases and the generator demands more reactive power that might not be acceptable under the actual grid codes. This demand for reactive power by the generator makes the voltage to recover slowly. The wind turbine system experiences speed oscillations at the drive train, which imply active and reactive power oscillations and consequently voltage oscillations induced into the grid.

A controller is designed to damp the oscillations and to prevent the generator rotor from accelerating to dangerous speed values and also to assist in voltage recovery. In active stall control, pitch angle regulation is a way of controlling the output power of the turbine and thus the transient fault controller is a pitch angle controller. As soon as a grid fault is detected, the controller reduces the aerodynamic power of the rotor to zero by changing the pitch angle with the maximum pitch rate. If the grid voltage has recovered and the generator speed is in normal range, the pitch angle returns to the value prior to the fault and the generator resumes to the normal operation.

(ii) *The Case of Pitch Angle-Controlled Wind Turbine*
The pitch angle control is employed in the case of variable-speed wind turbines, that is, doubly fed induction generator-based wind turbine system and full-scale converter connected wind turbine system. The ridding through fault procedures are different for the two types of wind turbine systems since in the case of the DFIG the stator is connected directly to the power grid [41].

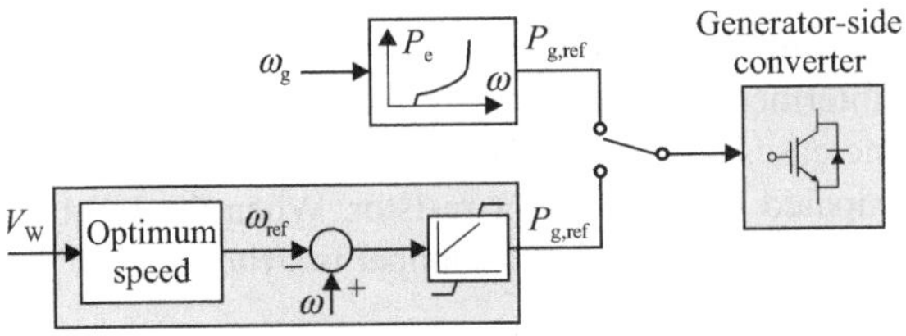

Figure 4.41. Torsional oscillations damping controller [10].

In the case of a DFIG, the fault ride-through procedure involves both the pitch angle control to prevent overspeeding and the so-called crowbar to protect the converter against overcurrents and the generator rotor and the DC-link against overvoltages. In normal operation, the rotor-side converter controls the output active and reactive powers, with the reference for the active power being given by the MPT characteristic, whereas when a grid fault is detected the active power reference is provided by a damping controller. Because the stator is connected directly to the power grid, when a fault occurs in the grid, the terminal voltage drops and the active power drops too. This produces an imbalance between the mechanical power of the turbine and the output electrical power resulting in acceleration of the turbine and generator, and the drive train starts to oscillate. The blade pitch angle control prevents the turbine from overspeeding and damps low-frequency oscillations but it is not able to damp the fast torsional oscillations. Therefore, a damping controller is employed to damp any torsional oscillations of the drive train under grid fault conditions (Figure 4.41).

A wind turbine system connected to the power grid via full-scale frequency converter can easily ride through fault since the synchronous generator is not directly connected to the power grid. Moreover, the generator-side converter can easily fulfill its task to control the DC-link voltage and the stator voltage without being influenced by the power grid disturbance.

A multipole PSMG has no inherent damping and any small speed perturbation can cause high mechanical stress on the drive train that can lead to instability, unless a damping controller is employed to damp the torsional oscillations. During a grid fault, the grid-side converter does not exchange active power with the power system, and the surplus of active power produced by the generator is dissipated in the resistance of the chopper module. On the other hand, the pitch angle controller rapidly limits the power extracted from the wind in order to limit the active power produced by the generator [41].

REFERENCES

[1] Clark, K., Miller, N., Sanchez-Gasca, J. *Modeling of GE wind turbine-generator for grid studies*, General Electric International, USA, April 16, 2010.

[2] *Wind turbine generator system. Technical documentation.* GE 2.5xl, General Electric, 2005.

[3] Ackermann, T. (editor). *Wind power in power systems*, John Wiley & Sons, Ltd, Chichester, 2009.

[4] Betz, A. *Wind-Energie und ihre Ausnutzung durch Windmühlen*, Vandenhoeck und Ruprecht, Göttingen, 1926.

[5] Anaya-Lara, O., Jenkins, N., Ekanayake, J., Cartwright, P., Hughes, M. *Wind energy generation. Modelling and control*, John Wiley & Sons, Ltd., Chichester, 2009.

[6] Heier, S. *Grid integration of wind energy conversion systems*, 2nd edition, John Wiley & Sons, Ltd., Chichester, 2006.

[7] Laverdure, N., Roye, D., Bacha, S., Belhomme, R. *Technologie des systèmes éoliens—Intégration dans les réseaux électrique*, Revue 3EI, 2004.

[8] Perdana, A. *Dynamic models of wind turbines. A contribution towards the establishment of standardized models of wind turbines for power system stability studies*, Ph.D. Thesis, Chalmers University of Technology, Göteborg, Sweden, 2008.

[9] Niiranen, J. *Voltage dip ride-through of a doubly fed generator equipped with and active crowbar*, Proceeding of Nordic Wind Power Conference, Gothenburg, Sweden, March 2004.

[10] Hansen, A., Iov, F., Sørensen, P., Cutululis, N., Jauch, C., Blaabjerg, F. *Dynamic wind turbine models in power system simulation tool DIgSILENT*, Risø-R-1400(ed.2)(EN), Technical University of Denmark, Aug. 2007.

[11] Grauers, A. *Design of a direct driven permanent magnet generators for wind turbines*, Ph.D. Thesis, Chalmers University of Technology, Report No. 292L, 1996.

[12] Chen, Z., Spooner, E. Grid interface options for variable-speed permanent magnet generators, *IEEE Proc. Electric Power Appl.*, Vol. 145, No. 4, pp. 273–283, 1998.

[13] Muyeen, S.M., Tamura, J., Murata, T. *Stability augmentation of a grid connected wind farm*, Green Energy and Technology, Springer-Verlag, London, 2009.

[14] Akhmatov, V., Nielsen, A.H., Pedersen, J.K., Nymann, O. Variable-speed wind turbines with multi-pole synchronous permanent magnet generators, Part I: Modelling in dynamic simulation tools, *Wind Eng.*, Vol. 27, pp. 531–548, 2003.

[15] Mihet-Popa, L., Blaabjerg, F., Boldea, I. Wind Turbine Generator Modeling and Simulation Where Rotational Speed is the Controlled Variable, *IEEE Trans. on Industry Applications*, Vol. 40, No. 1, Jan./Feg. 2004.

[16] Pierik, J.T.G., Morren, J., Wiggelinkhuizen, E.J., De Haan, S.W.H., Van Engelen, T.G., Bozelie, J. *Electrical and control aspects of offshore wind farms II* (ERAO II, Vol. 1: Dynamic models of wind farms), Jun. 2004.

[17] Wasynczuk, O., Man, D., Sullivan, J. Dynamic behaviour of a class of wind turbine generator during random wind fluctuations, *IEEE Trans. Power App. Syst.*, Vol. PAS-100, No. 6, pp. 2837–2845, Jun. 1981.

[18] Burton, T., Sharpe, D., Jenkins, N., Bossanyi, E. *Wind energy handbook*, John Wiley & Sons, Ltd., Chichester, 2001.

[19] Slootweg, J.G. *Wind Power: Modelling and Impact on Power System Dynamics*, PhD Thesis, Delft University of Technology, Netherlands, 2003.

[20] Răzusi, P.C., Eremia, M. *Prediction of Wind Power by Artificial Intelligence Techniques*, IEEE 16th International Conference on Intelligent System Applications to Power Systems – ISAP, Hersonissos, Crete, Greece, 25–28 September 2011.

[21] Andersen, P.M., Bose, A. Stability simulation of wind turbine systems, *IEEE Trans. Power App. Syst.*, Vol. 102, No. 12, pp. 3791–3795, 1983.

[22] Li, H., Zhao, B., Yang, C., Chen, H.W., Chen, Z. Analysis and estimation of transient stability for a grid-connected wind turbine with induction generator, *Renewable Energy*, Elsevier, No. 36, pp. 1469–1476, 2011.

[23] Jauch, C. *Stability and control of wind farms in power systems*, Riso Ph.D. dissertation, Aalborg University, Riso National Laboratory, Roskilde, Denmark, Oct. 2006.

[24] Kundur, P. *Power system stability and control*, McGraw-Hill, Inc., New York, 1994.

[25] Akhmatov, V. *Analysis of dynamic behaviour of electric power systems with large amount of wind power*, Ph.D. Thesis, Technical University of Denmark, Kgs. Lyngby, Denmark, 2003.

[26] Slootweg, J.G., Polinder, H., Kling, W.L. *Dynamic modelling of a wind turbine with doubly feed induction generator*, 2001 IEEE Power Engineering Society Summer Meeting, pp. 644–649, 2001.

[27] Holdsworth, L., Wu, X., Ekanayake, J.B., Jenkins, N. Comparison of fixed speed and doubly-fed induction wind turbines during power system disturbances, *IEE Proc. Generat. Transm. Distrib.*, Vol. 150, pp. 343–352, 2003.

[28] Akhmatov, V. *Modelling of a variable-speed wind turbines with doubly-fed induction generators in short-term stability investigations*, The 3rd International Workshop on Transmission Networks for Offshore Wind Farms, Stockholm, Sweden, 2002.

[29] Boldea, I. *Variable speed generators*, CRC Taylor & Francis Group, 2006.

[30] Pena, R.S., Clare, J.C., Asher, G.M. Doubly fed induction generator using back-to-back PWM converters and its application to variable wind-energy generation, *IEEE Proc. Electrical Power Appl.*, Vol. 143, No. 3, pp. 231–241, 1996.

[31] Müller, S., Deicke, M., De Doncker, R.W. Doubly fed induction generator systems for wind turbines, *IEEE Ind. Appl. Mag.*, Vol. 8, No. 3, pp. 26–33, 2002.

[32] Hughes, F.M., Anaya-Lara, O., Jenkins, N., Strbac, G. Control of DFIG-based wind generation for power network support, *IEEE Trans. Power Syst.*, Vol. 20, No. 4, pp. 1958–1966, 2005.

[33] Harnefors, L., Nee, H.P. Model-based current control of a.c. machines using the internal model control method, *IEEE Trans. Ind. Appl.*, Vol. 34, No. 1, pp. 133–141, 1998.

[34] Xu, L., Cartwright, P. Direct active and reactive power control of DFIG for wind energy generation, *IEEE Trans. Energy Conv.*, Vol. 21, No. 3, pp. 750–758, 2006.

[35] Wang, Z., Sun, Y., Li, G., Ooi, B.T. Magnitude and frequency control of grid-connected doubly-fed induction generator based on synchronised model for wind power generation, *IET Renewable Power Generat.*, Vol. 4, No. 3, pp. 232–241, 2010.

[36] Ekanayake, J.B., Holdsworth, L., Jenkins, N. Control of doubly fed induction generator (DFIG) wind turbine, *IEE Power Eng.*, Vol. 17, No. 1, pp. 28–32, 2003.

[37] Laverdure, N. *Sur l'intégration des generateurs éoliens dans les reseaux faible ou insulaires*, Ph.D. These, Institut National Polytechnique de Grenoble, 2005.

[38] Cristea, C., Peças-Lopes, J.A., Eremia, M., Toma, L. *The control of isolated power systems with wind generation*, 2007 IEEE Lausanne PowerTech, Lausanne, Switzerland, 1–5 Jul. 2007.

[39] Petersson, A. *Analysis, modelling and control of doubly-fed induction generators for wind turbines*, Licentiate thesis Chalmers University, Göteborg, Sweden, 2003.

[40] Feddersen, L. *Circuit arrangement and methods for use in a wind energy installation*, US Patent No. 7,102,247, Vestas Wind System A/S, Sep. 2006.

[41] Hansen, A., Cutululis, N., Markou, H., Sørensen, P., Iov, F. *Grid fault and design-basis for wind turbines*, Risø-R-1714(EN), Technical University of Denmark, Jan. 2010.

5

SHORT-CIRCUIT CURRENTS CALCULATION

Nouredine Hadjsaid, Ion Triştiu, and Lucian Toma

5.1 INTRODUCTION

A short circuit is defined as any accidental contact—direct or through an impedance—between two points of a power system that normally have different voltage, for example, between the conductors of two phases of a line or transformer, or between a conductor and earth, or between a conductor and permanently grounded components, such as the ground wires of overhead lines and the frames of electrical machines.

The analysis of short-circuit currents and related voltages as well as the power flows is useful for

- adequately setting up protection devices, which detect the fault and actuate the circuit breakers of the faulted component (line, transformer, generator, motor) thereby minimizing short-circuit consequences;
- checkup (determining) the breaking capacity of circuit breakers;
- changing the network structure, where possible, so as to make the short-circuit currents less severe.

Handbook of Electrical Power System Dynamics: Modeling, Stability, and Control. Edited by Mircea Eremia and Mohammad Shahidehpour.

5.1.1 The Main Types of Short Circuits

There are various types of short circuits that can occur in electrical installations [1]:

- Phase-to-earth (80% of faults).
- Phase-to-phase (15% of faults). This type of fault often degenerates into a three-phase fault.
- Three phase (only 5% of initial faults).

These different short circuits (abnormal connections) and their corresponding currents are presented in Figure 5.1 [2]. The primary characteristics of a short circuit are as follows:

- Duration (self-extinguishing, transient, and steady state).
- Origin:
  - lightning or switching overvoltages;
  - insulation breakdown due to heat, humidity, or a corrosive environment;
  - mechanical (break in a conductor, accidental electrical contact between two conductors via a foreign conducting body such as a tool or an animal).
- Location (inside or outside a machine or an electrical switchboard).

In Figure 5.1 the following notations have been used:

$\underline{I}''_{k1}$ is the initial phase-to-earth short-circuit current;
$\underline{I}''_{k2}$ is the initial phase-to-phase short-circuit current;
$\underline{I}''_{k2Ea}$, $\underline{I}''_{k2Eb}$ are the initial phase-to-phase short-circuit currents, flowing to earth from phase a or b, respectively;
$\underline{I}''_{kE2E}$ is the initial phase-to-phase short-circuit current flowing to earth;
$\underline{I}''_{k3}$ is the initial three-phase short-circuit current.

Note: It is necessary to distinguish between the short-circuit current at the short-circuit location and the partial short-circuit currents in the branches at any point of the network.

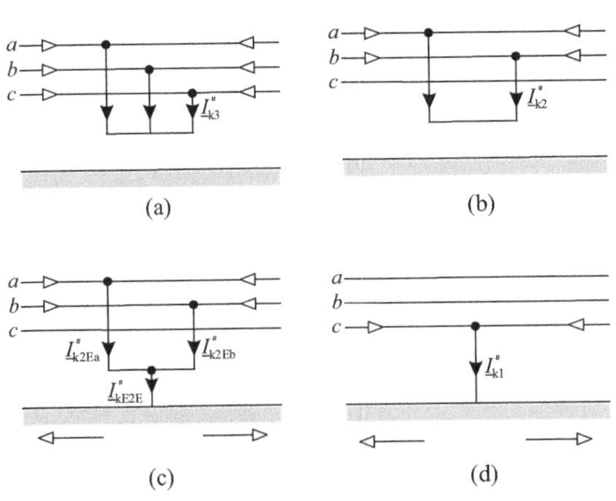

Figure 5.1. Different types of short circuits and their currents [2]: (a) symmetrical three-phase short circuit; (b) phase-to-phase short circuit clear of earth; (c) phase-to-phase-to-earth short circuit; (d) phase-to-earth short circuit. ➤ short-circuit current ▷ partial short-circuit currents in conductors and earth return

5.1.2 Consequences of Short Circuits

The consequences are variable depending on the type and the duration of the fault, the point in the installation where the fault occurs, and the short-circuit power. According to the work [2] of Schneider Electric, the consequences include

- at the fault location, the presence of electrical arcs, resulting in:
  - damage of insulation;
  - welding of conductors;
  - fire and danger to life;
- on the faulty circuit:
  - electrodynamic forces, resulting in: deformation of the busbars, disconnection of cables;
  - excessive temperature rise due to an increase in Joule losses, with the risk of damage to insulation;
- on other circuits in the network or in nearby by networks:
  - voltage dips during the time required to clear the fault, ranging from a few milliseconds to a few hundred milliseconds;
  - shutdown of a part of the network, the extent of that part depending on the design of the network and the discrimination levels offered by protection devices;
  - dynamic instability and/or the loss of machine synchronization;
  - disturbances in control/monitoring circuits, and so on.

As a consequences of these abnormal connections, the structure of the power system impedance (or admittance) network suddenly changes, entailing the passage of abnormal currents both through the accidental connection represented by a short circuit itself and through the different components of the power system; such currents are called short-circuit currents [3].

Usually, the maximum currents that may flow in the various components of the power system are the short circuit ones. These currents, which may reach very high values, have particularly adverse effects on the different components of the system (i.e., not only those directly involved by the fault) and on the operation of the system itself. Among these effects, the main ones (which explain why short-circuit currents should be known) are the following:

- *Thermal effects*, that is, increased temperature of conductors due to the increase of losses by the Joule effects; this effect obviously depends on both intensity and duration of the current.
- *Electrodynamic stresses* between the conductors of the same components (line, transformer, motors, generators) also depending on their spacing.
- *Variations of power flows on links* and, thus, of the powers generated by alternators, jeopardizing the stability of the system. In the effect, a short circuit represents a large perturbation and can cause loss a synchronism of one or more generators opening of links, separation of the network in two or more parts, nonsupply of all or part of the load.

5.2 CHARACTERISTICS OF SHORT-CIRCUIT CURRENTS

As it was shown in subsection 2.1.3.7, the short-circuit current contains a subtransient period, a transient, and a steady state one. The maximum value of the short-circuit current during the transient period occurs when the fault happens as the voltage passes through zero (Figure 2.18a), the minimal value occurs when the fault happens at the peak value of the voltage, in which case the aperiodical component disappears.

The transient and subtransient periods of short-circuit current are not identical on the three phases of the network. The three-phase stabilized short-circuit currents are, however, identical in their effective values on all the three phases.

A complete calculation of short-circuit currents should give the currents as a function of time at the short-circuit location from the initiation of short circuit up to its end, corresponding to the instantaneous value of the voltage at the beginning of short circuit [1].

If three-phase generators are short-circuited, either directly or through system impedances, this results in short-circuit currents that normally start with a high peak value but which decay to a steady-state value in accordance with the time constants of the circuits. The curve showing this process is represented in Figures 5.2a and b.

The transient conditions, prevailing while the short-circuit current develops, differ depending on the distance between the fault location and the generator. This distance is not necessarily physical, but means that the generator impedances are less than the link impedance, between the generator and the fault location [2].

- *Far-from-generator short circuit.* Short circuit during which the magnitude of the symmetrical AC component of the prospective (available) short-circuit current remains essentially constant.
- *Near-to-generator short circuit.* Short circuit to which at least one synchronous machine delivers an initial symmetrical short-circuit current, which is more than twice the machine's rated current, or a short circuit to which synchronous or asynchronous machines contribute more than 5% of the initial symmetrical short-circuit current (I''_k) without motors.

In the case of a far-from-generator short circuit (Figure 5.2a), the short-circuit current can be considered as the sum of two components:

- The AC component with constant amplitude during the whole short circuit.
- The aperiodic DC component beginning with an initial value A decaying to zero.

In case of a near-to-generator short circuit (Figure 5.2b), the short-circuit current can be considered as the sum of the following two components:

- The AC component with decaying amplitude during the short circuit.
- The aperiodic DC component beginning with an initial value A decaying to zero.

In practical applications, the following specific quantities present a major interest:

- *The initial symmetrical short-circuit current, I''_k, is the effective value of the AC symmetrical component* and serves as a basis for the calculation of the peak asymmetrical short-circuit current, i_p, as well as the breaking current and capacity.

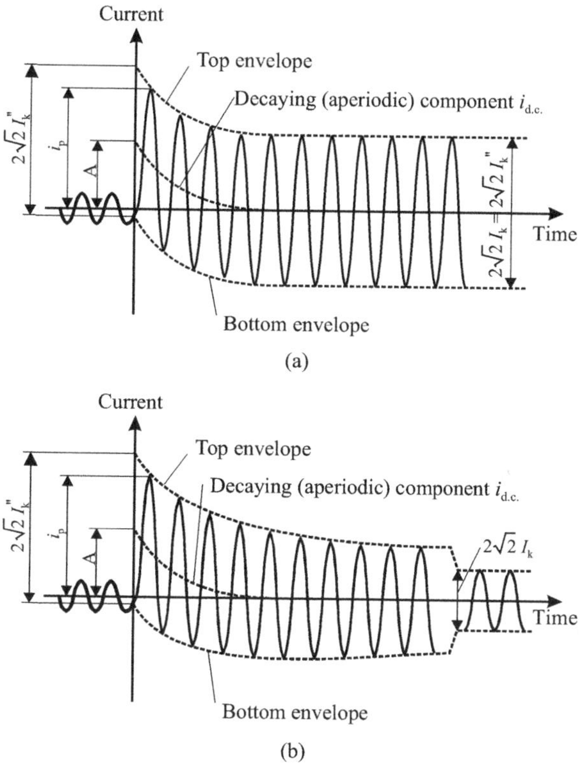

Figure 5.2. Typical wave of a short-circuit current: (a) far-from-generator short circuit with constant AC component; (b) near-to-generator short circuit with decaying AC component. (Reprinted with permission from IEC 60909. [1])

- *The peak asymmetrical short-circuit current*, i_p, is the maximum instantaneous value of this current during initial short-circuit conditions and is indicated as a peak value. In addition to the AC component, it includes the DC component $i_{d.c.}$, which occurs during a sudden short circuit. The value i_p determines the dynamic stresses in the components of the installation and also the load carried by the switchgear.
- *The breaking current*, I_b, is the effective value of the AC component of the short-circuit current at the instant of contact separation in the circuit breaker;
- *Initial symmetrical short-circuit power* S_k''. Fictive value determined as a product of the initial symmetrical short-circuit current I_k'' and the rated system voltage $U_r(S_k'' = \sqrt{3}U_rI_k'')$.
- *DC component* $i_{d.c.}$ of the short-circuit current and its initial value A.
- *The steady-state short-circuit current*, I_k, is rms value of a symmetrical current, which remains after the decay of the transient phenomena; it is quoted as an effective value.

In most practical cases, a determination like this is not necessary. Depending on the application of the results, it is of interest to know the rms value of the symmetrical AC component and the peak value i_p of the short-circuit current following the occurrence of the short circuit. The highest value i_p depends on the time constant of the decaying aperiodic component and the frequency f, that is, on the ratio R/X or X/R of the short-circuit impedance $\underline{Z}_k$, and is reached if the short circuit starts at zero voltage. The current i_p also depends on the decay of the symmetrical AC component of the short-circuit current (see 5.4.2).

In meshed networks, there are several direct current time constants. That is why it is possible to give an easy method for calculating i_p and $i_{d.c.}$. Special methods to calculate i_p with sufficient accuracy are given in Standard IEC 60909 [1].

In the case of a *near-to-generator short circuit* (Figure 5.2b), the variation in the impedance of the generator, in this case, the dominant impedance damps the short-circuit current.

The transient current-development conditions are complicated by the variation in the source voltage (electromotive force) resulting from the short circuit. For simplicity, the electromotive force is assumed to be constant and the internal reactance of the machine is variable [2].

The reactance develops in three stages:

- Subtransient (the first 10–20 milliseconds of the fault).
- Transient (up to 500 milliseconds).
- Steady state (or synchronous reactance).

Note that in the indicated order, the reactance acquires a higher value at each stage, that is, the subtransient reactance is less than the transient reactance, itself less than the steady-state reactance.

The successive effect of the three reactances leads to a gradual reduction in the short-circuit current, which is the sum of four components (Figure 5.3):

- The three alternating components (subtransient, transient, and steady state).
- The aperiodic component resulting from the development of the current in the circuit (inductive).

Note that the decrease in the generator reactance is faster than that of the aperiodic component. This is a rare situation that can cause saturation of the magnetic circuits and interruption problems because several periods occur before the current passes through zero [2].

Practically speaking, information on the development of the short circuit is not essential:

- In a LV installation, due to the speed of the breaking devices, the value of the subtransient short-circuit current, denoted I''_k, and of the maximum asymmetrical peak amplitude i_p is sufficient when determining the breaking capacities of the protection devices and the electrodynamic forces.
- In LV power distribution and in HV applications, however, the transient short-circuit current is often used if breaking occurs before the steady-state stage, in which case it becomes useful to use the short-circuit breaking current, denoted I_b, which determines the breaking capacity of the time-delayed circuit breakers. I_b is the value of the short-circuit current at the instant at which interruption is effective, that is, following a time t after the development of the short circuit, where $t = t_{min}$. Time t_{min} (minimum time delay) is the sum of the minimum operating time of a protection relay and the shortest opening time of the associated circuit breaker, that is, the

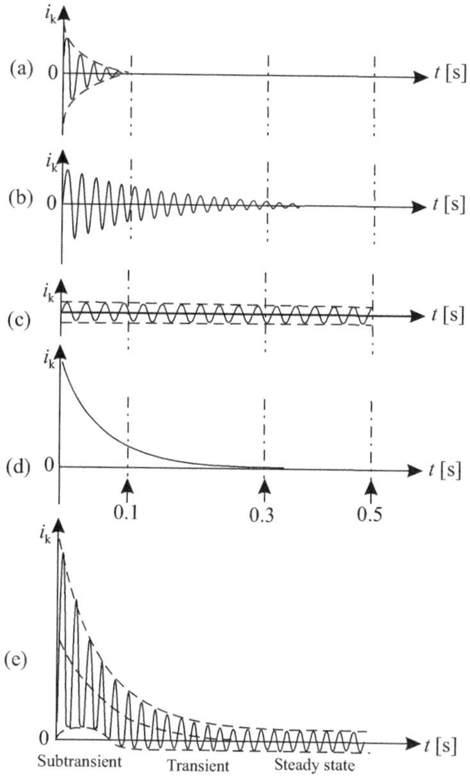

Figure 5.3. Total short-circuit current i_k and contribution of its components [2]: (a) subtransient reactance; (b) transient reactance; (c) steady-state reactance; (d) aperiodic component.

shortest time between the appearance of the short-circuit current and the initial separation of the pole contacts on the switching device.

Figure 5.4 presents the various currents of the short circuits defined above.

In the calculation of the short-circuit currents in systems supplied by generators, power station units and motors (near-to-generator and/or near-to-motor short circuits), it is of

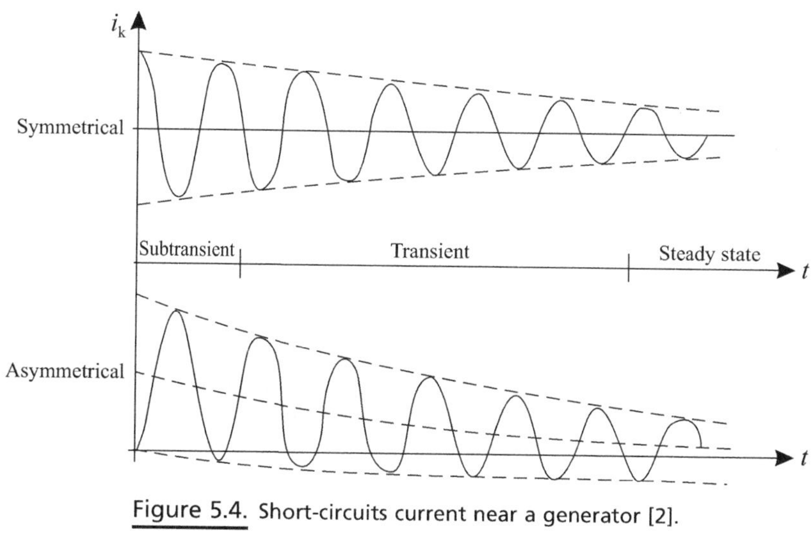

Figure 5.4. Short-circuits current near a generator [2].

interest not only to know the initial symmetrical short-circuit current I''_k and the peak short-circuit current i_p, but also the symmetrical short-circuit breaking current I_b and the steady state short-circuit current I_k. In this case, the symmetrical short-circuit breaking current I_b is smaller than the initial symmetrical short-circuit current I''_k. Normally, the steady-state short-circuit current I_k is smaller than the symmetrical short-circuit breaking current I_b.

For the calculation of the initial symmetrical short-circuit current I''_k, the symmetrical short-circuit breaking current I_b, and the steady-state short-circuit current I_k at the short-circuit location, the system may be converted by network reduction into an equivalent short-circuit impedance $\underline{Z}_k$ seen from the short-circuit location. This procedure is not allowed when calculating the peak short-circuit current i_p; in this case, it is necessary to distinguish between networks with and without parallel branches.

While using fuses or current-limiting circuit breakers to protect substations, the initial symmetrical short-circuit current is first calculated as if these devices were not available. From the calculated initial symmetrical short-circuit current and characteristic curves of the fuses or current-limiting circuit breakers, the cut-off current is determined, which is the peak short-circuit current of the downstream substation.

5.3 METHODS OF SHORT-CIRCUIT CURRENTS CALCULATION

5.3.1 Basic Assumptions

To simplify the short-circuit currents calculations, a number of assumptions are required. These impose limits for which the calculations are valid but usually provide good approximations, facilitating comprehension of the physical phenomena, and consequently the short-circuit current calculations. They nevertheless maintain a fully acceptable level of accuracy, "erring" systematically on the conservative side.

IEC 60909 is applicable for calculation of short-circuit currents [1]

- in low voltage three-phase AC systems;
- in high voltage three-phase AC systems.

Systems at highest of 550 kV and above with long transmission lines need special considerations.

The assumptions used are as follows [1,2]:

- The short-circuit current, during a three-phase short-circuit, is assumed to occur simultaneously on all three phases.
- During the short circuit, the number of phases involved does not change, that is, a three-phase fault remains three-phase and a phase-to-earth fault remains phase-to-earth.
- For the entire duration of the short circuit, the voltages responsible for the flow of the current and the short-circuit impedance do not change significantly.
- Transformer regulators or tap changers are assumed to be set to a medium position (if the short circuit occurs away from the generator, the actual position of the transformer regulator or tap changers does not need to be taken into account).
- Arc resistances are not taken into account.
- All line capacitances are neglected.

- Load currents are neglected.
- All zero sequence are taken into account.

When calculating short-circuit currents in systems with different voltage levels, it is necessary to transfer impedance values from one voltage level to another, usually to that voltage level at which the short-circuit current is to be calculated. For per unit or other similar unit systems no transformation is necessary, if these systems are coherent, that is, $U_{\mathrm{rTHV}}/U_{\mathrm{rTLV}} = U_{\mathrm{nHV}}/U_{\mathrm{nLV}}$ for each transformer in the system with partial short-circuit currents. $U_{\mathrm{rTHV}}/U_{\mathrm{rTLV}}$[1] is normally not equal to $U_{\mathrm{nHV}}/U_{\mathrm{nLV}}$[2] [1].

- U_{rTHV} and U_{rTLV} are the rated voltages of the transformer on the high voltage or low voltage side;
- U_{nTHV} and U_{nTLV} are the nominal system voltages of the transformer on the high voltage or low voltage side.

The impedances of the equipment in superimposed or subordinated networks are to be divided or multiplied by the square of the rated transformation ratio N_{r}. Voltages and currents are to be converted by the rated transformation ratio N_{r}.

5.3.2 Method of Equivalent Voltage Source

One way of calculating the short-circuit current assumes the introduction of an equivalent voltage source at the short-circuit location. The equivalent voltage source is the only active voltage of the system. All network feeders, synchronous and asynchronous machines are replaced by their internal impedance.

In all cases, it is possible to determinate the short-circuit current at the short-circuit location F with the help of an equivalent voltage source. Operational data and the load of consumers, tap changer position of transformers, excitation of generators, and so on, are dispensable; additional calculations about all the different possible load flows at the instant of short circuit are superfluous.

Figure 5.5 shows an example of the equivalent voltage source at the short circuit location F as the only active voltage of the system fed by a transformer without

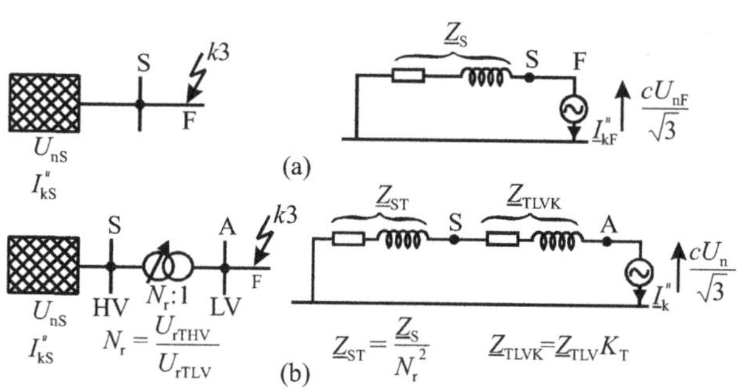

Figure 5.5. System diagram and equivalent circuit diagram for network feeders: (a) without transformer; (b) with transformer. (Reprinted with permission from IEC 60909 [1].)

[1] U_r denotes the rated phase-to-phase voltage;

[2] The nominal system voltage (U_n) is the voltage (phase-to-phase) at which a system is designated, and to which certain operating characteristics are referred.

or with on-load tap changer. All other active voltages in the system are assumed to be zero.

- If a three-phase short circuit is fed from a network in which only the initial symmetrical short-circuit current I''_{kS} at feeder connection point S is known (Figure 5.5a), then the equivalent impedance $\underline{Z}_S$ of the network (positive-sequence short-circuit impedance) at the feeder connection point S should be determined by

$$Z_S = \frac{cU_{nS}}{\sqrt{3}I''_{kS}} \tag{5.1}$$

If R_S/X_S is known, then X_S shall be calculated by

$$X_S = \frac{Z_S}{\sqrt{1 + (R_S/X_S)^2}} \tag{5.2}$$

- If the short circuit is fed by a transformer (Figure 5.5b) from a medium- or high-voltage network in which only the initial symmetrical short-circuit current I''_{kS} at the feeder connection point S is known, then the positive-sequence equivalent short-circuit impedance Z_{ST} referred to the low-voltage side of the transformer is to be determined by

$$Z_{ST} = \frac{cU_{nS}}{\sqrt{3}I''_{kS}} \cdot \frac{1}{N_r^2} \tag{5.3}$$

where

U_{nS} is the nominal system voltage at the feeder connection point S;

I''_{kS} is the initial symmetrical short-circuit current at the feeder connection point S;

c is the voltage factor (see Table 5.1) for the voltage U_{nS};

N_r is the rated transformation ratio at which on-load tap changer is in the main position.

- In the case of high-voltage feeders with nominal voltage above 35 kV fed by overhead lines (Figure 5.6), the equivalent impedance $\underline{Z}_S$ may be considered in many cases as a reactance, that is, $\underline{Z}_S \cong jX_S$. In other cases, if no accurate value is known for the resistance R_S of network feeders, we may substitute $R_S = 0.1\,X_S$, where $X_S = 0.995\,Z_S$.

TABLE 5.1. Voltage Factor c (Reprinted from IEC 60909-0 [1])

	Voltage Factor c for the calculation of	
Nominal Voltage U_n	Maximum Short-Circuit Currents c_{max}[a]	Minimum Short-Circuit Currents c_{min}
Low voltage: 100–1000 V	1.05[b] 1.10[c]	0.95
Medium voltage: 1–35 kV **High voltage[d]:** >35 kV	1.10	1.00

[a] $C_{max}U_n$ should not exceed the highest voltage U_m for equipment of power systems.
[b] For low-voltage systems with a tolerance of +6%, for example systems renamed from 380 V to 400 V.
[c] For low-voltage systems with a tolerance of +10%.
[d] If no nominal voltage is defined, $c_{max}U_n = U_m$ or $c_{min}U_n = 0.9U_m$ should be applied.

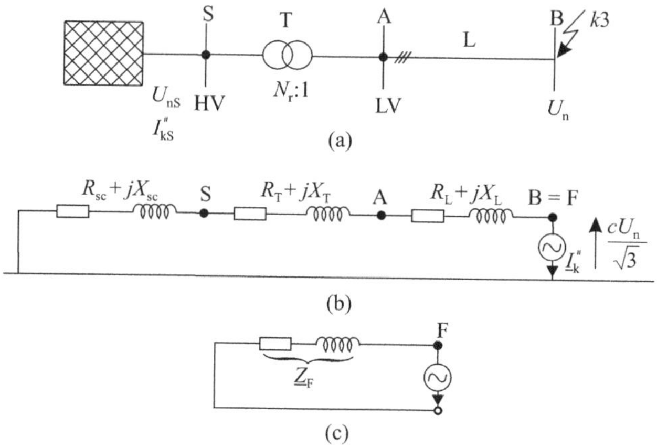

Figure 5.6. Illustrations for calculating the initial symmetrical short-circuit current $\underline{I}''_k$ in compliance with the procedure for the equivalent voltage source: (a) system diagram; (b) equivalent diagram of the positive-sequence system; (c) equivalent diagram with short-circuit impedance. (Reprinted with permission from IEC 60909 [1])

The initial symmetrical short-circuit currents $I''_{kS\,max}$ and $I''_{kS\,min}$ on the high-voltage side of the transformer shall be given by the supply company or by an adequate calculation according to the IEC 60909 standard.

Shunt admittances (e.g., line capacitances and passive loads) are not to be considered when calculating short-circuit currents (Figure 5.6b).

Note: If there are no national standards, it seems adequate to choose a voltage factor c according to Table 5.1, considering that the *highest voltage* in a normal system does not differ, on average, by more than approximately +5% (some LV systems) or +10% (some HV systems) from the nominal system voltage U_n [1].

5.3.3 Method of Symmetrical Components

5.3.3.1 General Principles. The majority of faults in a power system are not symmetrical three-phase short circuits, but occur between one phase and earth or less commonly occur between phases where they may also involve earth. In addition, there is usually a finite fault impedance rather than a short circuit. Thus, instead of a completely balanced three-phase circuit, which can be solved by calculating the current flowing in one of the three phases, an unbalanced three-phase circuit must now be solved. This can most conveniently be done by using the symmetrical components of an unbalanced system of currents or voltages [4].

In 1918, one of the most powerful tools for dealing with unbalanced polyphase circuit was discussed by C.L. Fortescue at a meeting of the American Institute of Electrical Engineers [5]. The method of symmetrical components can be applied to any polyphase system containing any number of phases, but the three-phases system is the only of interest here.

According the Fortescue's theorem, three unbalanced phasors of a three-phases system can be resolved into three fictitious balanced systems of phasors called symmetrical components of the original phasors (Figure 5.7).

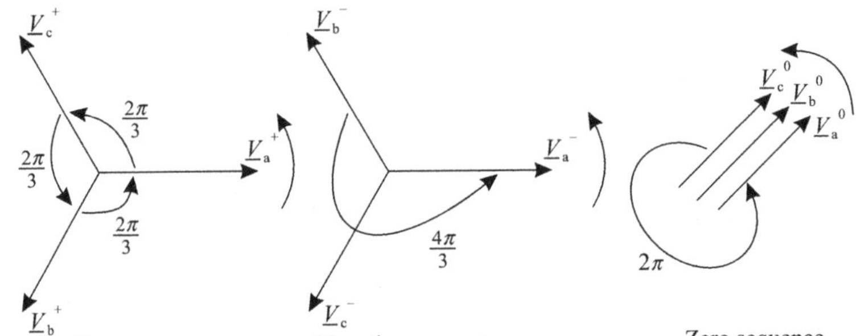

Figure 5.7. Three sets of balanced phasors that form the symmetrical components of three unbalanced phasors.

The three balanced systems of phasors are as follows [6]:

(i) A positive-sequence system consisting of three components (voltages or currents) equal in magnitude, shifted from each other by $2\pi/3$, with the phase sequence same to that of the original phasors a, b, c and denoted by $\underline{I}_a^+$, $\underline{I}_b^+$, $\underline{I}_c^+$ or $\underline{V}_a^+$, $\underline{V}_b^+$, $\underline{V}_c^+$.

(ii) A negative-sequence system consisting of three components (voltages or currents) equal in magnitude, shifted from each other by $2\pi/3$, with the phase sequence opposite to that of the original phasors a, b, c and denoted by $\underline{I}_a^-$, $\underline{I}_b^-$, $\underline{I}_c^-$ or $\underline{V}_a^-$, $\underline{V}_b^-$, $\underline{V}_c^-$.

(iii) A zero-sequence system consisting of three components (voltages or currents) equal in magnitude and with zero phase shifting from each other, and denoted by $\underline{I}_a^0$, $\underline{I}_b^0$, $\underline{I}_c^0$ or $\underline{V}_a^0$, $\underline{V}_b^0$, $\underline{V}_c^0$.

In order to represent the phase shifting of $2\pi/3$ between the components of the positive- and negative-sequence systems, it is convenient to use a rotation operator denoted by $\underline{a}$. This operator causes a rotation of $2\pi/3$ in the trigonometrical direction. It is a complex number of unit magnitude with an angle of $2\pi/3$ and is defined as follows:

$$\underline{a} = 1\underline{|2\pi/3} = 1 \cdot e^{j2\pi/3} = -\frac{1}{2} + j\frac{\sqrt{3}}{2}$$

If the operator $\underline{a}$ is applied to rotate a phasor twice in succession, the phasor is rotated by $2 \cdot 2\pi/3$, which is equivalent to $-2\pi/3$, and

$$\underline{a}^2 = 1\underline{|4\pi/3} = -\frac{1}{2} - j\frac{\sqrt{3}}{2}$$

Three successive applications of $\underline{a}$ rotate the phasor by 2π, thus

$$\underline{a}^3 = 1\underline{|2\pi} = 1\underline{|0^\circ} = 1$$

Therefore,

$$1 + \underline{a} + \underline{a}^2 = 0$$

5.3.3.2 The Symmetrical Components of Unsymmetrical Phasors. Since each of the original unbalanced phasors is the sum of its components, the original phasors expressed in terms of theirs components are [6]

$$\underline{V}_a = \underline{V}_a^0 + \underline{V}_a^+ + \underline{V}_a^- \tag{5.4a}$$

$$\underline{V}_b = \underline{V}_b^0 + \underline{V}_b^+ + \underline{V}_b^- \tag{5.4b}$$

$$\underline{V}_c = \underline{V}_c^0 + \underline{V}_c^+ + \underline{V}_c^- \tag{5.4c}$$

In order to resolve the three unsymmetrical phasors ($\underline{V}_a$, $\underline{V}_b$, and $\underline{V}_c$) into their symmetrical components, we reduce the number of unknown quantities by expressing each component of $\underline{V}_b$ and $\underline{V}_c$ in terms of the operator $\underline{a}$ and the components of $\underline{V}_a$.

Inspection of Figure 5.7 allows us to write:

$$\begin{array}{lll} \underline{V}_a^0 = \underline{V}^0 & \underline{V}_a^+ = \underline{V}^+ & \underline{V}_a^- = \underline{V}^- \\ \underline{V}_b^0 = \underline{V}^0 & \underline{V}_b^+ = \underline{a}^2\ \underline{V} & \underline{V}_b^- = \underline{a}\ \underline{V} \\ \underline{V}_c^0 = \underline{V}^0 & \underline{V}_c^+ = \underline{a}\ \underline{V}^+ & \underline{V}_c^- = \underline{a}^2\ \underline{V}^- \end{array} \tag{5.5}$$

Substituting the equations (5.5) into the equations (5.4) gives:

$$\underline{V}_a = \underline{V}^0 + \underline{V}^+ + \underline{V}^- \tag{5.6a}$$

$$\underline{V}_b = \underline{V}^0 + \underline{a}^2\ \underline{V}^+ + \underline{a}\ \underline{V}^- \tag{5.6b}$$

$$\underline{V}_c = \underline{V}^0 + \underline{a}\ \underline{V}^+ + \underline{a}^2\ \underline{V}^- \tag{5.6c}$$

or expressing in matrix form

$$\begin{bmatrix} \underline{V}_a \\ \underline{V}_b \\ \underline{V}_c \end{bmatrix} = \begin{bmatrix} 1 & 1 & 1 \\ 1 & \underline{a}^2 & \underline{a} \\ 1 & \underline{a} & \underline{a}^2 \end{bmatrix} \begin{bmatrix} \underline{V}^0 \\ \underline{V}^+ \\ \underline{V}^- \end{bmatrix} \tag{5.7}$$

Figure 5.8 illustrates the construction of the phase voltages $\underline{V}_a$, $\underline{V}_b$, and $\underline{V}_c$ in terms of the sequence components $\underline{V}^0$, $\underline{V}^+$, and $\underline{V}^-$, based on equations (5.5) and (5.6).

Let us denote by

$$[\mathbf{T}] = \begin{bmatrix} 1 & 1 & 1 \\ 1 & \underline{a}^2 & \underline{a} \\ 1 & \underline{a} & \underline{a}^2 \end{bmatrix} \tag{5.8}$$

a matrix with constant (complex or real) and independent time coefficients. It allows expressing the unbalanced phasors in terms of balanced components. In this way each of

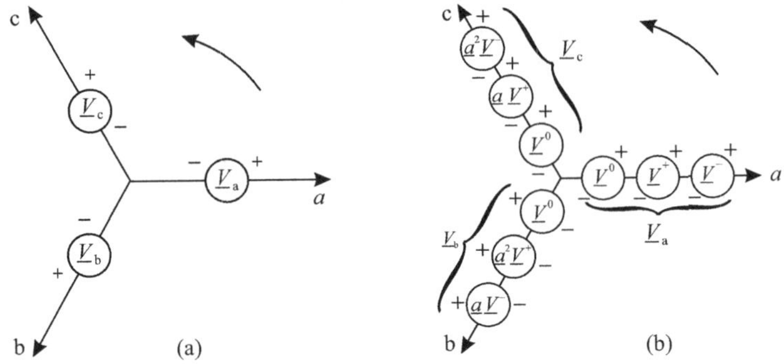

Figure 5.8. Representation of the voltage phasors in terms of sequence voltages.

the unbalanced phase components (current or voltage) can be expressed as a linear combination of positive-, negative-, and zero-sequence components.

Equation (5.7) can thus be written as follows:

$$[\underline{\mathbf{V}}_{abc}] = [\mathbf{T}]\left[\underline{\mathbf{V}}^{0+-}\right] \tag{5.7'}$$

The balanced components can be also expressed in terms of the unbalanced components. It may be verified easily that

$$[\mathbf{T}]^{-1} = \frac{1}{3}\begin{bmatrix} 1 & 1 & 1 \\ 1 & \underline{a} & \underline{a}^2 \\ 1 & \underline{a}^2 & \underline{a} \end{bmatrix} \tag{5.8'}$$

Premultiplying both sides of equation (5.7) by $[\mathbf{T}]^{-1}$ achieve

$$\begin{bmatrix} \underline{V}^0 \\ \underline{V}^+ \\ \underline{V}^- \end{bmatrix} = \frac{1}{3}\begin{bmatrix} 1 & 1 & 1 \\ 1 & \underline{a} & \underline{a}^2 \\ 1 & \underline{a}^2 & \underline{a} \end{bmatrix}\begin{bmatrix} \underline{V}_a \\ \underline{V}_b \\ \underline{V}_c \end{bmatrix} \tag{5.9}$$

This matrix equation can be written as separate equations in ordinary fashion:

$$\underline{V}^0 = \frac{1}{3}(\underline{V}_a + \underline{V}_b + \underline{V}_c) \tag{5.10a}$$

$$\underline{V}^+ = \frac{1}{3}(\underline{V}_a + \underline{a}\,\underline{V}_b + \underline{a}^2\,\underline{V}_c) \tag{5.10b}$$

$$\underline{V}^- = \frac{1}{3}(\underline{V}_a + \underline{a}^2\,\underline{V}_b + \underline{a}\,\underline{V}_c) \tag{5.10c}$$

The sequence components $\underline{V}_b^0$, $\underline{V}_b^+$, $\underline{V}_b^-$, $\underline{V}_c^0$, $\underline{V}_c^+$, and $\underline{V}_c^-$ can be found from equation (5.5).

Equation (5.10a) shows that no zero-sequence component exists if the sum of unbalanced phasors is zero. In a perfectly symmetrical and balanced three-phase system, the sum of the voltage phasors is always zero, and contains no zero-sequence components [6].

Similar to voltages, the currents phasors can be resolved using the theory of symmetrical components. They may be represented either analytically or graphically. The phase currents can thus be written in terms of their sequence components and the rotation operator as

$$\underline{I}_a = \underline{I}^0 + \underline{I}^+ + \underline{I}^- \tag{5.11a}$$

$$\underline{I}_b = \underline{I}^0 + \underline{a}^2\, \underline{I}^+ + \underline{a}\, \underline{I}^- \tag{5.11b}$$

$$\underline{I}_c = \underline{I}^0 + \underline{a}\, \underline{I}^+ + \underline{a}^2\, \underline{I}^- \tag{5.11c}$$

or

$$\underline{I}^0 = \frac{1}{3}(\underline{I}_a + \underline{I}_b + \underline{I}_c) \tag{5.12a}$$

$$\underline{I}^+ = \frac{1}{3}(\underline{I}_a + \underline{a}\, \underline{I}_b + \underline{a}^2\, \underline{I}_c) \tag{5.12b}$$

$$\underline{I}^- = \frac{1}{3}(\underline{I}_a + \underline{a}^2\, \underline{I}_b + \underline{a}\, \underline{I}_c) \tag{5.12c}$$

If a neutral wire is added to the three-phase system, the current that returns through the neutral $\underline{I}_n$ is equal the sum of the phase currents, that is

$$\underline{I}_a + \underline{I}_b + \underline{I}_c = \underline{I}_n \tag{5.13}$$

Substituting equation (5.13) in equation (5.12a) yield:

$$\underline{I}_n = 3\, \underline{I}^0$$

If no return path through neutral exists in the three-phase system, then $\underline{I}_n = 0$, and the phase currents contain no zero-sequence components. A Δ-connected load or transformer provides no return path, and thus the phase currents can contain no zero-sequence components.

Decoupling the Sequence Impedances. The relationship between voltages and currents on a three-phase electrical line can be written as [7, 13]

$$[\underline{\mathbf{V}}_{abc}] = [\underline{\mathbf{Z}}_{abc}][\underline{\mathbf{I}}_{abc}] \tag{5.14}$$

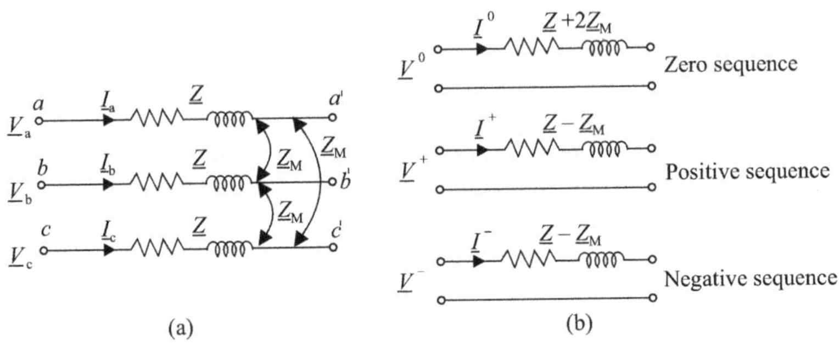

Figure 5.9. Representation of an electrical line: (a) phase components model with mutual coupling; (b) sequence components model.

where, $[\underline{\mathbf{Z}}_{abc}]$ is the impedance matrix of the line (Figure 5.9a), given as:

$$[\underline{\mathbf{Z}}_{abc}] = \begin{bmatrix} \underline{Z}_a & \underline{Z}_{ab} & \underline{Z}_{ac} \\ \underline{Z}_{ba} & \underline{Z}_b & \underline{Z}_{bc} \\ \underline{Z}_{ca} & \underline{Z}_{cb} & \underline{Z}_c \end{bmatrix} \tag{5.15}$$

where, $\underline{Z}_a$, $\underline{Z}_b$, and $\underline{Z}_c$ are the self-impedances of the phases a, b, and c, respectively, whereas $\underline{Z}_{ab} = \underline{Z}_{ba}$, $\underline{Z}_{ac} = \underline{Z}_{ca}$, and $\underline{Z}_{bc} = \underline{Z}_{cb}$ are the mutual impedance between corresponding phases.

Assuming that the three-phase line is constructed perfectly symmetrical, the self-impedances are equal, that is, $\underline{Z}_a = \underline{Z}_b = \underline{Z}_c = \underline{Z}$, and all the mutual impedances are equal between them and equal to $\underline{Z}_M$. Under these conditions, the impedance matrix becomes

$$[\underline{\mathbf{Z}}'_{abc}] = \begin{bmatrix} \underline{Z} & \underline{Z}_M & \underline{Z}_M \\ \underline{Z}_M & \underline{Z} & \underline{Z}_M \\ \underline{Z}_M & \underline{Z}_M & \underline{Z} \end{bmatrix} \tag{5.15'}$$

The above matrix is now fully symmetrical. In order to simplify the calculations, the phase components impedance matrix from (5.15′) can be decoupled using the theory of symmetrical components. Therefore, by appropriate transformation, and taking into account the expression $1 + \underline{a} + \underline{a}^2 = 0$, the sequence components impedance matrix is achieved [7]:

$$\begin{aligned} [\underline{\mathbf{Z}}^{0+-}] &= \frac{1}{3}\begin{bmatrix} 1 & 1 & 1 \\ 1 & \underline{a} & \underline{a}^2 \\ 1 & \underline{a}^2 & \underline{a} \end{bmatrix} \begin{bmatrix} \underline{Z} & \underline{Z}_M & \underline{Z}_M \\ \underline{Z}_M & \underline{Z} & \underline{Z}_M \\ \underline{Z}_M & \underline{Z}_M & \underline{Z} \end{bmatrix} \begin{bmatrix} 1 & 1 & 1 \\ 1 & \underline{a}^2 & \underline{a} \\ 1 & \underline{a} & \underline{a}^2 \end{bmatrix} = \\ &= \begin{bmatrix} \underline{Z} + 2\,\underline{Z}_M & 0 & 0 \\ 0 & \underline{Z} - \underline{Z}_M & 0 \\ 0 & 0 & \underline{Z} - \underline{Z}_M \end{bmatrix} \end{aligned} \tag{5.16}$$

The sequence components impedance matrix is thereby diagonal, the nondiagonal terms being equal to zero, which allow decoupling the sequence components. Figure 5.9a shows the representation of an electrical line in the three-phase components, whereas Figure 5.9b illustrates the three sequence networks, with no coupling between them.

Assuming now that the original three-phase system in not symmetrical, that is, the impedances $\underline{Z}_{\mathrm{a}}$, $\underline{Z}_{\mathrm{b}}$, and $\underline{Z}_{\mathrm{c}}$ are not equal, and neglecting the mutual impedances between phases, the impedance matrix from (5.15) becomes [7]

$$[\underline{\mathbf{Z}}''_{\mathrm{abc}}] = \begin{bmatrix} \underline{Z}_{\mathrm{a}} & & \\ & \underline{Z}_{\mathrm{b}} & \\ & & \underline{Z}_{\mathrm{c}} \end{bmatrix} \tag{5.17}$$

Under these conditions, transformation into sequence components give:

$$\begin{aligned}[\underline{\mathbf{Z}}^{0+-}] &= \frac{1}{3}\begin{bmatrix} 1 & 1 & 1 \\ 1 & \underline{a} & \underline{a}^2 \\ 1 & \underline{a}^2 & \underline{a} \end{bmatrix}\begin{bmatrix} \underline{Z}_{\mathrm{a}} & & \\ & \underline{Z}_{\mathrm{b}} & \\ & & \underline{Z}_{\mathrm{c}} \end{bmatrix}\begin{bmatrix} 1 & 1 & 1 \\ 1 & \underline{a}^2 & \underline{a} \\ 1 & \underline{a} & \underline{a}^2 \end{bmatrix} = \\ &= \frac{1}{3}\begin{bmatrix} \underline{Z}_{\mathrm{a}} + \underline{Z}_{\mathrm{b}} + \underline{Z}_{\mathrm{c}} & \underline{Z}_{\mathrm{a}} + \underline{a}^2\,\underline{Z}_{\mathrm{b}} + \underline{a}\,\underline{Z}_{\mathrm{c}} & \underline{Z}_{\mathrm{a}} + \underline{a}\,\underline{Z}_{\mathrm{b}} + \underline{a}^2\,\underline{Z}_{\mathrm{c}} \\ \underline{Z}_{\mathrm{a}} + \underline{a}\,\underline{Z}_{\mathrm{b}} + \underline{a}^2\,\underline{Z}_{\mathrm{c}} & \underline{Z}_{\mathrm{a}} + \underline{Z}_{\mathrm{b}} + \underline{Z}_{\mathrm{c}} & \underline{Z}_{\mathrm{a}} + \underline{a}^2\,\underline{Z}_{\mathrm{b}} + \underline{a}\,\underline{Z}_{\mathrm{c}} \\ \underline{Z}_{\mathrm{a}} + \underline{a}^2\,\underline{Z}_{\mathrm{b}} + \underline{a}\,\underline{Z}_{\mathrm{c}} & \underline{Z}_{\mathrm{a}} + \underline{a}\,\underline{Z}_{\mathrm{b}} + \underline{a}^2\,\underline{Z}_{\mathrm{c}} & \underline{Z}_{\mathrm{a}} + \underline{Z}_{\mathrm{b}} + \underline{Z}_{\mathrm{c}} \end{bmatrix}\end{aligned} \tag{5.18}$$

The resulting matrix shows that under asymmetrical construction of the three-phase system, the sequence matrix cannot be decoupled. Due to the network asymmetry, unbalanced currents on the three phases occur. The degree of asymmetry is, in general, small, and thereby in practical calculations the original three-phase system is considered symmetrical. When a fault occurs in the power system, except for the case of a symmetrical fault, the network impedances are no longer equal and the voltages and currents are unbalanced; the greatest unbalance occurs at the fault point [7].

Characteristics of the Symmetrical Voltages. Taking into account the representations from Figures 5.7 and 5.8, the characteristics of the sequence voltages are as follows [7]:

- The phasors $\underline{V}^0$ are called *zero-sequence voltage components*, are equal in magnitude in all phases (a, b, and c) and are cophasial, that is, all components are varying together.
- The phasors $\underline{V}^+$ are called *positive-sequence voltage components*, are equal in magnitude in all phases and form a system with the same phase sequence as the original three-phase system, that is, phase b lags phase a by $2\pi/3$, phase c lags phase b, and phase a lags phase c; the positive-sequence is also referred to as abc sequence.
- The phasors $\underline{V}^-$ are called *negative-sequence voltage components*, are equal in magnitude in each phase and are shifted also by $2\pi/3$; the sequence of voltages is changed as compared to the positive sequence, that is, phase b leads phase a rather than lagging, while phase c lags phase a; the order of sequence it thus acba.

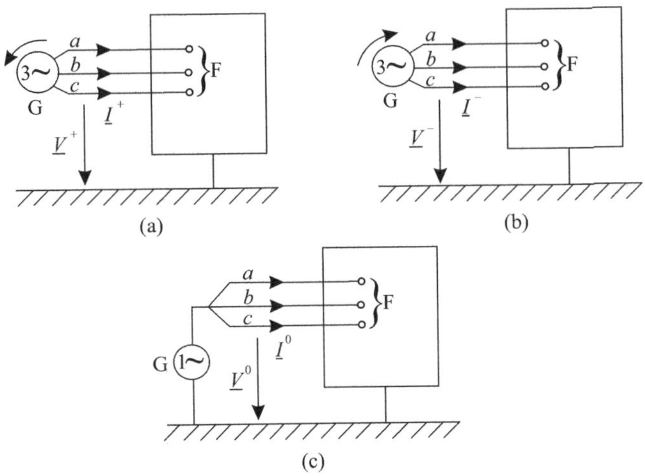

Figure 5.10. Three-phase AC system representation for calculation of the short-circuit impedances at the short-circuit location F: (a) positive-sequence short-circuit impedance; (b) negative-sequence short-circuit impedance; (c) zero-sequence short-circuit impedance. (Adopted from IEC 60909 [1])

- The order of sequence of the voltage components corresponds to the order in which each voltage attain the maximum value; the rotation of phasors is counterclockwise for all three sets of sequence components, as was assumed for the original unbalanced phasors (Figure 5.7).

These sequence currents are related to the sequence voltages by the corresponding sequence impedances ($\underline{Z}^+$, $\underline{Z}^-$, and $\underline{Z}^0$).

For the purpose of Standard IEC 60909, one has to make a distinction between *short-circuit impedances at the short-circuit location* F and *the short-circuit impedances of individual electrical equipment* [1].

- The positive-sequence short-circuit impedance $\underline{Z}^+$ at the short-circuit location F is obtained when a symmetrical system of voltages of positive-sequence phase order is applied to the short-circuit location F, and all synchronous and asynchronous machines are replaced by their internal impedances (Figure 5.10a):

$$\underline{Z}^+ = \frac{\underline{V}^+}{\underline{I}^+} \tag{5.19$'$}$$

- The negative-sequence short-circuit impedance $\underline{Z}^-$ at the short-circuit location F is obtained when a symmetrical system of voltages of negative-sequence phase order is applied to the short-circuit location F (Figure 5.10b):

$$\underline{Z}^- = \frac{\underline{V}^-}{\underline{I}^-} \tag{5.19$''$}$$

Note: The value of $\underline{Z}^+$ and $\underline{Z}^-$ impedances can differ from each other only in the case of rotating machines. When far-from-generator short circuits are calculated, it is generally allowed to take $\underline{Z}^- = \underline{Z}^+$.

- The zero-sequence short-circuit impedance $\underline{Z}^0$ at the short-circuit location F is obtained if an AC voltage is applied between the three short-circuited phase conductors and the joint return (e.g., earthing systems, neutral conductor, earth wires, cable sheaths, cable armoring) (Figure 5.10c):

$$\underline{Z}^0 = \frac{\underline{V}^0}{\underline{I}^0} \tag{5.19'''}$$

Note: The capacitances of overhead lines and cables of low-voltage networks may be neglected.

5.3.3.3 Sequence Impedance of Network Components. Decoupling the sequence impedances, as shown by equation (5.16), makes possible constructing independent sequence networks. A sequence network is a single-phase equivalent circuit consisting of a sequence impedance and eventually a voltage source. The voltage drop caused by the flow of the sequence current through the corresponding sequence network depends on the equivalent impedance of that network [6]. The impedance of a sequence network may be different from the impedances of the other sequence networks. The value of an impedance depends on the type of equipments from the analyzed part of the electrical network, for example, transmission lines, transformers, generators, and so on.

In the case of nonrotating equipment like transformers, the negative-sequence impedance is equal to the positive-sequence impedance, whereas in the case of rotating equipments, the two impedances are different. In general, for all type of equipments the zero-sequence impedance is different from both positive- and negative-sequence impedance (see Table 5.3). The sequence impedances are calculated using either equipment data from manufacturers or estimated data [7].

(i) *Construction of sequence networks* Three types of sequence networks are defined: positive, negative, and zero. The sequence networks are designed as viewed from the point of fault or unbalance, denoted by F, or F^+, F^- and F^0, respectively [7]. The sequence voltages are measured from the point of fault to the neutral point, marked by N, or N^+, N^- and N^0, respectively for each sequence network (Figure 5.11). Each network is designed in the form of a dipole through which sequence current flows.

The voltages applied to the network terminals are taken as phase-to-neutral voltages. The electrical generators are designed to supply balanced voltages and

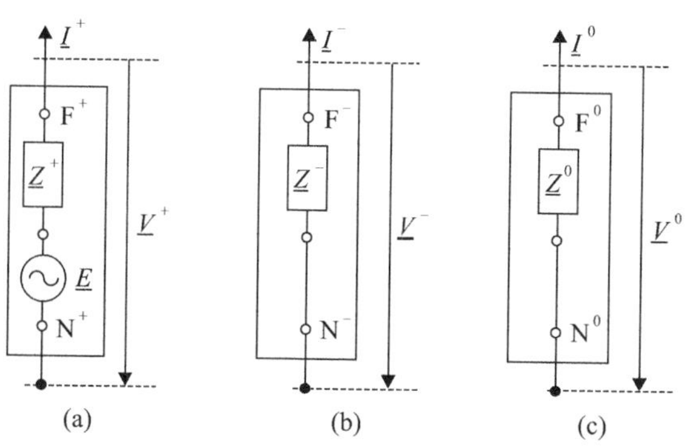

Figure 5.11. Sequence networks representation: (a) positive sequence; (b) negative sequence; (c) zero sequence.

thus an emf $\underline{E}$ is represented on the positive-sequence network only. Phase "a" voltage is taken as reference voltage and the voltages of phases b and c are expressed in terms of phase a voltage. The outward direction of the current flow in sequence networks is the positive direction (see Figure 5.11). Zero-sequence currents will flow through the network only if return path exists through which a completed circuit is provided.

The sequence networks carrying the sequence currents are interconnected to represents unbalance fault conditions [6].

A voltage source, $\underline{E}$, which is the Thévenin equivalent (Figure 5.11a), is represented on the positive-sequence network only. Applying the voltage Kirchhoff's law on each sequence network we may write:

$$\begin{aligned} \underline{V}^+ &= \underline{E} - \underline{I}^+ \underline{Z}^+ \\ \underline{V}^- &= \quad - \underline{I}^- \underline{Z}^- \\ \underline{V}^0 &= \quad - \underline{I}^0 \underline{Z}^0 \end{aligned} \tag{5.20}$$

or in matrix form

$$\begin{bmatrix} \underline{V}^0 \\ \underline{V}^+ \\ \underline{V}^- \end{bmatrix} = \begin{bmatrix} 0 \\ \underline{E} \\ 0 \end{bmatrix} - \begin{bmatrix} \underline{Z}^0 & 0 & 0 \\ 0 & \underline{Z}^+ & 0 \\ 0 & 0 & \underline{Z}^- \end{bmatrix} \begin{bmatrix} \underline{I}^0 \\ \underline{I}^+ \\ \underline{I}^- \end{bmatrix} \tag{5.21}$$

Equation (5.21), together with the equation that describes the conditions at the fault location, is used to derive $\underline{I}^+$ in terms of $\underline{E}$, $\underline{Z}^+$, $\underline{Z}^-$, and $\underline{Z}^0$, for each type of fault.

(ii) *Sequence impedance of electrical lines*

- *Positive- and negative-sequence impedances* An electrical line is a passive element. Assuming that the three phases of the line are symmetrical, the sequence of the phase voltages applied, that is, abc (positive sequence) or acb (negative sequence) makes no difference as regards the voltage drops produced. Therefore, the positive- and negative-sequence impedances of a symmetrical three-phase electrical line are identical. Their values are equal to the electrical line impedance used for steady-state calculations [8]. Note that for long transmission lines without phase transposition, the phase impedances and thus the sequence impedances may have slightly different values.

- *Zero-sequence impedance* The zero-sequence impedance of an electrical line is different by the positive-sequence and negative-sequence impedances because it also involves the paths to/through the ground and thus it accounts for the ground and the protection wires(s) impedances. The zero-sequence currents flowing through the phases of a three-phase electrical line are equal and cophasial, and their sum returns through the paths to/through the ground.

The magnetic field produced by the flow of the zero-sequence currents differs from that produced by the flow of the positive- or negative-sequence currents, which are dephased by $2\pi/3$, and the return current is zero. The resulting effect of the flow of the zero-sequence current is a zero-sequence reactance several times greater than the positive-sequence reactance, approximately by a factor [9]:

- Single-circuit overhead lines without ground wires or with steel ground wires 3.5
- Single-circuit overhead lines with copper or aluminum ground wires 2
- Double-circuit overhead lines without ground wires or with steel ground wires 5.5
- Double-circuit overhead lines with copper or aluminum ground wires . . . 3
- Three-phase cables . . . . 3 to 5
- Single-phase cables 1

The zero-sequence mutual impedance between parallel circuits of the same towers or even on different towers placed on the same right-of-way can also be important. Since the grounding impedance depends on the type of soil simplifying assumptions should be made.

(iii) *Sequence impedances of transformers*

- *Positive- and negative-sequence impedances* Similar to transmission lines, the transformers are passive elements, having no moving parts. Assuming that the transformer is symmetrical, the phase sequence has no effect on the winding reactances. Therefore, the equivalent circuits and the associated impedances are identical for both the positive-sequence and the negative-sequence networks. For fault analysis, only the series leakage impedance is considered.

- *Zero-sequence impedance* The zero-sequence impedance and the equivalent circuit depend upon the winding connection, method of neutral grounding, and so on.

Table 5.2 illustrates the sequence circuits of the two-winding and three-winding transformers for different winding connections.

The grounding impedance. The neutral point of Y-connected winding of the transformer may be grounded either solidly or through a neutral impedance $\underline{Z}_n$. Under symmetrical and balanced conditions, the potential of the neutral point is zero and no current will flow through the grounding impedance. However, if asymmetry or unbalance occurs, the potential of the neutral point will be different by zero and a current $\underline{I}_n = 3\,\underline{I}^0$ will flow through the impedance $\underline{Z}_n$ causing a voltage drop $3\,\underline{Z}_n\,\underline{I}^0$, which is added to all the phase-to-ground voltages and hence to the zero-sequence voltages [10]. In order to determine the zero-sequence current $\underline{I}^0$, which flows through the zero-sequence network, an impedance $3\,\underline{Z}_n$ must be added in this network in series with the transformer impedance [9].

Let us define the transformer impedance by $\underline{Z}_T^0$ as the sum of the perphase linkage reactances and resistances of the high (H) and low (L) voltage windings, that is, $\underline{Z}_T^0 = \underline{Z}_H^0 + \underline{Z}_L^0$, which may be about equal to the positive-sequence

TABLE 5.2. The Sequence Equivalent Circuits of Transformers for Different Winding Connections

Case	Winding Connections	Zero-Sequence Circuit	Positive- or Negative- Sequence Circuit
1	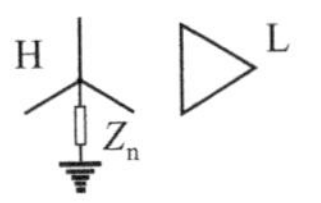	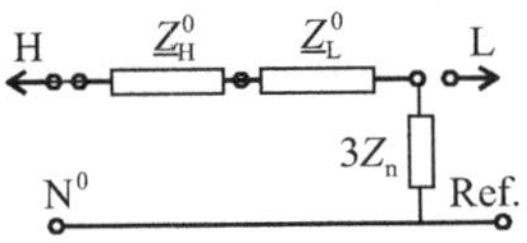	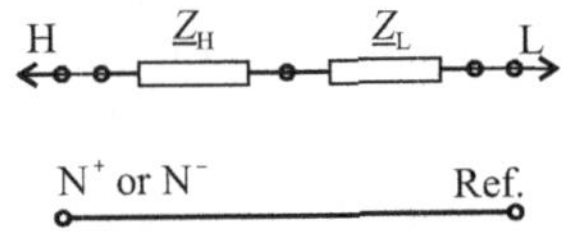
2	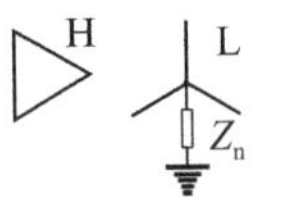	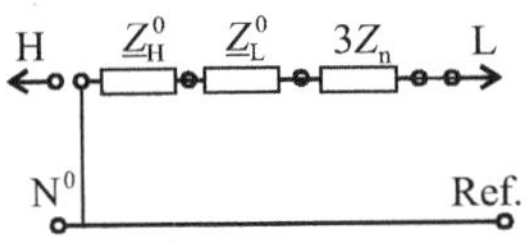	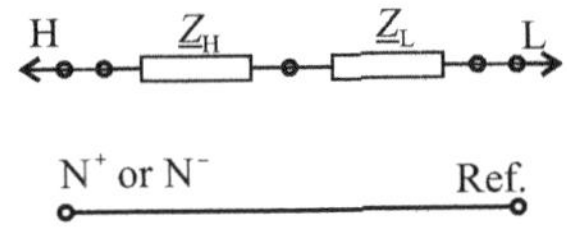
3	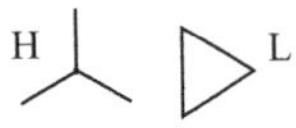	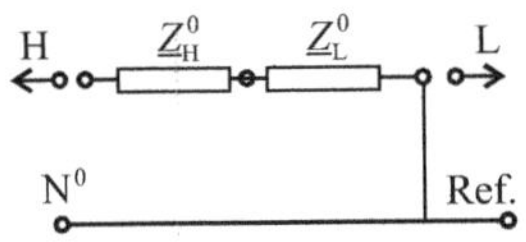	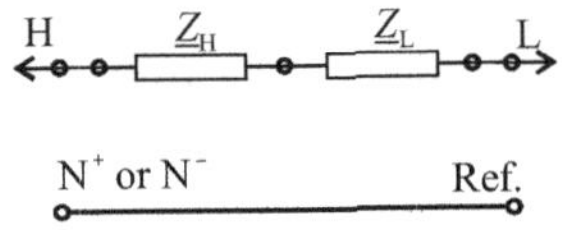
4		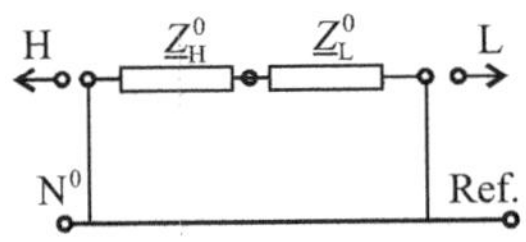	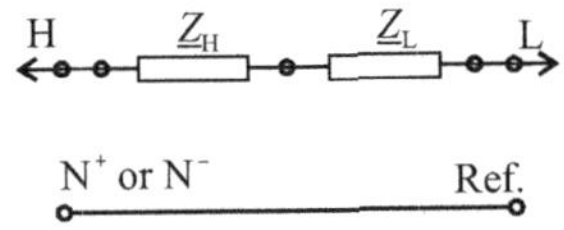
5	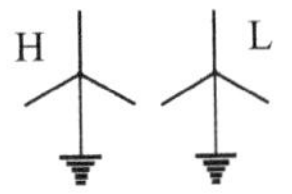	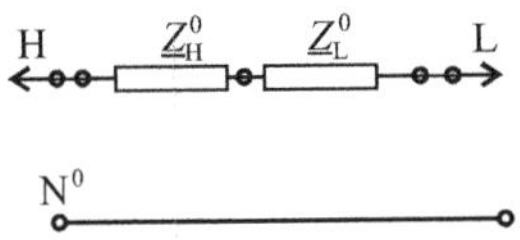	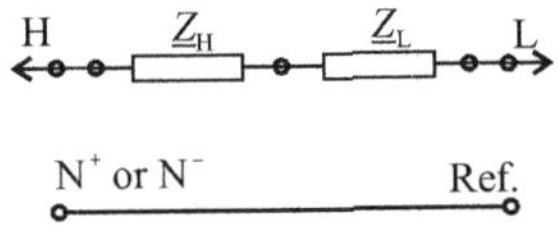
6	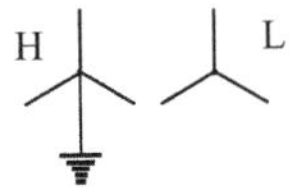	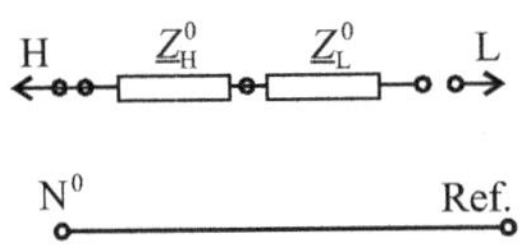	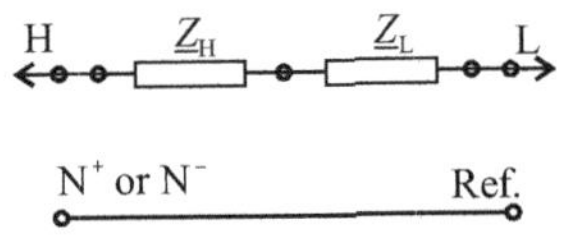
7	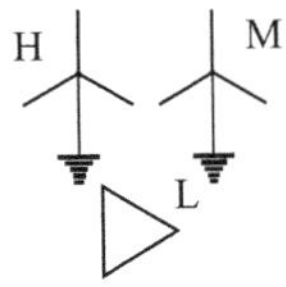	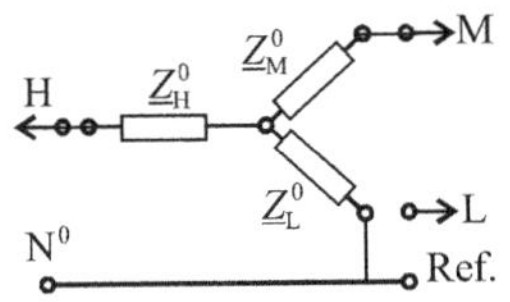	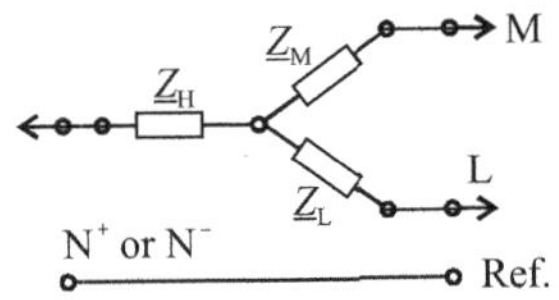
8	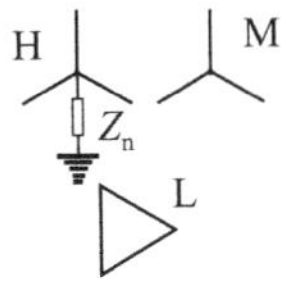	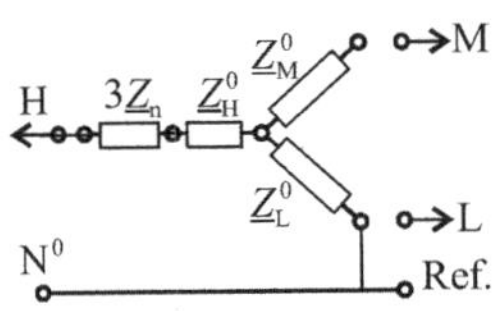	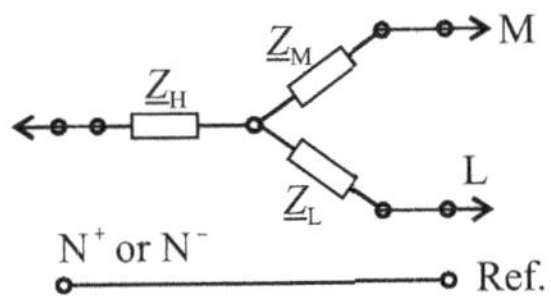
9	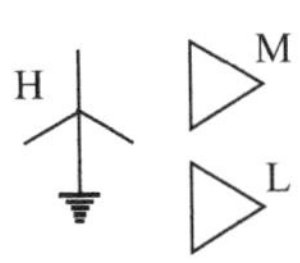	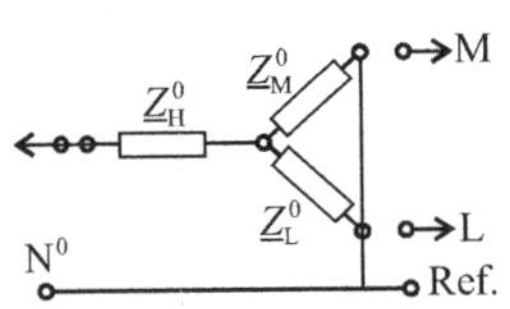	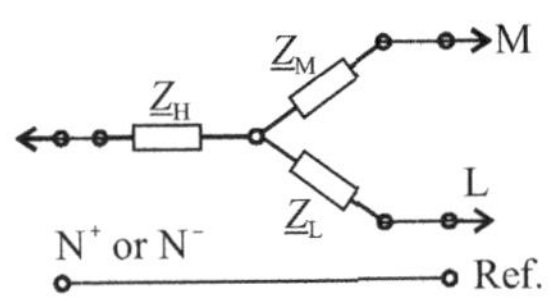
10	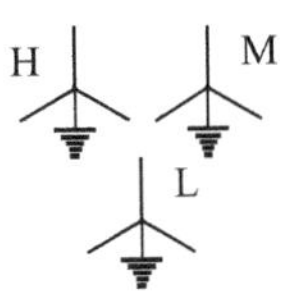	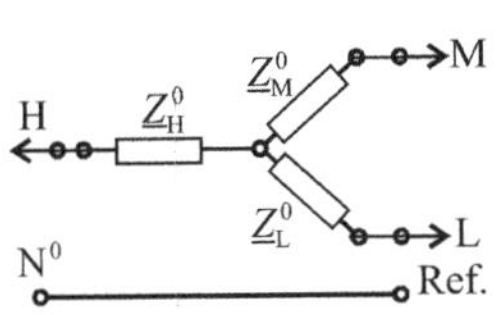	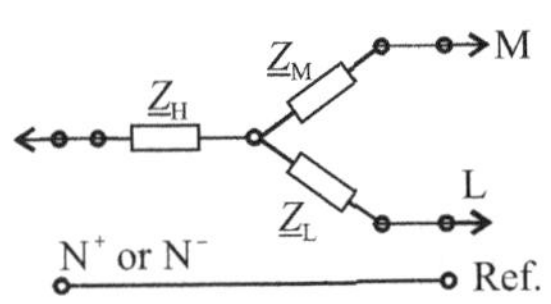

impedance. If the neutral of the Y-connected primary winding is grounded through an impedance $\underline{Z}_n$, the total zero-sequence impedance as seen from the primary side is $\underline{Z}^0 + 3\,\underline{Z}_n$. Since a Δ-connected winding provides no path for zero-sequence currents flowing in the line, the impedance seen from the secondary side is infinite.

Zero-sequence currents will flow through windings on one side of the transformer and into the connected lines only if a return path exists through which a completed circuit is provided. Also, a path for the corresponding current in the other windings must exist [11].

In the *case 1* presented in Table 5.2, on the primary windings side a return path for the zero-sequence currents is provided by the grounding system of the Y-connection, whereas on the secondary windings side a closed circuit is provide by the Δ-connection that serves to balance the currents in the primary windings. Of course, no zero-sequence currents will flow in the lines connected to the secondary winding. In the zero-sequence network an open circuit must be represented between the line and the reference bus on the Δ side.

If the neutral point of the Y-connected windings on one side is ungrounded, as in the *cases 3 and 6* given in Table 5.2, an open circuit exists in the zero-sequence network and no zero-sequence currents can flow between the two parts of the system connected by the transformer. Thus, the zero-sequence impedance seen from either the primary side or the secondary side is infinite. Since a Δ connection provides no return path for zero-sequence currents, no zero-sequence current can flow into a Δ-Δ bank, as shown in the *case 4*, although it can circulate within the Δ windings. The Y-Y transformer with both neutrals grounded (*case 5*) is a case in which zero-sequence currents can flow in both windings and outside the transformer on both sides. In this case, the zero-sequence network is similar to the positive-sequence network.

The cases 7–10 given in Table 5.2 illustrate the sequence networks for three-winding transformers. The effects of the leakage reactances and resistances of the three-windings are represented by a star-connected equivalent. Connection of this equivalent to the primary, secondary, and tertiary terminals, and the reference bus (Ref.) depends on the winding connections.

(iv) *Sequence impedances of synchronous machines*

- *Positive-sequence impedances* Detailed representation of the synchronous machines in the positive-sequence network applicable for stability studies was discussed in Chapter 2. This representation is also valid for short-circuit currents calculation for the entire study period, including the subtransient, transient, and steady-state conditions.

Figure 5.12 illustrates the simplified positive-sequence network of a synchronous machine. Thus, following the occurrence of a fault at the machine terminals,

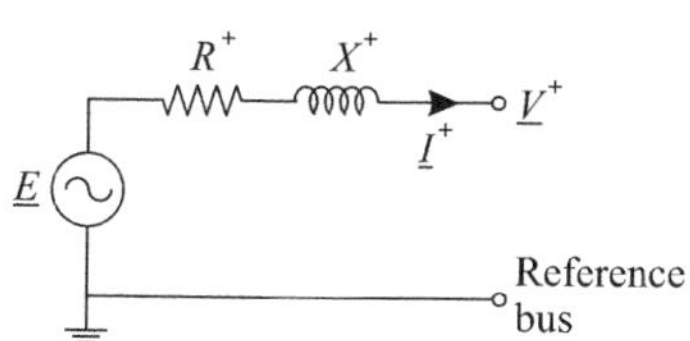

Figure 5.12. positive-sequence network of a synchronous machine.

the armature fluxes and the currents change. Three positive-sequence reactances X^+ are defined depending on the time frame of interest: subtransient, transient, and steady state. The positive-sequence resistance R^+ is the armature resistance.

The sequence network from Figure 5.12 was represented as a single-phase circuit under the assumption of symmetrical construction of the electrical machine. The reference bus is the neutral of the generator.

Neglecting the saliency effects, the positive-sequence reactance is: $X^+ = X''_d$, for the subtransient conditions; $X^+ = X'_d$, for the transient conditions; $X^+ = X_d$, for steady-state conditions.

- *Negative- and zero-sequence impedances* The negative-sequence and zero-sequence impedances are specified by the manufacturers and are determined by tests.

A synchronous machine is designed to normally produce positive-sequence voltages and currents. When an unbalance is created, for example, due to a short-circuit occurring near the generator, to single-phase loads or to open conductors, negative-sequence currents flow in the armature winding that produce a field rotating in opposite direction to the positive-sequence field created by the rotor winding and at double synchronous speed with respect to the rotor speed. The rotor is thus subjected to double frequency flux pulsation [12]. The existence of the negative-sequence currents results in rotor losses in the form of heating as well as small armature losses. The field produced by the negative-sequence currents passes alternatively through the direct axis and the quadrature axis, and thus the negative-sequence reactance is:

$$X^- = (X''_d + X''_q)/2$$

where X''_d and X''_q are the direct-axis and quadrature-axis subtransient reactances.

The negative-sequence resistance R^- is greater than the positive-sequence resistance and depends on the damper winding, since the double frequency currents are induced in the rotor field and damper winding.

Figure 5.13 illustrates the equivalent circuit of a synchronous machine appropriate for use in negative-sequence networks. No emf is represented, since the synchronous machine produces positive-sequence voltages only. The reference bus of the equivalent circuit is the neutral of the generator.

When zero-sequence currents flow through the armature, their instantaneous values in the three phases are equal and the resultant air gap flux is zero. Therefore, the zero-sequence reactance is approximately equal to the leakage reactance.

The flow of the zero-sequence currents in the three phases and the neutral of the synchronous machine is shown in Figure 5.14a. The zero-sequence equivalent circuit is illustrated in Figure 5.14b.

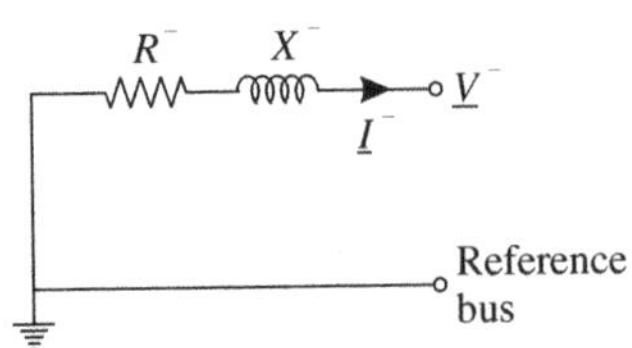

Figure 5.13. Negative-sequence network of a synchronous machine.

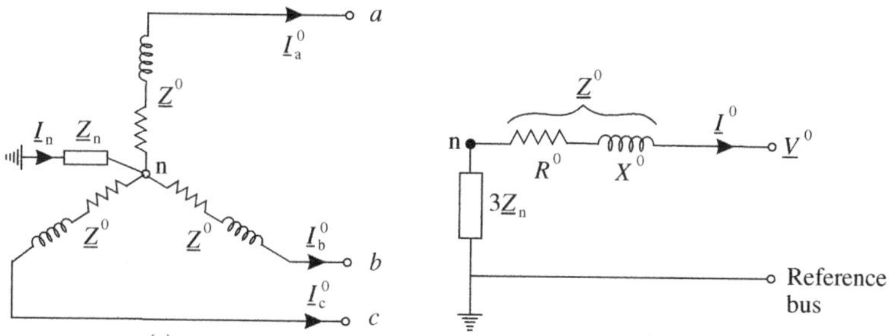

(a) (b)

Figure 5.14. Representation of the synchronous machine for zero sequence: (a) currents paths; (b) equivalent circuit (single phase).

The zero-sequence currents flowing through the three phases return through the grounding impedance $\underline{Z}_n$ as a sum $\underline{I}_n = \underline{I}_a^0 + \underline{I}_b^0 + \underline{I}_c^0 = 3\underline{I}_a^0$. This is why a term $3\underline{Z}_n$ is introduced in the equivalent circuit of Figure 5.14b. If the neutral is isolated from the ground, then $Z_n = \infty$ and $\underline{I}_n = 0$. The grounding impedance of the synchronous machine is not a part of either the positive- or negative-sequence network because neither positive- nor negative-sequence current can flow through this impedance [6]. The reference bus of the equivalent circuit is the ground at the generator.

Since all the neutral points of a symmetrical three-phase system through which balanced currents flow are at the same potential, all the neutral points have the same potential when either positive-sequence or negative-sequence currents flow [9].

Table 5.3 shows the zero-sequence characteristics of the various elements in an electrical network.

5.3.3.4 Unsymmetrical Fault Calculations.

In the sections 5.3.3.2 and 5.3.3.3, construction of the three sequence networks as seen from the fault point in terms of the characteristics of the power system elements has been presented. For unsymmetrical fault current calculations, the three separate networks can be connected in a certain manner, depending on the type of the fault [7].

TABLE 5.3. Zero-Sequence Characteristic of the Various Elements in an Electrical Network

Elements		$\underline{Z}^0$
Transformer (seen from secondary winding)		
No neutral		∞
Yyn or Zyn	free flux	∞
	forced flux	10 to $15X^+$
Dyn or YNyn		X^+
primary D or Y + Zn		0.1 to 0.2 X^+
Machine		
Synchronous		$\approx 0.5\ Z^+$
Asynchronous		≈ 0
Electrical Line		$\approx 3\ Z^+$

Shunt Faults of Zero Impedance. The following cases of faults to ground can be defined, assuming zero fault impedance [12]:

(i) Single phase-to-earth—on the a phase

$$\begin{aligned} \underline{I}_b &= 0 \\ \underline{I}_c &= 0 \\ \underline{V}_a &= 0 \end{aligned} \tag{5.22}$$

(ii) Phase-to-phase—between b and c phases

$$\begin{aligned} \underline{I}_a &= 0 \\ \underline{I}_b &= -\underline{I}_c \\ \underline{V}_b &= \underline{V}_c \end{aligned} \tag{5.23}$$

(iii) Phase-to-phase-to-earth—between b and c phases

$$\begin{aligned} \underline{I}_a &= 0 \\ \underline{V}_b &= 0 \\ \underline{V}_c &= 0 \end{aligned} \tag{5.24}$$

(iv) Three phase

$$\begin{aligned} \underline{I}_a + \underline{I}_b + \underline{I}_c &= 0 \\ \underline{V}_a &= \underline{V}_b \\ \underline{V}_b &= \underline{V}_c \end{aligned} \tag{5.25}$$

(v) Single-phase open-circuit fault

$$\begin{aligned} \underline{I}_a &= 0 \\ \underline{v}_b &= 0 \\ \underline{v}_c &= 0 \end{aligned} \tag{5.26}$$

Notice from the above that three equations that define the fault conditions can be written.

(i) *Single phase-to-earth fault* Consider a fault defined by equations (5.22) and by Figure 5.15a.

The conditions at the fault are expressed by the following equations:

$$\underline{I}_b = 0 \quad \underline{I}_c = 0 \quad \underline{V}_a = 0$$

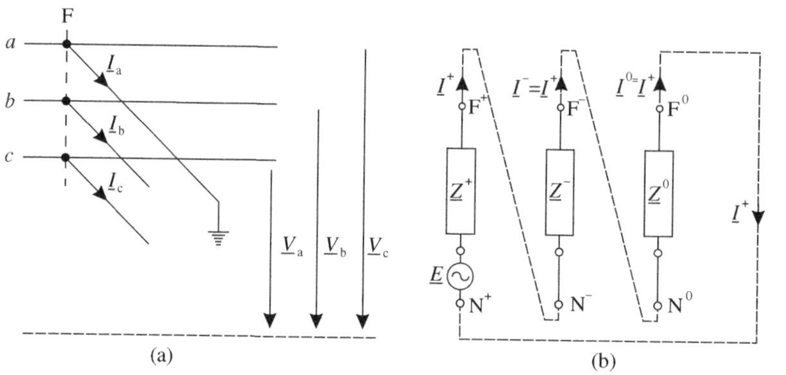

Figure. 5.15 Single phase-to-earth fault at F: (a) definition of fault; (b) equivalent circuit.

With $\underline{I}_b = 0$ and $\underline{I}_c = 0$, the symmetrical components of the current from (5.12 a,b, and c), are given by

$$\begin{bmatrix} \underline{I}^0 \\ \underline{I}^+ \\ \underline{I}^- \end{bmatrix} = \frac{1}{3} \begin{bmatrix} 1 & 1 & 1 \\ 1 & \underline{a} & \underline{a}^2 \\ 1 & \underline{a}^2 & \underline{a} \end{bmatrix} \begin{bmatrix} \underline{I}_a \\ 0 \\ 0 \end{bmatrix} \tag{5.27}$$

so that $\underline{I}^0$, $\underline{I}^+$, and $\underline{I}^-$ each equal $\underline{I}_a/3$ and

$$\underline{I}^+ = \underline{I}^- = \underline{I}^0 = \frac{1}{3}\underline{I}_a \tag{5.28}$$

Substituting $\underline{I}^+$ for $\underline{I}^-$ and $\underline{I}^0$ in equation (5.21), we obtain:

$$\begin{bmatrix} \underline{V}^0 \\ \underline{V}^+ \\ \underline{V}^- \end{bmatrix} = \begin{bmatrix} 0 \\ \underline{E} \\ 0 \end{bmatrix} - \begin{bmatrix} \underline{Z}^0 & 0 & 0 \\ 0 & \underline{Z}^+ & 0 \\ 0 & 0 & \underline{Z}^- \end{bmatrix} \begin{bmatrix} \underline{I}^+ \\ \underline{I}^+ \\ \underline{I}^+ \end{bmatrix} \tag{5.29}$$

Performing the indicated matrix multiplication and subtraction an equality of two column matrices is obtained. Premultiplying both column matrices by the row matrix $[1 \quad 1 \quad 1]$ gives

$$\underline{V}^0 + \underline{V}^+ + \underline{V}^- = -\underline{I}^+\underline{Z}^0 + (\underline{E} - \underline{I}^+\underline{Z}^+) - \underline{I}^+\underline{Z}^- \tag{5.30}$$

Since $\underline{V}_a = \underline{V}^0 + \underline{V}^+ + \underline{V}^- = 0$, we solve equation (5.30) for $\underline{I}^+$ and obtain:

$$\underline{I}^+ = \frac{\underline{E}}{\underline{Z}^+ + \underline{Z}^- + \underline{Z}^0} \tag{5.31}$$

Equations (5.28) and (5.31) are particular equations for a single phase-to-earth fault. They are used with equation (5.21) and the symmetrical-components relations to determine all the voltages and currents at the fault.

If the three sequence networks are connected in series (Figure 5.15b), we see that the currents and voltages resulting therefrom satisfy the equations above, and the three sequence impedances are in series with the voltage $\underline{E}$. With the sequence networks so connected, the voltage across each sequence network is the symmetrical component of $\underline{E}$ of that sequence.

If the neutral of the generator is not grounded, the zero-sequence network is open circuited and $\underline{Z}^0$ is infinite. Since equation (5.31) shows that $\underline{I}^+$ is zero when $\underline{Z}^0$ is infinite, $\underline{I}^-$ and $\underline{I}^0$ must be zero. Thus no current flows in phase a since $\underline{I}_a$ is the sum of its components, all of which are zero.

The fault current $\underline{I}_a$ is:

$$\underline{I}_a = 3\,\underline{I}^0 = \frac{3\,\underline{E}}{\underline{Z}^+ + \underline{Z}^- + \underline{Z}^0} \tag{5.32}$$

The voltage of phase b to ground under fault conditions is:

$$\underline{V}_b = \underline{a}^2\,\underline{V}^+ + \underline{a}\,\underline{V}^- + \underline{V}^0 = \underline{V}_a \frac{\underline{Z}^-(\underline{a}^2 - \underline{a}) + \underline{Z}^0(\underline{a}^2 - 1)}{\underline{Z}^+ + \underline{Z}^- + \underline{Z}^0} \tag{5.33}$$

Similarly, the voltage of phase c can be calculated.

(ii) *Phase-phase fault* Consider a fault defined by equation (5.23) and by Figure 5.16.

The conditions at the fault are expressed by the following equations:

$$\underline{I}_a = 0 \quad \underline{I}_b = -\,\underline{I}_c \quad \underline{V}_b = \underline{V}_c$$

With $\underline{V}_b = \underline{V}_c$, the symmetrical components voltage from equation (5.9) are given by:

$$\begin{bmatrix} \underline{V}^0 \\ \underline{V}^+ \\ \underline{V}^- \end{bmatrix} = \frac{1}{3}\begin{bmatrix} 1 & 1 & 1 \\ 1 & \underline{a} & \underline{a}^2 \\ 1 & \underline{a}^2 & \underline{a} \end{bmatrix}\begin{bmatrix} \underline{V}_a \\ \underline{V}_b \\ \underline{V}_b \end{bmatrix}$$

from which we find

$$\underline{V}^+ = \underline{V}^- \tag{5.34}$$

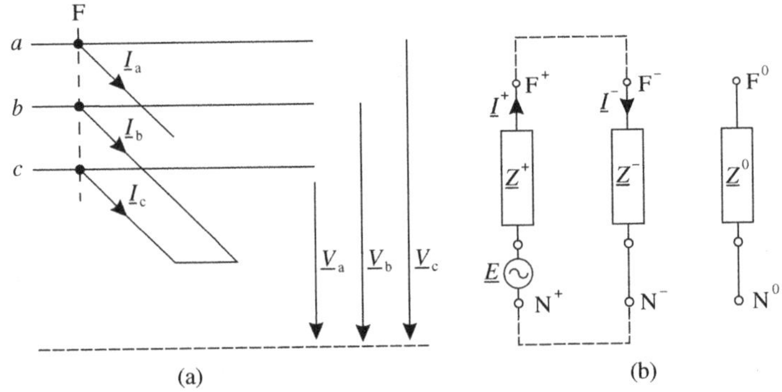

Figure 5.16. Phase-phase-earth fault at F: (a) definition of fault; (b). equivalent circuit.

Since $\underline{I}_b = -\underline{I}_c$ and $\underline{I}_a = 0$, the symmetrical components of current are given by equation:

$$\begin{bmatrix} \underline{I}^0 \\ \underline{I}^+ \\ \underline{I}^- \end{bmatrix} = \frac{1}{3}\begin{bmatrix} 1 & 1 & 1 \\ 1 & \underline{a} & \underline{a}^2 \\ 1 & \underline{a}^2 & \underline{a} \end{bmatrix}\begin{bmatrix} 0 \\ -\underline{I}_c \\ \underline{I}_c \end{bmatrix} = \frac{1}{3}\begin{bmatrix} 0 \\ -\underline{a} + \underline{a}^2 \\ -\underline{a}^2 + \underline{a} \end{bmatrix}[\underline{I}_c]$$

and therefore

$$\underline{I}^0 = 0 \tag{5.35$'$}$$

$$\underline{I}^- = -\underline{I}^+ \tag{5.35$''$}$$

Since $\underline{I}^0$ is zero,

$$\underline{V}^0 = 0 \tag{5.36}$$

Equation (5.29), with the substitutions allowed by equations (5.34), (5.35′), (5.35″), (5.36), becomes:

$$\begin{bmatrix} 0 \\ \underline{V}^+ \\ \underline{V}^+ \end{bmatrix} = \begin{bmatrix} 0 \\ \underline{E} \\ 0 \end{bmatrix} - \begin{bmatrix} \underline{Z}^0 & 0 & 0 \\ 0 & \underline{Z}^+ & 0 \\ 0 & 0 & \underline{Z}^- \end{bmatrix}\begin{bmatrix} 0 \\ \underline{I}^+ \\ -\underline{I}^+ \end{bmatrix} \tag{5.37}$$

Performing the indicated matrix operations and premultiplying the resulting matrix equation by the row matrix $[1 \quad 1 \quad -1]$ gives

$$0 = \underline{E} - \underline{I}^+\underline{Z}^+ - \underline{I}^+\underline{Z}^-$$

or

$$\underline{E} = \underline{I}^+(\underline{Z}^+ + \underline{Z}^-)$$

and solving for $\underline{I}^+$ yields

$$\underline{I}^+ = \frac{\underline{E}}{\underline{Z}^+ + \underline{Z}^-} \tag{5.38}$$

Equations (5.34), (5.35′), (5.35″), and (5.38) are the special equations for a phase-to-phase fault. They are used with equation (5.29) and the symmetrical-component relations to determine all voltages and currents at the fault.

From equation (5.11b) we have

$$\underline{I}_b = \underline{a}^2\,\underline{I}^+ - \underline{a}\,\underline{I}^+ = (\underline{a}^2 - \underline{a})\,\underline{I}^+ = -j\sqrt{3}\,\underline{I}^+ \tag{5.39}$$

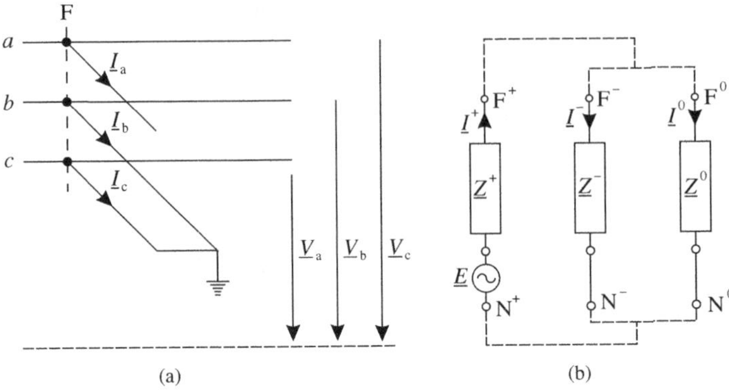

Figure 5.17. Phase-phase-earth fault at F: (a) definition of fault; (b). equivalent circuit.

The fault current is

$$\underline{I}_b = -\underline{I}_c = \frac{-j\sqrt{3}\,\underline{E}}{\underline{Z}^+ + \underline{Z}^-} \tag{5.40}$$

The constraints imposed by equations (5.35′) and (5.34) indicate that there is no zero-sequence network connection in the equivalent circuit and that the positive- and negative-sequence networks are connected in parallel. Figure 5.16b shows the equivalent circuit satisfying the above equations.

(iii) *Phase-phase-earth fault* Consider a fault defined by equation (5.24) and by Figure 5.17.

The condition at the fault are expressed by the following equations:

$$\underline{V}_b = 0 \quad \underline{V}_c = 0 \quad \underline{I}_a = 0$$

With $\underline{V}_b = 0$ and $\underline{V}_c = 0$, the symmetrical components of voltage are given by equation

$$\begin{bmatrix} \underline{V}^0 \\ \underline{V}^+ \\ \underline{V}^- \end{bmatrix} = \frac{1}{3}\begin{bmatrix} 1 & 1 & 1 \\ 1 & \underline{a} & \underline{a}^2 \\ 1 & \underline{a}^2 & \underline{a} \end{bmatrix}\begin{bmatrix} \underline{V}_a \\ 0 \\ 0 \end{bmatrix}$$

Therefore $\underline{V}^0$, $\underline{V}^+$, and $\underline{V}^-$ equal $\underline{V}_a/3$, and

$$\underline{V}^+ = \underline{V}^- = \underline{V}^0 \tag{5.41}$$

Substituting $\left(\underline{E} - \underline{I}^+ \underline{Z}^+\right)$ for $\underline{V}^+$, $\underline{V}^-$, and $\underline{V}^0$ in equation (5.29) and premultiplying both sides by $[\underline{\mathbf{Z}}]^{-1}$, where

$$[\underline{\mathbf{Z}}]^{-1} = \begin{bmatrix} \underline{Z}^0 & 0 & 0 \\ 0 & \underline{Z}^+ & 0 \\ 0 & 0 & \underline{Z}^- \end{bmatrix}^{-1} = \begin{bmatrix} \dfrac{1}{\underline{Z}^0} & 0 & 0 \\ 0 & \dfrac{1}{\underline{Z}^+} & 0 \\ 0 & 0 & \dfrac{1}{\underline{Z}^-} \end{bmatrix}$$

give

$$\begin{bmatrix} \frac{1}{\underline{Z}^0} & 0 & 0 \\ 0 & \frac{1}{\underline{Z}^+} & 0 \\ 0 & 0 & \frac{1}{\underline{Z}^-} \end{bmatrix} \begin{bmatrix} \underline{E} - \underline{I}^+ \underline{Z}^+ \\ \underline{E} - \underline{I}^+ \underline{Z}^+ \\ \underline{E} - \underline{I}^+ \underline{Z}^+ \end{bmatrix} = \begin{bmatrix} \frac{1}{\underline{Z}^0} & 0 & 0 \\ 0 & \frac{1}{\underline{Z}^+} & 0 \\ 0 & 0 & \frac{1}{\underline{Z}^-} \end{bmatrix} \begin{bmatrix} 0 \\ \underline{E} \\ 0 \end{bmatrix} - \begin{bmatrix} \underline{I}^0 \\ \underline{I}^+ \\ \underline{I}^- \end{bmatrix} \tag{5.42}$$

Premultiplying both sides of equation (5.42) by the row matrix $[1 \quad 1 \quad 1]$ and recognizing that

$$\underline{I}^+ + \underline{I}^- + \underline{I}^0 = \underline{I}_a = 0$$

we have

$$\frac{\underline{E}}{\underline{Z}^0} - \underline{I}^+ \frac{\underline{Z}^+}{\underline{Z}^0} + \frac{\underline{E}}{\underline{Z}^+} - \underline{I}^+ + \frac{\underline{E}}{\underline{Z}^-} - \underline{I}^+ \frac{\underline{Z}^+}{\underline{Z}^-} = \frac{\underline{E}}{\underline{Z}^+} \tag{5.43}$$

and upon collecting terms we obtain

$$\underline{I}^+ \left(1 + \frac{\underline{Z}^+}{\underline{Z}^0} + \frac{\underline{Z}^+}{\underline{Z}^-}\right) = \frac{\underline{E}}{\underline{Z}^0} + \frac{\underline{E}}{\underline{Z}^-} = \frac{\underline{E}\left(\underline{Z}^- + \underline{Z}^0\right)}{\underline{Z}^- \underline{Z}^0} \tag{5.44}$$

and

$$\underline{I}^+ = \frac{\underline{E}\left(\underline{Z}^- + \underline{Z}^0\right)}{\underline{Z}^+ \underline{Z}^- + \underline{Z}^+ \underline{Z}^0 + \underline{Z}^- \underline{Z}^0} = \frac{\underline{E}}{\underline{Z}^+ + \underline{Z}^- \underline{Z}^0 / \left(\underline{Z}^- + \underline{Z}^0\right)} \tag{5.45}$$

Equations (5.41) and (5.45) are the special equations for a double phase-to-earth fault. They are used with equation (5.29) and the symmetrical-component relations to determine all the voltages and currents at the fault.

Equations (5.41) indicates that the sequence networks should be connected in parallel (Figure 5.17b), since the positive-, negative-, and zero-sequence voltages are equal at the fault. Examination of Figure 5.17b shows that all the conditions derived above for the double-phase-to-earth fault are satisfied by this connection.

The diagram of network connections shows that the positive-sequence current $\underline{I}^+$ is determined by the voltage $\underline{E}$ impressed on $\underline{Z}^+$in series with the parallel combination of $\underline{Z}^-$ and $\underline{Z}^0$. The same relation is given by equation (5.45).

(iv) *Three-phase fault* Consider a symmetrical three-phase fault defined by equation (5.25) and by Figure 5.18.

The conditions at the fault are expressed by the following equations:

$$\underline{I}_a + \underline{I}_b + \underline{I}_c = 0$$
$$\underline{V}_a = \underline{V}_b = \underline{V}_c$$

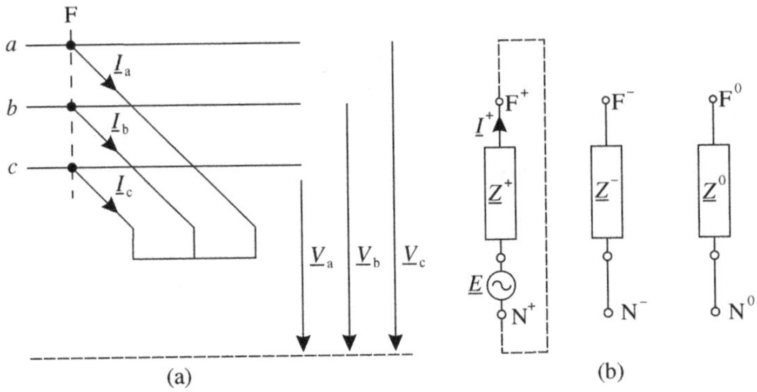

Figure 5.18. Three-phase fault at F: (a) definition of fault; (b) equivalent circuit.

In terms of symmetrical components

$$\underline{I}^0 = 0 \tag{5.46}$$

$$\underline{V}^+ = \underline{V}^- = \underline{V}^0 = 0 \tag{5.47}$$

There are no zero-sequence or negative-sequence emf values, and the voltages $\underline{V}^0$ and $\underline{V}^-$ at the fault point F are each equal to zero. Therefore, zero- and negative-sequence currents do not flow anywhere in the system. As expected, only the positive-sequence network is involved. The fault is simulated by shorting the fault point F^+ and the reference bus $N^+$of the positive-sequence network (Figure 5.18b).

(v) *Single-phase open circuit fault* The single-phase open circuit fault is shown in Figure 5.19. Let $\underline{v}_a$, $\underline{v}_b$, and $\underline{v}_c$ be series voltages drop in phases between F_1 and F_2 in phases a, b, c, respectively.

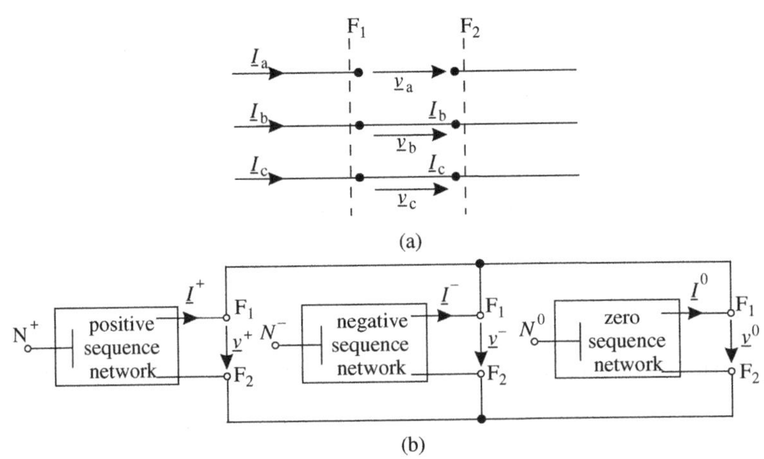

Figure 5.19. Open circuit on phase "a": (a) circuit diagram; (b) equivalent circuit.

At the fault point, the boundary conditions are (Figure 5.19a) [12]

$$\underline{I}_a = 0$$
$$\underline{V}_b = \underline{V}_c = 0 \tag{5.48}$$

Hence, from equation (5.9):

$$\underline{V}_a^0 = \frac{1}{3}\underline{V}_a$$
$$\underline{V}_a^+ = \frac{1}{3}\underline{V}_a$$
$$\underline{V}_a^- = \frac{1}{3}\underline{V}_a$$

and therefore

$$\underline{V}_a^+ = \underline{V}_a^- = \underline{V}_a^0 = \frac{1}{3}\underline{V}_a \tag{5.49}$$

From equation (5.11a), for $\underline{I}_a = 0$

$$\underline{I}_a = \underline{I}^+ + \underline{I}^- + \underline{I}^0 = 0 \tag{5.50}$$

From equation (5.49), it can be concluded that the sequence networks are connected in parallel, as shown in Figure 5.19b.

Faults Through Impedance. All the faults discussed in the preceding sections consisted of direct short circuits between phases and from one or two phases to earth. Most faults are the result of insulator flashovers, where the impedance between the phase and earth depends on the resistance of the arc, of the tower itself, and of the tower footing if ground wires are not used. Tower-footing resistances form the major part of the resistance between phase and earth and depend on the soil conditions [6].

The effect of impedance in the fault is found by deriving equations similar to those for faults through zero impedance. Connections of the hypothetical stubs for faults through impedance are shown in Figure 5.20.

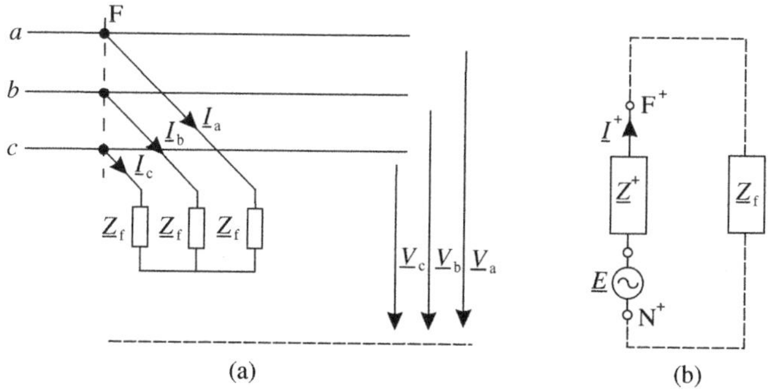

Figure 5.20. Connection of sequence network to simulate a three-phase fault through impedance at point F.

A balanced system remains symmetrical after the occurrence of a three-phase fault having the same impedance between each phase and a common point. Only positive-sequence currents flow.

With the fault impedance $\underline{Z}_f$ equal in all phases (Figure 5.20a), the voltage at the fault locations is

$$\underline{V}_a = \underline{I}_a \underline{Z}_f$$

and since only positive-sequence currents flow,

$$\underline{V}^+ = \underline{I}^+ \underline{Z}_f = \underline{V}_f - \underline{I}^+ \underline{Z}^+$$

and

$$\underline{I}^+ = \frac{\underline{V}_f}{\underline{Z}^+ + \underline{Z}_f} \tag{5.51}$$

The sequence-network connection is shown in Figure 5.20b.

A formal derivation can be made for the single phase-to-earth and double phase-to-earth faults through impedance shown in Figures 5.21a and b [6].

- *For a single phase-to-earth fault through* $\underline{Z}_f$ (Figure 5.21a):

$$\underline{I}^+ = \underline{I}^- = \underline{I}^0$$

$$\underline{I}^+ = \frac{\underline{V}_f}{\underline{Z}^+ + \underline{Z}^- + \underline{Z}^0 + 3\underline{Z}_f} \tag{5.52}$$

- *For a double phase-to-earth fault through impedance* $\underline{Z}_f$ (Figure 5.21b):

$$\underline{V}^+ = \underline{V}^-$$

$$\underline{I}^+ = \frac{\underline{V}_f}{\underline{Z}^+ + \underline{Z}^-\left(\underline{Z}^0 + 3\underline{Z}_f\right)/\left(\underline{Z}^- + \underline{Z}^0 + 3\underline{Z}_f\right)} \tag{5.53}$$

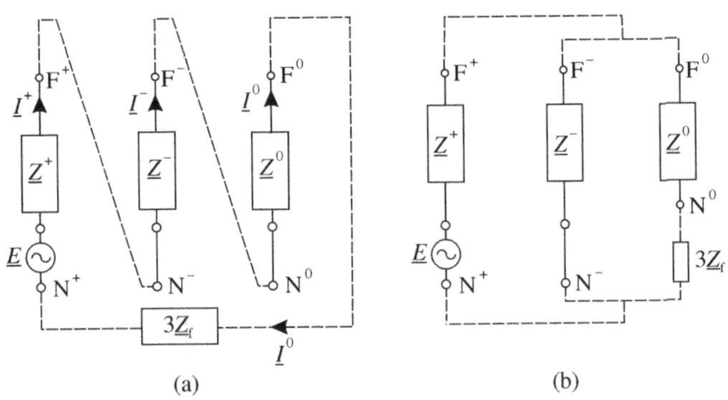

Figure 5.21. Connections of the sequence networks for: (a) single phase-to-earth fault through $\underline{Z}_f$; (b) double phase-to-earth fault through $\underline{Z}_f$.

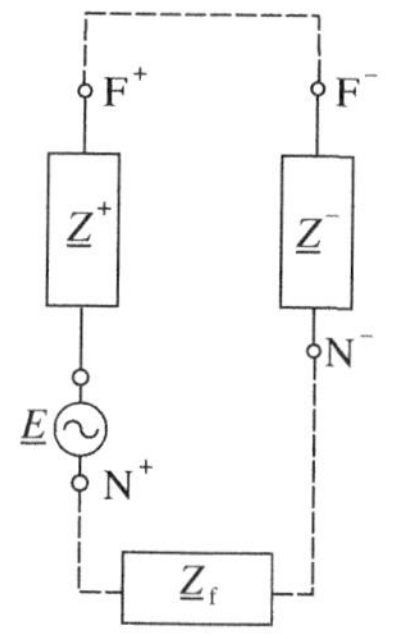

Figure 5.22. Connection of the sequence networks to simulate a phase-to-phase fault through impedance.

- *A phase-to-phase (phases b and c) fault through impedance* $\underline{Z}_f$ is shown in Figure 5.22.

The conditions at the fault are

$$\underline{I}_a = 0; \quad \underline{I}_b = -\underline{I}_c; \quad \underline{V}_c = \underline{V}_b - \underline{I}_b \underline{Z}_f$$

$\underline{I}_a$, $\underline{I}_b$ and $\underline{I}_c$ bear the same relations to each other as in the phase-to-phase fault without impedance. Therefore $\underline{I}^+ = -\underline{I}^-$

The sequence components of voltage are given by

$$\begin{bmatrix} \underline{V}^0 \\ \underline{V}^+ \\ \underline{V}^- \end{bmatrix} = \frac{1}{3} \begin{bmatrix} 1 & 1 & 1 \\ 1 & \underline{a} & \underline{a}^2 \\ 1 & \underline{a}^2 & \underline{a} \end{bmatrix} \begin{bmatrix} \underline{V}_a \\ \underline{V}_b \\ \underline{V}_b - \underline{I}_b \underline{Z}_f \end{bmatrix} \tag{5.54}$$

or

$$3\,\underline{V}^+ = \underline{V}_a + (\underline{a} + \underline{a}^2)\,\underline{V}_b - \underline{a}^2\,\underline{I}_b\,\underline{Z}_f \tag{5.55$'$}$$

$$3\,\underline{V}^- = \underline{V}_a + (\underline{a} + \underline{a}^2)\,\underline{V}_b - \underline{a}\,\underline{I}_b\,\underline{Z}_f \tag{5.55$''$}$$

therefore

$$3(\underline{V}^+ - \underline{V}^-) = (\underline{a} - \underline{a}^2)\,\underline{I}_b\,\underline{Z}_f = j\sqrt{3}\,\underline{I}_b\,\underline{Z}_f \tag{5.56}$$

Since $\underline{I}^+ = -\underline{I}^-$

$$\underline{I}_b = \underline{a}^2\,\underline{I}^+ + \underline{a}\,\underline{I}^- = (\underline{a}^2 - \underline{a})\,\underline{I}^+ = -j\sqrt{3}\,\underline{I}^+$$

and, upon substituting $\underline{I}_b$ in (5.56), we obtain

$$\underline{V}^+ - \underline{V}^- = \underline{I}^+\,\underline{Z}_f \tag{5.57}$$

Equation (5.57) requires the insertion of $\underline{Z}_f$ between the fault points in the positive- and negative-sequence networks to fulfill the required conditions for the fault.

5.4 CALCULATION OF SHORT-CIRCUIT CURRENT COMPONENTS[3]

In the short-circuit currents calculation in a node, the maximum and minimum values are of interest if the thermal and mechanical stresses and the power quality are involved, respectively.

5.4.1 Initial Symmetrical Short-Circuit Current I''_k

Considering two cases [1]:

For the common case when $\underline{Z}^0 > \underline{Z}^+ = \underline{Z}^-$, the highest initial short-circuit current will occur for the three-phase short circuit.

For short circuits near transformers with low zero-sequence impedance, $\underline{Z}^0$ may be smaller than $\underline{Z}^+$. In that case, the highest initial short-circuit current I''_{kE2E} will occur for a phase-to-phase short circuit with earth connection (see Figure 5.23) for $\underline{Z}^- / \underline{Z}^+ = 1$ and $\underline{Z}^- / \underline{Z}^0 > 1$, where $\underline{Z}^- = \underline{Z}^+$.

5.4.1.1 Three-Phase Short Circuit. In general, the initial symmetrical short-circuit current I''_k shall be calculated using equation (5.58) with the equivalent voltage source $\left(cU_n/\sqrt{3}\right)$ at the short-circuit location and the short-circuit impedance $(\underline{Z}_k = R_k + jX_k)$:

$$I''_k = \frac{cU_n}{\sqrt{3}Z_k} = \frac{1}{\sqrt{3}} \cdot \frac{cU_n}{\sqrt{R_k^2 + X_k^2}} \tag{5.58}$$

The equivalent voltage source shall be introduced at the short-circuit location (see Figure 5.6) with the factor c according to Table 5.1.

(i) *Single-fed short circuits* For a far-from-generator short circuit fed from a single source (see Figure 5.23a) the short-circuit current is calculated with equation

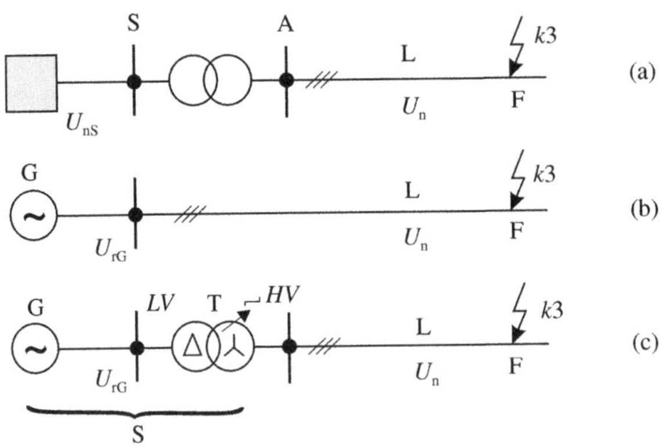

Figure 5.23. Examples of single-fed short circuits: (a) short circuit fed from a network feeder via a transformer; (b) short circuit fed from one generator (without unit transformer); (c) short circuit fed from one power station unit generator and unit transformer with or without on-load tap changer,. (Reprinted with permission from IEC 60909 [1])

[3] This section is partially reproduced from the IEC 60909 Standard with permission by International Electrotechnical Commission [1].

(5.58), with

$$R_k = R_{St} + R_{TK} + R_L$$

$$X_k = X_{St} + X_{TK} + X_L$$

where

R_k and X_k are the sum of series-connected resistances and reactances of the positive-sequence system, respectively (Figure 5.23a);

R_L is the line resistance for a conductor temperature of 20°C, when calculating the maximum short-circuit currents;

$\underline{Z}_{TK} = R_{TK} + jX_{TK} = K_T(R_T + jX_T)$ is the corrected transformer impedance;

K_T is the impedance correction factor, $K_T = 0.95 \cdot \frac{c_{max}}{1 + 0.6 \cdot x_T}$, where x_T is the relative reactance of the transformer $x_T = \frac{X_T}{(U_{rT}^2/S_{rT})}$.

Resistances R_k less than $0.3 \cdot X_k$ may be neglected. The impedance of the network feeder $\underline{Z}_{Sk} = R_{St} + jX_{St}$ is referred to the voltage of the transformer side connected to the short-circuit location (F).

For the examples in Figures 5.23b and c, the initial symmetrical short circuit is calculated with the corrected impedances of the generator and the power station unit in series with a line impedance $\underline{Z}_L = R_L + jX_L$. The impedances of generator (G), network transformer (T), and power station units (S) shall be multiplied with the impedance correction factors K_G, K_T,and K_S, respectively, K_{SO} when calculating short-circuit currents with the equivalent voltage source at the short-circuit location according to this Standard (See IEC 60909-4).

The short-circuit impedances for these examples are given by the following equations:

$$\underline{Z}_k = \underline{Z}_{Gk} + \underline{Z}_L = K_G(R_G + jX''_d) + \underline{Z}_L \quad \text{for Figure 5.23b} \qquad (5.59')$$

$$\underline{Z}_k = \underline{Z}_S + \underline{Z}_L = K_S(N_r^2\,\underline{Z}_G + \underline{Z}_{THV}) + \underline{Z}_L \quad \text{for Figure 5.23c} \qquad (5.59'')$$

where $\underline{Z}_{Gk}$ shall be determined from equation

$$\underline{Z}_{Gk} = K_G\,\underline{Z}_G = K_G(R_G + jX''_d) \qquad (5.60)$$

with

$$K_G = \frac{U_n}{U_{rG}} \cdot \frac{c_{max}}{1 + X''_d \sin\varphi_{rG}},$$

$\underline{Z}_S$ from equation

$$\underline{Z}_S = K_S(N_r^2\,\underline{Z}_G + \underline{Z}_{THV}) \qquad (5.61')$$

$$\underline{Z}_{SO} = K_{SO}(N_r^2\,\underline{Z}_G + \underline{Z}_{THV}) \qquad (5.62')$$

with

$$K_S = \frac{U_{nG}^2}{U_{rG}^2} \cdot \frac{U_{rTLV}^2}{U_{rTHV}^2} \cdot \frac{c_{max}}{1 + |X_d'' - X_T| \sin\varphi_{rG}} \quad (5.61'')$$

$$K_{S0} = \frac{U_{nS}}{U_{rG}(1 + p_G)} \cdot \frac{U_{rTLV}}{U_{rTHV}} \cdot (1 \pm p_T) \cdot \frac{c_{max}}{1 + X_d'' \sin\varphi_{rG}} \quad (5.62'')$$

where

p_G is the range of generator voltage regulation and p_T is the range of transformer voltage adjustment;

c_{max} is the factor voltage for the calculation of maximum short-circuit currents.

The generator impedance shall be transferred to the high-voltage side using the rated transformation ratio N_r. The unit transformer impedance $\underline{Z}_{THV} = R_{THV} + jX_{THV}$ according to the equations (5.63′) and (5.63″) without K_T is referred to the high-voltage side:

$$Z_T = \frac{u_{kr}}{100\%} \cdot \frac{U_{rT}^2}{S_{rT}} \quad (5.63')$$

$$R_T = \frac{u_{Rr}}{100\%} \cdot \frac{U_{rT}^2}{S_{rT}} \quad (5.63'')$$

$$X_T = \sqrt{Z_T^2 - R_T^2} \quad (5.63''')$$

where

u_{Kr} is the short-circuit voltage at rated current in percent;

u_{Rr} is the rated resistive component of the short-circuit voltage in percent;

U_{rT} is the rated voltage of the transformer on the high-voltage or low-voltage side;

S_{rT} is the rated apparent power of the transformer.

(ii) *Short circuits fed from nonmeshed networks* When there is more than one source contributing to the short-circuit current, and the sources are unmeshed, as shown in Figure 5.24, the initial symmetrical short-circuit current I_k'' at the

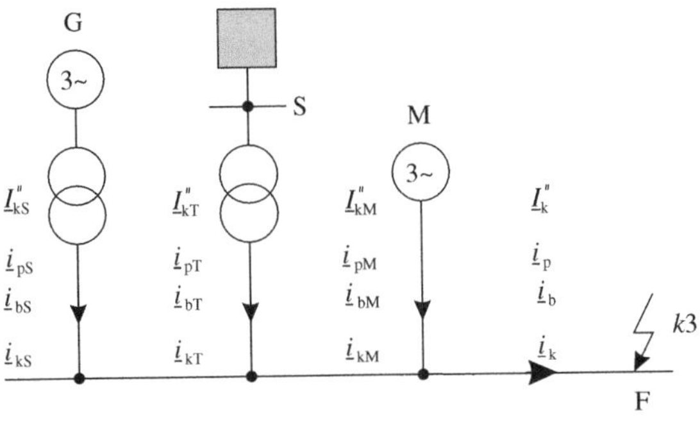

Figure 5.24. Example of a nonmeshed network.

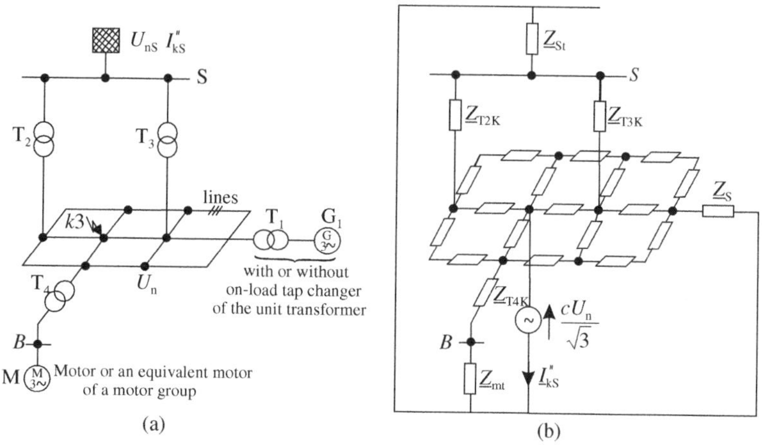

Figure 5.25. A meshed network fed from several sources: (a) system diagram; (b) equivalent circuit diagram for the calculation with the equivalent voltage source $cU_n/\sqrt{3}$ at the short-circuit location (Reprinted with permission from IEC 60909 [1]).

short-circuit location F is the sum of the individual branch short-circuit currents. Each branch short-circuit current can be calculated as an independent single-source three-phase short-circuit current in accordance with equation (5.58).

The initial short-circuit current at the short-circuit location F is the phasor, which is sum of the individual partial short-circuit currents [1]:

$$I''_k = \sum_i I''_{ki} \tag{5.64}$$

The short-circuit current at the short-circuit location F is considered, with good accuracy, as being the sum of the absolute values of the individual partial short-circuit currents.

(iii) *Short circuits in meshed networks* In meshed networks, such as those shown in Figure 5.25, it is generally necessary to determine the short-circuit impedance $\underline{Z}_k = \underline{Z}^+$ by network reduction (series connection, parallel connection, and delta-star transformation, for example) using the positive-sequence short-circuit impedances of electrical equipment.

The impedances in systems connected through transformers to the system, in which short circuit occurs, have to be transferred by the square of the rated transformation ratio. If there are several transformers with slightly differing rated transformation ratios (N_{rT1}, N_{rT2}, N_{rTn}), in between two systems, the arithmetic mean value can be used.

The initial symmetrical short-circuit current shall be calculated with the equivalent voltage source $cU_n/\sqrt{3}$ at the short-circuit location using equation (5.58).

5.4.1.2 Phase-to-Phase Short Circuit. In the case of a phase-to-phase short circuit, according to Figure 5.1b the initial short-circuit current shall be calculated by

$$I''_{k2} = \frac{cU_n}{|\underline{Z}^+ + \underline{Z}^-|} = \frac{cU_n}{2|\underline{Z}^+|} = \frac{\sqrt{3}}{2} I''_k \tag{5.65}$$

During the initial stage of the short circuit, the negative impedance is approximately equal to the positive-sequence impedance, independent of whether the short circuit is a near-to-generator or a far-from-generator short circuit. Therefore in equation (5.65), it is possible to introduce $\underline{Z}^- = \underline{Z}^+$.

Only during the transient of the steady-state stage, the short-circuit impedance $\underline{Z}^-$ is different from $\underline{Z}^+$ if the short circuit is a near-to-generator short circuit [1].

5.4.1.3 Phase-to-Phase Short Circuit with Earth Connection. To calculate the initial symmetrical short-circuit currents it is necessary to distinguish between the currents $\underline{I}''_{\mathrm{k2Ea}}$, $\underline{I}''_{\mathrm{k2Eb}}$ and $\underline{I}''_{\mathrm{kE2E}}$ (see Figure 5.1c).

For far-from-generator short circuits, $\underline{Z}^-$ is approximately equal to $\underline{Z}^+$. If in this case $\underline{Z}^0$ is less than $\underline{Z}^-$ the current $\underline{I}''_{kE2E}$ in the phase-to-phase short circuit with earth connection generally is the largest of all initial symmetrical short-circuit currents $\underline{I}''_{\mathrm{k}}$, $\underline{I}''_{\mathrm{k2}}$, $\underline{I}''_{\mathrm{k2E}}$, and $\underline{I}''_{\mathrm{k1}}$.

The equations (5.66′) and (5.66″) are given for the calculation of $\underline{I}''_{\mathrm{k2Ea}}$ and $\underline{I}''_{\mathrm{k2Eb}}$ in Figure 5.1c:

$$\underline{I}''_{\mathrm{k2Ea}} = -jcU_{\mathrm{n}}\frac{\underline{Z}^0 - a\,\underline{Z}^-}{\underline{Z}^+\,\underline{Z}^- + \underline{Z}^+\,\underline{Z}^0 + \underline{Z}^-\,\underline{Z}^0} \tag{5.66′}$$

$$\underline{I}''_{\mathrm{k2Eb}} = -jcU_{\mathrm{n}}\frac{\underline{Z}^0 - a^2\,\underline{Z}^-}{\underline{Z}^+\,\underline{Z}^- + \underline{Z}^+\,\underline{Z}^0 + \underline{Z}^-\,\underline{Z}^0} \tag{5.66″}$$

The initial short-circuit current $\underline{I}''_{\mathrm{kE2E}}$ flowing to earth and/or earthed wires, (Figure 5.1c), is calculated by

$$\underline{I}''_{\mathrm{kE2E}} = -\frac{\sqrt{3}\,cU_{\mathrm{n}}\underline{Z}^0}{\underline{Z}^+\,\underline{Z}^- + \underline{Z}^+\,\underline{Z}^0 + \underline{Z}^-\,\underline{Z}^0} \tag{5.66‴}$$

For a far-from-generator short circuit with $\underline{Z}^- = \underline{Z}^+$, these equations lead to the absolute values:

$$I''_{\mathrm{k2Ea}} = cU_{\mathrm{n}}\frac{|\,\underline{Z}^0/\,\underline{Z}^+ - \underline{a}|}{|\,\underline{Z}^+ + 2\,\underline{Z}^0|} \tag{5.67′}$$

$$I''_{\mathrm{k2Eb}} = cU_{\mathrm{n}}\frac{|\,\underline{Z}^0/\,\underline{Z}^+ - \underline{a}^2|}{|\,\underline{Z}^+ + 2\,\underline{Z}^0|} \tag{5.67‴}$$

$$I''_{\mathrm{kE2E}} = \frac{\sqrt{3}\,cU_{\mathrm{n}}}{|\,\underline{Z}^+ + 2\,\underline{Z}^0|} \tag{5.67‴}$$

5.4.1.4 Phase-to-Earth Short Circuit. The initial phase-to-earth short-circuit current $\underline{I}''_{\mathrm{k1}}$ in Figure 5.1a shall be calculated by

$$\underline{I}''_{\mathrm{k1}} = \frac{\sqrt{3}\,cU_{\mathrm{n}}}{\underline{Z}^+ + \underline{Z}^- + \underline{Z}^0} \tag{5.68}$$

For a far-from-generator short circuit with $\underline{Z}^{+} = \underline{Z}^{-}$, the absolute value is calculated by

$$I''_{k1} = \frac{\sqrt{3}\,cU_{n}}{|2\,\underline{Z}^{+} + \underline{Z}^{0}|} \tag{5.68'}$$

If $\underline{Z}^{0}$ is less than $\underline{Z}^{-} = \underline{Z}^{+}$, the initial phase-to-earth short-circuit current I''_{k1} is larger than the three-phase short-circuit current I''_{k}, but smaller than I''_{kE2E}. However, I''_{k1} will be the highest current to be interrupted by a circuit breaker if $1.0 > \underline{Z}^{0}/\,\underline{Z}^{+} > 0.23$ [1].

5.4.2 Peak Short-Circuit Current i_p

5.4.2.1 Three-Phase Short Circuit.

(i) *Short circuits in nonmeshed networks* For three-phase short circuits fed from nonmeshed networks as in Figures 5.23 and 5.24 the contribution to the peak short-circuit current from each branch can be expressed by

$$i_{p} = \kappa\sqrt{2}I''_{k} \tag{5.69}$$

The factor κ for the R/X or X/R ratio shall be obtained from Figure 5.26 or calculated by the following expression [1]:

$$\kappa = 1.02 + 0.98\,e^{-3R/X} \tag{5.70}$$

Equations (5.69) and (5.70) presume that the short circuit starts at zero voltage, and that i_p is reached approximately after one half-cycle. For a synchronous generator use R_{Gf} (fictions resistance $R_{Gf} = (0.05...0.15)x''_{d}$ in function of U_{rG} and S_{rG}).

The peak short-circuit current i_p at a short-circuit location F, fed from sources that are not meshed with one another, (Figure 5.24), is the sum of the branch peak short-circuit currents:

$$i_{p} = \sum_{i} i_{pi} \tag{5.71}$$

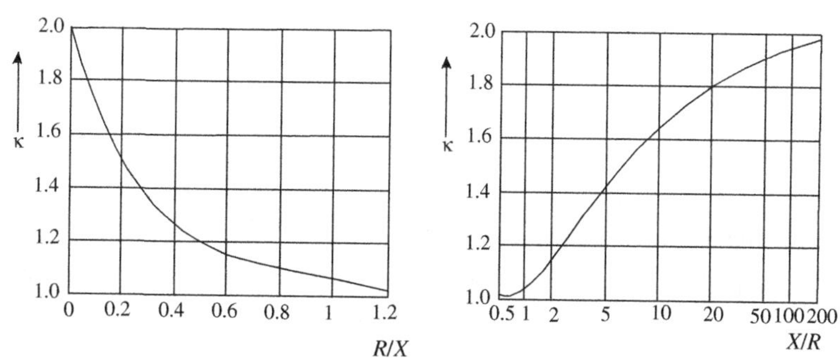

Figure 5.26. Factor κ for series circuit as a function of ratio R/X or X/R.

For the example from Figure 5.24 results:

$$i_{\mathrm{p}} = i_{\mathrm{pS}} + i_{\mathrm{pT}} + i_{\mathrm{pM}} \tag{5.72}$$

(ii) *Short circuits in meshed networks* When calculating the peak short-circuit current i_{p} in meshed networks, equation (5.69) shall be used with κ determined using one of the following methods (a, b, or c) [1].

(a) *Uniform ratio R/X or X/R*

For this method the factor κ is determined from Figure 5.26 taking the smallest ratio of R/X or the largest ratio of X/R of all branches of the network, having values between 1.6 and 1.9 in high voltage networks and 1.3 and 1.5 in low voltage networks.

It is only necessary to choose the branches, which carry partial short-circuit currents at the nominal voltage corresponding to the short-circuit location and branches with transformers adjacent to the short-circuit location. Any branch may be a series combination of several impedances.

(b) *Ratio R/X or X/R at the short-circuit location*

For this method the factor κ is multiplied by a factor 1.15 to cover inaccuracies caused by using the ratio R/X from a network reduction with complex impedances:

$$i_{\mathrm{p(b)}} = 1.15\,\kappa_{\mathrm{b}}\sqrt{2}\,I_{\mathrm{k}}'' \tag{5.73}$$

As long as R/X remains smaller than 0.3 in all branches, it is not necessary to use the factor 1.15. It is not necessary for the product $1.15 \cdot \kappa_{\mathrm{b}}$ to exceed 1.8 in low-voltage networks or to exceed 2.0 in medium- and high-voltage networks.

The factor κ_{b} is found from Figure 5.26 for the ratio $R_{\mathrm{k}}/X_{\mathrm{k}}$ given by the short-circuit impedance $\underline{Z}_{\mathrm{k}} = R_{\mathrm{k}} + jX_{\mathrm{k}}$ at the short-circuit location F, calculated for frequency $f = 50\,\mathrm{Hz}$ (or 60 Hz).

(c) *Equivalent frequency f_{c}*

An equivalent impedance $\underline{Z}_{\mathrm{c}}$ of the system as seen from the short-circuit location is calculated assuming a frequency $f_{\mathrm{c}} = 20\,\mathrm{Hz}$ (for a nominal frequency of $f = 50\,\mathrm{Hz}$) or $f_{\mathrm{c}} = 24\,\mathrm{Hz}$ (for a nominal frequency of $f = 60\,\mathrm{Hz}$). The R/X or X/R ratio is then determined according to equation (5.74):

$$\frac{R}{X} = \frac{R_{\mathrm{c}}}{X_{\mathrm{c}}} \cdot \frac{f_{\mathrm{c}}}{f} \tag{5.74 a}$$

$$\frac{X}{R} = \frac{X_{\mathrm{c}}}{R_{\mathrm{c}}} \cdot \frac{f}{f_{\mathrm{c}}} \tag{5.74 b}$$

where

$\underline{Z}_{\mathrm{c}} = R_{\mathrm{c}} + jX_{\mathrm{c}}$ is the equivalent impedance of the system as seen from the short-circuit location for the assumed frequency f_{c};

R_{c} is the real part of $\underline{Z}_{\mathrm{c}}$ (R_{c} is generally not equal to the R at nominal frequency);

X_{c} is the imaginary part of $\underline{Z}_{\mathrm{c}}$ (X_{c} is generally not equal to the X at nominal frequency).

The factor κ is found from Figure 5.26 using the R/X or X/R ratio from equation (5.74a), or with equation (5.74b).

Method (c) is recommended in meshed networks. When using this method in meshed networks with transformers, generators, and power station units, the impedance correction factors K_T, K_G, and K_S, respectively, K_{SO} shall be introduced with the same values as for the 50 Hz (or 60 Hz) calculations.

5.4.2.2 Phase-to-Phase Short Circuit. For a phase-to-phase short circuit the peak short-circuit current can be expressed by

$$i_{p2} = \kappa\sqrt{2}I''_{k2} \tag{5.75}$$

The factor κ shall be calculated according to 5.4.2.1, case (*i*) or to 5.4.2.2, case (*ii*), depending on the system configuration. For simplification, it is permitted to use the same value of κ as for the three-phase short circuit.

When $\underline{Z}^+ = \underline{Z}^-$, the phase-to-phase peak short-circuit current i_{p2} is smaller than the three-phase peak short-circuit current i_p as shown in equation:

$$i_{p2} = \frac{\sqrt{3}}{2} i_p \tag{5.76}$$

5.4.2.3 Phase-to-Phase Short Circuit with Earth Connection. For a phase-to-phase short circuit with earth connection, the peak short-circuit current can be expressed by

$$i_{p\,2E} = \kappa\sqrt{2}I''_{k2E} \tag{5.77}$$

The factor κ shall be calculated according to 5.4.2.1, case (*i*) or to 5.4.2.1, case (*ii*), depending on the system configuration. For simplification, it is permitted to use the same value of κ as for the three-phase short circuit.

It is only necessary to calculate i_{p2E}, when $\underline{Z}^0$ is much less than $\underline{Z}^+$ (less than about 1/4 of $\underline{Z}^+$) [1].

5.4.2.4 Phase-to-Earth Short Circuit. For a phase-to-earth short circuit, the peak short-circuit current can be expressed by

$$i_{pl} = \kappa\sqrt{2}I''_{kl} \tag{5.78}$$

The factor κ shall be calculated according to 5.4.2.1, case (*i*) or to 5.4.2.1, case (*ii*), depending on the system configuration. For simplification, it is permitted to use the same value of κ as for the three-phase short circuit [1].

5.4.3 DC Component of the Short-Circuit Current

The maximum decaying aperiodic component $i_{d.c.}$ of the short-circuit current as shown in Figures 5.2a and 5.2b may be calculated with sufficient accuracy by equation

$$i_{d.c.} = \sqrt{2}I''_k\, e^{-2\pi ft\, R/X} \tag{5.79}$$

where

I''_k is the initial symmetrical short-circuit current;

f is the nominal frequency;

t is the time;

R/X is the ratio according to 5.4.2.1 or the ratios according to the methods (a) and (c).

TABLE 5.4. Ratio of equivalent and nominal frequencies for various fxt values

$f \times t$	<1	<2.5	<5	<12.5
f_c/f	0.27	0.15	0.092	0.055

The correct resistance R_G of the generator armature should be used and not R_{Gf}.

For meshed networks, the ratio R/X or X/R is to be determined by the method (c) in 5.4.2.1, case (*ii*). Depending on the product $f \times t$ where f is the frequency and t is the time, the equivalent frequency f_c should be used (Table 5.4) [1].

5.4.4 Symmetrical Short-Circuit Breaking Current I_b

The breaking current at the short-circuit location consists in general of symmetrical current I_b and an aperiodic current $i_{d.c.}$ at the time t_{min} according to equation (5.79).

Note: for some near-to-generator short circuits the value of $i_{d.c.}$ at t_{min} may exceed the peak value of I_b and this can lead to missing current zeros [1].

5.4.4.1 *Far-from-Generator Short Circuit.* For far-from-generator short circuits, the short-circuit breaking currents are equal to the initial short-circuit currents [1]:

$$I_b = I''_k \tag{5.80a}$$

$$I_{b2} = I''_{k2} \tag{5.80b}$$

$$I_{b2E} = I''_{k2E} \tag{5.80c}$$

$$I_{b1} = I''_{k1} \tag{5.80d}$$

5.4.4.2 *Near-to-Generator Short Circuit.*

SINGLE-FED THREE-PHASE SHORT CIRCUIT. For a near-to-generator short circuit, in the case of a single fed short circuit as in Figure 5.23b and 5.23c or from nonmeshed networks as in Figure 5.24, the decay to the symmetrical short-circuit breaking current is taken into account by the factor μ (according to equation (5.82)):

$$I_b = \mu I''_k \tag{5.81}$$

The factor μ depends on the minimum time delay t_{min} and the ratio I''_{kG}/I_{rG}, where I_{rG} is the rated generator current. The values of μ in equation (5.82) apply if synchronous machines are excited by rotating exciters or by static converter exciters (provided, for static exciters, the minimum time delay t_{min} is less than 0.25 s and the maximum excitation voltage is less than 1.6 times rated load excitation voltage). For all other cases take $\mu = 1$, if the exact value is unknown [1].

When there is a unit transformer between the generator and the short-circuit location, the partial short-circuit current I''_{kS} at the high-voltage side of the unit transformer (in Figure 5.23c) shall be transferred by the rated transformation ratio to the terminal of the

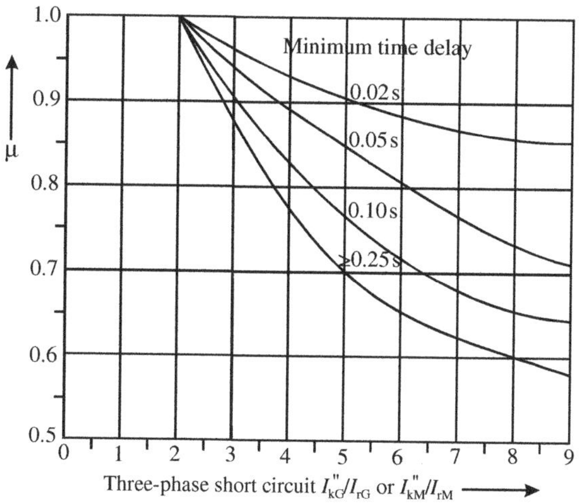

Figure 5.27. Factor μ for calculation of short-circuit breaking current I_b [1].

generator $I''_{kG} = N_r I''_{kS}$ before calculating μ, using the following equations [1]:

$$\begin{aligned} \mu &= 0.84 + 0.26e^{-0.26 \cdot I''_{kG}/I_{rG}} && \text{for } t_{min} = 0.02\,\text{s} \\ \mu &= 0.71 + 0.51e^{-0.30 \cdot I''_{kG}/I_{rG}} && \text{for } t_{min} = 0.05\,\text{s} \\ \mu &= 0.62 + 0.72e^{-0.32 \cdot I''_{kG}/I_{rG}} && \text{for } t_{min} = 0.10\,\text{s} \\ \mu &= 0.56 + 0.94e^{-0.38 \cdot I''_{kG}/I_{rG}} && \text{for } t_{min} \geq 0.25\,\text{s} \end{aligned} \tag{5.82}$$

If I''_{kG}/I_{rG} is not greater than 2, apply $\mu = 1$ for all values of the minimum time delay t_{min}. The factor μ may also be obtained from Figure 5.27. For other values of minimum time delay, linear interpolation between curves is acceptable.

5.4.5 Steady-State Short-Circuit Current I_k

The calculation of the steady-state short-circuit current I_k is less accurate than the calculation of the initial short-circuit current I''_k.

5.4.5.1 Three-Phase Short Circuit of One Generator or One Power Station Unit. For near-to-generator three-phase short circuits fed directly from one synchronous generator or one power station unit only, according to Figures 5.23b or 5.23c the steady-state short-circuit current I_k depends on the excitation system, the voltage regulator action, and saturation influences.

Synchronous machines (generators, motors, or compensators) with terminal-fed static exciters do not contribute to I_k in the case of a short circuit at the terminals of the machine, but they contribute to I_k if there is an impedance between the terminals and the short-circuit location. A contribution is also given if, in case of a power station unit, the short circuit occurs on the high-voltage side of the unit transformer (Figure 5.23c).

- *Maximum steady-state short-circuit current.* For the calculation of the maximum steady-state short-circuit current, the synchronous generator may be set at the maximum excitation:

$$I_{k\max} = \lambda_{\max} I_{rG} \tag{5.83}$$

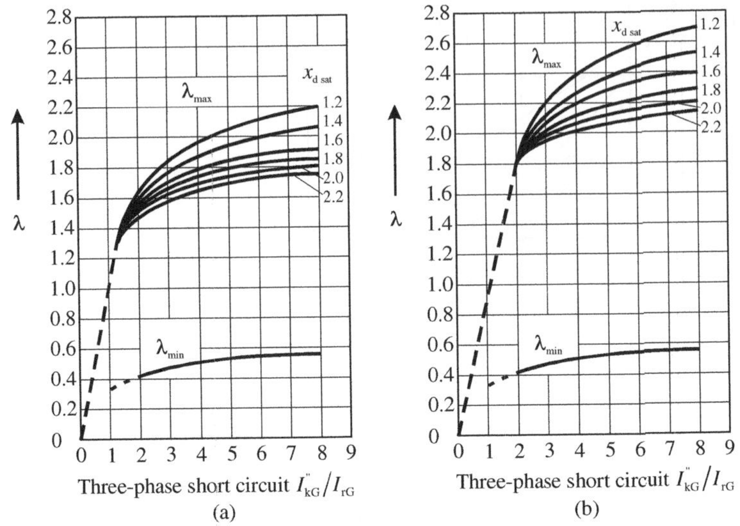

Figure 5.28. Factors λ_{min} and λ_{max} for cylindrical rotor generators: (a) series 1; (b) series 2.

For static excitation systems fed from the generator terminals and a short circuit at the terminals, the field voltage collapses as the terminal voltage collapses, therefore take $\lambda_{min} = \lambda_{max} = 0$ in this case.

- λ_{max} may be obtained from Figures 5.28 or 5.29 for cylindrical rotor generators or salient-pole generators. The saturated reactance $X_{d\text{sat}}$ is the reciprocal of the saturated no-load short-circuit ratio;
- λ_{max}—curves of series 1 are based on the highest possible excitation voltage according to either 1.3 times the rated excitation at rated apparent power and power factor for cylindrical rotor generators (Figure 5.28a) or 1.6 times the rated excitation voltage for salient-pole generators (Figure 5.29a);

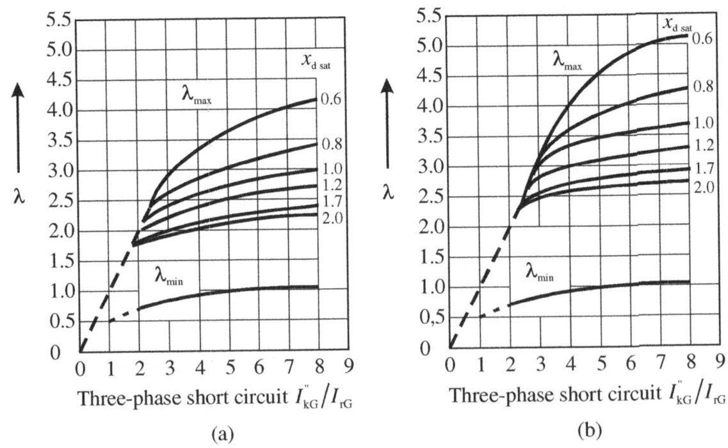

Figure 5.29. Factors λ_{min} and λ_{max} for salient-pole rotor generators: (a) series 1; (b) series 2.

- λ_{max}—curves of series 2 are based on the highest possible excitation voltage according to either 1.6 times the rated excitation at rated apparent power and power factor for cylindrical rotor generators (Figure 5.28b), or 2.0 times the rated excitation voltage for salient pole generators (Figure 5.29b).

Note: When calculating maximum short-circuit currents, it is necessary to introduce the following conditions:

- Voltage factor c_{max} according to Table 5.1 shall be applied for the calculation of maximum short-circuit currents in the absence of a national standard.
- Choose the system configuration and the maximum contribution from power plants and network feeders, which lead to the maximum value of short-circuit current at the short-circuit location, or for accepted sectioning of the network to control the short-circuit current.
- When equivalent impedances $\underline{Z}_S$ are used to represent external networks, the minimum equivalent short-circuit impedance shall be used, which corresponds to the maximum short-circuit current contribution from the network feeders.
- Motors shall be included.
- Resistance R_L of overhead lines and cables are to be introduced at a temperature of 20°C.

• *Minimum steady-state short-circuit current.* For the minimum steady-state short-circuit current in the case of a single-fed short circuit from one generator or one power station unit according to Figures 5.23b and 5.23c, constant no-load excitation (voltage regulator not being effective) of the synchronous machine is assumed:

$$I_{k\,min} = \lambda_{min} I_{rG} \tag{5.84}$$

λ_{min} may be obtained from Figures 5.28 and 5.29. In the case of minimum steady-state short circuit introduce $c = c_{min}$ according to Table 5.1.

The calculation of the minimum steady-state short-circuit current in the case of a near-to-generator short circuit, fed by one or several similar and parallel working generators with compound excitation, is made as follows:

$$I_{k\,min} = \frac{c_{min} U_n}{\sqrt{3}\sqrt{R_k^2 + X_k^2}} \tag{5.85}$$

For the effective reactance of the generators, introduce

$$X_{dP} = \frac{U_{rG}}{\sqrt{3} I_{kP}} \tag{5.86}$$

where I_{kP} is the steady-state short-circuit current of a generator at a three-phase terminal short circuit. The value should be obtained from the manufacturer.

Note: When calculating minimum short-circuit currents, it is necessary to introduce the following conditions:

- Voltage factor c_{max} for the calculation of minimum short-circuit currents has to be applied according to Table 5.1.

- Choose the system configuration and the minimum contribution from power stations and network feeders, which lead to a minimum value of short-circuit current at the short-circuit location.
- Motors are to be neglected.
- Resistances R_L of overhead lines and cables, line conductors, and neutral conductors, are to be introduced at a higher temperature.

$$R_L = [1 + \alpha(\theta_e - 20\,°C)] \cdot R_{L20}$$

where

- R_{L20} is the resistance at a temperature of 20°C.
- θ_e is the conductor temperature in degree Celsius at the end of the short-circuit duration.
- α is a factor equal to 0.004/K, valid with sufficient accuracy for most practical purposes for copper, aluminum, and aluminum alloy.
- λ_{max} is curves of series 1 or 2 may also be applied in the case of terminal-fed static exciters, if the short circuit is at the high-voltage side of the unit transformer of a power station unit or in the system, and if the maximum excitation voltage is chosen with respect to the partial breakdown of the terminal voltage of the generator during the short circuit.

5.4.5.2 Three-Phase Short Circuit in Nonmeshed Networks. In the case of a three-phase short-circuit in nonmeshed networks, as in Figure 5.24, the steady-state short-circuit current at the short-circuit location can be calculated by the summation of the individual steady-state short-circuit current contributions:

$$I_k = \sum I_{ki} \tag{5.87}$$

For example from Figure 5.24 we obtained:

$$I_k = I_{kS} + I_{kT} + I_{kM} = \lambda I_{rGt} + I''_{kT} \tag{5.88}$$

where λ (λ_{max} or λ_{min}) is found from Figures 5.28 and 5.29. I_{rGt} is the rated current of the generator transferred to the high voltage side of the unit transformer in Figure 5.24.

In the case of network feeders or network feeders in series with transformers (see Figure 5.24), $I_k = I''_k$ is valid (far-from-generator short circuit).

5.4.5.3 Three-Phase Short Circuit in Meshed Networks. In meshed networks with several sources, the steady-state short-circuit current may be calculated approximately by

$$I_{k\,max} = I''_{k\,max\,M} \tag{5.89}$$

$$I_{k\,min} = I''_{k\,min} \tag{5.90}$$

$I''_{k\,max} = I''_k$ is found according to (5.89) and $I''_{k\,min}$ according to (5.90).

Equations (5.89) and (5.90) are valid in the case of far-from-generator and in the case of near-to-generator short circuits.

5.4.5.4 Unbalanced Short Circuits. In all cases for steady-state unbalanced short circuits, the flux decay in the generator is not taken into account and the following equations should be used:

$$I_{k2} = I''_{k2} \tag{5.91a}$$

$$I_{k2E} = I''_{k2E} \tag{5.91b}$$

$$I_{kE2E} = I''_{kE2E} \tag{5.91c}$$

$$I_{k1} = I''_{k1} \tag{5.91d}$$

In the case of minimum steady-state short circuits introduce $c = c_{min}$ according to Table 5.1.

5.4.6 Applications

Application 1. Single-fed short circuit

Given the radial electrical network from Figure 5.30, fed from only one source, determine the rms and the peak values of the short-circuit current caused by the symmetrical three-phase fault indicated by $k3$. The equipment parameters are also presented in Figure 5.30.

Solution

Since a three-phase fault is concerned, the positive-sequence network is used. The equivalent positive-sequence network of the network from Figure 5.30 is presented in Figure 5.31.

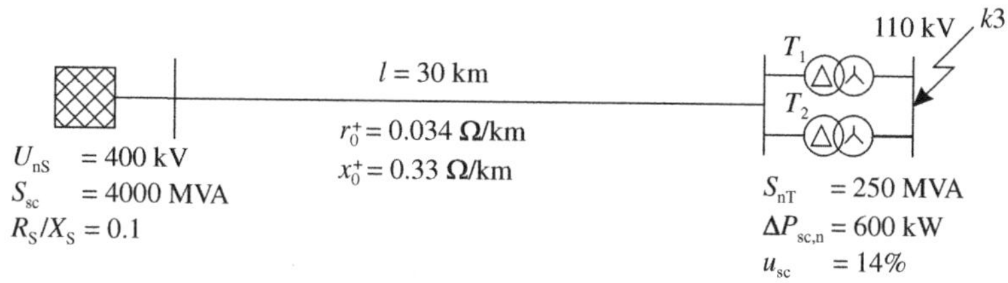

Figure 5.30. Radial electric network fed from one source.

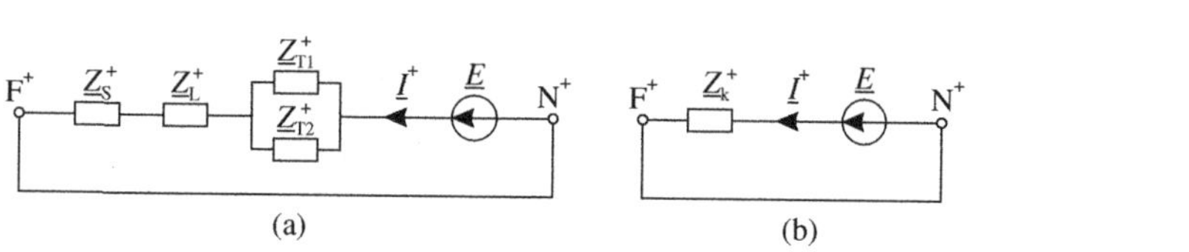

Figure 5.31. Equivalent circuit: (a) positive-sequence network; (b) reduced positive-sequence network.

Calculation can be performed either in absolute units or in per units.

Calculation using absolute units

All parameters are referred to the nominal voltage of the network in which the fault occurs, that is $U_{\mathrm{nK}} = 110\,\mathrm{kV}$.

(a) Calculation of the network parameters

– network feeder parameters

$$X_{\mathrm{S}}^{+} = \frac{U_{\mathrm{nS}}^2}{S_{\mathrm{sc}}} \cdot \left(\frac{U_{\mathrm{nK}}}{U_{\mathrm{nS}}}\right)^2 = \frac{U_{\mathrm{nK}}^2}{S_{\mathrm{sc}}} = \frac{110^2}{4000} = 3.025\,\Omega$$

$$R_{\mathrm{S}}^{+} = \frac{R_{\mathrm{S}}}{X_{\mathrm{S}}} \cdot X_{\mathrm{S}}^{+} = 0.1 \cdot 3.025 = 0.3025\,\Omega$$

– electrical line parameters

$$R_{\mathrm{L}}^{+} = r_0^{+} \cdot l \cdot \left(\frac{U_{\mathrm{nK}}}{U_{\mathrm{nL}}}\right)^2 = 0.034 \cdot 30 \cdot \left(\frac{110}{400}\right)^2 = 0.0771\,\Omega$$

$$X_{\mathrm{L}}^{+} = x_0^{+} \cdot l \cdot \left(\frac{U_{\mathrm{nK}}}{U_{\mathrm{nL}}}\right)^2 = 0.33 \cdot 30 \cdot \left(\frac{110}{400}\right)^2 = 0.7487\,\Omega$$

– transformers parameters

$$R_{T1,2}^{+} = \frac{\Delta P_{sc,n} \cdot U_{\mathrm{nK}}^2}{S_{\mathrm{nT}}^2} = \frac{600 \cdot 110^2}{250^2} \cdot 10^{-3} = 0.1162\,\Omega$$

$$Z_{T1,2}^{+} = \frac{u_{sc}}{100} \cdot \frac{U_{\mathrm{nK}}^2}{S_{\mathrm{nT}}} = \frac{14}{100} \cdot \frac{110^2}{250} = 6.776\,\Omega$$

$$X_{T1,2}^{+} = \sqrt{\left(Z_{T1,2}^{+}\right)^2 - \left(R_{T1,2}^{+}\right)^2} = \sqrt{6.776^2 - 0.1162^2} = 6.775\,\Omega$$

(b) Calculation of the short-circuit impedance

$$\underline{Z}_k^{+} = \underline{Z}_S^{+} + \underline{Z}_L^{+} + \frac{\underline{Z}_{T1}^{+} \cdot \underline{Z}_{T2}^{+}}{\underline{Z}_{T1}^{+} + \underline{Z}_{T2}^{+}} = (0.3025 + j3.025) + (0.0771 + j0.7487) +$$

$$+ \frac{(0.1162 + j6.775) \cdot (0.1162 + j6.775)}{(0.1162 + j6.775) + (0.1162 + j6.775)} = (0.4377 + j7.1612)\,\Omega$$

$$Z_k^{+} = \sqrt{\left(R_k^{+}\right)^2 + \left(X_k^{+}\right)^2} = \sqrt{0.4377^2 + 7.1612^2} = 7.1746\,\Omega$$

(c) Calculation of the initial symmetrical short-circuit current $I_k'' = I^{+}$ (5.58)

$$I_{\mathrm{k}}'' = \frac{cU_{\mathrm{nK}}}{\sqrt{3}Z_{\mathrm{k}}^{+}} = \frac{1.1 \cdot 110}{\sqrt{3} \cdot 7.1746} = 9.737\,\mathrm{kA}$$

(d) Calculation of the peak short-circuit current i_p

– factor κ (5.70)

$$\kappa = 1.02 + 0.98\mathrm{e}^{-3R_k/X_k} = 1.02 + 0.98\mathrm{e}^{-3\cdot\frac{0.4377}{7.1612}} = 1.836$$

– peak short-circuit current (5.69)

$$i_\mathrm{p} = \kappa\sqrt{2}I''_\mathrm{k} = 1.836\cdot\sqrt{2}\cdot 9.737 = 25.28\,\mathrm{kA}$$

Calculation in per units

For calculation in per-unit, definition of the base quantities is necessary:

- base voltage (nominal voltage of the network at the fault location) $U_\mathrm{b} = U_\mathrm{nK} = 110\,\mathrm{kV}$;
- base apparent power: $S_\mathrm{b} = 100\,\mathrm{MVA}$.

The base current and base impedance are therefore achieved:

$$I_\mathrm{b} = \frac{S_\mathrm{b}}{\sqrt{3}U_\mathrm{b}} = \frac{100}{\sqrt{3}\cdot 110} = 0.525\,\mathrm{kA}$$

$$Z_\mathrm{b} = \frac{U_\mathrm{b}^2}{S_\mathrm{b}} = \frac{110^2}{100} = 121\,\Omega$$

(a) Calculation of the network parameters

– network feeder parameters

$$x_\mathrm{S}^+ = \frac{S_\mathrm{b}}{S_\mathrm{sc}} = \frac{100}{4000} = 0.025\,\mathrm{p.u.}$$

$$r_\mathrm{S}^+ = \frac{R_\mathrm{S}}{X_\mathrm{S}}\cdot x_\mathrm{S}^+ = 0.1\cdot 0.025 = 0.0025\,\mathrm{p.u.}$$

– electrical line parameters

$$r_\mathrm{L}^+ = r_0^+\cdot l\cdot\frac{S_\mathrm{b}}{U_\mathrm{nL}^2} = 0.034\cdot 30\cdot\frac{100}{400^2} = 0.00064\,\mathrm{p.u.}$$

$$x_\mathrm{L}^+ = x_0^+\cdot l\cdot\frac{S_\mathrm{b}}{U_\mathrm{nL}^2} = 0.33\cdot 30\cdot\frac{100}{400^2} = 0.00619\,\mathrm{p.u.}$$

– transformers parameters

$$r_{\mathrm{T}1,2}^+ = \Delta P_{\mathrm{sc,n}}\cdot\frac{S_\mathrm{b}}{S_\mathrm{nT}^2}\cdot 10^{-3} = 600\cdot\frac{100}{250^2}\cdot 10^{-3} = 0.00096\,\mathrm{p.u.}$$

$$z_{\mathrm{T}1,2}^+ = \frac{u_\mathrm{sc}}{100}\cdot\frac{S_\mathrm{b}}{S_\mathrm{nT}} = \frac{14}{100}\cdot\frac{100}{250} = 0.056\,\mathrm{p.u.}$$

$$x_{\mathrm{T}1,2}^+ = \sqrt{\left(z_{\mathrm{T}1,2}^+\right)^2 - \left(r_{\mathrm{T}1,2}^+\right)^2} = \sqrt{0.056^2 - 0.00096^2} = 0.05599\,\Omega$$

(b) Calculation of the short-circuit impedance

$$\underline{z}_k^+ = \underline{z}_S^+ + \underline{z}_L^+ + \frac{\underline{z}_{T1}^+ \cdot \underline{z}_{T2}^+}{\underline{z}_{T1}^+ + \underline{z}_{T2}^+} = (0.0025 + j0.025) + (0.00064 + j0.00619) +$$

$$+ \frac{(0.00096 + j0.05599) \cdot (0.00096 + j0.05599)}{(0.00096 + j0.05599) + (0.00096 + j0.05599)} = (0.00362 + j0.05919)\,\text{p.u.}$$

$$z_k^+ = \sqrt{(r_k^+)^2 + (x_k^+)^2} = \sqrt{0.00362^2 + 0.05919^2} = 0.05929\,\Omega$$

(c) Calculation of the initial symmetrical short-circuit current $i_k'' = i^+$ (5.58)

$$i_k'' = \frac{c}{z_k^+} = \frac{1.1}{0.05929} = 18.552\,\text{p.u.}$$

$$I_k'' = i_k'' \cdot I_b = 18.552 \cdot 0.525 = 9.737\,\text{kA}$$

(d) Calculation of the peak short-circuit current i_p

– factor κ (5.70)

$$\kappa = 1.02 + 0.98\,e^{-3r_k/x_k} = 1.02 + 0.98\,e^{-3\cdot\frac{0.00362}{0.05919}} = 1.836$$

– peak short-circuit current (5.69)

$$i_p = \kappa \cdot \sqrt{2} \cdot I_k'' = 1.836 \cdot \sqrt{2} \cdot 9.737 = 25.28\,\text{kA}$$

Application 2. Short circuit fed from nonmeshed network

Given the nonmeshed electrical network from Figure 5.32, fed from two sources, determine the rms and the peak values of the short-circuit current caused by the symmetrical three-phase fault indicated by $k3$. The equipment parameters are also presented in Figure 5.32

Solution

Similar to the previous case, the positive-sequence network will be used (Figure 5.33).

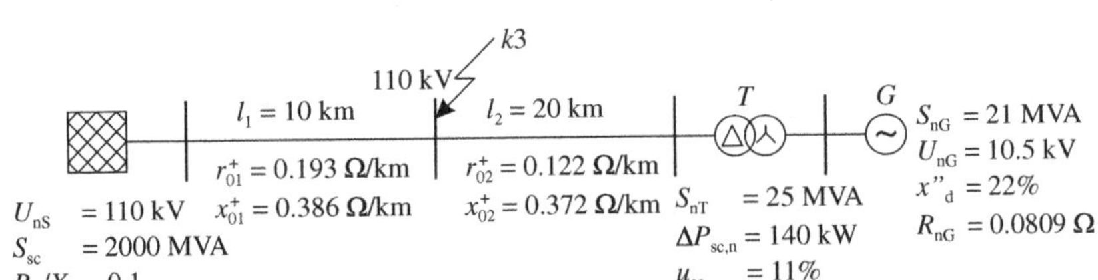

Figure 5.32. Nonmeshed electrical network fed from two sources.

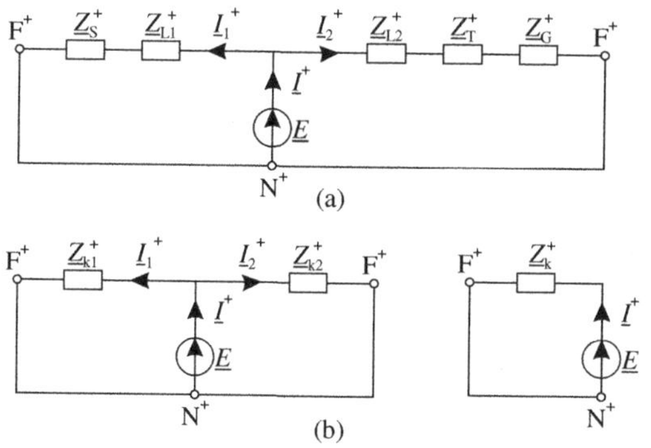

Figure 5.33. Equivalent circuit of the nonmeshed electric network from Figure 5.32: (a) positive-sequence network; (b) reduced positive-sequence network.

The calculation is performed in absolute units. Therefore, all parameters are calculated referred to the nominal voltage of the network in which the fault occurs, that is $U_{nK} = 110\,\text{kV}$.

(a) Calculation of the network parameters

– network feeder parameters

$$X_S^+ = \frac{U_{nS}^2}{S_{sc}} \cdot \left(\frac{U_{nK}}{U_{nS}}\right)^2 = \frac{U_{nK}^2}{S_{sc}} = \frac{110^2}{2000} = 6.05\,\Omega$$

$$R_S^+ = \frac{R_S}{X_S} \cdot X_S^+ = 0.1 \cdot 6.05 = 0.605\,\Omega$$

– generator parameters

$$R_G^+ = R_{nG} \cdot \left(\frac{U_{nk}}{U_{nG}}\right)^2 = 0.0809 \cdot \left(\frac{110}{10.5}\right)^2 = 8.8788\,\Omega$$

$$X_G^+ = \frac{X_d''}{100} \cdot \frac{U_{nK}^2}{S_{nG}} = \frac{22}{100} \cdot \frac{110^2}{21} = 126.7619\,\Omega$$

– electrical lines parameters

$$R_{L1}^+ = r_{01}^+ \cdot l_1 = 0.193 \cdot 10 = 1.93\,\Omega$$

$$X_{L1}^+ = x_{01}^+ \cdot l_1 = 0.386 \cdot 10 = 3.86\,\Omega$$

$$R_{L2}^+ = r_{02}^+ \cdot l_2 = 0.122 \cdot 20 = 2.44\,\Omega$$

$$X_{L2}^+ = x_{02}^+ \cdot l_2 = 0.372 \cdot 20 = 7.44\,\Omega$$

– transformer parameters

$$R_T^+ = \frac{\Delta P_{sc,n} \cdot U_{nK}^2}{S_{nT}^2} = \frac{140 \cdot 110^2}{25^2} \cdot 10^{-3} = 2.7104\,\Omega$$

$$Z_T^+ = \frac{u_{sc}}{100} \cdot \frac{U_{nK}^2}{S_{nT}} = \frac{11}{100} \cdot \frac{110^2}{25} = 53.24\,\Omega$$

$$X_T^+ = \sqrt{(Z_T^+)^2 - (R_T^+)^2} = \sqrt{53.24^2 - 2.7104^2} = 53.171\,\Omega$$

(b) Calculation of the short-circuit impedances

$$\underline{Z}^{+}_{k1} = \underline{Z}^{+}_{S} + \underline{Z}^{+}_{L1} = (0.605 + j6.05) + (1.93 + j3.86) = (2.535 + j9.91)\,\Omega$$

$$Z^{+}_{k1} = \sqrt{\left(R^{+}_{k1}\right)^2 + \left(X^{+}_{k1}\right)^2} = \sqrt{2.535^2 + 9.91^2} = 10.2291\,\Omega$$

$$\underline{Z}^{+}_{k2} = \begin{array}{l} \underline{Z}^{+}_{G} + \underline{Z}^{+}_{T} + \underline{Z}^{+}_{L2} = (8.8788 + j126.7619) + (2.7104 + j53.171) + \\ +(2.44 + j7.44) = (14.0292 + j187.3729)\,\Omega \end{array}$$

$$Z^{+}_{k2} = \sqrt{\left(R^{+}_{k2}\right)^2 + \left(X^{+}_{k2}\right)^2} = \sqrt{14.0292^2 + 187.3729^2} = 187.8973\,\Omega$$

$$\underline{Z}^{+}_{k} = \frac{\underline{Z}^{+}_{k1} \cdot \underline{Z}^{+}_{k2}}{\underline{Z}^{+}_{k1} + \underline{Z}^{+}_{k2}} = \frac{(2.535 + j9.91) \cdot (14.0292 + j187.3729)}{(2.535 + j9.91) + (14.0292 + j187.3729)} = (2.3209 + j9.4268)\,\Omega$$

$$Z^{+}_{k} = \sqrt{\left(R^{+}_{k}\right)^2 + \left(X^{+}_{k}\right)^2} = \sqrt{2.3209^2 + 9.4268^2} = 9.7083\,\Omega$$

Calculation of the initial symmetrical short-circuit currents (5.58)

$$I''_{k1} = I^{+}_{1} = \frac{cU_{\mathrm{nK}}}{\sqrt{3}Z^{+}_{k1}} = \frac{1.1 \cdot 110}{\sqrt{3} \cdot 10.2291} = 6.829\,\mathrm{kA}$$

$$I''_{k2} = I^{+}_{2} = \frac{cU_{\mathrm{nK}}}{\sqrt{3}Z^{+}_{k2}} = \frac{1.1 \cdot 110}{\sqrt{3} \cdot 187.8973} = 0.372\,\mathrm{kA}$$

$$I''_{k} = I^{+} = \frac{cU_{\mathrm{nK}}}{\sqrt{3}Z^{+}_{k}} = \frac{1.1 \cdot 110}{\sqrt{3} \cdot 9.7083} = 7.196\,\mathrm{kA}$$

Calculation of the peak short-circuit current

– factor κ (5.70)

$$\kappa_1 = 1.02 + 0.98\mathrm{e}^{-3R_{k1}/X_{k1}} = 1.02 + 0.98\,\mathrm{e}^{-3 \cdot \frac{2.535}{9.91}} = 1.475$$

$$\kappa_2 = 1.02 + 0.98\mathrm{e}^{-3R_{k2}/X_{k2}} = 1.02 + 0.98\,\mathrm{e}^{-3 \cdot \frac{14.0292}{187.3729}} = 1.803$$

– peak short-circuit currents (5.69)

$$i_{p1} = \kappa_1 \cdot \sqrt{2} \cdot I''_{k1} = 1.475 \cdot \sqrt{2} \cdot 6.829 = 14.245\,\mathrm{kA}$$

$$i_{p2} = \kappa_2 \cdot \sqrt{2} \cdot I''_{k2} = 1.803 \cdot \sqrt{2} \cdot 0.372 = 0.948\,\mathrm{kA}$$

$$i_p = i_{p1} + i_{p2} = 9.616 + 0.67 = 15.193\,\mathrm{kA}$$

Application 3. Short circuit in meshed network

Given the meshed electrical network from Figure 5.34, determine the rms and the peak values of the short-circuit current caused by a symmetrical three-phase fault indicated by

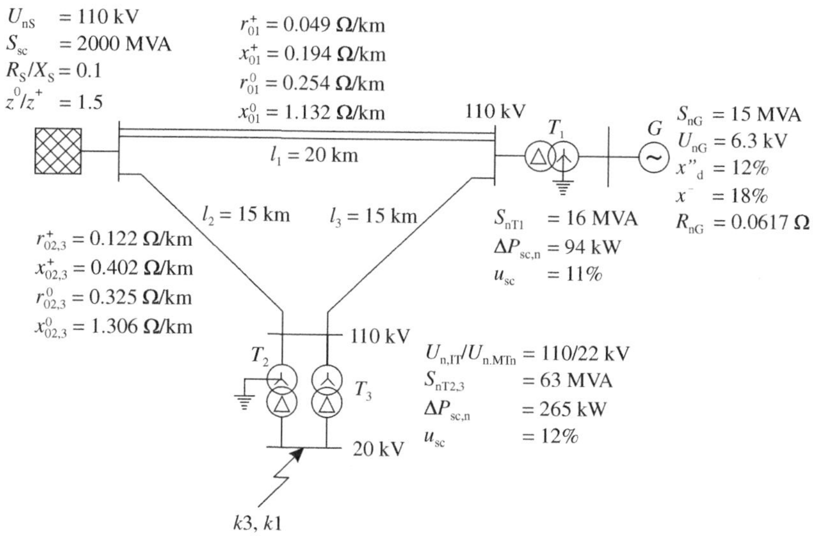

Figure 5.34. Short circuit fed from a meshed electrical network.

$k3$ and a single-phase fault indicated by $k1$, respectively. The equipment parameters are also shown in Figure 5.34

Solution

Figure 5.35 shows the positive-, negative-, and zero-sequence network of the meshed electrical network, used for three-phase and single-phase short-circuit current calculation.

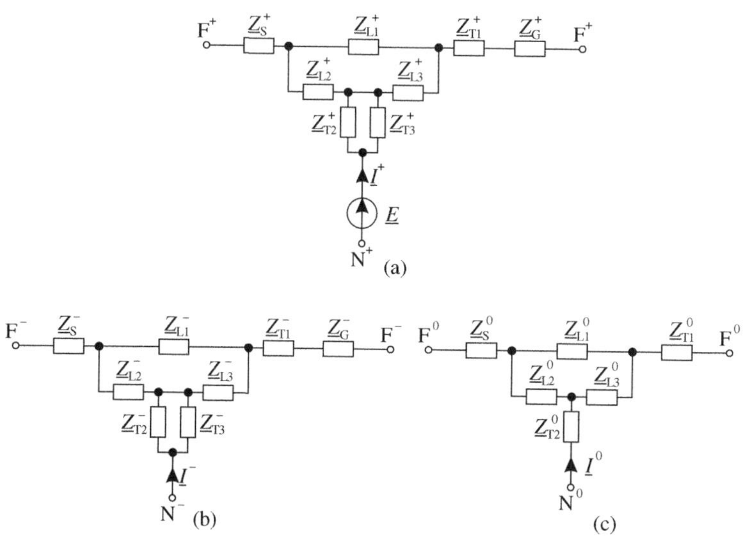

Figure 5.35. Equivalent circuit of the meshed electrical network: (a) positive-sequence network; (b) negative-sequence network; (c) zero-sequence network.

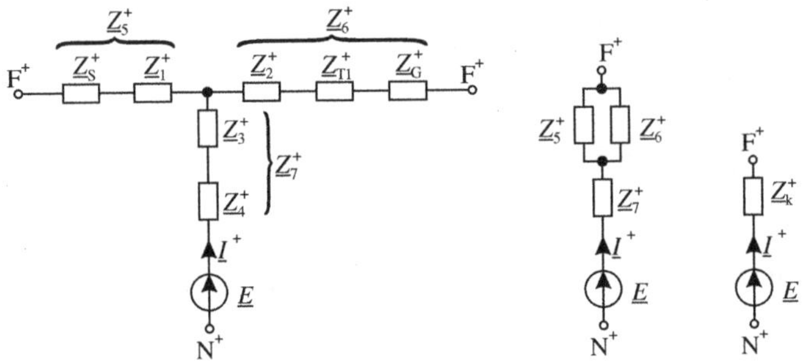

Figure 5.36. Reduced equivalent positive-sequence network.

The calculation is performed in absolute units. Therefore, all parameters are referred to the nominal voltage of the network in which the fault occurs, that is $U_{nK} = 110\,\text{kV}$.

(i) Reducing the positive-sequence network

Figure 5.36 illustrates the reduced equivalent positive-sequence network.

(a) Calculation of the network parameters

– network feeder parameters

$$X_S^+ = \frac{U_{\text{nS}}^2}{S_{\text{sc}}} \cdot \left(\frac{U_{\text{nK}}}{U_{\text{nS}}}\right)^2 = \frac{U_{\text{nK}}^2}{S_{\text{sc}}} = \frac{20^2}{2000} = 0.2\,\Omega$$

$$R_S^+ = \frac{R_S}{X_S} \cdot X_S^+ = 0.1 \cdot 0.2 = 0.02\,\Omega$$

– generator parameters

$$R_G^+ = R_{nG} \cdot \left(\frac{U_{\text{nk}}}{U_{\text{nG}}}\right)^2 = 0.0617 \cdot \left(\frac{20}{10.5}\right)^2 = 0.2239\,\Omega$$

$$X_G^+ = \frac{X_d''}{100} \cdot \frac{U_{\text{nK}}^2}{S_{\text{nG}}} = \frac{12}{100} \cdot \frac{20^2}{15} = 3.2\,\Omega$$

– electrical lines parameters

$$R_{L1}^+ = r_{01}^+ \cdot l_1 \cdot \left(\frac{U_{\text{nK}}}{U_{nL1}}\right)^2 = 0.049 \cdot 10 \cdot \left(\frac{20}{110}\right)^2 = 0.0162\,\Omega$$

$$X_{L1}^+ = x_{01}^+ \cdot l_1 \cdot \left(\frac{U_{\text{nK}}}{U_{nL1}}\right)^2 = 0.194 \cdot 10 \cdot \left(\frac{20}{110}\right)^2 = 0.0641\,\Omega$$

$$R_{L2,3}^+ = r_{02,3}^+ \cdot l_{2,3} \cdot \left(\frac{U_{\text{nK}}}{U_{nL2,3}}\right)^2 = 0.122 \cdot 15 \cdot \left(\frac{20}{110}\right)^2 = 0.0605\,\Omega$$

$$X_{L2,3}^+ = x_{02,3}^+ \cdot l_{2,3} \cdot \left(\frac{U_{\text{nK}}}{U_{nL2,3}}\right)^2 = 0.402 \cdot 15 \cdot \left(\frac{20}{110}\right)^2 = 0.1993\,\Omega$$

– transformers parameters

$$R_{T1}^{+} = \frac{\Delta P_{sc,n1} \cdot U_{nK}^2}{S_{nT1}^2} = \frac{94 \cdot 20^2}{16^2} \cdot 10^{-3} = 0.1469\,\Omega$$

$$Z_{T1}^{+} = \frac{u_{sc1}}{100} \cdot \frac{U_{nK}^2}{S_{nT1}} = \frac{11}{100} \cdot \frac{20^2}{16} = 2.75\,\Omega$$

$$X_{T1}^{+} = \sqrt{\left(Z_{T1}^{+}\right)^2 - \left(R_{T1}^{+}\right)^2} = \sqrt{2.75^2 - 0.1469^2} = 2.7461\,\Omega$$

$$R_{T2,3}^{+} = \frac{\Delta P_{sc,n2,3} \cdot U_{nK}^2}{S_{nT2,3}^2} = \frac{265 \cdot 20^2}{63^2} \cdot 10^{-3} = 0.0267\,\Omega$$

$$Z_{T2,3}^{+} = \frac{u_{sc2,3}}{100} \cdot \frac{U_{nK}^2}{S_{nT2,3}} = \frac{12}{100} \cdot \frac{20^2}{63} = 0.7619\,\Omega$$

$$X_{T2,3}^{+} = \sqrt{\left(Z_{T2,3}^{+}\right)^2 - \left(R_{T2,3}^{+}\right)^2} = \sqrt{0.7619^2 - 0.0267^2} = 0.7614\,\Omega$$

(b) Calculation of the short-circuit impedances

$$\underline{Z}_1^{+} = \frac{\underline{Z}_{L1}^{+} \cdot \underline{Z}_{L2}^{+}}{\underline{Z}_{L1}^{+} + \underline{Z}_{L2}^{+} + \underline{Z}_{L3}^{+}} = \frac{(0.0162 + j0.0641) \cdot (0.0605 + j0.1993)}{(0.0162 + j0.0641) + (0.0605 + j0.1993) + (0.0605 + j0.1993)} =$$
$$= (0.0072 + j0.0276)\,\Omega$$

$$\underline{Z}_2^{+} = \frac{\underline{Z}_{L1}^{+} \cdot \underline{Z}_{L3}^{+}}{\underline{Z}_{L1}^{+} + \underline{Z}_{L2}^{+} + \underline{Z}_{L3}^{+}} = \frac{(0.0162 + j0.0641) \cdot (0.0605 + j0.1993)}{(0.0162 + j0.0641) + (0.0605 + j0.1993) + (0.0605 + j0.1993)} =$$
$$= (0.0072 + j0.0276)\,\Omega$$

$$\underline{Z}_3^{+} = \frac{\underline{Z}_{L2}^{+} \cdot \underline{Z}_{L3}^{+}}{\underline{Z}_{L1}^{+} + \underline{Z}_{L2}^{+} + \underline{Z}_{L3}^{+}} = \frac{(0.0605 + j0.1993) \cdot (0.0605 + j0.1993)}{(0.0162 + j0.0641) + (0.0605 + j0.1993) + (0.0605 + j0.1993)} =$$
$$= (0.0267 + j0.0859)\,\Omega$$

$$\underline{Z}_4^{+} = \frac{\underline{Z}_{T2}^{+} \cdot \underline{Z}_{T3}^{+}}{\underline{Z}_{T2}^{+} + \underline{Z}_{T3}^{+}} = \frac{(0.0267 + j0.7614) \cdot (0.0267 + j0.7614)}{(0.0267 + j0.7614) + (0.0267 + j0.7614)} = (0.0134 + j0.3807)\,\Omega$$

$$\underline{Z}_5^{+} = \underline{Z}_S^{+} + \underline{Z}_1^{+} = (0.02 + j0.2) + (0.0072 + j0.0276) = (0.0272 + j0.2276)\,\Omega$$

$$\underline{Z}_6^{+} = \underline{Z}_G^{+} + \underline{Z}_{T1}^{+} + \underline{Z}_2^{+} = (0.2239 + j3.2) + (0.1469 + j2.7461) + (0.0072 + j0.0276) =$$
$$= (0.3779 + j5.9737)\,\Omega$$

$$\underline{Z}_7^{+} = \underline{Z}_3^{+} + \underline{Z}_4^{+} = (0.0267 + j0.0859) + (0.0134 + j0.3807) = (0.04 + j0.4666)\,\Omega$$

$$\underline{Z}_k^{+} = \underline{Z}_7^{+} + \frac{\underline{Z}_5^{+} \cdot \underline{Z}_6^{+}}{\underline{Z}_5^{+} + \underline{Z}_6^{+}} = (0.04 + j0.4666) + \frac{(0.0272 + j0.2276) \cdot (0.3779 + j5.9737)}{(0.0272 + j0.2276) + (0.3779 + j5.9737)} =$$
$$= (0.0657 + j0.6859)\,\Omega$$

$$Z_k^{+} = \sqrt{\left(R_k^{+}\right)^2 + \left(X_k^{+}\right)^2} = \sqrt{0.0657^2 + 0.6859^2} = 0.689\,\Omega$$

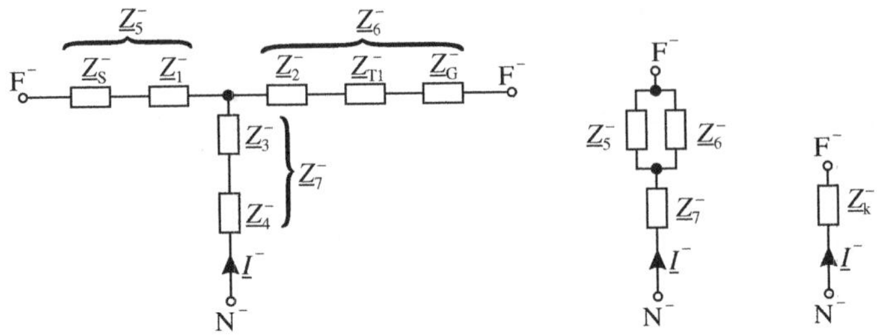

Figure 5.37. Reduced equivalent negative-sequence network.

(ii) Reducing the negative-sequence network

Figure 5.37 shows the reduced equivalent negative-sequence network.

(a) Calculation of the network parameters

– network feeder parameters

$$R_S^- = R_S^+ = 0.02\,\Omega$$

$$X_S^- = X_S^+ = 0.2\,\Omega$$

– generator parameters

$$R_G^- = R_G^+ = 0.2239\,\Omega$$

$$X_G^- = \frac{X^-}{100} \cdot \frac{U_{nK}^2}{S_{nG}} = \frac{18}{100} \cdot \frac{20^2}{15} = 4.8\,\Omega$$

– electric lines parameters

$$R_{L1}^- = R_{L1}^+ = 0.0162\,\Omega$$

$$X_{L1}^- = X_{L1}^+ = 0.0641\,\Omega$$

$$R_{L2,3}^- = R_{L2,3}^+ = 0.0605\,\Omega$$

$$X_{L2,3}^- = X_{L2,3}^+ = 0.1993\,\Omega$$

– transformers parameters

$$R_{T1}^- = R_{T1}^+ = 0.1469\,\Omega$$

$$X_{T1}^- = X_{T1}^+ = 2.7461\,\Omega$$

$$R_{T2,3}^- = R_{T2,3}^+ = 0.0267\,\Omega$$

$$X_{T2,3}^- = X_{T2,3}^+ = 0.7614\,\Omega$$

(b) Calculation of the short-circuit impedances

$$\underline{Z}_1^- = \frac{\underline{Z}_{L1}^- \cdot \underline{Z}_{L2}^-}{\underline{Z}_{L1}^- + \underline{Z}_{L2}^- + \underline{Z}_{L3}^-} = \frac{(0.0162 + j0.0641) \cdot (0.0605 + j0.1993)}{(0.0162 + j0.0641) + (0.0605 + j0.1993) + (0.0605 + j0.1993)} = (0.0072 + j0.0276)\,\Omega$$

$$\underline{Z}_2^- = \frac{\underline{Z}_{L1}^- \cdot \underline{Z}_{L3}^-}{\underline{Z}_{L1}^- + \underline{Z}_{L2}^- + \underline{Z}_{L3}^-} = \frac{(0.0162 + j0.0641) \cdot (0.0605 + j0.1993)}{(0.0162 + j0.0641) + (0.0605 + j0.1993) + (0.0605 + j0.1993)} = (0.0072 + j0.0276)\,\Omega$$

$$\underline{Z}_3^- = \frac{\underline{Z}_{L2}^- \cdot \underline{Z}_{L3}^-}{\underline{Z}_{L1}^- + \underline{Z}_{L2}^- + \underline{Z}_{L3}^-} = \frac{(0.0605 + j0.1993) \cdot (0.0605 + j0.1993)}{(0.0162 + j0.0641) + (0.0605 + j0.1993) + (0.0605 + j0.1993)} = (0.0267 + j0.0859)\,\Omega$$

$$\underline{Z}_4^- = \frac{\underline{Z}_{T2}^- \cdot \underline{Z}_{T3}^-}{\underline{Z}_{T2}^- + \underline{Z}_{T3}^-} = \frac{(0.0267 + j0.7614) \cdot (0.0267 + j0.7614)}{(0.0267 + j0.7614) + (0.0267 + j0.7614)} = (0.0134 + j0.3807)\,\Omega$$

$$\underline{Z}_5^- = \underline{Z}_S^- + \underline{Z}_1^- = (0.02 + j0.2) + (0.0072 + j0.0276) = (0.0272 + j0.2276)\,\Omega$$

$$\underline{Z}_6^- = \underline{Z}_G^- + \underline{Z}_{T1}^- + \underline{Z}_2^- = (0.2239 + j4.8) + (0.1469 + j2.7461) + (0.0072 + j0.0276) = (0.3779 + j7.5737)\,\Omega$$

$$\underline{Z}_7^- = \underline{Z}_3^- + \underline{Z}_4^- = (0.0267 + j0.0859) + (0.0134 + j0.3807) = (0.04 + j0.4666)\,\Omega$$

$$\underline{Z}_k^- = \underline{Z}_7^- + \frac{\underline{Z}_5^- \cdot \underline{Z}_6^-}{\underline{Z}_5^- + \underline{Z}_6^-} = (0.04 + j0.4666) + \frac{(0.0272 + j0.2276) \cdot (0.3779 + j7.5737)}{(0.0272 + j0.2276) + (0.3779 + j7.5737)} = (0.0659 + j0.6876)\,\Omega$$

$$Z_k^- = \sqrt{(R_k^-)^2 + (X_k^-)^2} = \sqrt{0.0659^2 + 0.6876^2} = 0.6907\,\Omega$$

(iii) Reducing the zero-sequence network

Figure 5.38 depicts the reduced equivalent zero-sequence network.

(a) Calculation of the network parameters

– network feeder parameters

$$R_S^0 = \frac{Z^0}{Z^+} R_S^+ = 1.5 \cdot 0.02 = 0.03\,\Omega$$

$$X_S^0 = \frac{Z^0}{Z^+} X_S^+ = 1.5 \cdot 0.2 = 0.3\,\Omega$$

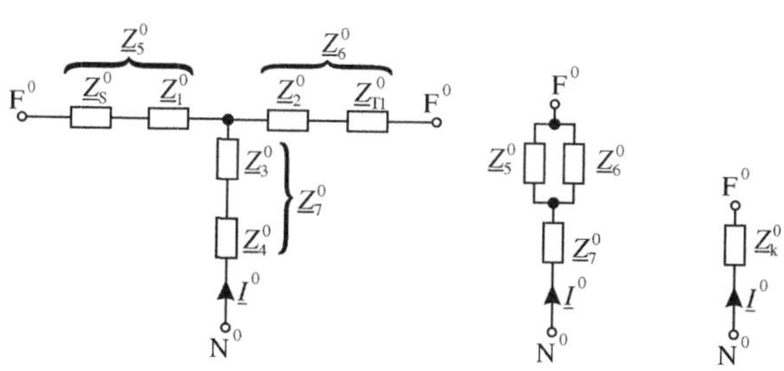

Figure 5.38. Reduced equivalent zero-sequence network.

– generator parameters

$$X_G^0 = \infty \quad ; \quad R_G^0 = \infty$$

– electrical lines parameters

$$R_{L1}^0 = r_{01}^0 \cdot l_1 \cdot \left(\frac{U_{nK}}{U_{nL1}}\right)^2 = 0.254 \cdot 10 \cdot \left(\frac{20}{110}\right)^2 = 0.084\,\Omega$$

$$X_{L1}^0 = x_{01}^0 \cdot l_1 \cdot \left(\frac{U_{nK}}{U_{nL1}}\right)^2 = 1.132 \cdot 10 \cdot \left(\frac{20}{110}\right)^2 = 0.3742\,\Omega$$

$$R_{L2,3}^0 = r_{02,3}^0 \cdot l_{2,3} \cdot \left(\frac{U_{nK}}{U_{nL2,3}}\right)^2 = 0.325 \cdot 15 \cdot \left(\frac{20}{110}\right)^2 = 0.1612\,\Omega$$

$$X_{L2,3}^0 = x_{02,3}^0 \cdot l_{2,3} \cdot \left(\frac{U_{nK}}{U_{nL2,3}}\right)^2 = 1.306 \cdot 15 \cdot \left(\frac{20}{110}\right)^2 = 0.6476\,\Omega$$

– transformers parameters

$$R_{T1}^0 = R_{T1}^+ = 0.1469\,\Omega \quad R_{T2}^0 = R_{T2}^+ = 0.0267\,\Omega \quad R_{T3}^0 = \infty$$
$$X_{T1}^0 = X_{T1}^+ = 2.7461\,\Omega \quad X_{T2}^0 = X_{T2}^+ = 0.7614\Omega \quad X_{T3}^0 = \infty$$

(b) Calculation of the short-circuit impedances

$$\underline{Z}_1^0 = \frac{\underline{Z}_{L1}^0 \cdot \underline{Z}_{L2}^0}{\underline{Z}_{L1}^0 + \underline{Z}_{L2}^0 + \underline{Z}_{L3}^0} = \frac{(0.084 + j0.3741) \cdot (0.1612 + j0.6476)}{(0.084 + j0.3741) + (0.1612 + j0.6476) + (0.1612 + j0.6476)} =$$
$$= (0.0334 + j0.1452)\,\Omega$$

$$\underline{Z}_2^0 = \frac{\underline{Z}_{L1}^0 \cdot \underline{Z}_{L3}^0}{\underline{Z}_{L1}^0 + \underline{Z}_{L2}^0 + \underline{Z}_{L3}^0} = \frac{(0.084 + j0.3741) \cdot (0.1612 + j0.6476)}{(0.084 + j0.3741) + (0.1612 + j0.6476) + (0.1612 + j0.6476)} =$$
$$= (0.0334 + j0.1452)\,\Omega$$

$$\underline{Z}_3^0 = \frac{\underline{Z}_{L2}^0 \cdot \underline{Z}_{L3}^0}{\underline{Z}_{L1}^0 + \underline{Z}_{L2}^0 + \underline{Z}_{L3}^0} = \frac{(0.1612 + j0.6476) \cdot (0.1612 + j0.6476)}{(0.084 + j0.3741) + (0.1612 + j0.6476) + (0.1612 + j0.6476)} =$$
$$= (0.0639 + j0.2512)\,\Omega$$

$$\underline{Z}_4^0 = \underline{Z}_{T2}^0 = (0.0267 + j0.7614)\,\Omega$$

$$\underline{Z}_5^0 = \underline{Z}_S^0 + \underline{Z}_1^0 = (0.03 + j0.3) + (0.0334 + j0.1452) = (0.0634 + j0.4452)\,\Omega$$

$$\underline{Z}_6^0 = \underline{Z}_{T1}^0 + \underline{Z}_2^0 = (0.1469 + j2.7461) + (0.0334 + j0.1452) = (0.1802 + j2.8913)\,\Omega$$

$$\underline{Z}_7^0 = \underline{Z}_3^0 + \underline{Z}_4^0 = (0.0639 + j0.2512) + (0.0267 + j0.7614) = (0.0906 + j1.0126)\,\Omega$$

$$\underline{Z}_k^0 = \underline{Z}_7^0 + \frac{\underline{Z}_5^0 \cdot \underline{Z}_6^0}{\underline{Z}_5^0 + \underline{Z}_6^0} = (0.0906 + j1.0126) + \frac{(0.0634 + j0.4452) \cdot (0.1802 + j2.8913)}{(0.0634 + j0.4452) + (0.1802 + j2.8913)} =$$
$$= (0.1414 + j1.3987)\,\Omega$$

$$Z_k^0 = \sqrt{\left(R_k^0\right)^2 + \left(X_k^0\right)^2} = \sqrt{0.1414^2 + 1.3987^2} = 1.4058\,\Omega$$

(c) Calculation of the initial short-circuit currents (5.58)

$$Z_{k3} = Z_k^+ = 0.689\,\Omega$$

$$I''_{k3} = I^+ = \frac{cU_{nK}}{\sqrt{3}Z_{k3}} = \frac{1.1 \cdot 20}{\sqrt{3} \cdot 0.689} = 18.435\,\text{kA}$$

$$\underline{Z}_{k1} = \underline{Z}_k^+ + \underline{Z}_k^- + \underline{Z}_k^0 = (0.0657 + j0.6859) + (0.0659 + j0.6876) + (0.1414 + j1.3987) =$$
$$= (0.2731 + j2.7722)\,\Omega$$

$$Z_{k1} = \sqrt{(R_{k1})^2 + (X_{k1})^2} = \sqrt{0.2731^2 + 2.7722^2} = 2.7856\,\Omega$$

$$I^+ = I^- = I^0 = \frac{cU_{nK}}{\sqrt{3} \cdot Z_{k1}} = \frac{1.1 \cdot 20}{\sqrt{3} \cdot 2.7856} = 4.56\,\text{kA}$$

$$I''_{k1} = 3 \cdot I^+ = 3 \cdot 4.56 = 13.679\,\text{kA}$$

(d) Calculation of the peak short-circuit current

– factor κ (5.70)

$$\kappa_{k3} = 1.02 + 0.98\text{e}^{-3R_{k3}/X_{k3}} = 1.02 + 0.98\text{e}^{-3 \cdot \frac{0.0657}{0.6859}} = 1.755$$

$$\kappa_{k1} = 1.02 + 0.98\text{e}^{-3R_{k1}/X_{k1}} = 1.02 + 0.98\text{e}^{-3 \cdot \frac{0.2731}{2.7722}} = 1.749$$

– peak short-circuit currents (5.69)

$$i_{pk3} = \kappa_{k3} \cdot \sqrt{2} \cdot I''_{k3} = 1.755 \cdot \sqrt{2} \cdot 18.435 = 45.757\,\text{kA}$$
$$i_{pk1} = \kappa_{k1} \cdot \sqrt{2} \cdot I''_{k1} = 1.47 \cdot \sqrt{2} \cdot 13.679 = 33.841\,\text{kA}$$

REFERENCES

[1] IEC 60909:0-4. *Short-circuit current calculation in three-phase AC systems*, 2002.

[2] de Metz-Noblat, B., Dumas, F., Poulain, C. – *Calculation of short-circuit currents*, Cahier technique No. 158, Schneider Electric, June 2000.

[3] Marconato, R. *Electric power systems*. Vol. 2, *Steady-state behaviour, controls, short-circuits and protection systems*, CEI-Italian Electrotechnical Committee, Milano, 2004.

[4] Guille, A.E., Paterson, W. *Electrical power system*, Vol. 1, 2nd edition (SI/Metric Units), Pergamon Press, 1979.

[5] Fortescue, C.L. Method of symmetrical coordinates applied to the solution of polyphase networks, *AIEE Trans.*, Vol. 37, pp. 1027–1140, 1918.

[6] Stevenson, W.D. *Elements of power system analysis*, 4th edition, McGraw-Hill, Inc., 1982.

[7] Das, J.C. *Power systems analysis: short-circuit load flow and harmonics*, CRC Press, New York, Basel, 2012.

[8] Eremia, M., et al. (editors). *Electric power systems. Vol. I. Electric networks*, Publishing House of the Romanian Academy, Bucharest, 2006.

[9] Kimbark, E.W. *Power system stability. Vol. I. Elements of stability calculations*, John Wiley & Sons, Inc., New York, London, 1961.

[10] Kundur, P. *Power system stability and control*, McGraw-Hill, Inc., New York, 1994.

[11] Anderson, P.M. *Analysis of faulted power systems*, Iowa State University Press, Ames, Iowa, 1973.

[12] Rush, P. (Coordinator) *Network protection & automation: Guide*, Alstom, 2002.

[13] Eremia, M., Crişciu, H., Ungureanu, B., Bulac, C. *Computer aided analysis of the electric power system states*, Technical Publisher, Bucharest, 1985. (in Romanian).

6

ACTIVE POWER AND FREQUENCY CONTROL

Les Pereira

6.1 INTRODUCTION

Electric power systems throughout the world are basically synchronously operated interconnected systems where alternating current generators are connected together in parallel by the transmission system to serve the load at the same frequency.[1] This chapter describes the details of "active power and frequency control" of the electric power system. This is also known as "load frequency control," notably in UCTE[2] [1], which operates in Europe supplying electricity to over 450 million customers.

At every instant in time, there should be a match of generation and load (plus losses), otherwise there will be a frequency deviation in the interconnected system. This deviation will be small if there are small mismatches (example random load fluctuations) and large if there are large mismatches (example a large generator or plant tripping out). The deviation could be positive or negative related to the nominal frequency depending upon whether there is excess or deficient generation, respectively, in the interconnected system with respect to load (plus losses).

The difference in energy is stored in the rotating masses in the system, both generation and dynamic load (motors). Typically, a load drop will result in excess mechanical energy

[1] The transmission systems are predominantly alternating current or AC. However, there could also be large high-voltage direct current (HVDC) links operating within the AC systems.

[2] UCTE is the former "Union for the Co-ordination of Transmission of Electricity: in Europe" [1]. *Note: Although UCTE has been included in the newly created ENTSO-E (European Network of Transmission System Operators for Electricity), we will refer to the former UCTE only.*

Handbook of Electrical Power System Dynamics: Modeling, Stability, and Control. Edited by Mircea Eremia and Mohammad Shahidehpour.

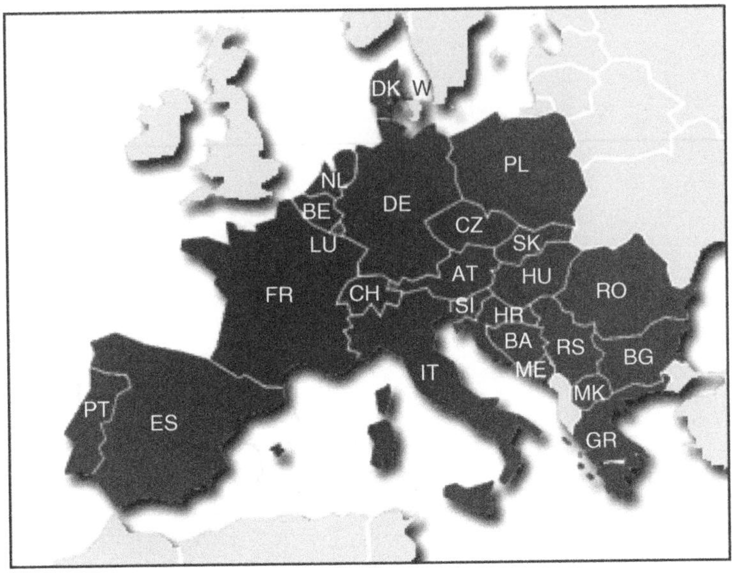

Figure 6.1. Map of UCTE system in Europe.

from the prime movers (turbines) that will be stored in the generator inertias and result in acceleration of the masses to increase frequency.

In order to understand how load frequency and control works in a practical interconnected system, it is necessary to understand the relationship between the size of the system (large or small MW systems), generation inertia (H, in MW-sec/MVA), and the nature and magnitude of the disturbances that cause frequency deviations. Two large interconnected systems will be discussed: the UCTE in Europe and the NERC[3] [2] interconnections in the United States. However, the discussion can be generally applied to large and small systems in other countries or regions.

The UCTE system is a very large interconnected electric system shown in the map in Figure 6.1, approximately 600,000 MW peak, with an annual electricity consumption of 2300 TWh supplied by generators through a highly interconnected transmission system comprising 200,000 km of 400 kV and 220 kV lines.

In comparison, the electricity consumption in the United States was 4055 TWh in 2005 [3], but is supplied by three interconnected systems operating independent of each other, namely the Western Interconnection (WECC)[4] of approximately 160,000 peak MW [4], the ERCOT region in Texas (about 70,000 peak MW), and the rest comprising the Eastern Interconnection (about 660,000 peak MW). The map of the NERC regions is shown in Figure 6.2.

The U.S. interconnections are electrically connected to Canada in the north and Baja Mexico in the south, but data presented herein is from the US government Energy Information Administration (EIA) for the United States part only.

[3] NERC web site is www.nerc.com, previously North American Electric Reliability Council, now North American Electric Reliability Corporation. [2].

[4] WECC is the Western Electricity Coordinating Council [4].

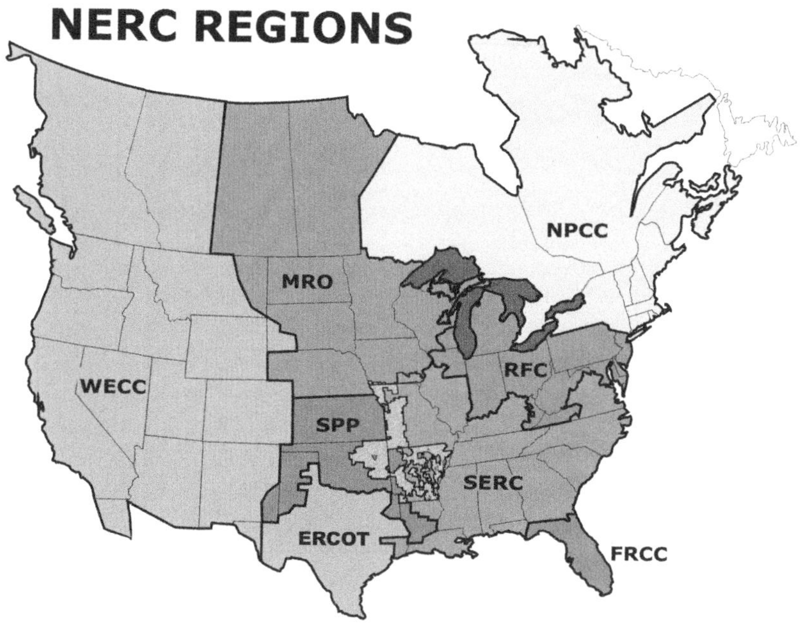

Figure 6.2. NERC regions in the United States and Canada.

6.2 FREQUENCY DEVIATIONS IN PRACTICE

6.2.1 Small Disturbances and Deviations

Typical small frequency deviations in the UCTE system are seen in Figure 6.3 [1], which illustrates that during the space of approximately 4 min, the deviation from the nominal 50 Hz operation has proceeded from negative to positive, that is, undergeneration to overgeneration.

These small deviations may be due to a variety of mismatch reasons including random differences in load, mismatch due to incorrect load forecasts versus system dispatch, slow or fast ramps not in synchronism with load deviations, AGC (automatic generation control) actions resulting in differentials that will eventually be corrected, and so on.

6.2.2 Large Disturbances and Deviations

Large deviations may be due to generator trips or load trips, remedial system actions (RAS) also known as special protection system (SPS) actions that result in generator or load

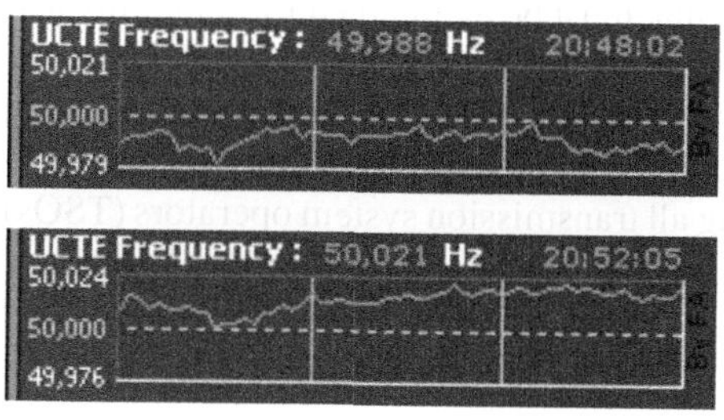

Figure 6.3. Typical small frequency deviation responses in UCTE.

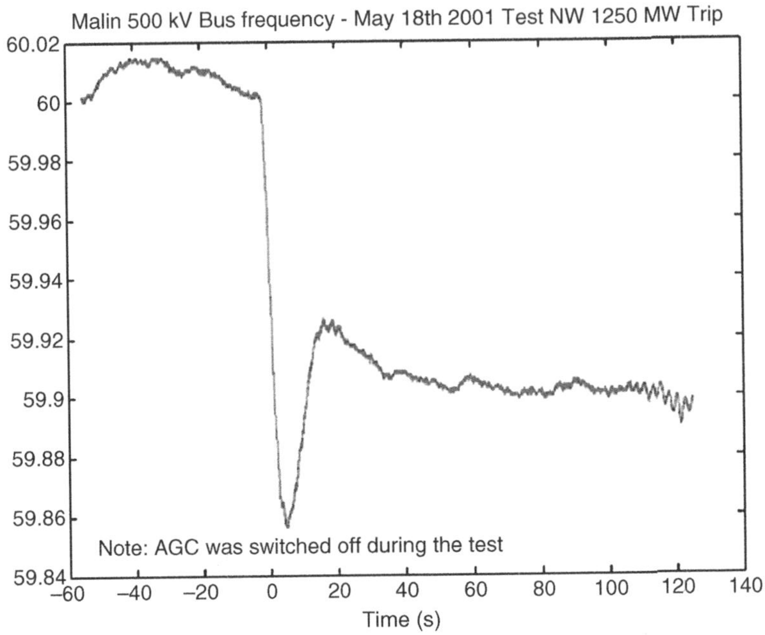

Figure 6.4. Large deviation frequency response resulting from a generation trip of 1250 MW in the WECC on May 18, 2001.

tripping, and so on. It should be apparent that a 1000 MW generating plant trip would create a very large frequency deviation in a relatively smaller system such as ERCOT, but will result in a smaller frequency deviation in WECC and an even smaller frequency deviation in the Eastern Interconnection.

Examples of a 1250 MW generation trip in the WECC and resulting frequency deviation is shown in Figure 6.4; these resulted from a staged test in WECC for the purpose of determining governor response and modeling, and are discussed in detail in the following sections below. Note that the nominal frequency in the United States and Canada is 60 Hz as opposed to 50 Hz in Europe.

6.3 TYPICAL STANDARDS AND POLICIES FOR "ACTIVE POWER AND FREQUENCY CONTROL" OR "LOAD FREQUENCY CONTROL"

6.3.1 UCTE Load Frequency Control

Load frequency control (LFC) is described in the "UCTE Operation Handbook" [1] as the continuous balance between supply and demand that must be maintained for reliability and economic operational reasons. The "UCTE Operation Handbook" is a manual including operation policies for generation control, performance monitoring and reporting, reserves, security criteria, and special operational measures. The basic objective of the Operation Handbook is to ensure the interoperability among all transmission system operators (TSOs) connected to the synchronous areas. Balance quality can be derived from system frequency, which should not vary significantly from its set point of 50 Hz. LFC is split into the following five controls:

A. Primary control
B. Secondary control
C. Tertiary control
D. Time control
E. Measures for emergency conditions

Control actions are performed in different successive steps, each with different characteristics and qualities, and all depending on each other:

- *Primary control* starts within seconds as a joint action of all undertakings involved;
- *Secondary control* replaces primary control after minutes and is put into action by the *responsible undertakings*;
- *Tertiary control* restores secondary control reserve by rescheduling generation and is put into action by the responsible undertakings;
- *Time control* corrects global time deviations of the synchronous time in the long term as a joint action of all undertakings.

6.3.1.1 Primary Control is by Governors. The objective of primary control is to maintain a balance between generation and consumption (demand) within the synchronous area, using turbine speed or turbine governors. The time for starting the action of primary control is in practice a few seconds starting from the incident (although there is no intentional time delay for governor pickup), the deployment time for 50% or less of the total primary control reserve is at most 15 s and from 50% to 100% the maximum deployment time rises linearly to 30 s.

To avoid calling up of primary control in undisturbed operation at or near nominal frequency, the frequency deviation should not exceed ±20 mHz. This reduces wear and tear of the governors due to too frequent operation and results in operation beyond the dead band of the governor. Load shedding schemes start at a frequency of 49 Hz and below; hence, the instantaneous frequency should not fall below 49.2 Hz. The maximum dynamic frequency should not exceed 50.8 Hz. Each control area should contribute to primary control reserves. Similar parameters exist for the 60 Hz systems in the United States.

6.3.1.2 Secondary Control by Automatic Generation Controls (AGCs). Secondary control maintains a balance between generation and consumption (demand) within each control area/block as well as the system frequency within the synchronous area, taking into account the control program, without impairing the primary control that is operated in the synchronous area in parallel but by a margin of seconds.

Secondary control makes use of a centralized automatic generation control, modifying the active power set points/adjustments of generation sets in the time frame of seconds to typically 15 min. Secondary control is based on secondary control reserves that are under automatic control. Adequate secondary control depends on generation resources made available by generation companies to the transmission system operators. Secondary control must be performed in the corresponding control center by a single automatic secondary controller that needs to be operated in an online and closed-loop manner. In order to have no residual error, the secondary controller must be of PI (proportional–integral) type. The integral term must be limited in order to have a nonwindup control action, able to react immediately in case of large changes or a change of the sign of the area control error (ACE). Within each control area/block, the individual ACE needs to be controlled to zero on a continuous basis. The ACE is calculated as the sum of the power control error and the frequency control error $(G = \Delta P + K \cdot \Delta f)$.[5]

6.3.1.3 Tertiary Control.
Tertiary control uses tertiary reserve {15 min reserve} that is usually activated manually by the TSOs after activation of secondary control to free up the secondary reserves. Tertiary control is typically operated in the responsibility of the TSO.

6.3.1.4 Self-Regulation of the Load.
The self-regulation of the load in all synchronous areas cannot be mandated by regulations. It is generally assumed to be 1%/Hz; that means a load *decrease* of 1% occurs in case of a frequency drop of 1 Hz.

6.3.2 NERC (U.S.) Standards

The U.S. system comprises numerous control areas (Balancing Authorities—BAL) independently controlled or via a ISO (Independent System Operator). Hence reliability standards spell out the rules of operation rather than detailed operation requirements and procedures as in UTCE's Handbook. NERC does not spell out "primary control" and "secondary control" but it is widely recognized and practiced that governors are the primary control and AGC systems are the secondary control. NERC mandates that the governors shall pick up with a 5% droop characteristic. However, as will be described under Section 6.4, governors, the pickup performance of governors for thermal units as primary control devices in practice has been eroding as more and more units are currently operated under power controllers with the resulting degradation of frequency response during disturbances. WECC, the Western Interconnection, notably has taken a lead in investigating the unresponsiveness of units. This is discussed further in Section 6.5.5.

The relevant NERC standards for power and frequency control currently are as follows:

- *BAL-001 Real Power Balancing Control Performance.* The purpose is to maintain Interconnection steady-state frequency within defined limits by balancing real power demand and supply in real time.
- *BAL-002 Disturbance Control Standard (DCS).* The purpose of this is to ensure the Balancing Authority is able to utilize its Contingency Reserve to balance resources and demand and return Interconnection frequency within defined limits following a reportable disturbance. Because generator failures are far more common than significant losses of load and because Contingency Reserve activation does not typically apply to the loss of load, the application of DCS is limited to the loss of supply and does not apply to the loss of load.
- *BAL-003 Frequency Response and Bias.* This standard provides a consistent method for calculating the frequency bias component of ACE.
- *BAL-004 Time Error Correction.* The purpose of this standard is to ensure that time error corrections are conducted in a manner that does not adversely affect the reliability of the Interconnection.
- *BAL-005 Automatic Generation Control.* This standard establishes requirements for Balancing Authority Automatic Generation Control necessary to calculate Area Control Error and to routinely deploy the Regulating Reserve. The standard also ensures that all facilities and load electrically synchronized to the Interconnection are included within the metered boundary of a Balancing Area so that balancing of resources and demand can be achieved.

[5] Further discussion of ACE is presented in Section 6.6.3.

6.3.3 Other Countries' Standards

Standards in other countries generally follow the same definitions and approach for rules as stated in UCTE in Europe and NERC in the United States. Local system conditions and the relative size of the system in particular play a large part in determining their standards and emphasizing one or other aspects of primary and secondary controls and reserves.

6.4 SYSTEM MODELING, INERTIA, DROOP, REGULATION, AND DYNAMIC FREQUENCY RESPONSE

To understand the load frequency control problem, the basic "swing equation" for a single generator operating with a governor droop is developed and then extended to multiple generators operating in parallel representing an "area." Two such areas connected by a tie-line are modeled and the load frequency control is developed for a two-area model with paralleled generators in each area. The basic form of AGC control is described using the two-area model. The concept of "system inertia," "system droop," and system frequency response to a disturbance are discussed in relation to the "area" models.

6.4.1 Block Diagram of the System Dynamics and Load Damping

When a disturbance occurs in the system and the system frequency deviates, each generator experiences an accelerating or decelerating torque C_a as discussed[6] in Chapters 2 and 10 as well as [5,6].

If consider the electromechanical model of the generator, developed in Section 2.1.2, given by equation (2.10)

$$\begin{aligned} 2H\frac{d\omega}{dt} + D\omega &= C_m - C_e \approx P_m - P_e \\ \frac{d\delta}{dt} &= \omega_0\omega \end{aligned} \quad (2.12)$$

where C_m is the turbine mechanical torque; C_e is the electrical torque; H is the inertia constant, $2H = M$; M is mechanical starting time.

Note: The self-regulation of the load (termed D) in all synchronous areas is usually assumed to be 1%/Hz; that means a load *decrease* of 1% occurs in case of a frequency *drop* of 1 Hz. Hence $D = 1$ in the equation if load damping is taken.

The block diagram of the system dynamics and load damping is shown in Figure 6.5.

Simplifying the block diagram, the two blocks can be combined into a single forward block using $1/(2Hs + D)$.

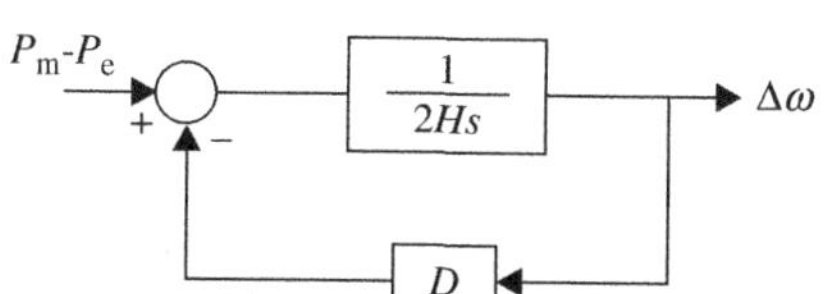

Figure 6.5. Block diagram of the system dynamics plus load damping.

[6] In practice, the generator inertia is much greater than the turbine inertia.

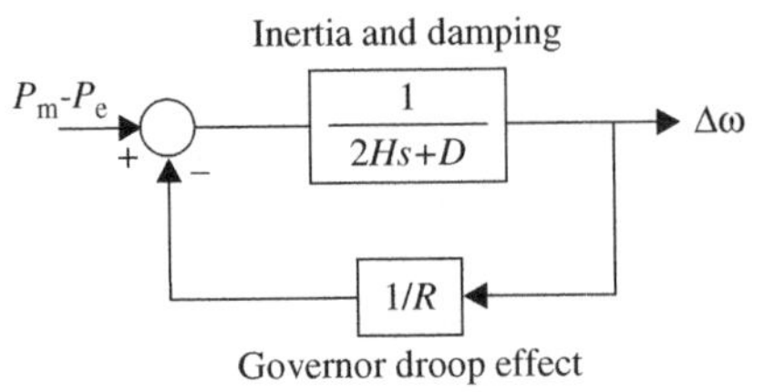

Figure 6.6. Block diagram of the system dynamics, load damping, and governor droop.

6.4.2 Effect of Governor Droop on Regulation

The regulation of the unit is defined as "droop" R and equals

$$R = \frac{\Delta\omega}{\Delta P} \tag{6.1}$$

The influence of the droop on the frequency regulation is shown in the block diagram in Figure 6.6.

The regulation of a unit with 4% droop is shown in Figure 6.7 for a unit loaded at 50%[7] and 100% frequency. This unit will then operate at zero load with a 2% overfrequency and at 100% load at 98% frequency or 2% underfrequency.

6.4.3 Increasing Load by Adjusting Prime Mover Power

By increasing the speed changer setting (also known as the load reference set point), the unit gets loaded at a higher output of 90% as shown in Figure 6.8, the system frequency remaining the same at 100%.

6.4.4 Parallel Operation of Several Generators

It is assumed that two units with different per unit settings of their speed changers both operating at the same 100% system frequency will take up load according to their settings as shown in Figure 6.9.

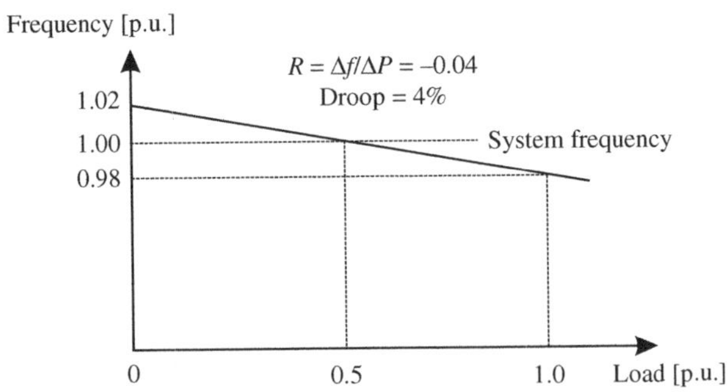

Figure 6.7. Governor droop operation with 50% load at 100% frequency.

[7] Note the range 50% from a practical standpoint is on the low side and is illustrative only.

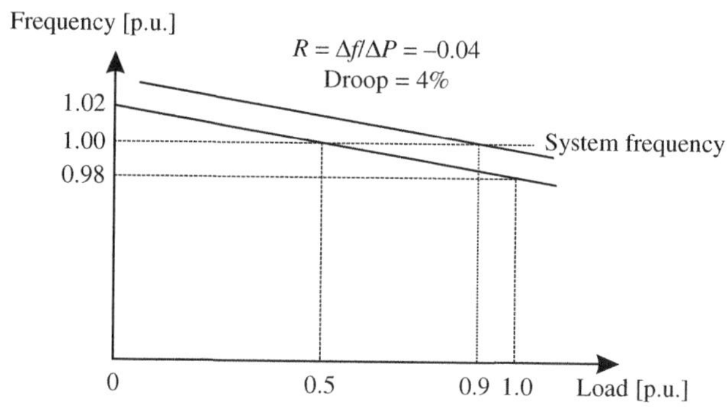

Figure 6.8. Governor droop operation with 90% load at 100% frequency.

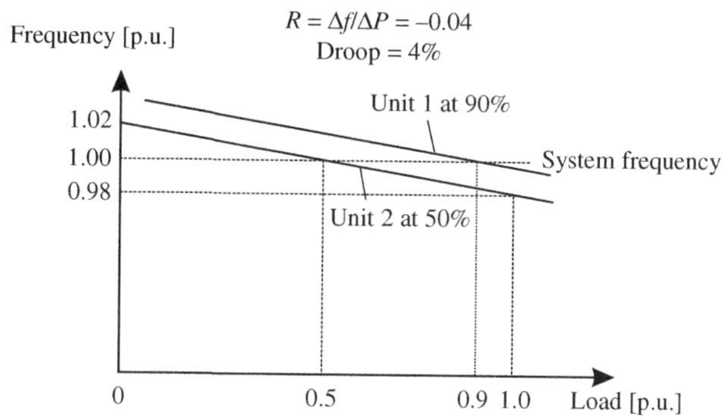

Figure 6.9. Parallel operation of two generators with different speed changer settings at 100% system frequency.

The block diagram of Figure 6.6 is modified as presented in Figure 6.10 to show parallel generators with different droops, R_1, R_2, and R_3. Hence, the inertia constant H and the damping D represent the system inertia and damping.

Isochronous operation with a zero droop is unacceptable for interconnected operation because the droop characteristic is essential for load sharing. An isolated system may, however, have isochronous operation.

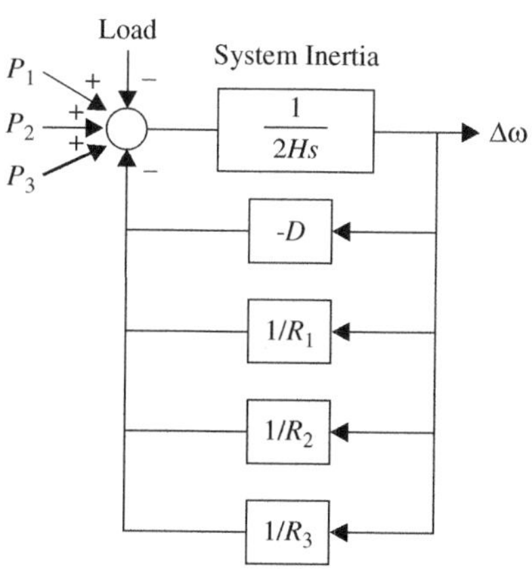

Figure 6.10. Block diagram showing parallel operation of generators.

An example with different speed changer settings and different droops will make parallel operation of generators in an isolated system more clear.

Application 1 [5]

Two generating units of rating 500 MVA and droop 6%, and 200 MVA and droop 4%, respectively, are operating in parallel in an isolated system. They share a load of 700 MW at 100% system frequency. Unit 1 supplies 500 MW, and unit 2 supplies 200 MW. If the load decreases by 80 MW, find the steady-state frequency and generation of each unit. Assume load varies 1% for every 1% change of frequency.

Solution

Using a base MVA of 1000 MVA and the relation (6.1), it results

- droops:

$$R_1 = (0.06)/(500/1000) = 0.12\,\text{p.u.}$$

$$R_2 = (0.04)/(200/1000) = 0.2\,\text{p.u.}$$

- per unit load change: $\Delta P_L = -80/1000 = -0.08$ p.u.
- damping $D = 1$ p.u.

Since the final operation is at steady state, transients due to the $2Hs$ factor can be ignored. The droops of individual units are given by

$$R_1 = \frac{\Delta\omega}{\Delta P_1} \quad \text{and} \quad R_2 = \frac{\Delta\omega}{\Delta P_2}$$

where the change in frequency is the same for both units. This is substituted in the following relationship where the pickup of each generated sums up to the load change:

$$\Delta P_1 + \Delta P_2 = \Delta P_L$$

and gives the frequency deviation of the isolated system as follows:

$$\frac{\Delta\omega}{R_1} + \frac{\Delta\omega}{R_2} = \boxed{\Delta P_L = \frac{\Delta\omega}{R_{sys}}}$$

where R_{sys} is the "equivalent system droop" and is given by

$$\frac{1}{R_{sys}} = \frac{1}{R_1} + \frac{1}{R_2}$$

Adding the effect of load damping, D, as seen in Figure 6.10, the frequency deviation is given by

$$\Delta\omega = \frac{\Delta P_L}{(1/R_1) + (1/R_2) + D} = \frac{-0.08}{(1/0.12) + (1/0.2) + 1.0} = -0.00558\,\text{p.u.}$$

and

$$\Delta f = -0.00558 \text{ p.u.} \times 60\,\text{Hz} = -0.3349\,\text{Hz}$$

Under these conditions, the active power deviation at unit 1 is

$$\Delta P_1 = \frac{\Delta\omega}{R_1} \cdot 1000 = -46.51\,\text{MW}$$

and the active power deviation at unit 2 is

$$\Delta P_2 = \frac{\Delta\omega}{R_2} \cdot 1000 = -27.91\,\text{MW}$$

Therefore, under the new conditions unit 1 supplies 453.5 MW and unit 2 supplies 172.1 MW at the new operating frequency of 59.6651 Hz.

The load damping is $D = (-0.00558 \cdot 1.0) \cdot 1000 = -5.58\,\text{MW}$

Under the initial load, the equivalent system droop was

$$R_{\text{sys}} = \frac{1}{(1/0.12) + (1/0.2) + 1} = 0.0697$$

while after the load deviation occurrence it increases to

$$R_{\text{sys}} = \frac{\Delta\omega}{\Delta P_{\text{L}}} = \frac{0.00558}{0.08} = 0.0725$$

Using the machine base and droop, it gives the same value[8] for pickup as given by formula $\Delta P_1 = \Delta\omega/R_1$, that is

- for unit 1 $\Delta P_1 = -0.00558/0.06 \cdot 500 = -46.51\,\text{MW}$, and
- for unit 2 $\Delta P_2 = -0.0056/0.04 \cdot 200 = -27.91\,\text{MW}$

Hence *generator sustained pickups by governors in the system are approximately linearly proportional to their own machine base and droop for a given frequency deviation* using the formula

$$\Delta P = \frac{\Delta\omega}{R}$$

6.4.5 Isolated Area Modeling and Response

Following the above discussion, it should be evident that an isolated area or interconnection can be approximately modeled using the concept of an equivalent *system inertia H* in seconds (MW-s/MVA), an equivalent *system droop R_{sys}*, and *area damping D*.

[8] Any small differences are due to the effect of damping, which is neglected in this calculation.

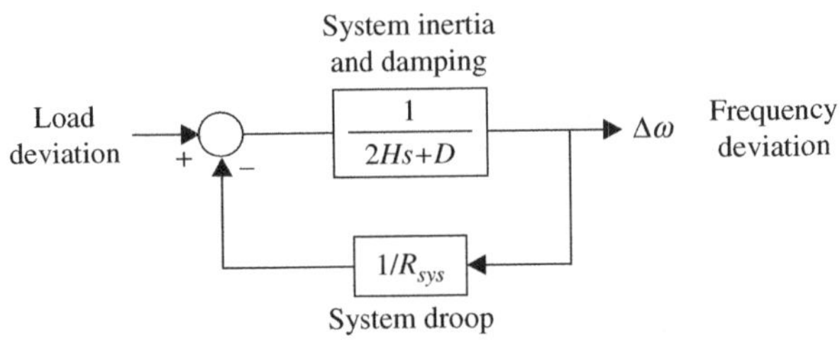

Figure 6.11. Block diagram of system frequency response, inertia, droop, and damping.

The equivalent formula for droop R_{sys} was developed in the previous section. H_{sys} can be similarly developed from first principles of stored energy in the system as distributed among generator rotating masses and is simply the summation of the MW-s of all generators divided by the MVA sum. System inertia H values are typically in range from 2 to 10 s.

The model of Figure 6.6 reproduced below in Figure 6.11 with some changes in definitions can be thus used in principle for an area, since the frequency deviation is one that is defined for the whole interconnection. The formula applicable for system droop is

$$\frac{1}{R_1} + \frac{1}{R_2} + \cdots = \frac{1}{R_{\text{sys}}}$$

Both the system droop and the system inertia can be calculated approximately from disturbance recordings after a known event such as a large generator or plant trip. As system generation, load, and spinning reserves vary, the responses and the value of these parameters will vary. The value of system inertia is a factor to be considered in the permissible operating transfer capacity nomograms (SCIT) in studies for establishing operating limits, as in the case of Southern California in WECC.

The concept of system frequency response and system droop arising from a large generator trip was central in the formulation and analysis from system trip tests performed in WECC that led to the development of an accurate "New Thermal Governor Model" (Section 6.5.5). This work that included system tests with all AGCs switched off in order to get pure governor responses only and led to accurate governor modeling is now the basis of development of new frequency response reserves standards and operating practices in the WECC.

An extension of the above system area concept including a PI controller for AGC is included in Section 6.6, which describes the basic concepts of AGC modeling.

6.5 GOVERNOR MODELING

As stated in Section 6.3, turbine governors are the "primary control" of frequency in interconnected systems and hence its performance and modeling are of paramount importance for operational and planning studies. It is important to understand the mix of generation types in an interconnection and hence what can be expected from the response of each type of governor during large frequency disturbances. In the United States as a whole, the mix in peak generation capacity in 2005, from EIA sources [3], is given in Table 6.1.

TABLE 6.1. Generation Capacity in United States

Natural gas	39%
Coal	32%
Nuclear	10%
Hydroelectric	8%
Hydro pumped storage	2%
Petroleum	6%
Other	3%

From system response recordings, it has been established that hydro governors have the most sustained response to frequency deviations. Nuclear governors are block loaded and unresponsive to frequency, and coal and gas units are largely not responsive in a sustained manner. Since the latter constitute the largest percentage of units, primary control responsive is largely diminished. These issues are discussed in [6]. Also the recent wind and solar generation are unresponsive to frequency. "Secondary control" by AGC therefore increasingly plays a larger part in the generation pickup following a large generation trip.

However, in the Western Interconnection (WECC), the generation mix is quite different; hydroelectric plants including pumped storage make up about 31% of the installed capacity and thermal plants (coal and natural gas) comprise 61%, with nuclear 5%. In the WECC, hydro plants dominate in the Northwest and, as shown later in Section 6.5.5, bear the brunt of the sustained generation pickup in the WECC following a large generation trip in the system.

6.5.1 Response of a Simple Governor Model with Droop

A new model (Figure 6.12) was developed based on the simulation building blocks shown in Figures 6.5 and 6.6 following the implementation of the "swing equation" $2H \times d\omega/dt = P_m - P_e$ and the droop equation $R = \Delta\omega/\Delta P$. The additional transfer function blocks have a simple single time constant (0.5 s) for a nonreheat steam turbine and for the governor valve. The other parameters of the model are the inertia constant $H = 5$ s and damping $D = 0.8$. The load step in the model is 20%. The model assumes operation in an isolated system.

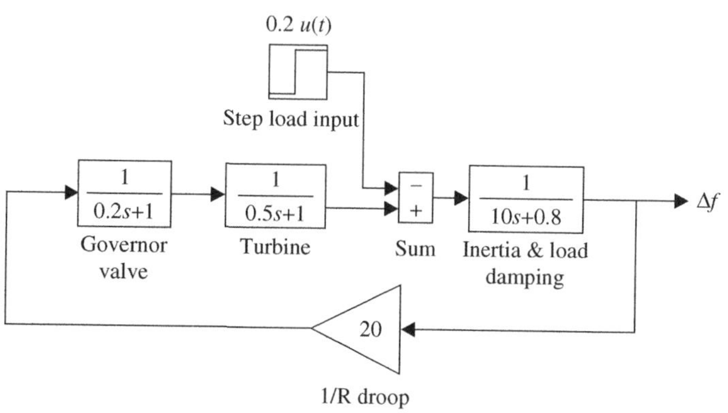

Figure 6.12. Simple governor droop model in isolated operation.

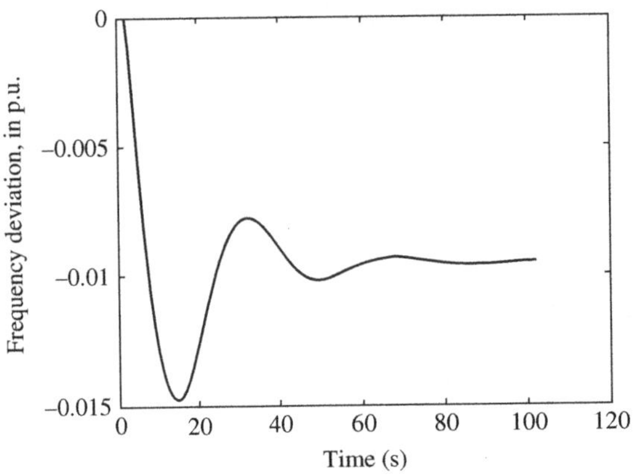

Figure 6.13. Frequency response of a simple governor droop model in isolated operation.

The sustained frequency deviation after transients dies away for a 20% load step, a 5% droop governor, and 0.8 damping can be also easily calculated, that is $\Delta\omega = \frac{\Delta P_L}{1/R_1 + D} = \frac{-0.2}{1/0.5 + 0.8} = -0.0096$ p.u.

Figure 6.13 shows the simulated response over a 100 s run of a simple turbine governor with step load.

6.5.2 Hydraulic Governor Modeling[9]

6.5.2.1 Hydraulic Turbines. A model for a hydraulic turbine can be formulated starting from the hydrodynamic properties, which govern the turbines behavior [6–8]. A block diagram of the model is shown in Figure 6.14, mentioning that the model is nonlinear (see also Figure 3.39).

The variables in Figure 6.14 are defined as follows:

g is gate position from governor;

A_t converts the actual turbine gate position g to the effective gate position G;

G is effective gate position;

q_* is water flow rate, p.u.;

q_{nl*} is water flow rate at no load, in p.u.;

h_* is head, in p.u.;

h_{0*} is initial head, in p.u.;

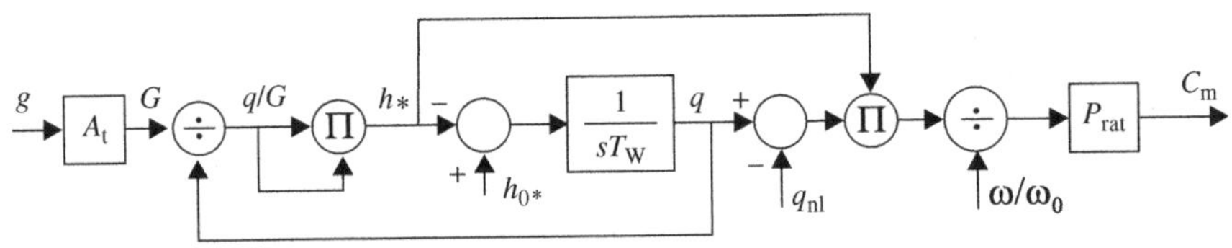

Figure 6.14. Hydraulic turbine.

[9] Additional information is provided in Chapter 3, Section 3.6.2.

T_W is water starting time;
ω/ω_o is per unit generator speed;
P_{rat} is ratio of turbine rating to generator rating;
C_m is per unit generator torque.

The actual gate position is converted to the effective gate position by

$$G = \frac{g}{g_{fl} - g_{nl}} = A_t g \tag{6.2}$$

where g_{fl} is the gate position at full load and g_{nl} is the gate position at no-load.

The turbine head h_*, per unit, is given by

$$h_* = \left(\frac{q_*}{G}\right)^2 \tag{6.3}$$

and the water velocity q_* is

$$\frac{dq_*}{dt} = -\frac{(h_* - h_{0*})}{T_W} \tag{6.4}$$

The turbine output power is

$$P_m = (q_* - q_{nl*})h_* \tag{6.5}$$

and the generator torque C_m is

$$C_m = \frac{P_m}{\omega/\omega_0} P_{rat} \tag{6.6}$$

The water flow at zero load is obtained from

$$q_{nl*} = A_t g_{nl} \sqrt{h_{0*}} \tag{6.7}$$

The initial head h_{0*} varies seasonally for most hydraulic turbines. Under rated conditions h_{0*} is unity. The coefficient P_{rat}, converts the turbine torque to the corresponding generator per unit torque.

6.5.2.2 Hydraulic Governors. The block diagram of Figure 6.15 shows a typical mechanical governor for a hydraulic turbine. The signal from the gate position "g" will feed into the hydraulic turbine model shown in Figure 6.14. The variables are defined in Table 6.2.

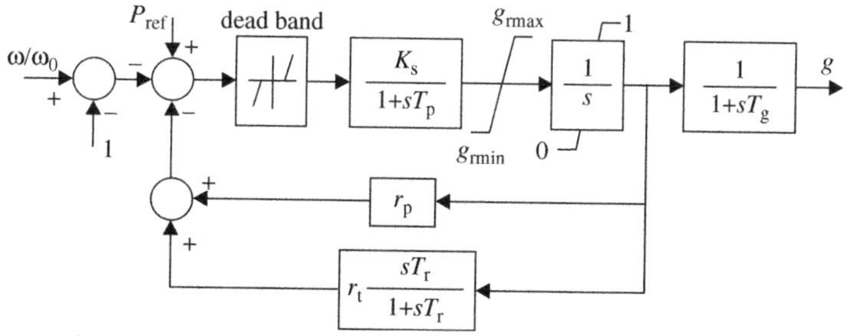

Figure 6.15. Mechanical governor for hydraulic turbine.

TABLE 6.2. The Variables of Hydraulic Governors

Parameter	Description	Typical Value	Range
r_p	Permanent droop	0.05	0.04–0.06
r_t	Temporary droop	0.3	0.2–1.0
T_r	Reset time	5	2.5–25.0
K_s	Gate servo gain	5	2–8
T_p	Pilot servo time constant	0.04	0.03–0.05
g_{rmax}	Maximum rate of change of gate position		
g_{rmin}	Minimum rate of change of gate position		
T_g	Gate power servo time constant	0.2	0.2–0.4
g	Gate position		

Typically, $T_r = 5T_W$ and $r_t = 2.5T_W/2H$, where H is the inertia constant of the generator-turbine unit; $T_W = \sum LV_{water}/9.81\,h_*$is water inertia time constant; L is the length of the penstock, V_{water} is the water velocity, h_* is the total head and the constant 9.81 represents the acceleration of gravity in m/s^2 [7,9].

The temporary droop (or rate feedback) is required to stabilize the turbine/governor generator system when the generator is not synchronized to the power system. This control may be taken out-of-service in some units once the generator is synchronized.

6.5.2.3 Hydraulic Turbine Model. The model shown in Figure 6.14 is for a detailed nonlinear model, where signal from the gate position g feeds into the hydraulic turbine model. However, for many of the smaller hydro plants, the nonlinear model can be replaced with a simpler model as shown in Figure 6.16 [9]. The output is the mechanical power P_{mech} that is the input to the generator model in stability programs.

The block diagram of a typical mechanical hydraulic actuator can be found in Figure 3.45. This may not be modeled completely in detail, and for simulations, the model shown in Figure 6.16 is used.

6.5.2.4 PID Governor. The primary damping adjustments for a PID type of governing controller are the proportional gain (K_P), the integral gain (K_I), and the derivative gain (K_D) (Figure 6.17). A typical range of adjustment for the proportional gain is from 0 to 20, respectively, for the integral gain is from 0 to 10 s^{-1}. A typical range of

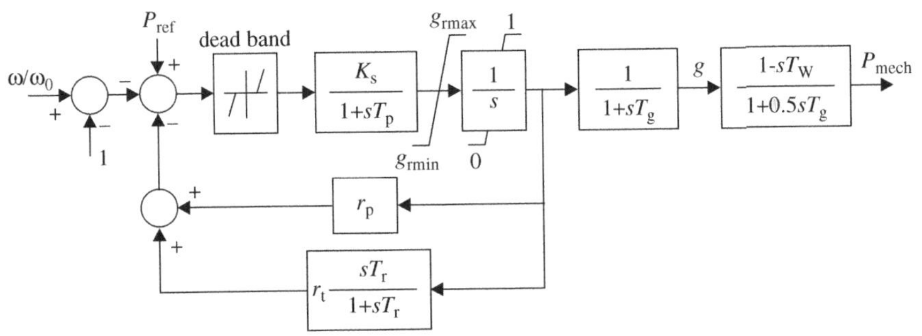

Figure 6.16. Mechanical governor and turbine hydro model.

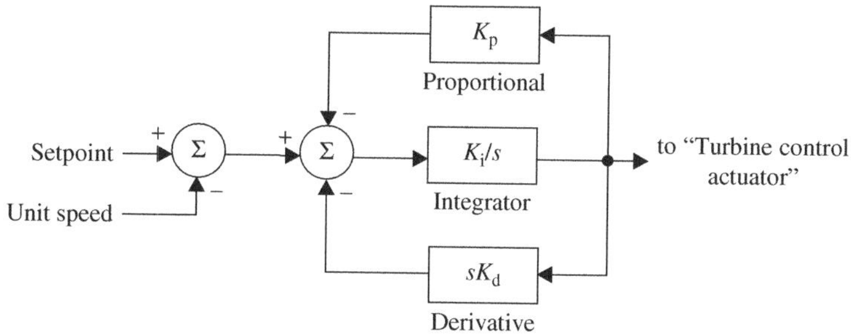

Figure 6.17. PID governor model.

adjustment for the derivative gain is from 0 to 5 s. The actual range of adjustment needed for any particular installation may vary from these guidelines.

6.5.3 Performance of Hydrogovernors with Parameter Variations

6.5.3.1 Isolated System Governor Simulations. Simulations were performed with a mechanical temporary droop hydro governor model, *varying only one parameter at a time*, to show the sensitivity of that parameter on the governor response of a hydro unit in an isolated system[10] [10].

An isolated system in this model consists in one generating unit supplying one load. The model includes the temporary droop mechanical governor shown in Figure 6.18, the turbine-penstock hydraulics, and the unit inertia. It does not model generator electrical effects.

The base case model parameters in the simulations are as follows:

- Permanent speed droop $b_p = 0.05$ (5%).
- Inertia constant $H = 4.75$.
- Mechanical starting time $M = 2 \cdot H = 9.5$ s.
- Water inertia time constant $T_W = 1.24$ s.

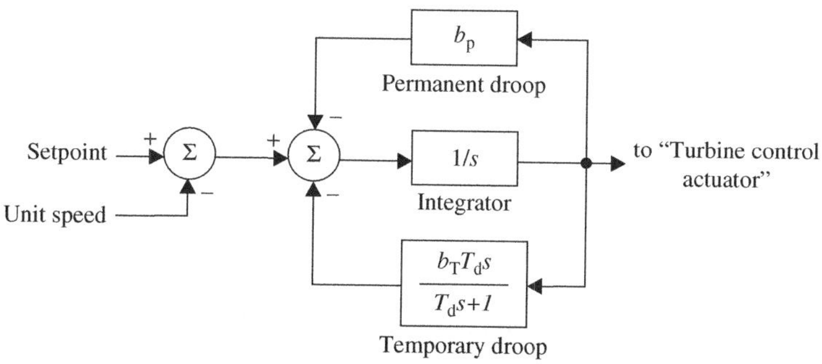

Figure 6.18. Temporary droop governor.

[10] The simulations shown in this section were prepared by the author L. Pereira for the IEEE Task Force report P1247 "Guide for the Application of Turbine Governing Systems for Hydroelectric Generating Units" [10].

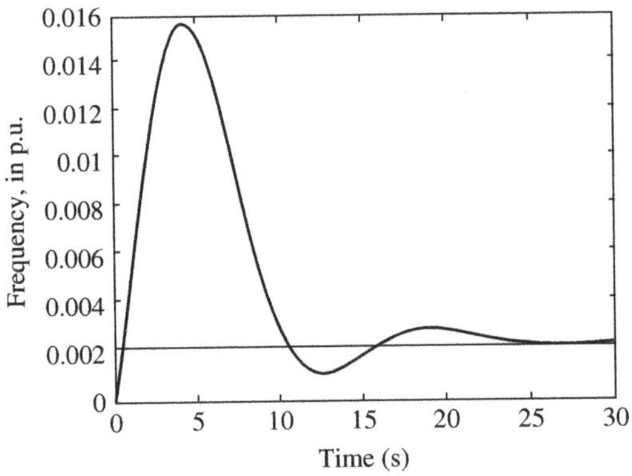

Figure 6.19. Base case: speed (frequency) versus time response for a 5% load change step.

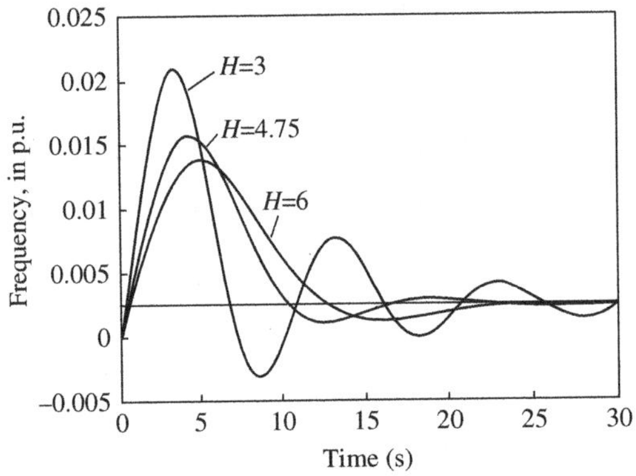

Figure 6.20. The effect of varying inertia constant H.

- Damping (reset) time constant $T_d = 5$.[11]
- Temporary speed droop $b_t = 0.27$ (27%).[12]

Figure 6.19 shows the frequency response versus time for the temporary droop governor shown in Figure 6.18.

Decreasing inertia constant H from 4.75 to 3.0, the slope of the frequency response is steeper and less damped for the smaller unit inertia. In increasing H from 4.75 to 6, as it can be seen in Figure 6.20, the slope of the response is less steep and the damping is greater.

Figure 6.21 shows the time response to a 5% load change for different values of T_W.

The transient response peak decreases with a smaller T_W by varying water inertia time constant from 1.24 to 0.8 s. The opposite effect is achieved when the water inertia time constant is increased to 1.6 s.

Figure 6.22 shows the time response to a 5% load change for different values of the temporary drop b_t.

Decreasing temporary droop from 27% to 12% increases the speed of response and increases oscillations. Increasing temporary droop from 27% to 40% decreases speed of response, improves damping, and decreases oscillations.

[11] Empirical formula for $T_d = 4–5$ times T_W.

[12] Empirical formula for $b_t = 2–2.5$ times T_W/C_m ratio.

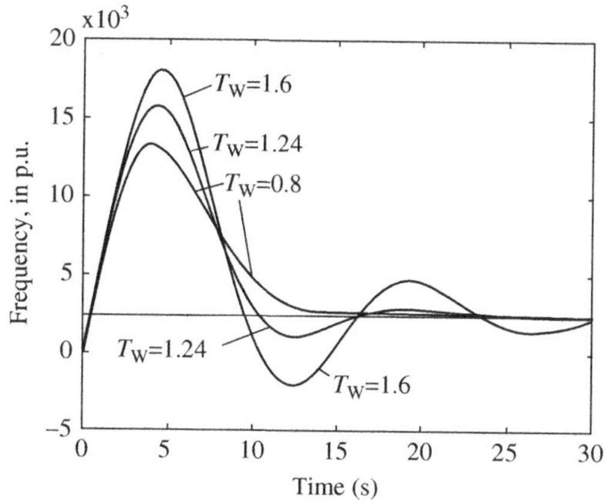

Figure 6.21. The effect of varying water inertia time constant T_W.

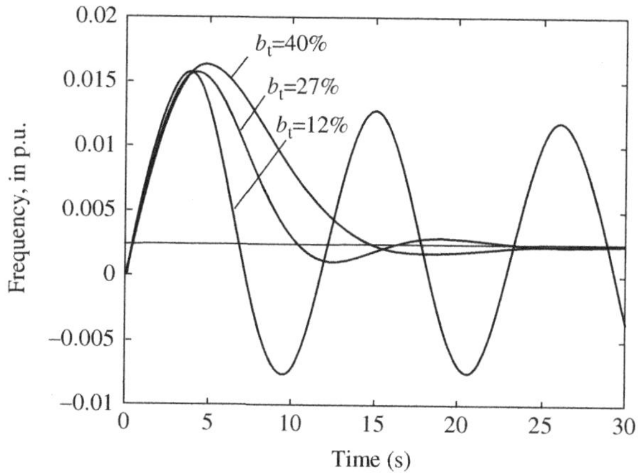

Figure 6.22. The effect of varying temporary droop (b_t).

Figure 6.23 shows the time response to a 5% load change for different values of the damping time constant T_d.

Decreasing T_d from 5 to 3 s decreases damping. Increasing T_d from 5 to 10 s increases damping; however, the setting of T_d is considered optimal in this case. Note that the peak of the speed transient remains the same in all three simulations and only damping is affected.

Figure 6.24 shows the time response to a 5% load change for different values of the speed drop b_p.

The final settling frequency deviation after the initial transient effects decay is the product of the speed droop and the 5% load change. It is 0.15% for a 3% permanent speed droop, 0.25% for a 5% b_p, and 0.3% for a 6% b_p. Note that the last response is essentially isochronous operation with a very small permanent speed droop of 0.0005% giving a final frequency deviation of almost zero.

6.5.3.2 Interconnected System Governor Simulations. To study the governor response of a hydro unit into an interconnected system, simulations were performed with a stability program (GE PSLF [11]) on a 120 MW unit in a small 2000 MW interconnected system. The interconnected system was modeled by two 120 MW hydro generators

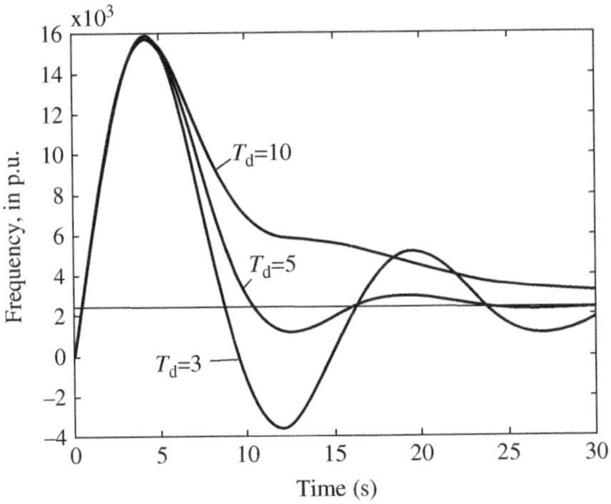

Figure 6.23. The effect of varying T_d, the damping constant.

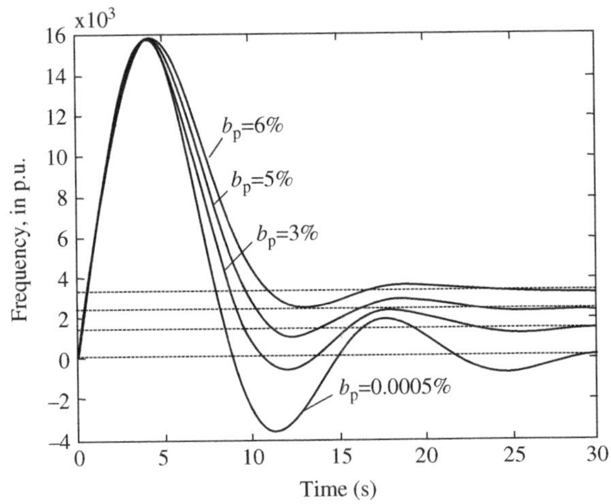

Figure 6.24. The effect of varying permanent speed droop (b_p) for a 5% step load change.

connected to a 2000 MW synchronous generator representing the system through a double circuit line. All machines are modeled with IEEE model representations for turbines, governors, generators, and excitation systems. The governor is modeled with a mechanical temporary droop hydro governor model similar to the isolated system example case. A 100 MW load, connected at the hydro plant, is tripped off to represent a 5% load trip in the system (Figure 6.25).

Simulations were performed *varying only one parameter at a time*, to show the sensitivity of that parameter on the interconnected system performance of the hydro unit.

The base case turbine governor model parameters are as follows:

- Permanent speed droop $b_p = 0.05$ (5%).
- Inertia constant $H = 4.75$.
- Damping (Reset) time constant $T_d = 5$ s.
- Temporary speed droop $b_t = 0.27$ (27%).
- Water inertia time constant $T_W = 1.24$ s.

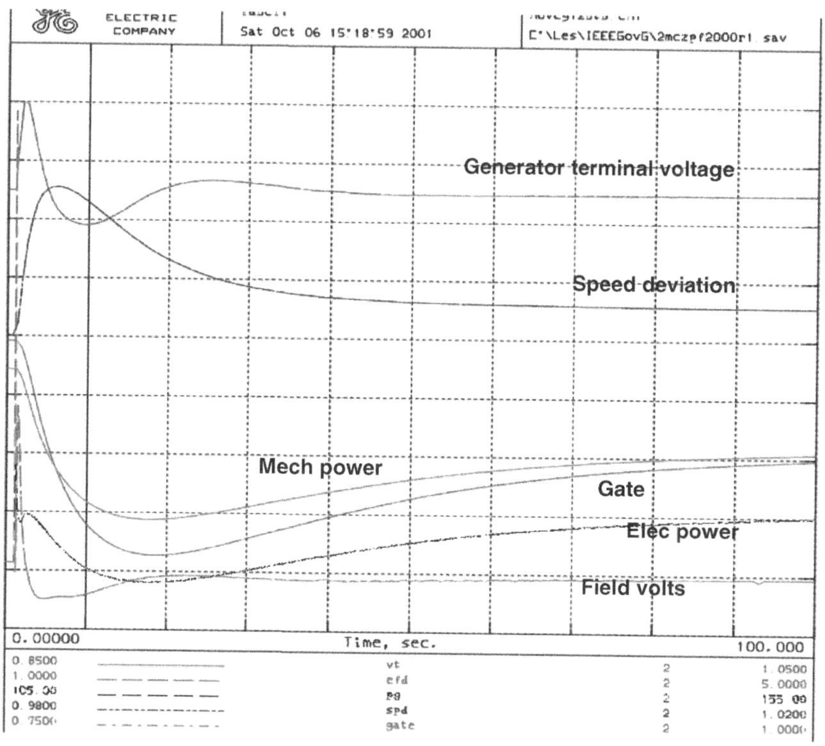

Figure 6.25. Interconnected system base case: speed (frequency) versus time response for a 5% load change step.

Each generator drops 5% of its rating in accordance with the speed droop characteristic, the 120 MW dropping 6 MW in the final steady state. The system frequency deviation is 0.25% (0.05 × 0.05 × 100).

Simulation results are shown in Figure 6.25. Note that in contrast with the isolated system case in Section 6.5.3.1, the first peak of the frequency transient is less in magnitude and subsequent oscillatory behavior in the interconnected system is more damped. The final settling frequency of 0.25% speed deviation is achieved in a longer period of 100 s.

The responses for a 5% step load change in a 120 MW hydro unit in a 2000 MW system are shown in Figure 6.26 where the only parameter changed was the temporary droop to show the sensitivity of that parameter on the interconnected system performance of the hydro unit. The 12% temporary speed droop change shows the fastest response as seen in the mechanical power response, which gets progressive slower as the temporary droop setting is increased. Responses are not oscillatory in comparison with the single unit isolated system model responses.

6.5.4 Thermal Governor Modeling

6.5.4.1 General Steam System Model. A general model of a steam turbine is shown in Figure 6.27 (see also Figure 3.16). Interpretation of the parameters used and typical values can be found in Report [7].

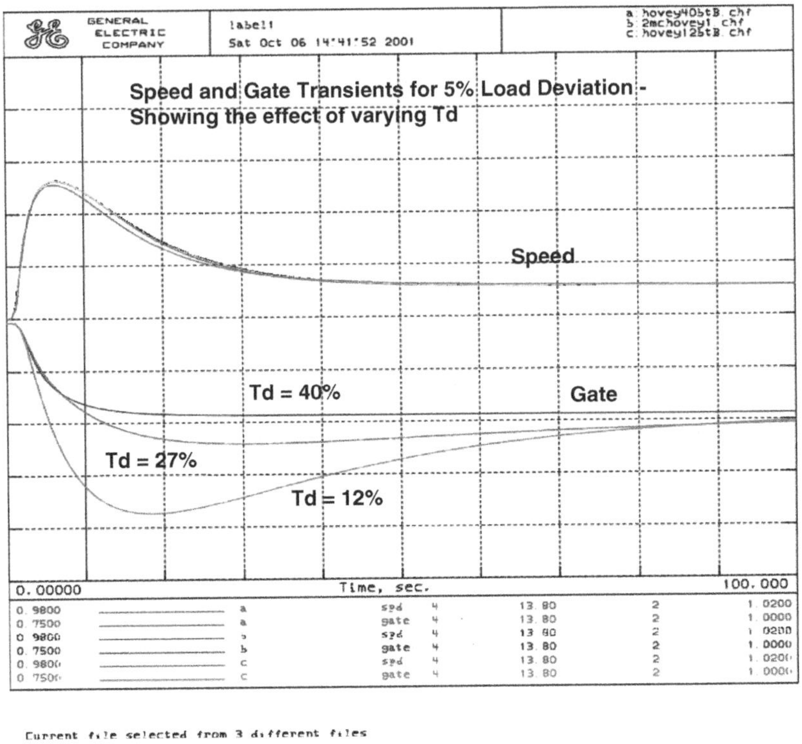

Figure 6.26. Speed and power responses for a 5% load step in an interconnected system: Showing the effect of varying temporary speed droop.

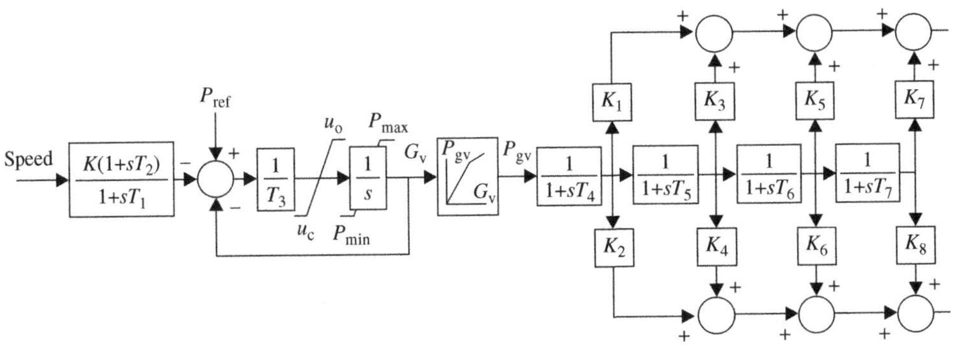

Figure 6.27. General steam turbine model.

Governor rate limits are nominally 0.1 p.u./s except for mechanical hydraulic system where the closing rate is 1.0 p.u./s. The nonlinear gain between gate position G_v and power P_v is usually input with multiple points.

The list of parameters and description is given in Table 6.3.

6.5.4.2 *Gas Turbine Model.* The “ggov1 model” shown in Figure 6.28 is a GE PSLF copyrighted model, which is extensively used in the WECC to model primarily gas turbine and single-shaft combined-cycle turbines. It can also be used to represent a variety

TABLE 6.3. Parameters of the Model from Figure 6.27

Parameter	Description
K	Governor gain (reciprocal of droop), p.u.
T_1	Governor lag time constant, seconds range 0.2–0.3 s
T_2	Governor lead time constant, seconds = 0
T_3	Valve positioner time constant, 0.1 s
u_o	Maximum valve opening velocity, p.u./s
u_c	Maximum valve closing velocity, p.u./s (< 0)
P_{max}	Maximum valve opening, p.u. of MWcap
P_{min}	Minimum valve opening, p.u. of MWcap
T_4	Inlet piping/steam bowl time constant, in seconds
K_1	Fraction of HP shaft power after first boiler pass
K_2	Fraction of LP shaft power after first boiler pass
T_5	Time constant of second boiler pass, in seconds
K_3	Fraction of HP shaft power after second boiler pass
K_4	Fraction of LP shaft power after second boiler pass
T_6	Time constant of third boiler pass, in seconds
K_5	Fraction of HP shaft power after third boiler pass
K_6	Fraction of LP shaft power after third boiler pass
T_7	Time constant of fourth boiler pass, in seconds
K_7	Fraction of HP shaft power after fourth boiler pass
K_8	Fraction of LP shaft power after fourth boiler pass

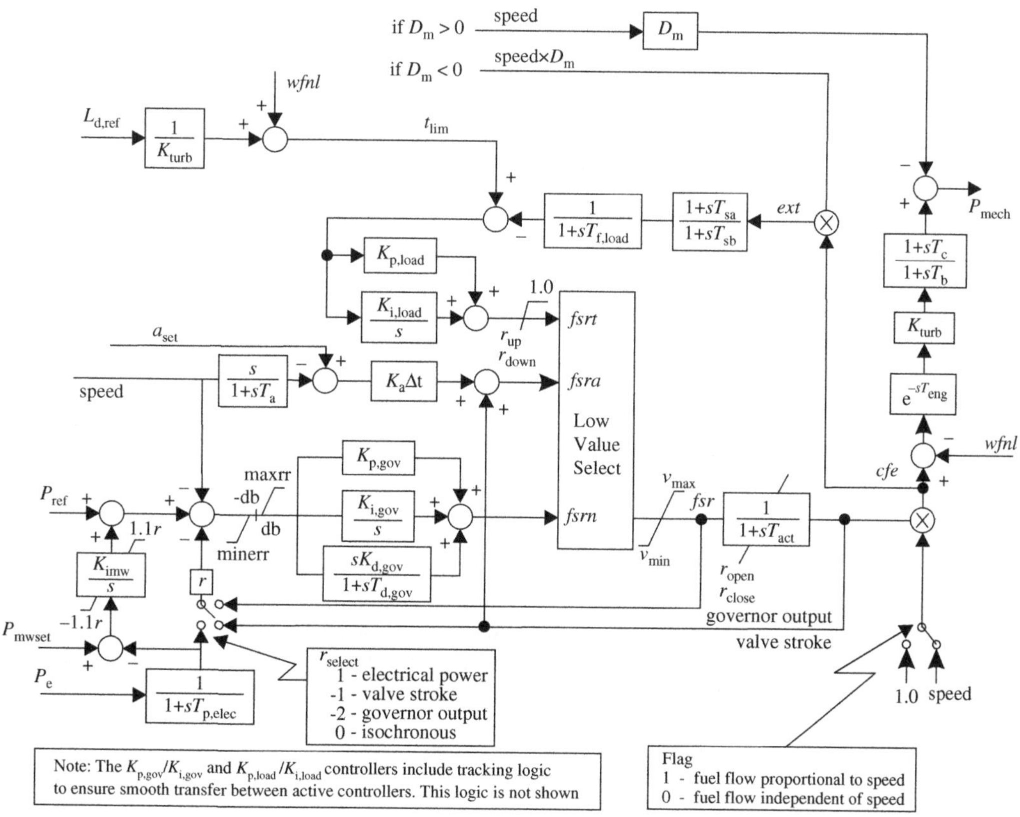

Figure 6.28. General thermal turbine governor model ggov1 (GE-PSLF stability program) described in Section 6.5.5 for steam and gas turbines.

of prime movers controlled by PID governors such as diesel engines with modern electronic or digital governors. It is also suitable, for example, for representation of steam turbines where steam is supplied from a large boiler drum or a large header whose pressure is substantially constant over the period under study. Parameters are given in Table 6.4, and

TABLE 6.4. Parameters of the Block Diagram from Figure 6.28

Parameter	Default Value	Description
r	0.04	Permanent droop, p.u.
r_{select}	1	Feedback signal for droop = 1 selected electrical power = 0 none (isochronous governor) = −1 fuel valve stroke (true stroke) = −2 governor output (requested stroke)
T_{pelec}	1.0	Electrical power transducer time constant, in seconds
max_{err}	0.05	Maximum value for speed error signal
min_{err}	−0.05	Minimum value for speed error signal
K_{pgov}	10.0	Governor proportional gain
K_{igov}	2.0	Governor integral gain
K_{dgov}	0.0	Governor derivative gain
T_{dgov}	1.0	Governor derivative controller time constant
v_{max}	1.0	Maximum valve position limit
v_{min}	0.15	Minimum valve position limit
T_{act}	0.5	Actuator time constant
K_{turb}	1.5	Turbine gain
$wfnl$	0.2	No load fuel flow, p.u
T_b	0.5	Turbine lag time constant
T_c	0.0	Turbine lead time constant
$Flag$	1.0	Switch for fuel source characteristic = 0 for fuel flow independent of speed = 1 fuel flow proportional to speed
T_{eng}	0.0	Transport lag time constant for diesel engine
T_{fload}	3.0	Load Limiter time constant
K_{pload}	2.0	Load limiter proportional gain for PI controller
K_{iload}	0.67	Load limiter integral gain for PI controller
L_{dref}	1.0	Load limiter reference value, p.u.
D_m	0.0	Speed sensitivity coefficient, p.u.
r_{open}	.10	Maximum valve opening rate, p.u./s
r_{close}	−0.1	Minimum valve closing rate, p.u./s
K_{imw}	0.002	Power controller (reset) gain
P_{mwset}	80.0	Power controller set point, MW
a_{set}	0.01	Acceleration limiter set point, p.u./s
K_a	10.0	Acceleration limiter gain
T_a	0.1	Acceleration limiter time constant, in seconds
d_b	0.0	Speed governor dead band
T_{sa}	4.0	Temperature detection lead time constant, in seconds
T_{sb}	5.0	Temperature detection lag time constant, in seconds
r_{up}	99.0	Maximum rate of load limit increase
r_{down}	−99.0	Maximum rate of load limit decrease

the block diagram is shown in Figure 6.28. Further details of the model can be obtained from the GE PSLF website [11].

The application of the "ggov1 model" in WECC is described in detail in Section 6.5.5.

6.5.5 Development of a New Thermal Governor Model in the WECC

6.5.5.1 The New Thermal[13] Governor Model. The development of a new thermal governor model [12,13] is described in some detail here because it is felt that the model, its tests, and its applications have attained far-reaching results and consequences in the Western Interconnection in the United States and Canada and provide a critical understanding of governor responses or the lack thereof in practical system operation.

Over the years, WECC have not been able to accurately simulate the frequency response in the Western Interconnection when large generators and plants trip. Comparisons of disturbance monitoring recordings with the computer simulations have indicated a wide discrepancy in both the "initial transient dips" and in the "settling" frequencies. Preliminary calculations indicated very high "system droops" (see Section 6.4.5 for definition) indicating high unresponsiveness of governors. Assessment of the first transient dip is important for load shedding while the subsequent time response is a measure of the responsiveness of the "primary control", that is, the turbine-governors, and the sustained settling frequency is a measure of the effectiveness of the "secondary control", that is, the AGCs in the system.

To further the governor investigation as to why governor modeling did not correspond to actual system performance, two separate generation trip tests, one in the Southwest and the other in the Northwest, were performed on May 18, 2001 in the WECC. During the generation trip tests, all AGCs were switched off throughout the WECC; hence, the resulting system frequency response observed was of governors only.

In the first test, 750 MW was tripped in the Hoover power plant in the Southwest. After the disturbance monitoring recordings were taken, the AGCs were made operational again to stabilize the frequency, and about 20 min later, 1250 MW was tripped in three power plants in the Northwest. Figure 6.29 shows the frequency response of the second 1250 MW test in the Northwest.

Figure 6.30 is a composite plot of both tests and shows a comparison of simulation results using existing (incorrect) governor models versus the disturbance recordings of the tests and how incorrect was the existing governor modeling.

A simple calculation using the formula below [12] indicates that only about 40% of the governors effectively respond in the real system in the settling time of about 60–100 s. If all the governors were responsive in the May 18, 2001 1250 MW trip test, out of a WECC generation base capacity of 91,000 MW online during the test, the calculated generation pickup for governors with a 5% droop, for a 0.1 Hz frequency deviation, would be 3185 MW. Since the actual pickup was only 1250 MW, the percentage of "responsive" governors would be only (1250/3185) or 39%.

The following parameters were chosen:

- Droop (regulation), $r = 0.05$ p.u.
- Damping, $D = 0$.
- Settling frequency deviation = 0.105 Hz.

[13] "Thermal" plants include conventional fired steam, nuclear steam, simple-cycle gas turbine, and combined-cycle gas turbine plants.

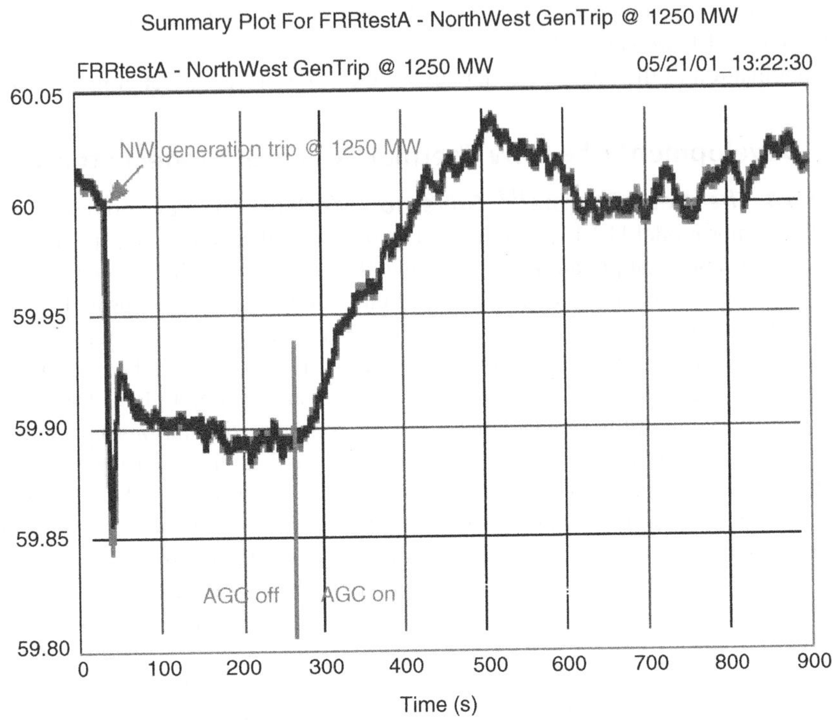

Figure 6.29. WECC NW 1250 MW trip test on May 18, 2001 shows governing response only with AGC switched off. Note the pickup of system frequency by AGC "after" the test.

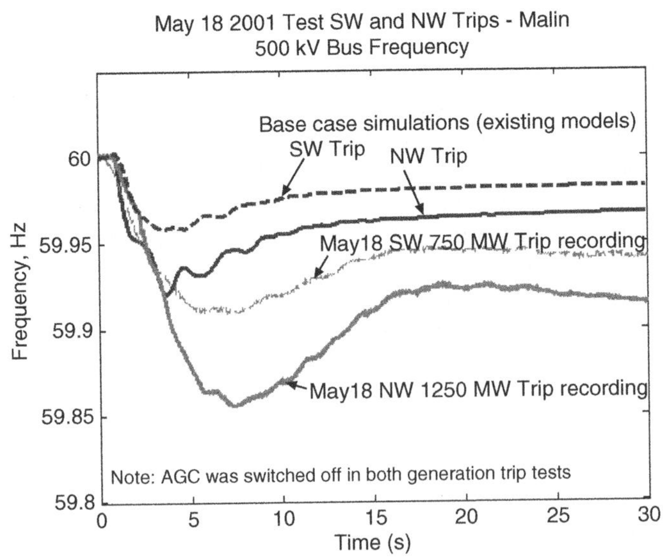

Figure 6.30. The discrepancy between existing model simulations and system frequency recordings (all AGCs were switched off during the tests).

- Frequency deviation, $\Delta\omega = 0.105/60 = 0.00175$ p.u.
- Generation pickup $\Delta P = 91000$ MWbase $= 3185$ MW, calculated from:

$$\frac{\Delta\omega}{\Delta P} = -\left(\frac{1}{1/r + D}\right)$$

Note that this calculation is a somewhat simplistic first approach to a complex response of units in the system because load damping, and the effect of redistributed losses due to different flow patterns in the system after the trips, is neglected. However, it does give a broad understanding of governor response in the system.

While the two tests hence indicated that only 40% of the expected governor response in the system actually occurred as a result of the initiating generation trip, the existing modeling practice (as in 2001) assumed that 100% of governors respond in accordance with the 5% speed droop governor characteristic. This resulted in a significant difference between simulations and actual recorded system responses as seen in Figure 6.30.

The principal reason for this large discrepancy was that base-loaded and load-limited generators and units operating with load controllers (also known as MW power-controllers) are not properly modeled. These are primarily "thermal" units, a classification that include conventional fired steam, nuclear steam, simple cycle gas turbine, and combined-cycle gas turbine plants. Analysis of the test recordings showed that the hydro governors were largely responsive. Investigations indicated that other effects such as nonlinear gate movement, dead band, and so on have some impact on simulation results, but a relatively minor one in comparison. In the modeling of governors, the "base-load" and "load-controller" operation of units was clearly the dominant effect.

Figure 6.31 compares frequency simulations of the "new" (correct) model with the "existing" (incorrect) model. The existing (incorrect) modeling assumes that 100% of governors respond in accordance with its 5% speed droop governor characteristic. The

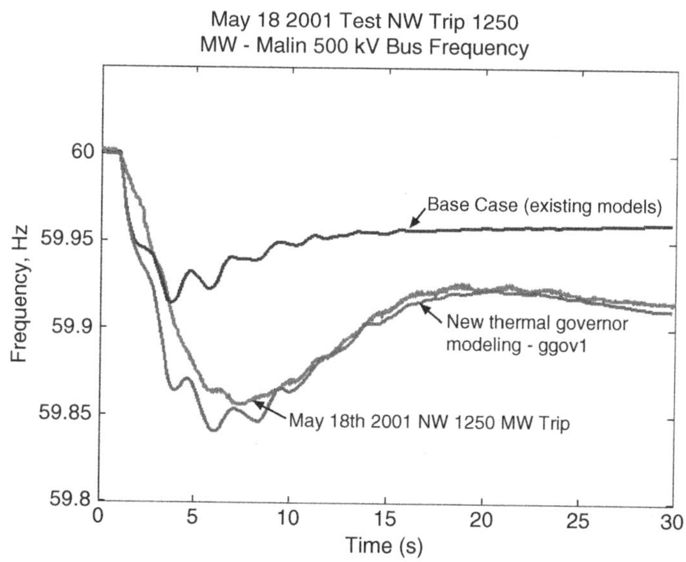

Figure 6.31. Simulations with the new thermal (ggov1) governor model compared with May 18, system test recordings for the NW 1250 MW trip, with AGC switched off.

comparison is with frequency recordings during system tests on May 18, 2001 when 1250 MW was tripped in the Northwest with AGC switched off system wide. It is clear that the "new" modeling is very accurate.

References [12,13] describe the development, validation, and verification of this "new thermal governor" model. The work included the creation of a WECC-wide system database based on disturbance monitoring and SCADA recordings of staged tests.

The new modeling approach has been extensively validated against recordings from three WECC system tests and several large disturbances. The model is now in use for all operation and planning studies in the WECC. The significance is not simply that frequency responses match accurately; it is that the units' pickup of generation is accurately modeled, with the result that the modeled and recorded frequencies now match accurately.

6.5.5.2 Analysis of Test Data: Thermal Versus Hydro Units. The analysis of the test data and recordings from both SCADA and disturbance monitors of the May 18, 2001 test showed that most of the hydro units were largely very responsive to frequency deviations and it was concluded, therefore, that the unresponsiveness was due mainly to the thermal units. A map of the location of hydro versus thermal generation in the Western Interconnection is shown in Figure 6.32.

The analysis of test data was a huge effort because the total thermal generation online during the May 18, 2001 test was about 67,000 MW out of a total WECC generation of 91,000 MW. A power flow base case was created to specifically model system conditions during the tests.

Each of the 1100 thermal governor units that were analyzed was given a code. About 60% of this thermal generation was "base" loaded. Typical responses for units that were

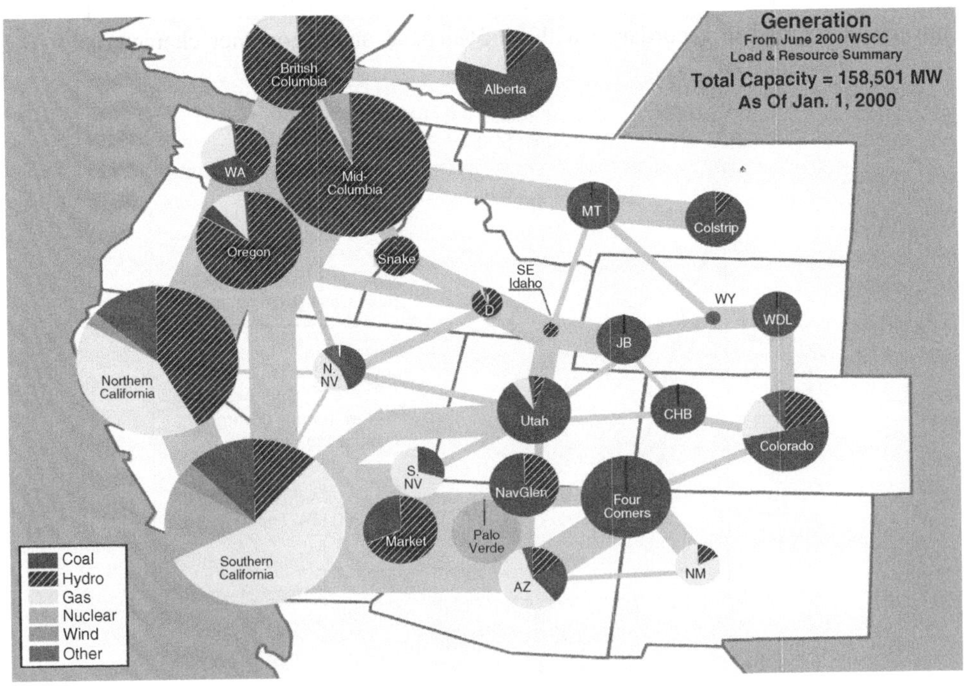

Figure 6.32. The location of hydro and thermal generation in the WECC.

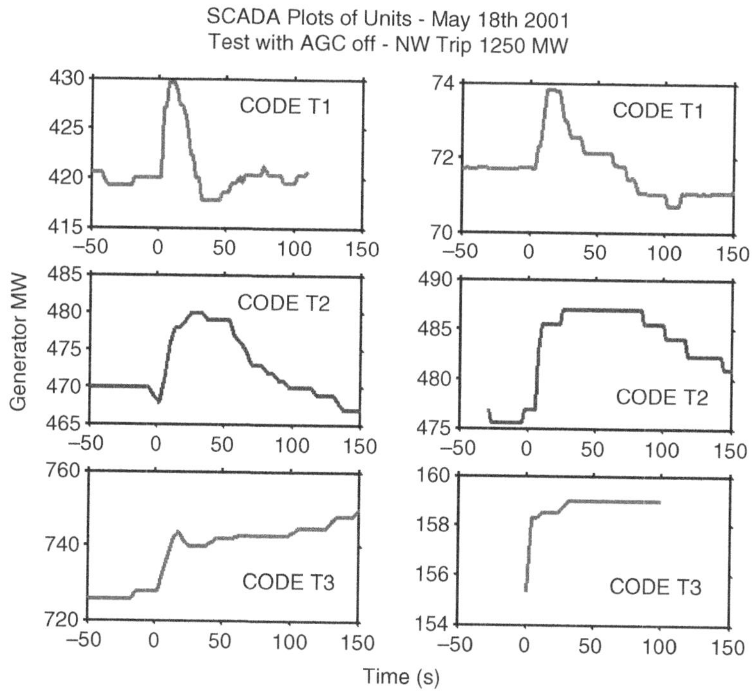

Figure 6.33. SCADA recordings of typical thermal units, Coded T1, T2, and T3 to denote fast, slow, and sustained governor responses, respectively, during the May 18, 1250 MW trip test (for which the system frequency response is shown in Figure 6.29).

coded T1 to T3 are shown in Figure 6.33 depicting the generator "responsiveness" in varying degrees.

Steam thermal units were coded T1 to T3, and corresponding gas turbine units coded G1 and G2 for fast and slow controller responses. Code T1 represents base loaded units. Code T3 units are the responsive 5% droop units, and Code T2 units only initially responsive.

Where SCADA data was not available for a specific unit, information obtained from a survey of owners/control areas, regarding the base loading or responsiveness of their units, was utilized in the selection of the turbine-governor code.

A simple block diagram of the thermal plant governor and controls is shown in Figure 6.34 [12,13]. The governor is a Proportional–Integral–Derivative (PID) type with the classic permanent droop feedback, typically 5%. The turbine is represented by a typical lag-lead transfer function.[14]

In the new governor model, "base" load operation is simulated by setting the limiters to limit the turbine power to a preset value. An additional MW power (load) controller is included to model Code T1-T3 and Code G1-G3 units. This is a simple reset (PI) controller with its gain (K_{imw}) typically having values of 0.01–0.02 per unit for "fast" controllers, 0.001–0.005 per unit for "slow" controllers, and 0 for no load controller action (i.e., a fully responsive 5% droop unit). The detailed block diagram of the thermal governor model (ggov1) is shown in Figure 6.35 [12] for further details of the model and the parameters.

[14] See Section 6.5.4.

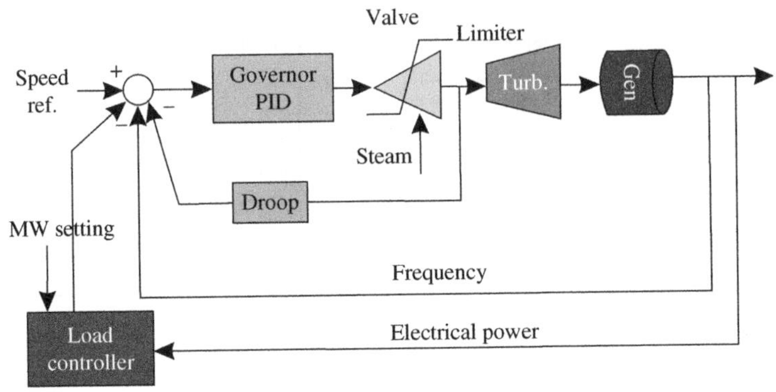

Figure 6.34. Block diagram of the new thermal turbine governor showing "Base" Load/Limiter and MW Load controller features.

Numerous sensitivity studies were performed to determine the effect of varying parameters in the dynamic database before the final selection of the parameters for each unit. The two new models, developed by GE for use in WECC studies, are the "ggov1" and the "lcfb1" models. The "ggov1 model" (Fig. 6.28) is a generic thermal governor/turbine model that incorporates both base loading and load controller effects [12,13]. The parameters of "ggov1 model" are presented in Section 6.5.4. Alternatively, thermal units may be represented by the general-purpose thermal governor model from IEEE Committee Report [7] such as shown in Figure 6.27 augmented by the *load controller model* (lcfb1) in Figure 6.35 [13]. The load controller model is a Proportional Integral or PI controller, and is similar in structure to the load controller portion of the ggov1 model. This is shown in Figure 6.35. Note that K_i in the lcfb1 model corresponds to K_{imw} in the ggov1 model, and is given identical values. The frequency bias, f_b, and K_p, proportional gain, are maintained as zero. This PI controller model may be simply added to existing thermal governor models shown in Figure 6.27.

Upon initialization, base-loaded units and load controllers are assigned MW set point values in the ggov1 and lcfb1 models equal to the generator dispatched value specified in the power flow data. The output of the unit will reset to the MW setting value of the load controller at a speed controlled by the value of the gain K_{imw} in the ggov1 model (or K_i in the load-controller model lcfb1).

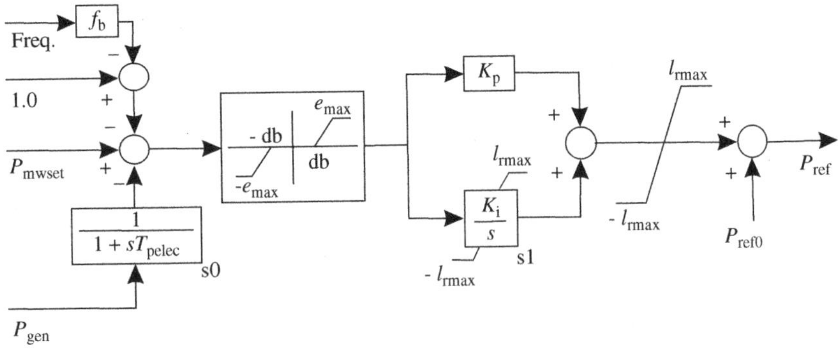

Figure 6.35. Block diagram of the load controller model lcfb1.

TABLE 6.5. Principal Parameters of the New Thermal turbine Govenor Model ggov for the Various Designated Codes [12,13]

Code		r	T_b	T_c	P K_{pgov}	I K_{igov}	D K_{dgov}	K_{imw}
T1	Fast load controller	0.05	10	2	10	2	0	0.005– 0.02
T2	Slow load controller	0.05	10	2	10	2	0	0.001–0.003
T3	No load controller	0.05	10	2	10	2	0	0
G1	With load controller	0.05	0.5	0	10	2	0	0.01–0.02
G2	No load controller	0.05	0.5	0	10	2	0	0

The major validation effort was in selecting the correct load-controller characteristic for each unit, that is, the gain K_{imw} so that the simulated MW response of the unit to the frequency deviation corresponded closely to the recorded MW response during the test. Also sensitivity studies were conducted for each of the parameters as described in [13].

The values set for the ggov parameters are presented in Table 6.5.

The principal parameters of the model are as follows:

R the permanent speed droop, p.u.
T_b is turbine lag time constant, s.
T_c is turbine lead time constant, s.
K_{pgov} is governor proportional gain, p.u.
K_{igov} is governor integral gain, p.u.
K_{dgov} is governor derivative gain, p.u.
K_{imw} is load (power) controller gain, p.u.

(i) *Model Validation with May 18, 2001 Test Data*
The results of the simulations for validation and verification of the "new" model compared with the real-time event frequency recordings from disturbance monitors are shown in Figures 6.36–6.42.

Simulations performed with the incorrect existing models are also shown for comparison. The existing (incorrect) modeling assumes that 100% of governors respond in accordance with its 5% speed droop governor characteristic. The corresponding SCADA, or disturbance monitoring unit, responses are identified through the word "recording". The accuracy of the new modeling WECC system frequency response is evident in Figures 6.36–6.38 for the May 18, 2001 trip tests. Many more comparative plots are given in references [12,13].

(ii) *Model Verification with Random System Trip Data*
Figures 6.39 and 6.40[15] show simulations of two typical large "random" system disturbances performed for the "Verification" of the new model (plots indicated by "new thermal governor ggov1 simulation") compared with disturbance monitoring frequency recordings (plots recognized through the word "recording"). The plots

[15] The reason that the new model simulation in Figure 6.40 differs from the disturbance recording in the 20–30 s range is that units picking up on AGC were not properly modeled due to insufficient AGC information.

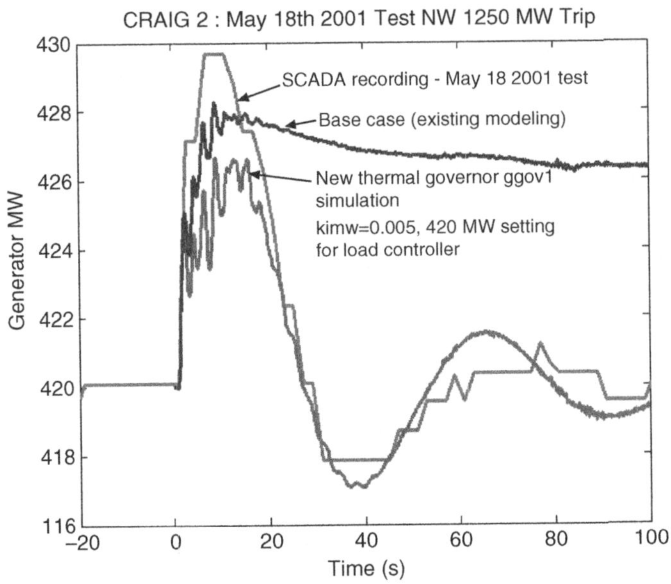

Figure 6.36. Simulations with the new turbine-governor model of a thermal unit with a "fast" load controller of Code T1 compared with its May 18, Test SCADA recordings.

indicated by "Base case" are the existing incorrect 5% droop simulations. Many more system disturbances were also verified that are not shown in this section. Note that it is required to use a power flow base case that represents closely the system existing conditions during the disturbance event in order to get a verifiable simulation.

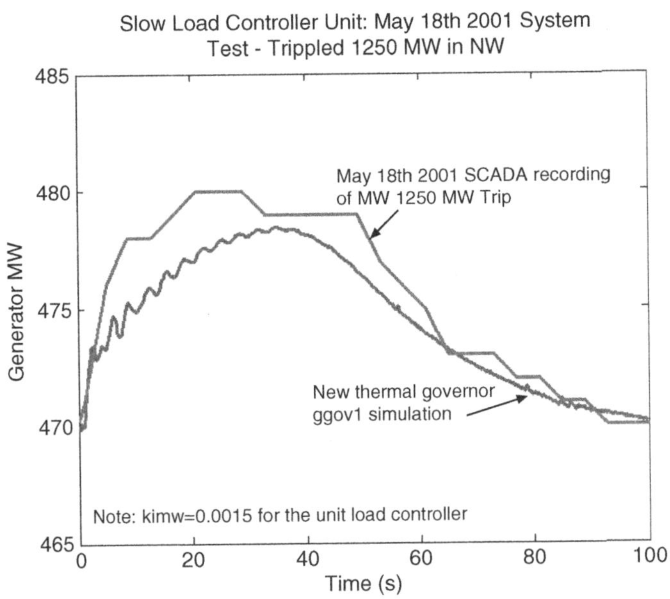

Figure 6.37. Simulations of a thermal unit with a "slow" load controller of Code T2 compared with its May 18, Test SCADA recordings.

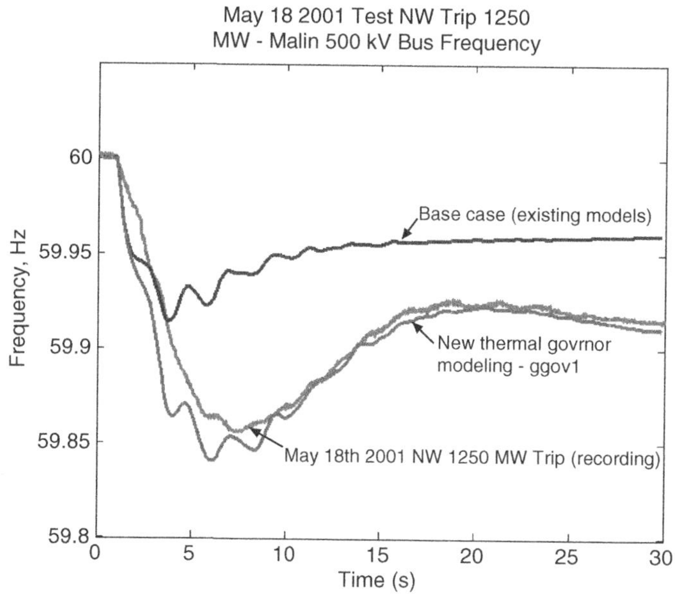

Figure 6.38. System frequency response simulations with the new governor model compared with May 18, Test recordings for the NW 1250 MW trip, all AGCs switched off.

The new governor model went through an intensive approval and validation process in the WECC. An intensive and coordinated effort was launched in the WECC to obtain "validated" governor model data from the generator owners to replace the "developmental" data created from the May 18, 2001 tests for validation studies of the new thermal governor model. This effort included two WECC Workshops, issue of Guidelines for selecting and validating new governor models,

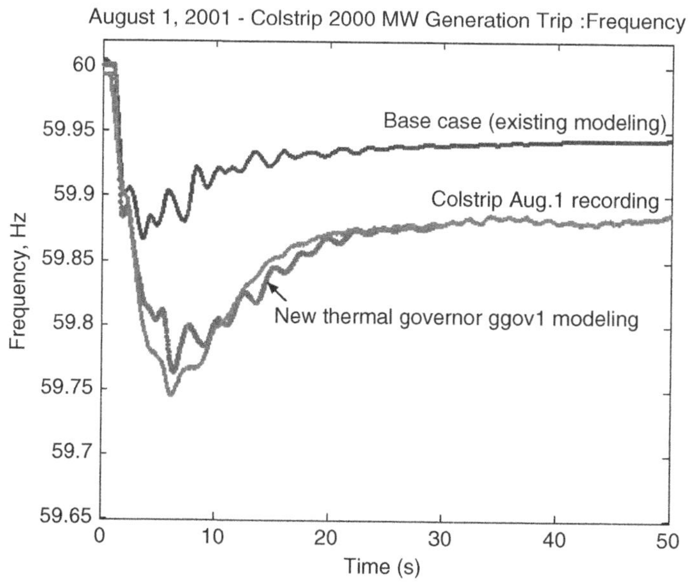

Figure 6.39. Governor model verification, 2000 MW Colstrip trip in Montana on Aug. 1, 2001.

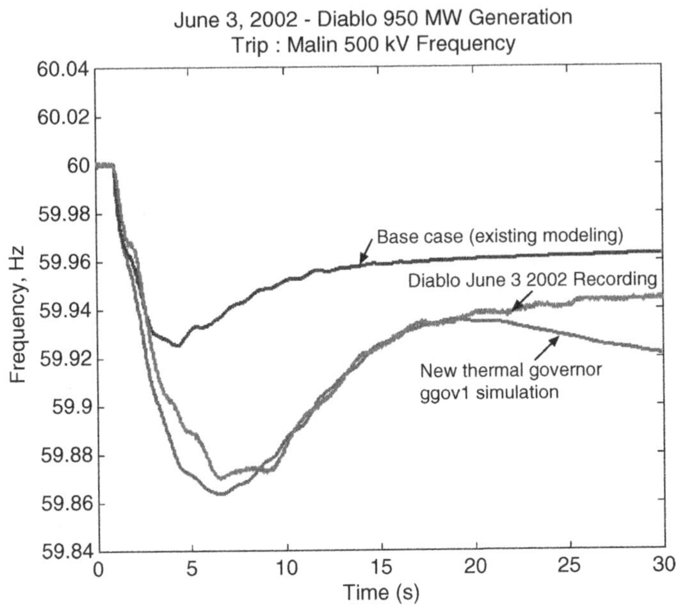

Figure 6.40. Governor model verification—950 MW Diablo generation trip in California on June 3, 2002.

and issue of new techniques and tutorial-type modeling programs for Model validation and Methodologies for assisting in the process of selecting model parameters and validating it.

To assist in the selection of the appropriate model, and the governor parameters, the generator owners were encouraged to answer typical questions with reference to the unit response diagram in Figure 6.41 to describe the response that best characterized their unit's electrical power response as recorded by disturbance recorders or SCADA. The initial electrical response "AB" is "inertial" and is common for all responses. The responses BC will end up in one of four 'boxes' characterizing "base-loaded", fast or slow controller, or 'responsive' operation of the unit.

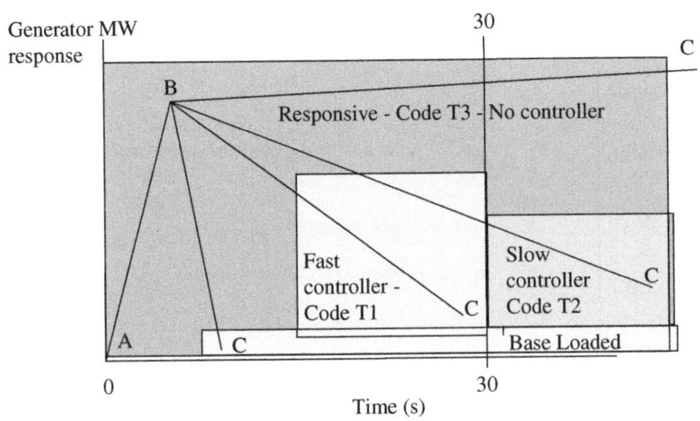

Figure 6.41. Unit electrical power response diagram and code classification [13].

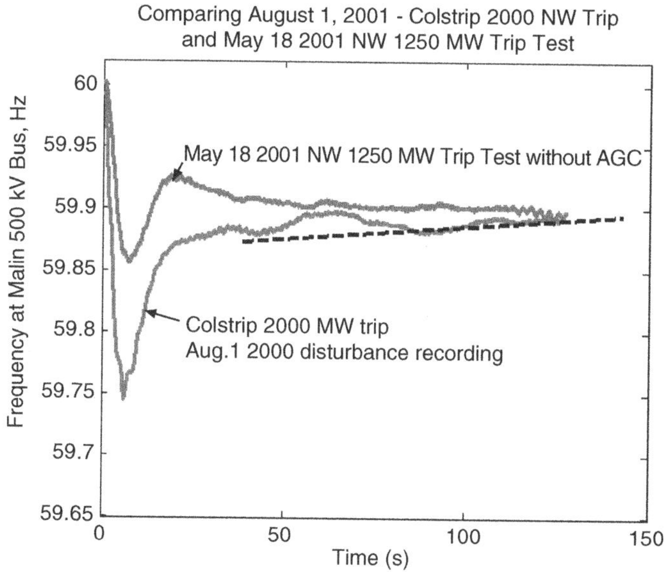

Figure 6.42. Disturbance monitoring recordings comparing two real-time recordings with and without AGC.

Effects of AGC. For studies extending to long periods, such as for system oscillations and dynamic voltage stability, it is desirable to model AGC. Comparison of the system recordings of the May 18, 2001 Test when all AGCs were switched off, and the system recording of a random trip the Colstrip 2000 MW plant on August 1, 2000 clearly indicates that AGC does make a difference in the frequency response of the system. This is illustrated in Figure 6.42.

Improved Hydro Plant Responses. An important finding of the impact of the thermal "unresponsiveness" on modeling was the greater importance of the "responsiveness" of hydro modeling, and hence the greater demands on more accurate hydro modeling [14]. The improved modeling of thermal plant response that results in a lower pickup of thermal plants also results in a corresponding increase of pickup by frequency 'responsive' hydro plants. Figure 6.43 shows the greater pickup of a typical hydro generator in the May 18, 2001 test simulation. Because thermal plants in the WECC are predominantly located in the South, and hydro generation is predominantly in the Northwest (see Fig. 6.32), the improved simulation of unit MW pickup and power flows across the system, particularly in intertie flows between the Northwest and the South, is critically significant in operation and planning studies.

System Simulation Impacts of the New Thermal Governor Modeling. It is important to realize that the correct modeling of the governing responsiveness of units resulted in significant system simulation improvements in the WECC in several areas. The following are some of the important impacts of the new thermal governor modeling on major system operation and planning are as

- system frequency responses can be predicted more accurately for large generation trips;
- accurately simulates whether units pickup or not by governor action;
- improved modeling of hydro versus thermal generation responses is achieved;

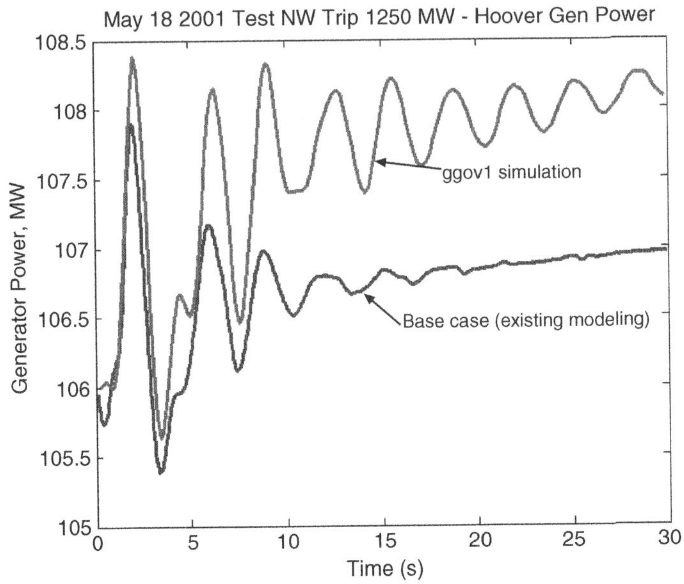

Figure 6.43. Hydro plant responses: improving the accuracy of the new thermal governor modeling (ggov1) increases the generator pickup of a typical frequency "responsive" hydro unit.

- improved underfrequency and load shedding studies, involving large generation trips and/or system islanding;
- a more accurate prediction of critical intertie flows and dynamic limits is obtained for operation in a system such as WECC where responsive hydro generation is located in the north and largely thermal generation is in the south;
- improved assessment of system oscillations and damping;
- more accurate assessment of Frequency Responsive Reserves (FRR) and Spinning Reserves to assist in new FRR standards for the industry;
- more accurate posttransient ("governor") power flow studies involving large generation trips.

Figure 6.44 shows the improvement in the simulations of intertie flow and oscillations resulting from the new thermal governor modeling.

More Accurate Posttransient (Governor) Power Flow Studies. The principles of the new thermal governor modeling approach for dynamic studies have also been implemented in the WECC in the "blocking" of *base loaded* generators in posttransient (or "governor") power flow studies. This is of significance in studies of constrained systems where system generation response to a large generation plant outage could create a voltage stability concern. For all posttransient power flow studies, units that are designated as "base-load" units are "blocked out" in the program so that those units do not pickup for a power flow run when generation is tripped.

Summary and Conclusions. A new thermal turbine governor modeling approach, based on improved simulation of base-loaded units and load-controlled units, has been

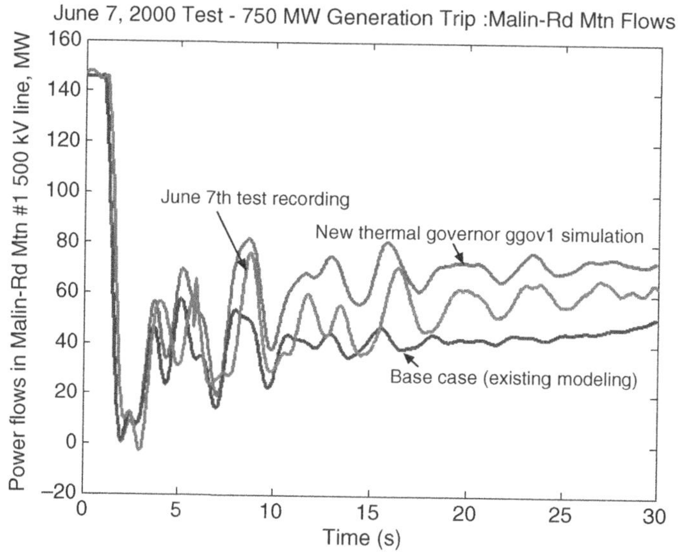

Figure 6.44. The difference in intertie flows and oscillations between the new governor modeling and existing modeling.

developed in the WECC. The development of this model went through an extensive study process that included validation to staged WECC system tests and verification with respect to numerous large system disturbances. After approval of the model for use in WECC, an intensive and coordinated effort was launched in the WECC to obtain validated governor model data from the generator owners. The new thermal governor modeling approach is being currently used in all operation and planning studies in the WECC.

Frequency Response in Other Interconnected Systems. While the interest of the WECC Governor Modeling Task Force as described in this section was specifically to governing relating to the WECC, the Western Interconnection in North America, the general principles of the new thermal governor modeling approach clearly apply to all large and small interconnections.

Reference [15] by an IEEE Task Force states that in recent years, power system operating and planning personnel have become increasingly aware of the fact that power plant governing response is considerably less than expected and planned [16]. Earlier observations in the Eastern Interconnection were similar to that reported in the WECC and thermal units appear to be mainly the unresponsive ones just as later demonstrated in the WECC.[16] This issue was addressed even earlier, from the operations side, by the (then) NERC Operating Committee and its Performance Subcommittee, who initiated an EPRI project [17]. Personnel in the system operating components of utilities conduct "regulation tests"that are initiated by and coordinated by the NERC Operations Subcommittee. The committee reached a broad consensus that approximately one-quarter to one-third of the expected governing response is found in analyses of the recorded power system frequency.

The Task Force concludes that the key points are that governing is a primary function within interconnected (and islanded) power systems. Today in many power systems, there

[16] It was also confirmed that hydro governors in the East were more responsive just as in the West.

is a discrepancy between actual system frequency response to a generation/load imbalance event and what is predicted using standard power system simulation tools and simulating the event based on the expected governing response of units. The report is devoted to providing insight on what is the cause of this discrepancy and how more refined modeling practices may be adopted to bridge this gap between actual and simulated system response. The Task Force also agrees that in smaller systems, such a disparity between actual and expected primary governing could be quite disastrous. It should be added that current trends indicate a declining primary governing response, and that if the declining trend continues, all interconnections will eventually be put at risk for inadequate security.

Use by Other Interconnections of the Methodology Created by the WECC in its New Thermal Governor Modeling [12,13]. The methodology described in this section can be very easily adapted by other interconnections. It may not be necessary to repeat the system trip tests performed by WECC on May 18, 2001 and described in Section 6.5.5 to prove whether individual governors in the interconnection are responsive or not in primary control.

The following recommended procedures at the "system" level and "unit" level could be followed:

a. Record the system frequency response of the interconnection with disturbance recorders during a large generation trip, for example, 1000 MW and more in systems up to 100,000 MW and larger trips in larger systems. Comparing with the simulations from a large-scale stability program of the generation trip will indicate whether there is a general modeling problem or not. A computer run of about 100 s is required.
b. Record the actual response of individual generating units during the disturbance using disturbance recorders or the unit's own data recording systems example PI or SCADA systems. The sampling intervals should be not greater than 2 s for a good match [13].
c. Simulate the response of individual generating units for the same disturbance by a large-scale stability program modeling the system with the known trip using existing dynamic models.
d. Alternatively simulate the response using a two-machine equivalent system with the frequency recording "played back" into the model as described in [13] using either a stability program or a Matlab/Simulink program or equivalent.
e. Compare the simulated response with the actual response of the individual generating units during the disturbance with the actual response recordings. This will indicate whether the existing dynamic models and the expected governor droop response is correct or not.

6.6 AGC PRINCIPLES AND MODELING

The purpose of this section is to describe the principles of automatic generation control or AGC, which is termed "secondary control" in the "UCTE Operation Handbook." A review of the literature indicates numerous publications on the many facets of AGC; however, there is no attempt to present anything more than a basic introduction as presented in numerous text books and published material [18–21].

AGC as a "secondary control" has been used for several decades to meet the objective of maintaining or bringing up the system frequency to its nominal value with its actions deliberately slower than the "primary control" that is provided by turbine governors. In any event, the first seconds of frequency dip and recovery after a major generator trip would necessarily be accomplished by governor control.

When the power system's self-regulation is insufficient to establish a stable state, the system frequency will continue to decay until it is arrested by automatic underfrequency load shedding (UFLS) to reestablish the load-generation balance within the time constraints necessary to avoid system collapse. Initial underfrequency load shedding relay settings are typically 59.3 Hz in the U.S. systems and 49 Hz in UCTE. Note that, starting with July 2009, the UCTE area is part of the new created ENTSO-E (European Network of Transmission System Operators for Electricity).

6.6.1 AGC in a Single-Area (Isolated) System

AGC may not be required for smaller isolated systems, which may operate well with governor control only with the backup provided by manual control. However, with droop governors, frequency deviations will result in a permanent steady-state deviation given by the equation $R = \Delta\omega/\Delta P$. The larger the disturbance, the larger will be the steady-state deviation. AGC with a PI controller, as shown in Figure 6.45, will bring the steady-state deviation to zero. The AGC signal is fed into the governor summing point along with the droop signal as shown. The integral gain K_i must be adjusted for an optimal response. The example shown in Figure 6.45 is taken from [5]. The isolated system single-area model concepts were discussed in Section 6.4.5.

Figure 6.46 shows the frequency response of AGC system in an isolated or single-area system.

6.6.2 AGC in a Two-Area System, Tie-Line Control, Frequency Bias

A simple model of a two-area system is shown in Figure 6.47. The two areas are each modeled exactly as in Figure 6.45 with an equivalent system inertia, droop and damping for

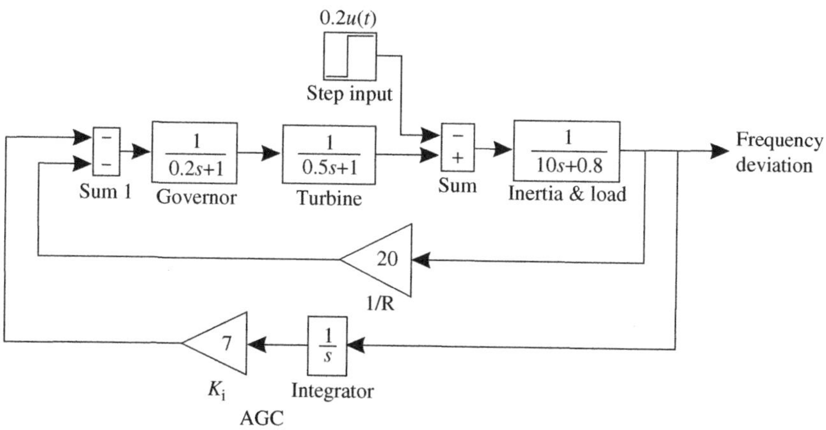

Figure 6.45. Block diagram of AGC for an isolated or single-area system.

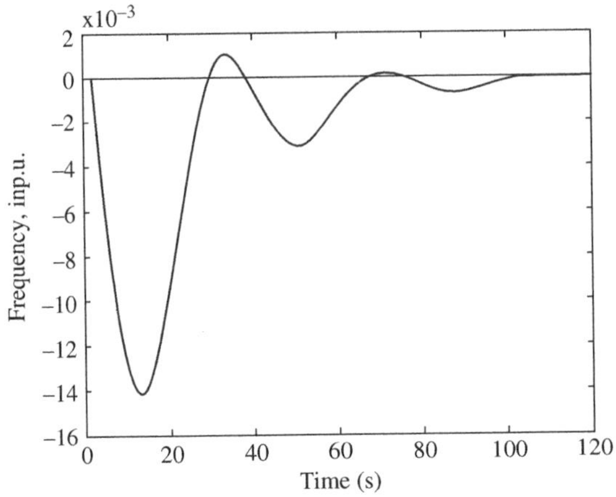

Figure 6.46. Frequency response of AGC in an isolated or single-area system.

each area, but connected together by a tie-line. An AGC with a PI controller is provided for each area and will bring the steady-state deviation to zero. The AGC signal is fed into the governor summing point along with the droop signal as shown. The integral gains K_I for each AGC are adjusted for an optimal response. The simulation of this model is shown in Figures 6.48 and 6.49. The model is taken from reference [5].

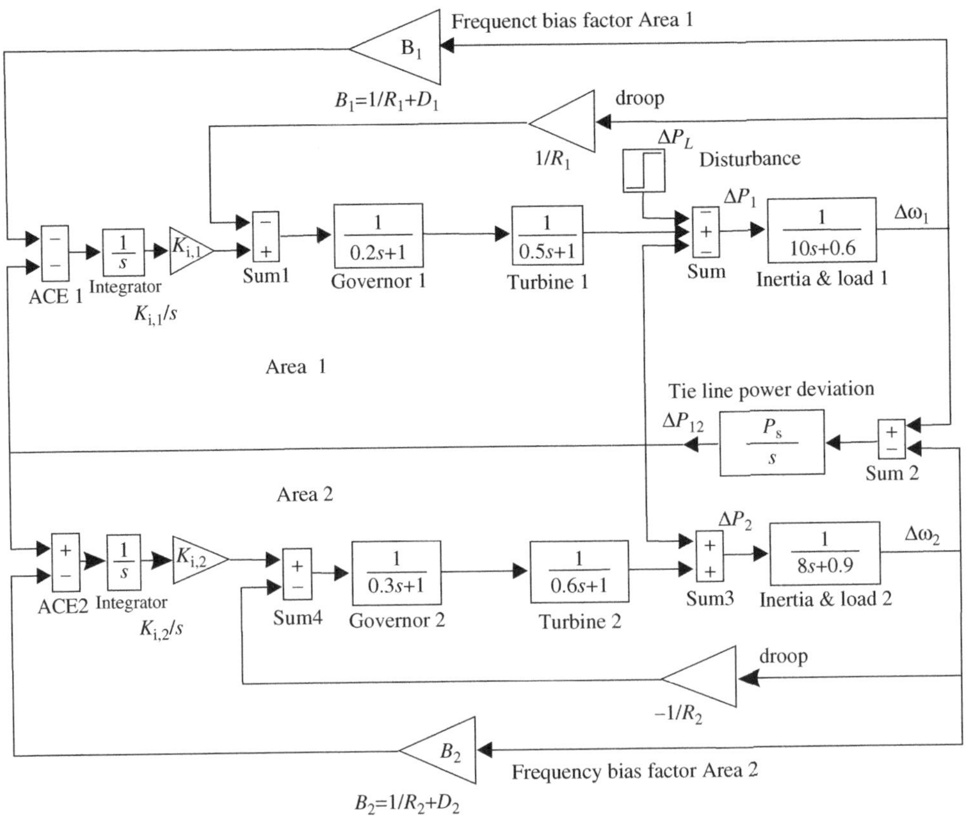

Figure 6.47. Two-area AGC tie-line model.

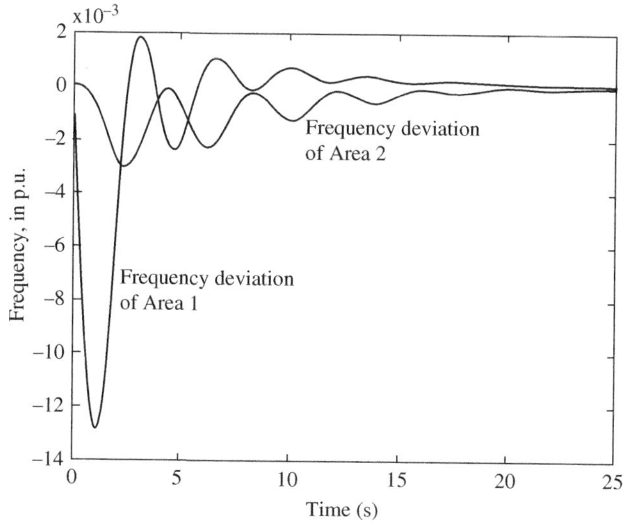

Figure 6.48. Simulated frequency response of the two-area AGC tie-line model shown in Figure 6.47.

The primary equations are as follows:

- Swing equation, same as in a single-area system.
- Turbine governor model, same as in a single-area system.
- Droop equation, same as in a single-area system.
- Tie-line power.
- Frequency bias factors.

The power transfer through the tie-line is given by the product of the synchronizing power P_s and the difference of the phase angles, that is, $\Delta P_{12} = P_s(\Delta\delta_1 - \Delta\delta_2)$.

Note if ΔP_{12} is assumed positive in one direction for one area, it will be in the opposite sign for the other area.

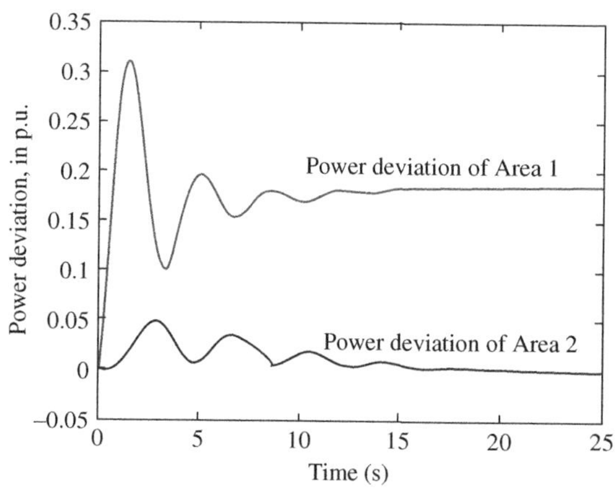

Figure 6.49. Simulated power deviation response of the two-area AGC tie-line model shown in Figure 6.47.

From the droop equation, $R = \Delta\omega/\Delta P$, mechanical powers in areas 1 and 2 are

$$\Delta P_1 = \frac{-\Delta\omega}{R_1} \quad \text{and} \quad \Delta P_2 = \frac{-\Delta\omega}{R_2} \tag{6.8}$$

In each area, the power deviations should balance between generation plus imports versus loads plus exports. In the Two Area diagram shown, these consist of the mechanical power from the rotating inertias, the load damping, and the disturbance step.

Hence in Area 1

$$\Delta P_1 - \Delta P_{12} - \Delta P_{L1} = \Delta\omega \cdot D_1$$

and in Area 2

$$\Delta P_2 + \Delta P_{12} = \Delta\omega \cdot D_2$$

Substituting the droop equations (6.8) into these power balance equations and solving for $\Delta\omega$, where

$$\Delta\omega = \frac{-\Delta P_{L1}}{(1/R_1 + D_1) + (1/R_2 + D_2)}$$

The denominators are known as the "Frequency Bias Factors" for Areas 1 and 2, respectively, as defined below:

$$B_1 = (1/R_1) + D_1$$

$$B_2 = (1/R_2) + D_2$$

From Figure 6.47, it is seen that the Area Control Error (ACE) for each area comprises the area's summated power mismatch and the bias factor multiplied by the frequency deviation:

$$ACE_1 = \Delta P_{12} + B_1\Delta\omega$$

$$ACE_2 = \Delta P_{21} + B_2\Delta\omega$$

It will be seen from Figure 6.49 that tie-line bias control forces the area with the disturbance to meet its own power mismatch (disturbance change) with the other area contributing to the transient condition of the system as a function of the frequency deviation and its bias factor.

6.6.3 AGC in Multiarea Systems

The basic scheme is in principle essentially the same as described for each control area. Each area will have its own centralized AGC. Within each area, the telemetered information required includes the MW output of each generator, the flow over each tie-line to the adjacent control areas, and the system frequency. The output from the AGC is transmitted to each selected generating unit's governor. Not all units are selected for AGC control.

At the AGC central controls, the difference between the measured frequency and the frequency standard is multiplied by the frequency bias factor for the control area. This signal is added to the summation of the tie-lines and the difference between the actual interchange and scheduled interchange to produce the ACE or Area Control Error. Unit output errors are added to ACE to form a composite error signal that drives the entire control system logic.

(i) *Area Control Error (ACE)*

In the previous section, the simple Area 2 model illustrated the function of ACE. ACE as used in control areas in practice is described in several papers in greater detail in references [18–21]. The paper in [21] gives AGC and ACE descriptions, which give a good understanding of practical system implementation.

$$\text{ACE} = T_\text{n} - T_\text{o} - 10 \cdot B_\text{n} \cdot (f - f_\text{o}) + \text{corr}_\text{n} \tag{6.9}$$

where

T_n is actual area net interchange, MW.

T_o is scheduled area net interchange, MW.

B_n is area frequency bias, –MW/0.1 Hz.

f is system frequency, Hz.

f_o is scheduled system frequency, Hz.

corr_n is corrective control such as inadvertent interchange reduction.

The equation shows this AGC to have an external or perimeter view of the control area. Control is based on the summation of tie-line telemetering with other control areas, the value of T_n. This tie-line telemetering can be noisy, have delays, and contain interconnected system interference, all of which become part of the ACE signal for generation control.

(ii) *AGC with processed ACE*

AGC thus requires a process or filter to reduce ACE noise, which delays AGC response for area load and interchange schedules. The lack of response becomes an imposition on and source of control interference in the interconnected system.

The power balance in each control area is

$$T_\text{n} = G_\text{c} + G_\text{b} - L_\text{n} \tag{6.10}$$

where

G_b is base load generation, MW.

L_n is control area load, MW.

G_c is generation on AGC, MW.

Substituting (6.10) in the (6.9) gives

$$\text{ACE} = G_\text{c} + G_\text{b} - L_\text{n} - T_\text{o} - 10 \cdot B_\text{n} \cdot (f - f_\text{o}) + \text{corr}_\text{n}$$

Without going into details, further refinements of the equation are necessary for practical applications and lead to the ACE as currently recommended by WECC [22]. Other interconnections and control areas have similar derivations.

The ACE signal as applied practically should not become too large. Absolute value of ACE in WECC should not exceed a control area's largest probable power or load loss contingency or absolute value of ACE is greater than 300 MW and system frequency deviation is less than 0.025 Hz. When ACE is greater than 300 MW and the system frequency deviation is less than 0.025 Hz, it is highly probable that the ACE data is incorrect. In WECC, depending on system loading, a sudden change of 1000 MW has been shown to cause a deviation in frequency of approximately 0.1 Hz.

(iii) *Typical frequency time responses in the WECC of large generation trips showing AGC action*

Figure 6.50 shows several typical responses following generation trips in the WECC.[17]

The responses differ from trip to trip and vary according to the particular control area in which the disturbance occurred, the special AGC control actions of that control area, and the AGCs of the remaining control areas via their frequency bias controls, time of day of the occurrence, system conditions at that time including inertia, load magnitude, load damping, and so on. Some control areas are significantly deficient in generation while others are surplus in relation to their loads. In most cases, the initial "settling" frequency after the first swing is about half the maximum initial dip, unless AGC action is sooner. AGC action seems to generally apply within 100 s in most plots, in some cases sooner.

(iv) *Suspension of AGC*

In many control areas, the automatic generation control (AGC) is suspended for large deviations of frequency. WECC rules are that AGC suspension should be considered when certain circumstances exist including system conditions that could be worsened by AGC, or if data required for ACE calculation is erroneous or missing, or for loss of any equipment that provides control input data to AGC. The control area with the disturbance would remain on AGC and be required to meet the NERC DCS (Disturbance Control Standard) criteria. AGC, if suspended, should be restored immediately after the system frequency disturbance has been mitigated.

Unplanned loss of a critical transmission facility that causes remaining transmission facilities to overload could result in suspension of AGC if continued AGC operation further aggravates the overloads on the remaining transmission facilities. Controlling a generation source that has been separated from its control area could aggravate system conditions.

The responsible dispatcher has authority to suspend AGC if control problems arise without discernible cause. This is a decision that has to be made on a case-by-case basis by the generation dispatcher. During AGC suspension, generating units should remain on governor control or local set point control. AGC should be restored to service as soon as possible after the condition that causes the suspension has been corrected or the appropriate mitigating action has taken place.

[17] While the WECC has several sophisticated disturbance monitoring equipment installed at various locations, the frequency recordings shown in Figure 6.50 have been taken from recordings made by Professor Grady of University of Texas with relatively simple laboratory made equipment. In reference [23] the results have been checked and compared closely with WECC's disturbance recordings.

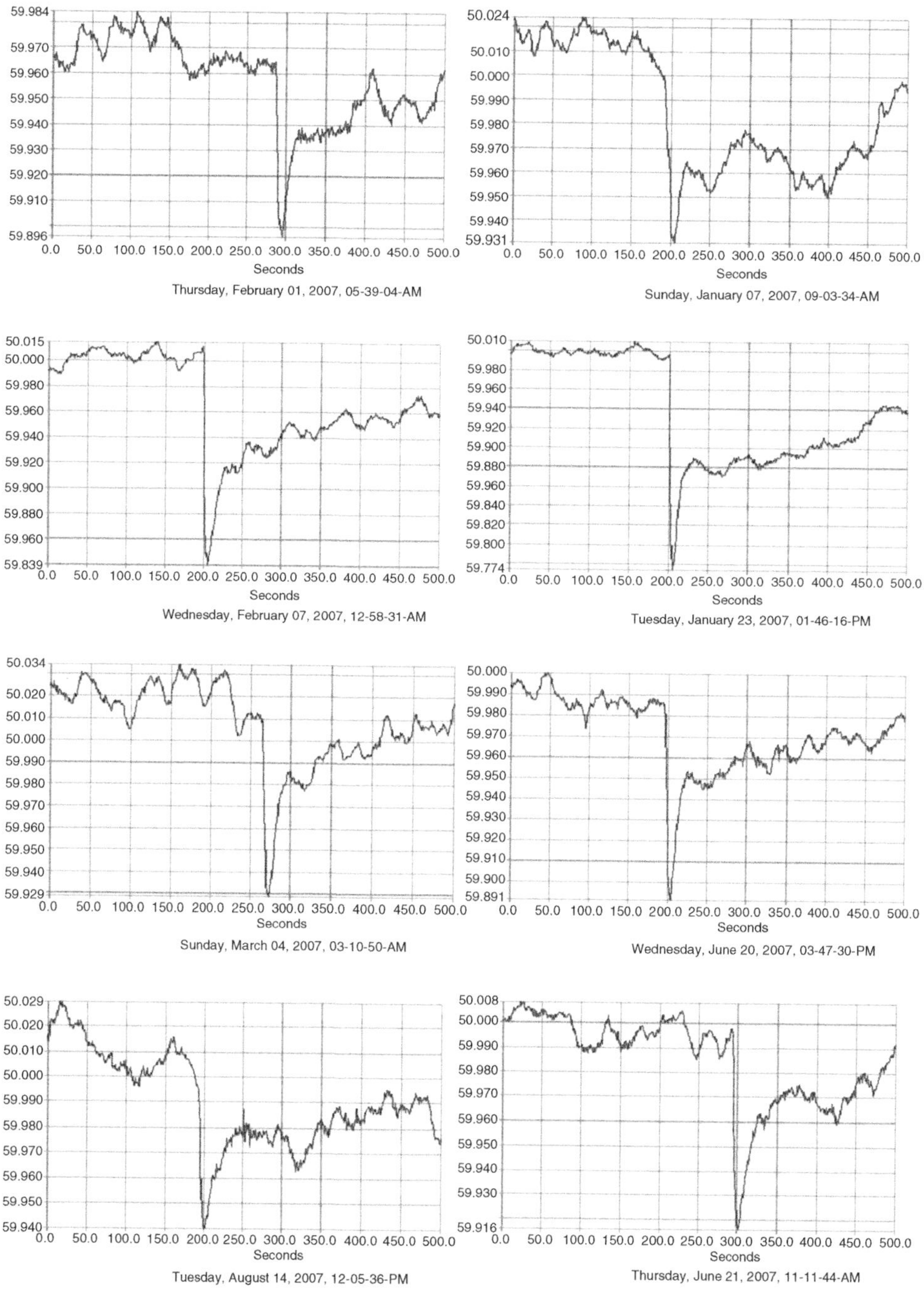

Figure 6.50. Typical frequency time responses in the WECC of large generation trips.

6.7 OTHER TOPICS OF INTEREST RELATED TO LOAD FREQUENCY CONTROL

A brief discussion follows relating to certain topics that have a bearing on the successful operation of load frequency control during normal operation and emergencies. These include the following:

a. Spinning reserves.
b. Underfrequency load shedding.
c. Operation of an interconnected system broken into electrical islands.
d. Blackstart operation.

6.7.1 Spinning Reserves

Spinning reserves are those available online and ready to supply power in seconds or minutes following a disturbance such as large generation unit or plant trip. This implies "primary" or "secondary" frequency controls, meaning governors, or AGCs. There is also an interpretation of spinning reserve that is of practical use to the system in disturbances and what is contracted for in the market. The everyday understanding of *spinning reserves* in market terms assumes that if, for example, a 200 MW generator is dispatched at 100 MW, the *spinning reserve* immediately available in that unit is 100 MW when a disturbance occurs. This is obviously technically incorrect because a 200 MW unit picks up only 6.6 MW for a 0.1 Hz deviation of system frequency, assuming a 5% droop governor action.[18] The remaining part of the "available" spinning reserve can only be automatically obtained if an AGC signal is sent to the unit. In addition, not *all* units with spinning reserves are on AGC. Hence only the selected units that are on AGC at a given moment could be technically capable of *automatically* delivering 100 MW of spinning reserve.

Given the loose market interpretation of spinning reserves, it is apparent that the "planned availability" of spinning response within the control area (or "contracted" spinning reserves from outside the control area) will clearly always tend to be over-optimistic per current market definitions. This could lead to potentially critical consequences if the generators are expected to suddenly supply "spinning" reserves during the immediate critical time period in a cascading or islanding situation, and obviously cannot.

6.7.2 Underfrequency Load Shedding and Operation in Islanding Conditions

UFLS is designed to assist in maintaining the load-generation balance by dropping sufficient load to match the generation. As frequency drops, the first stage of UFLS is triggered as the frequency reaches 59.3 Hz or 59.5 Hz in NERC regions. In UCTE, slightly lower settings of 49 Hz are used for the first stage. Since the initial dip in frequency is about twice the settling frequency in a few minutes before AGC action after a large generation trip, a 0.5 Hz frequency deviation means that a 7500–12,500 MW trip would have occurred in a fully interconnected WECC depending upon system conditions as shown in Figure 6.51.

[18] See the calculations in Section 6.5.5.

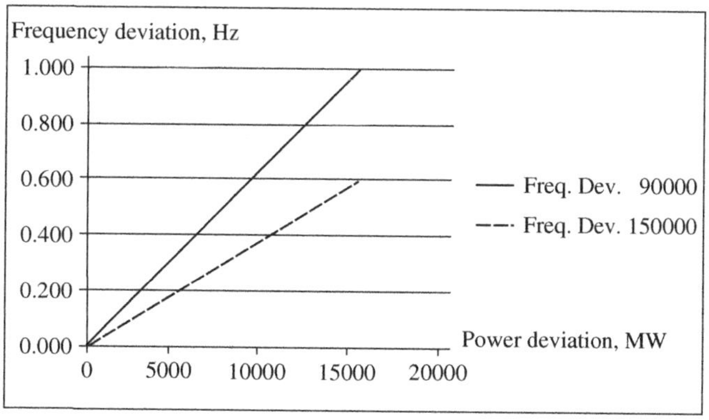

Figure 6.51. Load shedding versus frequency deviation for differing WECC system conditions.

The formula used for this illustrative calculation is the same as used in Section 6.5.5 except that the initial dip is twice the settling frequency for 5% droop governors:

$$\Delta P = \frac{\Delta\omega_{\text{pk}}/2}{R} \text{ in p.u.}$$

terms that translates into

$$\Delta P = \frac{\Delta\omega_{\text{pk}}/2/60 \cdot \text{System\_MVA}}{R}$$

from which for a given frequency deviation, the power deviation can be calculated and vice versa.

If this exercise is extended to the Eastern Interconnection (See Figure 6.52), it would be clear that a system-wide UFLS trip for 59.3 Hz or 59.5 Hz is highly unlikely and it is likely only after islanding (Figure 6.52).

The question of islanding brings up a whole different perspective to load frequency control. It should be clear that AGC would have been suspended previously after large system deviations in frequency and power occur. Hence any strategy based on AGC is highly suspect during emergency conditions. The focus then is entirely on the primary control by governors.

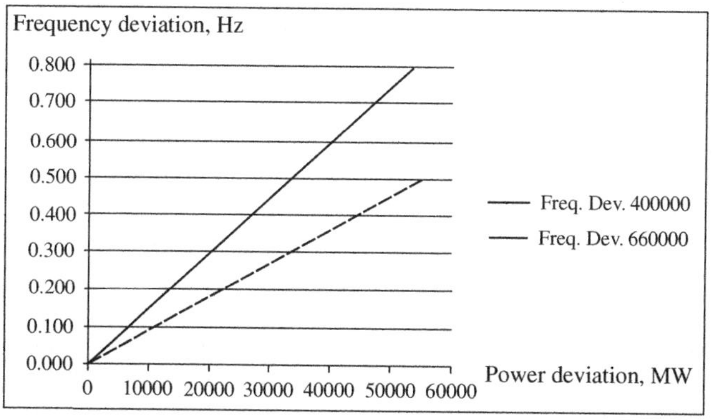

Figure 6.52. Load shedding versus frequency deviation for differing Eastern Interconnections system conditions—illustrating that system wide UFLS for 59.3 Hz or 59.5 Hz is unlikely.

If a large number of governors are unresponsive in operation to frequency deviation as dictated by the unit's load controllers as discussed in Section 6.5.5, it is a foregone conclusion that cascading will result in islands, which themselves would be subject to massive load shedding, and eventually more generator shedding, as the generator-load mismatch continues until a stable point is reached to minimize both the mismatch and control frequency by governors. In some islands, excessive generation led to high frequencies even approaching 63 Hz.

In the Eastern Interconnection blackout of August 14, 2003, at least 265 generating plants with more than 500 individual generating units shutdown [24]. Within less than an hour, UFLS load shedding was about 30,000 MW. At the end of the day, 62,000 MW of load was shed. Restoration is yet another matter for consideration; again the importance of governor response is critical for recovery, as generators have to operate stably controlling frequency as each section of load is added.

REFERENCES

[1] ENTSO-E, *UCTE Operation Handbook*, European Network of Transmission System Operators for Electricity 2004, www.entsoe.eu

[2] NERC, previously North American Electric Reliability Council, since 2006 North American Electric Reliability Corporation. Web site is www.nerc.com.

[3] EIA, US Government Energy Information Administration at www.eia.doe.gov.

[4] WECC, the Western Electricity Coordinating Council was formerly the WSCC (Western Systems Coordinating Council) is the Western Interconnection of the NERC regions in the USA and Canada.

[5] Sadaat, H. *Power System Analysis*, 2nd Edition, McGraw Hill, 2002.

[6] Kundur, P. *Power System Stability and Control*, McGraw-Hill, New York, 1994.

[7] IEEE Committee Governors, 1973.

[8] Rogers, G. *Cherry tree scientific software*, email: cherry@eagle.ca.

[9] Hovey, L.M. Optimum Adjustment of Governors on Manitoba Hydro System, *AIEE Trans*, Vol. PAS-81, pp. 581–587, 1962.

[10] IEEE *Guide for the Application of Turbine Governing Systems for Hydroelectric Generating Units*, IEEE Task Force report P1207, 2005.

[11] General Electric *PSLF Simulation program reference manual*, General Electric Company http://www.gepower.com/energyconsulting/en_us/pdf/pslf_manual.pdf.

[12] Pereira, L., Undrill, J., Kosterev, D., Davies, D., Patterson, S. A new thermal governor modeling approach in the WECC, *IEEE Transactions on Power Systems*, Vol. 18, No. 2, pp. 819–829, May 2003.

[13] Pereira, L., Kosterev, D., Davies, D., Patterson, S. New thermal governor model selection and validation in the WECC, *IEEE Transactions on Power Systems*, Vol. 19, No. 1, pp. 517–523, Feb. 2004.

[14] Patterson, S. *Importance of hydro generation response resulting from the new thermal modeling—and required hydro modeling improvements*, IEEE-PES General Meeting, Denver, July, 2004.

[15] IEEE *Interconnected power system response to generation governing: present practice and outstanding concerns. Final Report*, Task Force on Large Interconnected Power Systems Response to Generation Governing, of the Power System Stability Subcommittee, of the Power System Dynamic Performance Committee, of the IEEE Power Engineering Society, April 2007.

[16] Schulz, R.P. Modeling of governing response in the Eastern Interconnection, *Proceedings of the IEEE PES Winter Meeting, Symposium on Frequency Control Requirements, Trends and Challenges in the New Utility Environment*, 1999.

[17] Virmani, S. *Impacts of governor response changes on the security of North American Interconnections*, EPRI Report no. TR-101080, October1992.

[18] Wood, A.J., Wollenberg, B.F. *Power generation, operation, and control*, 2nd edition, Wiley, New York, 1994.

[19] Elgerd, O.I., Fosha, C. Optimum megawatt-frequency control of multiarea electric energy systems, *IEEE Trans. Power App. Syst.*, Vol. PAS-89, No. 4, pp. 556–563, April 1970.

[20] Jaleeli, N., Ewart, D.N., Fink, L.H. Understanding automatic generation control, *IEEE Transaction on Power Systems*, Vol. 7, No. 3, pp. 1106–1122, Aug. 1992.

[21] Schulte, R.P. *Generation control for deregulated electric power systems*, IEEE Conference publications, Summer Power Meeting 2000, Seattle, WA, USA.

[22] WECC Operations Committee Handbook http://www.wecc.biz/documents/library/publications/OC/OC_Handbook_Complete.pdf.

[23] Grady, M. *Course notes*, University of Texas, Austin, TX. http://users.ece.utexas.edu/~grady/.

[24] NERC-DOE *Final Report on the August 14, 2003 Blackout in the United States and Canada: Causes and Recommendations*, U.S.-Canada Power System Outage Task Force, April 5, 2004. August 14, 2003 Blackout Report. https://reports.energy.gov/BlackoutFinal-Web.pdf.

7

VOLTAGE AND REACTIVE POWER CONTROL

Sandro Corsi and Mircea Eremia

Open access and industry restructuring into generation, transmission, and distribution companies increased the need to sustain the grid by power plants voltage and reactive power control. In fact, the new trends of electricity market accompanied by more stressed conditions are leading the power systems toward higher risk operational states. The problem of an effective and automatic voltage and reactive power control in large and complex electric systems began to be seriously considered since 1980 and its solution is now demanding greater and greater effort for the system engineers in the definition and in the realization of sophisticated control schemes able to increase system security and operation efficiency. The utilities and transmission system operators (TSO) are certainly interested in improving reliability, security, and quality of supply with an effective solution having minimum impact on the investments. However, due to the subject novelty, they are also monitoring the other companies trying to get advantage from the first incoming on field experiences. Unfortunately this too prudent approach often stalls the decisions and delays the application of new advanced and already available wide area control solutions.

The reactive power–voltage control is indispensable in power systems either under normal or in emergency conditions. During the normal operation it ensures the transmission of electrical energy at the required voltage quality and in the most convenient conditions for the suppliers and users. Under emergency conditions, the role of voltage control is to increase system security by enlarging the margin with respect to the system voltage instability limits, therefore ensuring continuity in the system operation and proper operating conditions for the largest number of consumers.

The voltage regulation and reactive power compensation problems generally require a different approach when considering the transmission or distribution level. In the first case, the high-voltage (HV) network can benefit of the voltage–reactive power support offered by the largest generators through which controlling the overall grid while, at the

Handbook of Electrical Power System Dynamics: Modeling, Stability, and Control. Edited by Mircea Eremia and Mohammad Shahidehpour.

distribution level, the voltage control generally concerns each distribution area independently: one area representing a small separate part of the overall distribution system. Lastly, different dispatching centers and operators control the transmission and distribution levels and, while the extra high-voltage (EHV) system operation has a strong impact on the distribution area voltages, the vice versa is less relevant. Moreover, while the transmission networks are characterized by $R \ll X$ and large generators, on the contrary, the distribution networks have high load density, radial structure with $R < X$, and host few and small generators. Due to these differences, the objective and modalities of the $V - Q$ control can have different approach in the two overlapped networks even in case of the future Smart Grid with increased amount of distributed generators.

On *the transmission grid*, the main voltage control objectives are as follows:

1. Continuously maintain an appropriate high-voltage profile.
2. Minimize the power system losses.
3. Increase the system voltage stability margin.

To achieve these objectives it is necessary to have at disposal, on the transmission level, both

a. Adequate controllable reactive power reserves, to also face contingencies;
b. An effective and automatic wide area voltage control system.

On *the distribution networks*, the main voltage control objectives are as follows:

1. Maintain the voltage at the consumers terminals in acceptable range.
2. Minimize the system local losses.
3. Increase the distribution area voltage stability margin.

The achievement of these objectives would require, at each distribution area:

a. A strong voltage support by the local transmission network (high voltages and as much as possible maintained at constant value).
b. Areas with high-voltage controllability by on-load tap changers (OLTC).
c. Adequate compensating equipments, well located, to face the extreme load conditions.
d. An effective and automatic distribution area voltage regulation system coordinating the distributed generators (when available) and OLTCs controls and operating on the local compensating equipments only when needed—at the time their switching is of real voltage support—by also minimizing the number of their maneuvers.

From the above considerations, the voltage control in the transmission and distribution networks clearly appears as two distinct solutions to be achieved separately. Importance should be given to the great advantage the distribution area takes by an effective transmission network voltage control, due to the opportunity to minimize the MV level control effort as regards OLTCs and the number of maneuvers at the compensating equipments. Furthermore, the distribution losses minimization as well as the distribution area voltage stability increase can be easily achieved under the support of a transmission grid automatic voltage control.

Generally speaking, the voltage control problem is strongly influenced by the actual operating conditions of the power systems that continuously change in a way the dispatching operator is not able to manually track. In fact, the operator recovers the voltage lowering with some delay and with a clear difficulty given by a not coordinated and in some cases discretionary manual controls. Moreover, the tendency to exploit the electrical lines near the loadability limit determines a system voltage vulnerability increase.

The use of an automatic voltage control system, able to coordinate all the control variables and the available reactive power resources in the amount and at the time they are needed, is therefore the way to give important improvements both at the transmission and at the distribution levels.

Until now all the given considerations are based on the assumption that the power system main objective is the loads feeding in all the possible operating conditions. Therefore, the load shedding for voltage control was not considered inside the normal operation but against heavy contingency only, to protect and save part of the system in front of a real voltage instability risk.

This obvious consideration does not find in practice coherent generalized examples also because the short cut of the voltage control by load shedding is often proposed or justified as the only available solution. The writing authors are against this unnatural practice unless for protection needs. They fully agree with one of the Charles Concordia's [1,2] last comment on the subject: "too easy but not professional by power system engineers controlling grid voltages by mainly operating load shedding"!

The way the load shedding practice is operated around the world, under the push of the energy market liberalization, often put in evidence the system operator adaptation, without criticism, to the energy market rules instead of improving at the best performances of its voltage control system, thus minimizing the customer's interruptions vulnerability. This objective is also overlooked in most of the electrical energy monopolistic regimes, where the lack of innovating voltage controls is also relevant.

The voltage control strategy is obviously strongly influenced by the defined operating rules and available control structure in a power system as well as by the commercial supplier–consumer relationship. Therefore,

- the tendency to exploit the electrical lines near their loadability limits,
- the insufficient interconnection lines between neighboring systems, and
- the increasing power quality requirements by the customers

contribute to the system vulnerability in the voltage plan, that is, in the energy interruptions to the consumers as well as in the violation increase of power quality requirements.

7.1 RELATIONSHIP BETWEEN ACTIVE AND REACTIVE POWERS AND VOLTAGE

7.1.1 Short Lines

In order to establish the relationship between active and reactive powers and voltage, the one-line diagram and the phasor diagram of a short line are considered (Figure 7.1) [3,47].

In Figure 7.1, $\underline{V}_1$ and $\underline{V}_2$ are the phase voltages, while $\underline{I}_1$ and $\underline{I}_2$ are the currents at the line ends.

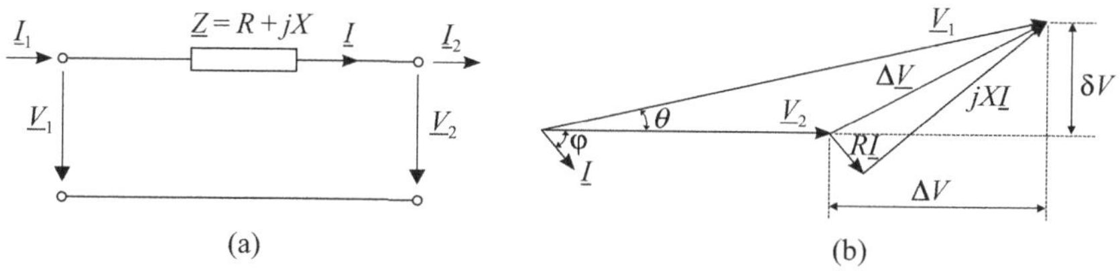

Figure 7.1. Short line model: (a) one-line diagram; (b) phasor diagram.

The absence of the shunt admittance leads to $\underline{I} = \underline{I}_1 = \underline{I}_2$ in all points along the line. Denoting by φ the angle between $\underline{I}$ and $\underline{V}_2$, then $I_a = I\cos\varphi$ and $I_r = I\sin\varphi$ are the active and reactive components of the current $\underline{I}$. Assuming that the voltage $\underline{V}_1$ is constant and $\underline{V}_2$ is taken as phase origin, the complex voltage drop $\Delta\underline{V} = \underline{Z}\,\underline{I} = \Delta V + j\,\delta V$ has two components:

$$\begin{aligned} \Delta V &= RI_a + XI_r \\ \delta V &= XI_a - RI_r \end{aligned} \tag{7.1}$$

where $\underline{I} = I_a - jI_r$ for inductive loads, and ΔV and δV are the longitudinal and transversal components of the voltage drop.

Let $\underline{S} = 3\underline{S}_0$ and $\underline{S}_0 = V_2(I_a + jI_r) = P_{02} + jQ_{02}$ denote the three-phase complex power and the single-phase complex power, respectively. Introducing the active power P_{02} and the reactive power Q_{02} in equation (7.1), we obtain

$$\begin{aligned} \Delta V &= \frac{RP_{02} + XQ_{02}}{V_2} \\ \delta V &= \frac{XP_{02} - RQ_{02}}{V_2} \end{aligned} \tag{7.2}$$

Because generally $R \ll X$ for transmission lines, it results

$$\begin{aligned} \Delta V &\approx \frac{XQ_{02}}{V_2} \\ \delta V &\approx \frac{XP_{02}}{V_2} \end{aligned} \tag{7.3}$$

Therefore, the voltage drop ΔV is mainly due to the reactive power flow on the line, so the magnitude difference between $\underline{V}_1$ and $\underline{V}_2$ depends mainly on the reactive power transits, while the active power substantially affects the phase difference. In order to reduce the voltage drop, the flow of reactive power should be avoided. In practice, this can be possible by reactive power generation near the consumption area.

Starting from relation (7.3), and taking into account the fact that V_1/V_2 is close to 1, we can write

$$\frac{\Delta V}{V_2} \approx \frac{\Delta V}{V_1} \approx \frac{XQ_{02}}{V_1^2} \approx \frac{Q_{02}}{S_{02cc}} \tag{7.4}$$

where $S_{02cc} = V_1^2/X$ is the short-circuit power at node 2.

The characteristic *V–Q* of the system is given by expression

$$V_2 \approx V_1\left(1 - \frac{Q_{02}}{S_{02cc}}\right) \tag{7.5}$$

This means that the voltage at the node 2 depends on the amount of the reactive power as well as on the weakness of the node.

The minimization of reactive power transfer is also motivated by the reduction of the Joule losses on the line. These losses are expressed by the equation

$$\Delta P_j = 3RI^2 = 3R\frac{S_2^2}{(\sqrt{3}U_2)^2} = R\frac{P_2^2 + Q_2^2}{U_2^2} \tag{7.6}$$

Due to the thermal limit defined for any network element by the admissible current, the reactive power transit also reduces the active power transmission possibilities. The losses minimization does, therefore, involve the reactive power compensation as well as the system operation at the highest voltage values.

From the above considerations it clearly appears the following:

- The voltage magnitudes are substantially determined by the reactive power flows.
- Reactive power flow on the line reduces the voltage at the receiving bus.
- Reactive power flow on the line produce increased losses.
- The stronger the short-circuit power in a given bus, the less the reactive power flow delivered to that bus reduces its voltage value.

From the point of view of the load voltages further considerations are required.

Referring to Figure 7.2, it shows two different types of reactive loads, one of inductive type $(+jX)$ and the second of capacitive type $(-jX)$.

Considering the inductive load (Figure 7.2), we have

$$\underline{V}_2 = jX(-jI_2) = XI_2 = V_2$$

$$V_2 = XI_2 = X\frac{Q_{02}}{V_2}$$

then

$$V_2^2 = XQ_{02}$$

In case of small variations

$$2V_2\,\Delta V_2 = X\,\Delta Q_{02}$$

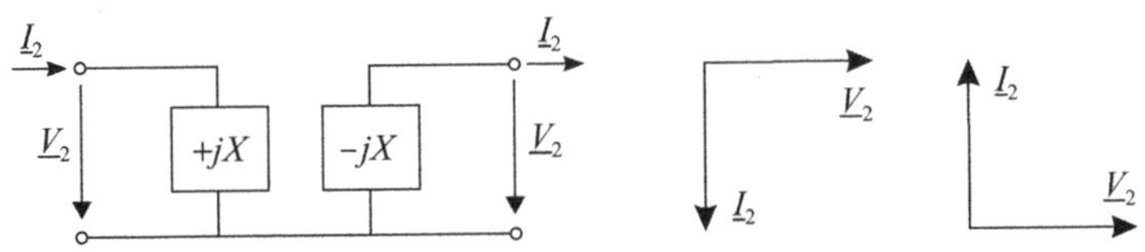

Figure 7.2. One-line diagram of inductive load $(+jX)$ and capacitive load $(-jX)$.

then

$$\Delta V_2 = \frac{X\,\Delta Q_{02}}{2V_2}$$

or

$$\frac{\Delta V_2}{V_2} = \frac{X\,\Delta Q_{02}}{2V_2^2} \tag{7.7}$$

According to (7.7), it clearly appears that the voltage of an inductive load increases when injecting reactive power into it. Obviously, the higher is the short-circuit power of the bus the less is the effect in terms of voltage increase, for a given amount of reactive power absorbed by the load.

Analogously, by considering the capacitive load (Figure 7.2), but with the reactive power delivered by the load

$$V_2 = (-jX)(jI_2) = XI_2$$

$$V_2 = XI_2 = \frac{XQ_{02}}{V_2}$$

then

$$V_2^2 = XQ_{02}$$

Therefore,

$$2V_2\,\Delta V_2 = X\,\Delta Q_{02}$$

$$\Delta V_2 = \frac{X\,\Delta Q_{02}}{2V_2}$$

$$\frac{\Delta V_2}{V_2} = \frac{X\,\Delta Q_{02}}{2V_2^2} \tag{7.8}$$

According to (7.8), it clearly appears that the voltage of a capacitive load increases when the load delivers reactive power to the grid.

In conclusion, the reactive power injection into a load bus increases or reduces the bus voltage depending on the inductive or capacitive nature of the load seen by the bus.

In a predominant inductive grid, the understanding of the positive or negative effect of the reactive power injection on the load bus voltage would require the comparison of the line voltage drop effect with the load bus voltage increase.

Because the transmission lines reactance is very small in comparison with the loads seen by the transmission busses and due to the fact that real loads are of reactive nature basically, the obvious conclusion is the correctness of the transmission network voltage control by reactive power injection into the load busses from the nearest resources. Furthermore, the higher is the voltage in the transmission grid the lower is the control effort to sustain voltages, this because of the capacitive effect of the lines.

7.1.2 Taking into Account the Shunt Admittance

Examining the medium/long length line case we may say that the shunt admittance can no longer be neglected, like in a short line case. Refer to the line π scheme, where the shunt parameters are modeled by a concentrated capacity C.

Again, denoting by $\underline{S} = 3\underline{S}_0$, the complex power at a given bus of voltage V, the current I is equal to the ratio S_0/V. The reactive power balance between the amount produced by the line and the amount absorbed by the line reactance is

$$Q_0 = \omega C V^2 - \omega L \frac{S_0^2}{V^2}$$

Notice that there exists an active power $P_{0,N}$ that makes the reactive balance zero:

$$P_{0,N} = V^2 \sqrt{\frac{C}{L}} = \frac{V^2}{Z_C}$$

Under these conditions, the power is transmitted on the line at constant voltage magnitude and unitary power factor. If the transmitted power S_0 is higher than the natural power $P_{0,N}$, like for high loaded overhead lines, the line absorbs reactive power. For cables, the term $\omega C V^2$ is preponderant and reactive generation always exists. For cable lines, the admissible maximum thermal power is always lower than the natural power.

In conclusion, the shunt capacitive susceptance (B_c) gives, in general, a significant contribution to the voltage support, partially compensating the local loads, unless of particular operating conditions. This reactive power contribution has to be undoubtedly considered in power system voltage analysis but only in terms or reactive power resources determining the operating point, and not to be used for the real-time voltage control. In other terms, the voltage control is not generally performed by continuously switching on/off the electrical lines.

7.1.3 Sensitivity Coefficients

In large networks, considering the active and reactive powers injection into the system busses, the corresponding effects are in general seen in terms of voltage and frequency variations. Referring to the voltage variations (vector dV), they are usually described in terms of the differential matrices $\partial V/\partial P$ and $\partial V/\partial Q$, otherwise called sensitivity matrices, denoted by S_{VP} and S_{VQ}, respectively. The voltage variation in a network point corresponds to a reactive power flow change on concurrence lines in that point. So we can write

$$\mathrm{d}V = \frac{\partial V}{\partial P}\mathrm{d}P + \frac{\partial V}{\partial Q}\mathrm{d}Q = S_{VP}\,\mathrm{d}P + S_{VQ}\,\mathrm{d}Q \tag{7.9}$$

The coefficients of these matrices obviously depend on the loads and lines characteristics and show at each bus, for a given injection, the resultant effect of the local changes in either reactive power flows or line losses or loads. Numerically speaking and also in terms of voltage control, the most important matrix is $\partial V/\partial Q$, whose coefficients indicate for each bus the reactive power amount necessary to produce a desired voltage variation in the same or in the other busses.

Section "Dynamic Model of the Regional Control System" provides further information on the linearized system model and its use in the voltage control system design.

7.2 EQUIPMENTS FOR VOLTAGE AND REACTIVE POWER CONTROL

From the previous considerations, the practical way to perform the voltage regulation in power systems substantially requires the control of the generated and consumed reactive powers and their flows at different levels (transmission and distribution) in the power system.

The synchronous generators are the main equipments in the power system able to deliver or absorb a significant amount of reactive power. The automatic voltage regulator (AVR) controls the generator excitation in order to maintain the voltage at the stator terminals at the set point value. Because this local priority control is mainly concerned with the generator voltage at LV level, it does not use at the best all the generators available with reactive power resources to cover the real voltage control needs at the HV load busses.

Other equipments contributing to support the system voltages are the compensating equipments, generally installed into the substations. These equipments can be classified as

- reactive power sources or loads: this category include shunt capacitors, shunt reactors, synchronous compensators, and static compensators;
- equipments providing compensation of the lines inductive reactance–the fixed or switched series capacitors;
- tap changing transformers.

The shunt reactors and capacitors as well as the series capacitors are passive compensation devices. They can be permanently connected or switched. In the first case, they have been designed as part of the basic grid to be controlled. In the second case, they represent part of the control resources supporting the basic grid voltages by recovering their variations. Hereinafter, we mainly refer to them as switchable and therefore controllable reactive power resources. Their stepping control is usually of manual, local, or remote type.

The synchronous and the static compensators are continuous and closed loop compensation devices. The reactive power absorbed or generated by them is automatically adjusted in such a manner to maintain constant the voltage level of the buses where they are connected. These equipments, similarly to the generators, maintain the controlled bus voltage at the set point value. In terms of voltage control they do not differ from the real generators.

7.2.1 Reactive Power Compensation Devices

Shunt capacitors are used for increasing lagging power factor contribution, whereas reactors are needed when leading power factor corrections are required such as in case of lightly loaded cables. In both cases, the effect is to supply/absorb reactive power to recover the voltage values around the nominal one. Unfortunately, lowering the voltage, the VArs produced by shunt capacitors or absorbed by reactors decrease, thus when needed most of their effectiveness, reduces unless they are controlled before the voltage significantly decreases. On the contrary, with light loads the voltages are high and the reactive powers produced by the capacitors or absorbed by the reactors are larger than the nominal values and their effectiveness is increased unless correctly controlled.

7.2.1.1 Shunt Capacitors. Capacitors are connected directly either to a bus bar or to the tertiary winding of a main transformer and are disposed along the route to minimize the losses and voltage drops. They compensate locally the reactive power of the consumers and are distributed throughout the system. The main advantages of the shunt capacitors are the low cost and the flexibility of installation and operation. Their principal disadvantage is the reactive power output reduction at low voltages being proportional to the voltage square. Moreover, the switching reduces their life.

The largest application of this compensation device is generally required by the distribution systems to supply the reactive power as close as possible to the point where it is consumed, that is, at the load buses.

Compensation schemes include both fixed and switched capacitors banks. In the case of transmission systems, the shunt capacitors are used to compensate for the XI^2 losses and to ensure satisfactory voltage levels during heavy load conditions. The capacitor banks are switched either manually or automatically by a voltage relay. Their location in the field is defined after detailed power flows, contingencies analysis, and transient studies. Switching on/off the capacitor banks provides the conventional mean of controlling the system voltages to recover large voltage deviation, typically due to load difference between night and day or after large contingencies. They cannot contribute to the continuous real-time voltage control due to the limitation on the switching maneuvers amount.

7.2.1.2 Shunt Reactors. Generally, the shunt reactors are used to compensate the line capacitance effects by limiting the voltage rise at open circuit or with light load. They are often used for EHV overhead lines longer than 150–200 km where the capacitive line-charging current flowing through the high value inductive reactance causes a voltage rise with the highest values at the sending end of the line. The shunt reactors can be connected directly to the electric line or through the tertiary windings of a transformer installed in the terminal station (Figure 7.3).

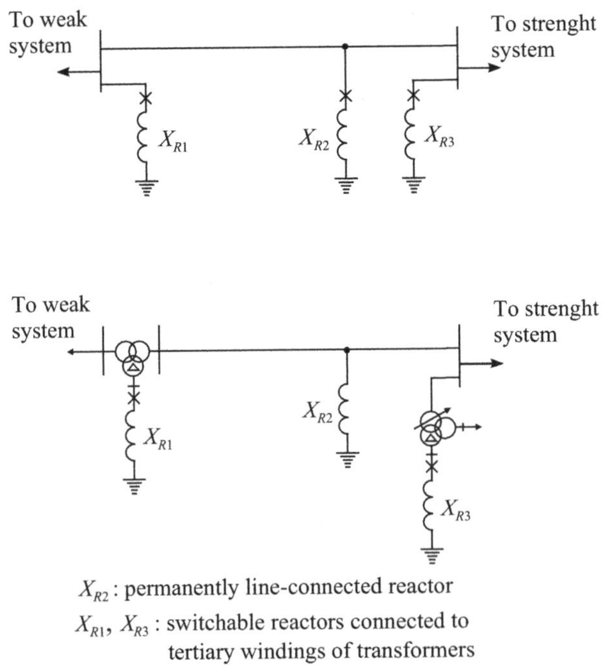

Figure 7.3. Shunt reactors possible connection configurations.

In the case of robust systems, the shunt reactors are permanently connected to the long electrical lines to limit temporary (lasting less than 1 s) or switching overvoltages up to 1.5 p.u. Additional shunt reactors can also be used on the electrical lines to limit the lightning over-voltages. During heavy load conditions the shunt reactors must be disconnected and, to this purpose, they are equipped with switching devices. On short lines supplied by weak systems only, switchable reactors are used. Shunt reactors cannot contribute to the real-time continuous voltage control due to the limitation in the number of the switching maneuvers.

7.2.2 Voltage and Reactive Power Continuous Control Devices

7.2.2.1 Synchronous Generators. Synchronous generators are primary voltage control devices and primary sources of spinning reactive power reserve. To meet reactive power demand, generators can be controlled inside their operating over- and underexcitation limits. Due to the time constant values associated with the generator rotor and stator thermal dynamics, a short time overload capability is allowed and can be usefully utilized through transient overvoltage and overexcitation limiting circuits. On a continuous basis, a generator will be operated successfully inside its voltage limits (usually between +5% and −5%) and its operating over- and underexcitation limits that define the available reactive power field. The generator capability curves are voltage dependant, therefore the overexcitation limit (OEL) and underexcitation limit (UEL) change dynamically. More precisely, if the generator voltage increases the overexcitation limit reduces the deliverable reactive powers while the underexcitation limit increases the absorbable reactive powers.

Figure 7.4 shows the limiting curves of the reactive power available at the generator in terms of the active load and terminal voltage at a given power factor.

The range area subtended by the armature current and the field current limits represents the available operating points of the generator in overexcitation.

By decreasing the active power generated by the machine, a larger reactive power reserve is created until the field current limit is reached. The diagonal dotted line represents the points at which the armature current and the field current of the generator have the same limiting values.

From Figure 7.4, it also results that if the terminal voltage decreases the available reactive power reserve of the generator increases (even instantaneously).

The generators provide voltage regulation in their admissible ranges. When a surplus of reactive power reserve exists in the system, the generators should absorb reactive power operating in the underexcitation domain until it reaches the underexcitation limit.

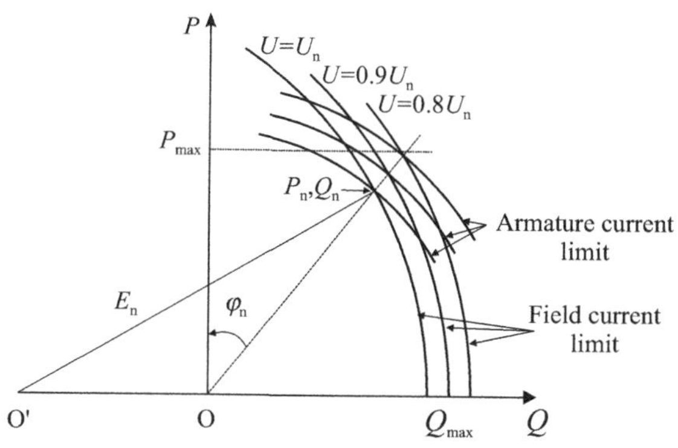

Figure 7.4. Variation of the armature current and the field current limits at a given power factor.

Nevertheless, because the turbine limit P_{max} is more restrictive than the armature limit, the possible decrease of the active power produced by the generator, in front of a lack of reactive power reserve, obtains an increase of reactive power availability mainly dependant on the shape of the field current limit curve. Therefore, the possibility to enlarge the reactive power limit by reducing the active power production does exist but it is not relevant and its use becomes reasonable only when critical voltage values are reached, that is, only after the system operator has used all the other available reactive power resources in the field.

In conclusion, the reactive power produced or absorbed by an electric generator mainly depends on the generated active power and the terminal voltage [3].

At constant generator voltage, the generator reactive power decreases if the grid voltage increases and conversely it increases when the grid voltage decreases. This has a stabilizing effect on the grid voltage. The increased or decreased reactive power output acts in a way to partially counteract the grid voltage variation.

7.2.2.2 Synchronous Compensators. A synchronous compensator/condenser is a synchronous machine running without a mechanical load and, depending on the value of excitation, it can absorb or generate reactive power in the same way as the synchronous generator. Even if its losses are considerable compared to static capacitors, its power factor is nearby zero, that is, it can operate at the maximum delivery or absorption of reactive power, according to the operating limits (Figure 7.5).

When the compensator operates in the overexcitation domain it injects reactive power into the network but absorb a field current that leads the voltage at the connection bus bar to 90°. When the compensator operates in the underexcitation domain it absorbs reactive power from the network and absorbs a field current that again leads the voltage to 90°. Acting on the machine excitation, the reactive power injected/absorbed from the network by the synchronous compensator can hereby be controlled.

When used with a voltage regulator, the compensator can automatically run overexcited at high loads and underexcited at light loads but its amount of supplied/absorbed reactive power will strongly depend on the operating voltage set point value. The compensator is run up as an induction motor in 2.5 min and then synchronized. The ratio between the maximum absorbed reactive power Q_k^{under} and the maximum supplied reactive power Q_k^{over} is an important characteristic of the synchronous compensator. Due to the fact that in under-excitation operation the compensator becomes unstable at low field currents, the synchronous compensators are usually designed with a ratio $Q_k^{under}/Q_k^{over} = 0.5 - 0.65$. If this ratio is low with respect to the required Q_k^{under} value, one of the following solutions can be adopted:

- Increase the ratio defined earlier up to the value 1 by enlarging the compensator air gap, with the disadvantage of increased cost.
- Combine the operation of the synchronous compensator with shunt reactors.

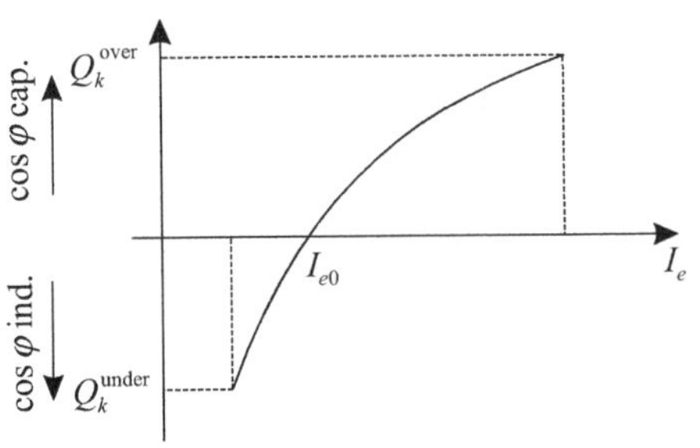

Figure 7.5. Variation of the reactive power supplied/absorbed by the synchronous compensator in terms of the field current.

A great advantage is a flexible operation at all load conditions. Even if the cost of such installations is high, it is justified in some circumstances, such as at the bus bar receiving end of a long high-voltage line, where the transmission at less then unity power factor cannot be tolerated. Maintenance costs of this machine are usually high.

7.2.2.3 Static VAr Controllers and FACTS[1]. SVC is a FACTS fully dedicated to voltage support. SVCs operate the electronic soft switching of their own shunt reactors and/or capacitors achieving a continuous reactive power variation. They are ideally suited to control the varying reactive power demand of large fluctuating loads and the overvoltages dynamics due to load rejection. They are also used in HVDC converter station and where the fast control of voltage and reactive power flow are required.

Compensation along the electrical lines requires a midpoint dynamic shunt. With a dynamic compensator at the midpoint, the symmetrical line behavior is achieved. The midpoint voltage will vary with the load, and an adjustable midpoint susceptance is a way to maintain constant voltage magnitude, therefore, the advantage of the SVC use is evident. With rapidly varying loads, the reactive power demand can be speedily corrected by SVC, with small overshoots and voltage rising. The power system oscillation damping can also be obtained by rapidly changing the output of the SVC from capacitive to inductive so as to counteract the acceleration or deceleration of interconnected machines.

The GTO technological advance, with a controlled on/off capability at high power level, opened new possibilities for power electronic equipments permitting better management of transmission grid through rapid, continuous, and flexible control of reactive and active power flows.

The high switching frequency of IGBT allows extremely fast control, which can be used in areas such as mitigation of voltage flicker caused by electric arc furnaces.

The STATic COMpensator (STATCOM) is another static VAR controller having characteristics similar to the synchronous compensator, but as an electronic device, it has no inertia and many advantages: better dynamics, a lower investment cost, and lower operating and maintenance costs. It can be seen as a voltage source behind a reactance. It provides reactive power generation as well as absorption purely by means of electronic processing of voltage and current waveforms in a voltage source converter. This means that capacitor banks and shunt reactors are not needed for generation and absorption of reactive power, facilitating a compact design and size. With the advent of STATCOM, still better performance can be reached in areas such as dynamic and steady-state voltage control in transmission and distribution systems and simultaneous control for both active and reactive powers.

Generally speaking, voltage source converters connected in "back-to-back configuration" between two AC bus bars enable active power transfer between two AC grids (synchronous or asynchronous or even with different frequencies), simultaneously providing reactive power support to the AC networks.

Lastly the unified power flow control (UPFC), incorporating a static synchronous series compensator (SSSC) as the series part and a STATCOM as the shunt element, is also appropriate for carrying out the following functions simultaneously: transient stability improvement, power swing damping, and voltage stability improvement.

[1] FACTS/Static devices are developed in the next volume of the editors titled "Advanced techniques and technologies in power systems: FACTS and Artificial Intelligence."

7.2.3 On-Load Tap Changing Transformers

7.2.3.1 Generalities. A transformer tap is a connection point along a transformer winding that allows the number of turns to be selected. By this means, a transformer with a variable turns ratio is obtained, enabling voltage regulation of the secondary side. Selection of the tap in use is made via a tap changer mechanism.

The tap changing transformers are efficient elements to regulate the voltage at one side sustained by the voltage value at the other side. The introduction of a supplementary longitudinal voltage is made in the direction of voltage increasing or decreasing, decreasing or increasing the turns ratio by changing the number of turns of a winding. The taps are usually made on the higher voltage, or lower current, side of the transformer in order to minimize the current handling during the switchovers. By changing the turns ratio the voltage in the secondary circuit is varied and the voltage control is obtained. This constitutes the most popular and widespread form of voltage control at all voltage levels.

Figure 7.6 is a schematic diagram of a tap changer placed in the secondary only or in the primary only: in both cases, the difference between the voltage to be regulated and the desired voltage value is transmitted to a motor, which actuates the changer. Contrary to the diagram of Figure 7.6, the turns to be disconnected or connected are not located at the ends of the windings: for the structural reasons, they are to be placed in a central position of the windings or symmetrically distributed between the two ends.

Figure 7.7 is indicating the equivalent circuit in Γ of a two-winding transformer if the tap changer is located in the secondary (Figure 7.6a).

- If *the connections of the primary and secondary are identical* (e.g., star–star or delta–delta), we can write the equation

$$\underline{V}'_{\mathrm{s}} = N\underline{V}_{\mathrm{s}}$$
$$\underline{I}'_{\mathrm{s}} = \frac{1}{N}\underline{I}_{\mathrm{s}} \tag{7.10}$$

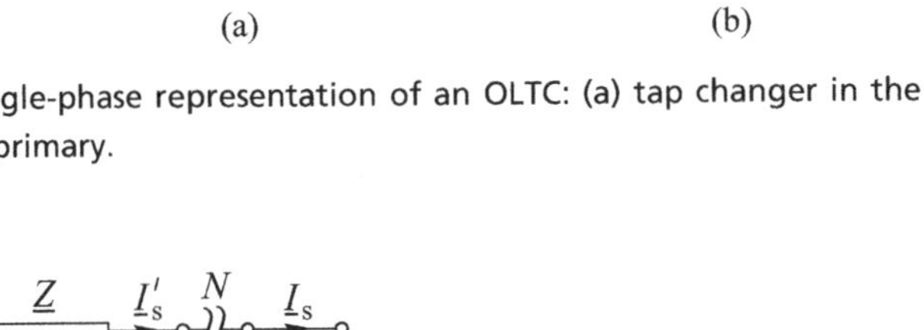

Figure 7.6. Single-phase representation of an OLTC: (a) tap changer in the secondary; (b) tap changer in the primary.

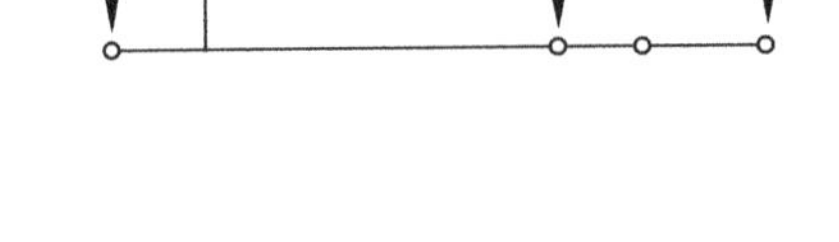

Figure 7.7. Single-phase equivalent circuit in Γ of a two-winding transformer.

N being the actual turns ratio, usually different from the nominal one N_{nom} [2]

$$N_{nom} = \frac{V_{n1}}{V_{n2}} = \frac{I_{n2}}{I_{n1}} \tag{7.11}$$

where V_{n1} and V_{n2} are the nominal phase-to-earth voltages of the primary and secondary, while I_{n1} and I_{n2} represent the nominal phase currents of the two windings.

From equation (7.10), we can infer that [4]

$$\begin{aligned} \frac{\underline{V}'_s}{V_{n1}} &= N\frac{V_{n2}}{V_{n1}}\frac{\underline{V}_s}{V_{n2}} \\ \frac{\underline{I}'_s}{I_{n1}} &= \frac{1}{N}\frac{I_{n2}}{I_{n1}}\frac{\underline{I}_s}{I_{n2}} \end{aligned}$$

that is, in p.u.

$$\begin{aligned} \underline{v}'_s &= m\underline{v}_s \\ \underline{i}'_s &= \frac{1}{m}\underline{i}_s \end{aligned} \tag{7.12}$$

where

$$m \triangleq \frac{N}{N_{nom}} \tag{7.13}$$

is the turns ratio in p.u.

Also, we can write the relation between the voltages and currents, respectively, at the ends of transformer:

$$\begin{aligned} \underline{v}_p &= \underline{z}'\underline{i}'_s + \underline{v}'_s \\ \underline{i}_p &= \underline{y}'\underline{v}'_p + \underline{i}'_s \end{aligned} \tag{7.14}$$

where

$$\begin{aligned} \underline{z}' &\triangleq \underline{Z}\frac{I_{n1}}{V_{n1}} \\ \underline{y}' &= \underline{Y}_0\frac{V_{n1}}{I_{n1}} \end{aligned} \tag{7.15}$$

Equations (7.12) and (7.14) completely describe the behavior of an OLTC in p.u., for each value of m. These equations may be easily associated with the single-phase equivalent circuit of Figure 7.8.

- *The primary and secondary windings are connected in a different way* (e.g., star–delta or delta–star configurations) [4].

[2] In literature, the turns ratio is also defined as a:1, where $a = N_{ik}$, and 1:a, where $1/a = N_{ki}$.

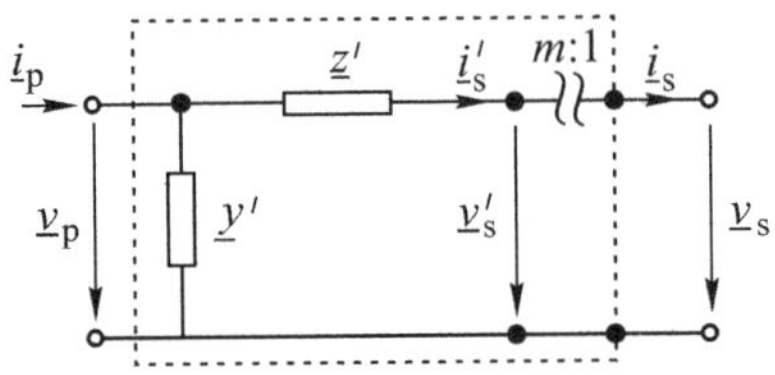

Figure 7.8. Single-phase equivalent circuits, in p.u., of an OLTC.

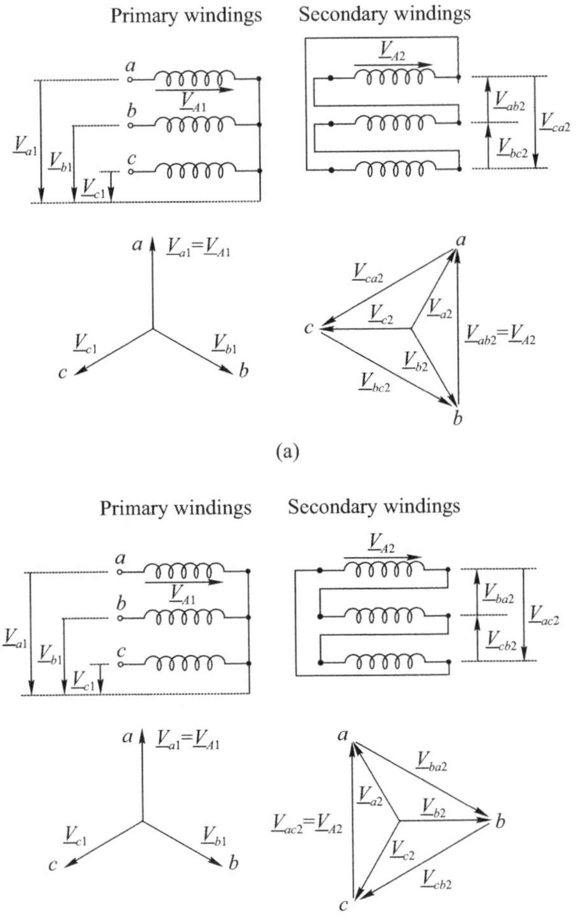

Figure 7.9. Star–delta connections of a two-winding three-phase transform and phasor diagrams of positive-sequence voltages [4].

If we consider the star–delta connection, which may produce the two situations shown in Figure 7.9 (the polarity of the delta winding is different).

With regard to the positive sequence, let us call $\underline{V}_{A1}$, $\underline{V}_{A2}$ the winding voltages, with the real turns ratio of ideal transformer

$$N = \frac{\underline{V}_{A1}}{\underline{V}_{A2}}$$

$\underline{V}_{A1}$ and $\underline{V}_{A2}$ are, of course, in phase. The phase voltages and winding voltages for the primary coincide, whereas the winding voltages of the secondary (see Figure 7.9a) are orderly equal to the phase-to-phase voltages[3] $\underline{V}_{ab,2}$, $\underline{V}_{bc,2}$, $\underline{V}_{ca,2}$.

Hence, the voltage of phase "a" of the primary leads the voltage of the same phase of secondary by 30°, and similarly for phases "b" and "c". More precisely

$$\underline{V}_{a2} = \frac{\underline{V}_{A2}}{\sqrt{3}} e^{-j30^\circ} \quad \text{or} \quad \underline{V}_{A2} = \sqrt{3}\underline{V}_{a2}\, e^{j30^\circ} \tag{7.16a}$$

that is,

$$\underline{V}_{a1} = \underline{V}_{A1} = N\underline{V}_{A2} = \sqrt{3}N\underline{V}_{a2}\, e^{j30^\circ} \tag{7.16b}$$

In the single-phase equivalent circuit of Figure 7.9b, the real turn ratio N of the ideal transformer becomes $\sqrt{3}N\, e^{j30^\circ}$ for the voltages, that is, complex.

For the currents, as the ideal transformer has remained such and the two complex powers should always be equal, that is, the equation

$$\underline{V}'_s \underline{I}'^*_s = \underline{V}_s \underline{I}^*_s \tag{7.17}$$

must be satisfied, we get

$$\frac{\underline{I}'_s}{\underline{I}_s} = \left(\frac{\underline{V}_s}{\underline{V}'_s}\right)^* = \left(\frac{\underline{V}_{a2}}{\underline{V}_{a1}}\right)^* = \frac{1}{\sqrt{3}N} e^{j30^\circ} \tag{7.18}$$

So, also the phase currents of the primary lead the currents of the secondary by 30°. In other terms, the turns ratio of the currents is equal to the conjugate inverse of the turns ratio of the voltages. In the case shown in Figure 7.9b, the phase variables of the primary lag (do not leg) the phase variables of the secondary by 30°. In conclusion, in star–delta (and delta–star) connection, the primary and secondary phase variables have a positive-sequence phase difference of 30° lead or lag.

Three-phase transformers with different configurations of their primary and secondary connections are transformers with complex turns ratio.

7.2.3.2 Switching Technologies.

Off-Load Designs. In low power, low-voltage transformers, the tap point can take the form of a connection terminal, requiring a power lead to be disconnected by hand and connected to the new terminal. Alternatively, the process may be assisted by means of a rotary or slider switch.

Because the different tap points are at different voltages, the two connections should not be made simultaneously, as this short circuits a number of turns in the winding and would result in an excessive circulating current. This therefore demands that the power to the load be physically interrupted during the switchover time. Off-load

[3] Note that in Europe, the phase-to-phase voltage is mainly denoted by U and the phase-to-earth (or neutral) voltage is denoted by V, and their relationship is $U = \sqrt{3}V$.

tap changing is also employed in high-voltage transformer designs, though it is only applicable to installations in which loss of supply can be tolerated.

On-Load Designs. Because interrupting the supply is usually unacceptable for a power transformer, these are often fitted with a more expensive and complex on-load tap changing mechanism. OLTC may be mechanical, or as electronic. Further, the basic switching technologies are presented and compared.

Mechanical Tap Changers. The more common type of mechanical OLTC has two parts: a *tap selector* to select the tap position and a *diverter* to transfer the current from one tap position to another. During tap changing, neither the load current must be broken nor the taps be short circuited. For this reason, an impedance or resistor or reactor is connected between the taps during the changeover to minimize the circulating current. The tap selector and the diverter are typically oil immersed.

A mechanical tap changer physically makes the new connection before releasing the old, but avoids the high current from the short-circuited turns by temporarily placing a large diverter resistor (sometimes an inductor) in series with the short-circuited turns before breaking the original connection. This technique overcomes the problems with open or short-circuit taps. The changeover nevertheless must be made rapidly to avoid overheating of the diverter. Powerful springs are wound up, usually by a low power motor, and then rapidly released to affect the tap changing operation. To avoid arcing at the contacts, the tap changers are filled with insulating transformer oil. Tapping normally takes place in a separate compartment to the main transformer tank to prevent contamination of its oil.

One possible design of on-load mechanical tap changer is shown in Figure 7.10 [5].

Let us assume that the tap is in position 2, with load supplied directly via the right-hand connection. Diverter resistor A is short circuited; diverter B is unused. In order to move from tap 2 to tap 3, the following sequence occurs:

1. Switch 3 closes, an off-load operation.
2. Rotary switch turns, breaking one connection and supplying load current through diverter resistor A.

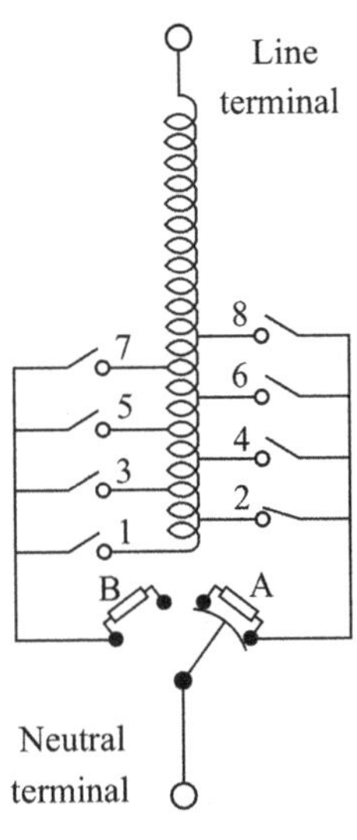

Figure 7.10. A mechanical on-load tap changer design. "A" and "B" are the diverter resistors [5].

3. Rotary switch continues to turn, connecting between contacts A and B. Load now supplied via diverter resistors A and B, winding turns bridged via A and B.
4. Rotary switch continues to turn, breaking contact with diverter A. Load now supplied via diverter B alone, winding turns no longer bridged.
5. Rotary switch continues to turn, shorting diverter B. Load now supplied directly via left-hand connection. Diverter A is unused.
6. Switch 2 opens, an off-load operation.

In the case of mechanical tap changers, the preferred location of tap changer is the neutral end of the higher voltage winding. The diverter duty is minimized with the lower current on the high-voltage winding. In the case of mechanical contacts in oil, the higher dielectric strength needed for the higher voltage stresses resulting from choosing the high-voltage winding can be obtained without significant cost penalty.

The main disadvantages of the mechanical OLTC, which should be overcome by other switching technologies are as follows:

- Breaking the current in the diverter leads to arcing at contacts. This leads to wear and erosion of the contacts, requiring appropriate maintenance.
- The power loss of transition resistors is raising the oil temperature inside the diverter switch compartment during commutations. Consequently, continuous operation is normally not allowed to keep the oil temperature within admissible limits.
- Contact arcing contaminates the oil and so oil replacement is needed on the occasion of regular inspection intervals.
- Special mechanism has to be installed at the OLTC to prevent disastrous consequences in case of failures with regard to shafting.
- The tap change operation, adequate for normal steady-state voltage regulation, is relatively slow: the speed of the diverter switch is in the range of 50–100 ms. The tap selector switch is slower and could take minutes to traverse the full tap range.

Figure 7.11 presents typical sectional views of a "diverter selector" and "selector switch," respectively [6].

Vacuum Switched Tap Changers. Some of the above disadvantages could be overcome by the use of vacuum switches to replace the mechanical contacts in oil, either in the diverter alone or in the diverter and the tap selector. In comparison with the mechanical switching technology, the vacuum switching technology improves the regulating performance due to the following features:

- Friendly to the environment: no oil carbonization (no arcing in the insulating oil), no oil filter unit, and expended lifespan of the insulating oil.
- Possibility to use alternative insulating media.
- No substantial effects of the contact wear on the life of the OLTC.
- Free of maintenance for up to 300,000 operations or even more for almost all network applications.
- Increased transformer availability.
- Significant reduction of life cycle costs.

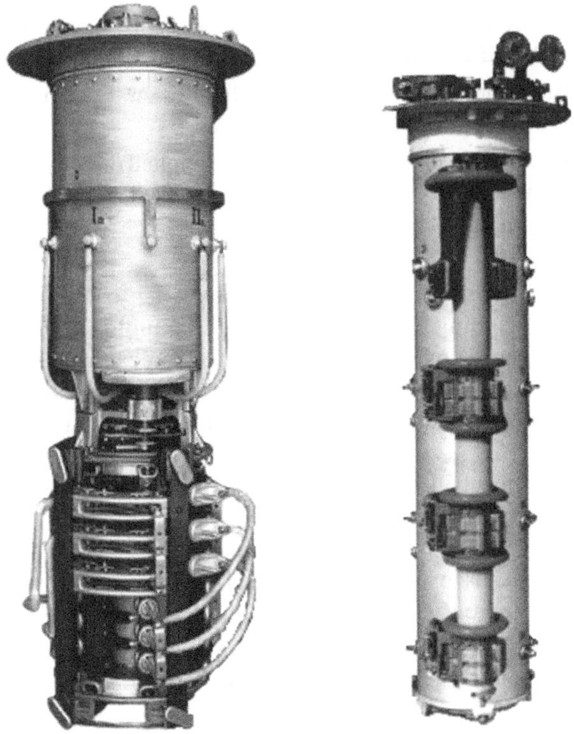

(a) (b)

Figure 7.11. Typical sectional views of diverter selector (a) and selector switch (b). (Courtesy of Easum-MR [6].)

These features allow this technology to be more and more implemented in power systems nowadays. Moreover, the vacuum switching technology has clear advantages in regard of the following points: maximum step voltage; maximum step power; long-term resistance in alternative insulating media; overall complexity and low risk of failures. In spite of these advantages, this technology does not improve the power system performances with respect to system stability or subsynchronous resonance.

In order to expose the operation principle of this technology, a particular case was chosen: the RMV-II Load Tap Changer (Figure 7.12) [7]. The load tap changer is used in conjunction with oil-immersed power transformers, regulators and phase shifting transformers (PSTs) to change taps under load, thereby controlling voltage magnitude or phase angle. The tap changer works on the preventive autotransformer (reactor) switching principle with vacuum interrupters to accomplish the tap change. Vacuum interrupters are used to interrupt the circuit within a half cycle. The interruption takes place in a vacuum of approximately 10^{-6} Torr instead of the usual arcing under oil. Thus, oil contamination is eliminated.

As regards the design, the RMV-II load tap changer consists of an oil compartment (Figure 7.12a) containing tap and changeover selectors, vacuum interrupters and bypass switches, a separately housed drive mechanism switch assembly, and other accessories depending on the on-site conditions. Each phase consists of a tap and changeover selectors positioned on an epoxy molded terminal board and a vacuum interrupter and bypass switch on

(a)

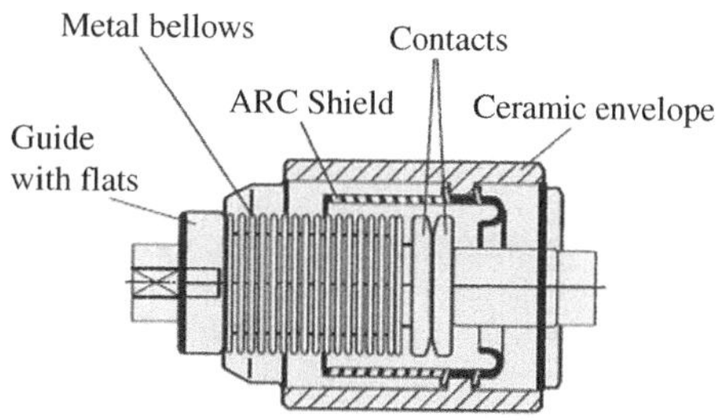

(b)

Figure 7.12. RMV-II load tap changer: (a) tap changer oil compartment; (b) schematic layout of vacuum interrupter (Courtesy of Maschinenfabrik Reinhausen) [7].

a separate vertical insulating panel mounted rigidly from the top of the tap changer oil compartment. The vacuum interrupter drive assembly is a cam action spring driven mechanism, which impacts an operating rod, when opening and closing. The operating rod is connected to the vacuum interrupter moving contact. The vacuum interrupter consists of a stationary and a moving contact enclosed in a vacuum-tight ceramic insulating envelope (Figure 7.12b). The moving contact is sealed through a flexible metal bellows protected from the arc by a shield. A metal shield surrounds the contacts forming an arc chamber and condensing surface to collect vaporized contact material that arises during arcing.

The tap changer operation (Figure 7.13) is divided into three major functions:

- Arc interruption and reclosing by use of the vacuum interrupters in conjunction with the associated bypass switches.
- Selection of the next tap position by the tap selector assemblies in proper sequence with the operation of the vacuum interrupters and bypass switches.
- Operation of reversing or coarse/fine selector in order to double the number of tap positions.

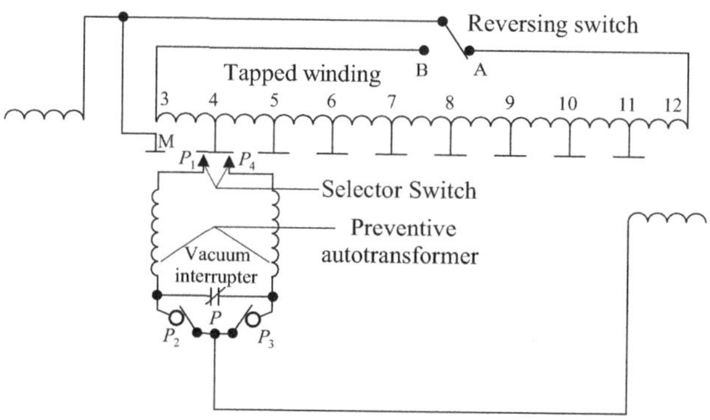

Figure 7.13. The load tap changer switching principle [7].

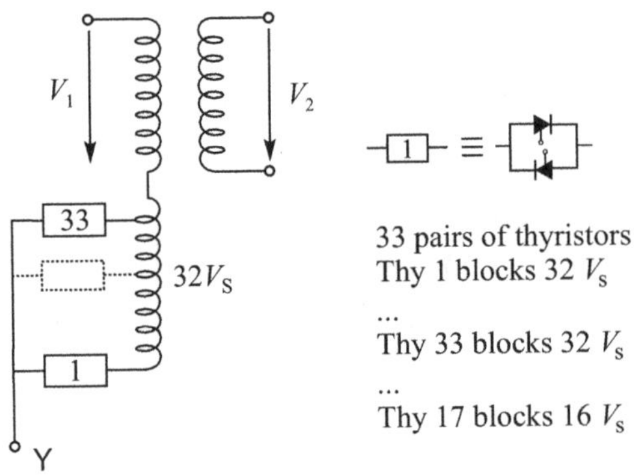

Figure 7.14. Thyristor substitution of mechanical contacts with tap selector of OLTC [8].

When moving from one tap position to the next, one set of bypass switch contacts opens, while the second set stays closed, routing the current through the vacuum interrupter just prior to its operation. The vacuum interrupter opens by a spring-operated mechanism before the tap selector moving contact selects the next tap. The vacuum interrupter then closes under spring force and locks in place followed by the bypass switch reclosing to shunt the vacuum interrupter, thus completing the tap change operation.

Static Switched Tap Changers. With the development of semiconductor devices, particularly thyristors, of relatively large current and voltage ratings, the possibility of developing static on-load tap changers is of special interest. There are different solutions for static OLTCs but not all of them are feasible or present a major advantage in comparison with mechanical or vacuum switches. In the case of thyristor tap changers, the voltage stresses would be the dominant consideration in determining the number of thyristors. From this point of view, in most cases, the preferred location of thyristor-based tap changers would be the lower voltage winding.

One of the possibilities is an *all-static thyristor OLTC* in which back-to-back connected thyristor valves replace the mechanical contacts of the conventional OLTC. This solution has a major disadvantage that makes it not a viable practical proposition: the number of thyristors. In order to illustrate this, Figure 7.14 shows a tap changer arrangement for ± 16 steps, assuming all the mechanical contacts are replaced by thyristors [8].

The voltage stress across each of the thyristor valves depends on its tap position. The valves at the end, 1 and 33, will have the maximum steady-state voltage stress, while the middle one, 17, will have the lowest. The maximum voltage across the other valves will vary, increasing progressively with the distance of the tap from the middle one. If the rms value of the step voltage is V_S, the sum of the steady-state voltage stresses on the all back-to-back connected thyristors valves in one phase of the tap changer with 32 steps will be approximately $800V_S$. If the step size is in the range from 0.5 to 1.0 kV, as a minimum, the number of thyristors needed for the tap changer should have a total voltage withstand capability of 2400–4800 kV. At 2 kA rating, so many thyristors will be enough to build thyristor valves for 1000–2000 MW AC/DC converter.

Alternative solutions pursued fall into three types [8]:

- Hybrid OLTCs with thyristor-assisted diverter switch.
- Hybrid OLTCs with GTO-assisted diverter switch.
- Fully static OLTCs with thyristor switched tap changer windings.

(i) *The Hybrid Thyristor-Type OLTC* consists of a thyristor-type diverter switch and a mechanically operated tap selector.

The thyristor-type diverter switch is generally equipped with several mechanical auxiliary switches, such as

- shunt contacts in order not to dimension the thyristors for circuit current and to reduce losses on the operating positions;
- contacts connected in series before the thyristors valves to eliminate the transient voltage stresses occurring in the step winding;
- if necessary, additional changeover switches to minimize the number of thyristors and to effect communication of the load current first to a transition resistor and then to disconnect the circulating current through the transition resistor by one back-to-back connected thyristor value.

Generally, there are two different arrangements:

- Hybrid thyristor-type diverter switch commutating without transition resistor.
- Hybrid thyristor-type diverter switch with transition resistor.

- The diverter of a commutating diverter switch *without transition resistor* operates—without the circulating current across a transition impedance, which is typical for an OLTC—by directly commutating the load current from the thyristor path of the current carrying diverter switch side (e.g., n) to the thyristor path of the preselected diverter switch side (e.g., $n + 1$) at zero current. The function of the commutating diverter switch components that are not suitable for operating in hot insulating oil, in particular: interlocking electronics to prevent simultaneous triggering of both thyristor paths, thus preventing a short circuit between steps to occur; measuring equipment (for the voltage increase at the thyristor groups after they have entered into their blocked state); commutating capacitors.
- On the other hand, the diverter switch *with transition resistor* (Figure 7.15) can be designed in a much simpler way, particularly the interlocking electronics and the commutating capacitors can be omitted.

In consequence, this type of diverters can be realized in the design and size of a conventional OLTC for in-tank installation. We used a basic connection diagram with only one antiparallel thyristor pair performing a transfer switching operation by first commutating the load current to the transition resistor and in the second switching process interrupting the circulating current flowing across the resistor.

The hybrid thyristor-type OLTC technology has only been implemented to a very small extent so far, and it can also be expected that it will not catch on to a

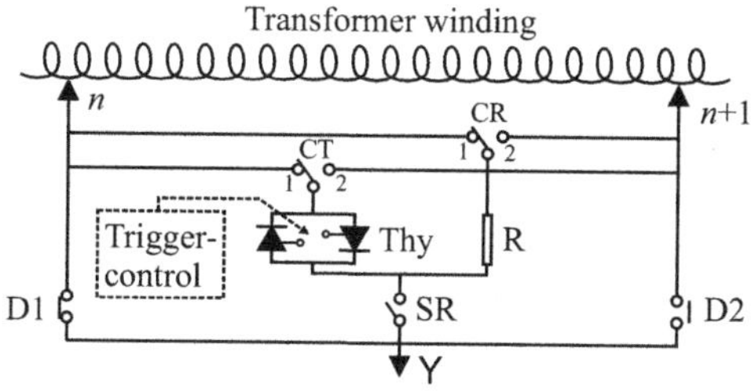

Figure 7.15. Hybrid thyristor-type diverter switch with transition resistor [8]: D_1, D_2 shunt contact; SR auxiliary; CT auxiliary; R transition.

wider extent in the future either as the vacuum switching technology clearly offers better performance.

(ii) *Hybrid GTO-Type OLTC*

The commutation problem associated with thyristor-based tap changer can be overcome by use of GTOs or IGBTs in the place of thyristors. With maximum voltage ratings of these devices lower than those of the thyristors, their use to replace mechanical contacts in the tap changer would be even more costly than in the case of thyristors. Hence, their use is also considered only in diverters with mechanically operated tap selector.

(iii) *Static OLTC with Thyristor Switched Windings*

Even from 1986, a fully static single-phase OLTC was installed in the 16.5 kV, 16 2/3 Hz systems of Norwegian State railways [9]. There has been no reported problem in its operation over many years.

Although this is in a distribution system, there is no reason why it cannot be applied to transmission system of higher voltages. It should be possible to extend the concept to a greater number of tap positions also.

7.2.3.3 Determination of the Current Operating Tap. In order to determine the operating tap of the transformer, consider the simple configuration of a load supplied from a source through a transmission line and a step-down transformer (Figure 7.16a). Neglecting the no-load losses of the transformer, its series impedance can be included together with the line impedance in the equivalent impedance $\underline{Z}$ of the circuit (Figure 7.16b) [10].

For a given operating point we know $\underline{V}_1, P_{20}$, and Q_{20}. The equation between voltages is

$$\underline{V}_1 = \underline{V}_{2'} + \Delta\underline{V}_{12} \tag{7.19}$$

Neglecting the transversal component of the voltage drop and considering the voltage at the load as phase reference, the relation (7.19) becomes [3]

$$\underline{V}_1 = V_{2'} + \Delta V = V_{2'} + \frac{RP_{20} + XQ_{20}}{V_{2'}} \tag{7.19$'$}$$

From equation (7.19′), it results a second-order equation with the unknown variable $V_{2'}$:

$$V_{2'}^2 - V_1 V_{2'} + RP_{20} + XQ_{20} = 0$$

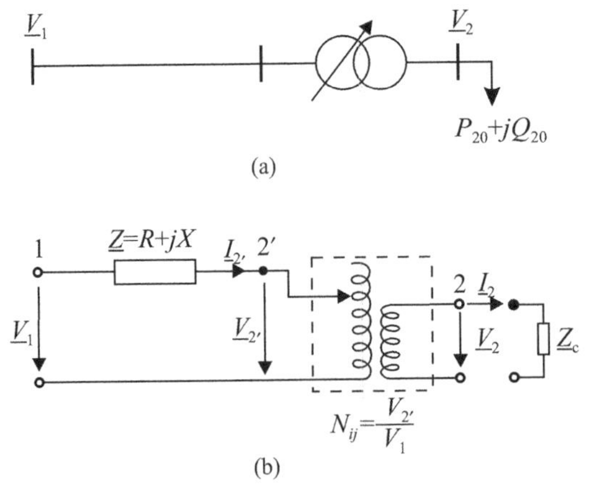

Figure 7.16. Simple circuit source–line–transformer–load: (a) one-line diagram; (b) equivalent circuit.

of which the solution is

$$V_{2'} = \frac{V_1 + \sqrt{V_1^2 - 4(RP_{20} + XQ_{20})}}{2} \tag{7.20}$$

Calculating the turns ratio $N_{ij} = V_{2'}/V_2$ and replacing $V_{2'}$ in expression (7.20), gives

$$N_{ij} = \frac{V_1 + \sqrt{V_1^2 - 4(RP_{20} + XQ_{20})}}{2\,V_2} \tag{7.20$'$}$$

It can be seen that the turns ratio N_{ij} is a function of the nominal voltage of the transformer, the percentage tap voltage ΔV_t and the number of the current tap n_p:

$$N_{ij} = \frac{V^{HV}}{V^{MV}}\left(1 + n_p \frac{\Delta V_t}{100}\right) \tag{7.21}$$

Starting from an operating state, characterized by $\underline{V}_1$, P_{20}, and Q_{20}, and also specifying a certain scheduled voltage value at the load $V_2 = V_{2sch}$, the number of operating tap should be

$$n_p = \frac{100}{\Delta V_t}\left[\frac{V_n^{MV}}{V_n^{HV}} \cdot \frac{V_1 + \sqrt{V_1^2 - 4(RP_{20} + XQ_{20})}}{2\,V_{2sch}} - 1\right] \tag{7.22}$$

The obtained value has to be rounded off to the nearest integer value.

7.2.3.4 Static Characteristic of the Transformer. On-load tap changing is used to keep the voltage V_2 on the secondary winding of the transformer close to a reference value V_{2sch} by modifying the turns ratio N.

In order to analyze the influence of the on-load tap changing, let us consider the example from Figure 7.16b, where the impedance $\underline{Z}_c$ represents the load. The following relationship between voltages and currents can be written as

$$\begin{aligned}\underline{V}_1 &= \underline{V}_{2'} + \underline{Z}\,\underline{I}_{2'}\\ \underline{V}_2 &= \underline{Z}_c\underline{I}_2\end{aligned} \tag{7.23}$$

To simplify the calculation, the resistances of both the line and the transformer are neglected, that is $\underline{Z} \cong jX$, and the load is considered resistive, that is $\underline{Z}_c = R_c$. Therefore, the load power is expressed as $P_2 = V_2^2/R_c$.

Choosing the voltage at the load terminals as phase origin, that is, $\underline{V}_2 = V_2$ and taking into account the expression of the turns ratio

$$N = \frac{V_{2'}}{V_2} = \frac{I_2}{I_{2'}} \tag{7.24}$$

it results then

$$\begin{aligned}\underline{V}_1 &= \underline{V}_{2'} + jX\underline{I}_{2'}\\ \underline{V}_2 &= R_c\underline{I}_2\end{aligned} \tag{7.23$'$}$$

Taking into account (7.24), expression (7.23′) can be written as

$$\underline{V}_1 = NV_2 + j\frac{X}{N}\frac{V_2}{R_c} = NV_2\left(1 + j\frac{X}{R_c}\frac{1}{N^2}\right) \tag{7.25}$$

and the absolute value is

$$V_1 = NV_2\sqrt{1 + \left(\frac{X}{R_c}\right)^2\frac{1}{N^4}} \tag{7.25′}$$

resulting

$$V_2 = \frac{V_1}{N\sqrt{1 + (X/R_cN^2)^2}} = \frac{V_1}{X/R_cN\sqrt{1 + (R_cN^2/X)^2}}$$

or

$$V_2 = f(N) = \frac{NR_cV_1}{X}\frac{1}{\sqrt{1 + (R_c/X)^2N^4}} \tag{7.26}$$

Expression (7.26), which defines the dependency of the voltage at the load terminals in terms of the turns ratio, where the source voltage V_1 and the resistance R_c are the equation parameters, defines the *static characteristic of the on-load tap changing transformer* (Figure 7.17) [10].

It can be seen that the function $V_2 = f(N)$ has a maximum, which can be determined by equating to zero the derivative $\frac{dV_2}{dN} = 0$. Denoting by

$$a = \frac{R_cV_1}{X}; \quad b = \frac{R_c^2}{X^2} \tag{7.27}$$

in (7.26), it results

$$V_2 = \frac{Na}{\sqrt{1 + bN^4}} \tag{7.28}$$

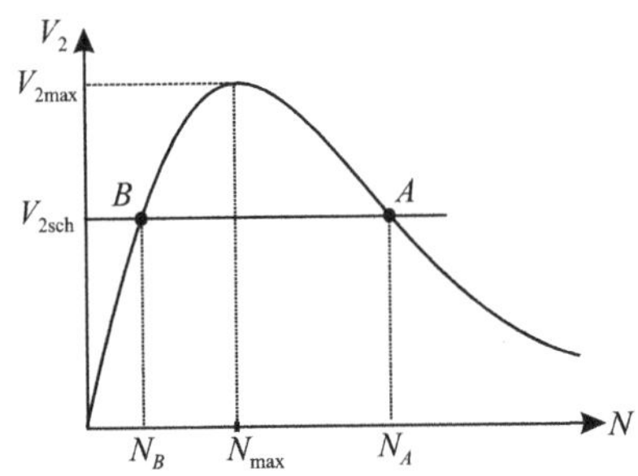

Figure 7.17. The static characteristic of the on-load tap changing transformer $V_2 = f(N)$.

Therefore,

$$\frac{dV_2}{dN} = \frac{a\sqrt{1+bN^4} - aN\left(4bN^3/\left(2\sqrt{1+bN^4}\right)\right)}{\left(\sqrt{1+bN^4}\right)^2} = \frac{a(1+bN^4) - 2abN^4}{\left(\sqrt{1+bN^4}\right)^3}$$

for which we impose

$$\frac{dV_2}{dN} = \frac{a\left(1 - bN_{max}^4\right)}{\left(\sqrt{1+bN_{max}^4}\right)^3} = 0 \qquad (7.29)$$

and it results that

$$1 - bN_{max}^4 = 0$$

that is

$$N_{max} = \sqrt[4]{\frac{1}{b}} = \sqrt{\frac{X}{R_c}} \qquad (7.30)$$

The maximum value of the voltage at the load terminals is obtained from (7.26), where we replace $R_c/X = 1/N_{max}^2$, resulting

$$V_{2max} = \frac{V_1}{\sqrt{2}N_{max}} = \frac{V_1}{\sqrt{2}\sqrt{(X/R_c)}} = V_1\sqrt{\frac{R_c}{2X}} \qquad (7.31)$$

From Figure 7.17 it can be seen that, for a scheduled voltage $V_{2sch} < V_{2max}$, two operating points A and B exist, while for $V_{2sch} > V_{2max}$ no operating point exist. This demonstrates the fact that it cannot be obtained any value V_{2sch} at the load by changing the turn ratio.

Figure 7.18 illustrates schematically the on-load tap changing mechanism and a simplified bloc diagram for turns ratio modification by an OLTC. The tap changer decrease the turns ratio when $V_2 < V_{2sch} - \varepsilon$, increase the turns ratio when $V_2 > V_{2sch} + \varepsilon$, or it will take no action when the voltage has values in admissible limits $[V_{2sch} - \varepsilon; V_{2sch} + \varepsilon]$ (Figure 7.18a).

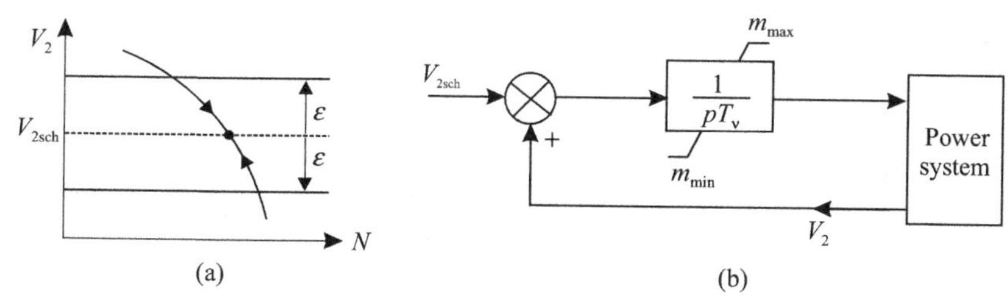

Figure 7.18. Commutation logic of the on-load tap changing (a); block diagram of turns ratio modification (b).

In Figure 7.18b, the integrator with the integration time T_v (equal to about 10 s at the least) corresponds to the motor that acts on the changer, whereas V_2 is the voltage to be regulated, V_{2sch} is the desired voltage, and m designates the turns ratio in p.u., comprised in the following range [4]:

$$m_{min} \leq m \leq m_{max} \tag{7.32}$$

The action of the tap changer is supposed to be continuous, even if it is actually stepped (high number of steps and small variation of m in the next two steps). Under steady-state conditions, the integrator ensures a zero voltage error if the required change of m lies within the specified range ($m_{min} \leq m \leq m_{max}$). This regulation is slow, as against other types of voltage regulation, and ensures the desired voltage value only under steady-state operating conditions.

The tap changing can be made automatically, by voltage automatic regulators, or manually. The transformers have to be able to provide regulation under normal conditions, such as load variations according to the daily load curve, as well as in emergency situations, when the voltage takes values outside the admissible operating limits, which can lead to the voltage instability phenomenon.

It is, therefore, desirable that the number of on-load commutations be as small as possible. In order to avoid the tap changer operation in case of transient voltage variations, the on-load tap changer is blocked during a delay time before the first commutation and between two successive commutations. Usually, if the voltage exceeds the admissible limits, the first commutation has a greater delay comparing with the next commutation. Furthermore, for cascading located transformers, the greater the voltage level is, the smaller is the delay of the first commutation.

7.2.3.5 Various Applications of the OLTC Transformers for Voltage and Reactive Power Control.

COMBINED USE OF OLTC AND REACTIVE POWER INJECTIONS IN TRANSMISSION NETWORKS [11].

- *First Case*. One of the common practice for reactive flow control in transmission networks is to use the tertiary of three-winding transmission transformer for the reactive power injection via synchronous compensators, or capacitor/reactor banks, as shown in Figure 7.19a.

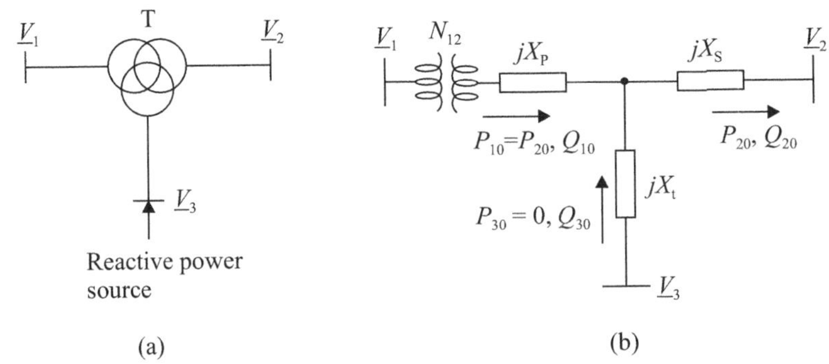

Figure 7.19. Three-winding OLTC transformer with synchronous compensator connected to its tertiary: (a) one-line diagram; (b) equivalent impedance diagram.

For a given load condition, it is necessary to find the transformer turns ratio for some specified reactive generation/consumption on the tertiary bus. By representing the three-winding transformer with its equivalent star (or Y) connection and by neglecting the winding resistances and transformer shunt losses, the impedance diagram of the system under consideration is shown in Figure 7.19b.

For a given secondary load P_{20}, Q_{20}, assuming $P_{30} \approx 0$, the voltage drop between buses 1 and 2, when the transversal component is neglected, is

$$|\Delta \underline{V}| \approx \Delta V = \frac{V_1}{N_{12}} - V_2 = \frac{X_P Q_{10} + X_S Q_{20}}{V_2} = \frac{X_P(Q_{20} - Q_{30}) + X_S Q_{20}}{V_2} \tag{7.33}$$

or

$$N_{12}V_2^2 - V_2 V_1 + N_{12}[(X_P + X_S)Q_{20} - X_P Q_{30}] = 0 \tag{7.33'}$$

Equation (7.33′) gives the relationship among V_1, V_2, and Q_{30}. Then, for V_1 and V_2 known, and for specific Q_{30}, from (7.33′) the required turns ratio is given by

$$N_{12} = \frac{V_1 V_2}{V_2^2 + (X_P + X_S)Q_{20} - X_P Q_{30}} \tag{7.34}$$

or, the value of reactive power injection Q_{30} for a specified turns ratio is determined by the relation

$$Q_{30} = \frac{N_{12}V_2 - V_1 V_2 + N_{12}(X_P + X_S)Q_{20}}{X_P N_{12}} \tag{7.35}$$

An usual arrangement is to apply the manual control of the OLTC turns ratio and the automatic control of synchronous compensator excitation.

- *Second Case.* The voltage at the load terminals cannot be maintained at the scheduled value by tap changing, and additional reactive power is required [12].

  In order that the rating of reactive power compensation device to be minimum, tap changing possibilities are used at the extreme limits. Therefore, the turns ratio has the minimum value N_{min} for maximum load, and the maximum value N_{max} for minimum load.

  If no reactive power compensation device is used, considering the voltage V_1 constant and neglecting the transversal component of the voltage drop, obtain

  – for maximum load:

$$V_1 = V_{2'max} + \frac{P_{20}^{max} R + Q_{20}^{max} X}{V_{2'max}} \tag{7.36}$$

  – for minimum load:

$$V_1 = V_{2'min} + \frac{P_{20}^{min} R + Q_{20}^{min} X}{V_{2'min}} \tag{7.36'}$$

subject to the following condition:

$$V_{2max} < V_{2sch} < V_{2min}$$

In order to achieve the scheduled voltage, the synchronous compensator has to provide a reactive power Q^{gen}_{comp}, in overexcitation, for the maximum load regime and to absorb a reactive power Q^{abs}_{comp}, in underexcitation, for the minimum load regime. By introducing the synchronous compensator, we need to modify the equations of voltage:

– for maximum load:

$$V_1 = V_{2'sch} + \frac{P_{20}^{max}R + \left(Q_{20}^{max} - Q^{gen}_{comp}\right)X}{V_{2'sch}} \tag{7.37}$$

– for minimum load:

$$V_1 = V_{2'sch} + \frac{P_{20}^{min}R + \left(Q_{20}^{min} + Q^{abs}_{comp}\right)X}{V_{2'sch}} \tag{7.37'}$$

Equating the expressions (7.36) and (7.37), the equation of the single-phase reactive power generated by the compensator when it is overexcited is obtained:

$$Q^{gen}_{comp} = \frac{V_{2'sch}}{X}\left(V_{2'sch} - V_{2'max} - \frac{P_{20}^{max}R + Q_{20}^{max}X}{V_{2'max}} + \frac{P_{20}^{max}R + Q_{20}^{max}X}{V_{2'sch}}\right) \tag{7.38}$$

Similarly, from expressions (7.36′) and (7.37′) it results the single-phase reactive power absorbed by the compensator when it is underexcited:

$$Q^{abs}_{comp} = \frac{V_{2'sch}}{X}\left(V_{2'min} - V_{2'sch} + \frac{P_{20}^{min}R + Q_{20}^{min}X}{V_{2'min}} - \frac{P_{20}^{min}R + Q_{20}^{min}X}{V_{2'sch}}\right) \tag{7.38'}$$

Neglecting the difference of the latter terms in expressions (7.38) and (7.38′), the simplified equations of the reactive power generated/absorbed by the synchronous compensator are obtained:

$$Q^{gen}_{comp} = \frac{V_{2'sch}(V_{2'sch} - V_{2'max})}{X} = \frac{N^2_{min}V_{2sch}(V_{2sch} - V_{2max})}{X} \tag{7.39}$$

$$Q^{abs}_{comp} = \frac{V_{2'sch}(V_{2'min} - V_{2'sch})}{X} = \frac{N^2_{max}V_{2sch}(V_{2min} - V_{2sch})}{X} \tag{7.39'}$$

If the inequality

$$Q^{abs}_{comp} \leq (0.5\ldots0.65)\cdot Q^{gen}_{comp} \tag{7.40}$$

holds, the rated power Q_n of the compensator results from $Q_n \geq Q^{gen}_{comp}$ condition. Otherwise, if the condition in (7.40) is not verified, the best solution is obtained by comparing two or more possibilities:

(a) Synchronous compensator of which rated power is determined from condition $Q_n \geq Q^{gen}_{comp}$, associated to the compensation reactor.
(b) Synchronous compensator of greater rated reactive power, which results from condition $Q_n \geq Q^{abs}_{comp}/(0.5\ldots0.65)$.

The general problem of placement and sizing the reactive power compensation sources in electrical networks is solved by technical–economical optimization calculations.

If the goal is to maintain the voltage only by modifying the generated/absorbed reactive power by the local source, the turns ratio being constant, the values $N_{\max} = N_{\min} = N$ are introduced in expressions (7.39) and (7.39′), and the turns ratio N is determined from (7.40), to which the restriction $N_{\min} < N < N_{\max}$ is imposed. The rated reactive power of the synchronous compensator is obtained following the above-presented algorithm.

In order to use all regulation range of the compensator (for instance, at maximum load to provide $Q_{\text{comp}}^{\text{gen}}$ and during minimum load $Q_{\text{comp}}^{\text{abs}} = 0.6 \cdot Q_{\text{comp}}^{\text{gen}}$), we obtain the expression of the scheduled voltage for both regimes. Therefore, we can write

$$V_1 = V_{2'\text{sch}} + \frac{P_{20}^{\min} R + (Q_{20}^{\min} + 0.6 \cdot Q_{\text{comp}}^{\text{gen}})X}{V_{2'\text{sch}}} = V_{2'\text{sch}} + \frac{P_{20}^{\max} R + (Q_{20}^{\max} - Q_{\text{comp}}^{\text{gen}})X}{V_{2'\text{sch}}}$$

obtaining

$$Q_{\text{comp}}^{\text{gen}} = \frac{R(P_{2\max} - P_{2\min}) + X(Q_{2\max} - Q_{2\min})}{1.6X} \tag{7.41}$$

Reactive Flow Control Between Two High-Voltage Networks Connected Through an OLTC Transformer. Transmission networks of different voltage levels are usually interconnected through OLTC transformers (Figure 7.20a) [11]. If such networks are infinite bus networks, the OLTC transformers are used as means for the reactive flow control between the two interconnected networks.

By using the equivalent impedance diagram for such a system, shown in Figure 7.20b, assuming that the taped winding is placed on the transformer primary side, and by neglecting transformer resistances and shunt losses, the following equation can be written (all quantities are expressed in p.u.):

$$|\Delta \underline{V}| \approx V_1 - NV_2 = \frac{X_T Q_0}{NV_2} \tag{7.42}$$

or

$$V_2^2 - \frac{V_2 V_1}{N} + \frac{X_T Q_0}{N^2} = 0 \tag{7.42′}$$

Equation (7.42′) gives the mutual relationship among V_1, V_2, Q_0, and N, for a known X_T. If, for example, the zero reactive flow between two networks for specific values of

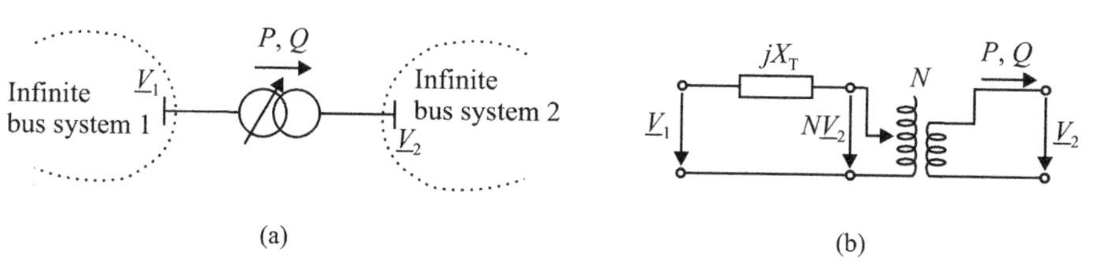

Figure 7.20. Interconnection of two strong networks through an OLTC: (a) one-line diagram; (b) equivalent impedance diagram.

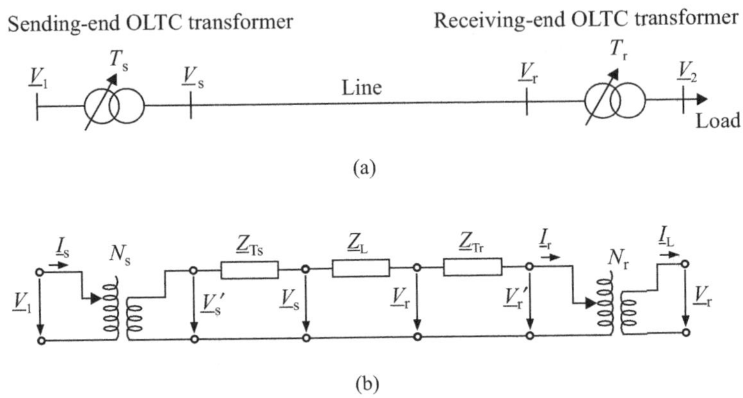

Figure 7.21. Radial transmission/distribution system with two cascaded OLTC transformers: (a) one-line diagram; (b) impedance diagram.

voltage magnitudes V_1 and V_2 is required, the transformer turns ratio is

$$N = \frac{V_1}{V_2}$$

that is, it must exactly match the actual voltages of two connected strong networks.

Radial Transmission/Distribution System with Two Cascaded OLTC Transformers. Very often, OLTC transformers are connected in series with the feeder line, in both the transmission and distribution networks, as shown in Figure 7.21 [11].

Let N_s and N_r be the turns ratios of OLTC transformers on the sending and receiving ends, respectively. It is of interest to determine the turns ratios N_s and N_r that will retain some specified relationship between steady-state magnitudes of voltages V_1 and V_2. With all quantities expressed in p.u., and considering the impedance diagram from Figure 7.21(b), the following relations can be written:

$$\underline{V}'_s = \frac{\underline{V}_1}{N_s}; \quad \underline{V}'_r = N_r \underline{V}_2$$

$$\underline{Z} = R + jX = \underline{Z}_{Ts} + \underline{Z}_L + \underline{Z}_{Tr}$$

The voltage drop in the entire system is

$$\Delta \underline{V} = \frac{\underline{V}_1}{N_s} - N_r \underline{V}_2 = \underline{Z}\,\underline{I}_r = (R + jX)\underline{I}_r = (R + jX)\frac{P_0 - jQ_0}{N_r \underline{V}_2^*} \tag{7.43}$$

Assuming $\underline{V}_2 = V_2 \angle 0°$ and neglecting the transversal component of the voltage drop, expression (7.43) becomes

$$|\Delta \underline{V}| \approx \Delta V = \frac{V_1}{N_s} - N_r V_2 = \frac{RP_0 + XQ_0}{N_r V_2} \tag{7.43'}$$

Rearranging the terms in equation (7.43′), gives

$$V_2^2 - \frac{V_1}{N_r N_s} V_2 + \frac{RP_0 + XQ_0}{N_r^2} = 0 \tag{7.44}$$

The solution of equation (7.44) is

$$V_2 = \frac{1}{2N_r N_s}\left(V_1 \pm \sqrt{V_1^2 - 4(RP_0 + XQ_0)N_s^2}\right) \tag{7.45}$$

Equation (7.45) has four variables (V_1, V_2, N_s, N_r) for a known load (P_0, Q_0) and a value of the system impedance ($R + jX$). Its application requires the additional specifications in view of variables involved. For example, if the requirement for equal magnitude of voltages $\underline{V}_1$ and $\underline{V}_2$ is specified (i.e., the complete compensation of the voltage drop in the system is required), it yields ($V_1 = V_2 = V$)

$$2VN_sN_r = V \pm \sqrt{V^2 - 4(RP_0 + XQ_0)N_s^2}$$

or

$$\frac{RP_0 + XQ_0}{V^2} = \frac{N_r}{N_s}(1 - N_sN_r) \tag{7.46}$$

then, for known R, X, P_0, Q_0, and V, the relationship between turns ratios N_s and N_r could be easily found.

7.2.4 Regulating Transformers

7.2.4.1 In-Phase Regulating Transformer (IPRT). A first type of regulating transformer is represented by "in-phase regulating transformer" or "booster," which are used to control the voltage amplitude (Figure 7.22).

Each phase has a winding that is series connected to the bus of which voltage is to be controlled, while the other winding is supplied by the same bus via an auxiliary transformer with variable turns ratio. Varying the turns ratio of the auxiliary transformer, the amplitude ΔV varies: the in-phase control action therefore varies the voltage amplitude, leaving the phase unaltered, as in the case of OLTCs.

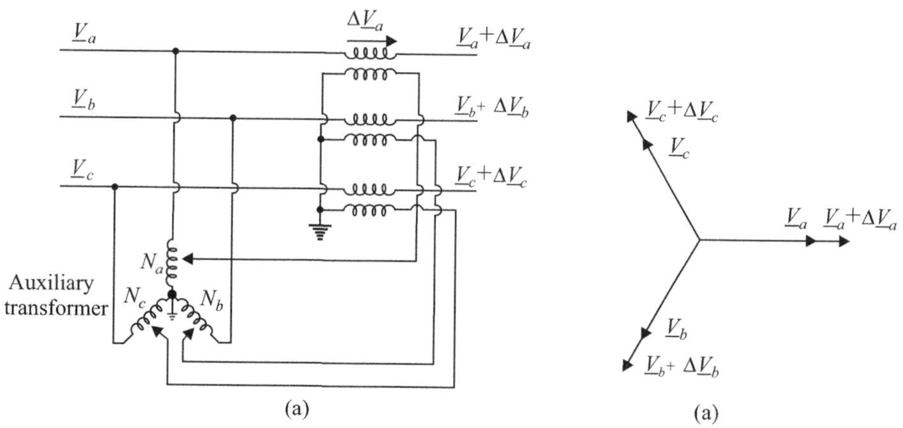

Figure 7.22. Basic booster scheme (a) and voltage phasor diagram (b).

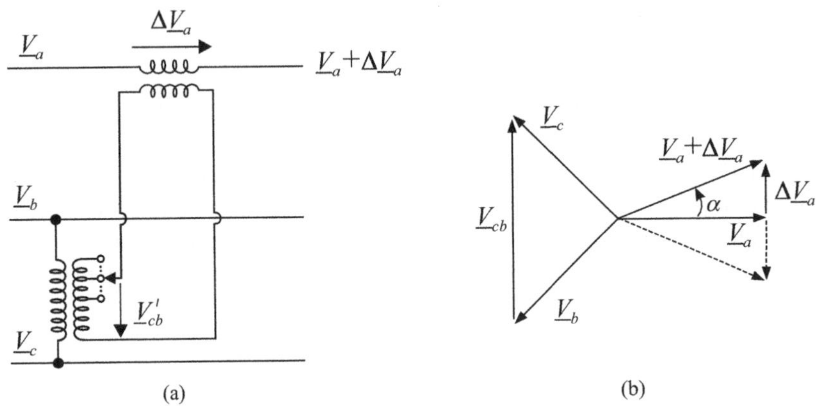

Figure 7.23. Basic scheme (phase "a") of a phase shifter and related voltage phasor diagram.

7.2.4.2 Phase Shifting Transformers. There are two types of phase shifters, corresponding to the schemes of Figures 7.23 and 7.24 (showing only phase "a", for convenience) [13].

(i) *In-Quadrature Regulating Transformers.* In the first type, the turns ratio variation of the auxiliary transformer causes a ΔV variation and thus a voltage phase variation.

This variation induces a phase shifting α (positive or negative) and a variation of the voltage amplitude (the smaller is the variation, the smaller is the phase shifting) between the voltages upstream and downstream of the transformer.

Therefore, these are complex-ratio transformers, similar to the ordinary three-phase transformers but with different connection modes of the primary and secondary windings and where $\alpha = \pm 30°$ (see Section 7.2.3.1).

From the phasor diagram of Figure 7.23b, we also deduce that the shifting α is approximately proportional to the variation ΔV of the voltage introduced by the transformer.

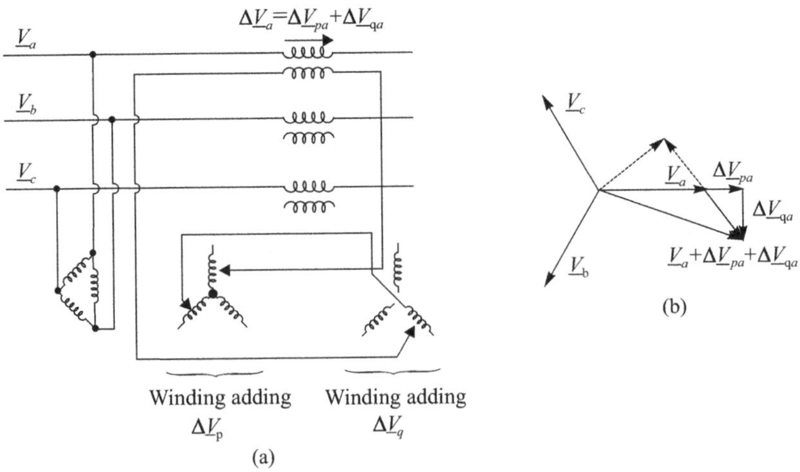

Figure 7.24. Basic scheme of regulating transformer for controlling voltage amplitude and phase [4].

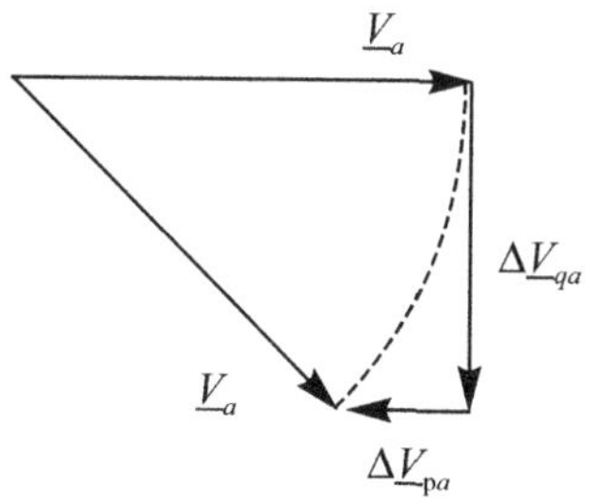

Figure 7.25. Phasor diagram of phase angle regulator.

(ii) *In-Phase and In-Quadrature Regulating Transformers (QBT).* This last type of regulating transformer controls the voltage amplitude and phase. It clearly embodies the previous two as exemplified in Figure 7.24 [13].

From the latter type, with appropriate measures, a purely phase angle regulator (PAR) may be derived, if the voltage amplitude is kept constant (Figure 7.25, phase "a" only).

Regulating transformers—with mechanically switches—which are series regulating transformers (see Figures 7.22–7.24), introduce an additional voltage between two nearby nodes at the same nominal voltage, that is, create an amplitude and phase voltage variation.

TABLE 7.1. Summary of Results (Adapted from Ref. 13)

IPRT	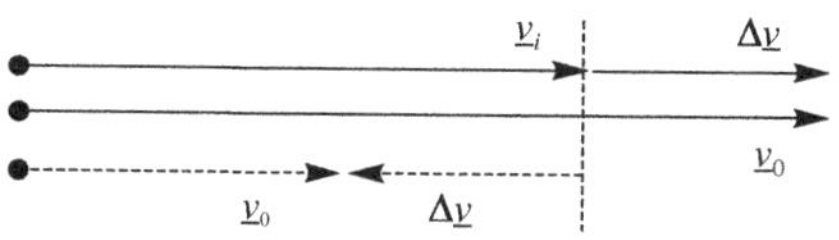	$v_0 \neq v_i$ $\alpha = \beta = 0$	Control the voltage in the point where they are installed and thus especially the reactive power flow
QBT	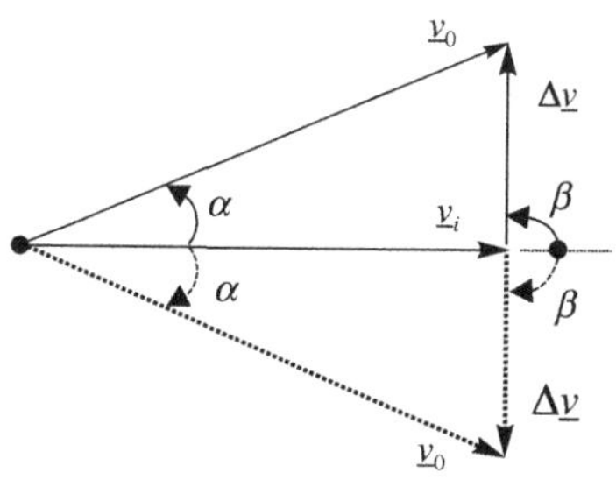	$v_0 \neq v_i$ $\beta = \pm\pi/2$ Variable α	Control of reactive and real power flows (due to a shift $\alpha \neq 0$)
PAR	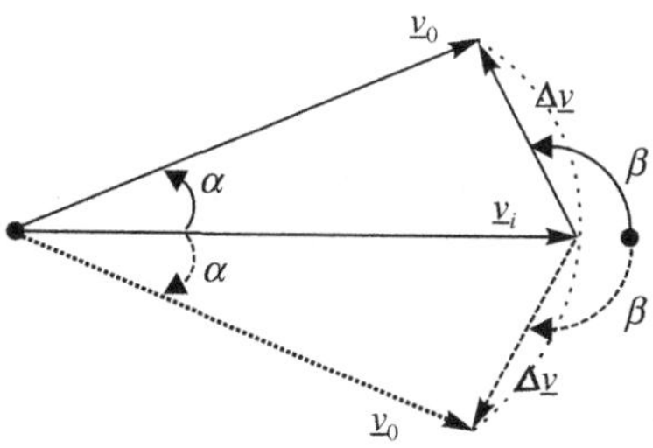	$v_0 = v_i$ variable β and α	

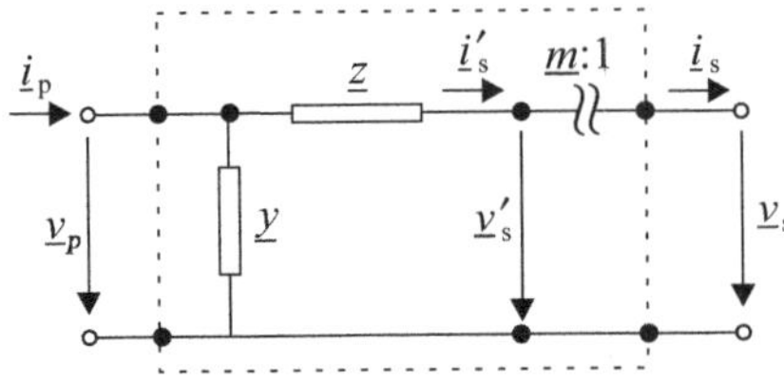

Figure 7.26. Single-phase equivalent circuit, in p.u., of a complex-ratio transformer.

Let us denote by

i, o the input and output terminals of the regulating transformer;
$\underline{v}_i$, $\underline{v}_o$ are the input and output voltage, in p.u.;
$\Delta\underline{v}$ is the additional voltage, in p.u.;
α is the phase displacement of $\underline{v}_o$ versus $\underline{v}_i$;
β is the phase displacement of $\Delta\underline{v}$ versus $\underline{v}_i$.

We, therefore, obtain in Table 7.1 the phase diagrams and the characteristics for the three types of regulating transformers.

Based on the above description, a regulating transformer may be represented by the single-phase equivalent circuit in p.u. (Figure 7.26), where the turns ratio, in p.u., is a complex ratio [13]:

$$\underline{m} = m\,e^{j\alpha} \tag{7.47}$$

The following values are appropriate:

- IPRTs: $m \neq 1$; $\alpha = 0$ as for OLTCs
- OBTs: $m \neq 1$; $\alpha \neq 0$
- PARs: $m = 1$; $\alpha \neq 0$

For three-phase transformers with different primary and secondary connection configurations, we have

$$m = 1;\ \alpha = \pm 30^\circ$$

7.3 GRID VOLTAGE AND REACTIVE POWER CONTROL METHODS

7.3.1 General Considerations

The control of grid voltages and reactive powers has become more critical in recent years due to the general trend by system operators and electrical utilities to operate the transmission networks as close as possible to their maximum capacity. The need for suitable control solutions capable to deal, in always tighter and meshed networks, with increased power loads and losses, possible grid contingencies and voltage collapse risk, has therefore grown. Nevertheless, the lack of real-time and closed loop "automatic" coordination of reactive power resources appears, in the common practice for network voltage control, persistent as much as unjustified.

Basically, the ways the system operator controls the grid voltages are of "manual" or "automatic" types or a combination of the two. Moreover, the transmission system operator controls the voltages on the EHV substations while the distribution system operator controls the voltages in the medium voltage and below networks. These two operators have different and complementary tasks as far as it concerns voltage and reactive power control.

In principle, the transmission system operator has the objective to impose the optimal voltage profile in the EHV system to achieve high transfer capability, minimum losses, and high-voltage stability. This objective can be achieved by controlling the available reactive power resources at that voltage level: injection/absorption of reactive power by the synchronized generators, switching on/off the compensating equipments, setting the voltage set points of the SVCs, STATCOMs, and OLTCs, blocking the OLTCs when the risk increases, and lastly paralleling the hot reserves, opening/re-energizing lines, shedding loads up to the very unconventional employment of the other system components, like for instance the phase shifters or UPFC used for voltage support.

The distribution system operator has, in principle, the objective to guarantee an adequate voltage level at the load side, again by controlling the reactive power resources available at the LV level: switching on/off the LV compensating equipments and setting the LV OLTC set points. Rarely, switching on/off the lines is used for this purpose.

The reactive power reserve in shunt static compensation and dynamic FACTS devices to improve the power factor of large loads is largely used to reduce reactive power flows from remote areas. Nevertheless, this solution is expensive and cannot be considered for the overall grid nodes, but only where strictly necessary. Therefore, the cheapest solution should be minimizing the reactive power injection distances and coordinating at the best the local reactive power resources. In this way, the effect of voltage drop due to the transmitting power at nonunity power factors is minimized, the losses reduced and the transmissible active power increased, with respect to the case of larger distance reactive power transfer. The ideal case, which allows the minimum losses, is obviously given by the artificial injection of reactive power at the loads themselves, by compensating at 100% the local reactive power demand. Even in this unrealistic situation, the compensating equipments (a very large amount) have to be continuously and widely controlled via a complex and therefore critical coordinating way. Coming back to a realistic situation, the compensating equipments are only a few in the transmission grid, while those applied at the distribution level are scheduled according to long-term load forecasting and switched on/off mostly by hands. When seen by the EHV grid, these compensating equipments are considered as part of the load and not as control variables. Apart that, these compensating equipments represent reactive power injections/absorptions at the voltage levels they are applied and the possible ways of their control is very important information for any grid voltage control proposal.

On the load shedding linked with voltage problems, this widely used practice is often overindulged while it should be linked to system protection needs, that is to be used in case of high system security risk only. Therefore, this should be the last control action to be actuated after having used at the best the other mentioned control variables.

Coming back to the transmission and distribution control centers, the way they are organized, their available control systems, their coordination, and the real ability to control the process, differ from one power system to another. Nevertheless the common understanding should be, in principle, on line with the following basics:

- The transmission system should take the major rule on the overall system voltage control because it is able to strongly affect the voltages of its fed distribution subgrids, while each of the low-voltage grids weakly affects the voltages on the transmission network. Therefore, the transmission grid should guarantee a high-quality and stable voltage control on the EHV and HV busses by an adequate timely and coordinated operation of the available reactive power resources at the amount and where required.

- Having robust and solid voltages on the transmission grid, it will be relatively simple to guarantee a high-quality voltage control at the load sides, by minimizing the control effort (switching) on the LV reactive resources and OLTC. Evidently, each distribution dispatcher should guarantee the fixed compensations, agreed with the transmission dispatcher at the planning stage.

According to that, the most advanced coordinated and fast control of the system voltages should be achieved by operating at the transmission level. This is the understandable reason why the voltage control has to distinguish the transmission network solution by the distribution network solution.

Without the transmission network voltage control, the distribution network has serious problems linked with voltage frequent variations; the OLTCs frequent (in few cases not sufficient) control to regulate by themselves the voltages at their low-voltage sides; the useless reactive power recirculation among the OLTC transformers operating in the same distribution area; the too slow OLTC dynamics with respect to the voltage decay speed; the OLTCs negative effects on the load voltages when approaching a voltage instability condition.

Hereinafter, we mainly refer to the transmission system voltage control due to its crucial importance.

Any equipment considered in Section 7.2 allows the voltage control at the bus it is installed. That is, each of the considered equipment is only able to control its local voltage or it is able to inject/absorb reactive power at the bus to which it is connected.

Considering the transmission network voltage control, the feasible objective is not to sustain the voltage of a single bus by itself, but the voltages in a given network area to increase the area operation security and efficiency and voltage quality. According to that, the grid voltage control does necessarily consists in a possible area voltage control by coordinating the area available equipments with the aim to sustain and optimize the main area load voltages in front of load variations and possible contingencies.

The grid control in general and the area voltage regulation in particular is a task of the grid dispatcher who defines his control strategy according to his organization, the available resources on field, the available system monitoring, and control solutions. Until today and notwithstanding a lot of operation problems are generally linked with voltage value changes, most of the system operators control the power system voltages by hands (manual control: written instructions, telephone calls, and remote telecontrols). Starting from this minimal system operation controls, there are round the world a wide variety of mixed manual and automatic solutions up to reach to the full automatic voltage control system.

The main issue of this section is the presentation of some of these area automatic voltage control systems.

Another relevant aspect to put in evidence when referring to an automatic voltage control is the difference between

- continuous voltage and reactive power control, also called voltage regulating system;
- discontinuous, extreme shut, voltage and reactive power control, also called voltage-protecting system.

The continuous voltage control operates at any instant to sustain the system voltages by controlling the system reactive power resources with continuity at the amounts required to maintain the voltages at the desired values. According to its nature, this control is of automatic type. The available resources allowing the continuous control are the generators, the SVCs, and the FACTs in general. The other components like the compensating equipments or the OLTCs do allow a discrete control because of their switching actuation but not continuous-discrete due to the limited number of the switching maneuvers. This discrete control can be combined with the continuous control of the generators, SVCs, STATCOMs, and so on [14], to realize the new advanced wide area voltage regulating systems (V-WAR) [15–23].

The discontinuous step control of the voltage and reactive power at a given section of the network is often a drastic control determining the loosing of the continuous feeding of a given load or a generator tripping or lines opening in front of a given sequence of events. Usually it is a protecting control tailored to a given event in a given part of the network, requiring fast voltage/reactive power measurements to recognize on time a dangerous incoming decay. This kind of protecting automatic control is always ready but never works unless given thresholds are overcome. These new advanced protecting controls are also called special protection scheme (SPS) or remedial control scheme (RCS) or wide area protection (WAP) when related to a given system area [24,25].

The regulating and protecting voltage control schemes are not alternative solutions but they complement each other when operating on the same grid area:

- The regulating control continuously works to maintain the system far from the voltage instability and with a voltage plan optimized to the current working conditions.
- The protecting control operates only when the regulating system has performed its maximum effort by using all the available reactive power resources to reach its saturation. It determines the loosing of part of the system/load to achieve the security of the remaining process maintained into operation.

Note: The protective voltage control is outside the scope of this chapter.

7.3.2 Voltage–Reactive Power Manual Control

The "manual" practice for grid voltage control, until now largely used by system operators worldwide, typically consists of the TSO or ISO control centers dispatching the forecasted reactive powers of the generating units, scheduling the power plants high-side voltages, switching the shunt capacitors or reactors banks, and setting the voltage set points of OLTC and FACTS controllers (usually by written rules). This conventional approach for solving the network voltage control problem does include also the operator real-time decisions based on the system monitoring and thresholds alarms and operated through commands sent by telephone or by telecommands, in a kind of "manual coordination" of the available reactive power resources and bus voltage controllers (those mentioned in Section 7.2).

This grid voltage control is nowadays considered quite unsatisfactory because

- units reactive power dispatching and plants high-side voltage scheduling are based on off-line forecasting study: actual network operating conditions are often different then their forecasted values and unpredictable;
- voltage set point coordination is often operated through written requirements or requested by system operator only when strongly needed; therefore, untimely or

inadequate control actions may occur during most of the system dynamic phenomena;

- the skills and ability of the operator to timely recognize and correctly face the grid problems have to be very high and continuously put to a severe test.

It often happens that the operators are familiar with repeating phenomena they face through tested countermeasures but in front of novelties they often fail. Moreover the operators, in general, are more worried about the system security with respect to the system efficiency and optimization, which are overlooked.

7.3.2.1 Manual Voltage Control by Reactive Power Flow. The possible ways the system operator can control the reactive power flowing through a given line are only few and of approximate effect. Basically the manual control has to modify the voltage difference at the ends of the considered line or equivalently modifying the reactive power delivery/absorption nearby one or both the ends of the line. This can be achieved by operating at the line ends by

- changing the transformation ratios of OLTCs;
- switching on/off the compensating equipments;
- changing the voltage set point of the local generators/synchronous compensators or their reactive power delivery/absorption;
- changing the voltage set point of the local FACTS.

Due to the different operating conditions and the interaction of the considered line with the surrounding grid, the results corresponding to a given control may differ, often requiring an iterative process to approximate the desired result.

Usually, the manual tracking is too slow with respect to the system needs.

Generally speaking, the control of reactive power flow is needed in more than one line and the multiflows control problem is very complex and with low probability of success when faced manually.

7.3.2.2 Manual Voltage Control by Network Topology Modification. An extreme way to control voltages, possible but generally not used, does consists in network topology modification based on the increasing or reducing the capacity effect provided by the electrical lines. In this case, the control strategy basically consists in switching-off some low charged lines in an area with high voltages and conversely switching-on lines in case of low voltage, after using all the other local reactive power resources. As before, the result corresponding to a given control switching may change in accordance with the specific grid operating conditions, therefore this control goes in the expected direction but its effect is not easily predictable.

7.3.3 Voltage–Reactive Power Automatic Control

The classic automatic voltage controls operating in the power systems are those provided by the generators' AVRs that maintain the stator voltages at the set point values by a fast closed loop control of the generators excitation. Therefore, changing the load seen by the generator or changing the AVR voltage set point value, the reactive power delivered or absorbed by the generator changes inside the limiting field defined by the OEL and UEL limits. This classic control, having a generalized application around the world, mainly contributes to the generator safe and stable operation but not enough to the EHV grid voltage support, even if the generator available reactive power resources could allow more.

For improving voltage control in transmission grids, the utilities and system operators have followed many approaches and many projects have been developed around the world.

Basically, most of them do consider neither the voltage–reactive power automatic control nor the continuous operator control of the generator AVRs due to the practical difficulties through this way. In most cases, the adopted approach is limited to the power factor correction based on off-line planning studies, by increasing the installed power in shunt capacitors or reactors banks, with large related investments. The switching of these components, when allowed, could be required by an automatic control system under many constrains on the amount of the maneuvers needed by the components life preservation.

Availability, in some cases, of unit step-up transformers equipped with OLTC, represents an additional chance for network voltage control, provided that their regulation system supports plant EHV side instead of generator stator terminals. Nevertheless their control is usually manual as well as stepping and slow when automated.

The solution of automatically supporting the power plant high-side voltage through AVR line drop compensation is also commonly used. This practice increases grid voltage support but introduces destabilizing interactions between primary voltage regulators (PVRs).

Use of FACTS controllers for automatic network voltage support, mainly SVC and STATCOM, has been seriously considered in the last years, even if the related costs do not always justify their choice and, if extensively applied, they require a coordinated control system similar to that described in Section 7.4.

Recently, under the push of the on-going market liberalization, some AVRs manufacturers offer AVR including the unit reactive power control or power factor control; in some cases, this is for the plant high-side voltage control.

The generator AVR controls and its evolution up to the high-side voltage control is presented hereinafter, being the most representative, together with FACTS, of a true, effective, fast, and continuous bus voltage–reactive power automatic control.

Moving from the single bus to the area voltage–reactive power control, Section 7.4 largely develops this new subject on grid automatic voltage regulation.

7.3.3.1 Automatic Voltage Control of the Generator Stator Terminals. The voltage regulation of the synchronous generator involves controlling the component voltage with the major objective of its correct and secure operation. This control (see Figure 7.27) is obtained by an AVR, based on local voltage fast measure, aiming to sustain the voltage at the set point value automatically, with dynamic performances characterized by a dominant time constant value within a few hundred of milliseconds up to one second, depending on the exciter characteristics.

Obviously the AVR control impacts with the transmission network, mainly at the bus the generator is connected, by sustaining the local medium voltages during normal and perturbed operating conditions, particularly during short circuits.

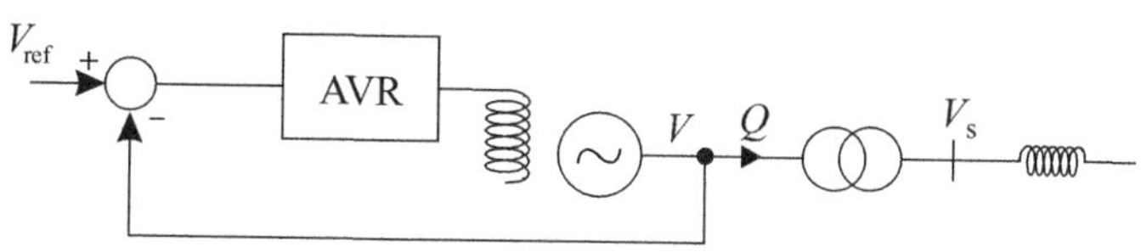

Figure 7.27. Schematic diagram of the generator voltage control loop.

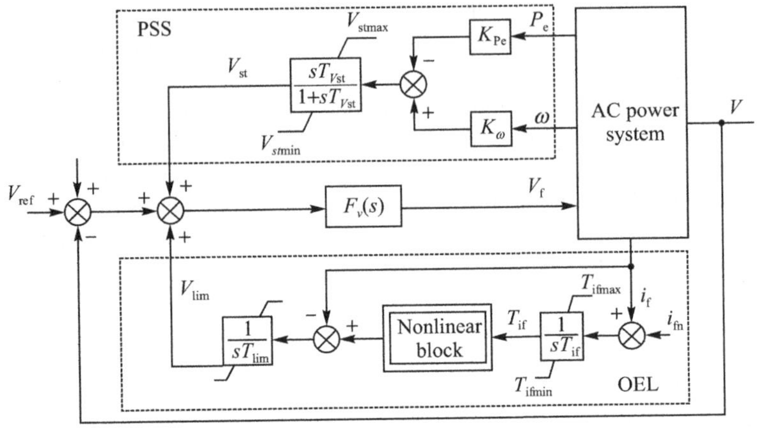

Figure 7.28. Block diagram of the AVR with evidence of voltage regulator control block $F_v(s)$, the overexcitation limit and the power system stabilizing feedbacks.

The AVR regulates the voltage at the generator's terminal by controlling the field excitation voltage V_f (Figures 7.27 and 7.28) [4,13,19].

The AVR realizes the primary voltage control of a generator. The AVR apparatus does also generally include the OEL and UEL limits, the stabilizing feedbacks PSS (power system stabilizing feedbacks) and the line drop compensation feedback (see Section 7.3.3.2).

In Figure 7.28, the main signals that appear in the control loop are as follows:

V_{ref} is the primary control loop's voltage reference value;
V is the measured value of the voltage at the generator's terminal;
V_{lim} is the control signal provided by the overexcitation limiting loop;
V_{st} is the additional signal provided by the PSS.

In normal operating conditions the OEL and PSS are open loop feedbacks (they do not reclose the correspondent V_{lim} and the V_{st} signal feedbacks), therefore the operating model of AVR is given by the $F_v(s)$ transfer function alone.

To simplify the analysis, the machine's excitation system is considered energized from an independent supply. Under this circumstance, the AVR's dynamics can be simply determined by a $F_v(s)$ control law with high frequency zeros $(1/Tz)$ and poles $(1/Tp)$ that allow the controlling range to be very wide, while high values of the amplification factor μ_0 at low frequencies allow a quasi-null steady-state regime error. $F_v(s)$ is of the

- third order: $F_v(s) = \mu_0(1+sTz_1)/(1+sTp_2)(1+sTp_3)(1+sTp_4)$ for systems with exciting dynamo or with alternator and rotating diodes. Rough values for the primary voltage control parameters are given in Table 7.2 [19].
- second order: $F_v(s) = \mu_0(1+sTz_1)/(1+sTp_2)(1+sTp_3)$ for systems with exciting dynamo and modern electronic voltage regulator. Rough values for the primary voltage control parameters are given in Table 7.2.
- first order: $F_v(s) = \mu_0(1+sTz_1)/(1+sTp_3)$ for static excitation systems, in absence of excitation current feedback. Rough values for the primary voltage control parameters are given in Table 7.2.

TABLE 7.2. The Parameters of the Primary Voltage Control Scheme

Parameter	$F_v(s)$ Order: First	$F_v(s)$ Order: Second	$F_v(s)$ Order: Third	Measure Unit
μ_0	400	3000	500	p.u./p.u.
Tz_1	0.8–1.5	2	2	s
Tp_2	–	0.05	0.05	s
Tp_3	5–20	200	20	s
Tp_4	–	–	0.02	s
T_{vst}			3	s
K_{pe}	0.15	0.15	0.15	p.u./p.u.
K_ω	15	15	15	p.u./p.u.
V_{stmin}	−0.05	−0.05	−0.05	p.u.
V_{stmax}	0.05	0.05	0.05	p.u.

LINEAR ANALYSIS OF THE GENERATOR VOLTAGE CONTROL LOOP. The $F_v(s)$ control parameters values are defined to achieve given voltage control loop steady-state accuracy as well as a proper dynamic behavior in terms of stability and response speed. A good accuracy requires a high static gain while the loop cutoff frequency is representative of the control speed.

The standard analysis of the generator voltage control loop refers to the case of a generator feeding an infinite bus or a load Z_C through the step-up transformer of reactance X_T and a line of reactance X_L. The equivalent scheme in Figure 7.29 represents the described system.

Considering the generator model referred to the d- and q-axes, according to the Park transform:

$$\begin{aligned} v_q &= a(s)v_f - x_d(s)i_d \\ v_d &= x_q(s)i_q \end{aligned} \tag{7.48}$$

Let us now refer to the Z_C load case and suppose that the load is linear, purely reactive and with impedance $Z_C \cong jX_C$, meaning that the machine operates as a compensator (delivered real power equal to zero). So, the stator voltage is entirely on the quadrature axis, that is

$$\begin{aligned} V_d &= 0; \quad V_q = V \\ I_d &= I; \quad I_q = 0 \end{aligned}$$

As a consequence, the operational reactance $x_q(s)$ from

$$V_d = x_q(s)I_q$$

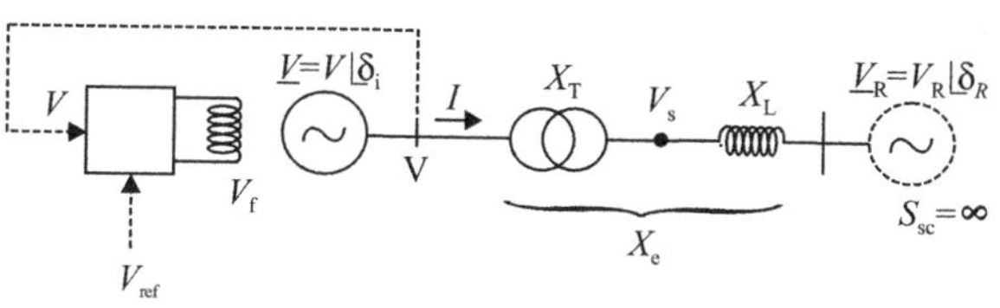

Figure 7.29. Generator connected to an infinite power bus.

is noninfluential. Conversely, as regards $x_d(s)$, let us assume that only the field circuit acts along the circuit axis.

The relationship between the stator current, voltage, and the excitation voltage is thus defined by (see also [13]):

$$V_q = \frac{1}{1+sT'_{d0}} V_f - x_d \frac{1+sT'_d}{1+sT'_{d0}} I_d \tag{7.49}$$

From (7.49), the linearized model is obtained:

$$\Delta V = \Delta V_q = \frac{1}{1+sT'_{d0}} \Delta V_f - x_d \frac{1+sT'_d}{1+sT'_{d0}} \Delta I_d \tag{7.50}$$

From $\underline{I} = (\underline{V} - \underline{V}_R)/jX_e$, it results

$$I_d = \frac{V_q - V_R \cos(\delta_R - \delta_i)}{X_e}$$

For $V_R =$ constant and $\delta_R - \delta_i \cong 0$, gives

$$\Delta I_d = \frac{\Delta V_q}{X_e} = \frac{\Delta V}{X_e} \tag{7.51}$$

Furthermore, taking into consideration equation (2.144) (see Section 2.1.6.1)

$$x'_d \triangleq x_d \frac{T'_d}{T'_{d0}} \tag{2.144'}$$

from equation (7.50) it results

$$\Delta V = \frac{x_e}{x_e + x_d} \cdot \frac{1}{1 + ((x_e + x'_d)/(x_e + x_d))sT_{d0}} \Delta V_f \tag{7.52}$$

or

$$\Delta V = \frac{h_e}{1+sT_e} \Delta V_f \tag{7.53}$$

with

$$\begin{aligned} h_e &\triangleq \frac{x_e}{x_e + x_d} \\ T_e &\triangleq T'_{d0} \frac{x_e + x'_d}{x_e + x_d} \end{aligned} \tag{7.54}$$

The block diagram from Figure 7.30 represents equation (7.53).

It should be noted that a similar expression to (7.53) can be obtained also when the generator feeds a load that is also resistive (and thus the real power delivered is different

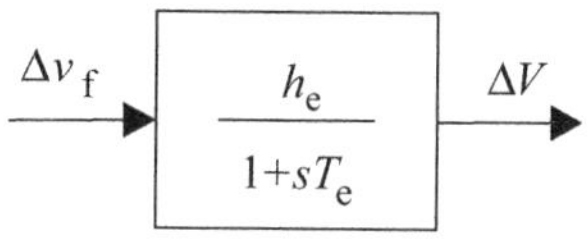

Figure 7.30. Linearized block diagram linking the generator and the excitation field voltages.

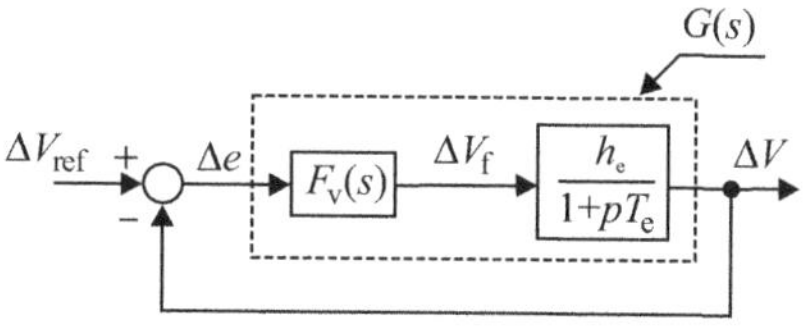

Figure 7.31. Block diagram of the primary voltage control loop of an on-load generating unit.

than zero) and when the generator is connected to an infinite bus system [13]. Taking into account equation (7.53), the block diagram of the voltage control loop of a generating unit is as displayed in Figure 7.31, whose forward transfer function is

$$G(s) = F_v(s) \cdot \frac{h_e}{1 + sT_e} \tag{7.55}$$

For static exciter, $F(s) = \mu_T = \mu_0 Tz_1/Tp_3$ (static gain of the voltage regulator amplifier) is a good approximation. This control loop usually designed to have a cutoff frequency of about 4/5 rad/s achieved with a high μ_T value that also guarantee a very low steady-state error.

Steady-State Operation. Under steady-state operating conditions ($s = 0$), by indicating the values of the various quantities with superscript "°", equation (7.52) becomes

$$\Delta V^\circ = x_e/(x_e + x_d) \cdot \Delta V_f^\circ$$

while

$$\Delta V_f^\circ = F_v^\circ(s)\left(\Delta V_{ref}^\circ - \Delta V^\circ\right) = \mu_0\left(\Delta V_{ref}^\circ - \Delta V^\circ\right)$$

therefore, the error e° in Figure 7.31 assumes the following value:

$$e^\circ = \left(\Delta V_{ref}^\circ - \Delta V^\circ\right) = \Delta V^\circ (x_e + x_d)/x_e \mu_0$$
$$e^\circ(\text{p.u.}) = \left(\Delta V_{ref}^\circ - \Delta V^\circ\right)/\Delta V^\circ = (x_e + x_d)/x_e \mu_0$$

To obtain a high steady-state accuracy on the voltage control loop, e° (p.u.) has to be lower than a very small quantity ε:

$$(x_e + x_d)/x_e \mu_0 < \varepsilon$$

Wishing to obtain a steady-state voltage error lower than 0.5% when passing from no-load to full load operation ($\varepsilon < 0.005$) and assuming $x_e \cong 1.0$, the static gain μ_0 assumes values of about

- $\mu_0 > 200\,\text{p.u./p.u.}$ for hydro units;
- $\mu_0 > 400\,\text{p.u./p.u.}$ for thermal units.

It is worth noting that the high static gain in the voltage regulators provides, at low frequencies, an effect similar to an integral control law. Therefore, in the absence of additional effects provided by the PSS, line drop compensation, OEL, or UEL, the generator terminal voltage is practically equal to its set point value.

Dynamic Behavior. Using the classic Bode diagrams of amplitudes and phases of the open loop transfer function, the stability and speed of response of the voltage control loop can be easily achieved.

As an example, reference can be made to the *hydraulic unit* with

$$x_d = 1.0\,\text{p.u.};\quad x'_d = 0.3\,\text{p.u.};\quad T'_{d0} = 7.0\,\text{s}$$

Therefore, $h_e = 0.5$ p.u. while $T_e = 4.55$ s.

In case of rotating exciters with modern voltage regulator, the second-order $F_v(s)$

$$F_v(s) = \mu_0(1 + sTz_1)/(1 + sTp_2)(1 + sTp_3)$$

with static gain $\mu_0 = 1000$ p.u./p.u., and the time constant values in Table 7.2 determines a control margin of about 90° and a cutoff frequency near to 2 rad/s.

In the case of static exciters of modern turboalternators, no one stability problem is recognized as here after shown.

As a different example, reference is now made to the *turboalternator* unit with

$$x_d = 2.0\,\text{p.u.};\quad x'_d = 0.3\,\text{p.u.};\quad T'_{d0} = 7.5\,\text{s}$$

Therefore, $h_e = 0.33$ p.u. while $T_e = 3.3$ s. Such quantities are thus included in the ranges:

$$0.33 < h_e < 1.0;\quad 3.2 < T_e < 7.5$$

where the upper bounds concern the no-load condition while the lower ones refer the full load operation. As a consequence the time constant T_e is always higher than the time constant Tz_1 and lower than the time constant Tp_3 of the first order $F_v(s)$:

$$F_v(s) = \mu_0(1 + sTz_1)/(1 + sTp_3)$$

As the cutoff frequency is certainly greater than $1/Tz_1$ and thus much greater than $1/T_e$, use of the following high-frequency approximation of $G(s)$ can be done:

$$G(s) = \mu_0(Tz_1/Tp_3)h_e/sT_e$$

which means that the loop control margin is of the order of 90°.

The loop cutoff frequency ω_v can be directly derived from

$$|G(s)| = 1.0 \cong \mu_T h_e/\omega_v T_e$$

which gives $\omega_v = 1.0 \cong \mu_T h_e/T_e$.

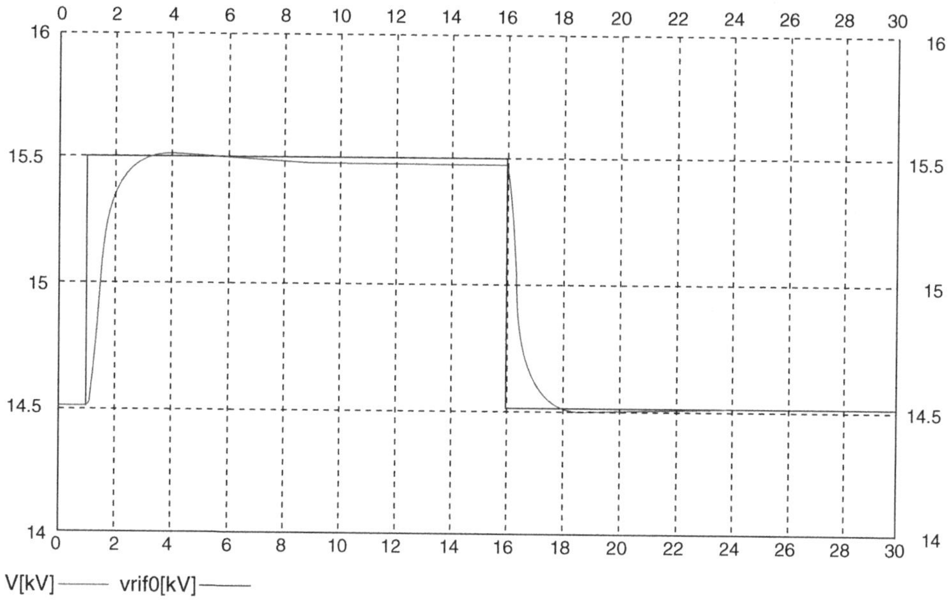

Figure 7.32. Generator voltage control loop transients following AVR set point steps.

The transient gain μ_T, also called "dynamic gain", is thus responsible for the response speed of the AVR control loop. The higher is μ_T the faster is the control loop. Taking into account that the usual value of T'_{d0} is 7.0 s, the transient gain μ_T should be close to 50.0 p.u./p.u. According to that, the parameters of the static exciters should have the following values:

$$\mu_0 \cong 400.0\,\text{p.u./p.u.}; \quad Tz_1 \cong 1.5\,(\text{s}); \quad Tp_3 \cong \mu_0/4T'_{d0}\,(\text{s})$$

Based on these values, in response to a step variation of the voltage set point, the generator voltage reaches the new imposed value, with near zero steady-state error, in an aperiodic way after about 1–2 s, as shown in Figure 7.32.

7.3.3.2 Automatic Voltage Control by Generator Line Drop Compensation.

Objective of the Compounding. This closed loop automatic control, at the generator level, consists of a reactive power feedback (α_c, called compound factor) on the input of the generator AVR (Figure 7.33). The objective is to increase the support of the local HV bus bar voltage (V_s). Therefore, it is an automatic voltage control more in favor of the grid that changes the principle of the AVR totally dedicated to the generator voltage. The analysis of

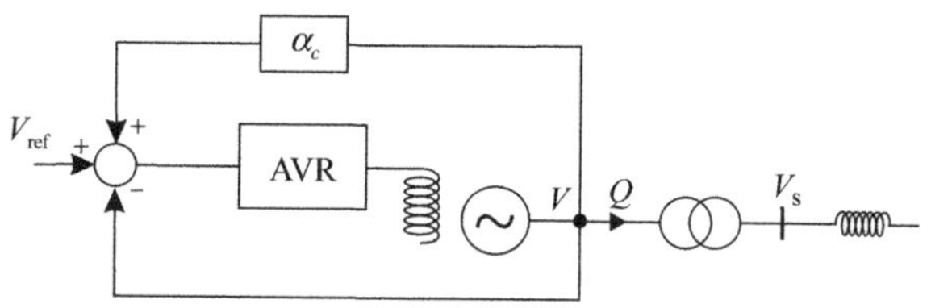

Figure 7.33. Schematic control diagram of the generator line drop compensation.

this additional feedback, overlapping the classic generator voltage control loop, requires the transfer function model of the generator under the AVR control, before described.

LINK BETWEEN VOLTAGE AND REACTIVE POWER. Consider the equivalent scheme in Figure 7.29 of a generator connected to an infinite bus.

The reactive power provided by the generator is given, with obvious meaning of the symbols, by

$$Q = \frac{V^2 - VV_R\cos(\delta_R - \delta_i)}{X_e}$$

Now assuming $P = 0$ (the machine operates as a reactive power compensator) and $\delta_R = \delta_i$:

$$Q = \frac{V^2 - VV_R}{X_e}$$

and with $V \cong V_R \cong 1$ p.u. in the considered operating point achieve (Figure 7.34):

$$\Delta Q \cong \frac{\Delta V}{X_e} \tag{7.56}$$

This relation is a good linear approximation also when $P \neq 0$ and $V \approx V_R \approx 1$ p.u. (Figure 7.34).

LINE DROP COMPENSATION (COMPOUNDING). As seen in Figure 7.33, the compounding is a reactive power feedback on the generator voltage control loop represented by the block diagrams from Figures 7.30 and 7.31. This simplified model and the considered hypotheses on it can also be correctly assumed for the analysis of the line drop compensation control loop.

Therefore, the following block diagram gives the understanding of the dynamic link of the generator voltage set point with the reactive power delivered by the generator (Figure 7.35).

The term $\alpha_c\,\Delta Q$ is the compound signal, while the coefficient α_c represents the compound factor. Moreover, when $\alpha_c > 0$ the compounding is "positive" while it is "negative" with $\alpha_c < 0$.

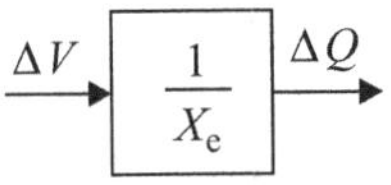

Figure 7.34. Linearized block model linking the generator voltage and reactive power variations.

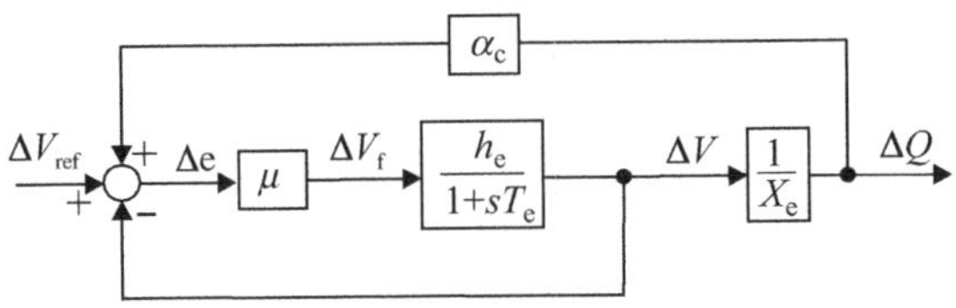

Figure 7.35. Block diagram of the line drop compensation control loop.

With enough high value of μ, in steady state, the error $\Delta e \approx 0$, therefore

$$\alpha_c \Delta Q + \Delta V_{ref} - \Delta V = 0$$

With constant voltage set point $\alpha_c \Delta Q - \Delta V = 0$, then the coefficient α_c is

$$\alpha_c = \Delta V / \Delta Q \tag{7.57}$$

Clearly α_c has the dimension of a reactance. So, under steady-state conditions, the generator stator voltage is given by

$$V = V_{ref} + \alpha_c Q \tag{7.58}$$

The equivalent circuit in Figure 7.36a represents equation (7.58) while Figure 7.36b represents the generator connected through the step-up transformer to the HV bus bar. Hence, Figure 7.36c is obtained from Figure 7.36a and b, by assuming that α_c is lower than the reactance x_T. Under steady-state conditions, the point where the voltage amplitude is kept constant at the V_{ref} value is thus inside the step-up transformer.

In this real hypothesis and upon the variations of the reactive power Q, the voltage drop in a portion of the step-up transformer is completely offset, just as if its reactance had diminished from x_T to $x_T - \alpha_C$. Of course, if $\alpha_C = x_T$, the entire drop in the transformer will be offset. In other words, the controlled voltage will be V_{HV} at the HV terminals of the transformer, instead of the voltage at the generator terminals as in the case without compound action. To be pointed out from now on, the V_{HV} compounding control cannot be achieved in practice because, as will be shown later, it generally corresponds to an unstable operating point.

According to the positive or negative value of α_c, equation (7.58) provides the linear voltage–reactive power steady-state characteristic (the solid line in Figure 7.37) that gives a clear view of the line drop compensation effect.

With line drop compensation the reference voltage of the generator AVR is therefore the voltage at the stator terminals under no-load conditions (zero reactive load) only. By analogy

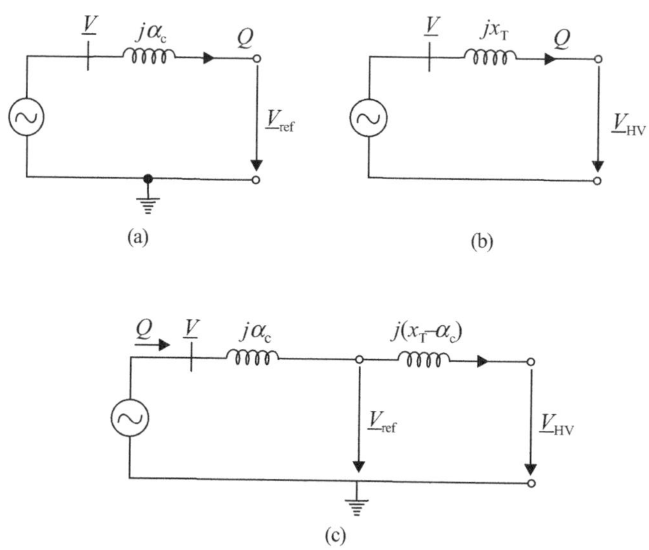

Figure 7.36. Equivalent circuit with compound (a) and in the presence of the step-up transformer with compound (b and c).

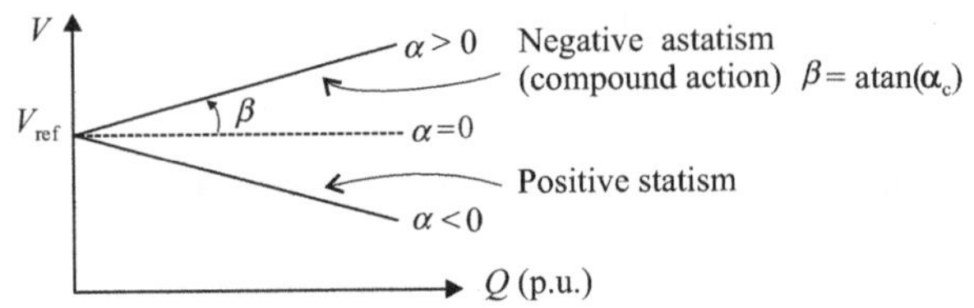

Figure 7.37. Voltage–reactive power steady-state characteristics with line drop compensation.

with the turbine steady-state characteristics (governor speed–real power), the compound effect is equivalent to the introduction of a negative drop or "statism" in the voltage regulator, since the presence of a negative compounding (positive statism) would mean decreasing stator voltage while the load increases, that is, assuming a negative α_c value. Vice versa, a positive value of α_c provides a positive compounding (negative statism).

As to the values of α_c, since $V_{max} = 1.05 - 1.1$ p.u. and assuming $\underline{V}_{ref} = 1.0$:

- at full real power ($Q_{max} \approx 0.5$ p.u.): $\alpha_c \approx 0.1 - 0.2$ p.u.;
- at the technical minimum ($Q_{max} \approx 0.7$ p.u.): $\alpha_c \approx 0.07 - 0.14$ p.u.

Hence, even a x_T reactance value greater than ≈ 0.13 p.u. might be offset. By adopting the most conservative condition ($\alpha_c \approx 0.07 - 0.14$ p.u.), the offsetting will be anyhow between $x_T/2$ and x_T and, as will see hereinafter, large offsetting will compromise the generator voltage control loop stability.

On the contrary, the consequence of the x_T offsetting is an increase in the short-circuit power of the power plant HV buses where the compound action is introduced. This increase makes the generator node stronger.

Line Drop Compensation and Stability. If the reactance x_T of the step-up transformer would completely offset ($\alpha_c = x_T$), this action would be equivalent to a high-side voltage control, where the generator would regulate the power plant HV bus voltage.

Unfortunately the line drop compensation does not allow the high-side voltage control. To better understand this point we have to refer to the possible power station configurations given in Figure 7.38.

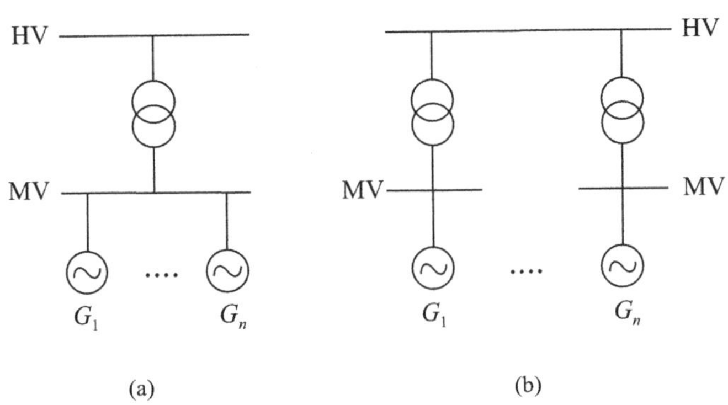

Figure 7.38. Generators parallel–connected to (a) MV bus; (b) HV bus.

The first case: In power plants with low-capacity units, all the units may be parallel connected to a single MV bus (Figure 7.38a). In this instance, since the voltage to be controlled is the same for all the units and assuming the horizontal voltage–reactive power steady-state characteristic (zero statism), than the compounding is out of operation and there will be "n" AVRs working in parallel to regulate the same MV bus bar voltage. This working condition is unstable for "real pole." Moreover, the distribution of reactive power among the various generators is indeterminate under steady-state conditions, but this consideration is not relevant in practice due to the mentioned instability. Similarly in the case of positive compound, because there will be "n" AVRs working in parallel to regulate the voltage of the same internal point of the MV/HV transformer, the high-side voltage control is not allowed. In conclusion, the stability in this case compulsorily requires the use of a negative compound at each AVR, in the way each generator regulates the voltage at an internal point of the generator stator windings reactance. This determines, with appropriate offset value of that reactance, enough electrical distance among the points inside the generators, having voltages automatically regulated by the plant AVRs.

Conclusion: $\alpha < 0$ required by generators with stator edges in parallel.

The second case: In power plants with high-capacity units, the generators are usually parallel connected to a single HV bus (Figure 7.38b).

In this scheme, the previous stability drawback is overcome, since each voltage regulator is sensitive to the corresponding voltage of its MV bus bar and the reactances of step-up transformers provide an adequate electrical distance among the MV busses. In this instance, the positive compound is now possible but the step-up transformers reactances should not be fully offset (high-side voltage control not allowed), because otherwise V_{HV} would be kept constant at the value of V_{ref} by the "n" AVRs with horizontal steady-state characteristics (V_{HV}, Q) and this determine voltage instability due to parallel voltage control loops. Obviously, with partial offset, the higher is the positive compounding the lower is the stability margin. This is also the reason why the compound factors should be within the range mentioned before: $\alpha_c \approx 0.07 - 0.14\,\text{p.u.}$

For assessing the compound effect on the stability of a generator primary voltage control loop, when operating alone at a power station, the case of a unit feeding a reactive load is considered (refer to Figure 7.29), having in Figure 7.35 the corresponding dynamic model of the line drop compensation control loop. In this case, the equivalent external reactance seen by the generator stator terminal is considered: $X_e = X_T + X_L + X_c$.

The block diagram in Figure 7.35 becomes as shown in Figure 7.39a or in its equivalent representation in Figure 7.39b.

Thus, for $1 - \alpha_c/x_e < 0$ the feedback becomes positive with the consequent instability (due to real pole) of the voltage control loop. On the contrary, with the feedback gain positive but lower than 1.0, the speed of the voltage control loop is reduced due to the loop lower static gain.

In practice, with $\alpha > 0$, the used compensation is usually less than 50% of the transformer reactance. Therefore, the compounding does not allow the real voltage control of the HV bus side, while it contributes to its objective.

Often, the case illustrated in Figure 7.38b shows, in practice, a lot of negative compounding! This is done for improving the AVR stability margins despite of supporting the local HV side bus voltages, but so doing the compounding fails its original objective being less effective than the classic generator voltage control loop on sustaining V_{HV}.

LINE DROP COMPENSATION SIMPLIFIED FEEDBACK. From equation (7.58), the compounding feedback comes from the reactive power. Often, in practice, the feedback comes from

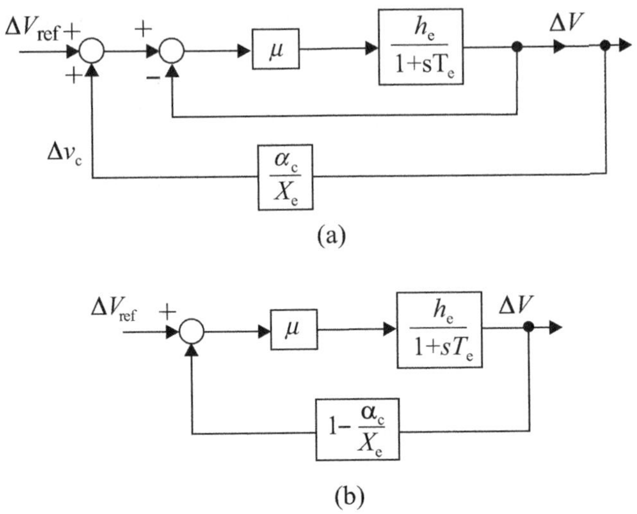

Figure 7.39. Block diagram of the generator voltage control loop of a unit with compound action (a) and the equivalent (b).

$I \sin \varphi$, therefore that equation becomes

$$V = V_{ref} + \alpha_c \frac{Q}{V} \tag{7.59}$$

or

$$V^2 - V_{ref} V - \alpha_c Q = 0$$

then

$$V = \frac{V_{ref}}{2} + \sqrt{\left(\frac{V_{ref}}{2}\right)^2 + \alpha_c Q} \tag{7.60}$$

In agreement with (7.58), this equation shows that if the reference voltage V_{ref} and the compound factor α_c are constant for all operating conditions, then the voltage at the alternator terminals increases with the delivered reactive power.

If the compound signal is

$$V_C = \alpha_c \cdot \frac{Q}{V} \tag{7.61}$$

and

$$Q = \frac{V^2}{x_e}$$

then

$$V_C = \frac{\alpha_c}{x_e} \cdot V \tag{7.62}$$

Linearizing, it can be obtained

$$\Delta V_C = \frac{\alpha_c}{x_e} \cdot \Delta V \tag{7.63}$$

This link is the same as in Figure 7.39a. Therefore, the dynamic analysis results and the stability conclusions reached with the feedback from Q are still valid when the feedback comes from $I \sin \varphi$.

7.3.3.3 Automatic High-Side Voltage Control at a Power Plant.

PRINCIPLE SCHEME. An innovating power plant automatic voltage control (HSVC) of the local, HV side bus bar, is achieved (Figure 7.40) through a nonconventional power station control able to coordinate the reactive powers of the operating generators in the plant.

This closed loop control is able to maintain the voltage V_s at the imposed value $V_{s,ref}$ by a continuous coordinated control of the reactive powers of the operating generators in the plant. The analysis of the control scheme requires an appropriate model of the process.

MODEL OF THE POWER PLANT. Considering a power plant with "n" generators in parallel on the same EHV bus, connected to a prevailing power bus through an equivalent reactance X_e, as represented below.

In Figure 7.41, the following notations are used:

S_{ni} is the MVA apparent nominal power of the ith unit;
S_n is the MVA apparent nominal power of the power station;

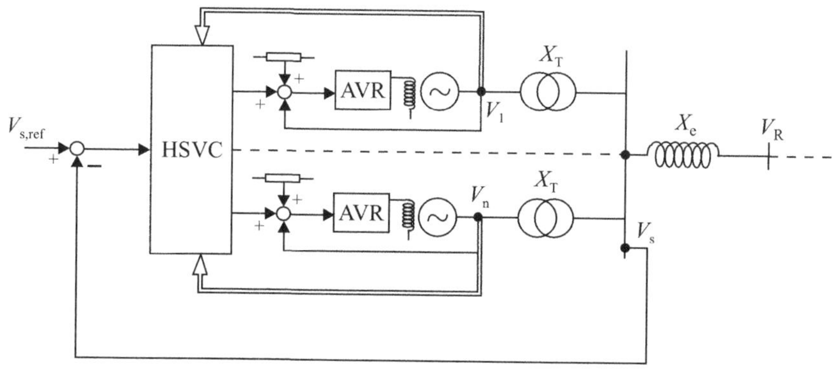

Figure 7.40. Principle scheme of the HSVC control.

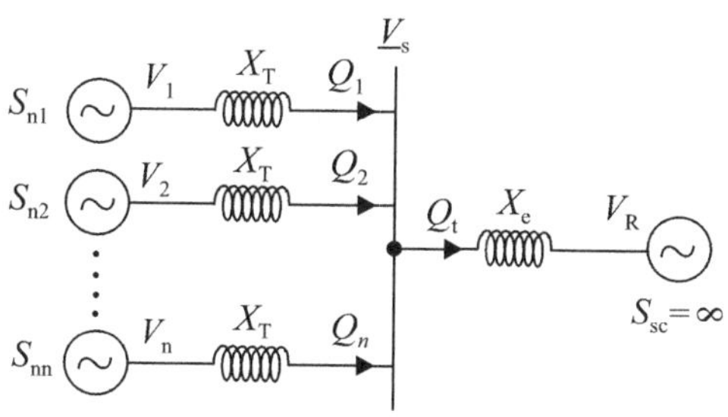

Figure 7.41. Power station connected to an equivalent representation of the grid.

V_n is the kV nominal voltage of the power station;
V_i is the voltage of the ith generator, in p.u. of V_{ni};
Q_i is the reactive power of the ith unit, in p.u. of S_{ni};
Q_t is the total reactive power of the power station, in p.u. of S_n;
V_s is the voltage of power station EHV bus bar, in p.u. of V_n;
V_r is the voltage of the grid bus bar, in p.u. of V_n;
X_T is the reactance of the unit transformer, in p.u. of $(V_{ni})^2/S_{ni}$;
X_e is the equivalent reactance of the external grid, in p.u. of $(V_n)^2/S_n$;

and

$\underline{V}_i = V_i e^{j\theta_i}, \quad i = 1, 2, \ldots, n$
$\underline{V}_s = V_s e^{j\theta_s}$
V_i and V_s are in p.u.
$\delta_i = \theta_i - \theta_s, \quad i = 1, 2, \ldots, n$

Now referring to the reactive power of each unit:

$$Q_i(\text{p.u.}) = \frac{V_i^2 - V_i V_s \cos\delta_i}{X_T}, \qquad i = 1, 2, \ldots, n$$

By the linear approximation around the operating point, under the usual assumed hypotheses

$$V_{i(0)} = V_{s(0)} \quad \text{and} \quad \delta_{i(0)} = 0$$

It can be written that

$$\Delta Q_i = \frac{V_{s(0)}}{X_T}(\Delta V_i - \Delta V_s), \quad i = 1, 2, \ldots, n$$
$$\Delta Q_t = \frac{V_{s(0)}}{X_e}(\Delta V_s - \Delta V_r) \tag{7.64}$$

Moreover, under the hypothesis of negligible reactive power losses on X_T

$$\Delta Q_t = \frac{1}{S_n}\sum_{k=1}^{n} \Delta Q_k S_{nk} = \frac{V_{s(0)}}{X_e}(\Delta V_s - \Delta V_r) \tag{7.65}$$

From equations (7.64) and (7.65), they are respectively achieved:

$$\Delta V_s = \Delta V_i - \frac{X_T}{V_{s(0)}}\Delta Q_i, \quad i = 1, 2, \ldots, n$$

$$\Delta V_s = \frac{X_e}{V_{s(0)}}\frac{1}{S_n}\sum_{k=1}^{n} \Delta Q_k S_{nk} + \Delta V_r$$

From the last two equations, the following result can be deduced:

$$\Delta V_i - \Delta V_r = \frac{X_T}{V_{s(0)}}\Delta Q_i + \frac{X_e}{V_{s(0)}}\frac{1}{S_n}\sum_{k=1}^{n}\Delta Q_k S_{nk}, \quad i = 1, 2, \ldots, n$$

$$S_n(\Delta V_i - \Delta V_r) = \frac{X_T S_n + X_e S_{ni}}{V_{s(0)}}\Delta Q_i + \frac{X_e}{V_{s(0)}}\sum_{k=1, k\neq i}^{n}\Delta Q_k S_{nk}, \quad i = 1, 2, \ldots, n$$

In matrix form

$$\begin{bmatrix} \Delta V_1 - \Delta V_r \\ \cdots \\ \Delta V_n - \Delta V_r \end{bmatrix} = \begin{bmatrix} a_{11} & \cdots & a_{1n} \\ \cdots & A & \cdots \\ a_{n1} & \cdots & a_{nn} \end{bmatrix} \begin{bmatrix} \Delta Q_1 \\ \cdots \\ \Delta Q_n \end{bmatrix}$$

where:

$$[A] = [a_{ii}] = \left\{ \begin{array}{l} a_{ii} = \dfrac{X_T S_n + X_e S_{ni}}{V_{s(0)} S_n} \\ a_{ij} = \dfrac{X_e S_{nj}}{V_{s(0)} S_n}, \text{ with } i \neq j \end{array} \right\}, \quad \text{with } i, j = 1, 2, \ldots, n \tag{7.66}$$

In a compact form, the result can be written as

$$\begin{aligned} &[\Delta V - \Delta V_r] = [A][\Delta Q] \\ &[\Delta Q] = [A]^{-1}[\Delta V - \Delta V_r I_{col}] \end{aligned} \tag{7.67}$$

where

$$[\Delta V] = \begin{bmatrix} \Delta V_1 \\ \Delta V_2 \\ \cdots \\ \Delta V_n \end{bmatrix}, \quad \text{while } [I_{col}] = \begin{bmatrix} 1 \\ 1 \\ \cdots \\ 1 \end{bmatrix} = (I_{row})^T$$

Therefore, the model result is a matrix with coefficients dependent on the operating point and on X_e and X_T. Both $[A]$ and $[A]^{-1}$ are full matrices; for this reason the reactive power of each unit depends on the voltages of all the other units inside the power station. Figure 7.42 shows the block diagram of the found linear model. In the scheme it is also assumed that $\Delta V_{ref} = \Delta V$ under the hypothesis that the primary voltage control loops have negligible dynamics with respect to the control under study.

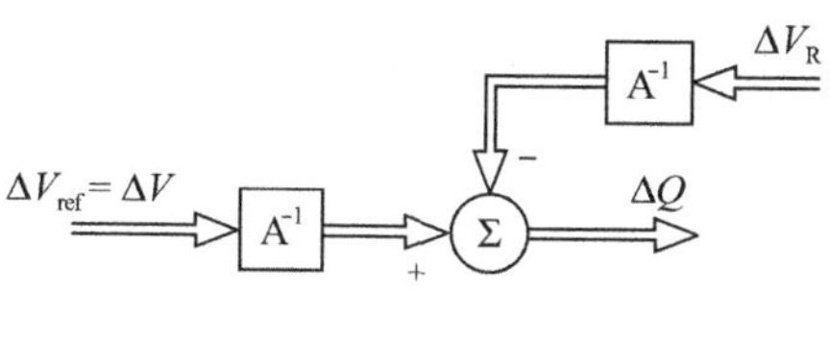

Figure 7.42. Linear model of a power plant with n generators connected to an equivalent grid.

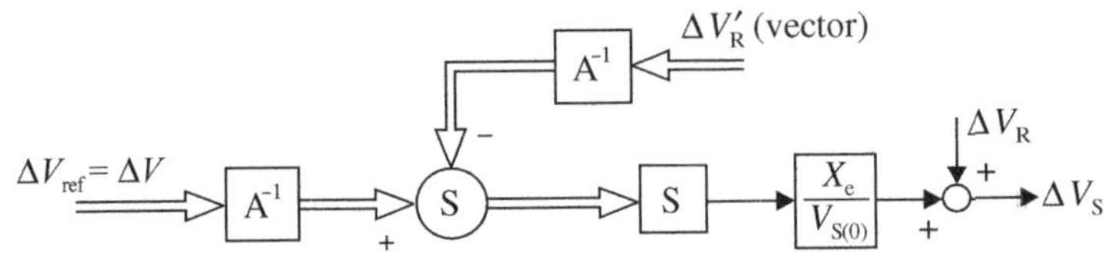

Figure 7.43. Linear equivalent model of a power plant with *n* generators, connected to an equivalent grid: link between the generators and the EHV bus bar voltages.

In Figure 7.42, ΔQ represents the overall reactive power delivered/absorbed by all together the power station units and any change of it determines a bus bar voltage variation according to

$$\Delta V_{\mathrm{s}} = \frac{X_{\mathrm{e}}}{V_{\mathrm{s}(0)}} \frac{1}{S_{\mathrm{n}}} \sum_{k=1}^{n} \Delta Q_k S_{\mathrm{n}k} + \Delta V_{\mathrm{r}} \tag{7.68}$$

The scheme from Figure 7.43, showing the link between the generators voltages at the power station and the local EHV bus bar voltages, comes from equation (7.68).

High-Side Voltage Regulator. Starting from the simplified system model shown in Figure 7.43, hereinafter the power station reactive power and voltage control loops are introduced by assuming that they are slower with respect to AVR voltage control loop (the dominant time constant of the AVR control loop is assumed of the order of 0.5 s, see Section "Linear Analysis of the Generator Voltage Control Loop" from Section 7.3.3).

The generator reactive power control loop, not common in practice, should in principle regulates the reactive power of its generator, without taking into account the reactive powers of the remaining generators operating at the same power station, so negatively interacting with them. Even in the case that all the generators in the power station are provided with autonomous reactive power control loops, their dynamic interaction generally determine poor oscillations damping performance at the power station level. Nevertheless, the interest to control the reactive powers of all the generators in a power station by avoiding reactive power dynamic interaction between them is high and becomes a real need in some plants with a large number of generators.

A possible solution to the mentioned problem is a "power station centralized control," corresponding to the block diagram in Figure 7.44.

A centralized, not-interacting control law, of integral type, allows the dynamic decoupling among the unit reactive power control loops as well as the reactive power absorption/delivery of each generator in accordance with the set point values. The following ΔV_{ref}, ΔQ_{ref}, ΔQ are vectors of variables while the control matrix A is given by equation (7.66).

The resulting diagonal control, when X_{e} properly identified, allows a first-order dynamics at each reactive power control loop, each characterized by a time constant T_{Q} of the order of 5 s (see Figure 7.45).

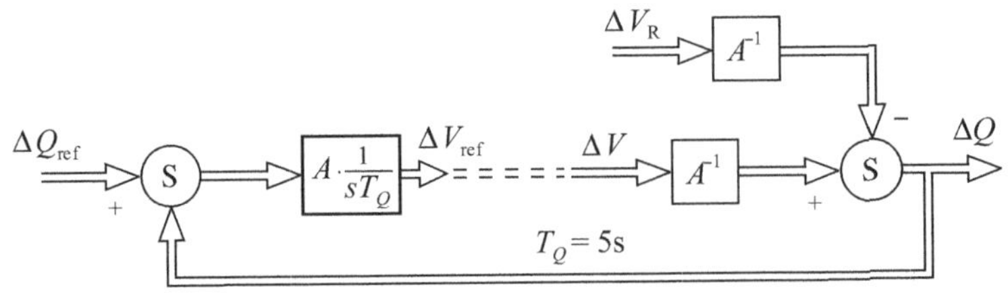

Figure 7.44. Block diagram of a power station reactive power control.

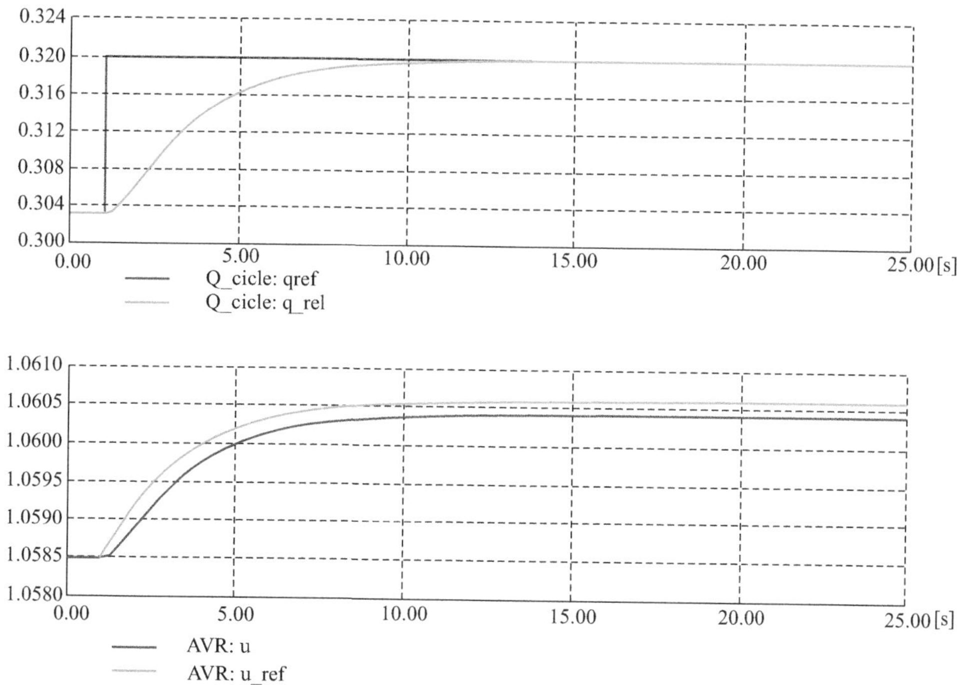

Figure 7.45. Set point step response of a generator reactive power control loop.

This centralized reactive power control accordingly moves the reactive powers of the generators in the power station by controlling the set point vector ΔQ_{ref}. This useful control, able to move together the operating points of the plant generators from the over Q^{+}_{lim} up to the under Q^{-}_{lim} excitation limits, can be effectively used in practice to pursue the objective of the EHV bus bar voltage regulation.

Being the automatic EHV voltage regulation, the true objective of a power system operation, a possible control scheme for the power station high-side voltage regulation, is given in Figure 7.46.

An external voltage control loop (Figure 7.46) overlaps the reactive power loops previously considered with a slower dynamics. A proportional-integral (PI) control law characterizes the proposed control scheme. The proportional contribution is a need to cover the large perturbations in the grid. The output of the control block Δq is called "reactive power level", ranging from $+1$ to -1, in p.u., for the generator's over- and under-excitation limits given by the column matrix linking Δq to ΔQ_{ref} in the block diagram. Equation (7.68) gives the shown link between ΔV_{s} and ΔQ.

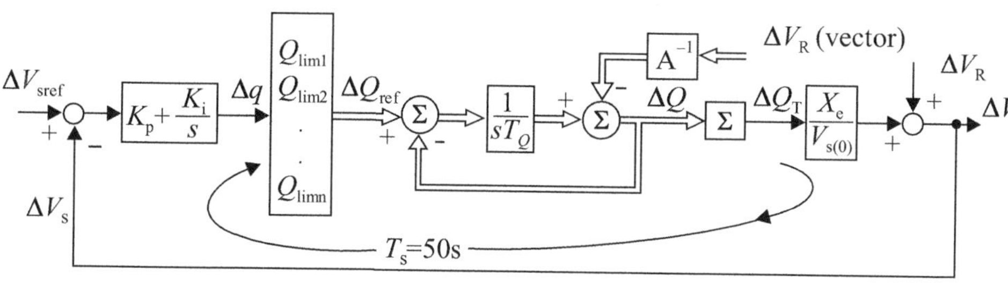

Figure 7.46. Block diagram of a power station EHV bus bar voltage regulation.

The external voltage control loop to be dynamically decoupled with respect to the inner reactive power control loops must have a dominant time constant of about $T_s = 50$ s. To this end, a proper design of the K_p and K_i coefficients of the PI control law is required.

The K_p coefficient can be computed by assuming the instantaneous compensation between the ΔQ variation determined by a grid perturbation ΔV_r at $t = 0^-$ and the ΔQ_{ref} provided at $t = 0^+$ by the ΔV_s variation corresponding to ΔV_r.

From this assumption comes out the following value:

$$K_p = \frac{V_{s(0)}}{X_T} \frac{1}{(1/S_n)\sum_{k=1}^{n} Q_{\lim k} S_{nk}} \tag{7.69}$$

with X_T in p.u. of the power station (equivalent transformer).

When considering all the units of the same size (S_{ni}), then

$$K_p = \frac{V_{s(0)}}{X_T} \frac{1}{(1/n)\sum_{k=1}^{n} Q_{\lim k}} \tag{7.70}$$

Hereinafter, the computing of the K_p coefficient is shown.

To simplify the analysis, reference is done to the equivalent generator of the power station in Figure 7.41. One generator of S_n size and stator voltage V is connected through an equivalent X_T to the local HV bus bar (V_s) that see the remaining network by the equivalent reactance X_e connected to an infinite bus (V_R).

Equations (7.64) and (7.65) describe the reactive power variation in this simplified equivalent scheme. From these equations it comes out

$$\Delta V_s = \Delta V - \frac{X_T}{V_{s(0)}} \Delta Q$$

$$\Delta V_s = \frac{X_e}{V_{s(0)}} \Delta Q + \Delta V_r$$

From them, the system model linking V, V_r, V_s is achieved:

$$(\Delta V - \Delta V_r) = \frac{X_T + X_e}{V_{s(0)}} \Delta Q$$

$$\Delta V_s = \frac{X_e}{V_{s(0)}} \Delta Q + \Delta V_r$$

Now, referring to the control system in Figure 7.46, its simplified equivalent representation, coherent with the hypotheses of very fast inner control loops with respect to the outer loop on V_s, is here after shown ($Q_{\lim}$ is in p.u. of S_n).

In fact, assuming the instantaneous compensation between the ΔQ variation determined by a grid perturbation ΔV_r at $t = 0^-$ and the ΔQ_{ref} provided at $t = 0^+$ by the ΔV_s variation corresponding to ΔV_r, that is

$$\Delta V_i(0) = 0$$

$$\Delta Q_i(0) = \Delta Q_{ref}(0)$$

$$\Delta Q_i(0) = -\frac{V_{s(0)}}{X_T + X_e} \Delta V_r$$

$$\Delta Q_{ref\,i}(0) = -Q_{\lim} K_p \left[\Delta V_r + \frac{X_e}{V_{s(0)}} \left(-\frac{V_{s(0)}}{X_T + X_e} \Delta V_r \right) \right]$$

Then the value of the proportional coefficient is obtained:

$$K_{\mathrm{p}} = \frac{V_{\mathrm{s}(0)}}{X_{\mathrm{T}}} \frac{1}{Q_{\mathrm{lim}}} \tag{7.71}$$

Considering the n generators as in (7.70):

$$K_{\mathrm{p}} = \frac{V_{\mathrm{s}(0)}}{X_{\mathrm{T}}} \frac{1}{(1/S_n) \sum_{k=1}^{n} Q_{\mathrm{lim}\,k} S_{\mathrm{n}k}}$$

Coming to the coefficient K_{i}, it must be able to impose a first-order dynamics to the outer control loop of Figure 7.47, with a dominant time constant $T_{\mathrm{s}} = 50\,\mathrm{s}$, when K_{p} assumes the above imposed value (7.71).

Therefore,

$$K_{\mathrm{i}} = \frac{K_{\mathrm{p}} + (V_{\mathrm{s}(0)}/X_{\mathrm{e}})\left(1/(1/S_n)\sum_{k=1}^{n} Q_{\mathrm{lim}\,k} S_{\mathrm{n}k}\right)}{T_{\mathrm{s}}} = \frac{V_{\mathrm{s}(0)}}{T_{\mathrm{s}}} \frac{1}{(1/S_n)\sum_{k=1}^{n} Q_{\mathrm{lim}\,k} S_{\mathrm{n}k}} \frac{X_{\mathrm{e}} + X_{\mathrm{T}}}{X_{\mathrm{e}} X_{\mathrm{T}}} \tag{7.72}$$

In fact, referring to Figure 7.47, the "pole" of the closed loop is given by

$$s = -\frac{Q_{\mathrm{lim}} K_{\mathrm{i}} (X_{\mathrm{e}}/V_{\mathrm{s}(0)})}{1 + Q_{\mathrm{lim}} K_{\mathrm{p}} (X_{\mathrm{e}}/V_{\mathrm{s}(0)})} = -\frac{1}{T_{\mathrm{s}}}$$

Then

$$K_{\mathrm{i}} = \frac{K_{\mathrm{p}} + (V_{\mathrm{s}(0)}/X_{\mathrm{e}} Q_{\mathrm{lim}})}{T_{\mathrm{s}}} = \frac{V_{\mathrm{s}(0)}}{T_{\mathrm{s}} Q_{\mathrm{lim}}} \frac{X_{\mathrm{e}} + X_{\mathrm{T}}}{X_{\mathrm{e}} X_{\mathrm{T}}}$$

Considering the n generators instead of the equivalent, equation (7.72) is achieved.

In conclusion and differently from the line drop compensation control, with the high-side voltage regulator it is possible to regulate, at the set point value, the voltage of the local

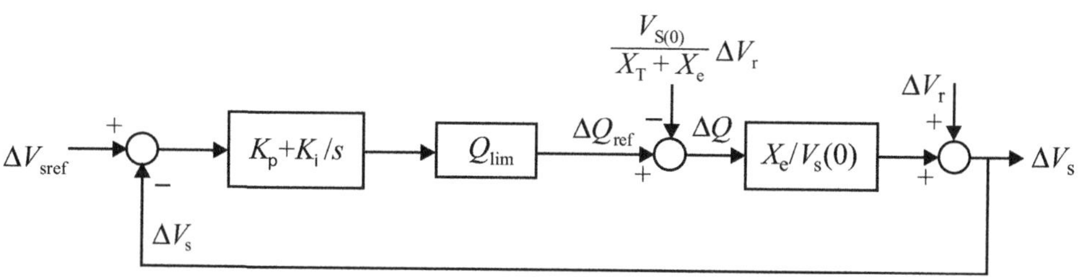

Figure 7.47. Simplified block diagram of EHV bus bar voltage control loop at $t = 0^-$, with a power station represented by an equivalent generator.

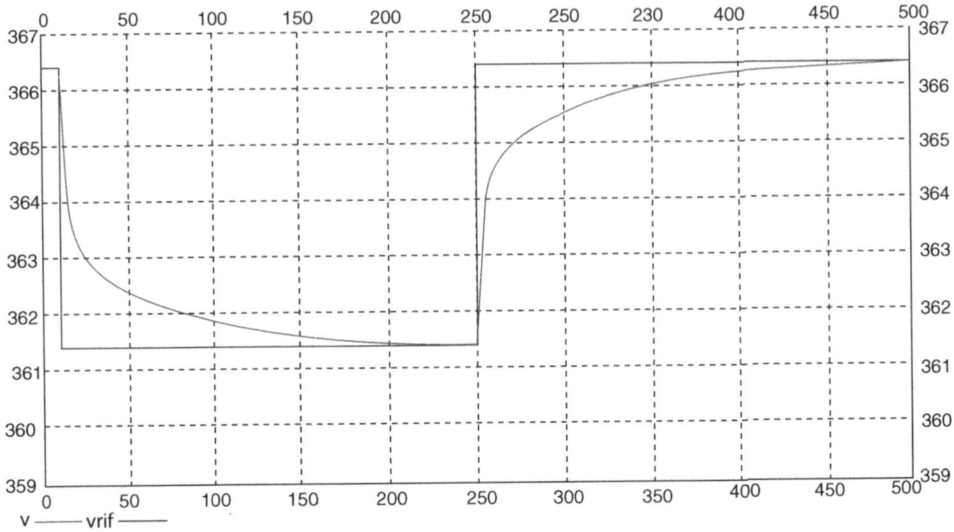

Figure 7.48. Transients of the power plant high-side bus bur voltage V_s following step variations of the HSVC voltage set point.

HV bus bar with a PI control law to characterize the HV transient by a dynamics with high-speed response in the first part and slower response in the second part, as shown in Figures 7.48–7.50.

Lastly, with the high-side voltage regulator it is also possible to increase the voltage stability during the black start-up maneuver, mainly at the initial voltage launch phase [26].

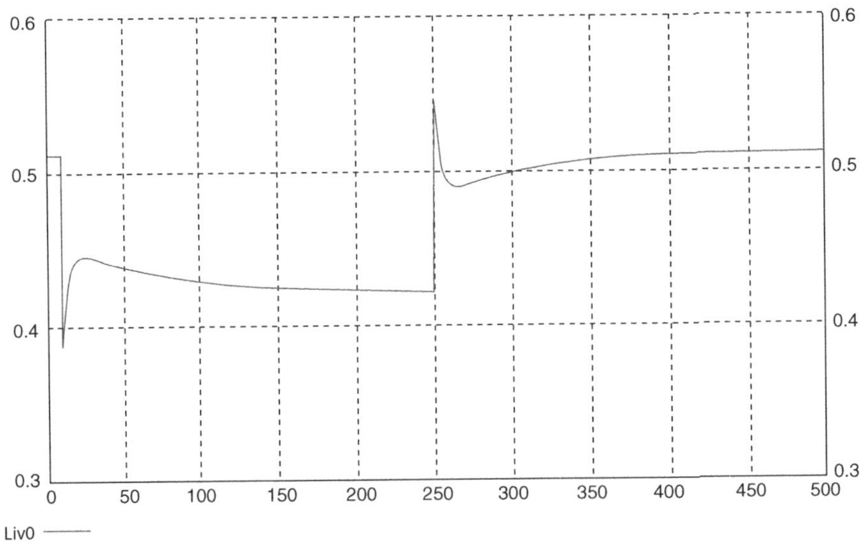

Figure 7.49. Transients of the HSVC output (q level) following step variations of the HSVC voltage set point.

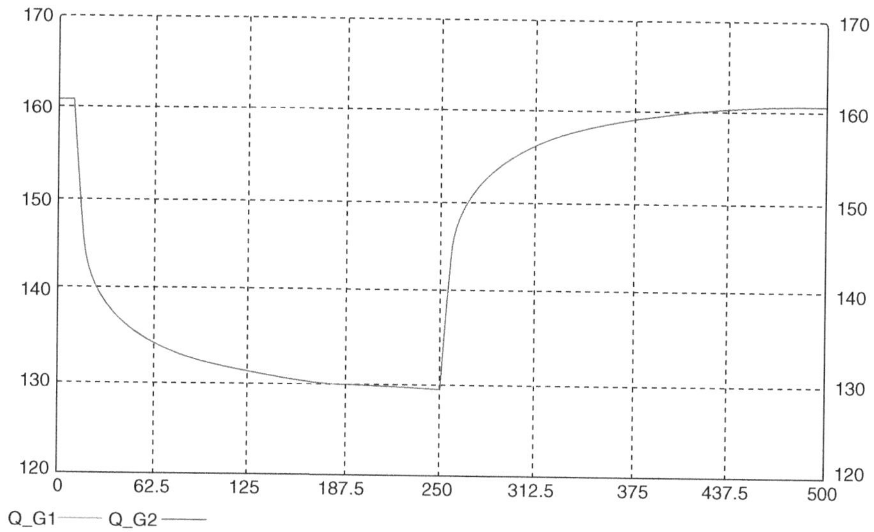

Figure 7.50. Power plant with two generators: transients of the generator reactive powers following step variations of the HSVC voltage set point.

7.4 GRID HIERARCHICAL VOLTAGE REGULATION

7.4.1 Structure of the Hierarchy

7.4.1.1 Generalities. Hierarchical systems, based on grid subdivision into areas and automatic coordination of each area reactive power resources to control the local voltages, have been first investigated in Europe (mainly in Italy and France) from 1980 and named coordinated voltage regulation (CVR) as well as secondary voltage regulation (SVR) and tertiary voltage regulation (TVR). Reference studies and applications come from Italy [15–18,27,28], France [20,21,29], followed by Belgium [30,31], and Spain [22,32], and more recently by United States, Brazil, Taiwan, South Korea, Romania, and South Africa [23,33–39]. An international CIGRE Task Force recently investigated the subject publishing an extensive report [19].

Some of the European applications in Italy and France operate in real systems and are extended at national level. Changing European utilities organization and facing-related energy markets liberalization, hierarchical voltage control systems (HVCSs) are being more appreciated and strengthened [25,26,40–42]. System operators, in fact, recognize that SVR and TVR permit, at the same time, to simplify the automatic control of the overall transmission network voltages by increasing the system efficiency and stability as well as to simply and correctly distinguish the contributions of different participants to the voltage ancillary service.

In North America, the interest for SVR and TVR or more simplified power plant solutions, like high-side voltage control, is growing [33]. At BPA, a wide area voltage control/protection is under test on site, based on the coordination of generators or loads tripping, reactive power switching, TCSC/SVC modulation, power plant high-side voltage schedule and OLTC tap changing [34]. SVR concepts are also under consideration in Brazil [35] where voltage control is proposed for critical distribution areas too, based on a methodology providing a guideline for OLTC assessment and coordination [36].

Progress and trends in transmission network voltage control require, at the beginning of new millennium, an important evolution moving from the largely still operating "manual controls" to the innovating "automatic controls", by using simple, effective, and closed loop regulating systems, managed and supervised directly by the dispatching centers. The cost/benefit analyses strongly support this innovation [43]. Moreover, because voltage control is a prevailing local problem, the feasible solutions must consider the automatic coordination of local reactive power resources, not only those of generators and compensators but also shunt capacitors and reactors, OLTCs, SVCs, and STATCOMs. For this reason, the goals of the voltage ancillary service (quality and security improvements in network operation) can be pursued through a decentralized voltage control system coordinating resources at each power system area/region. Such coordination requires data and signals exchanges between regional dispatcher and local plants/substations: the more data are exchanged in real time, according to power system dynamics, the more voltage control system can improve performances and effectiveness.

The benefit of the network voltage control in terms of grid efficiency is instead more strongly linked with interarea coordination, requiring an effective data and signals exchange among regional dispatchers and central/national system operator. In addition, the exchange of measurements with the neighboring utilities (e.g., edge-bus voltages and tie-line reactive power flows), as well as the coordination of mutual control actions, are also very important for reducing system losses.

The online and real-time monitoring of actual EHV control system performances represents also a challenging opportunity for a proper and indubitable recognition of the power plants contributions to the voltage service [40], in the framework of energy sector liberalization and ancillary market competition.

The main reasons, which definitely lead to a coordinated "automatic" real-time voltage regulation, can be then summarized as follows:

- The quality of power system operation is improved, in terms of reduced variations around the defined voltages profile across the overall transmission network.
- The security of power system operation is enhanced, in terms of larger reactive power reserves kept available by generating units for facing emergency conditions [44].
- The transfer capability of power system is improved, in terms of increased transmissible active power levels, with reduced voltage instability and collapse risks [45].
- The efficiency of power system operation is enhanced, in terms of active losses minimization, reduction of reactive flows, and better utilization of reactive resources.
- The controllability and measurability of voltage ancillary service is simplified, in terms of functional requirements definition and performance monitoring criteria [40].

The voltage and reactive power control of an electrical grid requires a geographical and temporal coordination of many on field components and control functions achievable by a hierarchical control structure. A real-time and automatic voltage control system can be, in fact, basically structured on three hierarchical levels: primary (component control), secondary (area control) and tertiary (power system control) levels. Figure 7.51 gives a preliminary spatial view of the three overlapped hierarchical levels of a voltage–reactive control system. It also puts in evidence the interaction between the tertiary level with the not-real time and off-line forecasting level based on state estimation and OPF. Distinguish

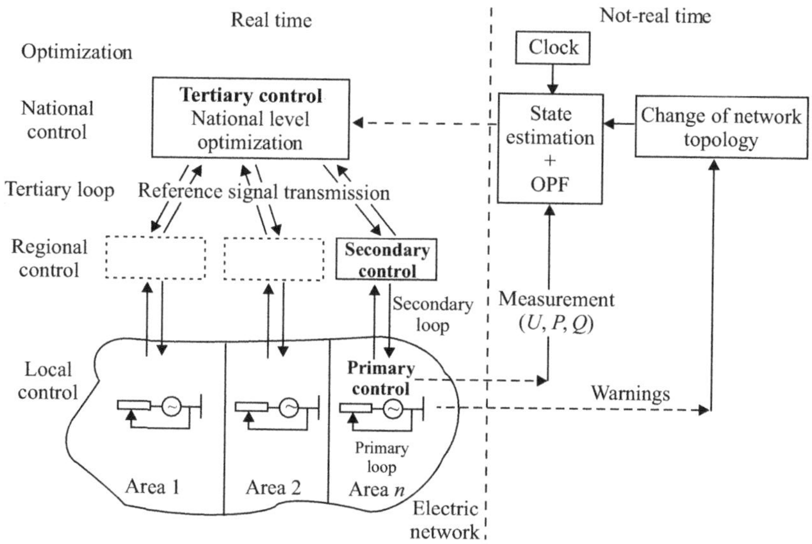

Figure 7.51. The structure of the hierarchical voltage–reactive power control.

the real-time levels with automatic closed loop voltage and reactive power controls from the day-before or short-term optimal forecasting computing, necessarily delayed with respect to the real-time power system operating conditions, is a fundamental clearing up that helps in recognizing relevant differences.

Often happens that the tertiary voltage control is wrongly confused with the static optimization problem of voltage–reactive power, which has to be considered (due to its high delay with respect to the system operating conditions) an open loop control or an off-line forecasting related to the system operation scheduling.

The off-line minimization of power losses is the most employed OPF objective function, which forecasts the generators reactive power scheduling in order to maintain the appropriate voltage levels within the power system.

Obviously, different performances are obtained with an automatic, closed loop, voltage control that minimizes losses in real time (TVR), in comparison with a system operation simply based on a forecast computation, linked to past working points (OPF). In fact, the present grid operating conditions could be, some time, very different from the forecasted one, mainly when facing critical operating conditions.

The off-line OPF voltage–reactive power issue can be an input to initialize the TVR computing, as shown in Figure 7.51.

7.4.1.2 Basic SVR and TVR Concepts. The basic concepts from which the SVR grew in Europe are here summarized to allow the understanding of the proposed control system and the reasons of its structure, performances, and advantages:

a. The idea to automatically control in real time hundreds of transmission bus voltages is too complex, very critical, not reliable and therefore unrealistic and uneconomic.
b. The generating units reactive power is, obviously, the main resource already available on field, at low cost, simple to control for network voltage support;

c. A realistic simple voltage control system should consider the dominant buses only (a small amount) so allowing a suboptimal but feasible and reliable control solution.
d. Dominant bus "pilot node" idea becomes solid assuming as joint buses those having high electrical coupling and voltages close to each other within a "regulation area."
e. The control structure, depending on the grid subdivision into "regulated areas," automatically and as much as possible independently, regulates each pilot node voltage.
f. The control resource is essentially based on the reactive powers of the largest units in the area ("control plants"), which mainly influence the local pilot node voltage.

The basic idea of TVR comes from the need of increasing the system operation security and efficiency through a centralized coordination of the SVR decentralized structure.

g. The pilot nodes voltage set points must be adequately updated and coordinated with dynamics slower than SVR, considering the real condition of the overall grid and avoiding useless and conflicting interarea control efforts.
h. The pilot nodes voltage set points can be computed and updated in real time, considering the global control system structure and its real-time measurements.
i. The pilot nodes voltage set points have to be optimized for minimizing grid losses always preserving control margins. That can be achieved by updating the optimal forecasted plan according to the real-time system working condition.

It has to be pointed out that, notwithstanding the objective of minimizing control system complexity, the effort for achieving an effective hierarchical control system is anyway relevant when a large transmission network is involved, as confirmed by the already undertaken experiences and existing applications. On the one hand, a new power plant apparatus is needed for controlling the reactive power production of generating units, as well as of synchronous compensators and FACTS devices, according to the local bus bar or remote pilot node voltage regulator and taking into account the instantaneous available capability of the plant generators or compensators. On the other hand, a specific regional dispatcher regulator [28] is required for automatically maintaining the pilot nodes voltages at their scheduled values, controlling by fast telecommunications the new power plant apparatuses, turning on/off reactors banks and shunt capacitor, ordering OLTCs and FACTS controllers set points.

Lastly, a new voltage and reactive power optimizing regulator is required at the national/utility control level for coordinating and updating, online and in real time, all the pilot nodes voltages set points (see also Figure 7.52). All these special not conventional control apparatuses do require a specific design. Moreover, the telecommunication speed for data exchange among the primary, secondary, and tertiary levels is high, of the order of 1 s delay, should require specific/dedicated telecommunication apparatuses and media.

7.4.1.3 Primary Voltage Regulation. Primary regulation involves controlling the local voltage of synchronous generators, synchronous/static compensators with the major objective to allow the correct and security operation of these equipments. Obviously the primary voltage control impacts with the transmission network, mainly at the MV busses they are connected, by sustaining the local medium voltages during

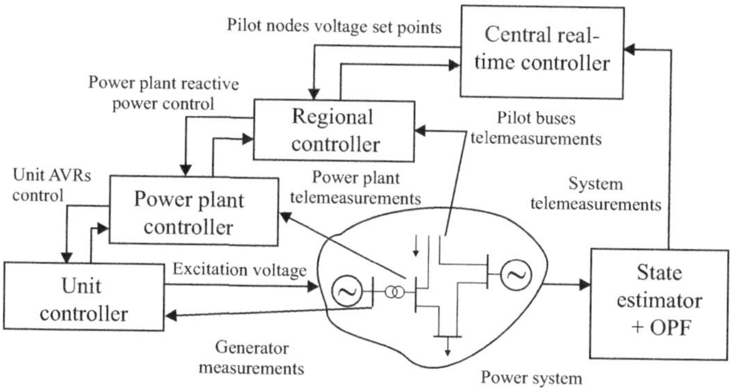

Figure 7.52. A hierarchical structure for transmission network voltage control.

normal and perturbed operating conditions. The control actions are based on local measures and aim to bring out the voltage at the set point value automatically, with a dynamics characterized by a dominant time constant value within a few hundred of milliseconds up to one second: this compensating control has to be considered in the high-speed voltage regulation.

The AVR realizes the primary voltage control of a generator (in Figure 7.52 unit controller). The AVR regulates the voltage at the generator's terminal by controlling the field excitation voltage V_f (Figure 7.53). This AVR basic functionality is already widely described in Section 7.3.3.1.

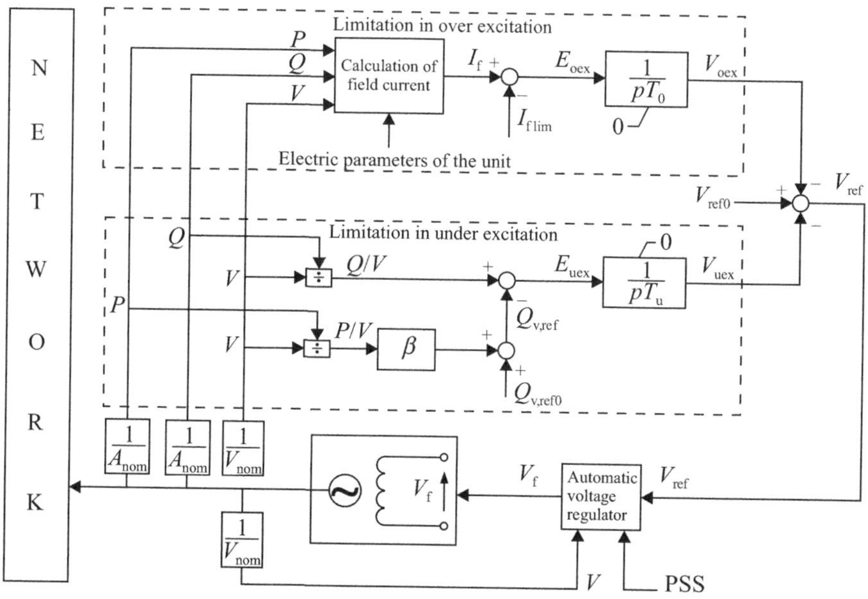

Figure 7.53. Structure of generator AVR including over- and underexcitation limits, respectively, OEL and UEL.

The synchronous machine operating at one of the limits (OEL or UEL) closes the correspondent V_{lim} (V_{oex} or V_{uex}) signal feedback. The signal is obtained from an integral regulator:

- The OEL input compares the actual excitation current I_f with its maximum value I_{flim}. Under steady-state operation I_f is lower than I_{flim} and the integral regulator does not participate at the primary voltage control. When I_f is greater than I_{flim}, the integral regulator generates a V_{lim} negative value in such a way to reduce the excitation voltage of the synchronous generator.

  A certain transient overload level of the excitation windings is allowed based on the slow transient of the thermal phenomena. According to that, the stator current limiting value is transiently increased. In this manner the excitation current I_f can reach very high values for a short time being limited by the generator thermal limits (rotor and winding heating, etc.).

- The UEL input compares the reactive power Q with a reference $Q_{v,ref}$. Under steady-state operation Q is greater than $Q_{v,ref}$ and the integral regulator does not participate at the primary voltage control. When Q is lower than $Q_{v,ref}$, the integral regulator generates a V_{lim} negative value in such a way to increase the excitation voltage of the synchronous generator.

  The control actions are based on local measures and aim to avoid operating points overcoming the limits, with a dynamics characterized by a dominant time constant value within a few seconds up to one or maximum two tens of seconds.

In a hierarchical automatic voltage control system, the role played by the OEL and UEL limits is very important and their curves shape and dynamics must be carefully reconstructed and taken into account by the power station regulator controlling the generator reactive powers. In fact, the generator operating point must be maintained inside the operating limits during normal and perturbed operating conditions, so avoiding any generator thermal stress and useless control efforts due to possible differences between the real and the reconstructed AVR limits.

An example of the OEL and UEL limiting curves is shown in Figure 7.54 that puts in evidence the voltage dependence of the OEL curve, shifting to the right while voltage reduces.

Rough values for the OEL and UEL parameters are given in Table 7.3. *Note*: Other aspects related to OEL and UEL are discussed in Section 2.1.8.4.

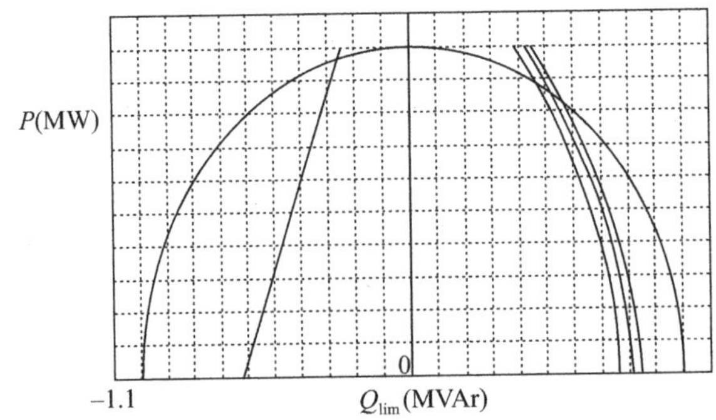

Figure 7.54. Curves of the over- and underexcitation generator limits, respectively, at the right-hand and left-hand sides.

TABLE 7.3. The Parameters of the OEL and UEL

Parameter	Value	Measure Unit
$T_{\text{lim}} = T_{\text{o}}; T_{\text{u}}$	10	s
OEL-$V_{\text{lim min}}$	0	–
UEL-$V_{\text{lim max}}$	0	–
β	Drop of the limit	

7.4.1.4 Secondary Voltage Regulation: Architecture and Modeling.

PRINCIPLE OF THE SECONDARY VOLTAGE REGULATION. The SVR has, as first objective, the voltage control of the system main transmission busses, which is the most important load busses, by controlling the main available reactive power resources on site. Therefore, the primary (see Sections 7.3.3.1. and 7.4.1.3) and secondary voltage controls have different objectives and in some cases are operated with opposite aims.

The secondary voltage control plays an important role during normal operating conditions as well as in front of contingencies.

- In normal grid operation, it ensures
  - maintaining the network voltages at a specified value and reducing their variations;
  - increasing dispatch control efficiency;
  - coordinating real-time controls of the reactive power resources;
  - assuring a dynamics of the first-order type, with a dominant time constant of about 50 s, to the HV voltage transients.
- Under disturbed conditions, the secondary voltage control
  - timely controls the generated/absorbed reactive powers in the perturbed area;
  - speedily recovers the perturbed area voltage level;
  - imposes a first-order dynamic response to the voltage transients, according to the PI control law, with a dominant time constant of about 50 s (I effect) as well as a fast recovery of most of the peak variations (due to large perturbation) during the first seconds of heavy transients (P effect).

Section "High-Side Voltage Regulator" is a useful reference in understanding the above statements and Figure 7.46 helps when considering the pilot node as the controlled bus.

The SVR basic principle is the voltage control of a small amount of grid buses, the most important ones, each of them able to determine the voltages in the surrounding buses, so each defining its area of influence. SVR, therefore, requires splitting the transmission network into "theoretically noninteracting areas," within which the voltage is controlled in the main bus, called "*Pilot Node*." A regional regulator (that controls the pilot nodes and therefore the areas in the region) separately coordinates the generators of a given area by automatically adjusting the area generators reactive powers to regulate the voltage of the area pilot node. Analogously to the high-side voltage regulation, the pilot node voltage regulation consists in a closed loop control of the pilot node voltage by controlling with a PI law, and therefore by an area reactive power level "q", the reactive powers of all the control

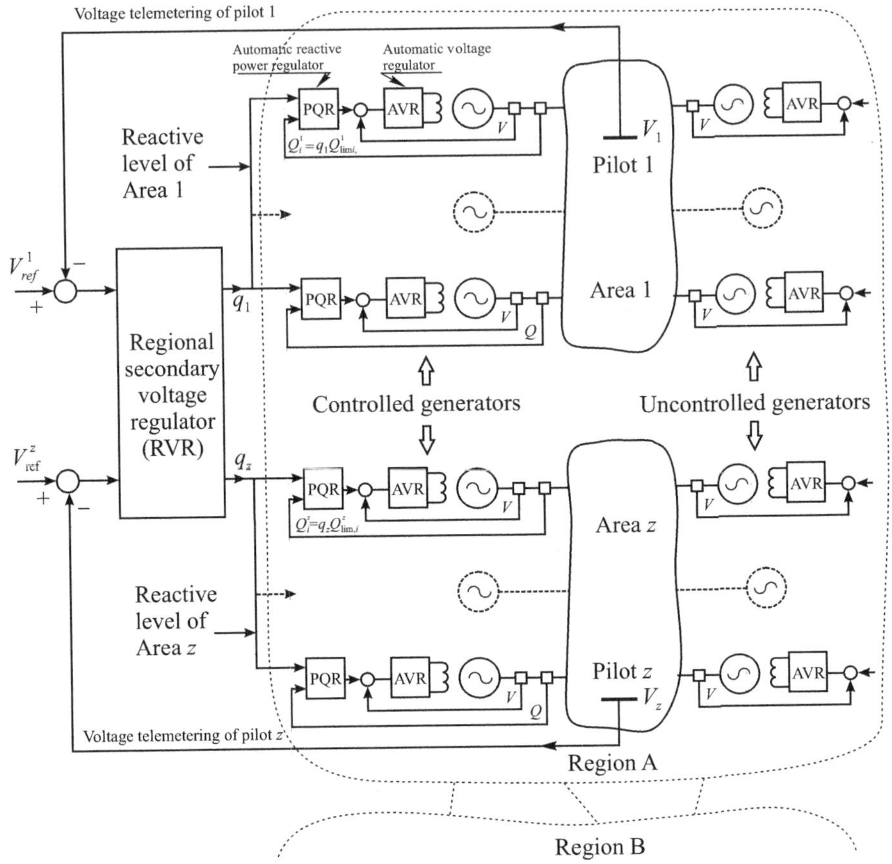

Figure 7.55. Schematic diagram of the hierarchical voltage control system.

power plants in the area. The secondary voltage regulator inputs the instantaneous voltage measure of the area pilot bus, compares it with the pilot node voltage set point, and determines instant by instant the reactive power level to be sent to the control power plants in the area. The reactive power level "q" therefore determines the alignment of each area generating units contributing, proportionally to their capabilities, to the total area reactive power.

The automatic voltage and reactive power control of the transmission network considers the hierarchical structure shown in Figure 7.55 where the control devices are now put in evidence:

- In this control structure, the first hierarchical level (primary level) consists of the conventional generator voltage regulators (AVR). These voltage regulators make it possible to take fast control action in the face of local perturbations (for instance, short circuits near a generator) and collectively determine the "primary" voltage regulation of the network.
- The second hierarchical level consists of power station regulators PQR, which achieve the reactive powers required by the regional voltage regulator (RVR) at a

higher hierarchical level (regional controllers) by operating on the reference of the primary voltage control loops.

- The third hierarchical level consists of slower RVR, one or few if the grid is subdivided in more than one region (regional dispatchers), which regulates in an integral way the voltage of the pilot nodes in the region by operating on the reactive powers of the controlled power stations considered in the second hierarchical level.

The switching of the compensating equipments such as capacitor banks and shunt reactors, as well as the blocking of the OLTC tap changers is one part of the SVR control actions. SVR operates at each area on the local switching resources, only when needed, in accordance with the area control margin value given by the difference of the real-time value of the area reactive power level "q" with respect to its $+1$ or -1 limits. Proper thresholds on the "q" value habilitate the area "on" and "off" switching according to predefined sequences. As far as the area OLTC automatic blocking is concerned, it can also be linked with SVR in case that the used "voltage instability indicator" is based on SVR trend in the area [24,46].

The bus set of the network in which the voltages, for normal perturbation, are close to the voltage of a pilot node defines an area. The RVR receives the voltage telemeasurements of its pilot nodes, and sends the area reactive level signals to the control power plants in each area, separately. The reactive power level signal is the reference of the reactive power regulators (PQR) of the controlled power stations in the area, and in accordance with this reference signal, the power stations deliver/absorb reactive power proportionally to their reactive capability limits; in this way, all the control generators of one area have the same reactive power margin with respect to the reactive bounds. The combined control actions of the RVRs and PQRs determine the "secondary" regulation of the regional network voltages.

As well known, the success of the above-described control scheme mainly depends on the way of choosing the pilot nodes and the control generators and on coordination of the RVRs set points by a possible Central Controller.

The criterion used for locating the pilot nodes is based on the intuitive assumption that such nodes must be chosen from among the strongest ones, that is the nodes that are able to impose the voltages of the load nodes electrically close to them. The criterion includes a constraint on the electrical coupling between pilot nodes: such coupling must be lower than a preestablished limit. In this way, in addition to avoiding problems of dynamic interaction between the secondary voltage control loops, it is possible to prevent excessive exchanges of reactive power due to differences, even small, between the voltage values imposed by the regulation system in adjacent pilot nodes.

From the above considerations, it clearly appears that the starting point of the proposed control scheme is the relatively low electrical coupling between two pilot nodes of adjacent areas.

As far as it concerns the control generators, they are chosen from among those inside each area strongly affecting the voltage of the local pilot node. In this way, each area would be enough autonomous in terms of resources to be used for the local control needs, also because of the electrical decoupling with the other surrounding areas. A good selection of areas and control generators should also put in evidence possible areas needed investments for installation of new reactive power resources.

Dynamic Model of the Regional Control System. Considering an *islanded system*, which is a system with connections with the neighboring grids represented by equivalents, its linear model around an operating point can be written as follows:

$$\begin{bmatrix} \Delta P_1 \\ \Delta Q_1 \\ \cdots \\ \Delta P_g \\ \Delta Q_g \\ \Delta P_{g+1} \\ \Delta Q_{g+1} \\ \cdots \\ \Delta P_n \\ \Delta Q_n \end{bmatrix} = \begin{bmatrix} \frac{\partial P_1}{\partial V_1} & \frac{\partial P_1}{\partial \theta_1} & \cdots & \frac{\partial P_1}{\partial V_g} & \frac{\partial P_1}{\partial \theta_g} & \frac{\partial P_1}{\partial V_{g+1}} & \frac{\partial P_1}{\partial \theta_{g+1}} & \cdots & \frac{\partial P_1}{\partial V_n} & \frac{\partial P_1}{\partial \theta_n} \\ \frac{\partial Q_1}{\partial V_1} & \frac{\partial Q_1}{\partial \theta_1} & \cdots & \frac{\partial Q_1}{\partial V_g} & \frac{\partial Q_1}{\partial \theta_g} & \frac{\partial Q_1}{\partial V_{g+1}} & \frac{\partial Q_1}{\partial \theta_{g+1}} & \cdots & \frac{\partial Q_1}{\partial V_n} & \frac{\partial Q_1}{\partial \theta_n} \\ \cdots & \cdots & \cdots & \cdots & \cdots & \cdots & \cdots & \cdots & \cdots & \cdots \\ \frac{\partial P_g}{\partial V_1} & \frac{\partial P_g}{\partial \theta_1} & \cdots & \frac{\partial P_g}{\partial V_g} & \frac{\partial P_g}{\partial \theta_g} & \frac{\partial P_g}{\partial V_{g+1}} & \frac{\partial P_g}{\partial \theta_{g+1}} & \cdots & \frac{\partial P_g}{\partial V_n} & \frac{\partial P_g}{\partial \theta_n} \\ \frac{\partial Q_g}{\partial V_1} & \frac{\partial Q_g}{\partial \theta_1} & \cdots & \frac{\partial Q_g}{\partial V_g} & \frac{\partial Q_g}{\partial \theta_g} & \frac{\partial Q_g}{\partial V_{g+1}} & \frac{\partial Q_g}{\partial \theta_{g+1}} & \cdots & \frac{\partial Q_g}{\partial V_n} & \frac{\partial Q_g}{\partial \theta_n} \\ \frac{\partial P_{g+1}}{\partial V_1} & \frac{\partial P_{g+1}}{\partial \theta_1} & \cdots & \frac{\partial P_{g+1}}{\partial V_g} & \frac{\partial P_{g+1}}{\partial \theta_g} & \frac{\partial P_{g+1}}{\partial V_{g+1}} & \frac{\partial P_{g+1}}{\partial \theta_{g+1}} & \cdots & \frac{\partial P_{g+1}}{\partial V_n} & \frac{\partial P_{g+1}}{\partial \theta_n} \\ \frac{\partial Q_{g+1}}{\partial V_1} & \frac{\partial Q_{g+1}}{\partial \theta_1} & \cdots & \frac{\partial Q_{g+1}}{\partial V_g} & \frac{\partial Q_{g+1}}{\partial \theta_g} & \frac{\partial Q_{g+1}}{\partial V_{g+1}} & \frac{\partial Q_{g+1}}{\partial \theta_{g+1}} & \cdots & \frac{\partial Q_{g+1}}{\partial V_n} & \frac{\partial Q_{g+1}}{\partial \theta_n} \\ \cdots & \cdots & \cdots & \cdots & \cdots & \cdots & \cdots & \cdots & \cdots & \cdots \\ \frac{\partial P_n}{\partial V_1} & \frac{\partial P_n}{\partial \theta_1} & \cdots & \frac{\partial P_n}{\partial V_g} & \frac{\partial P_n}{\partial \theta_g} & \frac{\partial P_n}{\partial V_{g+1}} & \frac{\partial P_n}{\partial \theta_{g+1}} & \cdots & \frac{\partial P_n}{\partial V_n} & \frac{\partial P_n}{\partial \theta_n} \\ \frac{\partial Q_n}{\partial V_1} & \frac{\partial Q_n}{\partial \theta_1} & \cdots & \frac{\partial Q_n}{\partial V_g} & \frac{\partial Q_n}{\partial \theta_g} & \frac{\partial Q_n}{\partial V_{g+1}} & \frac{\partial Q_n}{\partial \theta_{g+1}} & \cdots & \frac{\partial Q_n}{\partial V_n} & \frac{\partial Q_n}{\partial \theta_n} \end{bmatrix} \cdot \begin{bmatrix} \Delta V_1 \\ \Delta \theta_1 \\ \cdots \\ \Delta V_g \\ \Delta \theta_g \\ \Delta V_{g+1} \\ \Delta \theta_{g+1} \\ \cdots \\ \Delta V_n \\ \Delta \theta_n \end{bmatrix} \tag{7.73}$$

where Q_i is the reactive power at node i, V_i is the voltage magnitude at node i, n is the number of nodes, g is the number of generator nodes, and $m = n - g$ is the number of load nodes.

The linearized model can also be written as

$$[\Delta V] = [S_Q][\Delta Q] + [S_P][\Delta P]$$

where ΔV is the vector of the voltage modules changes in the region, ΔQ is the vector of the reactive power injection changes, ΔP is the vector of the active power injection changes, n is the number of HV buses in a region, and S_Q, S_p are the sensitivity matrices: $S_Q \in R^{nxn}$ and $S_p \in R^{n(n-1)}$.

Taking into account the decoupling between the active and reactive powers and separating the equations describing the dependence, in the power system, of the reactive powers from the voltages, it comes out from the system equation (7.73):

$$\begin{bmatrix} \Delta Q_1 \\ \cdots \\ \Delta Q_g \\ \Delta Q_{g+1} \\ \cdots \\ \Delta Q_n \end{bmatrix} = \begin{bmatrix} \frac{\partial Q_1}{\partial V_1} & \cdots & \frac{\partial Q_1}{\partial V_g} & \frac{\partial Q_1}{\partial V_{g+1}} & \cdots & \frac{\partial Q_1}{\partial V_n} \\ \cdots & \cdots & \cdots & \cdots & \cdots & \cdots \\ \frac{\partial Q_g}{\partial V_1} & \cdots & \frac{\partial Q_g}{\partial V_g} & \frac{\partial Q_g}{\partial V_{g+1}} & \cdots & \frac{\partial Q_g}{\partial V_n} \\ \frac{\partial Q_{g+1}}{\partial V_1} & \cdots & \frac{\partial Q_{g+1}}{\partial V_g} & \frac{\partial Q_{g+1}}{\partial V_{g+1}} & \cdots & \frac{\partial Q_{g+1}}{\partial V_n} \\ \cdots & \cdots & \cdots & \cdots & \cdots & \cdots \\ \frac{\partial Q_n}{\partial V_1} & \cdots & \frac{\partial Q_n}{\partial V_g} & \frac{\partial Q_n}{\partial V_{g+1}} & \cdots & \frac{\partial Q_n}{\partial V_n} \end{bmatrix} \cdot \begin{bmatrix} \Delta V_1 \\ \cdots \\ \Delta V_g \\ \Delta V_{g+1} \\ \cdots \\ \Delta V_n \end{bmatrix} \tag{7.74}$$

or, separating the EHV nodes between the "g" called generation buses from the "m" called load busses, it is possible to write

$$\begin{aligned}[\Delta Q] &= -\{[B_{GG}][\Delta V_G] + [B_{GL}][\Delta V_L]\} \\ [\Delta Q_L] &= -\{[B_{LG}][\Delta V_G] + [B_{LL}][\Delta V_L]\}\end{aligned} \tag{7.75}$$

With the obvious meaning of the symbols

$$[\Delta Q] = -[B][\Delta V] \tag{7.76}$$

where Q and V represent the reactive powers vector and the voltages vector of the overall system, respectively. This matrix equation allows a simplified but enough precise analysis of the links between the voltage variations on the busses and the injected reactive powers.

Denoting by $\widehat{n}$ the total number of the grid buses, then

$[\Delta Q]$: $(\widehat{n} \times 1)$ is the vector of injected reactive powers;
$[\Delta V]$: $(\widehat{n} \times 1)$ is the vector of the voltages in the grid busses;
$[B]$: $(\widehat{n} \times \widehat{n})$ is the symmetric matrix of the grid nodes susceptances, in p.u. (including lines and transformers).

The matrix $-[B]$ represents the sensitivity of the injected reactive powers with respect to the voltages.

Substituting $[\Delta V_L]$, it is possible to obtain

$$\begin{aligned}[\Delta V_L] &= -[B_{LL}]^{-1}[\Delta Q_L] - [B_{LL}]^{-1}[B_{LG}][\Delta V_G] \\ [\Delta Q_G] &= -\{[B_{GG}] - [B_{GL}][B_{LL}]^{-1}[B_{LG}]\}[\Delta V_G] + [B_{GL}][B_{LL}]^{-1}[\Delta Q_L]\end{aligned}$$

Therefore, the system model becomes

$$\begin{cases}[\Delta V_L] = [H][\Delta V_G] + [X_{CC}][\Delta Q_L] \\ [\Delta Q_G] = -[B_{eq}][\Delta V_G] + [D][\Delta Q_L]\end{cases} \tag{7.77}$$

where the introduced matrices are defined as follows:

$$\begin{aligned}&[X_{CC}] = -[B_{LL}]^{-1} \\ &[H] = -[B_{LL}]^{-1}[B_{LG}] = [X_{CC}][B_{LG}] \\ &[D] = [B_{GL}][B_{LL}]^{-1} = -[B_{GL}][X_{CC}] = -[H]^{T} \\ &[B_{eq}] = [B_{GG}] - [B_{GL}][B_{LL}]^{-1}[B_{LG}] = [B_{GG}] + [B_{GL}][H] \cdots \stackrel{\text{def}}{=} -[C]\end{aligned} \tag{7.78}$$

From (7.76) it is possible to deduce

$$[\Delta V] = [S_Q][\Delta Q]$$

Without loss of generality, the power system buses can be simply divided into generation buses (Q_G, V_G) and load buses (Q_L, V_L):

$$\Delta V = [S_{QG} \vdots S_{QL}] \cdot \begin{bmatrix}\Delta Q_G \\ \Delta Q_L\end{bmatrix}$$

Furthermore, it is necessary to distinguish between the generator buses allocated as control power plant buses (Q_{GC}, V_{GC}) and the others representing the uncontrolled power plant buses (Q_{GU}, V_{GU}):

$$\Delta V = \left[S_{QC} \vdots S_{QU} \vdots S_{QL} \right] \cdot \begin{bmatrix} \Delta Q_{GC} \\ \Delta Q_{GU} \\ \Delta Q_{L} \end{bmatrix} \tag{7.79}$$

The uncontrolled plants are under primary voltage regulation only, therefore

$$\Delta Q_{GU} = [K_{U}] \cdot [\Delta V_{GU}] = [K] \cdot [\Delta V] = [0 \quad K_{U} \quad 0] \cdot \begin{bmatrix} \Delta V_{GC} \\ \Delta V_{GU} \\ \Delta V_{L} \end{bmatrix}$$

By substituting, equation (7.79) can be rewritten as follows:

$$\Delta V = \left[S_{C} \vdots S_{L} \right] \cdot \begin{bmatrix} \Delta Q_{GC} \\ \Delta Q_{L} \end{bmatrix} \tag{7.80}$$

with obvious meaning of the symbols.

SVR CONTROL STRUCTURE. Let us denote by z the number of the pilot nodes in the region. From equation (7.80) it can be obtained

$$\begin{bmatrix} \Delta V_{p}^{1} \\ \Delta V_{p}^{2} \\ \Delta V_{p}^{z} \end{bmatrix} = \left[S_{Cp} \vdots S_{Lp} \right] \begin{bmatrix} \Delta Q_{GC}^{1} \\ \Delta Q_{GC}^{2} \\ \Delta Q_{GC}^{z} \\ \cdots \\ \Delta Q_{L} \end{bmatrix}$$

where ΔQ_{GC}^{k} represents the vector of the reactive power variation of the control power plants from area k.

More precisely, because the controlled power plant must operate as an equivalent unit

$$\Delta Q_{GC}^{k} = \text{Diag}\{\alpha_i\} \left| \Delta V_{\text{refC}}^{k} - \Delta V_{GC}^{k} \right| \tag{7.81}$$

where $\alpha_i \triangleq (V_{GCi}^{k})_0 / x_i$ is a constant that depends on the operating point and the unit transformer reactance.

From equation (7.80)

$$\Delta V_{GC}^{k} = S_{C}^{k} \Delta Q_{GC}^{k} + S_{L}^{k} \Delta Q_{L} \tag{7.82}$$

By substituting equation (7.82) in equation (7.81) it is possible to obtain, for the kth area, the dependence of the control power plant reactive powers from the set points of the

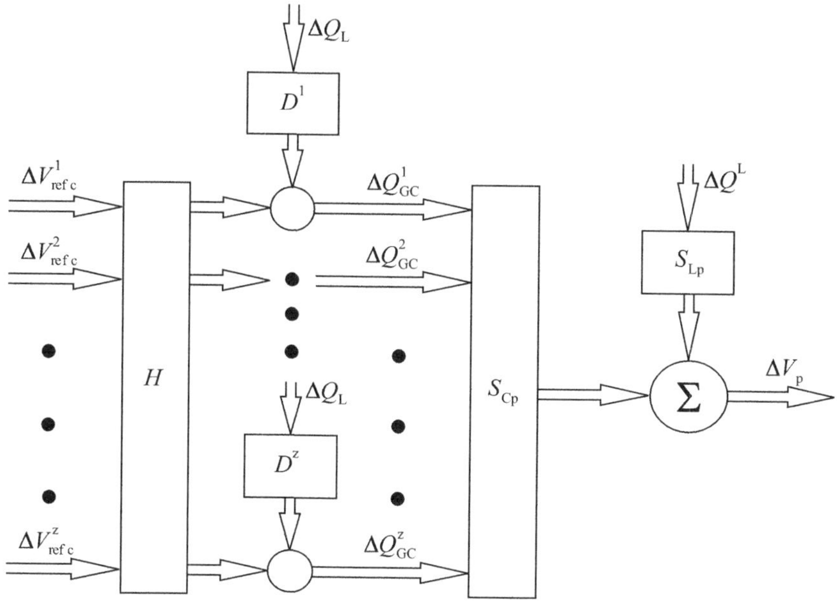

Figure 7.56. Schematic representation of the linearized dependence of the pilot node voltages with the load variation and power plant control.

corresponding AVRs and the load variations:

$$\Delta Q^k_{GC} = \text{Diag}\{\alpha_i\} \cdot \left|\Delta V^k_{refC} - S^k_C \Delta Q^k_{GC} - S^k_L \Delta Q_L\right|$$
$$\left[\Delta Q^k_{GC}\right] = \left[H^k\right]\left[\Delta V^k_{refC}\right] + \left[D^k\right]\left[\Delta Q_L\right]$$

For the overall region, the equation is

$$[\Delta Q_{GC}] = [H] \cdot [\Delta V_{refC}] + [D][\Delta Q_L]$$

A schematic representation of the system under control is shown in Figure 7.56, where the matrices S_{Cp} and H are in general diagonal dominant blocks.

Now, consider that the control scheme of the secondary voltage control it does include the following:

1. The power plant reactive power regulator (PQR) that provides the inner control loop to achieve the desired reactive power at the controlled plant.
2. The RVR that provides the outer control loop with the objective to achieve a desired voltage profile across the region through regulating the voltage of the pilot nodes.

A schematic representation of the secondary voltage regulation control structure is given in Figure 7.57 under the acceptable simplifying assumption to neglect the dynamics of the primary voltage control loops.

Hereinafter, g_K stands for the number of control generators associated with the kth pilot node, and $g = \sum g_K$ the total number of the controlled generators.

In the generic kth area, the reactive power regulator of the ith control generator is assumed purely of integral type. Its reference signal is proportional to the reactive power

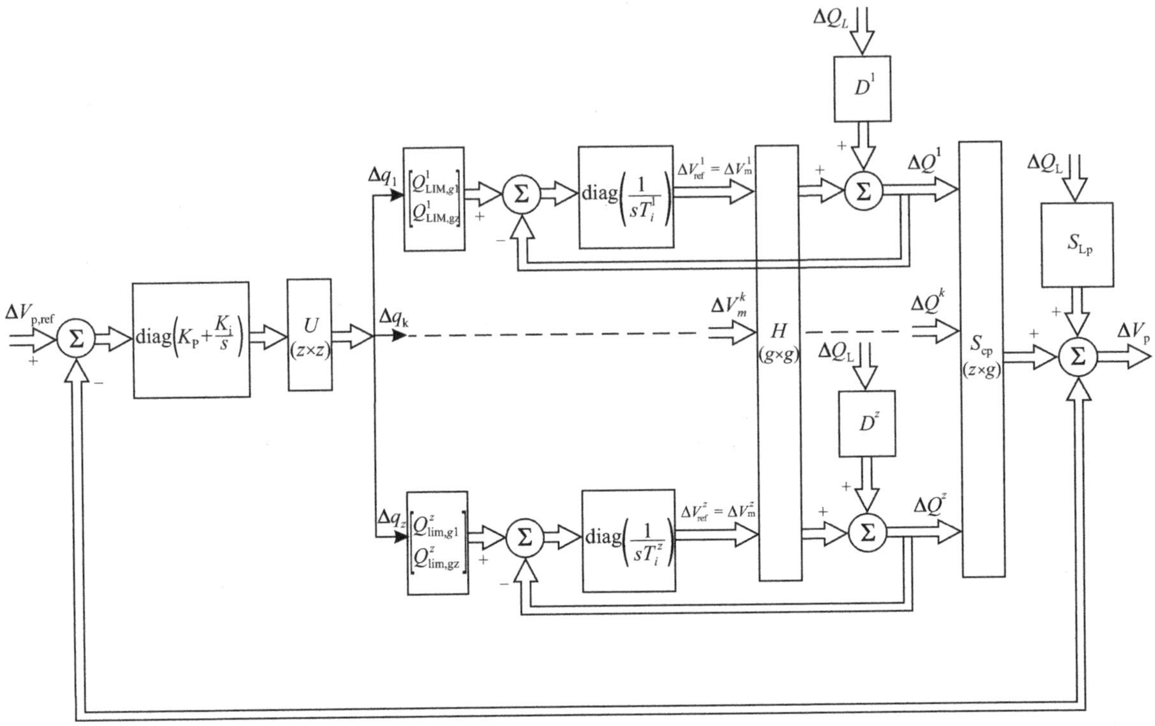

Figure 7.57. Block diagram of the secondary voltage regulation.

level q_k of the area considered. The proportionality coefficient is given by the reactive capability limit $Q^k_{\lim,i}$ of the ith generator. The reactive power level q_K, supplied by the pilot node voltage regulator, is defined by a proportional-integral law applied to a linear combination of the differences between the secondary voltage references and the corresponding pilot node voltages.

The dynamic design of the control system consists on choosing the time constants T^k_i of the integrators in the power plant reactive power control loops as well as the coefficients of the control matrix $\boldsymbol{U}$. From a practical point of view, the dynamic behavior of the controlled system is simplified by making the superimposed control loops dynamically time decoupled. This means that the response time constant of a control loop must be dominant with respect to those of its internal loops (time decomposition). In this connection, the response time constant T_Q of the power plant reactive power control loop must be chosen sufficiently higher than that of the primary voltage control loops and sufficiently lower than the desired dynamic response (time constant) of the secondary voltage regulation.

Bearing in mind the chosen time decoupling, the analysis of the slower modes associated with the main loop of the secondary voltage control system may be made on the basis of the block diagram in Figure 7.58.

This diagram results from Figure 7.57 by ignoring the response time constants of the power plant reactive power control loops and denoting by $Q_{\lim}$ $(g \times z)$ the block diagonal matrix of the reactive capability limits of the control generators.

It is important to realize that the assumption $\Delta q(\text{desired}) = \Delta q(\text{obtained})$, without considering in the model additional dynamics, is based on the following hypothesis: *the*

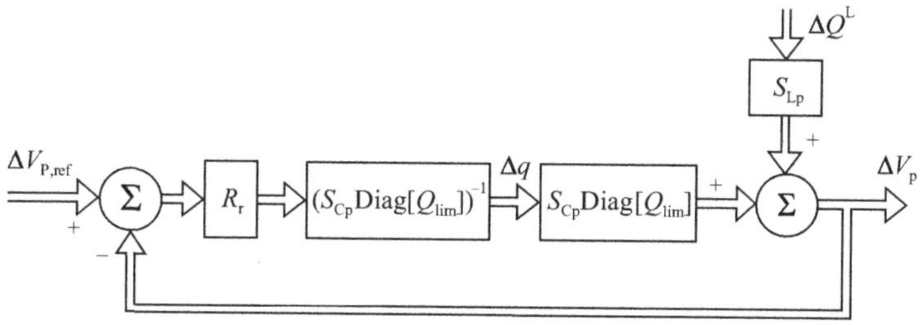

Figure 7.58. Pilot node voltage control loops in the region.

reactive power loops at PQR level are faster than the pilot node voltage loop and are also dynamically decoupled, thanks to the PQR control law; basically, R_r is a diagonal matrix.

Therefore, the synthesis of the control matrix $\boldsymbol{U}$ must be made with the aim of ensuring a reduced dynamic interaction between the voltage regulation of the individual pilot nodes: each pilot node dynamics should be characterized by only one dominant time constant.

Reference Transients. The transients of two separate tests are

- test on the PQR reactive power control loops following a q level step variation;
- test on the RVR pilot node voltage control loop after a voltage V_{Pref} set point step variation.

These overlapped control loops are clearly shown in Figure 7.55.

The PQR and RVR transients jointly represent the full SVR dynamic characteristics, the same seen for the HSVC, being designed with the same criteria.

- *Test on the PQR Reactive Power Control Loops.* Considering a power station with four generators under PQR control, the test results in Figure 7.59 show the transients following the step variations on the reactive power level q under the hypothesis of open pilot node voltage control loop.

  In the PQR presence, all the generators operating at the power station track the level q request with the imposed dynamics: first order with 5 s dominant time constant. It can be noticed that the four generators reactive powers and voltages move concordantly aligned, tracking the reference control signal step variations, in the figure top. Moreover, under PQR, no one dynamic interaction among the under test control loops are evidenced. The last transient at the figure bottom shows the pilot node voltage variation consequent to the considered power station reactive power delivery increasing (at the beginning) followed by its reduction.
- *Test on the RVR Pilot Node Voltage Control Loops.* Considering a power system with 12 pilot nodes under RVR control, the test results in Figure 7.60 show the transients following the step variations of the Area 2 pilot node voltage set point.

  The test emphasizes the proper selection of the pilot nodes with a light dynamic interaction among their voltage control loops. In fact, the step variations of the Area 2 voltage set point determine significant changes in the correspondent pilot node voltage and small transient effects on the others (Figure 7.60a). The Area 2 reactive power control level confirms this result significantly changing with respect to the other areas control levels that conversely remain practically constant.

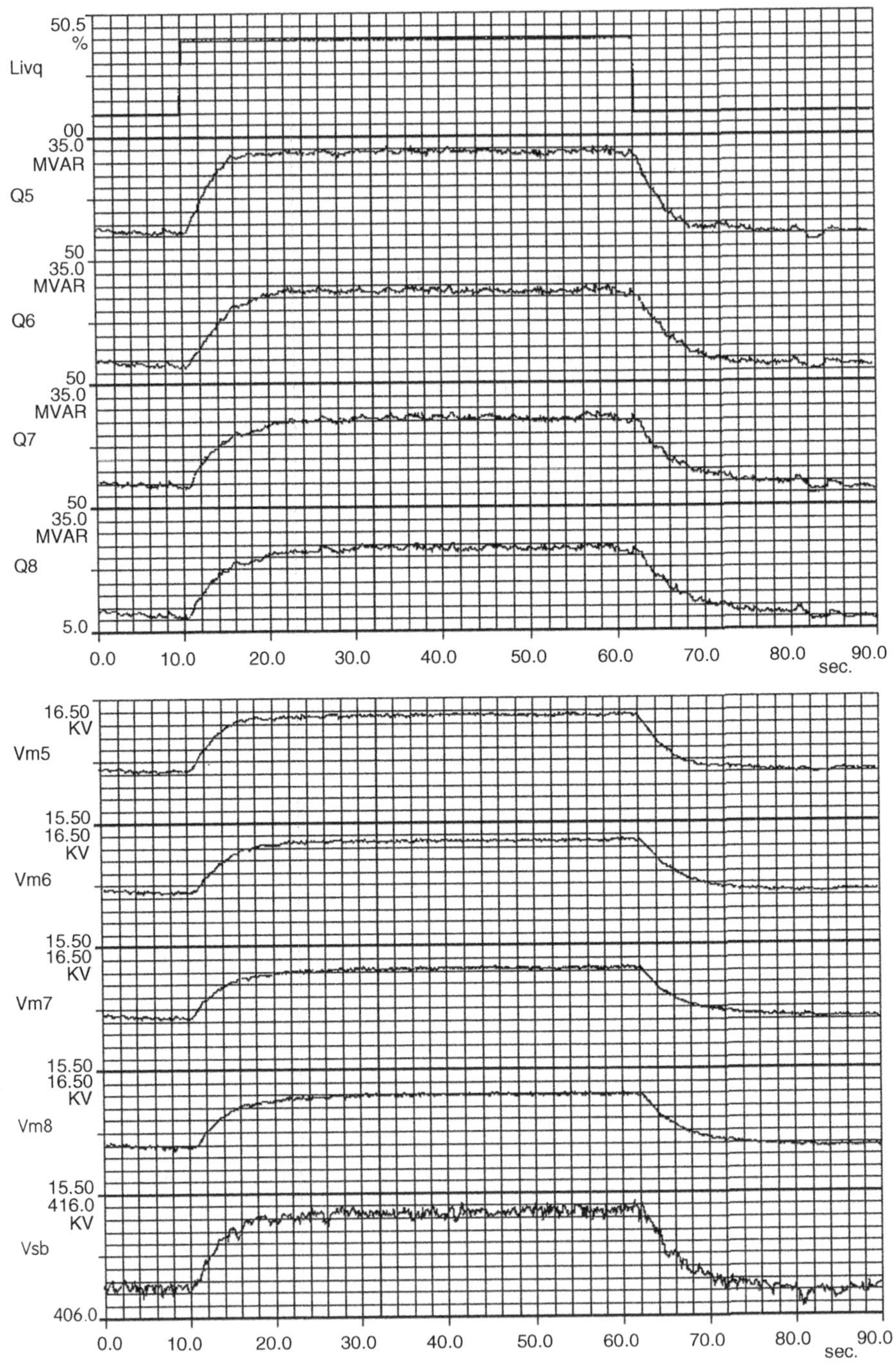

Figure 7.59. Step response of a power plant reactive power control loops showing a dominant time constant of about 5 s. V_{Sb} represents the local HV pilot bus bar voltage transient.

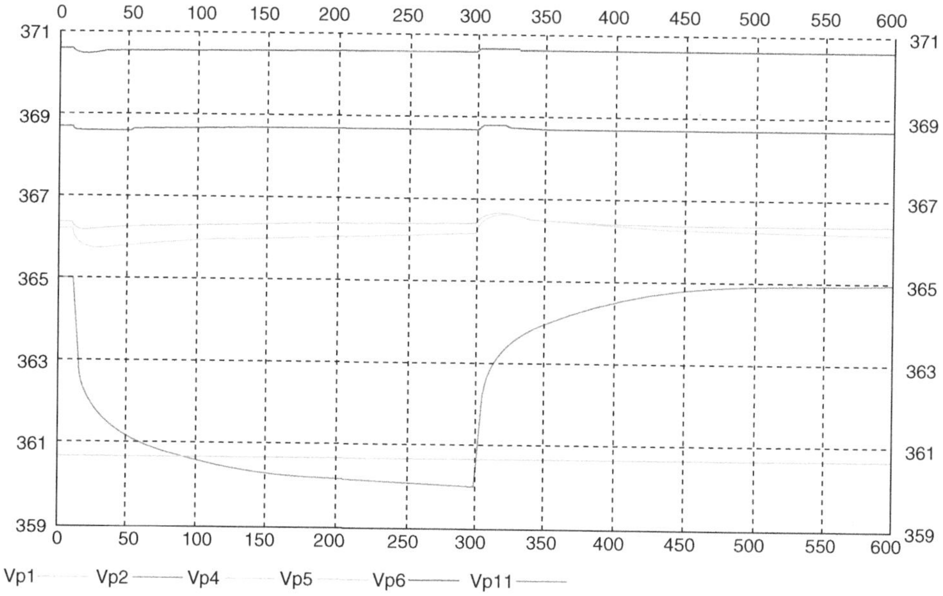

(a)

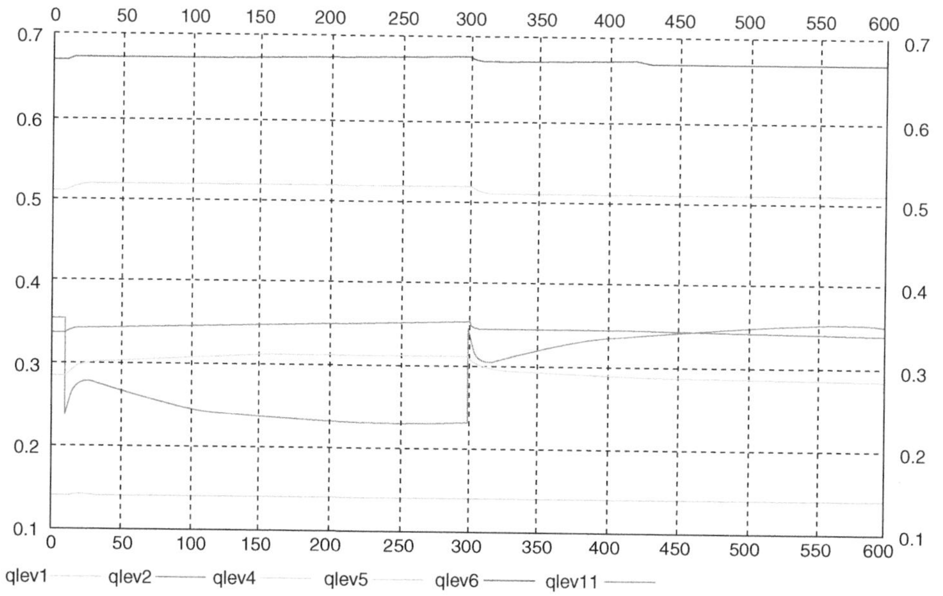

(b)

Figure 7.60. (a) Dynamic response of the pilot node voltages *Vp,i* following the set point V_{ref} step variations at Area 2 only. (b) The correspondent area reactive power control levels.

Again the PQRs impose to all the power system control generators the tracking of the corresponding control level q with a dynamics of 5 s dominant time constant. In its turn, RVR imposes to all the PQRs the tracking of the corresponding pilot node voltage set point with a dynamics of 50 s dominant time constant.

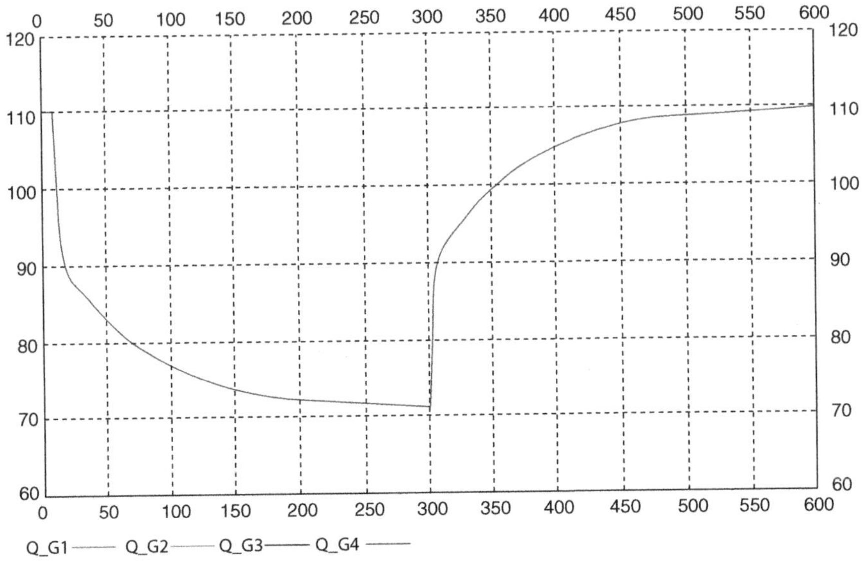

Figure 7.61. The overlapped transients of four generators allowing the Area 2 pilot node voltage set point tracking, following a $V_{ref,2}$ step variation, at a power station under control.

The shown test results clearly evidence that the appropriate operation of the secondary voltage control within each area mostly depends on the way the pilot nodes and control areas are selected.

A control area is correctly defined if the following conditions are met:

- With the pilot node voltage maintained constant, the voltages at the other nodes in the area have small variations even if the local load has significant changes.
- Voltage control within a control area does not influence significantly the voltage in the other areas.
- Control devices in each control area should be, in principle, able to maintain the voltage constant at the pilot node, under normal and disturbed conditions as well.

The combined effects of the two regulators give, as result in Area 2, the four generators reactive powers in Figure 7.61 that move concordantly aligned, tracking the reference control signal q_2 variations shown in Figure 7.60b. Lastly, under RVR, no dynamic interaction among the under test pilot nodes voltage control loops is evidenced (Figure 7.60a).

Identification of control areas and pilot nodes can be achieved by using electrical distance-based methods. The proposed algorithm, later described, is based on the following steps:

(i) Choose the network pilot nodes by selecting those with the largest short-circuit power.

(ii) For each selected pilot node, determine the corresponding area according to an electrical distance method.

(iii) Verify, by using the reactive power balance, if the area reactive power sources can supply the area demand.

(iv) Verify if the voltage variation at the pilot node is representative for the voltage variations in the other area busses.

(v) Verify that:

- distances between the pilot node and the other nodes in the area are confirmed in the presence of SVR;
- electrical distances between the pilot node of the area in study and the neighboring areas pilot nodes are considerable large.

7.4.1.5 Tertiary Voltage Regulation. The basic idea of TVR comes from the need to increase the system's operating security and efficiency through centralized real-time coordination of the decentralized SVR structure:

- The pilot nodes voltage set points must be adequately updated and coordinated online and in real time with dynamics slower than SVR, considering the real conditions of the overall grid and avoiding useless and conflicting SVR interarea control efforts.
- The pilot nodes voltage set points can be computed and updated in real time, considering the global control system structure and its real-time measurements.
- The pilot nodes voltage set points have to be optimized in real time to effectively minimize grid losses while still preserving control margin.

The third control level, related to the tertiary voltage regulation, is therefore aimed to optimize the nationwide voltage map *in real time*. This involves determining time by time the voltage set points for the pilot nodes in order to achieve safe and economic system operation.

The tertiary loop represents the global coordination in the voltage plan through automatic control actions. One of the objectives is to manage the reactive power flow between power system areas at a low value, minimizing the power losses into the system, and increasing the system controllability and stability. Since the tertiary control is currently performed manually, poor results are achieved with respect to the real time. If automated, becoming online and real time (as in Figures 7.51 and 7.52), it would have a dominant time constant of around 5–10 min.

The vector of the pilot node voltage set-pints $|V_{\mathrm{P,ref}}(t)|$ may be alternatively provided by: the remote RVR automatic systems (with the TVR control function out of service), in terms of the local daily trend; the regional dispatcher operator (RVR manual setting, with the TVR control function out of service); the RVR voltage set point real-time optimization tool (with the TVR control function in service).

The TVR real-time optimization determines the most appropriate pilot node voltage set points $|V_{\mathrm{P,ref}}(t)|$ update for secure/efficient operation, on the basis of an integral law (7.83) of the real-time optimized vector $|\Delta V_{\mathrm{P,ref}}(t)|$ representing the best increment to be actuated on $|V_{\mathrm{P,ref}}(t)|$ according to the minimization of the TVR objective function (7.84) [46]:

$$|V_{\mathrm{P,ref}}(t)| = \frac{1}{T_{\mathrm{T}}}\left[\left|\int_0^t \Delta V_{\mathrm{P,ref}}(\tau)\mathrm{d}\tau\right|\right]_{V_{\mathrm{P\,min}}}^{V_{\mathrm{P\,max}}} + |V_{\mathrm{P}}(0)| \qquad (7.83)$$

where T_{T} is the gain of the integral regulator, fixing the TVR closed loop dominant time constant at 5–10 min; S is the sensitivity matrix between area reactive power levels Δq_{LEV}

and pilot node voltages $\Delta V_{\mathrm{P,ref}}$:

$$[\Delta V_{\mathrm{P,ref}}] = [S][\Delta q_{\mathrm{LEV}}]$$

Equation (7.83) imposes the dynamics of the TVR control loop by integrating the result of the *TVR objective function minimization: Min* (*OF*), being *OF* based on the actual network state estimation and the forecasted optimal voltages and reactive powers plan:

$$\begin{aligned} OF = {} & \left[V_{\mathrm{P}} + \Delta V_{\mathrm{P,ref}} - V_{\mathrm{P}}{}^{0}\right]^{\mathrm{T}} Q^2 \left[V_{\mathrm{P}} + \Delta V_{\mathrm{P,ref}} - V_{\mathrm{P}}^{0}\right] \\ & + \left[q_{\mathrm{LEV}} + S^{-1}\Delta V_{\mathrm{P,ref}} - q_{\mathrm{LEV}}^{0}\right]^{\mathrm{T}} R^2 \left[q_{\mathrm{LEV}} + S^{-1}\Delta V_{\mathrm{P,ref}} - q_{\mathrm{LEV}}^{0}\right] \end{aligned} \tag{7.84}$$

where $[V_{\mathrm{P}}]$ and $[q_{\mathrm{LEV}}]$ are the vectors of the real-time measurements of the pilot node voltages and area reactive power levels; $[V_{\mathrm{P}}^{0}]$ and $[q_{\mathrm{LEV}}^{0}]$ are the vectors of the optimal forecasted pilot node voltages and area reactive power levels (coming from the "State Estimation and OPF" block); Q^2 and R^2 are weight matrices whose selection allows bestowing a privilege on pilot node voltage differences, rather then on the effort of control area reactive power levels.

In conclusion, equations (7.83) and (7.84) together represent the TVR control functionality that can be computed in real time because substantially dependent on real-time measurements. (In principle, $[V_{\mathrm{P}}^{0}]$, $[q_{\mathrm{LEV}}^{0}]$ could remain unchanged at given constant values if state estimation and/or OPF not working. In fact, their updated forecasting is not mandatory but simply a help.)

The compromise reached by TVR, when the available optimal forecasted plan does not fit well the real situation (obviously that more or less regularly happens), should properly consist in the achievement of the highest voltage plan consistent with real operating conditions, which minimize network losses as much as feasible. To achieve this result it is necessary to preserve system controllability, even if close to the limits, in such a way as to avoid the disastrous consequences of open loop operation. In this condition, in fact, the uncontrolled voltages determine undesired heavy reactive power flows, which increase system losses and worsen the operation efficiency. *OF* achieves the objective to preserve controllability; therefore, the proposed TVR is the correct and necessary completion of the hierarchical automatic real-time voltage control system.

Moving from the TVR control level to the higher voltage level where the OPF is computed, the real-time control is necessarily lost because OPF requires the state estimation and its correct updating that, even if achieved every 5 min, is too delayed to be able to track, by OPF, the power system voltage dynamics.

In conclusion, above the TVR level, the voltage control can be of forecasting type only.

7.4.2 SVR Control Areas

7.4.2.1 Procedure to Select the Pilot Nodes and to Define the Control Areas. The adopted criterion to choose the pilot nodes is based on the intuitive concept they have to be selected among the strongest ones. In fact, they must be able to impose the voltage to their surrounding buses also in front of the normal perturbations. The criterion includes a limit on the electrical coupling among the selected pilot nodes.

Evidence is given to the sensitivity matrix $[X_{\mathrm{CC}}]$ (see equations (7.77) and (7.78)).

$[X_{\mathrm{CC}}]$ is the sensitivity matrix of the voltages $[\Delta V_{\mathrm{L}}]$ of the EHV buses with respect to the reactive powers $[\Delta Q_{\mathrm{L}}]$ injected into the same load buses, when the generators voltages are maintained constant $[\Delta V_G] = 0$. This is a diagonal dominant matrix with negative coefficients.

- The diagonal coefficients are of the type

$$(X_{CC})_{hh} = \left(\frac{\Delta V_{L_h}}{\Delta Q_{L_h}}\right)_{\begin{cases}\Delta Q_{L_s} = 0\\ s \neq h\end{cases}}, \quad (h, s = n+1, \ldots, n+N) \tag{7.85}$$

- Those outside the main diagonal are of the type

$$(X_{CC})_{hk} = \left(\frac{\Delta V_{L_h}}{\Delta Q_{L_k}}\right)_{\begin{cases}\Delta Q_{L_s} = 0\\ s \neq k\end{cases}}, \quad (h, k, s = n+1, \ldots, n+N) \tag{7.86}$$

It is important to notice that the generic diagonal coefficient $(X_{CC})_{hh}$ is the equivalent reactance seen by the "h" bus: $\Delta V_{L_h} = (X_{CC})_{hh} \Delta Q_{L_h}$.

The $[X_{CC}]$ matrix plays a fundamental role for *the selection of the pilot nodes and related areas*.

The analytical procedure consists in the subsequent reordering of the $[X_{CC}]$ matrix, according to the following steps:

1. The rows and columns of the matrix are reordered in a way to satisfy the following condition:

$$\begin{cases}(X_{CC})^{(1)}_{1\,1} < (X_{CC})^{(1)}_{r\,r} \\ (X_{CC})^{(1)}_{1\,1} > (X_{CC})^{(1)}_{2\,1} > (X_{CC})^{(1)}_{3\,1} > \cdots > (X_{CC})^{(1)}_{N\,1}\end{cases} \quad \text{with} \quad r = 2, \ldots, N$$

2. They are computed the $(N-1)$ rations:

$$\beta_{ij} = \frac{(X_{CC})^{(1)}_{ij}}{(X_{CC})^{(1)}_{jj}} \quad \text{with} \quad \begin{cases} i = 1, 2, \ldots, N \\ j = 1, 2, \ldots, Z\end{cases}$$

where $0 \leq \beta_{ij} \leq 1$.

3. Establishing the lower limit of the "electrical distance" among the pilot nodes, they are excluded from the subsequent selections of those buses related to the first N_1 rows of the $[X_{CC}]^{(1)}$ reordered matrix having coupling coefficient β_{i1} with bus "1" greater then ε_P:

$$\varepsilon_P < \frac{(X_{CC})^{(1)}_{\eta,1}}{(X_{CC})^{(1)}_{1\,1}} \leq 1; \qquad \eta = 1, \ldots, N_1$$

4. The remaining $(N - N_1) = n_1$ rows and columns of the $[X_{CC}]^{(1)}$ matrix are reordered in such a way the new matrix $[X_{CC}]^{(2)}$ satisfy the following inequalities:

$$(X_{CC})_{11}^{(2)} < (X_{CC})_{rr}^{(2)}$$
$$(X_{CC})_{11}^{(2)} > (X_{CC})_{21}^{(2)} > (X_{CC})_{31}^{(2)} > \cdots > (X_{CC})_{N1}^{(2)}$$

where $[X_{CC}]^{(2)} \in R^{n_1 \times n_1}$; $r = 2, \ldots, n_1$. This corresponds to put the $(N - N_1)$ remaining buses in order to the electrical vicinity with the one, among them, of highest power.

5. Analogously to step 3, the first N_2 of the n_1 nodes (that is, the first N_2 rows of the reordered matrix $[X_{CC}]^{(2)}$), with coupling coefficient with bus "1": $(X_{CC})_{\eta 1}^{(2)}/(X_{CC})_{11}^{(2)}$ greater than ε_P, are no longer considered in the following steps:

$$\varepsilon_P < \frac{(X_{CC})_{\eta 1}^{(2)}}{(X_{CC})_{11}^{(2)}} \le 1; \qquad \eta = 1, \ldots, N_2$$

6. The reordering procedure of the matrices is repeated starting from the $(n_1 - N_2) = n_2$ remaining nodes, according to the indicated procedure up to the $(Z+1)$th reordering, that is, when among the $n_{(Z-1)} - N_Z \stackrel{\text{def}}{=} n_Z$ remaining buses, the coefficient $(X_{CC})_{11}^{(Z+1)}$ of the reordered matrix $[X_{CC}]^{(Z+1)}$ is greater than a predefined value $1/\gamma$ that represents the minimum admissible value of the short-circuit power for a pilot node:

$$(X_{CC})_{11}^{(Z+1)} > \frac{1}{\gamma}$$

7. The Z pilot nodes are those corresponding to the first row of the matrices:

$$[X_{CC}]^{(1)}; [X_{CC}]^{(2)}; [X_{CC}]^{(3)}; \cdots; [X_{CC}]^{(Z)}$$

After having defined the pilot nodes, the coupling parameter β_{ij} for each bus of the grid is computed as follows:

$$\beta_{ij} = \frac{(X_{CC})_{ij}}{(X_{CC})_{jj}}, \quad \text{with} \quad \begin{cases} i = 1, 2, \ldots, N \\ j = 1, 2, \ldots, Z \end{cases}$$

where $0 \le \beta_{ij} \le 1$.

These are the coefficients of the (N, Z) sensitivity matrix $[B_{RL}]$ that represent the sharing of the N grid buses among the Z areas in which the grid has been subdivided.

The ith bus is linked to the area j if it has the highest coupling coefficient with the jth *pilot node*. That is, the ith bus is associated to the area j if $\beta_{ij} > \beta_{ik} \forall k \neq j$.

Other formulations for the electrical distance between two buses, based on $[X_{CC}]$ matrix, are also possible.

7.4.2.2 Procedure to Select the Control Generators. After selecting the pilot nodes and the corresponding areas, it is necessary to select the control generators of each area, that is, the generators participating to the area pilot node voltage control. These control generators are obviously those able to mostly affect the voltage of the considered pilot node. The proposed analysis can be referred to the system model (7.77) adding the additional simplifications of reciprocal network.

Considering a reciprocal network and because $[B]$, $[B_{LL}]$, and therefore $[X_{CC}]$ and $[C]$ are symmetrical matrices, it becomes

$$[H]^T = [B_{GL}][X_{CC}] = -[D]$$

then

$$[B_{eq}] = [B_{GG}] - [D][B_{LG}] \tag{7.87}$$

The simplified system of equations become

$$\begin{cases} [\Delta V_L] = -[D]^T[\Delta V_G] + [X_{CC}][\Delta Q_L] \\ [\Delta Q_G] = -\left[B_{eq}\right][\Delta V_G] + [D][\Delta Q_L] \end{cases} \tag{7.88}$$

$$\begin{cases} [X_{CC}] = -[B_{LL}]^{-1} \\ [D] = -[B_{GL}][X_{CC}] \\ \left[B_{eq}\right] = [B_{GG}] - [D][B_{LG}] \cdots \overset{\text{def}}{=} -[C] \end{cases} \tag{7.89}$$

From these equations it is also possible to obtain the following relationships representing the voltage variations with respect to the injected reactive powers:

$$\begin{cases} [\Delta V_L] = -[S_{LG}][\Delta Q_G] + [S_{LL}][\Delta Q_L] \\ [\Delta V_G] = -[S_{GG}][\Delta Q_G] + [S_{LG}]^T[\Delta Q_L] \end{cases} \tag{7.90}$$

Having defined

$$\begin{aligned} &[S_{GG}] = -[C]^{-1} \\ &[S_{LG}] = -[H][S_{GG}] \\ &[S_{LL}] = [X_{CC}] - [S_{LG}][D] \end{aligned} \tag{7.91}$$

The control generators selection is based on the matrix $[S_{LG}]$ that represents the sensitivity of the EHV load bus voltages vector $[\Delta V_L]$ with respect to the vector $[\Delta Q_G]$ of the reactive powers injected by the generators. The procedure is based on the reordering of the submatrix $[S_{LG}]$ by considering the z rows correspondent to the pilot nodes and the n columns correspondent to the generation buses. This submatrix, called $[S_{RG}]$ represents the sensitivity of the pilot nodes voltages with respect to the injected reactive powers. The procedure selects for each $[S_{RG}]$ column the highest coefficient and reorders the n columns in such a way the first n_1 are all those having the highest coefficient at the first row, that is, those satisfying the following inequalities:

$$(S_{RG})_{1j} \geq (S_{RG})_{kj}, \quad k = 1, 3, \ldots, z; \quad j = 1, \ldots, n_1$$

The second n_2 columns are those having the highest coefficient in the second row, that is, those satisfying the following inequalities:

$$(S_{\text{RG}})_{2j} \geq (S_{\text{RG}})_{kj}, \quad k = 1, 3, \ldots, z; \quad j = 1, \ldots, n_2$$

The procedure continues up to the n_Z columns $(n_1 + n_2 + \cdots + n_Z = n)$. The grouping of the $n_1, n_2, \ldots, n_Z$ columns selects the generation buses linked with the pilot nodes "1", "2", . . . , "z". If A^k_{nj} stands for the generator nominal power at the jth bus of the kth area, the term $(S_{\text{RG}})_{kj} A^k_{nj}$ represents the real generator capacity to affect the pilot node voltage in the considered area. The control generators of the pilot node voltage of the kth area are selected from those inside the area satisfying the following inequality:

$$(S_{\text{RG}})_{kj} A^k_{nj} > \alpha^k_c$$

where α^k_c is the allowed minimum control capability in the considered kth area.

7.4.3 Power Flow Computation in the Presence of the Secondary Voltage Regulation

To obtain the generalized load flow mathematical model in the presence of secondary voltage control, a proportional loading of all the controlling generators of each area with respect their capability is required. Therefore, it is necessary to include in the system model a number of n_c equations that describe the constraints due to the secondary voltage control:

$$\frac{Q^j_1}{Q^j_{1,\max}} = \cdots = \frac{Q^j_{n^j_c}}{Q^j_{n^j_c,\max}}, \quad j = 1, \ldots, n_a \tag{7.92}$$

where n_p is the number of pilot buses, n_c is the total number of controlling generator buses, n_a is the number of SVR areas, and n^j_c is the number of controlling generators of area j.

The subscript "max" denotes the upper capability limit.

Under these considerations, the new mathematical model will be

$$\begin{cases} f_{Pi} = P^{\text{sp}}_i - P_i(V, \theta) = 0, & i = 1, 2, \ldots (n_u + n_g + n_c + n_p) \\ f_{Qi} = Q^{\text{sp}}_i - Q_i(V, \theta) = 0, & i = 1, 2, \ldots (n_u + n_g + n_p) \\ f_i = Q^j_{i+1,\max} \cdot Q^j_i - Q^j_{i,\max} \cdot Q^j_{i+1}, \; j = 1, \ldots, n_a; & i = 1, 2, \ldots, n^j_c \end{cases} \tag{7.93}$$

where n_u is the number of load buses and n_g is the number of generator buses.

The mathematical model is still a nonlinear system of equations and for its solution, a combination of the Gauss–Seidel and Newton–Raphson iterative methods may be used.

Linearizing the system (7.93) gives

$$\begin{bmatrix} \Delta P_i \\ \Delta Q_i \\ f^j_i \end{bmatrix} = \begin{bmatrix} \dfrac{\partial P_i(V,\theta)}{\partial \theta} & \dfrac{\partial P_i(V,\theta)}{\partial V} & \dfrac{\partial P_i(V,\theta)}{\partial Q} \\ \dfrac{\partial Q_i(V,\theta)}{\partial \theta} & \dfrac{\partial Q_i(V,\theta)}{\partial V} & \dfrac{\partial Q_i(V,\theta)}{\partial Q} \\ \dfrac{\partial f^j_i}{\partial \theta} & \dfrac{\partial f^j_i}{\partial V} & \dfrac{\partial f^j_i}{\partial Q} \end{bmatrix} \cdot \begin{bmatrix} \Delta \theta_i \\ \Delta V_i \\ \Delta Q_i \end{bmatrix} \tag{7.94}$$

which allows determining the unknown quantities V and θ, and the adjustments of the reactive power ΔQ_i at the controlling generators, through iterative computation.

7.5 IMPLEMENTATION STUDY OF THE SECONDARY VOLTAGE REGULATION IN ROMANIA

7.5.1 Characteristics of the Study System

The proposed methodology on the SVR pilot nodes and control areas selection is applied to the Romanian power system in order to evaluate the suitability of the secondary voltage control implementation. The case study considers the 220 and 400 kV transmission system: 254 buses, including some 110 kV buses, 280 lines and 63 generating units with a total installed power of about 20,000 MW. The 2008 winter time having a peak load of 7900 MW was studied.

The Romanian power system is operated and coordinated by the National Dispatching Center with the support of five Territorial Dispatching Centers (TDCs), chosen according to geographical and administrative criteria (Figure 7.62).

7.5.2 SVR Areas Selection

As far as the SVR areas partitioning is concerned, three scenarios (5 or 6 or 7 areas) are analyzed to link the different values for the electrical distance thresholds among pilot node

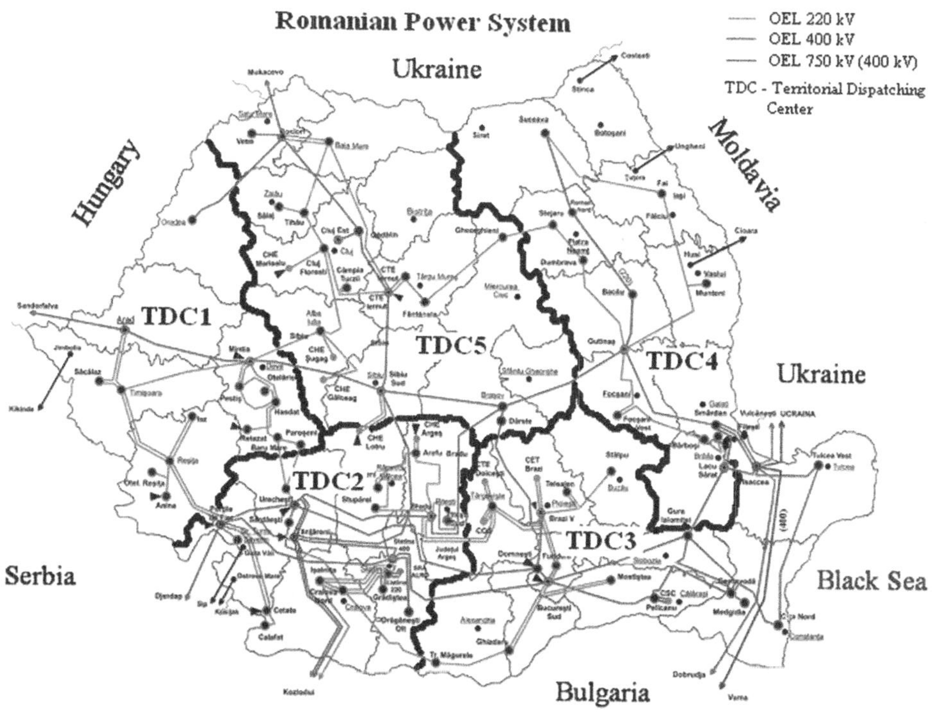

Figure 7.62. The Romanian power grid partitioning into five dispatching regions.

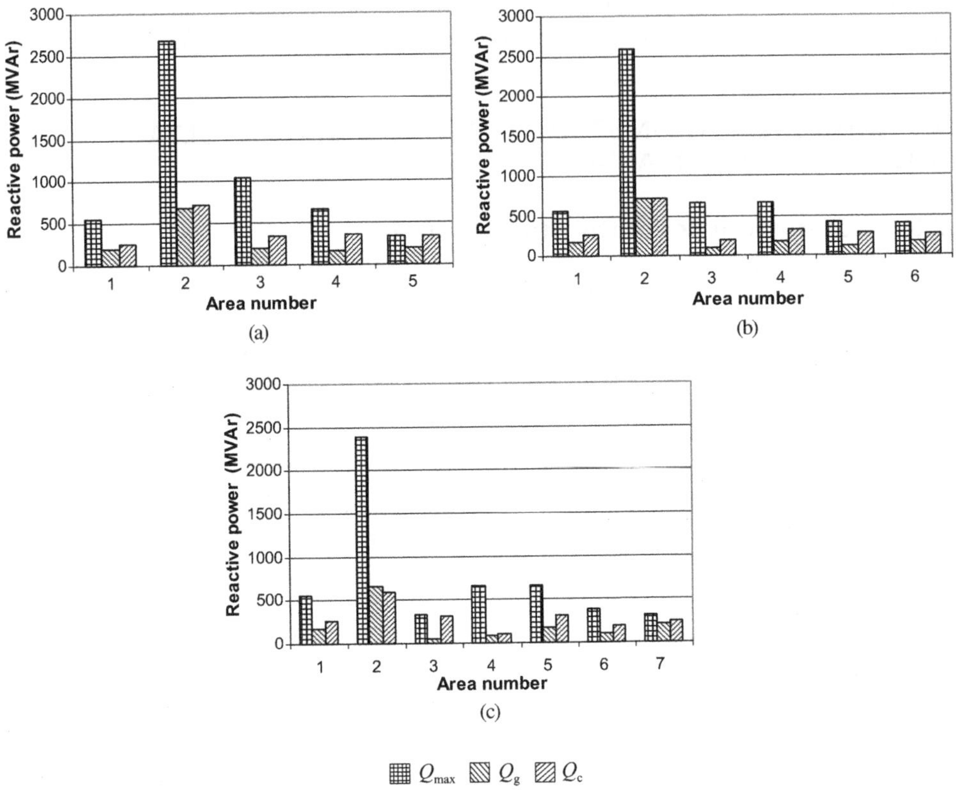

Figure 7.63. Comparisons of reactive powers in partitioned areas in different cases: (a) with 5 SVR areas; (b) with 6 SVR areas; (c) with 7 SVR areas.

buses (values that anyhow guarantee enough electrical distance) with the achievable control margin at each area.

In order to compare the impact of the different subdivisions into areas with the reactive power balance in each area, Figure 7.63a, b, and c shows the values of the reactive power Q_c required by each area loads, the reactive power Q_g produced by each area generators, and the maximum area reactive power reserve Q_{max}, for the examined cases with 5, 6, and 7 SVR areas.

From the ratios of each area reactive powers involved, it comes out that the case with 6 areas shows the largest control reserve at each area, when considering the already on field reactive power resources.

Figure 7.64 provides the ratio Q_g/Q_{max}, in (%), for the 6 areas case, from where the reserve margin of about 60% or more appears as a general result, with the exception of area 6 having a margin of about 25%. A contingency analysis is required to check whether these margins are enough to support voltages.

Figure 7.65 refers to the Romanian power system subdivided into 6 SVR areas, with the following 400 kV pilot buses: Mintia (Area 1), Tantareni (Area 2), Domnesti (Area 3), Lacu Sarat (Area 4), Gutinas (Area 5), and Iernut (Area 6).

The pilot nodes set point values, defined for the considered LF case, assume the following: 1 p.u. at Area 1, 1 p.u. at Area 2, 1 p.u. at Area 3, 1.02 p.u. at Area 4, 1.03 p.u. at Area 5, and 1.05 p.u. at Area 6.

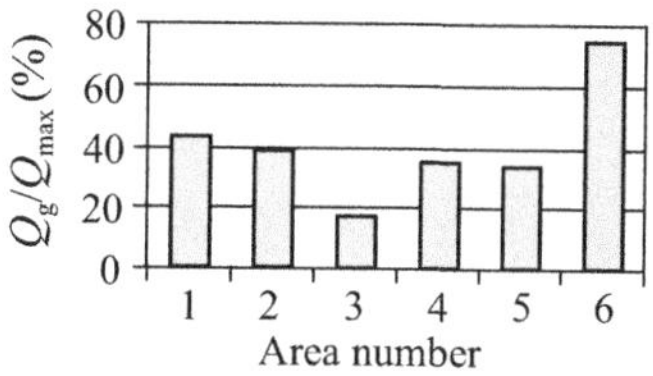

Figure 7.64. The area ratio $Q_g/Q_{max,}$ in the case with 6 SVR areas.

After 1990, the peak load of the Romanian system has decreased constantly from 11,500 to 7900 MW. Currently, the Romanian power system is not particularly stressed as the lines are loaded under the natural power value. Because of that, usually, the lines generate a large amount of reactive power and the control effort by the generators to support voltages is low.

In order to check the efficiency and robustness of the selected secondary voltage control scheme (six areas), the behavior of Romanian power system in front of lines and generators tripping are hereinafter analyzed with two contingency cases.

Case 1: *Line Outage.* The first study case considers as a contingence the tripping of the 220 kV line: Cluj Floreşti-Tihau, in Area 6 (Figure 7.66) [23].

The interest for this area is due to the fact it is directly interconnected with UCTE grid from which and the neighboring areas it receives 102.8 MVAr.

Figure 7.67 shows the voltage profile (in p.u.) of the most representative buses in Area 6. Four scenarios are analyzed: LF—load flow base case; $n - 1$ LF—load flow

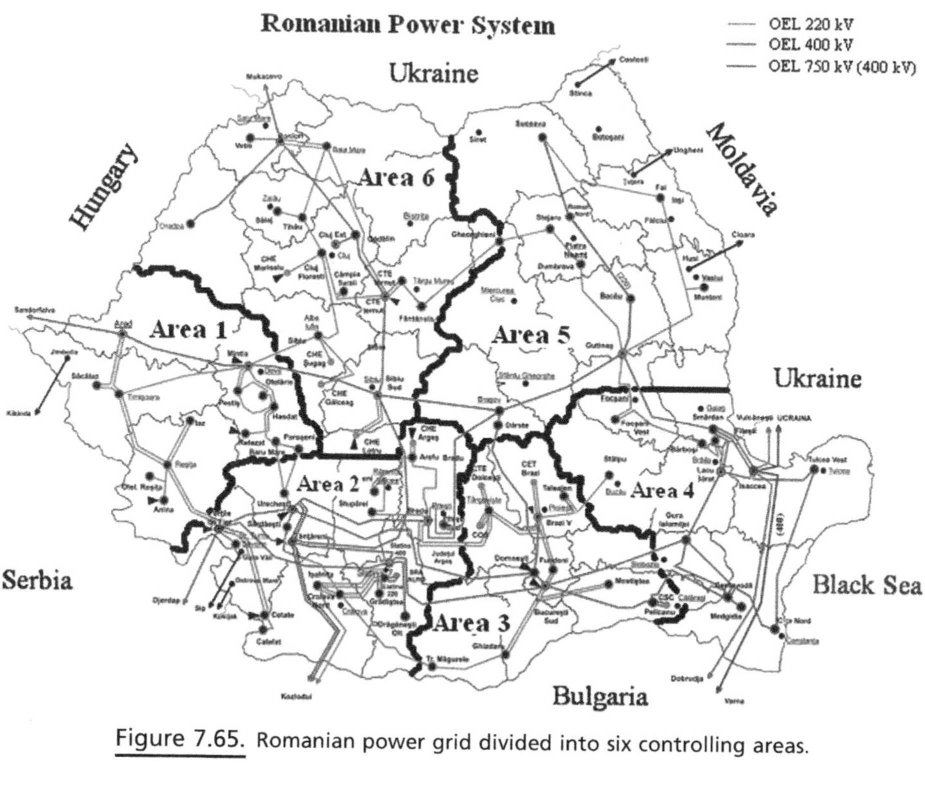

Figure 7.65. Romanian power grid divided into six controlling areas.

Figure 7.66. Area 6 interconnection lines with the neighboring areas and the tripped line in the test.

with the chosen 220 kV line tripping; SVCS—load flow with secondary voltage control system included; $n-1$ SVCS—load flow with contingency in the presence of the secondary voltage control system. The fact that the voltages at the SVCS case are lower than the LF case is simply due to the reduced voltage chosen for the Iernut pilot node with respect to the original LF case value. Usually happens the opposite because SVCS can provide higher voltage values than the basic LF. Anyhow this voltage difference is not relevant for the objective of the considered test.

As it can be seen, the line outage significantly lowers the area voltages, Vetis2 being the most affected bus with a voltage reduction to 0.9 p.u. Figure 7.67 also shows that the presence of secondary voltage control succeeds in maintaining a very good voltage control also in the presence of the contingency.

Furthermore, under SVR the two controlling generators of Area 6 (Mărişel, Iernut2) have the same reactive output (in p.u.) with respect to their maximum

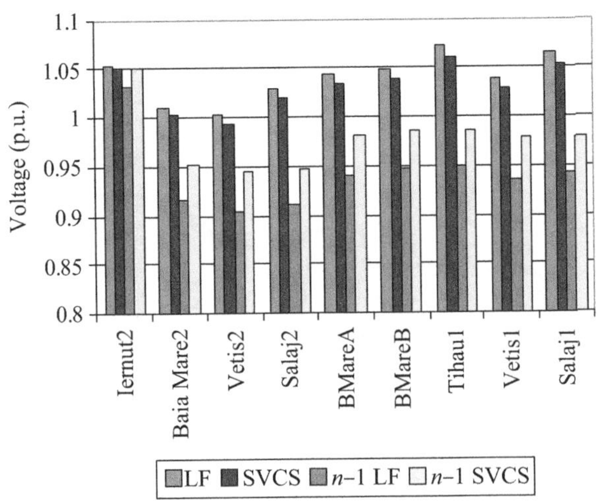

Figure 7.67. Area 6 voltage profiles with and without SVR and in the presence of the 220 kV line outage.

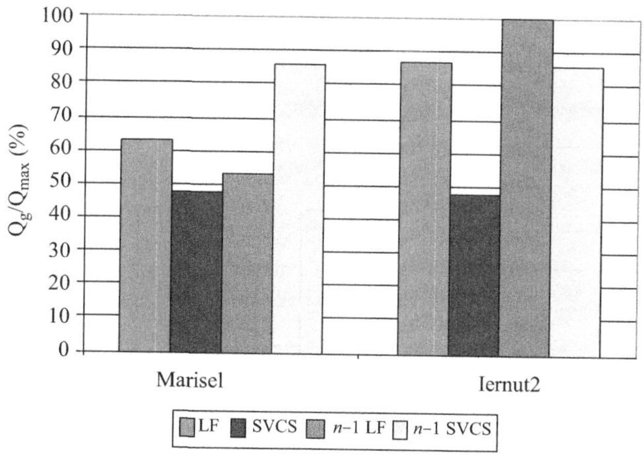

Figure 7.68. Area 6 reactive power delivery by the control generators before and after the line contingence.

capability limits, either with or without contingency (loaded at 47% in the case without contingency and at 85% with contingency). Therefore, the generator units under SVR maintain the same distance from their capability limits (Figure 7.68).

This figure also emphasizes a very different generators behavior in various Area 6 scenarios: in the base case, without SVCS and without contingency, the reactive output of the two generators is already unbalanced and higher than with SVR. The situation is much worse when the contingency occurs: generator Iernut2 reaches its maximum reactive capability reducing the reactive margin of the area.

In synthesis, the SVR control links the area generators to have the same reactive loading (in p.u.) and in case of contingency the lowered reactive output by generator Iernut2 results in a larger reactive margin (15% in the case with secondary voltage control and contingency, as compared to 1% without SVR) due to a better reactive powers coordination in the area.

The control generators of the other system areas are not significantly affected by the 220 kV line Cluj Floreşti-Tihau tripping (the reactive power outputs of these generators do not appreciably change).

Case 2: *Generator Outage.* In the second scenario, the tripping of the largest generating unit (700 MW) of the Romanian power grid is simulated [23]. Figure 7.69

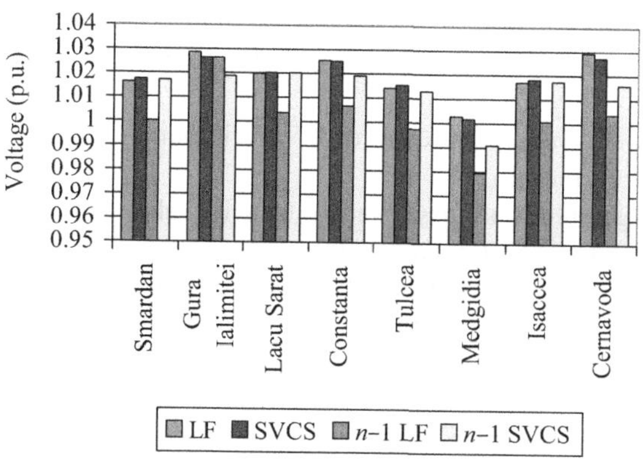

Figure 7.69. Voltage profile in case with the generator outage.

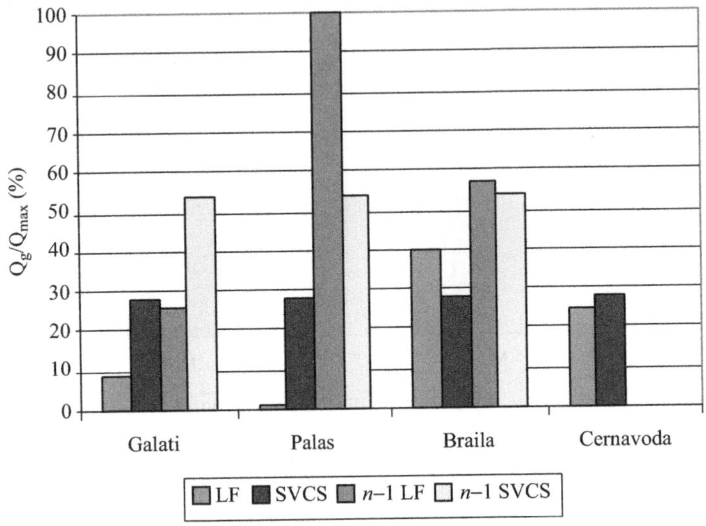

Figure 7.70. Area 4 reactive power delivery by the control generators before and after the generator tripping.

shows the voltage profile (p.u.) of the Area 4 most representative buses (Lacu Sarat is the pilot node), before and after the generator tripping as well as with and without SVR control.

Also in this case the SVR significantly contains the voltages lowering in all buses, with respect to the $n-1$ LF case, so achieving a more robust system with respect to the perturbation. Without SVCS, the closest buses (Medgidia, Cernavoda) to the area with contingency are more influenced than the others and their voltages become lower (of 0.02 p.u.) than in the case with SVCS.

Figure 7.70 shows the reactive power outputs of the Area 4 controlling generators, before and after the contingency. The controlling generator to be tripped has a reactive power output equal to 86.1 MVAr in the LF case and 97.7 MVAr in the SVCS case. Without SVCS we can see that the generating units (Palas, Braila) closer to the affected area provides more reactive power then the others.

The test results show the importance of SVR under steady-state conditions, but mostly at the contingency occurrence. Comparing the results of the power flow simulations without and in the presence of SVR it is possible to state that the secondary voltage control recovers at the best the voltages in the areas affected by contingencies while improves the reduction of the active power losses both under steady-state conditions and during transients.

Figure 7.71 shows the SVR losses reduction result of 2.5% without contingency and by 3% upon a contingency occurrence.

Better results in terms of voltage value and losses should be achievable with SVR voltage set point value maintained at the SVCS case value.

In conclusion, the results of the performed static analysis on the Romanian power system show the importance of proper selection of the pilot buses and reactive power control sources defining the secondary voltage control scheme. In the considered tests, the improvements gained by SVR (increase of reactive power control margins, voltages sustain, and losses reduction) are appreciable and allow extrapolating their

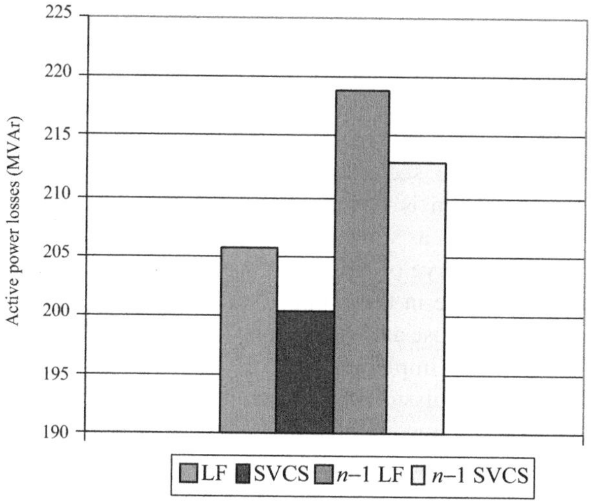

Figure 7.71. Active power losses of the system.

more evident impact in front of critical network situations. Other SVR advantages on system stability and security increase requiring dynamic studies are not here shown.

7.6 EXAMPLES OF HIERARCHICAL VOLTAGE CONTROL IN THE WORLD

7.6.1 The French Power System Hierarchical Voltage Control

7.6.1.1 General Overview. The coordinated voltage control of the French EHV grid operates at three different levels, which are temporally and spatially independent. Temporal independence means that the three controls do not significantly interact. If they did, the risk of oscillations or instability would increase.

Primary control involves keeping generator stator voltages at their set point values according to Sections 7.3.3.1 and 7.4.1.3. This performs partial automatic correction, within a few seconds, to compensate against rapid random variation in the EHV voltage.

The secondary voltage regulation is basically the one presented at Section 7.4.1.4 with some kind of differences when referring to the original dated SVR solution and other kind of differences when considering the more recent SVR control system, called coordinated secondary voltage regulation (CSVR). Secondary regulation involves splitting up the network into theoretically noninteracting zones, within them the voltage is controlled individually. The original SVR control system automatically adjusts the reactive power of certain generating units to control the voltage at a specific point (known as the pilot point) in the zone, this being considered representative for the voltages at all points inside the zone. The recent CSVR has been used in western France and operates by automatically and directly adjusting the AVR voltage set points of the Control Generators optimizing the pilot nodes voltages.

At the highest level, the tertiary regulation is another optimization function applied to the nationwide voltage map. This involves determining voltage set points for the pilot nodes in order to achieve safe and economic system operation. Tertiary regulation still not automated but if it were, it would have a dominant time constant of around 20 min.

Automatic control of the HV capacitors becomes necessary when the amount of MVArs required is high. Automatic control can be carried out at a local level—according to a voltage criterion for instance, or centrally. The local approach may result in insufficient use of all the reactive power sources available in case of an incident, or even due to functional incompatibilities. For this reason, actions had been oriented toward integrating HV capacitor control into the secondary voltage regulation system. As described in Section 7.4.1.4, the integration is governed by the following principle: the capacitors are switched on a priority basis as soon as the need to increase reactive power generation arises. In this way, a large reserve of reactive power can be maintained at generator level, which is immediately available in the event of an incident. Capacitors are progressively switched on, beginning with those at the lowest voltage level. In France, secondary voltage regulation began to be widely implemented in 1979 [20].

At present, France's transmission network comprises about 35 control zones including about 100 thermal generators (conventional fuel and nuclear) and 150 hydraulic generators. Total reactive power capacity available to perform voltage control is estimated at more than 30,000 MVAr.

7.6.1.2 Original Secondary Voltage Regulation. The original secondary voltage regulation system regulates the voltage profile in each zone by distributing reactive power from the various regulating generators. A control system (see also Section 7.4.1.4) comprising two distinct regulation loops is superimposed on the primary loop (AVR) of the regulating generators (Figure 7.72).

A proportional-integral law is used to calculate a control signal N, also termed the "level" of the zone, from the difference between the set point value at the pilot node and the voltage effectively measured at the given instant. The level thus indicates the zone's reactive power requirement:

$$N = \alpha \int_0^t \frac{V_{pp,sp} - V_{pp}}{V_n} dt + \beta \frac{V_{pp,sp} - V_{pp}}{V_n} \tag{7.95}$$

where α and β are the integral and proportional gains; V_{pp}, $V_{pp,sp}$, and V_n are the measured, set point, and nominal voltages, respectively; $V_{s,sp} = V_{s,sp0} + \Delta V_{s,sp}$ is the AVR set point.

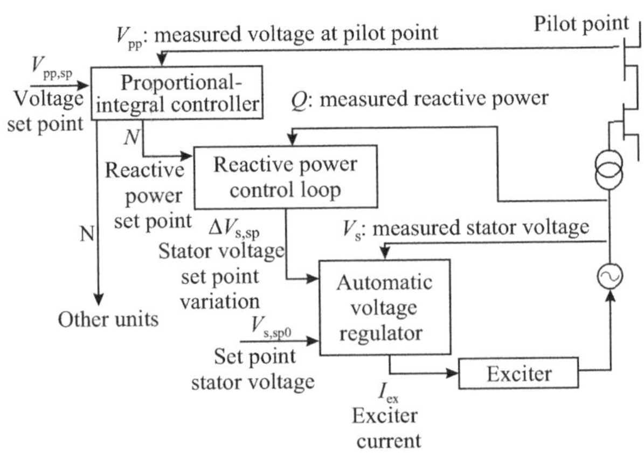

Figure 7.72. Secondary voltage regulation block diagram [19].

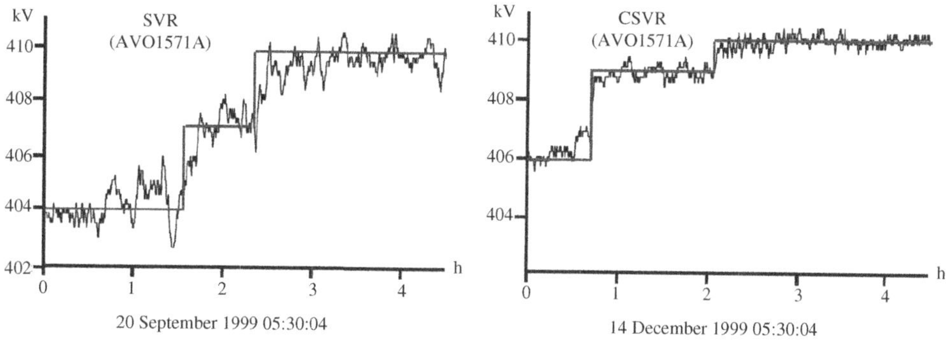

Figure 7.73. Voltage set point and voltage response with SVR and CSVR at EDF.

This original SVR control scheme that has been in operation since the early 1980s confirms the one shown in Figures 7.51 and 7.52 but its design and performances cannot be compared to those given in Section 7.4.1.4, due to the following limitations:

- The level of each pilot node is calculated by a dedicated microcomputer located in the zone's regional dispatching center. Therefore, each pilot node has a dedicated RVR regulator device computing its N level, eliminating any possible dynamic interaction among the pilot node controls of the edges zones.
- Very slow dynamics: in Figure 7.73 a dominant time constant of about 30 min is shown. It is very slow with respect to the declared 3 min or the 50 s indicated in Section 7.4.1.4.
- Abnormal transients due to the PQR initialization and standby period of 5 min to allow reactive power alignment among generator sets.
- Need of corrective actions on the control level by the operators because of the transients induced in the primary voltage control systems.

Other SVR French system limitations are declared structural [19], even if they appear fully linked with the choices made in the use of the control structure in Figure 7.71. Hereinafter, some comments on the reasons of the declared limitations are provided:

- In case the couplings between theoretically independent zones easily change, as a result of grid development subsequent to the already implemented SVR, which confirms the weakness of the pilot nodes selection that conversely is a mandatory issue. From this point of view the choice of 35 pilot nodes in the French system appears a very high and critical number, easily compromising the robustness of the zones selection in front of small system changes.
- If SVR, as suboptimal control, while requiring reactive power alignment of the generating units involved, does not allow excessive demands that might be made on certain units as a result of differences in physical proximity; this impediment becomes real and consistent mainly when the selection of the SVR control generators is not properly done, otherwise the alignment would not determine a so relevant renouncing.

- In case the internal reactive power control loops at generating-unit level become a destabilizing factor, amplifying the initial disturbance in the first few instants following certain incidents (e.g., generator drop-out), this could happen only when the proportional coefficient of the SVR area control law is not correctly computed or the telecommunications between RVR and PQR are delayed.

Other limitations are hardware and software design related:

- If the system makes only partial allowance for operating constraints (e.g., it does not fully integrate monitoring of permissible voltage limits or generating set operating limits), this is against the consolidated indication the PQR has to show to the power station operator as well as to send to the RVR operator real-time information on generator voltage and reactive power permissible limits together with its operating state under SVR.
- If thc use of fixed control loop parameters precludes optimum allowance for operating conditions, the solution of PQR and RVR adaptive control laws surely maintain their closed loops dynamics at the designed values; anyhow, the adaptive control is not a mandatory issue and can be a need in special cases only.

7.6.1.3 Coordinated Secondary Voltage Regulation. This new control system has been operating in western France since 1998. It is called coordinated secondary voltage regulation system because control signals for neighboring zones are no longer calculated on an independent basis, as is the case in the original SVR system [21].

The design of the CSVR is based on a layout similar to that used in the SVR with the additional goal of eliminating some practical limitations described above. The basic principle governing the coordinated secondary voltage regulation system continues to be that of regulating pilot node voltages at set point values. However, the control signal is calculated for a "region" comprising several pilot nodes and the effects of individual generators on all pilot nodes are correctly taken into account, analogously to the control scheme in Figure 7.55. The first relevant difference is the fact that CSVR directly computes the set point updates of the generator unit primary voltage controls by minimizing a multivariable quadratic function [21]. This direct control of the AVR set points is obviously less precise on field than the reactive power control because of the AVR set points offset in the electronic circuits and the small range of the voltage controllability ($\pm 5\%$) with respect to the large reactive power range between the over- and underexcitation limits. Therefore, the generators' alignment in a power station is more critical as this increases the complexity of the generator coordination and the risk of reactive power recirculation among the power plant generators. On the contrary, through this way, on one hand, the design problem of a correct computing of the PI control law is skipped but, on the other hand, the criticism related to the choices of the weights to be used in the following quadratic control function is introduced.

Through CSVR, the generators set point values are obtained by minimizing the following multivariable quadratic function:

$$\min\{\lambda_{\mathrm{v}}\left\|\alpha(V_{\mathrm{pp,sp}}-V_{\mathrm{pp}})-C_{\mathrm{v}}\Delta V_{\mathrm{s,sp}}\right\|^{2}+\lambda_{\mathrm{q}}\left\|\alpha(Q_{\mathrm{ref}}-Q)-C_{\mathrm{q}}\Delta V_{\mathrm{s,sp}}\right\|^{2}$$
$$+\lambda_{\mathrm{u}}\left\|\alpha(V_{\mathrm{s,sp}}-V_{\mathrm{s}})-\Delta V_{\mathrm{s,sp}}\right\|^{2}\}$$

where α is the control gain; V_{pp}, $V_{\mathrm{pp,sp}}$ are the measure and set point voltage values at pilot nodes; Q, Q_{ref} are the measure and set point reactive power values at generating units;

$V_{\rm s}, V_{\rm s,sp}$ are the measured and set point stator voltage values; $\Delta V_{\rm s,sp}$ is the vector of stator voltage variation, $\Delta V_{\rm s,sp} = V_{\rm s,sp} - V_{\rm s}$; $\lambda_{\rm v}, \lambda_{\rm q}, \lambda_{\rm u}$ are the weights for terms in the objective function: pilot node voltage, reactive power, and generator unit stator voltage; $C_{\rm v}$ is the sensitivity matrices linking variations in pilot node voltage to variations in stator voltage (network is modeled by sensitivity matrices for coordination between generating sites); and $C_{\rm q}$ is the sensitivity matrices relating variations in reactive power to variations in stator voltage.

Network and units constraints are taken into account at each computation step using the following equations:

$$\begin{aligned}
&\|\Delta V_{\rm s}\| \le \Delta V_{\max} \\
&a\left(Q + C_{\rm q}\Delta V_{\rm s}\right) + b\,\Delta V_{\rm s} \le c \\
&V_{{\rm pp}_{\min}} \le V_{\rm pp} + C_{\rm v}\Delta V_{\rm s} \le V_{{\rm pp}_{\max}} \\
&V_{{\rm ps}_{\min}} \le V_{\rm ps} + C_{\rm vs}\Delta V_{\rm s} \le V_{{\rm ps}_{\max}} \\
&V_{{\rm THT}_{\min}} \le V_{\rm THT} + C_{\rm v}\Delta V_{\rm s} \le V_{{\rm THT}_{\max}}
\end{aligned}$$

where *a*, *b*, and *c* are the coefficients of straight lines representing operating diagrams for generator units (*P*, *Q*, *V*); these diagrams depend on the active power output by the generator unit. $V_{\rm pp}, V_{{\rm pp}_{\min}}, V_{\rm pp}^{\max}$ are the measure, minimum, and maximum voltage at pilot nodes; $V_{\rm ps}, V_{{\rm ps}_{\min}}, V_{\rm ps}^{\max}$ are the measure, minimum, and maximum voltage at sensitive points; and $V_{\rm THT}$ is the voltage computed at generator unit EHV output.

The control system monitors the voltage at a limited number of network nodes, or "sensitive nodes," that are nodes at which the voltage must be kept between upper and lower limits and not controlled by an integral control law tracking the set point values like for the classic pilot nodes. The control function to be minimized apparently seems an improvement on the subject of voltage control: the weightings in the objective function may be adjusted to suit different control policies, giving priority to keeping pilot node voltages at reference values (e.g., high-voltage values), or to keep the reactive power generation close to the lower limit in order to gain reactive power margins. In practice, this is a minor facility for many reasons:

(i) Substituting the integral control with an optimal control law computed achieving a compromise between voltages and reactive powers, through fixed control weightings and matrices, does not allow minimizing the system losses at the best by renouncing to a full voltage support in normal and perturbed working conditions.

(ii) In front of system changes the available control matrices and weightings (off-line defined) cannot be adequate for the new unpredicted operating conditions, with consequent suboptimal control. To properly change these parameters, a state estimation update has to be waited that is not compatible with the SVR required dynamics.

(iii) In the voltage–reactive power control problem, the buses voltages are the regulated variables through the generators reactive powers control. No reason and some risk for any kind of balance between voltage and reactive power values mainly when the control parameters are not adequate to the new operating condition after contingencies.

(iv) In practice, the EHV voltages weighting choices, at Electricité de France, are higher than those for the other two terms. This confirms that the voltages are the main objective (see Figure 7.73) and they have to be controlled by the reactive powers at a changeable amount depending on the operating state. This also confirms the low usefulness, at the SVR level, of controlling voltages and reactive powers at the same time because the generators reactive powers must be free to timely move between the over- and underexcitation limits, mainly when the control matrices and weightings are not of adaptive type.

In conclusion, CSVR should have a performance very similar to the (7.95) control law and different from TVR (7.84) that notwithstanding the two optimizing functionalities have very similar structure. This also because only a very slow dynamics linked with the state estimation update should allow the correct computing and update of the full CSVR control parameters and this is not easily achievable also for a closed loop TVR with a dominant time constant of about 10–20 min. Therefore, CSVR should have different from zero only one weight (λ_V, reasonably) to be enough faster and less critically linked to the system state estimation.

7.6.1.4 Performances and Results of Simulations. After 5 years under full-time operation in western France, CSVR has gained local operator's confidence and showed its many advantages that are comparable with the reference performances given in Section 7.4.1.4.

The experimental CSVR system shows a better dynamic response (Figure 7.73) with respect to the very slow, original SVR in France.

Voltage Control in Case of Failures and Load Variation. In case of failures (unit or line tripping), primary voltage regulators contribute to enhance voltage regulation but sometimes this is not effective and sufficient. Consequently, the network remains weakened. After the AVR actions, the CSVR allows to maintain and restore the voltage profile by mobilizing and coordinating reactive generations. It is, therefore, possible to help on preventing voltage collapse on the CSVR control area.

Figure 7.74 shows the rapid restoration of pilot node voltage by the CSVR after a drop of 5 kV caused by a tripping of units, in less than 3 min.

Figure 7.75 compares the network voltage performance with primary control only and with CSVR for a very severe load increase situation (30%/h, 60,000 MW of initial load), without modifying the generator operating schedules. This example shows that the network

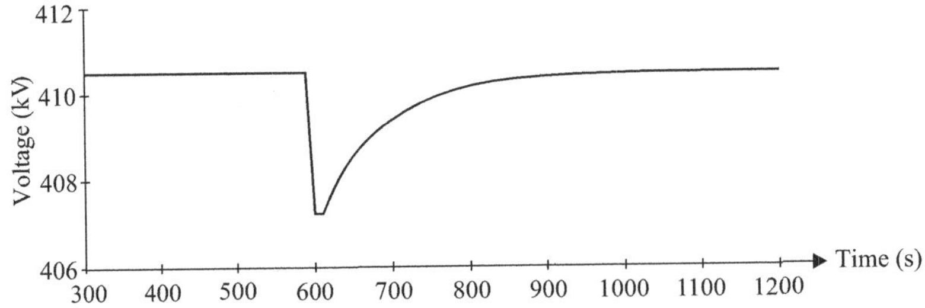

Figure 7.74. Tripping of units, pilot point voltage.

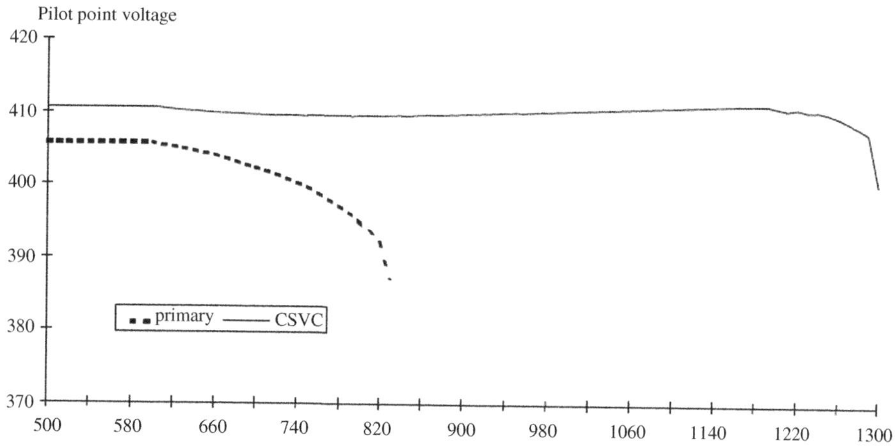

Figure 7.75. Pilot node voltage.

controlled with CSVR can be loaded 3000 MW more when compared to the control option only relying on AVRs.

7.6.1.5 Conclusion on the French Hierarchical Voltage Control System. Even if the CSVR secondary voltage regulator system appears as a solution overcoming the SVR level (no reactive power control loops but direct control of the primary set points; no pilot node integral control law but optimization of pilot nodes voltages and reactive power control margins), therefore deceiving on the possibility to merge together the second and third hierarchical levels by a secondary including the tertiary voltage control, in practice this is not the correct understanding. The control dynamics required to CSVR to realize a true secondary voltage control loop impose to the optimization function to correctly refer to pilot node voltages only (as in Figure 7.74). Otherwise, the inclusion of the optimal control margin objective asks for the continuously updating of the control weightings and matrices, according to the grid real operating conditions. This cannot be achieved without waiting for the state estimation update. This is an unavoidable conclusion mainly when the number of the pilot nodes is high and the limits on the sensitive nodes narrow. Moreover, because the optimization of voltages and control margins has to be computed for the overall network instead of each region separately, it is confirmed this task being proper for the TVR level.

7.6.2 The Italian Hierarchical Voltage Control System

7.6.2.1 General Overview. The Italian coordinated voltage control system is characterized by a hierarchical control structure (Figure 7.55), where each level generates the reference set points for the inner control level [16,17].

According to the reference description in Section 7.4.1.4:

- The primary level includes the classical AVR units already operating in the power plants.
- The secondary level includes the power plant voltage and reactive power regulators (REPORT, in figure called PQR), able to operate autonomously as an "advanced high-side voltage regulators" or in a coordinated way under the control of the RVR to achieve the SVR.

- The tertiary level includes the centralized TVR that updates in real time all the pilot nodes voltage set points (SVR set points) defined by the solution of an optimization problem with an objective function that represents a compromise between security and economy (Section 7.4.1.5).

The mentioned three hierarchical levels are real time, overlapped closed control loops that require a well-defined design, and investigation of the dynamics and stability.

The studies for selection of the pilot nodes and control power plants have resulted in the Italian power system (55,000 MW peak) subdivision into 18 control areas. This plan involves the largest thermal and hydro power plants connected to the 400 and 230 kV grids, for a total reactive capacity of about 20,000 MVAr. Figure 7.76 shows the subdivision into

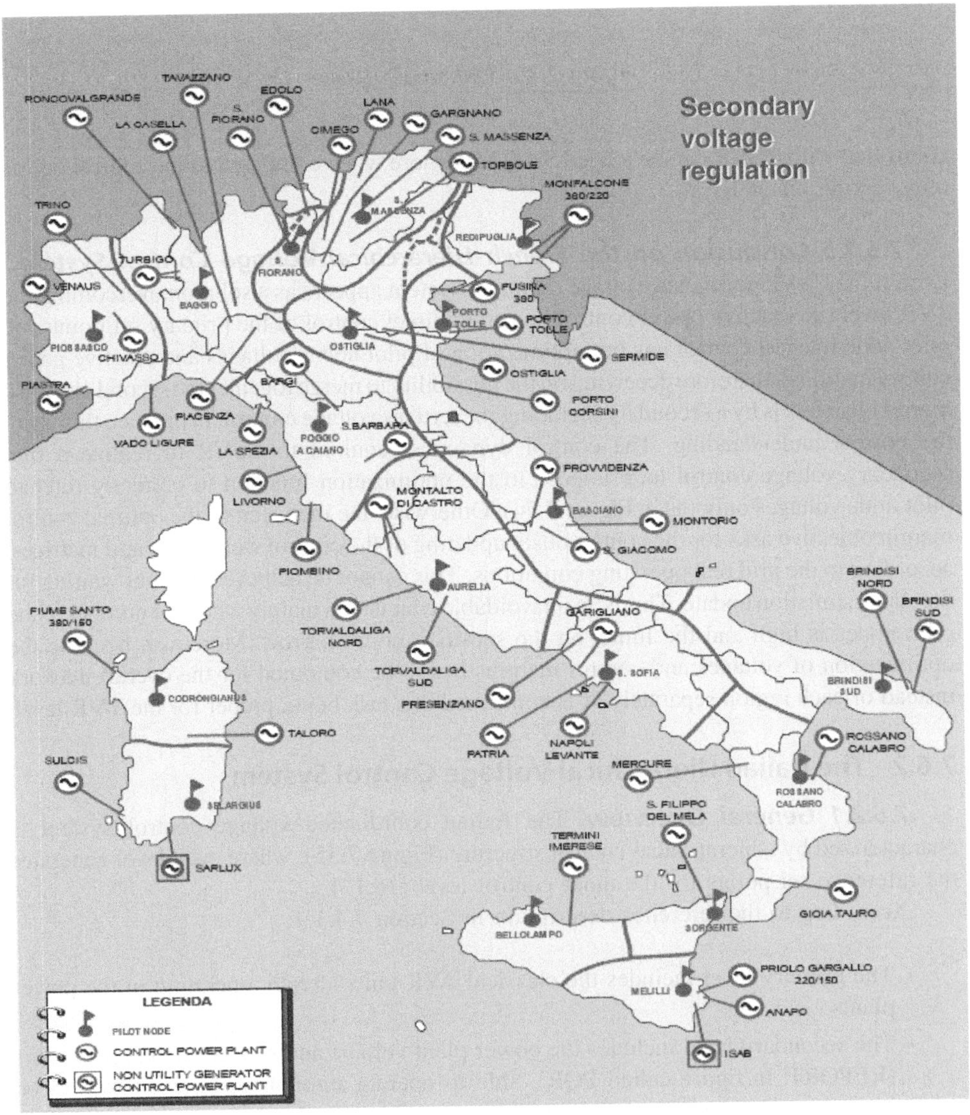

Figure 7.76. Application of the hierarchical SVR in the Italian grid.

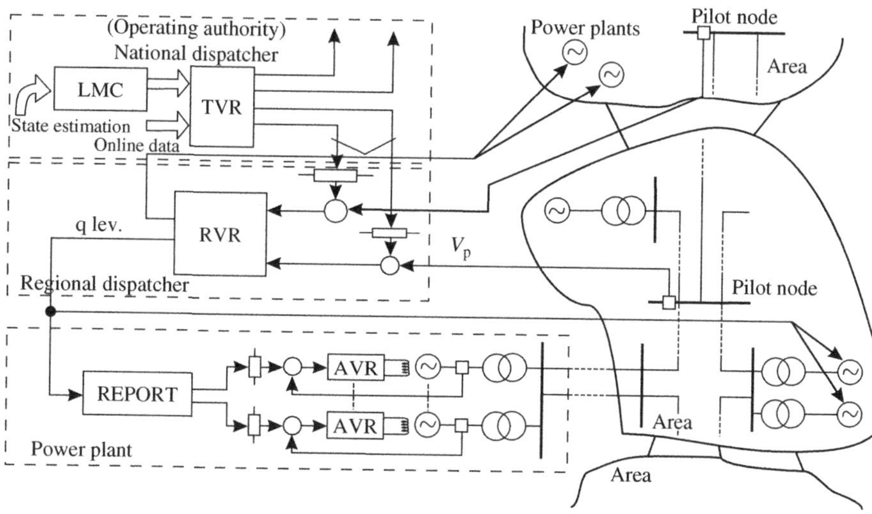

Figure 7.77. Schematic diagram of the Italian hierarchical voltage control system.

areas of the Italian network, pointing out the pilot nodes and the corresponding control power plants according to the SVR application plan.

The Italian hierarchical voltage control system (Figure 7.77) regulates the voltages of the main high-voltage buses (pilot buses) in a closed loop through real-time control of the reactive resources that mostly influences those buses. This permits secure transmission network operation, very close to the highest voltage limits, through rapid control of the main generators (control plants), coordinated by a reactive power level within the same control area and automatically forced to their limits only when needed. The regional voltage regulators close the control loops of the pilot node voltages, providing each area with a specific reactive power level that controls the local power plant's voltage and REPORTs). In turn, the REPORT closes the reactive power control loops of the plant units, directly acting on the set points of the generators' AVRs. RVR also controls capacitors banks, shunt reactors, OLTCs, and SVCs to avoid saturation of the area generators. AVR rapid control is referred to as PVR. The combination of REPORT and RVR implements the SVR. At the highest hierarchical control level, a tertiary voltage regulator coordinates the RVRs in a real-time closed loop. It establishes on the basis of the actual field measurements, the current pilot node voltages, which achieve the minimum feasible grid losses, by slow RVR set point correction, keeping the system under control at all times. To achieve this further aim, an optimal reactive power flow (ORPF) for losses minimization control (LMC) computes, in short (the day ahead) or very short terms (minutes ahead), the forecasted optimal voltages and reactive levels, starting from the forecasted/current state estimation. Therefore, TVR minimizes the differences between the actual field measurements and the optimal forecasted references. This computed "compromise" represents the maximum tenable voltages plan at any instant. The combination of TVR and LMC forms the national voltage regulator (NVR), which so links ORPF forecasting with the real-time optimization of the SVR set points.

The hierarchical voltage control system has different operation modes, according to its implementation progress, maintenance interventions and transient or persistent failures:

- Without plant telecommunications, or when the RVR is not operating, REPORT automatically regulates the local EHV bus voltage (high-side voltage regulation), according to defined daily trends or the plant operator's voltage set points, as agreed by phone with the regional dispatcher.
- Without TSO telecommunications or when TVR is not operating, the RVR autonomously regulates the pilot node voltages of its controlled areas, according to stored daily trends or the regional dispatcher's choices.
- When the LMC is not operating, the TVR autonomously coordinates the RVRs, assuming, as a reference for the optimization of pilot node voltages and reactive power margins, the available long-term forecasted optimal plan or the national control center operator's manual reference.

In the framework of the voltage control service, the Italian TSO, having completed the application of REPORT apparatuses on all the main power plants, as well as of RVR systems in the regional dispatchers' control rooms, can defines proper voltage service rules in connection with the operating SVR.

7.6.2.2 Improvements in the Power System Operation.

(i) *Voltage Control System Dynamics.*
Hereinafter, some examples of the SVR different on-site control loop dynamics, recorded during commissioning tests, are given.

At "La Casella" power station, the REPORT control recloses reactive power control loops on the units 3 and 4. These control loops are tested in front of their reactive power set points contemporarily change. The step variations of the reactive power level q show (Figure 7.78) transients on the unit reactive powers and voltages with a dominant time constant of about 5 s.

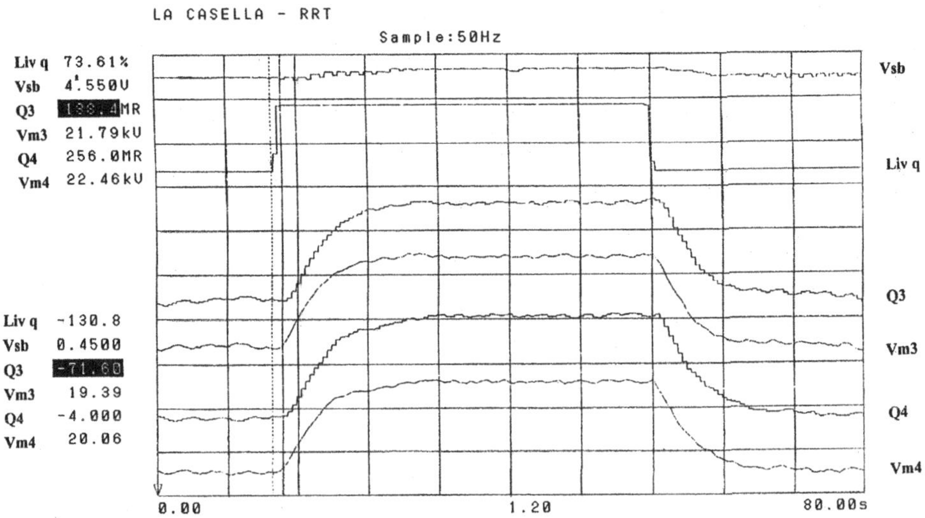

Figure 7.78. Transients of generator reactive power control loops following a step-up/ step-down variations on the reactive power level.

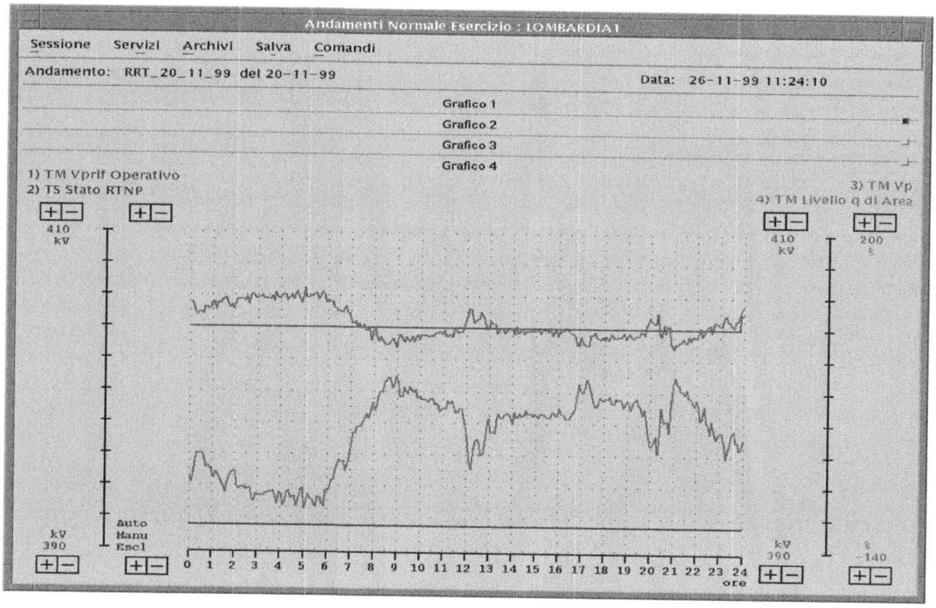

Figure 7.79. Baggio pilot node, daily voltage profile under the remote control of the Milan RVR by the reactive powers of Piacenza, Turbigo, and Tavazzano control power stations.

On "Baggio" pilot node, the following trace shows the daily recorded voltage profile determined by the remote control from the RVR at the Milan Dispatcher Control Room. The on-site test (Figure 7.79) refers to a constant voltage set point value (horizontal segment) for the full day: not optimized voltage trend. This set point choice requires a "compounded" pilot node voltage control loop allowing differences with respect to the set point value when the control effort is high (top trace around the constant set-point). The RVR control output q (bottom trend) represents the Baggio area control effort.

Figure 7.80 refers to the correspondent reactive power productions of the three power stations controlling the Baggio voltage. They show a concordant alignment with q as due. The higher control effort happens in underexcitation at the first 6 h of the day. The SVR voltage control loop dynamics with a dominant time constant of 50 s is too fast to be recognized from these daily traces.

(ii) *Voltage Stability Limit Increase.*

Suitable static and dynamic analyses show that SVR and TVR increase the overall loadability of the transmission system. The study case, the results of which are presented in Figure 7.81, involves a load ramp increase at some buses in the Rome control area. From this case it is possible to compare the simulation results, based on a system dynamic model, obtained with SVR only and with the TVR in service.

The *P–V* trajectory reveals the expected stability improving effect, in terms of both voltage profile and load margins (200 and 300 MW margin are increased at "Roma Nord" and "Roma Sud" buses, respectively).

Such increased overall loadability is also demonstrated by computing suitable off-line and online voltage stability indicators [24,28,44].

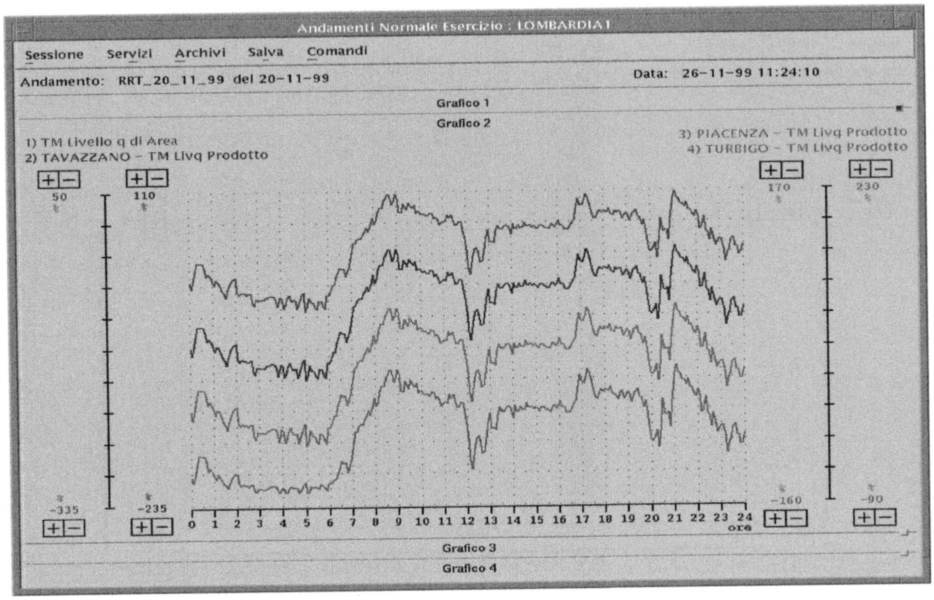

Figure 7.80. Daily reactive power profiles of Piacenza, Turbigo, and Tavazzano control power stations under the remote control of the Milan RVR to regulate the Baggio pilot node as in Figure 7.79.

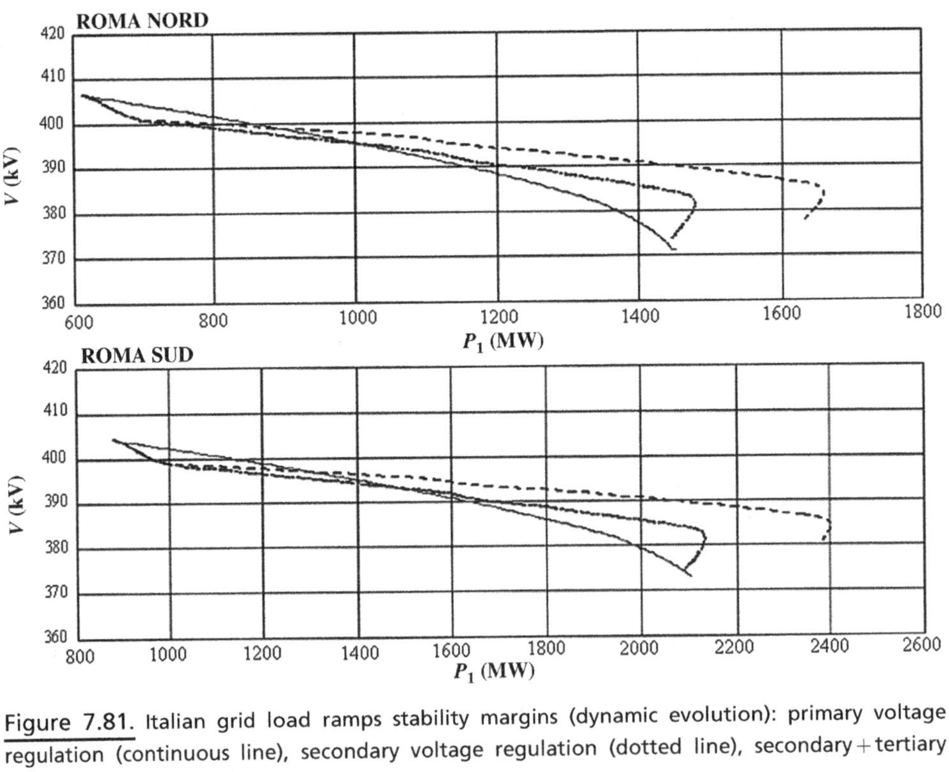

Figure 7.81. Italian grid load ramps stability margins (dynamic evolution): primary voltage regulation (continuous line), secondary voltage regulation (dotted line), secondary+tertiary voltage regulation (dashed line).

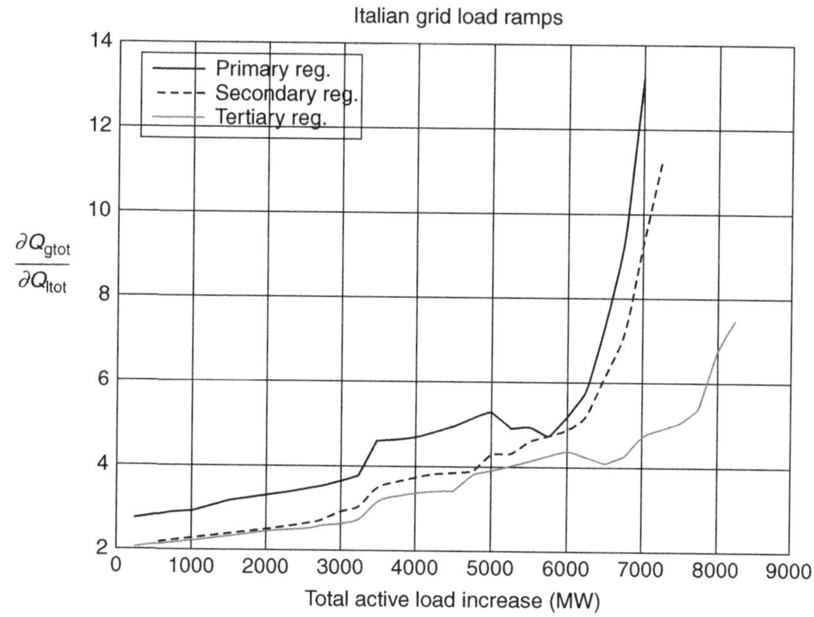

Figure 7.82. Italian grid load ramps stability margins (static evaluation): the $\partial Q_{gtot}/\partial Q_{ctot}$ voltage stability indicator represents the sensitivity of total reactive production with respect to total reactive consumption.

The study case, whose results are presented in Figure 7.82, concerns the overall Italian network and involves an increase in load ramps at all the buses.

In this particular simulation, very short-term reactive power redispatching by LMC has also been simulated: TVR uses, in fact, four different optimal pilot node voltages and area reactive levels, computed at every 2000 MW of total load increase. Due to the presence of SVR and TVR, the largest amount of load margin increase achieved is 1500 MW for the overall Italian grid.

(iii) *Network Losses Reduction.*

The main objective of LMC and TVR is the achievement of minimum losses in the grid, by short-term optimization of the values of network voltages (LMC) in order to update, in quasi-real time, the TVR reference to continuously reduce the overall

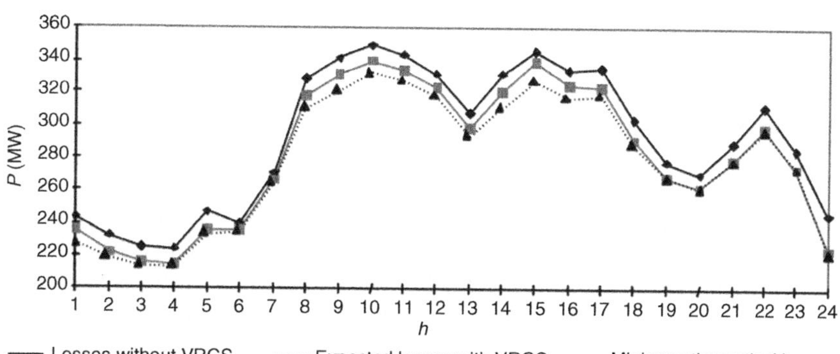

Figure 7.83. Expected loss reduction in the Italian grid with SVR and TVR.

system operation costs. Many static analyses conducted on the overall Italian network in recent years (Figure 7.83) have demonstrated that the application of the multilevel control system for grid voltages and reactive power regulation (VRCS = SVR + TVR) allows a reduction in transmission losses of about 4–6%. Such a control system also achieves better service to the final user in terms of operation quality (i), security (ii), and reliability (iii).

7.6.2.3 Conclusions on the Italian Hierarchical Voltage Control System. The hierarchical voltage control system has been proven to operate with success in the Italian power system since 1985, contributing to simplify and improve the network voltage operation. The SVR and the TVR allow grid operators to achieve the full exploitation of the transmission networks transfer capabilities, as required by the restructured and liberalized energy markets.

In the framework of the ancillary services market, data made available by the proposed control system also allows simple, correct recognition of the real contribution of each generator to the voltage service [43].

The Italian experience started with experimental applications in Florence area and in Sicily, which revealed significant benefits. The control system grew step by step, with plants first operating with REPORT (high-side voltage control) and then, with RVR, participating in SVR. The very satisfactory results inspired TERNA-GRTN to promote widespread application of SVR and begin the development of TVR-LMC [15].

7.6.3 The Brazilian Hierarchical Voltage Control System

7.6.3.1 General Overview. In the Brazilian power system the implementation of the HVCS is currently under study. Up to now, preliminary investigations related to the prospective application of the hierarchical coordinated voltage control to parts of the Brazilian EHV network have been made. The results are satisfactory and further investigations are expected [35].

The analyzed system is the Rio de Janeiro (Rio) grid, which is an energy importing area. This area, which is a part of the Brazilian Southeastern system, has a peak load of approximately 5000 MW in summer time (from January to March). The Rio Area equivalent system model consists of 387 AC buses, 678 AC transmission lines and transformers, 30 power plants, and 5 synchronous compensator units. Figure 7.84 illustrates the main transfer corridors that lead to this Area. The power flowing into Rio Area comes through four transmission corridors, identified as F1, F2, F3, and F4, in Figure 7.84. The main sources of reactive support within the Rio Area are 2 × 200 MVAr synchronous condenser (SC) at Grajau station and the Santa Cruz thermal station. The other reactive sources of interest are located in the transmission system around the Rio Area: Marimbondo, Furnas, and L.C. Barreto power station, and Ibiuna SC. The Rio Area is subdivided in four subsystems: Furnas (124 buses), Light (127 buses), Cerj (57 buses), and Escela (79 buses).

As regards the setup of the secondary voltage control system (SVR), the pilot bus set point error is sent to the generator and synchronous condenser units that participate in the SVR. At each unit the error is weighted by the participation factor K_i and integrated. The integrated output signal modulates the AVR set point such that to remotely regulate the pilot bus voltage. With respect to the reference Section 7.4.1.4, the first relevant difference is the direct computing of the set point updates of the generator unit primary voltage controls by skipping the generator reactive power control loop. As said before, this

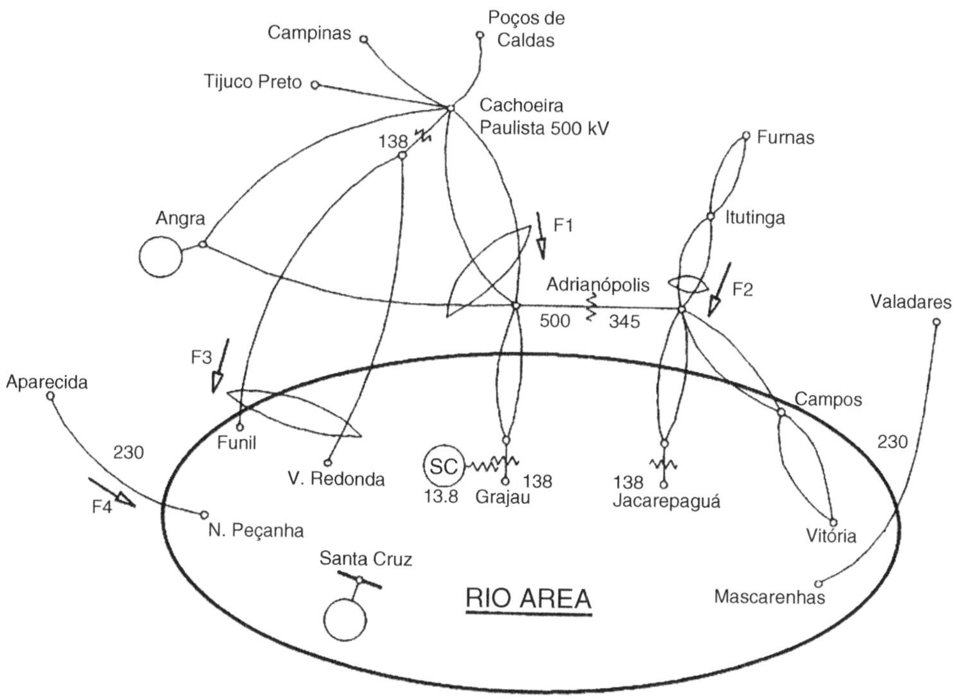

Figure 7.84. Main transmission corridors to the Rio Area (numbers denote voltage levels in kilovolt).

direct control of the AVR set points is obviously, from the engineering viewpoint, less precise than the reactive power control because of the AVR set points offsets and the small range of the voltage controllability (±5%) with respect to the large reactive power range between the over- and underexcitation limits. Therefore, the generators' alignment in a power station is more critical in practice as this increases the complexity of the generator coordination and the risk of reactive power recirculation among the generators. Besides, the generators over- and underexcitation limits are approximately and indirectly considered by the participation factors that also affect the dynamic interaction among the control generators in the area. Figure 7.85 shows the inner (PVR) and outer (SVR) voltage control loops setup; this scheme is useful for studies rather than for practical applications.

7.6.3.2 Results of the Study Simulations. The Rio Area is fully represented together with the transmission corridors through which the power flows into this area. The remaining parts of the Brazilian South-Eastern System are modeled with static equivalents. All participating factors from the control loops are set equal to one and the integrator time constants are set to 100 s. The AVR's steady-state gains are set equal to 50 p.u./p.u, in all generators and SCs.

(i) *SVR Step Response.*
In order to assess the SVR closed loop time response, a step increase of 5% in the reactive load is applied in the Light subsystem, which contains the majority of the loads at the Rio Area. The pilot bus is the Jacarepagua 138 kV bus depicted in Figure 7.84. The power stations participating in the SVR scheme are Furnas,

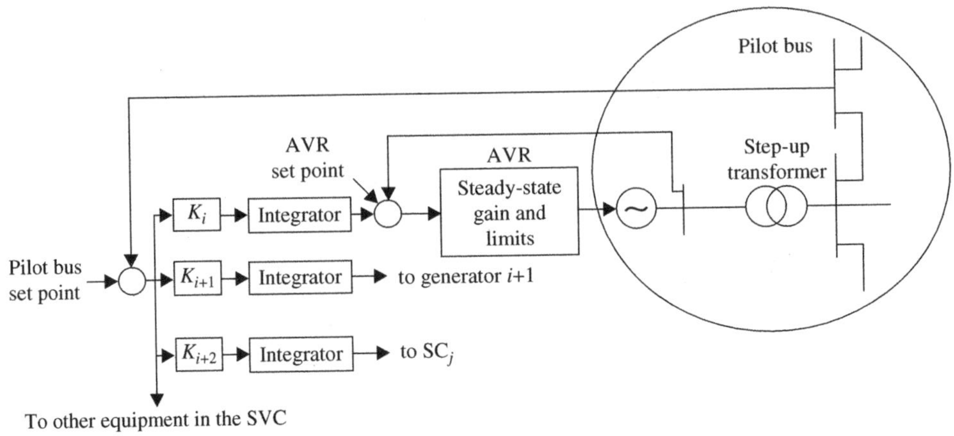

Figure 7.85. SVR setup in the Rio system studies.

Marimbondo, and Santa Cruz while Grajau and Ibiuna contribute with SCs. In Figure 7.86, the voltage with and without the SVR scheme for the step disturbance is shown. To be observed that in the presence of SVR the voltage of the pilot bus returns to its initial value (the value before the step increase is applied) and that the closed loop time response is over damped with a time constant of about 100 s.

(ii) *Load Variation.*

Starting from the base case (heavy load condition), a load variation simulation is performed as follows: (a) from 0 to 300 s: 5% ramp increase in the active and reactive loads of the light subsystem; (b) from 300 to 900 s: loads remain constant at the final values of the previous step; (c) from 900 to 1200 s: loads are reduced to the initial values through a ramp.

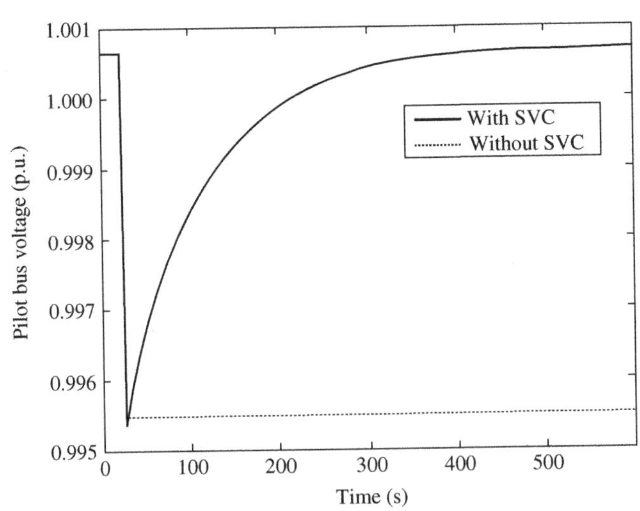

Figure 7.86. Jacarepagua bus voltages following a step increase in the Light subsystem reactive load (with and without SVR).

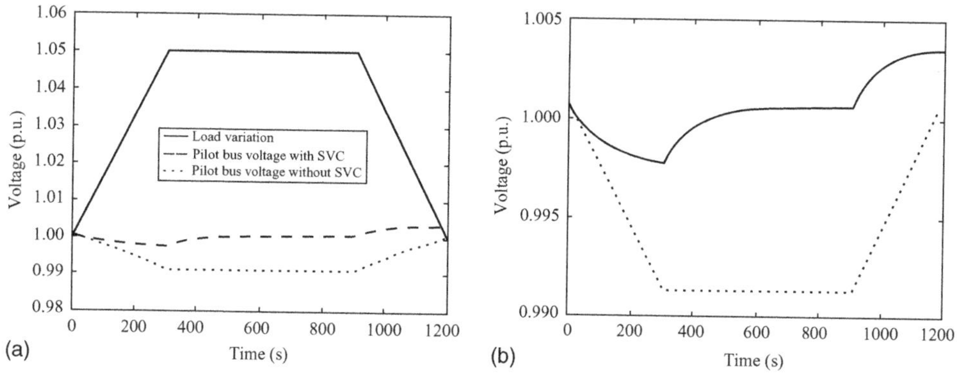

Figure 7.87. Pilot bus voltage response at a trapezoidal load variation: (a) pilot bus voltages and trapezoidal-shaped load variation; (b) detailed variation of the pilot bus voltages.

The load variation is shown in Figure 7.87a, together with the voltages at the pilot bus (Jacarepagua 138 kV) with and without the SVR scheme, while Figure 7.87b only shows the pilot bus voltage. As expected, when no SVR scheme is implemented the system voltages have an upside down trapezoidal shape. With the considered SVR control scheme, the pilot node voltage is characterized by a steady-state error.

(iii) *Single Contingency Case.*

In order to see the effect of the SVR in the case of a contingency, the outage of the 500 kV Angra-Adrianopolis transmission line (TL) is studied (Figure 7.88). The voltage at the pilot bus is 1 p.u. before the contingency and 0.958 p.u. after the contingency. The considered objective of the voltage control system is to maintain the pilot bus voltage at 0.98 p.u., after this contingency. In the absence of the SVR system clearly the objective is not reached.

Further, three SVR schemes, with different number of voltage control equipment regulating the pilot bus voltage, that is, participating in the SVR, are analyzed: (a) only the Jacarepagua OLTC; (b) Jacarepagua OLTC and Grajau SC; and (c) Jacarepagua OLTC, Grajau SC, and Santa Cruz thermal units.

Figure 7.88 shows the voltage at the pilot bus for all the three cases. The curves put in evidence that the control objective is only achieved in the third case. In the first and

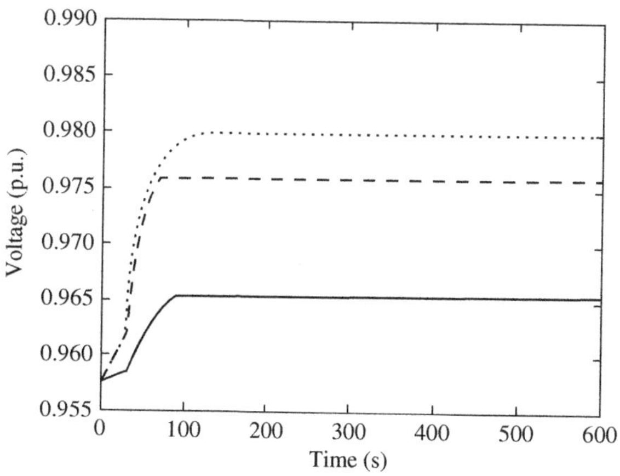

Figure 7.88. Pilot bus voltage: solid line (case a); dashed line (case b); dotted line (case c).

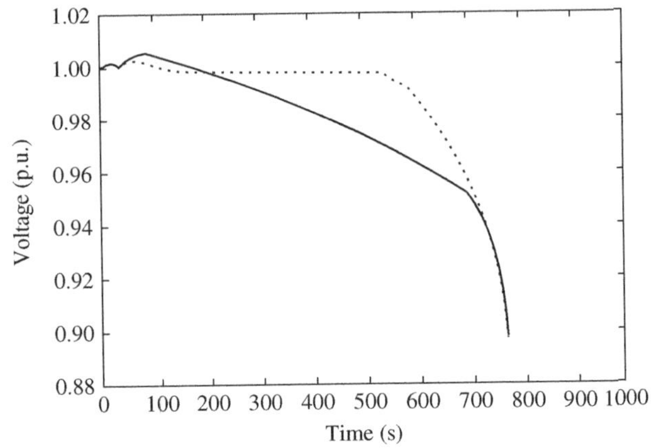

Figure 7.89. Pilot bus voltage: solid line (case a); dotted line (case c).

second cases, the control system steady-state errors, owning to the OLTC reaching its maximum tap limit in the former case and to the Grajau SC reaching its overexcitation limit in the later case. In the third case, a lower steady-state error allows achieving the waited result. More ambitious objective can be reached by using control scheme (Section 7.4.1.4 and integral control law.

(iv) *Loading the Light Subsystem.*
This simulation has the objective of showing not only the benefits on the overall system voltage profile but also the gains in loading margins when using SVR scheme in the Rio Area.

A load ramp is applied to the light subsystem consisting in a 30% load increase in 1000 s at constant power factor. The loads at the remaining subsystems are held constant. Only the first and the third cases described earlier are analyzed. The objective of the SVR is to regulate the pilot bus voltage at 1 p.u.

Figure 7.89 compares the pilot bus voltage for the two cases. Voltage instability is seen to occur shortly before 800 s of simulations.

In the first case (solid line), the voltage deteriorates as the system is loaded. On the other hand, in the third case (dotted line), the voltage at the pilot bus is held constant as long as the reserve of reactive power generation exists. One can note that when the last resource of reactive power hits its limit, a sharp decrease in the voltage at the pilot bus is observed. However, the utilization of the SVR scheme does not increase the maximum loadability of the system for these two cases as both voltage curves collapse at the same time instant. This is due to the limited resources under SVR.

In order to investigate the ability of the SVR system to increase the maximum loadability of the power system, another case (the forth—d) was studied: the same equipment considered in the third case plus Furnas and Marimbondo power plants and Ibiuna SC. Figures 7.90 and 7.91 show the pilot bus voltage for the simulations of the first case (solid line) and this last case (dotted line). It can be clearly seen the increase in the maximum loadability limit with the SVR system adopted in the forth case.

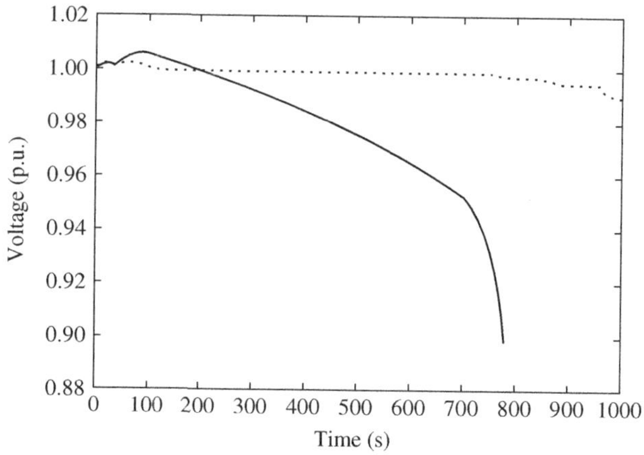

Figure 7.90. Pilot bus voltage: solid line (case a); dotted line (case d).

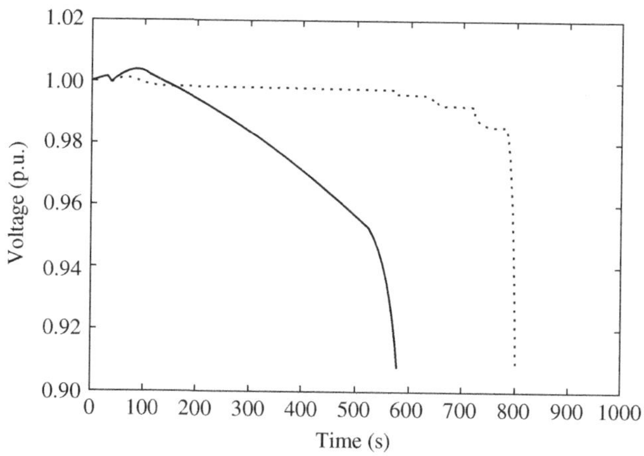

Figure 7.91. Pilot bus voltage: solid line (case a); dotted line (case d).

7.6.3.3 Conclusions on the Brazilian Voltage Control System. The preliminary results involving the use of SVR in the Rio Area showed the benefits gained regarding voltage profile and security. The results also showed the importance of properly selecting the pilot bus and the reactive power sources participating in the SVR scheme. The simulations also indicate that regulation at higher voltage levels leads to a better overall control performance and showed the benefits of including some remote reactive sources in the SVR scheme to control the voltage profile and increase the loading margin.

REFERENCES

[1] Concordia, C. Steady state stability of synchronous machines as affected by voltage-regulator characteristics, *Trans. Am. Inst. Electr. Eng.*, Vol. 63, No. 5, pp. 215–220, 1944.

[2] Concordia, C. *Synchronous machines: Theory and performance*, Wiley, 1951.

[3] Eremia, M., et al. (editors). *Electric power systems. Volume 1: Electric Networks*, Publishing House of the Romanian Academy, Bucharest, 2006.

[4] Marconato, R. *Electric power systems. Volume 1: Background and basic components*, 2nd edition, CEI—Italian Electrotechnical Committee, Jan. 2002.

[5] *Tap (transformer)*, Wikipedia, Online: http://en.wikipedia.org/wiki/Tap_(transformer).

[6] Easum-MR, *On-load tap changer type M – Product Manual*, Easum-MR Tap Changers (P) Ltd., India, 2006.

[7] MR, *On-load tap-changer RMV-II. Operating Instructions*, Product Manual, Maschinenfabrik Reinhausen, Germany, 2010.

[8] Vithayathil, J., et al. *Thyristor controlled voltage regulators*, Brochure CIGRE W.G. B4.35, Feb. 2004.

[9] Faester, A., Goransson, H. *Electronic tap-changers for railway power supplies*, ABB Review, April 1990.

[10] Bulac, C., Eremia, M. *Power system dynamics*, Printech Publishing House, Bucharest, 2006 (in Romanian).

[11] Calovic, M.S. *The use of under-load tap-changing transformer in voltage and VAr flow controls*, Publikacije Elektrotehnickog Fakulteta, Serija Elektroenergetika, No. 88-91, Belgrade, 1982.

[12] Taylor, C.W. *Power system voltage stability*, McGraw-Hill, 1994.

[13] Marconato, R. *Electric power systems. Volume 2: Steady-state behaviour, controls, short-circuits and protection systems*, 2nd edition, CEI—Italian Electrotechnical Committee, Jan. 2004.

[14] Miller, T.J.E. *Reactive power control in electric systems*, Wiley, 1982.

[15] Corsi, S. *The secondary voltage regulation in Italy*, Panel Session-2000 IEEE PES Summer Meeting, Seattle, Jul. 2000.

[16] Corsi, S., Pozzi, M., Sabelli, C., Serrani, A. The coordinated automatic voltage control of the Italian transmission grid, Part I: Reasons of the choice and overview of the consolidated hierarchical system, *IEEE Trans. Power Syst.*, Vol. 19, No. 4, Nov. 2004, ISSN 0885-8950.

[17] Corsi, S., Pozzi, M., Sforna, M., Dell'Olio, G. The coordinated automatic voltage control of the Italian transmission grid, Part II: Control apparatus and field performance of the consolidated hierarchical system, *IEEE Trans. Power Syst.*, Vol. 19, No. 4, Nov. 2004, ISSN 0885-8950.

[18] Corsi, S., Chinnici, R., Lena, R., Bazzi, U., et al. *General application to the main ENEL's power plants of an advanced voltage and reactive power regulator for EHV network support*, CIGRE Conference, 1998.

[19] Corsi, S., Martins, N. (Convenors). *Coordinated voltage control in transmission systems*, CIGRE Technical Brochure, Task Force 38.02.23, Jun. 2005.

[20] Paul, J.P., Leost, J.Y., Tesseron, J.M. Survey of secondary voltage control in France: Present realization and investigations, *IEEE Trans. Power Syst.*, Vol. 2, May pp. 505–511, 1987.

[21] Lefebvre, H., Fragnier, D., Boussion, J.Y., Mallet, P., Bulot, M. *Secondary coordinated voltage control system: Feedback of EdF*, Proceedings of the IEEE PES Summer Meeting, Jul. 2000.

[22] Sancha, J.L., Fernandez, J.L., Cortes, A., Abarca, J.T. Secondary voltage control: Analysis, solutions, simulation results for the Spanish transmission system, *IEEE Trans. Power Syst.*, Vol. 11, No. 2, pp. 630–638, 1996.

[23] Erbasu, A., Berizzi, A., Eremia, M., Bulac, C. *Implementation studies of secondary voltage control on the Romanian power grid*, IEEE PowerTech Conference, St. Petersburg, Russia, Jun. 27–30, 2005.

[24] Corsi, S., Cappai, G., Valadè, I. *Wide Area Voltage Protection*, CIGRE, Paper B5-208, Paris, 2006.

[25] Corsi, S. *Wide area voltage regulation & protection*, IEEE PowerTech Conference, Bucharest, Romania, Jun. 28–Jul. 2, 2009.

[26] Corsi, S., Pozzi, M. A multivariable new control solution for increased long lines voltage restoration stability during black startup, *IEEE Trans. Power Syst.*, Vol. 18, Aug. 2003.

[27] Arcidiacono, V. *Automatic voltage reactive power control in transmission system*, CIGRE-IFAC Survey Paper E, Florence, Sep. 1983.

[28] Corsi, S., Arcidiacono, V., Bazzi, U., Chinnici, R., Mocenigo, M., Moreschini, G. *The regional voltage regulator for ENEL's dispatchers*, CIGRE Conference, Paris, 1996.

[29] Lagonotte, P., Sabonnadiere, J.C., Leost, J.Y., Paul, J.P. Structural analysis of the electrical system: Application to the secondary voltage control in France, *IEEE Trans. Power Syst.*, Vol. 4, No. 2, pp. 479–486, May 1989.

[30] Piret, J.P., Antoine, J.P., Stubbe, M., et al. *The study of a centralized voltage control method applicable to the Belgian system*, CIGRE, 1992.

[31] Van Hecke, J., Janssens, N., Deude, J., Promel, F. *Coordinated voltage control experience in Belgium*, CIGRE Conference, Paris, 2000.

[32] Layo, L., Martin, L., Álvarez, M. *Final implementation of a multilevel strategy for voltage and reactive control in the Spanish electrical power system*, PCI Conference, Glasgow, 2000.

[33] Ilic, D.M., Liu, X., Leung, G., Athans, M., Vialas, C., Pruvot, P. *Improved secondary-new tertiary voltage control*, IEEE WM '95, NY, 1995.

[34] Taylor, C.W., Venkatasubramanian, V., Chen, Y. *Wide area stability and voltage control*, VII SEPOPE, Curitiba, Brazil, May 21–26, 2000.

[35] Taranto, G., Martins, N., Martins, A.C.B., Falcao, D.M., Dos Santos, M.G. *Benefits of applying secondary voltage control schemes to the Brazilian system*, Proceedings IEEE/PES SM, Seattle, Jul. 2000.

[36] Lemons, F.A.B., Feijo, Jr, W.L., Werberich, L.C., da Rosa, M.A. *Assessment of a transmission and distribution system under coordinated secondary voltage control*, PSCC 2002, Sevilla, Spain, 2002.

[37] Eremia, M., Petricica, D., Simon, P., Gheorghiu, D. *Some aspects of hierarchical voltage-reactive power control*, 2001 IEEE PES Summer Meeting, Vancouver, Canada, Jul. 15–19, 2001.

[38] Corsi, S., De Villiers, F., Vajeth, R. *Secondary voltage regulation applied to South Africa transmission grid*, 2010 IEEE PES General Meeting, Minneapolis, Jul. 2010.

[39] Ilea, V., Bovo, C., Merlo, M., Berizzi, A., Eremia, M. Reactive power flow optimisation in the presence of secondary voltage control, IEEE PowerTech Conference, Bucharest, Romania, Jun. 28–Jul. 2, 2009.

[40] Corsi, S., Pozzi, M., Biscaglia, V., Dell'Olio, G. *Fiscal measure of the generators support to the network voltage and frequency control in the ancillary service market environment*, CIGRE Meeting, Paris, 2002.

[41] Corsi, S., Marannino, P., Losignore, N., Moreschini, G., Piccini, G. Coordination between the reactive power scheduling and the hierarchical voltage control of the EHV ENEL system, *IEEE Trans. Power Syst.*, Vol. 10, pp. 686–694, 1995.

[42] Berizzi, A., Sardella, S., Tortello, F., Marannino, P., Pozzi, M., Dell'Olio, G. *The hierarchical voltage control to face market uncertainties*, Bulk Power Systems Dynamics and Control, IREP V Conference, Onomichi, 2001.

[43] Corsi, S., Arcidiacono, V., Cambi, M., Salvaderi, L. *Impact of the restructuring process at Enel on the network voltage control service*, IREP, Santorini, Greece, Aug. 1998.

[44] Corsi, S., Pozzi, M., Marannino, P., Zanellini, F., Merlo, M., Dell'Olio, G. *Evaluation of load margins with respect to voltage collapse in presence of secondary and tertiary voltage regulation*, Bulk Power Systems Dynamics and Control, IREP V Conference, Onomichi, 2001.

[45] Marannino, P., Zanellini, F., Berizzi, A., Medina, D., Merlo, M., Pozzi, M. *Steady state and dynamic approaches for the evaluation of the loadability margins in the presence of the secondary voltage regulation*, MedPower Conference, Athens, Nov. 4–6, 2002.

[46] Corsi, S., Pozzi, M., Bazzi, U., Mocenigo, M., Marannino, P. *A simple real-time and on-line voltage stability index under test in Italian secondary voltage regulation*, CIGRE, Paper 38-115, 2000.

[47] Eremia, M., Trecat, J., Germond, A. *Réseaux électriques. Aspects actuels*, Technical Publisher, Bucharest, 2000 (in French).

PART II

POWER SYSTEM STABILITY AND PROTECTION

Courtesy of Romanian Power Grid Company TRANSELECTRICA

8

BACKGROUND OF POWER SYSTEM STABILITY

S.S. (Mani) Venkata, Mircea Eremia, and Lucian Toma

8.1 INTRODUCTION

The power systems are the largest and most complex manmade dynamic systems. Similar to any dynamic system, the power system is continuously subjected to perturbations and experiences transitions from one operating state to another in the form of oscillations. A primitive condition for the power system to maintain the stability requires that the oscillations be damped.

Power system stability has been an important preoccupation for engineers since the 1920s [1,2]. However, blackouts caused by instability still occur despite the significant advancements in the control and protection technology that have been made. There has been also a continuing growth in interconnections between regional power systems aiming to improve the stability and control conditions, but the stability problem has gained new dimensions. Voltage stability, frequency stability, and interarea oscillations have become greater concerns than in the past. A clear understanding of different types of instability and how they are interrelated is essential for the satisfactory design and operation of power systems [3].

8.2 CLASSIFICATION OF POWER SYSTEMS STABILITY

A IEEE/CIGRE Joint Task Force on Stability Terms and Definitions has defined the power system stability as "the ability of an electric power system, for a given initial operating condition, to regain a state of operating equilibrium after being subjected to a physical

Handbook of Electrical Power System Dynamics: Modeling, Stability, and Control. Edited by Mircea Eremia and Mohammad Shahidehpour.

disturbance, with most system variables bounded so that practically the entire system remains intact" [3]. This definition refers mainly to an interconnected power system as a whole and is essentially a single problem. However, the various types of stability problems of a power system cannot be properly understood and effectively dealt with by treating them as a single problem. Furthermore, because of high dimensionality and complexity of the power system, which involves a large number of variables, simplifying assumptions are made in order to allow the analysis of specific types of problems with satisfactorily accuracy. Stability analysis is greatly facilitated by classification of stability into appropriate categories.

Classification of the power system stability by focusing mainly on only one variable (i.e., voltage, frequency, or rotor angle) is a meaningful practice. Such approach is also known in literature as *partial stability* [4–6]. Figure 8.1 gives the overall picture of the power system stability problem, based on the *dynamics of the phenomenon*, identifying its categories and subcategories that will be described in the following paragraphs [3,7]. This classification is based on the following considerations [3]:

- The physical nature of the resulting mode of instability as indicated by the main system variable in which instability can be observed.
- The size of the disturbance considered, which influences the method of calculation and prediction of stability.
- The devices, processes, and the time span that must be taken into consideration in order to assess stability.

Another possible classification of the power system stability can be done in terms of the duration of the phenomenon (*time domain*) and the *components that influence the phenomenon*, as illustrated in Table 8.1.

8.2.1 Rotor Angle Stability

Rotor angle stability refers to the ability of synchronous machines in a power system to remain in synchronism after being subjected to a perturbation. However, distinction should be made between types and severity of perturbations because the synchronous generator has the ability to withstand only some of them.

The rotor shaft of a synchronous generator rotates by application of a *mechanical torque* from a primary machine (turbine). As explained in Chapter 2, energization of the rotor winding from a DC source creates a stationary magnetic field in the rotor winding, which in turn induces alternating voltages in the stator windings when the generator rotor is driven by the primary machine (mover). When the generator is connected to the power grid,

TABLE 8.1. Classification of Power System Stability Based on the Time Domain and Influencing Component

<table>
<tr><th>Time Domain</th><th colspan="2">Generator Influenced Phenomenon</th><th>Load Influenced Phenomenon</th></tr>
<tr><td rowspan="2">Short term</td><td colspan="2">Rotor angle stability</td><td rowspan="2">Small-disturbance voltage stability</td></tr>
<tr><td>Small-disturbance stability</td><td>Transient stability</td></tr>
<tr><td>Long term</td><td colspan="2">Frequency stability</td><td>Large-disturbance voltage stability</td></tr>
</table>

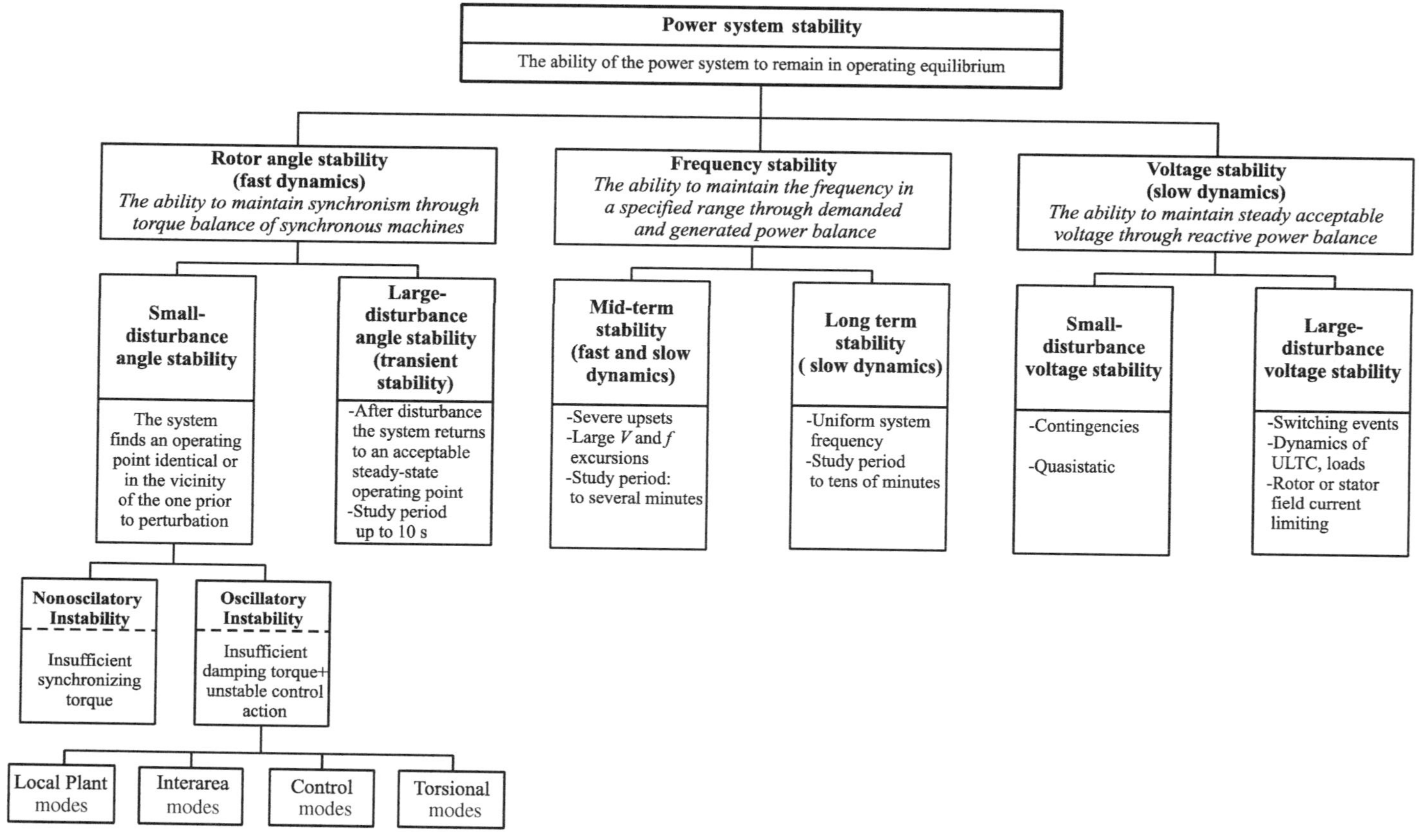

Figure 8.1. Classification of power system stability based on the dynamics of the phenomenon. (Adapted from [7].)

balanced three phase currents of frequency equal to the power system frequency will flow in the stator circuit, which will generate a rotating magnetic field. The stator rotating magnetic field interacts with the rotor magnetic field producing the *electromagnetic torque* that opposes the rotation of the rotor. Therefore, mechanical torque is required to support the rotor revolving at desired speed. In steady-state conditions, the two opposite torques are equal and the rotor speed is maintained at a constant value. In this state the generator is said to be in equilibrium. At nominal frequency, the rotor speed is equal to the synchronous speed for which the machine was optimized in the design stage. Therefore, the amount of mechanical torque necessary to be produced by the turbine is directly related to the amount of current flowing in the stator. Any change in the stator current requires a change in the mechanical torque.

At no load, when no power is transferred from rotor to stator, the stator magnetic field is aligned with the rotor magnetic field. As the load increases and more mechanical torque is produced by the turbine, there is an angular separation between the two fields. This angle can increase up to a maximum value beyond which the synchronous generator losses the synchronism. To determine this maximum value, let us consider a simple system consisting of a generator, with the electromotive force (emf.) $\underline{E}_g$ and the terminal voltage $\underline{V}_1$, supplying power to a power system through a network represented by the equivalent reactance X_{ech}. The voltage at the power system bus is denoted by $\underline{V}_2$. (Figure 8.2a).

The rotor angle, δ_g, also called load angle, of the synchronous generator represents the phase shift between the emf developed by the rotor, $\underline{E}_g$, and the voltage at the generator terminals, $\underline{V}_1$, and reflects the load of the generator. The more current is carried in the stator windings, the larger is the rotor angle and more stressed is the generator. In open circuit, the rotor angle is zero and the terminal voltage is equal to the emf $\underline{E}_g$.

In a simple form, the power transferred from the generator to the power system is given by

$$P_e = \frac{E_g V_2}{X_g + X_{ech}} \sin \delta \tag{8.1}$$

Thus, it can be seen that this power is a highly nonlinear function of δ, the phase shift between the emf developed by the rotor, E_g, and the voltage at the power system node, V_2. The network reactance can change if the network topology changes, whereas the generator reactance, X_g, depends on the purpose of the study. The steady-state reactance of the generator is much larger than the reactance in transient state (see Table 2.2).

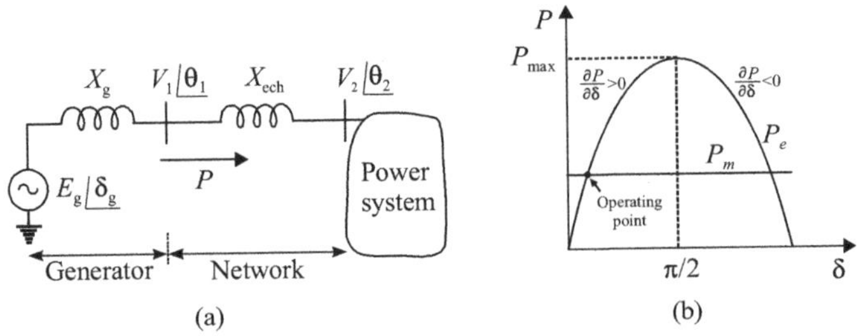

Figure 8.2. Illustration of a power transfer: (a) one-line diagram; (b) power-angle characteristic.

The relationship between the power transferred, P_e, and the phase angle, δ, is illustrated in Figure 8.2b. The power transferred increases as the angle increases and reaches a maximum value for an angle $\pi/2$, beyond which the power tends to decrease, which it not desirable. On the other hand, from equation (8.1) it can be seen that the higher is the network reactance, the less power can be transferred while keeping the generator in synchronism. Large network reactance values occur because of long distances or weak interconnections. Since the network is involved, the maximum value that can be reached for a rotor angle is less than $\pi/2$. There is also a phase difference between the terminal voltage of the generator and power system bus voltage. Usually, for stability reasons, the power transferred on an electrical line is limited to a phase difference of 30°. Therefore, in steady-state conditions, the maximum value of the rotor angle is less than $\pi/2$.

The power–angle characteristic illustrated in Figure 8.2b is a measure for evaluating the static (steady-state) stability. The steady-state stability limit corresponds to the point in which, if the receiving system demands a small additional load ΔP, the generator becomes unable to deliver it without a change in the field excitation. This is the case when $\partial P/\partial \delta = 0$. Therefore, the steady-state stability condition is satisfied for positive values of the derivative $\partial P/\partial \delta$.

To understand how the stress is exerted on the generator in terms of the rotor angle δ_g, let us assume the analogy between the voltage diagram of the network from Figure 8.2a and a mechanical system (Figure 8.3) [8,9].

The mechanical system consists of one fixed arm representing the voltage $\underline{V}_2$, one mobile arm representing the emf $\underline{E}_g$, and a spring representing the vector $(X_d + X_{ech})\underline{I}$, joining the extremities of the two arms. The lengths of the arms, representing the magnitudes of the two voltages, are considered to be constant. But, the length of the spring is propositional to the applied tensile force, that is, the current. A rotational torque $C_0 = q \cdot r$ is exerted at the fixed end of the arm E_g caused by the weight q, representing the load. In steady state, the torque C_0 is balanced by the torque C exerted by the spring on the arm E_g. The torque C is defined as the product of the length of the arm, E_g, the tensile force of the spring, f, and the sine of the angle, α, between the arm and the spring. The tensile force is the product of the spring elongation, l, and the elasticity coefficient, $k = 1/(X_d + X_{ech})$ [8].

Drawing the segment ab as a perpendicular from point b on the arm E_g, we find that $l \sin\alpha = V_2 \sin\delta$. Therefore, we find that the torque C has an expression similar to that given in equation (8.1). The torque C and the stress in the spring are at a maximum when the angle δ is 90°.

At equilibrium, the input mechanical torque equals the output electromagnetic torque at any individual generator in the power system. Any sudden change occurring in the power

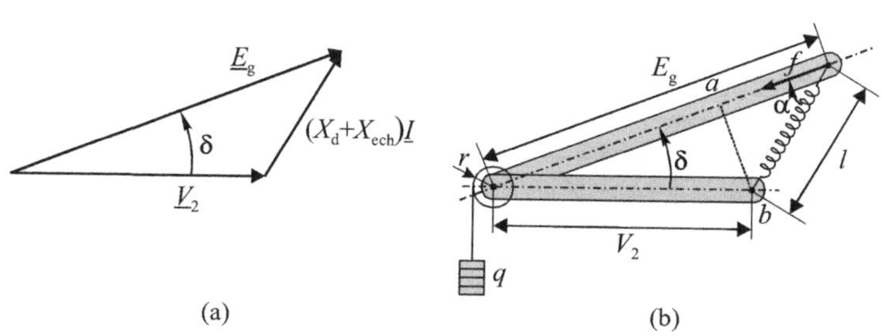

Figure 8.3. Voltage diagram of the power system: (a) electrical; (b) mechanical.

system causes unbalance between the network angular frequency and the angular speed of the rotor, and hence misalignment of the stator and rotor magnetic fields and unbalance between the two torques. Naturally the generators tend to align the two magnetic fields resulting in a change in the rotor speed. Therefore some generators may run faster than the others. The change in the rotor speed will result in a change in the load angle and hence the slower machines transfer part of the load to the faster machines. In accordance with equation (8.1), an increase in the phase angle beyond 90° results in decrease in the power transfer, which may lead to instability. For this case $\partial P/\partial \delta < 0$. The rate of change of the rotor speeds of individual generators is highly dependent on the rotor inertia and the amount of power deviation. The change in share of power production among different generators causes changes in the power flows on the network branches. Types of changes may be load change, generation change, limitation of power transfer from one or more generators to the loads due to line disconnection, severe voltage variations caused by short-circuits, and so on.

The control systems attempt to restore the power system to steady-state conditions and balance the mechanical torque with the electromagnetic torque of individual generators so that the rotor speeds and hence the load angles are stabilized. The turbine governor adjusts the mechanical torque and the voltage regulator attempts to restore the voltage. The transition to the new operating state, involving the rotor speeds, the voltages, or the power flows, is oscillatory in its nature and is usually called electromechanical oscillations. Of interest is the first oscillation of the rotor angle, which may indicate whether or not the generator will remain in synchronism.

An analogy of the electromechanical oscillations phenomena with a mechanical system is illustrated in Figure 8.4. The balls of different weights (e.g. inertia and installed power), representing the generators, are connected to each other by a set of strings, representing the transmission lines. The points of connection of the springs represent the substations.

In steady state the springs maintain constant elongation. However, if the weight of one ball is changed, changes in the tensile force in the spring occur and the spring oscillates causing the other springs to oscillate, which in turn affect the other balls. In this way, the entire interconnected mechanical system experiences oscillations. If the spring sustaining a certain ball experiences too large oscillations, that ball may break away becoming instable. But, using some damping solutions, the oscillations may die away and the system may returns to a stable state.

Taking a single generator as an example, its capability to return to a stable operating point after being subjected to perturbations can be analyzed using Figure 8.5 [10].

A ball, representing the operating point, is located in a potential hole. The heights of the hole show the strength of the generator with its control systems. This ball is maintained in equilibrium at the point A with the support of the prime mover providing the mechanical torque C_m that equals the electromagnetic torque C_e. Any perturbation reflects the effort

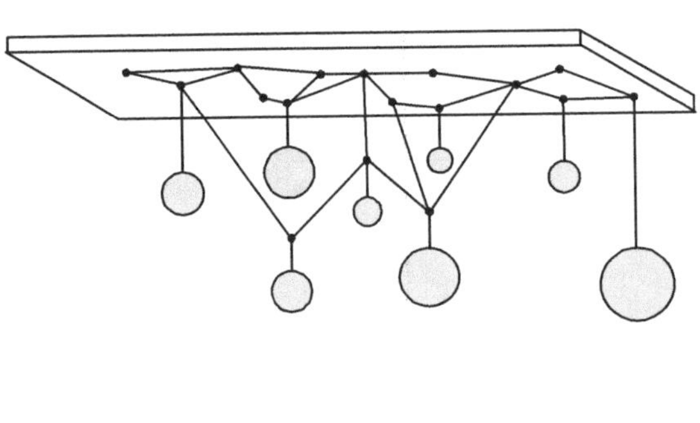

Figure 8.4. Mechanical analogy for the electromechanical oscillations.

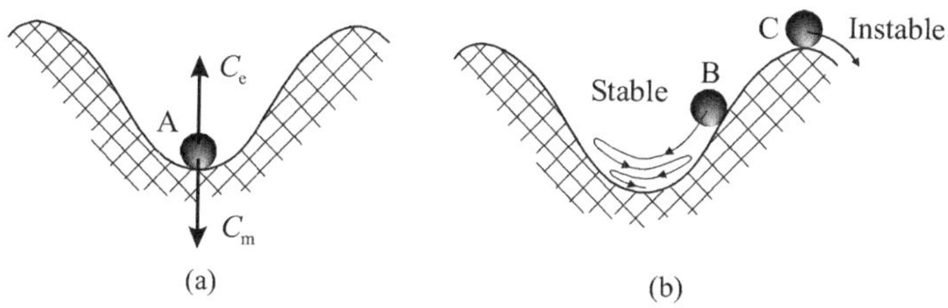

Figure 8.5. Generator states: (a) steady-state equilibrium; (b) perturbed operation.

required to pull the ball out of the hole. If a perturbation moves the ball at the point B, the generator is capable to return to a stable state, but after several oscillations, which should be damped as quickly as possible. However, if the perturbation moves the ball at the point C, the generator may become instable very easily because any additional small perturbation pushes the ball very quickly out of the hole.

Rotor angle oscillations may also appear in the power system any time because of weak electrical connection between generators or between generators and loads due to long distances (large reactances), because of uncoordinated fast voltage regulators or other types of controls. In large interconnected systems, local generators tend to attach to each other and in some situations to oscillate against other groups of generators. If not damped quickly, the rotor angle oscillations may cause damage to the power plants and other equipment in the power system.

The change in the electromagnetic torque of a synchronous generator following a perturbation can be resolved into two components [7]:

$$\Delta C_e = C_S \Delta\delta + C_D \Delta\omega \tag{8.2}$$

where $C_S\Delta\delta$ is the *synchronizing torque* component, which is in phase with the rotor angle deviation $\Delta\delta$; C_S is a synchronizing torque coefficient; $C_D\Delta\omega$ is the *damping torque* component, which is in phase with the speed deviation $\Delta\omega\,(= d\delta/dt)$; C_D is the damping torque coefficient.

The system stability is conditioned by the existence of both components of torque for each synchronous generator. Insufficient synchronizing torque will lead to *aperiodic* or *nonoscillatory* instability, whereas insufficient damping torque results in *oscillatory instability*.

According to [11], the synchronizing torque is the torque that acts on the shaft of a synchronous machine when the rotational speed of the rotor deviates from the synchronous speed. This keeps the machine in synchronism. In other words, it is torque that brings the rotor speed back to the synchronous speed. This can be done by contribution of the speed governor, the excitation system, and other internal control loops attached to the generator. External actions, for example, from FACTS devices, can also contribute to improvement of the synchronizing torque.

The power system is characterized by relatively small damping torque, which is mainly provided by the damper windings of the synchronous machines and by some load types. The damping capability is approximated by a damping term, which is included in the swing equation, given in Section 2.1. However, in the attempt to improve the synchronizing torque, the fast acting excitation systems can significantly weaken the damping torque thereby causing oscillatory instability. This effect is referred to as artificial negative damping. This problem can be corrected by using an additional signal to the excitation system loop called a *power system stabilizer*.

The synchronous generator, through its components and control systems, has the capability to restore the synchronous speed of the rotor after some disturbances. However, in some cases, the cause of perturbation must be quickly eliminated. To gain useful insight into the nature of stability problems, it is convenient to classify the rotor angle stability phenomena into the following two major categories: *small-disturbance rotor angle stability* and *large-disturbance rotor angle stability* [7].

8.2.1.1 Small-Disturbance (or Small-Signal) Rotor Angle Stability. Small-disturbance rotor stability is concerned with the ability of the power system to maintain synchronism under small disturbances, like small variations in load and generation [3]. The small disturbances are those changes occurring in the power system for which the rotor angle presents an almost linear variation allowing linearization of the system equations around the equilibrium point without encountering errors. The process following the disturbance depends on a number of factors, including the initial operating conditions, the transmission system strength, and the excitation systems performance.

When a small disturbance occurs in the power system, the instability may result in two forms [7] (Figure 8.6):

- Increase in rotor angle due to insufficient synchronizing torque; *nonoscillatory instability* may result when the automatic voltage regulators of the synchronous machines maintain constant field voltage.
- Rotor oscillations of increasing amplitude due to insufficient damping torque; *oscillatory instability* through oscillations of increasing amplitude may result when automatic voltage regulators act by continuously changing the field voltages.

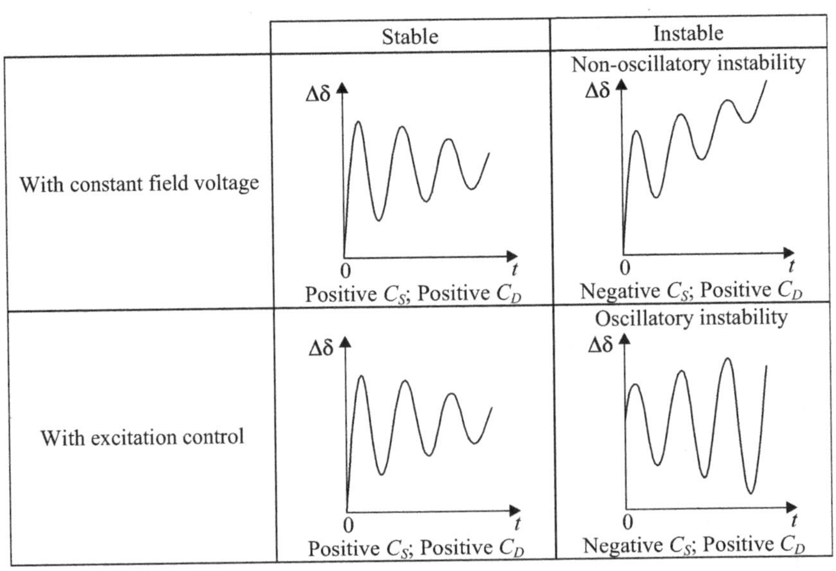

Figure 8.6. Nature of small-disturbance response.

Depending on the generators and the states involved in the unstable oscillations, the stability of the following types of oscillations is of concern [7]:

- *Local plant mode oscillations* are associated with the swinging of one generator or one power plant against the rest of the power system. The frequency of oscillation ranges from 1 to 2 Hz, depending on the machine characteristics and the power system operating conditions.
- *Interarea mode oscillations* are associated with the swinging of one geographically isolated group of synchronous machines against other machines. If the interarea oscillations become instable, the groups of generators can loss their synchronism and one part of the power grid can be islanded following transmission lines disconnection by the protection system. The frequency of this mode of oscillations ranges from 0.1 to 1 Hz.
- *Control mode oscillations* can be due to poorly tuned control systems attached to field exciters, speed governors, HVDC links, SVC, and so on.
- *Torsional modes* refer to the torsional vibrations induced in the generator-turbine rotational components due to sporadic changes either on the turbine side or on the power grid side. The torques applied on the rotor shaft in opposite directions create twisting of the shaft. Such phenomena can cause shaft break or turbine blade failures especially at the low-pressure cylinder of the thermal power plants. All perturbations are characterized by time, frequency, and amplitude. Perturbations in the power grid can be switching in the network or sudden large load changes such as arc furnaces operation or motor starting. A particular case of torsional mode is the subsynchronous resonance (SSR), which is due to the interaction between series capacitors and nearby power plants with long and flexible shafts [12]. Instability of torsional modes may be also caused by interaction with excitation controls and speed governors.

8.2.1.2 Large-Disturbance Rotor Angle Stability or Transient Stability. The large-disturbance rotor angle stability, commonly referred to as transient stability, is concerned with the ability of the power system or a synchronous generator to maintain synchronism when subjected to a severe disturbance. The resulting system response involves large excursions of the generator rotor angles emphasizing the nonlinear power–angle relationship. As a consequence, *the systems equations can no longer be linearized* as in the case of small perturbations, and the rotor angles variations can be analyzed using integration numerical methods. Examples of large perturbations are short-circuits on transmission lines, disconnection of large power plants, and disconnection of large loads. Stability depends on both the initial operating conditions of the power system and the severity of the disturbance. The most severe large perturbations involve changes in the network topology. For instance, a short circuit can be eliminated by disconnection of the affected element.

When a short circuit occurs on a transmission line close to a synchronous generator, the terminal voltage drops significantly and the generators' capability to produce electrical power is reduced. During the brief period following the perturbation occurrence, the excitation system, with fast response and high ceiling voltage, can contribute to maintain the stability of the generator by field forcing thereby boosting the terminal voltage. For such a perturbation, the first condition to maintain the stability is to quickly eliminate the short circuit by line disconnection. If the generator "survived," after fault elimination it enters into oscillations, which must be damped very quickly. The damping of the

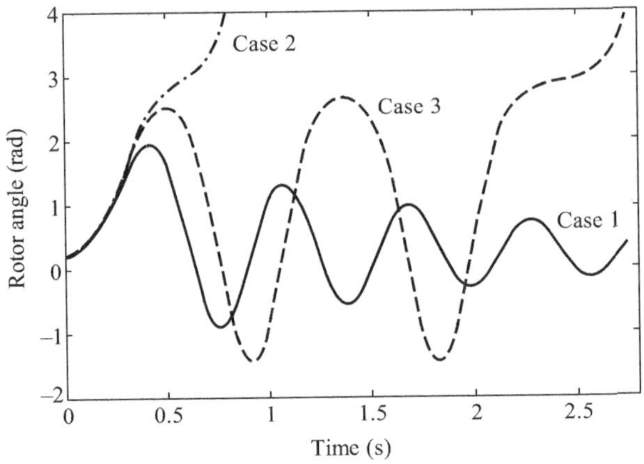

Figure 8.7. Rotor angle responses to large perturbations [7].

oscillations is then conditioned by the small-perturbation performance of the power system. The automation and protection systems attempt to reconnect the transmission line and restore the initial operating conditions. However, if the oscillations are not properly damped, a fast acting reclosure can cause larger oscillations and therefore greater stress on the generator shaft.

Figure 8.7 illustrates the rotor angle oscillations experienced by a synchronous machine indicating stability or instability of the machine (see Chapter 10 for more details). Case 1 shows that the rotor angle oscillations are damped (decrease in amplitude) and the rotor angle tends to stabilize to a constant value. Cases 2 and 3 indicate that the rotor angle can increase significantly in amplitude and the generator losses synchronism. In the Case 2, synchronism is lost at the first oscillation. This form of instability is known as *first swing instability* and is caused by insufficient synchronizing torque. In the Case 3, the machine maintains synchronism in the first oscillations, but the oscillations are growing in amplitude and synchronism is lost after a few oscillations. This form of instability generally occurs because the generator is not "small-disturbance stable" due to insufficient damping and/or synchronizing torques and conflicting actions of the control systems, even if appropriate actions are taken in time to eliminate the causes of the perturbation.

The time of interest for large perturbations stability ranges from 3 to 5 s and may extend to 10 s for very large power systems with weak interconnections between remote generators.

8.2.2 Voltage Stability

Voltage stability is the ability of a power system to maintain steady acceptable voltages at all buses in the system under normal operating conditions and after being subjected to a disturbance. The voltage may become instable when there is an unbalance between load demand and load supply, mainly an *imbalance of reactive power* [7,13–18]. This may happen for two reasons: either due to a sudden change in the load demand, such as loss of load in an area, or due to limitation in the load supply capacity due to tripping of a transmission line. The *voltage instability is* initially *a local phenomenon*. If appropriate measures are not taken in due time, the system voltages will deteriorate progressively with a widespread impact. Actions that may help voltage restoration can be load shedding, transformer tap changing, generator excitation forcing, and so on. Failure in voltage restoration will result in dangerous low voltages leading to voltage collapse and finally system blackout.

The most effective way to control the voltage is by reactive power support. The voltage stability conditions are met if the reactive power injection increase in one bus will lead to voltage increase in the same bus. In other words, a system is voltage stable if the V–Q sensitivity is positive for every bus and voltage unstable if the V–Q sensitivity is negative for at least one bus [19] (see also Section 11.3.2).

One way of explaining the voltage instability phenomenon is by considering a simple power system, consisting of one load center, represented by an equivalent impedance $\underline{Z}_c$, supplied through a transmission line and a transformer, represented by an equivalent impedance $\underline{Z}$, from an infinite power source, represented by the equivalent voltage $\underline{E}$, as shown in Figure 8.8b.

Denoting by β the angle of the equivalent impedance $\underline{Z}$ and by φ the angle of the load impedance $\underline{Z}_c$, the expression of the line current $\underline{I}$ may be written as [14]:

$$I = \frac{I_{\text{sc}}}{\sqrt{1 + 2(Z_c/Z)\cos(\beta - \varphi) + (Z_c/Z)^2}} \tag{8.3}$$

where $I_{\text{sc}} = E/Z = V_1/Z$ is the short-circuit current magnitude at the load bus.

The expressions for the load bus voltage and the delivered active power are

$$V_2 = |\underline{Z}_c \underline{I}| = \frac{I_{\text{sc}} Z_c}{\sqrt{1 + 2(Z_c/Z)\cos(\beta - \varphi) + (Z_c/Z)^2}} \tag{8.4}$$

$$P_2 = V_2 I \cos\varphi = \frac{I_{\text{sc}}^2 Z_c}{1 + (Z_c/Z)^2 + 2(Z_c/Z)\cos(\beta - \varphi)} \cos\varphi \tag{8.5}$$

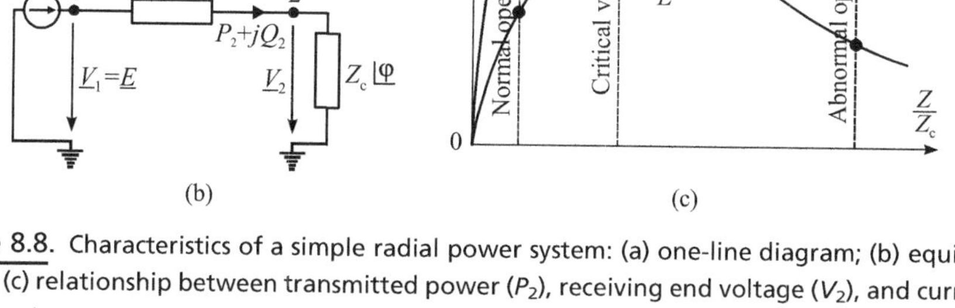

Figure 8.8. Characteristics of a simple radial power system: (a) one-line diagram; (b) equivalent circuit; (c) relationship between transmitted power (P_2), receiving end voltage (V_2), and current (I) for $I_{sc} = E/Z$.

Normalized diagrams for I, V_2, and P_2 as functions of the Z/Z_c ratio are suggestively shown in Figure 8.8c.

Under no-load conditions ($Z_c = \infty$) the current value is zero and the load bus voltage, V_1, is equal with the source bus voltage, E. As the load impedance decreases and the current increases asymptotically toward I_{sc}, the load bus voltage V_2 decreases due to the voltage drop on the equivalent impedance Z.

Analysis of characteristics from Figure 8.8c reveals the following:

- For values $Z_c > Z$, the increase in current is the dominant phenomenon and, thus, increase in the active power P_2 delivered to load center is possible.
- For values $Z_c < Z$, the voltage drop increase and the voltage decrease are dominant, which cause increase in the power losses, and in consequence the power P_2 that can be transferred to the load decreases.
- Maximum power transfer (point C) is achieved when the load impedance is equal to the network impedance ($Z_c = Z$).

Replacing for $Z_c = Z$ and $I_{sc} = V_1/Z$ in equation (8.5) gives the expression for the maximum power transfer:

$$P_{2\max} = \frac{V_1^2 \cos\varphi}{2Z[1 + \cos(\beta - \varphi)]} = \frac{V_1^2 \cos\varphi}{4Z \cos^2[(\beta - \varphi)/2]} \tag{8.6}$$

for which the critical voltage is obtained by

$$V_{2\mathrm{cr}} = \frac{I_{sc} Z}{\sqrt{2[1 + \cos(\beta - \varphi)]}} = \frac{V_1}{2\cos[(\beta - \varphi)/2]} \tag{8.7}$$

The operating point C, for which $V_{2\mathrm{cr}}$ and $P_{2\max}$ are achieved, is called *critical point* (Figure 8.8c). This point corresponds to the *maximum power transfer* achieved in theory, given that "the power delivered to a load through a dipole from a constant voltage source is maximum when the source impedance is equal with the load impedance".

The maximum power transfer corresponds to the normal operation limit, situated to the left of this value, as shown in Figure 8.8c.

For a power delivered smaller than the maximum power (for instance, $P_2 = 0.8$ p.u. $< P_{2\max}$), there are two possible operating points, A and B, obtained for two different values of the load impedance:

- *Point A*, placed to the left of the critical point, characterized by low current value and high voltage value, is a normal operating point for the power system.
- *Point B*, placed to the right of the critical point is an abnormal operating point and is characterized by large current and small voltage values, which causes high power losses. In this case, when the OLTCs supplying the load attempt to restore a normal voltage level, the apparent impedance "seen" by the transmission network will decrease. This will cause additional increase of the current and amplification of the voltage level degradation. For this reason, point B is considered an unstable operating point.

From equations (8.4) and (8.5) we see that the load power factor has a significant effect on the power–voltage characteristics of the system. This is to be expected since the voltage drop in the transmission line is a function of active as well as reactive power transfer. Voltage stability, in fact, depends on the relationship between P, Q, and V. Of special interest is the relationship between P_2 and V_2 (see Figure 11.4). The normal operation corresponds to the upper side of the characteristic, while the abnormal operation corresponds to the lower side. More aspects of the V–I characteristics for different values of the load power factor or different values of the source voltage E are developed in Chapter 11.

The above-presented aspects for a simple power system provide a basic overview on the voltage stability problem. However, in large power systems the voltage (in)stability depends on all factors influencing the voltage such as network topology and parameters, types of loads, network loading, and characteristics of reactive power compensating devices.

Classification of voltage stability in terms of the disturbance severity helps understanding and correctly analyzing the voltage variations in the power system [41]. Therefore, the voltage stability may be conveniently classified into large-disturbance voltage stability and small-disturbance voltage stability.

Large-disturbance voltage stability is concerned with the ability of a power system to maintain steady voltages within acceptable limits following large disturbances such as system faults or disconnection of a transmission line, a transformer or a synchronous generator. The voltage recovery process after the disturbance depends on the load characteristics and network characteristics as well as the actions of various protective and control systems. The mathematical analysis requires simulations over a period of time extended from a few seconds to tens of minutes [21].

Small-disturbance voltage stability refers to the ability of a power system to control the voltage when small changes occur in the power system, mainly incremental changes in load demand. The processes taking place in this case are slow and can be associated with steady-state conditions. Therefore, static analysis methods, also referred to as load-flow of steady-state analysis, can be applied, with good accuracy, to determine the stability margins and the indices showing the strength of the system in respect to voltage. The operating conditions are dependent on the characteristics of load and the control actions at the instant of time for which the analysis is performed [22].

In terms of the severity of disturbance and the operating conditions of the power system, voltage stability assessment may require analysis on various time frames. Therefore, voltage stability can also be classified into

- *Short-term voltage stability* involves the system dynamics, mainly of the fast acting load components capable to recover (e.g., induction motors, especially residential air conditioners, and heat pumps) during the brief period after a major disturbance occurrence. Instability may occur because of disturbance severity and that the discrete control systems are too slow. The study period of interest is of several seconds, and analysis requires solution of appropriate system differential equations.
- *Long-term voltage stability* deals with slower acting equipment such as tap changing transformers located close to loads (tens of seconds) or generator overexcitation limiters (many minutes) [23–25]. Long-term instability may be caused by controlled loads (e.g., constant energy loads, such as thermostatically controlled heating loads). Instability is due to the loss of long-term equilibrium, postdisturbance steady-state operating point being small-disturbance unstable, or a lack of attraction toward the stable postdisturbance equilibrium [17,18]. In many cases, static analysis

[19,24,26,27] can be used to estimate stability margins, identify factors influencing stability, and screen a wide range of system conditions, and a large number of scenarios. However, when timing of control actions is important, this should be complemented by quasi-steady-state time-domain simulations [17,23].

Voltage stability problems may be experienced in a power system when disturbances occur, for example, transmission line outage, which may reduce the reactive power transfer capability of the transmission network to the load. Let us consider the classical system consisting of a load center supplied through a radial network from the power source. It is assumed that voltage dynamics in the supply system are faster than the aggregated dynamics of the load center. Therefore, the network can be represented by three distinct quasi-steady-state *V–Q* characteristics for the three stages of the disturbance: predisturbance, postdisturbance, and postdisturbance with reactive power support (Figure 8.9).

In normal conditions, the "load-network" system operates in point **a** situated at the intersection between the *predisturbance characteristic* and the steady-state load characteristic $Q_d(V)$. Assuming that the network experiences a disturbance (e.g., a line outage) that reduces its reactive power supply capability, a new system characteristic is formed, called *postdisturbance characteristic*, and the operating point jumps to point **b** on this characteristic. The load center will respond instantaneously with its transient characteristic trying to restore the reactive power demand, characterized by lower voltage value. The load dynamics can be described through the generic model [20]:

$$T_q \frac{dy}{dt} = Q_s(V) - Q_d(V) \tag{8.8}$$

where T_q is the load time constant and y is a state variable. The load demand can be expressed as:

$$Q_d(V) = yQ_t(V) = yV^{\beta}$$

where $Q_t(V) = V^{\beta}$ is the transient load characteristic.

The increase in the reactive power demand can be modeled by increasing the state variable y. The load is modeled as an impedance if $\beta = 2$, whereas when $\beta = 1$ the load is modeled as a current.

Since at point **b** the voltage is lower than the nominal value, any increase in reactive power can be done by increasing the current which, in turn, will cause greater voltage drops on the network impedance. The voltage continues to decrease monotonically and the operating point moves from point **b** to point **c** or further to point **f** as long as the network reactive power supply, $Q_s(V)$, is lower than the reactive power demand, $Q_d(V)$.

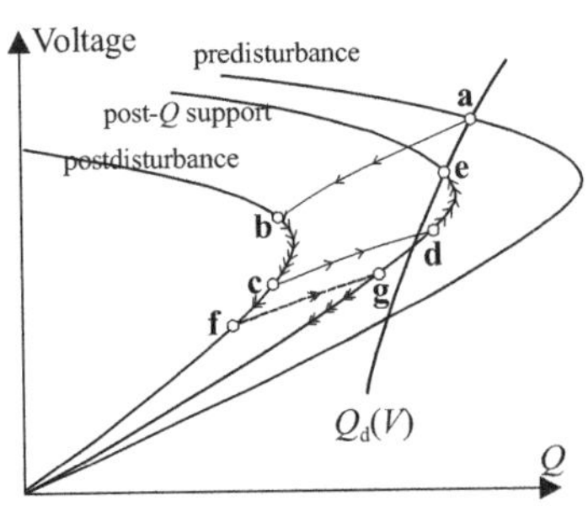

Figure 8.9. Voltage dynamics as viewed from the V–Q plane [20].

A new system characteristic is formed, called *post-Q support characteristic*, if a local reactive power source is connected to supply the load or remote reactive power support is initiated to boost the supply voltage. If the reactive power support is activated in a timely manner, the operating point jumps from point **c** to point **d** on the post-Q support characteristic. Since at point **d** the power supply, $Q_s(V)$, is greater than the load demand, $Q_d(V)$, the voltage begins to increase until the operating point reaches an equilibrium, at point **e**, at the intersection between the new system characteristic and the reactive power demand characteristic. However, if there is a large delay (slow response) in activating appropriate reactive power support and the system operation reaches the point **f** on postdisturbance characteristic, after reactive power support activation, the operating point jumps to point **g**, and the voltage begin to collapse monotonically.

8.2.3 Frequency Stability

Frequency stability refers to the ability of a power system to maintain steady frequency following a severe system upset resulting in a significant unbalance between generation and load [3].

Any unbalance between generation and load causes deviation of the system frequency with respect to the nominal value, and hence changes in the power grid angular frequency since their relationship is $\omega = 2\pi f$. Because the system frequency and the angular frequency are global quantities, any unbalance between generation and load affects the operation of all synchronous machines of the power system. Any change in the angular frequency results in change in the electromagnetic torque and finally unbalance between the electromagnetic torque and the mechanical torque of each synchronous machine. The change in the angular frequency is given by $\Delta\omega = \Delta P/2H_{sys}$, where ΔP is the active power unbalance and H_{sys} is the system inertia calculated as the sum of inertias of all turbine generators in the power system. The inertia of a turbine generator refers to the turbine generator's resistance to changes in the rotor speed. In large power systems with many synchronous generators connected, for example, UCTE, the system inertia in rotating masses is very large. Therefore, a significant change in the angular frequency needs a significant active power unbalance. However, the way in which the angular frequency is affected depends on the size of the power system and the severity of perturbation. Isolated power systems are more vulnerable to the disconnection of large generators [28].

The power systems are designed with automatic and manual control systems that are used for frequency control. As explained in Chapter 6, in Europe the frequency control is performed on three levels. The primary and the secondary controls are automatic systems and are able to respond very quickly to frequency deviations. The third control level consists in manual deployment of power reserves upon the call of the dispatching center. The primary control is designed to stabilize the frequency in the first seconds after perturbation occurrence, whereas the secondary control is designed to automatically balance the generation and load and bring the system frequency within predefined limits. The third control level has the purpose to replace the secondary reserve so that sufficient automatic active power reserve is maintained in the system in order to counteract inadvertent power unbalances and frequency deviations.

The national grid codes require that appropriate power reserve be maintained available in the power system for both upward and downward regulation in order to face the inadvertent disconnection of either the largest power plant or the largest load center. Furthermore, the neighboring power systems can contribute to restoring the equilibrium between generation and load. However, danger may also come from faults that can affect the network integrity. Disconnection of one transmission line or one transformer may cause overloading of other network branches, weakening the transfer capacity on some network corridors, which in the long run can create severe local unbalance between generation and load. The Italian blackout in 2003 [29] and the UCTE network splitting in 2006 [30] originated from disconnection of one transmission line, and the first process was related to frequency problems.

Severe system upsets resulting in large excursions of frequency also cause power flow changes, large voltage variations, as well as significant changes in other system variables. Voltage magnitude changes, which may be higher in percentage than frequency changes, also affect the load-generation balance. High voltage may cause undesirable generator tripping by poorly designed or coordinated loss of excitation relays or Volts/Hertz relays. In an overloaded system, low voltage may cause undesirable operation of impedance relays. Therefore, it may be said that frequency instability is not an independent phenomenon.

The involved corrective actions of processes, controls, and protection are not modeled in conventional transient stability or voltage stability studies. These processes may be very slow, such as boiler dynamics, or only triggered for extreme system conditions, such as Volts/Hertz protection tripping of generators [3].

Large frequency deviations are due to the inability of the synchronous generators to maintain/restore equilibrium between system generation and load. Instability that may result occurs in the form of sustained frequency swings leading to tripping of generating units and/or loads. For this reason, timely and optimal actions must be taken in order to minimize the amount of interrupted load.

Hydro generators may run almost unaffected when large drops in frequency occur, even up to 10%. However, thermal power plants are very sensitive to frequency drops, even by 5%, because their motor-driven auxiliaries such as boiler feedwater pumps, coal pulverizing and feeding equipment, and draft fans are high order frequency dependent. A small drop in the system frequency causes a large reduction in the output power to the auxiliaries; the energy input to the turbine decreases affecting the generator output, which in turn produces additional drop in the system frequency as a cascading effect. Furthermore, the turbines are damaged due to the operation at low frequency.

In large interconnected power systems, large power unbalance may cause weakening of the coupling between generators, and groups of generators may run at different frequencies leading to the network splitting into islands. In such cases, the question is whether the frequency in the islands can reach a steady value. It depends on the coordination of control and protection equipment. In order to prevent the collapse of the islands, underfrequency load shedding relays are activated in order to maintain the balance between generation and load.

The characteristic time of the processes evolving during frequency excursions can range from few seconds, corresponding to the responses from the generator control and protection devices and the underfrequency load shedding protection, to several minutes, characteristic to the prime mover reaction time or load voltage regulators [3]. However, the time characteristic for triggering the frequency instability depends on the power system size. Therefore, as also shown in Figure 8.1, the power system frequency stability can be either a *mid-term* phenomenon or a *long-term* phenomenon. Mid-term frequency instability and collapse can be observed in islands created after network splitting with insufficient generation resources and inappropriate control equipment [31], while long-term frequency stability problems, that can also lead to system collapse, can be observed when the steam turbine overspeed controls or boiler/reactor protection and controls experience disoperation [32,33].

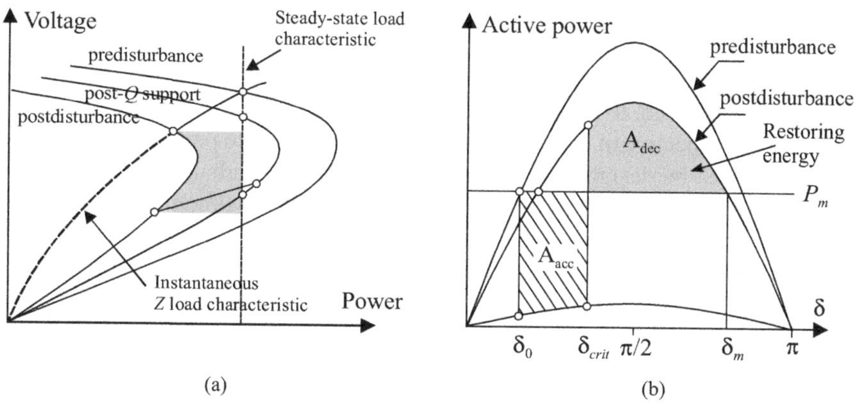

Figure 8.10. Parallelism between voltage stability and angle stability [20].

8.3 PARALLELISM BETWEEN VOLTAGE STABILITY AND ANGULAR STABILITY

Voltage instability does not always occur in its pure form. Often the angle and voltage instability go "hand in hand." One may lead to the other and the distinction may not be clear. Frequency stability may also be strongly connected to either of them. However, a distinction between angle stability and voltage stability is important for understanding the underlying causes of the problems in order to develop appropriate design and operating procedures.

More insight about the voltage stability limit of a load center can be obtained by comparing with the angle stability of a generation plant using graphical representation. Figure 8.10 presents a parallelism between the minimum voltage criteria for voltage stability (Figure 8.10a) with the equal area criteria for large-disturbance angle stability (Figure 8.10b).

The equal area criterion deals with excessive energy generation and the critical clearing time (see Chapter 10), whereas minimum voltage criterion deals with excessive energy demand and critical reactive power support (see Chapter 11) [20]. Excessive generation causes acceleration of the generators' rotor speed, which may lead to angular instability and finally loss of synchronism, while excessive load demand causes severe voltage drops and load reduction, which may lead to voltage collapse.

The shadowed surface shown in Figure 8.10b indicated as decelerating area (A_{dec}), represents the restoring energy available that allows the generator to stabilize after a large disturbance, and it can easily be calculated using the equal area criterion. In Figure 8.10a, the shadowed surface may theoretically indicate the minimum energy necessary to be activated immediately in order to help the power system survive to voltage stability problems. However, it is quite difficult to estimate exactly the restoring energy in this case.

8.4 IMPORTANCE OF SECURITY FOR POWER SYSTEM STABILITY

Evaluating the operating state of a power system is mandatory for the system operator in taking appropriate decisions and keeping the electrical quantities within acceptable limits.

Furthermore, analysis of the system performance under various conditions and phenomena helps designing the control systems.

The power system can be seen as a "black box," with a well-defined function, to which consumers are connected. It should ensure the continuity of supply with electrical energy to the consumers irrespective of its operating conditions. In order to perform this function, the power system must be designed so that to withstand any disturbance, for example, short circuits followed by a line tripping, tripping of any element without a fault, and so on.

The number of possible events that theoretically may occur is infinite and thus designing the power system to withstand all possible events or combination of events is impossible. The practice is to evaluate the power system state using one or more indexes for all possible events, and then rank the events in descending order of the combined index values. The indices penalize violation of transmission capacity limit, voltage limits, stability limits, and so on. Defense plans or countermeasures are designed for those disturbances that the power system cannot withstand. The defense plans are designed to counteract all dangerous disturbances starting from the top of the list.

The *reliability* of a bulk power system is its capability of continuously delivering electrical energy to any consumer at accepted quality standards and in the demanded amount, both in normal and disturbed conditions. The degree of reliability may be measured by the frequency, duration, and magnitude of adverse effects on consumer service [34].

The term *security* is used to indicate if the power system is able to withstand disturbances. The power system *security* may be defined as the ability to withstand any kind of disturbance without interruption of the power supply service. It can be said that the power system is fully reliable if it is secure all the time. Security may also be associated with the term *robustness*, which reflects the way in which the power system can withstand disturbances.

The power system security is conditioned by both physical security and cyber security. The *physical security* refers to the integrity and operation at normal parameters of the primary and secondary circuits of the electrical network, on one side, and of the generators and loads, on the other side. All computer, communication and digital control devices installed in control centers, power plants, substations, and so on form the cyber system, which is the backbone of the power system. Correct operation of the cyber system ensures the *cyber security* of the power system [35].

TCP/IP (Transmission Control Protocol/Internet Protocol) based cyber infrastructures have been developed and expanded significantly in the last decade. The SCADA (Supervisory Control And Data Acquisition) system is the main infrastructure used today in power systems for data acquisition and transmission, monitor and control, which includes TCP/IP protocols for remote access, remote maintenance, and so on. TCP/IP based communication is also used within the electricity markets. The remote access makes the power systems vulnerable to malicious attacks, sometimes with catastrophic consequences.

8.4.1 Power System States

The national/regional grid codes stipulate that the state quantities must remain within acceptable ranges for any disconnection of an element (line, transformer, generator, etc.). This is known as the N-1 security criterion. Consecutive outages may occur in a power system and thus it is highly recommended that the power system be designed so that to withstand double disconnections, known as the N-2 security criterion. Unfortunately, due

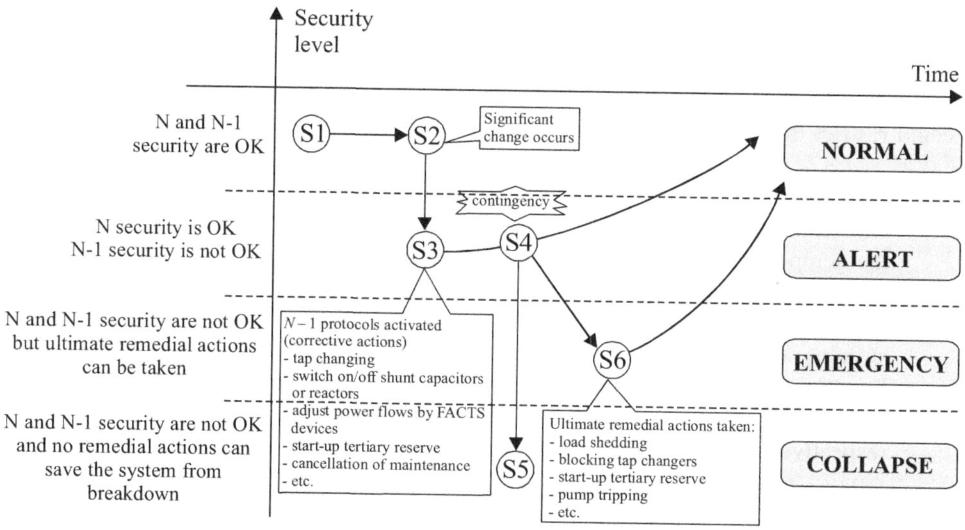

Figure 8.11. Power system states [36].

to the large financial investments required, many of the power systems may have problems to comply with the N-2 security criterion.

The operating conditions vary continuously and the power system moves from one state to another, as suggestively indicated by S in Figure 8.11. Transition to one state or another depends on the random events that may occur or on the decision taken by the system operator. In NORMAL state, all parameters are within acceptable ranges and the power system is stable and secure. Furthermore, from this state, disconnection of any element can bring no harm to the power system. However, significant changes such as large load increase or extreme weather conditions make the system vulnerable to disconnection of an additional element and the power system may enter in the ALERT state. Figure 8.11 shows also a classification of possible states of the power system depending on the event that may occur.

When the power system enters in the alert state, immediate corrective actions must be taken in order to restore the normal operation. The restoration process may take shorter or longer time depending on the dynamics of the corrective actions. If during this transition a new contingency occurs, the system can enter in an EMERGENCY state, in which there is a large number of bus voltage limits violations or exceeding of branch ampacity. In this state, ultimate (extreme) remedial actions can still be taken and system restoring to a normal operation is possible. If the contingency is too severe, the power system may become instable and finally COLLAPSEs.

Violation of an operating constraint may not necessarily mean that the power system becomes instable. Due to high loading, the voltage in the system nodes can become too low and the loads demand more current that, in time, may overload the transmission lines. This will finally jeopardize the electrical network integrity, limits the transmission capacity and lead to instability conditions.

Following the major incidents that have occurred in the European interconnected power system [29,30], UCTE has issued in 2009 a new policy for operation security policy [36]. This policy defines clear responsibilities for each power system operator for any operating conditions.

8.4.2 Power Flow Security Limits

Maintaining the power systems in a secure operating state assumes that some electrical quantities should be maintained within admissible limits. The most important limits that could restrict the power transfer in a transmission grid are as follows:

- *Thermal Limit* The electrical current flowing through an electrical line must not exceed the admissible value in order to avoid overloading and therefore increasing the temperature of line conductors. Exceeding this limit leads to conductors dilatation followed by increasing in the sag in conductors toward the earth. If there is too much sag, a short circuit will occur, which, when detected by the protection systems, will cause the line tripping. Unscheduled disconnection of a transmission line (or a transformer) can have severe consequences on the power system operation, that is, the redirection of the power flows and overloading of other branches, respectively, which will produce a chain reaction leading to blackout.
- *Voltage Limit* The electrical equipment is designed to operate at voltage values situated within an admissible range. A higher voltage level will cause insulation flashover, while a lower voltage level will cause the dynamic phenomenon know as "voltage collapse." On the other hand, the voltage exceeding outside the admissible limits will affect directly the consumers through the quality of the power supplied.
- *Stability Limit* The interconnected power systems undergo various electric and magnetic interactions due to the random behavior of the consumers and sometimes of the power plants, which lead to power, voltage, or frequency oscillations. In order to avoid power system instability, some measures should be taken by which sufficient stability reserves are ensured. Generator instability may cause damage to the attached equipment as well as to consumers through interruption in power supply.

The transmission capacity of a power grid can change in time due to changes of one of the above-defined limits (Figure 8.12). In particular, the total transfer capacity (TTC) denotes the maximum exchange program between two areas or between two nodes compatible with operational security standards applicable at each system if future network conditions, generation and load patterns were perfectly known in

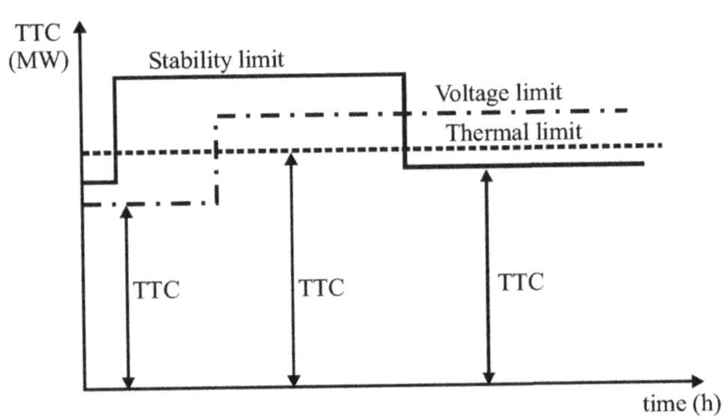

Figure 8.12. Limits of total transfer capacity [37,38].

advance [37]. In other words, the total transfer capacity is given by the most restrictive limit, that is:

$$\text{TTC} = \text{Minimum}\,\{\text{thermal limit, voltage limit, stability limit}\}$$

also including the case of the most severe contingency.

Accurate determination of the three limits is impossible due to the high number of unknown variables. For this reason, the system operators define a security margin (capacity not used) for each interconnection line or transmission corridor, and the remaining allowed transmission level is called Net Transfer Capacity (NTC).

The factors changing the security limits are multiple. For instance, the thermal limit is dependent on weather conditions, the stability limit depends on the power reserves, on the existence and performances of some regulation and control systems (PSS, AVR, "fast valving", speed governor, etc.), on the existence of advanced devices (FACTS, HVDC links, etc.), and the voltage limit can be influenced by the performance of the automatic voltage regulators.

Exceeding one of the security limits defined above denotes a network congestion that may jeopardize the power system operation. However, the congestion's related problems are managed by some technical or economical mechanisms. In the long term, the congestions management is seen as an economical problem because of the bilateral agreements involving power exchanges, which represents a great part of the total system load. However, in the short term or in real-time, the problem of congestions becomes a pure technical issue.

In real-time, network congestion may emerge as a consequence of load forecast errors and, thus, of unscheduled power flows, as well as by technical reasons such as unscheduled disconnection of a generator, an electrical line, or a transformer. The system operator has, therefore, to predict the risky situations that may emerge in any moment and adapt to the network conditions by taking the appropriate measures. Real time problems are solved by the system operators using the security services.

8.4.3 Services to Meet Power System Security Constraints

The services provided within a power system are classified in terms of destination, commercial feature, and the entity that provides them. From the commercial point of view, the power system services can be classified in two categories [39]:

- *System services* are those services provided by the system operator and transmission operators by some functions or means designed to monitor and control the continuous operation of the power system necessary to support the electrical energy supply to consumers while ensuring the secure conditions at the lowest costs.
- *Ancillary services* are those services procured by the system operator from the network users, producers or consumers, by means of which the system operator is carrying out its functions to coordinate the transport of the electrical energy from producers to consumers. The procurement of these services is usually performed by means of specific competitive markets, and the providers are financially remunerated.

In others words, power system services refer to those functions or services by means of which some electrical parameters of the network are controlled. In the following, we will focus mainly on those services that can make the object of commercial agreements, that is, the ancillary services.

From the point of view of controlling the electrical parameters, the following commercial services are defined:

- frequency control,
- voltage control,
- active power loss replacement,

and in case of blackouts

- black start capability.

Maintaining the power system stability should be added to these services, which is performed by the system operator by means of specific functions requiring short-, medium-, or long-term studies based on simulations. Although this service is classified as system service, it has to be mentioned due to its importance for the power system security.

The way in which the four ancillary services are performed may differ more or less from one control area to another, which are part of large synchronous areas, such as UCTE (Europe), the interconnected power systems from North America, or the interconnected systems from any part of the world. The necessity of implementing a certain ancillary service depends to a great extent on the geographical spread of the power system, the average consumption, the transmission capacity, the existence of advanced devices, and so on.

8.4.4 Dynamic Security Assessment

With the advancements in the computer and telecommunication technologies, more reliable and powerful tools have been implemented in power system control centers to assist the power system operators in their current activity such as real-time monitoring, on-/off-line analysis, state estimation, reporting, and so on. A particular tool is the Dynamic Security Assessment (DSA) system by means of which the system operator is able to evaluate if the system security conditions are met in both steady-state and transient conditions [21,26,40,41].

Off-line DSA tools perform stability analysis for all possible contingencies and operating conditions, including network topology, generation-load pattern, regulators settings, and so on. The analysis is performed in order to determine if the power system maintains the stability for the simulated operating conditions, also using time-domain simulations. On the other hand, calculations are performed to determine operating limits such as fault critical times that are references for relay settings, power transfer limits that allow the bus voltages to be maintained within security limits, loading levels on network sections that allow evaluating the operational risk for future changes within the power system, and so on. Calculations performed by off-line DSA tools might involve time consuming methods.

In the real-time operation, on-line DSA tools are used to perform fast stability evaluation based on the current data obtained from measurement systems and the state estimator. For real-time analysis, this evaluation requires simulations only for a limited number of contingencies, unusually the most dangerous cases from the list of possible contingencies generated by the off-line DSA system. Real-time simulations assume application of direct methods for transient stability assessment or calculation of local/global indices for evaluating the risk of triggering the voltage instability.

The DSA tools are part of the Energy Management System (EMS), which includes all tools and functions implemented in a control center for supervision, control, optimization and management. EMS receives measurements from Supervisory Control And Data Acquisition (SCADA) and, most recently, from Phasor Measurement Units (PMU). The EMS system is completed with modeling functions, archive, and visualization components. The results obtained by the DSA or other computation tools of the EMS system are sent to the tools dealing with real-time control or remedial plans.

Classical DSA simulations are performed using measurements acquired by the SCADA system at different time instants and sent to the dispatching center with a delay of 2–4 s. This may cause some additional errors in the results obtained by the DSA tool. With the PMU technology, synchronized measured data can be sent to the dispatching center much faster, in the order of milliseconds. PMU technology may therefore significantly improve the results of any transient stability analysis because they directly measure angles, present in the swing equations, as compared to the SCADA system where the angles are estimated.

REFERENCES

[1] Steinmetz, C.P. Power control and stability of electric generating stations, *AIEE Trans.*, Vol. XXXIX, Part II, pp. 1215–1287, Jul. 1920.

[2] AIEE Subcommittee on interconnections and stability factors. First report of power system stability, *AIEE Trans.*, pp. 51–80, 1926.

[3] Kundur, P., Paserba, J., Ajjarapu, V., Andersson, G., Bose, A., Canizares, C., Hatziargyriou, N., Hill, D., Stankovic, A., Taylor, C., Van Cutsem, T., Vittal, V. Definition and classification of power system stability, IEEE/CIGRE Joint Task Force on Stability Terms and Definitions, *IEEE Trans. Power Syst.*, Vol. 19, No. 2, pp. 1387–1401, May 2004.

[4] Vorotnikov, V.I. *Partial stability and control*, Birkhauser, Cambridge, MA, 1998.

[5] Rumyantsev, V.V. and Osiraner, A.S. *Stability and stabilization of motion with respect to a part of the variable*, Nauka, Moscow, Russia, 1987.

[6] Rouche, N., Habets, P., Laloy, M. *Stability theory by Liapunov's direct method*, Springer, New York, 1977.

[7] Kundur, P. *Power system stability and control*, McGraw-Hill, New York, 1994.

[8] Kimbark. E.W. *Power System Stability. Volume I. Elements of stability calculations*, Wiley-IEEE, 1995.

[9] Zhdanov, P.S. *Power system stability* (in Russian), 1948.

[10] *Etude simplifie de quelque régimes anormaux de fonctionnement des alternateurs: faux-couplage des alternateurs*, Journée d'Information sur la stabilité, Electricité de France, 1972.

[11] *The Great Soviet Encyclopedia*, Russian Academy of Sciences, Moscow, 2004.

[12] Anderson, P.M., Agrawal, B.L., Van Ness, V.E. *Subsynchronous resonance in power systems*, IEEE Press, New York, 1990.

[13] Eremia, M., Trecat, J., Germond, A. *Réseau électriques—Aspects actuels*, Technical Publishing House, Bucharest, 2000.

[14] Barbier, C. Barret, J.P. An analysis of phenomena of voltage collapse on transmission systems, *Revue Generale d'Electricite*, pp. 672–690, Oct. 1980.

[15] CIGRE TF 38-01-03. Planning against voltage collapse, *Electra*, No. 111, pp. 55–75, Mar. 1987.

[16] IEEE Special Publication 90TH0358-2-PWR. *Voltage stability of power systems: concepts, analytical tools, and industry experience*, New Jersey, 1990.

[17] Van Cutsem, T. Vournas, C. *Voltage stability of electric power systems*, Kluwer Academic Publisher, Norwell, MA, 2001.

[18] Taylor, C.W. *Power system voltage stability*, McGraw-Hill, New York, 1994.

[19] Gao, B., Morison, G.K., Kundur, P. Toward the development of a systematic approach for voltage stability assessment of large-scale power systems, *IEEE Trans. Power Syst.*, Vol. 11, pp. 1314–1324, Aug. 1996.

[20] Xu, W., Mansour, Y. Voltage stability analysis using generic dynamic load models, *IEEE Trans. Power Syst.*, Vol. 9, No. 1, 1994.

[21] Grigsby, L. *Power system stability and control*, CRC Press, 2007.

[22] Gao, B., Morison, G.K. Kundur, P. Voltage stability evaluation using modal analysis, *IEEE Trans. Power Syst.*, Vol. 7, No. 4, pp. 1529–1542, Nov. 1992.

[23] Van Cutsem, T. Voltage instability: Phenomenon, countermeasures and analysis methods, Proc. IEEE, Vol. 88, pp. 208–227, 2000.

[24] Morison, G.K., Gao, B., Kundur, P. Voltage stability analysis using static and dynamic approaches, *IEEE Trans. Power Syst.*, Vol. 8, pp. 1159–1171, Aug. 1993.

[25] Barbier, C., Carpentier, L., Saccomanno, F. CIGRE SC32 Report "*Tentative classification and terminologies relating to stability problem of power systems*", *Electra*, No. 56, 1978.

[26] Săvulescu, S. (editor). *Real time stability assessment in modern power system control*, IEEE Press Series on Power Engineering and John Wiley & Sons, USA, 2009.

[27] Lof, P.A., Smed, T., Andersson, G., Hill, D. J. Fast calculation of a voltage stability index, *IEEE Trans. Power Syst.*, Vol. 7, pp. 54–64, Feb. 1992.

[28] Hatziargyriou, N., Karapidakis, E., Hatzifotis, D. Frequency stability of power system in large islands with high wind power penetration, *Bulk Power Syst. Dynamics Control Symp.—IV Restructuring*, Vol. PAS-102, Santorini, Greece, Aug. 24–28, 1998.

[29] Berizzi, A. *The Italian 2003 blackout*, IEEE-PES General Meeting, Denver, USA, Jun. 6–10, 2004.

[30] UCTE, *System Disturbance on 4 November 2006. Final Report*, Union for the Co-ordination of Transmission of Electricity (currently ENTSO-E), Nov. 2007.

[31] CIGRE Task Force 38.02.14 Report. *Analysis and modeling needs of power systems under major frequency disturbances*, Jan. 1999.

[32] Chow, Q.B., Kundur, P., Acchione, P.N., Lautsch, B. Improving nuclear generating station response for electrical grid islanding, *IEEE Trans. Energy Conv.*, Vol. EC-4, pp. 406–413, Sep. 1989.

[33] Dy Liacco, T.E. *Control of power systems via the multi-level concept*, PhD Thesis Case Western Reserve University, 1968.

[34] *NERC Planning Standards*, Sep. 1997. [Online]. Available: http://www.nerc.com/.

[35] Ericsson, G.N. Cyber Security and Power System Communication—essential parts of a smart grid infrastructure, *IEEE Trans. Power Deliv.*, Vol. 25, No. 3, pp. 1501–1507, Jul. 2010.

[36] *UCTE Policy 3: Operational security*, Mar. 2009.

[37] ETSO. *Definitions of transfer capacities in liberalised electricity markets*, European Transmission System Operators, 2001.

[38] NERC *Available transfer capability definitions and determination: A framework for determining available transfer capabilities of the interconnected transmission networks for a commercially viable electricity market*, NERC, June 1996.

[39] Eurelectric, *Connection rules for generation and management of ancillary services*, Ref: 2000-130-0003, 2000.

[40] Machowski, J., Bialek, J., Bunchy, J. *Power system dynamics and stability*, John Wiley & Sons, 1997.

[41] Concordia, C. Voltage instability, *Electrical Power Energy Syst.*, Vol. 13, No. 1, Feb. 1991.

9

SMALL-DISTURBANCE ANGLE STABILITY AND ELECTROMECHANICAL OSCILLATION DAMPING

Roberto Marconato and Alberto Berizzi

9.1 INTRODUCTION

Nowadays, angle stability following small disturbances or variations (also called steady-state angle stability or, improperly, small-signal angle stability) is mostly related to damping of electromechanical oscillations, that is, to oscillations between synchronous generator rotors connected through the transmission grid (lines, transformers, etc.).

Up to about 1960, the typical primary voltage control time constants were in the range of some seconds and therefore the voltage control loop was much slower than the electromechanical oscillations, especially with respect to the local ones, characterized by periods of 0.4–0.7 s. Consequently, in the electromechanical loop, the operating conditions were the same as if the voltage regulator was in manual operating mode (i.e., excitation voltage constant) and the electromechanical loop and the primary voltage control loop were basically decoupled. The only issue potentially dangerous of the small-disturbance angle stability was the aperiodic instability, due to insufficient synchronizing power coefficients.

At present, thanks to the upgrading of transmission systems (characterized by a higher number of lines, an increased meshing, and higher nominal transmission voltage levels, all factors increasing short-circuit powers), the synchronizing power coefficients are no longer such as to induce aperiodic instability.

By contrast, the use of very fast modern voltage regulators, characterized by time constants much lower than in the past (about 0.2 s), gives rise to a coupling between the electromechanical and voltage loops, which may result in problems of periodic instability, that is, of insufficient (in some instances even negative) damping of electromechanical oscillations.

Handbook of Electrical Power System Dynamics: Modeling, Stability, and Control. Edited by Mircea Eremia and Mohammad Shahidehpour.

The problem of the aperiodic instability may still occurs nowadays (even in the presence of modern voltage regulators) in systems very large from the geographical point of view but poorly meshed and/or with too low transmission voltage levels (e.g., 150–220 kV rather than 400 kV, or at 400 kV but operated at 220 kV), that is, characterized by relatively small short-circuit powers. Conversely, in such systems, the problem of the damping of electromechanical oscillations may not arise, for reasons that will be covered in the following sections. An example of such systems will be given in Section 9.6.

In order to solve damping oscillation problems, the modal analysis and the synthesis of control actions (e.g., the introduction of power system stabilizers (PSSs) into excitation control systems) are performed. They require

- the determination of the dynamic matrix $[A]$ associated with the model of the system: this model is linearized around one or multiple operating points, defined by power flow (PF) computations;
- the evaluation of the eigenvalues of $[A]$ that are the poles of the system: a stable system requires that all the poles have negative real part;
- the evaluation of the eigenvectors of $[A]$ in order to fully describe the *modal characteristics* of the power system.

Therefore, in order to determine such features it is necessary to linearize the various equations that describe the dynamic behavior of the system and to find the dynamic matrix $[A]$.

Nevertheless, to evaluate the factors affecting electromechanical oscillations from a qualitative point of view, that is, to understand the phenomena involved, use may be made of the model of a power plant (true or equivalent to a system area) connected to an infinite system. This model is helpful to investigate both local or power plant oscillations and interarea electromechanical oscillations, if the problems affect the overall system.

In any case, the damping of electromechanical oscillations is strongly dependent on the dynamics of the various components. Hence, the behavior of such components and of the overall system should be described in sufficient detail, taking into account

- the structure of synchronous generators (field circuit, damper windings, or additional rotor circuits);
- the primary voltage control;
- the primary frequency control;
- the over/underexcitation limiting circuits of units;
- and so on.

9.2 THE DYNAMIC MATRIX

9.2.1 Linearized Equations

As it is well known, the overall model of a power system can be described by a system of nonlinear differential and algebraic equations. Such equations can be written as follows:

$$\begin{aligned} [\dot{x}] &= f([x],[w],[u]) \\ 0 &= g([x],[w],[z]) \end{aligned} \tag{9.1}$$

where

$[x]$ is the vector of the state variables; its size is n, if n is the dynamic order of the system;

$[w]$ is the vector of the voltage amplitudes and phases in the n_g generation nodes, its size is $2n_g$;

$$[w] = \begin{bmatrix} [v_G] \\ --- \\ [\theta_G] \end{bmatrix} \tag{9.2}$$

$[z]$ is the vector of the variations of the voltage amplitudes and phases in the n_L load nodes (HV nodes), its size is $2n_L$;

$$[z] = \begin{bmatrix} [v_L] \\ --- \\ [\theta_L] \end{bmatrix} \tag{9.3}$$

$[u]$ is the vector of the inputs, that is, of the set points of controls (at least primary voltage and frequency controls should be included).

For example, the set f of equations can include, if for each generator the fourth-order model is adopted (see Section 2.1.6.1), the differential equations (2.10), (2.155′), and (2.140′). Correspondingly, for the same generator, the set g will include equations (2.135′) and (2.149″), together with the power flow equations and, if necessary, additional algebraic equations.

In particular, the vector $[x]$ holds at least the following variables (core of the power system):

- Absolute rotor angles (δ).
- Absolute[1] angular speeds of rotors (Ω), which, of course, corresponds to ω_r used in previous chapters.
- Electromagnetic state variables (e''_d, e''_q, e'_d, e'_q) of rotors.
- State variables of excitation and primary voltage control systems (x_v).
- State variables of supply and primary frequency control systems (x_Ω).

The algebraic equations in (9.1) can be written ordering first the equations depending on the state variables $[x]$.

The linearization of such system of equations around an operating point (defined by a power flow computation), characterized by the superscript ^{o}, yields

$$\begin{cases} [\Delta\dot{x}] = [F_{xx}][\Delta x] + [F_{xw}][\Delta w] + [B][\Delta u] \\ 0 = [G_{wx}][\Delta x] + [G_{ww}][\Delta w] + [G_{wz}][\Delta z] \\ 0 = [G_{zw}][\Delta w] + [G_{zz}][\Delta z] \end{cases} \tag{9.1'}$$

[1] In this chapter, capital letters will be used for the absolute values of variables, while small letters will be used for relative (p.u.) values, according to the following table.

$U \leftrightarrow v$	$P \leftrightarrow p$	$Q \leftrightarrow q$	$E \leftrightarrow e$	$I \leftrightarrow i$

where $[F_{xx}], \ldots, [G_{zz}]$ are the matrices containing the suitable partial derivatives.

For example, matrix $[F_{xx}]$ is built as follows:

$$[F_{xx}] = \begin{bmatrix} \dfrac{\partial f_1}{\partial x_1} & \dfrac{\partial f_1}{\partial x_2} & \cdots & \dfrac{\partial f_1}{\partial x_n} \\ \dfrac{\partial f_2}{\partial x_1} & \dfrac{\partial f_2}{\partial x_2} & & \dfrac{\partial f_2}{\partial x_n} \\ \vdots & & \ddots & \vdots \\ \dfrac{\partial f_n}{\partial x_1} & \dfrac{\partial f_n}{\partial x_2} & \cdots & \dfrac{\partial f_n}{\partial x_n} \end{bmatrix}^0$$

and the other matrices are built in the same manner.

Accordingly, the so-called *complete system matrix* can be built, based on equations (9.1′)

$$[A_c] \triangleq \left[\begin{array}{c|cc} [F_{xx}] & [F_{xw}] & [0] \\ \hline [G_{wx}] & [G_{ww}] & [G_{wz}] \\ [0] & [G_{zw}] & [G_{zz}] \end{array}\right] \tag{9.4}$$

- This matrix has the structure depicted in Figure 9.1.

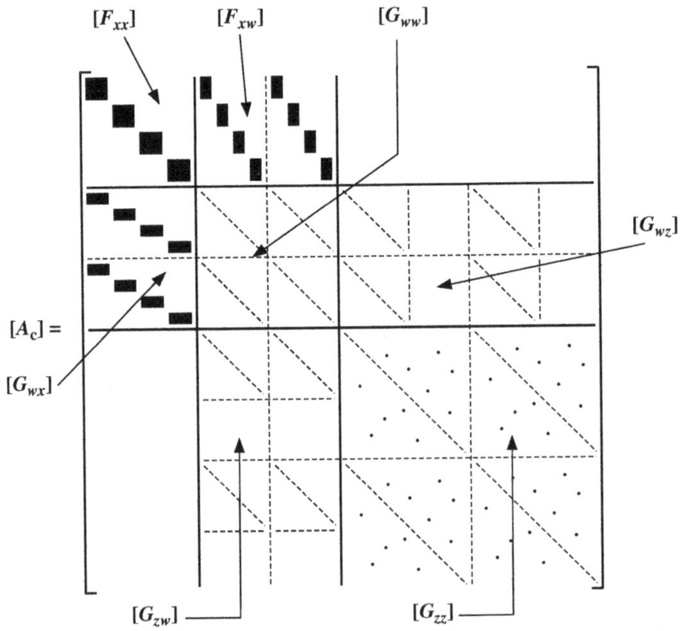

Figure 9.1. Structure of the complete system matrix.

9.2.2 Building the Dynamic Matrix

In equation (9.4), the submatrix

$$[J] \triangleq \begin{bmatrix} [G_{ww}] & [G_{wz}] \\ [G_{zw}] & [G_{zz}] \end{bmatrix} \tag{9.5}$$

is the complete Jacobian matrix, in polar coordinates, holding the partial derivatives of the PF equations with respect to all the algebraic variables and describes the linearization of all the system algebraic equations. They are always valid in analyzing the system steady state and dynamic behavior, except when dealing with electrical phenomena.

However, equations (9.1′) are not in the normal form, that is,

$$\begin{cases} [\overset{\bullet}{\Delta x}] = [A][\Delta x] + [B][\Delta u] \\ [\Delta y] = [C][\Delta x] + [D][\Delta u] \end{cases} \tag{9.6}$$

where $[y]$ represents all the outputs, that is, $[y]^T = [[w]^T[z]^T]$.

Equations (9.6) are obtained by removing the vectors $[\Delta w]$ and $[\Delta z]$ from equations (9.1′). For this purpose, from the third of equations (9.1′), we get

$$[\Delta z] = -[G_{zz}]^{-1}[G_{zw}][\Delta w] \tag{9.7}$$

and substituting into the second of equations (9.1′),

$$[0] = [G_{wx}][\Delta x] + \left\{[G_{ww}] - [G_{wz}][G_{zz}]^{-1}[G_{zw}]\right\}[\Delta w] \tag{9.8}$$

that is, putting

$$[H] \triangleq -\left\{[G_{ww}] - [G_{wz}][G_{zz}]^{-1}[G_{zw}]\right\}^{-1} \tag{9.9}$$

we obtain

$$[\Delta w] = [H][G_{wx}][\Delta x] \tag{9.10}$$

Substituting into the first of equations (9.1′), we get

$$[\overset{\bullet}{\Delta x}] = \{[F_{xx}] + [F_{xw}][H][G_{wx}]\}[\Delta x] + [B][\Delta u] \tag{9.11}$$

Finally, the *dynamic matrix associated with the linearized model* is obtained:

$$[A] \triangleq [F_{xx}] + [F_{xw}][H][G_{wx}] \tag{9.12}$$

whereas the outputs are in the normal form (9.6) with $[D] = 0$ and

$$[C] \triangleq \begin{bmatrix} [H][G_{wx}] \\ \hline -[G_{zz}]^{-1}[G_{zw}][H][G_{wx}] \end{bmatrix} \tag{9.13}$$

If the state variables are ordered as follows:

$$[x] \triangleq \left[[\delta]^T \ \vdots \ [\Omega]^T \ \vdots \ [e]^T \ \vdots \ [x_v]^T \ \vdots \ [x_\Omega]^T \right]^T \tag{9.14}$$

then the *dynamic matrix associated with the linearized model of the power system core* becomes

$$[A] = \begin{bmatrix} [0] & [I] & [0] & [0] & [0] \\ [A_{\Omega\delta}^{(*)}] & [A_{\Omega\Omega}] & [A_{\Omega e}^{(*)}] & [0] & [A_{\Omega x_\Omega}] \\ [A_{e\delta}] & ([A_{e\Omega}]) & [A_{ee}] & [A_{ex_v}] & [0] \\ [A_{x_v\delta}^{(*)}] & ([A_{x_v\Omega}]) & [A_{x_v e}^{(*)}] & [A_{x_v x_v}] & [0] \\ [0] & [A_{x_\Omega\Omega}] & [0] & [0] & [A_{x_\Omega x_\Omega}] \end{bmatrix} \tag{9.15}$$

Moreover, assuming $[u]^T = [[\Omega_{\text{ref}}]^T[v_{\text{ref}}]^T]$, where $[\Omega_{\text{ref}}]$ holds the speed set points of the primary frequency control systems and $[v_{\text{ref}}]$ the voltage set points of the primary voltage control systems, $[B]$ has the following structure:

$$[B] = \begin{bmatrix} [0] & [0] \\ [B_{\Omega\Omega}] & [0] \\ [0] & [B_{ev}] \\ [0] & [B_{x_v v}] \\ [B_{x_\Omega\Omega}] & [0] \end{bmatrix} \tag{9.16}$$

Dealing with $[A]$, it may be worth noting that

- if speed governors are not modeled, the corresponding rows and columns have to be erased (in particular, $[A_{\Omega\Omega}]$, which gives a contribution to the damping of electromechanical oscillations) while the remaining submatrices are unaffected;
- any PSSs derived from speed will affect only the matrices $[A_{e\Omega}]$ and $[A_{x_v\Omega}]$ (which would otherwise not be present);
- any PSSs derived from electrical real power will affect the matrices $[A_{\Omega\delta}]$, $[A_{\Omega e}]$, $[A_{x_v\delta}]$ and $[A_{x_v e}]$, marked with an asterisk in equation (9.15).

As the system poles are the eigenvalues of $[A]$, all the information about the stability of the system following small perturbations and, in particular, the periods and dampings of electromechanical oscillations can be evaluated. The number of these eigenvalues equals the size of the dynamic matrix.

It is worth emphasizing that, unlike in linear systems, the eigenvalues of $[A]$ also depend on the operating point, owing to the linearization that has been made.

9.3 A GENERAL SIMPLIFIED APPROACH

In order to take a simplified but sufficiently accurate approach, let us suppose that each generator is represented either by one of the third-order dynamic models, namely by the one having $x_q = x'_d = x''_d = x_i$, or by one of the second-order models; the internal emf $\underline{e}$ ($\underline{e}'$ or $\underline{e}''$), given by

$$\underline{e} = e(\cos\delta + j\sin\delta) \tag{9.17}$$

has a variable amplitude and can thus be regarded as an input or control variable.

Regarding the electromechanical oscillations, it should be noted that

- the damping of electromechanical oscillations is determined by the system characteristics in terms of masses (generators) and springs (network) with very low losses; therefore, it is smaller (0.1–0.2 at the most) than the value usually adopted for the control loops (e.g., voltage and frequency control loops), which may reach 0.7 or even higher values;
- the damping problem arises when it is positive but small (e.g., 0.03) or even negative (e.g., −0.04).

Hence, the use of a second-order dynamic model for the synchronous generators (implying zero damping) represents a good approximation and therefore provides very useful information when the computation of damping is not required. Of course, the determination of the damping requires the use of a more accurate dynamic model (e.g., at least fourth-order one) for synchronous generators.

9.3.1 Inertia and Synchronizing Power Coefficients

Assuming the second-order model (2.12) with $D = 0$, the differential equations are, in absolute values

$$\begin{cases} \dot{\delta}_k = \Omega_k - \Omega_n \\ \dot{\Omega}_k = \dfrac{\Omega_n}{T_{ak}S_{nk}}(P_{mk} - P_{ek}) \end{cases} \quad \text{with } k = 1, \ldots, N \tag{9.18}$$

where N is the number of the generators in the system, T_{ak} is the mechanical start-up time ($T_a = 2H$, being H the inertia constant) of the kth generator, S_{nk} is the nominal apparent power of the kth generator, Ω_n is the nominal angular speed (which, of course, corresponds to ω_0 used in previous chapters).

If, for the sake of simplicity and without losing generality, loads are assumed to be static and linear (i.e., complex but constant admittances and impedances), then the load impedances and the impedances jx_i of the generators may be embedded into the passive network, including the two-port networks of lines and transformers.

Therefore, the nodal admittance matrix can be reduced to the internal emf machine nodes, eliminating all other network nodes and resulting in a reduced matrix $[\underline{Y}]$:

$$[\underline{I}\,] = [\underline{Y}\,][\underline{E}\,] \tag{9.19}$$

where[$\underline{E}$] is the vector of the emfs of elements of (9.17) and $[\underline{I}\,]$ is the vector of the currents that generators inject into the network.

For the generic kth generator, we will have

$$\underline{I}_k = \sum_{i=1}^{N} \underline{Y}_{ki}\underline{E}_i \tag{9.20}$$

and a generated complex power

$$\underline{S}_k = P_{ek} + jQ_k = \underline{E}_k \sum_{i=1}^{N} \underline{Y}_{ki}^* \underline{E}_i^* \tag{9.21}$$

with

$$\underline{Y}_{ki} = G_{ki} + jB_{ki} \tag{9.22}$$

Therefore, the generated real and reactive powers are

$$\begin{cases} P_{ek} = E_k \sum_{i=1}^{N} E_i[B_{ki}\sin(\delta_k - \delta_i) + G_{ki}\cos(\delta_k - \delta_i)] \\ Q_k = E_k \sum_{i=1}^{N} E_i[G_{ki}\sin(\delta_k - \delta_i) - B_{ki}\cos(\delta_k - \delta_i)] \end{cases} \tag{9.23}$$

In particular, we will get

$$P_{ek} = f_k\{\delta_1, \ldots, \delta_N, E_1, \ldots, E_N\} \tag{9.24}$$

which is a nonlinear function of the angles $[\delta]$ and of the amplitudes $[E]$ of emfs.

The linearization of equations (9.18) around the operating point gives

$$\begin{cases} [\dot{\Delta\delta}] = [\Delta\Omega] \\ [\dot{\Delta\Omega}] = [M]^{-1}\{[\Delta P_m] - [\Delta P_e]\} \end{cases} \tag{9.25}$$

with

$$[\Delta P_e] \triangleq [K][\Delta\delta] + [F][\Delta E] \tag{9.26}$$

$$[M] \triangleq \mathrm{diag}\{M_k\} = \mathrm{diag}\left\{\frac{T_{ak}S_{nk}}{\Omega_n}\right\} \tag{9.27}$$

as shown in Figure 9.2a. The matrix $[M]$ is the *matrix of inertia coefficients or inertias*, while

$$[K] = \begin{bmatrix} K_{11} & \ldots & K_{1N} \\ \vdots & & \vdots \\ K_{N1} & \ldots & K_{NN} \end{bmatrix} \tag{9.28}$$

is the *matrix of synchronizing power coefficients*, which holds the sensitivities of P_{ek} with respect to δ_i

$$K_{ki} \triangleq \left(\frac{\partial P_{ek}}{\partial \delta_i}\right)^o \tag{9.29}$$

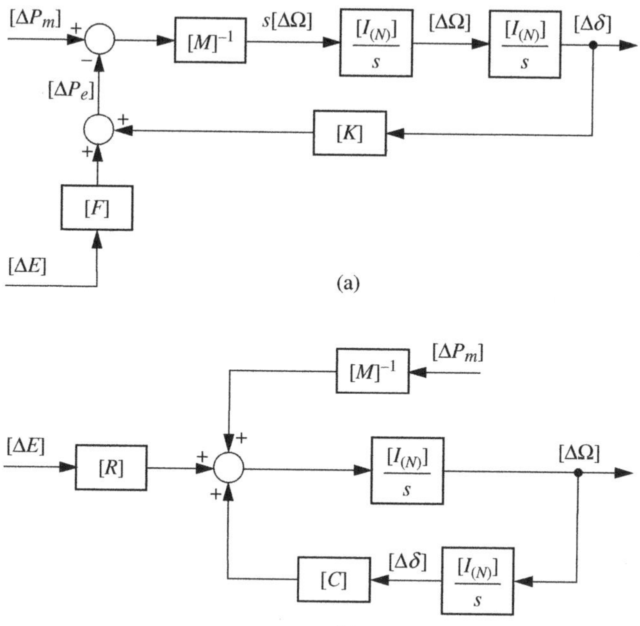

Figure 9.2. Block diagram of a system with second-order generator model.

Moreover,

$$[F] = \begin{bmatrix} F_{11} & \dots & F_{1N} \\ \vdots & & \vdots \\ F_{N1} & \dots & F_{NN} \end{bmatrix} \tag{9.30}$$

with

$$F_{ki} \triangleq \left(\frac{\partial P_{ek}}{\partial E_i}\right)^o \tag{9.31}$$

Equations (9.25) may also be written as

$$\begin{bmatrix} [\Delta\dot{\delta}] \\ [\Delta\dot{\Omega}] \end{bmatrix} = \begin{bmatrix} [0] & [I] \\ [C] & [0] \end{bmatrix} \begin{bmatrix} [\Delta\delta] \\ [\Delta\Omega] \end{bmatrix} + \begin{bmatrix} [0] & [0] \\ [R] & [M]^{-1} \end{bmatrix} \begin{bmatrix} [\Delta E] \\ [\Delta P_m] \end{bmatrix} \tag{9.32}$$

(as shown in Figure 9.2b), with

$$\begin{cases} [C] \triangleq -[M]^{-1}[K] \\ [R] \triangleq -[M]^{-1}[F] \end{cases} \tag{9.33}$$

Therefore, they are now in the normal form

$$[\Delta\dot{x}] = [A][\Delta x] + [B][\Delta u] \tag{9.34}$$

with

$$\begin{cases} [A] \triangleq \begin{bmatrix} [0] & [I] \\ [C] & [0] \end{bmatrix} \\ [B] \triangleq \begin{bmatrix} [0] & [0] \\ [R] & [M]^{-1} \end{bmatrix} \end{cases} \tag{9.35}$$

9.3.2 Electromechanical Oscillations

9.3.2.1 Oscillation Modes. Disregarding controls, that is, assuming $[\Delta E] = [\Delta P_m] = 0$, the system eigenvalues are the roots of the characteristic equation

$$\det|\lambda I - A| = 0 \tag{9.36}$$

that is, owing to the particular structure of $[A]$,

$$\det|\lambda I - A| = \det\left|\lambda^2 I - C\right| = 0 \tag{9.37}$$

The latter relationship shows that the eigenvalues of $[A]$ are given by the square roots of the eigenvalues of $[C]$.

In this respect, it should be noted that the determinant of $[C]$ is zero. Actually, the generic P_{ek} depends on the differences between the angular position δ_k of the kth machine and the angular positions of the other machines

$$P_{ek} = f_k\{\delta_1 - \delta_k, \ldots, \delta_N - \delta_k, E_1, \ldots, E_N\} \tag{9.38}$$

$$\begin{aligned} \Delta P_{ek} = {} & \left(\frac{\partial f_k}{\partial \delta_{1k}}\right)^o \Delta\delta_1 + \cdots + \left(\frac{\partial f_k}{\partial \delta_{Nk}}\right)^o \Delta\delta_N \\ & - \left[\left(\frac{\partial f_k}{\partial \delta_{1k}}\right)^o + \cdots + \left(\frac{\partial f_k}{\partial \delta_{Nk}}\right)^o\right]\Delta\delta_k + \sum_{i=1}^{N}\left(\frac{\partial f_k}{\partial E_i}\right)^o \Delta E_i \end{aligned} \tag{9.39}$$

where $\delta_{ik} = \delta_i - \delta_k$

The sum of the elements of each row of $[K]$ is zero ($k = 1, \ldots, N$):

$$\sum_{i=1}^{N} K_{ki} = 0 \tag{9.40}$$

As $[M]$ is diagonal, it follows that

$$\det[K] = \det[C] = 0 \tag{9.41}$$

Letting L_h be the generic eigenvalue of $[C]$ ($h = 1, \ldots, N$), in general

$$\det[C] = \prod_{h=1}^{N} L_h \tag{9.42}$$

Therefore, from (9.41), at least one eigenvalue of $[C]$ lies in the origin; by convention, it is assumed to be the last one ($L_N = 0$).

Since we are interested to problems of periodic-type stability, we assume that all the poles of $[A]$ are on the imaginary axis. Therefore, the other eigenvalues L_h ($h = 1, \ldots, N-1$) of $[C]$ must be real, distinct, and negative. Each L_h ($h = 1, \ldots, N-1$) corresponds to the following pair of eigenvalues of $[A]$:

$$\underline{\lambda}_h^{(1,2)} = \pm\sqrt{L_h} \tag{9.43}$$

that is,

$$\begin{cases} \underline{\lambda}_h = j\omega_{oh} \\ \underline{\lambda}_h^* = -j\omega_{oh} \end{cases} \tag{9.44}$$

with

$$\omega_{oh} = \sqrt{-L_h} \quad h = 1, \ldots, N-1 \tag{9.45}$$

The $(N-1)$ pairs of conjugate imaginary poles (9.44) correspond to $(N-1)$ *electromechanical oscillation modes*.

The previous approach refers to absolute rotor angles (Figure 9.3a). As the electrical power depends only on angular differences, we can model the mechanical part as in

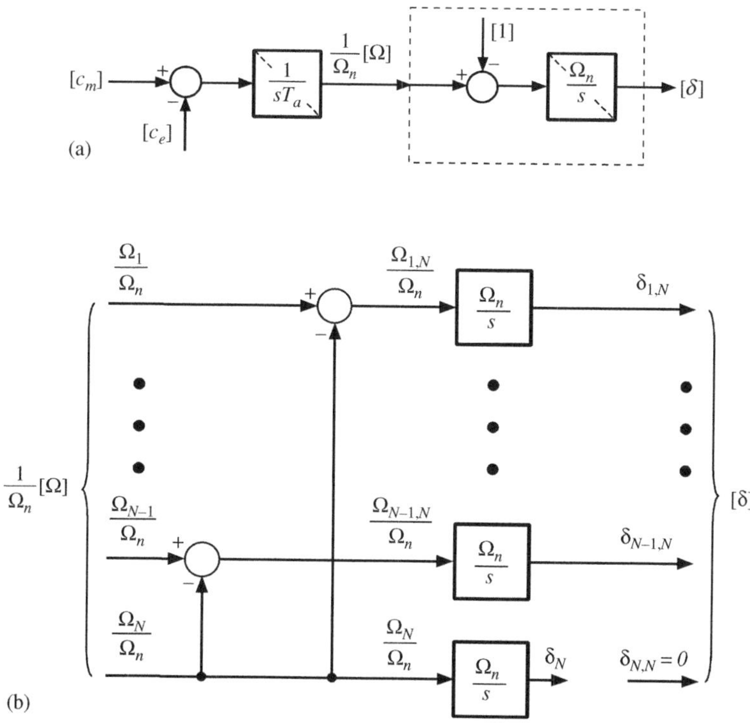

Figure 9.3. Mechanical part of the units: (a) absolute angles; (b) relative angles.

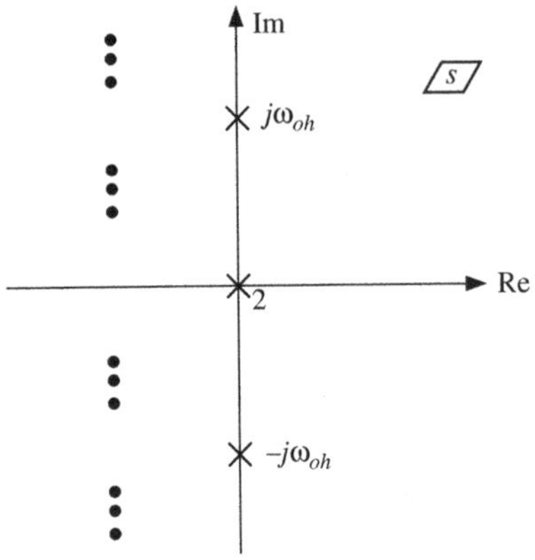

Figure 9.4. Position of eigenvalues of the dynamic matrix [*A*] with the simplified model.

Figure 9.3b. Therefore, one of the two poles of [*A*] lying in the origin is associated with the integrator that does not interact with the system, while the other one is associated with the *common or mean motion between machines,* corresponding to the *mean frequency of the network* (assuming that there is no speed control). This feature is represented in Figures 9.4 and 9.5 and will be explained in Section 9.3.2.2.

Actually, under the above-mentioned assumptions, a step $1/s$ of mechanical power corresponds to $1/s^2$ of frequency change that is to a ramp variation of frequency. The mean speed Ω_m of the system is the speed of the so-called *center of inertia.*

It is interesting to observe the formal analogy between the electromechanical oscillation modes and the torsional oscillation modes of units. In the latter case, the masses are those distributed among the turbine stages and the alternator and the springs are the torsional elastic constants of the shaft. Conversely, in the case of electromechanical oscillations, the masses are those relevant to the single generators and the springs are the synchronizing power coefficients (associated with the electrical connections between the synchronous generators).

Hence, electromechanical oscillations in a system of N machines are $N-1$ and, owing to the dynamic model that has been assumed (second-order synchronous generators), their damping is zero. Therefore, the second-order model of synchronous machines provides *only indications about the frequencies* ω_{oh} *of electromechanical oscillations* or about the related periods; the periods of electromechanical oscillations observed and studied up to now (including local and interarea oscillations) range from about 0.4 to 20 s.

However, as this approach does not provide any information about the damping of each mode, the latter should be determined by modeling the components (machines, regulators, and governors) with a sufficient level of detail, considering the state variables listed in Section 9.2.1.

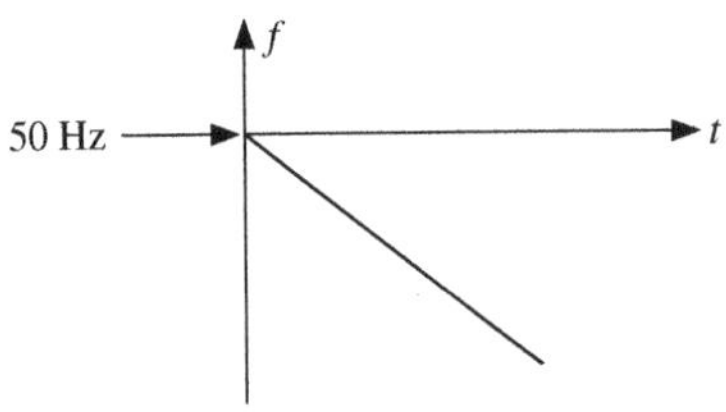

Figure 9.5. Mean motion of generators, without speed control, for a stepwise decrease in mechanical power.

9.3.2.2 Oscillation Amplitudes and Participation Factors. From equations (9.32), that is, from Figure 9.2, it is easy to obtain, in terms of Laplace transforms under null initial conditions

$$\begin{cases} s[\Delta\delta(s)] = [\Delta\Omega(s)] \\ s[\Delta\Omega(s)] = [C][\Delta\delta(s)] + [R][\Delta E(s)] + [M]^{-1}[\Delta P_m(s)] \end{cases} \tag{9.46}$$

that is the matricial equation

$$s^2[\Delta\delta(s)] = [C][\Delta\delta(s)] + [R][\Delta E(s)] + [M]^{-1}[\Delta P_m(s)] \tag{9.47}$$

or

$$[\Delta\delta(s)] = [s^2 I - C]^{-1}\left\{[R][\Delta E(s)] + [M]^{-1}[\Delta P_m(s)]\right\} \tag{9.48}$$

Then, if $[L]$ and $[G]$ are the matrices of the eigenvalues and eigenvectors of $[C]$, respectively

$$[L] = \mathrm{diag}\{L_h\} = \mathrm{diag}\{-\omega_{oh}^2\} \tag{9.49}$$

(with $L_N = 0$), we get, based on the properties of eigenvalues and eigenvectors

$$[\Delta\delta(s)] = [G][s^2 I - L]^{-1}[G]^{-1}\left\{[R][\Delta E(s)] + [M]^{-1}[\Delta P_m(s)]\right\} \tag{9.50}$$

and thus also

$$[\Delta\Omega(s)] = [G]s[s^2 I - L]^{-1}[G]^{-1}\left\{[R][\Delta E(s)] + [M]^{-1}[\Delta P_m(s)]\right\} \tag{9.51}$$

Similar expressions also hold for any other output variable $[Y(s)]$ of the linearized model. As a result, the simplified dynamic behavior of the system following small variations is completely defined by $[L]$ and $[G]$.

Dealing with $[G_N]$, the eigenvector corresponding to the zero eigenvalue,

$$[G_N] \triangleq [G_{1N} G_{2N} \cdots G_{NN}]^T \tag{9.52}$$

it results

$$G_{1N} = G_{2N} = \cdots = G_{NN} \tag{9.53}$$

If we define

$$\begin{cases} [W] \triangleq [G]^{-1}[R] = -[G]^{-1}[M]^{-1}[F] \\ [D] \triangleq [G]^{-1}[M]^{-1} \end{cases} \tag{9.54}$$

then the contribution associated with the eigenvalue $L_N = 0$, in all the $\Delta\Omega_k(s)$ of equation (9.51), can be written as

$$\frac{1}{s} G_{1N} \sum_{j=1}^{N} [W_{Nj} \Delta E_j(s) + D_{Nj} \Delta P_{mj}(s)] \tag{9.55}$$

and it is independent of the particular machine k ($k = 1, \ldots, N$), that is, it is the same for all the system generators. By identifying the variations of such speed with $\Delta\Omega_m$

$$\Delta\Omega_m = \Omega_m - \Omega_n \tag{9.56}$$

we will have

$$\Delta\Omega_m(s) = \frac{1}{s} G_{1N} \sum_{j=1}^{N} [W_{Nj} \Delta E_j(s) + D_{Nj} \Delta P_{mj}(s)] \tag{9.57}$$

If one of the power system generators is an infinite system, the eigenvalue L_N disappears. In such condition

$$\Delta\Omega_m = 0 \tag{9.58}$$

that is, as is intuitive, the mean network speed coincides with the nominal or synchronizing speed.

In the general case ($k = 1, \ldots, N$), from equation (9.51)

$$\Delta\Omega_k(s) = \sum_{h=1}^{N-1} \frac{s}{s^2 + \omega_{oh}^2} G_{kh} \left[\sum_{j=1}^{N} W_{hj} \Delta E_j(s) + \sum_{j=1}^{N} D_{hj} \Delta P_{mj}(s) \right] + \Delta\Omega_m(s) \tag{9.59}$$

where from equations (9.54), the elements W_{hj} and D_{hj} are

$$\begin{cases} W_{hj} = -\displaystyle\sum_{i=1}^{N} \frac{(G^{-1})_{hi}}{M_i} F_{ij} \\ D_{hj} = \dfrac{(G^{-1})_{hj}}{M_j} \end{cases} \tag{9.60}$$

Large or small structural perturbations (opening of lines, disconnection of generators) are equivalent to the application of mechanical power steps with amplitudes different from generator to generator. Therefore, in the absence of voltage and speed controls and within the framework of small variations, if P_{wk} is the amplitude of the equivalent perturbation for the kth machine, we will obtain

$$[\Delta P_m] = \frac{1}{s} [P_w] = \frac{1}{s} [P_{w1} \cdots P_{wN}]^T \tag{9.61}$$

Apart from the contribution of the pole in the origin, equation (9.59) becomes

$$\Delta\Omega_k(s) = \sum_{h=1}^{N-1} \frac{G_{kh}}{s^2 + \omega_{oh}^2} \sum_{j=1}^{N} D_{hj} P_{wj} \tag{9.62}$$

that is, in the time domain

$$\Delta\Omega_k(t) = \sum_{h=1}^{N-1} A_{kh}\sin(\omega_{oh}t) \tag{9.63}$$

where A_{kh} is the *participation factor* of the generator k to the mode h.

This expression shows that the speed of any generator of the network (and likewise for all other variables: angles, powers, voltages, etc.) is a *linear combination of the various oscillation modes*. The *amplitudes* A_{kh} are

$$A_{kh} = G_{kh}\sum_{j=1}^{N}\frac{(G^{-1})_{hj}}{\omega_{oh}M_j}P_{wj} \tag{9.64}$$

where, for each oscillation mode, the summation is independent of the particular machine k considered.

As a consequence, we may define

$$\alpha_h \triangleq \sum_{j=1}^{N}\frac{(G^{-1})_{hj}}{\omega_{oh}M_j}P_{wj} \tag{9.65}$$

a (real) coefficient identifying the hth oscillation mode and the (small) perturbation being investigated. Therefore,

$$A_{kh} = G_{kh}\alpha_h \tag{9.66}$$

In other terms, for the kth machine, the amplitude of the sinusoid pertaining to the hth mode is proportional to the corresponding element of the eigenvector $[G_h]$ of $[C]$.

For each mode h, α_h depends in particular on the oscillation period of the hth mode and on the value of the power steps applied to generators. In contrast, the *oscillation amplitudes pertaining to any mode*, that is, the amplitudes of sinusoids with frequency ω_{oh} (appearing in expressions (9.63) for the speeds of generators k, i, with $k, i = 1, \ldots, N$) are given by

$$\frac{A_{kh}}{A_{ih}} = \frac{G_{kh}}{G_{ih}} \tag{9.67}$$

which is the *ratio* between the elements of the eigenvector $[G_h]$ pertaining to the machines k and i. They do not depend on type, location, or extent of the perturbation.

On the contrary, for each h, the coefficient α_h takes account of the impact of the considered perturbation on the hth oscillation mode and in particular on the oscillation amplitudes A_{kh}. If the perturbation is such that $\alpha_h = 0$, then the mode h is not excited by that particular perturbation.

Of course, similar considerations also apply to angle variations and to the linearized model output variables, for example variations of currents and power flows on lines, of bus voltages, and so on, provided that the algebraic relationship between state variables and output variables is known.

Now, the matrix $[C]$ has real elements and real eigenvalues $[L]$: consequently, also the matrix of the eigenvectors $[G]$ and its inverse $[G]^{-1}$ have real elements. Thus, for *each oscillation mode*, generators *may be divided into three categories* (see equation (9.67)):

- Generators oscillating in phase between them, at generally different amplitudes; the elements G_{kh} of these generators have the same sign (e.g., positive).
- Generators oscillating in phase between them, at generally different amplitudes, in phase opposition with the generators of the first category; the elements G_{kh} of these generators have an opposite sign (e.g., negative) with respect to the generators of the first category.
- Generators not oscillating together with the generators of either the first category or the second category; these generators have $G_{kh}=0$, as if they were (for that oscillation mode) infinite systems.

It is worth stressing that, among the $N-1$ oscillation modes of a power system, the dominant ones (i.e., with significant amplitudes) are usually very few (2–4 at the most), even for large interconnected systems with 500 or more oscillation modes; this finding is also confirmed by worldwide experimental recording of perturbations. These dominant oscillation modes are also the slowest ones, in that the *zeros* actually overlap the fastest poles (see Figure 9.6), thereby canceling them.

In particular, if the eigenvectors of the matrix $[C]$ are known, then information may be obtained about the *characteristics of the various electromechanical oscillations in terms of local or interarea oscillations*. In the first instance, the elements G_{kh} will be large (in absolute values) only for some generators. In the second instance, a set of generators will have significant speed amplitudes of a given sign and oscillate with respect to another group of generators that either are characterized by equally significant amplitudes of an opposite sign, or practically do not oscillate. An example of the above is given in Figure 9.7: the oscillation mode may be interpreted as the oscillation between generators 1, 2, 3 and generators 5, 6.

The properties indicated above, rigorously valid if a second-order generator model is adopted, are extremely useful in practice for many analyses of the dynamic behavior of a network. Indeed, the conclusions of the previous simplified analysis are still valid with

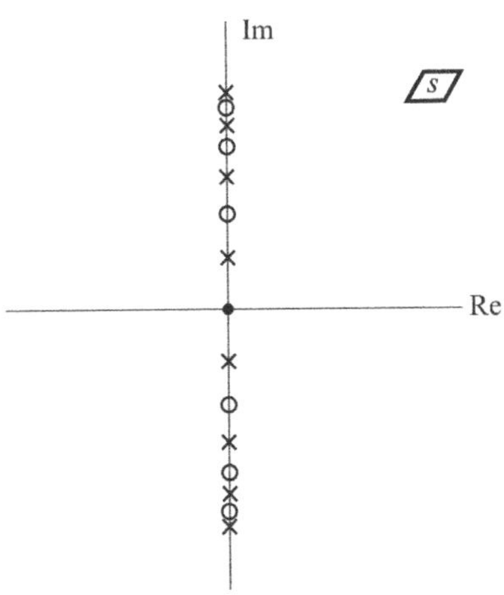

Figure 9.6. Poles and zeros associated with electromechanical oscillations.

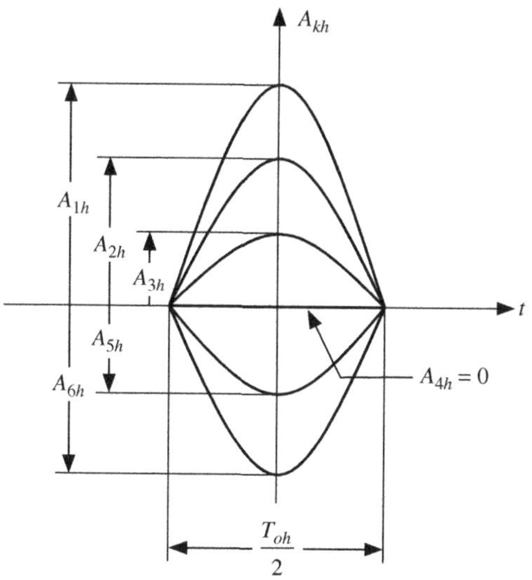

Figure 9.7. Example of oscillation amplitudes for a generic mode and for a 6-generator system.

sufficient approximation compared to the result of the eigenvalue analysis on the matrix $[A]$ built starting from more accurate models (e.g., those of the fourth-, fifth- or sixth order), as demonstrated by many practical cases.

In summary, the simplified analysis of the dynamic behavior of a system, based on the second-order model for synchronous machines, makes it possible to assess (through eigenvalues) the periods of the $(N-1)$ electromechanical oscillations between the N system generators

$$T_{oh} = \frac{2\pi}{\omega_{oh}} \tag{9.68}$$

and, through eigenvectors, the ratios between the amplitudes for the various oscillation modes and for the various generators, and thus also between the various points of the network and the actual oscillation amplitudes in response to perturbations (through the inverse matrix of eigenvectors, equation (9.64)).

Of course, if the *modal characteristics* of the state variables are known, it is immediate to determine the behavior of any output variable (voltages, real power flows, etc.). By contrast, the simplified analysis does not provide any information about the damping of electromechanical oscillations, since they are null by definition, owing to the dynamic model that has been adopted for generators.

9.3.3 Numerical Examples

In this section, some numerical examples are provided in order to demonstrate the properties of the simplified approach to study the dynamic phenomena associated to the electromechanical oscillations.

The first example shows the application of the approach to a simple two-area test system, while the second example is related to a three-area system representing a simplified model of the European Power System.

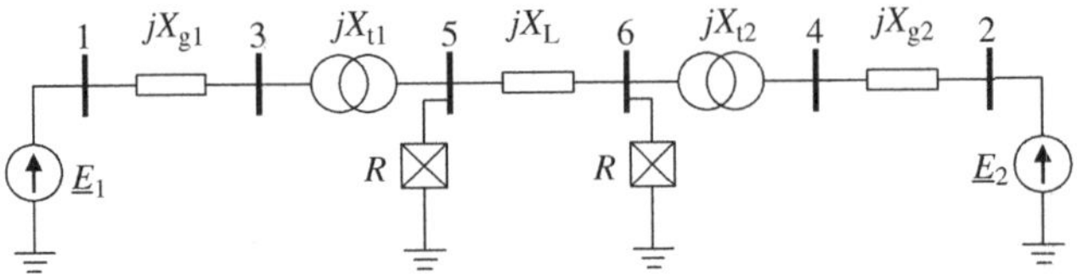

Figure 9.8. Two-area test system.

TABLE 9.1. Data of the Two-Area Test System

Generators	V_n (kV)	S_n (MVA)	x_g (p.u.)	T_a (s)
1	20	50,000	0.27	10
2	20	50,000	0.27	10
Lines	**Length (km)**		**x_L (Ω/km)**	
	100		0.3616	
Transformers	**n (kV/kV)**		**x_t (p.u.)**	
1	20/400		0.13	
2	20/400		0.13	

9.3.3.1 Application 1: Two-Area Test System. The test system used in this section (Figure 9.8) is characterized by two large areas connected by a transmission link. Each area, well meshed, has its own generation and load, quite balanced, and normally exchanges a low level of real and reactive power with the other area. It is, therefore, suitable to represent each area by an equivalent generator (a constant emf behind a reactance X_g), a transformer and a local load, represented as a resistance connected on the HV level. This system allows some interesting comments on the parameters that influence the electromechanical oscillations. The data of the test system are reported in Table 9.1.

The resistances R_1 and R_2 represent a real load of 40,000 MW each, at the rated voltage 400 kV; therefore, their value is 4 Ω.

The initial operating conditions of the test system are computed by a power flow, assuming that the voltage at the load busses is 400 kV and that the two areas are autonomous, that is, the real and the reactive power flowing through the interconnection line is approximately zero. The power flow results are reported in Table 9.2.

TABLE 9.2. Power Flow Results

Bus #	V (kV)	Phases (rad)
1	420.00	0.2061
2	420.00	0.2061
3	402.18	0
4	402.18	0
5	400.00	−1.1037
6	400.00	−1.1037

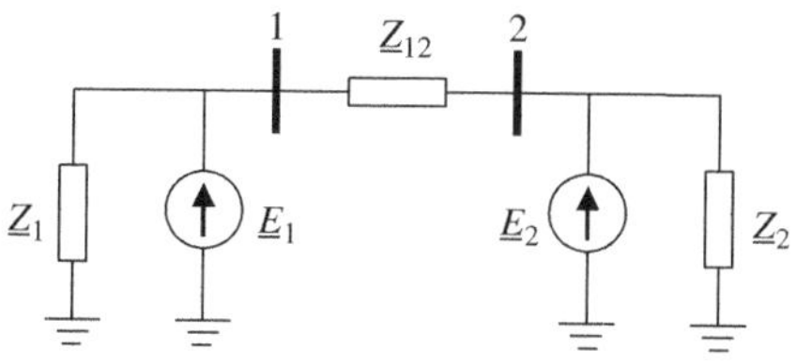

Figure 9.9. Equivalent circuit of the two-area test system.

The two-area test system is reduced, in order to determine its equivalent (Figure 9.9) as seen from nodes 1 and 2, where $\underline{E}_1$ and $\underline{E}_2$ are the emf behind the transient reactance of the generators.

Accordingly, the (6×6) admittance matrix of the system in Figure 9.8 can be manipulated and reduced to a (2×2) matrix $[\underline{Y}_{eq}]$ relevant to the network in Figure 9.9.

$$[\underline{Y}_{eq}] = [G_{eq}] + j[B_{eq}] = \begin{bmatrix} \underline{Y}_{11} & \underline{Y}_{12} \\ \underline{Y}_{21} & \underline{Y}_{22} \end{bmatrix} = \begin{bmatrix} (0.2133 - j0.0924) & (0.0135 + j0.0198) \\ (0.0135 + j0.0198) & (0.2133 - j0.0924) \end{bmatrix} \tag{9.69}$$

The matrix of synchronizing power coefficients $[K]$ is then

$$[K] = \begin{bmatrix} K_{11} & K_{12} \\ K_{21} & K_{22} \end{bmatrix} = \begin{bmatrix} 3490 & -3490 \\ -3490 & 3490 \end{bmatrix} \quad (\text{MW/rad}) \tag{9.70}$$

where

$$\begin{aligned} K_{11} &= E_1 E_2 [B_{eq12} \cos(\delta_{12}) - G_{eq12} \sin(\delta_{12})] \\ K_{12} &= -E_1 E_2 [-B_{eq12} \cos(\delta_{12}) + G_{eq12} \sin(\delta_{12})] \\ K_{22} &= E_2 E_1 [B_{eq21} \cos(\delta_{12}) - G_{eq21} \sin(\delta_{12})] \\ K_{21} &= -E_2 E_1 [-B_{eq21} \cos(\delta_{12}) + G_{eq21} \sin(\delta_{12})] \end{aligned} \quad (\text{MW/rad}) \tag{9.71}$$

The matrix of inertia coefficients is, according to (9.27)

$$[M] = \begin{bmatrix} M_1 & 0 \\ 0 & M_2 \end{bmatrix} = \begin{bmatrix} 1591 & 0 \\ 0 & 1591 \end{bmatrix} \quad (\text{MVA s}^2/\text{rad}) \tag{9.72}$$

and the matrix $[C]$ of equation (9.33)

$$[C] = -[M]^{-1}[K] = \begin{bmatrix} -\dfrac{K_{11}}{M_1} & +\dfrac{K_{11}}{M_1} \\ +\dfrac{K_{22}}{M_2} & -\dfrac{K_{22}}{M_2} \end{bmatrix} = \begin{bmatrix} -2.193 & 2.193 \\ 2.193 & -2.193 \end{bmatrix} \quad (\text{s}^{-2}) \tag{9.73}$$

The eigenvalues and eigenvectors of $[C]$ can be computed solving

$$\det \begin{bmatrix} L + \dfrac{K_{11}}{M_1} & -\dfrac{K_{11}}{M_1} \\ -\dfrac{K_{22}}{M_2} & L + \dfrac{K_{22}}{M_2} \end{bmatrix} = L\left[L + \left(\frac{K_{11}}{M_1} + \frac{K_{22}}{M_2}\right)\right] = 0 \tag{9.74}$$

The solution of the above equation

$$\begin{cases} L_1 = -\dfrac{K_{11}}{M_1} - \dfrac{K_{22}}{M_2} = -4.3855 \\ L_2 = 0 \end{cases} \tag{9.75}$$

shows that the only one electromechanical oscillation present is characterized by

$$T_o = \frac{2\pi}{\sqrt{-L_1}} = 3\,\text{s} \tag{9.76}$$

typical of interarea oscillations. In order to evaluate the ratio between the oscillation amplitudes, it is necessary to compute the eigenvector associated to L_1, held in the first column of the eigenvector matrix $[G]$:

$$[G] = \begin{bmatrix} -0.7071 & 0.7071 \\ 0.7071 & 0.7071 \end{bmatrix} \tag{9.77}$$

The ratio between the amplitudes of the oscillations of the two machines is

$$\frac{G_{11}}{G_{21}} = -1 \tag{9.78}$$

It is worth noting that this ratio is negative (this occurs always, in the case of two machines): this indicates that the generators oscillate always in phase opposition. Moreover, in this case, the ratio is -1, which indicates that the two amplitudes are equal, which is reasonable, as the two machines are equal.

In order to evaluate the accuracy of the simplified approach, some time-domain simulations have been carried out. The perturbation considered is a three-phase switching-on/-off, lasting 50 ms, of a solid connection to ground at bus bar 3. Since the duration is very short, this type of disturbance may be considered as a small perturbation.

In the simulation, all the controls have been switched-off in order to highlight the electromechanical oscillation only.

Figure 9.10 shows the oscillatory behavior of the system in terms of rotor speeds. The period of the electromechanical oscillation is approximately 3 s and the oscillation of the two rotors is equal in amplitude but in phase opposition, as expected. As the second-order model is used for generators, the oscillations are not damped. The same figure also highlights the mean speed variation of the system that is different from zero because the loads are assumed voltage dependent.

Figure 9.11 shows the simulation obtained in response to the same perturbation, assuming a fourth-order model for the generators: the period of electromechanical oscillations is practically the same as in the previous case, and the oscillation is well damped, thanks to the high positive contribution of the generator structure (see Section 9.4.3.2).

Figure 9.12 shows the results of a time-domain simulation carried out on the same perturbation, with the second-order model for generators, where the inertia of $G1$ has been increased doubling the start-up time. In this case, the simplified approach results in a period, for the electromechanical oscillation, of 3.46 s, while the amplitude of the speed variation of $G2$ is doubled with respect to the one of $G1$; this is confirmed by the time-domain simulation.

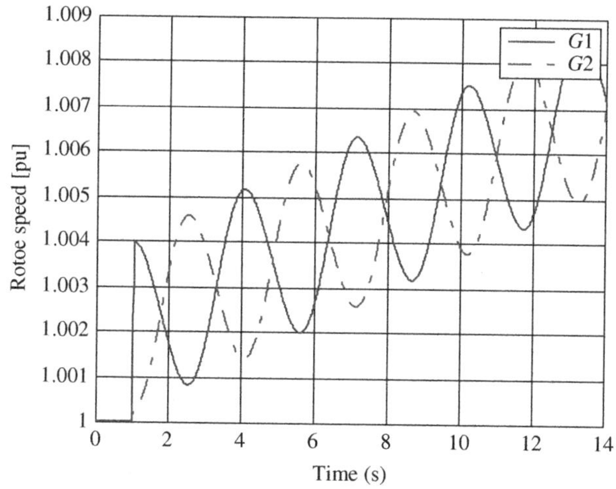

Figure 9.10. Behavior of the rotor speed of two equal generators, in case of a small perturbation (second-order dynamic model).

9.3.3.2 Application 2: Three-Area Test System. The second test system (Figure 9.13) is a simplified model of the European Power System (see Section 9.6), consisting of three cascade-connected areas (whose data are provided in Table 9.3); loads are assumed purely resistitive, and the areas are autonomous, in that the loads are local and the power exchanges are zero.

Local loads are assumed 80% of the relevant area generation capability, that is, area 1 and area 2 loads are 40,000 MW, while the load of area 3 is 200,000 MW, at the rated voltage 400 kV. The loads are represented by three resistances $R_1 = R_2 = 4\,\Omega$ and $R_3 = 0.8\,\Omega$.

The initial operating conditions are computed by a power flow, assuming that the power exchange among areas is zero and that voltages at load busses are 400 kV (Table 9.4).

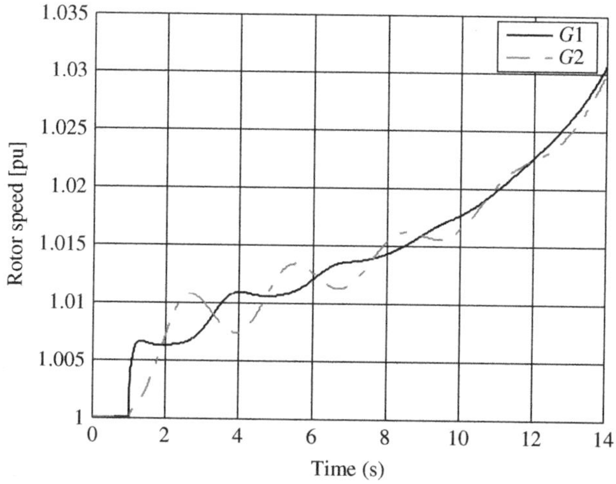

Figure 9.11. Behavior of the rotor speed of two equal generators, in case of a small perturbation (fourth-order dynamic model).

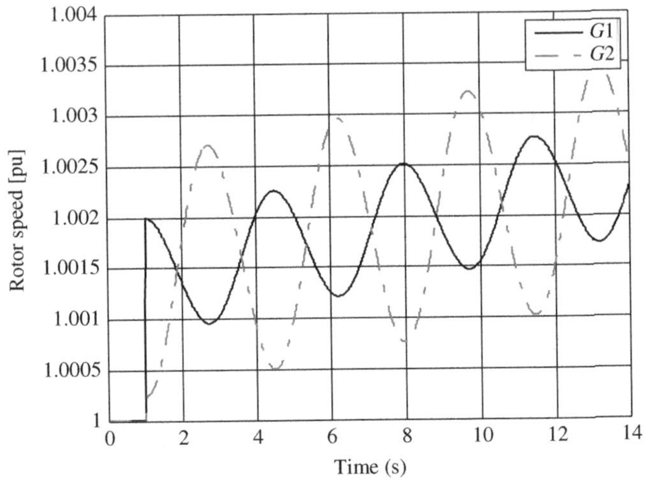

Figure 9.12. Behavior of the rotor speed of two generators (second-order dynamic model): case of doubled inertia of generator G1.

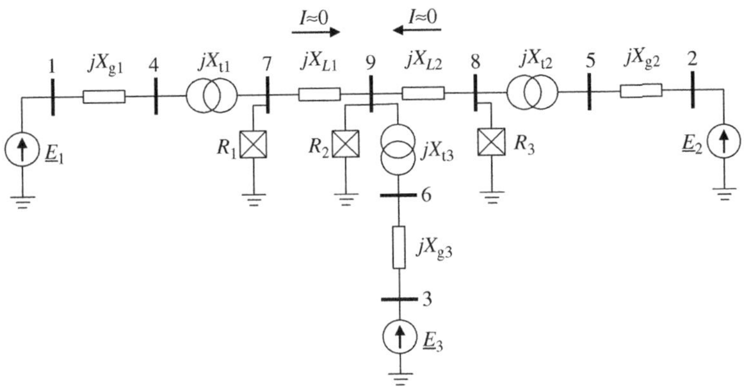

Figure 9.13. Three-area test system.

TABLE 9.3. Data of the Three-Area Test System

Generators	V_n (kV)	S_n (MVA)	X_g (p.u.)	T_a (s)
1	20	50,000	0.27	10
2	20	50,000	0.27	10
3	20	250,000	0.27	10

Lines	Length (km)	X_L (Ω/km)
7–9	120	0.36
8–9	80	0.28

Transformers	S_n (MVA)	n (kV/kV)	X_t (p.u.)
1	50,000	20/400	0.13
2	50,000	20/400	0.13
3	250,000	20/400	0.13

TABLE 9.4. Power Flow Results for the Three-Area Test System

Bus #	V (kV)	Phases (rad)
1	419.98	0.2060
2	419.98	0.2060
3	429.02	0.2079
4	402.16	0
5	402.16	0
6	402.16	0
7	400.00	−0.1047
8	400.00	−0.1047
9	400.00	−0.1047

The test system is then reduced in order to find its equivalent circuit at the terminals of the internal emf of generators, namely busses 1, 2, and 3.

Accordingly, the (9×9) admittance matrix of the system in Figure 9.13 can be manipulated and reduced to a (3×3) matrix $[\underline{Y}_{eq}]$ relevant to the network in Figure 9.14.

$$[\underline{Y}_{eq}] = [G_{eq}] + j[B_{eq}] = \begin{bmatrix} (0.2129 - j0.0924) & (0.0002 + j0.0001) & (0.0136 + j0.0197) \\ (0.0002 + j0.0001) & (0.2049 - j0.1044) & (0.0216 + j0.0317) \\ (0.0136 + j0.0197) & (0.0216 + j0.0317) & (1.0987 - j0.4142) \end{bmatrix} \tag{9.79}$$

The matrix of synchronizing power coefficients $[K]$ is determined starting from this admittance matrix and from the rotor angles. It results

$$[K] = \begin{bmatrix} 3503 & -27 & -3475 \\ -27 & 5631 & -5603 \\ -3466 & -5589 & 9055 \end{bmatrix} \quad (\text{MW/rad}) \tag{9.80}$$

The matrix of inertia coefficients is, according to (9.27)

$$[M] = \begin{bmatrix} 1592 & 0 & 0 \\ 0 & 1592 & 0 \\ 0 & 0 & 7961 \end{bmatrix} \quad (\text{MVA s}^2/\text{rad}) \tag{9.81}$$

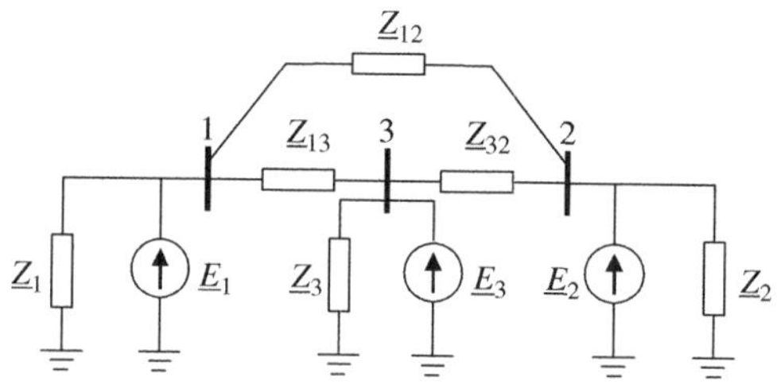

Figure 9.14. Equivalent of the three-area test system.

and the matrix $[C]$ of equation (9.33)

$$[C] = -[M]^{-1}[K] = \begin{bmatrix} -2.1998 & 0.0173 & 2.1825 \\ 0.0173 & -3.5363 & 3.5189 \\ 0.4353 & 0.7020 & -1.1373 \end{bmatrix} \quad (s^{-2}) \tag{9.82}$$

The eigenvalues and eigenvectors of $[C]$ are

$$[L] = \begin{bmatrix} -2.4730 & 0 & 0 \\ 0 & -4.4004 & 0 \\ 0 & 0 & 0 \end{bmatrix} \tag{9.83}$$

$$[G] = \begin{bmatrix} -0.9263 & -0.2241 & -0.5773 \\ 0.3593 & -0.9461 & -0.5773 \\ 0.1131 & 0.2334 & -0.5773 \end{bmatrix} \tag{9.84}$$

The eigenvalue analysis shows two electromechanical oscillation modes; their period is respectively,

$$T_{o1} = \frac{2\pi}{\sqrt{-L_1}} = 4\,s \tag{9.85}$$

$$T_{o2} = \frac{2\pi}{\sqrt{-L_2}} = 3\,s \tag{9.86}$$

Consider mode #1: the eigenvector relevant to the first column of $[G]$ shows that the amplitude of the oscillation for $G1$ is much higher than for $G2$ and $G3$, which oscillate in phase to each other. In particular, the amplitude of the oscillation for $G3$ is lower due to its larger inertia. The situation is represented in Figure 9.15.

For mode #2, the analysis of the relevant eigenvector (second column of $[G]$) is depicted in Figure 9.16. Generators $G1$ and $G2$ oscillate in phase, but $G2$ is characterized by a higher amplitude. $G3$ is in phase opposition with a lower amplitude.

Interesting results can be obtained increasing the inertia of $G3$, which means that area 3 is of infinite power. Assume that $T_{a3} = 30$ s. The eigenvalue analysis results in

$$T_{o1} = \frac{2\pi}{\sqrt{-L_1}} = 4.12\,s \tag{9.87}$$

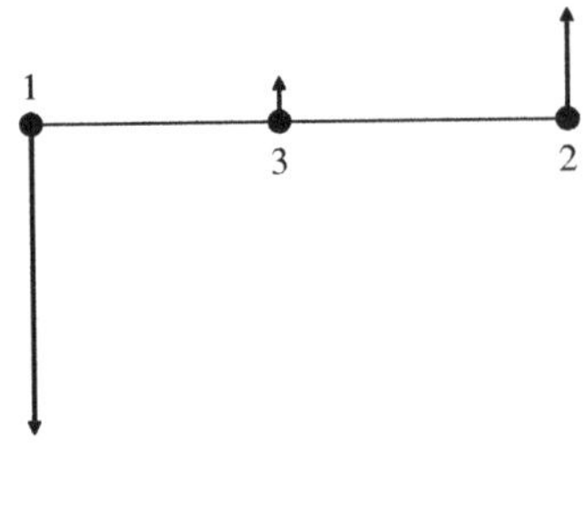

Figure 9.15. Speed oscillation amplitudes for oscillation mode #1.

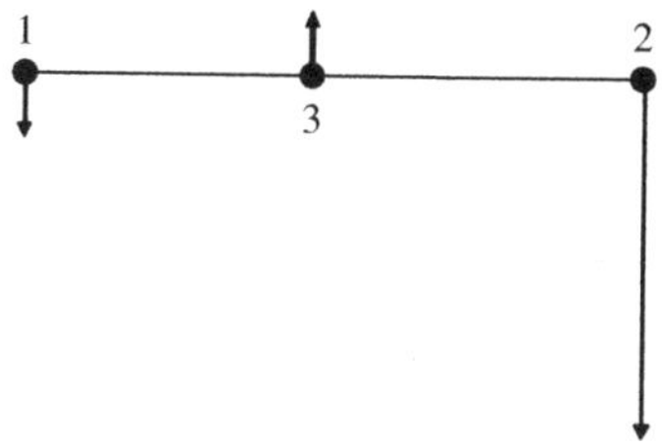

Figure 9.16. Speed oscillation amplitudes for oscillation mode #2.

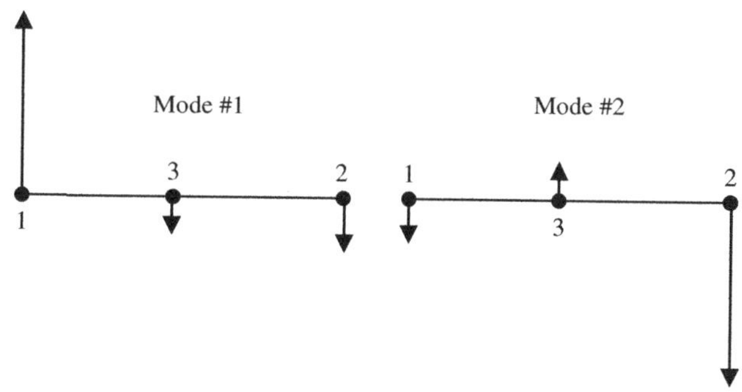

Figure 9.17. Speed oscillation amplitudes assuming area 3 of infinite power.

$$T_{o2} = \frac{2\pi}{\sqrt{-L_2}} = 3.23 \text{ s} \tag{9.88}$$

with the eigenvector matrix

$$[G] = \begin{bmatrix} 0.9873 & -0.0879 & -0.5773 \\ -0.1481 & -0.9935 & -0.5773 \\ -0.0558 & 0.0719 & -0.5773 \end{bmatrix} \tag{9.89}$$

It is worth noting that the assumption made results in (see Figure 9.17)

- a remarkable reduction of the amplitude of the oscillations for area 3, for both modes #1 and #2;
- a strong decoupling between areas 1 and 2, connected through the infinite bus. Basically, mode #1 describes the oscillation of $G1$ with respect to areas 2 and 3 together, while mode #2 describes the oscillation of $G2$.

9.4 MAJOR FACTORS AFFECTING THE DAMPING OF ELECTROMECHANICAL OSCILLATIONS

9.4.1 Introduction

Nowadays, the main stability problem that arises is the insufficient damping ζ of electromechanical oscillations, that is, oscillations between alternator rotors that fall within the following ranges in terms of oscillation frequencies and oscillation periods

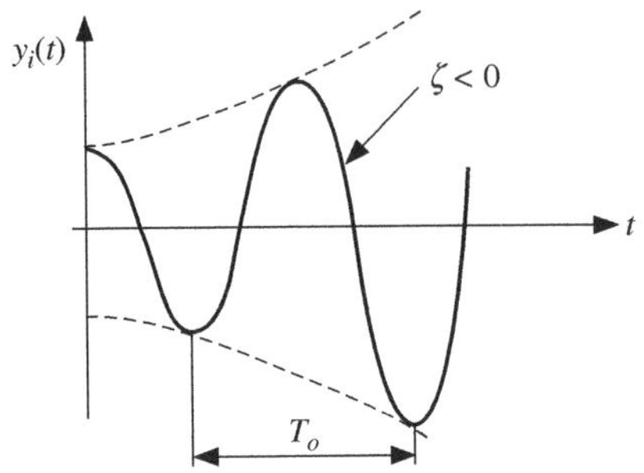

Figure 9.18. Undamped electromechanical oscillation.

$$\begin{cases} \omega_o = 0.3 - 15\,\text{rad/s} \\ f_o = 0.05 - 2.5\,\text{Hz} \\ T_o = 0.4 - 20\,\text{s} \end{cases} \tag{9.90}$$

positive but small or negative values of ζ (Figure 9.18) may occur in systems characterized by a particular structure (long lines, longitudinal configuration, etc.) and are usually due to the modern excitation systems of units, which are very fast.

The problem may be as follows:

- *Local*. Electromechanical oscillations have large amplitudes on the variables (p_e, Ω, δ, f) of either a single generator or power plant or a few power plants very close to each other. Local oscillations are characterized by

$$\begin{cases} \omega_o = 6 - 15\,\text{rad/s} \\ T_o = 0.4 - 1\,\text{s} \end{cases} \tag{9.91}$$

- *Between Areas* (or *Low-Frequency Oscillations*). Amplitudes are large on the variables of many power plants belonging to one or more zones of the system that are geographically far from each other. Such oscillations are characterized by

$$\begin{cases} \omega_o = 0.3 - 6\,\text{rad/s} \\ T_o = 1 - 20\,\text{s} \end{cases} \tag{9.92}$$

Anyhow, in both cases, the electromechanical oscillations are present on all the system variables: the difference is only due to the amplitudes.

The analysis to be developed has the *fundamental aim* to understand the phenomenon in qualitative terms. Consequently, reference can be made to the simplest scheme (Figure 9.19) of a generator (or area) connected to an infinite system; this scheme is valid in case of both local and interarea oscillations.

In case of *local oscillations* (actual power plant, actual load, etc.)

$$\begin{cases} p_e \lesseqgtr 0 \\ p_L \geq 0 \\ p_T \lesseqgtr 0 \quad , \quad \text{limit case}\, p_T = p_e \end{cases} \tag{9.93}$$

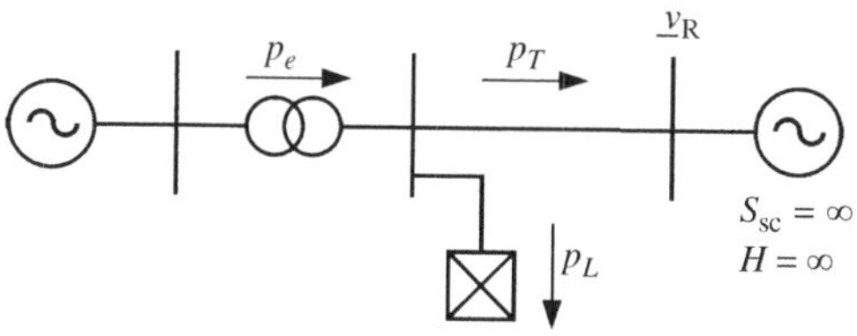

Figure 9.19. Reference case: generator or area connected to an infinite system.

In case of *interarea oscillations* (equivalent generator, equivalent load, etc.)

$$\begin{cases} p_e > 0 \\ p_L > 0 \\ p_T \gtreqless 0 \quad , \quad \text{small portion of } P_e \end{cases} \tag{9.94}$$

It is worth noting that the same scheme allows the analysis of the angle stability following large perturbations, always in qualitative terms.

9.4.2 Single Machine-Infinite Bus System: A Simplified Approach

As it is well known, in the case of a generator or an area connected to an infinite – system, the variations of the generated real power is a function of the rotor angle variations and of the excitation voltage variations:

$$\Delta p_e = h(s)\,\Delta v_f + k(s)\,\Delta\delta \tag{9.95}$$

Adopting the general dynamic model of generators depicted in Figure 9.20, that is, in p.u., the same as Figure 2.30, in terms of Laplace transforms we can obtain

$$h(s) \triangleq \left[\frac{\Delta p_e}{\Delta v_f}(s)\right]_{\delta=\text{const.}} = v_R^o \sin\delta^o \frac{a(s)}{x_e + x_d(s)} \tag{9.96}$$

$$\begin{aligned} k(s) \triangleq \left[\frac{\Delta p_e}{\Delta \delta}(s)\right]_{v_f=\text{const.}} &= \\ - v_R^o \sin\delta^o \left[i_q^o - \frac{v_R^o \sin\delta^o}{x_e + x_d(s)}\right] &+ v_R^o \cos\delta^o \left[i_d^o + \frac{v_R^o \cos\delta^o}{x_e + x_q(s)}\right] \end{aligned} \tag{9.97}$$

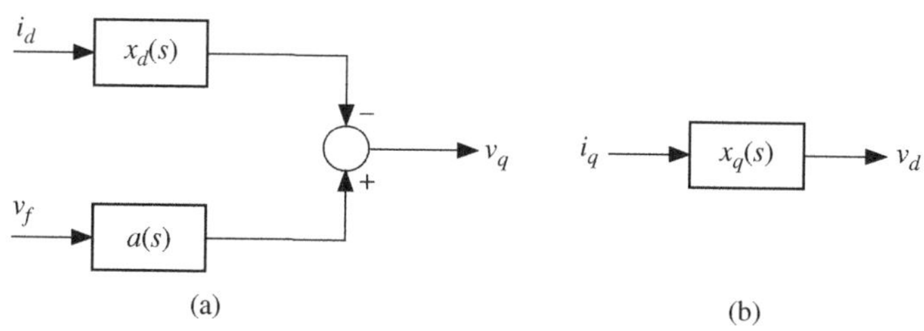

Figure 9.20. General dynamic model of the electrical part of a generator.

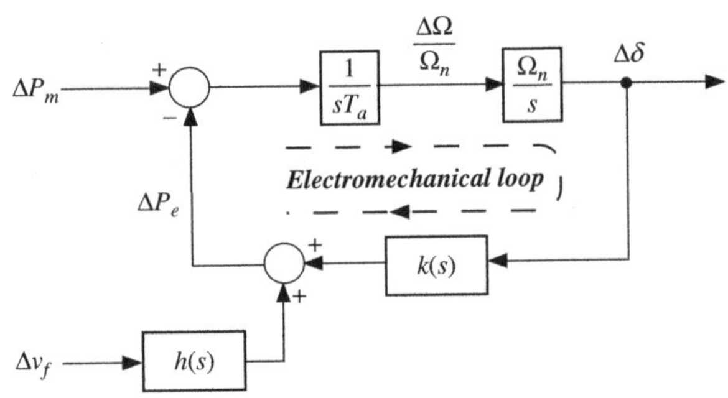

Figure 9.21. Linearized block diagram of a generator connected to an infinite system.

In particular, whatever the rotor circuit number (field circuit and additional circuits), the transfer function $k(s)$ may be put into the following form:

$$k(s) = q_{\mathrm{R}}^o + \left(v_{\mathrm{R}}^o\right)^2 \left[\frac{\sin^2 \delta^o}{x_e + x_d(s)} + \frac{\cos^2 \delta^o}{x_e + x_q(s)}\right] \tag{9.98}$$

where q_{R} is the reactive power entering into the infinite-system node having voltage $\underline{v}_{\mathrm{R}}$.

Such transfer function depends on the reactances $x_d(s)$ and $x_q(s)$ of the generator, on the reactance x_e (transformer plus line) connecting such generator to the infinite system, on the voltage of such system (v_{R}^o) and on the operating point of the generator, in terms of generated real and reactive powers (p_e^o and q^o), terminal voltage (v^o), that is, q_{R}^o, and rotor angle (δ^o).

The linearized system block diagram is shown in Figure 9.21, highlighting both the dynamic relationship between $\Delta\delta$ and Δp_e expressed by $k(s)$, and the inputs Δp_m (supply and speed control systems) and Δv_f (excitation and voltage control systems).

As a first approach to the problem of damping of electromechanical oscillations, let us consider the second-order generator model, characterized by

$$x_q'' = x_d'' = x_q' = x_i \tag{9.99}$$

With reference to the simplified circuit in Figure 9.22a the output of the generator is

$$p_e = f\{e, v_{\mathrm{R}}, \delta\} = \frac{e v_{\mathrm{R}} \sin \delta}{x_e + x_i} \tag{9.100}$$

For constant v_{R} and in the absence of voltage control (i.e., assuming the excitation voltage and thus the internal emf constant), the following equation would hold for small variations around the operating point:

$$\Delta p_e = \left[\frac{\partial p_e}{\partial \delta}\right]^o \Delta\delta \triangleq k\Delta\delta \tag{9.101}$$

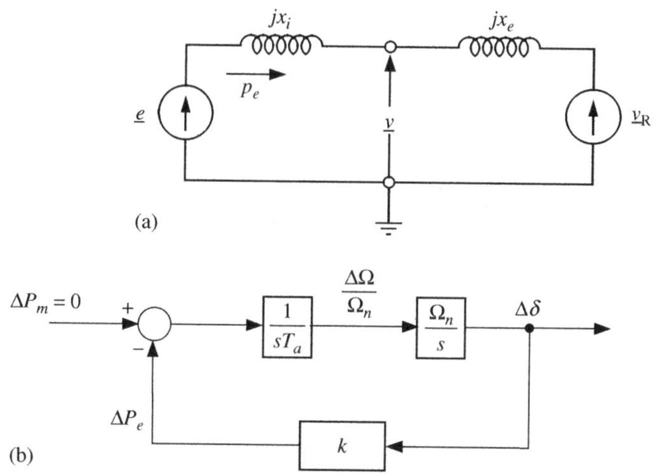

Figure 9.22. Second-order synchronous generator connected to an infinite system: (a) equivalent circuit; (b) linearized block diagram.

with

$$k = \frac{(e v_R \cos\delta)^o}{x_e + x_i} = q_R^o + \frac{(v_R^o)^2}{x_e + x_i} \tag{9.102}$$

Under the above assumptions, the modified *electromechanical loop* block is depicted in Figure 9.22b. The relevant characteristic equation is

$$s^2 + \frac{k\Omega_n}{T_a} = 0 \tag{9.103}$$

and its roots are, assuming that electromechanical oscillations are present,

$$\underline{\lambda}_{1,2} = \pm j\sqrt{\frac{k\,\Omega_n}{T_a}} \triangleq \pm j\omega_o = \pm j\sqrt{\frac{\Omega_n}{T_a}\frac{(e v_R \cos\delta)^o}{x_e + x_i}} \tag{9.104}$$

Therefore, the frequency of the electromechanical oscillation is inversely proportional to the square root of the total reactance $(x_i + x_e)$. In particular, such frequency decreases as the link (represented by x_e) to the infinite system becomes weak, or in case of interarea oscillation. It follows that *small* x_e gives rise to *local oscillations,* whereas *high values of* x_e bring about *interarea or low-frequency electromechanical oscillations.*

The second-order model considered for generators, in line with the general result of Section 9.3, provides no information on damping, which is zero by definition. The presence of the oscillation implies that

$$k > 0 \tag{9.105}$$

or

$$\delta^o < 90^\circ \tag{9.106}$$

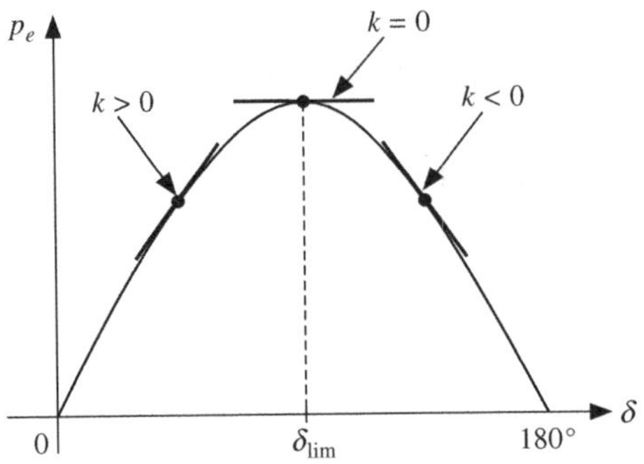

Figure 9.23. Power-angle steady-state characteristic and aperiodic stability limit.

otherwise the *instability* would be *of aperiodic type*, as the two roots would be real, equal, and of opposite sign. Therefore, the *aperiodic instability limit condition*, that is for a real pole, is given by

$$\begin{cases} k = 0 \\ \delta^o = \delta^o_{\text{lim}} = 90° \end{cases} \tag{9.107}$$

and thus (see Figure 9.23)

$$p_e = p_{e\,\max} = \frac{e v_{\text{R}}}{x_e + x_i} \tag{9.108}$$

In other terms, the *aperiodic or real-pole instability is due to insufficient synchronizing power coefficient k.*

Application: As an illustrative example, the situation depicted in Figure 9.24 is considered.

From Figure 9.24, we can write

$$\begin{cases} v_{\text{R}} = 1\,\text{p.u.} & q_{\text{R}} = 0 \\ x = x_e + x_i = 0.8\,\text{p.u.} & e^o \cos\delta^o = v_{\text{R}} = 1\,\text{p.u.} \end{cases} \tag{9.109}$$

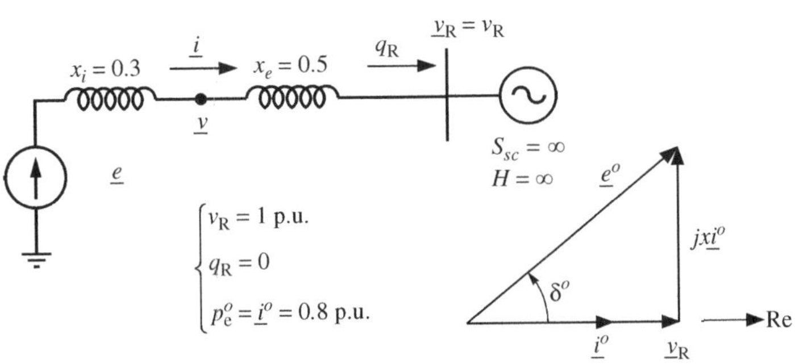

Figure 9.24. Example of computation of local electromechanical oscillation period.

If the alternator start-up time is taken to be $T_a = 7.5$ s and the nominal frequency to be 50 Hz, then

$$\begin{cases} k = \dfrac{e^o v_R}{x} \cos \delta^o = 1.25 \\ \omega_o = \sqrt{\dfrac{1.25 \times 314}{7.5}} \simeq 7.3\,\text{rad/s} \\ f_o = \dfrac{\omega_o}{2\pi} \simeq 1.15\,\text{Hz} \\ T_o = 0.86\,\text{s} \end{cases} \tag{9.110}$$

A few considerations are to be made about the simplified model adopted for the qualitative analysis. Actually, taking into account a more accurate model of the generator (damper windings, field circuit) and the presence of speed governors and voltage regulators, damping is not zero; but, even if it is positive, it remains usually very small as compared to the values normally required in control loops of typical regulations. For the damping of electromechanical oscillations, a value of 0.10 is usually regarded as good and 0.20–0.25 as excellent.

9.4.3 Single Machine-Infinite Bus System: A More Accurate Approach

9.4.3.1 Introduction. The general block diagram of Figure 9.25a can represent any case of a generator (or area) connected to an infinite system (with or without intermediate load and whatever its model, excitation system and voltage control, supply system and speed control), depending on $k(s)$, the transfer function between the angle variations $\Delta\delta$ and the decelerating power variations, given by

$$\Delta p_d = -\Delta p_a = \Delta p_e - \Delta p_m \tag{9.111}$$

where p_a is the accelerating power.

Generally, $k(s)$ takes into account the electromagnetic part of the generator, as well as its excitation system and voltage control, supply system and frequency control. Therefore, such function should not be mistaken for (9.98), which concerns the generator structure only. The characteristic equation of the system shown in Figure 9.25a is

$$1 + \frac{\Omega_n}{s^2 T_a} k(s) = 0 \tag{9.112}$$

that is

$$s^2 + \frac{\Omega_n}{T_a} k(s) = 0 \tag{9.113}$$

Equation (9.113) has a number of solutions, depending on $k(s)$; however, two of them must be associated with the electromechanical oscillation. Calling $(\underline{\lambda}_1, \underline{\lambda}_2)$ the relevant pair of (conjugate) poles:

$$\begin{cases} \underline{\lambda}_1 = \sigma_o + j\omega_o \\ \underline{\lambda}_2 = \sigma_o - j\omega_o \end{cases} \tag{9.114}$$

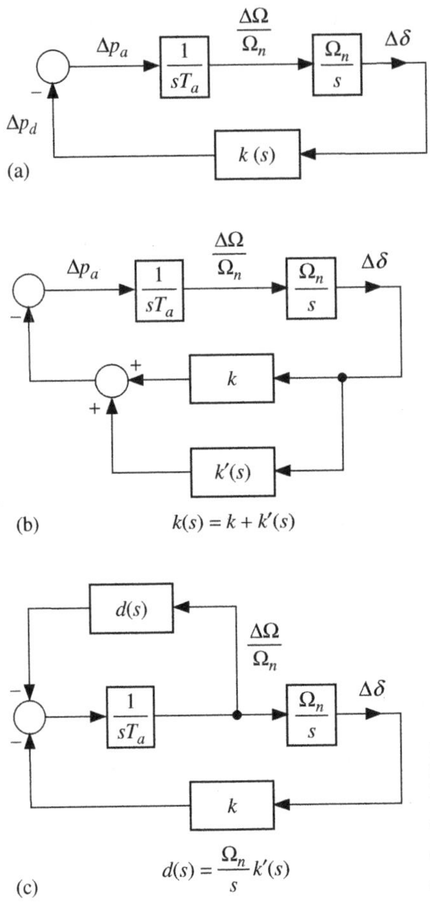

Figure 9.25. Different block diagrams suitable to represent the electromechanical loop to take into account the dynamic effects of the components (generator structure, load, excitation system and voltage control, supply system and frequency control).

on the basis of equation (9.113) it gets

$$-\underline{\lambda}_1^2 = \frac{\Omega_n}{T_a} k(\underline{\lambda}_1) \tag{9.115}$$

that is,

$$\omega_o^2 - \sigma_o^2 - 2j\sigma_o\omega_o = \frac{\Omega_n}{T_a}[\mathrm{Re}\{k(\underline{\lambda}_1)\} + j\,\mathrm{Im}\{k(\underline{\lambda}_1)\}] \tag{9.116}$$

$$\begin{cases} \omega_o^2 - \sigma_o^2 = \dfrac{\Omega_n}{T_a}\mathrm{Re}\{k(\underline{\lambda}_1)\} \\ -\dfrac{\sigma_o}{\omega_o} = \dfrac{1}{2\omega_o^2} \cdot \dfrac{\Omega_n}{T_a}\mathrm{Im}\{k(\underline{\lambda}_1)\} \end{cases} \tag{9.117}$$

Since ζ is anyway and always small (even upon local oscillations), we get

$$\sigma_o^2 \ll \omega_o^2 \tag{9.118}$$

$$k(\underline{\lambda}_1) \simeq k(j\omega_o) = \mathrm{Re}\{k(j\omega_o)\} + j\mathrm{Im}\{k(j\omega_o)\} \tag{9.119}$$

and from (9.117), we obtain

$$\omega_o = \sqrt{\frac{\Omega_n}{T_a}\mathrm{Re}\{k(j\omega_o)\}} \tag{9.120}$$

$$\zeta \simeq -\frac{\sigma_o}{\omega_o} \simeq \frac{1}{2\omega_o^2}\frac{\Omega_n}{T_a}\mathrm{Im}\{k(j\omega_o)\} = \frac{1}{2}\cdot\frac{\mathrm{Im}\{k(j\omega_o)\}}{\mathrm{Re}\{k(j\omega_o)\}} = \frac{1}{2}\tan \angle k(j\omega_o) \tag{9.121}$$

To understand whether the effect of $k(s)$ on damping is positive or negative, let us represent the electromechanical loop like in Figure 9.25b where the effect of $k(s)$ is split into two contributions:

$$\Delta p_d = k(s)\,\Delta\delta \tag{9.122}$$

with

$$k(s) = k + k'(s) \tag{9.123}$$

where k is the synchronizing power coefficient defined by (9.101) and (9.102), as if the generator were a second-order one; as mentioned above, it is not responsible for damping; and $k'(s)$ takes into account the actual structure of the generator and its controls; it is responsible for damping.

Now, as two integrators are associated with the mechanical part, the phase margin γ_c of the electromechanical loop is

$$\gamma_c = \angle[k + k'(j\omega_c)] \tag{9.124}$$

where ω_c indicates the cutoff frequency of the loop.

As, for small dampings, the following equation holds:

$$\omega_c \simeq \omega_o = \sqrt{\frac{k\,\Omega_n}{T_a}} \tag{9.125}$$

we may state that (see also Figure 9.26)

- the effect of $k'(s)$ is stabilizing, if $k'(j\omega)$ has a positive phase at the oscillation frequency ω_o, that is, if

$$0 < \angle k'(j\omega_o) < \pi \tag{9.126}$$

- conversely, the system is unstable, that is, the effect of $k'(s)$ is destabilizing, if

$$-\pi < \angle k'(j\omega_o) < 0 \tag{9.127}$$

Now, equations (9.119), (9.122), and (9.123) infer

$$\begin{aligned}\Delta p_d &= [k + \mathrm{Re}\{k'(j\omega_o)\} + j\mathrm{Im}\{k'(j\omega_o)\}]\Delta\delta \\ &= [k + \mathrm{Re}\{k'(j\omega_o)\}]\Delta\delta + \frac{\Omega_n}{\omega_o}\mathrm{Im}\{k'(j\omega_o)\}\frac{\Delta\Omega}{\Omega_n}\end{aligned} \tag{9.128}$$

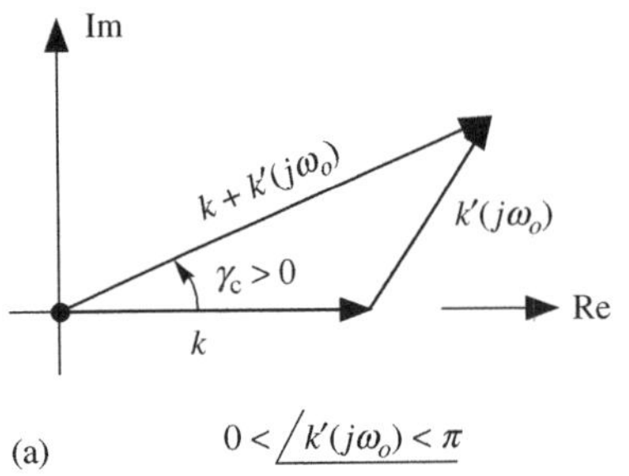

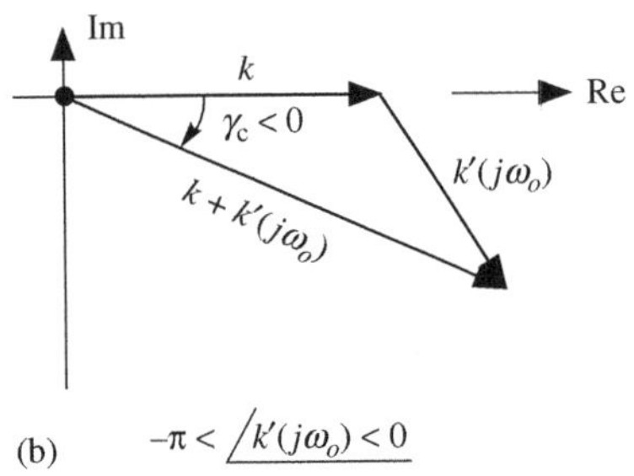

Figure 9.26. Stabilizing and destabilizing effects of the transfer function $k'(s)$: (a) stabilizing effect; (b) destabilizing effect.

If now we define

$$\begin{cases} \Delta p_S \triangleq [k + \mathrm{Re}\{k'(j\omega_o)\}]\Delta\delta \\ \Delta p_D \triangleq \dfrac{\Omega_n}{\omega_o}\mathrm{Im}\{k'(j\omega_o)\}\dfrac{\Delta\Omega}{\Omega_n} \end{cases} \tag{9.129}$$

the block diagram of Figure 9.27 results, where Δp_S and Δp_D are called *synchronizing power (or torque)* and *damping or braking power (or torque)*, respectively, while

$$\begin{cases} \hat{k} = k + \mathrm{Re}\{k'(j\omega_o)\} = \mathrm{Re}\{k(j\omega_o)\} \\ \hat{d} = \dfrac{\Omega_n}{\omega_o}\mathrm{Im}\{k'(j\omega_o)\} = \dfrac{\Omega_n}{\omega_o}\mathrm{Im}\{k(j\omega_o)\} \end{cases} \tag{9.130}$$

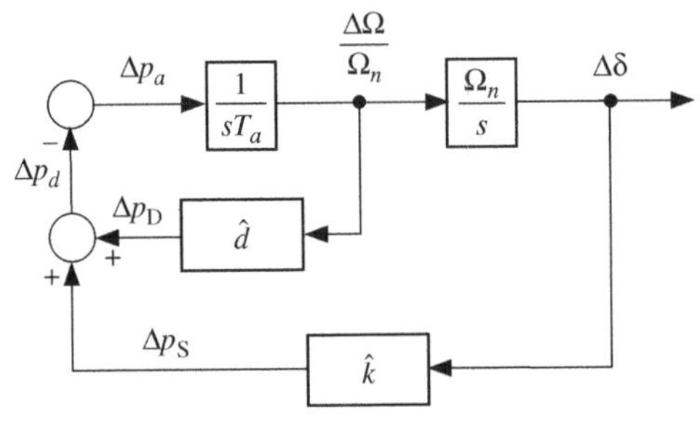

Figure 9.27. Electromechanical loop with damping or braking power (or torque) and synchronizing power (or torque).

The synchronizing power is proportional—in a static way—to angle variations, while the damping power is proportional—always in a static way—to speed variations. The characteristic equation of the electromechanical loop becomes

$$1+\frac{\Omega_n \hat{k}}{s(sT_a+\hat{d})}=0 \tag{9.131}$$

or

$$s^2+s\frac{\hat{d}}{T_a}+\frac{\Omega_n \hat{k}}{T_a}=0 \tag{9.132}$$

which is of the type

$$s^2+2\zeta\omega_n s+\omega_n^2=0 \tag{9.133}$$

where ζ is the damping and ω_n is the natural angular frequency, equal to the amplitude of the two poles associated to ω_o.

This equation has a pair of complex and conjugate poles; such pair is associated with the electromechanical oscillation and is characterized by

$$\omega_n=\sqrt{\frac{\Omega_n \hat{k}}{T_a}};\quad \zeta=\frac{1}{2\omega_n}\cdot\frac{\hat{d}}{T_a};\quad \omega_o=\omega_n\sqrt{1-\zeta^2} \tag{9.134}$$

and, as damping is small,

$$\begin{cases} \omega_o \simeq \sqrt{\Omega_n \dfrac{\hat{k}}{T_a}} \\ \zeta \simeq \dfrac{1}{2\omega_o T_a}\hat{d} \end{cases} \tag{9.135}$$

As a result, $\hat{k}$ can be considered responsible for the oscillation period and $\hat{d}$ for damping.

A *third* and last *form of representation* (equivalent to the previous ones) of the electromechanical loop to understand the effect of $k'(s)$ on damping is shown in Figure 9.25c, where

$$d(s) \triangleq \frac{\Omega_n}{s}k'(s) \tag{9.136}$$

Since damping is small, the following equation will hold (in analogy with equation (9.119)):

$$d(\underline{\lambda}_1) \simeq d(j\omega_o)=\mathrm{Re}\{d(j\omega_o)\}+j\mathrm{Im}\{d(j\omega_o)\} \tag{9.137}$$

and thus also

$$d(j\omega_o)=\frac{\Omega_n}{j\omega_o}k'(j\omega_o)=\frac{\Omega_n}{j\omega_o}[\mathrm{Re}\{k'(j\omega_o)\}+j\mathrm{Im}\{k'(j\omega_o)\}] \tag{9.138}$$

From (9.138)

$$\begin{cases} \operatorname{Re}\{d(j\omega_o)\} = \dfrac{\Omega_n}{\omega_o}\operatorname{Im}\{k'(j\omega_o)\} \\ \operatorname{Im}\{d(j\omega_o)\} = -\dfrac{\Omega_n}{\omega_o}\operatorname{Re}\{k'(j\omega_o)\} \end{cases} \tag{9.139}$$

Hence, three equivalent forms of representation of the electromechanical loop have been identified to obtain information about the various contributions to damping. They will be referred to in the following sections.

Furthermore, it is worth pointing out that

- equation (9.121) implies

$$\zeta = \frac{1}{2}\cdot\frac{\operatorname{Im}\{k(j\omega_o)\}}{\operatorname{Re}\{k(j\omega_o)\}} = \frac{1}{2}\cdot\frac{\operatorname{Im}\{k'(j\omega_o)\}}{k+\operatorname{Re}\{k'(j\omega_o)\}} \tag{9.140}$$

- since $\omega_o > 0$, from equations (9.135) and (9.130), we get

$$\hat{k} = k + \operatorname{Re}\{k'(j\omega_o)\} = \operatorname{Re}\{k(j\omega_o)\} > 0 \tag{9.141}$$

although small;

- moreover, equation (9.135) also shows that the sign of ζ depends on the sign of $\hat{d}$;
- as a first approximation, the damping value can be supposed to be the sum of various contributions:

$$\zeta = \zeta_S + \zeta_V + \zeta_{PF} + \cdots \tag{9.142}$$

where ζ_S is contribution to damping due to generator structure ($x_d(s)$ and ($x_q(s)$)), ζ_V is contribution to damping due to primary voltage control, ζ_{PF} is contribution to damping due to primary frequency control, and so on.

9.4.3.2 Contribution to Damping Due to Generator Structure. The above description suggests that second-order generator models cannot provide information about the damping of electromechanical oscillations. Indeed, account should be taken of the actual generator structure, represented by the operational reactances $x_d(s)$ and $x_q(s)$. A good dynamic model for the generator (see Figure 9.20) is the fourth-order one (also described by equations (2.88′), (2.89′), and (2.92) in Section 2.1.6.2)

$$\text{“}d\text{-axis”}\quad \begin{cases} x_d(s) = x_d\dfrac{1+sT'_d}{1+sT'_{do}} \\ a(s) = \dfrac{1}{1+sT'_{do}} \end{cases} \tag{9.143}$$

$$\text{“}q\text{-axis”}\quad x_q(s) = x_q\frac{1+sT''_q}{1+sT''_{qo}} \tag{9.144}$$

These equations are suitable to model the generator in the full range of (local and interarea) electromechanical oscillations.

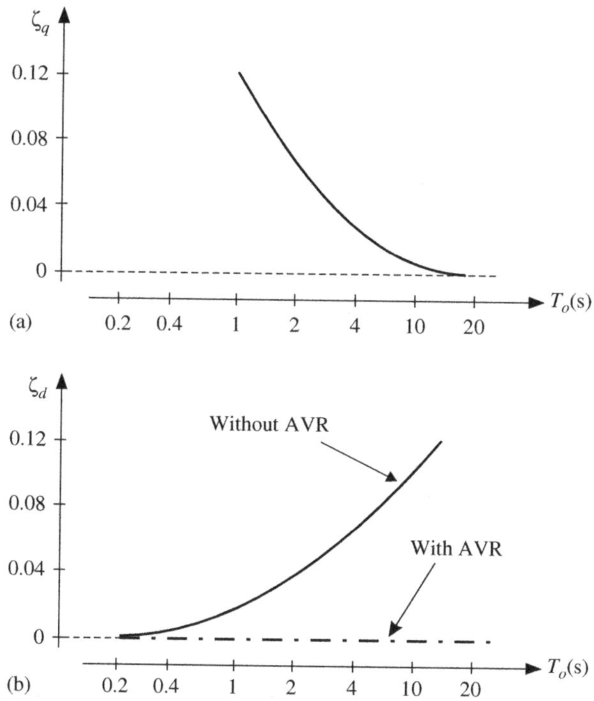

Figure 9.28. Contribution to damping of the generator structure along the q- and d-axes.

The contributions due to generator structure (i.e., to $x_d(s)$ and $x_q(s)$) are depicted in Figure 9.28.

As regards the contribution of the generator structure on the q-axis, the following comments can be offered:

- ζ_q is always positive, as intuitive, because it is due to the dynamic pole-zero characteristics of $x_q(s)$.
- When the oscillation period T_o rises, the dynamic effect tends to zero, because $x_q(s)$ tends to a constant value (x_q), so much so that the dynamic effect vanishes.
- ζ_q is sufficiently high ($\cong$ 0.10–0.15) for *local oscillations*, while it *radically drops* in case of *low-frequency or interarea oscillations*.

Dealing with the contribution of generator structure on the d-axis, with reference to Figure 9.28b, it is possible to conclude that

- ζ_d is always positive and this is intuitive, because it is due to the pole-zero dynamics along the d-axis;
- the behavior of ζ_d is opposite to the one of ζ_q, since the d-axis is phase shifted by 90° versus the q-axis;
- ζ_d is sufficiently high ($\cong$ 0.08–0.12) for *interarea oscillations*, while it *radically drops* in case of *local oscillations*.

9.4.3.3 Contribution of the Primary Voltage Control.

DAMPING OF OSCILLATIONS. The presence of the Primary Voltage Control affects the damping of electromechanical oscillations. The overall effect ζ_V of the Primary Voltage Control is given by two contributions:

$$\zeta_V = \zeta_F + \zeta_{PV} \tag{9.145}$$

The first contribution ζ_F, destabilizing, is due to the interaction with the structure of generators. Figure 9.28b shows also the contribution to damping in the presence of automatic voltage regulator (AVR). When the AVR is in operation, the cutoff frequency of the primary voltage control loop is only dependent on the transient reactance x'_d, that is, it is completely independent of the synchronous reactance x_d. This means that the voltage regulator has the effect of nullifying the dynamics of $x_d(s)$ that are responsible for the *positive effect* (ζ_d) *that the excitation circuit alone has on damping when there is no voltage control.* In other words, it can be stated that the presence of voltage control extends the validity of the approximations of $a(s)$ and $x_d(s)$ (along the d-axis only), which are strictly valid only at high frequency (i.e., only in the usual range of local oscillations), to the overall range of electromechanical oscillations. This result is described in Figure 9.29.

The primary voltage control loop gives rise to another contribution to the damping, called ζ_{PV}. In this connection it should be pointed out that, under the assumption

$$x'_d = x''_d = x''_q = x'' = x_i \tag{9.146}$$

the equivalent circuit of a generator (or area) connected to an infinite system, is depicted in Figure 9.30a. The corresponding phasor diagram is reported in Figure 9.30b where the internal emf $\underline{e}$ is variable in amplitude (e) and phase (δ). Consequently, the delivered real power p_e and the voltage v at the generator terminals, which are output variables, can be put

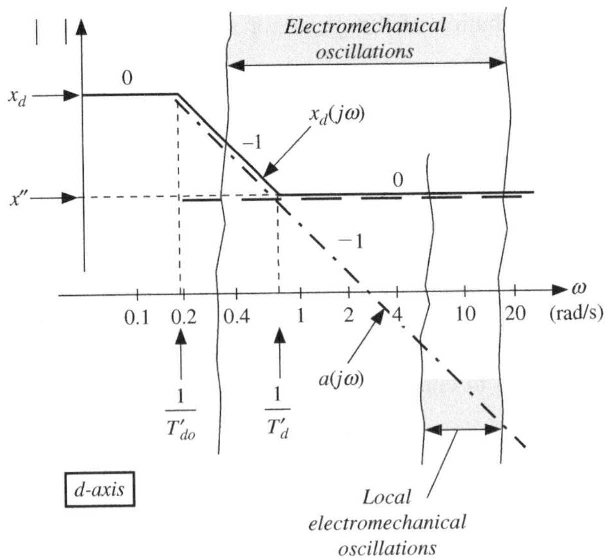

Figure 9.29. Transfer functions along the direct axis, local electromechanical oscillations, and overall range of electromechanical oscillations.

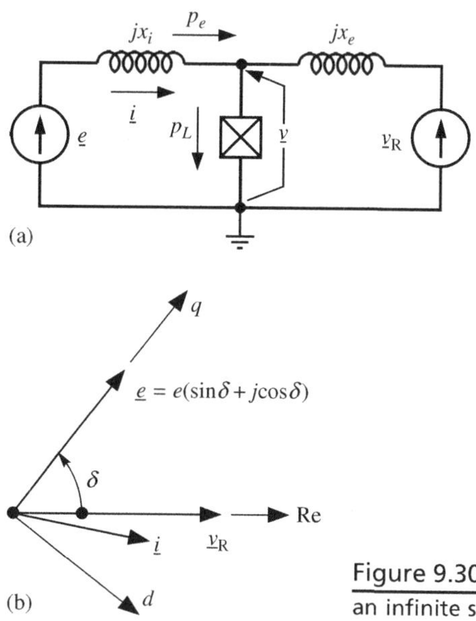

Figure 9.30. (a) Equivalent circuit of a generator connected to an infinite system; (b) relevant phasor diagram.

into the form

$$\begin{cases} p_e = f\{e, v_R, \delta\} \\ v = g\{e, v_R, \delta\} \end{cases} \tag{9.147}$$

If v_R is constant, then

$$\begin{cases} \Delta p_e = k\,\Delta\delta + h\,\Delta e \\ \Delta v = h_1\,\Delta\delta + h_2\,\Delta e \end{cases} \tag{9.148}$$

with

$$\begin{cases} k \triangleq \left(\dfrac{\partial p_e}{\partial \delta}\right)^o \\ h \triangleq \left(\dfrac{\partial p_e}{\partial e}\right)^o \end{cases} \tag{9.149}$$

and

$$\begin{cases} h_1 \triangleq \left(\dfrac{\partial v}{\partial \delta}\right)^o \\ h_2 \triangleq \left(\dfrac{\partial v}{\partial e}\right)^o \end{cases} \tag{9.150}$$

Therefore, the electromechanical and the voltage control loops can be described by the block diagram depicted in Figure 9.31.

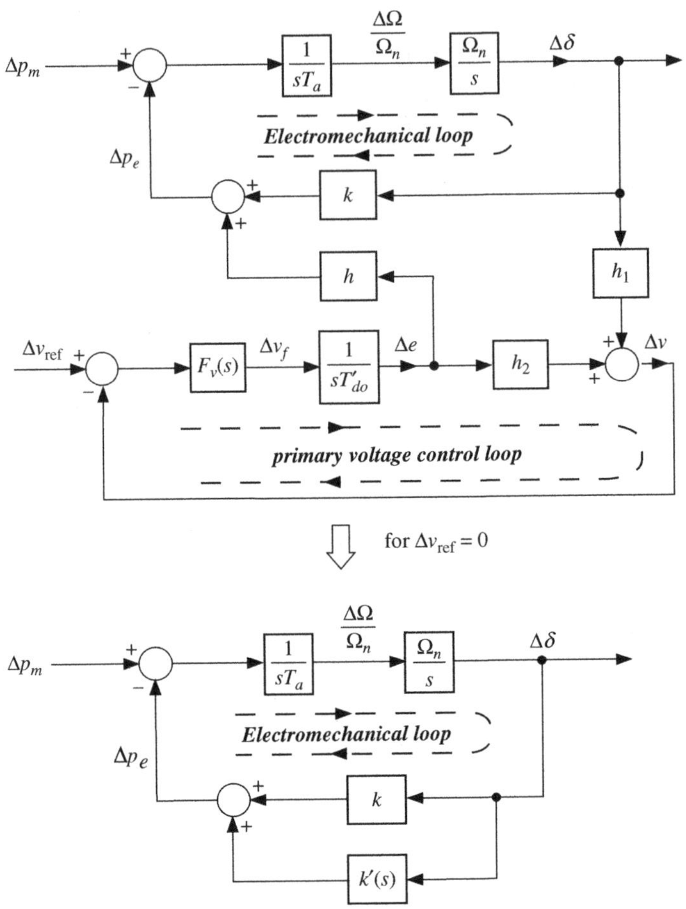

Figure 9.31. Block diagrams for analyzing the contribution ζ_{PV} due to primary voltage control.

Then

$$k'(s) = -\frac{hh_1}{h_2} g_v(s) \tag{9.151}$$

where $g_v(s)$ is the closed-loop transfer function of the primary voltage control loop at constant angle δ:

$$g_v(s) \triangleq \left[\frac{\Delta v}{\Delta v_{ref}}(s)\right]_{\Delta\delta=0} = \frac{h_2(F_v(s)/sT'_{do})}{1 + h_2(F_v(s)/sT'_{do})} \tag{9.152}$$

or

$$g_v(s) = \frac{1}{1 + s(T'_{do}/(h_2 F_v(s)))} \tag{9.153}$$

In the range of the electromechanical oscillations, it is usual to approximate the AVR transfer function $F_v(s)$ by its transient or dynamic gain μ_t (this assumption is strictly valid for static excitation systems, and to a first approximation also for rotating excitation systems):

$$F_v(s) \simeq \mu_t \tag{9.154}$$

From equation (9.153), it follows that

$$g_v(s) = \frac{1}{1 + sT_v} \tag{9.155}$$

with

$$T_v = \frac{1}{\omega_v} = \frac{T'_{do}}{\mu_t h_2} \tag{9.156}$$

where ω_v is the cutoff frequency of the primary voltage control loop.

At this point, it is possible to evaluate the contribution ζ_{PV} taking into account the general equation (9.121). Since

$$\text{Im}\{k'(j\omega_o)\} = -\frac{hh_1}{h_2}\text{Im}\{g_v(j\omega_o)\} \tag{9.157}$$

we obtain (see also equation (9.155))

$$\zeta_{PV} = -\frac{\Omega_n}{2T_a\omega_o^2}\frac{hh_1}{h_2}\text{Im}\{g_v(j\omega_o)\} = \frac{\Omega_n}{2T_a\omega_o^2}\frac{hh_1}{h_2}\frac{\omega_o/\omega_v}{1+(\omega_o/\omega_v)^2} \tag{9.158}$$

Based on equation (9.158), the contribution ζ_{PV} depends only on the sign of hh_1/h_2:

$$\begin{cases} \dfrac{hh_1}{h_2} < 0 & \text{destabilizing effect} \quad \zeta_{PV} < 0 \\ \dfrac{hh_1}{h_2} > 0 & \text{stabilizing effect} \quad \zeta_{PV} > 0 \end{cases} \tag{9.159}$$

It can be verified that h_2 is always positive, as it is intuitive; conversely, the signs of h and h_1 depend on the structure of the two-port network connecting the generator to the infinite system and on the operating conditions (power flow).

In this respect, it is necessary to examine both the situation of a generator or an area directly connected to the system (see Figure 9.24) and the situation with an intermediate load, as they may give rise to different effects.

In the case of a generator-network connection without intermediate load, we immediately deduce that

$$\begin{cases} \underline{e} - \underline{v} = jx_i\underline{i} \\ \underline{v} - \underline{v}_R = jx_e\underline{i} \end{cases} \tag{9.160}$$

from which we may derive the voltage v at the generator terminals. As a matter of fact, from

$$\frac{\underline{e} - \underline{v}}{jx_i} = \frac{\underline{v} - \underline{v}_R}{jx_e} \tag{9.161}$$

with

$$\begin{cases} \underline{e} = e(\cos\delta + j\sin\delta) \\ \underline{v}_{\mathrm{R}} = v_{\mathrm{R}} \end{cases} \tag{9.162}$$

we infer

$$\underline{v} = \frac{x_i x_e}{x_i + x_e}\left[\frac{\underline{e}}{x_i} + \frac{\underline{v}_{\mathrm{R}}}{x_e}\right] = \frac{x_i x_e}{x_i + x_e}\left[\left(\frac{e\cos\delta}{x_i} + \frac{\underline{v}_{\mathrm{R}}}{x_e}\right) + j\frac{e\sin\delta}{x_i}\right] \tag{9.163}$$

$$v = \frac{x_i x_e}{x_i + x_e}\sqrt{\left(\frac{e}{x_i}\right)^2 + \left(\frac{v_{\mathrm{R}}}{x_e}\right)^2 + \frac{2ev_{\mathrm{R}}\cos\delta}{x_i x_e}} \tag{9.164}$$

By differentiating, we obtain the expressions for h_1 and h_2:

$$\begin{cases} \left(\dfrac{\partial v}{\partial \delta}\right)^o = -\dfrac{x_i x_e}{(x_i + x_e)^2}\cdot\dfrac{e^o v_{\mathrm{R}}\sin\delta^o}{v^o} = -\dfrac{x_i x_e}{x_i + x_e}\left(\dfrac{p_e}{v}\right)^o \\ \left(\dfrac{\partial v}{\partial e}\right)^o = \left(\dfrac{x_i x_e}{x_i + x_e}\right)^2 \dfrac{1}{x_i v^o}\left(\dfrac{e^o}{x_i} + \dfrac{v_{\mathrm{R}}}{x_e}\cos\delta^o\right) \end{cases} \tag{9.165}$$

As an alternative, in order to obtain the expression of v as a function of e and δ, we can consider the phasor diagram of Figure 9.32, and obtain equivalent expressions for the coefficients h_1 and h_2:

$$\begin{cases} e\cos(\delta - \theta) = v + x_i i\sin\phi = v + x_i\dfrac{q}{v} \\ e\sin(\delta - \theta) = x_i i\cos\phi = x_i\dfrac{p_e}{v} \end{cases} \tag{9.166}$$

where

$$\begin{cases} q = \dfrac{v}{x_e}(v - v_{\mathrm{R}}\cos\theta) \\ p_e = \dfrac{v}{x_e}v_{\mathrm{R}}\sin\theta \end{cases} \tag{9.167}$$

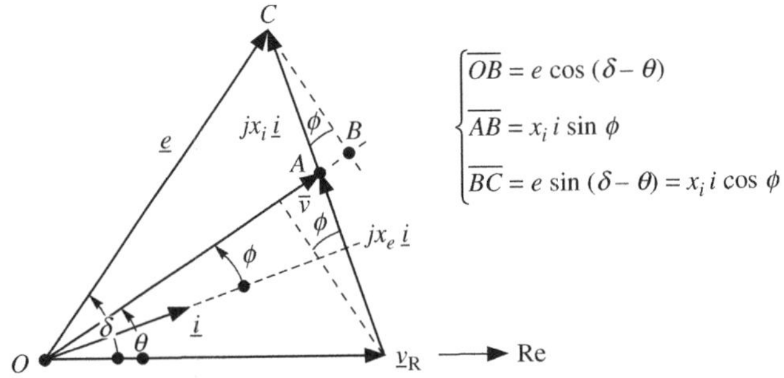

Figure 9.32. Phasor diagram of a generator connected to an infinite system.

By substitution of p_e and q into (9.166), we can see that the second equation is independent of v, so that the first equation relates v to e and enables to compute h_1 and h_2 univocally.

Indeed, it follows from the first of equations (9.166) and (9.167) that

$$e\cos(\delta-\theta) = v + \frac{x_i}{x_e}v - \frac{x_i}{x_e}v_R\cos\theta = v\frac{x_i+x_e}{x_e} - \frac{x_i}{x_e}v_R\cos\theta \tag{9.168}$$

$$v = \frac{x_e}{x_i+x_e}e\cos(\delta-\theta) + \frac{x_i}{x_i+x_e}v_R\cos\theta \tag{9.169}$$

$$\begin{cases} h_1 = \left(\dfrac{\partial v}{\partial \delta}\right)^o = -\dfrac{x_e}{x_i+x_e}e^o\sin(\delta-\theta)^o \\ h_2 = \left(\dfrac{\partial v}{\partial e}\right)^o = \dfrac{x_e}{x_i+x_e}\cos(\delta-\theta)^o \end{cases} \tag{9.170}$$

where h_2 is always >0, and also

$$\frac{h_1}{h_2} = -e^o\tan(\delta-\theta)^o \tag{9.171}$$

With regard to h_2, it is maximum at no load ($\delta-\theta=0$), when it is equal to $x_e/(x_i+x_e)$, and then decreases with increasing delivered real power (Figure 9.24).

Even in the case of a generator connected to an infinite system, the cutoff frequency ω_v of such loop proves to be maximum at no load and to diminish—although slightly—with increasing load.

As from equation (9.100), we have

$$\begin{cases} \left(\dfrac{\partial p_e}{\partial \delta}\right)^o = k = \dfrac{e^o v_R}{x_i+x_e}\cos\delta^o \\ \left(\dfrac{\partial p_e}{\partial e}\right)^o = h = \dfrac{v_R\sin\delta^o}{x_i+x_e} = \left(\dfrac{p_e}{e}\right)^o \end{cases} \tag{9.172}$$

we finally obtain ($-\pi/2 < \delta^\circ < \pi/2$), taking into account also equations (9.165) and (9.170)

$$\begin{cases} k = \dfrac{e^o v_R\cos\delta^o}{x_i+x_e} = q_R^o + \dfrac{v_R^2}{x_i+x_e} > 0 \\ h = \dfrac{v_R\sin\delta^o}{x_i+x_e} = \left(\dfrac{p_e}{e}\right)^o \\ h_1 = -\dfrac{x_i x_e}{x_i+x_e}\left(\dfrac{p_e}{v}\right)^o = -\dfrac{x_e}{x_i+x_e}e^o\sin(\delta-\theta)^o \\ h_2 = \left(\dfrac{x_i x_e}{x_i+x_e}\right)^2\dfrac{1}{x_i v^o}\left(\dfrac{e^o}{x_i} + \dfrac{v_R}{x_e}\cos\delta^o\right) \\ \quad = \dfrac{x_e}{x_i+x_e}\cos(\delta-\theta)^o > 0 \end{cases} \tag{9.173}$$

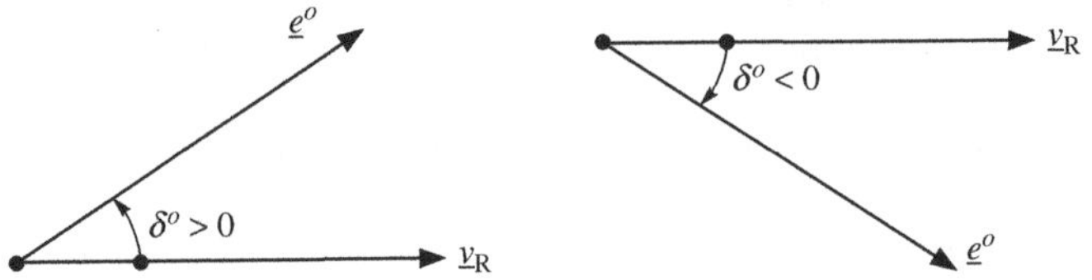

Figure 9.33. Unit in generation mode (a) and in pumping mode (b).

and

$$\frac{h h_1}{h_2} = -p_e^o \tan(\delta - \theta)^o \tag{9.174}$$

Therefore, the following *first major conclusions* can be drawn:

The signs of h and h_1 depend on the sign of δ^o, that is, on the sign of p_e^o; so, if the unit is in generation mode ($p_e^o > 0$):

$$\begin{cases} h > 0 \\ h_1 < 0 \end{cases} \tag{9.175}$$

If, instead, it is in pumping mode ($p_e^o < 0$), then (see Figure 9.33),

$$\begin{cases} h < 0 \\ h_1 > 0 \end{cases} \tag{9.176}$$

Thus, as $h_2 > 0$ always,

$$\frac{h h_1}{h_2} < 0 \tag{9.177}$$

when the generator is directly connected to the infinite system, the voltage control loop effect is always destabilizing (negative contribution to the damping of electromechanical oscillations). In other words, the voltage control contribution is doubly negative in that, on one hand, it nullifies the positive contribution of the field circuit ζ_F and, on the other hand, it introduces a negative contribution ζ_{PV}.

If the *generator* is *unloaded* ($p_e^o = 0$), as in the case of *synchronous compensators*

$$h = h_1 = 0 \tag{9.178}$$

the voltage loop is completely decoupled from the electromechanical one (see Figure 9.31) and therefore the *voltage control loop does not have any effect on the electromechanical oscillations*. Therefore, reducing the loading of the generators has a stabilizing effect: *in the absence of PSSs, the only action available to increase damping is the reduction of the generated power.*

As far as the aperiodic limit is concerned, it is possible to show that the presence of the voltage regulator makes the limit angle δ^o_{lim} lie above 90°; therefore, the voltage control loop increases the aperiodic stability region.

In general, for ζ_{PV}, we can derive the following expression, from (9.158) and (9.174):

$$\begin{aligned}\zeta_{PV} &= \frac{1}{2\omega_o^2} \cdot \frac{\Omega_n}{T_a} p_e^o \tan(\delta - \theta)^o \mathrm{Im}\{g_v(j\omega_o)\} \\ &= -\frac{\Omega_n}{2T_a\omega_o^2} p_e^o \tan(\delta - \theta)^o \frac{\omega_o/\omega_v}{1 + (\omega_o/\omega_v)^2}\end{aligned} \tag{9.179}$$

Hence, the following *second major conclusions* can be drawn:

- For $\omega_o \to \infty$ or $\omega_o/\omega_v \to \infty$ or $\omega_o \gg \omega_v$, $\zeta_{PV} \to 0$. In the past (1960s), the AVR had no effect on damping of electromechanical oscillations, because it was very slow (large T_v, small ω_v, $\omega_v \ll \omega_o$). A negative but small ζ_{PV} was largely compensated by ζ_q.
- When the speed of response of the primary voltage control loop grows, that is, when ω_v gets closer to ω_o, as in the case of modern static exciters, the contribution of ζ_{PV} tends to become increasingly negative. This occurs when

  –x_e increases (ω_o small), that is, the generator (area) is weakly connected to the infinite system;

  – p_e^o increases, in case of either delivered or absorbed power (when p_e^o reverts, also $\tan(\delta - \theta)^\circ$ reverts);

  –the internal angle $(\delta - \theta)^\circ$ increases, that is, at a given real power, the delivered reactive power q^o is smaller (remembering in particular that, as q tends toward the underexcitation limit, the internal angle tends to 90°);

  –T_a decreases, that is, the (true or equivalent) generator inertia is smaller.

ABOUT OSCILLATION FREQUENCY. As shown in the previous sections, the general expression for the oscillation frequency is given by

$$\omega_o^2 = \frac{\Omega_n}{T_a} \mathrm{Re}\{k(j\omega_o)\} \tag{9.180}$$

with

$$\mathrm{Re}\{k(j\omega_o)\} = k - \frac{h\,h_1}{h_2} \mathrm{Re}\{g_v(j\omega_o)\} \tag{9.181}$$

Under the assumption

$$k \gg \frac{h\,h_1}{h_2} \mathrm{Re}\{g_v(j\omega_0)\} \tag{9.182}$$

the oscillation frequency can be approximated to

$$\omega_o \simeq \sqrt{\frac{\Omega_n k}{T_a}} \tag{9.183}$$

The above approximation depends on k, h, h_1, and h_2, that is, on the structure of the connecting two-port network and on the operating point. In particular, since

$$\mathrm{Re}\{g_v(j\omega_o)\} = \frac{1}{1 + (\omega_o/\omega_v)^2} \tag{9.184}$$

$$0 \leq \mathrm{Re}\{g_v(j\omega_o)\} \leq 1 \tag{9.185}$$

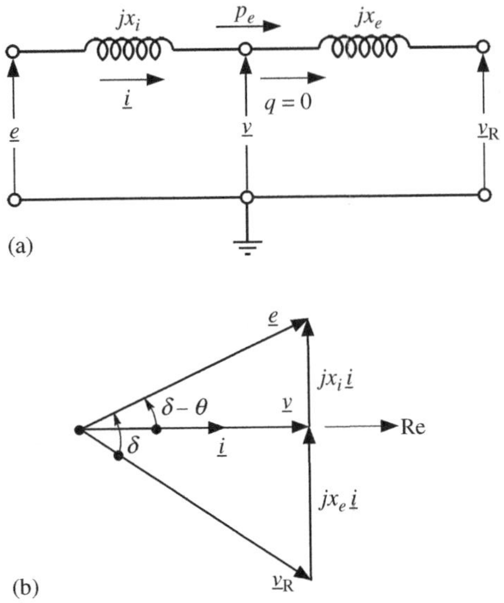

Figure 9.34. Example of evaluation of approximations when computing oscillation frequency.

we can state that, for each assigned p_e^o, if x_e decreases (and this better models local oscillations), then δ^o decreases, k and ω_o increase, h_2 and ω_v decrease, ω_o/ω_v increases, $\text{Re}\{g_v(j\omega_o)\}$ decreases; therefore, equation (9.182) is more and more true, and equation (9.183) is suitable for local oscillations more than for interarea oscillations.

Application: As an example, reference can be made to Figure 9.34 with the following parameters

$$\begin{cases} T_a = 10\,\text{s}; \quad T'_{do} = 7.5\,\text{s} \\ \mu_t = 50\,\text{p.u./p.u.}; \quad x_i = 0.2\,\text{p.u.} \\ x_e = 1\,\text{p.u.} \end{cases} \tag{9.186}$$

and the operating point

$$\begin{cases} v^o = 1\,\text{p.u.} \\ p_e^o = i^o = 0.8\,\text{p.u.} \end{cases} \tag{9.187}$$

The result is

$$\begin{cases} \underline{v}_R^o = 1 - j0.8; \; \underline{e}^o = 1 + j0.16 \\ v_R^o = 1.28; \qquad e^o = 1.0127 \\ (\delta - \theta)^o \simeq 9^\circ; \; \delta^o \simeq 48^\circ \end{cases} \tag{9.188}$$

and then

$$\begin{cases} k = 0.723; \quad \dfrac{h\,h_1}{h_2} = -0.128; \quad h_2 = 0.833 \\ \omega_o \simeq \sqrt{\dfrac{k\,\Omega_n}{T_a}} = 4.76; \quad \omega_v = 5.55 \\ \dfrac{\omega_o}{\omega_v} = 0.86; \quad \text{Re}\{g_v(j\omega_o)\} = 0.576 \\ \dfrac{h\,h_1}{h_2}\text{Re}\{g_v(j\omega_o)\} = -0.074 \ll k \end{cases} \tag{9.189}$$

In this case, the condition (9.182) is well satisfied.

When this approximation fails, ω_o can be calculated from (9.180), (9.181), and (9.184):

$$\omega_o^2 = \frac{\Omega_n}{T_a}\left(k - \frac{h h_1}{h_2}\frac{1}{1+(\omega_o/\omega_v)^2}\right) \tag{9.190}$$

that is, from the biquadratic equation

$$\omega_o^4 + \left(\omega_v^2 - \frac{k\,\Omega_n}{T_a}\right)\omega_o^2 - \frac{\Omega_n\,\omega_v^2}{T_a}\left(k - \frac{h h_1}{h_2}\right) = 0 \tag{9.191}$$

whose real and positive single solution

$$\omega_o = \sqrt{\frac{((k\,\Omega_n/T_a) - \omega_v^2) + \sqrt{((k\,\Omega_n/T_a) - \omega_v^2)^2 + 4(\Omega_n/T_a)\omega_v^2(k - (h h_1/h_2))}}{2}} \tag{9.192}$$

provides the frequency of the electromechanical oscillation.

Oscillation Frequency and Damping Contribution of the Primary Voltage Control on Local and Interarea Oscillations. The expression of ω_o depends on k, h, h_1, and h_2 or on k only. It is general, that is, without and with intermediate load between the generator and the infinite system.

Further, let us suppose that the investigated subset of electromechanical oscillations is expressed in terms of ω_o/ω_v (this is meaningful, because ω_v is practically always ranging between 4 and 6 rad/s). So, we can generalize some conclusions based on some particular cases:

(i)

$$\frac{\omega_o}{\omega_v} \simeq 1 \tag{9.193}$$

or $\omega_o \cong 4$–6 rad/s, $T_o \cong 1$–1.5 s, that is, in practice, the *borderline between local and interarea oscillations*:

$$\begin{cases} \text{Re}\{g_v(j\omega_o)\} \simeq \dfrac{1}{2} \\ \text{Im}\{g_v(j\omega_o)\} \simeq -\dfrac{1}{2} \end{cases} \tag{9.194}$$

$$\omega_o^2 \simeq \frac{\Omega_n}{T_a}\left(k - \frac{h h_1}{2h_2}\right) \tag{9.195}$$

Moreover, equation (9.182) being satisfied:

$$\omega_o \simeq \sqrt{\frac{k\Omega_n}{T_a}} \tag{9.196}$$

we obtain from (9.179)

$$\zeta_{PV} = \frac{1}{2\omega_o^2} \cdot \frac{\Omega_n}{T_a} p_e^o \tan(\delta - \theta)^o \operatorname{Im}\{g_v(j\omega_o)\} = -\frac{1}{4k} p_e^o \tan(\delta - \theta)^o \tag{9.197}$$

independent of the dynamic gain μ_t *of the AVR.*

(ii)

$$\frac{\omega_o}{\omega_v} \geq 2 \tag{9.198}$$

or $\omega_o \geq 8$–12 rad/s, $T_o \leq 0.5$–0.8 s, that is, in the *domain of local oscillations,*

$$\begin{cases} \operatorname{Im}\{g_v(j\omega_o)\} \simeq -\dfrac{\omega_v}{\omega_o} \\ \operatorname{Re}\{g_v(j\omega_o)\} \simeq \left(\dfrac{\omega_v}{\omega_o}\right)^2 \end{cases} \tag{9.199}$$

so that the real part tends to zero more rapidly than the imaginary part.

As a consequence, equation (9.182) is certainly satisfied and thus

$$\omega_o \simeq \sqrt{\frac{k\Omega_n}{T_a}} \tag{9.200}$$

$$\zeta_{PV} = -\frac{1}{2\omega_o^2} \cdot \frac{\Omega_n}{T_a} \cdot \frac{\omega_v}{\omega_o} p_e^o \tan(\delta - \theta)^o = -\frac{1}{2k\omega_o} \omega_v p_e^o \tan(\delta - \theta)^o \tag{9.201}$$

In this case, all other parameters being equal, the negative contribution ζ_{PV} increases when ω_v or the dynamic gain μ_t increases.

(iii)

$$\frac{\omega_o}{\omega_v} \leq \frac{1}{4} \tag{9.202}$$

or $\omega_o \leq 1$–1.5 rad/s, $T_o \geq 4$–6 s, that is, in the *domain of interarea oscillations,*

$$\begin{cases} \operatorname{Im}\{g_v(j\omega_o)\} \simeq -\dfrac{\omega_o}{\omega_v} \\ \operatorname{Re}\{g_v(j\omega_o)\} \simeq 1 \end{cases} \tag{9.203}$$

$$k \text{ and } h\,h_1/h_2 \text{ comparable} \tag{9.204}$$

$$\omega_o \simeq \sqrt{\frac{\Omega_n}{T_a}\left(k - \frac{h\,h_1}{h_2}\right)} \tag{9.205}$$

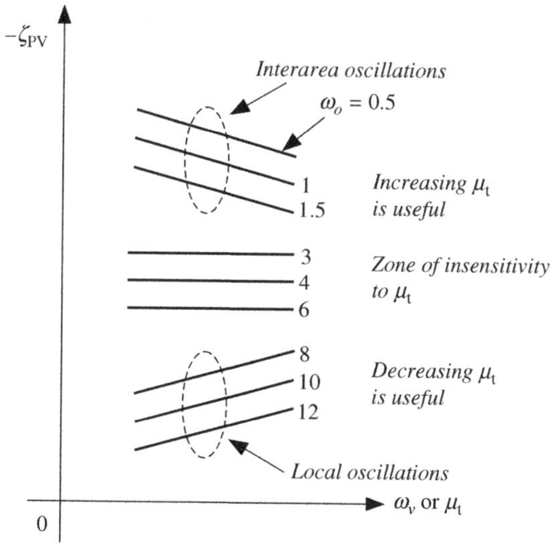

Figure 9.35. Damping contribution of AVR versus its dynamic gain.

$$
\begin{aligned}
\zeta_{PV} &= -\frac{1}{2\omega_o^2} \cdot \frac{\Omega_n}{T_a} \cdot \frac{\omega_o}{\omega_v} p_e^o \tan(\delta - \theta)^o \\
&= -\frac{1}{2\omega_o} \cdot \frac{\Omega_n}{T_a} \cdot \frac{1}{\omega_v} p_e^o \tan(\delta - \theta)^o
\end{aligned}
\tag{9.206}
$$

So, in this case, all other parameters being equal, the *negative contribution* ζ_{PV} *increases when* ω_v *or the transient gain* μ_t *decreases.*

The above considerations are summarized in qualitative terms in Figure 9.35.

Damping Improvement in the Absence of PSS. The three conclusions of the previous section justify why, when no appropriate PSSs are available, the following control action is suitable in order to damp any type of electromechanical oscillation:

- Reduction of delivered real power and limitations of reactive power in under-excitation operation, with consequent reduction of generator utilization with respect to thermal and turbine limits (see Figure 9.36).

  Besides, where no appropriate PSSs are available, the following additional control actions can be taken (depending on the type of electromechanical oscillations), at a given operating point
- In case of local oscillations, reduction of the dynamic AVR gain μ_t; however, this deteriorates the speed of response of the primary voltage control loop.
- In case of relatively slow interarea oscillations ($T_o \geq 4$ s, $\omega_o \leq 1.5$ rad/s), increase of μ_t.

For oscillations in the range of 4–6 rad/s ($T_o = 1$–1.5 s), that is, in case of interarea oscillations having a small oscillation period, ζ_{PV} is practically unaffected by variations of μ_t.

All the conditions described so far, just as those concerning the signs of h and h_1, remain valid also for a more general model of $g_v(s)$, remembering that $g_v(j\omega)$ has a negative phase in the range of 0–180°.

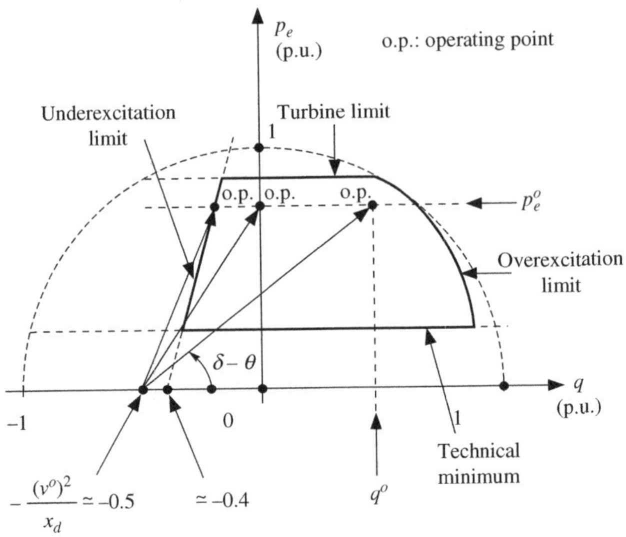

Figure 9.36. Generator capability curves.

In the case of the generator on the infinite system without intermediate load, under manual control ($\Delta v_f \cong 0$), the aperiodic-type stability limit of the rotor angle was found to be 90°. The above conclusions show that the negative contribution ζ_{PV} may result in a more restrictive constraint: in order to ensure a reasonable oscillatory-type stability margin, δ^o can be limited, in the absence of PSSs, to below 90°. Accordingly, the limit value for the real delivered power may be lower than

$$\left(p_e^o\right)_{\delta^o=90^\circ} = \frac{e^o v_R}{x_i + x_e} \tag{9.207}$$

Figure 9.37 depicts the qualitative curves of aperiodic and oscillatory stability limits with and without AVR. The figure stresses, as of now, the benefits deriving from the use of appropriate stabilizing signals.

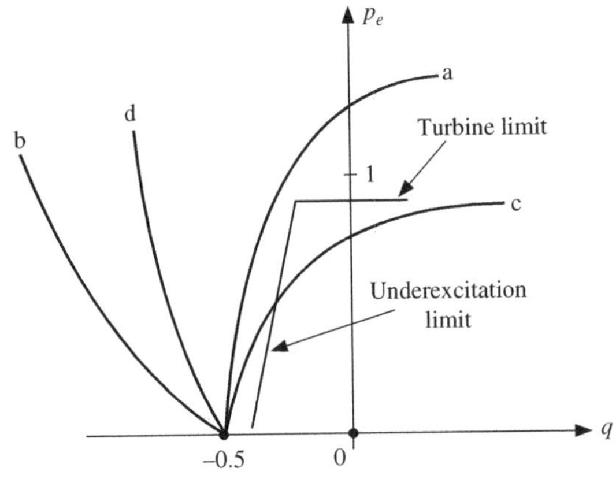

Figure 9.37. Qualitative curves of stability limits following small perturbations: (curve a) aperiodic without AVR; (curve b) aperiodic with AVR; (curve c) oscillatory with AVR and without PSSs; (curve d) oscillatory with AVR and with PSSs.

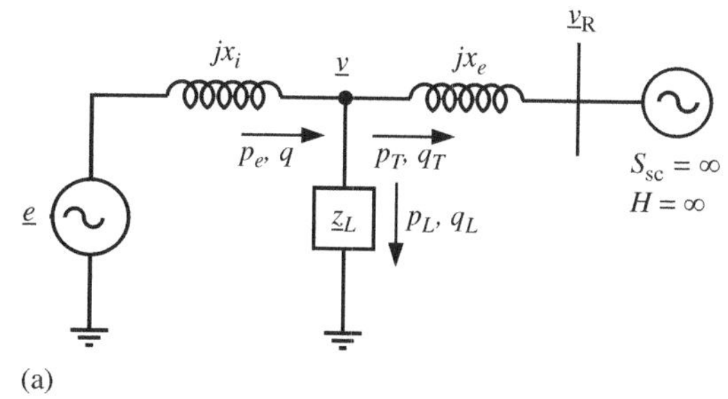

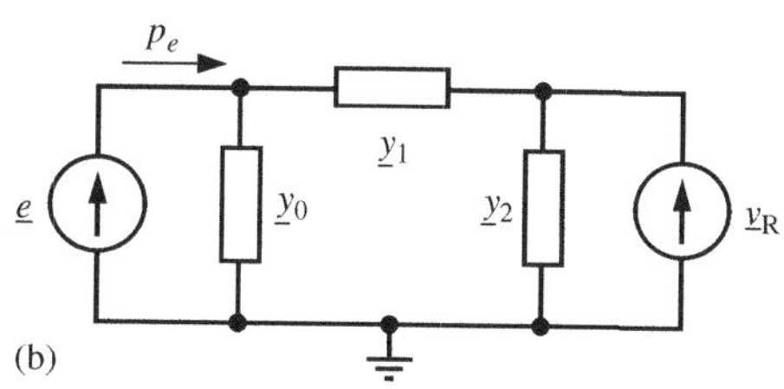

Figure 9.38. Generator connected to an infinite system with intermediate load: (a) electrical scheme; (b) equivalent circuit.

IMPACT OF LOADS AND POWER FLOWS ON THE DAMPING. The case of an intermediate load between the generator and the infinite system is certainly a more significant case from the practical point of view. In particular, it can also model the case of a system area, with its generators (represented by an equivalent generator) and loads (represented by an equivalent load), connected to the rest of the system.

For the sake of simplicity, it is suitable to assume the load static, linear and directly connected to the generator terminals (Figure 9.38a).

The following equation applies

$$\frac{\underline{e} - \underline{v}}{jx_i} = \frac{\underline{v} - \underline{v}_R}{jx_e} + \frac{\underline{v}}{\underline{z}_L} \tag{9.208}$$

that is

$$-j\left[\frac{\underline{e}}{x_i} + \frac{\underline{v}_R}{x_e}\right] = \underline{v}\left[\frac{1}{jx_e} + \frac{1}{jx_i} + \frac{1}{\underline{z}_L}\right] \tag{9.209}$$

By putting

$$\frac{1}{\underline{z}} \triangleq \frac{1}{\underline{z}_L} + \frac{1}{jx_e} + \frac{1}{jx_i} \tag{9.210}$$

$$\begin{cases} \underline{v}_R = v_R \\ \underline{e} = e(\cos\delta + j\sin\delta) \end{cases} \tag{9.211}$$

we infer

$$\underline{v} = -j\underline{z}\left[\left(\frac{e\cos\delta}{x_i} + \frac{v_R}{x_e}\right) + j\frac{e\sin\delta}{x_i}\right] \tag{9.212}$$

from which

$$v = g\{e, v_{\mathrm{R}}, \delta\} = z\sqrt{\left(\frac{e}{x_i}\right)^2 + \left(\frac{v_R}{x_e}\right)^2 + \frac{2ev_{\mathrm{R}}\cos\delta}{x_i x_e}} \tag{9.213}$$

quite similar to equation (9.164). As a result (z takes the place of $x_i x_e/(x_i + x_e)$), we obtain

$$\begin{cases} \left(\dfrac{\partial v}{\partial \delta}\right)^o = -\dfrac{z^2}{x_i x_e} \cdot \dfrac{e^o v_{\mathrm{R}}}{v^0} \sin\delta^o \\ \left(\dfrac{\partial v}{\partial e}\right)^o = \dfrac{z^2}{x_i v^o}\left[\dfrac{e^o}{x_i} + \dfrac{v_{\mathrm{R}}}{x_e}\cos\delta^o\right] \end{cases} \tag{9.214}$$

To derive the expression of p_e as a function of e, δ, and v_{R}, it is useful to apply the star-delta transformation, as shown in Figure 9.38b, obtaining

$$\begin{cases} \underline{y}_1 = -\underline{z}\dfrac{1}{x_i x_e} \triangleq g_1 + jb_1 \\ \underline{y}_0 = -j\underline{z}\dfrac{1}{x_i \underline{z}_L} \triangleq g_0 + jb_0 \end{cases} \tag{9.215}$$

As

$$p_e = \mathrm{Re}\{\underline{e}(\underline{y}_0\underline{e})^* + \underline{e}[\underline{y}_1(\underline{e} - \underline{v}_{\mathrm{R}})]^*\} \tag{9.216}$$

we have

$$\begin{aligned} p_e &= g_0 e^2 + g_1(e^2 - ev_{\mathrm{R}}\cos\delta) - b_1 ev_{\mathrm{R}}\sin\delta \\ &= (g_0 + g_1)e^2 - ev_{\mathrm{R}}(g_1\cos\delta + b_1\sin\delta) \end{aligned} \tag{9.217}$$

and by differentiating

$$\left(\frac{\partial p_e}{\partial \delta}\right)^o = [ev_{\mathrm{R}}(g_1\sin\delta - b_1\cos\delta)]^o \tag{9.218a}$$

$$\left(\frac{\partial p_e}{\partial e}\right)^o = [2e(g_0 + g_1) - v_{\mathrm{R}}(g_1\cos\delta + b_1\sin\delta)]^o \tag{9.218b}$$

Hence, in lieu of equations (9.173), we get

$$\begin{aligned} k &= [ev_{\mathrm{R}}(g_1\sin\delta - b_1\cos\delta)]^o \\ h &= [2e(g_0 + g_1) - v_{\mathrm{R}}(g_1\cos\delta + b_1\sin\delta)]^o \\ &= \left[\frac{p_e}{e} + e(g_0 + g_1)\right]^o \\ h_1 &= -\frac{z^2}{x_i x_e}\left[\frac{ev_{\mathrm{R}}}{v}\sin\delta\right]^o \\ h_2 &= \frac{z^2}{x_i}\left[\frac{1}{v}\left(\frac{e}{x_i} + \frac{v_{\mathrm{R}}}{x_e}\cos\delta\right)\right]^o > 0 \end{aligned} \tag{9.219}$$

where z, g_1, b_1, and $(g_0 + g_1)$ only depend on the electrical parameters of the system (x_e), generator (x_i), and load (r_L, x_L) and do not depend on the transit p_T^o or on the operating point.

Based on equations (9.219), the following remarks may be made:

1. As the contribution of the primary voltage control loop to damping is always provided by

$$\zeta_{PV} \simeq -\frac{1}{2\omega_o} \cdot \frac{\Omega_n}{T_a} \cdot \frac{h h_1}{h_2} \text{Im}\{g_v(j\omega_o)\} \tag{9.220}$$

the approximation

$$\omega_o \simeq \sqrt{\frac{k\,\Omega_n}{T_a}} \tag{9.221}$$

is subject to the same considerations made for the case without intermediate load.

2. To have $k > 0$, it is required that

$$\tan \delta^o > b_1/g_1 \tag{9.222}$$

and thus the condition $k = 0$ is no longer defined by $\delta^o_{\text{lim}} = \pm 90°$ as in the case without intermediate load $(g_1 = 0)$ but rather by the more restrictive

$$\tan \delta^o_{\text{lim}} = b_1/g_1 \tag{9.223}$$

3. The sign of h not only depends on the sign of p_e, as in the case without intermediate load but also on the value of $(g_0 + g_1)$. In this connection, let us assume that the impedance $\underline{z}_L$ of the load is purely resistive

$$\underline{z}_L = r_L \tag{9.224}$$

This assumption does not prevent from generality, as the case of an resistive–inductive load can be studied using Thévenin's Theorem, as in Figure 9.39.

The star-delta transformation provides the following values of b_0, b_1, g_0, g_1

$$\begin{cases} b_0 = -\dfrac{1}{x_i}\alpha & b_1 = -\dfrac{1}{x_i + x_e}\alpha \\ g_0 = \dfrac{1}{r_L} \cdot \dfrac{x_e}{x_i + x_e}\alpha & g_1 = -\dfrac{1}{r_L} \cdot \dfrac{x_p}{x_i + x_e}\alpha \end{cases} \tag{9.225}$$

with

$$\begin{cases} x_p = \dfrac{x_i x_e}{x_i + x_e} \\ \alpha = \dfrac{1}{1 + (x_p/r_L)^2} \end{cases} \tag{9.226}$$

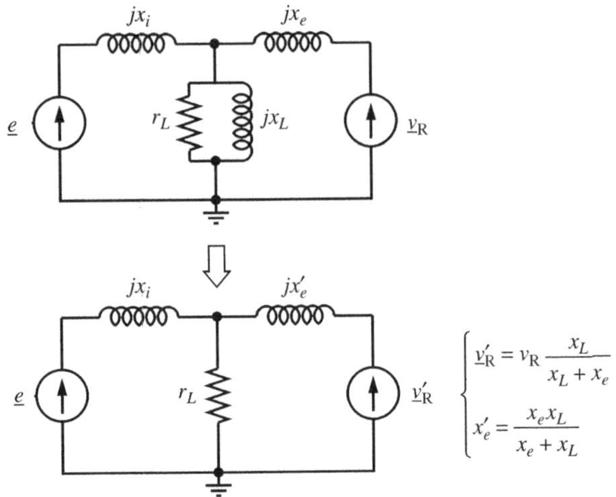

Figure 9.39. Application of Thévenin's Theorem to the case of a resistive-inductive load.

Now, note that

$$g_0 + g_1 = \frac{1}{r_L} \cdot \frac{x_e - x_p}{x_i + x_e} \alpha = \frac{1}{r_L} \left(\frac{x_e}{x_i + x_e} \right)^2 \frac{1}{1 + (x_p/r_L)^2} > 0 \tag{9.227}$$

which is always satisfied, whatever the value of x_p or x_e. It follows that

– if the unit is in generation mode ($p_e^o > 0$), then

$$h > 0 \tag{9.228}$$

– If the unit is unloaded ($p_e^o = 0$), then

$$h > 0 \tag{9.229}$$

– If the unit is in pumping mode ($p_e^o < 0$), then

$$h \gtrless 0 \tag{9.230}$$

depending on

$$p_e^o \gtrless -(e^o)^2 (g_0 + g_1) \tag{9.231}$$

Thus, the coefficient h becomes negative for

$$p_e^o < -(e^o)^2 (g_0 + g_1) \tag{9.232}$$

4. The sign of h_1 depends only on the sign of δ^o, as in the case without load. However, now δ^o may be negative, even if the unit or the area is in generation mode and this depends on the load and on the real power transit p_T^o. Figure 9.40 qualitatively

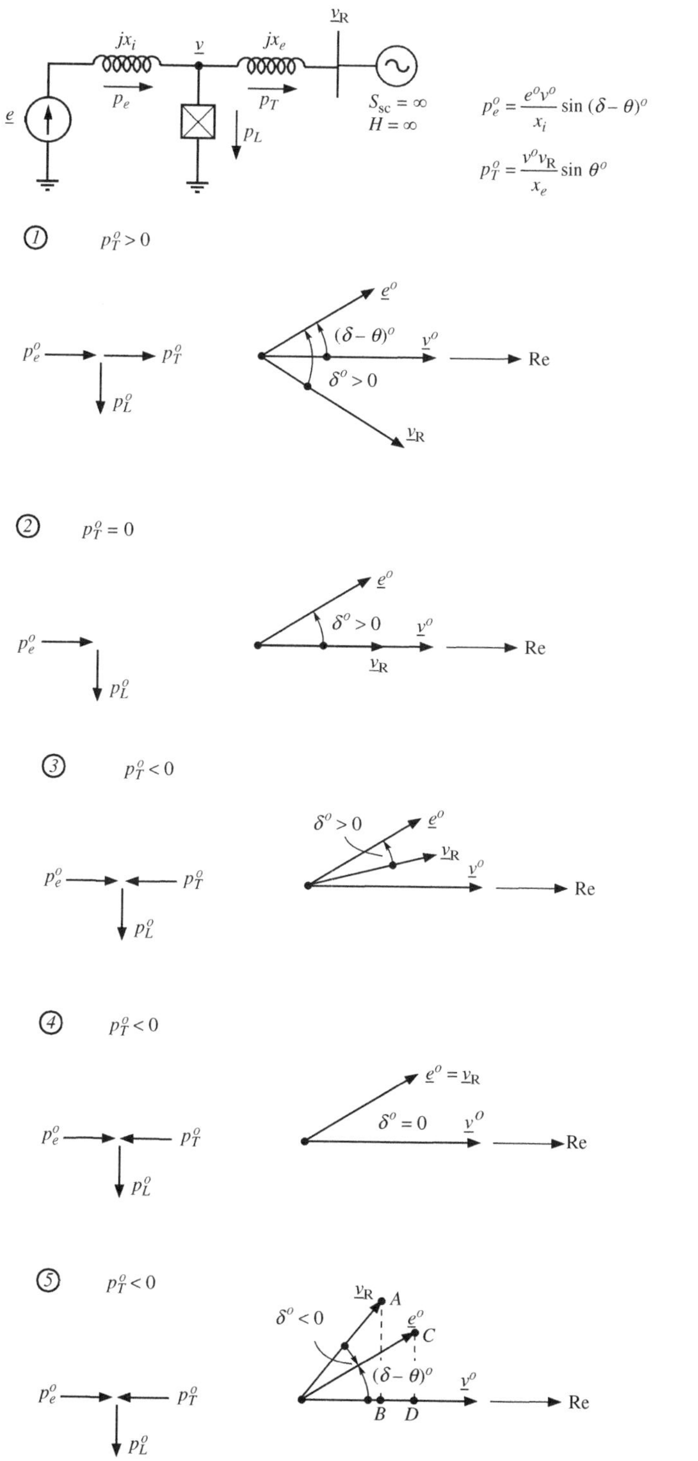

Figure 9.40. Unit in generation mode in the same operating point (e^o, v^o, $(\delta - \theta)^o$, p_e^o are const.): effect of the decrease in real transit p_T^o with the increase in real load p_L^o.

describes this situation: it is worth noting that the generator operating conditions (e^o, v^o, $\sin(\delta-\theta)^o$, p_e^o) are the same in all cases.

Since

$$\begin{cases} p_e^o = \dfrac{e^o v^o}{x_i} \sin(\delta-\theta)^o \\ p_T^o = \dfrac{v^o v_R}{x_e} \sin\theta^o \end{cases} \tag{9.233}$$

we deduce (see situation 5)

$$\begin{cases} \bar{AB} = v_R \sin\theta^o = \dfrac{x_e p_T^o}{v^o} \\ \bar{CD} = e^o \sin(\delta-\theta)^o = \dfrac{x_i p_e^o}{v^o} \end{cases} \tag{9.234}$$

and the condition $\delta^o = 0$ occurs when the load is also supplied by the network, that is, $p_T^o < 0$.

In particular, if we assume

$$e^o \simeq v_R \tag{9.235}$$

we get (point A matches C, situation 4 in Figure 9.40)

$$x_e |p_T^o| = x_i p_e^o \tag{9.236}$$

whereas in situation 3

$$|p_T^o| < \frac{x_i}{x_e} p_e^o \tag{9.237}$$

and in situation 5

$$|p_T^o| > \frac{x_i}{x_e} p_e^o \tag{9.238}$$

Therefore, δ^o is negative when the approximated condition (9.238) is fulfilled. Additionally, for a load also supplied by the network, we have

$$p_L^o = p_e^o + |p_T^o| \tag{9.239}$$

Therefore, the condition (9.238) may also be written as

$$p_L^o > \left(1 + \frac{x_i}{x_e}\right) p_e^o \tag{9.240}$$

5. From (9.219), we have

$$-hh_1 = \frac{z^2}{x_i x_e} \left[\frac{e v_R}{v} \sin\delta\right]^o \left[\frac{p_e}{e} + e(g_0 + g_1)\right]^o \tag{9.241}$$

Neglecting the case $p_e^o < 0$ (unit in pumping mode), the sign of $-hh_1$ depends on the sign of the angle δ^o. Hence, again under the assumption (9.235) and taking in mind that $h_2 > 0$, we may conclude that the effect of the voltage control loop, when a load is present, is (due to condition (9.159))

- destabilizing if

$$|p_T^o| < \frac{x_i}{x_e} p_e^o \qquad (9.242)$$

- stabilizing if

$$|p_T^o| > \frac{x_i}{x_e} p_e^o \qquad (9.243)$$

In particular, in the case of generator at no load ($p_e^o = 0$ like in the case of *synchronous compensators*), as $h > 0$, $h_2 > 0$, and $h_1 < 0$ for $\delta^o < 0$, the effect of AVR is positive (it was null in the case of load far from the generator).

6. Therefore, the presence of a real load always has stabilizing effects; this is depicted in Figure 9.41 (cases numbered as in Figure 9.40), which shows the qualitative shift of the electromechanical mode as the power flows from the infinite system to the area increases.

Some of the previous results may be simplified in practice, in particular equation (9.243). In effect, for both local and interarea oscillations, we get

$$\begin{cases} |b_0| \gg g_0 \\ |b_1| \gg g_1 \end{cases} \qquad (9.244)$$

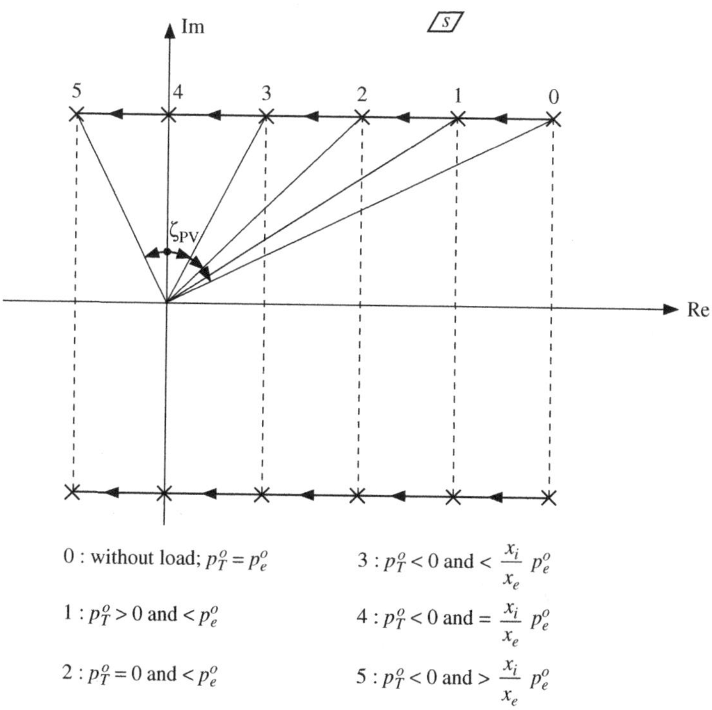

Figure 9.41. Qualitative shift of electromechanical oscillation following transit decrease.

and also

$$\alpha \simeq 1 \tag{9.245}$$

For instance, if $x_i \cong 0.3$ p.u., $r_L \cong 1.25$ p.u. ($p_L^o \cong 0.8$ p.u.), we obtain (all parameters are in p.u.)

- for local oscillations with $x_e \cong 0.7$ p.u., $T_o \cong 1$ s

$$\begin{cases} x_p = 0.21 \quad \alpha = 0.97 \\ b_0 = -3.24 \; b_1 = -0.97 \simeq -1/(x_i + x_e) \\ g_0 = 0.543 \;\; g_1 = -0.163 \end{cases} \tag{9.246}$$

- for interarea oscillations with $x_e = 9.7$ p.u., $T_o \cong 3.5$ s

$$\begin{cases} x_p = 0.29 \quad \alpha = 0.95 \\ b_0 = -3.16 \; b_1 = -0.098 \simeq -\dfrac{1}{x_i + x_e} \\ g_0 = 0.74 \quad g_1 = -0.022 \end{cases} \tag{9.247}$$

Consequently, from (9.225)

$$\begin{cases} \underline{y}_0 \simeq g_0 = \dfrac{1}{r_L} \cdot \dfrac{x_e}{x_i + x_e} \\ \underline{y}_1 \simeq jb_1 = -j\dfrac{1}{x_i + x_e} \end{cases} \tag{9.248}$$

In this expression, the larger x_e with respect to x_i (case of low-frequency oscillations), the closer will be g_o to $g_L(= 1/r_L)$. The equivalent circuit of Figure 9.38b becomes as shown in Figure 9.42, as if the *load* were *directly supplied by the internal emf of the generator.*

Therefore, the expressions of k, h_1, and h_2 are those applying to the situation without load with p_T^o in place of p_e^o, in that

$$\frac{e^o v_R \sin \delta^o}{x_i + x_e} = p_T^o \tag{9.249}$$

Therefore, from (9.219), we get

$$h_1 = -\frac{x_i x_e}{x_i + x_e} \left(\frac{p_T}{v}\right)^o \tag{9.250}$$

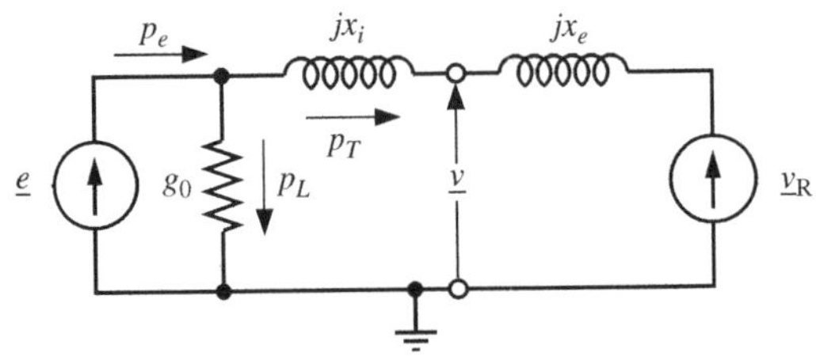

Figure 9.42. Approximation of the equivalent circuit of Figure 9.38b.

$$h = \left(\frac{p_e}{e} + g_0 e\right)^o \tag{9.251}$$

Then, g_o being positive and excluding the pumping-mode case, we obtain

$$-hh_1 < 0 \tag{9.252}$$

if

$$p_T^o < 0 \tag{9.253}$$

replacing equation (9.243). Hence, to a first approximation, the mere fact that the power flow comes from the infinite system (p_T^o negative) will cause the primary voltage control loop to have a stabilizing effect.

Based on the above considerations on the real load dependence on voltage, it may be stated that

- At no load ($p_e^o = 0$), the effect of the voltage control loop is positive; conversely, in case of absence of load or of load far from the generator, such effect was null.
- The presence of a *real load always has a stabilizing effect*; indeed, at a given p_e^o, it reduces the rotor angle δ^o: if equation (9.243) is not satisfied, the value of ζ_{PV} remains negative, but it is lower in absolute terms than in absence of load.
- Thus, *braking resistors*, shunt connected to the generator terminals (see Figure 9.43) and fully equivalent to a linear real load, represent a solution—expensive—to improve the damping of electromechanical oscillations.
- If equation (9.243) is satisfied, then the *effect* of the primary voltage control loop is *positive*, that is, the stabilizing effect of load is such as to offset completely the effect, which is otherwise always negative, of the voltage loop. Such feature depends on x_e. Therefore, in terms of system characteristics, such situation may be different, in practice, depending on whether *fast or local oscillations* or *slow or interarea oscillations* are involved. Given the fact that equation (9.243) requires a negative p_T^o, that is, a load also supplied by the network, positive damping is obtained:
  - for fast electromechanical oscillations (relatively small x_e), if p_T^o accounts for a significant percentage of p_e^o or is comparable to p_e^o;
  - for slow electromechanical oscillations (relative large x_e), if p_T^o accounts for an even small percentage of p_e^o, for example 5–10%.
- The previous considerations, along with those made in the preceding section, clearly show that the contribution ζ_{PV} and thus the *problem of damping* of electromechanical oscillations is a *problem of real power transmission*.

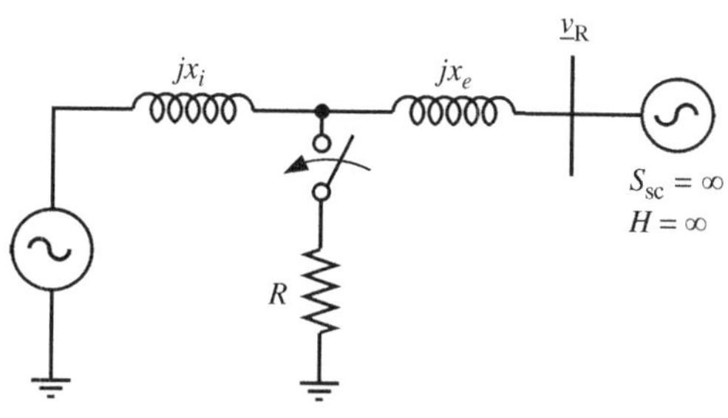

Figure 9.43. Switching-on of braking resistors at generator terminals.

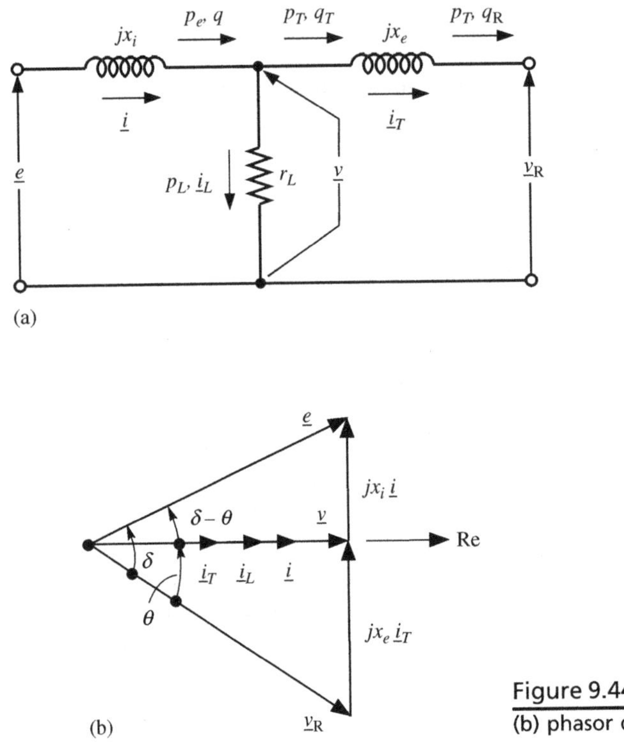

Figure 9.44. Evaluation of ζ_{PV}: (a) equivalent circuit; (b) phasor diagram.

Major Conclusions on the Influence of the Primary Voltage Control. To get an idea of the value that ζ_{PV} may take on with and without intermediate load, the realistic situation of a generation and possibly load area will be considered. The area may export power ($p_T^o > 0$), import power ($p_T^o < 0$), or have zero power exchange ($p_T^o = 0$). To simplify computations, let us also suppose that $q_L^o = 0$ (purely real load) and that (see Figure 9.44a).

$$q_T^o = q = 0 \tag{9.254}$$

Consequently, v_R and q_R will adapt to the various operating conditions. At $v^o = 1$, we will thus have

- for each value of p_e^o

$$\begin{cases} \underline{i}^o = p_e^o = p_L^o + p_T^o \\ e^o \simeq 1 \\ \delta - \theta = \text{const.} \end{cases} \tag{9.255}$$

- for each value of p_T^o

$$\begin{cases} v_R = \sqrt{1 + (x_e p_T^o)^2} \\ \tan\theta = x_e p_T^o \end{cases} \tag{9.256}$$

- for each value of load and of x_e, the values of z, g_1, b_1, and $(g_0 + g_1)$ are constant.

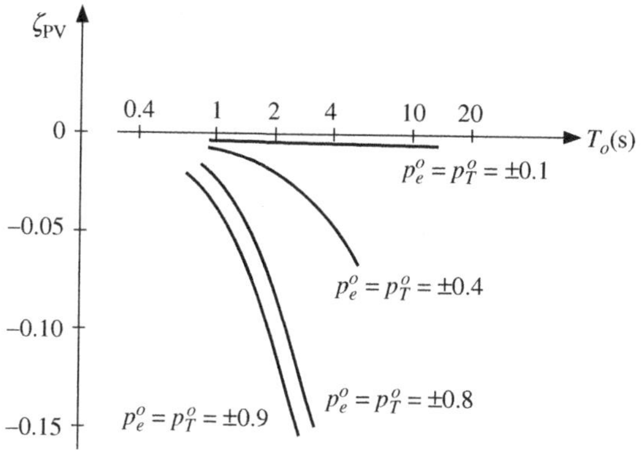

Figure 9.45. Trend of ζ_{PV} without intermediate load.

Figures 9.45 and 9.46 show the behavior of ζ_{PV} as a function of x_e, that is, of the oscillation period T_o, and of the power flows. They are based on (9.173), without intermediate load, and on equations (9.219) in the presence of load ($p_L^o = 0.8\,\text{p.u.}$, $r_L = 1.25\,\text{p.u.}$, $q_L^o = 0$). In particular, the curves are obtained by assuming

$$\begin{cases} x_i = 0.2\,\text{p.u.} & T_a = 10\,\text{s} \\ T'_{do} = 7.5\,\text{s} & \mu_t = 50\,\text{p.u./p.u.} \end{cases} \tag{9.257}$$

These results can be summarized as follows (see Figure 9.46):

- If the area exports power, $\zeta_{PV} < 0$.
- If the area imports power, $\zeta_{PV} > 0$ is likely to occur.

9.4.3.4 Effect of Primary Frequency Control.

Basic Considerations. Let $-d(s)$ be the transfer function of the primary speed governor of the unit, of the related supply system and of the turbine:

$$-d(s) \triangleq \left[\frac{\Delta p_m}{\Delta\Omega/\Omega_n}(s)\right] \tag{9.258}$$

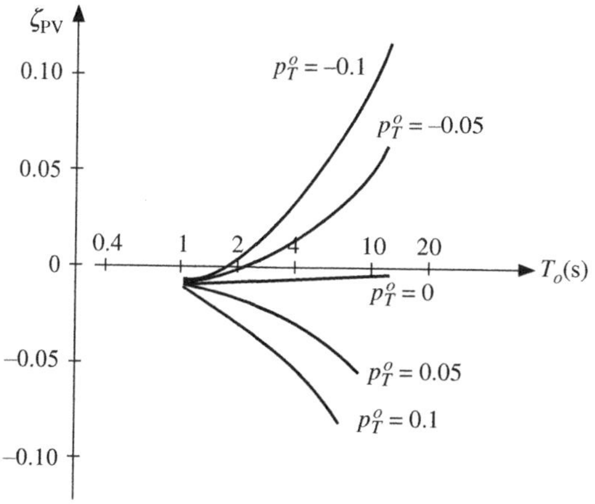

Figure 9.46. Trend of ζ_{PV} with intermediate load.

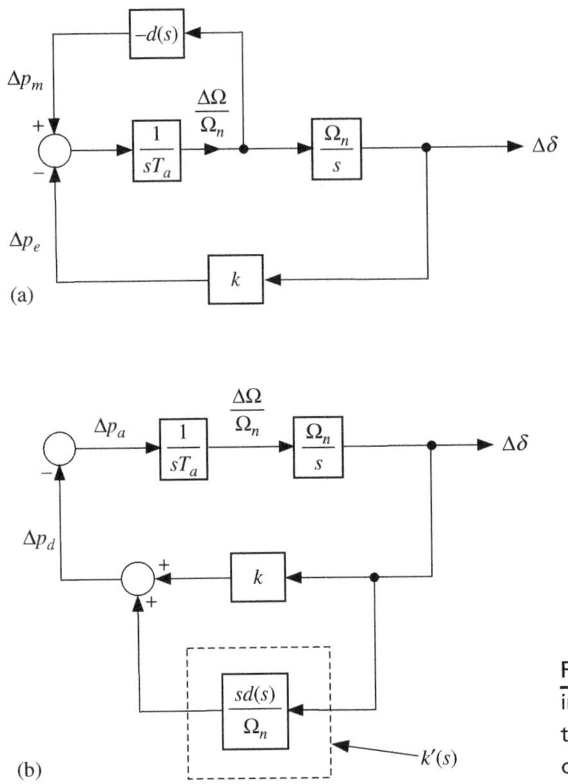

Figure 9.47. Block diagrams for assessing the effect of primary frequency control on the damping of electromechanical oscillations.

Figure 9.47a shows the block diagram for assessing the effect on electromechanical oscillations. This diagram may be transformed into the one shown in Figure 9.47b, where

$$k'(s) \triangleq \frac{s}{\Omega_n} d(s) \tag{9.259}$$

and thus

$$k'(j\omega_o) = j\frac{\omega_o}{\Omega_n} d(j\omega_o) = j\frac{\omega_o}{\Omega_n}[\mathrm{Re}\{d(j\omega_o)\} + j\mathrm{Im}\{d(j\omega_o)\}] \tag{9.260}$$

that is

$$\mathrm{Im}\{k'(j\omega_o)\} = \frac{\omega_o}{\Omega_n}\mathrm{Re}\{d(j\omega_o)\} \tag{9.261}$$

So, from equation (9.121), we obtain

$$\zeta_{\mathrm{PF}} \simeq \frac{1}{2\omega_o T_a}\mathrm{Re}\{d(j\omega_o)\} \tag{9.262}$$

Therefore, the effect of primary frequency control loop is stabilizing only if

$$-\frac{\pi}{2} < \angle d(j\omega_o) < \frac{\pi}{2} \tag{9.263}$$

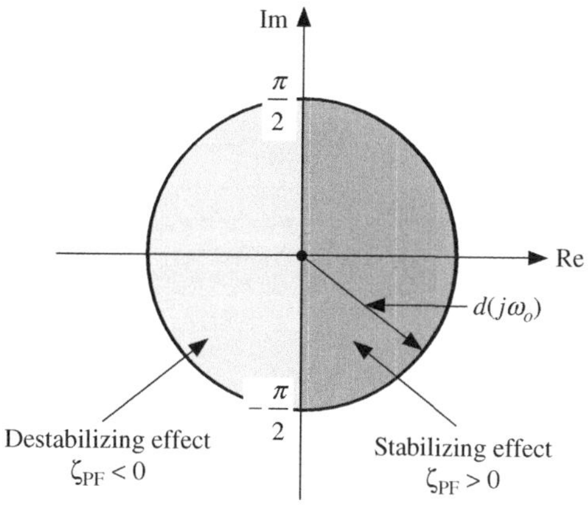

Figure 9.48. Stabilizing and destabilizing effects of primary frequency control loop.

as shown Figure 9.48, or

$$\begin{cases} \mathrm{Re}\{d(j\omega_o)\} > 0 \Rightarrow \zeta_{\mathrm{PF}} > 0 \\ \mathrm{Re}\{d(j\omega_o)\} < 0 \Rightarrow \zeta_{\mathrm{PF}} < 0 \end{cases} \tag{9.264}$$

Now, to evaluate the actual contribution of primary frequency control, it is necessary to study the details of the dynamic characteristics of $d(s)$, due to differences in speed governors, supply systems and turbines. It is suitable to evaluate at least the main features of the primary frequency control of

- conventional thermal units;
- gas-turbine and combined-cycle units;
- hydro units.

Conventional Thermal Units. Assume that the effect of the slowest control loops is neglected, as they do not interact with electromechanical oscillations. The following transfer function for the frequency control loop is suitable for the analysis:

$$d(s) = \frac{1}{b_{\mathrm{p}}} \cdot \frac{1}{(1 + sT_{\mathrm{sc}})^2} \cdot \frac{1 + s\alpha T_{\mathrm{R}}}{1 + sT_{\mathrm{R}}} \tag{9.265}$$

where b_{p} is the permanent droop, T_{sc} is the steam-chest time constant, T_{R} is the reheater time constant, and α is the fraction of power produced by high-pressure stages.

For these parameters, typical data are

$$\begin{cases} b_{\mathrm{p}} = 0.05\,\mathrm{p.u./p.u.} & \alpha = 0.3\,\mathrm{p.u./p.u.} \\ T_{\mathrm{sc}} = 0.3\,\mathrm{s} & T_{\mathrm{R}} = 10\,\mathrm{s} \end{cases} \tag{9.266}$$

Therefore,

$$d(s) = 20\frac{1}{(1 + s0.3)^2} \cdot \frac{1 + s3}{1 + s10} \tag{9.267}$$

and the relevant Bode plots are shown in Figure 9.49.

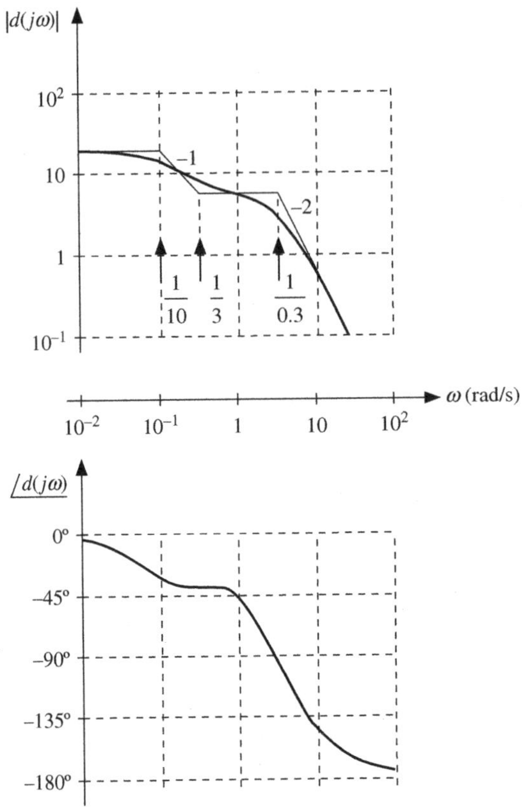

Figure 9.49. Typical Bode plots of primary frequency control loop transfer function d(*s*) for a thermal unit.

Based on equation (9.262), Figure 9.50 shows the approximate damping contribution that occurs upon the variation of the electromechanical oscillation period T_o and for a start-up time $T_a = 8$ s.

It is worth noting that

- in case of local oscillations ($T_o = 0.4$–1 s), ζ_{PF} is negative but very small, because $\mathrm{Re}\{d(j\omega_o)\}$ is very small and can be neglected. Moreover, $\Delta p_m \cong 0$ because the amplitude of $d(j\omega_o)$ is very small too;
- in case of interarea oscillations ($T_o \geq 2$ s), ζ_{PF} becomes positive and high ($\Delta p_m \neq 0$), so high as to certainly compensate $\zeta_{\mathrm{PV}} < 0$ for $T_o \geq 4$ s;
- if, T_a remaining equal, b_{p} is higher than assumed (areas with regulating and nonregulating units), for example, $b_{\mathrm{p}} = 0.15$ p.u./p.u. (as shown in Figure 9.50), then any effect will be reduced proportionally;
- if, b_{p} remaining equal, the start-up time is greater than 8 s (e.g., $T_a = 12$ s, in case of areas with many rotating loads), the effects will be similarly reduced;
- good frequency control (well-damped control loop and presence of many regulating units) provides a strongly positive contribution to damping of very slow electromechanical oscillations (whose frequency lies within the frequency range controlled by the regulation). This is another good reason for having good frequency control in large interconnected power systems that experience interarea oscillations.

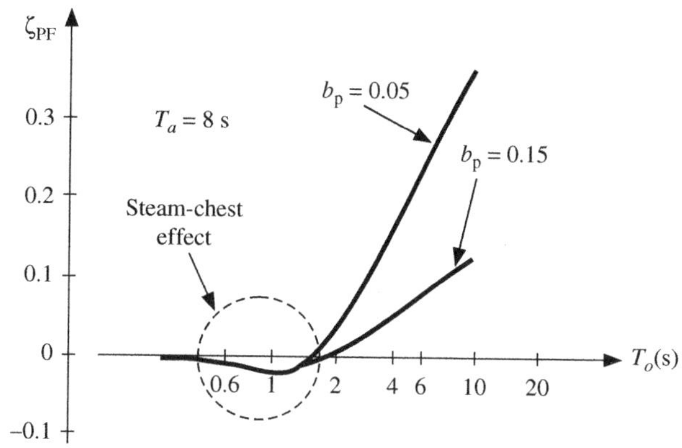

Figure 9.50. Qualitative contribution of primary frequency control loop of conventional thermal units to damping of electromechanical oscillations.

Gas Turbines and Combined-Cycle Power Plants. In case of gas turbines, the following expression holds for $d(s)$:

$$d(s) = \frac{1}{b_p} \cdot \frac{1}{(1+sT_t)} \cdot \frac{1}{(1+sT_f)^2} \cdot \frac{1+s\alpha T_m}{1+sT_m} \tag{9.268}$$

where b_p is the permanent droop, T_t is the throttle time constant, T_f is the fuel equivalent time constant, T_m is the engine time constant, α is the engine transient gain.

With reference to industrial engines, the following typical parameter values may be assumed

$$\begin{cases} b_p = 0.05\,\text{p.u./p.u.} & T_t = T_f = 0.1\,\text{s} \\ T_m = 1\,\text{s} & \alpha = 0 \end{cases} \tag{9.269}$$

and then

$$d(s) = \frac{20}{(1+s0.1)^3(1+s)} \tag{9.270}$$

Bearing in mind that

$$T_a \simeq 20\,\text{s} \tag{9.271}$$

the diagram of Figure 9.51 is obtained.

The remarks previously made for conventional thermal units will apply. In addition, it can be observed that

- in comparison with conventional thermal units, the reheater time constant (equal to about 10 s) is missing and, accordingly, ζ_{PF} is higher in the range of interarea electromechanical oscillations.

In case of combined-cycle units, the results are intermediate between those depicted in Figures 9.50 and 9.51.

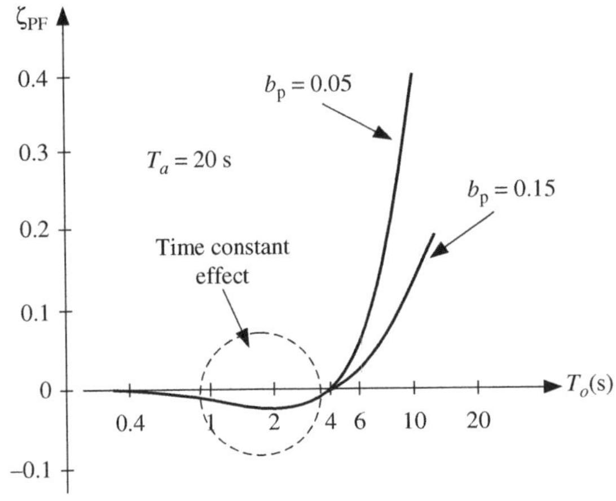

Figure 9.51. Qualitative contribution of primary frequency control loop of gas turbines to damping of electromechanical oscillations.

Hydro Units. To highlight the effect of the penstock and therefore also of the operating point, it is useful to consider separately the speed governor (represented by $g_r(s)$) and of the water supply system (represented by $g_a(s)$):

$$d(s) = g_r(s)g_a(s) \tag{9.272}$$

$$\begin{cases} |d(j\omega)| = |g_r(j\omega)| \cdot |g_a(j\omega)| \\ \angle d(j\omega) = \angle g_r(j\omega) + \angle g_a(j\omega) \end{cases} \tag{9.273}$$

Assuming that (i) the speed governor is equipped with an accelerometer and thus has no transient feedback and (ii) the penstock and the turbine are described by the low-frequency approximation, we obtain

$$\begin{cases} g_r(s) = \dfrac{1}{b_p} \cdot \dfrac{1 + sT_{ac}}{1 + sT_b} \cdot \dfrac{1}{1 + s(T_s/b_p)} \\ g_a(s) = \dfrac{1 - s(T_{wn}q^o)}{1 + s(T_{wn}q^o/2)} \end{cases} \tag{9.274}$$

where b_p is the permanent droop, T_{ac} is the accelerometer time constant, T_b is the tachometer time constant ($\cong 0.3$ s, not negligible, like in the case of thermal units), $1/T_s$ is the control servomotor gain, T_{wn} is the penstock nominal start-up time, and q^o is the water flow rate in the operating point.

Assuming the typical parameter values

$$\begin{cases} b_p = 0.05\,\text{p.u./p.u.} & T_{ac} = 3\,\text{s} \\ T_b = 0.3\,\text{s} & T_s = 0.5\,\text{s} \end{cases} \tag{9.275}$$

for the governor (supposing that the unit is equipped with a *Pelton turbine*) and $T_{wn} = 1.25$ s for the penstock, with the unit at full load

$$q^o = 0.8\,\text{p.u.} \tag{9.276}$$

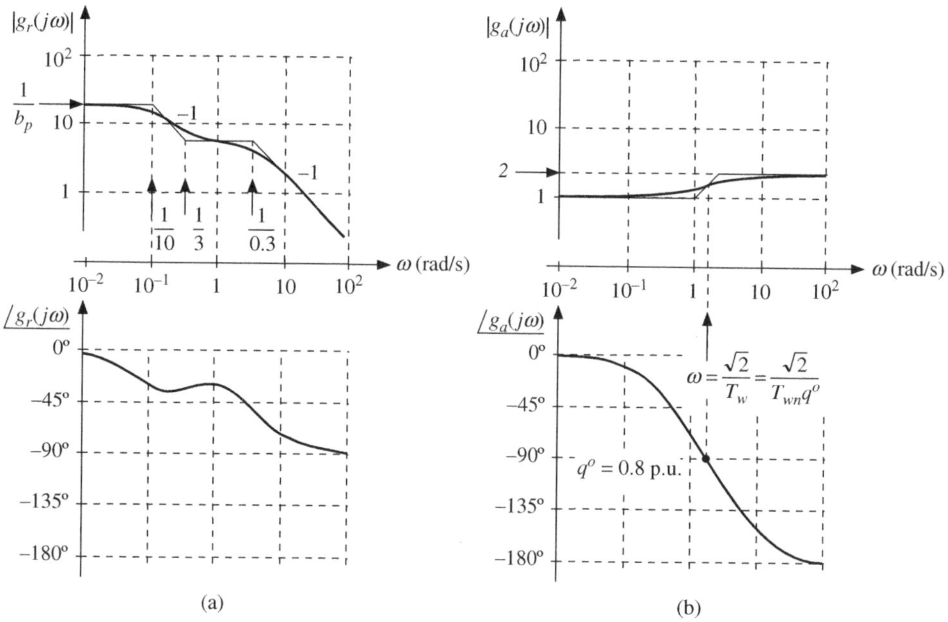

Figure 9.52. Typical Bode plots of the transfer functions of (a) speed governor; (b) water supply system of hydro unit equipped with Pelton turbine and $q^o = 0.8$ p.u. (full load).

A realistic form of $d(s)$ can be obtained by

$$\begin{cases} g_r(s) = 20\dfrac{1 + s3}{(1 + s0.3)(1 + s10)} \\ g_a(s) = \dfrac{1 - s}{1 + s0.5} \end{cases} \tag{9.277}$$

whose Bode plots are reported in Figure 9.52.

In order to compare the results with thermal units, let us assume $T_a = 8$ s (instead of $\cong 6$ s, typical value for hydro units). Figure 9.53 shows the damping contributions of the speed governor alone, that is, of the unit at no load ($q^o = 0$ and $g_a(s) = 1$) and of the governor-supply system set at full load and at half load ($q^o = 0.4$ p.u.).

It can be observed that

- at no load, only the speed governor has an effect, which is stabilizing;
- the effect of the penstock is destabilizing: the greater the load, the more destabilizing the effect for a given T_o;
- at high load, the overall effect is destabilizing and to a significant extent ($\Delta p_m \neq 0$) in the range of interarea electromechanical oscillations, whereas it is small ($\Delta p_m \cong 0$) in the range of local oscillations, as in the case of thermal units;
- at low load, the effect of the penstock tends to disappear ($g_a(s) \to 1$) and the overall effect may also be stabilizing, but still very small in the range of local electromechanical oscillations. Conversely, it is certainly stabilizing and to a significant extent for interarea oscillations;
- for $T_o < 0.6$ s, the overall effect is $\cong 0$;

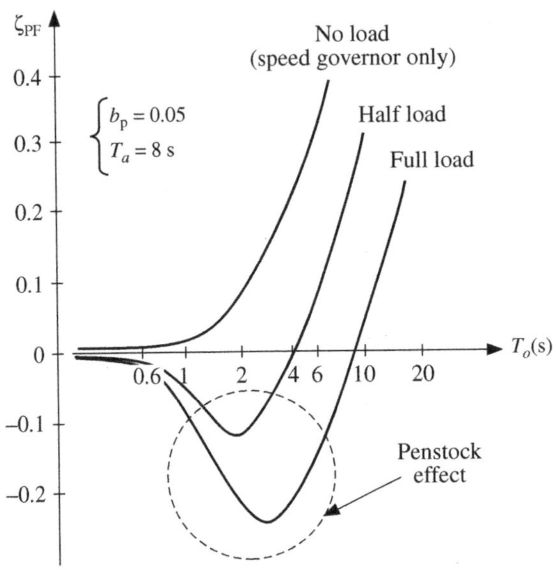

Figure 9.53. Qualitative contribution of primary frequency control loop of Pelton-turbine hydro unit to damping of electromechanical oscillations.

- for $T_o > 10\,s$, the overall effect is highly stabilizing;
- for b_p and T_a, the same considerations as those made about thermal units will apply.
- in case of Francis and Kaplan turbines (Figure 9.53 refers to Pelton turbines), that is for hydro units with low water-head, the penstock effect tends to disappear ($T_w \rightarrow 0$) in the overall range of electromechanical oscillations.

MAJOR CONCLUSIONS ON THE INFLUENCE OF THE PRIMARY FREQUENCY CONTROL. According to the analysis developed above, the following conclusions on the contribution of ζ_{PF} in interconnected power systems can be drawn:

- It is generally highly dependent on *type of the units in operation, actual regulating capacities* and *load of hydro units.*
- It can be disregarded ($\Delta p_m \cong 0$) in case of local oscillations involving thermal units and low water-head hydro units.
- It must be taken into account and can have positive or negative effects ($\Delta p_m \neq 0$) in case of interarea oscillations.

9.4.3.5 Outline of Other Contributions. Other possible issues could affect the damping of electromechanical oscillation in electric power systems. For the sake of brevity, we summarize the conclusions for each of such factors:

- The nonlinear dependence of a real load on voltage has a stabilizing effect, as in the linear case. The higher the exponent k_{pv} of the voltage dependence of the load, the higher will the positive contribution of the load be.
- As the load regulating energy is very small compared to the one of the units, the dependence of the frequency on a real load is stabilizing but generally small.

- Reactive loads, shunt reactors, and shunt capacitors have *indirect* effects: positive in the case of capacitors banks and capacitive loads; negative in the case of inductive loads and shunt reactors. However, such effects are negligible, as they are related to the reactive power.
- The compound action and the over/underexcitation limiting circuits worsen ζ_{PV} but in a negligible way, because the related loops are very slow.
- Secondary voltage control (thanks to the reactive power control loop) provides a positive contribution to interarea oscillations only. However, also this contribution is very small.
- Secondary frequency control (or LFC), just as boiler controls, has no effect on damping.
- FACTS controllers, such as SVCs and TCPARs, provide negative but negligible direct contributions, since they have very fast controls. Thanks to their voltage support action, SVCs provide a stabilizing indirect contribution; indeed, by decreasing the electromechanical oscillation periods, they increase the contribution ζ_q of the generator structure along the q-axis.
- HVDC links improve ζ_{PV}; indeed, if they are in power control mode, they reduce real power flows on AC links. In frequency control mode, they provide a strongly stabilizing contribution, thanks to the very small value of the permanent droop (0.01 p.u./p.u. instead of 0.05 p.u./p.u.).

9.4.4 Summary of the Major Factors Affecting the Damping of Electromechanical Oscillations

The stability, and in particular the damping of electromechanical oscillations is dependent on a number of factors: the characteristics of generators, the controls and the physical process of the energy conversion. The main conclusions of this section are as follows:

- The structure of generators introduces a highly positive contribution ζ_S; in particular, the quadrature axis has a highly positive ζ_q effect on local oscillations, whereas the direct axis (in the absence of the AVR) has a highly positive effect ζ_d on interarea oscillations.
- Primary voltage control nullifies the positive contribution ζ_d. Moreover, with no load, it introduces a second negative contribution ζ_{PV}; the slower the oscillations, the larger the latter contribution.
- The presence of the real load and its dependence on voltage has a stabilizing effect and, depending on the directions of real power flows in the system, it may be such as to make ζ_{PV} strongly positive.
- Primary frequency control (on conventional thermal, gas-turbine and combined-cycle units) provides a negative and very small contribution ζ_{PF} to local oscillations. By contrast, such contribution becomes positive upon interarea oscillations and so high as to compensate ζ_{PV}.
- In the case of hydro units with high water-head (Pelton turbines), ζ_{PF} is strongly variable and depends on the delivered power. At low load, it is always stabilizing; at medium-high load, it is strongly negative, especially upon interarea oscillations; conversely, upon very slow oscillations, it may become strongly positive.

- Instead, in the case of hydro units with low water-head (Francis and Kaplan turbines), ζ_{PF} is actually always positive.
- Thus, the contribution ζ_{PF} is highly dependent on the type of the units in operation, actual regulating capacities and load of hydro units. For thermal units and low water-head hydro units, it can be disregarded in the case of local oscillations. In case of interarea oscillations, it must be taken into account.

9.5 DAMPING IMPROVEMENT

9.5.1 Introduction

The concepts described in the previous section highlight that modern voltage regulators are among the main responsible factors for negative damping of electromechanical oscillations. Interestingly, if we consider the block diagram of Figure 9.31 at constant set point, we obtain (with or without intermediate load)

$$\Delta e = -\frac{F_v(s)}{sT'_{do}}\,\Delta v = -\frac{F_v(s)}{sT'_{do}}(h_2\,\Delta e + h_1\,\Delta\delta) \tag{9.278}$$

that is,

$$\Delta e = -h_1\frac{F_v(s)/sT'_{do}}{1 + h_2(F_v(s)/sT'_{do})}\Delta\delta \tag{9.279}$$

Taking into account that

$$\begin{cases} \Delta\delta = \dfrac{\Omega_n}{s}\cdot\dfrac{\Delta\Omega}{\Omega_n} \\ g_v(s) = \dfrac{h_2(F_v(s)/sT'_{do})}{1 + h_2(F_v(s)/sT'_{do})} \end{cases} \tag{9.280}$$

we also derive

$$\Delta e = -\frac{h_1\,\Omega_n}{h_2 s}\,g_v(s)\frac{\Delta\Omega}{\Omega_n} \tag{9.281}$$

As around the electromechanical oscillation frequency ω_o

$$g_v(j\omega) = \frac{1}{1 + j(\omega/\omega_v)} = \frac{1 - j(\omega/\omega_v)}{1 + (\omega/\omega_v)^2} \tag{9.282}$$

we obtain

$$h\,\Delta e = \frac{h\,h_1}{h_2}\cdot\frac{\Omega_n}{\omega\left[1 + (\omega/\omega_v)^2\right]}\left(\frac{\omega}{\omega_v} + j\right)\frac{\Delta\Omega}{\Omega_n} \tag{9.283}$$

representing real power changes due to changes in the internal emf as a result of voltage control:

$$h\,\Delta e = \left(\frac{\partial p_e}{\partial e}\right)^o \Delta e \tag{9.284}$$

The total variation of electrical power is

$$\Delta p_e = k\,\Delta\delta + h\,\Delta e = k\frac{\Omega_n}{s}\cdot\frac{\Delta\Omega}{\Omega_n} + h\,\Delta e = \Delta p_e' + \Delta p_e'' \tag{9.285}$$

which, around ω_o, becomes

$$\Delta p_e = \left[-j\frac{k\Omega_n}{\omega} + \frac{h\,h_1}{h_2}\cdot\frac{\Omega_n}{\omega\left[1+(\omega/\omega_v)^2\right]}(\omega/\omega_v + j)\right]\frac{\Delta\Omega}{\Omega_n} \tag{9.286}$$

Hence, the imaginary part of Δp_e depends on both $\Delta p_e'$ and on the imaginary part of $\Delta p_e''$, while the real part of Δp_e depends only on the real part of $\Delta p_e''$.

On the other hand, the conditions for stability are, as shown in (9.159)

$$\begin{cases} \dfrac{h\,h_1}{h_2} < 0 & \Rightarrow \quad \zeta_{PV} < 0 \\ \dfrac{h\,h_1}{h_2} > 0 & \Rightarrow \quad \zeta_{PV} > 0 \end{cases} \tag{9.287}$$

This means that, when $\zeta_{PV} < 0$, electrical power variations have a *component that is due to internal emf variations and that is in phase opposition to speed variations,* that is, when speed increases the generated power decreases and vice versa, thus resulting in a destabilizing contribution; therefore, this is the component responsible for *the negative damping induced by the voltage control.*

Figure 9.54 shows that, from the physical point of view, the situation is similar to the case of a speed governor with negative gain ($1/b_p < 0$), that is, a negative braking torque (or power).

Therefore, the compensation of this destabilizing effect can be realized by *modifying the excitation control structure*, by introducing a suitable transfer function between the speed variation and the AVR set point:

$$\frac{\Delta v_{ref}}{\Delta\Omega/\Omega_n}(s) \triangleq k_{PSS}(s) \tag{9.288}$$

that is termed *Additional Feedbacks* (AFs) or *additional signals* (ASs) or PSSs. Consequently, the diagram of Figure 9.31 turns into the one of Figure 9.55.

Moreover, the goal of the AFs is that, around the oscillation frequency ω_o, the harmonic transfer function k_{PSS} ($j\omega$) causes the emf variations Δe to be in phase with the speed variations $\Delta\Omega/\Omega_n$

$$\frac{\Delta e}{\Delta\Omega/\Omega_n}(j\omega) = \hat{k}_\Omega \tag{9.289}$$

In that case, the block diagram of Figure 9.55 turns into the one in Figure 9.56, whose characteristic equation is

$$\frac{k\Omega_n}{s}\cdot\frac{1}{sT_a + h\hat{k}_\Omega} + 1 = 0 \tag{9.290}$$

that is

$$s^2T_a + sh\hat{k}_\Omega + k\Omega_n = 0 \tag{9.291}$$

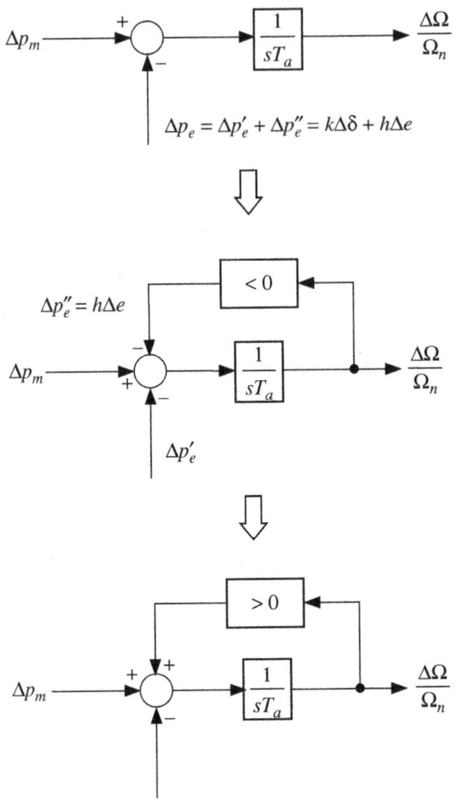

Figure 9.54. Physical interpretation of negative contribution of primary voltage control to damping.

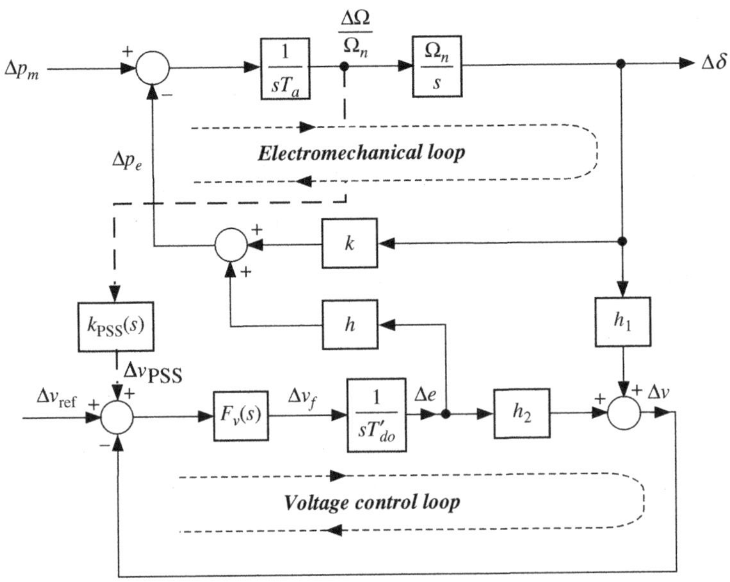

Figure 9.55. Block diagram of a unit connected to an infinite system (with or without intermediate load) in the presence of stabilizing signals.

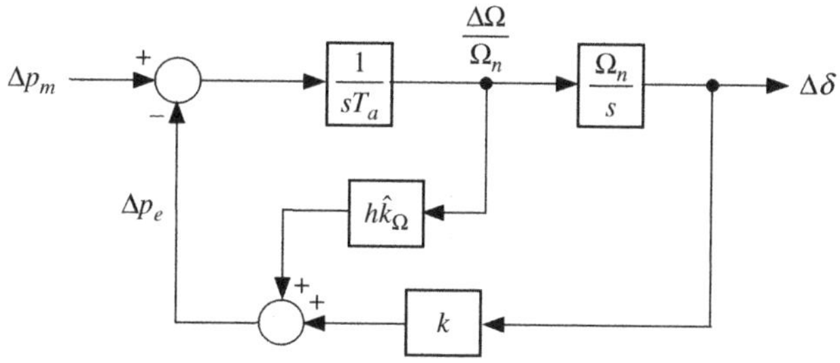

Figure 9.56. Block diagram equivalent to the one of Figure 9.55, in which internal emf variations are in phase with speed variations.

Thus, the system poles have the form

$$\underline{\lambda}_{1,2} = -\zeta\omega_n \pm j\omega_n\sqrt{1-\zeta^2} = -\zeta\omega_n \pm j\omega_o \tag{9.292}$$

with

$$\begin{cases} \omega_n = \sqrt{\dfrac{k\Omega_n}{T_a}} \\ \zeta = \dfrac{1}{2\omega_n} \cdot \dfrac{h\hat{k}_\Omega}{T_a} \\ \omega_o \triangleq \omega_n\sqrt{1-\zeta^2} \end{cases} \tag{9.293}$$

where ω_n coincides with ω_o if $h\Delta e = 0$ (e.g., if there is no AVR).

In practice, as damping values are small ($\zeta \cong 0.1$–0.2 at the most), we may always assume that

$$\omega_n = \sqrt{\frac{k\Omega_n}{T_a}} \simeq \omega_o \tag{9.294}$$

and thus also

$$\zeta \simeq \frac{1}{2\omega_o} \cdot \frac{h\hat{k}_\Omega}{T_a} \tag{9.295}$$

With regard to damping, we may observe that

- if $k_{PSS}(s)$ is selected in such a way that

$$h\hat{k}_\Omega > 0 \tag{9.296}$$

  then, the additional feedback makes the damping of oscillations positive;
- in the case of a generator connected to an infinite system without intermediate load, we should remember that (see (9.175) and (9.176))

$$\begin{cases} h > 0 & \text{if} \quad p_e^o > 0 \\ h = 0 & \text{if} \quad p_e^o = 0 \\ h < 0 & \text{if} \quad p_e^o < 0 \end{cases} \tag{9.297}$$

As a result,

- the effect of additional signals is zero if the generated real power is zero, as in the case of synchronous compensators;
- if the generator can operate in both generation and pumping modes (like modern hydro units), then the *sign of the additional feedback should be made dependent on the sign of the generated real power* in order to make the contribution to the damping always positive.

9.5.2 Modal Synthesis Based on the Theory of Small Shift Poles

The normally required increase $\Delta\zeta$ in the damping of electromechanical oscillations is fairly small ($\Delta\zeta$ in the range of 0.2 is excellent), which involves a small shift of poles. Consequently, the *theory of small shifts of poles based on eigenvalue sensitivity to additional feedback* can be adopted.

To this end, a generic system with one input $U(s)$ and one output $Y(s)$ will be considered (see Figure 9.57). In Figure 9.57, $G(s)$ is the forward transfer function and $H(s)$ is the feedback function (to be closed), with

$$H(s) = \varepsilon\, h(s) \tag{9.298}$$

where ε is the static gain and $h(s)$ has an assigned structure.

Moreover, let us assume that the poles of $G(s)$ and $H(s)$ are distinct, $\underline{\lambda}_1, \ldots, \underline{\lambda}_n$ are the poles of $G(s)$, and $\underline{C}_1, \ldots, \underline{C}_n$ the corresponding residues, so that

$$G(s) = G(\infty) + \sum_i \frac{\underline{C}_i}{s - \underline{\lambda}_i} = G(\infty) + \frac{\underline{C}_h}{s - \underline{\lambda}_h} + \sum_{i \neq h} \frac{\underline{C}_i}{s - \underline{\lambda}_i} \tag{9.299}$$

Now, let $\underline{\lambda}_h$ be the generic pole of $G(s)$ to be shifted: after the closure of the positive feedback, it will shift by $\Delta\underline{\lambda}_h$ while satisfying the characteristic equation

$$1 - G(s)H(s) = 0 \tag{9.300}$$

from which, for $s = \underline{\lambda}_h + \Delta\underline{\lambda}_h$, we get

$$1 - \varepsilon h(\underline{\lambda}_h + \Delta\underline{\lambda}_h)C(\underline{\lambda}_h + \Delta\underline{\lambda}_h) = 0 \tag{9.301}$$

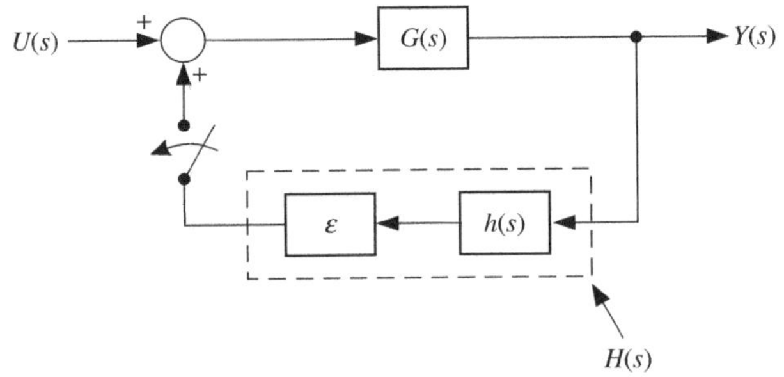

Figure 9.57. Closure of feedback for single-input and single-output systems.

$$1 - \varepsilon h(\underline{\lambda}_h + \Delta\underline{\lambda}_h)\left[\frac{\underline{C}_h}{\Delta\underline{\lambda}_h} + \sum_{i\neq h}\frac{\underline{C}_i}{\underline{\lambda}_h - \underline{\lambda}_i + \Delta\underline{\lambda}_h}\right] = 0 \tag{9.302}$$

and thus

$$1 - \varepsilon h(\underline{\lambda}_h + \Delta\underline{\lambda}_h)\sum_{i\neq h}\frac{\underline{C}_i}{\underline{\lambda}_h - \underline{\lambda}_i + \Delta\underline{\lambda}_h} = \varepsilon h(\underline{\lambda}_h + \Delta\underline{\lambda}_h)\frac{\underline{C}_h}{\Delta\underline{\lambda}_h} \tag{9.303}$$

that is,

$$\frac{\Delta\underline{\lambda}_h}{\varepsilon} = \frac{\underline{C}_h h(\underline{\lambda}_h + \Delta\underline{\lambda}_h)}{1 - \varepsilon h(\underline{\lambda}_h + \Delta\underline{\lambda}_h)\sum_{i\neq h}(\underline{C}_i/(\underline{\lambda}_h - \underline{\lambda}_i + \Delta\underline{\lambda}_h))} \tag{9.304}$$

Observing that (Figure 9.57) for $\varepsilon \to 0$, we also have $\Delta\underline{\lambda}_h \to 0$, we deduce

$$\lim_{\varepsilon\to 0}\frac{\Delta\underline{\lambda}_h}{\varepsilon} = \underline{C}_h h(\underline{\lambda}_h) \tag{9.305}$$

In other words, for *small values of static gain of the feedback*, we obtain the (small) shift of the pole h as a result of feedback closure

$$\Delta\underline{\lambda}_h = \underline{C}_h \varepsilon h(\underline{\lambda}_h) = \underline{C}_h H(\underline{\lambda}_h) \tag{9.306}$$

Now, let us define the pair of conjugate poles associated with the mode h of the electromechanical oscillation (whose damping is to be improved):

$$\begin{cases} \underline{\lambda}_h = \sigma_h + j\omega_{oh} \\ \underline{\lambda}_h^* = \sigma_h - j\omega_{oh} \end{cases} \tag{9.307}$$

Observing that (see (9.306))

$$|H(\underline{\lambda}_h)| = \frac{|\Delta\underline{\lambda}_h|}{|\underline{C}_h|} \tag{9.308}$$

then, as the hth residue does not depend on the shift, all the points of the circle centered in $\underline{\lambda}_h$ and radius $|\Delta\underline{\lambda}_h|$ are characterized by the same value of $|H(\underline{\lambda})_h|$ as shown in Figure 9.58. Since, for the mode to be damped, the following is certainly true

$$\sigma_h \ll \omega_{oh} \tag{9.309}$$

we also have

$$H(\underline{\lambda}_h) \simeq H(j\omega_{oh}) \tag{9.310}$$

$$\frac{|\Delta\underline{\lambda}_h|}{|\underline{C}_h|} \simeq |H(j\omega_{oh})| \tag{9.311}$$

In other words, all the points of the circumference of Figure 9.58 correspond to the same gain of the harmonic transfer function of the additional feedback. For a given $|H\,(j\omega_{oh})|$, the value (among all the possible shifts $\Delta\underline{\lambda}_h$) that corresponds to the *maximum increase of damping* is orthogonal to $\underline{\lambda}_h$, in the direction $j\,\underline{\lambda}_h$. As

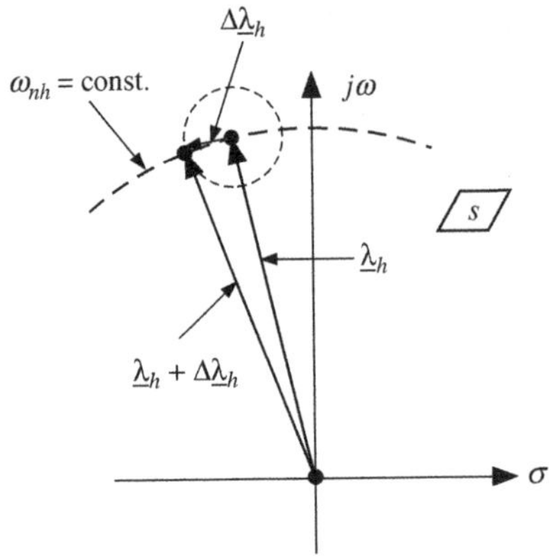

Figure 9.58. Small shift of the pole $\underline{\lambda}_h$.

$$\underline{\lambda}_h = \sigma_h + j\omega_{oh} \triangleq -\zeta_h\omega_{nh} + j\omega_{nh}\sqrt{1-\zeta_h^2} \tag{9.312}$$

the orthogonal shift of the pole $\underline{\lambda}_h$ implies that the amplitude of the same pole is taken to be constant, that is,

$$\omega_{nh} = \text{const.} \tag{9.313}$$

Therefore,

$$[\Delta\underline{\lambda}_h]_{\omega_{nh}=\text{const.}} = \left[-\omega_{nh} - j\frac{2\zeta_h\omega_{nh}}{2\sqrt{1-\zeta_h^2}}\right]\Delta\zeta_h = j\frac{\underline{\lambda}_h}{\sqrt{1-\zeta_h^2}}\Delta\zeta_h \tag{9.314}$$

In view of the above, we may develop a synthesis criterion that *maximizes damping, with a constant gain of the additional feedback transfer function*. Such criterion corresponds to selecting

$$H(\underline{\lambda}_h) = \frac{\Delta\underline{\lambda}_h}{\underline{C}_h} = j\frac{1}{\sqrt{1-\zeta_h^2}} \cdot \frac{\underline{\lambda}_h}{\underline{C}_h}\Delta\zeta_h \tag{9.315}$$

In other terms, on the basis of equation (9.309)

$$H(j\omega_{oh}) = -\frac{\omega_{oh}}{\underline{C}_h}\Delta\zeta_h \tag{9.316}$$

$$\begin{cases} |H(j\omega_{oh})| = \dfrac{\omega_{oh}}{|\underline{C}_h|}\Delta\zeta_h \\ \angle H(j\omega_{oh}) = \pi - \angle\underline{C}_h \end{cases} \tag{9.317}$$

and so also

$$\Delta\zeta_h = \frac{|\underline{C}_h H(j\omega_{oh})|}{\omega_{oh}} \tag{9.318}$$

The above equations clearly infer that the stabilizing effect on the hth electromechanical oscillation only depends on the amplitude and phase values of the harmonic transfer function of the additional feedback at the oscillation frequency ω_{oh}. These values may be calculated once the pole $\underline{\lambda}_h$ and its residue $\underline{C}_h$ have been determined and the increase $\Delta\zeta_h$ has been established.

9.5.3 PSSs on Excitation Control

9.5.3.1 Base Case and Theory. With reference to equation (9.288), let us determine the residue $\underline{C}$ of the transfer function (in the absence of PSSs) between the input Δv_{ref} and the output $\Delta\Omega/\Omega_n$. From Figure 9.55, for

$$k_{\text{PSS}}(s) = 0 \tag{9.319}$$

we get, first of all,

$$\Delta e = -\frac{h_1}{h_2} g_v(s)\,\Delta\delta + \frac{1}{h_2}\, g_v(s)\,\Delta v_{\text{ref}} \tag{9.320}$$

On the other hand, for $\Delta p_m = 0$, the following equation holds

$$\frac{\Delta\Omega}{\Omega_n} = -\frac{1}{sT_a}\,\Delta p_e = -\frac{1}{sT_a}(k\,\Delta\delta + h\,\Delta e) \tag{9.321}$$

and, finally,

$$\frac{\Delta\Omega/\Omega_n}{\Delta v_{\text{ref}}}(s) = \frac{-s(1/T_a)\cdot(h/h_2)g_v(s)}{s^2 + (\Omega_n/T_a)\,[k - (h\,h_1/h_2)\,g_v(s)]} \tag{9.322}$$

In the latter expression, the denominator corresponds to the characteristic equation of the electromechanical loop. Focusing on the electromechanical mode and neglecting the other poles, we get two poles with negligible damping, that is,

$$\begin{cases} \underline{\lambda}_1 \simeq j\omega_o \\ \underline{\lambda}_2 \simeq -j\omega_o \end{cases} \tag{9.323}$$

$$s^2 + \frac{\Omega_n}{T_a}\left[k - \frac{h\,h_1}{h_2}\, g_v(s)\right] \simeq s^2 + \omega_o^2 \tag{9.324}$$

and, accordingly,

$$\frac{\Delta\Omega/\Omega_n}{\Delta v_{\text{ref}}}(s) \simeq -\frac{s(1/T_a)\cdot(h/h_2)\,g_v(s)}{s^2 + \omega_o^2} \tag{9.325}$$

The residue $\underline{C}$ associated with $\underline{\lambda}_1$ is

$$\underline{C} = \left[(s - \underline{\lambda}_1)\,\frac{\Delta\Omega/\Omega_n}{\Delta v_{\text{ref}}}(s)\right]_{s=\underline{\lambda}_1} = -\frac{1}{2T_a}\cdot\frac{h}{h_2}\, g_v(j\omega_o) \tag{9.326}$$

Now, by (9.316), we get around ω_o

$$k_{\text{PSS}}(j\omega_o) = 2\omega_o T_a\,\frac{h_2}{h}\cdot\frac{1}{g_v(j\omega_o)}\,\Delta\zeta \tag{9.327}$$

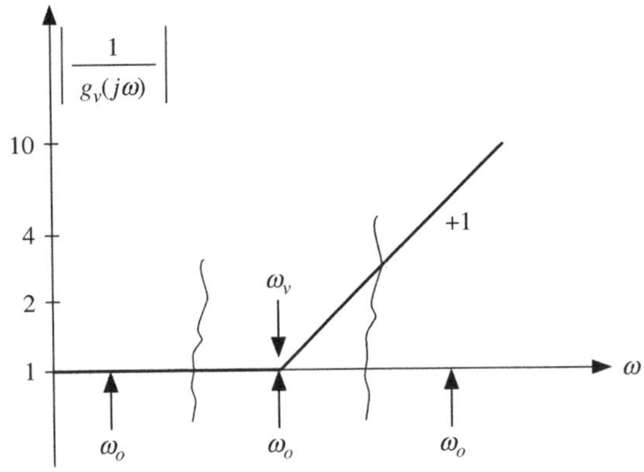

Figure 9.59. Asymptotic plot of the amplitude of the transfer function $1/g_v(s)$.

or

$$k_{\rm PSS}(j\omega_o) = 2\omega_o T_a \frac{h_2}{h} \Delta\zeta \left(1 + j\frac{\omega_o}{\omega_v}\right) \tag{9.328}$$

In other words, the dynamic characteristics of $k_{\rm PSS}(s)$ have to be opposite to the features of $g_v(s)$ (see Figure 9.59), which actually is the transfer function to be compensated by the power system stabilizers.

Therefore, we immediately realize that (see (9.328) and also Figure 9.59)

- if $\omega_o \ll \omega_v$, $1/|g_v(j\omega_o)|$ is constant, the stabilizing signal is required to be proportional to speed only;
- if $\omega_o \gg \omega_v$, $1/|g_v(j\omega_o)|$ increases with slope +1, the stabilizing signal is required to be proportional to the derivative of the speed only, that is, the acceleration;
- if ω_o is in a neighborhood of ω_v, both signals are required.

We can observe that no need arises for speed derivatives of an order higher than the first one because, in practice, modern excitation and voltage control systems are sufficiently fast, so that the phase of $1/g_v(j\omega)$, near ω_o never exceeds 90° significantly.

In the light of the above statements, an additional signal $\Delta v_{\rm PSS}$, capable of damping electromechanical oscillations of any frequency is of the type (see Figure 9.60a)

$$\Delta v_{\rm PSS} = k_\Omega \frac{\Delta\Omega}{\Omega_n} + k_c\left(\Delta p_m - \Delta p_e\right) \tag{9.329}$$

corresponding to the transfer function

$$\frac{\Delta v_{\rm PSS}}{\Delta\Omega/\Omega_n}(s) \triangleq k_{\rm PSS}(s) = k_\Omega \left[1 + \frac{k_c T_a}{k_\Omega} s\right] \tag{9.330}$$

where k_Ω and k_c are *gains of the additional signals* from speed and acceleration (accelerating power or torque), respectively.

Based on the above, we may also claim that

- k_c is suitable to damp local oscillations;
- k_Ω is suitable to damp interarea oscillations.

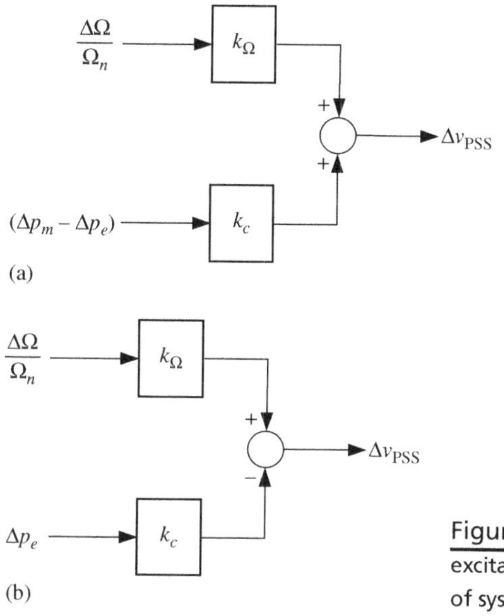

Figure 9.60. (a) General block diagram of PSSs on excitation control of units; (b) its simplification in case of systems with dominantly thermal generating units.

Owing to equation (9.327), we will have, around the frequency ω_o to be damped:

$$k_\Omega + jk_c T_a \omega = 2\frac{h_2}{h}\omega_o T_a \Delta\zeta \frac{1}{g_v(j\omega)} \tag{9.331}$$

Therefore, for $\omega = \omega_o$, the additional feedback gains are

$$\begin{cases} k_\Omega = \dfrac{2h_2}{h}\omega_o T_a \operatorname{Re}\left\{\dfrac{1}{g_v(j\omega_o)}\right\}\Delta\zeta \\ k_c = \dfrac{2h_2}{h}\operatorname{Im}\left\{\dfrac{1}{g_v(j\omega_o)}\right\}\Delta\zeta \end{cases} \tag{9.332}$$

that is,

$$\begin{cases} k_\Omega = \dfrac{2h_2}{h}\omega_o T_a \Delta\zeta \\ k_c = \dfrac{2h_2}{h}\cdot\dfrac{\omega_o}{\omega_v}\Delta\zeta = \dfrac{k_\Omega}{T_a\omega_v} \end{cases} \tag{9.333}$$

which enables the synthesis of the needed transfer function.

It should be pointed out that, for a system with dominant thermal generation (see also Section "Conventional Thermal Units" in 9.4.3.4),

- for local oscillations, that is, k_c important,

$$\zeta_{PF} \simeq 0 \Rightarrow \Delta p_m = 0 \tag{9.334}$$

and, consequently, the block diagram of Figure 9.60a becomes that of Figure 9.60b;

- for interarea oscillations, the gain k_c becomes less and less important (in particular, for very slow oscillations, ζ_{PF} is so positive and large as to completely compensate the negative contribution ζ_{PV} and therefore there is no need for PSSs).

In any case, the use of a suitable phase-shifting filter with the only speed signal as input can result in an output signal proportional both to the speed and to the acceleration, thereby facing any situation.

To have an idea of the numerical values that the gains k_Ω and k_c may take on, further let us refer to the example of equations (9.188) and (9.189), assuming $\Delta\zeta = 0.1$; we get the following typical values:

$$k_\Omega \simeq 10\,\text{p.u./p.u.}; \quad k_c \simeq 0.35\,\text{p.u./p.u.} \tag{9.335}$$

In conclusion, introducing additional feedbacks into excitation control can significantly improve the damping of electromechanical oscillations, that is, ensure an adequate oscillatory mode stability margin. This also allows the full utilization of the capability chart during the operation of the generator (see Figure 9.36).

9.5.3.2 Synthesis of PSSs on Excitation Control: General Case.

Choice of Generators on Which PSSs are to be Introduced. The most common means adopted to improve the damping of electromechanical oscillations is the use of local PSSs on the excitation control of units. In some instances, the limitations on the PSSs operation due to the excitation control may suggest their use in FACTS controllers (typically in SVCs and TCPARs) or in HVDC-link controls. In all cases, equation (9.316), is the basic relationship to be used.

In this section, we describe how to select the generators most suitable to damp electromechanical oscillations and how to determine the relevant best values of the gains k_Ω and k_c of the transfer functions for such generators, that is, of the local additional feedbacks.

Assuming that *additional feedbacks* are of *local type* (Figure 9.61) and namely of the type (9.330), that is, deriving from the speed of a unit and adding to the set point of the

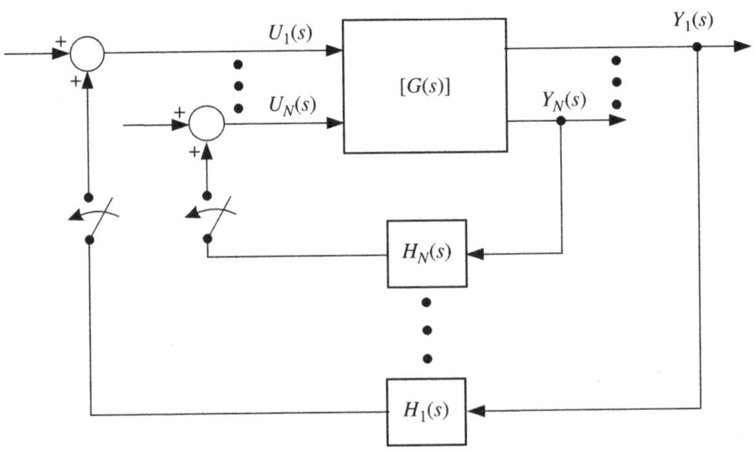

Figure 9.61. Local additional feedbacks.

voltage regulator of the same unit, then equations of the type (9.316) will apply. Therefore, for the kth machine ($k = 1, \ldots, N$)

$$H_k(j\omega_{oh}) = -\frac{\omega_{oh}}{\underline{C}_{kh}}\Delta\zeta_h \tag{9.336}$$

It is possible to compute the *damping sensitivity* to the gain $|H_k(j\omega_{oh})|$ of local additional feedbacks:

$$\mu_{kh} \triangleq \frac{\zeta_h}{|H_k(j\omega_{oh})|} = \frac{|\underline{C}_{kh}|}{\omega_{oh}} \tag{9.337}$$

The most effective generators are characterized by the highest values of μ_{kh}, that is, of the residues $|\underline{C}_{kh}|$.

To determine these residues, reference may be made to the dynamic matrix, adopting a sufficiently detailed model of the power system. Thus, if $[\underline{\Lambda}]$ and $[\underline{\Gamma}]$ are the matrices of the eigenvalues and eigenvectors of the dynamic matrix $[A]$, respectively, using the notation of (9.6), we get

$$[\Delta x(s)] = [\underline{\Gamma}]\,[sI - \underline{\Lambda}]^{-1}[\underline{\Gamma}]^{-1}\,[B]\,[\Delta v_{\text{ref}}(s)] \tag{9.338}$$

where the inputs of interest are only the set points of the voltage regulators.

As a consequence, the matrix $[B]$ (see equation (9.16)) is

$$[B] = \left\{[0]^T\,[0]^T\,[B_{ev}]^T\,[B_{x_v v}]^T\,[0]^T\right\}^T \tag{9.339}$$

where $[B_{ev}]$, $[B_{x_v v}]$ are square matrices of order equal to the number N of generators.

In particular, the transfer function between the input $\Delta v_{\text{ref},k}$ and the output $\Delta\Omega_k/\Omega_n$ is

$$\begin{aligned}\frac{\Delta\Omega_k/\Omega_n}{\Delta v_{\text{ref},k}}(s) = {} & \sum_h \frac{\underline{\Gamma}_{kh}^{(\Omega)}\underline{Q}_{hk}}{s - \underline{\lambda}_h} \\ & + \sum \begin{matrix}\text{Heaviside's partial-fraction expansion} \\ \text{relevant to nonelectromechanical complex poles}\end{matrix} \\ & + \sum \begin{matrix}\text{Heaviside's partial-fraction expansion} \\ \text{relevant to real poles}\end{matrix}\end{aligned} \tag{9.340}$$

where in the first summation the Heaviside partial-fraction expansion for electromechanical modes is highlighted; $\underline{\Gamma}_{kh}^{(\Omega)}$ identifies the (generally complex) kth element of the eigenvector $[\underline{\Gamma}_h]$, associated with the eigenvalue $\underline{\lambda}_h$ and pertaining to the state variable $\Delta\Omega_k/\Omega_n$; $\underline{Q}_{hk}$ is the (generally complex) element of the matrix.

$$[\underline{Q}] = [\underline{\Gamma}]^{-1}\,[B] \tag{9.341}$$

pertaining to the eigenvalue $\underline{\lambda}_h$ and to the input $\Delta v_{\text{ref},k}$. As a result, the residue is

$$\underline{C}_{kh} = \underline{\Gamma}_{kh}^{(\Omega)}\,\underline{Q}_{hk} \tag{9.342}$$

It is interesting to note that the residues $\underline{C}_{kh}$ may also be experimentally deduced.

Once all the residues are determined for all the system units, it is possible to select the power plants most suitable for the introduction of additional feedbacks.

At this point, assuming a given damping increase $\Delta\zeta_h$, equations (9.316) and (9.330) allow the determination of the values of the feedback gains relevant to the ith unit, $k_{\Omega i}$ and k_{ci}

$$\begin{aligned} H_i(j\omega_{oh}) &= \frac{\omega_{oh}}{|\underline{C}_{ih}|}\Delta\zeta_h(\cos(\pi - \angle\underline{C}_{ih}) + j\sin(\pi - \angle\underline{C}_{ih})) \\ &= k_{\Omega i} + j\omega_{oh}k_{ci}T_{ai} \end{aligned} \tag{9.343}$$

that is,

$$\begin{cases} k_{\Omega i} = \dfrac{\omega_{oh}}{|\underline{C}_{ih}|}\Delta\zeta_h\cos(\pi - \angle\underline{C}_{ih}) \\ k_{ci} = \dfrac{1}{T_{ai}|\underline{C}_{ih}|}\Delta\zeta_h\sin(\pi - \angle\underline{C}_{ih}) \end{cases} \tag{9.344}$$

from which also

$$\frac{k_{ci}}{k_{\Omega i}} = \frac{1}{\omega_{oh}T_{ai}}\tan(\pi - \angle\underline{C}_{ih}) \tag{9.345}$$

independent of the damping and only dependent on the system behavior.

Hence, the most effective power plants for improving the damping of a given electromechanical oscillation and the relevant synthesis of the additional feedback transfer function require the knowledge of the poles and residues of the appropriate transfer functions; this can be very burdensome for large interconnected power systems. However, it is worth noting that the most effective power plants practically depend on the structure of the network and on the operating point and not on voltage and frequency controls. This allows the simplification of the approach, also considering that, as already mentioned, the typical damping of electromechanical oscillations is very small; therefore, reference can be made to the second-order model for generators and to linear loads.

Simplified Synthesis. In analogy with the approach taken for the base case (bearing in mind, in particular, Figures 9.55 and 9.56), let us call $k_{\Omega k}$ the gain between the output $\Delta\Omega_k/\Omega_n$ and the input Δe_k, corresponding, in Figure 9.2, to the closure of the feedback between $[\Delta\Omega]$ and $[\Delta E]$. With a procedure similar to the one described above, we may derive the transfer functions between $\Delta\Omega_k/\Omega_n$ and ΔE_k, with a view to assessing the damping sensitivity. From equation (9.59) we get, neglecting the contribution due to the pole in the origin

$$\begin{aligned} \frac{\Delta\Omega_k/\Omega_n}{\Delta E_k}(s) &= \frac{1}{\Omega_n}\sum_h G_{kh}W_{hk}\frac{s}{s^2 + \omega_{oh}^2} \\ &= \frac{1}{2\Omega_n}\sum_h G_{kh}W_{hk}\left(\frac{1}{s - j\omega_{oh}} + \frac{1}{s + j\omega_{oh}}\right) \end{aligned} \tag{9.346}$$

implying that

$$\underline{C}_{kh} = \underline{C}_{kh}^* = \frac{1}{2\Omega_n}G_{kh}W_{hk} \tag{9.347}$$

and then, based on equation (9.336)

$$\hat{k}_{\Omega k} = -\frac{2\omega_{oh}\Omega_n}{G_{kh}W_{hk}}\Delta\zeta_h \tag{9.348}$$

while, based on equation (9.337), we get the *damping sensitivity in simplified form*:

$$\gamma_{kh} = \frac{|G_{kh}W_{hk}|}{2\Omega_n\omega_{oh}} \tag{9.349}$$

The above equations allow the determination of the most effective power plants for damping the hth mode of electromechanical oscillation, regardless of the model of voltage regulators and only taking into account the structure of the network (in particular, the location of power plants and the operating conditions). Therefore, the simplified analysis may help to determine the damping sensitivity with a more realistic model of the system (see (9.337)), that is, taking into account the actual structure of generators (e.g., via the fifth-order model) and the presence of voltage and speed regulators: the model can be made more complete only for the power plants whose simplified analysis has highlighted large sensitivities (9.349), that is, large amplitudes of the electromechanical oscillation to be damped.

ACTUAL STRUCTURE OF PSSS. The structure of PSSs depicted in Figure 9.60 is ideal. In practice, it should be taken into account that

- use can be made of the frequency Δf at the generator terminals, instead of the electrical angular speed $\Delta\Omega$;
- a *washout filter* is to be introduced, in order to eliminate the continuous components of the input signals.

Moreover, it should be observed that

- in many systems, Δv_{PSS} is obtained as a combination of signals derived both from speed (or frequency) and delivered power, like in Figure 9.60b;
- in other cases, the AF is derived from power only or from speed (or frequency) only: in such cases, we can obtain a complete effect, equivalent to k_Ω and k_c, that is, an effect from both speed and acceleration, by introducing a suitable *phase-shifting* filter, as already mentioned.

Therefore, apart from the necessary transducers, the general block diagram of PSSs on excitation control is that of Figure 9.62.[2]

In Figure 9.62

$$F_w(s) = \frac{sT_w}{1+sT_w} \tag{9.350}$$

[2] In Figure 9.62, one or two inputs will be actually present (in particular, either the frequency or the rotor speed), depending on the particular situation.

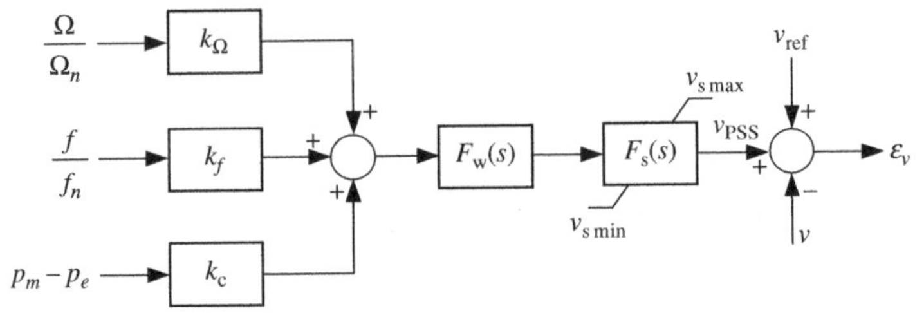

Figure 9.62. General block diagram of PSSs on excitation control.

is the transfer function of the washout filter, while

$$F_s(s) = k_s \frac{1 + sT_2}{1 + sT_1} \tag{9.351}$$

is the transfer function of the phase-shifting filter. Besides, the additional voltage v_{PSS} must be limited in order to avoid large variations in voltage loops after large perturbations (large variations of Ω, f, and acceleration), or to avoid angle stability problems resulting from large perturbations. Hence

$$v_{\mathrm{s\,min}} \leq v_{\mathrm{PSS}} \leq v_{\mathrm{s\,max}} \tag{9.352}$$

with

$$v_{\mathrm{s\,max}} = -v_{\mathrm{s\,min}} \simeq 0.06 - 0.08\,\mathrm{p.u.} \tag{9.353}$$

As to the phase-shifting filter

(a) if the additional signal is derived from acceleration, a phase-lag (pole-zero) transfer function can be introduced to also obtain the acceleration integral, that is speed, within a proper frequency range;

(b) if the additional signal is derived from speed, a phase-lead (zero-pole) transfer function can be considered to also have the speed derivative, within a proper frequency range.

Online Updating. With the above-illustrated approach, the values of the gains k_Ω and k_c can be determined under given conditions of structure and operation of the power system, so as to have a minimum acceptable damping for them, for instance, $\zeta_{\min} = 0.10$.

Sometimes, during the operation of the power system, the damping ζ tends to be smaller than $\zeta_{\min}$. In this case, the gains should be updated.

This may be done online, by using the measured values of p_e, q, v, and i at generator terminals, both to determine the initial values of the state variables (typically δ^o and e^o) and to estimate the external reactances x_e seen from the various generators involved, as if each of them were connected to an infinite system.

Once all the necessary data are collected, through real-time simulations of the power plant (and of its controls) connected to an infinite system, we may recompute online the residues $\underline{C}_{kh}$ and update the values of the PSS gains.

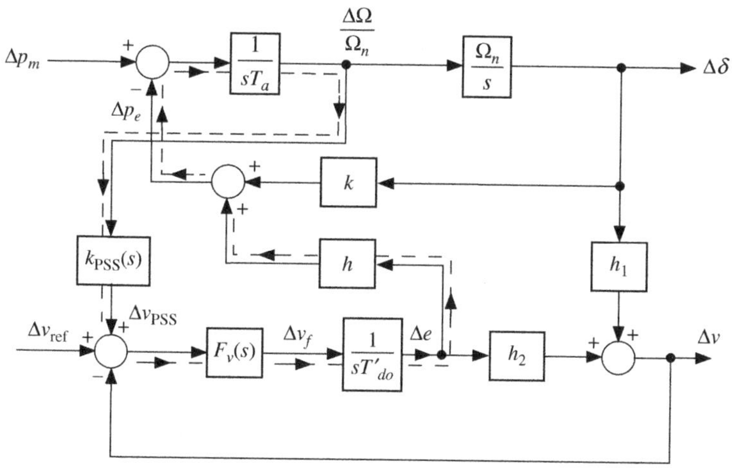

Figure 9.63. Effect of PSSs on voltage loop.

9.5.4 Limitation on PSS Gains

The introduction of PSSs changes the voltage loop, in that it closes a loop between Δe and Δv_{ref}, as shown by the dashed line of Figure 9.63.

On the usual assumption

$$F_v(s) = \mu_t \tag{9.354}$$

we immediately see that the voltage loop block diagram changes as depicted in Figure 9.64. The transfer function of the dashed-line block is

$$\frac{1}{sT'_{do}/\mu_t + (h/sT_a)k_{PSS}(s)} \tag{9.355}$$

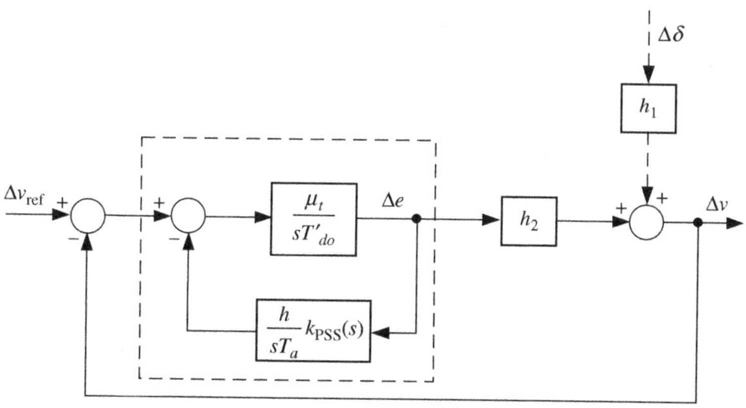

Figure 9.64. PSSs seen from the voltage loop.

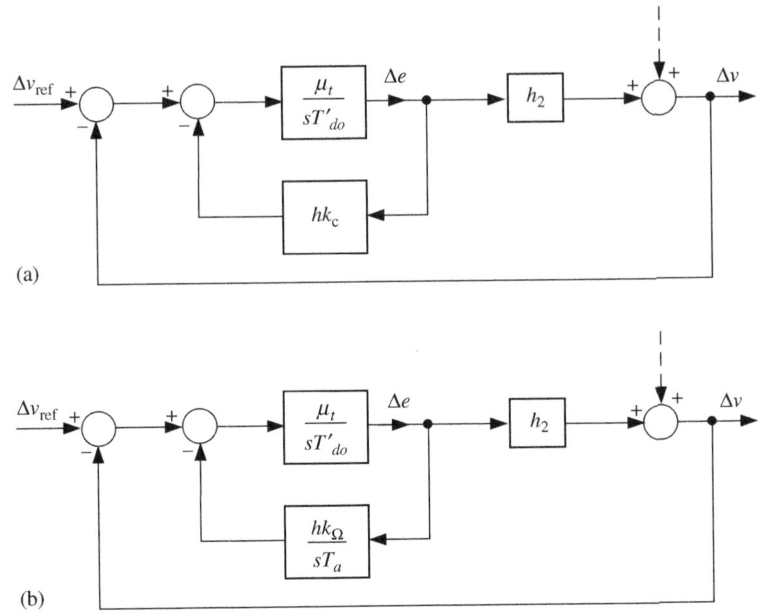

Figure 9.65. Effects of PSSs on voltage loop: (a) effect of k_c; (b) effect of k_Ω.

where

$$\frac{h}{sT_a} k_{PSS}(s) = \frac{h}{sT_a}(k_\Omega + sT_a k_c) = \frac{h k_\Omega}{sT_a} + hk_c \tag{9.356}$$

which enables to separately analyze the effects of k_c and k_Ω.

In the presence of k_c alone (see Figure 9.65a), the feed-forward transfer function is

$$h_2 \frac{1}{hk_c + (sT'_{do}/\mu_t)} = \frac{\mu_t h_2}{sT'_{do} + \mu_t hk_c} \tag{9.357}$$

and thus the characteristic equation of the voltage loop is

$$sT'_{do} + \mu_t(h_2 + hk_c) = 0 \tag{9.358}$$

The (real) pole is equal to

$$\underline{\lambda} = -\frac{\mu_t(h_2 + hk_c)}{T'_{do}} = -\omega'_v \tag{9.359}$$

Now, as

$$hk_c > 0 \tag{9.360}$$

(the stabilizing signal is subjected to the sign of h) and h_2 is always positive, k_c shifts the pole leftward, improving the voltage loop stability (Figure 9.66a).

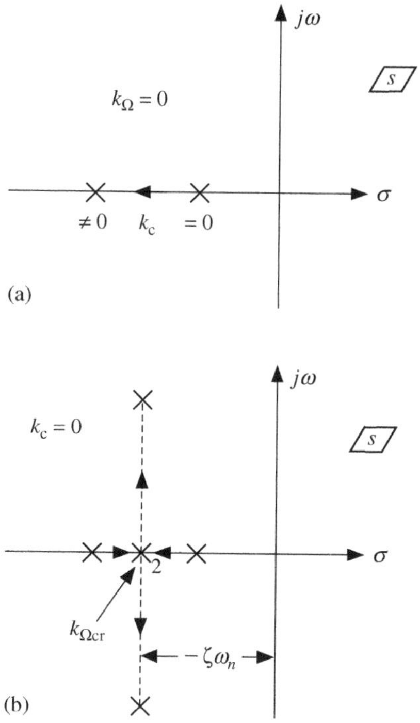

Figure 9.66. Pole shift of voltage control loop: (a) effect of k_c; (b) effect of k_Ω.

In the presence of k_Ω alone (Figure 9.65b), the feed-forward transfer function is

$$h_2 \frac{1}{sT'_{do}/\mu_t + (hk_\Omega/sT_a)} = \frac{sh_2T_a\mu_t}{s^2T_aT'_{do} + \mu_t hk_\Omega} \tag{9.361}$$

Hence, the characteristic equation of the loop is

$$s^2 + s\frac{h_2\mu_t}{T'_{do}} + \frac{\mu_t hk_\Omega}{T_aT'_{do}} = 0 \tag{9.362}$$

or in normal form

$$s^2 + 2\zeta\omega_n s + \omega_n^2 = 0 \tag{9.363}$$

Therefore, k_Ω has the effect of creating another pole. As

$$\begin{cases} \omega_n = \sqrt{\dfrac{\mu_t hk_\Omega}{T_aT'_{do}}} \\ \zeta = \dfrac{1}{2}h_2\sqrt{\dfrac{T_a\mu_t}{hk_\Omega T'_{do}}} \end{cases} \tag{9.364}$$

with

$$hk_\Omega > 0 \tag{9.365}$$

(the stabilizing signal is subjected to the sign of h) and h_2 is always positive, the damping (always remaining positive) decreases when k_Ω increases. Moreover, with the simplification

$$h_2 = \frac{x_e}{x_i + x_e} \cos(\delta - \theta)^o = \text{const.} \tag{9.366}$$

the amount

$$\zeta \omega_n = \frac{1}{2} \cdot \frac{h_2 \mu_t}{T'_{do}} \tag{9.367}$$

(independent of k_Ω) is constant and the root-locus upon the variation of k_Ω is as depicted in Figure 9.66b: when k_Ω increases, the poles are first real and distinct, then real and coincident when $\zeta = 1$, that is (from (9.364))

$$k_{\Omega\,\text{cr}} \triangleq (k_\Omega)_{\zeta=1} = \frac{1}{4} h_2^2 \frac{T_a \mu_t}{h T'_{do}} = \frac{1}{4} \cdot \frac{h_2}{h} T_a \omega_v \tag{9.368}$$

Assuming the typical values

$$\begin{cases} \omega_v \simeq 5\,\text{rad/s} \\ T_a \simeq 7.5\,\text{s} \end{cases} \tag{9.369}$$

$$h = \left(\frac{p_e}{e}\right)^o \simeq 0.8 \tag{9.370}$$

$$h_2 = 0.75 - 1 \tag{9.371}$$

(lower value for local oscillations, upper value for interarea ones), we get

$$k_{\Omega\,\text{cr}} \simeq 10 - 12\,\text{p.u./p.u.} \tag{9.372}$$

If

$$k_\Omega > k_{\Omega\,\text{cr}} \tag{9.373}$$

the poles become complex conjugate and an additional increase in k_Ω decreases the damping of the loop.

In conclusion, on the contrary of k_c, too high values of k_Ω may make the loop oscillatory, with a certainly positive but not too high damping: the stabilizing effect of k_Ω on electromechanical oscillations destabilizes the voltage loop.

This may be *undesired, because voltages at generator terminals, and thus also on the network, have an oscillatory behavior.*

Therefore, the value of k_Ω should be kept within an acceptable range in order to achieve a satisfactory response of the primary voltage control loop.

9.6 TYPICAL CASES OF INTERAREA OR LOW-FREQUENCY ELECTROMECHANICAL OSCILLATIONS

A *first typical case* (see Figure 9.67) is represented by a highly meshed area, with generators and loads, that is connected to a system with a nominal power much larger than the power of the area, so that it can be considered as an infinite system. The connection is

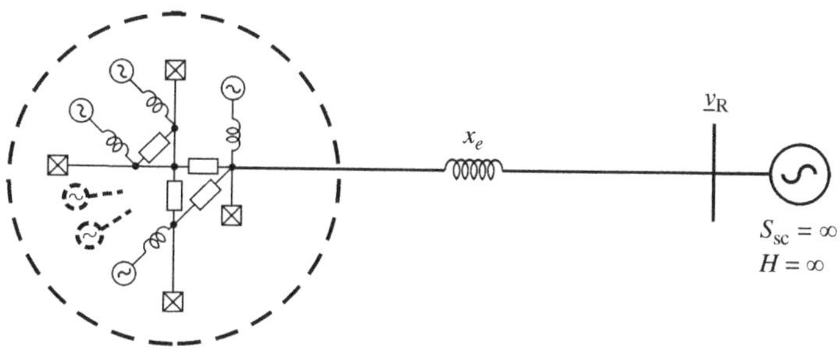

Figure 9.67. Highly meshed area connected to the rest of the system via a long line.

made up of an equivalent long line (weak connection, if x_e is large, T_o is large too). It is intuitive that, in such a system, a slow electromechanical oscillation mode arises between the various generators of the area—oscillating together and practically with the same amplitude—and the infinite system.

The *second typical case* is represented by systems with a longitudinal structure (connected to an infinite system) or systems extending mainly in a given geographic direction and having generators and loads distributed along this path in a more or less uniform way (see Figure 9.68).

In this case, there exists a slow oscillation mode (the slowest one), whose amplitude varies with distance from the infinite system: maximum at the free end, decreasing in the rest of the system and obviously null at the infinite bus. At least three examples of structures of this type can be made.

(i) The *first example* is relevant to the *Central-Southern Italian Power System* in the past (in the late 1960s, prior to the development of the 400 kV transmission

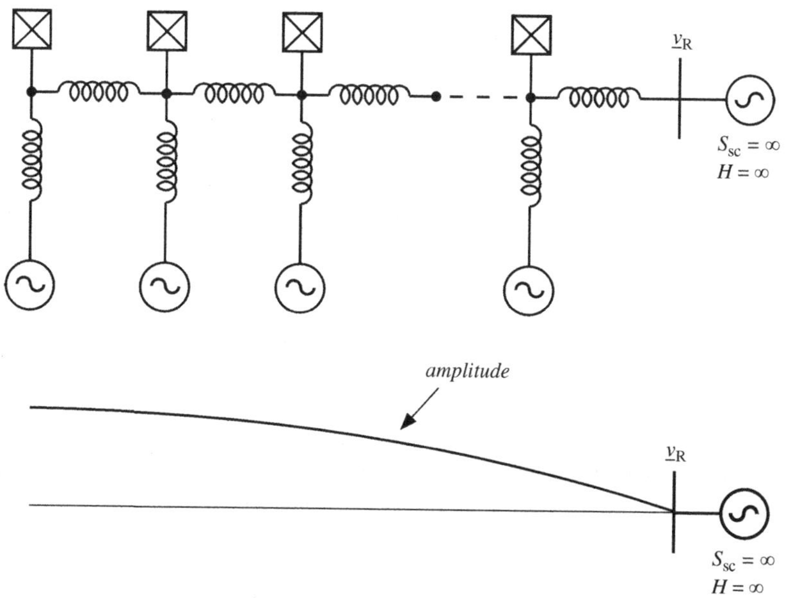

Figure 9.68. Systems with longitudinal structure.

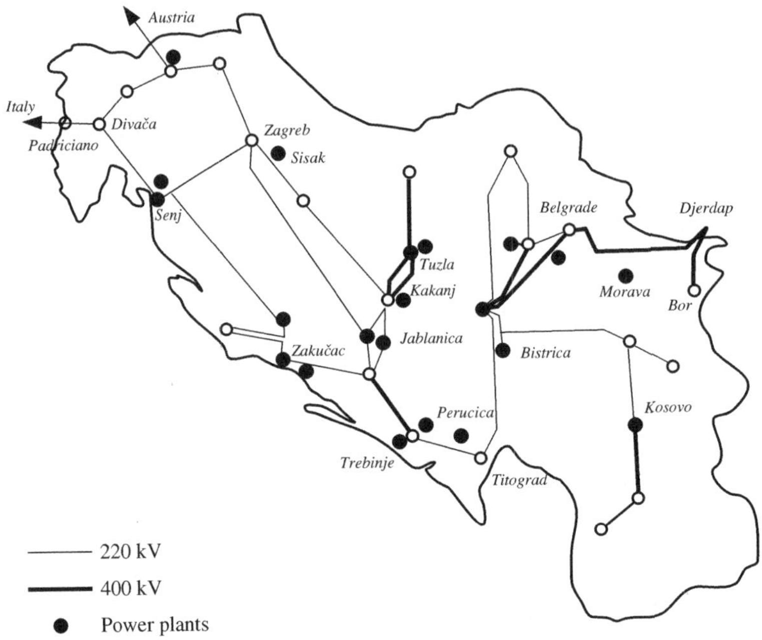

Figure 9.69. Structure of the power system of Yugoslavia in the years 1972–1973.

system), for which the North Italy and the European interconnected network represented a *de facto* infinite system. The slowest electromechanical oscillation, with a period $T_o \cong 4$ s, was characterized by a $\zeta_{PF} \cong 0$ (prevalently thermal units) and a strongly negative ζ_{PV}. The damping of such oscillation required the adoption of k_Ω having high values in Sicily (large amplitudes) and gradually decreasing values along the peninsula and toward North.

At present, the infinite system lies close to South Italy ($T_o \cong 2$–2.5 s) and so the situation is as in the *first typical case*.

(ii) The *second example* is represented by the former *Yugoslav Power System* (Figure 9.69) in its interconnected operation with the rest of the European network (years 1972–1973): the situation was similar to the previous one but with $T_o \cong 6$ s and maximum amplitudes toward the border with Romania.

In this case, it was necessary to increase the value of k_Ω in the hydro power plant of *Djerdap* (border with Romania), as such value had been set up for damping a slower oscillation in isolated operation ($\cong 2.9\,s$).

(iii) A *third example* is the interconnection of the *Greek Power System* with the European one via the former Yugoslav network (years 1974–1975). In this instance, the slowest mode had a period of $T_o \cong 8$–10 s with maximum amplitudes at the free end, that is, in Greece. In the Greek network, having a dominantly thermal generation, ζ_{PF} proved to be positive and sufficiently large with respect to a negative ζ_{PV}: no need for PSSs.

A *third typical case* is represented by small-sized systems (Figure 9.70) having a slow oscillation mode with variable amplitude: some generators oscillate with positive amplitude and other generators are in phase opposition to them (negative amplitude). An

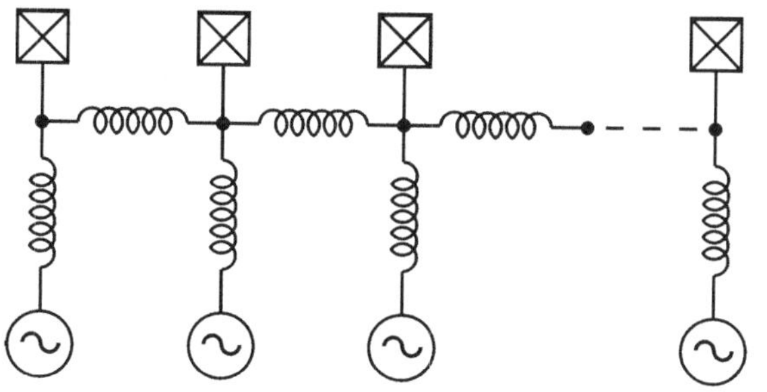

Figure 9.70. System having a slow oscillation mode and where some generators oscillate with positive amplitude and the remaining generators are in phase opposition to the first.

example was the *Central Peru Power System* (years 1978–1979) in which the slowest electromechanical oscillation was characterized by $T_o \cong 1$ s and negative damping. PSSs with k_c and k_Ω were installed in the *Mantaro* hydro power plant (the biggest one, on the Andes region) in order to damp this oscillation.

The *fourth typical case* is represented by systems with large geographic extension, having a cascade configuration with a structure of longitudinal type at both ends (Figure 9.71).

An example is the interconnection between *CENTREL* and the European system (UCTE) in 1997:

- Area A defines the western end of the European system (Portugal and Spain).
- Area B consists of the CENTREL Power System (Poland, Hungary, Czech Republic, Slovakia).
- The central area C identifies the rest of the UCTE system (highly meshed, of high nominal power and including France, Germany, Switzerland, Italy, etc.).

This system, which can be studied in a simplified way as presented in Section 9.3.3.2, had two slow oscillation modes (with periods of about 4 and 3 s) of area A with respect to C and of area B with respect to C (as if C were an infinite system), respectively. These oscillation modes, with a positive but very small damping, were damped by adopting appropriate values of k_Ω at the free end of area B (Poland) and by introducing braking resistors in area A.

The *fifth typical case* is represented by large interconnected systems characterized by a very wide geographic extension and a filiform structure (weak connection) along a long electrical backbone; an example is the planned *MEDRING* (Figure 9.72) around the Mediterranean Sea (ring structure, weak backbone along the southern-eastern coast). In North Africa and Middle East, most of the power plants are thermal (both conventional, with gas-turbine units and combined cycles): it follows that, thanks to the high stabilizing

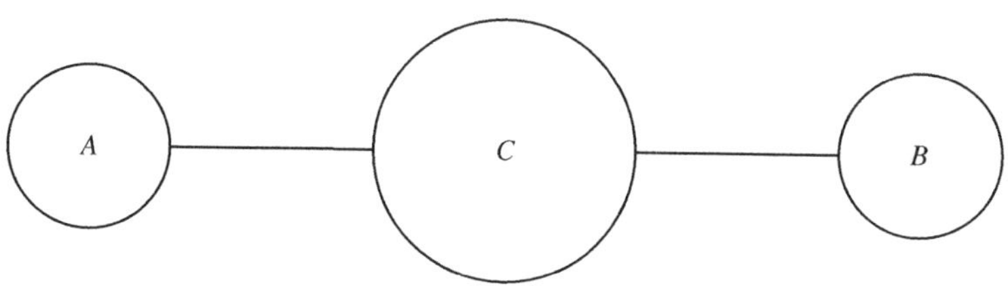

Figure 9.71. Cascade systems.

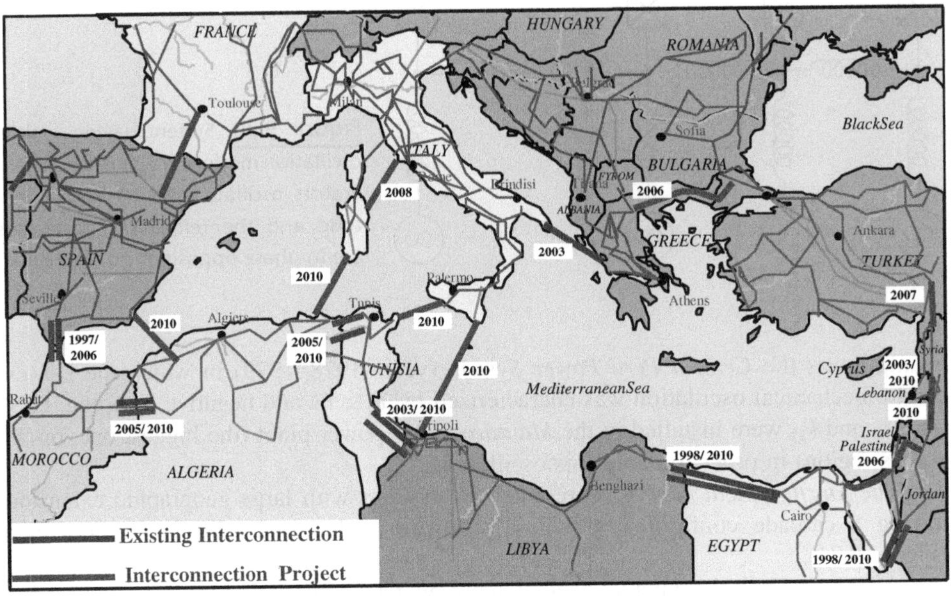

Figure 9.72. The MEDRING Power System.

contribution that is owed to the primary frequency control loop of such generating units, the slowest electromechanical oscillation (period of about 10 s with the ring closed and about 20 s with the ring open side Turkey) is well damped even in the presence of very fast modern voltage regulators. As a result, the adoption of PSSs to damp such oscillation does not appear to be necessary.

By contrast, owing to the poor meshing of the system together with not too high transmission voltage levels and thus, ultimately, low short-circuit powers (see also Section 9.1), problems of steady-state angle instability of aperiodic type may arise.

In conclusion, as practically experienced, it is worth mentioning that the problem of the damping of electromechanical oscillations in case of interconnection between two initially separate systems often lies in adapting the already installed PSSs, whose values are set up for damping oscillations in isolated operation. Interconnection makes the system *different*, changing its modal characteristics completely. Consequently, the most efficient power plants and the PSS gains are to be identified with reference to the new characteristics.

REFERENCES

[1] Crary S.B. *Power system stability*, Vol. 2, Wiley, 1947.

[2] Concordia C. *Synchronous machines*, Wiley, 1951.

[3] Kimbark E.W. *Power system stability*, Vol. 3, Wiley, 1956.

[4] Adkins B. *The general theory of electrical machines*, Chapman-Hall, 1962.

[5] Venikov V.A. *Transient phenomena in electrical power systems*, Pergamon Press, 1965.

[6] Weedy B.M. *Electric power systems*, 1st edition, Wiley, 1967.

[7] Marin R., Valtorta M. *Trasmissione ed Interconnessione*, CEDAM, 1971.

[8] Porter B., Crossley R. *Modal control: Theory and applications*, Taylor-Francis, 1972.
[9] Arcidiacono V., Marconato R. *Power system dynamics*, Enel-Consulting Services, 1978.
[10] Weedy B.M. *Electric power systems*, 3rd edition, Wiley, 1979.
[11] Marconato R. *Sistemi Elettrici di Potenza*, Vol. 2, CLUP, 1984–1985.
[12] Ferrari E. *Regolazione della Frequenza e Controllo dell'Eccitazione dei Gruppi Generatori*, CLUP, 1986.
[13] Marconato R. *Training course on power system dynamics and control*, 1st edition, Enel-Consulting Services, 1988.
[14] Marconato R. *Electric power systems*, Vol. 3, 2nd edition, CEI—Italian Electrotechnical Committee, 2002–2008.
[15] Arnoldi W.E. The principle of minimised iterations in the solution of the matrix eigenvalue problem, *Quart. Appl. Math.*, Vol. 9, pp. 17–29, 1951.
[16] Heffron W.G., Phillips R.A. Effects of modern amplidyne voltage regulator in underexcited operation of large turbine generators, *AIEE Trans.*, Vol. PAS-71, pp. 692–697, Aug. 1952.
[17] Francis J.G.F. The QR transformation—A unitary analogue to the LR transformation, Parts 1 and 2, *Comput. J.*, Vol. 4, pp. 265–271, 1961–1962.
[18] Hanson O.W., Goodwin C.J., Dandeno P.L. Identification of excitation and speed control parameters in stabilising inter-system oscillation, *IEEE Trans.*, Vol. PAS-87, pp. 1306–1316, May 1968.
[19] De Mello F.P., Concordia C. Concepts of synchronous machine stability as affected by excitation control, *IEEE Trans.*, Vol. PAS-88, pp. 316–329, Apr. 1969.
[20] Ferrari E., Floris R., Saccomanno F. *Limiti di Stabilità di Turboalternatori con Eccitazione Statica e Diverse Strutture del Controllo dell'Eccitazione. Parte I: Analisi della Stabilità per Piccole Variazioni*, Riunione Annuale AEI 1969, Paper No. 4.1.19.
[21] Saccomanno F. *Sensitivity analysis of the characteristic roots of a linear time-invariant dynamic system: Application to the synthesis of damping action in electric power systems*, IFAC, 1975.
[22] Arcidiacono V., Ferrari E., Marconato R., Brkic T., Niksic M., Kajari M. *Studies and experimental results on electromechanical oscillation damping in Yugoslav Power System*, IEEE PES Summer Meeting, San Francisco (USA), Jul. 1975, Paper No. A75 420-0.
[23] Kundur P., Dandeno P.L. *Practical application of eigenvalue techniques in the analysis of power system dynamic stability problems*, 5th edition PSCC, Cambridge, 1975.
[24] Arcidiacono V., Ferrari E., Marconato R., Saccomanno F. *Analysis of factors affecting the damping of low-frequency oscillation in multimachine systems*, CIGRE, 1976, Paper No. 32-19.
[25] Arcidiacono V., Brkic T., Epitropakis E., Ferrari E., Marconato R., Saccomanno F. *Results of some recent measurements of low-frequency oscillations in a European power system with longitudinal structure*, CIGRE, Session 1976. Paris, Paper No. 32-19A.

10

TRANSIENT STABILITY

Nikolai Voropai and Constantin Bulac

10.1 GENERAL ASPECTS

The transient stability is defined as the ability of a power system to maintain the generators in synchronism and reach acceptable steady-state operating conditions when subjected to large disturbances such as short circuits, loss of large generating units, loss of critical network branches, or large load variations.

The behaviors of a power system are well known described by a hybrid DAE model consisting of algebraic equations (A), modeling the electrical network, and nonlinear differential equations (D), modeling the generators and/or loads behavior:

$$\begin{cases} \dot{\mathbf{x}} = \mathbf{f}(\mathbf{x}, \mathbf{y}, \boldsymbol{\mu}) = D(\mathbf{x}, \mathbf{y}, \boldsymbol{\mu}) \\ 0 = g(\mathbf{x}, \mathbf{y}, \boldsymbol{\mu}) = A(\mathbf{x}, \mathbf{y}, \boldsymbol{\mu}) \end{cases} \tag{10.1}$$

where $\mathbf{x}$, $\mathbf{y}$ are state variables and $\boldsymbol{\mu}$ is the vector of parameters.

The solution of the nonlinear differential equations is obtained through step-by-step integration methods. In this regard, the transient state is decomposed into a series of consecutive instantaneous states separated by small enough time intervals so that the variation in any variable during an integration step is evaluated through its derivative. Therefore, in the DAE system of equations, the electrical variables are divided into two categories:

Handbook of Electrical Power System Dynamics: Modeling, Stability, and Control. Edited by Mircea Eremia and Mohammad Shahidehpour.

- *Inertial or slow variation variables*—**x**, the vector of inertial variables—are variables related to the rotor fluxes influencing the transient process. These variables remain constant in the first moments after the disturbance (the law of constant flux linkages). In the computation process, the values of the inertial variables at the beginning of an integration step are taken equal to their values at the end of the previous step, while their variations during the integration step are evaluated by the system of differential equations. In the transient stability studies, the inertial variables of interest are the inductive fluxes, the emfs proportional to these fluxes, the rotor angle δ, the torque, and so on.
- *Noninertial or fast variation variables*—**y**, the vector of noninertial variables—are variables related to the electrical network, which suddenly changes, in jumps, upon the occurrence of a disturbance; for instance, when a three phase-to-earth fault occurs on a bus-bar, the bus voltage is $U = U_0$ until $t = 0$, and becomes $U = 0$ for $t > 0$. In the computation process, these variables remain constant during an integration step but varies from one step to another. The noninertial variables are the currents, the voltages, and the powers generated.

For the noninertial variables, the existence conditions for the steady-state solutions are studied, while for the inertial variables, the conditions for stability of steady-state solutions are investigated. The symmetrical, balanced, and sinusoidal steady state is the solution (x_0, y_0) of the algebraic equations (*A*), which is verified through the nonlinear differential equations. The stability of the algebraic equations solution is given by the stability of the inertial variables (*D*); the steady state is stable if the dynamic elements of the power system are in stable equilibrium. The transient states of interest in the planning and operation of power systems are the transition states between two equilibrium states of the dynamic elements. We may say that "the transition is stable" if the transient state ends in a stable steady state.

Thus, in the transient stability analysis, three distinct phases are considered: the predisturbance or the initial steady state, the disturbance state, and postdisturbance or the final steady state.

During the disturbance, the transition of the power system from one equilibrium state to another may be a normal operation process or may be caused by a random event due to external factors (e.g., a short circuit on a transmission network, followed by the unsimultaneous line disconnection at the two ends and, eventually, automatic or manual reclosing).

One of the most used tools for the transient stability assessment is *the critical fault clearing time* defined as the maximum time period elapsed from the instant of fault occurrence to the instant of fault clearing and isolation of the affected section, necessary for the power system to maintain its ability to reach a postdisturbance steady state (after the fault clearance). Note that, the fault clearing time also includes the necessary time for protection system operation and breakers switching.

In practice, the most used approach for the transient stability analysis is the *time-domain* simulation of the inertial and noninertial variables after the occurrence of a large disturbance.

The most relevant information regarding the transient stability/instability is obtained by the examination of *oscillation curves*, which represents the evolution in time of the relative rotor angles. Three characteristic cases are distinguished

(i) *the stable case*, in which the rotor angle, after several damped oscillations, reaches a constant postdisturbance steady-state value;

(ii) *the unstable case*, at first swing, also called *first-swing instability*, in which case the rotor angle increases continuously in time;

(iii) *the case* in which the generator is *stable in the first swing*, but the next oscillations are growing in magnitude and the stability is lost; this form of instability generally occurs when the postdisturbance state does not fulfill small-signal stability conditions.

An important role in ensuring the transient stability is given by the fast, selective, and safe operation of the protection system, which must be capable to distinguish between the faulty operation, the stable oscillations, and instability conditions. Therefore, the protection systems must avoid false actions and undesired disconnection of network elements, which may worsen the stability conditions of the power system.

More simple methods were developed for transient stability assessment among which the most known is the equal area criterion.

10.2 DIRECT METHODS FOR TRANSIENT STABILITY ASSESSMENT

10.2.1 Equal Area Criterion

10.2.1.1 Fundamentals of Equal Area Criterion. The equal area criterion is a graphical–analytical method for fast assessment of the first-swing transient stability, applicable for the following cases:

(i) The synchronous machine is connected through a passive network to an infinite power bus.

(ii) Two synchronous machines of finite power interconnected through a passive network.

(iii) Multimachine system, but possible only by aggregating all generators in two equivalent synchronous machines, case for which the method was called extended equal area criterion.

In order to substantiate the principles of the equal area criterion, let us consider a *synchronous generator connected through a passive network to an infinite power system*, while the generator is modeled by the classical model (a constant emf E' behind the transient reactance). Therefore, the rotor angle δ from the swing equation is replaced with the emf angle $\delta'(\underline{E}' = E' e^{j\delta'})$. Furthermore, if the damper winding is neglected, that is $D = 0$, the swing equation (2.10) becomes

$$\frac{M}{\omega_0} \cdot \frac{\mathrm{d}^2\delta'}{\mathrm{d}t^2} = P_{\mathrm{m}} - P_{\mathrm{e}} = P_{\mathrm{a}} \tag{10.2}$$

where P_{a} is the accelerating power.

Multiplying by $2\dfrac{\omega_0}{M} \cdot \dfrac{\mathrm{d}\delta'}{\mathrm{d}t}$ both sides of equation (10.2) and given that $H = M/2$, yields

$$2\frac{\mathrm{d}\delta'}{\mathrm{d}t} \cdot \frac{\mathrm{d}^2\delta'}{\mathrm{d}t^2} = \frac{\omega_0}{H} P_{\mathrm{a}} \frac{\mathrm{d}\delta'}{\mathrm{d}t} \tag{10.3}$$

Given that

$$\frac{\mathrm{d}^2\delta'}{\mathrm{d}t^2} = \frac{\mathrm{d}}{\mathrm{d}t}\left(\frac{\mathrm{d}\delta'}{\mathrm{d}t}\right) = \left[\frac{\mathrm{d}}{\mathrm{d}\delta'}\left(\frac{\mathrm{d}\delta'}{\mathrm{d}t}\right)\right]\frac{\mathrm{d}\delta'}{\mathrm{d}t} = \frac{\mathrm{d}}{\mathrm{d}\delta'}\left(\frac{1}{2}\left(\frac{\mathrm{d}\delta'}{\mathrm{d}t}\right)^2\right)$$

the left side of (10.3) becomes

$$2 \cdot \frac{d\delta'}{dt} \cdot \frac{d}{d\delta'} \cdot \left[\frac{1}{2} \cdot \left(\frac{d\delta'}{dt}\right)^2\right] = \frac{d\delta'}{dt} \cdot \frac{d}{d\delta'} \cdot \left(\frac{d\delta'}{dt}\right)^2 = \frac{d}{dt} = \left(\frac{d\delta'}{dt}\right)^2$$

and therefore equation (10.3) becomes

$$\frac{d}{dt}\left(\frac{d\delta'}{dt}\right)^2 = \frac{\omega_0}{H} P_a \left(\frac{d\delta'}{dt}\right)$$

Integrating in time gives

$$\left(\frac{d\delta'}{dt}\right)^2 = \int \frac{\omega_0}{H} P_a d\delta'$$

The $\left(\frac{d\delta'}{dt}\right)$ derivative represents the angular velocity of the synchronous machine rotor related to the general reference system, which rotates with the synchronous speed ω_0. Under stable steady-state conditions, this angular velocity is zero but modifies if a disturbance occurs in the system (increases if the disturbance is a short circuit).

The first-swing stability of the generator is fulfilled if the angle δ', measuring the rotor position with respect to the reference axis, which rotates with the synchronous speed, does not continuously increase in time, that is, to reach the maximum swing at δ'_m and thereafter decrease (Figure 10.1). This condition is satisfied if the angular velocity $d\delta'/dt$ returns to zero after a certain time interval elapsed from the time instant of disturbance occurrence, when $\delta' = \delta'_m$. Therefore, the stability criterion becomes

$$\int_{\delta'_0}^{\delta'_m} \frac{\omega_0}{H} P_a d\delta' = \int_{\delta'_0}^{\delta'_m} \frac{\omega_0}{H} (P_m - P_e) d\delta' = 0 \quad (10.4)$$

where δ'_0 is the initial value of δ'.

The graphical representation from Figure 10.1 shows that $\int_{\delta'_0}^{\delta'_m} (\omega_0/H) P_a d\delta'$ is given by the algebraic sum of the shaded surfaces A_1 and A_2. The A_1 surface represents the

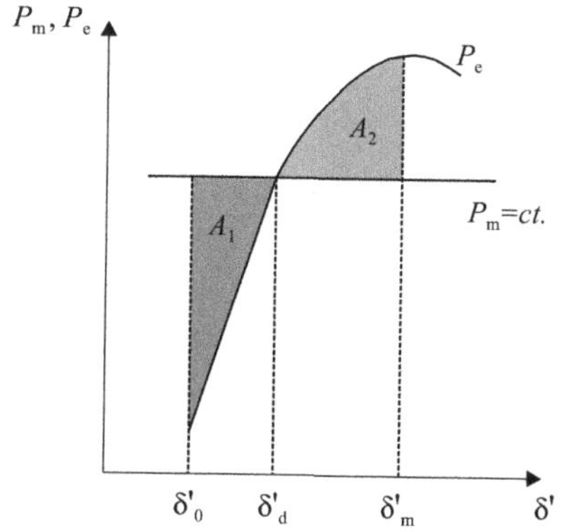

Figure 10.1. Basic concepts of the equal area criterion ($A_1 = A_2$).

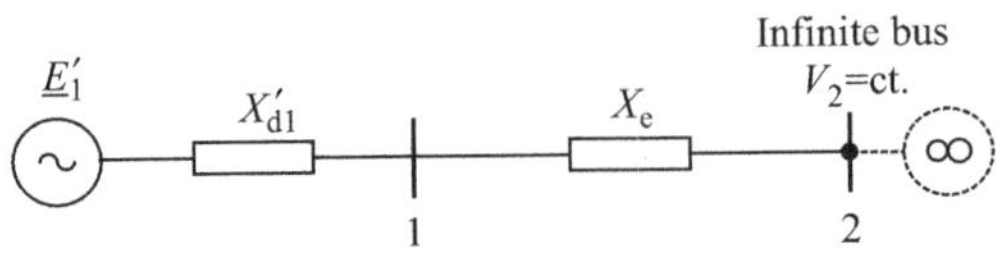

Figure 10.2. A single machine infinite bus (SMIB) system.

accelerating area when $P_m > P_e$, while A_2 is the decelerating area when, in this case, $P_m < P_e$.

The first-swing stability criterion is satisfied if the A_1 surface is smaller or at most equal to the A_2 surface ($A_1 \leq A_2$).

In practice, in order to assess the first-swing stability of the synchronous generator, the mechanical power is assumed constant ($P_m = \text{ct.}$) and the electromagnetic power is expressed in terms of the parameters of the network through which the generator is connected to the infinite power bus.

Let us consider the circuit from Figure 10.2, in which a single synchronous generator (bus 1) is connected through an external reactance X_e to the infinite power bus (bus 2), of which voltage $\underline{V}_2$, taken as phase reference, is constant, and the electromagnetic power is given by

$$P_{e1} = \frac{E'_1 V_2}{X'_{d1} + X_e} \sin \delta'_{12} \tag{10.5}$$

where δ'_{12} is the angle between the phasor of the emf $\underline{E}'_1 = E'_1 e^{j\delta'_{12}}$, behind the generator transient reactance, and the phasor of the infinite power bus voltage $\underline{V}_2 = V_2 e^{j0} = V$.

If loads are supplied from one of the connection network nodes, these are modeled as constant shunt admittances. All the passive nodes of the network are then eliminated by a Gauss procedure applied to the nodal admittances matrix $[Y'_{nn}]$, modified by addition to the diagonal terms of the admittances modeling the loads. The nodal admittance matrix reduced to the generator buses is thus obtained:

$$\left[\underline{Y}_{nn}^{red}\right] = \begin{array}{|c|c|} \hline \underline{Y}_{11} & \underline{Y}_{12} \\ \hline \underline{Y}_{21} & \underline{Y}_{22} \\ \hline \end{array}$$

For the considered case, the $\left[\underline{Y}_{nn}^{red}\right]$ matrix is constructed for two buses only, that is,

- bus 1, to which the reactance X'_{d1} is connected, having the voltage equal to the emf $\underline{E}'_1 = E'_1 e^{j\delta'_{12}}$;
- bus 2, of infinite power, having the voltage $\underline{V}_2 = V_2 e^{j0} = V$.

If $\underline{Y}_{11} = G_{11} + jB_{11}$ and $\underline{Y}_{12} = G_{12} + jB_{12}$ are the nodal admittances on the first row of the $\left[\underline{Y}_{nn}^{red}\right]$ matrix, then the expression of the electromagnetic power delivered by the synchronous generator connected to bus 1 is given by

$$\begin{aligned} P_{e1} &= \text{Re}(\underline{E}'_1 \underline{I}_1^*) = G_{11} E'^2_1 + E'_1 V_2 \left(G_{12} \cos \delta'_{12} + B_{12} \sin \delta'_{12}\right) \equiv \\ &\equiv Y_{11} E'^2_1 \sin \alpha_{11} + Y_{12} E'_1 V_2 \sin\left(\delta'_{12} + \alpha_{12}\right) \end{aligned} \tag{10.6}$$

where

$$Y_{11} = \sqrt{G_{11}^2 + B_{11}^2}; \quad \tan \alpha_{11} = \frac{G_{11}}{B_{11}}$$

$$Y_{12} = \sqrt{G_{12}^2 + B_{12}^2}; \quad \tan \alpha_{12} = \frac{G_{12}}{B_{12}}$$

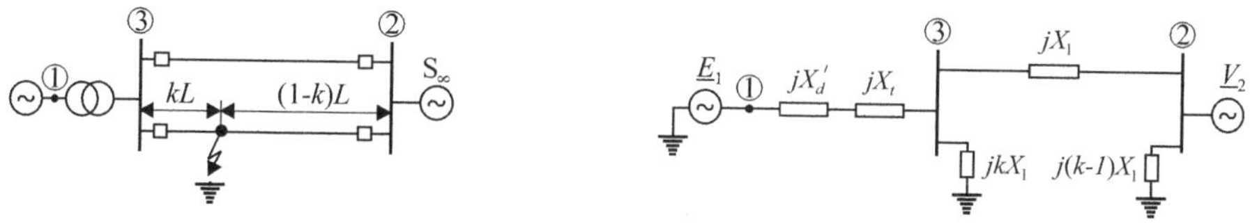

Figure 10.3. The SMIB system used for application of the equal area criterion (*a*) and its equivalent circuit (*b*).

10.2.1.2 Calculation of the Fault Clearing Time. In order to illustrate the application principles of the equal area criterion, let us consider the power system from Figure 10.3, which consists of a synchronous generator, connected through a transformer and a double-circuit line to the infinite power bus 2. A symmetrical three phase-to-earth short circuit is applied on one of the line circuits (Figure 10.3a).

The electromagnetic power at the generator terminals, in the three stages of the transient state, is expressed using equation (10.5) as follows:

(i) Prefault steady state (both line circuits are connected)

$$P^n_{e1} = P_{m1} = \frac{E'_1 V_2}{X^n_{12}} \sin \delta'_{12}$$

where $X^n_{12} = X'_d + X_t + \dfrac{X_1}{2}$

(ii) During fault conditions, when a three-phase short circuit occurs on one of the line circuits at the $k \cdot L$ distance from bus 3 (Figure 10.3b)

$$P^f_{e1} = \frac{E'_1 V_2}{X^f_{12}} \sin \delta'^f_{12}$$

where X^f_{12}, given by the expression

$$X^f_{12} = X'_d + X_t + X_1 + \frac{(X'_d + X_t)X_1}{kX_1}$$

is obtained by applying the $Y \rightarrow \Delta$ transformation to the Δ circuit formed by $X_{Y,1} = X'_d + X_t$, $X_{Y,2} = X_1$ and $X_{Y,3} = kX_1$ reactances. If the fault occurs close to bus 3, then $k = 0$ and therefore, $X^f_{12} \rightarrow \infty$ and $P^f_{e1} = 0$, respectively. This is the worst case because the accelerating power $P_a = P_{m1} - P^f_{e1} = P_{m1}$ is maximum.

(iii) Postfault steady state (the fault was cleared by simultaneous tripping of the breakers at both ends of the affected circuit)

$$P^{pf}_{e1} = \frac{E'_1 V_2}{X^{pf}_{12}} \sin \delta'^{pf}_{12}$$

where $X^{pf}_{12} = X'_d + X_t + X_1$.

The power–angle transient characteristics corresponding to the three stages of the disturbance are illustrated in Figure 10.4.

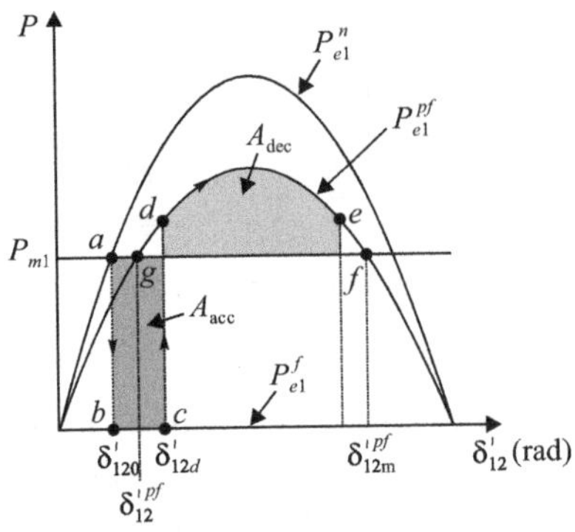

Figure 10.4. Application of the equal area criterion to the SMIB system.

The point a, situated at the intersection of the $P_{el}^n(\delta')$ curve with the mechanical power line P_{m1} = ct., to which the δ'_{120} angle corresponds, defines the initial state of the studied system for which stability conditions have to be ensured. At the instant of the fault occurrence, the operating point suddenly moves in the point b since $P_{el}^f = 0$ (since the angle δ' is an inertial variable it does not show sudden variations). Now, because $P_{m1} > P_{el}^f$, the generator rotor accelerates and the angle δ'_{12} increases up to δ'_{12d} (point c on the P_{el}^f characteristic), when the affected circuit is disconnected.

At the time instant t_d, corresponding to δ'_{12d}, when the fault is cleared, the operating point suddenly jumps from c to d and continues to move on the P_{el}^{pf} curve with negative acceleration since $P_{m1} < P_{el}^{pf}$, until the deceleration area A_{dec} becomes equal to the acceleration area A_{acc} and the relative speed of the rotor becomes zero (point e in Figure 10.4). The angle δ'_{12} reaches the maximum swing in the point e and thereafter decreases, indicating that the transient stability condition is satisfied. In the absence of damping, the rotor will continue to swing around δ'^{pf}_{12} (point g is the new postfault steady state point).

If the two areas does not become equal up to point f, the operating point continues to move on the P_{el}^{pf} characteristic and the angle increases, while the acceleration becomes again positive since $P_{m1} > P_{el}^{pf}$, and the synchronism is lost.

The equal area criterion is practically a method for determination of the maximum rotor angle swings and, therefore, it can be used for the transient stability assessment without analyzing the $\delta'(t)$ curve, that can be plotted by solving the swing equation.

If δ'^{pf}_{12m} is the maximum swing that the rotor angle can reach up to without losing the synchronism (the rotor angle corresponding to point f on the P_{el}^{pf} characteristic), then the two areas A_{acc} and A_{dec} are computed as follows:

$$A_{acc} = \int_{\delta'_{120}}^{\delta'_{12d}} P_a \mathrm{d}\delta'_{12} = \int_{\delta'_{120}}^{\delta'_{12d}} \left(P_{m1} - P_{el}^f\right) \mathrm{d}\delta'_{12} = \int_{\delta'_{120}}^{\delta'_{12d}} P_{m1} \mathrm{d}\delta'_{12} = P_{m1}\left(\delta'_{12d} - \delta'_{120}\right) \tag{10.7'}$$

$$A_{dec} = \int_{\delta'_{12d}}^{\delta'_{12m}} \left(P_{el}^{pf} - P_{m1}\right) \mathrm{d}\delta'_{12} = \int_{\delta'_{12d}}^{\delta'_{12m}} \left(\frac{E'_1 V_2}{X_{12}^{pf}} \sin\delta'_{12} - P_{m1}\right) \mathrm{d}\delta'_{12}$$

$$= \frac{E'_1 V_2}{X_{12}^{pf}} \left(\cos\delta'_{12d} - \cos\delta'_{12m}\right) - P_{m1}\left(\delta'_{12m} - \delta'_{12d}\right) \tag{10.7''}$$

For the considered disturbance, the transient stability limit corresponds to the situation in which the two areas are equal. The limit value of δ' is called *critical angle* (δ'_{crit}) and can be determined by solving the equation:

$$\int_{\delta'_{120}}^{\delta'_{crit}} P_{\text{m1}} \, \mathrm{d}\delta'_{12} = \int_{\delta'_{12d}}^{\delta'_{12m}} \left(P_{\text{e1}}^{pf} - P_{\text{m1}} \right) \mathrm{d}\delta'_{12} \tag{10.8}$$

resulted from the equality of the two areas ($A_{\text{acc}} = A_{\text{dec}}$). From expressions (10.7′) and (10.7″), given that $P_{\text{m1}} = P_{\max} \sin \delta'_{120}$, where $P_{\max} = \frac{E'_1 V_2}{X_{12}^{pf}}$, we get

$$P_{\max} \left(\cos \delta'_{\text{crit}} - \cos \delta'_{\text{12m}} \right) = P_{\max} \sin \delta'_{120} \left(\delta'_{\text{12m}} - \delta'_{120} \right) \tag{10.9}$$

and

$$\delta'_{\text{crit}} = \operatorname{a} \cos \left[\left(\delta'_{\text{12m}} - \delta'_{120} \right) \sin \delta'_{120} + \cos \delta'_{\text{12m}} \right] \tag{10.10}$$

The critical fault clearing time

Given the value of the critical angle δ'_{crit} we can determine the fault clearing time corresponding to this angle, called *critical fault clearing time*, t_{crit}, by solving the equation $\delta'(t) = \delta_{\text{crit}}$. The values of $\delta'(t)$ can be determined either by integrating the swing equation or by Taylor series expansion.

(i) ***Numerical Integration of the Swing Equation***

Since during the fault we have $P_{e1}^{f} = 0$, the swing equation becomes

$$\frac{\mathrm{d}^2 \delta'_{12}}{\mathrm{d}t^2} = \frac{\omega_0}{M} P_{\text{m1}}$$

and its solution is

$$\delta'_{12}(t) = \frac{\omega_0}{M} \cdot P_{m1} \cdot \frac{t^2}{2} + C_1 t + C_2 \tag{10.11}$$

The integration constants, C_1 and C_2, are determined from the initial conditions: $\left. \frac{\mathrm{d}\delta'_{12}}{\mathrm{d}t} \right|_{t=0} = \omega_0 \omega |_{t=0} = 0$ and $\delta'_{12}|_{t=0} = \delta'_{120}$, resulting $C_1 = 0$ and $C_2 = \delta'_{120}$, respectively.

Thus, $\delta'_{12}(t) = \frac{\omega_0}{M} \cdot P_{\text{m1}} \cdot \frac{t^2}{2} + \delta'_{120}$ and, setting $\delta'_{12}(t) = \delta'_{\text{crit}}$, the critical fault clearing time is

$$t_{\text{crit}} = \sqrt{\frac{2M}{\omega_0 P_{\text{m1}}} \cdot \left(\delta'_{\text{crit}} - \delta'_{120} \right)} \tag{10.12}$$

(ii) ***Taylor Series Expansion***

Expanding $\delta'(t)$ in Taylor series and keeping the first two terms only, yields

$$\delta'(t) \cong \delta'(t_0) + (t - t_0) \left. \frac{\mathrm{d}\delta'}{\mathrm{d}t} \right|_{t=t_0} + \frac{(t - t_0)^2}{2} \left. \frac{\mathrm{d}^2 \delta'}{\mathrm{d}t^2} \right|_{t=t_0} \tag{10.13}$$

At the instant of disturbance occurrence, $t_0 = 0$, in accordance with the swing equation, we have

$$\left.\frac{d\delta'_{12}}{dt}\right|_{t=0} = \omega_0\omega = 0$$

and

$$\left.\frac{d^2\delta'_{12}}{dt^2}\right|_{t=0} = \omega_0 \left.\frac{d\omega}{dt}\right|_{t=0} = \frac{\omega_0}{M}\left(P_{m1} - P^f_{e1}\big|_{t=0}\right) \tag{10.14}$$

Equating the right side of equation (10.13), in which angle derivatives are calculated with (10.14), to the critical angle δ'_{crit}, achieve a second-order equation, from where we get the critical time.

Note that, if $P_{m1} = \text{ct.}$ and the fault occurs near bus 3, that is $P^f_{e1} = 0$, then equation (10.13) is identical with (10.11).

The three possible cases for a Single Machine Infinite Bus (SMIB) system are illustrated in Figure 10.5. Analyzing the stable case, the transient stability margin, η, can be defined as follows:

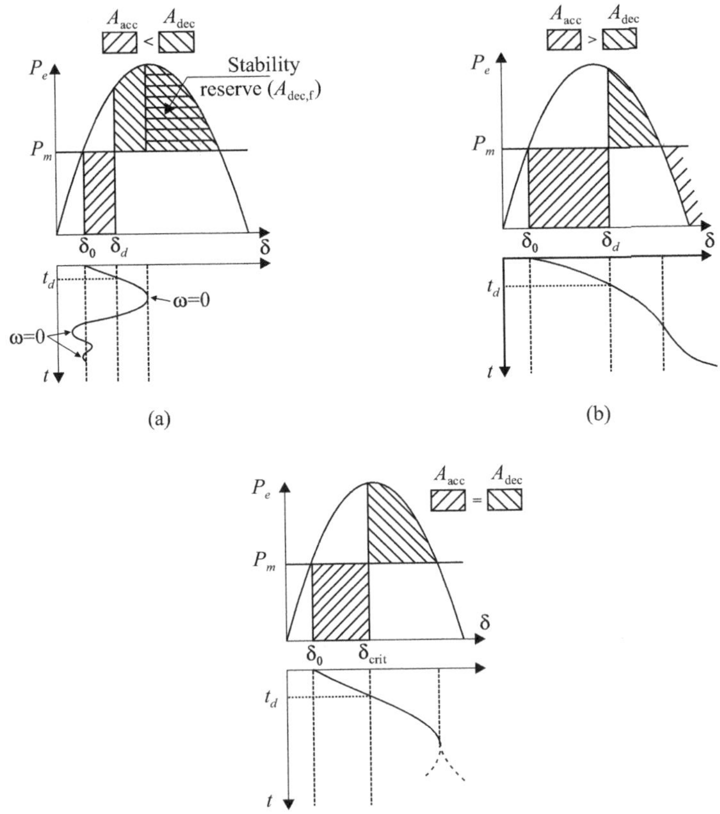

Figure 10.5. Equal area criterion applied for a SMIB system: (a) stable case; (b) unstable case; (c) critical case.

- In terms of *the fault clearing time*:

$$\eta_t = \frac{t_{crit} - t_d}{t_{crit}} \tag{10.15a}$$

- In terms of *the available deceleration area*:

$$\eta_a^{(1)} = \frac{A_{dec}}{A_{dec,f}} \tag{10.15b}$$

- In terms of *the difference between the deceleration and the acceleration areas*:

$$\eta_a^{(2)} = \frac{A_{dec} - A_{acc}}{A_{acc}} \tag{10.15c}$$

From (10.15c) we can conclude that η has positive values for stable system cases and negative values for unstable system cases.

Figure 10.6 shows the qualitative dependence of η in terms of the fault clearing time, t_d.

10.2.1.3 Two Finite Power Synchronous Generators. Let us consider two generators connected through a passive network that can be reduced to the simple SMIB system case. The swing equations of the two machines can be written as:

$$\frac{2H_1}{\omega_0} \cdot \frac{d^2\delta_1'}{dt^2} = P_{a1} = P_{m1} - P_{e1}$$

$$\frac{2H_2}{\omega_0} \cdot \frac{d^2\delta_2'}{dt^2} = P_{a2} = P_{m2} - P_{e2}$$

Subtracting the two expressions yields the swing equation of the equivalent synchronous machine:

$$\frac{2H}{\omega_0} \cdot \frac{d^2\delta_{12}'}{dt^2} = P_a = P_m - P_e \tag{10.16}$$

where $\delta_{12}' = \delta_1' - \delta_2'$;

$H = \dfrac{H_1 H_2}{H_1 + H_2}$;

$P_a = \dfrac{H_2 P_{a1} - H_1 P_{a2}}{H_1 + H_2}$ (valid also for P_m and P_e)

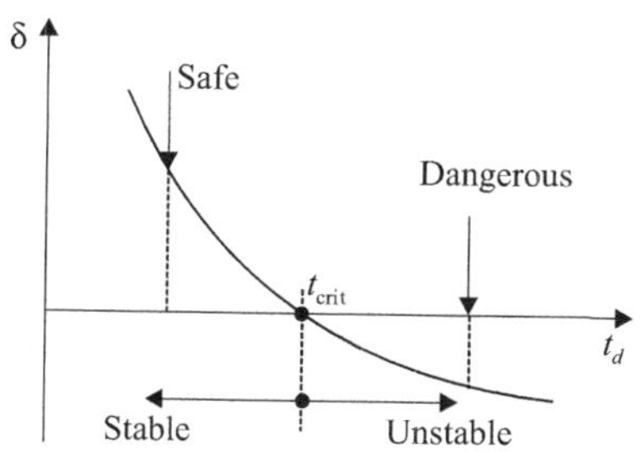

Figure 10.6. The plot of η against the fault clearing time t_d.

The electric powers produced by the two generators are given by

$$\begin{aligned} P_{e1} &= Y_{11}E_1'^2 \sin\alpha_{11} + Y_{12}E_1'E_2' \sin\left(\delta_{12}' + \alpha_{12}\right) \\ P_{e2} &= Y_{22}E_2'^2 \sin\alpha_{22} + Y_{21}E_2'E_1' \sin\left(\delta_{21}' + \alpha_{21}\right) \end{aligned} \quad (10.17)$$

where $\delta_{21}' = -\delta_{12}'$ and $\underline{Y}_{12} = \underline{Y}_{21}$.

The differential equation (10.16) has the same form as the swing equation of a synchronous machine connected to an infinite power bus. Therefore, the equal area criterion becomes

$$\int_{\delta_{120}'}^{\delta_{12m}'} (H_2 P_{a1} - H_1 P_{a2}) d\delta_{12}' = 0 \quad (10.18)$$

10.2.2 Extended Equal Area Criterion–EEAC

The Extended Equal Area Criterion—EEAC [1] evaluates the transient stability of a power system using an equivalent system consisting of a generator connected to an infinite power bus called Single Machine Infinite Bus System—SMIB. The method allows fast evaluation of the stability, since it uses an algebraic formulation for the stability index.

The principle of the EEAC method is illustrated in Figure 10.7. The A_{acc} and A_{dec} areas are the acceleration and deceleration areas of the equivalent generator, while $\delta_e = \delta(t_e)$ and $\delta_u = \delta(t_u)$ are its rotor angle values corresponding to the clearing time t_e and the time to instability t_u, respectively.

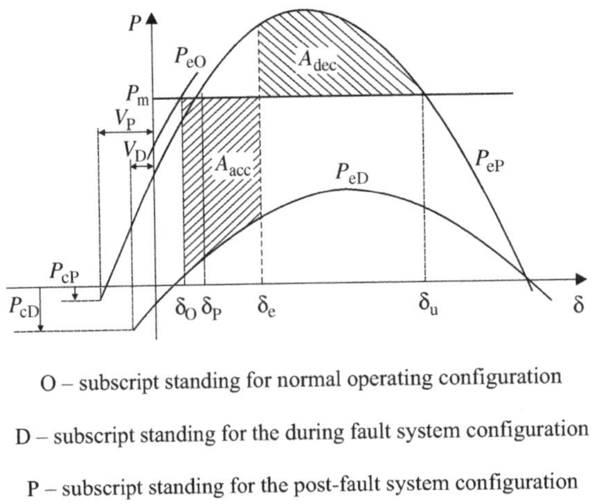

Figure 10.7. Basics of the EEAC method.

The critical fault clearing time $t_{\rm crit}$ represents the duration of fault for which the two areas are equal and maximum (for the given disturbance).

The EEAC method consists of two main steps:

Step 1: Determine the parameters of the equivalent SMIB system.

Step 2: Apply the equal area criterion to the SMIB system in order to determine the critical clearing time $t_{\rm crit}$.

If the second step does not raise problems, the equal area criterion being simple and fast to apply, the first step may be critical; it consist in a procedure involving the following operations:

(i) For a given fault, the electrical network is divided into two generator groups: the *group of weak disturbed generators* or *the noncritical generators*, identified by N, and *the group of strong disturbed generators* or *the critical generators*, identified by C.

(ii) *Reduce each group*, noncritical and critical, to one equivalent generator, each representing the center of inertia of the corresponding group.

(iii) *Reduce the resulted system*, consisting of two equivalent generators, to an *SMIB system*.

The groups of critical and noncritical generators are determined using the *Critical Machine Ranking* (CMR) [1] method. The CMR is based on the *Degree of Criticality (Involvement) of the Machine* (*DCM*) defined in terms of the rotor angles magnitude. In this regard, a simplified time simulation that uses Taylor's series expansion is performed to determine the time evolution of the rotor angles. The series expansion allows using a large integration step, which is reflected in less computation effort.

The algorithm of the CMR method requires the following steps:

1. *Determine the initial conditions*: identify an initial group of critical generators and compute the corresponding critical angle—$\delta_{\rm crit}$ and the unstable equilibrium angle—$\delta_{\rm u}$; this computation is performed using a modified version of the CMR method, that is the EEAC method.
2. *Compute $t_{\rm crit}$ and $t_{\rm u}$* corresponding to $\delta_{\rm crit}$ and $\delta_{\rm u}$ using *global type* Taylor's series expansion;
3. *Compute the rotor angles time evolution* (up to $t_{\rm u}$) using *individual type* Taylor's series expansion. The time intervals $[0, t_{\rm crit}]$ and $[t_{crit}, t_u]$ are divided into a number or subintervals established through desired accuracy.
4. *Rank the generators in descending order of the rotor angles magnitudes* in the $[t_{\rm crit}, t_{\rm u}]$ interval.
5. After ranking, the critical generators can be grouped into various possible groups by selecting $k = 1$, then $k = 2$ generators, and so on, starting from the top of the list. For each critical group compute the critical clearing time, $t^k_{\rm crit}$, using the EEAC criterion. The real critical group is that with the smallest critical clearing time;

The initial conditions are determined by following the next algorithm within the step 1 of the CMR method:

1.1 Determine the evolution of the rotor angles over a time window of 3 s, for the considered disturbance, using the individual type Taylor's series expansion. In order to obtain an unstable trajectory and to reduce the computation time, large values are chosen for the fault duration ($t_e \cong 1$ s) and for the step size ($\Delta t \cong 0.1$ s).

1.2 If for the considered fault duration ($t_e \cong 1$ s) the generators do not lose the synchronism, it can be said that the power system is very stable. Otherwise, the generators are ranked in descending order of the magnitudes of the rotor angles at the time instant in which the system becomes unstable.

1.3 Select the generators from the top of the list established at point 1.2 in order to set the critical group to determine the initial conditions of the CMR method.

The *advantage* of the EEAC method is that the method is reliable and accurate, and the stability index is computed very fast. The main *disadvantage* is that the method is applicable for the classical generator model only. Moreover, correction coefficients are required to eliminate the errors caused by the truncated Taylor's series expansion [1].

10.2.3 The SIME (SIngle - Machine Equivalent) Method

The research developed by Mania Pavella from Université de Liège regarding, on one hand, preserving the main advantages of the time-domain methods and of the equal area criterion and, on the other hand, the elimination of the main disadvantage of the EEAC method, that is the use of detailed methods for the synchronous generators, led to the creation of the hybrid method generically called *SIME* [3,44]. This method converts the trajectories of the multimachine system to a single trajectory of an equivalent system—OMIB. The SIME method combines *the step-by-step time-domain integration method applied to the multimachine system* and *the equal area criterion applied to the equivalent generator.* This combination requires two basic steps: *the identification of the critical generators* (i.e., the generators responsible for the loss of synchronism) and *the assessment of stability reserve.*

For a given unstable scenario, defined by a system operating point before the disturbance and by a contingency (type, location, sequence of events), the SIME method performs, in a first stage, the time-domain simulation "*during the disturbance*" period and, afterwards, the time-domain simulation for the "*after disturbance*" period. In the beginning of the last step, SIME builds the so-called OMIB candidates to which the equal area criterion is applied. The candidates are built using the data of the multimachine system given for each step of the time-domain simulation. The process stops when one of the candidates is declared unstable. In this moment SIME identifies the critical generators, declares that this OMIB candidate is the real one and computes the corresponding stability reserve.

SIME preserves the method accuracy and capacity to deal with the desired model and stability scenario by updating the OMIB parameters for every step of the time-domain simulation. Furthermore, the use of the OMIB and of the equal area criterion allows for the significant expansion of the method's time-domain simulation problem, adding the following possibilities:

(i) *Fast stability assessment.*
(ii) *Contingencies filtering* (the elimination of the riskless contingencies) and *classification–evaluation of the dangerous contingencies.*
(iii) *Sensitivity analysis.*
(iv) *Preventive control*, that is, the identification of the necessary measures to stabilize the system in case of dangerous contingencies.
(v) *Description*—the time evolution of the rotor angles, of the OMIB, and the OMIB's P-δ curves—and the multiform *physical interpretations* given by OMIB and the equal area criterion.

10.2.3.1 Method Formulation. The SIME method is based on the following two remarks:

a. No matter the complexity of the system, the transient instability phenomenon starts when the system generators split into two groups causing the irreversible loss of synchronism.
b. SIME substitutes the study of the multimachine system dynamics with the study of the OMIB, which is simpler and faster using the equal area criterion, by replacing each generators group trajectory with the equivalent generators trajectory and, further, with the trajectory of a single-generator system—OMIB.

In this regard, the parameters of the OMIB system are computed starting from the two equivalent generators system. Hence, the rotor angle (δ), the angular velocity (ω), the inertial coefficient (M), the mechanical power (P_m), the electrical power (P_e), and the acceleration power (P_a) are determined in the following ways:

- The OMIB rotor angle is computed with respect to the angle difference between the two generators:

$$\delta(t) = \delta_C(t) - \delta_N(t) \tag{10.19}$$

where

$$\delta_C(t) = M_C^{-1} \sum_{k \in C} \delta_k(t) M_k; \quad \delta_N(t) = M_N^{-1} \sum_{j \in N} \delta_j(t) M_j \tag{10.20}$$

and

$$M_C = \sum_{k \in C} M_k \quad M_N = \sum_{j \in N} M_j \quad M = \frac{M_C M_N}{M_C + M_N} \tag{10.21}$$

- The mechanical and the electrical powers are determined using the expressions

$$\begin{aligned} P_m(t) &= M\left[M_C^{-1} \sum_{k \in C} P_{mk}(t) - M_C^{-1} \sum_{j \in N} P_{mj}(t)\right] \\ P_e(t) &= M\left[M_C^{-1} \sum_{k \in C} P_{ek}(t) - M_C^{-1} \sum_{j \in N} P_{ej}(t)\right] \end{aligned} \tag{10.22}$$

- The acceleration power is:

$$P_a(t) = P_m(t) - P_e(t) \tag{10.23}$$

- the OMIB angular speed is:

$$\omega(t) = \omega_C(t) - \omega_N(t) \tag{10.24}$$

In the above equations, C represents the set of the critical generators, N represents the set of the noncritical generators and M is the OMIB inertial coefficient.

Thus, the differential equation describing the OMIB dynamics is:

$$M\frac{d^2\delta}{dt^2} = P_m(t) - P_e(t) = P_a(t) \tag{10.25}$$

We stress that the quantities P_m, P_e, P_a are obtained on the basis of the information provided by the time-domain simulation program, to which SIME is coupled, which considers all the system and generator controls activated and uses Park equations. The OMIB's trajectory is not affected by any simplifying hypothesis, except the ones used by the time-domain simulation program. These quantities are computed for every time step of the simulation program.

As an example, the results obtained by applying the SIME method to a three-machines power system (m_1, m_2, and m_3) [4] are illustrated in Figure 10.8. In Figure 10.8a, the rotor angle variation curves for the three machines and the rotor angle variation for the OMIB equivalent system are shown for a fault clearing time $t_e = 0.117$ s. Notice that the most advanced machine is m_2 and that the maximum angle deviation is in between m_1 and m_3 (126.6°). Hence, the generators m_2 and m_3 form a critical group while the generator m_1 forms a noncritical group.

On the other hand, Figure 10.8b, shows the P–δ characteristic for the OMIB and illustrates the equal area criterion. The results of the equal area criterion for a fault clearing angle of $\delta_e = 71.1°$ are: $\delta_u = 158°$ and $t_u = 0.458$ s.

As it is known (see subsection 10.2.1), the equal area criterion provides that the stability of a dynamic system described by a differential equation of type (10.25) depends on the sign of the stability reserve, η, defined by:

$$\eta = A_{dec} - A_{acc} \tag{10.26}$$

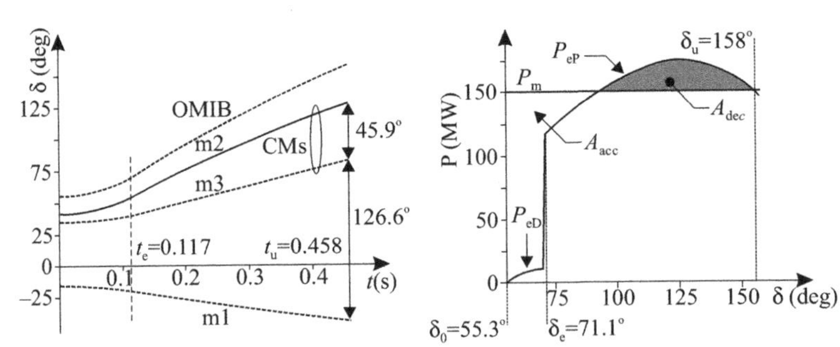

Figure 10.8. SIME method results for a three-generators system [2]: (a) the time evolution of the three generators and of the OMIB; (b) the $P-\delta$ characteristic of the OMIB and the equal area criterion.

The power system is stable if η is positive, unstable if η is negative, while for $\eta = 0$ the system is at its stability limit.

The expression of type (10.26), which assesses the OMIB case using the equal area criterion, contains easy-to-compute particular expressions, as it will result below.

10.2.3.2 Criteria and Degree of Instability. *An unstable OMIB trajectory reaches the unstable angle δ_u at the time instant t_u when*

$$P_a(t_u) = 0 \quad \text{and} \quad \left.\frac{dP_a}{dt}\right|_{t=t_u} > 0 \tag{10.27}$$

with $\omega > 0$ for $t > t_u$ (see Figure 10.8b).

The conditions (10.27) are "stopping conditions" for the time-domain simulation program provided by SIME. They mark the beginning of the loss of synchronism and, hence any further computation is useless, except when a particular research is desired.

At the instant of time $t = t_u$, the stability reserve η is defined by another, very easy to compute, expression:

$$\eta_u = -\frac{1}{2}M\omega_u^2 \tag{10.28}$$

10.2.3.3 Criteria and Corresponding Stability Reserve. *A stable OMIB trajectory reaches its "recovery angle" $\delta_r < \delta_u$ at the instant t_r when the OMIB's angle reaches its maximum value then decreases*, that is,

$$\omega(t_r) = 0 \quad \text{and} \quad P_a(t_r) < 0 \tag{10.29}$$

The conditions (10.29) are "the stopping conditions" of the SIME based time-domain simulation program. They show whether the system is stable—at least as regards the first oscillation—and any further computation is useless, unless the next oscillations are of interest.

At the instant $t = t_r$, the stability reserve η is given by (see Figure 10.9b):

$$\eta_{st} = \int_{\delta_r}^{\delta_u} |P_a| d\delta \tag{10.30}$$

It should be mentioned that, unlike the instability margin defined by (10.28), the stability reserve (10.30) can be determined only approximately. This happens because neither the angle δ_u nor the trajectory $P_a(\delta)$ with $\delta \in [\delta_r, \delta_u]$ are known since the OMIB trajectory "recovers", that is the angle of the equivalent system starts decreasing after the maximum value $\delta = \delta_r < \delta_u$ is reached.

The following two approximations are proposed in [2]:

a. *the triangle approximation*, TRI in Figure 10.9b:

$$\eta_{st} \cong \frac{1}{2}P_a(\delta_u - \delta_r) \tag{10.31}$$

b. *the least squares approximation* (weighted or not), denoted by WLS in Figure 10.9b, where the $P_a(\delta)$ curve is extrapolated on the interval $\delta \in [\delta_r, \delta_u]$.

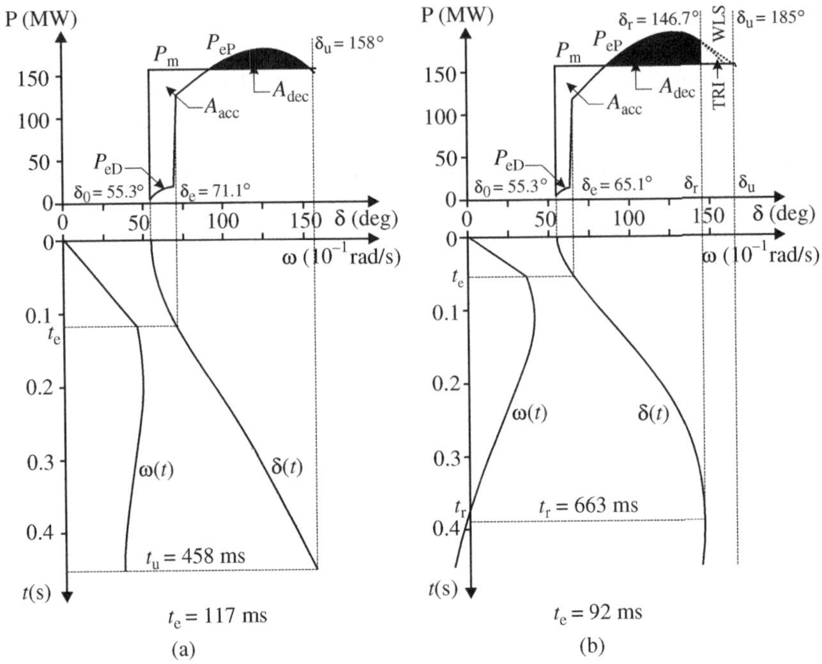

Figure 10.9. Time-domain and $P-\delta$ plane representation of the OMIB equivalent trajectory for the three-generators system case [2]: (a) the unstable trajectory; (b) the stable trajectory.

10.2.3.4 Identification of the OMIB Equivalent. This identification is based on the following observation: "no matter how complex the system is, the transient instability phenomenon is triggered when the generators separate into two groups and the irreversible loss of synchronism results".

The OMIB identification is carried out in the following manner:

(i) For every computation step, starting from t_e (the instant of time when the fault is cleared and the multimachine system enters the final configuration), SIME performs a ranking of the generators in descending order in terms of their rotor angles, and then considers the first "electrical distances" between the ranked generators (e.g., for the first 10 generators).

(ii) Each of these "distances" splits the generators into two groups on both sides of the considered distance; SIME computes the corresponding "OMIB candidates" and applies on it the test (10.27).

(iii) If one of the "OMIB candidates" satisfies the conditions (10.27), it is considered as the "real OMIB." The critical generators will be the ones situated above the distance of the real OMIB, while the noncritical generators will be the ones bellow. Once the real OMIB is identified, SIME stops the time-domain simulation and computes the stability reserve using the equation (10.28).

(iv) Otherwise, if the conditions (10.27) are not satisfied, SIME continues the time-domain simulation, proceeding to the next computation step, and repeating the steps (i) and (ii) until the conditions (10.27) are satisfied, after which the step (iii) is executed.

Observation: the OMIB identification is performed only for an unstable trajectory. Extending, one considers that the OMIB of the previously identified unstable trajectory is still valid for a stable trajectory close enough to this one (e.g., a trajectory obtained for a relatively close fault clearing time).

10.2.4 Direct Methods Based on Lyapunov's Theory

10.2.4.1 Lyapunov's Method. The direct methods for transient stability assessment based on Lyapunov stability theory deal with the autonomous nonlinear dynamic system:

$$\Delta\dot{x} = f(\Delta x) \tag{10.32}$$

where Δx is a vector of variables deviation from the equilibrium point (the origin of coordinates), that is $\Delta x = x - x_0$. The system of differential equations (10.32) is a particular case of the system (10.1) and represents a so-called classical model of the dynamics of the electric power system defined by (10.2).

All functions f and their first partial derivatives are supposed to be continuous and defined in any Δ-neighborhood of the origin of coordinates. This means that the system (10.32), under the initial conditions $\Delta x_0 = \{\Delta x_{10}, \Delta x_{20} \ldots, \Delta x_{\mathrm{uo}}\}$ in the Δ-neighborhood of the origin of coordinates, has a well (unique) defined solution (unique trajectory).

The Lyapunov stability is defined as follows:

- The equilibrium point $\{0, 0, 0, \ldots, 0\}$ of the system (10.32) is called stable, if for any positive number ε, no matter how small it can be, it is possible to indicate another positive number $\eta(\varepsilon)$ such that for all initial disturbances satisfying the condition

$$\sum_{i=1}^{n} \Delta x_{io}^2 \leq \eta \tag{10.33}$$

in further changes in the system the following inequality holds

$$\sum_{i=1}^{n} \Delta x_i^2(t) < \varepsilon \tag{10.34}$$

Instability of system equilibrium is understood as the absence of formulated stability properties.

- If the equilibrium is stable, and also the conditions

$$\lim_{t\to\infty} \Delta x_i^2(t) = 0 \tag{10.35}$$

are met, the equilibrium is called *asymptotically stable.*

For dynamic systems as described by (10.32), the second (direct) Lyapunov method, best suited to the problem of constructing the criteria of transient stability of electric power systems, is formulated as follows.

The equilibrium of the autonomous system (10.32) is stable "in the large," if there exists a continuous Lyapunov function $U(\Delta x_1, \Delta x_2, \Delta x_3, \ldots, \Delta x_n)$ that is determined in the phase space of variables $\{\Delta x_1, \Delta x_2, \Delta x_3, \ldots, \Delta x_n\}$, with continuous partial derivatives such that

1. $U(\Delta x_1, \Delta x_2, \Delta x_3, \ldots, \Delta x_n)$ is a positive definite function in the closed region Ω, including the origin of coordinates;
2. one of the surfaces of $U = \text{ct.}$ will be a boundary of the region Ω;
3. the Lyapunov function gradient—grad U—is not equal to zero in Ω, excluding the origin of coordinates and the boundary of the region Ω;
4. by virtue of equations (10.32), the Lyapunov function derivative dU/dt is a negative sign function in Ω or identically equal to zero.

Some explanations of the key notions applied in the considered theorem are given in the following:

- The function of multiple variables is called a constant sign function if, besides the zero values, it takes only values of one sign for all variables. The constant sign function is called sign definite if it vanishes only at the origin of coordinates.
- The system of coordinates $\{\Delta x_1, \Delta x_2, \Delta x_3, \ldots, \Delta x_n\}$ is called a phase space of the dynamic system (10.32).
- If in terms of the considered theorem dU/dt is a negative definite function in Ω, the asymptotic stability "in the large" of equilibrium of the dynamic system (10.32) is ensured.

Classical theorems of the Lyapunov method stability functions [5,6], in contrast to the indicated modified theorem [7–9], require that only conditions 1 and 4 be fulfilled in any vicinity of the origin of coordinates. The modified theorem according to LaSalle's Invariance Principle actually guarantees the existence of the attraction (stability) region Ω with the boundary $U = C = \text{ct.}$ The estimate of constant C, with which the region Ω will be a peculiar "trap" for the trajectories of the disturbed system motion, is a major problem of the Lyapunov method functions. Virtually, the region described by Lyapunov function at $U = C$ results in some approximations of the real attraction region of a dynamic system [10].

Figure 10.10 illustrates graphically the Lyapunov's method functions to estimate stability "in the large" for the second-order system.

The upper part of the figure presents a section of Lyapunov function along the axis Δx_1 and shows lines of the equal level of function U. Here, 0 stands for stable equilibrium and 1 for unstable equilibrium of the "saddle" form, which is seen from the topology of lines of the equal level of function U (see the lower part of Figure 10.10). The region outlined by line 2 is an estimate of the attraction region by using functions U. Line 3 is a real attraction region of the dynamic system. If the attraction region estimated using Lyapunov function U is smaller than the real attraction region of the dynamic system equilibrium, it means that the second method of Lyapunov gives sufficient stability conditions that are far from the necessary and sufficient ones.

Controlling the system stability "in the large" is performed by determining the coordinates of the points 0 and 1, calculation of Lyapunov functions U values for some disturbances (e.g., at point 4 that is situated in Figure 10.10 beyond the plane of axis Δx_1) and at the point of unstable equilibrium 1, and then by comparing different values of function U. If $U_4 < U_1$ the system is stable, if $U_4 > U_1$ the system is unstable, whereas for $U_4 = U_1$ a critical situation is reached.

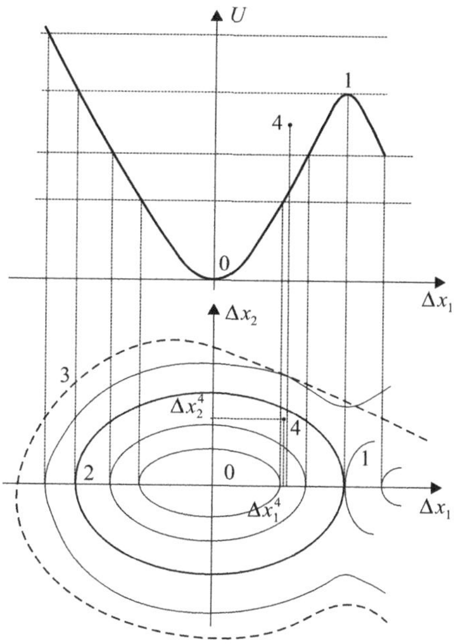

Figure 10.10. Illustration of the method of Lyapunov functions.

In the classical form, the direct Lyapunov's method proceeds from the fact that at the initial time instant the dynamic system is in a disturbed state and the system stability "in the large" is estimated under the assumption of its free motion from the disturbed state. Thus, the previous process that led to the disturbed state of the system was not taken into account.

The transient process evolving in the power system as a result of a disturbance is accompanied by changes in the system parameters at certain instants of time (e.g., a short circuit on a transmission line involves line tripping, automatic reclosing, and operation of emergency control devices) The coordinates of the system at the instant of the last change in its parameters represent a disturbed state from which its free motion toward equilibrium occurs. The equilibrium is stable if the system stability is maintained or unstable if stability is lost.

Based on the above-mentioned properties, the application of the second Lyapunov's method for the power system transient stability assessment requires that the following three problems be solved:

1. Calculation of the system motion trajectory in time up to the last change in its parameters, for example, by numerical integration of differential equations of the mathematical model of the system dynamics.
2. Construction of an appropriate Lyapunov function.
3. Determination of the stable and unstable system equilibriums coordinates.

The first problem does not cause any important difficulty. However, the extent to which the sufficient conditions for stability obtained by the Lyapunov function method are far from the necessary and sufficient ones will depend on how efficient the solution to the second and third problems is. In the next two sections attention will be given to the approaches for the Lyapunov function construction and the methods used to determine the system equilibrium coordinates.

10.2.4.2 Designing the Lyapunov Function. Construction of the Lyapunov function $U(x)$ is one of the key problems of this method. A well proven criterion to estimate the constructed Lyapunov function is the following requirement: the sufficient conditions obtained using this function in a nonlinear case should be also necessary conditions in the linear case [11]. For this purpose, the Routh–Hurvitz stability conditions for linear (linearized) system should be satisfied.

It should be noted that no general technique of constructing the Lyapunov function for nonlinear systems was defined so far. Construction of a good Lyapunov function is a matter of luck.

Lyapunov in [5] suggested the function U as a quadratic form of coordinates for the linear (linearized) system with constant coefficients

$$U = \Delta x^T \boldsymbol{P} \Delta x \qquad (10.36)$$

where $\boldsymbol{P}$ is a sought matrix. The derivative of Lyapunov function is defined as

$$\dot{U} = \Delta x^{\mathrm{T}} \boldsymbol{Q} \Delta x \qquad (10.37)$$

where $\boldsymbol{Q}$ is a given matrix.

The sought matrix $\boldsymbol{P}$ is determined from the Lyapunov matrix equation [9].

$$\boldsymbol{A}^{\mathrm{T}} \boldsymbol{P} + \boldsymbol{P} \boldsymbol{A} = \boldsymbol{Q} \qquad (10.38)$$

where $\boldsymbol{A}$ is a coefficient matrix of the linearized system (10.32) in the form

$$\Delta \dot{x} = \boldsymbol{A} \Delta x \qquad (10.39)$$

The main problems of this approach are related to setting the matrix $\boldsymbol{Q}$ (since there are no effective techniques to set it), and solving the Lyapunov matrix equation (10.38). Positive definiteness of Lyapunov function (10.36) is determined by meeting the Sylvester's criterion, which requires that all the determinants of the matrix $\boldsymbol{P}$ be positive.

Development of the considered approach is connected with a search for the derivative of Lyapunov function $\dot{U}$ by virtue of the initial nonlinear system (10.32) [10–12].

The experience shows that the best Lyapunov functions are obtained when they can be interpreted physically. In this context, it is interesting to mention the energy method for constructing Lyapunov functions. In an explicit form, this method has apparently been used since the "birth" of the analytical mechanics. For the conservative systems, we find the total energy H equal to the sum of kinetic and potential energies of the system, as a function of generalized coordinates. Then the elements that correspond to absorption and dissipation of mechanical energy are added to the system and for this system the sought function H will be Lyapunov function [9].

To illustrate the approach, let us consider the differential equation [9]:

$$\ddot{x} + \varphi(\dot{x}) + f(x) = 0 \qquad (10.40)$$

where $\varphi(0) = f(0) = 0$.

This equation is apparently equivalent to the system

$$\dot{x} = y, \quad \dot{y} = -f(x) - \varphi(y) \qquad (10.41)$$

Equation (10.40) allows simple interpretation in terms of mechanics, that is, it describes the oscillations of a particle affected by the nonlinear restoring force $f(x)$ in an environment with the impedance (damping) that depends nonlinearly on the speed y.

Taking into account that in accordance with (10.40) the mass of a particle equals unity, the total energy can be written in the form

$$U = \frac{y^2}{2} + \int_0^x f(x)\,\mathrm{d}x \tag{10.42}$$

where the first term in the right-hand side corresponds to the kinetic energy, and the second term to the potential energy.

If the environment has no impedance ($\varphi(y) = 0$), the system (10.41) would allow the first integral of U, which corresponds to the known law of energy conservation. However, since the mechanical energy in the process of oscillations, due to the impedance, is converted into thermal energy, the function U should decrease along the trajectory of the system described in (10.41). Indeed, it is easy to see that by virtue of system (10.41) we have $\dot{U} = -\varphi(y)y$.

If the condition U is positive definite, the inequality $\varphi(y)y > 0$ holds true for $y \neq 0$, then obtain $\dot{U} \leq 0$.

For the function U to be positive definite the equality $f(x)x > 0$ should hold true.

In order to apply the stability theorems, it is necessary to make sure that $\lim_{|x|\to\infty} \int_0^x f(x)\mathrm{d}x = \infty$ or impose some conditions to provide boundedness of the trajectories of the system (10.41).

In the end, it is also necessary to make sure that on the curve $y = 0$, where function $\dot{U}$ vanishes, there are no integral trajectories except for zero equilibrium. However, if along the trajectory $y = 0, \dot{y} = 0$ as well. It follows then from the second equation of the system (10.41) that $f(x) = 0$, but since by virtue of the condition $f(x)x > 0$, $x = 0$ is the only zero of function $f(x)$, we obtain $x = 0$.

The presented reasoning about the function U is typical for the case of a physical model of a differential system of equations under study, that is, for the electric power system as well.

Some known techniques of constructing the Lyapunov function as applied to the classical model of the multimachine power system dynamics described by (10.2) and (10.6) will be considered:

$$\frac{\mathrm{d}\delta_i}{\mathrm{d}t} = s_i$$

$$J_i \frac{\mathrm{d}s_i}{\mathrm{d}t} = P_{\mathrm{mi}} - E_i'^2 Y_{ii} \sin\beta_{ii} - \sum_{\substack{j=1 \\ j\neq i}}^{n} E_i' E_j' Y_{ij} \sin\left(\delta_i - \delta_j - \beta_{ij}\right) \tag{10.43}$$

where

δ_i is the synchronous machine rotor angle relative to an arbitrary synchronously rotating axis;

$s_i = \omega_i - \omega_0$ is the synchronous machine slip relative to a synchronously rotating axis;

$J_i = M_i/\omega_0$ is the inertia constant of the synchronous machine;

$\beta_{ii} = \pi/2 - \alpha_{ii}$ and $\beta_{ij} = \pi/2 - \alpha_{ij}$;

n is the number of synchronous machines.

Considering that $\beta_{ij} = 0$, from (10.43) a conservative model of the power system dynamics is obtained, for which the conditions

$$\frac{\partial f_k}{\partial x_m} = \frac{\partial f_m}{\partial x_k} \quad ; \quad k, m = \overline{1, n} \tag{10.44}$$

are met, where f stands for the right-hand sides of equations (10.43), and x are variables (angles, slips). The conditions (10.44) are necessary and sufficient for the existence of the first integral for the initial system of equations (at $\beta_{ij} = 0$) in the form of a sum of kinetic and potential energies, which in some neighborhood of the origin coordinates is a positive constant sign function and can be used as Lyapunov function.

Gorev in [13] suggested a technique for constructing the first integral of a conservative model of the power system dynamics. The technique implies a preliminary transformation of the initial equations. The transformation results in the total differential equation with separable variables. Then, Lyapunov function takes the following form:

$$\begin{aligned} U = & \frac{1}{2}\sum_{i=1}^{n} J_i s_i^2 - \sum_{i=1}^{n}\left(P_{mi} - E_i'^2 Y_{ii}\sin\beta_{ii}\right)(\delta_i - \delta_{oi}) - \\ & - \sum_{\substack{i=1, j=2 \\ i<j}}^{n} E_i' E_j' Y_{ij}\left(\cos\left(\delta_i - \delta_j\right) - \cos\left(\delta_{oi} - \delta_{oj}\right)\right) \end{aligned} \tag{10.45}$$

where $\delta_{oi}, i = \overline{1, n}$, is the coordinates of system equilibrium, whose stability is studied.

Analysis of function (10.45) shows that it is a positive constant sign function in the neighborhood of the origin of coordinates, which represents a sum of kinetic (the first term) and potential (the second and third terms) energies of the system in its disturbed motion. Its total time derivative is identically equal to zero.

The paper by Magnusson [14], published in 1947, suggested a transient energy method. The method is illustrated by the example of a three-machine power system. However, it can be generalized for a system with n machines. As shown by Ribbens-Pavella in [15], the transient energy function for the classical conservative model of the power system is equivalent to the Lyapunov function. The assumptions made are mostly the same as in [13] and the transient energy function is constructed in the form of a sum of kinetic and potential energies for the conservative model of the system. In doing so, the potential energy is sought relative to the considered steady state (equilibrium, whose stability is studied).

Kinnen and Chen [16] suggested the procedure of constructing Lyapunov function of the form (10.45) by deriving an auxiliary system of total differential equation similarly to Gorev [13]. The procedure implies selecting the new functions of variables taking into account the right-hand sides of the initial system, for which the conditions (10.44) are satisfied. Then it is possible to write the total differential equation and find its total integral.

The authors of [17] suggested the method of introducing additional terms in the right-hand side of the second equation of the system (10.43). The terms, depending on $\delta_i - \delta_j$, minimize the effect of the assumption $\beta_{ij} = 0$.

For the conservative model of the power system, Andreyuk [18], Aylett [19], Gless [20], and other authors suggested that applying the equations of mutual motion and construct the Lyapunov function in the form of the first (energy) integral of the initial system of equations. In a more general form of the motion model of the power system under disturbed conditions (with variables $\Delta\delta_{\text{i}} = \delta_{\text{i}} - \delta_{\text{oi}}$), the equations of mutual motion of the synchronous machine rotor were obtained by Putilova [21] by multiplying the equations of the i-th machine by $J_{\text{j}} \Big/ \sum_{i=1}^{n} J_{\text{i}}$ and the equations of the j-th machine by $J_{\text{i}} \Big/ \sum_{i=1}^{n} J_{i}$ and subtracting one equation obtained from the other.

The conservative model of an electric power system is obtained as before for $\beta_{\text{ij}} = 0$. Then the first integral can be found by integrating the sum of the left- and right-hand sides of the obtained system of equations, previously multiplied by $\text{d}\Delta\delta_{\text{ij}}$ [21]. The obtained first integral is a Lyapunov function and its total derivative, by virtue of the conservative model equations, is identically equal to zero.

If the Lyapunov function is constructed in the same manner as for the conservative model, its derivative can be obtained by virtue of the initial system of equations. In this case, the total derivative of the Lyapunov function alternates in sign and is close to zero. Therefore, a so-called generalized Lyapunov function is obtained, which nonstrictly corresponds to the conditions of the modified Lyapunov theorem. Nevertheless, the studies show the practical acceptability of the obtained estimates [22].

An additional term, proportional to the synchronous machine slip, is quite often added to equations (10.43) in order to take into account the natural damping of the synchronous machine oscillations in the transient processes due to automatic voltage regulators and PSS and because of energy dissipation in the active impedances of the electric network. In the elementary case, the slip factor is set by the constant value D_i, then the indicated additional term $D_i s_i$ can be added to the left-hand side of the second equation of system (10.43).

For the obtained extended model, the Lyapunov function can be constructed on the basis of its conservative idealization, for example, in the form of (10.45) and its total derivative, taken by virtue of the extended model equations, will have the form $\dot{U} = -\sum_{i=1}^{n} D_{\text{i}} s_{\text{i}}^2$ [9]. It is quite obvious that the function $\dot{U}$ has a negative sign, that is, the conditions of the modified theorem relating the stability of dynamic systems are satisfied.

Podshivalov in [23], using recommendations from [6,9], considered the construction of the Lyapunov function for the Lienard vector equation with application to some modified extended model of electric power system dynamics. In the model, an additional term in the second equation of system (10.43) for the i-th synchronous machine is represented by $\sum_{k=1}^{n} D_{\text{ik}} s_{\text{k}}$. Thus, the constructed Lyapunov function and its total time derivative satisfy the conditions of the modified Lyapunov function (LaSalle's Invariance Principle).

Tavora and Smith in [24], on the basis of the concept of the center of mass from the classical mechanics, introduced the transformation of coordinates in the system of equations (10.43) with respect to the center of angles (the center of inertia) of the system. The same transformation was considered by Gorev in [13]. Athay, Rodmore, and Virmani in [25] constructed the energy Lyapunov function for the classical model of an electric power system, transformed in accordance with [24], with respect to the center of inertia. This approach was also used to develop the modified method for studying the transient stability on the basis of Lyapunov functions. The method is presented in Subsection 10.2.4.4. Therefore, the transformation of the classical model with respect to the center of inertia of a system and formulating the respective Lyapunov function are considered below in more detail.

Taking into account the fact that the classical models of electric power system dynamics are not universal, attempts were made to construct the Lyapunov functions for rather detailed mathematical models.

One of the approaches refers to determining Lyapunov function for a system model that includes the electric network structure. Tsolas, Arapostathis, and Varaiya [26] proposed such a Lyapunov function to represent the synchronous machines by the classical model (swing equation) and the loads by constant capacities. Alberto and Bretas [27] considered a more realistic model of the load by including a constant component, a linear dependence of active load on frequency, and nonlinear dependence of reactive load on voltage.

Another approach is represented by the attempts to construct Lyapunov function for detailed models of synchronous machines within the classical model of a reduced electric network. The general method implies constructing the Lyapunov function in a quadratic form of all coordinates of the linearized system of differential equations [8,9]. Other methods are related to determination of the approximated Lyapunov function components, which represent additional variables of the detailed model of the synchronous machine with respect to the classical model of synchronous machine [28–30]. The most promising method is the one with Lyapunov function constructed as a sum of energy integral for the conservative model of the system and a quadratic form of coordinates that are not considered in the conservative model or are considered as constant values [22,31,32].

10.2.4.3 Determination of Equilibrium. Application of the Lyapunov functions method to solve transient stability problems for multimachine power systems is of particular interest because it gives an answer to the question about stability or instability of the considered dynamic transition. Furthermore, it helps finding the conditions to be satisfied by the initial disturbances and parameters of the swing equations for the transition to be stable, that is, constructs the criterion of stability "in the large."

In the phase space of variables, this criterion separates a stability (an attraction) region that is bounded (for the above considered Lyapunov functions) by the separation surface passing through one of the saddle points of the Lyapunov functions. From Morse's theory [33], it is known that such a separation surface represents a surface passing through the saddle point of the Lyapunov function, at which the function takes a minimum value.

Thus, the study of transient stability of electric power systems by using the Lyapunov function method aims to determine the coordinates of the saddle point, through which the separation surface passes, and to calculate the Lyapunov function value in this point. Then the criterion of transient stability can be represented in the form [8]:

$$\begin{aligned} \Delta\delta &< \Delta\delta_{cr} \\ U &< U_{cr} \end{aligned} \tag{10.46}$$

where U is a Lyapunov function value at the initial disturbed state; U_{cr} is a Lyapunov function value at the criterion saddle point; $\Delta\delta_{cr}$ are the coordinates of the saddle point; $\Delta\delta$ are values of variables at the initial disturbed state.

The coordinates of system equilibrium, for which stability is estimated, and the coordinates of the saddle points (unstable equilibriums) in the neighborhood of the supposed stable equilibrium (postfault state) are determined from the systems of nonlinear algebraic equations that are obtained by equating to zero the right-hand sides of equations (10.43) or the derivatives of Lyapunov function of the form (10.45) for all the variables. In general, the number of unstable equilibrium points in the neighborhood of the postfault state can be no less than $(2^{n-1} - 1)$, where n is the number of machines.

The postfault state, for which stability is estimated, is determined by different general methods for solving nonlinear systems of algebraic equations. For this purpose, the Newton–Raphson method was applied in [34] and the steepest descent method was suggested in [35]. Podshivalov [23] proposed a modification of the controlled differential descent method by constructing an auxiliary system of differential equations on the basis of the derivatives of the Lyapunov function for all variables and by adding the artificial damping to the differential equations. The damping coefficients are chosen so that the asymptotic approach of the system trajectory to equilibrium is granted and a quite high convergence speed of the computational process is achieved. The values of variables in the prefault system state are considered as an initial approximation in the above-mentioned methods.

A considerably more complicated problem is to determine the coordinates of the critical saddle points. In this case, two approaches are possible. The first approach is to find the coordinates of all critical points of Lyapunov function and to compare them in terms of the Lyapunov function value at these points [34,36]. The second approach is characterized by the avoidance to determine coordinates of all critical points and directly search for a required saddle point of Lyapunov function [22,35,37].

In Refs. [34,38], the first approximation for identifying the saddle point of the Lyapunov function is found by reducing the multimachine system to $(n-1)$ two-machine systems considering the equilibriums between one basic machine and the rest of machines in the system. It is advisable to choose the machine with the highest inertia constant as reference. When constructing two-machine systems the angle difference of all other machines, excluding the reference and the considered ones, is neglected. In [36,39], it was proposed that similar two-machine systems should be arranged by considering each machine with respect to the remaining part of the system. The obtained estimates are taken as initial values of the variables for the steepest descent method that is applied to determine exact values of the coordinates of saddle points of the Lyapunov function.

In Ref. [35], it is noted that even if the suggested procedure for determining the first approximation in the steepest descent method is valid and provides convergence to the unstable equilibrium; it is still not clear whether this position will characterize a saddle point of the separation surface covering the stability region in the phase space of variables. The procedure proposed in Refs. [36,39] has also drawbacks, since the equilibrium can be violated not only between one machine and the rest of the system, but also between groups of machines.

The authors of Refs. [8,37] proposed an analytical procedure for determining the coordinates of the criterion saddle point by searching for a rigorous lower bound of the function in the form given by (10.45) on the multidimensional cube faces. Its essence is presented in the following.

The nature of variation in the components of Lyapunov function U that depends only on deviations of machine angles from equilibrium is studied in definite directions from the origin of coordinates (stable equilibrium). These directions are chosen so that the multidimensional problem is reduced to the one-dimensional. Then the maximum component of Lyapunov function and its location are determined based on the simple analytical expressions.

All the extreme values of the Lyapunov function components found in such a way are compared with one another, and the minimum ones among them are taken as the criterion constant U_{cr} (as a first approximation). The values of coordinates of the saddle point criterion of Lyapunov function are determined simultaneously.

If the choice of directions, in which the studies are performed on the components of the Lyapunov function, is related to the centers of multidimensional cube faces, the following analytical relations are obtained as a result for the criterion constant estimation [8,37]:

$$U_{\mathrm{cr}} = \min \omega_{\mathrm{i}} \tag{10.47}$$

$$\omega_{\mathrm{i}} = 2A_{\mathrm{i}} - |B_{\mathrm{i}}|\varepsilon_{\mathrm{i}} \tag{10.48}$$

$$\varepsilon_i = 2A \operatorname{atan}(A_i/|B_i|) \tag{10.49}$$

$$A_{\mathrm{i}} = \sum_{\substack{j=1 \\ j\neq i}}^{n} E'_{\mathrm{i}}E'_{\mathrm{j}}Y_{\mathrm{ij}}\cos\beta_{\mathrm{ij}}\cos\delta_{\mathrm{oij}} \tag{10.50}$$

$$B_{\mathrm{i}} = \sum_{\substack{j=1 \\ j\neq i}}^{n} E'_{\mathrm{i}}E'_{\mathrm{j}}Y_{\mathrm{ij}}\cos\alpha_{\mathrm{ij}}\sin\delta_{\mathrm{oij}} \tag{10.51}$$

Then the steepest descent method, the Newton–Raphson method or any other method can be applied to specify the coordinates of the saddle point criterion and the Lyapunov function value for it.

Vaiman in Ref. [40] suggested the search for the required saddle point of the Lyapunov function for the conservative models of electric power systems as a common point of the region boundary that is obtained when the generalized Routh–Hurwitz conditions meet the boundary of the region of admissible deviations of the phase variables in a multimachine system represented by the closed surface of the equal level of function U.

Zubov in Ref. [41] treats the general algorithm of searching for stability regions of the Lyapunov functions by using the conditions that hypersurfaces of the levels of Lyapunov function $U = C$ are in contact with the hypersurface $\mathrm{d}U/\mathrm{d}t = 0$.

The algorithms presented in [40,41] involve some computational difficulties.

10.2.4.4 Extension of the Direct Lyapunov's Method.

Let us consider the system of equations (10.43) in another form based on [25], representing self- and transfer-admittances of the reduced network, not in the polar but rectangular coordinate system. Then from (10.43) we obtain

$$\dot{\delta}_i = s_i \tag{10.52}$$

$$J_i\dot{s}_i = P_i - P^*_{ei}, \quad i = \overline{1,n} \tag{10.53}$$

where

$$P^*_{ei} = \sum_{\substack{j=1 \\ j\neq i}}^{n} \left(C_{ij}\sin(\delta_i - \delta_j) + D_{ij}\cos(\delta_i - \delta_j)\right) \tag{10.54}$$

$$P_i = P_{mi} - E_i'^2 G_{ii} \tag{10.55}$$

$$C_{ij} = E_i' E_j' B_{ij} \tag{10.56}$$

$$D_{ij} = E_i' E_j' G_{ij} \tag{10.57}$$

and G_{ij} and B_{ij} are the active and reactive components of the transfer-admittance of the reduced network.

In accordance with Ref. [24], we transform the coordinates of the model given by (10.52) and (10.53) to the coordinates of the system inertia center:

$$\delta_c = \sum_{i=1}^{n} J_i \delta_i / J_c; \quad J_c = \sum_{i=1}^{n} J_i \tag{10.58}$$

$$s_c = \sum_{i=1}^{n} J_i s_i / J_c \tag{10.59}$$

The equation of motion of the system inertia center will then have the following form:

$$\dot{\delta}_c = s_c \tag{10.60}$$

$$J_c \dot{s}_c = P_c - P_{ec}^* \tag{10.61}$$

where:

$$P_c = \sum_{i=1}^{n} P_i; \quad P_{ec}^* = \sum_{i=1}^{n} P_{ei}^* \tag{10.62}$$

The equation of motion of the synchronous machine i with respect to the inertia center will be as follows:

$$\dot{\delta}_{ic} = s_{ic} \tag{10.63}$$

$$J_i \dot{s}_{ic} = P_i - P_{ei}^* - J_i \dot{s}_c - \left(P_e - P_{ec}^*\right) J_i / J_e \tag{10.64}$$

where $\delta_{ic} = \delta_i - \delta_c$; $s_{ic} = s_i - s_c$.

In Ref. [25], Athay, Podmore, and Virmani construct the Lyapunov function by analogy with the manner in which Aylett did in [19] for the conservative model of an electric power system in mutual motion. They add to the expression of potential energy an extra component reflecting the dissipated energy, $D_{ij} \cos\left(\delta_i - \delta_y\right)$, in (10.54) as the integral along the system trajectory from the stable equilibrium to the current values of coordinates. Finally the Lyapunov function for the system of equations (10.63) and (10.64), based on [25], can be represented as

$$\begin{aligned} U = & \frac{1}{2} \sum_{i=1}^{n} J_i s_{ic}^2 - \sum_{i=1}^{n} P_i (\delta_{ic} - \delta_{oic}) - \\ & - \sum_{i=1}^{n-1} \sum_{j=i+1}^{n} \left[C_{ij} \left(\cos\left(\delta_{ic} - \delta_{jc}\right) - \cos\left(\delta_{oic} - \delta_{ojc}\right) \right) - \right. \\ & \left. - \int_{\delta_{oic}+\delta_{ojc}}^{\delta_{ic}+\delta_{jc}} D_{ij} \cos\left(\delta_{ic} - \delta_{jc}\right) \mathrm{d}\left(\delta_{ic} + \delta_{jc}\right) \right] \end{aligned} \tag{10.65}$$

Physically, the Lyapunov function components (the transient energy function) of (10.65) can be interpreted as given below [25]:

- The first component is the kinetic energy of the synchronous machine motion with respect to the system inertia center.

$$\frac{1}{2}\sum_{i=1}^{n} J_i s_{ic}^2 = \frac{1}{2}\sum_{i=1}^{n} J_i s_i^2 - \frac{1}{2} J_c s_c^2 \tag{10.66}$$

- The second component is part of the potential energy of the system that depends basically on the mechanical moments of turbines (see equation (10.42)) with respect to the system inertia center.

$$\sum_{i=1}^{n} P_i(\delta_{ic} - \delta_{oic}) = \sum_{i=1}^{n} P_i(\delta_i - \delta_{oi}) - \sum_{i=1}^{n} P_i(\delta_c - \delta_{oc}) \tag{10.67}$$

- The third component is the potential electromagnetic energy transmitted through the tie line ij.

$$\sum_{i=1}^{n-1}\sum_{j=i+1}^{n} C_{ij}\left(\cos\left(\delta_{ic} - \delta_{jc}\right) - \cos\left(\delta_{oic} - \delta_{ojc}\right)\right) \tag{10.68}$$

- The fourth component is the dissipated energy in the active conductance of tie line ij:

$$\sum_{i=1}^{n-1}\sum_{j=i+1}^{n} \int_{\delta_{oic}+\delta_{ojc}}^{\delta_{ic}+\delta_{jc}} D_{ij}\cos\left(\delta_{ic} - \delta_{jc}\right) \mathrm{d}\left(\delta_{ic} + \delta_{jc}\right) \tag{10.69}$$

An approximate expression for (10.69) after integration for the case of linear trajectory of the system from δ_o to δ_{cr} is presented in [25].

The obtained Lyapunov function (10.65), as noted in [25], enables the estimation of equilibrium stability in the coordinates of the synchronous machines with respect to the inertia center. However, it does not estimate stability of equilibrium in the inertia center coordinates, that is, for the equation of system frequency dynamics.

The Ref. [25] does not address the issue of the total time derivative of the Lyapunov function (10.65). Since the system of equations (10.52) and (10.53), from which equations (10.63) and (10.64) are obtained, is fully equivalent to the system of equations (10.43), it is believed that (10.65), by analogy with [22], is a generalized Lyapunov function and the conditions of negative definiteness of the total time derivative are met nonstrictly for it.

In subsection 10.2.4.3, it was noted that the study on transient stability of electric power systems by the method of Lyapunov functions is based on determining the coordinates of the saddle point, through which the separation surface passes, and calculating the Lyapunov function value at this point. In this case, the Lyapunov function at the saddle point criterion has a minimum value as compared to its values at other saddle points. The transient stability criterion from (10.46) constructed on this base is unique and universal for all possible disturbed states of the system and that is the reason why for many disturbed states it gives too conservative estimates of the stability region, which are far from the necessary and sufficient ones.

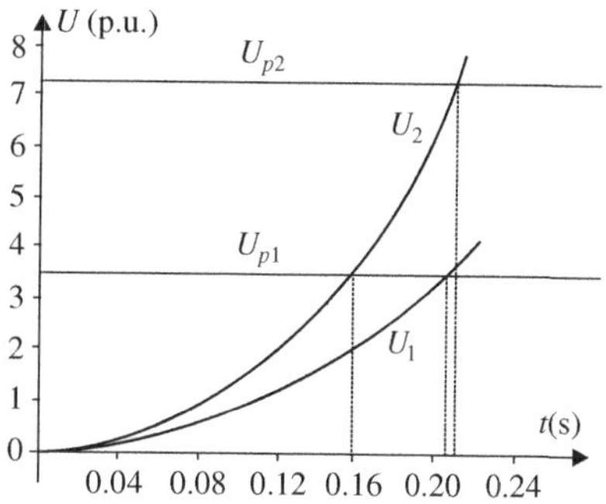

Figure 10.11. Estimation of critical clearing time by different ways of applying the Lyapunov function method.

The above-noted circumstance is illustrated by Figure 10.11 and Table 10.1 [25]. Figure 10.11 shows the total energies U_1 and U_2 as functions of time for two disturbances occurred at two different buses in a three-machine system. The critical levels of potential energy U_{p1} and U_{p2} at each of the saddle points are shown along the y axis.

The system trajectory approaches these saddle points in the considered two fault cases. As seen in Figure 10.11, if the saddle point with the minimum value of the Lyapunov function is used as criterion, the critical fault clearing time equals 0.16 s while the system trajectory in each case brings the system to its saddle point, and the estimates of the critical clearing times are essentially higher.

For the well-known IEEE test system consisting of 39 buses and 10 synchronous machines, Table 10.1 presents the values of critical fault clearing time at different buses. The values are obtained by simulating transient process and on the basis of the Lyapunov function method for the two cases. As seen in Table 10.1, the estimates of the critical clearing time on the basis of the Lyapunov function method, taking into account the system trajectories, practically coincide with the estimates obtained from simulations, whereas the classical method gives less accurate estimates.

In the bulk electric power systems, the trajectory of synchronous machines can be rather complex. The stability can be lost not during the first cycle of swings but during the subsequent ones. In Ref. [42] Gupta and El-Abiad provided an example of a complex system trajectory resulting from the initial disturbance and subsequent control action for the 225 kV CIGRE test system (Figure 10.12).

For this example, Figure 10.13 shows a diagram of change in the total energy of the system versus time along the trajectory of its motion. Figure 10.13 is analogous to Figure 10.11 except for the fact that the numbers along the lines of potential energy level are the numbers of critical generators in the system from the viewpoint of potential stability loss.

TABLE 10.1. Critical Fault Clearing Times

	Simulation		Lyapunov Function Method	
Disturbance Location	Stable	Unstable	Considering Motion Trajectory	Traditional
Bus 31	0.28	0.30	0.28	0.23
Bus 32	0.30	0.32	0.29	0.22
Bus 35	0.34	0.36	0.33	0.29
Bus 38	0.18	0.20	0.18	0.18

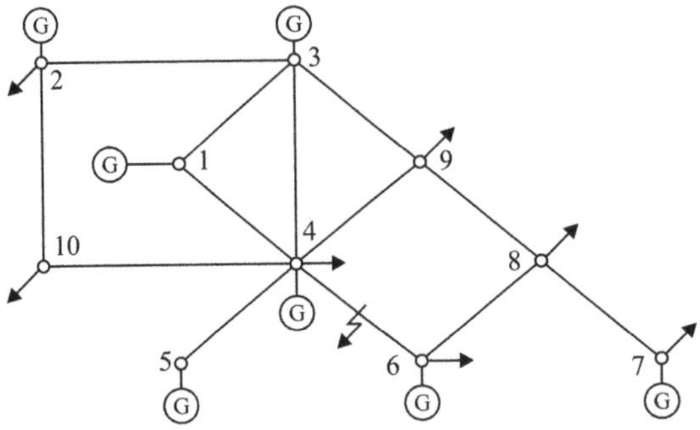

Figure 10.12. A 225 kV CIGRE test network.

In Ref. [42], the authors present an algorithm for determining the possible unstable equilibrium states of a complex system. At the initial stage, consideration is given to two-machine systems formed by each machine in relation to the electric equivalent of the rest of the system. The obtained coordinates of the saddle points criterion are specified for the entire postemergency network by the Newton–Raphson method. Since the method converges in a small number of iterations, its convergence in the given number of iterations is assumed as a criterion for the nonexistence of equilibrium. Thus, the existing saddle points are selected.

The next stage deals with the situation concerning possible loss of stability in a group of machines with respect to the remaining part of the system. For each situation, the system is divided into two groups of machines. The number of machines in the group that losses the stability with respect to the remaining part of the system cannot exceed $n/2$, where n is the number of machines in the system.

It is assumed that in the majority of practically important cases small groups of generators lose their stability with respect to the rest of the system. Hence, the analysis is started with such situations. Similar to the initial stage, the coordinates of the saddle points criterion are obtained for a two-machine electric equivalents, representing two considered groups of machines. Then, they are specified for the entire postemergency configuration by the Newton–Raphson method.

In the real emergency situations, the results obtained allow the saddle point to be taken as a criterion one in the direction of system motion trajectory to assess the power system stability.

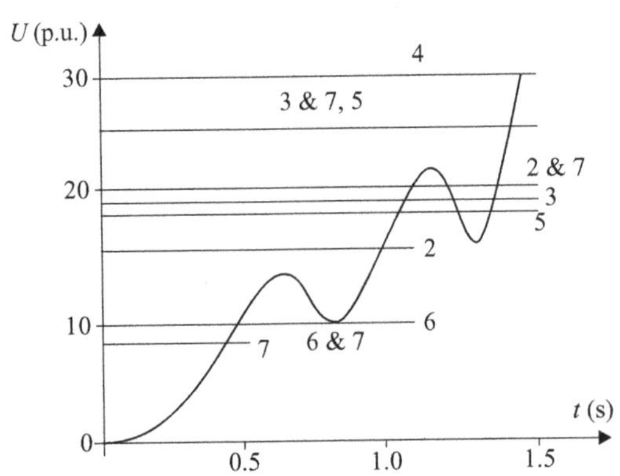

Figure 10.13. A change in the total energy along the system trajectory.

Thus, constructing the Lyapunov function for the dynamic equations of the synchronous machines with respect to the system inertia center and applying the saddle point as a criterion one in the direction of system motion trajectory results in acceptable estimates of transient stability of power systems on the basis of the second Lyapunov method. The estimates virtually coincide with the estimates obtained by simulation of transients.

A systematic description of the method of Lyapunov functions applied in transient stability study of power systems can be found in the technical literature [28,29,42,43].

10.2.4.5 New Approaches.

PEBS METHOD. The method of potential energy boundary surface (PEBS) was suggested by Kakimoto, Oshawa, and Hayashi in [45]. Vaiman names this boundary a dividing line [40]. The key advantage of the PEBS method is that there is no need to determine the coordinates of the unstable equilibriums and hence it is very fast and easy to be applied.

Let the electric power system be represented by the following differential equations in the matrix form [46]:

$$\dot{\delta} = s \tag{10.70}$$

$$J\dot{s} = P_m - P_e - Ds \tag{10.71}$$

The Lyapunov function is also expressed as the sum of the kinetic $U_k(s)$ and the potential $U_p(\delta)$ energies:

$$U = U_k(s) + U_p(\delta) \tag{10.72}$$

The potential energy can be represented as a bowl in the space of angles. The stable equilibrium of the system with a minimum value of potential energy is found on the bottom of this bowl. The points of local maximums and the saddle points lie on the bowl edge. The gradient of potential energy function at these points is equal to zero. The line connecting these points of unstable equilibrium is the potential energy boundary surface or the dividing line. The line is perpendicular to the lines of the equal level of potential energy function.

Let the fault-on trajectory of system $\{\delta(t), s(t)\}$ for determining the critical fault clearing time be directed to PEBS. The point in which the trajectory will cross the boundary is called an exit point with δ^* coordinates. The criterion value of Lyapunov function is determined in this point, that is $U_{cr} = U(\delta^*)$.

Chiang, Wu, and Varaiya have shown [47] that the PEBS method in some cases can give too optimistic estimates of the critical fault clearing time. Figure 10.14, presented also in [46], illustrates such a situation. It shows a fault-on trajectory to the exit point δ^* and a postfault trajectory (situation 2), which reaches the critical point with a potential energy value lower than $U_{er}(\delta^*)$.

BCU METHOD. The boundary controlling unstable (BCU) method was proposed by Chiang, Wu, and Varaiya [48]. The method is based on the concept of controlling unstable equilibrium points (CUEPs) of dynamic systems. Nowadays, this is the most efficient direct method for transient stability analysis in electric power systems.

Let the electric power system be represented by the following differential equations in the matrix form [46]:

$$\dot{\delta} = s \tag{10.73}$$

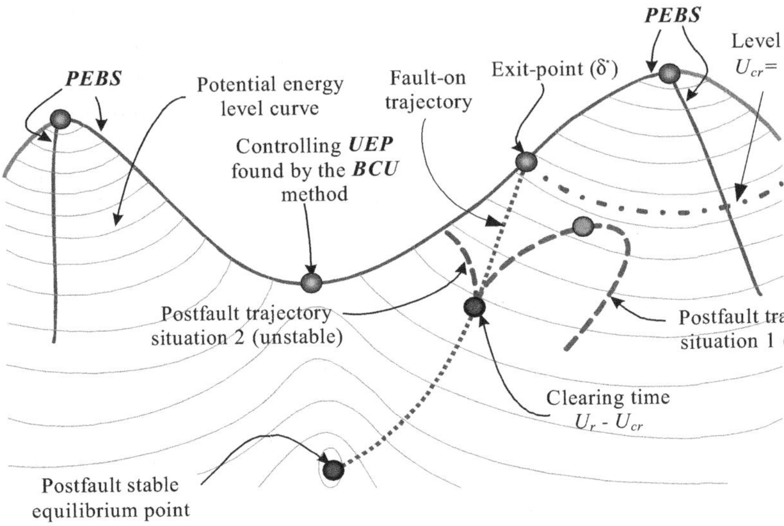

Figure 10.14. PEBS fails.

$$J\dot{s} = -\frac{\partial U_{\mathrm{p}}(\delta)}{\partial \delta} - Ds \tag{10.74}$$

Associated to (10.73) and (10.74), consider the following gradient system:

$$\dot{\delta} = -\frac{\partial U_{\mathrm{p}}(\delta)}{\partial \delta} \tag{10.75}$$

It can be seen that the PEBS method limits the attraction area of the associated gradient system (10.75).

The BCU method exploits the relationship between the systems (10.73), (10.74), and (10.75). Note that $\{\delta, 0\}$ is an equilibrium of the systems (10.73) and (10.74), if and only if δ is an equilibrium of system (10.75). Other interesting relationships between these systems can be found in Ref. [47].

Under quite reasonable hypotheses about the vector field, it is possible to show that the boundary of the attraction area is composed by the stable manifolds of unstable equilibriums, which belong to the boundary of the attraction area. The estimated exit point, therefore, is quite close to the stable manifold of some unstable equilibrium point on the boundary of the attraction area of the associated gradient system. In the BCU method, this point is defined as the controlling unstable equilibrium point that in essence is the saddle point (see Figure 10.14).

The mentioned explanation is the theoretical justification for the general case of the property noted by Tagirov [49] for a one-machine system. This property of system motion trajectories to leave the attraction area near the saddle point was implicitly applied in [25], [42], etc.] for multimachine systems.

The estimate of coordinates of the exit point of the fault-on trajectory δ^* by the PEBS method is taken as an initial approximation for the BCU method.

Further, the coordinates of the controlling unstable equilibrium point are determined by the gradient method. In contrast to the PEBS method, the BCU method does not yield unreasonably optimistic estimates of the critical fault clearing time.

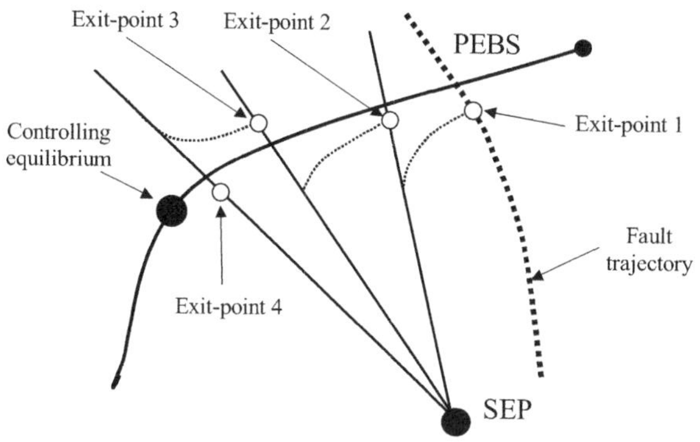

Figure 10.15. Illustration of shadowing method [51].

Shadowing Method. The studies in Ref. [50] show that in some cases the BCU method does not converge to the required CUEP, first, due to the fact that the PEBS method can obtain other exit point δ^*, and secondly even at correct estimation by the PEBS method the gradient algorithm of the second step of the BCU method can converge to the point other than the closest CUEP. Based on this Scruggs and Mili [50] suggested a dynamic method for determining the PEBS.

In order to improve the convergence of the method, Treinen, Vittal, and Klienman suggested the shadowing method [51], which is illustrated in Figure 10.15.

In this method the coordinates of the exit point δ^* are specified by the PEBS method through the fixed intervals of the conjugate gradient system trajectory. In consequence, each new exit point δ^* is a new initial approximation for the gradient algorithm until it converges to a sought CUEP.

Comprehensive Method. In order to efficiently use the positive properties of the above methods, Xue, Mei, and Xie [52] suggested a comprehensive method for determining the CUEP. The flowchart illustrating how the method works is presented in Figure 10.16.

In order to improve the convergence of the gradient method, a so-called reflected gradient system (RGS) is introduced in (10.75) and also in Ref. [52], under the form

$$\dot{\delta} = \left(I - 2\nu(\delta)\nu(\delta)^T\right)f(\delta) \tag{10.76}$$

where $f(\delta)$ is a right-hand side of (10.75), I is the unit matrix, $\nu(\delta)$ is the eigenvector corresponding to the largest eigenvalue (real) of the Jacobian matrix $Y(\delta) = \partial f/\partial\delta$.

In Ref. [52], it is proved that the theorems determine the correspondence of the unstable equilibrium points of system (10.75) with the stable equilibrium points of the reflected system (10.76).

10.3 INTEGRATION METHODS FOR TRANSIENT STABILITY ASSESSMENT

10.3.1 General Considerations

Application of the step-by-step methods, usually employed in transient stability assessment, needs some remarks:

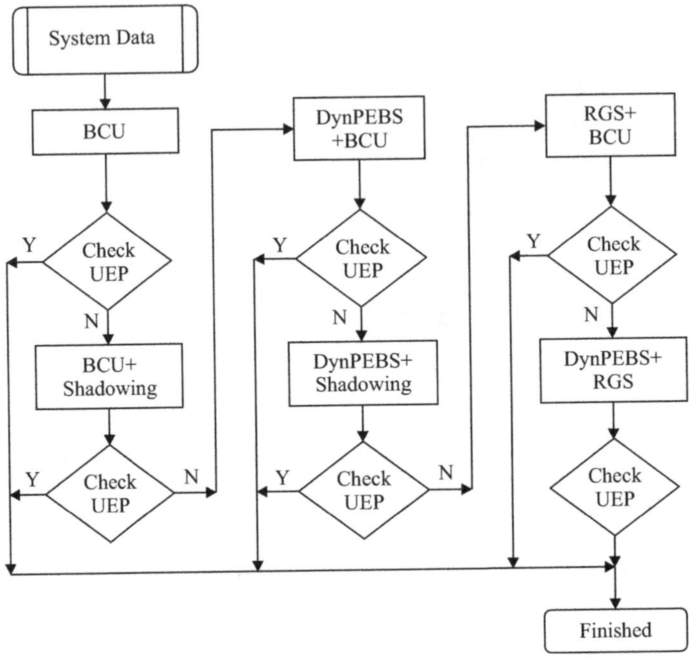

Figure 10.16. Flowchart of the comprehensive method [52].

(i) Two approaches can be used to solve the system of equations (10.1):

- *The sequential approach*, in which the differential (D) and the algebraic (A) systems of equations are solved sequentially.
- *The simultaneous approach*, in which the differential system of equations (D) is transformed into an algebraic system of equations, which is thereafter solved simultaneously with the algebraic equations.

(ii) The integration methods can be classified into two categories: explicit methods and implicit methods.

(iii) The integration methods can be classified into: Euler methods, Runge–Kutta methods, predictor–corrector methods, and so on.

(iv) If the integration step size (Δt) is considered, then we have

- *Constant-step methods*, applicable to transitory stability problems.
- *Variable-step methods*, used in general computational programs (for e.g., EUROSTAG, NEPLAN, NETOMAC, and EDSA) to analyze transient regimes (short-time) and medium and long-term dynamics regimes.

The explicit methods allow computing the state variables, $\mathbf{x}_{n+1}$, at the end of the n-th integration step, in terms of their values at the beginning of the integration step, $\mathbf{x}_n$. However, these methods can lead to numerical instability of the solution due to accumulation of errors, which amplify during the considered interval and cause the divergence of the solution. Therefore, in order to successfully employ an explicit method, it is necessary to set an integration step Δt smaller than the smallest time constants that are included in the differential system of equations. On this basis, the explicit methods (e.g., Runge–Kutta method) are used with good accuracy in computer programs for short-term transient stability assessment (the time interval analyzed is of few seconds).

In an *implicit method*, the state variables $\mathbf{x}_n$ at the end of the n-th integration step are determined in terms of both the values at the beginning of the integration step $\mathbf{x}_{n-1}$ and the actual values $\mathbf{x}_n$. The system of differential equations is thus transformed into a nonlinear system of algebraic equations, of which solution is determined by applying an iterative method at each integration step (e.g., the Newton–Raphson method). The implicit methods are numerically stable because the integration step size is less important than for the explicit methods. The most used implicit method in computation programs is based on the trapezoidal rule, which allows expanding the simulation interval to minutes.

Generally, the performances of a computer program designed to simulate the transient behavior over in the time domain of a dynamic system depend both on the nature of the modeled phenomena (fast and/or slow dynamic phenomena) and on the integration method employed.

Thus, the solution of a linear differential system of equations consists in a linear combination of exponential functions that describe the individual variation modes corresponding to the systems eigenvalues λ_k, that is

$$x_i(t) = \sum_{k=1}^{n} C_k e^{\lambda_k t} \quad i = 1, 2, \ldots, n \tag{10.77}$$

If the eigenvalues are spread over a large area of the complex plane then the solution (10.77) of the system of differential equations is a sum of fast-variation dynamic modes corresponding to the eigenvalues located far from the imaginary axis (the magnitudes of the real parts of eigenvalues are large) and slow-variation dynamic modes given by the eigenvalues positioned close to the imaginary axis (the magnitudes of the real parts of eigenvalues are small). From the integration method viewpoint, such a dynamic system is "*numerically difficult*". Moreover, a nonlinear dynamic system is numerically difficult if its linear approximation is a numerically difficult system.

In the power systems analysis, if besides the electromechanic equations (swing equations) the differential equations modeling the dynamics of the rotor flux (fast-dynamics processes), the AVR, the governor, the turbine, and so on are added in the DAE system, then the set of differential equations is a numerically difficult system of equations, and hence robust algorithms and integration methods are required to simulate the dynamic behavior.

In exchange, if the classic model is used to model the generator then the system of differential equations is no longer a numerically difficult system, and simple explicit methods, such as the Runge–Kutta methods, can be used for numerical integration. The basic features of the most employed integrations methods in power system simulations are presented in the following. Let us consider a simple nonlinear differential equation under the form

$$\dot{x} = \frac{\mathrm{d}x}{\mathrm{d}t} = f(x(t)) \tag{10.78}$$

for which the initial solution is $x(t_0) = x_0$.

The integration methods are employed to solve differential equations, for example equation (10.78), by calculating a series of values $(x_1, x_2, \ldots, x_n, \ldots)$ at different time instants $(t_1, t_2, \ldots, t_n, \ldots)$ that may estimate the dynamic behavior of a system with acceptable accuracy. The value x_{n+1} at the next step is determined in terms of the values obtained at the previous steps $(\ldots, x_{n-2}, x_{n-1}, x_n)$. Depending on the formulae used to calculate each value of x, the integration methods may be classified into two groups: the single-step explicit methods (e.g., Runge–Kutta methods) and multistep implicit methods or predictor–corrector methods.

The value of variable x at a certain step has normally an error with respect to the real value, on one hand, due to solution round off and, on the other hand, due to the integration method used. The errors may propagate from one computation step to the next and, if they do not amplify, it is said that the integration method is numerically stable; otherwise it is numerically unstable [44].

In the integration process the new value x_{n+1} of the state variable x can be determined either by integrating the function $f(x(t))$ over the time interval $[t_n, t_{n+1}]$, or by integrating the variable $x(t)$ over the interval $[x_n, x_{n+1}]$. In both cases, in order to determine the integration expression, extrapolation and interpolation polynomials are used, of which coefficients are calculated based on a set of previous consecutive r values $(f_{n-r+1}, \dots, f_{n-1}, f_n)$ and $(x_{n-r+1}, \dots, x_{n-1}, x_n)$, respectively, computed at the time instants $(t_{n-r+1}, \dots, t_{n-1}, t_n)$. The number of values r is called *the order of the integration method* [44]. Depending on the way in which a new value x_{n+1} is determined using the set of previous r values, the integration methods are classified in *Adams type methods* and *Gear type methods* [44,53].

In the ***Adams type methods*** the value of the state variable at the next step can be determined using the integration formula [44,115]:

$$x_{n+1} = x_n + \Delta t\left(\sum_{k=1}^{r} b_k f_{n+1-k} + b_0 f_{n+1}\right) \tag{10.79}$$

where $\Delta t = t_{n+1} - t_n$ is the integration step size.

For $b_0 = 0$, expression (10.79) provides the *explicit integration formulae*, known as *Adam–Bashforth formulae*. Otherwise, for $b_0 \neq 0$, expression (10.79) provides the *implicit integration formulae*, known as *Adam–Moulton formulae*.

Table 10.2 provides the Adams–Bashforth–Moulton integration formulae up to the third order [44,115]. The first-order integration formulae, for $r = 1$, are the Euler formulae, whereas the second-order integration formulae, for $r = 2$, provides the trapezoidal formulae.

In Adams type formulae, the error at step $(n + 1)$ is [44]:

$$\varepsilon_{n+1} = \varepsilon_0 \cdot x_n^{(r+1)}(\tau) \cdot \Delta t^{(r+1)} \tag{10.80}$$

TABLE 10.2. The Adams–Bashforth–Moulton Integration Formulae [44]

Type	Order	Integration Formulas	Error ε_0
Adams–Bashforth	1	$x_{n+1} = x_n + \Delta t \cdot f_n$	1/2
	2	$x_{n+1} = x_n + \frac{\Delta t}{2}(3f_n - f_{n-1})$	5/12
	3	$x_{n+1} = x_n + \frac{\Delta t}{12}(23f_n - 16f_{n-1} + 5f_{n-2})$	9/24
Adams–Moulton	1	$x_{n+1} = x_n + \Delta t \cdot f_{n+1}$	−1/2
	2	$x_{n+1} = x_n + \frac{\Delta t}{2}(f_{n+1} + f_n)$	−1/12
	3	$x_{n+1} = x_n + \frac{\Delta t}{12}(5f_{n+1} + 8f_n - f_{n-1})$	−1/24

where $x_n^{(r+1)}(\tau)$ is the $(r+1)$ order derivative of x in a point $\tau \in [t_{n-r}, t_{n+1}]$, whereas ε_0 is a constant that depends on the method order (see Table 10.2).

Notice that, for higher order formulae and the same integration step size, the implicit methods generate smaller error than the explicit methods. Another advantage of the implicit methods is that they have better numerical stability. However, their disadvantage resides in the fact that they do not allow direct computation of the value x_{n+1}. Indeed, considering that $f_{n+1} = f(x_{n+1})$ and denoting $\beta_n = x_n + \Delta t \sum_{k=1}^{r} b_k f_{n+1-r}$, from (10.79) the following nonlinear equation results:

$$x_{n+1} = \beta_n + \Delta t \cdot b_0 \cdot f(x_{n+1}) \tag{10.81}$$

where x_{n+1} is the unknown variable.

A *predictor–corrector* method is required to solve the equation (10.81). In the predictor step, an explicit formula is used to determine the initial value $x_{n+1}^{(0)}$. In the corrector step, the initial estimated value is iteratively corrected using the implicit integration formula. In this regard, the following expression is used:

$$x_{n+1}^{(m+1)} = \beta_n + \Delta t \cdot b_0 \cdot f(x_{n+1}^{(m)}) \tag{10.82}$$

where the exponent m is the iteration number.

This iterative computation process converges if [44]

$$\Delta t \cdot b_0 \cdot L < 1 \tag{10.83}$$

where $L = \sqrt{\lambda_{max}}$ is the Lipschitz's constant; λ_{max} is the largest eigenvalue of the matrix product $\mathbf{A}^T\mathbf{A}$, whereas $\mathbf{A}$ is the state matrix computed in the x_{n+1} point.

The smaller the product $\Delta t \cdot b_0 \cdot L$ the faster the process converges. Thus, one way of achieving convergence of numerically difficult systems of differential equations, with large eigenvalues, is to set very small integration steps sizes Δt. This is a drawback because it leads to large computation efforts, which, however, can be reduced by using the Newton method to solve the nonlinear equation (10.81) instead of the recursive formula (10.82).

In the ***Gear type methods*** extrapolation and interpolation are applied to approximate the variable $x(t)$ over the considered integration interval as compared to the Adams methods where variation of the function $f(x(t))$ is applied. In the Gear type method, the following integration formulae are applied [44,115]:

- *The explicit integration formula*

$$x_{n+1} = \sum_{k=0}^{r} a_k x_{n-k} + b_0 \cdot \Delta t \cdot f_n \tag{10.84}$$

- *The implicit integration formula*

$$x_{n+1} = \sum_{k=0}^{r} a_k x_{n-k} + b_0 \cdot \Delta t \cdot f_{n+1} \tag{10.85}$$

The main advantage of the Gear integration methods is that they have a larger numerical stability compared to the Adams methods [44,115].

When the implicit Gear method is employed to solve the nonlinear equation (10.85), similar to (10.81), the predictor–corrector method is used. Furthermore, in the case of numerically difficult differential equations, the initial value estimated by an expression

TABLE 10.3. Gear Type Integration Formulae [44]

Type	Order	Integration Formula	Error ε_0
Explicit	1	$x_{n+1} = x_n + \Delta t \cdot f_n$	
	2	$x_{n+1} = x_{n-1} + 2\Delta t \cdot f_n$	
	3	$x_{n+1} = -\frac{3}{2}x_n + 3x_{n-1} - \frac{1}{2}x_{n-2} + 3\Delta t \cdot f_n$	
Implicit	1	$x_{n+1} = x_n + \Delta t \cdot f_{n+1}$	$-1/2$
	2	$x_{n+1} = \frac{4}{3}x_n - \frac{1}{3}x_{n-1} + \frac{2}{3}\Delta t \cdot f_{n+1}$	$-2/9$
	3	$x_{n+1} = \frac{18}{11}x_n - \frac{9}{11}x_{n-1} + \frac{2}{11}x_{n-2} + \frac{6}{12}\Delta t \cdot f_{n+1}$	$-3/22$
Lagrange	1	$x_{n+1} = 2x_n - x_{n-1}$	
	2	$x_{n+1} = 3x_n - 3x_{n-1} + x_{n-2}$	
	3	$x_{n+1} = 4x_n - 6x_{n-1} + 4x_{n-2} - x_{n-3}$	

similar to (10.84) does not ensure a good approximation when a large integration step size is used. In order to avoid this, the initial value is directly determined by using the Lagrange approximation, that is,

$$x_{n+1}^{(0)} = \sum_{k=0}^{r} a_k x_{n-k} \tag{10.86}$$

The implicit and explicit Gear formulae and the Lagrange extrapolation polynomials are given in Table 10.3 [44].

When the dynamic behavior of a power system is simulated using numerical integration methods, due to the nonlinearity of the mathematical model equations, the state matrix and its eigenvalues are not constant and, therefore, the criteria that bounds the integration step size are changed at each iteration. Thus, choosing an optimal step that, on one hand, can ensure the numerical stability, the convergence, and the accuracy of the results and, on the other hand, that does not lead to large computation time, represents one of the fundamental aspects when developing simulation programs.

Depending on the objective, one of the following two measures are used:

(i) If only the simulation of the transient behavior is performed, then small order (1 or 2) implicit formulae, which have a good numerical stability, are used and a constant integration step of which size is limited to a value that guarantees a good convergence and minimizes the errors, is adopted.

(ii) Numerical methods that automatically modify the order of the method and the integration step size to reduce the computation time are used when designing integrated software that perform simulations both the transient response and the medium- and long-term dynamic behavior.

10.3.2 Runge–Kutta Methods

In the Runge–Kutta methods the solution x_{n+1} is approximated at the end of the actual step $n+1$ by Taylor's series expansion without being necessary for explicit evaluation of higher order derivatives. The contribution of the terms from the Taylor's series expansion containing higher order derivatives are included in the calculus of x_{n+1} by successive evaluation of the first-order derivatives. Different order Runge–Kutta methods exist depending on the number of the retained terms from the Taylor's series expansion.

The second-order Runge–Kutta method consists in successive application of the following relationship:

$$x_{n+1} = x_n + \Delta x_{n+1} = x_n + \frac{K_{1,n} + K_{2,n}}{2} \tag{10.87}$$

where

$n = 0, 1, 2, \ldots$ is the number of the integration step;
Δt is the integration step size;
$K_{1,n} = f(x_n)\Delta t$; $K_{2,n} = f\left(x_n + K_{1,n}\right)\Delta t$

Note that the adjustment $\Delta x_{n+1} = \left(K_{1,n} + K_{2,n}\right)/2$ is the arithmetic mean of the tangents to the variation curve evaluated at the beginning and at the end of the integration step.

The second-order Runge–Kutta method assumes considering only the first and the second-order derivatives of the Taylor's series expansion. In this case, the computation error for each step is proportional to Δt^3.

The fourth-order Runge–Kutta method approximates more accurately the solution x_{n+1} in a computation step by using the expression:

$$x_{n+1} = x_n + \frac{1}{6}\left(K_{1,n} + 2K_{2,n} + 2K_{3,n} + K_{4,n}\right) \tag{10.88}$$

where

$$\begin{aligned} K_{1,n} &= f(x_n)\Delta t & K_{2,n} &= f\left(x_n + 0.5 \cdot K_{1,n}\right)\Delta t \\ K_{3,n} &= f\left(x_n + 0.5 \cdot K_{2,n}\right)\Delta t & K_{4,n} &= f\left(x_n + K_{3,n}\right)\Delta t \end{aligned} \tag{10.89}$$

The adjustment of the variable x at the step $n+1$:

$$\Delta x_{n+1} = \frac{1}{6}\left(K_{1,n} + 2K_{2,n} + 2K_{3,n} + K_{4,n}\right) \tag{10.88}$$

represents a weighted mean of the slopes evaluated at the beginning, the mid-point, and the end of the integration step.

In the fourth-order Runge–Kutta method, the first four-order derivatives of the Taylor's series expansion are considered. The computation error for each step in this case is proportional to Δt^5.

10.3.3 Implicit Trapezoidal Rule

The implicit trapezoidal rule is actually the implicit second-order Adams method. The method is based on the assumption that the adjustment of the variable x at the actual computational step, that is,

$$\Delta x_{n+1} = x_{n+1} - x_n$$

is equal to the arithmetic mean of the exact values that the function f takes at the beginning and at the end of the step multiplied with the integration step size Δt:

$$\Delta x_{n+1} = x_{n+1} - x_n = \frac{\Delta t}{2}[f(x_n) + f(x_{n+1})] \tag{10.90}$$

The trapezoidal rule consists in transforming the differential equation (10.78) into a nonlinear algebraic equation, with respect to the unknown quantity x_{n+1}, which is solved by an adequate method such as the Newton's method.

If the trapezoidal rule is used to determine the transient states of a power system, the differential system of equations (10.1) is transformed into a nonlinear algebraic system of equation, which is solved simultaneously with the algebraic system of equations (10.1) using the Newton–Raphson method.

Application of the trapezoidal rule is illustrated in the following for obtaining the solution of the system of differential equations describing the behavior of a synchronous generator connected to an infinite power bus, using the classical model.

If the damper is neglected, the electromechanical equations are:

$$\frac{d\omega}{dt} = \frac{1}{M}\left(P_m - P_e^{max}\sin\delta\right)$$
$$\frac{d\delta}{dt} = \omega_0\omega$$

where

$P_e^{max} = \frac{E'U}{X'_{d,e}}$ is the maximum value of the electromagnetic power;
E' is emf behind the transient reactance X'_d;
$X'_{d,e}$ is equivalent reactance between the internal generator point, of voltage $E'\underline{|\delta}$, and the infinite power bus, of voltage $U\underline{|0}$.

Applying the implicit trapezoidal rule at the n-th integration step gives:

$$\begin{aligned}\omega_{n+1} - \omega_n &= \frac{\Delta t}{2M}\left(P_m - P_e^{max}\sin\delta_n\right) + \frac{\Delta t}{2M}\left(P_m - P_e^{max}\sin\delta_{n+1}\right)\\ \delta_{n+1} - \delta_n &= \omega_0\frac{\Delta t}{2}(\omega_n + \omega_{n+1})\end{aligned} \tag{10.91}$$

Given the values of ω_n and δ_n at the beginning of the integration step, by solution of the nonlinear algebraic system of equations (10.91) yields their values ω_{n+1} and δ_{n+1} at the end of the integration step. By simple transformations, the system of equations (10.91) becomes:

$$\begin{aligned}a_{11}\omega_{n+1} + a_{12}\sin\delta_{n+1} &= b_1\\ a_{21}\omega_{n+1} + a_{22}\delta_{n+1} &= b_2\end{aligned} \tag{10.92}$$

where

$$a_{11} = 1;\ a_{12} = \frac{\Delta t}{2M}P_e^{max};\ a_{21} = -\omega_0\frac{\Delta t}{2};\ a_{22} = 1$$
$$b_1 = \omega_n + \frac{\Delta t}{M}P_m - \frac{\Delta t}{2M}P_e^{max}\sin\delta_n \text{ and } b_2 = \delta_n + \omega_0\frac{\Delta t}{2}\omega_n$$

The algebraic system of equations (10.92) is solved using the Newton–Raphson method. Therefore, for every iteration p following the current iteration p-1, the system (10.92) is linearized resulting

$$\begin{aligned} a_{11}\Delta\omega_{n+1}^{(p)} + a_{12}\cos\delta_{n+1}^{(p-1)}\Delta\delta_{n+1}^{(p)} &= b_1 - a_{11}\omega_{n+1}^{(p-1)} - a_{12}\sin\delta_{n+1}^{(p-1)} \\ a_{21}\Delta\omega_{n+1}^{(p)} + a_{22}\Delta\delta_{n+1}^{(p)} &= b_2 - a_{21}\omega_{n+1}^{(p-1)} - a_{22}\delta_{n+1}^{(p-1)} \end{aligned} \quad (10.93)$$

from which the adjustments $\Delta\omega_{n+1}^{(p)}$ and $\Delta\delta_{n+1}^{(p)}$ are determined. Then new values of the variables are determined by

$$\omega_{n+1}^{(p)} = \omega_{n+1}^{(p-1)} + \Delta\omega_{n+1}^{(p)} \text{ and } \delta_{n+1}^{(p)} = \delta_{n+1}^{(p-1)} + \Delta\delta_{n+1}^{(p)} \quad (10.94)$$

The initial values of the iterative procedure are chosen equal to those from the beginning of the integration step, that is $\omega_{n+1}^{(0)} = \omega_n$ and $\delta_{n+1}^{(0)} = \delta_n$, and the iterative process is continued until the convergence test $\max\left\{\left|\Delta\omega_{n+1}^{(p)}\right|, \left|\Delta\delta_{n+1}^{(p)}\right|\right\} \le \varepsilon_{adm}$ is satisfied. When the convergence is achieved, set $\omega_{n+1} = \omega_{n+1}^{(p)}$ and $\delta_{n+1} = \delta_{n+1}^{(p)}$, then go of the integration step.

10.3.4 Mixed Adams-BDF Method

Researches have shown that the simulation of the transient, the medium-term, and the long-term processes of the power systems can be performed using integrated computation software. The numerical integration methods implemented in this software are robust and allow changing the order of the method and the integration step size. Such a method is Adams-BDF (backward differentiation formulae), which is implemented in the EUROSTAG software [56]. The method is based on the general Gear–Hindmarsh method used for solving hybrid DAE systems.

The general Gear–Hindmarsh method

Let $\mathbf{z} = [\mathbf{x}^T, \mathbf{y}^T]^T$ be the vector that groups the vector of inertial state variables $\mathbf{x}$ and the vector of noninertial state variables $\mathbf{y}$. The DAE system (10.1) can therefore be written as

$$\begin{cases} \dot{\mathbf{z}} = \mathbf{f}(\mathbf{z}(t)) \\ 0 = \mathbf{g}(\mathbf{z}(t)) \end{cases} \quad (10.95)$$

Given the vector $\mathbf{z}_n = \mathbf{z}(t_n)$ of state variables at the time instant t_n as well as their derivatives up to the order r (r is the order of the method), the problem of determining the vector $\mathbf{z}_{n+1} = \mathbf{z}(t_{n+1})$, which is the solution of the system of equations (10.95) at the time instant t_{n+1} arises. For this, the vector $\mathbf{z}(t)$ and the vectors of derivatives, $\mathbf{z}^{(m)}(t), m = 1, 2, \ldots, r$, are stored in the vector $\bar{\mathbf{z}}$, called the Nordsieck vector:

$$\bar{\mathbf{z}}(t) = \left[\mathbf{z}(t), h \cdot \mathbf{z}^{(1)}(t), \frac{h^2}{2!}\mathbf{z}^{(2)}(t), \ldots, \frac{h^r}{r!}\mathbf{z}^{(r)}(t)\right] \quad (10.96)$$

where $h = \Delta t$ is the integration step size.

The advantage of using the Nordsieck vector is that when changing the integration step size, from h to αh, the new vector is obtained by

$$\bar{\mathbf{z}}(t + \alpha h) = \mathbf{D}\bar{\mathbf{z}}(t + h) \quad (10.97)$$

where $\mathbf{D} = diag\{1, \alpha, \ldots, \alpha^r\}$ is a diagonal matrix, and α is the ratio between the new and the old step size. Also, using the Nordsieck vector, changing the order of the method is achieved by simply changing the dimension of the $\mathbf{D}$ matrix (addition or deletion of a line and of a column).

In the *Gear–Hindmarsh* method, determination of the new vector $\mathbf{z}_{n+1}$ is based on a predictor–corrector procedure.

In *the predictor step*, the initial values of the state variables $\tilde{\mathbf{z}}_{n+1}, \tilde{\mathbf{z}}_{n+1}^{(1)}, \ldots, \tilde{\mathbf{z}}_{n+1}^{(r)}$ as well as the Nordsieck vector at the time instant t_{n+1} are estimated by Taylor series expansion up to the r order and using the known values of the state variables $\mathbf{z}_n, \mathbf{z}_n^{(1)}, \ldots, \mathbf{z}_n^{(r)}$ and their derivatives at the time instant t_n:

$$\bar{\mathbf{z}}_{n+1}^{(0)} = \mathbf{A}\bar{\mathbf{z}}_n \tag{10.98}$$

where $\bar{\mathbf{z}}_n$ is the Nordsieck vector calculated at the previous step t_n, and $\mathbf{A}$ is the Pascal's triangle array of which terms are given by

$$a_{ik} = \begin{cases} \dfrac{k!}{(k-i)!i!} & \text{if} \quad i \le k \\ 0 & \text{if} \quad i > k \end{cases} \tag{10.99}$$

In *the corrector step*, the Nordsieck vector estimate $\bar{\mathbf{z}}_{n+1}^{(0)}$ is adjusted using to the expression [56]:

$$\bar{\mathbf{z}}_{n+1} = \bar{\mathbf{z}}_{n+1}^{(0)} + \mathbf{l}_{n+1}(\mathbf{z}_{n+1} - \tilde{\mathbf{z}}_{n+1}) \tag{10.100}$$

where $\mathbf{l}_{n+1} = [l_{0,n+1}, \ldots, l_{r,n+1}]^T$ is a vector of which components depend on the integration method and on its order. The vector of state variables $\mathbf{z}_{n+1} = \mathbf{z}(t_{n+1})$, which is solution for the DAE system at the time instant t_{n+1} and satisfy the relationships (10.95), is determined by solving the following system of nonlinear algebraic equations

$$\begin{aligned} &h_n\bar{\mathbf{z}}_{n+1}^{(1)} + \mathbf{l}_{1,n+1}(\mathbf{z}_{n+1} - \tilde{\mathbf{z}}_{n+1}) - h_n\mathbf{f}(\mathbf{z}_{n+1}) = 0 \\ &\mathbf{g}(\mathbf{z}_{n+1}) = 0 \end{aligned} \tag{10.101}$$

using the Newton method. This is performed by determining the vector of adjustment values $\Delta\mathbf{z}_{n+1} = \mathbf{z}_{n+1} - \tilde{\mathbf{z}}_{n+1}$ that satisfies the equations:

$$\begin{aligned} &h_n\tilde{\mathbf{z}}_{n+1}^{(1)} + \mathbf{l}_{1,n+1}\Delta\mathbf{z}_{n+1} - h_n\mathbf{f}(\tilde{\mathbf{z}}_{n+1} + \Delta\mathbf{z}_{n+1}) = 0 \\ &\mathbf{g}(\tilde{\mathbf{z}}_{n+1} + \Delta\mathbf{z}_{n+1}) = 0 \end{aligned} \tag{10.102}$$

In order to reduce the computation time, in the iterative process for solving the system of equations (10.102), the Jacobian matrix is computed only once (in the first iteration). Moreover, the Jacobian matrix is maintained constant even for several successive integration steps; the matrix is recalculated when the step size and the method order are changed.

Changing the Step Size and the Method Order. The method error given by (10.81) can be estimated using the Norsieck vector [56]. Therefore, from equation (10.100) achieves

$$\frac{h^r\mathbf{z}_{n+1}^{(r)}}{r!} - \frac{h^r\mathbf{z}_n^{(r)}}{r!} = l_{r,n+1}\Delta\mathbf{z}_{n+1} \tag{10.103}$$

Taking into account the expression (10.103) and that

$$\frac{h^{r+1}(\mathbf{z}(t_{n+1}))^{(r+1)}}{r!} = \frac{h^r\mathbf{z}_{n+1}^{(r)}}{r!} - \frac{h^r\mathbf{z}_n^{(r)}}{r!} \tag{10.104}$$

achieve the approximate expression of the error

$$\varepsilon_{n+1} \cong \varepsilon_0 l_{r,n+1} r! \|\Delta\mathbf{z}_{n+1}\| \tag{10.105}$$

where $\|\Delta\mathbf{z}_{n+1}\|$ is the norm of the vector of adjustment values $\Delta\mathbf{z}_{n+1}$ computed by solving the system of equations (10.102).

Comparing the error calculated using (10.105) with the admissible value ε_{adm} specified by the user according to the desired accuracy, the current integration step size is evaluated and a new value is determined using the expression:

$$h^{new} = h^{old}\left[\frac{\varepsilon_{\text{adm}}}{\varepsilon_0} l_r \|\Delta\mathbf{z}\|\right]^{\frac{1}{r}} \tag{10.106}$$

Also, the equation (10.106) is used to estimate the error when the method order is reduced to $(r-1)$ or increased to $(r+1)$. It is, therefore, possible to identify the time instant when the method order and/or the integration step size should be changed.

THE MIXED ADAMS-BDF METHOD. Although the Gear–Hindmarsh method is adequate to simulate the behavior of the dynamic systems governed by numerically difficult algebraic-differential equations (dynamic systems characterized by high ratio, around 10^4 or higher, between the largest and the smallest eigenvalue of the state matrix), the accuracy of the results and the computational effort, quantified by the CPU time, depend on the integration method used. When analyzing the disturbed states of a power system, in order to obtain a good simulation of the stable cases and to identify the unstable cases, the numerical stability domain of the integration method must include the left-side semiplane of the complex plane, where the eigenvalues of the state matrix are represented [44]. This condition is fulfilled by the implicit first- and second-order Adams methods. Implementation of the second-order Adams method, known as the trapezoidal rule, contributed to elaboration of robust software, for example, the EUROSTAG software, designed to simulate both the transient processes and the medium- and long-term dynamics.

Using the Nordsieck notation, the integration formula of the trapezoidal rule

$$\mathbf{z}_{n+1} = \mathbf{z}_n + \frac{h}{2}(\dot{\mathbf{z}}_n + \dot{\mathbf{z}}_{n+1}) = \mathbf{z}_n + \frac{h}{2}(\mathbf{f}(\mathbf{z}_n) + \mathbf{f}(\mathbf{z}_{n+1})) \tag{10.107}$$

provides the vector

$$\mathbf{l}_{ADAMS} = [l_0 = 0.5,\ l_1 = 1,\ l_2 = 0.5]^T \tag{10.108}$$

used in the general Gear–Hindmarsh method.

Although the trapezoidal rule has a good numerical stability, the tests have shown that there are situations when unstable cases cannot be identified.

Moreover, if the model takes into consideration the algebraic state variables in order to estimate the error and the accuracy of the results, requiring small values of the integration step size (automatically selected by the method), then excessively large computation times would be necessary. In order to eliminate this disadvantage, the BDF method was used in the EUROSTAG software for the algebraic variables. The mixed Adams–BDF method that has thus resulted uses the Adams method to deal with the inertial state variables and the BDF method for the algebraic variables.

The formula of the implicit BDF method is

$$\mathbf{z}_{n+1} = -\frac{1}{3}\mathbf{z}_{n-1} + \frac{4}{3}\mathbf{z}_{n+1} + \frac{2}{3}h_n\dot{\mathbf{z}}_{n+1} = -\frac{1}{3}\mathbf{z}_{n-1} + \frac{4}{3}\mathbf{z}_{n+1} + \frac{2}{3}h_n\mathbf{f}((\mathbf{z}_{n+1})) \tag{10.109}$$

which, in the Nordsieck notation, leads to the vector

$$\mathbf{l}_{BDF} = \left[l_0 = \frac{2}{3}, l_1 = 1, l_2 = \frac{1}{3}\right]^T \tag{10.110}$$

Implementing the mixed Adams–BDF method in the general Gear–Hindmarsh algorithm is very simple and consists in replacing the vector $\mathbf{l}_{n+1}$with the vector $\mathbf{l}_{ADAMS,\,n+1}$ when dealing with the inertial state variables, and with the vector $\mathbf{l}_{BDF,\,n+1}$ when dealing with the algebraic state variables.

10.4 DYNAMIC EQUIVALENTS

10.4.1 Generalities

The actual electric power systems are the most complex technical systems due to the large number of interdependent sets of elements, the various types of operating states (conditions) and the complicated processes that develop after disturbances occurrence. It is, therefore, necessary to investigate them, under the classical postulate of the systems theory, by utilizing simplifying methods. The formulation of dynamic equivalents, in fact, represents the science of reducing the dimension of the mathematical models of the systems [57].

Drawing upon the approaches employed in studying the transient stability of the electric power systems, the following stages of mathematical model simplification may be identified [58,59]: *a priori* idealization; simplification of power system mathematical description, and finding the adequate power system mathematical model.

An *a priori* idealization is based on the rich experience by theoretical research and practical calculations of transient processes of various durations in modern bulk power systems, which take into account the capabilities of computers and software tools used for simulating transient processes. Examples of such idealizations are: representing the system as one-line positive-sequence diagram, neglecting stator resistances, and ignoring the electromotive forces of transformers and the synchronous generator rotation and some others. This renders it reasonable to neglect some of the factors in mathematical description of power system transient processes. The resulting mathematical model of electric power system dynamics in the form of a design model and mathematical descriptions of elements can then be employed.

The simplification stage of the mathematical description of the power system dynamics implies simplifying the mathematical models representing classes of elements through neglecting less relevant features. A mixture of widely used expert approaches [60,61], the method of small parameters (a singular case) [60,62,63] and modal analysis [60,64] can be applied in this respect.

The next problem of the simplification process is to determine the impact of disturbances on the behavior of the electric power system elements [59,65]. The solution to this problem is based on a well-known property of large electric power systems that the impact of a disturbance is experienced less intense the further away from the originator the element is located due to energy dispersion, the presence of insensitivity zones in the generators' speed governors, and so on. Different familiar indices measuring the disturbance impact are based on the assessment of "electric distance" (transfer admittances) or "distance with respect to perception" (synchronizing powers) [65,67]. Also, these indices take into account the disturbance characteristics and the dynamic parameters of generators [58,60,68,69]. Further investigations on determining these complex indices are provided in Section 10.4.2.

Along with the impact degree of a disturbance, the level of detail necessary in modeling a power system element for a given disturbance is provided by assessing the significance of an element in the sense of its impact on the transient process. The elements of high significance (generators and loads) are characterized by a large capacity and strong ties with the system as a whole. The quantitative significance estimates of the electric power system elements are obtained from the analysis of system structure (see Subsection 10.4.3) [65,70] and by using the modal analysis [71]. Notice that the element significance estimates do not employ the disturbance characteristics and therefore are invariant with respect to them.

Based on the estimates of disturbance impact and of element significance, the level of detail in its mathematical model can be assessed. Finding such estimates at this stage for all elements allows a simplified mathematical description of the dynamics of electric power system behavior to be formulated for a given network configuration, operating conditions, and disturbance. This also provides the basis for identifying the power system subsystems in which electric network can be reduced.

The last stage is concerned with the simplification of the mathematical model for the electric power system dynamics by formulating an equivalent. It includes finding the solution of two problems: determining the subsystems to be represented by an equivalent and calculating the parameters of the power system equivalent model [58,65].

The subsystems to be represented by a single equivalent are determined on the basis of generator motion coherency. Originally this problem was solved using approximated, often empiric, characteristics: full symmetry of the subsystem represented by equivalent, equality of initial accelerations of generator rotors, equality of synchronizing powers of machines, meeting the stability conditions within a group [58,60,72–75,109]. In order to achieve a more accurate estimate of the coherency, a numerical calculation of the initial stage of transient process was employed in [76–78] either by using a nonlinear model or based on modal analysis of a linearized model. The analytical approach for the estimation of motion coherency of machine pairs on the basis of different indices is presented in Subsection 10.4.4.

In order to study long-term transient processes in electric power systems, the use of estimates of generator motion coherency at non-zero initial conditions, that are obtained on the basis of the second form of the stability criterion, was suggested by Gorev [79]. Subsection 10.4.4 considers the possibilities of obtaining and applying these estimates.

A new step forward was the introduction of local and global coherency concepts [80]. Local coherency is determined by the structural properties of the subsystem and the disturbance, whereas the global coherency is determined only by structural properties, that is, it is invariant with respect to disturbances. Global coherency is revealed on the basis of structural analysis methods using the indices of electric and dynamic connectivity (Subsection 10.5.4) [65,70,74,75,77,80], modal analysis [60,63,66,77,81,82,109,110], transformation of transfer functions of a linearized system [83], and analysis of the potential members of the Lyapunov function [84–86].

Essential progress in the area of local and global coherency was made on the basis of transformation of the electric power system coordinates to the *inertia center coordinates*:

$$\{\delta_i\} \rightarrow \{\delta_c, \delta_{ic}\}, \quad i = \overline{1, n}$$

where $\delta_{ic} = \delta_i - \delta_c$ is the angle of generator i with respect to the system (subsystem) inertia center.

The transformation was introduced by Gorev [79] and then used in [76,87]. It was proved that the inertia center of a coherent subsystem executes a slow motion, whereas the coordinates δ_{ic} vary considerably faster [63,72,88]. This observation provided the background for developing applications of the methods of singular disturbance theory to reveal the coherency and to develop the concept of "slow coherency" between subsystems (inertia centers of subsystems) that are invariant with respect to disturbances [63,89].

The solution to the simplification problems of mathematical description of the electric power system and the determination of the equivalent subsystems allows the division of the system under analysis in two parts–the subsystem where transient processes will be investigated and the external subsystem that can be represented by a simplified equivalent. These parts of the system (subsystems) are interconnected through the adjacent nodes (boundary nodes).

Computing the equivalent power system parameters require that certain criteria for transforming the initial system to the equivalent one must be fulfilled. For the non-transformed subsystem the equivalence criteria require the invariance of its behavior, which often reduces to the invariance of state variables at the boundary nodes. For the equivalent subsystem the equivalence criteria determines the relationships that are used to determine the parameters of the equivalent. Both groups of the criteria should be coordinated [90–92]. They are considered further in Subsection 10.4.5.

For the determination of the equivalent parameters in the classical mathematical model for electric power system dynamics, the technique of averaging the parameters [58,60,86,91], the Dimo-REI method [54], the modal analysis [60,63,66,80,92], and the small parameter method, in addition to the relationships generated by the equivalence criteria (Subsection 10.4.6.), were also used. The small parameter method was also used for transforming the coordinates with respect to the inertia center [73,88,89]. Consideration was given to the possibilities of determining the equivalent's parameters so that the voltage regulators and generators' speed governors in the determined subsystem [58,65,82] and also the structure of electric network [94] are taken into account.

Since no ideal coherency of generators motion is encountered in practice but rather an approximated coherency is observed, the need arises to take into account the noncoherency of motion in the parameters of the equivalent. Some elementary methods were suggested to solve this problem. These methods introduce weight coefficients for the initial parameters with the initial accelerations or active powers of equivalated generators, representing the equivalent of two or more generators [95–97]. The general approach to the consideration of the motion noncoherency in the equivalent generator parameters is based on averaging techniques (Subsection 10.4.6) [59,65].

Among the other methods for determining the parameters of the equivalent, an emphasis should be placed on the aggregation of linearized models [98], dispersion of the inertia constants of generators in the subsystem to be represented by equivalent with reduced networks [99], continual idealization [72], identification methods [100–102], and methods for functional equivalent formulation [103,104]. These methods, however, have not become widespread.

10.4.2 Simplification of Mathematical Description of a System

10.4.2.1 The Disturbance Impact Index. In transient stability studies, the impact of a disturbance upon the behavior of a considered power system element (generator, load) during the transient process is determined by the disturbance characteristics (magnitude, duration, etc.) and by the distance between the considered element and the disturbance point. The influence of the disturbance magnitude and duration are obvious, while the second factor is a vivid expression of the above—mentioned property of bulk power systems—the disturbance impact decays as the distance between the point of its application and the element increases.

Hence, the index that characterizes the disturbance impact at node i in the system is determined by two components: a coefficient measuring the electrical distance of node i from the disturbance point, k_{ia}, and the effect of disturbance at node i, Δx_i, that is,

$$\gamma_i = f(k_{ia}, \Delta x_i) \tag{10.111}$$

Here, k_{ia} is given by the configuration of the system and the initial state of the system, and Δx_i by the disturbance characteristics.

Let the boundary values, $\overline{\gamma}_1 > \overline{\gamma}_2 > \ldots > \overline{\gamma}_\ell$, be set such that: if $\gamma_i > \overline{\gamma}_1$, the element at node i should be represented in the mathematical model of the system by the most detailed mathematical description (model M_1); if $\overline{\gamma}_1 > \gamma_i > \overline{\gamma}_2$, the element at node i is introduced by a simpler model M_2; for $\overline{\gamma}_2 > \gamma_i > \overline{\gamma}_3$ a further simplified model M_3 of the element is used and so on.

For the disturbance impact index, as shown in [58], the so-called F coefficient [68] can be adopted, which is determined by the expression:

$$\gamma_i = Y_{ia}\Delta P_i t_a^2 / 2J_i \tag{10.112}$$

where t_a is the operation duration under emergency conditions, ΔP_i is power imbalance at node i under the emergency conditions; J_i is inertia constant of generator or (a)synchronous motor at node i; Y_{ia} is magnitude of the transfer admittance between the node i and the node where the disturbance is applied.

A cumbersome problem is caused by selecting the boundary admissible values $\overline{\gamma}_\ell$ for the disturbance impact index (10.112). There are no rigorous methods for solving this problem. Therefore, it is necessary to use the experience of research and control computations to compare the results obtained for the initial and for the simplified mathematical models of the electric power system.

10.4.2.2 The Study of the Disturbance Impact Index. Consider a specific power system example [58], of which configuration is shown in Figure 10.17. For illustration, only the problem of simplifying the mathematical description for the

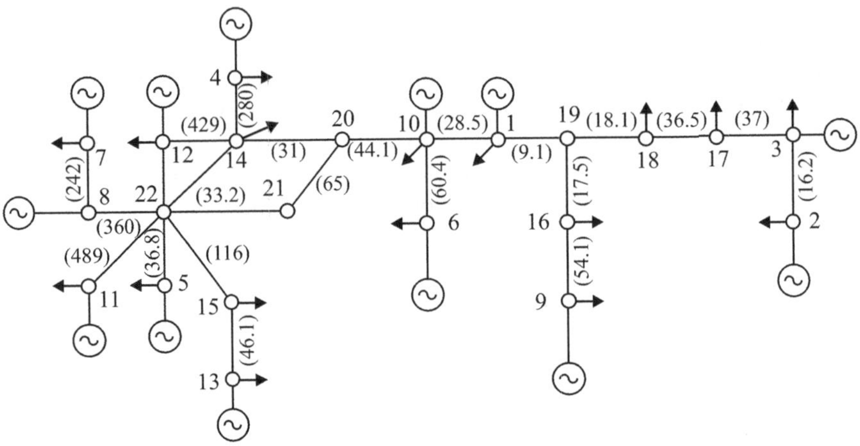

Figure 10.17. Configuration of the studied system.

generators will be considered. The reactances of the tie lines are shown in brackets, in ohms. The other necessary parameters of the system are given in Table 10.4, where P_g denotes the generation and P_ℓ denotes load. The classical model of an electric power system is considered as an initial approximation that takes into account the voltage regulation systems and the speed governors of generators.

A symmetrical three-phase short circuit at node 1 lasting for 0.2 s was considered as the disturbance for which the system response is analyzed. The transient process caused by this disturbance, assuming the initial mathematical model of the system, is shown in Figure 10.18.

Figures 10.19 and 10.20 illustrate the estimates of the electric distances of generators with respect to the disturbance point.

The regions of equal electric distance of generators (Figure 10.20) are determined in accordance with the values of the transfer admittances between generators and the node where the disturbance is applied. The regions are plotted on the number axis (Figure 10.19).

The regions obtained on the basis of electric distance will be compared with the regions that will then be formulated on the basis of the disturbance impact indices. The values of these indices are presented in Figure 10.21 and Table 10.5.

TABLE 10.4. Parameters of the Studied System [58]

Node	P_g [MW]	J[s]	P_ℓ[MW]	Node	P_g[MW]	J [s]	P_ℓ[MW]
1	1120	25.2	470	12	255	112.0	435
2	1515	53.5	1060	13	1145	216.0	700
3	1275	10.5	1300	14	–	–	602
4	155	19.5	313	15	–	–	880
5	798	28.8	140	16	–	–	800
6	310	37.2	255	17	–	–	190
7	755	106.0	677	18	–	–	124
8	175	49.5	–	19	–	–	–
9	780	16.8	76	20	–	–	–
10	130	47.6	155	21	–	–	–
11	380	64.0	480	22	–	–	420

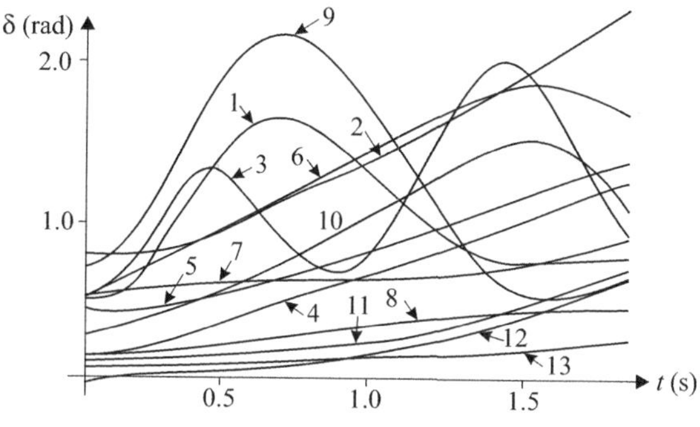

Figure 10.18. The rotor angle variations for a selected disturbance.

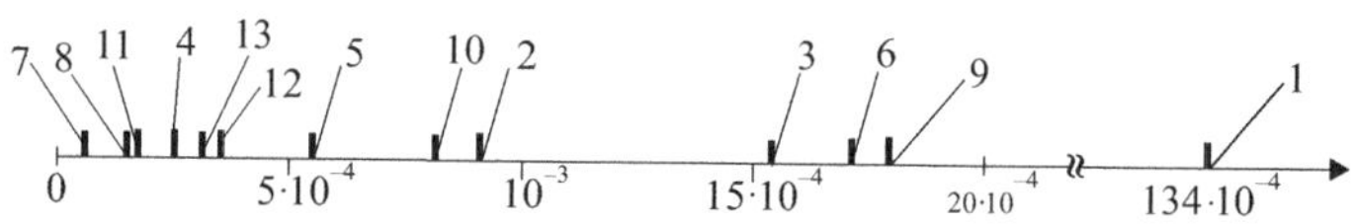

Figure 10.19. Location on the number axis of magnitudes of generator transfer admittances with respect to the node where the disturbance was applied.

Consider the response of the most critical generator connected at node 1, where the short circuit occurs. Figure 10.22 shows the variation of the rotor angle of this generator for various levels of detail considered in modeling the other generators, that is: 1—the initial model of the system; 2—generators 7, 11, 13 are modeled by the swing equation (SE), also called equation of motion, assuming $E' = \text{ct.}$ and $P_m = \text{ct.}$; 3—the same as 2 plus generators 4 and 8 modeled using SE only; 4—the same as 3 plus generator 12 modeled using SE only; 5—the same as 4 plus generators 5 and 10 modeled using SE only; 6—the same as 5, plus generators 2 and 6 modeled using SE only; 7—all generators except for the first one are modeled by SE only.

Based on the similarity of generator no. 1 rotor angle variation for the assumed initial representation of the other generators in the system and for the simplified models

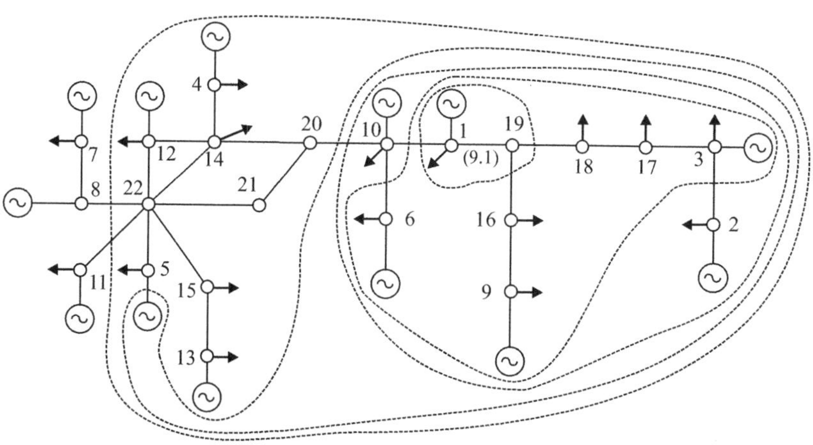

Figure 10.20. Decomposition of the system by the criterion of electric distance of generators with respect to the disturbance point.

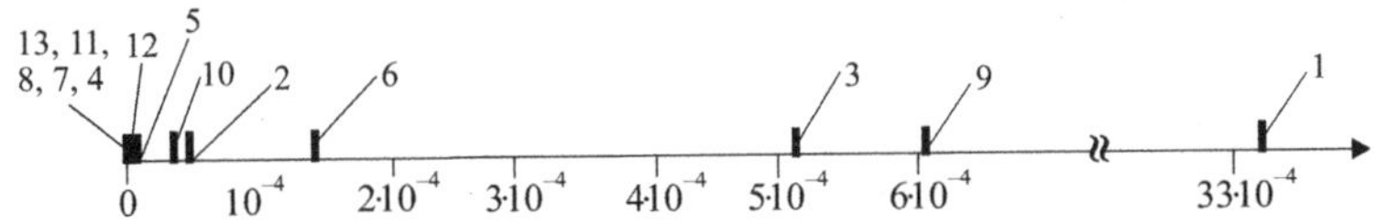

Figure 10.21. Location on the number axis of disturbance impact indices for generators.

TABLE 10.5. The Values of Disturbance Impact Index

Generator	1	2	3	4	5	6	7
$\gamma/10^{-4}$	33.4	0.42	5.2	0.011	0.075	1.45	0.005
Generator	8	9	10	11	12	13	–
$\gamma/10^{-4}$	0.01	6.1	0.33	0.001	0.037	0.0004	–

subsequently used for other generators it can be concluded, based on the results shown in Figure 10.22, that sufficient evidence exists for using the swing equation only to represent generators no. 7, 11, 13. It is less advisable to use the swing equation only for generators no. 4, 8 and, probably, no. 12.

In order to gain more confidence in determining the condition for using the simplified model of generators and to estimate a suitable threshold for $\overline{\gamma}$, a comparison of the critical fault clearing times from the transient stability point of view is performed in the following. For the cases presented in Figure 10.22 these are given in Table 10.6 in absolute and relative (to the critical clearing time for the initial model) units.

It results from Table 10.6 that if 5% is assumed an admissible simplification error, the swing equation can be used to represent generators no. 4, 7, 8, 11, 12, 13, with $\overline{\gamma} = 0.37 \cdot 10^{-5}$. However, if the simplification error is assumed less than 10% the simplified model can be used to represent generators no. 4, 5, 7, 8, 10, 11, 12, 13, with $\overline{\gamma} = 0.33 \cdot 10^{-4}$.

Taking into account the rather large difference between the first and second case (from 5% to 10%) and the fact that the error of 10% is hardly acceptable, the threshold value $\overline{\gamma}$ can be considered to be higher than $0.37 \cdot 10^{-5}$ but lower than $0.33 \cdot 10^{-4}$ (for example, 10^{-5}). The boundary separating the regions for transition from the initial complete model of system's generators to the simplified one for this case is represented in Figure 10.23 by the dashed line.

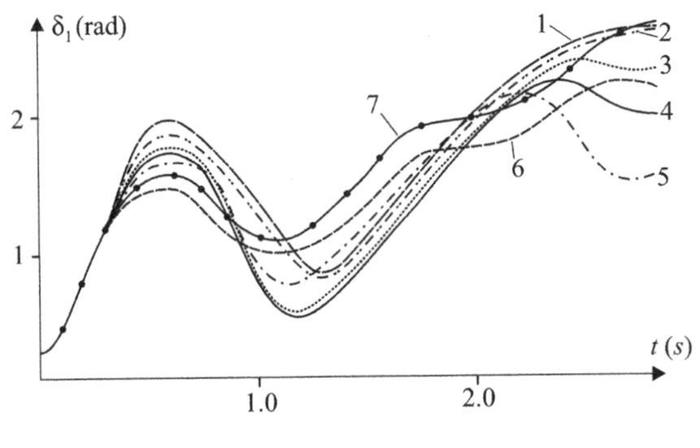

Figure 10.22. Generator 1 rotor angle variation assuming different mathematical modeling of the system elements.

TABLE 10.6 Critical Clearing Time

No	Type of Model	Clearing Time s	Clearing Time %
1	Complete system model	0.360	0
	Simplified models of generators:		
2	7, 11, 13	0.370	2.70
3	4, 7, 8, 11, 13	0.375	4.10
4	4, 7, 8, 11, 12, 13	0.378	4.90
5	4, 5, 7, 8, 10, 11, 12, 13	0.395	9.80
6	2, 4, 5, 6, 7, 8, 10, 11, 12, 13	0.425	18,05
7	All generators except for generator 1	0.450	25.00

Assumes that the loads of the studied power system are also represented by rather complete models (e.g., by static characteristics dependent on voltage and frequency and/or by dynamic equations in the case of asynchronous motors). The authors of [105] present the diagram for choosing a complex load model in terms of the level and duration of load decrease due to disturbance at the node where the load is connected. Using these recommendations, a threshold condition for the transition from the detailed load model to the simplified one can be determined, for the considered problem. The analysis shows [58] that for the studied disturbance all loads in the system can be represented by simplified models, $\dot{Y}_\ell = \text{ct}$. Thus, the threshold condition for transition from one type of model to another for generators and loads do not coincide in a general case.

As a result, a reasonably simplified mathematical model of the electric power system is obtained, which is used in the next stage of the model simplification.

10.4.3 Estimating the System Element Significance

10.4.3.1 Index of the System Structural Connectivity. The system structure is critical when carrying out studies on complex systems, such as the electric power system, because it reflects the most essential interrelations between the individual elements and their groups. These interrelations are quite invariant when perturbations occur in a system and they guarantee the system operation and its typical behavior. System integrity, nonadditivity, and intrinsic behavior are determined by its structure, that is, by the configuration, the strength of tie lines, and the interactions between elements and

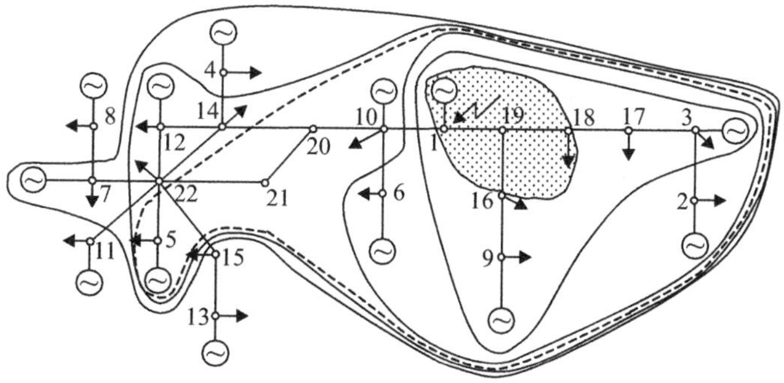

Figure 10.23. Decomposition of a system on the basis of disturbance impact index.

subsystems. These ties and interactions have quite definite quantitative estimates and it is important to know how to calculate these estimates.

The connectivity of the electric power system structure offers the opportunity to determine the subsystems that have strong ties between their elements and weak ties and cutsets between the determined subsystems. The connectivity is determined by the numerical measure of "distance" between generators in the system and expresses the extent of interaction between them under steady-state and transient conditions. The electric connectivity of generators plays an important role. The connectivity is very often estimated by the magnitude of transfer admittance Y_{ij} of a reduced network of the classical model of electric power system dynamics.

A more appropriate numerical measure of electric connectivity is given by

$$w_{\mathrm{ij}} = E'_{\mathrm{i}} E'_{\mathrm{j}} Y_{\mathrm{ij}} \tag{10.113}$$

which, besides the electric distance Y_{ij}, it takes into account the voltage levels at the nodes where the two generators i and j are connected, that are indirectly expressed through their electromotive forces [65,70].

The measure of electric connectivity w_{ij} based on (10.113) within an electric power system can differ greatly, by orders of five or more. These differences form the basis for determining strongly connected subsystems interconnected by weak tie lines.

It should be noted that although the electromotive forces of generators depend on the operating conditions, they vary within a comparatively narrow range, which is indicative of the robustness of electric connectivity estimates.

10.4.3.2 Significance of a System Element. The significance of a system element (a generator) is estimated through the analysis of the relationship between the individual components of the maximum generated power:

$$P_{\mathrm{ei}} = P_{\ell i} + w_{iI} + w_{jJ}, \quad i \in I, j \in J \tag{10.114}$$

where

$P_{\ell i}$ are loads connected to the generator node i (generator loads);

w_{iI} and w_{jJ} are the components of the power that can be used inside and outside the subsystem.

There can be three cases [65].

- case a:

$$P_{ei} \approx P_{\ell i}; \quad w_{iI} \approx w_{jJ} \approx 0 \tag{10.115}$$

This means that the power generated is practically completely used to supply the loads located close to the generator, and the connection between the generator and the system is rather weak. This generator has local influence and does not affect the system.

- case b:

$$P_{ei} \approx P_{\ell i} + w_{iI}, \quad w_{iJ} \approx 0 \tag{10.116}$$

In this case, the generator power is used not only to cover the local loads but also to supply remote loads within the subsystem I. This generator affects essentially the processes that occur within the subsystem I and practically does not affect the remaining part of the system.

- case c:

$$P_{ei} = P_{\ell i} + w_{iI} + w_{ij} \tag{10.117}$$

In this case, the ties of the considered generator with all other generators in the system are significant. This generator affects considerably the processes that occur in the entire system and is called a system generator.

The estimates of significance of the elements (generators) in the system supplement, to a certain extent, the estimates of the disturbance impact that were introduced in Section 10.4.2. Based on these estimates, it is possible to formulate recommendations about using the detailed modeling of system elements and to establish the general directions of suitable aggregation that may be identified in the system structure, which can be useful at the subsequent stages of simplification.

10.4.4 Coherency Estimation

10.4.4.1 Equation of the Mutual Motion of a Pair of Machines. Consider a classical model of an electric power system dynamics in the form of differential equations of generator motion as (see also §10.2.4.4):

$$\frac{d^2\delta_i}{dt^2} = \frac{1}{J_i}\left(P_{mi} - E_i'^2 G_{ii} - \sum_{\substack{j=1 \\ j\neq i}}^{n} (C_{ij}\sin(\delta_i - \delta_j) + D_{ij}\cos(\delta_i - \delta_j))\right) \tag{10.118}$$

where

$$C_{ij} = E_i' E_j' B_{ij} \tag{10.119}$$

$$D_{ij} = E_i' E_j' G_{ij} \tag{10.120}$$

where G_{ij} and B_{ij} are the active and reactive components of the transfer admittance of the reduced network and G_{ii} is the self-conductance of the reduced network.

The following notations are introduced:

$$P_i^* = \left(P_{mi} - E_i'^2 G_{ii}\right)/J_i \tag{10.121}$$

$$C_{ij}^* = C_{ij}/J_i \tag{10.122}$$

$$D_{ij}^* = D_{ij}/J_i \tag{10.123}$$

In order to define the equation of mutual motion of the generators i and j, a similar equation of (10.118) is derived for j. Hence, we obtain:

$$\frac{d^2\delta_{ij}}{dt^2} = P_{ij}^* - \left(F_{ij}\sin\delta_{ij} + Q_{ij}\cos\delta_{ij}\right) - A_{ij}; \quad i,j = \overline{1,n} \tag{10.124}$$

where

$$P_{ij}^* = P_i^* - P_j^* \tag{10.125}$$

$$F_{ij} = C_{ij}^* + C_{ji}^* \tag{10.126}$$

$$Q_{ij} = D_{ij}^* + D_{ji}^* \tag{10.127}$$

$$A_{ij} = \sum_{\substack{k=1 \\ k \neq i,j}}^{n} (C_{ik} \sin \delta_{ik} + D_{ik} \cos \delta_{ik}) + \sum_{\substack{k=1 \\ k \neq i,j}}^{n} (C_{jk} \sin \delta_{jk} + D_{jk} \cos \delta_{jk}) \tag{10.128}$$

$$\delta_{ij} = \delta_i - \delta_j \tag{10.129}$$

Assuming that F_{ij} and Q_{ij} are the components in the rectangular coordinates of some complex quantity with the magnitude R_{ij} and the phase φ_{ij}, equation (10.124) can be written as

$$\frac{d^2\delta_{ij}}{dt^2} = P_{ij}^* - R_{ij} \sin(\delta_{ij} - \varphi_{ij}) - A_{ij}; \quad i,j = \overline{1,n} \tag{10.130}$$

where

$$R_{ij} = \sqrt{F_{ij}^2 + Q_{ij}^2} \tag{10.131}$$

$$\varphi_{ij} = \text{atan}(F_{ij}/Q_{ij}) \tag{10.132}$$

The third term A_{ij} in (10.130) represents the impact of the motion of the remaining generators in the system (except the generators i and j) with respect to the considered generators i and j. This impact is essential for high values of A_{ij}, which is comparable in magnitude with the first two terms of (10.130). For low values of A_{ij}, the mutual motion of the generators i and j will be negligibly influenced by the motion of generators from the remaining part of the system. This dependence can be neglected by assuming $A_{ij} = \text{ct.}$ The latter will be valid if the generators i and j are strongly connected with each other and the connection of each with the rest of generators is weak, that is, $Y_{ij} \gg Y_{ik}(Y_{jk})$ or $w_{ij} \gg w_{ik}(w_{jk})$. Strong connectivity between the generators i and j, in turn, is a necessary condition for their coherency (necessary but not sufficient, since the coherency of generator motion is determined supplementary by the dynamic parameters of generators and the disturbance characteristics).

Thus, for low values of A_{ij}, assuming $A_{ij} = \text{ct.}$, the equation of mutual motion of the generators i and j can be obtained instead of equation (10.130) in the form

$$\frac{d^2\delta_{ij}}{dt^2} = P_{ij} - R_{ij} \sin(\delta_{ij} - \varphi_{ij}) \tag{10.133}$$

where

$$P_{ij} = P_{ij}^* - A_{oij}$$

and A_{oij} is the value of A_{ij} in the steady-state preemergency condition.

Since (10.133) is valid for a strong connectivity between the generators i and j in contrast to their ties with the remaining generators in the system and because this situation is a necessary coherency condition between the generators i and j, the coherency estimates calculated by equation (10.133) will be acceptable in accuracy. For incoherencies between the generators i and j, the effect of A_{ij} on their motion will be important, that is, the assumption $A_{ij} = \text{ct.}$ is not valid, and the numerical coherency estimates based on equation (10.133) will be inaccurate. The latter, however, is of no significance, since the interest is only in estimates that correspond to coherent motion of the generators i and j, and the accuracy limits for these were already established.

10.4.4.2 Coherency Indices. Accordingly to early notes in Section 10.4.4.1, the coherency estimates of generator motion can be obtained based on the structural characteristics of an electric power system, either by neglecting or by considering the disturbance characteristics. In a first case, the index of structural (electric) connectivity, w_{ij}, determined by (10.113) may be taken as a coherency index. In the second case, the Lyapunov function method (see Section 10.2.4) is applied to the equation of mutual motion of the generators i and j (10.133) [65].

Denoting

$$x_{ij} = \delta_{ij} - \varphi_{ij} \tag{10.134}$$

the equation (10.133) can be transformed into

$$\frac{\mathrm{d}x_{ij}^2}{\mathrm{d}t^2} = P_{ij} - R_{ij}\sin x_{ij} \tag{10.135}$$

For equation (10.135), the Lyapunov function is known in the form of energy integral (see Section 10.2.4). Based on the requirement to be positive defined, this type of Lyapunov function for (10.135), in terms of inverse transformation of the variables, will have the following form:

$$U_{ij} = \frac{1}{2}\left(\frac{\mathrm{d}\delta_{ij}}{\mathrm{d}t}\right)^2 + \left[P_{ij}(\delta_{oij} - \delta_{ij}) - R_{ij}(\cos(\delta_{oij} - \varphi_{ij}) - \cos(\delta_{oij} - \varphi_{ij}))\right] \tag{10.136}$$

If the system (10.135) defines the operation in a stable equilibrium point, then $U_{oij} = 0$. The coordinates of unstable equilibrium point for (10.135), considering inverse transformation of the variables, will be equal to $\mathrm{d}\delta_{ij}/\mathrm{d}t = 0$ and $\delta_{ij}^{cr} = \pi - \delta_{oij} + 2\varphi_{ij}$, where the superscript "cr" stands for the unstable equilibrium point. The Lyapunov function value at the unstable equilibrium point will be:

$$U_{crij} = P_{ij}\left(2\delta_{oij} - 2\varphi_{ij} - \pi\right) - 2R_{ij}\cos\left(\delta_{oij} - \varphi_{ij}\right) \tag{10.137}$$

The value of U_{crij} determines an attraction region for the system (10.133). The larger is the region, the greater is the dynamic connectivity between the generators i and j. Therefore, the index of dynamic connectivity between the generators i and j can be taken as a coherency index

$$v_{ij} = U_{crij} \tag{10.138}$$

Possible coherency indices based not only on the structural characteristics of the power system, but also on the disturbance characteristics, are analyzed in the following. Therefore, taking into consideration equation (10.133) of mutual motion of the generators i and j and considering the second form of Gorev's stability criterion, which is given by (10.135) in [65], then:

$$\int_{x_{oij}}^{x_{crij}} \left(P_{ij} - R_{ij}\sin x_{ij}\right)\mathrm{d}x_{ij} + \frac{1}{2}\left(\frac{\mathrm{d}x_{ij}}{\mathrm{d}t}\right)^2 \le 0 \tag{10.139}$$

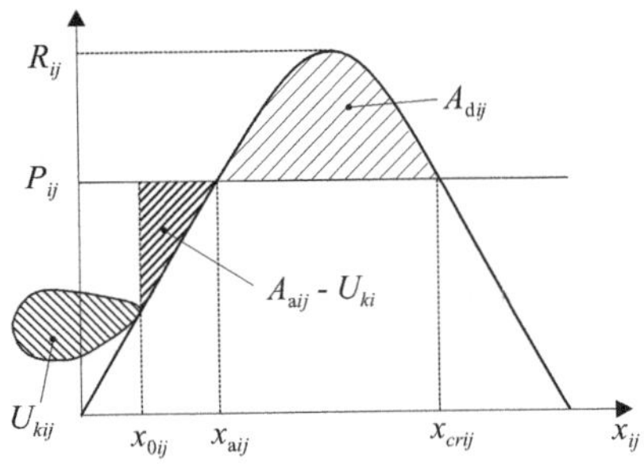

Figure 10.24. Illustration of the second form of Gorev's stability criterion.

where the first term represents the potential energy and the second is the kinetic energy of the system (U_{kij} in Figure 10.24).

After separating the integration interval in accordance with Figure 10.24, grouping of the terms after integration and reverse transformation to the initial variables δ_{ij} in accordance with (10.134), condition (10.139) is obtained as a relationship corresponding to the criterion of equal areas:

$$A_{aij} \leq A_{dij} \tag{10.140}$$

where A_{aij} is the acceleration area determined by

$$A_{aij} = \left|P_{aij}\left(\delta_{aij} - \delta_{oij}\right) - R_{aij}\left(\cos\left(\delta_{aij} - \varphi_{ij}\right) - \cos\left(\delta_{oij} - \varphi_{ij}\right)\right)\right| + \frac{1}{2}\left(\frac{d\delta_{ij}^{o}}{dt}\right)^{2} \tag{10.141}$$

and A_{dij} is the possible deceleration area determined by

$$A_{dij} = \left|P_{ij}\left(\delta_{aij} - \delta_{crij}\right) - R_{ij}\left(\cos\left(\delta_{aij} - \varphi_{ij}\right) - \cos\left(\delta_{crij} - \varphi_{ij}\right)\right)\right| \tag{10.142}$$

where $d\delta_{ij}^{o}/dt$ is the derivative of the mutual rotor angle between the generators i and j at the initial time interval that in general can be nonzero; P_{aij} and R_{aij} are the system parameters in the emergency condition; P_{ij} and R_{ij} are the parameters in the postemergency condition.

The value δ_{aij} that corresponds to the final stage of the emergency period can be obtained assuming the invariability of accelerating generator powers during the emergency conditions, since this condition is short-term and because the generators' rotors inertia does not allow them to considerably change their position. Subject to this assumption,

$$\delta_{ai} = \delta_{oi} + \frac{1}{2J_i}\Delta P_i t_{a}^{2} \tag{10.143}$$

where t_a is the duration of the emergency condition; ΔP_i is the value of accelerating (decelerating) power of generator i at the time of disturbance initiation. Hence,

$$\delta_{aij} = \delta_{ai} - \delta_{aj}$$

Based on the above observations, the motion coherency index for the generators i and j can be calculated with the expression:

$$\nu_{ij} = A_{aij}/A_{dij} \tag{10.144}$$

In Ref. [58], the coherency index analogous to (10.144) is determined subject to $d\delta_{ij}^{o}/dt = 0$. Table 10.7 presents estimates of this index for the electric power system

TABLE 10.7. Indices of Generators Coherency

13	5	11	8	7	12	4	6	10	1	9	2	3	Gen.
	0.0076	0.0008	0.0044	0.0001	0.0022	0.011	0.18	0.048	10^8	10^8	10^7	10^9	13
		0.042	0.0002	0.015	0.0031	0.0013	0.0031	0.0006	10^7	10^8	10^6	10^8	5
			0.0069	0.0038	0.0034	0.016	0.22	0.086	10^8	10^8	10^7	10^9	11
				0.0007	0.012	0.014	0.11	0.035	10^8	10^8	10^6	10^9	8
					0.0068	0.038	1.6	0.34	10^9	10^9	10^8	10^{10}	7
						0.0039	0.091	0.057	10^7	10^8	10^7	10^8	12
							0.0052	0.0045	10^7	10^8	10^6	10^8	4
								0.001	10^7	10^7	10^6	10^8	6
									10^6	10^7	10^5	10^8	10
										10^6	10^7	10^7	1
											1.9	1.6	9
												0.052	2
													3

presented in Figure 10.17 assuming that a short circuit is applied at node 1 as per conditions described in Section 10.4.2.2. The correlation between the estimates of coherency indices in Table 10.7 and the character of generator response in Figure 10.18, allows the conclusion that generators 13, 5, 11, 8, 7, 12, 4 (the first group) and 8, 10 (the second group) should be treated as sufficiently coherent. The coherency index values in this case do not exceed 0.038. This value can be assumed a threshold for defining the generator motion coherency.

Studies on many known coherency indices carried out in [106] show that the coherency indices determined based on various interpretations of the equal area criterion, including those presented in this section, result in rather valid estimates of generator motion coherency for short time intervals suitable for transient stability studies.

The long-term transient processes in a bulk power system are characterized by possible substantial structural changes in the system as a result of essential changes in powers balance in individual system parts and because of operation of emergency control systems during the transient process. The structural changes can cause significant changes in the motion of individual generators and subsystems with respect to one another. Moreover, the studies on long-term transients require that the power system model should take into account the equations of voltage regulators, the dynamics of generators' prime movers and their associated control systems, and also characteristics of loads that are frequency and voltage dependent.

The variations of voltages at the nodes and the frequency in the strongly tied subsystem, which includes generators in coherent motion, are analyzed in the following. The emf of generators will obviously have a profound effect on the behavior of voltages vector $\dot{V}_i(t)$ at the subsystem nodes. For $|\underline{\dot{E}}_i| = \text{ct.}$, $i = \overline{1, n}$, the variation of $\dot{V}_i(t)$ at different nodes will differ insignificantly in both magnitude and phase. The frequency in the subsystem will correspond to the motion of its center of inertia with only minor mutual swings of the generators. These conditions are sufficient to apply rather simple approaches for formulating the load equivalents in the study of long-term transient processes in electric power systems.

Overall, the application of the classical model of power system dynamics for estimating the generators' motion coherency during long-term transient processes can be a reasonable approach, as long as the assumptions for its application are valid $\left(P_m = \text{ct.}, E' = \text{ct.}, \dot{Y}_\ell = \text{ct.}\right)$. Hence, the coherency of generator motion that is obtained at the initial stage of the transient process should be tested periodically, later in the process, during certain time intervals for which the estimates from the classical model are valid. The verification of assumed hypothesis should be also carried out when switching off system elements by protection relays and emergency control devices and when large deviations in state variables are experienced, that may lead to variations in coherency conditions. The controlled variables during the transient process can be represented by mutual angles, generator rotor slips in the subsystem, the deviations of the mutual angles, and rotor slips with respect to the subsystem center of inertia, as well as the voltage deviations at the subsystem nodes. The coherency of generator motion is tested in this case by the indices (10.144) that take into account the nonzero initial conditions.

10.4.4.3 Clustering of Coherency Indices. Clustering of coherency indices of generator motion is required as a formal algorithm for determining the groups of generators that can be represented by an equivalent. Two clustering algorithms are presented in the following, which rely on using the indices v_{ij} determined from (10.144), although application of any other coherency indices is also possible.

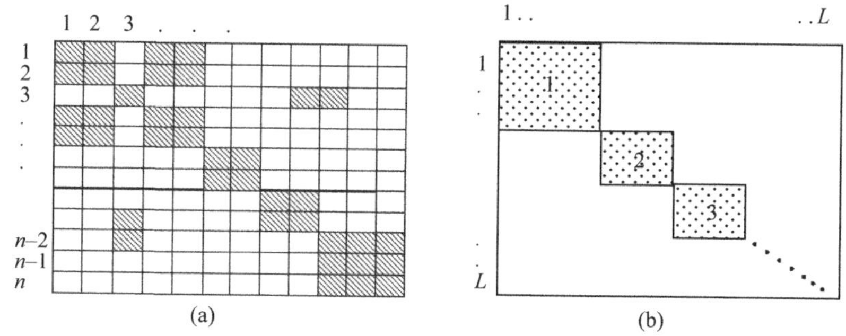

Figure 10.25. Illustration of the first algorithm application.

The use of the first algorithm is illustrated in Figure 10.25. In this case, a threshold for the coherency index is assumed $\overline{\nu}_{ij}$ in the sense that, if $\nu_{ij} \le \overline{\nu}_{ij}$, the generators i and j are considered to be coherent, and if $\nu_{ij} > \overline{\nu}_{ij}$, they are incoherent [58,65].

Under this assumption, all the components of the matrix $\{\nu_{ij}\}$, $i, j = \overline{1, n}$, for which the condition $\nu_{ij} \le \overline{\nu}_{ij}$ is satisfied are set to zero, that is if $\nu_{ij} \le \overline{\nu}_{ij}$ then $\nu_{ij} = 0$. After this operation, the matrix of the indices $\{\nu_{ij}\}$ will have the form as in Figure 10.25a, where the shaded cells correspond to $\nu_{ij} = 0$. The rows and columns in the matrix shown in Figure 10.25a can be rearranged to the form presented in Figure 10.25b where the shaded submatrices highlight the subsystems with coherent generators.

A more general second approach for grouping the coherency indices is based on the methods of cluster analysis [65,102,106].

Consider the simplest case of a generator grouping for one configuration assuming one operating scenario and one disturbance. The matrix $\{\nu_{ij}\}$ of coherency indices will be interpreted as a similarity matrix [107], in which each pair of objects is characterized by some degree of closeness ν_{ij}, with small values corresponding to closer objects. In this case, the generators are the objects to be clustered.

The algorithm divides an initial set of generators G, $(i = \overline{1, n}) \in G$, into k nonempty subsets of clusters (taxons), where $k < n$. In this algorithm, the final number of clusters can be either predefined or determined during clustering. Note that the closeness matrix $\{\nu_{ij}\}$ is an $n \times n$ symmetric matrix with zero diagonal elements and positive remaining elements, i.e. $\nu_{ij} > 0$ for $i \ne j$, $\nu_{ii} = 0$, $\nu_{ij} = \nu_{ji}$, where $i, j = \overline{1, n}$. The index ν_{ij} will be called an inner degree of closeness if the objects i and j belong to the same cluster or an outer degree of closeness if they belong to different clusters.

The outer degree of closeness between any two subsets G_I and G_J is assumed to be an average value of outer degrees of closeness between the objects in these clusters.

$$\nu(G_I, G_J) = \frac{1}{n_I n_J} \sum_{i \in G_I} \sum_{j \in G_J} \nu_{ij} \tag{10.145}$$

where n_I and n_J are the numbers of objects in the subsets G_I and G_J, respectively.

The inner degree of closeness in the subset G_I is assumed to be an average value of inner degree of closeness in this cluster

$$\nu(G_I, G_I) = \frac{2}{n_I(n_I - 1)} \sum_{i \in G_I} \sum_{\substack{j \in G_I \\ j > i}} \nu_{ij} \tag{10.146}$$

The clustering algorithm employs an iterative search process. Each iteration reduces by one unit, the initial number of subsets being is equal to the number of objects at the beginning of algorithm application. In each step, a pair of the subsets G_μ and G_η is sought, with minimal average value of the outer degrees of closeness. These subsets are combined into one cluster, resulting thus in a reduction of the number of subsets by one unit.

When the predefined number of clusters is obtained, the algorithm ends. If the number of clusters that was formed exceeds a predefined initial number, the procedure continues.

If the number of clusters is unknown, the following condition will be the criterion for cluster formation:

$$I_{in}^k / I_{out}^k \geq \varepsilon \tag{10.147}$$

where

$$I_{in}^k = \frac{1}{k} \sum_{i=1}^{k} \nu(G_i, G_i) \tag{10.148}$$

$$I_{out}^k = \frac{2}{k(k-1)} \sum_{i=1}^{k-1} \sum_{j=i+1}^{k} \nu(G_i, G_j) \tag{10.149}$$

and I_{in}^k is the average value of the inner degrees of closeness; I_{out}^k is the average value of the outer degrees of closeness; k is the number of clusters.

The second algorithm is illustrated by clustering the matrix of coherency indices that are given in Table 10.7. Figure 10.26 shows a dendogram of clustering. The minimum values of the outer degrees of closeness, following the formation of a new cluster from two previous ones, are indicated at each step. These results are in agreement with the results of expert analysis of possible identification for coherent generators based on the matrix of indices given in Table 10.7.

The generalized distance between the initial conditions k and ℓ is given by:

$$d(k, \ell) = \left\{ \frac{2}{m(m-1)} \sum_{k=1}^{m-1} \sum_{\ell=k+1}^{m} \left(d_{ij}^k - d_{ij}^\ell \right)^r \right\}^{\frac{1}{r}} \tag{10.150}$$

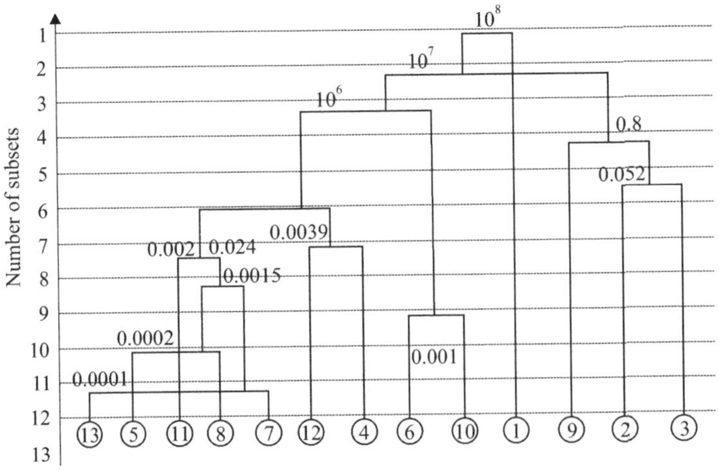

Figure 10.26. A dendrogram of clustering based on coherency indices.

where m is the number of initial conditions; $d_{ij}^k = 1/v_{ij}^k$ is the generalized distance between generators for the k-th case of initial conditions; r is a positive integer. The transformations of the obtained similarity matrix can be similar to those employed in the algorithm for generators clustering.

A similar algorithm can be applied for clustering the coherent generators for a set of configurations, operating conditions, and disturbances [65]. In this case, the objects to be clustered are represented in the m-dimensional space of classification features, where m equals the number of initial conditions. It differs from the second algorithm basically through the expressions used for the determination of outer and inner degrees of closeness among the objects that can be viewed as distances of appropriate metric (taxonometric, Euclidian and the like) [43].

Moreover, the last approach can be applied for grouping the initial conditions, rather than generators, into appropriate clusters. This would enable performing detailed calculations of the transient processes for only one operating condition that is representative for each cluster instead of calculating the transients for every operating condition [65]. This issue is discussed in detail in Section 10.5. In this case, the matrices of indices are the objects to be clustered. The dimension of the similarity matrix that indicates the closeness of the considered initial conditions corresponds to the number of these conditions. The appropriate values of closeness can be obtained by using the coefficients such as distances in the multidimensional space [43].

10.4.5 Equivalencing Criteria

The equivalencing criteria determine the relationship between parameters of the initial and the equivalent systems on the basis that the initial system and the equivalent one deliver quasi-identical transient responses. In this case, the external subsystem is equivalated and the subsystem under study remains in the original form.

The determination of the dynamic equivalent of an electric power system requires that the analysis of the transients for generators included in the subsystem under study can be carried out using the equation of motion of the classical model and still deliver identical responses for identical initial conditions. The necessary conditions for obtaining an equivalent response in this case emerge from the comparison between the generator equations of motion for the studied subsystem before and after the equivalencing of the external subsystem, namely [55,90]:

$$J_i \frac{\mathrm{d}^2 \delta_i}{\mathrm{d}t^2} = P_{mi} - P_{ei}, \quad i = \overline{1, n} \tag{10.151}$$

$$J_i^e \frac{\mathrm{d}^2 \delta_i^e}{\mathrm{d}t^2} = P_{mi}^e - P_{ei}^e, \quad i = \overline{1, n} \tag{10.152}$$

where the superscript "e" in (10.152) shows that this equation corresponds to the generator i in the studied subsystem assuming that a simplified external subsystem (an equivalent) was used; the absence of the superscript in (10.151) indicates that the equation corresponds to the generator prior to the external subsystem simplification and equivalencing; n is the number of generators in the studied subsystem.

The identity of processes in the studied subsystem for the initial and equivalent systems will be satisfied provided that:

$$\delta_i(t) = \delta_i^e(t) \tag{10.153}$$

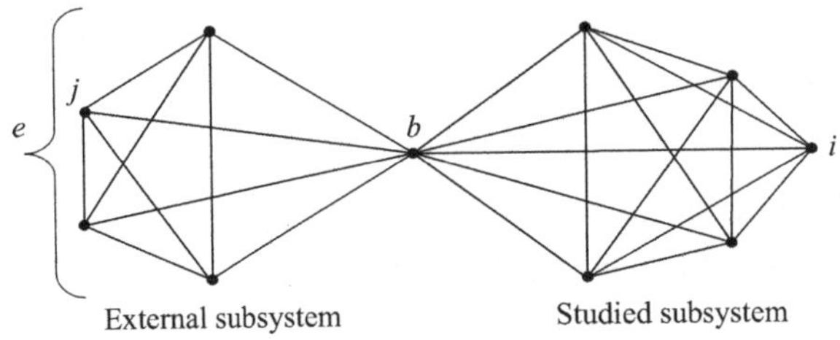

Figure 10.27. Illustrative system.

and hence

$$J_i \frac{d^2\delta_i}{dt^2} = J_i^e \frac{d^2\delta_i^e}{dt^2} \tag{10.154}$$

Since $J_i = J_i^e$ and $P_{mi} = P_{mi}^e$, therefore

$$P_{ei} = P_{ei}^e \tag{10.155}$$

Based on the requirement to obtain identical voltage values in the studied subsystem before and after the external subsystem equivalencing, equation (10.155) is extended to the total powers of generators, that is,

$$\dot{S}_{ei} = \dot{S}_{ei}^e \tag{10.156}$$

The simple system configuration shown in Figure 10.27, which includes only one adjoining node (boundary node) b, while the generators $j = \overline{1, m}$ of the external subsystem are combined into one equivalent generator, is considered for the subsequent computations.

The condition (10.156) suggests that the similar condition for the boundary node b is met, that is,

$$\underline{S}_b = \underline{S}_b^e \tag{10.157}$$

On the grounds that $\underline{S} = \underline{V}\,\underline{I}^* = \underline{V}\,\underline{Y}^*\underline{V}^*$, for the configuration shown in Figure 10.27, the conditions (10.157) can be written in detail in the form

$$\underline{V}_b \sum_{i=1}^{n} \underline{E}_i^* \underline{Y}_{ib}^* + \underline{V}_b \sum_{j=1}^{m} \underline{E}_j^* \underline{Y}_{ib}^* = \underline{V}_b \sum_{i=1}^{n} \underline{E}_i^* \underline{Y}_{ib}^* + \underline{V}_b \underline{E}_e^* \underline{Y}_{eb}^* \tag{10.158}$$

from where by canceling the identical components it follows that

$$\sum_{j=1}^{m} \underline{E}_j^* \underline{Y}_{jb}^* = \underline{E}_e^* \underline{Y}_{eb}^* \tag{10.159}$$

Here the subscript "e" corresponds to the equivalent generator representing an external subsystem.

For the equivalent generator of the external subsystem the following assumption is reasonable from a physical point of view:

$$P_m^e = \sum_{j=1}^{m} P_{mj}; \quad P_e^e = \sum_{j=1}^{m} P_{ej} \tag{10.160}$$

or

$$\underline{S}_e^e = \sum_{j=1}^{m} \underline{S}_{ej} \tag{10.161}$$

for a more general case because of the requirement to achieve identical voltage values at the boundary node.

Expanding (10.161) similarly to (10.157) and (10.158), the following expression is obtained:

$$\underline{E}_e \underline{V}_b^* \underline{Y}_{eb}^* = \sum_{j=1}^{m} \underline{E}_j \underline{V}_b^* \underline{Y}_{jb}^* \tag{10.162}$$

or eliminating $\underline{V}_b^*$ from both sides, obtain

$$\underline{E}_e \underline{Y}_{eb}^* = \sum_{j=1}^{m} \underline{E}_j \underline{Y}_{jb}^* \tag{10.163}$$

A comparison between (10.159) and (10.163) indicates that they are not identical and the values $\underline{Y}_{eb}^*$ obtained from both relations are different. A similar situation occurs in the more general case with several boundary nodes and several equivalent generators representing an external subsystem. This inconsistency will be discussed further in this section, but the conditions that determine the interrelation of the equations for motion of generators in the external subsystem and their equivalent generator are analyzed next.

From (10.160), the following expression can be written:

$$\sum_{j=1}^{m} J_j \frac{d^2\delta_j}{dt^2} = \sum_{j=1}^{m} P_{mj} - \sum_{j=1}^{m} P_{ej} \tag{10.164}$$

On the other hand, the equivalent generator of the external subsystem is represented by the equation of motion:

$$J_e \frac{d^2\delta^e}{dt^2} = P_m^e - P_e^e \tag{10.165}$$

Hence,

$$J_e \frac{d^2\delta^e}{dt^2} = \sum_{j=1}^{m} J_j \frac{d^2\delta_j}{dt^2} \tag{10.166}$$

from which

$$J_e = \left(\sum_{j=1}^{m} J_j \frac{d^2\delta_j}{dt^2} \right) \Bigg/ \frac{d^2\delta^e}{dt^2} \tag{10.167}$$

Since, in general $\frac{d^2\delta^e}{dt^2} \neq \frac{d^2\delta_j}{dt^2}$, $j = \overline{1, m}$, then

$$J_e \neq \text{ct.} \tag{10.168}$$

which contradicts normal expectations for an equivalent generator from physical point of view.

Assuming a simplified situation when $\frac{d^2\delta^e}{dt^2} = \frac{d^2\delta_j}{dt^2}$, which is obviously typical for the coherent motion of generators in the external subsystem, the following expression can be inferred from (10.167):

$$J_e = \sum_{j=1}^{m} J_j \tag{10.169}$$

which is compliant with normal expectations for an equivalent generator from physical point of view.

Moreover, for coherent motion of generators in the external subsystem, the values of $\underline{Y}^*_{eb}$ calculated from (10.159) and (10.163) prove to be practically in agreement, which is dictated by the relationship between the elements of the strongly tied external subsystem in contrast with the weak ties between the external subsystem and the subsystem under study [44]. The relations obtained provide the basis for determining the parameters of the equivalent. This stage is analyzed in detail in Section 10.4.6.

10.4.6 Center of Inertia. Parameters of the Equivalent

The parameters of the equivalent generators that present coherent swings in the external subsystem can be determined by the relations stemming from the accepted equivalencing criteria (see Section 10.4.5). Despite the same qualitative nature of the criteria that require identity of the processes in the studied subsystem, their specific interpretation in the form of formulas for calculating parameters of the equivalent can be different. This Subsection starts from conditions (10.157) and (10.161).

Let the initial external subsystem be described by the equation [65]:

$$\begin{aligned} J_j \frac{d^2\delta_j}{dt^2} = P_{mj} - E_j^2 G_{jj} - \sum_{\substack{\ell=1\\ \ell\neq j}}^{m} E_j E_\ell Y_{j\ell} \sin\left(\delta_{j\ell} - \beta_{j\ell}\right) - \\ - \sum_{\substack{i=1\\ i\neq j}}^{n} E_j E_i Y_{ji} \sin\left(\delta_{ji} - \beta_{ji}\right); \quad j = \overline{1, m} \end{aligned} \tag{10.170}$$

Based on equation (10.170), the transformation of coordinates with respect to the external subsystem's center of inertia coordinates, by analogy with Section 10.2.4.4, will result in:

$$\begin{aligned} J_j\left(\frac{d^2\delta_{jc}}{dt^2} + \frac{d^2\delta_c}{dt^2}\right) = P_{mj} - E_j^2 G_{jj} - \sum_{\substack{\ell=1\\ \ell\neq j}}^{m} E_j E_\ell Y_{j\ell} \sin\left(\delta_{jc} - \delta_{\ell c} - \beta_{j\ell}\right) - \\ - \sum_{\substack{i=1\\ i\neq j}}^{n} E_j E_i Y_{ji} \sin\left(\delta_{jc} + \delta_c - \delta_i - \beta_{ji}\right); \quad j = \overline{1, m} \end{aligned} \tag{10.171}$$

where:

$$\delta_c = \sum_{j=1}^{m} J_j \delta_j / J_c \tag{10.172}$$

Following the transformation above, the swings described by the inertia center of the external subsystem are governed by the equation:

$$\begin{aligned} J_c \frac{d^2\delta_c}{dt^2} = P_{mc} - \sum_{j=1}^{m} \left[E_j^2 G_{jj} + \sum_{\substack{\ell=1 \\ \ell \neq j}}^{m} E_j E_\ell Y_{j\ell} \sin(\delta_{jc} - \delta_{\ell c} - \beta_{j\ell}) \right] - \\ - \sum_{j=1}^{m} \sum_{\substack{i=1 \\ i \neq j}}^{n} E_j E_i Y_{ij} \sin(\delta_{jc} + \delta_c - \delta_i - \beta_{ji}) \end{aligned} \tag{10.173}$$

where J_c is determined from (10.169) and P_{mc} from (10.160); the subscripts "c" and "ℓ" are inter-changeable from this point forward and they designate quantities referring to the equivalent subsystem.

Similarly to (10.165), the equivalent generator of the external subsystem is represented by the equation:

$$J_c \frac{d^2\delta_c}{dt^2} = P_{mc} - E_c^2 G_{cc} - \sum_{\substack{i=1 \\ i \neq c}}^{n} E_c E_i Y_{ci} \sin(\delta_c - \delta_i - \beta_{ci}) \tag{10.174}$$

From (10.173) and (10.174) it follows that:

$$\begin{aligned} E_c^2 G_{cc} + \sum_{\substack{i=1 \\ i \neq c}}^{n} E_c E_i Y_{ci} \sin(\delta_c - \delta_i - \beta_{ci}) = \\ = \sum_{j=1}^{m} \left[E_j^2 G_{jj} + \sum_{\substack{\ell=1 \\ \ell \neq j}}^{m} E_j E_\ell Y_{j\ell} \sin\left(\delta_{jc} - \delta_{\ell c} - \beta_{j\ell}\right) \right] + \\ + \sum_{j=1}^{m} \sum_{\substack{i=1 \\ i \neq j}}^{n} E_j E_i Y_{ji} \sin\left(\delta_{jc} + \delta_c - \delta_i - \beta_{ji}\right) \end{aligned} \tag{10.175}$$

Equality (10.175) can be satisfied based on the values E_c, G_{cc}, Y_{ci} and β_{ci}, determined by the following relations [58,65]:

$$\underline{S}_i^c = \underline{S}_i, \text{ hence } \underline{E}_c \underline{Y}_{ic} = \sum_{\substack{j=1 \\ j \neq i}}^{m} \underline{E}_j \underline{Y}_{ij} \tag{10.176}$$

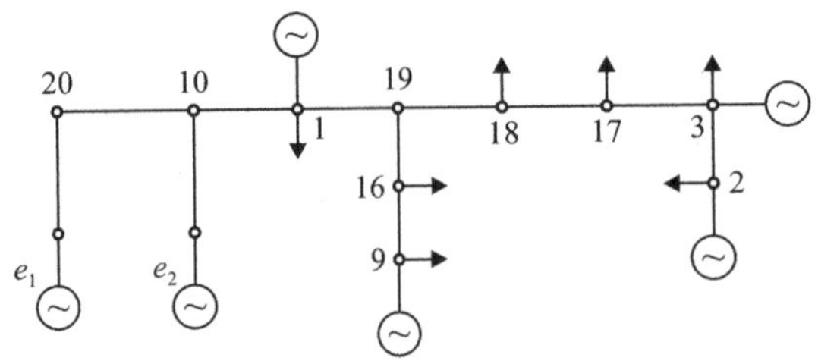

Figure 10.28. An equivalent system configuration.

as shown in (10.159), and also:

$$\sum_{i=1}^{n} \underline{Y}_{ic} = \sum_{i=1}^{n} \sum_{\substack{j=1 \\ j \neq i}}^{m} \underline{E}_j \underline{Y}_{ij} \tag{10.177}$$

$$E_c = \left| \sum_{i=1}^{n} \sum_{\substack{j=1 \\ j \neq i}}^{m} \underline{E}_j \underline{Y}_{ij} \right| \Bigg/ \left| \sum_{i=1}^{n} \sum_{\substack{j=1 \\ j \neq i}}^{m} \underline{Y}_{ij} \right| \tag{10.178}$$

$$\underline{Y}_{ic} = \sum_{\substack{j=1 \\ j \neq i}}^{m} \underline{E}_i \underline{Y}_{ij} / \underline{E}_c; \quad i = \overline{1, n} \tag{10.179}$$

Since the value δ_c is determined by the transformation of the coordinates with reference to the center of inertia, G_{cc} is determined uniquely from (10.175).

An equivalent system configuration obtained from the initial configuration of Figure 10.17 is used as an example of applying the described approach for determining the parameters of the equivalent in Figure 10.28.

According to the estimates of generator motion coherency (see Section 10.4.4) the external subsystem is replaced by two equivalent generators e_1 and e_2, representing groups of generators no. 13, 5, 11, 8, 7, 12, 4 and 6, 10 of the initial configuration, respectively.

Figure 10.29 illustrates the rotor angle swings of these equivalent generators during the transient process in the scenario that was considered in Section 10.4.4. In order to outline the consistency of the equivalent generators' formulation, Figure 10.29 preserve the same angle curves of the equivalated generators as in Figure 10.19, whereas the swings of equivalent generators are shown by dashed lines.

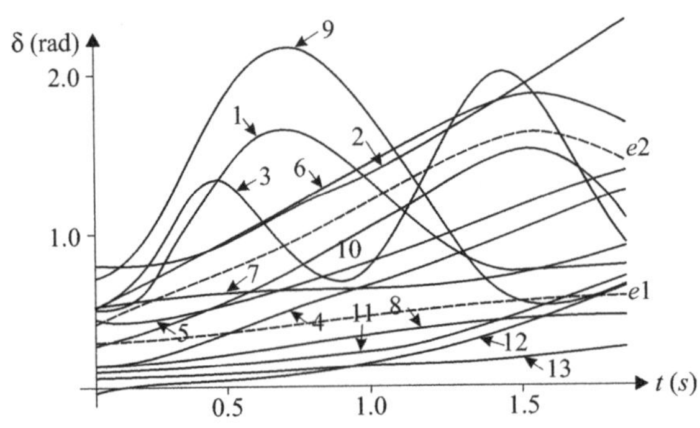

Figure 10.29. Illustration of the transient process in the equivalent system.

In the consistent system formed by (10.171) and (10.173), the first equation for coherent generator motion represents fast motions with respect to the center of inertia of the subsystem (see Section 10.4.6). Hence, it can be roughly considered with respect to the slow motion of the center of subsystem inertia at $\delta_c = \text{ct.}$ and $\mathrm{d}^2\delta_c/\mathrm{d}t^2 = 0$

$$J_j \frac{\mathrm{d}^2\delta_{jc}}{\mathrm{d}t^2} = P_{mj} - E_j^2 G_{jj} - \sum_{\substack{\ell=1 \\ \ell \neq j}}^{m} E_j E_\ell Y_{j\ell} \sin\left(\delta_{jc} - \delta_{\ell c} - \beta_{j\ell}\right) - \sum_{\substack{i=1 \\ i \neq j}}^{n} E_j E_i Y_{ji} \sin\left(\delta_{jc} - \delta_i - \beta'_{ji}\right), \quad j = \overline{1,m} \tag{10.180}$$

where $\beta'_{ji} = \beta_{ij} - \delta_c$.

As far as the condition $Y_{j\ell} \gg Y_{ji}$ is the necessary condition for coherent generator motion in the external subsystem (see Section 10.4.4), the theory of differential equations with small parameters [107] allows the inference that the second sum in (10.180) contains a small parameter ε that decreases towards zero. Furthermore, since in coherent motion the deviations $\Delta\delta_{jc} = \delta_{jc} - \delta_{ojc}$ are small, the use of the equation system (10.180) can be easily replaced by its linear approximation:

$$J_j \frac{\mathrm{d}^2\Delta\delta_{jc}}{\mathrm{d}t} = -\psi_j \Delta\delta_{jc} + \sum_{\substack{\ell=1 \\ \ell \neq j}}^{m} \psi_{j\ell} \Delta\delta_{\ell c}; \quad j = \overline{1,m} \tag{10.181}$$

The obtained equation system formed by (10.173) and (10.181) can be solved by using asymptotic methods from the theory of differential equations with small parameters. We consider Volosov's algorithm [107] that is applicable to the system $\dot{x} = f(x, z)$, $\varepsilon\dot{z} = cx + dz$, when the general solution of the subsystem of fast motions is known explicitly on the bases of the slow ones. Then, the average values of the fast variables z_i are determined from:

$$\bar{z}_i = \lim_{T \to \infty} \frac{1}{T} \int_{t_o}^{t_o+T} z_i(t) dt \tag{10.182}$$

where $t_o = \text{ct.}$ is an arbitrary variable, and the subsystem of slow motions is considered in the form $\dot{x} = f(x, \bar{z})$.

In order to apply Volosov's algorithm, a special case of (10.181) is first considered, assuming $\psi_{j\ell} = 0$; $\ell = \overline{1,m}$ – which is reasonable, since $\psi_j \gg \psi_{j\ell}$ for large values of m. Then, each independent equation from (10.181) has a solution in the form [65]:

$$\Delta\delta_{jc}(t) = \Lambda_j \sin\left(\Omega_j t + \chi_j\right) \tag{10.183}$$

where $\Omega_j = \sqrt{\psi_j / J_j}$, Λ_j and χ_j are determined from the initial conditions represented by $\Delta\delta_{jc} = \Delta\delta^o_{jc}$ and $\mathrm{d}\Delta\delta_{jc}/\mathrm{d}t = s^o_{jc}$. Assuming in this special case that $t_o = 0$, for some final interval T of averaging with (10.182) in terms of solution $\Delta\delta_{jc}(t)$, we get averaged values $\Delta\overline{\delta}_{jc}$ based on the following approximate relation:

$$\Delta\overline{\delta}_{jc} \approx \left(\cos\chi_j^o - \cos\left(\Omega_j \mathrm{T} + \chi_j^o\right)\right)\Lambda_j^o/\mathrm{T}\Omega_j \tag{10.184}$$

Thus, in order to take into account the motion incoherency in the equivalated subsystem, $\delta_{jc} = \delta_{jc}^o + \Delta\overline{\delta_{jc}}$ must be considered in (10.175) when determining the equivalent system parameters.

For a more general case, with $\Psi_{j\ell} \neq 0$, this system of equations should be reduced in (10.181) to the normal form by linear transformation of coordinates used in the theory of linear differential equations. Once the reduction is performed, then Volosov's algorithm can be applied in a similar way.

The approach described previously for calculating the long-term transient process in electric power systems makes possible the correction of parameters of the equivalent. The correction can be done by averaging on the time interval during which fast motions of generators occurs with respect to the slow motion of the equivalent generator of the external subsystem [65]; this time interval is identified by estimates of generator motion coherency (see Section 10.4.4).

10.5 TRANSIENT STABILITY ASSESSMENT OF LARGE ELECTRIC POWER SYSTEMS

10.5.1 Characteristics of Large Electric Power Systems

The development of electric power systems may refer to increasing the number of elements, to the geographical expansion, to the interconnection with other power systems in the same country or at continental/planetary level, and so on. More generation capacities are necessary to ensure power supply reliability, new transmission lines are needed due to the increased load demand and relocation of generation facilities, there is a need for more efficient tools for matching the generation and loads under time shifting load peaks conditions, improvements in the voltage and frequency control systems are required, and so on. Within this trend, the electric power systems infrastructure becomes more complex. This constitutes a prerequisite for ensuring the electrical energy supply service of any consumer located in any point, both at a required quality and an acceptable price.

On the other hand, with the increased share of distributed generation, the stability problems that used to be specific in the past only in the transmission systems are now a new subject in the distribution networks as well.

Therefore, robust tools are necessary for transient stability assessment of large-dimension power systems, consisting of thousands or tens of thousands of buses and branches, hundreds and thousands of generators, and so on. In other words, nowadays, the transient stability studies involve multidimensional power systems.

The multidimensionality of power systems results in some other specific characteristics related to the diversity of their topology (normal, repair, and postemergency) and the diversity of operating conditions that are determined by different load levels and curve profiles at different nodes and different loading of generators. The various operating conditions are overlapped on a large number of network topologies that have to be considered. Furthermore, for each set consisting of both the topology and the operating conditions, it is necessary to consider a set of relevant disturbances for which the transient stability assessment procedures have to be performed in order to decide for the most appropriate countermeasures to the power system problems.

Thus, a very large number of laborious computations result as necessary for the power systems transient stability assessment, and hence a very large number of sets of results for which it is practically impossible to conduct studies over some acceptable and often very limited periods of time. This problem is particularly critical for the design studies of power system, including transient stability issues, when several scenarios of the system expansion have to be considered.

A considerable contribution to solving the multidimensional problems of power systems is given by the dynamic equivalents. This Subsection presents additional features, which, in combination with building the power system dynamic equivalents, allow the simplification of the considered problem in terms of computational efforts. In fact, current interests in transient stability assessment techniques are to develop a hierarchical multi-stage approach, in which from one stage to the next, the system is represented in more detail but simultaneously the set of conditions to be studied (configuration, operating conditions, disturbances) is reduced [65,106].

10.5.2 Initial Conditions

Setting the initial conditions for the transient stability studies of large power systems is the first stage in the multistage approach. This stage is based on the a priori knowledge of states, events, and processes in the system, which has quite obvious physical grounds supported by the experience of studies that allow to:

- Limit the initial size of the studied system configuration.
- Update the studied system configuration according to maintenance schedules and operating conditions (e.g., the equipments maintenance are performed during yearly minimum load periods and are not performed during yearly maximum load periods).
- Reduce a priori the number of considered operating conditions, starting from a specific configuration of the system (e.g., in systems with an insignificant share of hydro power plants the operating conditions during flood periods on the rivers are not critical).
- Reduce the set of possible disturbances to the most severe ones (e.g., three-phase short circuits on the 400 kV–500 kV lines); severe but less important disturbances as compared to others may be eliminated (e.g., short circuits occurring at the middle of the line are less severe than the short circuits occurring at the line ends), and so on.

The number of initial conditions to be assumed for transient stability studies of large power systems depends on the experience and individual particularities of a researcher. Very optimistic assumptions in this stage can result in significant errors, both in the design and operation stages. Therefore, in setting the initial conditions, the researcher must consider some security margins, that may ensure the system stability also in case of unexpected events.

10.5.3 Standard Conditions for Transient Stability Studies

10.5.3.1 Studied Conditions and Disturbances. Besides the initial conditions assumed for the power system analysis, the experience of the researcher for real large-scale power systems is of great importance. Based on such experience, standard conditions can be formulated for the analysis of power system stability [112]. In this respect, the electric power system configurations can be divided into two categories: normal and maintenance. In the normal configuration, all network elements, for which the power system is sensitive from the stability point of view, are in operation. In the maintenance configuration, one or several critical elements of the electric network are disconnected, reducing the transmission capacity on some cutsets.

Regarding the loading level of the power system on the cutsets, the system operation can be classified in the following:

- Normal (the largest admissible flow is called maximum admissible).
- Forced (the largest admissible flow is called emergency admissible).

The forced loading conditions are allowed in order to prevent or minimize the power shortages to consumers and losses of energy resources under the need to save individual kinds of energy resources; with the unfavorable overlapping of planned and emergency maintenance of basic equipment in power plants and networks and during operation at minimum load, when it is impossible to reduce the flow due to insufficient maneuverability of nuclear power plants (except for the cutsets adjacent to the nuclear power plants).

For designing the loading conditions of the cutsets the steady states are divided in normal and heavy.

The loading conditions are considered to be heavy when characterized by unfavorable overlapping of the main equipment maintenance at power plants in the maximum and minimum load operation with the total duration of such operation within a year not exceeding 10%.

The most severe disturbances that are taken into account in the power system stability requirements, the so-called standard disturbances, are subdivided into three groups: I, II, and III [114]. The groups include the following disturbances:

a. *Short circuits (SCs) with disconnection of network elements.* Distribution of disturbances among the groups is presented in Table 10.8.

   The studied fault clearing times are presented in Table 10.9.

b. *Sudden emergency imbalance of active powers for any reasons: tripping of one generator or a group of generators connected through a common breaker, disconnection of a large substation, a large consumer, and so on.* The distribution of imbalances in terms of group of disturbance is presented in Table 10.10.

   Emergency imbalances of group III refer to the case when the power system stability is analyzed with respect to the tie lines between the interconnected systems.

   In addition, group III includes the following disturbances:

c. *Simultaneous disconnection of two overhead lines located in the same right of way on more than half the length of a shorter line*, as a result of disturbances classified in group I in accordance with Table 10.8.

d. Disturbances in the groups I and II *with tripping of a network element or generator which, as a result of maintenance of one of the breakers, leads to disconnection of another network element or generator that is connected to one and the same switchgear.*

If the self starting-up processes of a motor of a large consumer can cause considerable voltage drop to one substation busbar (by more than 15%), this event is considered as a disturbance and should be included in group I.

TABLE 10.8. List of Standard Disturbances [114]

Disturbances	Groups of Standard Disturbances Classified in Terms of Nominal Voltage, kV			
	110–220	330–500	750	1150
Disconnection of a network element by the main protection devices (or by back-up devices, but with slower operation) following a single-phase SC with successful automatic reclosing: single-phase automatic reclosing in 330 kV networks and higher and three-phase automatic reclosing in 110–220 kV networks	I	I	I	I
The same, but with failed automatic reclosing[b]	I	I	I[a], II	II
Disconnection of a network element by the main protection devices following a three-phase SC with successful or failed automatic reclosing[b]	II	–[c]	–	–
Disconnection of a network element by the back-up protection devices following a single-phase SC with successful or failed automatic reclosing[b]	II	–	–	–
Disconnection of a network element by the main protection devices following a phase-to-phase-to-ground SC with failed automatic reclosing[b]	–	II	III	III
Disconnection of a network element using redundant switch operator following a single-phase SC with one breaker failure; here disconnections of all neighboring network elements, including busbars, are accounted, related to the disconnection of adjacent circuit breakers	II	III	III	III
The same but following a phase-to-phase-to-ground SC	–	III	III	–
The same but following a three-phase SC	III	–	–	–
Disconnection of a busbar following a single-phase SC, which does not include disconnection of tie lines between the network nodes	I	I	II	II
The same but with disconnection of tie lines	III	III	–	–

[a]On the tie line between the nuclear power plant and the system.
[b]With blocking of automatic reclosing; in case of no arc extinction the failed automatic reclosing is not considered.
[c]A dash means that the disturbance is not considered.

10.5.3.2 Stability Margins. The static (aperiodic) stability margin with respect to active power in the cutset is calculated by the formula:

$$k_p = \frac{P_{\lim} - (P + \Delta P_{irr})}{P_{\lim}} \tag{10.185}$$

TABLE 10.9. Standard SC Durations

Nominal voltage, kV	110	220	330	500	750	1150
Clearing time of SC, s	0.18	0.16	0.14	0.12	0.10	0.08

TABLE 10.10. Standard Emergency Power Imbalances [114]

Emergency Power Imbalance	Group of Disturbances
Power of one generator or one block of generators connected to the network through a common circuit breaker. The power of two NPP generators connected to one reactor block	II
Power of generators connected to section (system) of buses or a switchgears of the same voltage, within a power plant	III

where

$P_{\lim}$ is the maximum active power flow in the considered cutset with respect to the aperiodic stability;

P is the actual active power flow in the cutset under the considered conditions;

ΔP_{irr} is the range of irregular active power fluctuations in the cutset; it is assumed that under the action of irregular fluctuations the flow varies in the range $P \pm \Delta P_{irr}$.

The range of irregular active power fluctuations is set for each cutset of the system based on measurements. In the absence of such measurements, the calculated range of the irregular active power fluctuations in the cutset can be determined by the expression:

$$\Delta P_{irr} = K\sqrt{\frac{P_{\ell 1} \cdot P_{\ell 2}}{P_{\ell 1} + P_{\ell 2}}} \tag{10.186}$$

where $P_{\ell 1}$ and $P_{\ell 2}$ represent total load powers in subsystems on the two sides of the considered cutest; the coefficient K is assumed equal to 1.5 for manual regulation and 0.75 for automatic regulation (limitation) of power flow in the cutset.

The rules of determining the maximum active power flow in a cutset by its successive loading is presented in [112].

The stability margin with respect to voltage relates the load nodes and is calculated by the formula:

$$k_V = \frac{V - V_{cr}}{V} \tag{10.187}$$

where V is the actual voltage at the node for the considered conditions, and V_{cr} is the critical voltage at the same node in terms of stability of motors.

The critical voltage at load nodes with nominal voltage of 110 kV and higher, in the absence of more accurate data, should be assumed equal to the largest value of the two values: 0.7 V_{nom} or 0.75 V_{norm}, where V_{nom} is the nominal voltage, and V_{norm} is the actual voltage at the considered load node under normal operating conditions of the system.

10.5.3.3 System Stability Requirements. The power system stability requirements should not be less restrictive than those given in Table 10.11. The dash means that for the given situation the system stability is not ensured. In [112], some specific features and details on applying the standards of power system stability requirements are presented.

TABLE 10.11. Standards for Stability Requirements [112]

Loading Conditions of the Cutset	Minimum Stability Margin with Respect to Active Power	Minimum Stability Margin with Respect to Voltage	Groups of Disturbances for which System Stability should be Ensured	
			Normal Configuration	Maintenance Configuration
Normal	0.20	0.15	I, II, III	I, II
Heavy load	0.20	0.15	I, II	I
Forced	0.08	0.10	–	–

The use of the standards presented in this Subsection helps reducing the set of conditions to be studied when analyzing the stability of large electric power systems.

10.5.4 Reducing the Studied Conditions by Structural Analysis

The quantitative estimates of structural heterogeneity as a fundamental property of large electric power systems and the efficiency of using these quantitative estimates to construct dynamic equivalents were demonstrated in Section 10.4. It turns out that these estimates can be successfully applied to reduce the number of studied conditions for the transient stability analysis of a large electric power system on the basis of structural analysis methods presented in Section 10.4 [65].

It is obvious that, in the considered approach, the (transient) stability is most likely to be lost, first of all, due to weak tie lines in the complex power system. Therefore, the approach is based on the quantitative estimation of electric power system connectivity and the determination of strongly connected subsystems and weak ties (cutsets) between them. This approach is illustrated in the following by grouping normal, maintenance, and postemergency system configurations in terms of the transfer capacity of weak cutsets.

Consider a small size electric network, as shown in Figure 10.30. Based on the structural analysis and using the indices of electrical connectivity w_{ij}, that are determined using (10.113), two strongly connected subsystems I and J and a weak cutset between them (Figure 10.30) can be identified. The transfer capacity of this cutset is determined by the expression:

$$w_{IJ} = \sum_{i=1}^{n} \sum_{j=1}^{m} w_{ij} \tag{10.188}$$

where n and m are the number of nodes in which the generator electromotive force is applied in subsystems I and J, respectively.

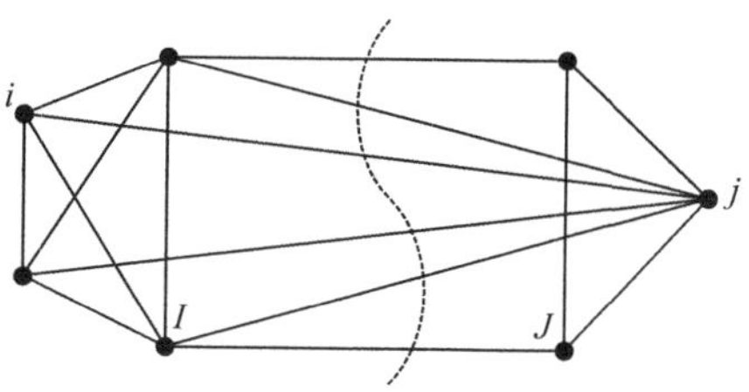

Figure 10.30. An example of a reduced network.

From the whole set of considered configurations, a basic one (for example, the normal operation) is determined, which is then compared with all the remaining configurations by the values of transfer capacities of the weak cutsets (in a general case there can be L of them) using the formula:

$$\Delta w_\ell^k = (w_\ell^b - w_\ell^k)/w_\ell^b \tag{10.189}$$

where $k = \overline{1, K}$ is the number of configurations, $\ell = \overline{1, L}$ is the number of weak cutsets, and the subscript "b" corresponds to the basic scheme.

Expression (10.189) shows how much the transfer capacity of the ℓ-th cutset in the considered k-th configuration differs from the transfer capacity of this cutset in the basic configuration. If the value Δw_ℓ^k is insignificant it means that, for example, disconnection of a certain tie line for maintenance in the initial system does not affect the value of transfer capacity of the ℓ-th cutset. In a general case with L weak cutsets, the assumed conditions should be met for all weak cutsets; it can be deducted that the configuration k is identical to the basic configuration in terms of transfer capacity of the weak cutsets, allowing the exclusion of configuration k from the analysis, by assigning it the results obtained for the basic configuration.

These considerations are the essence of the presented approach. In a general case with K configurations and L weak cutsets, the configurations should be grouped based on generalized distances in a multidimensional space of the form (10.150). Here, the number of initial conditions for equation (10.150) in this case is formed by the sets of K configurations and L weak cutsets in these configurations.

Out of each cluster of configurations, a representative configuration to be further studied is determined. The representative configuration reflects the most severe situation in the considered cluster.

Similarly, using the corresponding indices we can also group the studied operating conditions and disturbances and form the sets of representative operating conditions and disturbances, respectively [65]. Therefore, the set of conditions to be studied is reduced.

10.5.5 Using the Simplified Models and Direct Methods

The previous stages were focused on determining the reduced set of studied conditions (configurations, operating conditions, disturbances) for which the transient stability need to be estimated and the measures to ensure it should be chosen. Under these circumstances, the simplified models of power system dynamics (e.g., the classical model) and direct methods of analysis (e.g., the Lyapunov function method, and EEAC) can be applied. The use of the simplified models and direct methods will enable the revealing of combinations of configurations, operating conditions, and disturbances that are known to be secure in terms of stability. The other situations should be studied using more detailed models of electric power system dynamics taking into account voltage regulators and speed governors, load characteristics, and so on. The studies should employ simulations of transient processes in time domain by numerical methods.

For some combinations of configurations, operating conditions and disturbances, the system may happen to lose its stability and it is necessary to choose the control actions in order to ensure the system stability (e.g., control actions of emergency control devices). The process of choosing the control actions is normally iterative. At the initial iterations, in order to speed up the process, the stability can be estimated using the simplified models of system dynamics and direct methods. However, at the final stage of the iterative process a more detailed model can be employed to specify the values of control actions in order to ensure system stability.

The techniques considered in this section simplify and regularize the study on stability of large electric power systems and the choice of control actions to ensure stability.

10.6 APPLICATION

The following application examines the transient stability of a thermal power plant consisting of three 388 MVA, 24 kV, 50 Hz units ($G_1 \ldots G_3$) supplying power to an infinite bus (IB) through two transmission circuits (TC_1, TC_2) as shown in Figure 10.31.

The nominal voltage of the transmission network is $U_n = 400$ kV. The reactance of each line in per unit on $S_b = 100$ MVA and $U_b = U_n = 400$kV is $X_L = 0.0309$ p.u..

The generators are modeled as a single equivalent generator represented by the classical model with the following parameters expressed in per unit on the $S_{ng} = 3 \times 388 = 1164$ MVA and $U_b = 400$ kV base

$$X'_d = 0.364\,\text{p.u. and}\, H = 3.1\,\text{MW}{\times}\text{s/MVA}.$$

The transformers are modeled as a single equivalent transformer with an impedance of $X = 0.156$ p.u. on $S_{nt} = 3 \times 400 = 1200$ MVA and $U_b = 400$ kV, and off-nominal ratio of 1.0.

The initial system operating conditions with quantities expressed in per units on S_b and U_b are as follows:

- Generated active power $P_g = 8.5$;
- Generator terminals voltage $U_{LV} = U_1 = 1.0$;
- Infinite bus voltage $U_{IB} = U_2 = E = 1.0$.

Assume that at time instant $t_0 = 0$, a three-phase-to-earth short circuit occurs on the line circuit TC_2, in the point F, which is eliminated by disconnection of the affected circuit simultaneously at both ends at time instant $t = t_d$.

A. Determine the critical clearing angle δ_{crit} and the critical fault clearing time t_{crit}, using the equal area criterion.
B. Check the above value, using numerical integration.

Solution

A. *Determination of the critical clearing angle and the critical fault clearing time using the equal area criterion*

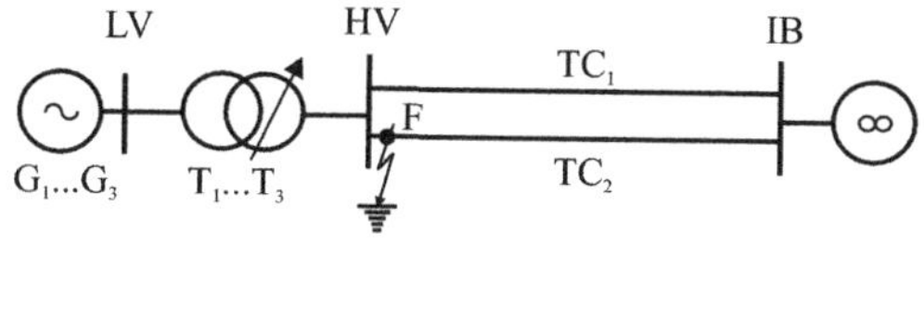

Figure 10.31. One-line diagram of the examined power system.

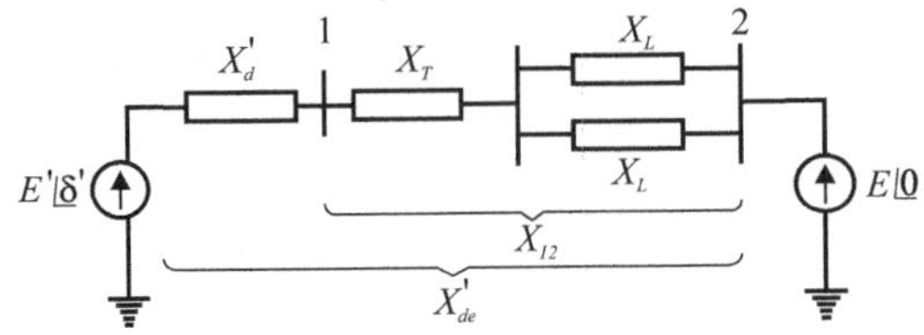

Figure 10.32. Equivalent circuit.

With the generator and the network reactance related to the HV of the step-up transformer, the system equivalent circuit is as shown in Figure 10.32.

In accordance with the considered scenario, the system undergoes the following three operating states:

(i) The *normal operation* state or *before fault* (bf) corresponding to the time interval before the fault occurrence $t < t_0$.

(ii) *Faulty* state or *during fault* (df), corresponding to the time interval between fault occurrence and fault clearance $t_0 \leq t < t_d$.

(iii) The *post fault* (pf) state, corresponding to the time period following the fault clearance $t \geq t_d$.

Denoting by X_{12} the equivalent reactance of the electrical network (between nodes 1 and 2) and by $X'_{de} = X'_d + X_{12}$ the equivalent reactance between the two emfs, the electromagnetic powers expressions, corresponding to the three operating states, based on which the transient characteristic $P - \delta$ from Figure 10.33 is plotted, are

$$P_e = \begin{cases} \dfrac{E'E}{X'^{(bf)}_{de}} \sin\delta & \text{for} \quad t < t_0 \\ 0 & \text{for} \quad t_0 \leq t < t_d \\ \dfrac{E'E}{X'^{(pf)}_{de}} \sin\delta & \text{for} \quad t \geq t_d \end{cases} \tag{10.190}$$

where

$\delta = \delta'$ is the angle of emf $\underline{E}'$used to define the rotor position with respect to the system of axes of the network;

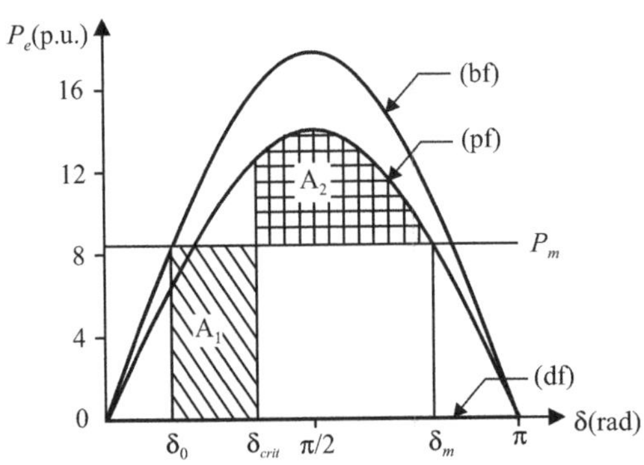

Figure 10.33. Transient characteristic.

$X_{de}^{\prime\,(bf)}$ is the equivalent reactance in the normal operating state;

$X_{de}^{\prime\,(pf)}$ is the equivalent reactance after the fault clearance.

Because the generator and transformer parameters are related to their rated values, they need to have a common base in S_b, resulting:

$$X_d' = X_d' \frac{S_b}{S_{ng}} = 0.364 \frac{100}{1164} = 0.0313\,\text{p.u.}$$

$$H = H \frac{S_{ng}}{S_b} = 3.1 \frac{1164}{100} = 36.084\,\text{MW}\times\text{s/MVA},\ M = 2H = 72.168$$

and

$$X_T = X_T \frac{S_b}{S_{nt}} = 0.156 \frac{100}{1200} = 0.013\,\text{p.u.}$$

respectively

$$X_{12}^{(bf)} = X_T + \frac{X_L}{2} = 0.0285\,\text{p.u.} \quad \text{and} \quad X_{de}^{\prime\,(bf)} = X_d' + X_T + \frac{X_L}{2} = 0.0598\,\text{p.u.}$$
$$X_{12}^{(pf)} = X_T + X_L = 0.0439\,\text{p.u.} \qquad X_{de}^{\prime\,(pf)} = X_d' + X_T + X_L = 0.0752\,\text{p.u.}$$

To calculate the transient emf $\underline{E}' = E'\lfloor\underline{\delta}'$, the quantities at the generator terminals are required. Therefore, given the generated active power $P_g = 8.5\,\text{p.u.}$ and the terminal voltage magnitude $U_g = U_1 = 1.0\,\text{p.u.}$, the voltage angle θ and the reactive power Q_g are computed using the relationships:

$$P_g = \frac{U_g E}{X_{12}^{(bf)}} \sin\theta$$
$$Q_g = \frac{U_g^2}{X_e^{(bf)}} - \frac{U_g E}{X_{12}^{(bf)}} \cos\theta \qquad (10.191)$$

From the first relationship gives $\theta = \text{asin}\frac{P_g X_{12}^{(bf)}}{U_g E} = 14.0914^{\circ}$, while from the second it results that $Q_g = 1.0451\,\text{p.u.}$. Therefore $\underline{S}_g = P_g + jQ_g = (8.5 + j1.0451)\,\text{p.u.}$, and $\underline{U}_1 = U_g e^{j\theta} = 0.9702 + j0.2423 = 1 \cdot e^{j14.0194^{\circ}}\,\text{p.u.}$

Next, calculate $\underline{E}' = \underline{U}_g + jX_d'\left(\underline{S}_g/\underline{U}_g\right)^{*} = (0.9375 + j0.5081)\,\text{p.u.}$ Thus, the transient emf magnitude, which remains constant, is $E' = 1.0663\,\text{p.u.}$, and the initial value of the rotor angle is $\delta_0 = \delta' = 0.4966\,\text{rad} = 28.4531^{\circ}$.

The maximum value of the rotor angle, after fault clearance, corresponds to the intersection point between the line $P_m = \text{ct.}$ and the $P - \delta$ post fault characteristic. Therefore, from $P_m = \frac{E'E}{X_{de}^{\prime\,(pf)}} \sin\delta$ gives:

$$\delta_m = \pi - \text{asin}\frac{P_m X_{de}^{\prime\,(pf)}}{E'E} = 2.4987\,\text{rad} = 143.165^{\circ}$$

According to the equal area criterion, the critical clearance is obtained by equating the accelerating area A_1 and the deceleration area A_2 (Figure 10.33). Thus:

$$\delta_{crit} = \text{acos}\left\{ P_m(\delta_m - \delta_0)\frac{X_{de}^{\prime\,(pf)}}{E'E} + \cos\delta_m \right\} = 1.1594\,\text{rad} = 66.4284^{\circ}.$$

Given the critical clearing angle, the critical fault clearing time is calculated using the expression

$$t_{crit} = \sqrt{\frac{2(\delta_{crit} - \delta_0)M}{\omega_0 P_m}} = 0.1893\,\text{s}$$

B. *Integration of Swing Equation.* Simulation of power system dynamic response requires numerical integration of the system of differential equations that characterize its behavior. When using the classical model of the synchronous generator, the only differential solutions present in the model are the electromagnetic state equations:

$$\begin{aligned} \frac{d\omega}{dt} &= \frac{1}{M}(P_m - P_{\max}\sin\delta - D\omega) = f_\omega(\omega,\delta) \\ \frac{d\delta}{dt} &= \omega_0\omega = f_\delta(\omega,\delta) \end{aligned} \tag{10.192}$$

in which:

$$P_{\max} = \begin{cases} \dfrac{E'E}{X'^{(bf)}_{de}} & \text{for} \quad t < t_0 \\ 0 & \text{for} \quad t_0 \le t < t_d \\ \dfrac{E'E}{X'^{(pf)}_{de}} & \text{for} \quad t \ge t_d \end{cases} \tag{10.193}$$

Next, for the numerical integration of parameters, the fourth-order Runge–Kutta method and the trapezoidal rule are used, given that ω and δ are inertial quantities, and the mechanical power P_m is constant.

B1. Runge–Kutta Method. Given the values for $\omega(t)$ and $\delta(t)$:

$$\begin{cases} \omega(t+\Delta t) = \omega(t) + \dfrac{1}{6}(K_{\omega,1} + 2K_{\omega,2} + 2K_{\omega,3} + K_{\omega,4}) \\ \delta(t+\Delta t) = \delta(t) + \dfrac{1}{6}(K_{\delta,1} + 2K_{\delta,2} + 2K_{\delta,3} + K_{\delta,4}) \end{cases} \tag{10.194}$$

where

$$\begin{cases} K_{\omega,1} = f_\omega(\omega(t), \delta(t)) \cdot \Delta t \\ K_{\delta,1} = f_\delta(\omega(t), \delta(t)) \cdot \Delta t \\ K_{\omega,2} = f_\omega(\omega(t) + 0.5K_{\omega,1}, \delta(t) + 0.5K_{\delta,1}) \cdot \Delta t \\ K_{\delta,2} = f_\delta(\omega(t) + 0.5K_{\omega,1}, \delta(t) + 0.5K_{\delta,1}) \cdot \Delta t \\ K_{\omega,3} = f_\omega(\omega(t) + 0.5K_{\omega,2}, \delta(t) + 0.5K_{\delta,2}) \cdot \Delta t \\ K_{\delta,3} = f_\delta(\omega(t) + 0.5K_{\omega,2}, \delta(t) + 0.5K_{\delta,2}) \cdot \Delta t \\ K_{\omega,4} = f_\omega(\omega(t) + K_{\omega,3}, \delta(t) + K_{\delta,3}) \cdot \Delta t \\ K_{\delta,4} = f_\delta(\omega(t) + K_{\omega,3}, \delta(t) + K_{\delta,3}) \cdot \Delta t \end{cases} \tag{10.195}$$

Therefore, considering a time step $\Delta t = 0.001$ s and a damping coefficient $D = 10$, to determine the values for $\omega(0.001)$ and $\delta(0.001)$ it is known that:

$$\begin{cases} \omega(0) = \omega(0_+) = \omega(0_-) = 0 \\ \delta(0) = \delta(0_+) = \delta(0_-) = \delta_0 = 0.4966\,\text{rad} \\ P_{\max} = 0 \end{cases}$$

then calculate the Runge–Kutta coefficients using the expressions (10.194) and (10.195):

– the first class of Runge–Kutta coefficients

$$K_{\omega,1} = \frac{1}{72.168}(8.5 - 0 \cdot \sin 0.4966 - 50 \cdot 0) \cdot 0.001 = 1.1778 \cdot 10^{-4}$$

$$K_{\delta,1} = 314.15 \cdot 0 = 0$$

– the second class of Runge–Kutta coefficients

$$K_{\omega,2} = \frac{1}{72.168}(8.5 - 0 \cdot \sin(0.4966 + 0.5 \cdot 0) - 50 \cdot 0) \cdot 0.001 = 1.1778 \cdot 10^{-4}$$

$$K_{\delta,2} = 314.15 \cdot (0 + 0.5 \cdot 1.1778 \cdot 10^{-4}) = 1.8501 \cdot 10^{-5}$$

– the third class of Runge–Kutta coefficients

$$K_{\omega,3} = \frac{1}{72.168}\left(8.5 - 0 \cdot \sin(0.4966 + 0.5 \cdot 1.8501 \cdot 10^{-5}) - 50 \cdot 0\right) \cdot 0.001$$

$$= 1.1778 \cdot 10^{-4}$$

$$K_{\delta,3} = 314.15 \cdot (0 + 0.5 \cdot 1.1778 \cdot 10^{-4}) = 1.8501 \cdot 10^{-5}$$

– the fourth class of Runge–Kutta coefficients

$$K_{\omega,4} = \frac{1}{72.168}\left(8.5 - 0 \cdot \sin(0.4966 + 0.5 \cdot 1.8501 \cdot 10^{-5}) - 50 \cdot 0\right) \cdot 0.001$$

$$= 1.1778 \cdot 10^{-4}$$

$$K_{\delta,4} = 314.15 \cdot (0 + 1.1778 \cdot 10^{-4}) = 3.7002 \cdot 10^{-5}$$

Finally obtain:

$$\omega(0.001) = 0 + \frac{1}{6}(1.1778 + 2 \cdot 1.1778 + 2 \cdot 1.1778 + 1.1778) \cdot 10^{-4} = 1.1778 \cdot 10^{-4}$$

$$\delta(0.001) = 0.4966 + \frac{1}{6}(0 + 2 \cdot 1.8501 + 2 \cdot 1.8501 + 3.7002) \cdot 10^{-5} \cong 0.4966 \text{ rad}$$

The rotor angle variation curves, for various values of the fault clearing time t_d, obtained with the Runge–Kutta method implemented in MATLAB, are presented in Figure 10.34.

B2. Trapezoidal Rule. Assume that the electromagnetic power is a noninertial quantity, that is $P_e(t + \Delta t) = P_e(t)$. Therefore, applying the trapezoidal rule to the electromechanic state equations (10.192) it results:

$$\begin{cases} \omega(t + \Delta t) = \dfrac{1}{1 + \dfrac{D}{M}\dfrac{\Delta t}{2}}\left\{[P_m - P_e(t)]\dfrac{\Delta t}{M} + \left[1 - \dfrac{D}{M}\dfrac{\Delta t}{2}\right]\omega(t)\right\} \\ \delta(t + \Delta t) = \delta(t) + \omega_0[\omega(t + \Delta t) + \omega(t)]\dfrac{\Delta t}{2} \end{cases} \tag{10.196}$$

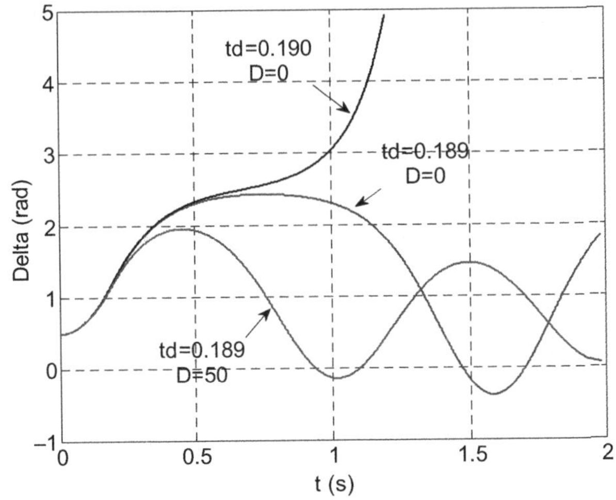

Figure 10.34. Simulation of rotor angle variation with Runge–Kutta method.

which help us to determine the angular speed and the rotor angle at the time instant $t + \Delta t$ in terms of the quantities determined at the time instant t. In this regard, the following steps are taken:

(i) Initialize the simulation:

– set: $t = 0_+$ (the time instant immediately after the fault occurrence);

$$\omega(0_+) = \omega(0_-) = 0$$

$$\delta(0_+) = \delta(0_-) = \delta_0 = 0.4966\,\text{rad}$$

– choose an integration time step $\Delta t = 0.001$ s, the fault clearing time as t_d and the simulation time $t_s = 2$ s.

(ii) Calculate the electromagnetic power at the time instant t using the appropriate expression from (10.193).

Therefore, for $t = 0_+ < t_d$ gives $P_e(t) = P_e(t + \Delta t) = 0$

(iii) Calculate $\omega(t + \Delta t)$ and $\delta(t + \Delta t)$ using the equations (10.195)

$$\omega(0.001) = \frac{1}{1 + \dfrac{0}{72.168}\dfrac{0.001}{2}}\left\{[8.5 - 0]\frac{0.001}{72.168} + \left[1 - \frac{0}{72.168}\frac{0.001}{2}\right]0\right\}$$
$$= 0.1178 \cdot 10^{-3}$$

$$\delta(0.001) = 0.4966 + 314.15\left[0.1178 \cdot 10^{-3} + 0\right]\frac{0.001}{2} \cong 0.4966\,\text{rad}$$

(iv) Set $t = t + \Delta t$ and go to step (ii)

The rotor angle variation curves, for various values of the fault clearing time t_d, obtained using the trapezoidal rule are presented in Figure 10.35.

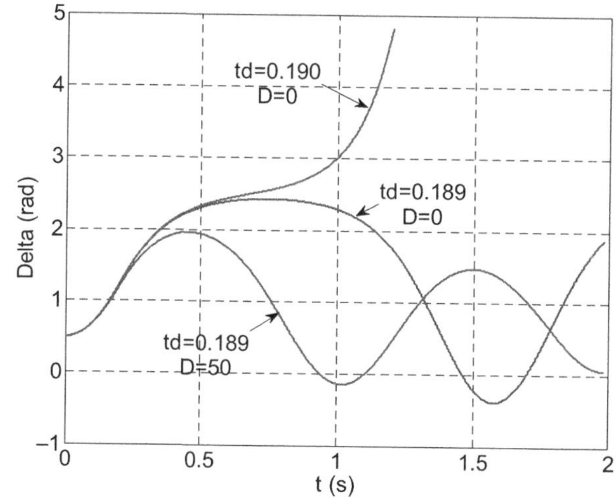

Figure 10.35. Simulation of rotor angle variation with trapezoidal method.

Note that the results obtained by dynamic simulation verify the critical fault clearing time value previously obtained using the equal area criterion. Therefore, neglecting the damper winding, for a value $t_d = 0.190$ s the stability is lost (the rotor angle increases, and the generator losses its synchronism with the infinite power system), and for a value $t_d = 0.189$ s the dynamic response is oscillatory and undamped, that is the generator maintains the synchronism. Furthermore, if the damper effect is considered, when the fault clearing times are smaller than t_{crit}, the oscillations are dampened.

REFERENCES

[1] Pavella, M., Murthy, P.G. *Transient Stability of Power Systems. Theory and practice*, John Wiley & Sons, New York, 1994.

[2] Pavella, M., Ernst, D., Ruiz-Vega, D. *Transient Stability of Power Systems. A Unified Approach to Assessment and Control*, Kluver Academic Publisher, Boston, 2000.

[3] Zhang, Y., Wehenkel, L., Pavella, M. SIME: A comprehensive approach to fast transient stability analysis, *IEE of Japan Proceedings*, Vol. 118-B, No. 2, pp. 127–132, 1998.

[4] Crappe, M. *Stabilité de sauvegarde des réseaux électriques*, Hermes Science Publications, Paris, 2003.

[5] Lyapunov, A.M. *General problem of moving stability*, Acad. of Sci. of the USSR, Moscow, 1956. (in Russian)

[6] LaSalle, J.P., Lefscetz, S. *Stability by Lyapunov's direct method, with applications*, Acad. Press, New York, 1961.

[7] Leondes, C.T. *Advances in control systems*, New York – London, 1965.

[8] Putilova, A.T., Tagirov, M.A. *Stability criteria of electric power systems*, VINITI, Moscow, 1971. (in Russian)

[9] Barbashin, E.A. *Lyapunov functions*, Nauka, Moscow, 1970. (in Russian)

[10] Rao, N.D. Routh-Hurwitz conditions and Lyapunov methods for transient stability problem, *Proc. Inst. Electr. Eng.*, Vol. 116 (C), No. 4, pp. 539–547, 1969.

[11] Korotkov, V.A. *On the study of power system stability with application of the quadratic Lyapunov functions*, Proc. of Siberian Electric Power Research Inst., Moscow, Energiya, Vol. 20, pp. 26–32, 1971 (in Russian).

[12] Natesan, T.R. Quadratic Lyapunov function and transient stability study, *Journ. Inst. Eng. (India), Electric Eng. Div.*, Vol. 52, No. 12, pp. 341–344, 1972.

[13] Gorev, A.A. *On the stability of parallel work of synchronous machines*, Proceedings of Leningrad Electromechanical Institute, Leningrad, Kubuch, pp. 75–136, 1934 (in Russian).

[14] Magnusson, P.C. The transient-energy method of calculating stability, *AIEE Trans.*, Vol. 66, pp. 747–755, 1947.

[15] Ribbens-Pavella, M. *Critical survey of transient stability studies of multimachine power systems by Lyapunov's direct method*, Proc. of the 9th Annual Allerton Conference on Circuits and Systems Theory, pp. 751–767, October 1971.

[16] Kinnen, E., Chen, C.S. *Lyapunov function from auxiliary exact differential equations*, NASA Contractor Report-799, Washington, NASA, 1967.

[17] Ambarnikov, G.A., Rudnitsky, M.P. *An approximate mathematical model of electric power system and study of its stability*, Proc. of the 2nd Seminar-Symposium on Application of the Method of Lyapunov Functions in Energy, Novosibirsk, Nauka, pp. 172–180, 1979. (in Russian)

[18] Andreyuk, V.A. *Derivation of sufficient stability conditions "in the large" for a system of synchronous machines*, Proc. of Direct Current Research Inst., Vol. 2, Leningrad, Gosenergoizdat, pp. 239–257, 1957. (in Russian)

[19] Aylett, P.D. *The energy-integral criterion of transient stability limits of power systems*, *Proc. Inst. Electr. Eng.*, Vol. 105 (C), No. 3, pp. 527–536, 1958.

[20] Gless, G.E. Direct method of Lyapunov applied to transient power system stability, *IEEE Trans. Power Appar. Syst.*, Vol. 85, No. 2, pp. 156–168, 1966.

[21] Putilova, A.T. *Lyapunov function for mutual motion equations for synchronous machines and its modification*, Proc. of the 2nd Seminar-Symposium on Application of the Method of Lyapunov Functions in Energy, Novosibirsk, Nauka, pp. 104–114, 1970. (in Russian)

[22] Putilova, A.T. *Analysis of stability of bulk electric power systems by Lyapunov criteria*, Proc. of the 2nd Seminar-Symposium on Application of the Method of Lyapunov Functions in Energy, Novosibirsk, Nauka, pp. 151–171, 1970. (in Russian)

[23] Podshivalov, V.I. *Construction of Lyapunov function for the Lienard vector equation*, Proc. of the 2nd Seminar-Symposium on Application of the Method of Lyapunov Functions in Energy, Novosibirsk, Nauka, pp. 68–78, 1970 (in Russian)

[24] Tavora, C.J., Smith, O.J.M. Characterization of equilibrium and stability in power systems, *IEEE Trans. Power Appar. Syst.*, Vol. 91, No. 3, pp. 1127–1130, 1972.

[25] Athay, T., Podmore, R., Virmani, S. A practical method for the direct analysis of transient stability, *IEEE Trans. Power Appar. Syst.*, Vol. 98, No. 2, pp. 573–584, 1979.

[26] Tsolas, N.A., Arapostathis, A., Varaiya, P.P. A structure preserving energy function for power system transient stability analysis, *IEEE Trans. Circuits Syst.*, Vol. 32, No. 10, pp. 981–989, 1985.

[27] Alberto, L.F.C., Bretas, N.G. *Energy function for power systems with frequency dependent loads*, Proceedings of the XII Brazilian Automatic Control Conference, San Paolo, Brazil, pp. 703–707, September 8–12, 1998. (in Portuguese)

[28] Kakimoto, N., Ohsawa, Y., Hayashi, M. Transient stability analysis of multimachine power systems with field flux decays via Lyapunov's direct method, *IEEE Trans. Power Appar. Syst.*, Vol. 99, No. 5, pp. 1819–1827, 1980.

[29] Pai, M.A. *Energy function analysis for power system stability*, Kluwer, Norwell, MA, 1989.

[30] Fouad, A.A., Vittal, V. *Power system transient stability analysis using the transient energy function method*, Prentice Hall, Englewood Cliffs, 1991.

[31] Kartvelishvili, N.A. *Energy system stability problems as the problems of general stability theory*, The Second Lyapunov's Method and Its Application in Energy. Proceedings of the Workshop-Symposium. Part II, Novosibirsk, Nauka, pp. 122–150, 1966. (in Russian)

[32] Venikov, V.A., Bampi, Yu.S. Opportunities, methodology and prospects for the electric system stability studies by the direct method of A.M. Lyapunov, *Electrichestvo*, No. 12, pp. 15–23, 1972. (in Russian)

[33] Morse, M. Relations between the critical points of a real function of independent variables, *Trans. Amer. Mathem. Soc.*, Vol. 27, No. 3, pp. 345–396, 1925.

[34] Ambarnikov, G.F., Rudnitsky, M.P. *Determination of equilibriums of a bulk power system*, Proc. of the 2nd Seminar-Symposium on Application of the Method of Lyapunov Functions in Energy, Novosibirsk, Nauka, pp. 96–103, 1970. (in Russian)

[35] El-Abiad, A.H., Nagapan, K. Transient stability regions of multimachine power systems, *IEEE Trans. Power Appar. Syst.*, Vol. 85, No. 1, pp. 169–179, 1966.

[36] Ribbens-Pavella, M. *Theorie generale de la stabilite transitoire de machines synchrones*, Univ Liege, Collect. des Publicat., No. 17, pp. 3–134, 1970.

[37] Zaslavskaya, T.B., Putilova, A.T., Tagirov, M.A. Lyapunov function as a criterion of synchronous transient stability, *Elektrichestvo*, No. 6, pp. 19–24, 1967. (in Russian)

[38] Teichgraeber, R.D., Harris, F.W., Johnson, L. New stability measure for multimachine power systems, *IEEE Trans. Power Appar. Syst.*, Vol. 89, No. 2, pp. 233–239, 1970.

[39] Prabhakara, F.S., El-Abiad, A.H. A simplified determination of stability regions for Lyapunov methods, *IEEE Trans. Power Appar. Syst.*, Vol. 94, No. 2, pp. 672–689, 1975.

[40] Vaiman, M.Ya. *Study of systems stable "in the large,"* Nauka, Moscow, 1981. (in Russian)

[41] Zubov, V.I. *Mathematical methods for the study of automatic control systems*, Sudpromgiz, Leningrad, 1959. (in Russian)

[42] Gupta, C.L., El-Abiad, A.H. Determination of the closest unstable equilibrium state for Lyapunov methods in transient stability studies, *IEEE Trans. Power Apar. Syst.*, Vol. 95, No. 5, pp. 1699–1712, 1976.

[43] Van Rysin, J. *Classification and clustering*, Academic Press Inc., New York, 1977.

[44] Machowski, J., Bialek, J.W., Bumby, J.R. *Power system dynamics and stability*, Wiley, New York, 1997.

[45] Kakimoto, N., Ohsawa, Y., Hayashi, M. Transient stability analysis of electrical power system via Lure type Lyapunov function, *IEE Jpn. Proc.*, Vol. 98, No. 5, pp. 31–36; No. 6, pp. 77–83, 1978.

[46] Alberto, L.F.C., Silva, F.H.J.R., Bretas, N.G. *Direct methods for transient stability analysis in power systems: state of the art and future perspectives*, 2001 IEEE Porto Power Tech Conference, Porto, Portugal, September 10–13, 2001.

[47] Chiang, H.D., Wu, F.F., Varaiya, P.P. Foundations of PEBS method for power system transient stability analysis, *IEEE Trans. Circuits Syst.*, Vol. 35, No. 6, pp. 675–682, 1988.

[48] Chiang, H.D., Wu, F.F., Varaiya, P.P. A BCU method for direct analysis of power system transient stability, *IEEE Trans. Power Syst.*, Vol. 9, No. 6, pp. 1194–1208, 1994.

[49] Tagirov, M.A. *On development of transient stability criteria for mathematic models of electric power systems (in Russian)*, Proceedings of the 2nd Workshop-Symposium on Application of the Lyapunov Function Method in Energy, Novosibirsk, Nauka, pp. 140–150, 1970.

[50] Scruggs, J.T., Mili, L. Dynamic gradient method for PEBS detection in power system transient stability assessment, *Electrical Power Energy Syst.*, Vol. 23, No. 2, pp. 155–165, 2001.

[51] Treinen, R.T., Vittal, V., Klienman, W. An improvement technique to determine the controlling unstable equilibrium point in a power system, *IEE Trans. Circuits Syst.*, Vol. 43, No. 4, pp. 313–323, 1996.

[52] Xue, A., Mei, S., Xie, B. *A comprehensive method to compute the controlling unstable equilibrium point*, The 3rd Int. Conf. on Electric Utility Deregulation and Restructuring and Power Technologies, Nanjing, China, April 6–9, 2008.

[53] Maria, G., Tang, C., Kim J. Hybrid stability analysis, *IEEE Trans. Power Syst.*, Vol. 5, No. 2, pp. 384–391, 1990.

[54] Dimo, P. *Nodal analysis of power systems*, Abacus Press, Kent, England, 1975.

[55] Zhukov, L.A., Stratan, I.P. Steady states of complex electric networks and systems: calculation techniques, Energiya, Moscow, 1979. (in Russian)

[56] EUROSTAG. *Release 4.2 Tractebel and Electricité de France*, Oct. 2002.

[57] Ashby, W.R. *An introduction to cybernetics*, Wiley, New York, 1956.

[58] Voropai, N.I. *Simplification of the mathematical models of power system dynamics*, Nauka, Novosibirsk, 1981. (in Russian)

[59] Voropai, N.I. *Simplification of the mathematical models of power systems in the dynamic processes of different length*, 8th PSCC, Helsinki, Finland, Aug. 19–24, 1984.

[60] Guseinov, F.G. *Simplifcation of calculated schemes of electric systems*, Energiya, Moscow, 1978. (in Russian)

[61] Venikov, V.A., Sukhanov, O.A. *Cybernetic models of electric systems*, Energoizdat, Moscow, 1982. (in Russian)

[62] Castro-Leon, E.G., El-Abiad, A.H. *Bibliography on power system dynamic equivalents and related topics*, IEEE PES 1980 Winter Meeting, New York, pp. 34.9/1–34.9/9, 1980.

[63] Chow, J.H., Winkelman, J.R., Pai, M.A., Sauer, P.W. *Application of singular perturbations theory to power system modeling and stability analysis*, Proc. Amer. Contr. Conf., Boston, Vol. 3, pp. 1401–1407, 1985.

[64] Perez-Arriaga, I.J., Verghese, G.C., Schweppe, F.C. Selective modal analysis with applications to electric power systems, *IEEE Trans. Power Appar. Syst.*, Vol. 101, No. 9, pp. 3117–3125, 1982.

[65] Abramenkova, N.A., Voropai, N.I., Zaslavskaya, T.B. *Structural analysis of electric power systems*, Nauka, Novosibirsk, 1990. (in Russian)

[66] Lee, S.T.Y., Schweppe, F.C. Distance measures and coherency recognition for transient stability equivalents, *IEEE Trans. Power Appar. Syst.*, Vol. 92, No. 5, pp. 1550–1558, 1973.

[67] Wu, F.F., Narasimhamurthi, N. Coherency identification for power system dynamic equivalents, *IEEE Trans. Circuits Syst.*, Vol. 30, No. 3, pp. 140–147, 1983.

[68] Chafurian, A., Cory, B.J. *Improvements in the calculation of transient stability in large interconnected power systems*, Proc. 5th Iran Conf. Elec. Eng., Shiraz, Vol. 1, pp. 360–370, 1975.

[69] McCauley, T.M. *Disturbance dependent electromechanical equivalent for transient stability studies,* IEEE Winter Power Meeting, New York, Jan. 1975.

[70] Abramenkova, N.A. Determination of electric power system structure for static stability analysis, *Izv. AS USSR. Energetika i Transport*, No. 3, pp. 33–40, 1985 (in Russian).

[71] Octojic, D. Identifikacija elektromehanickih oscilacija i analiza osetljivosti u slozenium elektroenergetskim sistemima, *Elektroprivreda (SFRY)*, T. 39, No. 7/8, pp. 277–284, 1986. (in Serbian)

[72] Kartvelishvili, N.A., Galaktionov, Yu.I. *Idealization of complex dynamic systems*, Nauka, Moscow, 1976. (in Russian)

[73] Grujic, L., Darwish, M., Fantin, J. Coherency, vector Lyapunov functions and large-scale power systems, *Int. J. Syst. Sci.*, Vol. 10, No. 3, pp. 351–362, 1979.

[74] Eremia, M., Palasanu, T. Criteria for electrical networks partitioning in building transient equivalents, *Bull. Inst. Polytech., Series Energ., Bucharest*, No. 5, pp. 65–74, 1983. (in Romanian)

[75] Machowski, J. Dynamic equivalents for transient stability studies of electrical power systems, *Elec. Power Energy Syst.*, Vol. 7, No. 4, pp. 215–224, 1985.

[76] Zhukov, L.A., Krug, N.K., Yarnykh, L.V. Quantitative estimation of equivalenting admissibility for transient stability calculations, *Izv. AS USSR. Energetika i Transport*, No. 4, pp. 21–28, 1971. (in Russian)

[77] Di Caprio, U., Barbier, C., Humphreys, P. The techniques and applications of power system dynamic equivalents at CEGB, EdF and ENEL, *Energ. Elect.*, Vol. 59, No. 5, pp. 498–506, 1982.

[78] Giri, G. Coherency reduction in the EPRI stability program, *IEEE Trans. Power Appar. Syst.*, Vol. 102, No. 5, pp. 1285–1293, 1983.

[79] Gorev, A.A. *Transient processes of a synchronous machine*, Gosenergoizdat, Leningrad, 1950. (in Russian)

[80] Dorsey, J., Schlueter, R.A. Global and local dynamic equivalents based on structural archetypes for coherency, *IEEE Trans. Power Appar. Syst.*, Vol. 102, No. 6, pp. 1793–1801, 1983.

[81] Pai, M.A., Angaonkar, R.P. Electromechanical distance measure for decomposition of power system, *Elec. Power Energy Syst.*, Vol. 6, No. 4, pp. 249–254, 1984.

[82] Guseinov, F.G., Guseinov, A.M. Mathematical modeling and equivalenting for control of power system operating conditions, *Izv. AS USSR. Energetika i Transport*, No. 3, pp. 158–161, 1986. (in Russian)

[83] Semlyen, A., Ruiz, G.A.I. *A building dynamic system equivalents for stability studies*, IEEE Electr. and Electron. Conf. and Exp., Toronto, Canada, pp. 42–43, 1979.

[84] Ohsawa, Y., Hayashi, M. Transient stability equivalents based on Lyapunov function, *Mem. Fac. Eng. Kyoto Univ.*, Vol. 48, No. 2, pp. 137–152, 1980.

[85] Jovanovic, S.M., Calovic, M.S. Contribution to the analysis of power system transient security, *Elec. Power Syst. Research*, Vol. 6, No. 4, pp. 259–267, 1983.

[86] Guseinov, F.G., Abdullaev, N.Sh., Efendiev, S.E. Identification of groups of coherent generators in electric power system, *Elektrichestvo*, No. 6, pp. 6–10, 1986. (in Russian)

[87] Stanton, K.M. Dynamic energy balance studies for simulation of power-frequency transient, *IEEE Trans. Power Appar. Syst.*, Vol. 91, No. 1, pp. 110–117, 1972.

[88] Pai, M.A., Sauer, P.W., Khorasani, K. Singular perturbations and large scale system stability, Proc. 23rd IEEE Conf. Decis. and Contr., Las Vegas, Vol. 1 pp. 173–178, December 12–14, 1984.

[89] Peponides, G., Kokotovic, P.V., Chow, J.H. Singular perturbations and time scales in nonlinear models of power systems, *Int. Symp. Circuits Syst., Rome*, Vol. 1, pp. 535–540, May 10–12, 1982.

[90] Venikov, V.A., Zhukov, L.A., Pospelov, G.E. *Electric systems. Operating conditions of electric systems and networks*, Vyschaya Shkola, Moscow, 1975. (in Russian)

[91] Di Caprio, U. Theoretical and practical dynamic equivalents in multimachine power systems, *Elec. Power Energy Syst.*, Vol. 4, No. 4, pp. 224–232; Vol. 5, No. 1, pp. 40–54, 1982.

[92] Nishida, S., Takeda, S. Derivation of equivalents for dynamic security assessment, *Elec. Power Energy Syst.*, Vol. 6, No. 1, pp. 15–23, 1984.

[93] Pires de Souza, E.J.S., Cardaso, E.N. *Dynamic equivalent utilizing by the REI approach to aggregate coherent generators*, IEEE Int. Symp. Circuits and Syst., Montreal, May 7–10, Vol. 2 pp. 622–625, 1984.

[94] Nath, R., Lamba, S.S. Development of coherency based time-domain equivalent model using structure constraints, *IEEE Proc.*, Vol. 133, Pt. C, No. 4, pp. 165–175, 1986.

[95] Zhukov, L.A. On transformations of complex electric systems for stability calculations, *Izv. AS USSR, Energetika i Transport*, No. 2, pp. 202–209, 1964 (in Russian)

[96] Kovalenko, V.P. Equivalent transformation of complex power systems, *Izv. AS USSR, Energetika i Transport*, No. 2, pp. 182–190, 1964. (in Russian)

[97] Stephen, D.D. Simulating multiple machines for stability studies, *Elec. Times*, Vol. 160, No. 10, pp. 36–38, 1971.

[98] Darwish, M., Fantin, J., Grateloup, C. On the decomposition-aggregation of large scale power systems, *Automat. Contr. Theory Appl.*, Vol. 5, No. 1, pp. 18–25, 1977.

[99] Brown, H.E., Shipley, R.B., Coleman, B., Nied, R.E. A study of stability equivalents, *IEEE Trans. Power Appar. Syst.*, Vol. 88, No. 3, pp. 200–206, 1969.

[100] Giri, Y., Bose, A. *Identification of dynamic equivalents for on line transient security assessment*, IEEE PES 1977 Summer Meeting, Mexico City, pp. 514.3/1–514.3/9, 1977.

[101] Rudnick, H., Hudhes, F.M., Brameller, A. *Identification of dynamic equivalents for power system transient studies*, 7th PSCC, Guildford, pp. 997–1001, 1981.

[102] Almutairi, A.M., Yee, S.K., Milanovic, J.V. *Identification of coherent generators using PCA and cluster analysis*, 16th PSCC, Glasgow, UK, Jul. 14–18, 2008.

[103] Orurk, I.A. *New methods for synthesis of linear and some nonlinear systems*, Nauka, Moscow, 1965. (in Russian)

[104] Shchedrin, N.N. *Simplification of electric systems for modeling*, Energiya, Moscow, 1966. (in Russian)

[105] Gurevich, Yu.E., Libova, L.E. *Studied load models for analysis of electric system stability*, Proc. of All-Union Research Institute of Electric Power Industry, issue 51, pp. 204–215, 1976. (in Russian)

[106] Voitov, O.N., Voropai, N.I., Gamm. A.Z., Golub, I.I., Efimov, D.N. *Analysis of inhomogenities of electric power systems*, Nauka, Novosibirsk, 1999. (in Russian)

[107] Volosov, V.M., Mogrugov, V.I. *An averaging method in the theory of nonlinear oscillatory systems*, Publ. House of Moscow State University, Moscow, 1971. (in Russian)

[108] Eremia, M. (Editor) *Electric power systems. Vol. 1. Electric networks*, Publishing House of the Romanian Academy, Bucharest, 2006.

[109] Bulac, C., Eremia, M. *Power system dynamics*, Printech Publisher, Bucharest, 2006. (in Romanian)

[110] Eremia, M., Trecat, J., Germond, A. *Réseaux électriques. Aspects actuels*, Technical Publisher, Bucharest, 2000. (in French)

[111] Surdu, C. *Transient stability assessment of power systems using hybrid methods,* PhD thesis, University "Politehnica" of Bucharest, 2005. (in Romanian)

[112] *Russian National Standardization Authority Methodological guidelines on stability of energy systems*, Approved by the order of Ministry of Energy of the RF of Jun. 30, 2003, No.277, Moscow, Publ. House of RC ENAS, 2004. (in Russian)

[113] Eremia, M., Crişciu, H., Ungureanu, B., Bulac, C. *Computer aided analysis of the electric power systems regimes*, Technical Publisher, Bucharest, 1985. (in Romanian)

[114] Gurevich, Y, Okin, A *Analysis of power system stability and emergency control implementation (in Russian)*, Energoatomizdat, Moscow, 1990.

[115] Chua, L.O., Lin, P.-M. *Computer-aided analysis of electronic circuits*, Prentice Hall, 1975.

11

VOLTAGE STABILITY

Mircea Eremia and Constantin Bulac

11.1 INTRODUCTION

Voltage stability is the power system ability to maintain the voltage level in acceptable limits in all the buses, under normal operating conditions as well as after perturbations.

An electrical power system enters a voltage instability state when a disturbance—load increase or change in the system topology—causes a progressive and uncontrollable voltage level degradation in a node, in an area or in the whole system. In the first moments the degradation process is slow, then it becomes more and more rapid, in general, if the system operates near its transmission capacity. This limit is much lower in case of unavailability of one generating unit or one component of the transmission network.

The main cause of the voltage instability phenomenon is the voltage drops due to changes in the power flows through the inductive elements of the transmission network following:

- load increase correlated with local or regional reactive power deficit;
- certain incidents that "weaken" either *the local voltage control* (tripping of some generating units, exceeding some generators reactive power limits) or *the transmission network* (tripping of some transmission lines, transformers or autotransformers, faults occurring on substations bus bars) or cause heavily loading of the transmission network (separation of network), and so on;
- *malfunction* of on-load tap changing transformers.

Handbook of Electrical Power System Dynamics: Modeling, Stability, and Control. Edited by Mircea Eremia and Mohammad Shahidehpour.

Although, in essence, the voltage instability is a local phenomenon, its consequences have a major impact on the system operation, sometimes initiating a voltage collapse process.

The voltage "avalanche" or "collapse" phenomenon is characterized by a succession of cascading events, associated with the instability phenomenon, that determine severe reduction of the voltage level in an area or in the whole system and finally the loss of the system angle stability.

Voltage collapse is typically associated with the reactive power demands of load not being met because of limitations on the production and transmission of reactive power:

- limitations on the production of reactive power include generator and SVC reactive power limits and the reduced reactive power produced by capacitors at low voltage;
- the primary limitations on the transmission of power are the high reactive power loss on heavily loaded lines, as well as possible line outages that reduce transmission capacity;
- reactive power demands of loads increase with load increases, motor stalling, or changes in load composition such as an increased proportion of compressor load.

The voltage stability can be classified into:

- small-disturbance voltage stability;
- large-disturbance voltage stability.

11.2 SYSTEM CHARACTERISTICS AND LOAD MODELING

11.2.1 System Characteristics

Let us consider a simple radial network consisting of a source, a transmission line and a load. Figure 7.1 shows the equivalent circuit and the phasor diagram. The characteristic of a system or transmission network "seen" from a load bus is the relationship between the voltage at that bus and the delivered active and reactive powers, independent of the static load characteristic.

The relationship between the voltages at both ends of the transmission line, $\underline{V}_1 = V_1\underline{|0}$ and $\underline{V}_2 = V_2\underline{|0}$, is obtained as follows:

$$\underline{V}_1 = V_2 + \frac{RP_2 + XQ_2}{V_2} + j\frac{XP_2 - RQ_2}{V_2} \tag{11.1}$$

where $R = Z\cos\beta$ and $X = Z\sin\beta$ are the resistance and reactance, respectively, corresponding to the equivalent impedance $\underline{Z}$ of the transmission line.[1]

$$\underline{S}_2 = P_2 + jQ_2 = S_2(\cos\varphi + j\sin\varphi) = V_2\,\underline{I}^*$$

[1] The European standards use letter "U" to denote the phase-to-phase voltage (called also line-to-line voltage); for compatibility with the North American standards, in this chapter we will use letter "V", and the expression $\underline{S} = P + jQ = \underline{V}\,\underline{I}^*$ stands for the three-phase complex power.

Taking into consideration the phasor diagram from Figure 7.1, by separation of the real and imaginary parts of (11.1) yields:

$$V_1 V_2 \cos\theta = V_2^2 + RP_2 + XQ_2 \tag{11.2a}$$

$$V_1 V_2 \sin\theta = XP_2 - RQ_2 \tag{11.2b}$$

To eliminate θ, the sum of squares of equations (11.2a) and (11.2b) is computed, resulting:

$$V_1^2 V_2^2 (\cos^2\theta + \sin^2\theta) = V_2^4 + 2V_2^2(RP_2 + XQ_2) + (RP_2 + XQ_2)^2 + (XP_2 - RQ_2)^2$$

or

$$f(V_2, P_2, Q_2) = V_2^4 + \left[2(RP_2 + XQ_2) - V_1^2\right]V_2^2 + Z^2 S_2^2 = 0 \tag{11.3}$$

where

$$(RP_2 + XQ_2)^2 + (XP_2 - RQ_2)^2 = Z^2(P_2^2 + Q_2^2) = Z^2 S_2^2$$

Equation (11.3) can be written as:

$$f(V_2, P_2, \varphi) = V_2^4 + \left[2S_2(R\cos\varphi + X\sin\varphi) - V_1^2\right]V_2^2 + Z^2 S_2^2 = 0 \tag{11.4}$$

which defines the implicit relationship between the receiving-end voltage of the transmission line V_2 and the transmitted powers P_2 and Q_2. Starting from this equation, the voltage V_2 can be expressed as an explicit function:

$$V_2 = g(P_2, Q_2) = g(S_2, \varphi) \tag{11.5}$$

that defines the system or transmission network characteristics represented by the surface from Figure 11.1.

The explicit function defined by expression (11.5) may exist only if the biquadratic equation (11.4) obeys the implicit functions theorem conditions, that is, to have real roots. This condition is fulfilled if the discriminant Δ is positive:

$$\left[2S_2(R\cos\varphi + X\sin\varphi) - V_1^2\right]^2 - 4Z^2 S_2^2 \geq 0 \tag{11.6}$$

At the limit, when the discriminant Δ is equal to zero, gives

$$2S_2(R\cos\varphi + X\sin\varphi) - V_1^2 = \pm 2ZS_2$$

or

$$2S_2(R\cos\varphi + X\sin\varphi \pm Z) = V_1^2$$

thus:

$$S_2 = \frac{V_1^2}{2(R\cos\varphi + X\sin\varphi \pm Z)} = \frac{V_1^2}{2Z[\cos(\beta - \varphi) \pm 1]} \tag{11.7}$$

Because $S_2 > 0$, the positive sign "+" at the nominator of equation (11.7) holds, therefore, the maximum loadability of the line:

$$S_{2\,\max} = \frac{V_1^2}{2Z[\cos(\beta - \varphi) + 1]} = \frac{V_1^2}{4Z\cos^2[(\beta - \varphi)/2]} \tag{11.7'}$$

thus resulting the same expression (8.6), previously computed following a different path (see Section 8.2.2), respectively:

$$P_{2\,\max} = \frac{V_1^2\cos\varphi}{4Z\cos^2[(\beta - \varphi)/2]} \tag{11.8}$$

11.2.2 Load Modeling

11.2.2.1 Load Characteristics. Voltage stability, also known as "load stability," is strongly influenced by the static characteristics and the dynamic response of the load [1].

In the power systems terminology, the term "load" can have different meanings, such as:

(i) a device connected to the power system that consumes active and/or reactive power;
(ii) the total active and/or reactive power consumed by all devices connected to the power system;

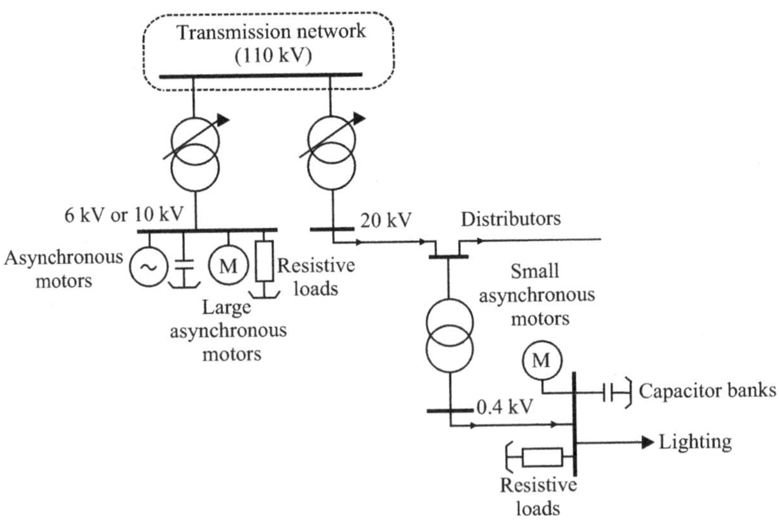

Figure 11.1. The load structure of a transmission network bus.

(iii) a part of the power system, which is not explicitly defined, but it is assigned as a device connected to the power system.

The driving force for voltage instability is usually the loads. After a disturbance, load power restoration is attempted by motor slip adjustment, tap changing, and thermostats.

Generally, the load modeling is an important issue, since in a power system the loads are aggregated, that is, they consist of different receivers. The main problem is the identification of the load components at a given instant and their mathematical representation. To understand the nature of the dependence between voltage stability and load dynamics, it is necessary to analyze the load in an individualistic approach. For this purpose, this section focuses on the dependence between voltage and load, especially on the properties of the exponential and polynomial load models.

In power system stability studies, in general, and in voltage stability, in particular, the meaning of load is taken as the definition (iii), representing a distribution network that receives power from the transmission network through a substation. Therefore, the load connected to a bus, also known as *complex load*, includes, besides individual consumers (synchronous and asynchronous motors, resistive loads, etc.), the feeders, distributors, transformers, and so on (Figure 11.1).

The *load characteristics* are a set of parameters, such as the power factor or variation of active and reactive powers with voltage and frequency, that characterize the specific behavior of the load in static or dynamic states. The *load mathematical model* can be formulated in the general form as Ref. [3,51,52]:

$$\begin{aligned} P &= K_{\mathrm{P}} \cdot P_0 \cdot f_{\mathrm{P}}(V,f) \\ Q &= K_{\mathrm{Q}} \cdot Q_0 \cdot f_{\mathrm{Q}}(V,f) \end{aligned} \tag{11.9}$$

where P_0, Q_0 are the active and reactive load powers, for values of voltage and frequency corresponding to the normal operating state (V_0, f_0); K_{P}, K_{Q} are the independent quantities, known as load demand coefficients; and $f_{\mathrm{P}}, f_{\mathrm{Q}}$ are the functions expressing the dependency of the active and reactive load powers on the voltage and frequency at a given time.

In steady state, the load characteristics, called *static load characteristics*, represent the dependency at a given time of the active and reactive powers on the connection bus voltage and the system frequency in quasi-static state. This state is characterized by very slow modifications of the operating conditions, and the transition from one state to another can be considered a series of steady states (a series of equilibrium points). The modeling of these characteristics involves only algebraic equations.

In dynamic state, the load characteristics, called *dynamic load characteristics*, also represent the dependency of the active and reactive powers on the connection bus voltage and the system frequency, but in transient state, when the operating conditions substantially changes from one instant of time to another.

As general simplification, for voltage stability studies, the frequency dependency of the active and reactive powers is neglected, and therefore the general load mathematical model can be written as:

$$\begin{aligned} P &= K_{\mathrm{P}} \cdot P_0 \cdot f_{\mathrm{P}}(V) \\ Q &= K_{\mathrm{Q}} \cdot P_0 \cdot f_{\mathrm{Q}}(V) \end{aligned} \tag{11.10}$$

Unlike the system characteristics, where the amounts of the active and reactive powers that can be delivered to the load zone, *the load characteristics* define the variation of the active and reactive powers in terms of the receiving-end voltage of the transmission line. For the considered configuration, this dependence can be written as:

$$\begin{aligned} P_2 &= f_P(V_2, P_{0,2}) \\ Q_2 &= f_Q(V_2, Q_{0,2}) \end{aligned} \tag{11.11}$$

where P_2 and Q_2 represent the demanded powers under certain operating conditions.

The last expressions define, in the (V_2, P_2, Q_2) space, a curve that, for a specific power demand, which does not exceed "the capability of the transmission network," intersects the surface defined by equation (11.5) in one or more points representing possible operating points. When the demand changes, these points move on the surface and their projection in the (V_2, P_2) plane represents the V_2–P_2 characteristic of the system (transmission network). Note that the system characteristic cannot be defined without specifying how the power delivered to the load varies. Therefore, the V_2–P_2 characteristic corresponding to a constant power factor represents *a particular case of system characteristic*.

11.2.2.2 Static Models. Exponential Model. This is one of the most used models, having the general form:

$$P = K_P \cdot P_0 \cdot \left(\frac{V}{V_0}\right)^{\alpha_P} \tag{11.12a}$$

$$Q = K_Q \cdot Q_0 \cdot \left(\frac{V}{V_0}\right)^{\alpha_Q} \tag{11.12b}$$

where K_P and K_Q are dimensionless load coefficients, being equal to 1 for the base case; V_0 is the reference voltage; and α_P, α_Q are the coefficients depending on the load type.

The terms $K_P \cdot P_0$ and $K_Q \cdot Q_0$ represent the absorbed active and reactive powers, respectively, at a voltage V equal to the reference voltage V_0, and correspond to the nominal load powers.

It is necessary to make distinction between the actually consumed power and the load demand. This distinction is important for understanding the basic instability mechanism, since as the load demand increases the consumed power may decrease, as a consequence of the voltage decrease.

Three particular cases for α_P and α_Q are usually considered as follows:

- constant impedance load: $\alpha_P = \alpha_Q = 2$
- constant current load: $\alpha_P = \alpha_Q = 1$
- constant power load: $\alpha_P = \alpha_Q = 0$

For lower voltage values the exponential model is not sufficiently precise, because, when the voltage drops below a certain value (e.g., $V \leq 0.6$ p.u.), many consumers are disconnected or their characteristics are completely altered.

Two of the most important properties of the exponential model are presented.

(i) To initialize the exponential model, any voltage level can be used as reference. To demonstrate this property $K_P = 1$ is considered, while the reference voltage V_0

and the powers P_0 and Q_0 can be arbitrarily specified without changing the characteristic. Thus, for a voltage level V_1, we have

$$P_1 = P_0 \cdot \left(\frac{V_1}{V_0}\right)^{\alpha_P}$$

If P_0 is expressed from the last equation then replaced in (11.12a) results:

$$P = P_1 \cdot \left(\frac{V}{V_1}\right)^{\alpha_P}$$

Thus, V_0 was replaced with V_1 and P_0 with P_1, and the new reference voltage becomes V_1.

(ii) The exponents α_P and α_Q used in the load exponential model determine the powers sensitivity to voltage change. Considering V_0 as reference voltage and P_0 the active power load corresponding to this voltage value, the power sensitivity to voltage change can be calculated as:

$$\frac{dP}{dV} = \alpha_P \cdot P_0 \cdot \left(\frac{V}{V_0}\right)^{\alpha_P - 1} \cdot \frac{1}{V_0}$$

Evaluating the sensitivity for $V = V_0$ obtain as follows:

$$\frac{dP/P_0}{dV/V_0} = \alpha_P \tag{11.13a}$$

respectively

$$\frac{dQ/Q_0}{dV/V_0} = \alpha_Q \tag{11.13b}$$

Therefore, the sensitivities have the same values irrespective of the voltage reference value. The expressions (equation 11.13a,b) can be written as:

$$\Delta P = \alpha_P \frac{\Delta V}{V_0} P_0$$

$$\Delta Q = \alpha_Q \frac{\Delta V}{V_0} Q_0$$

If the pervious equations are divided member by member, gives:

$$\frac{\Delta P}{\Delta Q} = \frac{\alpha_P}{\alpha_Q} \frac{P_0}{Q_0}$$

Given the relationship $Q = P \cdot \tan\varphi$ if $\tan\varphi = \tan\varphi_0 = \text{ct.}(\tan\varphi_0 = (Q_0/P_0))$ obtain as follows:

$$\Delta Q = \frac{Q_0}{P_0} \Delta P$$

meaning that for constant power factor, $\alpha_P = \alpha_Q$.

Polynomial Model. Since different components of a complex load present different voltage characteristics, an alternative load model obtained by summing up the load

TABLE 11.1. Polynomial Load Models [3]

Load Type	$f_P\left(\frac{V}{V_0}\right)$	$f_Q\left(\frac{V}{V_0}\right)$
Air conditioner	$2.97 - 4\left(\frac{V}{V_0}\right) + 2.03\left(\frac{V}{V_0}\right)^2$	$12.9 - 26.8\left(\frac{V}{V_0}\right) + 14.9\left(\frac{V}{V_0}\right)^2$
Asynchronous motors	$0.72 + 0.11\left(\frac{V}{V_0}\right) + 0.17\left(\frac{V}{V_0}\right)^{-1}$	$2.08 + 1.63\frac{V}{V_0} - 7.6\left(\frac{V}{V_0}\right)^2 + 4.89\left(\frac{V}{V_0}\right)^3$
Complex load	$0.83 - 0.3\left(\frac{V}{V_0}\right) + 0.74\left(\frac{V}{V_0}\right)^2$	$6.7 - 15.3\left(\frac{V}{V_0}\right) + 9.6\left(\frac{V}{V_0}\right)^2$

components, which presents the same (or almost the same) exponent, should be considered. When all exponents are integers, the load characteristic becomes a polynomial function of V.

A special case is the load model consisting of three components: constant impedance, constant current, and constant power. In this case, the characteristics of the active and reactive power, respectively, of the load model are given by the expression [2]:

$$
\begin{aligned}
P &= K_P \cdot P_0 \cdot \left[a_P \cdot \left(\frac{V}{V_0}\right)^2 + b_P \cdot \frac{V}{V_0} + c_P\right] \\
Q &= K_Q \cdot Q_0 \cdot \left[a_Q \cdot \left(\frac{V}{V_0}\right)^2 + b_Q \cdot \frac{V}{V_0} + c_Q\right]
\end{aligned}
\tag{11.14}
$$

where a_P, b_P, c_P, a_Q, b_Q, c_Q represent the weights of each component and satisfy the relationships $a_P + b_P + c_P = 1$ and $a_Q + b_Q + c_Q = 1$; and $K_P \cdot P_0$, $K_Q \cdot Q_0$ are active and reactive consumed powers, corresponding to the reference voltage V_0.

Similar to the exponential model, for reduced voltage values, the polynomial model cannot be used to appropriately model the loads.

Table 11.1 shows some polynomial models [3].

11.2.2.3 Dynamic Models. These models describe, through algebraic-differential equations, the dynamic behavior of the load given either by the intrinsic load characteristics or by the actions of some control devices such as the automatic on-load tap changer or the thermostat load control systems.

Because the use of physical dynamic models specific to some loads, such as large synchronous and asynchronous motors individually represented, is inadequate for complex load modeling, the so-called *generic dynamic models* have been developed. They describe the specific behavior of the complex load when subjected to sudden voltage variations in the connecting node, due to various disturbances occurring in the transmission network.

The measurements at the HV bus bars of some substations supplying the distribution network showed that the response of the complex load to sudden voltage variations is as illustrated in Figure 11.2 for a real power P [4]. This response reflects the collective effects of all the complex load components, from the on-load tap changer to the individual household loads (Figure 11.1). The time span for a load to recover to steady state is

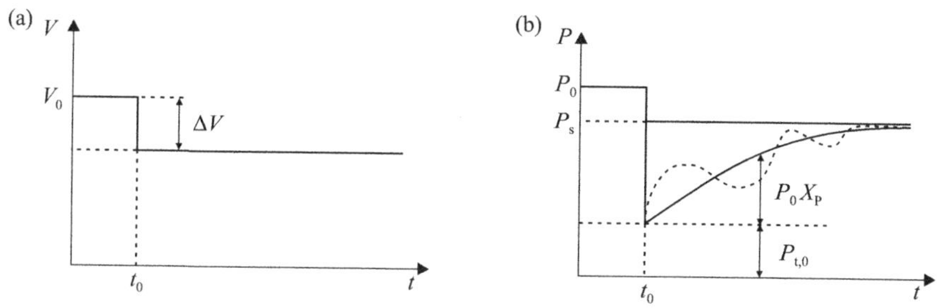

Figure 11.2. The dynamic active power response of a complex load (b) at sudden voltage variation in the connection node (a) [4].

normally in the range of several seconds to minutes, depending on the load composition. Responses for real and reactive power are qualitatively similar.

It is obvious that a sudden voltage variation determines an instantaneous change in the consumed power. This variation defines *the transient characteristic of the load* and can be modeled as constant impedance $\underline{Z}_c$, as used in the transient stability analysis tools, or, more general, as an exponential characteristic. A dynamic load restoration process follows characterized by an increase in the power demand up to a new voltage value on the steady-state characteristic.

On a longer timescale, the lower voltage tap changers and other control devices act to restore voltages and so the load. In fact, the aggregate effect of distribution-level restoration can be to cause load overshoot. With recovery time (T_p) on a timescale of a few seconds, this behavior captures the behavior of induction machines. On a timescale of minutes, the role of tap changes and other control devices is included. Over hours, the load recovery and possible overshoot may emanate from heating load [4].

The dynamic restoration process can be asymptotic or even oscillatory (Figure 11.2b) and can be modeled through a set of differential equations [2]:

$$\begin{aligned} T_P \frac{dX_P}{dt} &= \frac{1}{P_0}(P_s - P_t) \\ T_Q \frac{dX_Q}{dt} &= \frac{1}{Q_0}(Q_s - Q_t) \end{aligned} \tag{11.15}$$

where P_0, Q_0 are the active and reactive consumed powers in normal operating state characterized by the voltage V_0; P_s, P_t are the steady-state and transient load characteristics of the active power; Q_s, Q_t are the steady-state and transient load characteristics of the reactive power; X_p, X_Q are the generic dimensionless state variables associated with the load restoration dynamic process; and T_p, T_Q are the load restoration dynamic process.

The generic models are usually associated with the exponential load models. Hence, the steady-state characteristic is given by expressions (11.12) where $K_P = K_Q = 1$:

$$P_s = P_0 \cdot \left(\frac{V}{V_0}\right)^{\alpha_P}; \qquad Q_s = Q_0 \cdot \left(\frac{V}{V_0}\right)^{\alpha_Q} \tag{11.12$'$}$$

Depending on the way in which the generic variables X_P and X_Q intervene in the expressions of the transient load characteristics, there are two types of generic dynamic models: *the additive model* where the generic state variables are appended to the transient characteristic and *the multiplicative model* where the generic state variables multiply the transient characteristics.

(i) The additive generic model

The transient load characteristic is given by [2]:

$$P_{\mathrm{t}} = P_0 \cdot \left[\left(\frac{V}{V_0}\right)^{\alpha_{\mathrm{P,t}}} + X_{\mathrm{P}}\right]$$
$$Q_{\mathrm{t}} = Q_0 \cdot \left[\left(\frac{V}{V_0}\right)^{\alpha_{\mathrm{Q,t}}} + X_{\mathrm{Q}}\right] \tag{11.16}$$

where $\alpha_{\mathrm{P,t}}$ and $\alpha_{\mathrm{Q,t}}$ are exponents of the transient load characteristics.

Hence, the differential system of equations (11.15) describing the load restoration dynamic process becomes:

$$T_{\mathrm{P}}\frac{\mathrm{d}X_{\mathrm{P}}}{\mathrm{d}t} = \frac{1}{P_0}(P_{\mathrm{s}} - P_{\mathrm{t}}) = \left(\frac{V}{V_0}\right)^{\alpha_{\mathrm{P}}} - \left(\frac{V}{V_0}\right)^{\alpha_{\mathrm{P,t}}} - X_{\mathrm{P}}$$
$$T_{\mathrm{Q}}\frac{\mathrm{d}X_{\mathrm{Q}}}{\mathrm{d}t} = \frac{1}{Q_0}(Q_{\mathrm{s}} - Q_{\mathrm{t}}) = \left(\frac{V}{V_0}\right)^{\alpha_{\mathrm{Q}}} - \left(\frac{V}{V_0}\right)^{\alpha_{\mathrm{Q,t}}} - X_{\mathrm{Q}} \tag{11.17}$$

(ii) The multiplicative generic model

The transient load characteristic is given by:

$$P_{\mathrm{t}} = X_{\mathrm{P}} \cdot P_0 \cdot (V/V_0)^{\alpha_{\mathrm{P,t}}} \quad ; \quad Q_{\mathrm{t}} = X_{\mathrm{Q}} \cdot Q_0 \cdot (V/V_0)^{\alpha_{\mathrm{Q,t}}} \tag{11.16$'$}$$

and the differential system of equations (11.15) describing the load restoration dynamic process becomes:

$$T_{\mathrm{P}}\frac{\mathrm{d}X_{\mathrm{P}}}{\mathrm{d}t} = \frac{1}{P_0}(P_{\mathrm{s}} - P_{\mathrm{t}}) = (V/V_0)^{\alpha_{\mathrm{P}}} - X_{\mathrm{P}} \cdot (V/V_0)^{\alpha_{\mathrm{P,t}}}$$
$$T_{\mathrm{Q}}\frac{\mathrm{d}X_{\mathrm{Q}}}{\mathrm{d}t} = \frac{1}{Q_0}(Q_{\mathrm{s}} - Q_{\mathrm{t}}) = (V/V_0)^{\alpha_{\mathrm{Q}}} - X_{\mathrm{Q}} \cdot (V/V_0)^{\alpha_{\mathrm{Q,t}}} \tag{11.17$'$}$$

The following remarks are made regarding the two generic dynamic models [2]:

(i) the exponents $\alpha_{\mathrm{P,t}}$ and $\alpha_{\mathrm{Q,t}}$ have usually large values and thus the dependence on voltage of the load powers is greater in transient state than in steady state;

(ii) the generic state variables X_{P} and X_{Q} must range within acceptable limits, that is:

$$X_{\mathrm{P}}^{\min} \le X_{\mathrm{P}} \le X_{\mathrm{P}}^{\max}$$
$$X_{\mathrm{Q}}^{\min} \le X_{\mathrm{Q}} \le X_{\mathrm{Q}}^{\max} \tag{11.18}$$

(iii) the constant power type component introduced in the additive models may cause nonphysical singularities to the transient characteristic, which is not the case for the multiplicative model;

(iv) the simple model given in(11.12′) can be obtained if we take $X_P = X_Q = 0$ in the additive model, and $X_P = X_Q = 1$ in the multiplicative model, respectively;

(v) the exponents α_P and α_Q and the time constants T_P and T_Q can be determined using measurements of powers supplied to the distribution network either considering disconnection of a transformer or on-load tap changing in the HV/MV distribution substation leading to gradual modification of the voltage level Ref. [2].

11.3 STATIC ASPECTS OF VOLTAGE STABILITY

11.3.1 Existence of Steady-State Solutions

The projection of the points corresponding to the maximum transmissible power $S_{2\max}$ in the (P_2, Q_2) plan determines a curve that separates the plan into two zones (see Figure 11.3) [5]:

The hatched zone represents the domain of existence of solutions and corresponds to the possible operating points (A and B in Figure 11.4), for which the biquadratic equation (11.3) has two distinct real positive solutions.

$$V_2^4 + \left[2(RP_2 + XQ_2) - V_1^2\right]V_2^2 + Z^2\left(P_2^2 + Q_2^2\right) = 0 \tag{11.3}$$

Using the notations:

$$\begin{gathered} y = V_2^2 \quad ; \quad \alpha = V_1^2 - 2(RP_2 + XQ_2) \\ \Delta = \left[2(RP_2 + XQ_2) - V_1^2\right]^2 - 4Z^2\left(P_2^2 + Q_2^2\right) \end{gathered} \tag{11.19}$$

from (11.3) a quadratic equation in y is obtained as follows:

$$y^2 - \alpha y + Z^2 S_2^2 = 0$$

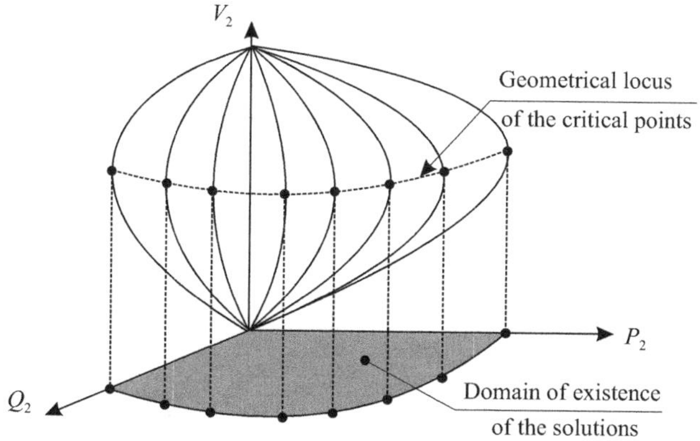

Figure 11.3. The dependence between the voltage V_2 and the powers P_2 and Q_2, for a simple generator-line-load configuration.

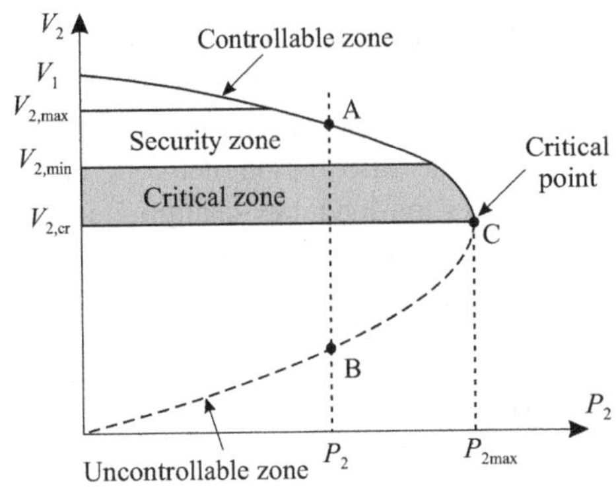

Figure 11.4. The V_2–P_2 system characteristic: operating points.

with the solutions $y_{1,2}$, which give the voltages corresponding to the two possible operating points A and B:

$$V_{2A} = \sqrt{\frac{\alpha + \sqrt{\Delta}}{2}}; \qquad V_{2B} = \sqrt{\frac{\alpha - \sqrt{\Delta}}{2}} \tag{11.20a,b}$$

- *The external zone* represents the domain of inexistence of solutions, thus no operating points are found in this area.
- The *critical points situated on the frontier* represent static bifurcation points for the system evolution as they are characterized by zero value of the discriminant from equation (11.19), $\Delta = 0$, of the biquadratic equation (11.3), resulting:

$$\left[2(RP_2 + XQ_2) - V_1^2\right]^2 = 4Z^2 S_2^2$$

If $\Delta = 0$ is considered in the expressions (11.20a and b), identical solutions V_{2A} and V_{2B} are obtained in the critical points, corresponding to the possible operating points:

$$V_{2A} = V_{2B} = V_{cr} = \sqrt{\frac{\alpha}{2}} = \sqrt{\frac{V_1^2 - 2(RP_{2max} + XQ_{2max})}{2}} \tag{11.21}$$

Note that in the critical points situated on the separation frontier of the domain of existence of solutions in the P_2–Q_2 plan, the Jacobian matrix of the Newton–Rhapson method-based load flow mathematical model become singular.

We know that:

$$P_i = G_{ii}V_i^2 + \sum_{\substack{k=1 \\ k \neq i}}^{n} V_i V_k GG_{ik}$$

$$\frac{\partial P_i}{\partial V_i} V_i = 2G_{ii}V_i^2 + \sum_{\substack{k=1 \\ k \neq i}}^{n} V_i V_k GG_{ik} = G_{ii}V_i^2 + G_{ii}V_i^2 + \sum_{\substack{k=1 \\ k \neq i}}^{n} V_i V_k GG_{ik} = G_{ii}V_i^2 + P_i$$

$$\frac{\partial P_i}{\partial V_i} = \frac{P_i + G_{ii}V_i^2}{V_i}$$

$$Q_i = -B_{ii}V_i^2 - \sum_{\substack{k=1 \\ k \neq i}}^{n} V_i V_k BB_{ik}$$

$$\frac{\partial Q_i}{\partial V_i} V_i = -2B_{ii}V_i^2 - \sum_{\substack{k=1 \\ k \neq i}}^{n} V_i V_k BB_{ik} = -B_{ii}V_i^2 + Q_i$$

$$\frac{\partial Q_i}{\partial V_i} = \frac{Q_i - B_{ii}V_i^2}{V_i}$$

where

$$GG_{ik} = G_{ik}\cos(\theta_i - \theta_k) + B_{ik}\sin(\theta_i - \theta_k)$$

$$BB_{ik} = G_{ik}\sin(\theta_i - \theta_k) - B_{ik}\cos(\theta_i - \theta_k)$$

The Jacobian matrix of the power system from Figure 7.1, considering that the bus 1 (the slack bus) voltage is taken as phase reference with $\underline{V}_1 = V_1 \underline{|\theta_1 = 0}$, and the bus 2 voltage is $\underline{V}_2 = V_2 \underline{|\theta_2}$, has the following form:

$$[J] = \begin{bmatrix} J_{P\theta} & J_{PV} \\ J_{Q\theta} & J_{QV} \end{bmatrix} = \begin{bmatrix} \dfrac{\partial P_2}{\partial \theta_2} & \dfrac{\partial P_2}{\partial V_2} \\ \dfrac{\partial Q_2}{\partial \theta_2} & \dfrac{\partial Q_2}{\partial V_2} \end{bmatrix}$$

where

$$\frac{\partial P_2}{\partial \theta_2} = -Q_2 - B_{22}V_2^2; \quad \frac{\partial Q_2}{\partial V_2} = \frac{Q_2 - B_{22}V_2^2}{V_2};$$

$$\frac{\partial Q_2}{\partial \theta_2} = P_2 - G_{22}V_2^2; \quad \frac{\partial P_2}{\partial V_2} = \frac{P_2 + G_{22}V_2^2}{V_2};$$

Thus, the determinant of the Jacobian matrix is

$$\begin{aligned} \det[J] &= \frac{\partial P_2}{\partial \theta_2} \cdot \frac{\partial Q_2}{\partial V_2} - \frac{\partial P_2}{\partial V_2} \cdot \frac{\partial Q_2}{\partial \theta_2} \\ &= \frac{\left(-Q_2 - B_{22}V_2^2\right)\left(Q_2 - B_{22}V_2^2\right)}{V_2} - \frac{\left(P_2 + G_{22}V_2^2\right)\left(P_2 - G_{22}V_2^2\right)}{V_2} \\ &= \frac{1}{V_2}\left[V_2^4\left(G_{22}^2 + B_{22}^2\right) - \left(P_2^2 + Q_2^2\right)\right] = \frac{1}{V_2}\left(V_2^4 Y_{22}^2 - S_2^2\right) = \frac{1}{V_2 Z^2}\left(V_2^4 - Z^2 S_2^2\right) \end{aligned}$$

Because in the critical point $V_2^4 = V_{2cr}^4 = Z^2 \cdot S_2^2$, it results:

$$\det[J] = 0$$

This property, valid for a simple generator-line-load configuration as well as for any complex network, is the core of voltage instability and collapse risk assessment, due to the inexistence of the equilibrium points (steady-state solutions) caused by load increase and/or due to the contraction of the domain of existence of the solutions after a contingency.

11.3.2 Operating Points and Zones

For analysis of the points inside the hatched zone (Figure 11.3), the dependence between P_2 and Q_2 through the power factor $\cos\varphi$ is considered. Thus, equation (11.4) can be written as:

$$V_2^4 + \left[2P_2(R + X\tan\varphi) - V_1^2\right]V_2^2 + Z^2\left(\frac{P_2}{\cos\varphi}\right)^2 = 0 \quad (11.22)$$

If V_1 and $\cos\varphi$ are considered constant, then for values of the active power $P_2 \in [0, P_{2\max}]$, an expression of the form $V_2 = f(P_2)$, called the V_2–P_2 "system characteristic," can be obtained (see Figure 11.4) [5].

Figure 11.4 shows that for a transmitted power $P_2 \le P_{2\max}$, two possible operating points are found as: point A, characterized by a high-voltage value, corresponds to the normal operating conditions being a stable operating point, whereas point B, characterized by a low-voltage value, corresponds to abnormal operating conditions.

To demonstrate why A is stable and B is unstable, the effects of voltage regulation by the control of the power factor at the transmission line receiving end and voltage variation at the transmission line sending-end will be analyzed.

(i) *The effect of compensating the transmitted reactive power*

Considering different values for $\cos\varphi$ and keeping constant the sending-end voltage of the transmission line ($V_1 = \text{ct.}$), the family of V_2–P_2 characteristics from Figure 11.5 is obtained.

For a given value $P_2 < P_{2\max}$, the V_{2A} voltage corresponding to the operating point A increases as the power factor changes from inductive to capacitive values. At the same time, in the point B, compensation of the reactive power has opposite

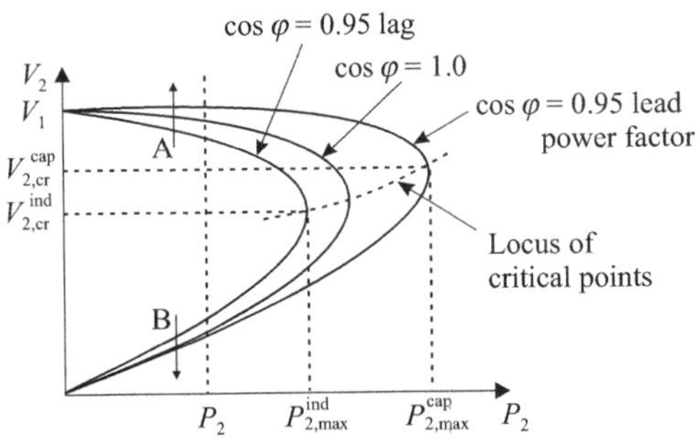

Figure 11.5. Effect of reactive power compensation on the V_2–P_2 characteristic.

effects since the voltage decreases as $\cos\varphi$ changes from inductive to capacitive values. Therefore, point A is a controllable operating point and thus stable; whereas point B is uncontrollable and thus unstable.

The locus of critical operating points is shown by dotted line in Figure 11.5. Normally, only the operating points above the critical points represent satisfactory operating conditions. Thus, a sudden reduction in the power factor (increase in Q_2) can cause the system to change from stable operating conditions to unsatisfactory, and possibly unstable, operating conditions represented by the lower part of a V_2–P_2 curve.

The influence of the reactive power characteristics of the equipments at the receiving end (loads and compensating devices) is more obvious in Figure 11.6 [6], which shows a family of curves applicable to the power system from Figure 11.4, each of which representing the relationship between V_2 and Q_2 for a fixed value of P_2.

For a complex system, these curves can be obtained for a load bus through repeated load flow calculations. In this regard, consider that a virtual reactive power source (without power limits) is connected to the bus. As a consequence, this bus becomes a PV bus where the voltage is imposed and the reactive power is calculated.

An example of this type of characteristics, in p.u., $q = QX/V_1^2$, is given in Figure 11.6 for a transmission line ($X \gg R$) having an active source as parameter ($p = PX/V_1^2$) and a switched capacitor bank.

The system is stable in the region where the derivative dQ_2/dV_2 is positive. The voltage stability limit (critical operating point) is reached when the derivative is zero. Thus, the zones on the q_2–u_2 curves to the right of the minima represent stable operation, and the zones to the left represent unstable operation. Stable operation in the region where dQ_2/dV_2 is negative can be achieved only with a continuously regulated reactive power equipment that has sufficient control range and high Q/V gain with polarity opposite to that of the normal.

If no capacitor bank is connected, the operating point is at the intersection of the V–Q characteristic with the horizontal axis (Figure 11.6, the a curve). If reactive power compensation is provided, a new operating point can be found at the intersection between the bus characteristic and the capacitor bank characteristic (Figure 11.6, curves b and d). Notice that for large values of the transferred active powers, if there is insufficient compensation, no operation point can be defined.

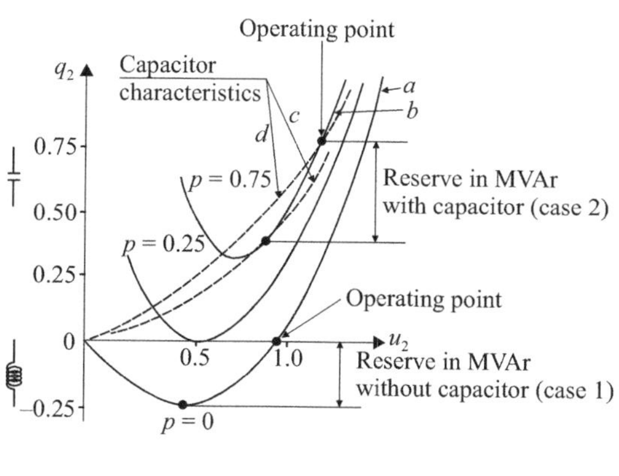

Figure 11.6. The q_2–u_2 curves for various real-power load [6, 51].

The reactive power reserve to maintain normal operation is the distance, in MVAr, from the operating point to the minimum of the bus characteristic (Figure 11.6, case 1), without capacitor, or to a point where the capacitor bank characteristic on the lowest step is tangent to the bus characteristic (Figure 11.6, case 2).

The slope of the characteristic in the operating point is the $\partial Q/\partial V$ sensitivity that shows the robustness or weakness of the analyzed bus as well as the effectiveness of reactive power compensation.

(ii) *The effect of controlling the transmission line sending-end voltage*

Considering different values for the transmission line sending-end voltage, $V_1^{(1)}$ and $V_1^{(2)}$, obtained by modifying the generator's field current, and constant power factor ($\cos\varphi = \text{ct.}$), the family of V_2–P_2 characteristics from Figure 11.7 is obtained.

In this case, for a constant value $P_2 < P_{2\max}$ of the transmitted active power, the point A is a normal operating point because the corresponding voltage (V_A) increases as the transmission line sending-end voltage increases. In exchange, the point B is an abnormal operating point because the sending-end voltage increases as the voltage corresponding to this point decreases.

Therefore, the control of V_1 voltage at the transmission line sending-end through change of the generator's field excitation current has the desired effect on the operating point A, while for the operating point B the effect is opposite.

Moreover, analyzing the characteristics from Figures 11.5 and 11.7, we can see that the critical voltage and the maximum transmissible power increase as the reactive power is compensated with the help of reactive power devices, or as the transmission line sending-end voltage increases. When operating near the limit, the two voltages corresponding to the possible operating points will have high and closed values. Therefore, the voltage attached to one operating point only is not an indicator of voltage instability proximity. Additional information such as voltage sensitivities to active and reactive powers variation is thus necessary.

(iii) *Voltage sensitivities to active and reactive powers variation*

The voltage sensitivities can be obtained by differentiating equations (11.4). To simplify the calculus, a small phase difference between the voltage phasors at the transmission line ends is assumed, that is, $\theta = (\theta_1 - \theta_2) \cong 0$. Therefore, using equations (11.2a) and (11.2b), since $\cos\theta \cong 1$ and $\sin\theta \cong 0$, equation (11.3) becomes:

$$V_2^2 - V_1 V_2 + RP_2 + XQ_2 = 0$$

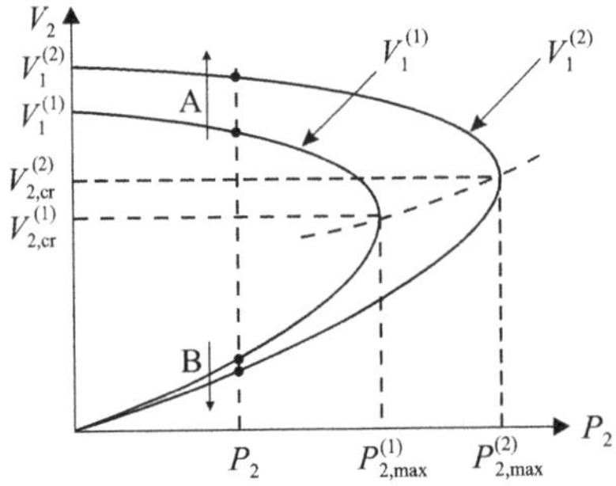

Figure 11.7. The effect of controlling the voltage V_1 on the V_2–P_2 characteristic.

The voltages corresponding to the possible operating points A and B are as follows:

$$V_{2A} \cong \frac{V_1 + \sqrt{V_1^2 - 4(RP_2 + XQ_2)}}{2} \tag{11.23a}$$

$$V_{2B} \cong \frac{V_1 - \sqrt{V_1^2 - 4(RP_2 + XQ_2)}}{2} \tag{11.23b}$$

A power system, seen from a bus i, is stable from the viewpoint of the bus voltage, if the addition of a conductance ΔG_i or a susceptance ΔB_i, respectively, eventually infinitesimal, determines the increase in the consumed active and reactive powers, and a simultaneous decrease in the voltage of the considered bus. Therefore, the necessary and sufficient condition for the power system to be controllable is [7]

$$\begin{aligned} &\left.\frac{dV_i}{dG_i}\right|_{B_i=\mathrm{ct}} < 0; \quad \left.\frac{dP_i}{dG_i}\right|_{B_i=\mathrm{ct}} > 0 \\ &\left.\frac{dV_i}{dB_i}\right|_{G_i=\mathrm{ct}} < 0; \quad \left.\frac{dQ_i}{dB_i}\right|_{G_i=\mathrm{ct}} > 0 \end{aligned} \tag{11.24}$$

The sensitivities [2] of voltage to the active and reactive powers in the A and B points can be calculated starting from the expressions (11.23a,b):

$$\left.\frac{\partial V_{2A}}{\partial P_2}\right|_{Q_2=\mathrm{ct}} = \frac{-4R}{2\cdot 2\sqrt{V_1^2 - 4(RP_2 + XQ_2)}} = \frac{-R}{\sqrt{V_1^2 - 4(RP_2 + XQ_2)}} < 0 \tag{11.25a}$$

$$\left.\frac{\partial V_{2A}}{\partial Q_2}\right|_{P_2=\mathrm{ct}} = \frac{-X}{\sqrt{V_1^2 - 4(RP_2 + XQ_2)}} < 0 \tag{11.25b}$$

$$\left.\frac{\partial V_{2B}}{\partial P_2}\right|_{Q_2=\mathrm{ct}} = \frac{R}{\sqrt{V_1^2 - 4(RP_2 + XQ_2)}} > 0 \tag{11.25c}$$

$$\left.\frac{\partial V_{2B}}{\partial Q_2}\right|_{P_2=\mathrm{ct}} = \frac{X}{\sqrt{V_1^2 - 4(RP_2 + XQ_2)}} > 0 \tag{11.25d}$$

[2] If $y = \sqrt{u(x)}$ then $y' = \frac{u'(x)}{2\sqrt{u(x)}}$, or $\frac{dy}{du} = \frac{u'(x)}{2\sqrt{u(x)}}$

According to the definitions from (11.24) and (11.25), point A is a controllable operating point, thus stable, and point B is an uncontrollable operating point, thus unstable.

Considering these aspects, the following operating zones can be defined on the V_2–P_2 characteristic (Figure 11.4):

- *The Controllable Zone*. Corresponds to the upper side of the characteristic and comprises the operating points for which $Z_c > Z$ and sensitivities are negative. In the technical literature, this zone is known as *the stable operating zone*. If the critical voltage is lower than the minimum admissible operating voltage (V_{min}), this zone can be divided into two subzones:
  - *The Security Zone*. Defined between the maximum and the minimum admissible voltages (V_{min} and V_{max}), characterized by negative and close to zero sensitivities;
  - *The Critical Zone*. Defined between V_{min} and V_{cr} voltages, characterized also by negative sensitivities but high in magnitude. Depending on the load characteristic, operation in this zone can trigger voltage instability phenomenon or voltage avalanche, phenomenon characterized by a voltage decrease toward V_{cr}, either due to an increase in the power demand or due to the on-load tap changing transformers action.
- *The Uncontrollable Zone*. Corresponds to the lower side of the characteristic and comprises the operating points characterized by $Z_c < Z$ and positive sensitivities. This zone is known in the technical literature as *the unstable operating zone*.

Note that the assumption of a small phase difference (θ small) is no longer possible to determine the critical voltage because, if in equations (11.23a,b) the zero condition of the discriminant would be imposed, the critical voltage is obtained $V_{crit} = V_1/2$, value that depends on the sending-end voltage but is independent of the power factor, which is in contradiction with the remark previously made.

11.4 VOLTAGE INSTABILITY MECHANISMS: INTERACTION BETWEEN ELECTRICAL NETWORK, LOADS, AND CONTROL DEVICES

In power systems operation, the maximum transmissible power and the critical voltage are only for theoretical importance since they cannot be precisely determined due to the phenomenon complexity, depending on a series of factors influencing each other (load behavior, keeping the voltage to the reference value in the generator nodes, the reactive power compensation devices action, etc.).

11.4.1 Interaction between Electrical Network and Load

There are two basic mechanisms that can lead to voltage instability and collapse phenomena in power systems. One of these mechanisms is the slow increase in the load demand, and the other one resides in the disturbance occurrence in the transmission network (disconnection of electrical lines, transformers, generators, etc.) that lead to the modification of the V_2–P_2 characteristic.

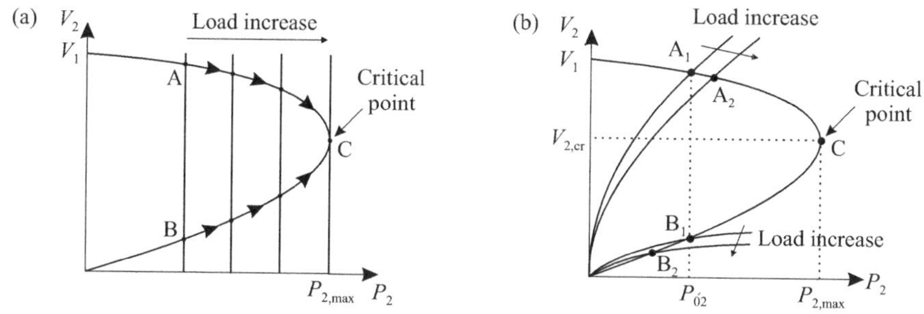

Figure 11.8. Voltage instability mechanism in terms of load demand increase: (a) $P = $ ct; (b) voltage-dependent load.

(i) *The voltage instability phenomenon mechanism, caused by the load increase, considering different static load characteristics*, is illustrated in Figure 11.8.

- Thus, for *a* $P = $ ct. *load characteristic*, as the demand power increases, due to the load characteristic shifting, the two equilibrium points approach each other, overlap in the critical point C (which represents a singular point coinciding, in this case, with the maximum transmissible power) and then disappear (Figure 11.8a);
- If *the load is represented by exponential model*, and the operating point is situated on the upper side, as the load demand increases, this point moves from A_1 to A_2 (Figure 11.8b) leading to voltage decrease simultaneously with increase in the transmitted power, which corresponds to a normal operation of the "transmission network-load" configuration. Instead, if the operating point is situated on the lower side, as load demand increases, the point moves from B_1 to B_2 causing a decrease both in the voltage level and in the transmitted power, which corresponds to an abnormal operation of the transmission network-load configuration.

(ii) *The voltage instability mechanism to a major disturbance*, caused by disconnection of one circuit of the transmission line, is shown in Figure 11.9. The disturbance occurrence causes in the first moments a decrease in voltage and therefore sudden decrease in the load demand, in accordance with the transient load characteristic, and the operating point moves from A_1, situated on the normal operation characteristic, to A_2, situated on the postdisturbance characteristic. As the transient process evolves, the point A_2 moves toward a new equilibrium point, given by the intersection between the static load characteristic and the postdisturbance characteristic of the transmission network. We see that, if *the load is of* $P = $ ct. *type* there is no intersection point between the two characteristics, which determines the voltage to collapse at the load terminals (Figure 11.9a).

Instead, if *the load is represented by exponential model*, the two characteristics intersect in point A′, which is the new operating point characterized by a reduced voltage level (Figure 11.9b—curve *a*) (see also Section 8.2.2). However, in this case, the voltage collapse can be triggered by the on-load tap changer actions, trying to restore the voltage to the reference value. The consequences of this action are the change of the load characteristic seen from the transformer primary and the increase in the consumed power from the system, which increase the risk for voltage instability and voltage collapse (Figure 11.9b—curve *b*).

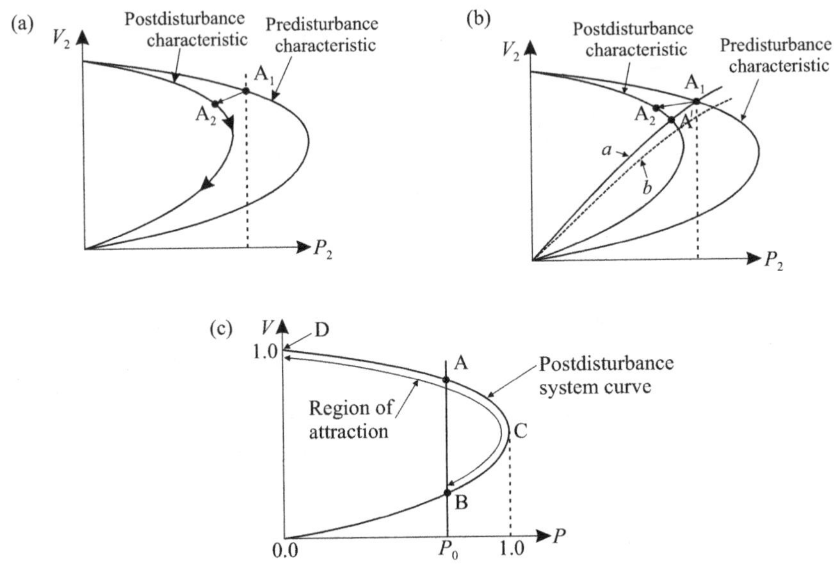

Figure 11.9. The voltage instability mechanism to major disturbances (a and b) and the region of attraction (c).

These mechanisms caused by the load behavior are strongly influenced by the on-load tap changer action, a reactive power deficit due to limitations in the generator's field and armature currents, as well as following the secondary voltage control system actions.

The portion of the *V–P* characteristic that allows restoring the system operation to a stable equilibrium point is called *region of attraction* (Figure 11.9c). Consider the simplified radial power system with constant sending-end bus voltage and unity power factor load. As shown before, there are two possible equilibrium points corresponding to $P_0 = \text{ct}$: point A situated on the upper section of the *V–P* curve, which is a stable equilibrium point, and point B situated on the lower section of the *V–P* curve, which is an unstable equilibrium point (Figure 11.9c). It can be shown that, after a disturbance, if the power system reaches an operating point situated anywhere on the DACB section of the *V–P* characteristic, appropriate measure can help increasing the voltage and thus moving the system operation in a stable equilibrium point corresponding to the demanded power. [8,9].

11.4.2 Influence of the On-Load Tap Changer

In some power system operating conditions, the on-load tap changer actions performed under an automatic voltage regulator (AVR) may enhance the risk for or even trigger the voltage instability process [10]. To analyze this phenomenon, let us consider the one-line diagram from Figure 11.10a of a system supplying through a short electrical line and an on-load tap changing (OLTC) transformer equipped with AVR.

11.4.2.1 Modeling the On-Load Tap Changing Dynamics. Neglecting the transformer no-load losses, including its series impedance in the equivalent impedance $\underline{Z}$ and modeling the load by the impedance $\underline{Z}_c$, the equivalent circuit from Figure 11.10b is obtained.

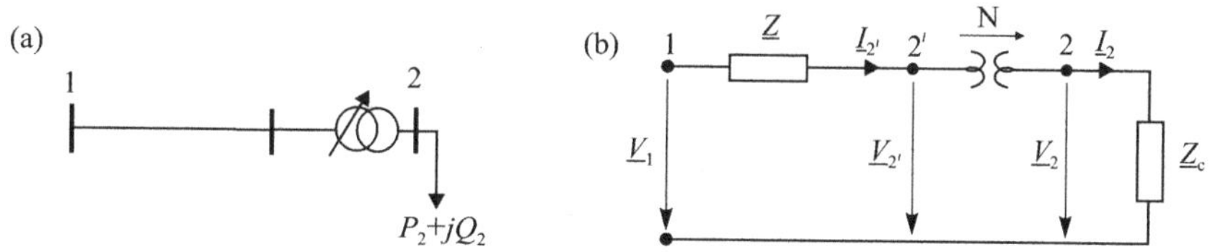

Figure 11.10. One-line diagram (a) and equivalent circuit (b) for the analysis of the on-load tap changing influence.

The turns ratio N is a real number given by:

$$N = \frac{V_{2'}}{V_2} = \frac{I_2}{I_{2'}} = \frac{V_{\text{np}}\left(1 \pm n_{\text{p}}\frac{\Delta V_{\text{p}}}{100}\right)}{V_{\text{ns}}} \tag{11.26}$$

where V_{np} is the primary winding nominal voltage (the nominal voltage of the high-voltage winding, equipped with tap changer); V_{ns} is the secondary winding nominal voltage; n_{p} is the actual tap position (actual operating tap); and ΔV_{p} is the percentage of voltage variation per tap (expressed in percents).

The turns ratio N varies within the admissible interval $[N_{\text{lo}}, N_{\text{up}}]$ given by the transformer's control capability (number of taps).

The role of AVR is to change on-load the ratio N so that the load side voltage V_2 is kept inside the interval $[V_{2,\text{sch}} - \varepsilon, V_{2,\text{sch}} + \varepsilon]$, called *the insensitivity domain of the regulator*, around the reference voltage $V_{2,\text{sch}}$, determined under stable-operating conditions. Therefore, if the load side voltage is lower than $V_{2,\text{sch}} - \varepsilon$ then, after a period called *delay period*, according to expression (11.26), written as $V_2 = V_{2'}/N$, the regulator will command the tap change with one position in the sense of turns ratio decrease, resulting the increase of V_2. Reverse, if $V_2 > V_{2,\text{sch}} + \varepsilon$ the regulator will command the tap movement with one position in the sense of turns ratio increase, resulting the decrease of V_2. The process will continue until either the voltage is brought back within the admissible interval or one of the limits, N_{lo} or N_{up}, is reached (the tap changer reaches one of its extreme positions).

The dynamic behavior of an on-load tap changing transformer equipped with AVR can be described using either a discreet model or a continuous model.

- The *discrete model* assumes that the regulator, activated whenever the load side voltage does not fulfill the condition $\left|V_2 - V_{2,\text{sch}}\right| \le \varepsilon$, will command the tap change at discrete time intervals t_k, $k = 0, 1, \ldots, n$, which satisfies the recursive formula: $t_{k+1} = t_k + \Delta T_k$ and $t_0 = 0$. In this expression, ΔT_k denotes the time delay and depends on the regulator characteristics and the voltage error. Generally, ΔT_k is not necessarily constant and may be determined using the expression [2]:

$$\Delta T_k = T_{\text{d}}\frac{\varepsilon}{\left|V_2 - V_{2,\text{sch}}\right|} + T_{\text{f}} + T_{\text{m}} \tag{11.27}$$

where T_{d} is the maximum time delay of the inverse time characteristic; T_{f} is a fixed value of the time delay; T_{m} is the mechanical time necessary to perform the tap change; and ε is half of the OLTC deadband.

The tap changing logic, called *direct* or *normal logic*, can be defined as:

$$N_{k+1} = \begin{cases} N_k - \Delta N & \text{if} \quad V_2 < V_{2\text{sch}} - \varepsilon \quad \text{and} \quad N_k > N_{\text{lo}} \\ N_k & \text{if} \quad |V_2 - V_{2\text{sch}}| \leq \varepsilon \\ N_k + \Delta N & \text{if} \quad V_2 > V_{2\text{sch}} + \varepsilon \quad \text{and} \quad N_k < N_{\text{up}} \end{cases} \tag{11.28}$$

where ΔN is the size of tap step and N_{up}, N_{lo} are the upper and lower tap limits.

- The *continuous model* assumes a continuous variation of the turns ratio $N(t)$ within the interval $[N_{\text{lo}}, N_{\text{up}}]$ and ignores the insensitivity domain of the regulator. Therefore, the tap changer dynamics can be modeled through the following differential equation:

$$\frac{\mathrm{d}N}{\mathrm{d}t} = \frac{1}{T_{\text{AVR}}}(V_2 - V_{2\text{sch}}) \quad ; \quad N_{\text{lo}} \leq N \leq N_{\text{up}} \tag{11.29}$$

where T_{AVR} is the time constant of the regulator.

Although the continuous model is less accurate than the discrete model, it is more frequently used in the technical literature because it allows the analytical investigation of the tap changing behavior and highlights the influence on the voltage stability [2].

11.4.2.2 The Effect of Automatic Tap Changing on the Possible Operating Points. It was demonstrated in Section 7.2.3.4 that the static characteristic $V_2 = f(N)$ of an on-load tap changing transformer can be obtained by using expression (7.26), as the curve shown in Figure 7.17:

$$V_2 = f(N) = \frac{NV_1R_c}{X}\left(1\Big/\sqrt{1 + \left(\frac{R_c}{X}\right)^2 \cdot N^4}\right)$$

The maximum load side voltage, $V_{2\max}$, (7.31) can be obtained for a turns ratio $N_{\max}$ (7.30):

$$N_{\max} = \sqrt{\frac{X}{R_c}} \tag{7.30}$$

$$V_{2\max} = V_1\sqrt{\frac{R_c}{2X}} \tag{7.31}$$

The equilibrium points defined by condition $\mathrm{d}N/\mathrm{d}t = 0$ are obtained by intersecting the characteristic $V_2 = f(N)$ defined by expression (7.26) with the horizontal line of $V_2 = V_{2,\text{sch}}$ (Figure 7.17).

Analysis of Figures 7.17 or 11.11 reveals that for a scheduled value $V_{2,\text{sch}} < V_{2,\max}$ two operating points, A and B, are found, whereas for $V_{2,\text{sch}} > V_{2,\max}$ no operating point is possible. This shows that no value of the load side voltage $V_{2,\text{sch}}$ can be reached through on-load tap changing and turns ratio control.

In order to determine the nature of the equilibrium points A and B, a virtual disturbance is assumed, such as an incremental change $\Delta V_{2,\text{sch}}$ of the load side scheduled voltage $V_{2,\text{sch}}$ (Figure 11.11), then the resulted transient state is analyzed. According to the

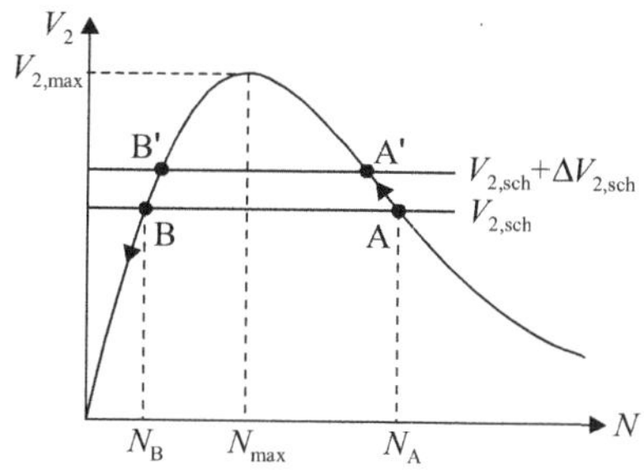

Figure 11.11. The effect of the on-load tap changing regulation on the possible operating points A and B.

normal control logic, the regulator will command the turns ratio to decrease so that V_2 increases to the new reference value $V_{2,sch} + \Delta V_{2,sch}$.

It is obvious that, if the operating point is A, the turns ratio decrease will result in the movement of this point toward the new equilibrium point A′ demonstrating that point A is a stable operating point. Instead, point B is an unstable point since a turns ratio decrease will cause movement away of the new equilibrium point B′. Therefore, the stable operation of the transformer's AVR corresponds to the points on the static characteristic located on the right side of N_{max}, characterized by negative values of the derivative V_2/dN. Instead, for the points located on the left side of the N_{max} value, characterized by $V_2/dN > 0$, the AVR operation is unstable.

11.4.2.3 Influence of On-Load Tap Changing on the Voltage Stability [18].

In order to establish the conditions under which the on-load tap changer actions can cause voltage instability, on one hand, the $V_2 = f(N)$ curves are analyzed in two situations influenced by various factors among which the sending-end voltage V_1 and the load power, and, on the other hand, the tap changer behavior in the possible operating points on the V_2–P_2 system characteristic. In this regard, assumes that the static characteristic from (7.26) is expressed in p.u., hence the nominal turns ratio is $N_{nom} = 1$.

(i) *The Constant Load Case ($R_c = ct.$).* For different values of the voltage V_1, the maximum point of the $V_2 = f(N)$ characteristic does not change ($N_{max} = \sqrt{X/R_c}$ remains constant!), while the maximum value of the receiving-end voltage, $V_{2max} = V_1\sqrt{R_c/2X}$, decreases simultaneously with V_1 (Figure 11.12).

Thus, as the V_1 voltage decreases, the equilibrium points A and B move closer to each other, overlap in the maximum point—which constitutes a static bifurcation point, then disappear. When the equilibrium points disappear, the tap changer dynamic behavior becomes unstable and, in its attempt to reach a new operating point, the AVR causes the voltage collapse phenomenon.

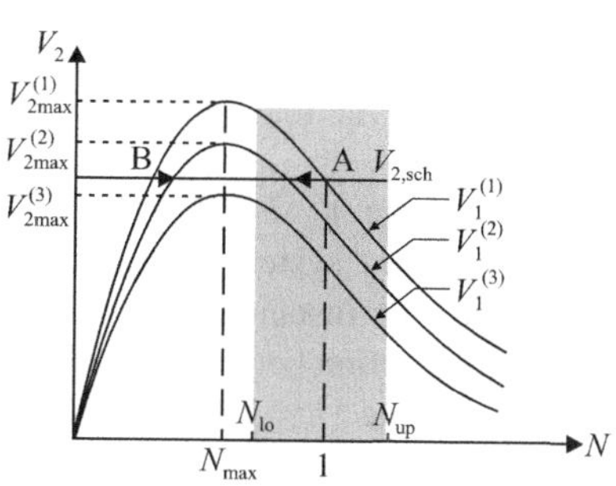

Figure 11.12. The evolution of the $V_2 = f(N)$ static characteristic with the sending-end voltage V_1.

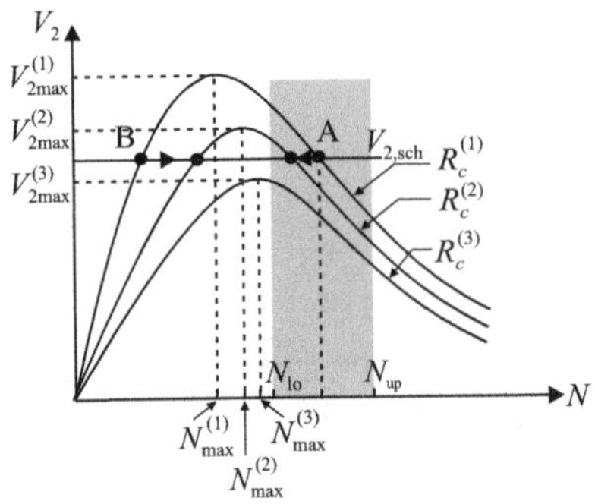

Figure 11.13. The evolution of the $V_2 = f(N)$ static characteristics with the load.

(ii) *The Constant Source Voltage Case* ($V_1 = \text{ct.}$). If the load increases from $P_2^{(1)}$ to $P_2^{(2)}$ and further to $P_2^{(3)}$, that is, the load resistance decreases ($R_c^{(3)} < R_c^{(2)} < R_c^{(1)}$), then the maximum point $N_{\max} = \sqrt{X/R_c}$ of the static characteristic moves to the right (Figure 11.13) since:

$$N_{\max}^{(3)} = \sqrt{\frac{X}{R_c^{(3)}}} > N_{\max}^{(2)} = \sqrt{\frac{X}{R_c^{(2)}}} > N_{\max}^{(1)} = \sqrt{\frac{X}{R_c^{(1)}}}$$

while, according to equation (7.31), the maximum receiving-end voltage decreases:

$$V_{2\max}^{(1)} = \frac{V_1}{\sqrt{2}}\sqrt{\frac{R_c^{(1)}}{X}} > V_{2\max}^{(2)} = \frac{V_1}{\sqrt{2}}\sqrt{\frac{R_c^{(2)}}{X}} > V_{2\max}^{(3)} = \frac{V_1}{\sqrt{2}}\sqrt{\frac{R_c^{(3)}}{X}}$$

Also in this case, a load increase ($P_2^{(1)} < P_2^{(2)} < P_2^{(3)}$ or $R_c^{(1)} > R_c^{(2)} > R_c^{(3)}$, respectively) determines the two operating points, A and B, to move closer to each other then to disappear in a static bifurcation point triggering the voltage collapse phenomenon.

The most severe situation appears when, during operation, the sending-end voltage decreases while the load increases simultaneously. Therefore, paradoxically, an AVR controlled transformer, in its action to keep the voltage constant, may cause the voltage collapse phenomenon!

(iii) *The Behavior of the On-Load Tap Changer in the Two Possible Operating Points of the V_2–P_2 System Characteristic*. In order to analyze the tap changer behavior in the two possible operating points on the system characteristic, assumes that, for some reasons (the transmitted power P_2 increases, the sending-end voltage V_1 decrease, etc.), the load side voltage has decreased below $V_{2,sch} - \varepsilon$. Under this assumption, in according with the normal control logic, after a time delay the AVR will initiate the turns ratio decreasing in order to restore the load side voltage.

For the operating point A, situated on the upper side of the V_2–P_2 characteristic specific for $R_c > X$ (Figure 11.14a), the automatic tap changing system has a stable behavior. Indeed, from Figure 11.14b, it can be "seen" that, for instance, as the load increases (R_c decreases), the operating point, situated at the intersection between the $V_2 = f(N)$ characteristic, which corresponds to $R_c^{(1)}$, and the horizontal line $V_2 = V_{2,sch}$, moves in the point situated on the $V_2 = f(N)$ characteristic, which corresponds to $R_c^{(2)}$. Therefore, as

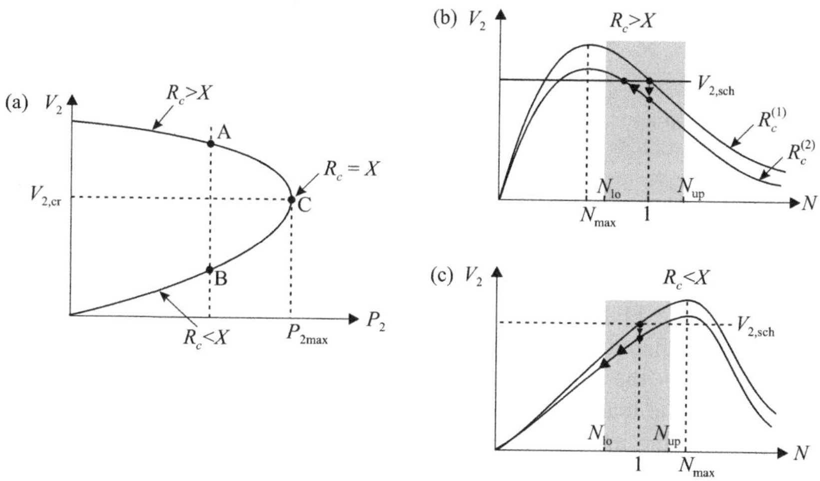

Figure 11.14. The behavior of the on-load tap changer in the two possible operating points.

the turns ratio decreases, the point moves on the new characteristic toward the final operating point, which corresponds to $V_2 = V_{2,sch}$.

For the operating point B, situated on the lower side of the V_2–P_2 characteristic (Figure 11.14a) specific for $R_c < X$, the AVR action is destabilizing. Indeed, from Figure 11.14c, it can be seen that, as the turns ratio N decreases, the operating point moves away from the new equilibrium point and the load side voltage decreases. This process will continue until the ratio reaches its lower limit N_{lo} or until the voltage collapses.

From physical viewpoint, when the load side voltage decreases, the tap changer acts to restore the voltage to its reference value. For this purpose, the AVR initiates the turns ratio decreasing no matter if R_c is higher or lower than X. Following this action, the load impedance value seen from the transformer primary decreases, and the primary current increases while additional voltage drop occurs. Consequently, the secondary voltage evolution is given, on one hand, by an increase caused by the control of the transformer ratio, and, on the other hand, a decrease caused by decrease in the primary voltage. Therefore:

- for $R_c > X$, the first effect is dominant and the load side voltage is restored to the specified value;
- for $R_c < X$, the second effect is dominant and the tap changing action cannot longer compensate the decrease in the secondary voltage but, on the contrary, it will accentuate the phenomenon of voltage decrease.

From the above analysis, it is obvious that the operating point must be situated on the upper side of the V_2–P_2 system characteristic. On this side, the tap changer can compensate the voltage fluctuations caused by changes in power flows or by disturbances that could modify the values of Z or V_1. For this reason, the points situated on the upper side of the V_2–P_2 characteristic are considered to be stable operating points and the ones situated on the lower side are considered to be unstable operating points.

Also, the turns ratio changing action indirectly restores the load power.

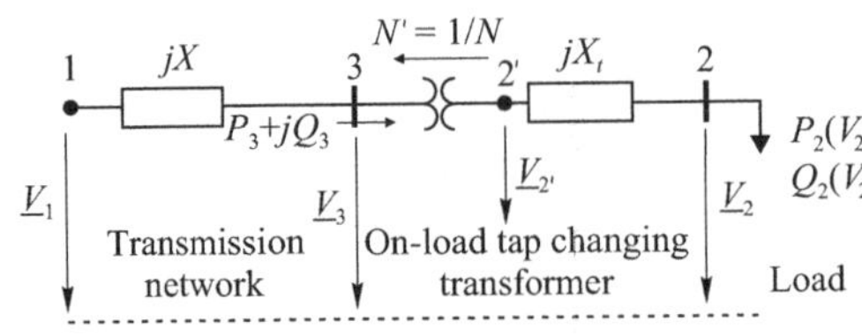

Figure 11.15. The equivalent circuit for analysis of load restoration through automatic on-load tap changing actions.

The *load restoration process through the automatic tap changing actions* and its impact on the voltage stability can be analyzed, considering again the simple radial network from Figure 11.10a and b. In this case, however, the transformer is explicitly modeled through its parameters referred to the medium voltage, being assigned to the load area. Thus, to analyze the interaction between the transmission network and the load, it is necessary to trace the load characteristic seen from the transformer primary.

For simplicity, the resistances and the shunt parameters are neglected resulting the equivalent diagram from Figure 11.15. Furthermore, consider that the load connected to bus 2 has a voltage-dependent characteristic, that is:

$$P_2 = K_P \cdot P_{0,2} \cdot (V_2/V_{0,2})^{\alpha_P} \quad \text{and} \quad Q_2 = K_Q \cdot Q_{0,2} \cdot (V_2/V_{0,2})^{\alpha_Q}$$

If the voltage $\underline{V}_2$ is taken as phase reference, from the equivalent circuit it results that:

$$\underline{V}_{2'} = N' \, \underline{V}_3 = V_3/N = V_2 + jX_t(P_2 - jQ_2)/V_2$$

and

$$\left(\frac{V_3}{N}\right)^2 = \left(V_2 + \frac{X_t Q_2}{V_2}\right)^2 + \left(\frac{X_t P_2}{V_2}\right)^2 \tag{11.30}$$

However, the active and reactive powers entering the transformer from the transmission network depend on the secondary voltage V_2:

$$\begin{aligned} P_3 &= P_2 = f_P(V_2) \\ Q_3 &= Q_2 + X_t \frac{P_2^2 + Q_2^2}{V_2^2} = f_Q(V_2) \end{aligned} \tag{11.31}$$

Using equation (11.30), P_3 and Q_3 can be expressed in terms of the V_3/N ratio:

$$\begin{aligned} P_3 &= P_2 = f_{\mathrm{P}}(V_3/N) \\ Q_3 &= Q_2 + X_{\mathrm{t}} \frac{P_2^2 + Q_2^2}{V_2^2} = f_{\mathrm{Q}}(V_3/N) \end{aligned} \tag{11.31'}$$

which, for a given turns ratio value N, they may represent the transient load characteristics seen from the transformer primary.

When the transformer is equipped with AVR aiming to maintain the voltage at bus 2 at a scheduled value $V_{2,\text{sch}}$, the load active and reactive powers are constant and given by equations (11.31) by replacing V_2 with $V_{2,\text{sch}}$. These powers do not depend on the

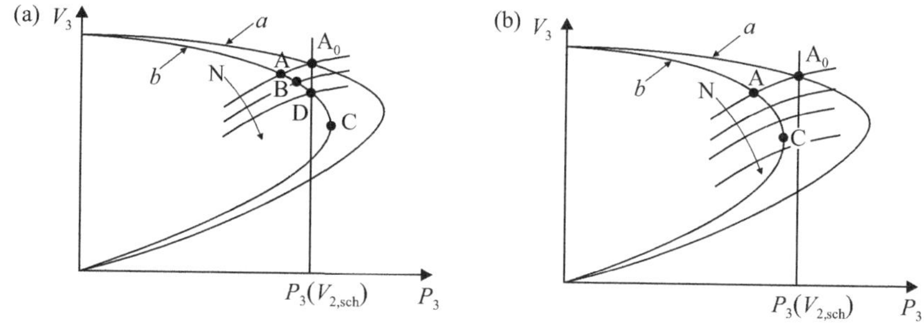

Figure 11.16. The load restoration process following the on-load tap changing action: (a) stable case; (b) unstable case.

transmission line receiving-end voltage V_3 and therefore the static characteristic of the load "seen" from the transformer primary is of constant power type.

In normal operating conditions, the operating point A_0 is situated at the intersection between the system characteristic (Figure 11.16 a, curve *a*) and the load characteristic. When a disturbance occurs in the transmission network (for instance, disconnection of one transmission line circuit), the operating point, characterized by the voltage $V_{2,sch}$, moves in the point *A* situated at the intersection between the postdisturbance characteristic of the system (Figure 11.16 a, curve *b*) and the transient load characteristic that corresponds to the actual turns ratio N_0.

Note that, in the new operating point, which is a short-term equilibrium point, the consumed power $P_2(V_2)$ is smaller than the power consumed in the point A_0 and $V_2 < V_{2,sch}$ since the turns ratio did not change. If the voltage is no longer within the admissible limits, that is, $V_2 < V_{2,sch} - \varepsilon$, the OLTC will start changing the tap in the sense of voltage V_2 increase. The operating point moves on the postdisturbance characteristic, in accordance with the new transient load characteristics (corresponding to the new turns ratio values), until V_2 returns inside the admissible band $V_{2,sch} - \varepsilon \leq V_2 \leq V_{2,sch} + \varepsilon$. A new operating point *D* is, therefore, obtained at the intersection between the static load characteristic and the postdisturbance system characteristic. However, although the load voltage (the secondary transformer voltage) is restored, the transmission line receiving-end voltage is lower in comparison with the case in which the transformer would have been equipped with AVR (Figure 11.16 a, point *A*).

Depending on the loading degree and the gravity of disturbance occurred within the transmission network, it is possible to have no equilibrium point anymore (the post-disturbance characteristic does not intersect the static load characteristic) and the on-load tap changing action triggers the voltage instability phenomenon (Figure 11.16b).

11.4.3 Effect of the Generated Reactive Power Limitation

Another important aspect of the voltage instability mechanism is *the loss of voltage control at a generator node*. In order to illustrate the effect of generator reactive power limitation due to limitations in the armature and field currents, let us consider the simple case of a load supplied from a synchronous generator through a short line (3-2). The generator is also interconnected with the infinite power system through a long line (1-3) (see Figure 11.17).

The generator is provided with an excitation system and is able to keep the voltage at the bus 3 at a constant value. However, the excitation voltage is constrained by the minimum and maximum limits. Because the bus 3 voltage is kept constant by the generator,

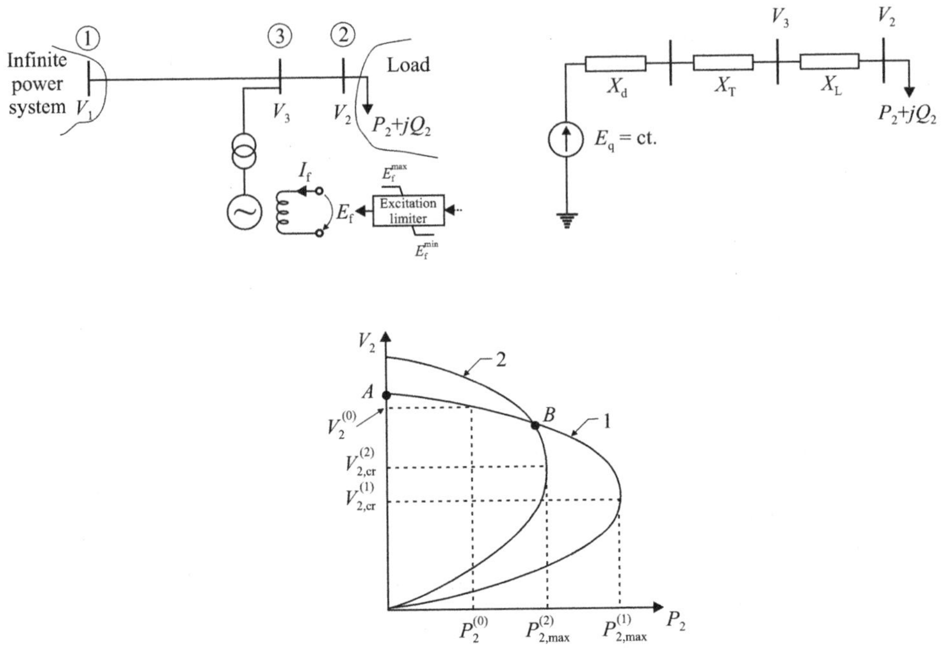

Figure 11.17. Impact of voltage control fail at an intermediate generator bus.

the voltage regulation capability of the infinite power system is neglected. On the other hand, because the line between buses 1 and 3 is long, the voltage control from the power system side is difficult.

- For normal operation, when the generator has sufficient reactive power reserve to keep the voltage V_3 constant, the voltage characteristic is presented in Figure 11.17c (case 1). A certain load voltage $V_2^{(0)}$ corresponds to a certain consumed power $P_2^{(0)}$. The maximum transmissible power $P_{2,max}^{(1)}$ and the critical voltage $V_{2,cr}^{(1)}$ can also be identified. Note that the generator is represented through a simple model.
- When the load power increases, the operating point moves in a new position. If the load power has increased too much, the generator might not be able to provide the required reactive power, and it is no longer capable to keep the voltage at bus 3 at the scheduled value. Hence, the voltage characteristic changes and the operating point is part of the new characteristic (case 2). In this case, the fixed voltage is the electromotive voltage E_q behind the generator reactance X_d. The maximum transmissible power $P_{2,max}^{(2)}$ is now smaller and the critical voltage is higher, as the equivalent impedance between the controlled voltage bus and the load bus is now higher and therefore the situation becomes critical. So, the voltage instability can be triggered for a very small load increase, and the danger comes from the generator reactive power limiters.

More reactive power can be provided by the generator in extreme conditions if the active power is reduced. However, this action can have two implications. The active power reduction is a power market issue since nonscheduled changes involves balancing reserves deployment. On the other hand, pushing the generator to operate beyond certain limits may cause undesired power losses and faster aging. If frequent, such situation can be avoided by installation of new reactive power devices.

With these being considered, accurate representation of the loading capability curves of the synchronous generators is needed, especially the reactive power limits for long-term and short-term operation. For this purpose, one option is to model the power system nodes, to which generators or compensators are connected, as *controlled voltage nodes* with *the reactive power limits Q_{min} and Q_{max} dependent on the terminal voltage*.

This approach, different by the classical one, in which Q_{min} and Q_{max} are fixed, has the following advantages:

- allows considering the total reactive power reserve, which increases as the generated active power decrease;
- allows considering the effect of field and armature currents limitation in the methods for steady-state voltage stability analysis.

To determine the expression of the maximum reactive power corresponding to a certain operating state of the synchronous machine, with a certain voltage Vat the generator terminals, the coefficients $K_S = I_S^{max}/I_{S,n}$ and $K_R = i_f^{max}/i_{f,n} = E_f^{max}/E_{f,n}$ are defined. These coefficients are used to determine the armature current limit, I_S^{max}, and the field current limit, i_f^{max}, in terms of their nominal values $I_{S,n}$ and $i_{f,n}$. In defining the coefficient K_R, the proportionality of the induced e.m.f. E_f to the field current was considered.

Therefore, the maximum reactive power corresponding to the maximum armature current is determined by specifying the condition $S \leq S_{max} = VI_S^{max}$ from which, at the limit, it results:

$$Q_{S,max} = \sqrt{\left(VI_S^{max}\right)^2 - P^2} \tag{11.32}$$

To establish the maximum reactive power corresponding to the maximum field current, we make use of the expressions of the generated active and reactive powers in terms of the induced e.m.f. E_f and the internal electrical angle δ_{int} under the assumption that $X_d = X_q$, that is:

$$P = \frac{VE_f}{X_d} \sin \delta_{int} \tag{11.33a}$$

$$Q = \frac{VE_f}{X_d} \cos \delta_{int} - \frac{V^2}{X_d} \tag{11.33b}$$

Eliminating δ_{int} from (11.33a,b) obtain $P^2 + \left(Q + V^2/X_d\right)^2 = V^2E_f^2/X_d^2$. Then, the expression of reactive power is achieved

$$Q = -V^2/X_d + \sqrt{(VE_f/X_d)^2 - P^2} \tag{11.34}$$

From (11.34), assuming that $E_f = E_f^{max}$, the reactive power limit corresponding to the maximum field current is obtained as follows:

$$Q_{R,max} = -V^2/X_d + \sqrt{\left(VE_f^{max}/X_d\right)^2 - P^2} \tag{11.35}$$

Finally, with $Q_{S,\max}$ and $Q_{R,\max}$ being defined, the maximum reactive power limit can be determined as follows:

$$Q_{\max} = \min\{Q_{S,\max}, Q_{R,\max}\} \tag{11.36}$$

The minimum reactive power limit, $Q_{\min}$, can be specified by the system operator based on static stability considerations, or can be calculated in terms of the terminal voltage V, the active power P, and the maximum rotor angle in under-excitation state, $\delta_{\max}$, by means of the expression [11]:

$$Q_{\min} = P/\tan\delta_{\max} - V^2/X_d \tag{11.37a}$$

or

$$Q_{\min} = P/\tan\delta'_{\max} - E^2_{ext}/(X_d + X_{ext}) \tag{11.37b}$$

when the purpose is to limit the difference between the generator rotor angle and the phase angle of an external node, considered of infinite power, modeled through the e.m.f. E_{ext} and the reactance X_{ext}.

11.4.4 The Minimum Voltage Criteria

Qualitative aspects regarding the mechanisms of voltage instability occurrence have been presented in Section 8.2.2. A minimum voltage criteria concept is presented in this section [12,13].

The dynamics of a power system following a large disturbance occurrence are influenced by various aspects, including the types of loads, the power transfer capacity, or the performances of the reactive power compensation equipments. The maximum time delay allowed before taking appropriate corrective measures, in order to avoid voltage collapse, determines the type of equipments recommended, for example, Static VAr Compensator, mechanically switched capacitors, and so on. This maximum time delay is determined at the limit when, following the activation of the reactive power support, the immediate operating voltage is equal to *the minimum voltage* ($V_{\min}$) *determined by the steady-state load characteristic and the V–Q post-Q support characteristic* (Figure 11.18).

Assumes that in steady state, the load is modeled by a constant power characteristic, that is, $Q_s(V) = Q_0$, and in transient state the load is modeled by a constant impedance characteristic, that is, $Q_t(V) = V^{\beta}$. Changes in the load demand are described by $Q_{demand} = y(t)Q_t(V)$. At the equilibrium point, that is, at the intersection of the steady-

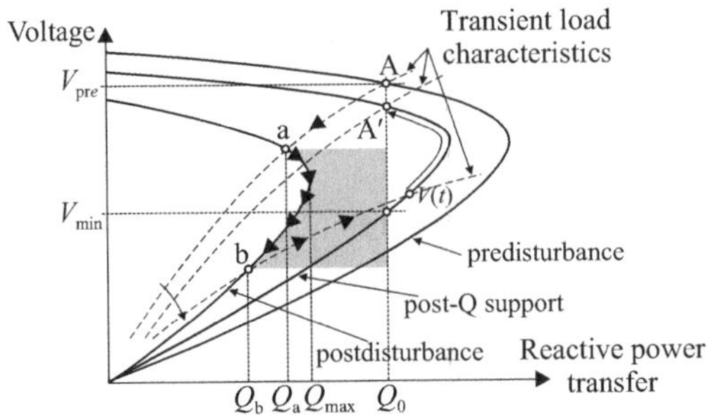

Figure 11.18. Determining the area of the minimum voltage criteria. *Source*: Adapted from [13].

state characteristic and the drifting transient load characteristic, we have [12]

$$Q_0 = y(t)Q_t(V) = y(t)V^{\beta} \tag{11.38}$$

or

$$V(t) = [Q_0/y(t)]^{1/\beta} \tag{11.38$'$}$$

Given that a contingency occurs at $t = 0$ and the reactive power support is activated at $t = \tau$, the *minimum voltage criteria* requires that

$$V(\tau) = [Q_0/y(\tau)]^{1/\beta} > V_{\min} \tag{11.39}$$

The state variable $y(t)$ can be determined from equation (8.8)

$$T_q \int_{y(0)}^{y(t)} dy = \int_0^t (Q_0 - Q^{\text{post}})dt \Rightarrow y(t) = y_0 + \frac{1}{T_q}\int_0^{\tau}(Q_0 - Q^{\text{post}})dt \tag{11.40}$$

where Q^{post} is the power defined by the postdisturbance curve and y_0 is the state variable at the predisturbance condition, whereas

$$Q_0 = y_0 Q_t(V_{\text{pre}}) \quad \text{or} \quad y_0 = Q_0 V_{\text{pre}}^{-\beta}$$

and V_{pre} is the predisturbance operating voltage.

The critical condition satisfying the minimum voltage criteria is then jointly defined by equations (11.39) and (11.40) as follows:

$$\frac{Q_0}{y_0 + \frac{1}{T_q}\int_0^{\tau}(Q_0 - Q^{\text{post}})dt} = V_{\min}^{\beta}$$

or

$$\frac{1}{T_q}\int_0^{\tau}(Q_0 - Q^{\text{post}})dt = Q_0\left(V_{\min}^{-\beta} - V_{\text{pre}}^{-\beta}\right) \tag{11.41}$$

The integration at the left-hand side has a clear meaning. It is the supply–demand energy mismatch.

However, since it is the integration of the area between the steady-state load curve and the postdisturbance V–Q curve, the upper and lower bounds of the integration can be estimated using the upper and lower bounds of the area. Condition $Q = Q_{\max}$ gives the lower bound whereas condition $Q = \min(Q_a, Q_b)$ gives the upper bound. The shaded area in Figure 11.18 shows the lower bound of the area.

Thus, equation (11.41) is simplified as

$$\frac{\tau}{T_q} = \frac{Q_0}{\Delta Q}(V_{\min}^{-\beta} - V_{\text{pre}}^{-\beta}) \tag{11.42}$$

where $\Delta Q = Q_0 - Q_{\max}$ or $Q_0 -$ *Minimum*(Q_a, Q_b) depending on whether lower or upper area bound is used: if lower bound is used, there is less supply–demand imbalance and optimistic (slow response) requirement for reactive power support is obtained; if upper bound is used, conservative result is obtained.

The significant features of equation (11.42) are as follows. The maximum time delay for voltage stability control strongly depends on the transient load characteristic parameter β, in addition to the load time constant T_q. The delay is inversely proportion to the percentage margin deficit $\Delta Q/Q_0$. Thus, for a smaller postcontingency margin deficit, slower voltage control response may be acceptable.

One concern for utilities is to switch very fast from the postcontingence curve to the post-Q support curve in order to avoid a voltage collapse. The minimum voltage criteria addressed by equation (11.42) can be a practical indicator for choosing the best equipments. Thus, SVC or STATCOM is preferred (see also Section 11.6.4).

11.5 VOLTAGE STABILITY ASSESSMENT METHODS

11.5.1 Overview of Voltage Collapse Criteria

As previously shown for a simple "generator-line-load" configuration, also summarized in Figure 11.19, for larger power systems, the mechanisms leading to voltage instability and collapse phenomena are very complex.

Voltage stability assessment methods are employed in order to identify the vulnerable sections of the system as well as the appropriate measures to enhance the system security. In this context, the assessment methods should be accurate and fast enough to help the system operator to take the appropriate decisions in proper time.

In terms of the simulation type and the test performed to identify possible voltage stability problems, the assessment methods are divided into steady-state and dynamic methods (Figure 11.20) [5,53].

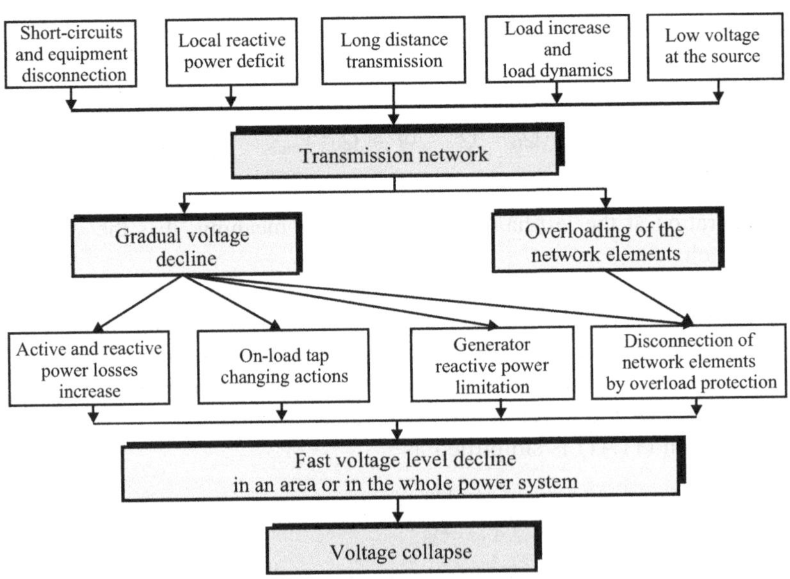

Figure 11.19. Overview of the voltage instability and collapse mechanisms in a power system.

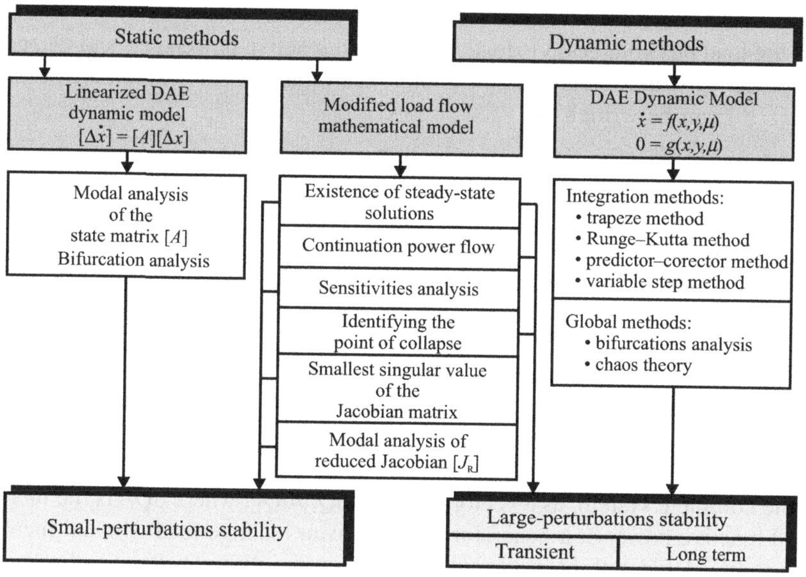

Figure 11.20. Voltage stability assessment methods [5].

Steady-State Methods. The steady-state methods are based on the fact that the main support for the voltage stability in a power system is its capability to transmit power to the load zones both under normal and disturbed operating conditions. This capability can be evaluated using steady-state methods based on the powers flow equations, adjusted so that to include the essential factors that may trigger the voltage instability mechanism, and on the quasistatic approximation[3] system evolution during mid- and long-term dynamic processes. Considering this aspect as well as the disadvantages of the dynamic methods, static methods for voltage stability assessment were developed.

These methods provide good results that are much faster than the dynamic methods and allow local or global voltage stability indices to be defined. These indices can be used in the operation activity to evaluate the risk for voltage instability both under small disturbances and during different instants of the dynamic evolution at large disturbances, as well as to identify the most efficient preventive or corrective actions.

One class of methods employed to evaluate if the voltage profile across the network ranges within acceptable limits from stability point of view is the *load flow feasibility* (LFF) [15,16]. These methods determine also the maximum power transfer capability or the VAr capability of the network by evaluating at the limit the existence of load flow solutions. Usually, the system state is evaluated by means of a scalar *performance index (PI)*, also known as Voltage Instability Proximity Indicator, load flow feasibility index, and so on, which is compared with a *threshold value* (TH), that is, [16]:

$$\begin{aligned} &\mathrm{PI}(z, z^c) \leq \mathrm{TH} \Rightarrow \{\text{Acceptable voltage profile exists}\} \\ &\mathrm{PI}(z, z^c) > \mathrm{TH} \Rightarrow \{\text{No acceptable voltage profile exists}\} \end{aligned} \tag{11.43}$$

[3] The Quasistatic approximation regards the parameters as variable inputs to the system whose dynamics are neglected. Thus, although the parameters can vary and pass through a value, the system dynamics are computed assuming that the parameter is fixed at that value. The quasistatic approximation holds when the parameter variation is slow enough compared to the dynamics of the rest of the system because then the parameters can be approximated as constants at the timescale of the dynamics of the rest of the system [14].

where $PI(z, z^c)$ is evaluated for a set of variables that form the operating point z, which includes the load bus voltages, V_L, load bus real and reactive powers, P_L and Q_L, generator reactive power, Q_G, and so on, and z^c is a reference value of z. In general, the performance index $PI(z, z^c)$ can be written as

$$PI(z, z^c) = \sum_i w_i f_i(z, z^c)$$

where $f_i(z, z^c)$ is the ith real-valued function of z and z^c, and $\{w_i\}$ is the positive weighting coefficients.

It is worth providing a literature overview describing some indices employed for voltage stability assessment [17].

Barbier and Barret [18] suggested an approximate method to calculate the critical value of bus voltage as a threshold value. Using a maximum transfer condition and reduced bus admittance matrices, bus voltage stability is checked in a static manner.

Carpentier et al. [19] defined a proximity indicator for voltage collapse for a bus, an area or the complete system, as a vector of ratios dQ_G/dQ_L, where dQ_G is the incremental generated reactive power at a generator when a given reactive load demand increases by dQ_L. When any element of this vector of ratios becomes infinite, voltage collapse is said to occur. Optimal power flow is proposed to evaluate these indicators.

Jarjis and Galiana [20] suggested a method for the analysis of voltage stability that does not rely on power flow or optimal power flow simulations. It is based on the concept of feasibility regions of power flow maps and the feasibility of margins. This is an exact method. However, the procedure does not lend itself to an application in larger systems due to its enormous computational requirements.

Kessel and Glavitsch [21] proposed a different type of indicator to express the risk of voltage collapse. The indicator uses information from a normal power flow and it can be obtained with reasonable computational effort. They used a hybrid model, where partial inversion of the bus admittance matrix for the load buses is necessary to generate a hybrid matrix.

Tiranuchit and Thomas [22] proposed the minimum singular value of the Jacobian of the descriptor network equations as a voltage security index. Instead of performing singular value decomposition for every change of system operating conditions, they established incremental linear relationship between parameters (dP's and dQ's) and increments of the minimum singular value.

Table 11.2 summarizes the above presented load flow feasibility methods for predicting the proximity to voltage collapse.

Another class of methods employed to evaluate the power system state is the so-called *Steady-State Stability* methods. This approach is concerned with determining if the steady-state dynamic equations are stable around the equilibrium point.

As an alternative to directly calculating the eigenvalues of the linearized dynamic equations and checking if they are in the left half of the complex plan, various indirect methods have been proposed to determine if the linearized dynamic equations are stable. These are based on calculating a *sensitivity matrix* $[S]$ either from the linearized dynamic equations [19] or the Jacobian matrix $[J]$ of the power flow equations [24–26] and check if certain matrix properties are satisfied, namely [27]:

$$\begin{aligned} [S] \in \boldsymbol{M}^{\mathrm{P}} &\Rightarrow \text{(system is voltage stable)} \\ [S] \notin \boldsymbol{M}^{\mathrm{P}} &\Rightarrow \text{(system is voltage unstable)} \end{aligned} \tag{11.44}$$

TABLE 11.2. Load Flow Feasibility Methods to Predict Voltage Collapse (Adapted from [15])

	Load Feasibility Criterion of $PI(z, z^c)$	Form of Performance Indices		$[u]$
Maximum power transfer Barbier and Barret [18]	$\mathrm{PI}_1 \leq \mathrm{TH}_1 = 1.0$	Critical load bus voltages calculated based on maximum power transfer limit	$\mathrm{PI}_1(u) = \max\limits_{i \in L} \left\| \dfrac{\lvert \underline{V}_{Li}^{\text{critical}} \rvert}{\lvert \underline{V}_{Li}(u) \rvert} \right\|$ where $\lvert \underline{V}_{Li} \rvert$ is the voltage magnitude at load bus i; $\lvert \underline{V}_{Li}^{\text{critical}} \rvert$ is the critical voltage magnitude at bus i	$\begin{bmatrix} V_G \\ \theta_G \\ P_L \\ Q_L \end{bmatrix}$
Carpentier [19]	$\mathrm{PI}_3 \leq \mathrm{TH}_3$ (selected by operators)	Voltage collapse sensitivity index based on maximum power transfer criteria and computed from an optimal power flow	$\mathrm{PI}_3(u) = \max\limits_{i \in L} \left\{ \left. \dfrac{\partial Q_G^{\text{tot}}}{\partial Q_L(u)} \right\|_* \right\}$ Note: Change in total reactive power generation for incremental change in $Q_{\mathrm{L}i}$ from optimum power flow solutions (*)	$\begin{bmatrix} V_G \\ \theta_G \\ P_L \\ Q_L \end{bmatrix}$
Load flow feasibility Galiana [20]	$\mathrm{PI}_4 \leq \mathrm{TH}_4$ (selected by operators)	At boundary of load flow feasibility region $\mathrm{TH} = \infty$	$\mathrm{PI}_4(u) = \dfrac{1}{\{\sin([u_\alpha], [u_1])\}^k}$ $u_\alpha = \alpha u_1 + (1 - \alpha) u_0$ where u_0 is the base case input vector; u_1 is the closest input vector boundary of load flow region from u_0; k is a positive integer; $\sin([u_\alpha], [u_1])$ is the sine of the angle between vectors $[u_\alpha]$ and $[u_1]$; and α is a variable scale $0 \leq \alpha \leq 1$	$\begin{bmatrix} P_G \\ V_G^2 \\ P_L \\ Q_L \end{bmatrix}$
	$\mathrm{PI}_5 \leq \mathrm{TH}_5 = 1.0$			

(*continued*)

TABLE 11.2 *(Continued)*

	Load Feasibility Criterion of $PI(z, z^c)$	Form of Performance Indices		$[u]$
Glavitsch and Kessel [21]		At boundary of load flow feasibility region TH = 1.0	$\mathrm{PI}_5(u) = \max_{i \in L} \left\lvert 1 - \frac{\sum_{i \in G} F_{ij} \lvert \underline{V}_{Gi} \rvert}{\lvert \underline{V}_{Lj} \rvert} \right\rvert < 1$ where $\lvert \underline{V}_{Li} \rvert$ is the load bus voltage magnitude of bus i; and F_{ij} is the elements of the matrix, $[\underline{F}] = -\underline{Y}_{LL}^{-1}\underline{Y}_{LG}$	$\begin{bmatrix} V_G \\ \theta_G \\ P_L \\ Q_L \end{bmatrix}$
Multiple load flow solutions Tamura [23]	if $\mathrm{PI}_2\{x^a(u^0 + \Delta u^0)\}$ $< \mathrm{PI}_2\{x^b(u^0 + \Delta u^0)\}$ $\Rightarrow x^a$ feasible or if $\mathrm{PI}_2\{x^b(u^0 + \Delta u^0)\}$ $< \mathrm{PI}_2\{x^a(u^0 + \Delta u^0)\}$ $\Rightarrow x^b$ feasible	Necessary condition on sensitivity matrix for voltage stability of multiple load flow solutions, x^a and x^b	$\mathrm{PI}_2(u) = \frac{1}{2}\sum_{i,j} \gamma_{ij}$ where $\gamma_{ij} = \mathrm{sign}\{[J(x^a)]\} - \mathrm{sign}\left\{[J(x^{BC})]\rvert_{ij}\right\}$ $[J(x^a)]_{ij} = ij$th element of the $[J]$ matrix	$\begin{bmatrix} P_G \\ V_G \\ P_L \\ Q_L \end{bmatrix}$
Singular value of Jacobian $\sigma(J)$ Thomas [22]	$\mathrm{PI}_6 \le \mathrm{TH}_6$ (selected by operators)	A security index for characterizing nearness of the system to instability	$\mathrm{PI}_6(u) = \frac{1}{\sigma(J)}$ where σ is a singular value of Jacobian matrix	$\begin{bmatrix} V_G \\ \theta_G \\ P_L \\ Q_L \end{bmatrix}$

Source: Adapted from [15]

where M^P *is a set of matrices having the property P such as positive definite, positive matrix,* M-matrix property [28], positive (or negative) value of the determinant of [S], *singular value greater than some prescribed value, eigenvalues in open left half of complex planes,* and so on.

Table 11.3 presents a summary of the steady-state stability methods implemented in some computer programs to predict voltage collapse.

Table 11.4 provides details of the expressions for sensitivity matrices.

All methods presented in Table 11.3, except for that of Brucoli [29], evaluate the steady-state stability by checking if a specific sensitivity matrix [S_i], for $\forall i$, satisfies some properties rather than testing if $\mathrm{Re}\{\lambda_i[L]\} \leq 0$. Thus, the effectiveness of these methods depends on: (*a*) the relationship between the sensitivity matrix [S] and the system matrix [A_{sys}]; (*b*) the matrix property that [S] must satisfy. Thus, the M-matrix theorem is used to judge the system stability: the matrix [S] must be positive definite [31–33]. Non-negative matrix condition (defined by $S \geq 0$) [34] requires that some matrices ([S_6], [S_7], and $\Sigma[S_8]_{ij}$) must be positive, which means that each element must satisfy the condition $S_{ij} > 0$.

Concluding, either scalar indices (*PI*) or matrix indices ([S] matrix) are used to evaluate the steady-state (static) performance of the power system. Local or global indices are determined based on different analysis methods such as sensitivity analysis,

TABLE 11.3. Steady-State Stability Methods to Predict Voltage Collapse

Reference	Stability Criterion of $[S]\in M^P$	Comments
Eigenvalues of linearized dynamics		
Brucoli [29]	$\mathrm{Re}\{\lambda_i[A_{\mathrm{Syst}}]\} < 0$	
Jacobian matrix [J]		
Venikov [30]	At two solutions x^a and x^b $\mathrm{sign}\{\det[J(x^a)]\} = \mathrm{sign}\{\det[J]\|_x\}_{\text{base case}}$ and $\mathrm{sign}\{\det[J(x^b)]\} = \mathrm{sign}\{\det[J]\|_x\}_{\text{base case}}$	• Sufficient condition for a periodic instability; • Criterion to predict static bifurcation of multiple load flow solution
Araposthatis [31]	S_2 is "Positive definite"	Necessary and sufficient condition for stability under power system model
Liu and Wu [32]	S_3 is "Positive definite"	Necessary and sufficient condition for stability under power system model
Abe [25]	S_4 is an M-matrix	Sufficient condition for dynamic stability
Kwatny [33]	S_3 is "Positive definite"	System model assumed to be "strictly casual" at solution
Schlueter (IEEE) [34]	PQ stability: [S_5] is a M matrix and $[S_6] \geq 0, \forall i, j$; *PV* stability: $[S_7] \geq 0, \forall i, j$, and $\sum_i [S_8]_{ij} \geq 0, \forall j$	Sufficient conditions on sensitivity matrices for voltage stability at both PQ and PV bus

M-matrix is a matrix which has all negative or zero off diagonal elements, is nonsingular, and has an inverse matrix that has all positive or zero elements. *Source*: Adapted from [15,16]

TABLE 11.4. Form of Sensitivity Matrix

Matrix	Comment
$[J]=\begin{bmatrix}\left[H\left(=\frac{\partial P}{\partial\theta}\right)\right] & \left[L\left(=\frac{\partial P}{\partial V}\right)\right]\\ \left[M\left(=\frac{\partial Q}{\partial\theta}\right)\right] & \left[N\left(=\frac{\partial Q}{\partial V}\right)\right]\end{bmatrix}$	Jacobian of AC load flow power equations in polar form
$[S_2]=[H_{GG}]$	For generator terminal buses only
$[S_3]=[H]-[L][N]^{-1}[M]$	For generator internal and slack buses
$[S_4]=\{[N'_{LL}]-[A_\theta][L'_{LL}]\}^{-1}\cdot\{[L'_{LL}]-[B_V][A_\theta][C_V]\}$ $[N'_{LL}]=[N_{LL}]\frac{1}{V_L^2};\ [L'_{LL}]=[L_{LL}]\frac{1}{V_L^2}$	For load busses
$[A_\theta]=-[M_{LL}]\frac{1}{V_L^2}\begin{bmatrix}[H_{LL}]\frac{1}{V_L^2}\\ \frac{\partial R}{\partial\theta_L}\end{bmatrix};\ [B_V]=\frac{\partial f_{BL}}{\partial V_L}$ $[C_V]=\begin{bmatrix}[H_{LL}]\frac{1}{V_L^2}-\frac{\partial f_{GL}}{\partial V_L}\\ \frac{\partial R}{\partial\theta_L}-\frac{\partial R}{\partial G_L}\frac{\partial f_{GL}}{\partial V_L}\end{bmatrix}$	f_{BL} and f_{GL} are load bus voltages characterized by B_L and G_L, respectively
$[S_5]=[N_{LL}]-[M_{LL}][H_{LL}]^{-1}[L_{LL}]$	Submatrices are partition of Jacobian
$[S_6]=[S_5]^{-1}[N_{LG}]-[M_{LL}][H_{LL}]^{-1}[L_{LG}]$	–
$[S_7]=[N_{GG}]-[M_{GL}][H_{LL}]^{-1}[L_{LG}][S_5]^{-1}$	–
$[S_8]=[S_7]-\left\{[N_{GG}]-[M_{GL}][H_{LL}]^{-1}[L_{LG}][S_5]^{-1}[S_6]\right\}$	–

Source: Adapted from [15]

bifurcations theory, loading margin, the smallest singular value method, or the modal analysis, which are analyzed in detail in the following section.

Dynamic Methods. The dynamic methods imply that the method is based on an ADE dynamic system model characterized by a set of nonlinear differential and algebraic equations. The dynamic methods are, generally, methods based on the numerical integration of the ADE dynamic model and make use of algorithms similar to those used for transient stability assessment, but completed with more complicated models for voltage regulators, for dynamic behavior of loads or special equipments such as on-load tap changers and FACTS devices (SVC, STATCOM, UPFC, etc.).

Analyzing the mechanisms triggering the voltage instability or voltage collapse, this approach is difficult, even inadequate, because, besides the fact that they require large computation effort, they present also some disadvantages, such as:

- do not provide information regarding the stability or instability level;
- do not provide quantitative information regarding the stability reserves;
- the results require appropriate analysis and interpretation that should be performed by specialists.

For these reasons, the dynamic methods are not suitable in the power system operation activity. They are used in the planning and design stages of the power system or in the examination–validation studies of the voltage stability steady-state assessment methods.

Performance indices to predict proximity to voltage collapse problems are of great interest for researchers and technical staff in power systems operation, as these indices could be used online or off-line to help operators determine how close the system is to collapse [3,35]. The objective of these indices is to define a scalar magnitude that can be monitored as system parameters change.

To this extent, the paper [14] can be considered as a reference regarding the voltage stability indices (VSI).

11.5.2 Sensitivities Analysis Method: Local Indices

In the sensitivity analysis approach, voltage stability is treated as a steady-state phenomenon. Sensitivity factors are well-known indices used in several utilities to detect voltage stability problems and to device corrective measures [35,36]. These indices were first used to predict voltage control problems in generator Q–V curves, and they may be defined as

$$\mathrm{VSF}_i = \max_i\left\{\frac{\mathrm{d}V_i}{\mathrm{d}Q_i}\right\}$$

where VSF stands for voltage sensitivity factor. As generator i approaches the bottom of its V–Q curve, the value of VSF_i becomes large and eventually changes sign, indicating an "unstable" voltage control condition.

According to the definition formulated in Section 11.3.2 and briefly presented through expressions (11.25), the voltage sensitivities to active and reactive load power variations are voltage stability local indices.

In a power system with n nodes, with node 1 as slack node, nodes 2,3, . . . ,n_g as generator nodes (PV type) and nodes n_g+1, . . . ,n as load nodes (PQ type), the sensitivities are calculated starting from the powers balance equations [5]:

$$\begin{aligned}&\sum_{k=1}^{n} V_i V_k[G_{ik}\cos(\theta_i-\theta_k)+B_{ik}\sin(\theta_i-\theta_k)]-P_i=0; \quad i=2,3,\ldots,n\\&\sum_{k=1}^{n} V_i V_k[G_{ik}\sin(\theta_i-\theta_k)-B_{ik}\cos(\theta_i-\theta_k)]-Q_i=0; \quad i=n_g+1,\ldots,n\end{aligned} \tag{11.45}$$

or written in compact form:

$$\boldsymbol{F}=([\theta],[V],[P],[Q])=0$$

The parameters of $\boldsymbol{F}$ are grouped in terms of the nodes type as [37]:

- state variables vector $[x] = [\theta_2, \theta_3, \ldots, \theta_n, V_{n_g+1}, \ldots, V_n]^T$;
- input variables vector $[s] = \left[P_{n_g+1}, \ldots, P_n, Q_{n_g+1}, \ldots, Q_n\right]^T$;
- control variables vector $[c] = [V_1, V_2, \ldots, V_{n_g}, P_2, \ldots P_{n_g}]^T$;
- output variables vector $[q] = \left[Q_1, \ldots, Q_{n_g}\right]^T$.

Taking into account this regrouping of the state variables, which define the power system at an instant of time, the last equation becomes:

$$\boldsymbol{F} = ([x], [c], [s]) = 0 \tag{11.46}$$

and the output variables vector can be written as:

$$[q] = f([x], [c]) \tag{11.47}$$

Expanding the equations (11.46) and (11.47) in Taylor series, it results:

$$[\Delta F] = \left[\frac{\partial F}{\partial x}\right][\Delta x] + \left[\frac{\partial F}{\partial c}\right][\Delta c] + \left[\frac{\partial F}{\partial s}\right][\Delta s] = 0 \tag{11.46$'$}$$

$$[\Delta q] = \left[\frac{\partial f}{\partial x}\right][\Delta x] + \left[\frac{\partial f}{\partial c}\right][\Delta c] \tag{11.47$'$}$$

where $[\partial F/\partial x]$ is the Jacobian matrix from the Newton–Raphson-based power flow computation method.

Assuming that $[\partial F/\partial x]$ is an invertible matrix, from expressions (11.46′) gives:

$$[\Delta x] = -\left[\frac{\partial F}{\partial x}\right]^{-1}\left(\left[\frac{\partial F}{\partial c}\right][\Delta c] + \left[\frac{\partial F}{\partial s}\right][\Delta s]\right) = [S_{xc}][\Delta c] + [S_{xs}][\Delta s] \tag{11.48}$$

where

$$[S_{xc}] = -\left[\frac{\partial F}{\partial x}\right]^{-1}\left[\frac{\partial F}{\partial c}\right] \tag{11.49a}$$

is the sensitivity matrix of the state variables to the control variables;

$$[S_{xs}] = -\left[\frac{\partial F}{\partial x}\right]^{-1}\left[\frac{\partial F}{\partial s}\right] \tag{11.49b}$$

is the sensitivity matrix of the state variables to the input variables.

Furthermore, because $-[\partial F/\partial s]$ is the unit matrix, it is obvious that $[S_{xs}]$ is the inverse of the Jacobian matrix.

From equations (11.47′) and (11.48) yields

$$[\Delta q] = [S_{\mathrm{qc}}][\Delta c] + [S_{\mathrm{qs}}][\Delta s] \tag{11.49c}$$

where $[S_{qc}]$ and $[S_{qs}]$ are the sensitivity matrices of the output variables to the control variables and to the input variables, respectively.

Although the voltage stability is affected by both the active and the reactive powers flow, considering the strong V–Q coupling, in this section, only the effects of the reactive power variations on the power system voltage stability will be studied.

(i) *Sensitivity of the load voltage to the reactive power load* ($\boldsymbol{S_{V_cQ_c}}$)

Analysis of the V_2–P_2 system characteristic (Figure 11.4), which for constant power factor represents the V_2–Q_2 dependency, at a different scale, reveals that the slope of the tangent in the normal operating point A is negative and close to zero. As the point approaches the critical point, the slope increases and tends to infinity, then changes the sign when the point moves on the lower side of the characteristic.

The main diagonal elements of the $[S_{xs}]$ matrix, corresponding to the voltage magnitudes and reactive power loads, represent the slopes of the tangents to the V_2–Q_2 curves, for the consumer nodes $c = n_g + 1, \ldots, n$, and therefore they may be used as voltage stability indicators.

According to definitions from Section 11.3.2, the power system "seen" from a load node c is controllable and, hence, stable if:

$$S_{V_cQ_c} = \frac{\partial V_c}{\partial Q_c} < 0 \tag{11.50a}$$

The magnitude of $S_{V_cQ_c}$ represents an indicator for the voltage instability risk: the greater the sensitivity $S_{V_cQ_c}$, the greater the voltage instability risk.

(ii) *Sensitivity of the reactive power generation to the reactive power load* ($\boldsymbol{S_{Q_gQ_c}}$)

When the operating point is located in the critical zone on the V_2–Q_2 characteristic, the reactive power losses increase very much and the generated reactive power does not match the load. If the reactive power in a load node c increases by $\Delta Q' = 1$ MVAr, and the control variables $[c]$ do not vary, that is, $\Delta c = 0$, then, from (11.49c) we see that the sum of the additional reactive powers supplied by the generators is equal to the sum of the elements situated on the column corresponding to node c in the $[S_{qs}]$ matrix. Therefore:

- the operating point is located in the stable zone if this sum is close 1;
- if the operating point is located in the critical zone, mathematically, the sum of the additional reactive powers supplied by the generators is very high, and tends to infinity in the critical point.

Hence, the value of:

$$S_{Q_gQ_c} = \sum_{k=1}^{n_g} S_{qs}(k, c) \tag{11.50b}$$

represents a voltage stability local indicator.

Therefore, the power system "seen" from a node is stable from voltage viewpoint if $S_{V_cQ_c} < 0$ and $S_{Q_gQ_c} \cong 1$, respectively.

11.5.3 Loading Margin as Global Index

The $S_{V_cQ_c}$ and $S_{Q_gQ_c}$ indices provide local information, their validity being limited only to the operating point around which the linearization was made. Their values significantly change if a disturbance causing a strong nonlinearity, similar to reactive power limitation at the generators, occurs in the system. For this reason, indices taking into account the load evolution as well as system nonlinearities under various disturbances are needed. Such an indicator is the loading margin which represents the additional power that may be transferred to the load, located in a node c or in a load zone, so that the power system, initially found in a stable state, moves to the final state that corresponds to the voltage stability limit.

A version of a loading margin measures the amount of power transferred between two areas when studying the transfer capability between areas.

The loading margin is used as a voltage collapse index because it has several advantages, such as [14]:

- it is simple, quick to calculate, and does not require a particular system model;
- it can be calculated using a static model only; a dynamic model can also be employed but its details do not affect the calculation [38];
- it is highly accurate because it takes into account the power system nonlinearities and limits such as reactive power capacity limits.

The loading margin index has disadvantage also [14]:

- it assumes considering points away from the current operating points, which requires large computation times; this is, the greatest disadvantage;
- it requires initialization of a direction of load increase, which is not always easily to define.

The loading margin is usually calculated starting from the current operating point and assumes small load increments, for which a new load flow calculation is performed, until the nose of the *V–P* curve, which is the critical point, is reached. The loading margin is the total load increment from the current operating point to the critical point. In practice, continuation or direct methods are used for loading margin calculation [14].

(a) *Direct methods*, also known in power system applications as point of collapse methods [27], where originally developed to compute singular bifurcation points of nonlinear systems [39].

The method consists in solving a set of equations to directly obtain the point of collapse. Computation on large power systems involves a large number of equations and therefore large computational effort. There are also other disadvantages of this method, for example, it cannot detect when reactive power control limits are reached and it can only determine a collapse point associated with system singularities (bifurcations) [14].

(b) *The continuation method* (*Voltage profiles*), also known as *V–Q*, *V–P* or nose curves, is used for determining proximity to collapse [14].

The method consists in calculating successive power flow solutions and tracing the voltage profiles, by gradually increasing the load, up to the collapse point (the maximum loading point), then determine the loading margin. This technique is used as an alternative to the direct method; the loading margin provides information regarding the voltage behavior in the system buses.

This method is currently in use at some system operators because it is reliable and very informative. It may help the operators to take timely preventive measures to avoid system voltage collapse. However, it requires large computation times, especially in large systems with multiple limits.

Three loading margins can be defined

- reactive loading margin: DQ_c (P_c = ct.);
- active loading margin: DP_c (Q_c = ct.);
- apparent loading margin: DS_c ($\cos\varphi$ = ct.).

Early evaluation of these margins was based on the continuation power flow technique. However, convergence problems were observed for this method in the vicinity of the critical point. To overcome this disadvantage, either the *optimal multiplier-based Newton–Raphson method* or the *local parameterization technique* can be used to detect the points situated on the critical surface.

As known, for the load flow calculation, the numerical solution of the nonlinear system of equations (11.45) is necessary. In this regard, the iterative Newton–Raphson method is used. Thus, for the $(p+1)$ iteration following the current (p) iteration, we need to go through the following steps [40]:

(a) linearization of the system of equations (11.45) in order to obtain the linear system of equations:

$$\left[J^{(p)}\right]\begin{bmatrix}\Delta\theta^{(p)} \\ \Delta V^{(p)}\end{bmatrix} = \begin{bmatrix}\Delta P^{(p)} \\ \Delta Q^{(p)}\end{bmatrix} \tag{11.51}$$

where

$$\left[J^{(p)}\right] = \begin{bmatrix}J_{P\theta} & J_{PV} \\ J_{Q\theta} & J_{QV}\end{bmatrix}$$

is the Jacobian matrix calculated at iteration (p);

(b) solve the linear system of equations (11.51) and adjust the vector of unknown variables using the expression:

$$\begin{bmatrix}\theta^{(p+1)} \\ V^{(p+1)}\end{bmatrix} = \begin{bmatrix}\theta^{(p)} \\ V^{(p)}\end{bmatrix} + \begin{bmatrix}\Delta\theta^{(p)} \\ \Delta V^{(p)}\end{bmatrix} \tag{11.52}$$

To avoid convergence problems due to singularity of the Jacobian matrix near the critical point and to detect the inexistence of solutions, the load flow calculation is performed using an optimization technique. In this regard, for adjusting the unknown variables vector, the following expression is used

$$\begin{bmatrix}\theta^{(p+1)} \\ V^{(p+1)}\end{bmatrix} = \begin{bmatrix}\theta^{(p)} \\ V^{(p)}\end{bmatrix} + \beta\begin{bmatrix}\Delta\theta^{(p)} \\ \Delta V^{(p)}\end{bmatrix} \tag{11.52$'$}$$

where β is the optimal multiplier and is determined through the minimization condition of the objective function $F(\beta) = \sqrt{(\Sigma \Delta P^2 + \Sigma \Delta Q^2)}$ (the Euclidian norm of the active and reactive powers adjustments vector). In case of convergence, the objective function tends to 0 and the β coefficients tend to 1. Instead, if there is no possible solution, the β coefficients tend to 0 and the objective function tends to a positive value.

The local parameterization technique is based on plotting the static bifurcation diagram of a dynamic system whose behavior depends on one parameter and derives from the general class of the so-called path-following methods [39].

For the implementing this method, besides computation of the loading margins between the current operating point and the critical point (the saddle-node bifurcation point), parameterization of load flow equations is required. Thus, in order to simulate the load increase, the demanded nodal powers are written as:

$$\begin{aligned} P_{C_i} &= P_{CO_i} + \alpha k_{CP_i} \cos \varphi_i \\ Q_{C_i} &= Q_{CO_i} + \alpha k_{CQ_i} \sin \varphi_i \end{aligned} \tag{11.53}$$

where P_{CO_i}, Q_{CO_i} are the demanded powers in the X_0 initial state; k_{CP_i}, k_{CQ_i} are the coefficients for selection of the nodes in which the load increase is simulated; and α is a parameter having the dimension of an apparent power and simulates the loads evolution.

If an increase in the demanded active power is also simulated, the load surplus $\Delta P = \sum \alpha k_{CP_i} \cos \varphi_i$ must be allocated among the system's power plants. The coefficients $k_{gP_{ii}}$ are used for load allocation, and the generated active power is adjusted by $P_{gi} = P_{g0_i} + k_{gP_i} \Delta P$, where P_{g0_i} is the active power supplied by the generator at the i node in the initial state X_0.

Thus, the system of equations representing the powers balance in the static model is written in compact form as:

$$\boldsymbol{F}(\boldsymbol{\theta}, \boldsymbol{V}, \alpha) = 0 \tag{11.54}$$

which is the parameterized mathematic model for the load flow computation.

Obviously, the system of equations (11.54) does not accept solutions for any value of the parameter α. There is a critical value α_{cr} corresponding to the critical point situated on the critical surface for which the two equilibrium points of the power system coincide in a so-called *saddle-node bifurcation point* (see Section 11.5.4.3). In this point, a significant change in the power system dynamic evolution is produced through new trajectories that lead to voltage instability. To determine α_{cr} a *predictor–corrector technique* is used.

In *the predictor step*, a new solution is predicted starting from a known solution and considering a movement along the tangent vector. Differentiating equation (11.54) gets:

$$\begin{bmatrix} \dfrac{\partial F}{\partial \theta} & \dfrac{\partial F}{\partial U} & \dfrac{\partial F}{\partial \alpha} \end{bmatrix} \begin{bmatrix} d\theta \\ dV \\ d\alpha \end{bmatrix} = 0 \tag{11.55}$$

To calculate the tangent vector $[\mathrm{TAN}] = [d\theta\, dV\, d\alpha]^t$ using equation (11.55), an additional equation is needed. This can be done by specifying the value of one component of the vector [TAN]. Thus, initially (at iteration $p = 0$) the condition $d\alpha = 1$ is imposed, then for $p = 1, 2, \ldots, n$ the component $\mathrm{TAN}_k^{(p)}$ is determined by:

$$\mathrm{TAN}_k^{(p)} = \max\left(d\theta^{(p-1)} \quad dV^{(p-1)} \quad d\alpha^{(p-1)} \right) \tag{11.56}$$

This selection technique of a component from the tangent vector is known as *local parameterization* and allows writing the system of equations (11.55) as:

$$\begin{bmatrix} \dfrac{\partial F}{\partial \theta} & \dfrac{\partial F}{\partial V} & \dfrac{\partial F}{\partial \alpha} \\ & e_k & \end{bmatrix} \begin{bmatrix} \mathrm{d}\theta \\ \mathrm{d}V \\ \mathrm{d}\alpha \end{bmatrix} = \begin{bmatrix} 0 \\ 0 \\ \pm 1 \end{bmatrix} \tag{11.55'}$$

In equation (11.55′), e_k is a row unit vector with all the elements equal to zero except for the kth one, which corresponds to the current continuation parameter, that is, $e_k = [0, \ldots, 1, \ldots, 0]$, and the values $+1$ or -1 are chosen so that to reflect the increasing or the decreasing trend of the specified component.

Next, knowing $\left[\mathrm{TAN}^{(p)}\right]$, a new solution is predicted through:

$$\left[x^{(p)}\right] = \left[x^{(p-1)}\right] + h\left[\mathrm{TAN}^{(p)}\right] \tag{11.57}$$

where $[x]$ is vector of unknown variables and h is the step length that can be determined at the limit, for instance, by specifying that a generator reaches its reactive power limits.

In the *corrector step*, the predicted solution is corrected by solving the system of equations:

$$\begin{cases} \boldsymbol{F}(\boldsymbol{\theta}, \boldsymbol{V}, \alpha) = 0 \\ x_k = x_k^{(p)} \end{cases} \tag{11.58}$$

where x_k is component of the vector $[x]$ chosen as parameter, of which value is taken equal to the predicted value.

This computation technique, briefly illustrated in Figure 11.21, allows complete plot of the *V–P* characteristic of a complex power system as well as accurate assessment of the loading margins.

The predictor–corrector technique can be used for loading margins computation, either for a certain "stress" path or to the closest critical point, by simulating different load increase scenarios. In practice, to calculate the loading margins considering a certain "stress" path, it is advantageous to use both methods in a complementary manner [35].

The step increase in load is first set, and a succession of steady states (equilibrium points) is then computed using the optimal multiplier-based Newton–Raphson algorithm, until the vicinity of the critical point is reached (until there is no possible solution). Then, to accurately determine the critical point, the local parameterization technique is employed starting from the last determined equilibrium point.

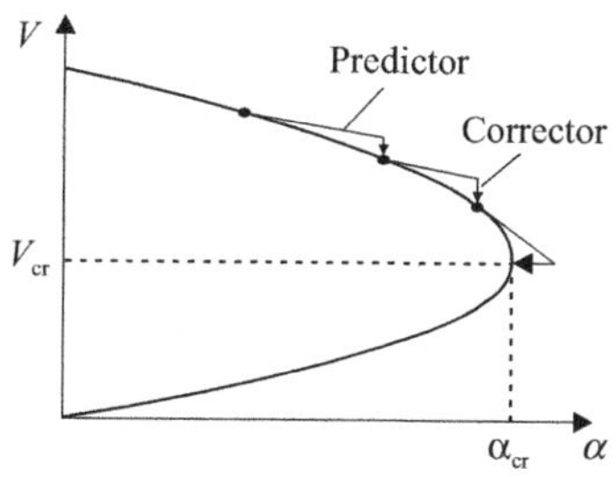

Figure 11.21. Predictor–corrector computation technique.

11.5.4 Some Aspects of the Bifurcations Theory

11.5.4.1 Generalities. A power system is characterized by a nonlinear dynamic behavior. Voltage collapse is, thus, a dynamic phenomenon that requires appropriate analysis techniques such as bifurcation theory [2,14,41].

In the bifurcation theory, slow changes in the system parameters are assumed then the trajectory that leads to instability is identified. The analysis is based on steady-state calculations, which include eigenvalue analysis at every system parameters μ and/or p change in the following set of differential-algebraic equations (DAEs) [42]:

$$\dot{x} = f(x, y, \mu, p) \tag{11.59a}$$

$$0 = g(x, y, \mu, p) \tag{11.59b}$$

where x is a vector of state variables associated with the dynamic power system components (generators, loads, controllers, FACTS devices, etc); y is a vector of algebraic variables associated with steady-state models of the power system components (loads, transmission lines, etc); μ is a set of controllable parameters such as controllers set points; p is a set of uncontrollable parameters, such as variations in the load powers; $f(\cdot)$ is the set of differential equations associated with the state variables x; and $g(\cdot)$ is the set of algebraic equations associated with the algebraic variables y.

Linearization of DAEs from (11.59) around an equilibrium point x^* gives

$$\begin{bmatrix} \Delta\dot{x} \\ 0 \end{bmatrix} = [J]\begin{bmatrix} \Delta x \\ \Delta y \end{bmatrix} \tag{11.60}$$

where $[J]$ is the system unreduced Jacobian determined by:

$$[J] = \begin{bmatrix} f_x = \left(\dfrac{\partial f}{\partial x}\right)\Big|_{x=x*} & f_y = \left(\dfrac{\partial f}{\partial y}\right)\Big|_{x=x*} \\ g_x = \left(\dfrac{\partial g}{\partial x}\right)\Big|_{x=x*} & g_y = \left(\dfrac{\partial g}{\partial y}\right)\Big|_{x=x*} \end{bmatrix} \tag{11.60$'$}$$

If the submatrix $[g_y]$ is nonsingular, extracting the vector of algebraic variables $[\Delta y]$ from the second equation of (11.60) and substituting in the first equation gives

$$[\Delta\dot{x}] = \left([f_x] - [f_y][g_y]^{-1}[g_x]\right)[\Delta x] = [A][\Delta x] \tag{11.61}$$

and the set of DAEs reduces to a set of ordinary differential equations (ODEs).

The state matrix $[A]$ of the linearized system equations, called also the reduced Jacobian:

$$[A] = \left[[f_x] - [f_y][g_y]^{-1}[g_x]\right] \tag{11.62}$$

is the Schur's complement of the submatrix $[g_y]$ of the unreduced Jacobian J [2].

The DAEs system is subjected to bifurcations similar to a simple ODEs system when the parameters p vary. Bifurcations can be detected by monitoring the eigenvalues of the state matrix $[A]$. Figure 11.22 illustrates some types of bifurcations identified by observing the eigenvalues.

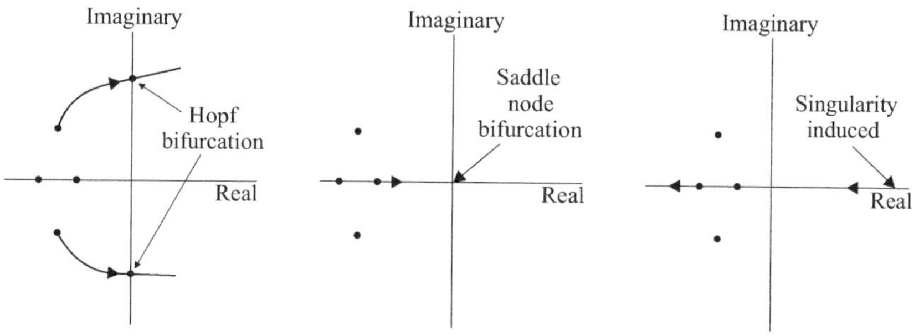

Figure 11.22. Locus of the eigenvalues close to three types of bifurcation to changes in the loading parameters [43].

In practice, the eigenvalues of [A] are obtained directly from the system Jacobian [J] because it is a sparse matrix, whereas [A] is a full matrix [44].

Interpretation of the eigenvalues:

- the equilibrium point x^* is asymptotically *stable* if the real part of all the eigenvalues of the state matrix [A] are negative, that is, $\text{Re}(\lambda_i) < 0, \forall i$;
- the equilibrium point x^* is *unstable* if the real part of at least one eigenvalue is positive;
- the unstable equilibrium is called *saddle* if some eigenvalues have positive real parts and all other have negative real parts.

In bifurcation theory, two different characteristics corresponding to two equilibrium points join, creating then a bifurcation. In the bifurcation points the Jacobian become singular, and therefore the implicit function theorem cannot be applied [2]. A bifurcation may occur at any point in the parameter space, for which the qualitative structure of the system (11.59) changes for a small variation of the parameter vector.

Definitions

- An *equilibrium point* x^* is a steady-state solution of a differential equation (transient stability) model. An equilibrium point is a load flow solution.
- A *bifurcation* is a qualitative change in the system, such as when an equilibrium disappears (saddle-node bifurcation) or the steady-state changes from equilibrium to an oscillation (Hopf bifurcation) [14].
- A *static bifurcation* is an operating condition for a transient stability model where there are multiple equilibrium points of the transient stability model for small perturbation in the transient model or the initial conditions. *A static bifurcation may exist if the Jacobian of the linear transient stability model is singular.* Multiple load flow solutions can occur at a static bifurcation [23].
- A *periodic orbit* is a steady-state oscillation. It is visualized in the state space as a closed loop, which the state traverses once every period [14].

Generally, the bifurcations are classified into local and global bifurcations.

- *Local bifurcations* occur when changes in critical parameters set within a small neighborhood cause qualitative changes in the properties of stability of an equilibrium. The small local changes are usually studied by analyzing changes of the eigenvalues in response to parameter variations [9].

Local bifurcations can be classified depending on the specific way in which the critical parameter conditions are defined and situation of the system operating point. The most important bifurcations in the power system models are as follows:

(i) *Saddle-Node Bifurcations*. When two equilibrium points coalesce and disappear as the loading parameter moves beyond a certain limit, then no equilibrium exists which means that the collapse occurs;

(ii) *Hopf Bifurcations*. When a conjugated complex pair of eigenvalues crosses the stability boundary as one critical parameter gradually changes, and stability is lost in the form of oscillations;

(iii) *Singularity Induced Bifurcations*. When singularity of the network differential-algebraic occurs, and system stability is lost characterized by a per se indeterminable behavior.

- *Global bifurcations* are characterized by changes in the dynamic properties of the power system models over a large domain of the state space. A dynamic system can have a solution that repeats itself in time thus exhibiting a periodic orbit. Such systems are called oscillators. The system can lose its stability when the periodic orbit disappears under parametric changes. Voltage stability problems occur when the operating periodic orbit vanishes similarly to the case when an operating equilibrium disappears at saddle node bifurcation.

Table 11.5 illustrates the main types of bifurcations.

11.5.4.2 Hopf Bifurcation. Consider the autonomous differential equation of a power system dynamics model:

$$\dot{x} = f(x, p) \tag{11.63}$$

TABLE 11.5. Main types of bifurcation

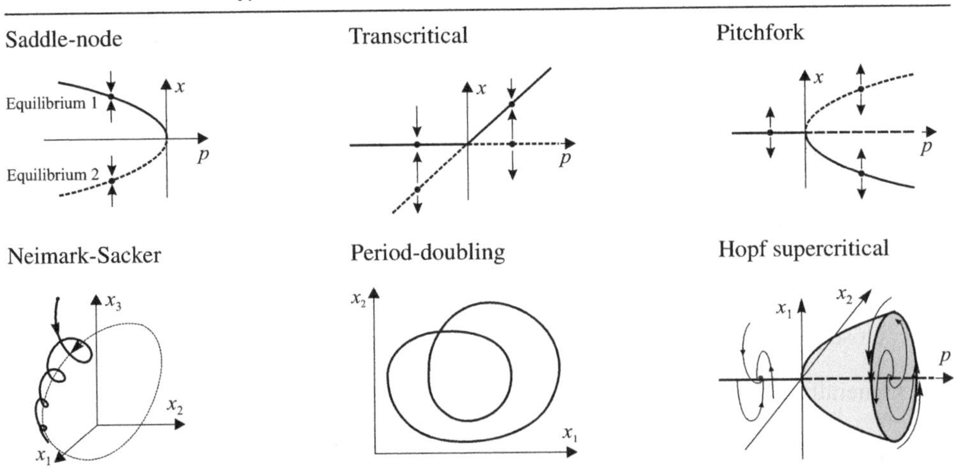

and vary the parameter p on a certain trajectory, then the state vector x and the associated with eigenvalues of the Jacobian evaluated on this trajectory change accordingly. In the vicinity of a fixed equilibrium point $x = x^*$, the left hand side term becomes zero, that is:

$$0 = f(x^*, p) \tag{11.63'}$$

equation that defines the position of the equilibrium point x^* in the surface given by variation of p. The power system operating point on this surface is asymptotically stable if the eigenvalues of the Jacobian have negative real parts at that point.

The power system state can become critical if the real part of at least one eigenvalue becomes zero or a pair of complex conjugate eigenvalues crosses the imaginary axis. As the pair of eigenvalues moves to the right half plane (Figure 11.22a), the system may start oscillating, either with stable oscillations or growing oscillatory transients. The onset of this oscillatory phenomenon is described by Hopf bifurcation theory [39,45]. Hopf bifurcations occur as periodic orbits, which are steady-state oscillations emerging around an equilibrium point [14].

At a Hopf bifurcation point, one pair of eigenvalues of the state matrix $[A]$ (see equation 11.61) is purely imaginary. The Hopf bifurcation point is reached for a particular set of values of the p parameters in the system of equations (11.59). A Hopf bifurcation is called supercritical if the periodic solutions are stable (Table 11.5), whereas it is called subcritical if the periodic solutions are unstable and voltage collapse phenomenon is induced.

Hopf bifurcations can be detected using several indications [14]:

(i) A power system, previously found in a stable equilibrium state begins to oscillate monotonically in a periodic orbit or has a growing oscillation transient as one parameter slowly change;

(ii) The system equilibrium persists as parameter slowly varies, but it changes from stable to oscillatory unstable;

(iii) The system Jacobian has *one purely imaginary pair of eigenvalues*, which means that *the system linearized about the equilibrium becomes unstable at a Hopf bifurcation as a pair of eigenvalues crossing the imaginary axis.*

11.5.4.3 Saddle-node Bifurcation. A saddle-node bifurcation of a power system operation state is defined as a point at which two power flow solutions merge then disappear as the loading parameter, that is, the load active power, is slowly increased. The disappearance of the power flow solutions in the mathematical calculation indicates that the power system steady state changes dynamically and it can fall into a voltage collapse [14,41]. The dynamics are first slow and then fast.

The saddle-node bifurcation is the point of maximum loading indicated by point C in Figure 11.4 and corresponds to a singularity of the state Jacobian f_x. At maximum loading, the two equilibrium points A (stable) and B (unstable) merge into point C. The necessary conditions for a saddle-node bifurcation are given by the equilibrium equation (11.63′) and the following singularity condition:

$$\det f_x(x^*, p) = 0 \tag{11.64}$$

However, these are not sufficient conditions that must be met by an operating point to become a saddle-node bifurcation. The points satisfying these conditions can be transcritical or pitchfork bifurcations [46] (see Table 11.5).

For a multivariable system, a saddle-node bifurcation is achieved when two equilibrium points, one stable (having real negative eigenvalues) and the other one unstable (having at least one real positive eigenvalue) coalesce when both eigenvalues become zero then disappear simultaneously and no equilibrium point can exist.

For a nonsingular matrix g_y, the Schur's determinant formula gives [2]:

$$\det J = \det g_y \cdot \det\left(f_x - f_y g_y^{-1} g_x\right) = \det g_y \cdot \det A \tag{11.65}$$

and the state matrix A and the unreduced state Jacobian become singular. Therefore, a saddle-node bifurcation in the power system operation occurs when the unreduced state Jacobian J is singular.

A saddle-node bifurcation is thus detected when the following conditions occur [14]:

(i) Two equilibrium points, one stable and one unstable, coalesce;
(ii) The sensitivity of a state variable to the loading parameter becomes infinite at the same time the solution curve exhibit infinite slope. On the particular case illustrated in Figure 11.4, the tangent to the nose in point C is infinite;
(iii) The system Jacobian has one zero eigenvalue and/or one zero singular value;
(iv) When voltage collapse is induced, the dynamics are first slow then fast.

11.5.4.4 Singularity Induced Bifurcation. A singularity induced bifurcation occurs when a state parameter changes slowly and the algebraic equations become singular [14]. The midterm dynamics of a power system can be described by the DAEs model:

$$\dot{x} = f(x, y) \tag{11.66a}$$

$$0 = g(x, y) \tag{11.66b}$$

The power system dynamics studied with the model from equation (11.66) are slow (<5 Hz) and thus the quasi-stationary or *quasi-steady-state assumption* may be employed. This allows representing the network by lumped parameters using the concept of quasi-stationary phasor representation [14]. For the cases in which the above condition is not sufficient, additional condition called the *singularity* must be defined in order to solve the phasor network equations (11.66b).

A power system is normally stable at small perturbations about an equilibrium point. Because the load and thus the generation changes continuously, there may be an infinite number of operating points. One way of verifying if a certain operating point, say (x_0, y_0), is small perturbations stable is by calculating the eigenvalues of the system matrix $[A]$ by linearizing the equations (11.66) at the studied equilibrium point (x_0, y_0), where

$$A = \left[\frac{\partial f}{\partial x} - \frac{\partial f}{\partial y}\left(\frac{\partial g}{\partial y}\right)^{-1}\frac{\partial g}{\partial x}\right]\Bigg|_{(x_0, y_0)} \tag{11.67}$$

which can be defined only if the matrix $[g_y]$ $\left(= \left.\frac{\partial g}{\partial y}\right|_{(x_0, y_0)}\right)$ has an inverse. When a matrix is singular, that is, its determinant is zero, it does not have an inverse. If $[g_y]$ is singular at the studied operating point (x_0, y_0), the system matrix $[A]$ cannot be defined. This condition of singularity of the matrix $[g_y]$ at the operating point is defined as *the singularity induced bifurcation* [14]. Singularity of matrix $[g_y]$ is a consequence of the operating point lying in the vicinity of an algebraic singularity of equations (11.66b). In this situation the network cannot be represented by quasi-stationary phasor as assumed earlier.

The singularity-induced bifurcation (Figure 11.22c) is detected when some eigenvalues of the system matrix $[A]$ are infinite (unbounded eigenvalues). When approaching such a bifurcation, some variables of the DAEs start to very fast. As one eigenvalue moves to the right half plane the power system become unstable at small perturbations.

Concluding, a singularity-induced bifurcation is characterized by the following mathematical attributes [14]:

(i) The Jacobian associated with the algebraic equations becomes singular for a certain set of values of the parameter y;
(ii) One eigenvalue of the matrix $[A]$ moves from the left to the right half plane or vice versa by diverging through infinity.

11.5.4.5 Global Bifurcations. While the stability of an equilibrium point is evaluated by calculation of eigenvalues at that point, in global bifurcations the stability is evaluated by means of the Floquet multipliers [4] of a periodic orbit [47]. The Floquet multipliers are similar to the eigenvalues, associated to a discrete map (Poincaré map), and are used to quantify the local stability properties of a point on the periodic orbit [14,39,45].

Given a set of values of p in equation (11.63′), the monodromy matrix $M(p)$ has n Floquet multipliers: $\eta_1(p)$, $\eta_2(p)$, . . . , $\eta_n(p)$, where n is the number of variables involved. In the Floquet theory, one of the multipliers is always equal to unity, and the remaining $n-1$ multipliers determine the stability. For a certain periodic solution $x(t, p_0)$ the system is asymptotically stable if all the $n-1$ multipliers are inside the unit circle, that is, $|\eta(p_0)| < 1$. If at least one multiplier is outside the unit circle, that is, $|\eta(p_0)| > 1$, then the system is unstable. The boundary of the stability domain is thus represented by the critical points p_{cr} for which the matrix $M(p)$ has multipliers on the unit circle, that is, $|\eta(p_{cr})| = 1$. When the multipliers of a periodic orbit cross the unit circle under changes of the parameter p, global bifurcations result.

Any change in the parameter p results in change in the multipliers $\eta(p)$, and some of them may cross the unit circle. There are two ways in which the system stability can be lost [48]:

- one real multiplier crosses outside the unit circle along the positive real axis, with $\eta(p_{cr}) = 1$ (Figure 11.23a) or one multiplier crosses outside the unit circle along the negative real axis, with $\eta(p_{cr}) = -1$ (Figure 11.23b); this is called parametric resonance;
- a complex conjugate pair of simple multipliers crosses the unit circle at the points $\exp(\pm j\omega)$, where $0 < \omega < \pi$, with $\mathrm{Im}(\eta(p_{cr})) \neq 0$ (Figure 11.23c); this is called combination resonance.

The system can be stabilized and reach a stable equilibrium by reversing the arrow and forcing the Floquet multipliers to enter in the unit circle.

[4] If the trajectories of the differential equation can be defined $x = \varphi(t,z)$, which is the periodic solution of $\dot{x} = f(x,p)$, the matrix $[\partial\varphi(T,z^*)/\partial z]$ is the *monodromy matrix*. The eigenvalues of the monodromy matrix are called Floquet multipliers or characteristic multipliers [47].

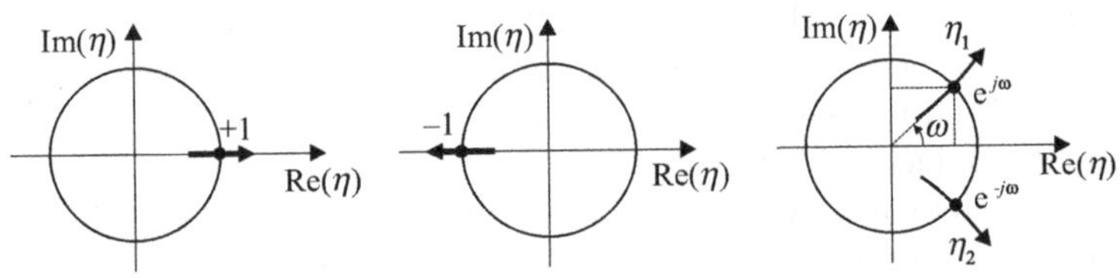

Figure 11.23. Three ways of losing stability [45].

There are three types of global bifurcations by which the loss of stability of a periodic orbit occurs [14]:

(i) *Cyclic fold bifurcation* occurs when a Floquet multiplier of the periodic orbit reaches the stability boundary at "+1" under parameter variations. In this case the global bifurcation is called saddle-node bifurcation of periodic orbits. A cyclic fold bifurcation is identified if the following attributes are observed:

- the system is operating initially at a stable periodic orbit with dampened oscillations;
- under parameter variations bifurcation may occur, the periodic orbit becomes poorly stable and the operating point starts moving away from the stable point on a trajectory that diverges while the periodic orbit disappears. The divergence is first slow then fast and oscillations are still observed as the trajectory diverges. The divergence results either in voltage collapse or in loss of synchronism.

(ii) *Period doubling bifurcation* occurs when a Floquet multiplier of the periodic orbit reaches the stability boundary at "−1" under parameter variations (see Table 11.5). This type of bifurcation is characterized by double period oscillations of the periodic orbit. A *supercritical* bifurcation occurs when a stable branch is created, whereas a *subcritical* bifurcation occurs when a branch of unstable period-doubled solution disappears at the point of bifurcation (see Table 11.5).

(iii) *Secondary Hopf (Neimark-Sacker) bifurcation* occurs when the moduli of one pair of complex Floquet multipliers reach the unit circle. Similar to the period doubling bifurcation, two subtypes of secondary Hopf bifurcations may occur, that is, subcritical and supercritical.

11.5.5 The Smallest Singular Value Technique. VSI Global Index

As shown in Section 11.3, in the critical points, located on the boundary between the domain of existence and the domain of inexistence of the steady-state solutions, the Jacobian of the system of equations (11.45) is a singular matrix. Since, from mathematical viewpoint, the singularity of an nth order matrix is observed when the singular value σ_n becomes null, it results that *the smallest singular value of the Jacobian matrix* is another way of calculating the loadability, from the current operating point to the critical point.[5]

[5] For a real symmetric matrix $[A](\equiv[J])$ the absolute values of the eigenvalues from an eigenvalue decomposition are identical to the singular values of the same matrix. This relation between the singular values obtained from a singular value decomposition and the eigenvalues of the same matrix comes from the fact that the inputs σ_i in the diagonal matrix Σ are the singular values of $[A]$, and by construction their squares are the eigenvalues of $[A]^T[A]$ (or $[A][A]^T$). If the minimum singular value is equal to zero, then the studied matrix is singular and no power flow solution can be obtained [11].

Applying *singular values decomposition* of the Jacobian matrix ($n \times n$), of the linear model (11.51) achieved by linearization of the system of equations (11.45) around the analyzed equilibrium point, gives [11]:

$$[J] = [U][\Sigma][V]^T = \sum_{i=1}^{n} \sigma_i [u_i][v_i]^T \tag{11.68}$$

where $[U]$ and $[V]$ are $n \times n$ orthonormal matrices, whose columns are the right and the left singular vectors $[u_i]$ and $[v_i]$, which are the ith columns of the unitary matrices $[U]$ and $[V]$ respectively, and $[\Sigma]$ is a diagonal matrix with

$$\Sigma(J) = \text{diag}\{\sigma_i(J)\},\ i = 1, 2, \ldots, n$$

where the singular values of $[J]$ are $\sigma_i \geq 0$ for all i. The diagonal elements in the matrix Σ are usually ordered so that $\sigma_1 \geq \sigma_2 \geq \cdots \geq \sigma_n \geq 0$.

If the $[J]$ matrix is not singular then its inverse is given by:

$$[J]^{-1} = [V][\Sigma]^{-1}[U]^T = \sum_{i=1}^{n} \sigma_i^{-1} [v_i][u_i]^T \tag{11.68'}$$

Therefore, according to equations (11.51) and (11.68′), the effect of modification of the active and reactive powers on the voltage magnitudes and angles can be assessed using the relationship:

$$\begin{bmatrix} \Delta\theta \\ \Delta V \end{bmatrix} = [J]^{-1} \begin{bmatrix} \Delta P \\ \Delta Q \end{bmatrix} = \sum_{i=1}^{n} \sigma_i^{-1} [v_i][u_i]^T \begin{bmatrix} \Delta P \\ \Delta Q \end{bmatrix} \tag{11.69}$$

The minimum singular value is a relative measure of how close the system is to the voltage collapse or singular point. Furthermore, near this collapse point, since σ_n is close to zero, equation (11.69) can be rewritten as [2]

$$\begin{bmatrix} \Delta\theta \\ \Delta V \end{bmatrix} \cong \sigma_n{}^{-1} [v_n][u_n]^T \begin{bmatrix} \Delta P \\ \Delta Q \end{bmatrix} \tag{11.69'}$$

Since σ_n^{-1} is very large, any finite ΔP, ΔQ will produce a large change in ΔV and $\Delta\theta$ where u_n gives the most sensitive direction of the change in nodal powers and v_n gives the most sensitive voltage magnitudes and angles.

If consider $\begin{bmatrix} \Delta P \\ \Delta Q \end{bmatrix} = [u_n]$ in the expression (11.69′) and taking into account that $[u]$ is orthonormal, that is, $[u_i]^T \cdot [u_i] = 1$ and $[u_i]^T \cdot [u_j] = 0$ for $i \neq j$, then:

$$\begin{bmatrix} \Delta\theta \\ \Delta V \end{bmatrix} = \sigma_n^{-1} [v_n] \tag{11.70}$$

According to equation (11.70), the following meanings were assigned to the smallest singular values $\sigma_{\min}(J)$ of the Jacobian matrix and their associated singular vectors [35]:

- $\sigma_{\min}(J)$ is a global index for the power system voltage stability, indicating how close is the current operating point to the critical operating point, characterized by the singularity of the Jacobian matrix [J];
- the components of the right singular vector $[v_n]$ denotes the sensitivities of the voltage magnitudes and angles to variations in the active and reactive powers;
- the components of the left singular vector $[u_n]$ denotes the most sensitive variation directions of the power.

Although, generally, in a power system the voltage stability is influenced by both the active and reactive powers, however, considering the tight *V–Q* coupling, instead of using the Jacobian matrix, a matrix reflecting this property, which may also be used for singular values decomposition and analysis, is of great importance. For this purpose, the $[J_{QV}]$ matrix from the linear model can be used:

$$\begin{bmatrix} \Delta P \\ \Delta Q \end{bmatrix} = \begin{bmatrix} J_{\mathrm{P}\theta} & J_{\mathrm{PV}} \\ J_{\mathrm{Q}\theta} & J_{\mathrm{QV}} \end{bmatrix} \cdot \begin{bmatrix} \Delta\theta \\ \Delta V \end{bmatrix}$$

However, given that under stressed conditions the *V–Q* coupling is critical, the use of the reduced Jacobian matrix is recommended [35]. This can be obtained starting from the above equation written as:

$$[\Delta P] = [J_{\mathrm{P}\theta}][\Delta\theta] + [J_{\mathrm{PV}}][\Delta V] \tag{11.71a}$$

$$[\Delta Q] = [J_{\mathrm{Q}\theta}][\Delta\theta] + [J_{\mathrm{QV}}][\Delta V] \tag{11.71b}$$

Considering $\Delta P = 0$ (weak *V–P* coupling) from (11.71*a*) it results that:

$$[\Delta\theta] = -[J_{\mathrm{P}\theta}]^{-1}[J_{\mathrm{PV}}][\Delta V] \tag{11.72}$$

Then, substituting [$\Delta\theta$] from (11.72) in (11.71b) gives:

$$[\Delta Q] = \left([J_{\mathrm{QV}}] - [J_{\mathrm{Q}\theta}][J_{\mathrm{P}\theta}]^{-1}[J_{\mathrm{PV}}]\right)[\Delta V] \tag{11.73}$$

or

$$[\Delta Q] = [J_{\mathrm{R}}][\Delta V] \tag{11.73$'$}$$

If the matrix $[J_R]$ is not singular, from (11.66) it results:

$$[\Delta V] = [J_{\mathrm{R}}]^{-1}[\Delta Q] \tag{11.74}$$

where

$$[J_{\mathrm{R}}] = [J_{\mathrm{QV}}] - [J_{\mathrm{Q}\theta}][J_{\mathrm{P}\theta}]^{-1}[J_{\mathrm{PV}}] \tag{11.75}$$

is the reduced Jacobian matrix obtained considering that $[J_{\mathrm{P}\theta}]$ is not singular, that is, there is no small-signal stability problem in the analyzed power system.

Thus, according to Schur's formula (11.65):

$$\det([J]) = \det([J_{P\theta}])\det([J_R]) \tag{11.65'}$$

we may conclude that the singularity of the Jacobian matrix is given by the singularity of the $[J_R]$ matrix, which is a proof of the voltage stability problems.

Therefore, *the smallest singular value of the reduced Jacobian is a global voltage stability index.*

Although the smallest singular value $\sigma_{min}(J_R)$ is a global index, indicating the proximity to voltage instability, it does not provide precise information regarding how close is the system to the critical point. In order to have global information on the position of the equilibrium point with respect to the critical surface, *the smallest singular value $\sigma_{0min}(J_R)$ for no-load operating conditions*, considered as the most stable state, is also calculated. Therefore, the global voltage stability index (VSI) is defined as:

$$\mathrm{VSI} = \frac{\sigma_{min}(J_R)}{\sigma_{0min}(J_R)} \tag{11.76}$$

Values close to zero of this index show that the system is situated in the vicinity of the critical point, while values close to 1 show that the system is far from the critical point.

11.5.6 Modal Analysis of the Reduced Jacobian Matrix

The modal analysis is a method applied for dynamic systems stability assessment based on linearization around an equilibrium point and eigenvalues and eigenvectors computation of the state Jacobian.

At a saddle-node bifurcation a real eigenvalue becomes zero and the equilibrium point disappears. The corresponding eigenvectors contain valuable information on the nature of the bifurcation, the response of the system and the effectiveness of control measures [2,49,50]:

- The right eigenvector shows the direction in state space along which the states will evolve due to the saddle-node bifurcation;
- The left eigenvector shows which states have a prominent effect on the zero eigenvalue, that is, which states are more effective in order to control the bifurcation.

Since voltage stability strongly depends on the dynamic behavior of the load and the *V–Q* response of the system, it can be shown that under favorable conditions for voltage instability, the singularity of the state matrix is given by the singularity of the Jacobian matrix of the Newton-Raphson method, while considering the effects of the static load characteristics, the armature and field limits, the on-load tap changing, the FACTS devices, and so on.

On the other hand, because the state matrix and its eigenvalues require a large computation effort, the reduced Jacobian matrix, as defined by expression (11.75) [44], is used for small-signal voltage stability assessment instead of the modal analysis.

This method is based on the quasi-symmetry of the $[J_R]$ matrix and uses the eigenvalues and eigenvectors decomposition of a matrix. Indeed, if the electrical lines and transformers resistances and the effect of phase-shifting transformers are neglected, that is, the nodal admittances matrix is symmetrical, then the $[J_R]$ matrix is also symmetrical. Therefore, all eigenvalues of this matrix are real, and the left and right eigenvectors corresponding to the same eigenvalue are identical.

11.5.6.1 The V-Q Variation Modes of the Power System. Eigenvalues decomposition of the Jacobian matrix $[J_R]$ gives:

$$[J_R] = [R][\Lambda][L] = \sum_{i=1}^{n} \lambda_i [r_i][l_i] \tag{11.77}$$

where $[R]$ is the right eigenvector matrix of $[J_R]$; $[L]$ is the left eigenvector matrix of $[J_R]$; and $[\Lambda]$ is the diagonal eigenvalue matrix of $[J_R]$.

From equation (11.77)

$$[J_R]^{-1} = [R][\Lambda]^{-1}[L] = \sum_{i=1}^{n} \lambda_i^{-1} [r_i][l_i] \tag{11.77'}$$

Substitution of (11.77′) in (11.74) yields the voltages variations in terms of the reactive powers demand variations $[\Delta Q]$:

$$[\Delta V] = [R][\Lambda]^{-1}[L][\Delta Q] = \sum_{i=1}^{n} \lambda_i^{-1} [r_i][l_i][\Delta Q] \tag{11.78}$$

Each eigenvalue λ_i and the associated eigenvectors $[r_i]$ and $[l_i]$, forms a *V–Q* variation mode of the power system.

Assuming that eigenvectors are normalized so that $[l_i]^T[v_i] = 1$, and that they are orthogonal to each other (for different eigenvalues), $[l_i]^T[v_j] = 0$, results $[L] = [R]^{-1}$.

Then, the equation (11.78) can be written as:

$$[L][\Delta V] = [\Lambda]^{-1}[L][\Delta Q]$$

or

$$[v] = [\Lambda]^{-1}[q] \tag{11.79}$$

where $[v] = [L][\Delta V]$ is the *voltage modal variations* vector; and $[q] = [L][\Delta Q]$ is the *reactive power modal variations* vector.

Matrix $[\Lambda]^{-1}$ from equation (11.79) is diagonal, as compared to matrix $[J_R]^{-1}$ from equation (11.74) which is not diagonal, and therefore the expression (11.79) forms an uncoupled first order system of equations.

The voltage variation in the ith mode is

$$v_i = \frac{1}{\lambda_i} q_i \quad i = 1, 2, \ldots, n_c \tag{11.80}$$

We can see that each voltage variation mode is equal to the ratio of the reactive power variation mode q_i to the eigenvalue λ_i. Hence, the eigenvalue λ_i represents an index for the degree of voltage stability, being in fact a sensitivity of the modal voltage variation to the reactive power variation.

Thus:

- if $\lambda_i > 0$, the system is stable in the *i*th *V–Q* variation mode because the modal voltage v_i and the modal reactive power q_i have the same variation direction, that is, a reactive power injection causes an increase in the voltage level;
- if $\lambda_i < 0$, the system is unstable in the *i*th *V–Q* variation mode because the modal voltage v_i and the modal reactive power q_i have opposite variation directions, that is, a reactive power injection causes a decrease in the voltage level;
- if $\lambda_i = 0$, the modal voltage v_i collapses because any modal reactive power modification q_i leads to an infinite variation of the modal voltage v_i.

Therefore, *the condition for voltage stability of a power system is satisfied if all the eigenvalues of the reduced Jacobian are positive.*

At the first inspection we might say that the above-mentioned remarks are not in accordance with the conclusion drawn in the sensitivities analysis method, that is, the system is stable if all $S_{V_cQ_c}$ sensitivities are negative. To eliminate this apparent ambiguity, the connection of the $S_{V_cQ_c}$ sensitivities with the eigenvalues of the matrix $[J_R]$ are further analyzed. For this, let us consider a 1 MVAr variation in the load reactive power in a *PQ* type node *k*. Since the nodal power is defined as the difference between the demanded and the generated powers, we can write as follows:

$$\Delta Q_k = \Delta Q_{gk} - \Delta Q_{ck} = -1\,\text{MVAr}$$

and, hence,

$$[\Delta Q] = [0,\ \ldots,\ -1,\ \ldots,\ 0]^t = -e_k$$

Substitution of $[\Delta Q] = -e_k$ in equation (11.78) gives $[\Delta V] = -\sum_{i=1}^{n} \frac{l_{ik}}{\lambda_i}[r_i]$, that is, $\Delta V_k = -\sum_{i=1}^{n} \frac{r_{ki} l_{ik}}{\lambda_i}$.

Therefore, the sensitivity of voltage at the *k* node to the reactive power demand variation, in the same node, is:

$$S_{V_k Q_{ck}} \cong \frac{\Delta V_k}{\Delta Q_{ck} = 1} = -\sum_i \frac{r_{ki} l_{ik}}{\lambda_i} \tag{11.81}$$

where r_{ki} is the *k*th element of the right eigenvector $[r_i]$; and l_{ik} is the *k*th element of the left eigenvector $[l_i]$.

Since the $[J_R]$ matrix is quasi-symmetrical, the eigenvectors $[r_i]$ and $[l_i]$ are identical and, hence, the $r_{ki} l_{ik}$ products are positive. Hence, the sensitivities are negative if the eigenvalues are positive and vice versa.

Note that according to(11.81) the sensitivity value provides information regarding the combined effect of all the *V–Q* variation modes and, in consequence, it cannot detect peculiar voltage collapse modes, which can result through modal analysis of the $[J_R]$ matrix only.

The eigenvalues amplitude provides a relative measure of the proximity to voltage instability rather than an absolute one, because of the problem nonlinearities. Analogy can be made with the use of the damping coefficient in the small oscillations angular stability studies, which, although it provides the degree of damping, it is not an absolute measure of the stability reserve.

The modal analysis of the reduced Jacobian is a technique used either individually for the analysis of a certain operating point or in combination with the loading margin computation methods. Therefore, in computing a loading margin on a certain stress path, the modal analysis is applied for every operating point on the calculated trajectory.

Application of modal analysis allows evaluating the degree of system stability, or in what amount the load or the transmitted power can increase under stable conditions, while in the critical situation when the system reaches the voltage stability critical point, it may help for identification of the critical areas and the elements connected to its nodes which contribute to voltage instability.

11.5.6.2 Definition of Participation Factors in Voltage Stability Analysis. In order to identify the weak points, from voltage viewpoint, of the power system and to establish the required preventive actions, the participation factors of *the buses*, *the branches* (electric lines and transformers) and *the generators* to the *V–Q* variation modes are then calculated.

- ***The bus participation factors*** The relative participation of the k bus/node to the ith *V–Q* variation mode is given by the participation factor [44]:

$$A_{ki} = r_{ki} l_{ik} \tag{11.82}$$

From equation (11.81) it results that A_{ki} determines the contribution of the λ_i eigenvalue to the *V–Q* sensitivity to the k node.

The bus participation factors determine the areas associated with each variation mode. The sum of all the bus participation factors for each variation mode is equal to unity since the left and right eigenvectors are normalized. The value of the participation factor of a node shows the effect of the corrective actions taken in that node in damping the *V–Q* variation mode.

Generally, two types of variation modes are defined [44]:

- *the localized variation mode*, characterized by the existence of few nodes with high-value participation factors, while all other nodes have participation factors close to zero. Such a localized mode appears if one node supplying a load zone is connected to a strong electrical grid through a long transmission line;
- *the dispersed mode*, characterized by the existence of a large number of nodes with small, but close, participation factors, while the other nodes have participations close to zero; this means that the variation mode is not localized. An unlocalized mode is identified when the load increases in an area of a large power system and there is no additional reactive power support available for that area.

To identify the weak points of the power system and to establish the required preventive actions, besides the node participation factors to the *V–Q* variation modes, it is necessary to also determine the branch and the generator participation factors.

In this regard, for the ith *V–Q* variation mode, all components of the modal reactive power variations vector are considered equal to zero except for the ith component considered equal to 1, that is, $[q] = [0, \ldots, 1, \ldots, 0]^T = e_i$ (the ith vector of the canonical base). Under these circumstances, from relationship $[q] = [L][\Delta Q]$, since $[L]^{-1} = R$, it results:

$$\left[\Delta Q^{(i)}\right] = [L]^{-1}[q] = [R][q] = [r_i] \tag{11.83}$$

where $[r_i]$ is the ith right eigenvector of the $[J_{\mathrm{R}}]$ matrix. Assumes that for all the right eigenvectors, we have $\sum_j r_{ji}^2 = 1$.

From equation (11.78) yields the voltage magnitude variations vector:

$$\left[\Delta V^{(i)}\right] = \frac{1}{\lambda_i}\left[\Delta Q^{(i)}\right] \tag{11.84}$$

and, finally, from expression (11.72) yields the voltage angle variations vector:

$$\left[\Delta\theta^{(i)}\right] = -[J_{\mathrm{P}\theta}]^{-1}[J_{\mathrm{PV}}]\left[\Delta V^{(i)}\right] \tag{11.85}$$

Given the variations $\left[\Delta\theta^{(i)}\right]$ and $\left[\Delta V^{(i)}\right]$ corresponding to the ith V–Q variation mode, the variations of the reactive power losses in each branch of the transmission network, as well as the variations of the reactive powers supplied by the generators, which did not reach their reactive power limit, can be determined.

- ***Branch participation factors***

The k–jth branch participation factor to the ith V–Q variation mode is defined as the ratio of the reactive power losses variation in the k–j branch, $\Delta Q_{kj}^{(i)}$, to the maximum variation in the reactive power losses in all the power grid branches, caused by a reactive power load variation in the system, that is:

$$A_{kj}^{(i)} = \frac{\Delta Q_{kj}^{(i)}}{\max\limits_{(i,j)\in l}\left\{\Delta Q_{kj}^{(i)}\right\}} \tag{11.86}$$

The $A_{kj}^{(i)}$ participation factors allow identification, for each ith V–Q variation mode, of the branches with the highest reactive power losses as response to an incremental variation in the reactive load. The branches of high participation factors are either weak, from electrical viewpoint, or highly loaded electrical lines. The branch participation factors can be used to establish the corrective actions in preventing voltage instability problems and for contingencies screening.

- ***Generator participation factors***

The relative participation of the mth generator to the ith V–Q variation mode is given by the participation factor:

$$A_{mi} = \frac{\Delta Q_m^{(i)}}{\max\limits_{g\in n_g}\left\{\Delta Q_g^{(i)}\right\}} \tag{11.87}$$

that is by the ratio of the reactive power variation of the mth generator, $\Delta Q_m^{(i)}$, to the maximum variation of the reactive powers produced by all the generators from the system to a reactive power load variation.

The A_{mi} participation factors allow identification, for each ith V–Q variation mode, of the way in which each generator participates in supplying additional reactive power load. These factors can be employed in the reactive power reserves planning, to maintaining an adequate voltage stability level.

For the practical implementation of the modal analysis method of the reduced Jacobian matrix $[J_R]$, the following remarks should be taken into account [44]:

- Equation (11.80) reveals that the most dangerous V–Q variation mode is obtained for the smallest eigenvalue, $\lambda_i^{\min}$, of the reduced Jacobian $[J_R]$. However, calculating the smallest eigenvalue only is not sufficient to have a correct overview of the power system stability. There can be several modes associated with other parts of the power system that can induce instability as the operating conditions are worsened;
- Since the reduced Jacobian matrix is quasi-symmetrical, its eigenvalues are identical with the associated singular values, and thus it is preferable to directly calculate the singular value, $\sigma_{\min}$, and its associated with the left and the right singular vectors rather than calculating the smallest eigenvalue, $\lambda_i^{\min}$, and the left and the right eigenvectors.

11.6 VOLTAGE INSTABILITY COUNTERMEASURES

Countermeasures against power system instability and in particular against voltage instability and collapse are taken from the design stage to the real-time operation. The aim is to ensure adequate voltage stability margin. In the planning activity or in real-time operation, this can be done by appropriate coordination of voltage control equipments, including scheduling of reactive power resources. By midterm and long-term design and planning activity, a power system should be able to meet the load supply and power transfer requirements under the actual economical power market conditions. However, if the available resources cannot ensure the required stability margin, technical measures must be taken such as limitation of power transfers or starting up additional generating units [44]. At the power system planning stage, transmission network reinforcement would be the first choice, but under the actual environmental constraints and control performance requirements the FACTS devices became more effective solutions.

Voltage instability and collapse countermeasures can include a large number of options for the real-time operation, for example, blocking of tap changers, transformers reverse control, switching of capacitor banks, generation redispatching, changing the reference voltage at generators and pilot buses, load shedding, or temporary reactive power overload of generators.

11.6.1 Some Confusions

The voltage instability is a nonlinear phenomenon that evolves in stages, associated with different physical and mathematical behavior. The effectiveness of the countermeasures mentioned earlier for the real-time operation is different in the different stages of voltage instability. One concern is the effectiveness of capacitor switching and load shedding when the operating point reaches the uncontrollable zone of the characteristic given in Figure 11.4, characterized by voltage values below the critical point, favorable for voltage collapse.

From physical point of view, one would expect the bus voltage magnitude to increase when a capacitor is switched on, or some load is shed, at that bus, under any operating condition [9]. As explained in Section 11.3.2, the compensation actions that normally lead to voltage increase when the operating point is situated in the controllable zone have adverse consequences on the voltage when the operating point is situated in the uncontrollable zone (see Figure 11.4).

A practice in the voltage stability analysis is to calculate the voltage sensitivities dV/dP and dV/dQ from the power flow Jacobian. These sensitivities change the sign when the operating point moves from the security zone to the uncontrollable zone. Therefore, the sensitivities obtained by load flow calculation for the normal operation are not valid when the power system encounters instability conditions. This is because a steady-state Jacobian calculation assumes static and constant MVA load model, which is not the case when the voltage starts to decrease. As shown in [8], in voltage instability analysis, representing constant MVA loads by a static model can lead to erroneous and, often, misleading results because a constant MVA load is not a static load.

Voltage instability is largely influenced by the load characteristics. In areas with large amounts of static loads, for example, resistive loads, operation in the low-voltage region, on characteristic from Figure 11.4, would be possible, and no corrective measures may be necessary if the voltage level is acceptable [9]. For loads with constant MVA (self-restoring) characteristics, advanced shunt compensation devices, for example, SVC (static VAr compensator) or STATCOM, can help the power system to operate in this region in stable conditions. Furthermore, in system with large amounts of such loads instability is characterized by slow dynamics and there is a significant time available for corrective actions to avoid a voltage collapse.

11.6.2 Load Shedding: An Emergency Measure

Load shedding/disconnection can be an effective corrective measure against voltage stability problems in power system operation. However, load shedding should be considered as an in extremis action, only when it is the last emergency corrective measure available to avoid voltage collapse. It is preferable to "sacrifice" a certain part of the load under predefined protection scheme and "save" the system from blackout, which can have a significant social and economical impact.

The following cases assume that a part of the load is of resistive type (static) and no special control devices (e.g., SVCs or STATCOMs) are available.

- *Case a.* Assume that, following a major change in the power system, the operating point moves on the low-voltage side of the system V–P characteristic (Figure 11.24a).

  The static load characteristic (solid line) intersects the V–P characteristic in two points, A and B. It can be shown that point A is stable and point B in unstable by analyzing the effect of load shedding. By shedding some load, a new load

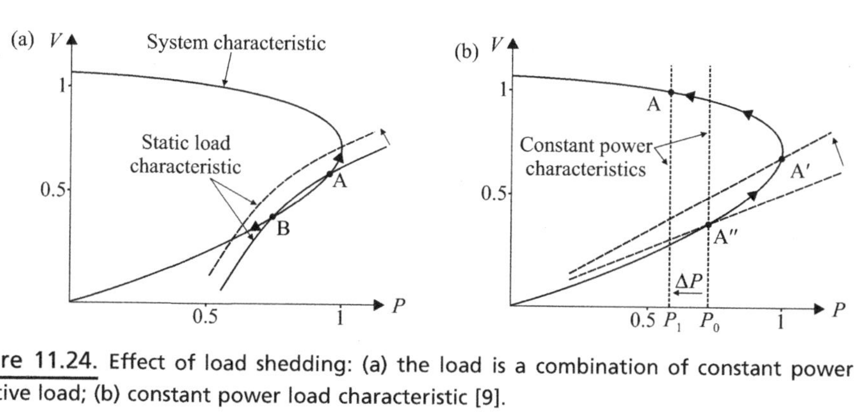

Figure 11.24. Effect of load shedding: (a) the load is a combination of constant power and resistive load; (b) constant power load characteristic [9].

characteristic is formed (dashed line), point A moves upward along the *V–P* characteristic in the region of attraction helping the voltage to increase, while point B moves downward leading to voltage collapse.

- *Case b.* Assumes that the *load* can be modeled by a *constant power characteristic* and the system is operating on the lower voltage side of the system *V–P* characteristic (Figure 11.24b). For a certain power transfer P_0 there is one possible operating point, indicated by A″. In the absence of special controls, the operating point will slide down on the postdisturbance *V–P* characteristic, except temporarily in the case of a resistive load, and the voltage may collapse. However, if an amount of load ΔP is timely shed, the operating point will move in point A′, located in the region of attraction (see Section 11.4.1; Figure 11.9), at the intersection with a new constant impedance characteristic. The system operation will finally settle in the stable equilibrium point A, corresponding to the new load P_1.
- *Case c.* Consider that the power system operates initially in the point A, which corresponds to a constant power load P_0. Following a large disturbance the power system reduces to the postdisturbance characteristic and the operating point moves instantly in point A' (Figure 11.25). Because of the low voltage in the point A', the consumer could try to maintain the constant power; this is done by demanding more current which in effect generates more voltage drop, and the operating point moves downward on the postdisturbance *V–P* characteristic. If no action is timely taken, the voltage will continue to decrease and finally to collapse [9].

The power system can be saved from a collapse if the operating point remains in the region of attraction. The maximum system loadability for the postdisturbance conditions is $P_1 < P_0$. To restore the constant power load, while bringing the power system into a stable equilibrium point, the minimum load to be shed is $(P_0 - P_1)$.

In order to keep the load shedding requirements to a minimum, load must be shed promptly. Any delay in triggering the load shedding would results in a larger amount of load necessary to be shed in order to bring the system in the region of attraction then in a stable equilibrium point. The rate of drop of voltage depends on the type of load (see Section 11.2.2.3). In systems with large amounts of static loads, the voltage drops at very slow rate; whereas in systems with loads characterized by fast response, for example, motor loads, the voltage drops at a much faster rate. The rate of drop of voltage would, therefore, serve as an indicator for an immediate need for load shedding.

For static load type, with characteristic similar to that from Figure 11.24b, the operating point can be at a new equilibrium point after the disturbance and load shedding may not be necessary. In this case, other measures can be taken in order to bring the voltage within acceptable limits if necessary.

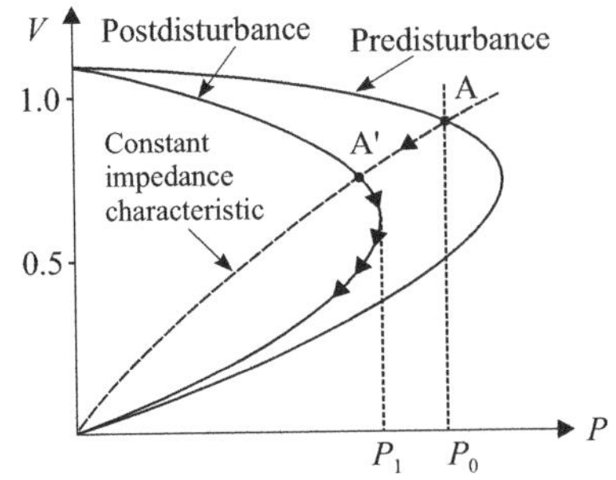

Figure 11.25. Minimum load needed to be shed after a disturbance in the case of constant MVA load [9].

11.6.3 Shunt Capacitor Switching

In power systems with large load areas and lack of synchronous generators, the shunt capacitors can be efficient solutions to the voltage stability problems. To avoid additional loading of transmission paths and also to efficiently use the synchronous generators for active power generation, the reactive power should be supplied locally with special compensation equipments. The switching shunt capacitors can successfully be employed to restore the system voltages after a significant change in the power system (Figure 11.26a).

Consider that the power system operates initially in point A, at the intersection between the V–P system characteristic and the constant power load characteristic P_0 (Figure 12.26b). Following a major disturbance, a postdisturbance system characteristic is formed and the operating point moves instantaneously to point M along the constant impedance characteristic (dashed line). The load tries to restore to constant power and the voltage starts to decrease. Because no intersection point exists between the postdisturbance characteristic and the initial load demand characteristic P_0, the system collapse will result if no action will be timely taken.

To bring out the system in a stable equilibrium point, appropriate reactive power compensation by switching shunt capacitors on at or near the load bus must be done. This action is not stringent if a large part of the load is static. However, if acting too late, no amount of reactive power can restore the system stability.

If timely contribution with appropriate amount of reactive power is provided by shunt capacitors, say in point A', the system operation will temporarily move in point B' in the region of attraction, at the intersection between the postdisturbance characteristic with capacitor and the new constant impedance characteristic. From here, because the load power is greater the load demand P_0, the system moves in the new stable equilibrium point B. However, if the reactive power support is delayed too much and the voltage become to low, for example, in point A," no amount of reactive power can prevent the system from voltage collapse [9].

11.6.4 Extending the Voltage Stability Limit by FACTS Devices

With appropriate reactive power support, larger power transfers can be possible as shown in Figure 11.5. However, when the system operates below the critical point of the V–P curve, stable operation is possible only by support of fast and continuous acting equipments, such as SVC or STATCOM devices [9].

In general, voltage and stability limits (see Section 8.4.2) are more restrictive than the thermal limit (MVA rating) for the power transfer on the transmission lines. Series

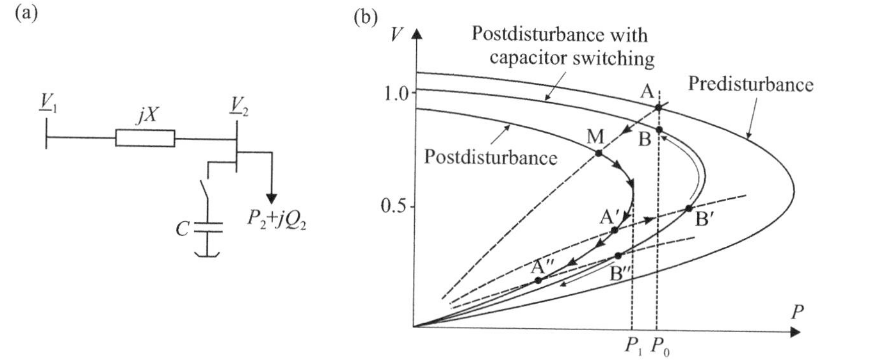

Figure 11.26. Restoration of system voltages by shunt (a) capacitors following a disturbance.

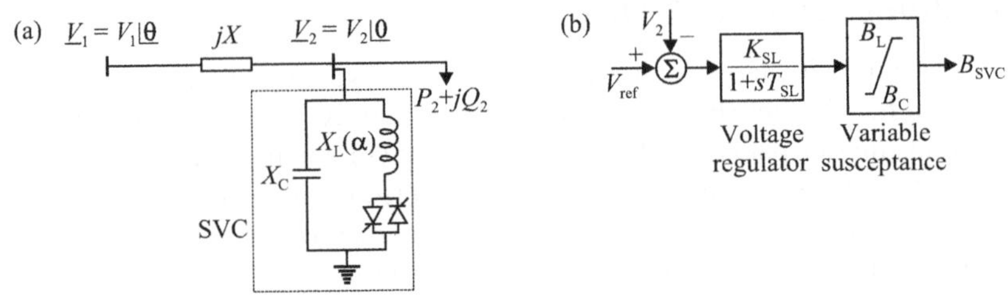

Figure 11.27. One-line diagram (a) and a simple first-order delay model (b) of a SVC.

compensation is much more effective than shunt compensation for increasing the voltage stability limit.

Shunt Compensation with SVC (Static VAr Compensator). To analyze the influence of an SVC device on the voltage stability limit let us consider again a radial transmission system supplying a unity power factor load, with voltage controlled by a SVC (Figure 11.27a). An SVC acts by increasing the capacitive susceptance as the voltage drops from a set value and vice versa [9].

The control logic of the SVC for voltage control may be described by a simple first-order delay model (Figure 11.27b).

$$T_Q \frac{dB_{SVC}}{dt} = V_{ref} - V_2 \tag{11.88}$$

where $B_{SVC} = B_L(\alpha) + B_C$ is the SVC susceptance; $B_L(\alpha)$ is the inductive component of the SVC device, controlled by means of the firing angle α; B_C is the capacitive component of the SVC device; $T_Q(= T_{SL})$ is a time constant.

For a unity power factor load, that is, $B_2 = 0$, the load model is given by

$$T_L \frac{dG_2}{dt} = P_0 - V_2^2 G_2 \tag{11.89}$$

where $G_2 = P_0/V_2^2$ is the load conductance; P_0 is the power set point; and T_L is the time constant of the load.

The powers transferred to the load bus are as follows:

$$P_2 = V_2^2 G_2 = \frac{V_1 V_2}{X} \sin\theta \tag{11.90a}$$

$$Q_2 = 0 = \frac{V_1 V_2}{X} \cos\theta - \frac{V_2^2}{X} + V_2^2 B_{SVC} \tag{11.90b}$$

Linearization of equations (11.88–11.90) and eliminating the nonstate variables, the state-space model is obtained as [9]:

$$\begin{bmatrix} \Delta\dot{B}_{SVC} \\ \Delta\dot{G}_2 \end{bmatrix} = [A_{SVC}] \begin{bmatrix} \Delta B_{SVC} \\ \Delta G_2 \end{bmatrix}$$

where

$$[A_{SVC}] = \begin{bmatrix} a_{11} = -\dfrac{V_2}{2T_Q G_2} \sin 2\theta & a_{12} = \dfrac{V_2}{T_Q G_2} \sin^2\theta \\ a_{21} = -\dfrac{V_2^2}{T_L} \sin 2\theta & a_{22} = -\dfrac{V_2^2}{T_L} \cos 2\theta \end{bmatrix}$$

The system is stable if the following conditions hold

$$-(a_{11}+a_{22})>0 \tag{11.91a}$$

and

$$\det(A_{\mathrm{SVC}})>0 \tag{11.91b}$$

The first condition (11.91a) is automatically satisfied, provided that the second condition (11.91b) is satisfied.

Starting from (11.91a), the stability condition can be rewritten as follows:

$$\frac{V_2}{2T_\mathrm{Q}G_2}\sin 2\theta+\frac{V_2^2}{T_\mathrm{L}}\cos 2\theta>0 \Rightarrow \frac{V_2}{2T_\mathrm{Q}G_2}\tan 2\theta+\frac{V_2^2}{T_\mathrm{L}}>0 \Rightarrow$$

$$V_2T_\mathrm{L}\tan 2\theta+2T_\mathrm{Q}G_2V_2^2>0 \Rightarrow \tan 2\theta>-\frac{2T_\mathrm{Q}G_2V_2}{T_\mathrm{L}} \tag{11.92}$$

Taking into account equations (11.90a) and (11.90b), yields

$$\left.\begin{array}{r}\sin\theta=\dfrac{V_2}{V_1}XG_2\\ \cos\theta=\dfrac{V_2}{V_1}(1-B_{\mathrm{SVC}}X)\end{array}\right\}\Rightarrow \tan\theta=\frac{XG_2}{1-B_{\mathrm{SVC}}X} \tag{11.93}$$

Using the formula to calculate the tangent of the double angle, equation (11.92) becomes

$$\frac{2\tan\theta}{1-\tan^2\theta}>-\frac{2T_\mathrm{Q}G_2V_2}{T_\mathrm{L}}$$

Replacing tan θ by the expression given in (11.93) gives the final stability condition, that is

$$1-\tan^2\theta>-\frac{T_\mathrm{L}}{T_\mathrm{Q}V_2}\frac{X}{1-B_{\mathrm{SVC}}X} \tag{11.94}$$

Since, in general, the time reaction of the SVC device is much smaller than the load restoration time, that is, $T_\mathrm{Q}\ll T_\mathrm{L}$, the voltage stability limit can be extended theoretically to $\theta=90°$, which may correspond to operation on the lower portion of the V–P characteristic.

The maximum active power that can be transferred to the receiving end of a transmission line (Figure 7.1) at a given power factor cos φ, in the absence of compensation, is as follows:

$$P_{\max}=\frac{V_1^2}{2X}\frac{1-\sin\varphi}{\cos\varphi} \tag{11.95}$$

to which the critical voltage it corresponds:

$$V_{2,\mathrm{cr}} = \frac{V_1}{\sqrt{2}}\sqrt{\frac{1-\sin\varphi}{\cos^2\varphi}} \tag{11.96}$$

For unity power factor, $\cos\varphi = 1$, and $\varphi = 0$, the above expressions become

$$P_{\max} = \frac{V_1^2}{2X} \tag{11.95$'$}$$

$$V_{2,\mathrm{cr}} = \frac{V_1}{\sqrt{2}} \tag{11.96$'$}$$

The active power transfer on a transmission line can also be written as:

$$P_e = \frac{V_1 V_2}{X}\sin\theta \tag{11.97}$$

At the limit, the receiving-end voltage is $V_2 = V_{2,\mathrm{cr}}$. From voltage stability point of view, the maximum voltage phase shift can be obtained by equating the powers from (11.95′) and (11.97), which gives

$$\frac{V_1^2}{2X} = \frac{V_1}{X}\frac{V_1}{\sqrt{2}}\sin\theta \Rightarrow \sin\theta = \frac{\sqrt{2}}{2} \Rightarrow \theta = 45^\circ$$

We may, thus, conclude that, for unity power factor load, the point of maximum loadability on the *V–P* characteristic, which is a static bifurcation point, corresponds to $\theta = 45^\circ$. For values $\theta > 45^\circ$, the operating point moves on the lower portion of the *V–P* characteristic where rapid reactive power support is required from a shunt device. However, if the load recovers very fast to constant power, that is, $T_L \rightarrow 0$, no additional improvement in the stability limit can be obtained, and instability occurs at $\theta = 45^\circ$.

For other power factors, $\cos\varphi \neq 1$, equating the expressions from equations (11.95) and (11.97), and taking into account (11.96), we achieve

$$\frac{V_1^2}{2X}\frac{1-\sin\varphi}{\cos\varphi} = \frac{V_1}{X}\left(\frac{V_1}{\sqrt{2}}\sqrt{\frac{1-\sin\varphi}{\cos^2\varphi}}\right)\sin\theta \Rightarrow \sin\theta = \sqrt{\frac{1-\sin\varphi}{2}}$$

In this case, for lagging power factors, the maximum loadability is achieved for $\theta < 45^\circ$ [9].

With appropriate reactive power support at the receiving end of the line, for example, by an SVC device, the maximum transmissible power can be extended to values corresponding to $\theta = 90^\circ$. In the absence of continuous voltage control, operation at values $\theta > 45^\circ$ (at unity power factor) is not desirable.

Figure 11.28 illustrates the influence of an SVC device on the *V–P* characteristic. In the absence of reactive compensation (Figure 11.28, curve *a*) the maximum power that can be transferred to the load is limited to $P_{1,\max}$. Any load demand beyond this value is associated with the voltage collapse. Provision of an SVC at the load bus will, in terms of

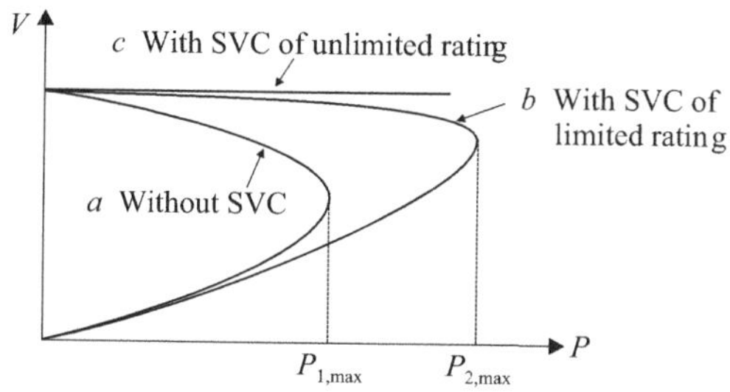

Figure 11.28. Influence of shunt compensation by SVC on the *V–P* characteristic.

SVC rating range, help smoothing the upper portion of the *V–P* characteristic (Figure 11.28, curve *b*), and thus easily maintaining the voltage within normal limits. In this case, the maximum transmissible power can be extended up to $P_{2,max}$. Theoretically, if the SVC has unlimited rating (Figure 11.28, curve *c*) it is possible to hold the voltage constant for any load demand.

Series Compensation with TCSC (Thyristor Controlled Series Capacitors). Voltage stability limit can also be expanded by *series compensation* using various capacitor-based devices. Mechanically switched capacitors have been classically used on long lines in order to allow more active power to be transferred while meeting the stability conditions. Their major disadvantages are the large reaction time, compensation in a discrete domain and the switching transients generates. With the advent of power electronics, new devices have been employed, for example, TCSC, which are capable to continuous control the active power flow on a transmission line to desired values. The capacitive reactance of TCSC compensates the inductive reactance of the transmission line thus allowing a larger amount of power flow.

To show how series compensation can contribute to extending the voltage stability limit, consider the classical example of a transmission line, with reactance X_L, supplying a load of impedance $\underline{Z}$, as shown in Figure 11.29a. A capacitor bank is series connected on the line, having the capacitive reactance X_C sized for a percentage of the line reactance X_L.

Figure 11.29c illustrates the *V–P* characteristic for a constant power factor load, one without series compensation, $X_C = 0$, and two levels of series compensation at 50% and 75%, respectively. The "nose-point," corresponding to the voltage stability limit, moves to the right as the compensation level increases. While mechanically switched capacitors provide compensations in steps, a TCSC (Figure 11.29b) can provide a continuous control

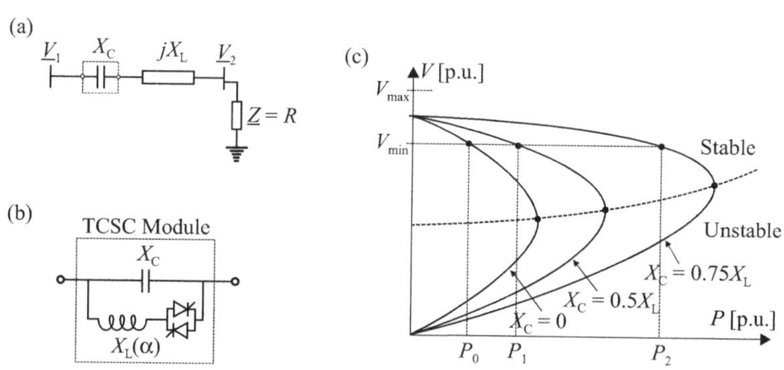

Figure 11.29. Extending the voltage stability limit of a radial transmission line by series capacitive compensation.

of the capacitive reactance. Furthermore, as illustrated in Figure 11.29c, compensation by TCSC helps maintaining the receiving-end voltage in acceptable limits and preventing voltage collapse [5]. Figure 11.29c is analogous to Figure 11.28, where compensation in the sample radial system is provided by shunt compensation. Both solutions can effectively increase the voltage stability limit, but in the case of shunt compensation this is done by reactive power supply, while in the series compensation the capacitance of the series device cancels a part of the inductive reactance of the transmission line.

11.6.5 Countermeasures Against the Destabilizing Effect of the Load Tap Changer

Taking into account the aspects discussed in Section 11.4.2.3 regarding the influence of on-load tap changing on the voltage stability, in the next lines are mentioned the most used methods in power system operation as countermeasures against the destabilizing effect of the on-load tap changing actions are as follows:

(i) *Blocking the Tap Changer on the Actual Tap Position.* This is one of the simplest and most frequently used actions in emergency situations and consists in blocking the transformer on the actual tap position, in case in which the voltage level in the transmission network decreases under a certain reference value. However, this method has the disadvantage that it cannot overcome other load restoration actions in progress, and the voltage level can continue to decrease both in the transmission and in the distribution systems.

(ii) *Switching on a Pre-Established Operation Tap.* This action consists in switching the transformer on a pre-established tap, based on simulations of contingencies with the highest probability of occurrence. A compromise has to be made, in this case, as regards choosing the operating tap, since only one position is not sufficient to withstand the various contingencies that may occur in power system operation.

(iii) *Decreasing the Reference Voltage Value.* To prevent voltage problems, the system operator may modify the reference value of the voltage $V_{2,\text{sch}}$. This method is identical with the previous one except for the difference that a voltage value is defined instead of an operating tap.

(iv) *Shifting to Inverse Logic.* This method assumes monitoring the voltage both on the primary and secondary sides of the transformer. The aim is to maintain an appropriate voltage level in the transmission network, which may allow the system operator to take the necessary measures to prevent the voltage instability phenomenon [2].

11.7 APPLICATION[6]

Perform the voltage stability analysis for the simple electric power system illustrated in Figure 11.30, consisting of a load area $(P_c + jQ_c)$ supplied from an infinite power system through a double-circuit transmission line and a step-down transmission substation.

The parameters of the power system components are as follows.

[6] Notice that V denotes the phase-to-phase voltage. In some figures, V may be represented as phase-to-ground quantity, but expression in per unit will make no difference to the phase-to-phase voltage.

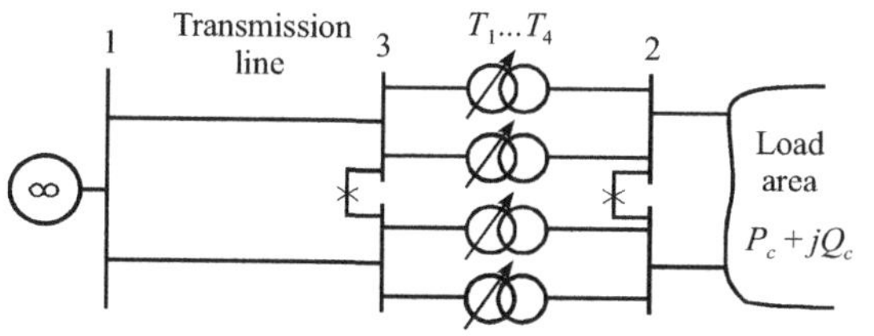

Figure 11.30. The one-line diagram of the electric power system.

- Transmission lines parameters: nominal voltage $V_n = 400\,\text{kV}$; resistance $R = 5.6\,\Omega$, reactance $X = 52.6\,\Omega$, conductance $G_0 \cong 0\,\mu S$; susceptance $B_0 = 526\,\mu S$, circuits. All these parameters are per one circuit.
- Substation transformers parameters:

Rated power	$S_n = 250\,\text{MVA}$
Rated low (fixed) voltage	$V_{nf} = 121\,\text{kV}$
Rated high (regulated) voltage	$V_{nr} = 400\,\text{kV}$
No-load (open circuit) test losses	$\Delta P_0 = 180\ \text{kW}$
No-load test current	$i_0 = 0.45\,\%$
Full-load (short circuit) test losses	$\Delta P_{sc}^{nom} = 780\,\text{kW}$
Short-circuit test voltage	$u_{sc} = 16\%$
Regulation	± 8 taps $\times\ \Delta V_t (= 1.56\,\%)$
Actual operating tap	Tap $= -2$

- Load demand: active power $P_c = 700\,\text{MW}$; reactive power $Q_c = 300\,\text{MVAr}$.

Solution

For simplicity, the calculations are performed in per unit, and the base power is $S_b = 100\,\text{MVA}$.

(1) The Equivalent Circuit. Representing the transmission line by Π quadripole, and the transformers by Γ quadripole, having the parameters referred to the regulated winding voltage (primary voltage) in series with the transformation operator N, obtain the equivalent circuit illustrated in Figure 11.31.

Since the substation parameters are represented on the transmission line side, all parameters are referred to the base impedance

$$Z_{b1} = \frac{1}{Y_{b1}} = \frac{V_{b1}^2}{S_b} = \frac{400^2}{100} = 1600\,\Omega$$

where the base voltage is taken equal to the transmission line voltage $V_{b1} = V_{n1} = 400\,\text{kV}$.

Parameters of the transmission line equivalent circuit

$$\underline{y}_{13} = \frac{1}{\underline{Z}_L/n_c} Z_{b1} = \frac{Z_{b1}}{(1/n_c)(R + jX)} = \frac{1600}{(1/2)(5.6 + j52.8)} = 6.3564 - j59.9319\,\text{p.u.}$$

$$\underline{y}_{130} = \underline{y}_{310} = n_c \frac{1}{2} \frac{\underline{Y}_{L0}}{Y_{b1}} = n_c \frac{1}{2}(G_0 + jB_0) Z_{b1} =$$
$$= 2\frac{1}{2}(0 + j526)10^{-6} \times 1600 = 0 + j0.8416\,\text{p.u.}$$

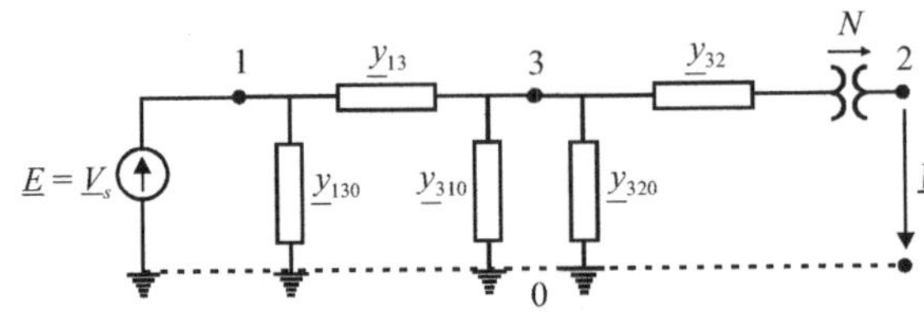

Figure 11.31. Equivalent circuit of the studied network.

where $n_c = 2$ is the number of parallel circuits of the transmission line.

Parameters of the power transformers and transmission substation equivalent circuit.

The actual tap of the step-down transformers is tap $= -2$, and all parameters are referred to the primary voltage, that is, the regulated voltage:

$$V_r = V_{nr}\left(1 + \text{tap}\frac{\Delta V_t}{100}\right) = 400 \times \left(1 - 2\frac{1.56}{100}\right) = 387.52\,\text{kV}$$

- Series parameters

$$R_T = \Delta P_{sc}^{nom}\frac{V_r^2}{S_{nT}^2}10^{-3} = 780\frac{387.52^2}{250^2}10^{-3} = 1.8741\,\Omega$$

$$Z_T = \frac{u_{sc}}{100}\frac{V_r^2}{S_{nT}} = \frac{16}{100}\frac{387.52^2}{250} = 96.1099\,\Omega$$

$$X_T = \sqrt{Z_T^2 - R_T^2} = 96.0916\,\Omega$$

$$\underline{Z}_T = R_T + jX_T = 1.8741 + j96.0916\,\Omega$$

- Shunt parameters

$$G_{T0} = \frac{\Delta P_0}{V_r^2}10^{-3} = \frac{180}{387.52^2}10^{-3} = 1.1986 \times 10^{-6}\,\text{S}$$

$$Y_{T0} = \frac{i_0}{100}\frac{S_{nT}}{V_r^2} = \frac{0.45}{100}\frac{250}{387.52^2} = 7.4914 \times 10^{-6}\,\text{S}$$

$$B_{T0} = \sqrt{Y_{T0}^2 - G_{T0}^2} = 7.3949 \times 10^{-6}\,\text{S}$$

$$\underline{Y}_{T0} = G_{T0} - jB_{T0} = (1.1986 - j7.3949) \times 10^{-6}\,\text{S}$$

- Transformer ratio

$$N = \frac{(V_{nr}/V_{b1})(1 + \text{tap} \times \Delta V_t/100)}{V_{nf}/V_{b2}} = \frac{400(1 - 2 \times 1.56/100)}{121}\frac{110}{400} = 0.8807$$

Notice that the secondary (fixed) voltage is referred to the base voltage $U_{b2} = U_{n2} = 110\text{kV}$.

In per unit, the per unit admittances of the transmission substation equivalent circuit, taking into account that the number of parallel transformers is $n_T = 4$, achieve

$$\underline{y}_{32} = \frac{Z_{b1}}{(\underline{Z}_T/n_T)} = \frac{1600}{(1/4)(1.874 + j96.1099)} = 1.2985 - j66.5778\,\text{p.u.}$$

$$\underline{y}_{320} = n_T \frac{\underline{Y}_{T0}}{Y_{b1}} = n_T\,\underline{Y}_{T0} Z_{b1} =$$

$$= 4(1.1986 - j7.3949) \times 10^{-6} \times 1600 = 0.0077 - j0.0473\,\text{p.u.}$$

(2) The Thévenin Equivalent Circuit. To simplify the calculations, the Γ quadripole of the transmission substation equivalent circuit is transformed into a Π (galvanic) quadripole, as shown in Figure 11.32. The equivalent circuit of the studied power system from Figure 11.31 can, therefore, be represented as shown in Figure 11.33a.

Finally, the equivalent circuit from Figure 11.33a is reduced to the equivalent circuit from Figure 11.33b by applying the Thévenin theorem.

Thus, the admittances from Figure 11.33a are as follows:

$$\underline{Y}_1 = \underline{y}_{13} = 6.3564 - j59.9319\,\text{p.u.}$$

$$\underline{Y}_2 = N\,\underline{y}_{32} = 0.8807 \times (1.2985 - j66.5778) = 1.1436 - j58.6368\,\text{p.u.}$$

and

$$\underline{Y}_{10} = \underline{y}_{130} = j0.8416\,\text{p.u.}$$

$$\underline{Y}_{20} = N(N-1)\,\underline{y}_{32} = -0.1364 + j6.9938\,\text{p.u.}$$

$$\underline{Y}_{30} = \underline{y}_{310} + \underline{y}_{320} + (1-N)\,\underline{y}_{32} = 0.1625 - j7.1466\,\text{p.u.}$$

- *Calculation of the Thévenin electromotive voltage*
  The Thévenin electromotive voltage $\underline{E}_{Th}$ is the voltage applied to the equivalent circuit "as seen" from the bus 2 under open-circuit conditions, that is, for $\underline{I}_2 = 0$.

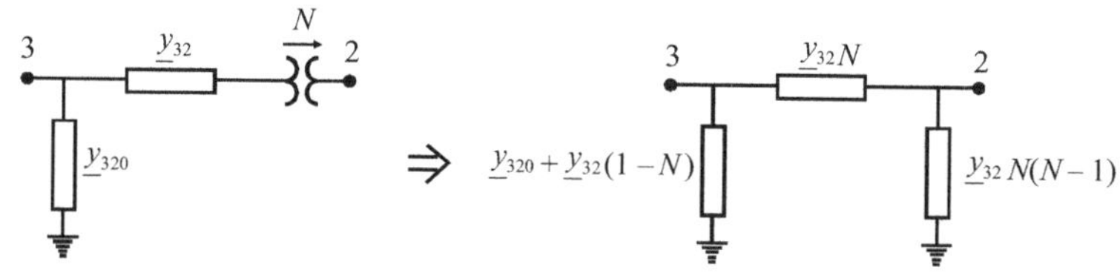

Figure 11.32. Transformation from a Γ quadripole into a Π (galvanic) quadripole for a transformer equivalent circuit.

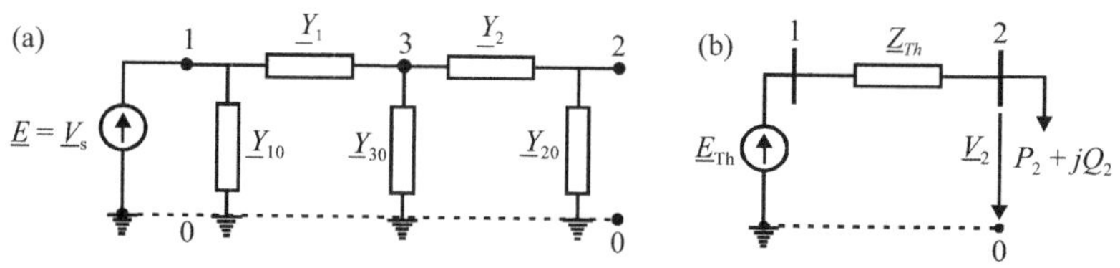

Figure 11.33. The reduced circuit of the studied power system. (a) Simplified equivalent circuit. (b) The Thévenin equivalent.

Given that the infinite power bus voltage is imposed, that is. $\underline{V}_1 = 1 + j0$ p.u., the nodal voltages matrix equation for the Thévenin equivalent shown in Figure 11.33b is as follows:

$$\begin{bmatrix} \underline{Y}_{11} = \underline{Y}_1 + \underline{Y}_{10} & 0 & \underline{Y}_{13} = -\underline{Y}_1 \\ 0 & \underline{Y}_{22} = \underline{Y}_2 + \underline{Y}_{20} & \underline{Y}_{23} = -\underline{Y}_2 \\ \underline{Y}_{31} = -\underline{Y}_1 & \underline{Y}_{32} = -\underline{Y}_2 & \underline{Y}_{33} = \underline{Y}_1 + \underline{Y}_2 + \underline{Y}_{30} \end{bmatrix} \begin{bmatrix} \underline{V}_1 = 1 + j0 \\ \underline{V}_2 = \underline{E}_{Th} \\ \underline{V}_3 \end{bmatrix} = \begin{bmatrix} \underline{I}_1 \\ \underline{I}_2 = 0 \\ 0 \end{bmatrix}$$

Thereby, the following system of equations is obtained as:

$$\begin{cases} \underline{Y}_{22}\,\underline{V}_2 + \underline{Y}_{23}\,\underline{V}_3 = 0 \\ \underline{Y}_{31}\,\underline{V}_1 + \underline{Y}_{32}\,\underline{V}_2 + \underline{Y}_{33}\,\underline{V}_3 = 0 \end{cases}$$

from where the Thévenin voltage results, that is:

$$\underline{E}_{\text{Th}} = \underline{V}_2 = \frac{\underline{Y}_{23}\,\underline{Y}_{31}}{\underline{Y}_{22}\,\underline{Y}_{33} - \underline{Y}_{23}\,\underline{Y}_{32}}\underline{V}_1 = 1.1505 - j0.0018\,\text{p.u.}$$

- *Calculation of the Thévenin Impedance*
  The Thévenin admittance is the equivalent admittance of the passivized circuit (the voltage sources are short circuited) for which the formulae for series and parallel circuits are applied. It results:

$$\underline{Y}_{\text{Th}} = \underline{Y}_{20} + \frac{\underline{Y}_1(\underline{Y}_1 + \underline{Y}_{30})}{\underline{Y}_1 + \underline{Y}_2 + \underline{Y}_{30}} = 1.6046 - j24.3402\,\text{p.u.}$$

and the Thévenin impedance is as follows:

$$\underline{Z}_{\text{Th}} = \frac{1}{\underline{Y}_{\text{Th}}} = R_{\text{Th}} + jX_{\text{Th}} = Z_{\text{Th}}\angle\beta = 0.0027 + j0.0409 = 0.0410\angle 86.2284^\circ$$

The Thévenin equivalent is used in the voltage stability study as follows.

(3) Analysis of the Possible Operating Points A and B. The load is modeled as constant power, $P = \text{const.}$ In per unit, the complex power demand is $\underline{S}_2 = (P_2 + jQ_2)/S_b = 7 + j3$ p.u.

Therefore, $S_2 = \sqrt{P_2^2 + Q_2^2} = 7.6158$ p.u., and $\cos\varphi = \dfrac{P_2}{S_2} = \dfrac{7}{7.6158} = 0.9191$

- *The voltages in the possible operating points, $V_{2\text{A}}$ and $V_{2\text{B}}$*
  Taking into account that $V_1 = E_{\text{Th}} = 1.1505$ p.u., $R_{\text{Th}} = 0.0027$ p.u., and $X_{\text{Th}} = 0.0409$ p.u., from equation (11.19) achieve:

$$\alpha = V_1^2 - 2(R_{\text{Th}}P_2 + X_{\text{Th}}Q_2) = 1.1505^2 - 2(0.0027 \times 7 + 0.0409 \times 3) = 1.0404$$

$$\Delta = \alpha^2 - 4Z_{\text{Th}}^2S_2^2 = 1.0404^2 - 4 \times 0.0410^2 \times 7.6158^2 = 0.6926$$

and the voltages in the possible operating points (see Figure 11.4) are determined according to equation (11.20), that is:

$$V_{2A} = \sqrt{\frac{\alpha + \sqrt{\Delta}}{2}} = \sqrt{\frac{1.0404 + \sqrt{0.6926}}{2}} = 0.9676\,\text{p.u.}$$

$$V_{2B} = \sqrt{\frac{\alpha - \sqrt{\Delta}}{2}} = \sqrt{\frac{1.0404 - \sqrt{0.6926}}{2}} = 0.3227\,\text{p.u.}$$

- *The voltage sensitivities*
  Equation (11.25) gives:

$$\frac{\partial V_{2A}}{\partial P_2} = \frac{-R_{Th}}{\sqrt{V_1^2 - 4(R_{Th}P_2 + X_{Th}Q_2)}} = \frac{-0.0027}{\sqrt{1.1505^2 - 4(0.0027 \times 7 + 0.0409 \times 3)}} = -0.0036$$

$$\frac{\partial V_{2A}}{\partial Q_2} = \frac{-X_{Th}}{\sqrt{V_1^2 - 4(R_{Th}P_2 + X_{Th}Q_2)}} = \frac{-0.0409}{\sqrt{1.1505^2 - 4(0.0027 \times 7 + 0.0409 \times 3)}} = -0.0540$$

and

$$\frac{\partial V_{2B}}{\partial P_2} = \frac{R_{Th}}{\sqrt{V_1^2 - 4(R_{Th}P_2 + X_{Th}Q_2)}} = 0.0036$$

$$\frac{\partial V_{2B}}{\partial Q_2} = \frac{X_{Th}}{\sqrt{V_1^2 - 4(R_{Th}P_2 + X_{Th}Q_2)}} = 0.0540$$

The negative values of the point A voltage sensitivities show stability, whereas the positive values of the point B voltage sensitivities show instability.

(4) The Transmission Network Characteristics. Assumes that the power factor is constant and the critical point is determined.

- *The maximum transmissible powers* The maximum powers that can be transmitted to the load area can be determined based on the relationship (11.7). Thus, for $\cos\varphi = 0.9191$ ($\sin\varphi = 0.3940$ and $\tan\varphi = 0.4286$), achieve as follows:

$$S_{2max} = \frac{1.1505^2}{2(0.0027 \cdot 0.9191 + 0.0409 \cdot 0.3940 + 0.0410)} = 11.1064\,\text{p.u.}$$

$$P_{2max} = S_{2max}\cos\varphi = 11.1064 \cdot 0.9191 = 10.2084\,\text{p.u.}$$
$$Q_{2max} = S_{2max}\sin\varphi = 11.1064 \cdot 0.3940 = 4.3750\,\text{p.u.}$$

and the distance in apparent power, from the actual operating point to the critical point (calculated at constant power factor), is

$$DS_c = S_{2max} - S_2 = 11.1064 - 7.6158 = 3.4906\,\text{p.u.}$$

- *The critical voltage* Taking into account that in the critical point $\Delta = 0$, the voltage value in this point is

$$V_{2\text{cr}} = \sqrt{\frac{V_1^2 - 2(R_{\text{Th}}P_{2\max} - X_{\text{Th}}Q_{2\max})}{2}} =$$
$$= \sqrt{\frac{1.1505^2 - 2(0.0027 \times 10.2084 - 0.0409 \times 4.3750)}{2}} = 0.6748\,\text{p.u.}$$

- *The effect of reactive power compensation* Three different values of the power factor are considered ($\cos\varphi = 0.9191$ inductive, $\cos\varphi = 1$ and $\cos\varphi = 0.98$ capacitive), for which the V_2–P_2 characteristics are plotted, considering the source voltage as constant ($V_s = 1$ p.u. and thus $V_1 = 1.1505$ p.u.). The maximum transmissible powers and the critical voltages are first calculated in the same way as above. The results are given in Table 11.6.

  The V_2–P_2 characteristics, corresponding to the three different power factors, are plotted for transmitted active power values, P_2, varying from 0 to $P_{2\max}$. These characteristics are achieved by successive calculations of the voltages $V_{2\text{A}}$ and $V_{2\text{B}}$ in the possible operating points. The three characteristics thereby obtained are illustrated in Figure 11.34a.
- *The effect of modifying the source voltage* Two different values of the source voltage are considered ($V_s = 1$ p.u. and $V_s = 1.05$ p.u.) for which cases the V_2–P_2 characteristics are plotted, assuming constant power factor ($\cos\varphi = 0.9191$ inductive). For a voltage value $V_s = 1.05$ p.u., the electromotive

TABLE 11.6. Maximum Transmissible Powers and Critical Voltages for Different Values of the Power Factor and $V_s = 1$ p.u.

		Maximum Transmissible Powers			
$\cos\varphi$	$\tan\varphi$	$S_{2\max}$ [p.u.]	$P_{2\max}$ [p.u.]	$Q_{2\max}$ [p.u.]	$V_{2\text{cr}}$ [p.u.]
0.9191	0.4286	11.1064	10.2084	4.3750	0.6748
1	0	15.1471	15.1471	0	0.7880
0.98	−0.2031	18.6437	18.2708	−3.7100	0.8742

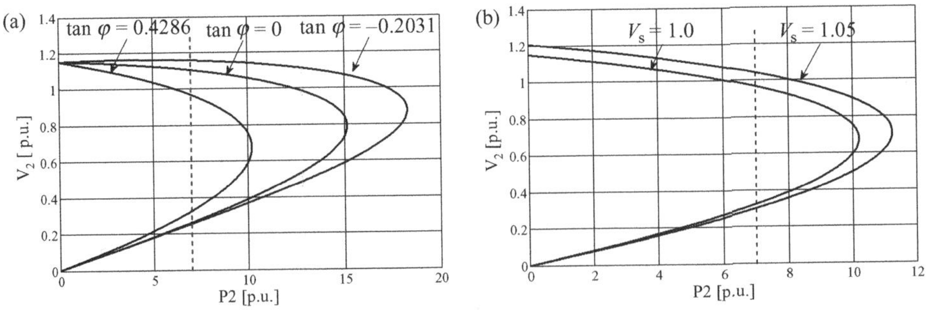

Figure 11.34. V_2–P_2 characteristics. (a) Constant source voltage and variable power factor. (b) Variable source voltage and constant power factor.

TABLE 11.7. Maximum Transmissible Powers and Critical Voltages for Different Source Voltages and Constant Power Factor $\cos\varphi = 0.9191$

		Maximum Transmissible Powers			
V_s [p.u.]	$V_1 = E_{Th}$ [p.u.]	S_{2max} [p.u.]	P_{2max} [p.u.]	Q_{2max} [p.u.]	V_{2cr} [p.u.]
1	1.1505	11.1064	10.2084	4.3750	0.6748
1.05	1.2080	12.2449	11.2548	4.8235	0.7085

voltage E_{Th} must be recalculated because the operating conditions have changed (the value of the impedance $\underline{Z}_{Th}$ does not change because the network topology does not change).

For the assumed cases, the maximum transmissible powers and the critical voltages are shown in Table 11.7.

The V_2–P_2 characteristics, for two voltage values at the source (the infinite power bus), assuming a constant factor, are illustrated in Figure 11.34b.

(5) Modal Analysis of the Reduced Jacobian. Consider the reduced circuit (Figure 11.33b), where bus 1 is the slack and bus 2 is a PQ bus. The balance equations at bus 2 are given by

$$\begin{cases} P_{2g} - P_{2c} = P_{2t} \\ Q_{2g} - Q_{2c} = Q_{2t} \end{cases}$$

Since $P_{2g} = Q_{2g} = 0$, $P_{2c} = P_2$, and $Q_{2c} = Q_2$, whereas

$$\begin{cases} P_{2t} = V_1 V_2 (G_{21}\cos\theta_2 + B_{21}\sin\theta_2) + G_{22}V_2^2 \\ Q_{2t} = V_1 V_2 (G_{21}\sin\theta_2 - B_{21}\cos\theta_2) - B_{22}V_2^2 \end{cases}$$

we achieve $P_{2t} = -P_2$ and $Q_{2t} = -Q_2$, and the nonlinear equation system becomes:

$$\begin{cases} P_2 + V_1 V_2 (G_{21}\cos\theta_2 + B_{21}\sin\theta_2) + G_{22}V_2^2 = 0 \\ Q_2 + V_1 V_2 (G_{21}\sin\theta_2 - B_{21}\cos\theta_2) - B_{22}V_2^2 = 0 \end{cases}$$

where P_2 and Q_2 are the active and reactive powers demanded by the load area.

Under these conditions, the system Jacobian is as follows:

$$[J] = \begin{bmatrix} J_{P\theta} = \dfrac{\partial P_{2t}}{\partial \theta_2} & J_{PV} = \dfrac{\partial P_{2t}}{\partial V_2} \\ J_{Q\theta} = \dfrac{\partial Q_{2t}}{\partial \theta_2} & J_{QV} = \dfrac{\partial Q_{2t}}{\partial V_2} \end{bmatrix} = \begin{bmatrix} -Q_{2t} - B_{22}V_2^2 & \dfrac{P_{2t} + G_{22}V_2^2}{V_2} \\ P_{2t} - G_{22}V_2^2 & \dfrac{Q_{2t} - B_{22}V_2^2}{V_2} \end{bmatrix}$$

and the reduced Jacobian is $J_R = J_{QV} - J_{Q\theta} \times J_{P\theta}^{-1} \times J_{PV}$.

To calculate the matrix J_R and to evaluate the voltage stability using the modal analysis, assumes that $V_1 = 1.1505$ p.u. and the subsequent steps are followed:

(i) *Calculation of the bus admittances matrix*

$$[\underline{Y}_{nn}] = \begin{bmatrix} \underline{Y}_{Th} & -\underline{Y}_{Th} \\ -\underline{Y}_{Th} & \underline{Y}_{Th} \end{bmatrix} = \begin{bmatrix} \underline{Y}_{11} & \underline{Y}_{12} \\ \underline{Y}_{21} & \underline{Y}_{22} \end{bmatrix}$$

$$= \begin{bmatrix} 1.6046 - j24.3402 & -1.6046 + j24.3402 \\ -1.6046 + j24.3402 & 1.6046 - j24.3402 \end{bmatrix}$$

Therefore $G_{22} = 1.6046$ p.u. and $B_{22} = -24.3402$ p.u.

(ii) *Calculation of voltage V_2*
Since the operating conditions are the same as those considered at point 3, the voltage V_2 in the two operating points are $V_{2A} = 0.9676$ p.u. and $V_{2B} = 0.3227$ p.u.

(iii) *Calculation of the reduced Jacobian for voltage stability assessment*
As shown earlier, in the two operating points we have $P_{2t} = -P_2$ and $Q_{2t} = -Q_2$, and thus

- for the operating point A, where $V_2 = V_{2A} = 0.9676$ p.u., the Jacobian terms are as follows:

$$J_{P\theta} = Q_2 - B_{22}V_2^2 = 3 - (-24.3402) \times 0.9676^2 = 25.7902$$

$$J_{PV} = \frac{-P_2 + G_{22}V_2^2}{V_2} = \frac{-7 + 1.6046 \times 0.9676^2}{0.9676} = -5.6815$$

$$J_{Q\theta} = -P_2 - G_{22}V_2^2 = -7 - 1.6046 \times 0.9676^2 = -8.5024$$

$$J_{QV} = \frac{-Q_2 - B_{22}V_2^2}{V_2} = \frac{-3 - (-24.3402) \times 0.9676^2}{0.9676} = 20.4521$$

and the reduced Jacobian is as follows:

$$J_{R,A} = J_{QV} - J_{Q\theta} \times J_{P\theta}^{-1} \times J_{PV} = 20.4521 - \frac{-8.5024 \times (-5.6815)}{25.7902} = 18.5791$$

Therefore, in this particular case, the reduced Jacobian is a 1×1 matrix. The eigenvalue of the $J_{R,A}$ matrix is equal to $\lambda = 18.5791 > 0$ and thus the operating state is stable from the voltage viewpoint. Furthermore, the minimum singular value is equal to the eigenvalue and thereby the global index VSI is the ratio of the eigenvalue calculated for the operating point A to the eigenvalue calculated under no-load.

- for the operating point B, where $V_2 = V_{2B} = 0.3227$, the elements of the Jacobian are as follows:

$$J_{P\theta} = Q_2 - B_{22}V_2^2 = 3 - (-24.3402) \times 0.3227^2 = 5.5339$$

$$J_{PV} = \frac{-P_2 + G_{22}V_2^2}{V_2} = \frac{-7 + 1.6046 \times 0.3227^2}{0.3227} = -21.1774$$

$$J_{Q\theta} = -P_2 - G_{22}V_2^2 = -7 - 1.6046 \times 0.3227^2 = -7.1670$$

$$J_{QV} = \frac{-Q_2 - B_{22}V_2^2}{V_2} = \frac{-3 - (-24.3402) \times 0.3227^2}{0.3227} = -1.4445$$

TABLE 11.8. Results of the Modal Analysis of the Matrix J_R and the Values of the Global Index VSI in Various Points of the P–V Characteristic

		Operating Point A		Operating Point B		
P_2 [p.u.]	Q_2 [p.u.]	V_{2A} [p.u.]	$\lambda = J_R$	V_{2B} [p.u.]	$\lambda = J_R$	VSI
0	0	1.1505	28.1247	0	–	1
1	0.4286	1.1321	27.2667	0.0394	−64.3456	0.9695
2	0.8571	1.1119	26.2905	0.0802	−57.9089	0.9348
3	1.2857	1.0896	25.1770	0.1228	−51.8190	0.8952
4	1.7143	1.0648	23.8998	0.1676	−45.9750	0.8498
5	2.1429	1.0369	22.4204	0.2151	−40.2790	0.7972
6	2.5714	1.0050	20.6796	0.2663	−34.6238	0.7353
7	3.0000	0.9676	18.5791	0.3227	−28.8714	0.6606
8	3.4286	0.9221	15.9316	0.3870	−22.8022	0.5665
9	3.8571	0.8617	12.2858	0.4658	−15.9375	0.4368
10	4.2857	0.7551	5.4725	0.5907	−6.0851	0.1946
P_{2max} 10.2084	Q_{2max} 4.3750	0.6748	0	0.6748	0	0

and the reduced Jacobian is as follows:

$$J_{R,B} = J_{QV} - J_{Q\theta} \times J_{P\theta}^{-1} \times J_{PV} = -1.4445 - \frac{-7.1670 \times (-21.1774)}{5.5339} = -28.8714$$

Therefore, the eigenvalue of the reduce Jacobian, $J_{R,B}$, is $\lambda = -28.8714 < 0$ and thus the operating state is unstable from the voltage viewpoint.

Table 11.8 presents the results of the voltage stability assessment using the modal analysis of the reduced Jacobian for various values of the power demand in the load area.

The results presented in Table 11.8 shows that as the transmitted power increases, the eigenvalues of the matrix J_R corresponding to the two operating points A and B get closer and become equal to zero in the critical point. Moreover, the global index VSI decreases from 1, under no-load conditions, to 0, in the critical point.

REFERENCES

[1] Concordia, C. Voltage instability, *Int. J. Elect. Power Energy Syst.*, Vol. 13, No. 1, pp. 14–20, 1991.

[2] Van Cutsem, T., Vournas, C.D. *Voltage stability of electric power systems*, Kluwer Academic Publishers, Boston, 1998.

[3] CIGRE Task Force 38-02-11. *Indices predicting voltage collapse including dynamic phenomena*, 1994.

[4] Hill, D.J. Nonlinear dynamic load models with recovery for voltage stability studies, *IEEE Trans. Power Syst.*, Vol. 8, No. 1, pp. 166–175, February 1993.

[5] Eremia, M., Trecat, J., Germond, A. *Réseaux electriques. Aspects actuels*, Technical Publishers, Bucharest, 2000.

[6] CIGRE Task Force 38-02-10. *Modeling of voltage collapse including dynamic phenomena*, CIGRE Brochure No. 75, 1993.

[7] Borremans, P. et al. Voltage stability: fundamental concepts and comparison of practical criteria, *CIGRE Proceeding paper 38-11*, August 1984.

[8] Pal, M.K. Voltage stability conditions considering load characteristics, *IEEE Trans. Power Syst.*, Vol. 7, No. 1, pp. 243–249, February 1992.

[9] Pal, M.K. *Lecture notes on power system stability*, Edison, New Jersey, June 2007.

[10] Weedy, B.M. *Electric power systems*, John Wiley & Sons, 3rd edition, 1979.

[11] Löf, P.A., Andersson, G., Hill, D.J. Voltage stability indices for stressed power systems, *IEEE Trans. Power Syst.*, Vol. 8, pp. 326–335, 1993.

[12] Xu, W., Mansour, Y., Harrington, P.G. Planning methodologies for voltage stability limited power systems, *Int. J. Elect. Power Energy Syst., Special issues "Voltage Stability and Collapse"*, Vol. 15, No. 4, pp. 221–228, 1993.

[13] Xu, W., Mansour, Y. Voltage stability analysis using generic dynamic load models, *IEEE Trans. Power Syst.*, Vol. 9, No. 1, pp. 479–493, 1994.

[14] Canizares, C. (Editor) *Voltage stability assessment: concepts, practices and tools*. IEEE-PES Power System Stability, Special publication, SP101PSS, 2003.

[15] Fischl, R. (Drexel University). *Performance indexes for predicting voltage collapse*, Final Report EPRI EL-6461, July 1989.

[16] Chow, I.C., Fischl, R., Yan, H. On the evaluation of voltage collapse criteria, *IEEE Trans. Power Syst.*, Vol. 5, No. 2, May 1990.

[17] EPRI EL-6183. *Proceedings: Bulk power system voltage phenomena – Voltage stability and security*, Project 2473-21, January 1989.

[18] Barbier, C., Barret, J.P. Analyse des phénomènes d'écroulement de tension sur un réseau de transport, *Rev. Générale d'Elect.*, Tome 89, No. 10, pp. 672–690, October 1980.

[19] Carpentier J. et al. *Voltage collapse proximity indicators computed from an optimal power flow, Proceedings of Systems Computation Conference*, Helsinki, Finland, pp. 671–678, September 1984.

[20] Jarjis, J. Galiana, F.D. Quantitative analysis of steady state stability in power networks, *IEEE Trans. Power Apparatus Syst.*, Vol. PAS-100, No. 1, pp. 318–326, January 1981.

[21] Kessel, P., Glavitsch, H. Estimating the voltage stability of a power system, *IEEE Trans. Power Deliv.*, Vol. PWRD-1, No. 3, pp. 346–354, July 1986.

[22] Tiranuchit, A. Thomas, R.J., A posturing strategy against voltage instability in electric powers systems, *IEEE Trans. Power Syst.*, Vol. 3, No. 1, pp. 87–93, February 1988.

[23] Tamura, Y., Mori, H., Iwamoto, S. Relationship between voltage instability and multiple load flow solutions in electric power systems, *IEEE Trans. Power Apparatus Syst.*, Vol. PAS-102, No. 5, pp. 1115–1125, 1983.

[24] Lachs, W.R. *Voltage collapse in EHV power systems*, Paper No. A78 057-2, 1978 IEEE/PES Winter Power Meeting, January, 1978.

[25] Abe, S. et al. Power system voltage stability, *IEEE Trans. Power Apparatus Syst.*, Vol. PAS-101, No. 10, pp. 3820–3839, October 1982.

[26] Costi, A., Shu, L., Schlueter, R.A. *Power system voltage stability and controllability, Proceedings of the 1986 IEEE ISCAS*, Vol. 3, San Jose, California, pp. 1023–1027, May 1986.

[27] Alvarado, F.L., Jung, T.H. *Direct detection of voltage collapse conditions, Proceedings of Bulk Power Systems Voltage Phenomena – Voltage stability and security*, EL-6183, EPRI, January 1989.

[28] Bergen, A.R., Vittal, V. *Power systems analysis*, Prentice Hall Inc., New Jersey, 2000.

[29] Brucoli, M. et al. A generalized approach to the analysis of voltage stability in electric power systems, *Elect. Power Syst. Res.*, Vol. 9, pp. 49–62, 1985.

[30] Venikov, V.A., Stroev, V.A., Idelchick, V.I., Tarasov, V.I. Estimation of electrical power system steady-state stability in load flow calculations, *IEEE Trans. Power Apparatus Syst.*, Vol. PAS-94, No. 3, pp. 1034–1041, May–June 1975.

[31] Arapostathis, A. et al. Analysis of power flow equation, *Elect. Power Energy Syst.*, Vol. 3, pp. 115–126, July 1981.

[32] Liu, C.C., Wu, F.F. *Steady-state voltage stability regions of power systems, Proceedings of the 3rd IEEE CDC*, Las Vegas, Nevada, pp. 488–493, December 1984.

[33] Kwatny, H.G., et al. Static bifurcations in electric power networks: loss of steady-state stability and voltage collapse, *IEEE Trans. CAS*, Vol. CAS-33, No. 10, pp. 981–991, October 1986.

[34] EPRI EL-5967 *Voltage stability and security assessment*, Final report on Advanced concepts for power systems engineering project RP – 1999-8, August. 1988.

[35] IEEE Special publication 93TH0620_SPWR. *Suggested techniques for voltage stability analysis*, 1993.

[36] CIGRE Task Force 38-02-12. *Criteria and countermeasures for voltage collapse*, Report, Electra, Vol. 162, October 1995.

[37] Flatabo, N., Ognedal, R., Carlsen, T. Voltage stability condition in a power transmission system calculated by sensitivity methods. IEEE Trans. Power Syst., Vol. PWRS-5, No. 4, pp. 1286–1293, 1990.

[38] Dobson, I. The irrelevance of load dynamics for the loading margin to voltage collapse and its sensitivities, *Proceedings on Power System Voltage Phenomena III – Voltage stability and security*, ECC Inc., Davos, Switzerland, pp. 509–518, August 1994.

[39] Seydel, R. *From equilibrium to chaos. Practical bifurcation and stability analysis*, Elsevier, New York, 1988.

[40] Ajjarapu, V. *Computational techniques for voltage stability assessment and control*, Springer, Power Electronics and Power Systems Series, 2006.

[41] Gomez-Exposito, A., Conejo, A.J., Canizares, C. *Electric energy systems. Analysis and operation*, CRC Press, 2008.

[42] Gonzales, J.M., Canizares, C.A., Ramirez, J.M. Stability modeling and comparative study of series vectorial compensators, *IEEE Trans. Power Deliv.*, Vol. 25, No. 2, April 2010.

[43] Ilić, M., Zabosszky, J. *Dynamics and control of large electric power systems*, John Wiley & Sons Inc., New York, 2000.

[44] Kundur, P. *Power systems stability and control*, McGraw Hill Inc., New York, 1994.

[45] Ajjarapu, V., Lee, B. Bifurcation theory and its application to nonlinear dynamical phenomena in electric power system, IEEE Trans. Power Syst., Vol. 7, No. 1, pp. 424–431, February 1992.

[46] Jing, Z., Xu, D., Chang, Y., Chen, L. Bifurcations, chaos, and system collapse in a three node power system, *Elect. Power Energy Syst.*, Elsevier, No. 25, pp. 443–461, 2003.

[47] Arnold, V. *Geometrical methods in the theory of ordinary differential equations*, Springer-Verlag, New York, 1983.

[48] Seyranian, A.P., Mailybaev, A.A. *Multiparameter stability theory with mechanical applications*, World Scientific Publishing Co. Pvt. Ltd., Singapore, 2003.

[49] Gao, B., Morison, G.K., Kundur, P. Voltage stability evaluation using modal analysis, *IEEE Trans. Power Syst.*, Vol. PWRS-7, No. 4, pp. 1529–1542, 1992.

[50] Dobson, I. Observations on the geometry of saddle-node bifurcation and voltage collapse in electric power systems, *IEEE Trans. Circuits Syst. I*, Vol. 39, No. 3, pp. 240–243, 1992.

[51] Taylor, C.W. *Power system voltage stability*, McGraw Hill Inc., New York, 1994.

[52] IEEE Task Force Report Standard load models for power flow and dynamic performance simulation, *IEEE Trans. Power Syst.*, Vol. 10: pp. 1302–1313, 1995.

[53] Kundur, P., Paserba, J., Ajjarapu, V., Andersson, G., Bose, A., Canizares, C., Hatziargyriou, N., Hill, D., Stankovic, A., Taylor, C., Van Cutsem, T., Vittal, V. Definition and classification of power system stability, IEEE-CIGRE Joint Task Force on Stability terms and definitions, *IEEE Trans. Power Syst.*, Vol. 19, No. 2, pp. 1387–1401, May 2004.

12

POWER SYSTEM PROTECTION

Klaus-Peter Brand and Ivan De Mesmaeker

12.1 INTRODUCTION

12.1.1 Motivation

A lot of protection books are published ([1–3]) indicating the importance of protection for power systems and discussing protection concepts and measures in all the details. The reports (brochures) of the working groups of the CIGRE SC B5 ("Protection and Automation") published by CIGRE [4] provide information about a lot of actual protection issues. Numerous details are published in many professional journals, which may not be listed here.

The book is focused mainly to power system stability and control, besides the basic concepts, the described protection functions are grouped according to the power system structure and the objects of the power system to be protected. Important features contributing to safety and stability of the power system will be highlighted. Details regarding the endless number of situations and the many sophisticated protection parameters will be left to the above-mentioned protection books.

Also, not the history of protection will be discussed. The state-of-the-art implementation of protection functions in IEDs (intelligent electronic devices) will be explained including the serial communication, which may replace nearly all parallel copper wires in the future. For this topic, references are made to the standard IEC 61850 defining not only the communication but also a comprehensive data model.

Handbook of Electrical Power System Dynamics: Modeling, Stability, and Control. Edited by Mircea Eremia and Mohammad Shahidehpour.

It should be noted already here that protection is the fastest automatic function available in the power system. Since protection is mostly implemented in substations it is seen as an important part of *substation automation* (SA) also.

12.1.2 The Task of Protection

The task of the protection in the electrical power system is

- to protect people against dangerous situations;
- to protect the power system against instabilities;
- to protect the assets of the power system (overhead lines, generator, power transformer, etc.) against malfunction and destruction.

In this respect, protection can be considered as an assurance: more assurance is needed when the protected object is more important.

Figure 12.1 shows the different states in the power system and where the protection acts. In the *normal* state, the power flow is dispatched at network level by the power system management system. If the power system or part of its deviates from the normal state resulting in the *alert* or *emergency* state, then the protection in the substation has to act. The goal is always to come back to normal as soon as possible. From *alert* state it may be done, for example, with some load relieving of an overloaded transformer by reallocation of the power flow. For transient faults, the way back from *emergency* it is done quickly after the trip, for example, the so-called autoreclosing of the tripped circuit breaker. Persistent faults have to be first isolated reliably in the *interrupted* state until the power supply is restored again. Applying system (wide area) protection, the allocation of states stays the

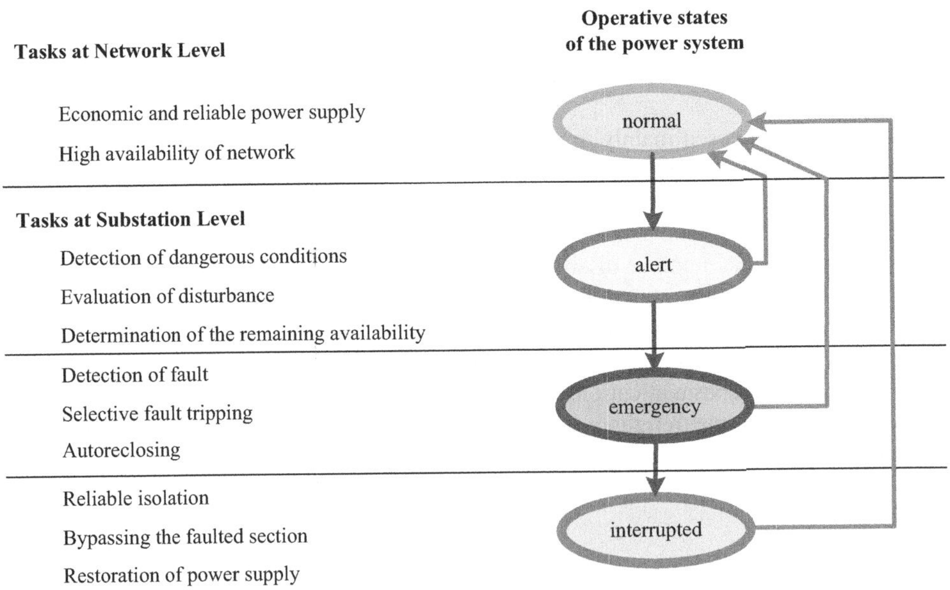

Figure 12.1. States in the power system and protective actions.

same but this protection is complementing the local one and residing in system protection IED(s) also outside the substation.

It should not be neglected that the protection actions resulting mostly in opening of circuit breakers may also have some negative impact on the power system, that is, they

- interrupt the power flow and, therefore, impact the power delivery;
- produce transients that may impact the system stability;
- cause deviations of current, voltage, and frequency from nominal values degrading the power quality.

Normally, these negative impacts are overruled by the necessary task to isolate the faulted component or part of the power system. The important task in the design of power systems is to respect the ($N - 1$) rule for power system stability also in case of protection operation, that is, the loss of one component caused by protection operation has to be tolerated by the power system without loosing its stability. Another point to be considered in protection and power system design is to minimize the risk of consequential faults causing the sequential loss of components and ending in a blackout. The related operative requirement of the protection is a *high selectivity*, which will be discussed below.

12.1.3 Basic Protection Properties and Resulting Requirements

For understanding protection, the basic properties have to be known and both the measuring and protection principles have to be understood. It is also important how protection is implemented in modern numerical relays commonly named IEDs. This allows to discuss the trends of protection in general and especially the evolution of protection from *local* protection (e.g., based on a single overcurrent measurement only) to *zone* or *object* protection (e.g., comparing values at the zone of an object like the currents per feeder for the busbar protection) and, finally, to *global* power system protection schemes.

One basic property is that the protection is *permanently* supervising the protected part of the power system but has to work, that is, to release trip command just for some *few short times* in its lifetime. This requires high reliability with impact both on the expected high MTTF (mean time to failure) of the IED, the requested comprehensive self-supervision and the maintenance strategy to be applied.

12.1.4 From System Supervision to Circuit Breaker Trip

The path from the sensor collecting power system data like voltage and current over the supervising and deciding function(s) in the protection IED to the actuator like circuit breaker acting on the process is schematically shown in Figure 12.2. All the links may be conventionally hardwired or realized as serial links. If not mentioned especially, in case of serial links we refer always to the standard IEC 61850 [5], which was developed for communication within substations but is now defined and used already for some applications beyond the substation or in other application domains of the power system automation.

The protection IED is connected today to higher level control systems like a substation automation or network control system (HMI, SA System) getting parameter or parameter set changes and providing information about faults (e.g., by time tagged events, disturbance/fault records) and the protection action taken. This information will end on displays of the HMI, in alarm and event list and in archives.

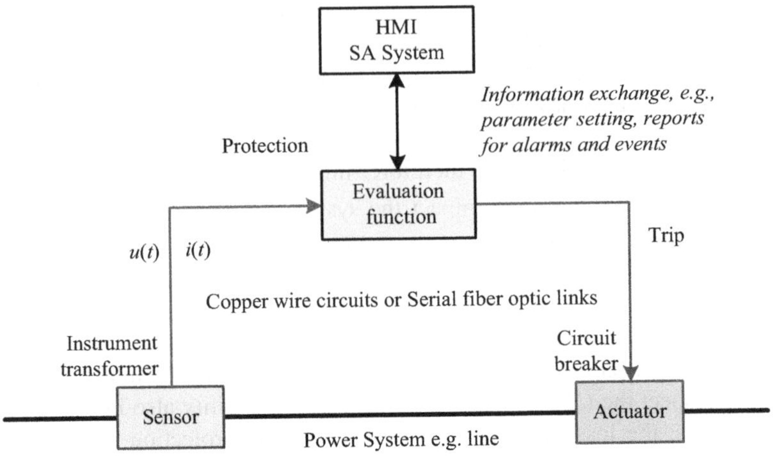

Figure 12.2. From sensor via supervision to actuator (circuit breaker).

12.1.5 Main Operative Requirements

The main requirements for the operation of protection are as follows:

- Selectivity
- Reliability
- Speed or performance
- Adaptation to changing power system conditions
- Backup protection

12.1.5.1 Selectivity. Selectivity is one of the main issues of a protection. In case of a fault in the power system, the protection should eliminate only the faulted part or component of the power system but not more. That means:

- Assure that the faulted part will be eliminated by the protection
- Assure that no healthy part will be eliminated by the protection

Selectivity has to be checked between all protection functions implemented in the IEDs of the protection schemes for the power system. This means a very careful coordination of all protection functions applied in the power system.

12.1.5.2 Reliability. *Reliability* according to IEV 448-12-05 [6] is the probability that a protection can perform the required function under given conditions for given time. This means also high security and dependability:

- *Security* according to IEV 448-12-06 [6] is the probability for protection of not having unwanted operations under given conditions for a given time interval.
- *Dependability* according to IEV 448-12-07 [6] is the probability for a protection not having a failure to operate under given conditions for a given time interval.

12.1.5.3 Speed and Performance. The protection has to eliminate quickly the faulty part of the power system in order not to jeopardize the stability and to minimize the destruction due to the short-circuit current. Generally, the total fault clearing time, that is, total time to eliminate the fault including also the operation time of the circuit breaker in the chain sketched in Figure 12.2 has to be considered in this respect. Typical clearing times are about 80 ms where 40 ms refer to the fault detection, trip decision, and all communication times and 40 ms to the mechanical moving and fault breaking action of the circuit breaker. At this instant, the fault is cleared from the power system point of view. To avoid uncontrolled reaction of protection after opening the breaker the reset time of the protection (about 20 ms) has to be considered.

12.1.5.4 Adaptation. If protection makes a wrong trip due to power system changing conditions, this will be a high risk for the power system stability in respect to the ($N-1$) conditions. Therefore, it is important that the protection behavior is adapted to changing power system conditions. Adaptation means that the protection characteristic and behavior are set in that way that the protection will react correctly even with changing network conditions, for example, dissymmetrical overload during single-phase recloser on parallel line.

12.1.5.5 Adaptive Protection. The need for adaptive protection is the same as for adaptation above. The difference is that adaptive protection means that not all situations are predefined but that the protection parameters are changed remotely from an instance having an overview over a wider part of the power system. In special cases, this instance may also be implemented in the protection device itself.

12.1.5.6 Backup Protection. Because of the high availability requested for protection, it is an imperative rule that backup protection functionality is available in case the main protection fails. This requirement may be solved by

- protection installed at a place more remote from the fault location resulting generally in a less selective and delayed trip ($\Delta t > 0$) as shown in Figure 12.3;
- Use of a second protection at the same place with no delayed trip ($\Delta t = 0$) as shown in Figure 12.4. In this case the selectivity is generally not impacted.

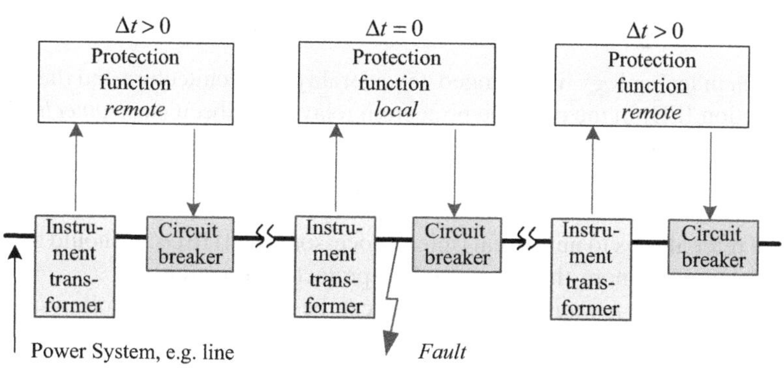

Figure 12.3. Concept of remote backup protection with time delays.

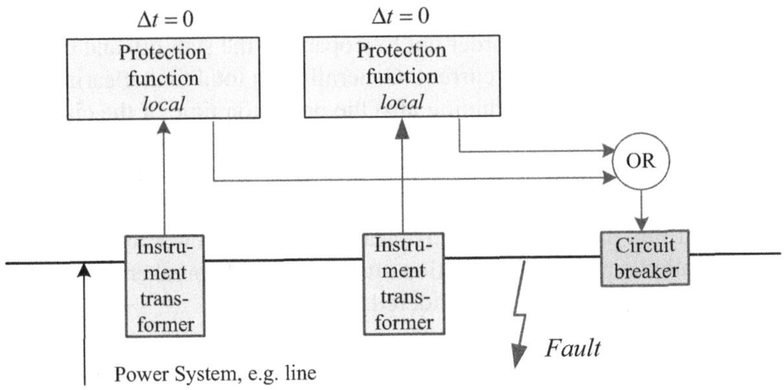

Figure 12.4. Concept of local backup protection without time delays.

To minimize also systematic protection errors, the investment for a second protection is done either by *another protection principle* or by a protection with the same principle but from *another manufacturer* resulting in the use of different algorithms. Several users decide for two identical protections because of maintenance reasons. These both protections are called Main 1 and Main 2. In some installations, there is in addition to Main 1 and Main 2 still an additional backup protection in a single IED or integrated in the control IED. It should be noted that in case of local redundant protection, each has its own sensor set (conventionally provided by separate, dedicated secondary windings in the instrument transformers (IT), by separate, dedicated trip coils in the circuit breaker drive, and by independent copper circuits or serial links. The logical OR for the trip shown in Figure 12.4 is realized by the circuit breaker itself. The important issue of Backup protection is analyzed very carefully in [7].

12.1.5.7 General Remarks About Features Like Performance, Reliability, and Availability. In such an availability analysis and the resulting protection system, design not only the protection function itself but also the whole chain according to Figure 12.2 including data acquisition by measuring transformers or sensors, the circuit breaker and the trip circuit, and the DC supply of the IED hosting the protection function have to be considered.

12.1.6 Advantages of State-of-the-Art Protection

The evolution in technology has changed considerably the architecture and the capabilities of the protection IED. In the past, the protection relays have been *electromechanical* ones. Later, they have been replaced or complemented by *solid state* relays with logical gates. Today, all protection relays are based on *numerical technology using microprocessors*. These numerical devices are commonly called IEDs. If not mentioned especially, this chapter will refer always to numerical (microprocessor based) IEDs. It should be noted that IEDs could comprise more than the classical protection functions.

The impact of the *numerical technology* can be summarized as follows:

- A/D conversion of analog inputs as currents and voltages.
- Use of microprocessors supporting complex calculations and decisions.

- Use of formulas similar as learned from the textbooks.
- Provision of serial communication facilities as add-on given by microprocessors.
- Provision of time tagging of events and alarms as add-on given by the microprocessor clock.
- Support of standardized data models and communication protocols like IEC 61850.
- Support of integration of several functions in one equipment (multifunctional relays).
- Use of hardware platform with a software library of functions.
- Support for system optimization of the whole secondary voltage equipment with functions like protection, control, and monitoring inside a substation called substation automation.

The advantages refer to the complete life cycle of the protection, that is, from specification, project execution, commissioning, testing, operation, and maintenance. It should be noted that the state-of-the-art protection has also an influence on the qualification needed both by providers and users of this equipment and these systems.

In Figure 12.5, the principal layout of a protection IED is shown. It contains analog inputs for currents and voltages (for all phases). The *input transformers* applied both adapt the values coming from the instrument transformers to the electronic world inside the IED and decouple both electrically. The resulting output are snapshot by a so-called *Sample and Hold* (S/H) function and provided *multiplexed* to the *analog digital (A/D) converter* to provide the numbers as needed by the microprocessor. Binary signals are collected by

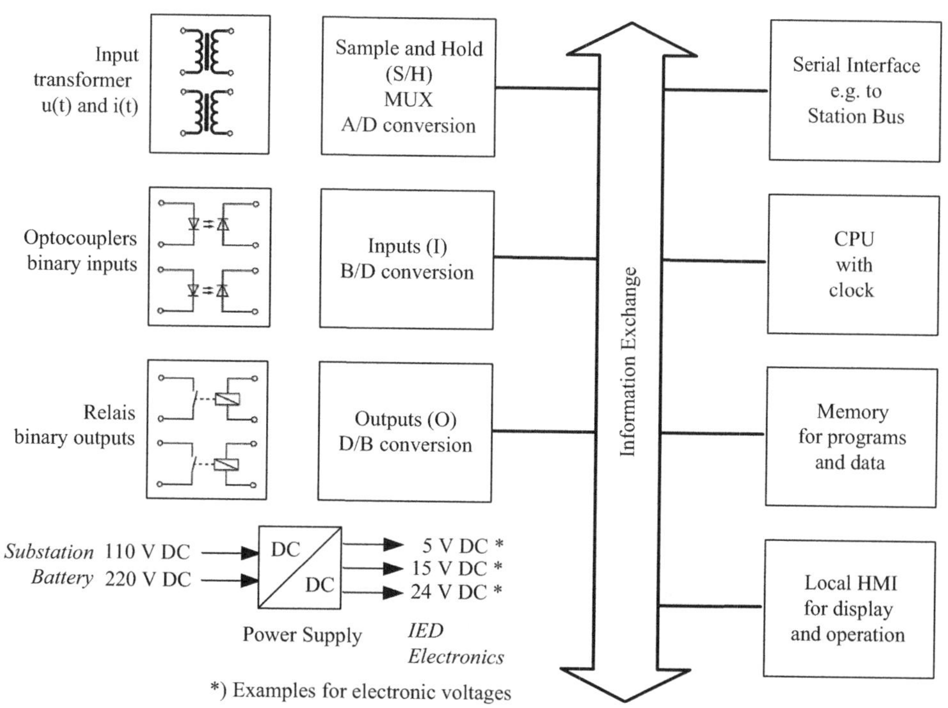

Figure 12.5. Principle diagram of a numerical protection device.

opto-couplers providing a high electric separation. The resulting noise free binary signals are transformed by the *Binary Digital (B/D) converter* into digital signals, that is, numbers as needed by the microprocessor. Binary signals are issued by signaling or heavy-duty *output relays* with galvanic separation.

The numerical signals are processed according to the function and the implemented algorithm by the *microprocessor*, which has by definition an integrated *clock* for the beat of the processor. The microprocessor clock within any IED may be used also for other purposes like time tagging of events and alarms. For time coherence between all the IEDs within the substation, all these clocks have to be synchronized by a SA master clock not shown here. Global time coherence is provided if all the master clocks refer to the time signals of GPS (Global Positioning System). For 1 ms time tagging for events, the synchronization is done by standard means. The *memory* inside the IED is mandatory for programs and operative data but is also used for historical fault data like disturbance records and events.

According to the need of the microprocessor, all the data inside the IED exist as digital numbers and, therefore, are ready for the *serial interface* to be exchanged over the station bus with all other IEDs of the substation automation systems. Nearly all protection IED have some kind of local HMI providing data by LEDs or a small display screen and allowing operation or setting changes, for example, by buttons.

Last not least the IED needs a *power supply*, which is fed by the substation battery (110 V DC or 220 V DC) and converted to the common electronic DC voltages.

If a serial link called *Process Bus* is used to collect current and voltage already as samples, the related serial interface is replacing both the input transformer and the complete A/D unit. For synchronized time coherent sampling with accuracy of the order of 1 μs, the synchronization needs special means as discussed below.

12.2 SUMMARY OF IEC 61850

For interoperable communication in substations, the standard IEC 61850 "Communication networks and systems in substations" [5] was created. It consists of a domain specific data model with data and services. The data are grouped object oriented in so-called Logical Nodes (LN). The name of the LN refers to the related function. The services define how in a standardized way the data are accessed. The services range from the simple *read* and *write service* to more complex ones as the *control service* for the operation of switchgear and supporting also SBO (select before operate commands). The operator gets automatically information with the help of reports started under well predefined conditions. Between IEDs time critical data (block, trip, release, new switch position) are exchanged by the so-called GOOSE messages (*Generic Object Oriented System Event, GOOSE service*) with the most demanding performance class of 4 ms. Analog samples are transmitted by the *Sampled Value(SV) service*.

The domain specific model is mapped on an ISO/OSI 7 layer stack, which defines the coding and decoding of data and services to and from bits to be transmitted in messages (telegrams) over the physical link like wires and fibers. The components of the stack have not been newly defined but taken from the main stream communication technology. For layers 1 and 2, *Ethernet* with 100 MB/s transmission rate and support of priorities is chosen. Ethernet is the most widely used protocol where the most money is invested extending it from office applications to any kind of application in industrial system ones. For layer 3, the *TCP* (Transport Control Protocol) and for layer 4 the IP (Internet Protocol) has been selected being the de facto standard for all communication networks including

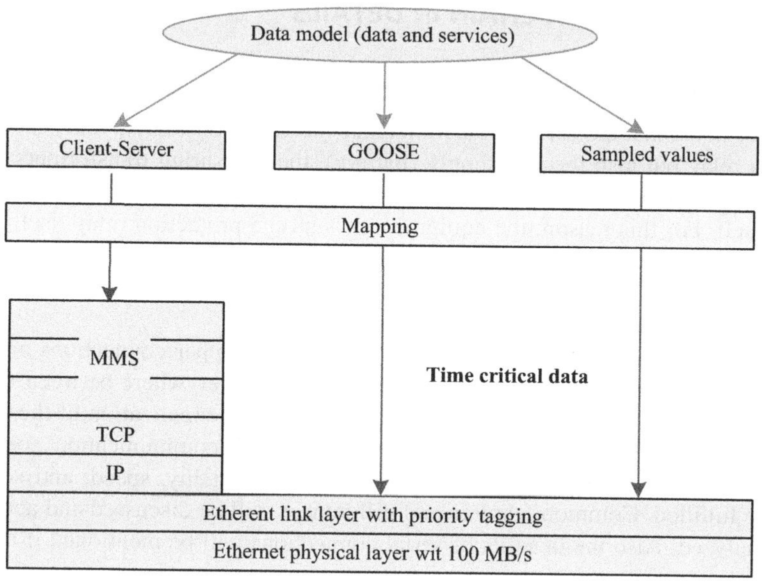

Figure 12.6. Stack as used in IEC 61850 and the relationship to the data model.

Internet. With some simplification we may say that layers 5–7 are covered by *MMS* (Manufacturing Message Specification). The result is a comfortable but in coding and decoding time consuming protocol for communication between HMIs as clients and protection IEDs as servers. Between IEDs there are two time critical services, that is, GOOSE and SV. These services are mapped directly to the link layer of Ethernet, that is, to layer 2 in the ISO/OSI model reaching the highest performance class of 4 ms for transmission time between IEDs, which is of outmost importance for the protection chain. The stack and the relationship to the data model are given in Figure 12.6.

Time synchronization for events requests accuracy of about 1 ms, which is provided between all IEDs in the system by SNTP (Simple Network Time Protocol) over Ethernet. Time synchronization for synchronized sampling with accuracy of 1 μs is provided either by 1 pps (pulse per second) over a dedicated link (Edition 1 of IEC 61850) or by IEEE 1588 [8] over Ethernet (Edition 2 of IEC 61850 or Amendment).

The goal of IEC 61850 is the interoperability but by its comprehensive standardization of all communication features with serial links the standard will finally result in replacement of all copper cables for signals by serial links, which have to be realized at least outside the cubicles for EMC reason as optical fibers.

According to IEC 61850, the source of voltage and current samples from one measuring point is the Merging Unit (MU) merging the currents and voltages (e.g., $3 \times I$, I_0, $3 \times U$, U_0) in one message (telegram). Depending on the inputs, the MU also performs the sampling (and resampling if necessary) and the following A/D conversions as the S/H and A/D unit in the IED structure shown schematically in Figure 12.5. If implementing the synchronization accuracy of 1 μs according to IEEE 1588 [8], it will possible in some future also to restrict the MU to a dedicated unit per measuring point, that is, in the extreme case to one for each current or voltage sensor if beneficial for the user.

More details are found in the documents of the standard IEC 61850 [5] and in a large number of publications. An early application example for protection is found in [9].

12.3 THE PROTECTION CHAIN IN DETAILS

12.3.1 Copper Wires vs. Serial Links

It should be noted that the whole protection chain has to be considered, that is, not only the protection relay but also the DC supply (battery), the measuring transformers (CTs and VTs), the wiring, the auxiliary relays, the tripping coils (of the relay and of the breaker), the breaker itself. For this reason, the equipment around the protection relay itself is shortly analyzed. As mentioned above, considering the backup needs also all these equipment has to be considered. To discuss all these issues, the simple scheme of Figure 12.2 is now used with more details as shown in Figures 12.7 and 12.8.

This chapter refers both to conventional hardwired copper connections and to fiber optic serial connection between the components, that is, everywhere between the sensor, protection, and the actuator. The implementation, that is, the realization of the protection chain may be totally different. Independently from the communication solution, the requirements of the protection function regarding functionality, speed, and supervision have to be fulfilled. Common features and differences will be discussed and actual trends will be analyzed. Also the benefits of serial connections will be mentioned if applicable.

12.3.2 Supervision

Because of the importance of protection chain for the reliability of the power system at all possible point steps in the chain supervision may take place:

- CT and VT (instrument transformers): Current and voltage check
  - phase symmetry check
  - $U_0 \times$ not I_0 or $U_2 \times$ not I_2 and/or other checks

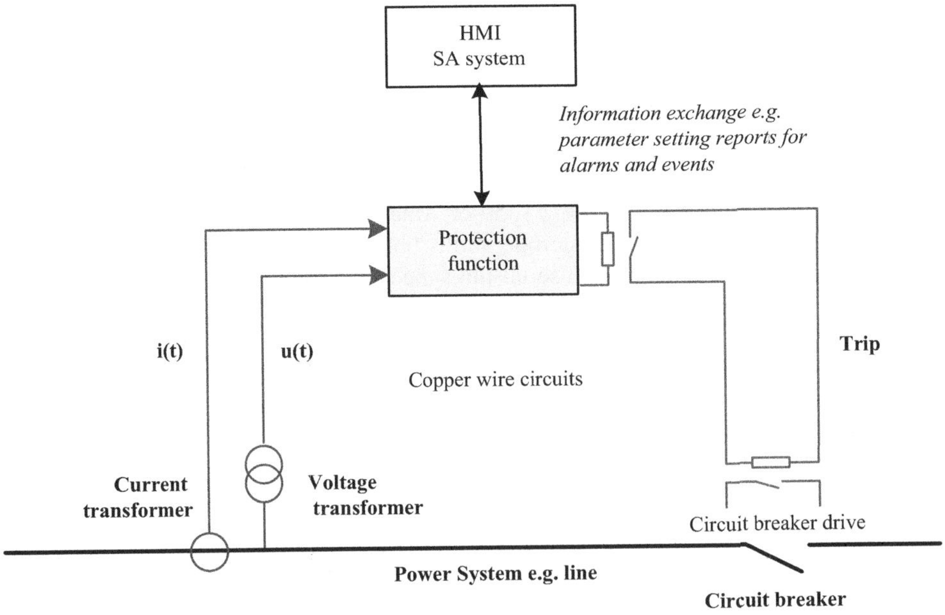

Figure 12.7. Protection chain hardwired.

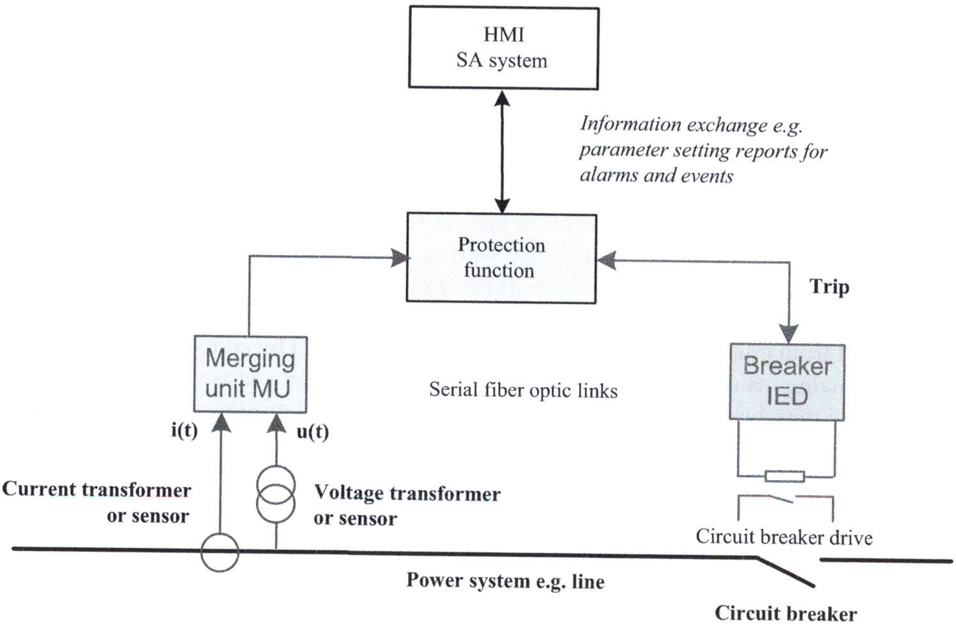

Figure 12.8. Protection chain with serial links.

- Data acquisition system for voltage and current
  - Hardwired connection allows plausibility checks for hardwired connection (see also CT and VT)
  - Serial connection allows checking the reception and errors of the telegrams
- Protection IED
  - Input transformers: see CT and VT above (current and voltage checks)
  - A/D conversion: continuous supervision of 2 reference signals
  - Program processing in CPU: Watchdog functions
  - Memory check: Read/write comparison & checksum function
  - Serial communication: Check the reception and errors of serial telegrams
  - Power supply: tolerance checks of the external and internal voltages
- Data sending system for issuing the protection trip
  - Hardwired connection allows check the trip circuit by a small test current
  - Serial connection allows checking the reception and errors of the telegrams
- Circuit breaker
  - Supervision of the isolation gas (SF_6) density
  - Opening time of contacts
  - Coherency of all the three-phase contacts
- Station battery
  - Battery supervision including earthing

12.3.3 Values Measured for Protection

12.3.3.1 Nonelectrical Values. One example for a measured nonelectrical value is the temperature inside the power transformer (hot spot) or the pressure inside a GIS volume, which shows a strong increase in case of an internal fault. Other examples are vibration inside the generator (rotor) or the degassing of the oil in a power transformer (Buchholz protection).

Generally, these types of protection have at least two limit values, that is, an *alarm* limit in order to inform the maintenance staff that the installation has reached an alert state and a *trip* limit in order to switch off the protected object that is not able to remain in service. An example is the insufficient isolation. This limit crossing may be calculated from continuously measured values or signalized by contacts of a dedicated sensor. Analog values continuously measured by dedicated sensors like temperature or pressure may be provided as mA values (e.g., 4–20 mA) or by a serial communication link.

In addition, several binary signals are needed in the environment of protection, for example, "drive ready for the breaker trip" or the "position of switches like circuit breakers and isolators." They are commonly collected by using contacts of auxiliary relays but may also be acquired internally by electronics is, for example, integrated in the switchgear and provided by serial communication link.

12.3.3.2 Electrical Values. Most of the protection is based on electrical values, that is, on measured *currents* and/or *voltages*. Relatively simple protection is based only on one of these values, that is, only on current(s) or only on voltage(s).

More sophisticated protection is based on calculated values, current and/or voltage measurement, for example, on the *difference* between the currents at both or many ends of the protected object and zone (differential protection), calculated *impedance* (distance protection scheme), *phasors* (e.g., for system protection schemes), *frequency* (e.g., for load shedding), or *power* (e.g., for over or und load protection) calculated out of current and voltage. The calculation could be done in protection devices.

12.3.4 Data Acquisition from Sensors

12.3.4.1 Sensors. Electrical measured values are collected via current and voltage IT or, more generally named, by current and voltage sensors. These instrument transformers can be conventional (CIT) or nonconventional (NCIT) ones. Some examples are shown in Figure 12.9.

They have to assure the *electric separation* between the primary (High voltage) and the secondary parts (LV, equipment for protection, control, monitoring, etc). They provide also the *transformation* of the primary process signal (U in kV, I in kA) to a level compatible for the protection equipment (e.g., 1A by conventional current transformers, $110/\sqrt{3}$ V by conventional voltage transformer). A first level of electric separation is done by the magnetic field of instrument transformer, a second one by the input transformer of the IED. Opto-couplers at the input of the IEDs improve the galvanic separation referring to the coupling by light. It should be noted that opto-couplers also represent a capacitance between input and output and, therefore, do not block completely high-frequency noise. The best electric separation is reached if the copper wires between the sensors and the protection IED are replaced by fiber optic links, but this implies always a transmission of analog values in telegrams.

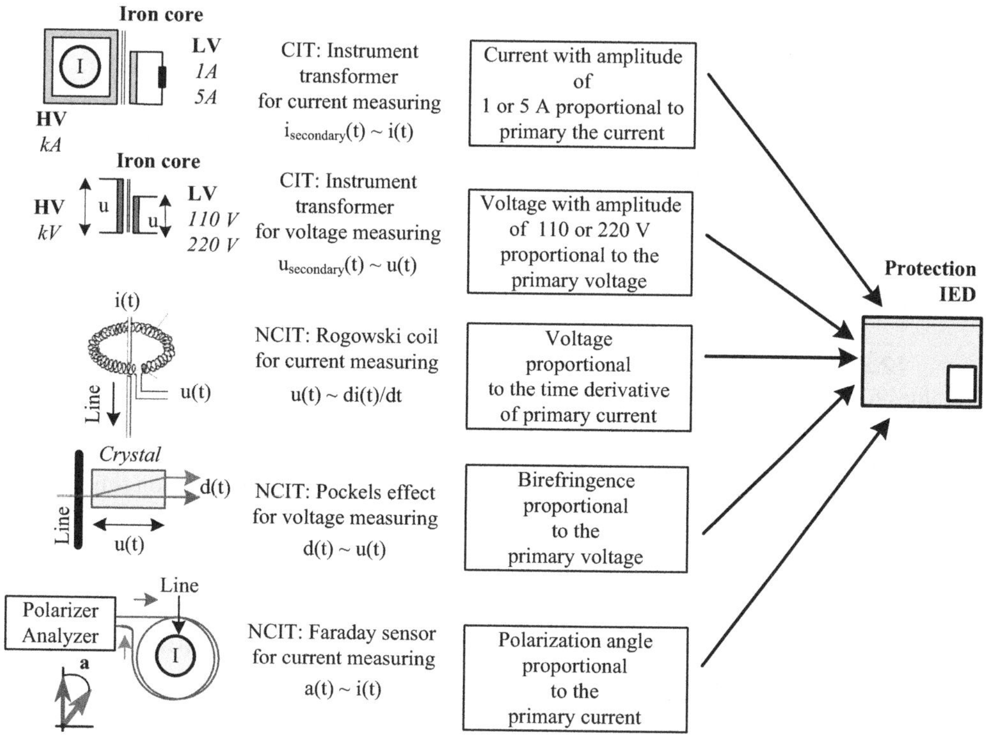

Figure 12.9. Examples for instrument transformer.

The *dynamic behavior* of the instrument transformers (CIT or NCIT), that is, the step response and the frequency response curve have to be considered carefully because it can influence dramatically the behavior of the protection itself.

Conventional transformers (CIT) are magnetic analog ones with iron core working like transformers; the secondary values are proportional to the primary ones at least in steady state case without saturation. The secondary values may deviate from the primary one by transformer *saturation* in case of high input currents or in case of transient phenomena. The output values are fed in the protection equipment using copper wires.

Nonconventional instrument transformers (NCIT) may be realized by electrical and nonelectrical measuring principles. The detailed discussion of all these principles goes beyond this chapter but some are indicated in Figure 12.9 and shortly described below.

- The electrical principle is used, for example, by current measuring with the Rogowksi coil having no iron core and, therefore, providing on secondary side a signal proportional to di/dt needing spike limitation and proper integration. An example for voltage transformers is the capacitive divider where the high primary voltage is divided down to the protection input voltage level by capacitances.
- Nonelectrical principles are base on opto-electrical effects. One example is the Faraday effect where the magnetic field of the current to be measured changes the angle of polarized light sent through a fiber enclosing the current. An example for voltage measuring is the Pockel's effect where the incoming light is split in two paths

differing according to the voltage applied. The secondary values are proportional to the current and voltage, respectively.

The secondary values of NCITs are of different nature and, therefore found no acceptance in the past. Today, all such analog values are converted to numbers and send in standardized telegrams over a serial link to the protection equipment. These telegrams are the common denominator for all instrument transformers or sensors and may be also used for CITs if their values are converted accordingly. The use of serial telegrams is strongly boosted with the definition for the transfer of samples in the standard IEC 61850 (so-called Process Bus as defined in part IEC 61850-9-2 [6,10,11]).

12.3.4.2 A/D Conversion and Merging Unit. Without NCIT and process bus the values for current and voltage a transferred by copper wires to the inputs of the protection IEDs (1 or 5 A, $110/\sqrt{3}$). The A/D conversion provides the samples for the numerical processor. This analog values for current and voltage may be supervised at more than one place but the decisive point is the output of the A/D converter. The supervision is based, for example, on the transformer ratio verification, on the properties of the three-phase system, and on the plausibility between current and voltage readings. The A/D conversion itself is often tested by the conversion of a reference signal.

In case of nonconventional instrument transformers and/or the use of process bus, the values are transferred as serial telegrams to the inputs of the protection IEDs and handled directly by the microprocessor.

The A/D conversion still needed is moved from the IED toward the measuring point (sensor). For this purpose, the IEC TC38 has defined a so-called MU, which was introduced by IEC TC57 into IEC 61850 as source of the standardized telegrams. The MU is collecting the current and voltage from the instrument transformers of any physical principle—restricted only by the input channels available in the MU—and provides these after the A/D conversion as telegrams according to IEC 61850. The term *merging unit* indicates that different values are merged into one telegram. In the common three-phase power system, it is convenient to merge all three or four currents and voltages from the one measurement point and the same instant in time into one telegram. This reduces not only the telegram overhead but also provides inherently *synchronized* values per merging unit allowing the calculation of *time coherent* phase-to-phase and phase-to-earth values. This coherency is also a prerequisite for protection functions, which calculate values from these samples like Z (impedance) in the distance protection. Note that the protection functions itself are described in section [6].

12.3.4.3 Time Synchronization. Currents and voltages to be time coherent not only for impedance (distance) calculation but also for the calculation of sums in nodes according to the Kirchhoff law (current differential protection) or for the transformation of the three phases to other systems like the common positive, negative, and zero sequence. Note that in conventionally wired protection IEDs with sample and hold (S/H) process and A/D conversion time coherence is of about 5 μs. In the MU, the order of 1 μs is provided.

If the samples to be coherent come from more than one IED or MU, which has to be assumed especially for differential protections (e.g., distributed busbar protection) these units have to be synchronized with each other to allow synchronized sampling with the order of 1 μs as defined as highest accuracy class in IEC 61850-5 [12]. The time synchronization method used today between different IEDs like two or more MUs is the 1 PPS method. In the near future, such synchronization will be possible over the Ethernet-based serial bus according to IEEE 1588 [8].

Comparing events or samples from different substations a global time reference is needed which may be provided most convenient by one master clock per substations using the time signals from the satellite-based GPS.

12.3.5 Protection Data Processing

12.3.5.1 General. Protection equipment based on electro*mechanical* or *solid state technology* can only process the acquired data in a limited way and deliver a limited amount of information only, that is, some few basic binary signals like start and trip in case of a fault, which phase is involved in the fault, and so on. These signals are given via binary contacts. *Numerical relays*, that is, microprocessor-based IEDs may easily do any kind of evaluation of the input data and provide much more information about the supervised values, especially also in case of a fault. The most important trip signal is only one of much information.

12.3.5.2 Trip Decision and Related Information. Information related to the fault itself is provided like as follows:

- Trip (command) to be sent the to the circuit breaker.
- Calculated values dedicated for the protection function (current, impedance, etc.).
- Crossing of a preset limit (protection parameter, characteristics) indicating a fault.
- Snapshot of the fault peak of the supervised value if applicable.
- Information about the fault and the trip to be sent to higher level HMI (with 1 ms time tag for alarm list and event list, for short report and for any other kind of display or storage).
- Number of trips with corresponding tripping current or more precise calculations providing the circuit breaker wear of the circuit breaker in fault clearing.
- Disturbance records with analog and binary values.

It should be noted that in one IED there could be many protection functions active (function library for a hardware platform).

12.3.5.3 Other Data Handling Features. In addition, more general information or functions not directly fault oriented may be also provided like

- reading of actual current and voltage values;
- management of parameter and parameter lists including parameter changes initialized from local or remote;
- self-supervision;
- operation modes (e.g., test mode) including mode changes initialized from local or remote.

12.3.6 Data Sending to the Actuators

The decision of the protection processing has to be communicated to the actuator, that is, to the circuit breaker.

This could be done conventionally by copper wires and relay contacts forming a trip circuit connecting the trip contact of the protection with the related trip coil in the breaker.

This trip circuit should be as short and direct as possible. The number of intermediate auxiliaries should be reduced as much as possible. The supervision is commonly done by injecting a small current below the tripping level of the coil (contact at the circuit breaker). Furthermore, some periodical tests of the tripping contact inside the protection IED itself may be done.

If a serial link is used between the protection IED and the circuit breaker, an IED commonly named Breaker IED (BIED) is needed integrated or nearby to the circuit breaker in order to operate the breaker and get the breaker response. The supervision of this communication link complements the self-supervision of the IEDs on both sides. All these supervision procedures are working continuously and providing a high reliable "trip circuit."

In some applications, a trip has to be maintained until the maintenance people are resetting the trip locally (lock-out relay feature). This is done to avoid any reclosing command, which would reenergize the faulted object without the guarantee that this object is again healthy. Examples are the power transformer and the busbar. An equivalent behavior has to be provided by the BIED also.

12.3.7 Process Interface

In case of hardwired connection between sensors and protection and between protection and actuator (circuit breaker), the process interface comprises all these copper circuits. To have all interactions and interferences under control, a clear separation between primary (switchgear) and secondary (protection, control) technology has to be provided. To allow easy engineering and maintenance, this process interface should be localized near to the switchgear. Input transformers provide a barrier for analog values like U and I, auxiliary contacts, input relays, and opto-couplers provide the same for binary indications and signals. The process interface contains the exchange of data in both directions.

If serial links are applied a maximum decoupling between the protection IED and sensors, and actuators is given by the fiber optic links. The process interface as such has an electronic part, that is, the IEDs both at sensor (MU) and actuator (BIED). Depending on the state-of-the-art of integrating electronics in switchgear, still some short wires, dedicated fibers or other means may be left between these IEDs and the switchgear.

12.3.8 Circuit Breaker

The circuit breaker has to be able to interrupt any short circuit, which is detected by the protection as fast as possible. Therefore, the breaking capability of the circuit breaker has to be carefully supervised.

The fast contact movement needed for the circuit breaker is enabled by proper energy storage for this high mechanical acceleration needed. The energy storage is realized either a spring or a pressurized nitrogen tank. This energy storage of the breaker drive is supervised to guarantee operation of the breaker in general and, especially at least one Opening (O) after closing by the operator or by the autorecloser. This switching capability is often called CO. If the trip (O) of the protection and autoreclosing is seen as one sequence the request may be also formulated as OCO.

The contact nozzles have to be also in good shape to providing contact, to allow for the pressure buildup needed for extinguishing the arc by a proper gas flow. But the supervision of the contact nozzle (breaker) wear is not so easy to perform.

Therefore, the first approach is to count the number of switching operation. This information is provided by the circuit breaker respectively the circuit breaker electronic only. The next level of approach is to sum up the number of switched amperes (fault currents). This can be done with existing means but needs in addition to the circuit breaker information also the peak fault current from the protection. More precise would be the integration of the arc power needing the arc voltage, the current, the contact opening time (start of the fault arc), and the time of the arc extinction. The integrand refers to the power and, therefore, if the arc resistance is known, it should be proportional to I^2. Since the arc and the arc length are not stable but being extended during the opening its resistance is not constant. Therefore, the exponent for the current is less than 2. The best guess is 1.8 stable for more than three decades and confirmed just in Ref. [13].

Since breaking the short circuit is the key task of the protection, the so-called *breaker failure protection* acts if the circuit breaker fails to open in case of a fault trip. This function will repeat in a first step the trip to this failed breaker (eventually to a second trip coil of the breaker) and in a second step sent a trip to all surrounding breakers in order to eliminate the fault in the power system. This increases the fault clearing time and decreases the selectivity but results in any case to switching off the fault current. Details will be discussed in Section 12.6.5.1 containing the protection principle.

12.3.9 Power Supply

The power supply is generally using the DC supply from the station battery (110 V DC or 220 V DC) and comprises of a DC–DC converter inside the IED providing the electronic DC voltages of some volts as needed. This DC power supply is not only used supplying the IEDs but also input and output circuits, for example, for position indications, control commands, and protection trips. The station battery has its own supervision system covering also its load status. The power supply supervision in the IED is based on measuring both the input voltage coming from the battery and the voltages provided to the electronics of the IED. Independent from the IEDs the battery itself is supervised on voltage and stored energy (load status) to be reloaded continuously or at least in time.

In some specific application, the power supply may be taken directly from the protected object (e.g., from the overhead line) but these systems have to respect the isolation level and need to have enough power buffer capacity for power-off times. Also the case of energizing both the protected line and the power supply of the protection IED has to be considered. The IED and, therefore, the readiness of the protection functions will lag always behind.

12.4 TRANSMISSION AND DISTRIBUTION POWER SYSTEM STRUCTURES

Traditionally, transmission and distribution power systems are classified according to the voltage level as seen in Table 12.1. It is seen that the voltage ranges are overlapping. In real power systems, we will find in rural areas power systems that have the function of transmission systems also if the voltage is at about 60 kV and below. On the other side, power systems with voltage up 220 kV distribute the power in heavy load centers like to the center of big cities. If we look at the voltage only, we get the impression that there is no obvious reason to use different protection functions at different voltage levels.

There is a difference in the value of the protected object, for example, HV transformers are much more expensive than MV ones. Another issue is the impact of the destruction of a component on the power system and supply. This may result in the request for a higher trip

TABLE 12.1. Allocation of Voltage Levels to Power System Types

Functional Name	Main Purpose	Voltage Level	Allocated Voltages
Power transfer power systems	Bulk power over large distances	EHV (extra high voltage)	400–750 kV
Transmission power systems	Bulk power over short and medium distances	HV (high voltage)	50–500 kV
Distribution power systems	Distribution of power	MV (medium voltage) Systems	3–170 kV

speed (interrupt higher fault power faster) and in more protection functions applied. There may be a difference only in the requested trip speed and in EMC.

If we like to discuss the differences in protection of transmission and distribution networks in more details, we have to consider differences in the structure of these power systems:

- *Distribution power systems* have to distribute power from one infeed point to many consumers. Therefore, they are of radial type or of small rings with one infeed (Figure 12.10). The protection may be done, for example, with a simple protection scheme out of overcurrent protection (Figure 12.18).
- *Transmission power systems* have to guarantee the provision of power to all infeed point of distribution power systems and control the complete power flow. Therefore, they have multiple infeed and as highly meshed structure (Figure 12.11). The protection has to refer to the more complex structure using, for example, distance protection (Figure 12.24) and high speed dedicated busbar protection.

It may be easily seen from Figures 12.10 and 12.11 where possible faults currents are flowing and where the chance for protection is to interrupt these currents and clear the

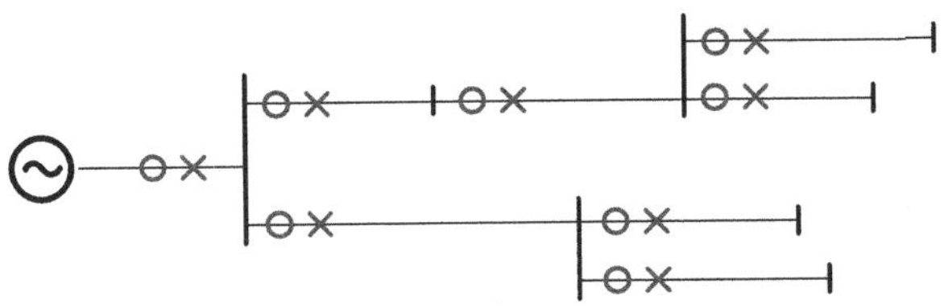

(a)

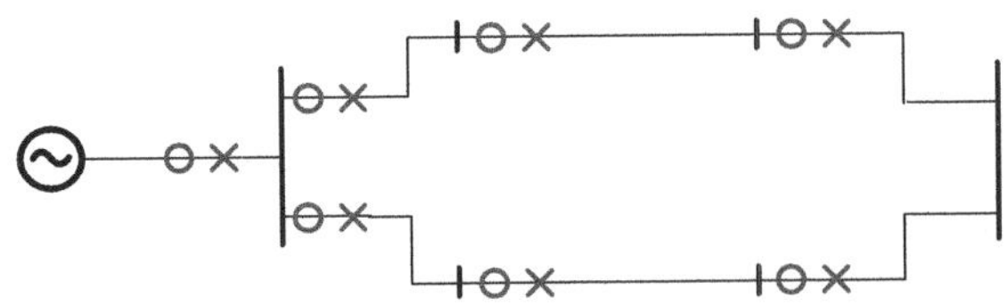

(b)

Figure 12.10. Principle structures of distribution power systems: (a) radial network; (b) ring network with single infeed.

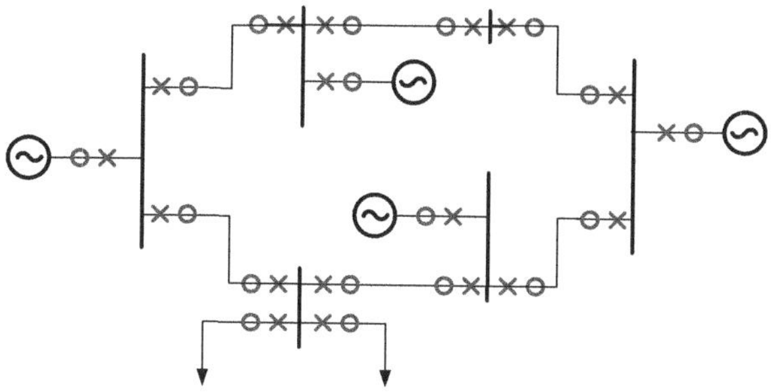

Figure 12.11. Principle structures of transmission power systems.

faults. In distribution power systems by opening one breaker (tree) or two breakers (ring), all downstream short circuits are cleared but with relatively low selectivity, that is, a lot of power users may be impacted. In transmission power systems for the clearance of a fault, two or more circuit breakers have to be opened to stop the fault current fed from different sources. But if the faulted part is selectively switched off, the power flow may have in the meshed power system structure alternative parts. The number of impacted power users may be minimized.

12.5 PROPERTIES OF THE THREE-PHASE SYSTEMS RELEVANT FOR PROTECTION

12.5.1 Symmetries

The three phases of the power system are generated symmetrically, that is, with the same *amplitude* (here I_0 and U_0) and *power frequency* $f = \omega/2\pi$ but with a *phase difference* of 120° respectively $2\pi/3$ (Figure 12.13). Between current and voltage there is a *phase difference* φ depending on the power system conditions, which will be discussed later.

$$
\begin{aligned}
u_a(t) &= U_0 \sin(\omega t - \varphi) & i_a(t) &= I_0 \sin(\omega t) \\
i_b(t) &= I_0 \sin(\omega t - 2\pi/3) & u_b(t) &= U_0 \sin(\omega t - 2\pi/3 - \varphi) \\
i_c(t) &= I_0 \sin(\omega t - 4\pi/3) & u_c(t) &= U_0 \sin(\omega t - 4\pi/3 - \varphi)
\end{aligned}
$$

Note: I_0 may have a different meaning in different power system equations. Here I_0 represents the amplitude of the sinusoidal current.

Because of the symmetry, it is convenient to connect all three phases in one star point. This is in line with the three-phase transformers at generator or other places where such star points are realized. The symmetry results in the equation

$$i_a(t) + i_b(t) + i_c(t) = 0$$

which means that not return conductor is needed. Therefore, the symmetric three-phase system consists of only three conductors, which handle all the power transfer. It should be noted that this is valid only for a symmetric three-phase system (Figures 12.12).

Using the Euler relationship

$$e^{jx} = \cos x + j \sin x$$

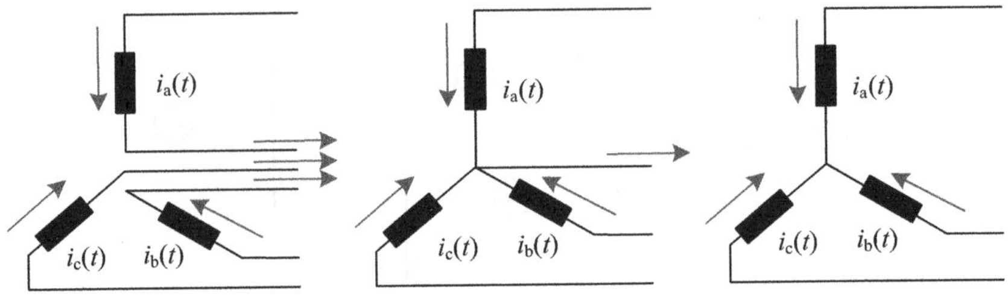

Figure 12.12. Three-phase system with star point connection: (a) the three-phase loops; (b) the three-phase loops combined in the (transformer) star point; (c) the three-phase currents add up zero in case of symmetry.

sinusoidal values for current and voltage may be transformed to *phasors*, that is, to vectors with amplitude and angle in the complex plane:

$$\underline{A}_a(t) = A_0 e^{\omega t}$$
$$\underline{A}_b(t) = A_0 e^{\omega t - 2\pi/3}$$
$$\underline{A}_c(t) = A_0 e^{\omega t - 4\pi/3}$$

The term ωt causes the rotation of the phasors (Figure 12.13). If we are interested only on relations between the three phases we my neglect this term or, mathematically more correct, transform the phasors in a coordination system rotating with ωt.

12.5.2 Unbalance

In the *normal* operation state of the power system, the three phases are generally symmetric but there may be unbalances if the different phases are not equally loaded or by unbalances in the transmission line. In addition, the symmetry is strongly broken in case of any nonsymmetrical fault, for example, single-phase-to-earth or phase-to-phase fault (*emergency* state). Such unbalanced faults are the most common ones. The impact of unbalance on the currents is shown in Figure 12.14.

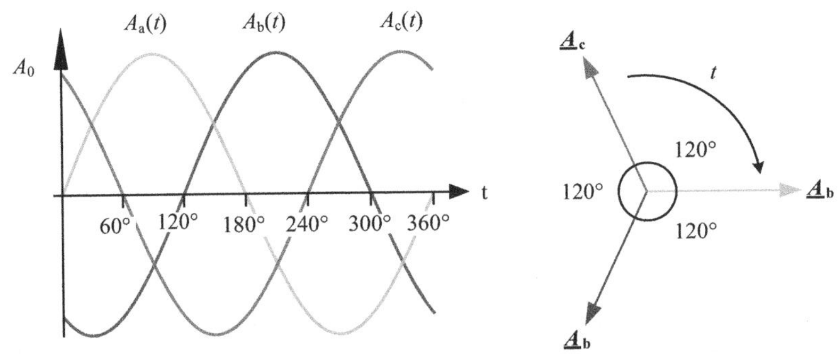

Figure 12.13. From sinusoidal representation to rotating phasors: (a) the degrees indicated the phase differences between the sinusoidal values (current or volatages); (b) representation as rotating phasors.

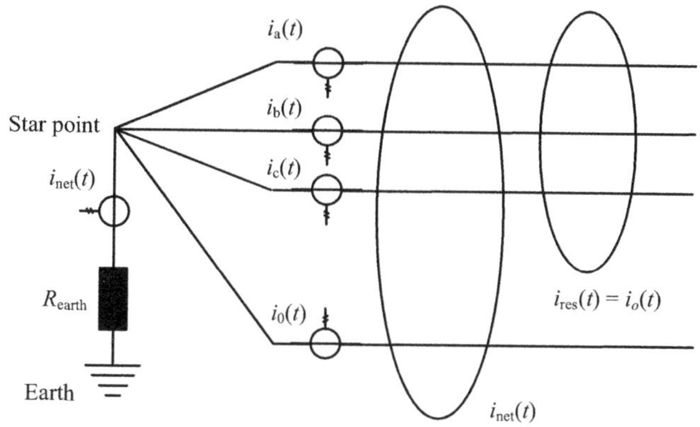

Figure 12.14. Currents caused by unbalance of faults in the three-phase systems.

If the system is symmetric (no fault) and not earthed, the sum of the phase currents is zero:

$$i_a(t) + i_b(t) + i_c(t) = 0$$

The star point of *transmission* systems is normally strongly earthed, that is, the potential of the "Star point (N)" is equal to "Earth." There exists no fourth conductor for closing the current flow at potential 0 unequal to earth. Nevertheless, there is the so-called "protective earth" (potential 0 = Ground) on top of overhead lines to screen the phase conductors against lightning strokes. It also helps to equalize and reduce the earth impedance, which may vary from tower to tower and to restrict the surges to the healthy phases. Today, it is also commonly used as container for fiber optic communication cables.

Because of the line-ground capacitance, very small capacitive currents may flow from the conductors to earth and return. Since all three phases contribute about the same to this effect it will have no impact on the unbalance and neglected by protection. If there is an asymmetry by uneven loads, unequal phase conductors, or faults, the exceeding return current comes back by both the earth and ground wire.

In *distribution* systems, the star point may be earthed alternatively by

- earth resistor R_{earth} with or without bypass switch;
- dynamic reactance (suppression or Petersen coil), which is tuned to keep the short-circuit current to a minimum by compensating the capacitance of the power system (grid);
- a combination of dynamic reactance and (switchable) resistor;
- not earthed at all.

In all theses cases, there is a potential difference between the star point (N) and Earth. Therefore, unbalance in the phases caused by a fault may create a net current to/from Earth i_{net}. The sum looks according to the Kirchhoff's node Law as:

$$i_a(t) + i_b(t) + i_c(t) + i_{net}(t) = 0$$

The current i_{net} may be calculated or more conveniently measured. If this current goes over some predefined level, a fault is assumed (detected) and the related breaker tripped.

In lower voltage *distribution* power systems, the star point may be not grounded at all and its potential 0 respectively the residual current named I_{res} or $\underline{I}_0$ may be transported via a fourth conductor not only as earth wire for overhead lines but also as conductor in cables. The Kirchhoff's law results in:

$$i_a(t) + i_b(t) + i_c(t) + i_{res}(t) = 0$$

respectively

$$i_a(t) + i_b(t) + i_c(t) + i_0(t) = 0$$

If also in this case the star point is grounded, the sum of all involved currents is zero:

$$i_a(t) + i_b(t) + i_c(t) + i_{net}(t) + i_{res}(t) = 0$$

If one of these currents, especially i_{net} and/or i_{res} goes over some predefined level, this event may be used both for a phase fault detection and short-circuit trip.

12.5.3 Symmetrical Components

Instead of measuring the additional components i_{res} and/or i_{net} as discussed in Section 12.5.2, the easy and everywhere measurable three phases are transformed into other alternative systems retaining the three-phase system properties but facilitating the asymmetry and fault detection and all derived protection functions, respectively.

The *asymmetric three-phase system* in phasor representation ($\underline{A}_a$, $\underline{A}_b$, $\underline{A}_c$), that is, amplitudes different in each phase and the angle between the phases unequal to 120° ($2/3\pi$) may be transformed in the so-called *symmetrical components* ($\underline{A}^+$, $\underline{A}^-$, $\underline{A}_0$) as seen in Figure 12.15. The *symmetrical components* consist of three symmetrical three-phase systems with different length represented each by one phasor, that is:

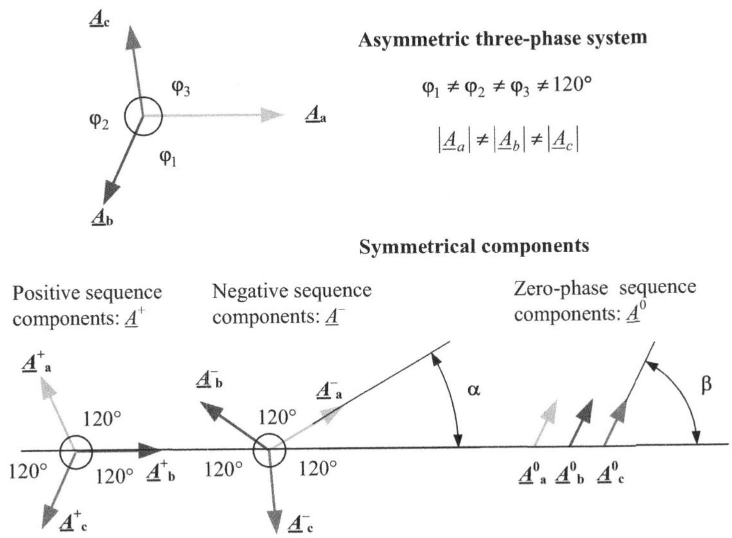

Figure 12.15. Transformation of the three phases into symmetrical components.

- Positive sequence components ($\underline{A}_a^+$, $\underline{A}_b^+$, $\underline{A}_c^+$) consisting of three symmetrical phasors, that is, with the same length and the angle difference 120° (2/3π) rotating in the same direction as the original system with angular velocity ω with the original phase sequence a → b → c.
- Negative sequence components ($\underline{A}_a^-$, $\underline{A}_b^-$, $\underline{A}_c^-$) consisting of three symmetrical phasors, that is, with the same length and the angle difference 120° (2/3π) rotating in the same direction as the original system with angular velocity ω but with the opposite phase sequence a → c → b.
- Zero sequence components ($\underline{A}_a^0$, $\underline{A}_b^0$, $\underline{A}_c^0$) consisting of three symmetrical phasors, that is, with the same length and the angle difference 0° (0π) rotating in the same direction as the original system with angular velocity ω where the phase sequence has no meaning.
- There is a fixed angle difference α between the positive sequence components, for example, $\underline{A}_a^+$ and the negative sequence components, for example, $\underline{A}_a^-$ as seen in Figure 12.15.
- There is a fixed angle difference β between the positive sequence components, for example, $\underline{A}_a^+$ and the zero sequence components, for example, $\underline{A}_a^0$ as seen in Figure 12.15.

The transformation equations are as follows [14]:

$$\begin{bmatrix} \underline{A}_a^+ \\ \underline{A}_a^- \\ \underline{A}_a^0 \end{bmatrix} = \frac{1}{3} \begin{bmatrix} 1 & \underline{a} & \underline{a}^2 \\ 1 & \underline{a}^2 & \underline{a} \\ 1 & 1 & 1 \end{bmatrix} \begin{bmatrix} \underline{A}_a \\ \underline{A}_b \\ \underline{A}_c \end{bmatrix}$$

The extension to the three phasors per symmetrical sequence reads as:

$$\begin{bmatrix} \underline{A}_a^+ \\ \underline{A}_b^+ \\ \underline{A}_c^+ \end{bmatrix} = \begin{bmatrix} 1 \\ -\underline{a} \\ -\underline{a}^2 \end{bmatrix} \begin{bmatrix} \underline{A}_a^+ \\ \underline{A}_a^+ \\ \underline{A}_a^+ \end{bmatrix} \text{ and } \begin{bmatrix} \underline{A}_a^- \\ \underline{A}_b^- \\ \underline{A}_c^- \end{bmatrix} = \begin{bmatrix} 1 \\ -\underline{a}^2 \\ -\underline{a} \end{bmatrix} \begin{bmatrix} \underline{A}_a^- \\ \underline{A}_a^- \\ \underline{A}_a^- \end{bmatrix} \text{ and } \begin{bmatrix} \underline{A}_a^0 \\ \underline{A}_b^0 \\ \underline{A}_c^0 \end{bmatrix} = \begin{bmatrix} 1 \\ 1 \\ 1 \end{bmatrix} \begin{bmatrix} \underline{A}_a^0 \\ \underline{A}_a^0 \\ \underline{A}_a^0 \end{bmatrix}$$

The transformation back using $\underline{A}_X = \underline{A}_X^+ + \underline{A}_X^+ + \underline{A}_X^0$

$$\begin{bmatrix} \underline{A}_a \\ \underline{A}_b \\ \underline{A}_c \end{bmatrix} = \begin{bmatrix} 1 & 1 & 1 \\ -\underline{a} & \underline{a}^2 & 1 \\ -\underline{a}^2 & \underline{a} & 1 \end{bmatrix} \begin{bmatrix} \underline{A}_a^+ \\ \underline{A}_a^- \\ \underline{A}_a^0 \end{bmatrix}$$

This transformation may be done both for voltages and currents. It is used mostly for *currents* to detect very sensitive the fault currents in the symmetrical sequence components $\underline{A}^-$ and $\underline{A}^0$ (I^- and I^0) deviating from zero.

12.6 PROTECTION FUNCTIONS SORTED ACCORDING TO THE OBJECTS PROTECTED

12.6.1 Protection Based on Limits of Locally Measured Values

If one or more parameters (current, voltage) are locally measured and these measurands provoke by limit crossing the decision to trip a local breaker then the fault is cleared if not a second infeed to the fault exists. This limits may graphically look like a simple boundary line or have complex structures, for example, in the complex impedance plane but they are called the *characteristic* of a protection function. In this chapter, some *characteristics* are mentioned but for the multitude of all the details the reader has refer to the fast amount of protection books or the data sheet of a dedicated protection IED.

This kind of local protection is very well suited for distribution networks with one infeed only as shown in Figure 12.10. A typical protection for this application is the *overcurrent protection.*

12.6.1.1 Overcurrent and Time Overcurrent Protection. If the protection trips immediately after crossing the preset limit, we have a Protection with an *Instantaneous Overcurrent Function.* If the trip is delayed, we have a Protection with a *Time Overcurrent function.* The delay is fix as soon the fault current has reached a certain limit. This protection is called *Definite Time Lag* (DTL) *function* ($I>$). Generally, this type of protection has a second set of settings with short time trip in case of very high currents ($I\gg$). If the overcurrent functions responds continuously faster to heavier currents, it is called *Inverse Definite Minimum Time Lag* (IDMT) *function.* Also other curves may be applied. These protection functions can be used as *main protection* in distribution networks, that is, mainly in radial networks with single infeed or as *backup protection* in transmission networks. To avoid that in case of high currents, the inverse overcurrent backup function trips faster than the main protection, always this minimum time lag (IDTM) has to properly set (Figure 12.16).

12.6.1.2 Overload Protection. The load, that is, temperatures inside the protected object may often not be measurable with reasonable effort but have to be simulated based on current measurement with appropriate model parameters inside the overload function. Any load bias is taken into account by this thermal replica in accordance with the heating and cooling curves. Alarm signals or trip commands are issued if in this thermal replica set temperature limits are exceeded. Such overload functions are used for objects that can overheat like transformers and motors, but less commonly for cables or overhead lines for which the thermal replica is more complex. The quality of this protection depends strongly on the accuracy of the thermal replica and the parameters used. An example according to

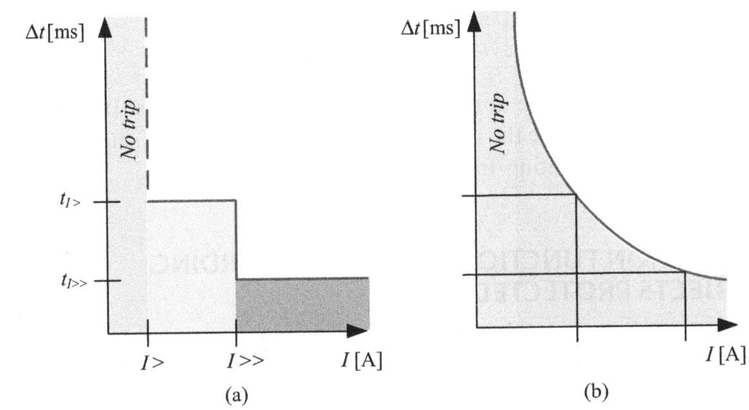

Figure 12.16. Characteristics of overcurrent protection: (a) two stage DTL; (b) IDMT function.

IEC 60255-8 [15] is given in the following equation:

$$\frac{d\Theta}{dt} + \frac{1}{\tau_{oil}} \times \Theta = \frac{1}{\tau_{oil}} \times \left(\frac{I}{I_{\max}}\right)^2 \qquad (12.1)$$

where

Θ is the oil temperature (°C) either at top-oil or hot-spot;
I is actual transformer load (A);
$I_{\max}$ is maximum permissible load current (A);
τ_{oil} is appropriate oil time constant, either top-oil or hot-spot.

12.6.1.3 Frequency Protection. The frequency is evaluated based on one local voltage measurement. The deviation of the frequency from the rated one is a good indication for an unbalance between produced and consumed active power (P) as seen from the P–f relationship valid in power system. Therefore, if the frequency (f) goes stepwise above ($f>f_{hl}$) or below set limits ($f<f_{Lim1} f_{Lim2}$) or decays (df/dt) at an unacceptable rate, quick action is needed.

In case of underfrequency load shedding, in case of overfrequency generator shedding is indicated. A more complex reaction is islanding, that is, the split of the power system in self-containing parts. Generally, load shed has to be not more than necessary to avoid unwanted interruption of power supply for the consumers. The best response is given if not only breakers are opened according to some predefined priorities but according to the actual measured load behind the breakers. This improves the classical frequency protection and load shedding to an optimized or adaptive load shedding (Figure 12.17).

This protection cannot be seen as a local one because local measurement (maybe complemented with decentralized load measurement) may act on the local and many remote breakers.

12.6.1.4 Voltage Protection. There are undervoltage and overvoltage protection functions because both overvoltages and undervoltages may cause malfunctions at user's site (power quality) and may damage the protected object. Examples are the overvoltage for power transformers and the undervoltage for generators.

The voltage used for protective decisions can be the directly measured phase-to-phase or phase-to-earth voltage separately per phase or some calculated voltages representing the three-phase system as such. This transformation is enabled by three-phase properties and the results are the so-called symmetrical components, that is, the positive, negative, and zero sequence voltage (Figure 12.15). It should be noted that the same is possible for the phase currents also.

12.6.1.5 Limit Supervision and Protection. Other protective devices used for dedicated objects in the substation include, for example, protection for interturn faults, for appearance of unwanted negative sequence values, and against reverse power flow for generators. Buchholz protection, temperature monitors, oil level indicators, oil and air flow indicators are used for power transformers. Insulation monitoring is common for all components on potential enclosed by an insulation medium like SF6, oil and dry insulation material. For gas (density) and for liquid (level) insulation, there exist at least two levels:

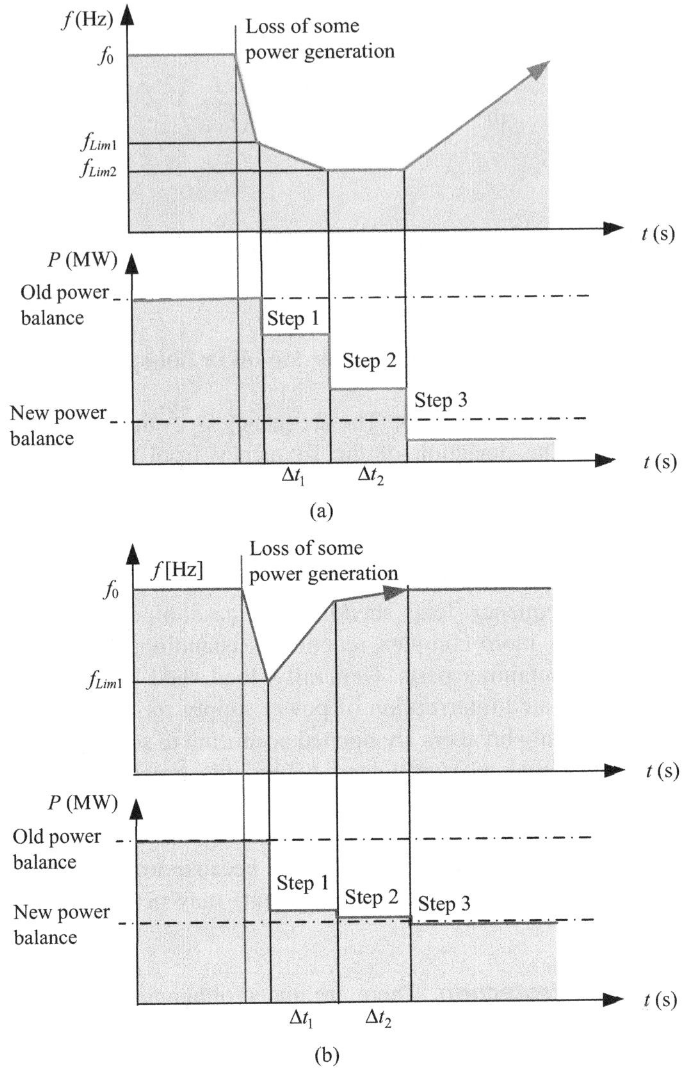

Figure 12.17. Classical frequency protection (a) and adaptive load shedding (b).

Level 1

Insulation still provided (> 100%) but medium below the start level

Supervision alarm and request for refilling the insulation medium!

Level 2

Insulation at the limit ($\leq$ 100%) or already below

Protection trip to remove voltage for expected danger!

12.6.1.6 Protection with Improvement of Selection by Time Delays. In a radial distribution power system with one infeed as shown in Figure 12.18, any fault should be cleared as downstream as possible to avoid that nonfaulted branches are switched off.

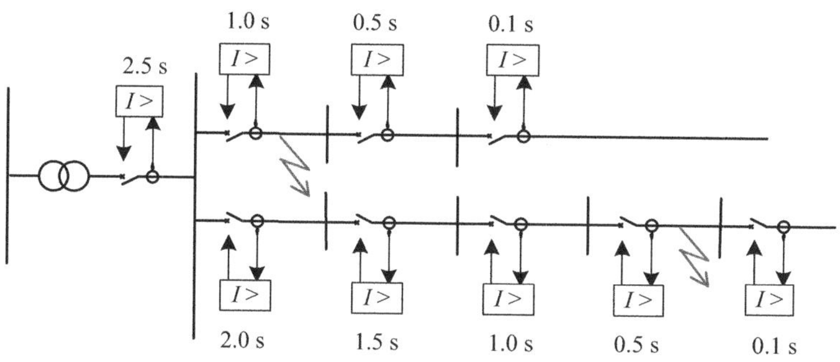

Figure 12.18. Grading (acceleration) of overcurrent delay times in radial networks.

The simple overcurrent protection used typically in such nets has neither fault direction detection nor a communication between the relays. Therefore, the time delay between start (fault detection) and trip is the only parameter to improve the selectivity. Grading this delay between and along the branches means shorter delays for more downstream located relays (acceleration). Protection IEDs near the end of the branches trip first in trying to minimize the impact on the branch. If not successful, then according to the longer delay the next upstream relays trips and so on until the protection next to the fault trips and stops the fault current. Typical faults locations are indicated as flashes in Figure 12.18. The only problem is that for upstream fault the clearing time may be reasonable long. Therefore, the proper grading of the delays is the key.

12.6.1.7 Protection with Improvement of Selection by Communication. Figure 12.19 represents a radial power system, typically in distribution level. The *over-current functions* are *time-graded*. That means the protection installed at the secondary side of the transformer feeding the busbar (upstream overcurrent function) is delayed compared to the ones installed in the feeders (see Figure 12.18). Therefore, a fault on the busbar will be cleared after an unwanted delay.

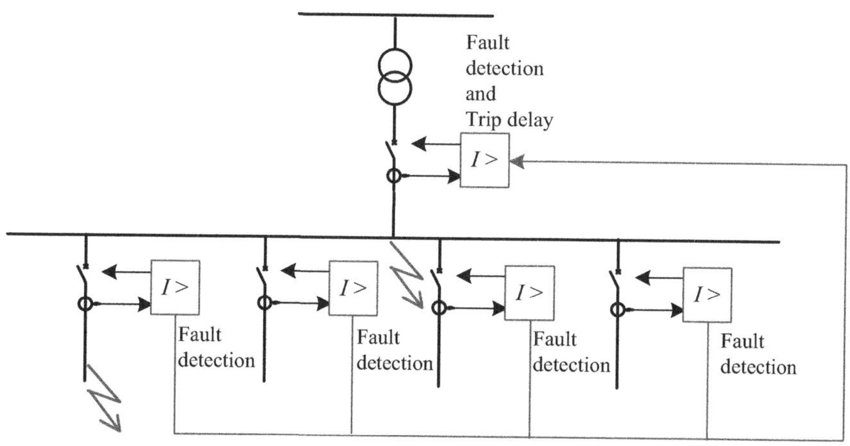

Figure 12.19. Reverse blocking of overcurrent relays in radial networks.

But it is possible to *accelerate* the tripping and fault clearance by using *communication* between the protection IEDs. If there is a fault at one of the distribution lines outgoing from the busbar (feeders), it will be detected both by the related feeder overcurrent protection and by the upstream overcurrent protection creating a start signal. If the fault is on the busbar, the upstream overcurrent function only will detect the fault.

If the feeder overcurrent protection has detected a fault, it will send its start signal to the upstream overcurrent protection. This protection will wait for tripping according to its time-graded delay giving the feeder overcurrent a chance to clear the fault. If the upstream overcurrent protection has detected the fault but within a reasonable short time no start signal comes from the feeder protection a fault at the busbar is assumed and the upstream overcurrent function will trip without further delay (protection or trip acceleration). This scheme is generally known as *reverse blocking* in the sense that the fault detection in the feeder is increasing the delay in the upstream relay, that is, blocking it for some time.

Also in case of loss of communication or of feeder protection failure, no acceleration takes place and the selectivity remains based on time-grading as in Figure 12.18.

12.6.2 Protection with Fault Direction Detection

12.6.2.1 Directional Protection. This type of protection measures locally *current* and *voltage* and is able to detect the *direction* of the fault. A fault changes the angle between the rotating phasors for current and voltage as seen in Figure 12.20. If the fault voltage phasor $\underline{U}_F$ is seen fixed at the vertical (imaginary) axis of the phasor plane, then the fault current phasor $\underline{I}_F$ will be in the first quadrant. If the fault is in forward direction and will be in the third quadrant, the fault is in the backward direction. The reason for this behavior is that the fault arc is a resistive element between phases or phase-to-earth changing the *line impedance* producing the mentioned phase angle difference. Therefore, we may see the fault direction also in the quadrant where the *fault impedance* $\underline{Z}_F$ calculated out of $\underline{U}_F$ and $\underline{I}_F$ is located.

$$\underline{Z}_F = \frac{\underline{U}_F}{\underline{I}_F} \tag{12.2}$$

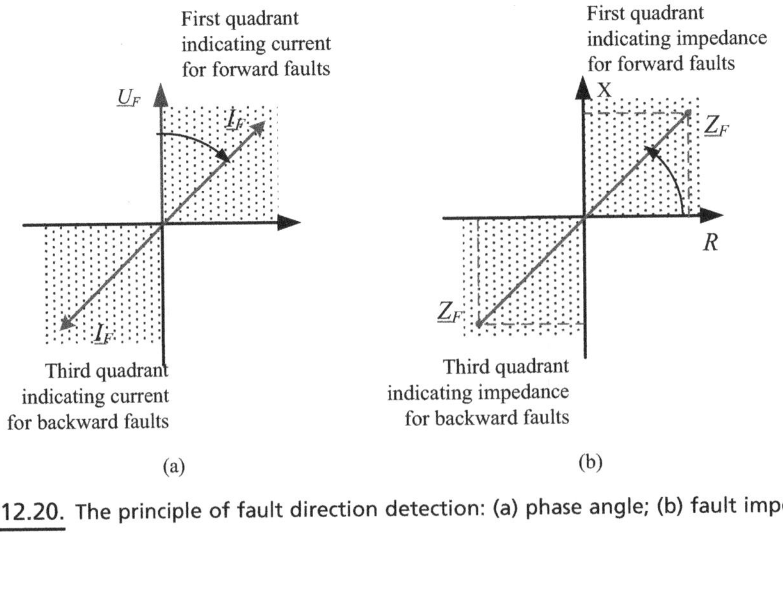

Figure 12.20. The principle of fault direction detection: (a) phase angle; (b) fault impedance.

As for all directional functions, current and voltage are used for calculations both have to be measured time coherently, that is, nearly exactly at the same time meaning a jitter of the order of 1 μs. This has to be considered especially if voltage and current are measured in different IEDs and being transmitted as samples in serial telegrams. It has to be regarded also that the voltage may be influenced by the short circuit. Especially in case of near fault in the power system, the voltage drops considerably and the directional measurement becomes difficult. Therefore, some so named "healthy voltages" are used, that is, voltages, which are not dropping due to the fault or which have been memorized before the fault has happened. Note that *forward* and *backward* have to be always defined.

The directional function can be added to the overcurrent function in order to build a so-called *Directional Overcurrent Protection*. IEDs with such a function are typically used for ring networks with single infeed (see Figure 12.10) or on the lower voltage sides of parallel operating transformers.

A special version is the *Directional Earth Fault Protection*, which is based on the neutral voltage $\underline{U}_0$ and neutral current $\underline{I}_0$.

12.6.2.2 Improvement of Directional Protection by Communication.

The directional protection (PDIR) gives the direction to the fault but not the line segment where the fault is located. Therefore, directional protection as such is used often at distribution level for rings with single infeed because the pure time delay grading alone is not suited for a ring since following a ring instead of a brunch we come to the infeed again. Starting with the same grading at both sides of the ring near to the source directional information will discriminate on what side of the ring the fault may be located.

If there is a communication between directional protections transmitting only the fault directions as seen by all related protections, the fault may be allocated to the faulted line segment as seen in Figure 12.21.

In case (a), the fault directions evaluated by the directional protections on both sides of the line show the opposite directions but both in direction of Line 12. Therefore, they agree that the fault is on the line segment between substation 1 and substation 2 (Line 12) and may trip the circuit breaker on each line end. In case (b), both directions to the fault are not

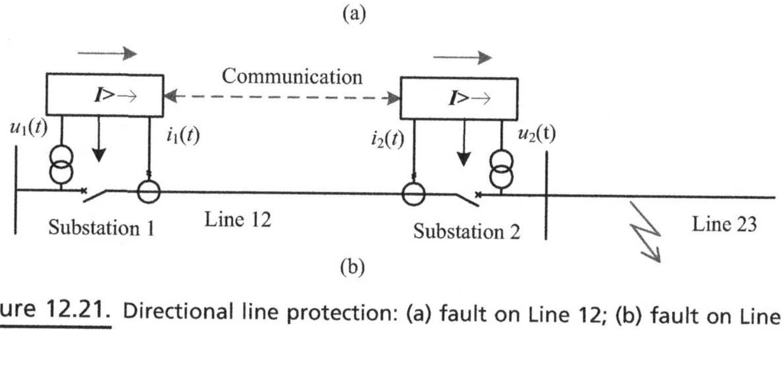

Figure 12.21. Directional line protection: (a) fault on Line 12; (b) fault on Line 23.

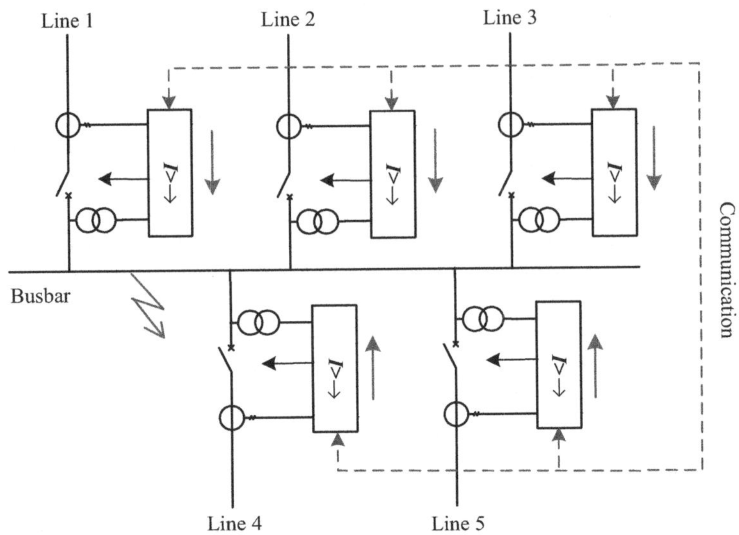

Figure 12.22. Directional protection being use as busbar protection.

opposite and point in direction of Line 23. Therefore, both protections may conclude the fault is not on their line but somewhere behind substation 2.

In Figure 12.22, there is a single busbar shown with 5 connected lines all equipped with directional protection. If in case of a fault the direction evaluation of directional protections points to the busbar, the fault must be on the busbar and all line breakers are tripped (opened). If there is one fault direction not in line with all other fault directions the fault is not on the busbar but on the deviating line.

Directional protection with its communications (only low transmission capacity is needed) has a high selectivity and is seen as *object protection* scheme because protecting the whole objects and only this (e.g., a line).

Note: Busbar protection based only on the directional criterion is used only in some particular distribution application and cannot be used in higher voltage level due to reliability aspects and more busbar arrangements.

12.6.3 Impedance Protection

12.6.3.1 Distance Protection. If the line impedance $\underline{Z}_L$ as seen at the voltage and current measuring point is continuously monitored a sudden change of this valued indicates that a fault has happened. The new value is the above mentioned fault impedance $\underline{Z}_F$ (Figure 12.23). By comparing $\underline{Z}_L$ and $\underline{Z}_F$ the *distance to fault* may be calculated. Therefore, this protection is called *distance protection.*

The distance protection is one of the most common protection types since it gives by measuring voltage and current at one point already a very selective information about the fault. Same as already mentioned for the directional protection in Section 12.6.2.1 a high time coherence between current and voltage in the order of 1 μs is needed for correct results.

The faulted object is to be protected, for example, a line with a given L (km) respectively impedance $\underline{Z}$ (0.2–0.4 Ohm/phase and km for overhead lines). Within this protected line, the calculated fault impedance may be allocated to the fault location. The

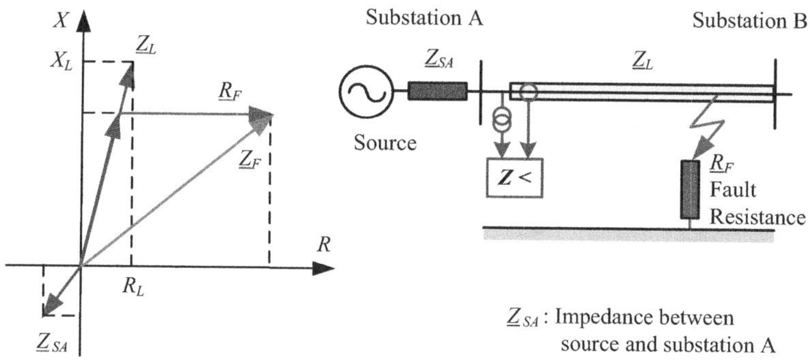

Figure 12.23. The principle of distance protection.

measured distance to fault may also be beyond the limit of the next substation, which may have a lot of lines or, more generally, connect the complexity of the power system to the protected line. Therefore, an allocation of a fault distance to the fault impedance is not possible in any case. Regarding the clearance of remote faults seen by the local protection will interfere with remote protections. Also, complete protection schemes have to solve these problems involving *delay times* and also *communication*.

The whole range of accessible fault impedances is grouped into so-called *protection zones* where the first zone refers to the line to be protected and higher zones to more and more remote faults. Since faults in backward direction are visible also backward zones may be defined. IEDs for distance protection comprise normally one to four forward zones and one backward zone. The tripping characteristic is represented in the impedance plane as complex polygons or circles. The first zone covers only about 80–85% of the line to get a selective decision if the fault is on the line clear also near the line end. Such a fault is tripped very quickly by the distance protection. Fault seen in other zones refer to the remote line end (the last 10–15%) and beyond. To keep the selectivity and give other distance protections the chance for fast tripping in their first zones, trips in higher zones are normally delayed. The overlapping zones in both directions and the time delays Δt are shown both in Figure 12.24. For better visibility, both the circuit breakers and the backward zones have been left out.

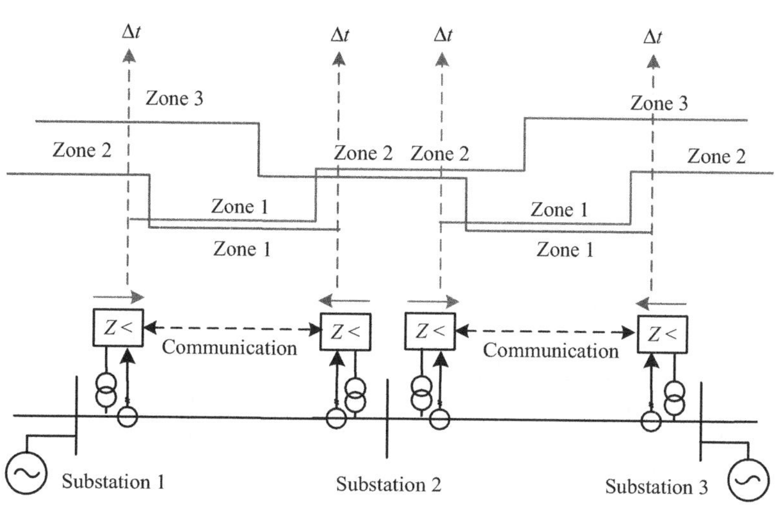

Figure 12.24. Example for a distance protection zones referring also to trip time delay Δt.

The optimal protection schemes with the best selectivity tripping only the faulted part needs a coordination of the finding of the local distance protection with the distance protection at the other side of the line. In the past, the most common *communication* method was the analog power line carrier with a very low bandwidth. Together with the supervision of the carrier only very few signals for blocking and releasing the protection at the other end of the line have been communicated. The protection schemes today are still referring to this state. Modern digital *communication* methods with high bandwidth allow transferring the complete view of the protection IEDs and more efficient protection schemes in the future. The arc in the top of the distance protection IED (Z<) shows the defined forward direction.

As further requirement the distance protection has to provide a clear indication, which phases are concerned by the power system fault (phase selectivity). This is very important in case single-phase tripping and single-phase autoreclosing is applied.

It is not discussed here that in case of parallel lines the mutual coupling of two lines is influencing the impedance and has to be considered in the equations and, therefore, in the protection settings.

12.6.3.2 Special Impedance-Based Functions.

Power Swing Blocking Function. This function is also based on impedance and generally used together with the distance protection function. In case of power swing, the distance protection should not give wrong trip. The evolution or the slow change of the measured impedance is monitored and used as criterion in order to block the distance protection trip. This slow change can be easily checked based on the change of the calculated quantity $\Delta U\cos\varphi$ (φ is the angle between voltage and current). In several applications, the slow change is just detected by the time needed to cross two impedance characteristics.

Pole Slipping Protection or Power System Islanding. This function is also based on impedance. In case of out of step due to a pole slip inside the generator or due to the fact that the synchronizing torque between two power system parts is not strong enough, it is necessary to trip. The pole slipping function is based on changes of the measured impedance where at least two main conditions have to be fulfilled. First, the impedance vector has to cross over from one quadrant to the other one in the complex impedance plane. Second, the slip has to be near to the device location, that is, the impedance vector has to be inside a certain impedance zone. Sometimes as third condition, it is added that the impedance vector has to leave again this impedance zone.

12.6.4 Current Differential Functions

12.6.4.1 Differential Protection. These functions use for an object (line, transformer, busbar) within incoming and outgoing current carrying lines one of the Kirchhoff's laws, that is, the sum of all these current must be zero, $\Sigma i_i(t) = 0$ (Figure 12.25). Since the measurements are made at the object boundaries and any deviation from zero must result from current loss by a short circuit, inside the object this protection function is by definition an object protection function. Since the algorithm to sum up currents is faster than the calculation of directions or impedances, differential protection is faster than the directional or distance protection.

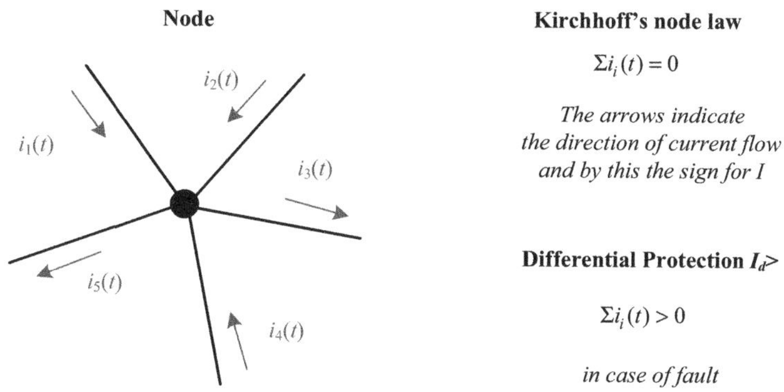

Figure 12.25. Kirchhoff's node law as basis for any current differential protection.

Since Kirchhoff's Law is valid in each instant of time, instantaneous currents have to be measured and summed up. This means that also here a high time coherence in the order of 1 μs of all involved current is requested. In addition, the distance between the current measuring points depends on the extension of the object, at least for line differential protection serial transmission of coherent current values is needed but today it is also convenient for busbar protection. The collection of the instantaneous currents can be done in a decentralized way (BU BBP) and the trip may be decided centrally (CU BBP) or distributed but normally the IEDs measuring or collecting the current on this distributed measuring points issue also the trip to the allocated circuit breakers.

For Transformer differential protection (Figure 12.26a) the impact of the transformer ratio on the currents has to be considered: $i_2(t) = (u_1(t)/u_2(t))i_1(t)$ for healthy transformer.

Because of its high selectivity and sensitivity, current differential protection schemes do not wait after detection of a fault until they trip, that is, these protections have no *start* but only a *trip*. To get a stable sum also in case of saturation in the current transformers, the

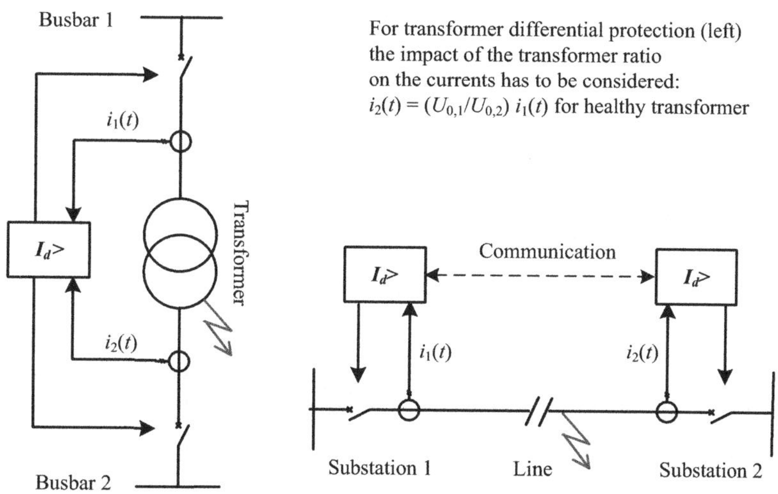

Figure 12.26. Transformer (a) and line differential (b) protection examples.

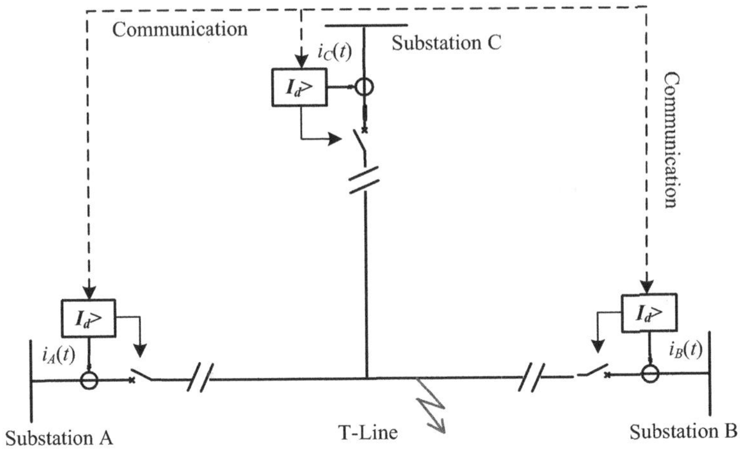

Figure 12.27. Line differential protection for a T-Line.

tripping condition according Kirchhoff's law is scaled by the sum of absolute values involved:

$$
\begin{aligned}
&\frac{\sum i_i(t)}{\sum |i_i(t)|} = \frac{\Delta i(t)}{\sum |i_i(t)|} < \frac{\Delta I_{Limit}}{\sum |i_i(t)|} \quad \text{without fault} \\
&\frac{\sum i_i(t)}{\sum |i_i(t)|} = \frac{\Delta i(t)}{\sum |i_i(t)|} > \frac{\Delta I_{Limit}}{\sum |i_i(t)|} \quad \text{with fault}
\end{aligned} \tag{12.3}
$$

12.6.4.2 Application Issues for Busbar Protection. The busbar in a substation is a node in the power system. According to Kirchhoff's law, the sum of all incoming and outgoing currents has to be zero. The busbar protection acquires and sums up all these

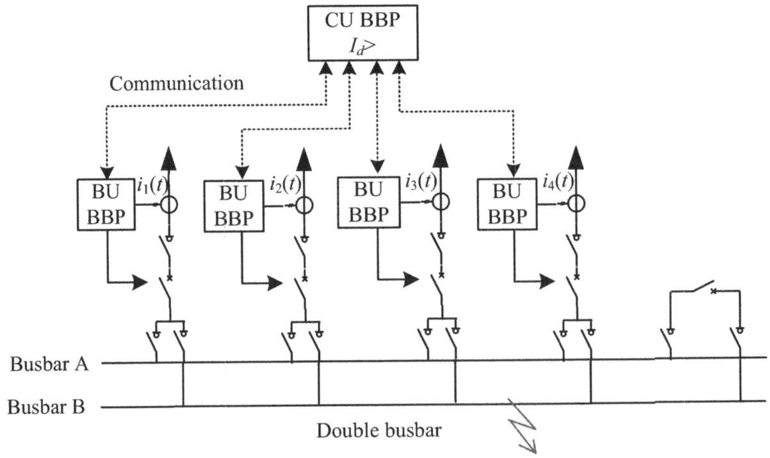

Figure 12.28. Busbar differential protection exampled (double busbar).

currents to the differential current ($\Delta i(t)$). If $\Delta I(t)$ exceeds the preset value ΔI_{Limit}, all connected feeders are tripped (Figure 12.28).

It has to be noted that the busbar is not just a simple node but may consist of multiple bus sections and protection zones. Common are $1\frac{1}{2}$ breaker and double busbar schemes. For double busbar schemes, each feeder may be connected alternatively to one of the busbars by busbar isolators. To identify the actual node configuration, a dynamic busbar image (topology) out of the status position of all isolators has to be created. This allows both summing up the correlated currents and tripping in case of a busbar fault only the faulted part.

Today, IED-based (numerical) busbar protection systems are installed. Because of the computation power provided today, other functions like end fault protection (elimination of fault between open breaker and the current transformer), breaker failure protection, timed-overcurrent protection, undervoltage protection, and phase discrepancy monitoring may be integrated.

Very commonly today, busbar protection consists of one *central* unit (CU BBP) to calculate the current difference and make the trip decision, and one decentralized unit per bay (BU BBP) for data acquisition and trip execution. In the BU, line protection functions may be integrated also, that is, for the main 2 protection or at least for the backup protection.

12.6.4.3 Application Issues for Line Differential Protection. The communication for transformer and busbar current differential protection is inside the substation, the communication for line current differential protection between the substations. Therefore, the line differential protection needs some special considerations (Figures 12.26b, 12.27).

The requested time coherence in the range of 1 μs for the data on both sides of the line is achieved historically by a proper "hand-shake" done measuring the propagation time and adjusting by this the local clocks for the synchronized sampling in the IEDs on both sides of the line. Today, more and more the synchronization is made with help of GPS-based clocks making the synchronization independently from changes in the propagation time and the compensation algorithms. More and more also the line differential protection should deal with communication network changes and rerouting/reconfiguration but it is very important to have both a reliable link, which is not provided by public networks and as well a backup protection function, which is independent from any communication link.

Note that IEC 61850 Ed.1 [5] as standard for communication inside the substation provides all means for serial communication of currents for transformer and busbar differential protection. IEC 61850 Ed.2 [16] (last part finalized beginning of 2013) will cover also the communication between substations and, therefore, provide the communication means for line differential protection.

The differential protection is a very selective *object* protection. The protected zone is defined by the location of the current transformers. For correct protection operation, the *communication link* has to be supervised properly. To cope with the loss of the communication link, line differential protection is generally complemented by another protection function like distance or directional overcurrent protection.

12.6.4.4 Comparative Protection as Simplified Differential Protection. For this protection, the variables measured at beginning and end of the protected object are not compared per sample (amplitude and phase angle comparison in vector representation) but as averages in a certain time window (e.g., for half-wave of sinusoidal values). Then the coincidence (phase comparison protection) or the equal signal direction (signal comparison) is checked. Such protection schemes require a much lower communication bandwidth between the IEDs at the ends compared to current differential protection (Figure 12.29).

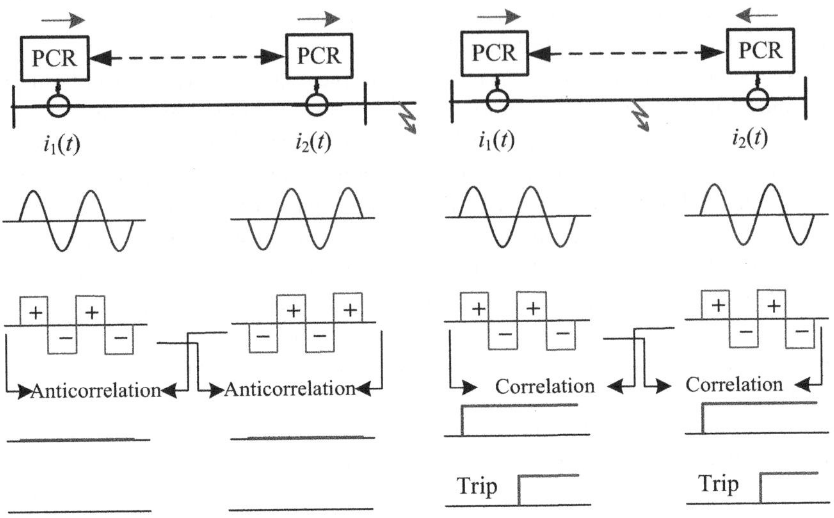

Figure 12.29. Comparative protection example.

12.6.5 Protection-Related Functions

12.6.5.1 Breaker Failure Protection. Since breaking the short circuit is the key task of the protection, the so-called breaker failure protection acts if the circuit breaker fails to open in response to the protection trip. This protection-related function will repeat in a *first step* the trip to breaker to the same or to a second trip coil of the breaker if applicable. If not successful the breaker failure protection sends in a *second step* a trip to all surrounding breakers in order to eliminate the faulted part in the power system. This results in any in fault clearing but with a delay, which should be kept acceptable. For this second step, the breaker failure protection needs to know the actual topology and, therefore, the breaker failure protection can be successfully implemented in the same IEDs as the busbar protection using the busbar image of the busbar protection.

12.6.5.2 Autoreclosing. In case of faults on overhead lines, the line protection (e.g., time-overcurrent, directional, or distance protection) interrupts one or all three phases to cut off the power infeed into the fault. Assuming a transient fault the line or respectively its power supply capacity should be switched on again as soon as possible. For this purpose, the protection-related function *autoreclosure* is used. This function provides normally a closing sequence of one fast step and sometimes also two or more additional slow ones. If the closing step is successful, the autoreclosure function is reset. If the fault persists, the protection will trip again and the next autoreclosing step is initiated until the number of pre-set autoreclosing steps is reached.

After the first unsuccessful step, which can be single after a single-phase trip or three-phase trip after a multiphase trip, tripping and autoreclosing are done normally in the next steps for all three phases independently from the first step. Also, the so-called evolving fault, that is, the development from a single phase to a three-phase fault before the start of the first reclosing operation has to be covered. There exist a big variety of autoreclosing schemes with marginal differences based on the users' philosophy. Autoreclosing assumes an appropriate communication between the protective IED(s) and the autoreclosing one.

12.6.5.3 Synchrocheck. The synchrocheck function is used as additional (protection related) function to the slow steps (> 1) of the autoreclosing functions to check the synchronisms before reclosing. Synchronism means that the differences of voltage, phase angle, and frequency from both sides of the circuit breaker are within a pre-set range. The synchrocheck function is also used in case of manual close commands by an operator.

12.7 FROM SINGLE PROTECTION FUNCTIONS TO SYSTEM PROTECTION

12.7.1 Single Function and Multifunctional Relays

There is a trend driven by cost issues and enabled by technology to reduce the number of protection IEDs by integrating more and more function within one IED. The enabling technology is microprocessors getting more powerful and memories getting larger. The discussion is given in more detail in Ref. [15].

As a result of integration, there has already been a substantial reduction in hardware in comparison with the one device per function approach of the past (Figure 12.30). In the boxes representing IEDs, the implemented functions are indicated by the Logical Nodes names according to IEC 61850 [5].

- CSWI: Control function object for circuit breaker and other switches; one instance per switch needed.
- PDIF_L: Line differential protection object; line indication _L not according to IEC 61850.
- PDIS: Distance protection object; one instance per zone needed.

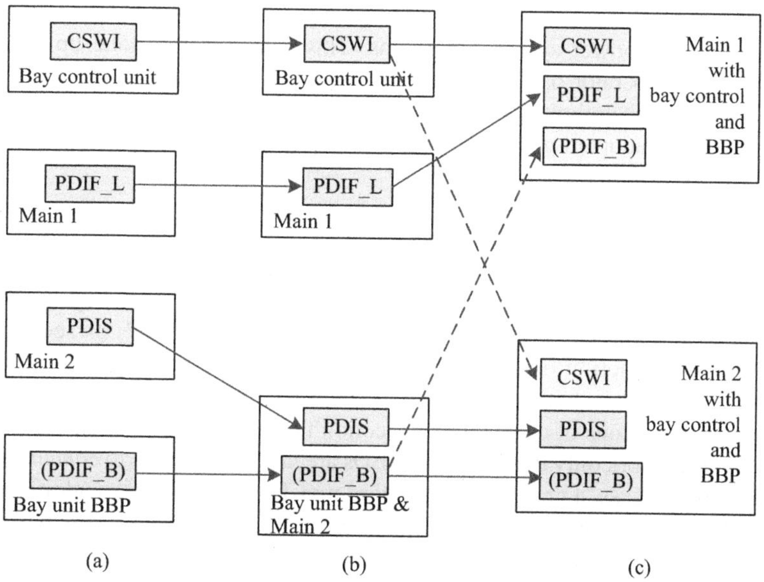

Figure 12.30. Trends to integrate protection functions in multifunctional relays.

PDIF_B: Busbar differential protection object; busbar indication _B not according to IEC 61850; brackets not according to IEC 61850 indicate the bay unit part of decentralized busbar protection.

A vast majority of utilities have accepted the advantages of function integration. In transmission, the highest accepted level of integration today is 3 or 4 IEDs for a typical Double Busbar *bay*, namely one unit for controlling and two units for feeder protection, that is, one each for Main 1 and Main 2, and may be a fourth unit for the bay unit of decentralized busbar protection or other functions needed, that is, (a) in Figure 12.30.

The next step already accepted partly today is to combine in the bay unit of the busbar protection also protection functions like Main 2 according to (b) in Figure 12.30.

The already mentioned ($N-1$) condition states the following requirement: Even in case that one of the secondary equipment is out of operation (under test, component failure, blocked), tripping in due time and with acceptable selectivity shall be possible. This also applies to a certain extent to CTs, VTs, and CBs. Since current/voltage transformers and circuit breakers are expensive, these are not duplicated. Instead, redundant cores and trip coils are made available. This must be considered for the use of NCIT and the process bus.

The consequence of the ($N-1$) condition for bay control and protection is that there have to be *at least two independent IEDs* and two independent communication paths also in the future. If in each of the two remaining IEDs also control functions may be implemented and then not only protection would be redundant but also control as seen in (c) in Figure 12.30. This is normally not requested if the control functions having its dedicated IED. The integration of the control function in the protection IED is already common at distribution level today resulting in one IED per bay. In case of two IEDs per bay at transmission level, the same integration may happen for Main 1 and Main 2 resulting in a redundant control also. This was not yet requested in case of dedicated control IED but is a challenge for the future.

12.7.2 Adaptive Protection

Protections have parameters combined commonly to parameter sets, which are switched over in case of changing power system conditions. This switchover is done up today nearly exclusively by the operator or a protection specialist. Some solutions imply a time-based switchover. To make an optimized use of protection in any situation, the parameter sets should be switched over or adopted *automatically* based on the *actual power system conditions.*

12.7.3 Distributed Protection

12.7.3.1 Differential Object Protection Functions. Functions like the busbar protection may be implemented by using more than one IED, that is, on bay unit (BU BBP) per bay/feeder and one central unit for the trip decision (CU BBP) as seen already in Figure 12.28. The function of this CU may be distributed in the future to all bay units sharing the trip decision as it was possible for interlocking implementations. It may be a benefit that part functions of distributed protection functions are realized process (switchgear) near but they work only if *all* included IEDs are *healthy* and able to *communicate* with each other. Nevertheless, a graceful degradation may be considered. An additional step in decentralization may be reached if in the future the careful designed process bus according to IEC 61850 is a commodity. Another distributed function besides the BBP is other current differential functions like this one for the line. Only if IEDs on both sides work together the protection function may be performed.

12.7.3.2 Directional Object Protection Functions. At first glance directional functions (directional overcurrent, distance) work standalone in one IED. If they are interconnected by communication to protect selectively, one object they may perform this task only *together*, that is, the form a distributed object function as seen, for example, in Figures 12.21 and 12.22. The key is again reliable *communication*.

The wide area protection as discussed below may be seen as a very special distributed function. The protected object is a part of the power system or the complete one.

12.7.4 Wide Area Protection

Protection just reacts by definition in case of a *fault* in the power system based on current and voltage measurement as discussed before. For these local protections, there is no global information about *system stability* available, which would allow counteracting in time against developing instabilities.

If there is global information, that is, from a wide area of the power system available, system-wide monitoring and protecting functions may be realized. Such wide area information is collected from all or at least many nodes in the power system by so-called *phasor measuring functions* implemented either in dedicated units (PMU) or in protection IEDs, which collect both current and voltage phasors with a *time coherence of 1 μs* (see Section 12.3.4.3) using GPS-based time synchronization. This phasor information is sent to a centralized function called *Wide Area Protection System (WAPS) or System Protection System (SPS)* implemented either in a stand-alone IED or as function in one of the IEDs (computers) in the network control centers (Figure 12.31).

It monitors all the collected phasors and may support getting a time snapshot of the power system named state estimation. The conventional state estimation is time consuming to overcome by many iterations the sensor specific response behavior and the bad time coherency. At least the time coherency of the data from the synchronized PMUs is much better and the iterations needed are drastically reduced. The *power system equations* (globally applied Kirchhoff's laws) are as function implemented in the WAPS/SPS. Solving this equations with the PMU data allow to decide if the power system is running stable or if an

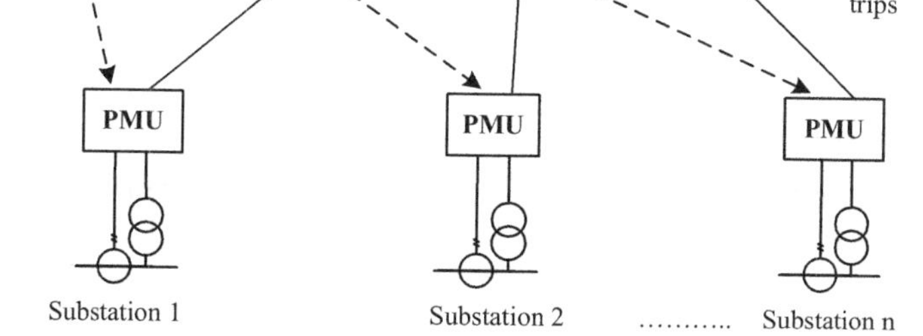

Figure 12.31. Example for the physical setup of a Wide Area Protection System.

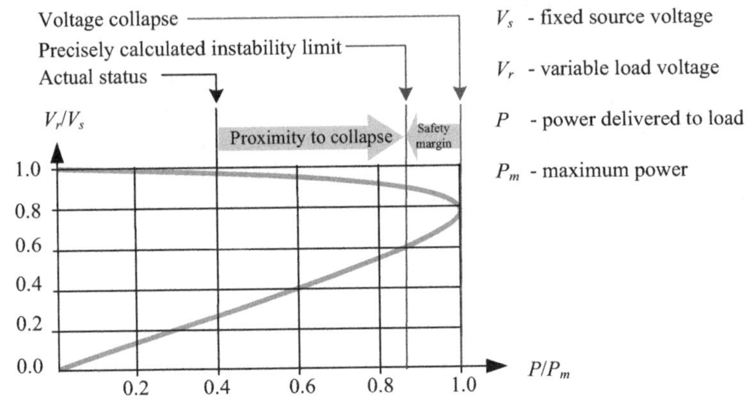

Figure 12.32. Example for power system stability: power transmission of a line and voltage.

instability is existing, which may cause after some time (typically from some 100 ms to few minutes mainly depending on the inertia of the feeding generators), for example, voltage collapse and, therefore, a blackout. The result may be some alarm for the operator similar to the start of local protection and—if implemented and accepted—after some delay and adapted to the power system situation trips (switch off circuit breakers) to initiate load shedding or islanding the system. The goal is to come back to power system stability by the actions of the operator or of the automatics. Therefore, the requirements for many time coherent data are high, for the requested response time low compared with local protection.

FACTS (Flexible AC Transmission Systems) devices allow to control the power flow continuously, for example, by changing the phase angle and may be also controlled by the wide area protection [17]. If these power electronic devices are available, the wide area protection may act very adaptive. It should be noted that *wide area protection* without and with FACTS has to be very carefully *coordinated* with all the more *local protection schemes* applied in the power system.

Monitoring and alarm functions are already implemented in some projects; the very complex protection functionality is still in the phase of development or collecting some first experience. It should be noted that global and local protection functions have to be carefully coordinated.

There are some instability modes in the power system. In Figure 12.32, the voltage collapse of a line caused by exceeding the power delivery capacity from source to load is given.

The parabolic type of curve is given by the solution of the nonlinear power system equations. To be on the safe side of the stability limit, some safety margin exists between the limit until the movement to the collapse is slow allowing counteractions and the absolute limit where the voltage collapse would happen very fast.

12.7.5 General Guide

12.7.5.1 General Recommendations for Protection Application. The protection is like an **assurance** for the power system and for the protected object. That means the whole protection system will cover and eliminate more risks in relation with fault or incidents in the power system. Generally, also more complicate networks will request more sophisticated protection systems.

For example, in low voltage radial power system simple overcurrent relays could be sufficient; the backup functionality is realized by the remote installed devices with time delay (Figures 12.3 and 12.18).

In higher voltage levels—transmission networks—the clearing time (no danger of stability loss in case of a fault) will be the main aspects to considered, independently of the fault in the power system and taking also into account a possible failure in the protection chain (condition $N-1$). This leads to the concept of "Main 1–Main 2" with high performance of the two protection schemes (Figure 12.4).

In this respect, the protection engineer has to consider the incidence of a fault in any location in the power system. Typical examples for better understanding are as follows:

- Fault in the last 10–15% of the line to be cleared with respect of the clearing time.
- Busbar fault to be eliminated according the clearing time requirements (protection given by the second zone of the remote distance protection to slow).
- Detection of fault between bus and bushing measuring current transformers in power transformer, between open breaker and current transformers in busbar, and so on.
- Detection of high impedance fault in overhead lines: need for specific function in order to fulfill this requirement. The use of the higher zones of the distance protection can lead to wrong trip (and can have a significant impact leading to blackout).

This last example shows that there is a specific protection function for any application (location and type of fault). Generally, trying to use one function in order to solve another application than its specific one leads to strong restrictions or even to dangerous mal operation. In this respect, the probability of some specific faults should be considered. A typical example is a fault between parallel lines in high voltage power system. This probability is not negligible and it should be avoided that all six phases (two three-phase lines) are tripped in case of a fault "one phase on line 1—another phase on line 2) if a single-phase autoreclosing is foreseen.

The $(N-1)$ condition for protection means

- existence of a backup protection in case the main protection chain fails; backup protection is generally delayed compared to the main protection and will trip slower and larger area in the power system (less selectivity, see Figure 12.3);
- to avoid the disadvantages mentioned above, it is necessary to install a second main protection (Figure 12.4).

As typical example, the case of the overhead line protected by a single distance protection can be mentioned. Generally, the higher zones realize the backup protection. These are slower and less selective. Sometimes also the higher zones do not cover the adjacent lines due to the parallel infeed at the remote bus. A correct Main 1–Main 2 concept consists in taking two full protection chains (separate DC supply, separate tripping circuit, and coil). The protection principle can be the same for both protection schemes (advantage for the maintenance; disadvantage risk of common mode failure) or can be based on two distinct principles (advantage no common mode failure, disadvantage maintenance, and knowledge needed for two different devices). Today for overhead lines that means, for example, one distance protection and one differential protection.

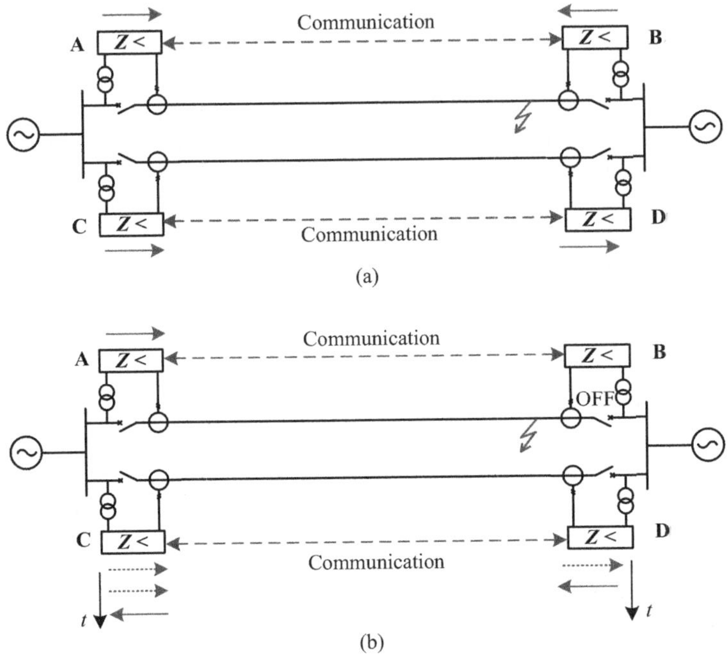

Figure 12.33. Impact of sequential tripping for a fault on one of two parallel lines.

Another important aspect is the **protection coordination**. The tripping time of all the protection functions as well as the protected area of each of the protection function should be carefully analyzed in order to avoid nonselective tripping. A typical example is the protection installed in the zero-to-earth connection of the star point of a power transformer (Figure 12.12): this protection should not trip quicker than all the protection installed on the secondary side of the power transformer respectively in the corresponding power system part. Any backup protection should first give the opportunity to the main protection to operate. This case shows that *protection coordination* is important to keep selectivity and avoid the loss of healthy parts. Extrapolated to the complete power system, local and global *protection coordination* (including also WAPS/SPS if applicable) is one of the main preventive measures against blackouts.

Also the **adaptability** of the protection to changing power system conditions should be analyzed. A typical example for this is a fault on one of two parallel lines (Figure 12.33) protected by *distance protections* ($Z<$) on both sides of each line. These protections communicate pairwise the fault direction seen to agree if the line is faulty and has to be tripped. The elimination of a fault on one line may change considerably the current and voltage on the other line. If there is a fault near point B as shown in case (a) of Figure 12.33, the protections on both sides will see the fault inside the line and agree that both breaker should open to eliminate the faulted line. This opening should happen at the about the same time. The healthy line shall not be impacted. A more detailed view shows a more complex situation. The protection at point B will see the fault in zone 1 and trip directly. The protection at point A may see the fault in the delayed zone 2 and not trip immediately. If the protection scheme (logics with communication) cannot compensate this delay or if the breaker at point A is delayed for some other reason, only the breaker at A is open (OFF) but the fault is not yet eliminated. The distance protections at C and D have seen the fault in the same direction outside the healthy line, maybe at C in zone 2 and at D in the backward zone and, therefore, no trip was in initiated. With the open breaker (OFF) at point B the fault current will change its direction (current reversal) and the fault should now seen by both protections, that is, at point C and D in the opposite direction as before. If the change in direction happens faster in the protection at point D, the protections at C and D communicating the assumed fault direction before the change in fault direction is completed will assume an internal fault and trip the healthy line also. The complete power transfer between the ends of the double line is interrupted.

12.7.6 Security and Dependability

Settings have to be done carefully: most wrong trips are due to wrong settings. Today, numerical protection has integrated fault recorder. This allows a much better postfault analysis than in the past and this contributes to reduce the repetition of wrong trips. Registered fault recordings can be reinjected easily in protection equipment in the laboratory and allow checking the reaction of the equipment with the given settings.

Due to the fact that protection is tripping very few times per year, periodic testing has been necessary with the previous technologies. The numerical technology has the advantage of built-in self-supervision, allowing a reduction of periodical tests down to maintenance on demand.

Tests on site during the lifetime of the equipment should only be done in the healthy state of the equipment. Today, the reading of current and voltages values and their comparison with the values indicated in other equipment or in control units will indicate if a big part of the equipment is running properly. It is important to check periodically the tripping circuit from the tripping contact of the protection until the breaker coil. In that sense, the periodical injection of analog values to the protection can be reduced or eliminated, avoiding also the risks inherent due to the tests: the protection is not set back in the same state as before the tests have been done (Table 12.2).

TABLE 12.2. Summary about protection methods and their impact on selectivity

Method	Result	Modes	Selectivity	Speed
Single measurement point	Limit value crossings	Direct trip Graded delay Communication	0 + ++	*High* *Low* *Medium*
Single measurement point	Limit values crossing and fault direction	Direct trip Graded delay Communication	+ +++ ++++	*High* *Low* *Medium*
Multiple measurement points (samples)	Deviation from zero according to Kirchhoff's current node law	Locally hardwired Distributed and serial communication	+++++ +++++	*High* *High*
Multiple measurement points (phasors)	Solution of power system equations based on Kirchhoff's laws for I and U	Distributed and serial communication	+++++ Goal: power system stability	*Low – High* Depends on situation and algorithm

12.7.7 Summary

- Protection is like an assurance.
- $(N-1)$ condition for protection and power system has to be fulfilled.
- Consideration of any location for faults and any type of fault and the corresponding protection function.
- Consideration security and dependability aspects with impact on maintenance.
- Protection coordination.

The relative selectivity is indicated by the number of +, the speed for the trip and, therefore, fault clearance speed is characterized by three classes High, Medium, and Low. The values for the phasor-based WAMP/SPS depend strongly on the actual situation and the decision algorithm implemented.

12.8 CONCLUSIONS

- The introduction of the *numerical technology* has opened new opportunities and leads to optimize the complete protection, monitoring, and control functions as part of a common system, which are not yet fully exploited.
- The increasing *communication facilities* based on the worldwide standard IEC 61850 allow getting access to much better information than in the past, to realize much better protection functionalities including testing and have much better fault analysis. Serial communication will dominate at all levels including the process bus.
- The *basic requirements* of the protection do not change and the general guidelines remain but protection can be integrated with control giving the possibility to buildup systems, which are better optimized and better automated. These will impact all levels of tasks inside the protection from specification to maintenance.
- *Wide area protection* will not replace but complement the more local protection functions used today.

Annex 12.1. IDENTIFICATION OF PROTECTION FUNCTIONS

A.12.1. General Remarks

In some fields of the table, there is more than one entry caused by not identical definitions in all three identification systems.

A.12.1.1. IEEE Device Numbers

The identification of protection with Device Numbers made by IEEE for ANSI was started 1928 as AIEE No.26 to describe converter substations. This standard was revised in 1937, 1945, 1956, 1962, 1979, 1996, and 2008 [18]. To each ***device number***, there is short descriptive ***text*** allocated. The version from 2008 contains also a reference to the Logical Nodes names from IEC 61850 Ed.1.

The positive side of his long history is that it contains a large amount of identifications used in many countries. It stands in an old tradition and always tried to enhance this standard to new technologies. Unfortunately, this continuity over different technologies leads to a very vaguely defined meaning, or reinterpretation of meaning at different places. Sometimes the device numbers mean devices, sometimes the same device numbers stand for a function.

A.12.1.2. IEC Designation

The IEC standard IEC 60617 [19] is available as database accessible by Internet and replaces the paper versions of the parts IEC 60617-2 to IEC 60617-13 (2001-11) allocates graphical symbols to a lot of electrotechnical item including protection functions. It allows also to compose missing symbols according to some basic rules. The easy access to this large database is the benefit of IEC 61850 but the free composition of missing symbols may result in some individual symbols. To use these symbols also in text documents, alphanumeric derivatives have been created also with some individuality.

A.12.1.3. Logical Nodes Names

The object oriented data model of IEC 61850 has on the highest standardized level the objects called Logical Nodes (LN) named by four letter acronyms, which refer to automation functions in power systems started in Edition 1 with the domain substation. The names of the Logical Nodes referring, for example, to protection functions start always with the letter P, to protection-related function with the letter R, and to control functions with the letter C. For more function groups and functions, see Ref. [20]. The letters 2–4 contain the mnemonics for the function allocated as seen, for example, in the LN PDIS (distance protection). Logical Nodes contain all standardized Data Objects, and the Data Objects themselves all Data Attributes which are needed for the data exchange of the function with other functions respectively primary objects represented also by LNs in the substation (for example, circuit breaker by the LN XCBR) or beyond.

Note that Logical Nodes describe functions and no IEDs. It may be seen as drawback that the LN identification refers to the core functionality of the function only but it provides as benefit the highest flexibility to realize any function allocation in IEDs today and in the future but always with standardized data. Therefore, the LN identification appears in any IEC 61850-based substation or power automation system.

A.12.2. Identification List

IEEE Std C37.2		IEC 60617		IEC 61850-7-4	
Text	Device number	Graphical symbols	Alpha-numeric	Logical Nodes	Comments
Transient earth fault				PTEF	
Directional earth fault wattmetric protection				PSDE	Sensitive earth fault protection
Checking or interlocking relay	3			CILO	
Over speed	12	$\omega >$			

(*continued*)

IEEE Std C37.2		IEC 60617		IEC 61850-7-4	
Text	Device number	Graphical symbols	Alpha-numeric	Logical Nodes	Comments
Zero speed and under speed	14	$\omega <$		PZSU	
Distance	21	$Z <$	$Z <$	PDIS PSCH	One instance of PDIS per zone Line protection logics (scheme)
Volts per Hz	24			PVPH	
Synchronism-check	25			RSYN	
Over temperature	26	$\theta >$	$\vartheta >$		
(Time) Under voltage	27	$U <$		PTUV	
Directional over power/reverse power	32	$\rightarrow$ $P >$		PDOP	Directional over power Reverse power modeled by PDOP plus additional mode “reverse”
Undercurrent	37	$I <$		PTUC	
Underpower	37	$P <$		PDUP	
Loss of field/Under excitation	40			PDUP	
Negative sequence relay	46	$I_2 >$	$I_2 >$	PTOC	Time overcurrent for sequence current
Negative sequence voltage relay	47	$U_2 >$		PTOV	Time overvoltage for sequence current
Motor start-up	49, 66, 48, 51LR			PMRI PMSS	Motor restart inhibition Motor starting-time supervision
Thermal overload	49	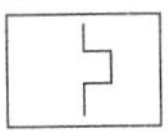	$\Theta >$	PTTR	
Rotor thermal overload	49R			PTTR	
Stator thermal overload	49S			PTTR	

IEEE Std C37.2		IEC 60617		IEC 61850-7-4	
Text	Device number	Graphical symbols	Alpha-numeric	Logical Nodes	Comments
Instantaneous overcurrent or rate of rise	50	$I>>$	$I\gg$	PIOC	
Instantaneous earth fault overcurrent relay	50N		$I_E\gg$		
AC time overcurrent	51 (Inverse time), 50TD (Definite time)	$I>$	$I>, t$	PTOC	
Inverse time earth fault overcurrent relay	51G	I ⏚ $>$		PTOC	
Definite time earth fault overcurrent relay	51N	I ⏚ $>$	$I_E>, t$	PTOC	
Voltage controlled/dependent time overcurrent	51V	$U\uparrow\ I>$		PVOC	
Circuit breaker	52			XCBR	
Power factor	55	$\cos\varphi>$		POPF PUPF	Over power factor Under power factor
(Time) Over voltage	59	$U>$	$U>$	PTOV	Both for dc and ac
Neutral point displacement relay	59N	$U_{rsd}>$	$U_0>$		
Voltage or current balance	60			PTOV PTUV	Over voltage Under voltage
Breaker failure protection	50BF or 62BF			RBRF	
Earth fault/ground detection	64	I ⏚ $>$		PHIZ	
Rotor earth fault	64R			PTOC PHIZ	Time overcurrent High impedance ground detector
Stator earth fault	64S			PTOC PHIZ	Time overcurrent High impedance ground detector
Inter-turn fault	64W			PTOC PHIZ	Time overcurrent High impedance ground detector

(continued)

IEEE Std C37.2		IEC 60617		IEC 61850-7-4	
Text	Device number	Graphical symbols	Alpha-numeric	Logical Nodes	Comments
AC directional overcurrent	67	→ $I>$	$I>$	PTOC	Time overcurrent
	67N	→ $I\stackrel{\perp}{=}>$	$I_E>$		Time overcurrent
Directional earth fault	67G[a]			PTOC RDIR	Time overcurrent Direction comparison
Power swing detection/blocking	68			RPSB	
DC overcurrent	76			PTOC	
Phase angle or out-of-step	78	$\varphi>$		PPAM	
AC auto reclosing	79	$O \rightarrow I$		RREC	
Underfrequency relay	81U	$f<$	$f<$	PUFP	Under frequency
Overfrequency relay	81O	$f>$	$f>$	POFP	Over frequency
Current Differential relay	87	$I_d>$	$I_d>, \Delta I>$	PDIF	
Phase Comparison	87P	$I_d>$	$I_d>, \Delta I>$	PDIF	
Differential Line	87L	$I_d>$	$I_d>, \Delta I>$	PDIF	
Differential Transformer	87T	$I_d>$	$I_d>, \Delta I>$	PDIF PHAR	Differential transformer Harmonic restraint
Differential Busbar	87B	$I_d>$	$I_d>, \Delta I>$	PDIF	Busbar differential
Busbar Protection	87B			PDIR	Direction comparison
Motor Differential	87M	$I_d>$	$I_d>, \Delta I>$	PDIF	
Generator Differential	87G	$I_d>$	$I_d>, \Delta I>$	PDIF	

REFERENCES

[1] Phadke, A.G., Horrowitz, S.H. *Power system relaying*, 3rd edition, Wiley & Sons, Inc., 2008.

[2] Blackburn. J.L., Domn, T.J. *Protective relaying—principles and applications*, 3rd edition, CRC Press, Taylor & Francis Group, 2008.

[3] Ungrad, H., Winkler, W., Wiszniewski, A. *Protection techniques in electrical energy systems*, 1st edition, Marcel Dekker Inc., 1995.

[4] www.cigre.org

[5] IEC 61850 – *Communication networks and systems in substations*, 1st edition, 2002–2005 (www.iec.ch)

[6] IEV International Electrotechnical Vocabulary—Part 448: *Power system protection/Reliability of protection* (www.iec.ch)

[7] CIGRE Brochure 140—*Reliable fault clearance and back-up protection*, 1998.

[8] IEEE 1588—*Precision clock synchronization protocol for networked measurement and control systems.*

[9] Brand, K.P., Brunner, C., de Mesmaeker, I. *How to use IEC 61850 in protection and automation*, Electra 222, pp. 11–21, Oct. 2005.

[10] IEC 61850-9-2—*Communication networks and systems in substations—Part 9-2: Specific communication service mapping (SCSM)*—Sampled values over ISO/IEC 8802-3.

[11] IEC 61850-9-2LE (Light edition)—*Implementation guideline for digital interface to instrument transformers using IEC 61850-9-2*, UCA International Users Group, www.ucainternational.org

[12] IEC 61850-5—*Communication networks and systems in substations—Part 5: Communication requirements for functions and device models.*

[13] Dalke, G., Horak J. Application of numeric protective relay circuit breaker duty monitoring, *IEEE Trans. Ind. Appl.*, Vol. 4, No. 41, pp. 1118–1124, 2005.

[14] Elmore, W.A. *Protection relaying*, Marcel Dekker, Inc., New York, 1994.

[15] Baass, W., Brand, K.P., Menon, A. *Acceptable function integration of protection and control at bay level*, CIGRE SC B5 Colloquium, Paper B5-217, Madrid, 2007.

[16] IEC 61850—*Communication networks for power utility automation*, 2nd edition, 2009–2013 (www.iec.ch)

[17] Zhang, X.P., Rehtanz, C., Pal, B. *Flexible AC transmission systems—modelling and control*, Birkhäuser (Springer), 2006.

[18] IEEE Std C37.2™-2008—IEEE Standard for Electrical Power System Device Function Numbers, Acronyms, and Contact.

[19] IEC 60617—*Graphical symbols for diagrams.*

[20] IEC 61850-7-4 Ed.1—*Communication networks and systems in substations—Part 7-4: Basic communication structure for substation and feeder equipment—Compatible logical node classes and data classes.*

[21] Rehtanz, C., Bertsch J. *Wide Area Measurement and Protection system for emergency voltage stability control*, Proceedings of IEEE Power Engineering Society Winter Meeting, Vol. 2, pp. 842–847, New York, Jan. 27–31, 2002.

[22] IEC 60255-8—*Thermal electrical relays.*

PART III

GRID BLACKOUTS AND RESTORATION PROCESS

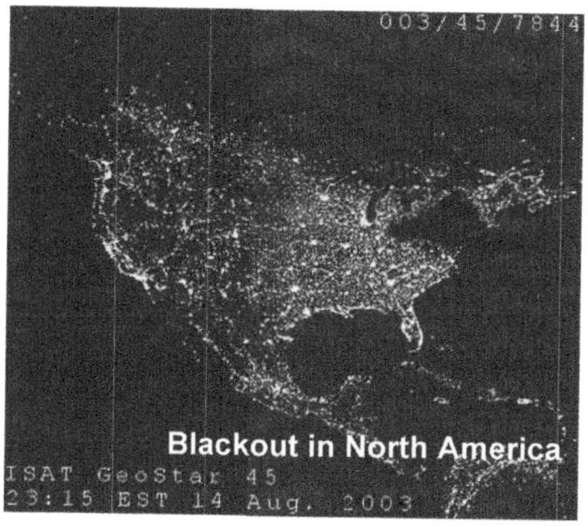

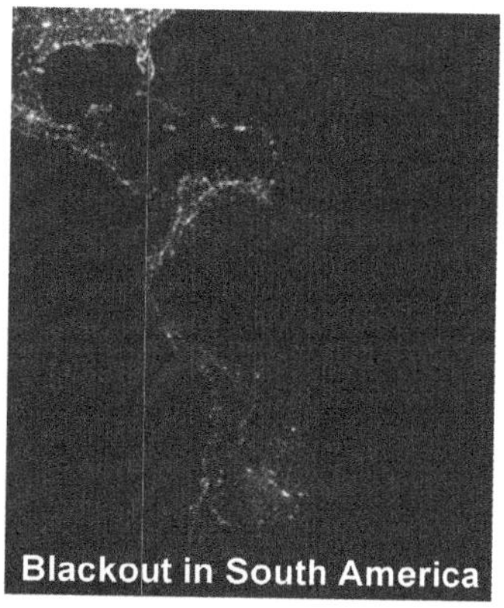

Handbook of Electrical Power System Dynamics: Modeling, Stability, and Control. Edited by Mircea Eremia and Mohammad Shahidehpour.

13

MAJOR GRID BLACKOUTS: ANALYSIS, CLASSIFICATION, AND PREVENTION

Yvon Besanger, Mircea Eremia, and Nikolai Voropai

13.1 INTRODUCTION

The power system's operators have to ensure the security of their grid and to transport stable energy to the customers. However, thousands of disturbances occur in the modern power systems every year in the world, and some of them can lead to blackouts. Of course, we all heard about some famous blackouts in medias or in the literature, and everybody knows what it is a blackout. Moreover, some of us have already experienced it (several times?). Nevertheless, let us try to give a definition and/or some characteristics to fix our minds.

First, blackouts are major incidents in the power systems. It sounds like evidence. We can also talk about generalized outages.

Then, we may say that a blackout is characterized by its geographical scale, depth, and duration. The depth is related to the number of not supplied customers, or in other words, to the geographical density of customers. It is then related to the load that is lost during the incident. The combination of geographical scale and depth determines the importance of the blackout: we often talk about small-scale blackouts and large-scale blackouts. The duration directly quantifies the severity of the incident and its consequences, particularly in terms of cost, which can be very high (several billions dollars for a large-scale blackout). It also gives indications on the difficulty for operators to restore their power system.

Next, an important point is that blackouts always result from a succession of conjunctural events. These events should be almost anodyne on the power system if they should be considered separately. However, in a particular operation context, they can be linked unfortunately from an initiating event, resulting in cascading outages of power

Handbook of Electrical Power System Dynamics: Modeling, Stability, and Control. Edited by Mircea Eremia and Mohammad Shahidehpour.

system equipments (lines, power plants, power stations, etc.). At the end of the process, regions or entire countries can be pulling down in black for some minutes or tens of minutes. More precisely, a blackout always results from an initiating event (sometimes more) and worsening factors. This conjunction is always verified, as modern power systems are generally operated using the N-1 criterion (the system can lost one major equipment without losing the stability). It means that, ordinarily, a single event is not able to lead to a catastrophic situation. We give, hereafter, a nonexhaustive list of initiating events and worsening factors:

- *Natural Factors.* Storm, geomagnetic storm, earthquake, lightning, contact between line and tree, animals, and so on. Note that the contact between line and tree can also be considered as a technical or a human reason (lack of trimming).
- *Technical Factors.* Short circuit, component failure, heavy load, maintenance of key equipments, breakdown with age, and so on.
- *Human Factors.* Switching mistakes, erroneous or inadequate communications between operators, sabotage, lack of training especially for emergency situations, and so on.

The combination of some of these factors can lead to a snowball (or domino) effect and then to a catastrophic situation. To avoid this, transmission system operators (TSOs) operate the power system according to different states, as shown in Figure 13.1.

The power system works under two principal types of constraints: constraint of providing electricity to customers and operating constraints. The first type imposes that all the customers must be supplied, and the second type requests that all the system variables (frequency, voltage, line currents, etc.) have to stay in their authorized range.

The system is in *normal state* if the constraint of providing electricity to customers and the operating constraints are satisfied.

The system is in *emergency state* if some operating constraints are violated. This can be provoked by a perturbation that leads some system variables to go out of the limits. If the intervention actions of the operator are able to bring back the variables in the limits, the system is preserved and goes to the alarm state.

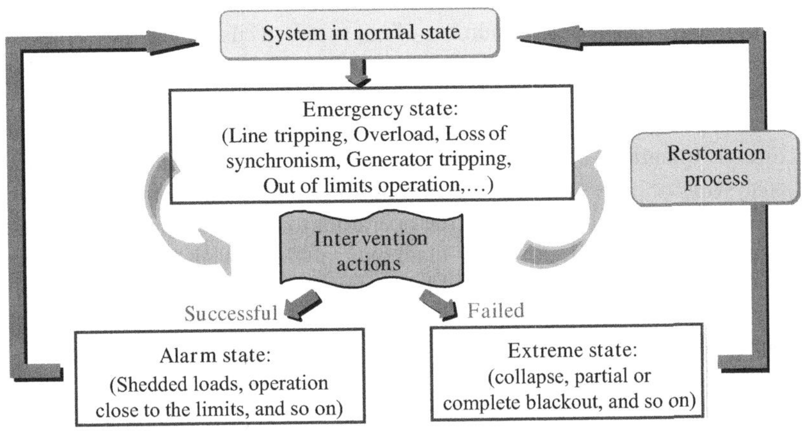

Figure 13.1. Operating states of the power system.

The system is in *alarm state* when the constraint of providing electricity to customers is not entirely satisfied (a part of the customers is not supplied).

The system is in *extreme state* if both types of constraints are violated.

The power system is generally sized to survive perturbations, which are considered as "probable," like the loss of a major component (line, generator, transformer, etc.). This is the N-1 criterion. During the normal system operation, only the probable perturbations are taken into account. On the opposite, that is, facing nonprobable perturbations, some emergency mechanisms are launched (action on protections, load shedding, large-scale defense plan, etc.) to keep the system safe. That is why blackouts generally do not result from a single critical event, but from an initial perturbation followed by a combination of events.

The initiating event provokes perturbations that cannot be eliminated by the operator (not enough time, bad evaluation of the severity level, etc.). Then, the occurrence and/or succession of supplementary events (more or less hazardous) lead to the degradation of the system state.

Of course, the system operator has the possibility (if he has enough time) to contact other operators in case of serious problem in order to launch coordinated actions (community of interests of the interconnected transmission grid actors).

Nevertheless, small-scale blackouts often happen, whereas large-scale blackouts rarely occur. Large-scale blackouts cause enormous economical and social damages, including the security of fragile people (old persons, hospitals, etc.). For example, large scale blackouts happened in the Northeast of the United States and in the Canada on August 14, 2003, and in Italy on September 28, 2003. The first one affected approximately 50 million people in eight U.S. states and two Canadian provinces. The economic losses were about $7–10 billion [1,2]. The second one affected about 57 millions people in Italy and 180 GW of load was interrupted [3–5]. This blackout in Italy is the largest one in Europe. This, added to the fact that customers tolerate less and less supply interruptions, makes that it is necessary to avoid, as much as possible, the appearance and the impacts of such large incidents. Analyzing and studying the reasons and the mechanisms of blackouts would be the first step for blackout prevention and so, we should review the previous large-scale incidents and draw their common characteristics. This is the purpose of this chapter.

This study investigated 39 blackouts that occurred throughout the world since 1965. Table 13.1 shows the corresponding list.

Of course, this list is not exhaustive. Others blackouts happened all around the word recently, for example: Iran, Finland, and Algeria in 2003; Australia, Jordan, Bahrain, and Libya in 2004, Florida (USA) in 2008, India in 2012 (world's larget blackout ever recorded) and so on. It seems that blackouts trend to become recurrent phenomena. Then, people are more and more sensible and attend to these major incidents, helped on this by medias. Thus, power systems are critical infrastructures for the welfare of modern societies as well as communication and computer networks. During the past several years, these three types of networks have become so intimately interlinked that it will be necessary to consider these infrastructures in an integrated framework in the future.

Section 13.2 of this chapter shows a brief description of some large incidents. In Section 13.3, we introduced the divisions of the whole process and divided the progression of blackouts into several sequential phases. According to these different phases, the phenomena of those cited large-scale incidents were compared, and it seems to exist some common characteristics. Then, we highlighted the common mechanisms of blackouts by analyzing their progression. Some details about economical and social impacts are given in Section 13.4. Some possible methods are suggested for the prevention of large-scale incidents in Section 13.5. General considerations about defense and restoration plans are discussed in Section 13.6.

TABLE 13.1. The Blackouts that were Analyzed

	Date	Country	Ref.
1	09/11/1965	10 states in Northeast of United States	[6]
2	1967	Pennsylvania, New Jersey, Maryland, United States	[7]
3	05/1977	Miami, United States	[7]
4	07/1977	New York City, United States	[7]
5	19/12/1978	France	[8,9]
6	01/1981	Idaho, Utah and Wyoming, United States	[7]
7	03/1982	Oregon, United States	[7]
8	27/12/1983	Sweden	[8]
9	23/07/1987	Japan	[8]
10	12/01/1987	Western France	[9]
11	13/03/1989	Québec, Canada	[8,10]
12	24/08/1994	Italy	[9]
13	14/12/1994	Arizona and Washington states, United States	[7]
14	17/01/1995	Japan	[11]
15	08/06/1995	Israel	[12]
16	12/03/1996	Florida, United States	[13]
17	16/04/1996	Southwestern of United States	[13]
18	02/07/1996	14 states in the United States	[13]
19	07/08/1996	Big Rivers Electric Corporation, United States	[13]
20	10/08/1996	California Pacific Northwest, United States	[13]
21	26/08/1996	New York City, United States	[13]
22	21/09/1996	Allegheny Power System, United States	[13]
23	30/10/1996	New York City, United States	[13]
24	01/1998	Canada, New York and New England	[7]
25	12/1998	San Francisco and California Bay Area, United States	[7]
26	07/1999	New York City, United States	[7]
27	11/03/1999	Brazilian power system	[14]
28	02/01/2001	India	[15]
29	12/01/2003	Southern part of Croatia and a part of Bosnia Herzegovina	[16]
30	14/08/2003	Northeast of United States and Canada	[1,2]
31	28/08/2003	South London	[17,18]
32	05/09/2003	The West Midlands, UK	[17,18]
33	23/09/2003	Eastern Denmark and Southern Sweden	[19]
34	28/09/2003	Italian power system	[3–5]
35	07/11/2003	Most of Chile	[20]
36	12/07/2004	The Athens and Southern Greece	[21]
37	14/03/2005	Queensland, NSW, Victoria and South Australia	[22]
38	25/05/2005	Moscow, Russia	[23,24]
39	04/11/2006	European power system	[25,26]

13.2 DESCRIPTION OF SOME PREVIOUS BLACKOUTS

Some large incidents are briefly summarized in this section, according to the data that are available. The descriptions are made in the same way for each blackout, considering the following sequential phases: precondition, initiating events, cascading events, and final state. A simple flowchart showing the mechanisms of the incidents is given for most cases.

13.2.1 August 14, 2003 Northeast United States and Canada blackout

13.2.1.1 Precondition.

(i) *Hot Weather, High Electric Demand, and Power Flow* August 14 was warm in the Midwest and Northeast. Temperatures were not exceptionally high, but above normal and there was very little wind. The weather was typical of a warm summer day. The warm weather caused electrical demand in northeastern Ohio to be high, especially high air-conditioning demand. This causes power system to consume high levels of reactive power in First Energy (FE: a system operator) control area, but electrical demand was not close to a record level. The flow of power through the ECAR (East Central Areas Reliability Coordination Agreement) region was heavy as a result of large transfers of power from the south (Tennessee, Kentucky, Missouri, etc.) and west (Wisconsin, Minnesota, Illinois, etc.) to the north (Michigan) and east (New York). The destinations for much of the power were northern Ohio, Michigan, and Ontario, Canada [1]. This causes the system to be operated near the security limit. There was a risk of overloaded lines.

(ii) *Facilities Out of Service* Several key generators were out of service going into August 14 (planned generation outages), but the unavailability of these generation units did not cause the blackout. Some unplanned outages of transmissions and generating units earlier in the day of August 14 affected the safety of the system:

a. 12:08, the Bloomington-Denois Creek 230-kV outage was not automatically communicated to the MISO (Midwest Independent System Operator) state estimator and the missing status of this line caused a large mismatch error that stopped the MISO's state estimator from operating correctly at about 12:15.

b. 13:31, the Eastlake 5 generating unit along the shore of Lake Erie tripped. This is a major source of reactive power support for the Cleveland area. The loss of this unit made voltage management in northern Ohio more challenging, and gave FE operators less flexibility in operating their system. With Eastlake 5 forced out of service, transmission line loadings were notably higher but well-below ratings.

c. 14:02, the Stuart-Atlanta 345-kV line in southern Ohio tripped. Since the line was not in MISO's control area, MISO operators did not monitor the status of this line and did not know that it had tripped out of service. That cause MISO's state estimator to continue to operate incorrectly, even after the previously mentioned mismatch was corrected [1].

The loss of key generator and the lack of awareness about the line tripping show that the system was in weak condition and the system operators were not able to take a right decision in the emergency condition.

(iii) *Voltage and Reactive Power Conditions Prior to the Blackout* FE operators began to address voltage concerns early in the afternoon of August 14. He noted to most of them that system voltages were sagging. FE was a major importer of power and northern Ohio was a net importer of reactive power.

From Figure 13.2, we can see that there was just a small reactive power reserve in Cleveland area of FE. Because the reactive power cannot flow from long distances, even if the neighboring areas had many reactive power reserves, the voltage decline continued in the Cleveland–Akron area.

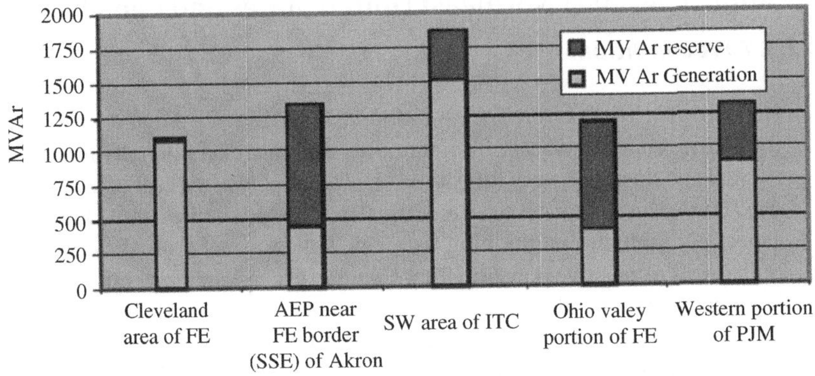

Figure 13.2. Reactive reserves of representative groups of generators at 04:00 p.m. on August 14, 2003. AEP, American Electric Power system operator.

The analysis showed that the declining voltages in the Cleveland–Akron area were strongly influenced by the increasing temperatures and loads in that area and minimally affected by transfers through FE to other systems. At the same time, FE did not have enough reactive production in the Cleveland–Akron area on August 14 to meet reactive power demands and to maintain a safe margin from voltage collapse [1].

P-Q and V-Q analysis by investigators determined that the low-voltages and low-reactive power margins in the Cleveland–Akron area on August 14 prior to the blackout could have led to a voltage collapse. This voltage collapse influenced the other parts of the system and caused the wide-area blackout. So, the monitoring of reactive power margins is important to prevent voltage collapse.

(iv) *System Frequency* Before the blackout, there were insufficient resources to control frequency for the existing scheduling practices.

(v) *Conclusion* In the precondition of the 2003 USA blackout, the system was near the steady-state limit. Power flows were heavy in the lines; there was a risk of line overloads. Due to the high demand of the air-conditioning, the reactive power demand was high. In the part of the system that lacked of reactive power support, the voltage declined. At the same time, voltage declining caused that the system needed more reactive power to support the voltage. This is the typical process of voltage collapse. Before the blackout, active power resources were also insufficient to support the frequency. On the other hand, due to the deregulation context, each ISO (independent system operator) could not know the condition of the others ISOs directly by its measure system. The information was transmitted between the two ISO just by phone calls, and that was not sufficient and efficient in the emergency condition. So, this caused that each ISO lacks information from the other parts of the power system. When the system became unstable, they could not take some effective actions to bring back the system in the safety condition. Generally, heavy load flow, lack of reactive power and active power, and unknowing the entire system condition of each ISO are factors that put the system into a danger condition, and the blackout happen is hardly avoidable.

13.2.1.2 Initiating Events. Due to the precondition that was presented in the previous section, the system was near the steady-state limit of stability. When an unforeseen big disturbance happened in the power system, the system has become unstable.

From 15:05:41 to 15:41:35, three 345 kV lines failed with power flows at or below each transmission line's emergency rating. Each trip and lockout was the result of a contact between an energized line and a tree that had grown so tall that it had encroached into the minimum safe clearance of the line. As each line failed, the corresponding power flow was shifted onto the remaining lines, that is, others transmission paths. As a consequence, the voltages on the rest of FE system degraded further.

The following key events occurred during this period:

1. *15:05:41:* The Chamberlin-Harding 345 kV line tripped, reclosed, tripped again, and locked out. (the line sagged into the tree and phase C fault-to-ground);
2. *15:31–33:* MISO called PJM (PJM Interconnection, LCC: AEP's reliability coordinator) to determine if PJM had seen the Stuart-Atlanta 345 kV line outage. PJM confirmed Stuart-Atlanta was out;
3. *15:32:03:* The Hanna-Juniper 345-kV line tripped, reclosed, tripped again, and locked out.

13.2.1.3 Cascading Events. Due to high reactive power output, FE's Eastlake unit 5 generator regulator tripped to manual because of over-excitation at 01:31 p.m.

The Chamberlin-Harding 345 kV line tripped at 15:05 FE's system due to tree contact. Due to the load flow change, Hanna-Juniper 345 kV and Star-Canton 345 kV lines were heavily loaded and tripped due to tree contact at 03:32 and 03:41 p.m., respectively. The lines tripped one by one.

After the tripping of Sammis-Star 345 kV line, due to zone 3 relay operating by overload, the widespread cascading happened in Ohio.

At 04:10 p.m., due to cascading loss of major interconnection lines in Ohio and Michigan, power flow started flowing counterclockwise from Pennsylvania, through New York and Ontario and finally into Ohio and Michigan. At this time, voltage collapsed due to extremely heavily loaded transmission, and cascading outage of several hundred lines and generators provoked a wide-area blackout.

In Figure 13.3, a simple flowchart of the mechanism of U.S. blackout is presented.

Figure 13.4 presents the time line of the blackout. It includes grid events, computer events, and human events. From this figure, we can see that, before the blackout, FE's computer system had problems. The system operator could not know the line tripping and could not receive alarms information. On the other hand, the system operators lack training for the emergency conditions, they were not aware of the fault of the computer and they did not take effective actions to stop cascading events.

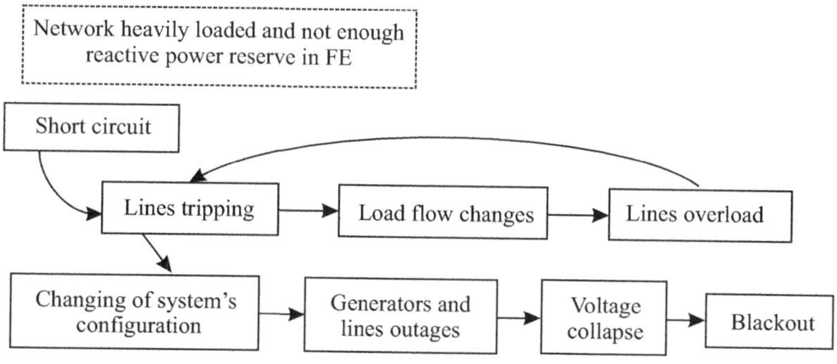

Figure 13.3. The mechanism of U.S. blackout.

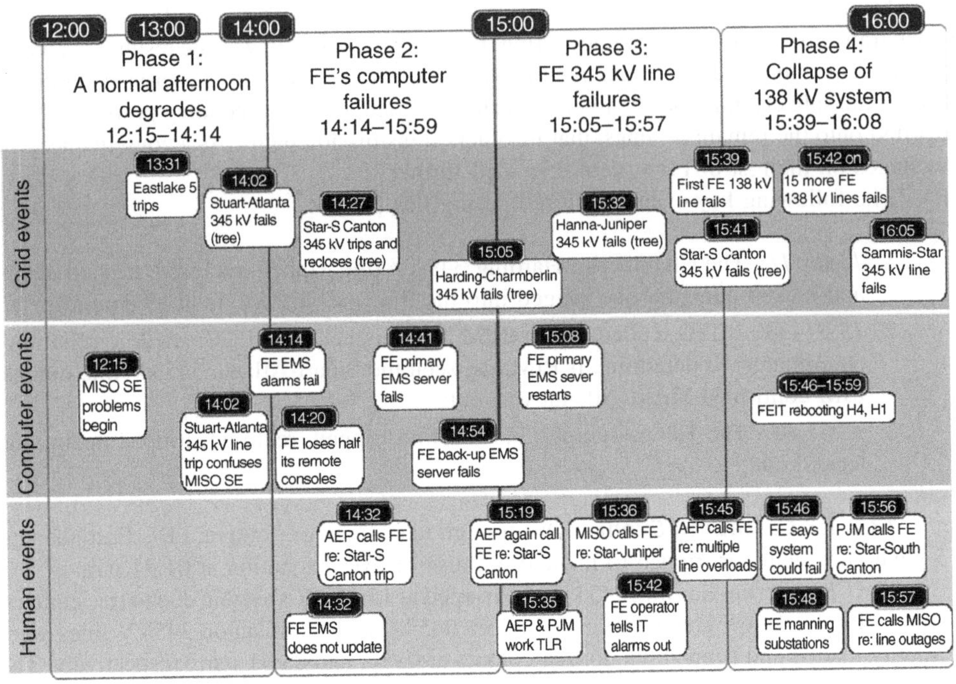

Figure 13.4. Time line: start of the blackout in Ohio [2].

To analyze the cascading events, to find critical lines and generators, and to understand why cascading spread, Table 13.2 presents the blackout time line.

In Table 13.2, there are 87 cascading events. These events can be divided into three periods: precondition, steady-state progression, and high-speed cascade. During the steady-state progression, the cascading events already start, but the system still keeps the steady-state stability: voltages and frequency are not at critical values. The high-speed cascading means that many lines and generators trip in a short time. At this point, the blackout cannot be stopped.

There are some important events that should be noticed:

1. The third event: at 01:31:34 p.m., the Eastlake unit 5 (rating: 597 MW) tripped. This generator is a major source of reactive power support for the Cleveland area and the loss of this unit made voltage management in northern Ohio more challenging and gave FE operators less flexibility in operating their system. So, it is a critical unit in the Cleveland area.
2. The fourth, fifth, and sixth events: Stuart-Atlanta, Harding-Chamberlain, Hanna-Juniper three 345 kV lines tripping, which were not known by the system operator due to the computer fault until 3:31–3:41 p.m., when the computer system was recovered. The Stuart-Atlanta 345 kV line tripped due to the result of a tree contact at 02:02 p.m. Because the line was not in Midwest ISO's (MISO) control area, MISO operators did not monitor the status of this line and did not know it has gone out of service. This led to a data mismatch that prevented MISO's state estimator (a key monitoring tool) from producing usable results later in the day at a time when system conditions in FE's control area were deteriorating. This problem was solved

TABLE 13.2. Time Line of the August 14, 2003 Northeast United States and Canada blackout

Period	No.	Time	Event	Equipment	Observation	Remark
Precondition	1	12:05:44	Conesville unit 5 (rating 375 MW)	G	O	
	2	01:14:04	Greenwood unit 1 (rating 785 MW)	G	O	
	3	01:31:34	Eastlake unit 5 (rating: 597 MW)	G	O	Critical unit
	4	02:02	Stuart-Atlanta 345 kV	L	NO	O:03:31–33
Steady-state progression about 1 h 5 min	5	03:05:41	Harding-Chamberlain 345 kV	L	NO	O:03:41
	6	03:32:03	Hanna-Juniper 345 kV	L	NO	O:03:41
	7	03:39:17	Pleasant Valley-West Akron 138 kV line tripped and reclosed	L	O	
	8	03:41:33	Star-South Canton 345 kV	L	O	Critical unit
	9	03:42:05	Pleasant Valley-West Akron 138 kV West line tripped and reclosed	L	O	
	10	03:42:49	Canton Central-Cloverdale 138 kV line tripped on fault and reclosed	L	O	
	11	03:42:53	Cloverdale-Torrey 138 kV line tripped	L	O	
	12	03:44:12	East Lima-New Liberty 138 kV line tripped from sagging into an underlying distribution line	L	O	
	13	03:44:32	Babb-West Akron 138 kV line tripped on ground fault and locked out	L	O	
	14	03:44:40	Pleasant Valley-West Akron 138 kV West line tripped and locked out	L	O	
	15	03:45:33	Canton Central-Tidd 345 kV	L	O	
	16	03:45:39	Canton Central-Cloverdale 138 kV line tripped on fault and locked out	L	O	
	17	03:45:40	Canton Central 345/138 kV transformer tripped and locked out due to 138 kV circuit breaker operating multiple times, which then opened the line to FE's Cloverdale station	T	O	
	18	03:51:41	East Lima-N. Findlay 138 kV line tripped, likely due to sagging line, and reclosed at East Lima end only	L	O	
	19	03:58:47	Chamberlin-West Akron 138 kV line tripped	L	O	
	20	03:59:00	West Akron 138 kV bus tripped, and cleared bus section circuit breakers at West Akron 138 kV	L	O	
	21	03:59:00	West Akron-Aetna 138 kV line opened	L	O	
	22	03:59:00	Barberton 138 kV line opened at West Akron end only. West Akron-B18 138 kV tie breaker opened, affecting West Akron 138/12 kV transformers no. 3, 4, and 5 fed from Barberton	L	O	
	23	03:59:00	West Akron-Granger-Stoney-Brunswick-West Medina opened	L	O	

(continued)

TABLE 13.2. (Continued)

Period	No.	Time	Event	Equipment	Observation	Remark
	24	03:59:00	West Akron-Pleasant Valley 138 kV East line (Q-22) opened	L	O	
	25	03:59:00	West Akron-Rosemont-Pine-Wadsworth 138 kV line opened	L	O	
	26	04:05:55	Dale-West Canton 138 kV line tripped due to sag into a tree, reclosed at West Canton only	L	O	
	27	04:05:57	Sammis-Star 345 kV tripped by zone 3 relay	L	O	Critical line
	28	04:06:02	Star-Urban 138-kV line tripped	L	O	
	29	04:06:09	Richland-Ridgeville-Napoleon-Stryker 138 kV line tripped on overload and locked out at all terminals	L	O	
Steady-state progression about 1 h 5 min	30	04:08:58	Galion-Ohio Central-Muskingum 345 kV on zone 3 relay	L	O	
	31	04:08:58	Ohio Central-Wooster 138 kV line tripped	L	O	
	32	04:09:06	East Lima-Fostoria Central 345 kV line tripped on zone 3 relay, causing major power swings through New York and Ontario into Michigan	L	O	Power swing
	33	04:09:08	Michigan Cogeneration Venture plant reduction of 300 MW (from 1263 to 963 MW)		O	
	34	04:09:17	Avon Lake 7 unit trips (82 MW)		O	
	35	04:09:17	Burger 3, 4, and 5 units trip (355 MW total)	G	O	
	36	04:09:30	Kinder Morgan units 3, 6, and 7 trip (209 MW total)	G	O	
	37	04:10	Harding-Fox 345 kV	L	O	
High-speed cascade about 3 min	38	4:10:04–4:10:45	Twenty generators along Lake Erie in northern Ohio (loaded to 2174 MW total) tripped	G	O	
	39	04:10:36	Argenta-Battle Creek 345 kV line tripped	L	O	
	40	04:10:36	Argenta-Tompkins 345 kV line tripped	L	O	
	41	04:10:36	Battle Creek-Oneida 345 kV line tripped	L	O	
	42	04:10:37	West-East Michigan 345 kV	L	O	
	43	04:10:37	Sumpter Units 1, 2, 3, and 4 units tripped on undervoltage (300 MW near Detroit)	G	O	
	44	04:10:37	MCV Plant output dropped from 963 to 109 MW on overcurrent protection	G	O	
	45	04:10:38	Midland cogeneration venture (loaded to 1265 MW)	G	O	
	46	04:10:38	Transmission system separates northwest of Detroit		O	
	47	04:10:38	Hampton-Pontiac 345 kV line tripped	L	O	
	48	04:10:38	Thetford-Jewell 345 kV line tripped	L	O	
	49	04:10:38	Perry-Ashtabula-Erie West 345 kV	L	O	

	50	04:10:38	Cleveland separated from Pennsylvania, flows reversed and a huge power surge flowed counterclockwise around Lake Erie		O
	51	04:10:38	Erie West-Ashtabula-Perry 345 kV line tripped at Perry	L	O
	52	04:10:38	Large power surge to serve loads in eastern Michigan and northern Ohio swept across Pennsylvania, New Jersey, and New York through Ontario into Michigan		O
	53	04:10:39	Bay Shore-Monroe 345 kV line	L	O
	54	04:10:39	Allen Junction-Majestic-Monroe 345 kV line	L	O
	55	04:10:40	Majestic-Lemoyne 345 kV line Majestic 345 kV substation: one terminal opened sequentially on all 345 kV lines	L	O
	56	04:10:40	Homer City-Watercure Road 345 kV	L	O
	57	04:10:40	Homer City-Stolle Road 345 kV	L	O
	58	04:10:40	Lakeshore unit 18 (156 MW, near Cleveland) tripped on underfrequency	G	O
	59	04:10:41	South Ripley-Dunkirk 230 kV	L	O
	60	04:10:41	Fostoria Central-Galion 345 kV	L	O
	61	04:10:41	Perry 1 nuclear unit (rated 1252 MW)	G	O
	62	04:10:41	Avon Lake 9 unit (rated 616 MW)	G	O
	63	04:10:41	Beaver-Davis Besse 345 kV line	L	O
High-speed cascade about 3 min	64	04:10:41	Underfrequency load shedding in Ohio: First Energy shed 1754 MVA load AEP shed 133 MVA load		O
	65	04:10:41	Eastlake 1, 2, and 3 units (304 MW total, near Cleveland) tripped on underfrequency	G	O
	66	04:10:41	Avon Lake unit 9 (580 MW, near Cleveland) tripped on underfrequency	G	O
	67	04:10:41	Perry 1 nuclear unit (1223 MW, near Cleveland) tripped on underfrequency	G	O
	68	04:10:41	Belle River unit 1 tripped (637 MW) on out-of-step	G	O
	69	04:10:41	St. Clair unit 7 tripped (221 MW, DTE unit) on high voltage	G	O
	70	04:10:42	Bay Shore Units 1–4 (551 MW near Toledo) tripped on over-excitation	G	O
	71	04:10:42	Ashtabula unit 5 (184 MW, near Cleveland) tripped on underfrequency	G	O
	72	04:10:42	Greenwood unit 1 tripped (253 MW) on low voltage, high current	G	O
	73	04:10:42	Trenton Channel units 7A, 8, and 9 tripped (648 MW)	G	O
	74	04:10:42	Campbell unit 3 (rated 820 MW) trips	G	O
	75	04:10:43	Keith-Waterman 230 kV	L	O
	76	04:10:43	West Lorain units (296 MW) tripped on undervoltage	G	O
	77	04:10:44	South Ripley-Erie East 230 kV, and South Ripley-Dunkirk 230 kV	L	O
	78	04:10:44	East Towanda-Hillside 230 kV	L	O

(continued)

TABLE 13.2. *(Continued)*

Period	No.	Time	Event	Equipment	Observation	Remark
	79	04:10:45	Wawa-Marathon 230 kV	L	O	
	80	04:10:45	Branchburg-Ramapo 500 kV	L	O	
	81	04:10:46–04:10:47	New York–New England transmission lines disconnect		O	
	82	04:10:48	New York transmission splits east-west		O	
	83	04:10:50	The Ontario system just west of Niagara Falls and west of St. Lawrence separates from New York		O	
	84	04:11:22	Long Mountain-Plum Tree 345 kV	L	O	
	85	04:11:22	Southwest Connecticut separated from New York City		O	
	86	04:11:57	Remaining transmission lines between Ontario and eastern Michigan separate		O	
	87	04:13	Cascading sequence essentially complete		O	

In this table, the fifth column presents the type of the component: L for a line, G for a generator, and T for a transformer; the sixth column presents whether the events could be observed by system operator: O is observable, NO is not observable.

by the system operator at 04:04 p.m., about 2 min before the start of the cascade. Because of this problem, the system operator lost a lot of time to prevent blackout.

Because of these computer faults, FE's system operators were not aware that their electrical system condition was started to degrade!

3. The possible load shedding to prevent blackout. There are three time points, in which if system operator took load shedding, the cascading events should be stopped:

   a. Before the eighth event, 03:41:33 p.m. Star-South Canton 345 kV tripping, if 500 MW of load would have been shed within the Cleveland-Akron area, this would have improved voltage at the Star bus from 91.7% up to 95.6%, pulling the line loading from 91% to 87% of its emergency current rating; an additional 500 MW of load would have had to be dropped to improve Star voltage to 96.6% and the line loading to 81% of its emergency current rating. It would have been possible to prevent the blackout by shedding load within the Cleveland-Akron area.

   b. Shedding 1500 MW of load in that area before the loss of the Sammis-Star line (27th event) at 04:05:57 p.m. might have prevented the cascade and blackout.

   c. After the East Lima-Fostoria Central 345 kV line tripping (32nd event) at 04:09:06 p.m., if automatic undervoltage load shedding had been in place in northeast Ohio, it might have been triggered at or before this point, and dropped enough load to reduce or eliminate the subsequent line overloads that spread the cascade. The East Lima-Fostoria Central 345 kV line tripping also caused major power swings through New York and Ontario into Michigan [2].

   From the analysis, at these time points, *the system operators could took some actions to stop cascading event and limit the blackout in a small area. But because of computer faults, lack of effective defense plan in the system (e.g., Special Protection Scheme, emergency load shedding) and lack of effective training of operators, the blackout could not be avoided.*

13.2.1.4 Final State. At the end of the blackout (16:13), most of the Northeast had blacked out. Some isolated areas of generation and load remained online for several minutes. These were some of those areas in which a close generation-demand balance could be maintained operational. There is 62,000 MW of load shedding and 531 generators at 261 plants that tripped out in this blackout.

13.2.1.5 What Stopped the Cascade Spreading?

1. The effects of a disturbance that travel over power lines and become damped the further they are from the initial point (such as the waves generated by a stone thrown into water). Thus, the voltage and current swings seen by relays on lines farther away from the initial disturbance are not as severe, and at some point they are no longer sufficient to cause lines to trip.
2. Higher voltage lines and more densely networked lines, such as the 500 kV system in PJM and the 765 kV system in AEP, are better able to absorb voltage and current swings and thus serve as a barrier to the spread of a cascade. The cascade progressed into western Ohio and then northward through Michigan through the areas that had the fewest transmission lines. Because there were fewer lines, each line absorbed more of the power and voltage surges and was more vulnerable to

tripping. A similar effect was seen toward the east as the lines between New York and Pennsylvania, and eventually northern New Jersey tripped. The cascade of transmission line outages became contained after the Northeast United States and Ontario were completely separated from the rest of the Eastern Interconnection and no more power flows were possible into the Northeast (except the DC ties from Québec, which continued to supply power to western New York and New England).

A solution against cascade spreading may be to implement FACTS (flexible AC transmission system) in some critical lines in the network. These power electronics equipments can represent a barrier against cascade spreading by damping voltage and power swings.

3. Line trips isolated some areas from the portion of the grid that was experiencing instability. Many of these areas retained sufficient online generation or the capacity to import power from other parts of the grid, unaffected by the surges or instability, to meet demand. As the cascade progressed, and more generators and lines tripped off to protect themselves from severe damage, some areas completely separated from the unstable part of the Eastern Interconnection. In many of these areas, there was sufficient generation to match load and stabilize the system. After the large island was formed in the Northeast, symptoms of frequency and voltage decay emerged. In some parts of the Northeast, the system became too unstable and shut down by itself. In other parts, there was sufficient generation, coupled with fast-acting automatic load shedding, to stabilize frequency and voltage. In this manner, most of New England and the Maritime Provinces remained energized. Approximately, half of the generation and load remained on in western New York, aided by generation in southern Ontario that split and stayed with western New York. There were other smaller isolated pockets of load and generation that were able to achieve equilibrium and remain energized.

From this point, a solution could be the design of an *emergency isolation plan*. When a blackout appears in an area and when the system operator cannot prevent the blackout spread to neighboring areas, an emergency isolation plan can split the system into some parts. This can stop the spreading of blackout and keep the other areas safety.

13.2.1.6 Causes of Blackout. There are three principal reasons for this blackout:

1. The loss of the Sammis-Star line in Ohio, following the loss of other transmission lines and weak voltages within Ohio, triggered many subsequent line trips, by power flow transfers.
2. The protection systems can save the equipments of the power system; they can also participate in the cascading events. They have been involved in about 70% of the blackout. Many of the key lines that tripped between 16:05:57 and 16:10:38 operated on zone 3 impedance relays (or zone 2 relays set to operate like zone 3s), which responded to overloads rather than faults on the protected facilities. The speed at which they tripped accelerated the spread of the cascade beyond the Cleveland-Akron area.
3. There are no effective remedial plans to prevent blackout. The evidence indicates that the relay protection settings for the transmission lines, generators, and underfrequency load shedding in the Northeast may not be sufficient to reduce the likelihood and consequences of a cascade, nor were they intended to do so.

There are some causes of blackout summarized by NERC [1]:

1. The NERC and ECAR compliance programs did not identify and resolve specific compliance violations before those violations led to a cascading blackout.
2. There are no commonly accepted criteria that specifically address safe clearances of vegetation from energized conductors.
3. Problems identified in studies of prior large-scale blackouts were repeated on August 14, including deficiencies in vegetation management, operator training, and tools to help operators to better visualize system conditions.
4. Reliability coordinators and control areas have adopted different interpretations of the functions, responsibilities, authorities, and capabilities needed to operate a reliable power system.
5. In ECAR, data used to model loads and generators were inaccurate due to a lack of verification through benchmarking with actual system data and field testing.
6. In ECAR, planning studies, design assumptions, and facilities ratings were not consistently shared and were not subject to adequate peer review among operating entities and regions.
7. Available system protection technologies were not consistently applied to optimize the ability to slow or stop an uncontrolled cascading failure of the power system.
8. FE was operating its system with voltages below critical voltages and with inadequate reactive reserve margins.
9. FE did not have an effective protocol for sharing operator information within the control room and with others outside the control room.
10. FE did not have an effective generation redispatch plan and did not have sufficient redispatch resources to relieve overloaded transmission lines supplying northeastern Ohio.
11. FE did not have an effective load reduction plan and did not have an adequate load reduction capability, whether automatic or manual, to relieve overloaded transmission lines supplying northeastern Ohio.
12. FE did not adequately train its operators to recognize and respond to system emergencies, such as multiple contingencies.
13. FE did not have the ability to transfer control of its power system to an alternate center or authority during system emergencies.
14. FE operational planning and system planning studies were not sufficiently comprehensive to ensure reliability because they did not include a full range of sensitivity studies based on the 2003 summer base case.
15. FE did not perform adequate hour-ahead operations planning studies after Eastlake 5 tripped off-line at 13:31 to ensure that FE could maintain a 30-min response capability for the next contingency.
16. FE did not perform adequate day-ahead operations planning studies to ensure that FE had adequate resources to return the system to within contingency limits following the possible loss of their largest unit, Perry1.
17. FE did not have or use specific criteria for declaring a system emergency.
18. ECAR and MISO did not precisely define "critical facilities" such that the 345 kV lines in FE that caused a major cascading failure would have to be identified as critical facilities for MISO.

19. MISO did not have additional monitoring tools that provided high-level visualization of the system.
20. ECAR and its member companies did not adequately follow ECAR Document 1 to conduct regional and interregional system planning studies and assessments.
21. ECAR did not have a coordinated procedure to develop and periodically review reactive power margins.
22. Operating entities and reliability coordinators demonstrated an over-reliance on the administrative levels of the TLR procedure to remove contingency and actual overloads, when emergency redispatch of other emergency actions were necessary.
23. Numerous control areas in the Eastern Interconnection, including FE, were not correctly tagging dynamic schedules, resulting in large mismatches between actual, scheduled, and tagged interchange on August 14.

13.2.1.7 Recommendations to Prevent Blackouts. In the deregulation context, the U.S. power system splitted into several independent system operators (ISOs), such as New York ISO, Midwest ISO. Each ISO should keep balance generations and loads in real time to maintain reliable operation and keep the safety of the system. On the other hand, ISOs are linked with each other through interconnection lines. The power exchange schedules should be done among the ISOs. Therefore, in these conditions, each ISO cannot know the entire condition of the system and they have not authority to control the generators of the other ISOs for the safety of their own system. It means that if a wide-area blackout appears, individual ISO does not have the ability to prevent it. We give, hereafter, some recommendations for this problem.

1. *Establish wide-area measurement system* (WAMS), this kind of system can get the effective information of the entire system and can help each ISO to know the system condition directly. This system also can be the database for the entire system models. Based on these models, system operators can do wide-area emergency control and wide-area system optimization, and then, keep the system running in optimal and safety conditions.
2. *Establish wide-area control system* (normal and emergency control), such as wide-area stability and voltage control system (WACS), wide-area fault tolerant control system (WAFTCS). These systems can ensure the entire system to run in safety condition.
3. *Use some new technologies to improve stability of the system, such as FACTS.* FACTS can be implemented in some critical lines. The first function of FACTS is to increase the capacity of lines and to change the power flows. When lines are overloaded, FACTS can modify their apparent impedance and then, can allow these lines to pass more power or transferring power on nonoverloaded lines. This can give more time for the system operator to control generators and loads to come back to normal condition. The second function of FACTS is to damp power oscillations in the system, using, for example, PSSs (power system stabilizers) or more sophisticated decentralized or coordinated controllers in their control loops, as well as in excitation control loop of some generators. The third function of FACTS is providing reactive power for voltage control. Such system is a reactive power resource and can be used in voltage control actions.
4. *Improve performance of the protection system* and u*se available system protection technologies* to slow and stop an uncontrolled cascading failure in the blackout.

5. *Improve the operator training,* especially in emergency condition.
6. *Use some effective tools for hour-ahead operations planning studies and day-ahead operations planning studies.*

13.2.2 September 28, 2003 Italy Blackout

13.2.2.1 Precondition.

(i) *History.* Due to historical reasons, the energy price in Italy is much higher than in the European market. Therefore, the pressure of Italian consumers (especially large industrial customers) for importing the maximum possible amount of cheap power from foreign countries is constantly increasing.

The interconnection between the Italian system and the other UCTE grids consists of six 380 kV lines and nine 220 kV lines (Figure 13.5). An additional 500 MW DC undersea cable between Italy and Greece was put in service during 2002. The most important interconnection lines are those to the French (three 380 kV and one 220 kV lines) and the Swiss (two 380 kV and six 220 kV lines) systems, while the capacity of the interconnections with Austria and Slovenia is lower and therefore less important [4].

(ii) *Energy Import.* The Italian ISO, called GRTN (Gestore della Rete di Transmissione Nazionale), defined the maximum secure levels for the power import on the different transmission grid sections. In September 2003, the net transfer capability (NTC) computed according to the (N-1) security rule was about 5400 MW and the transmission reliability margin (TRM) was 500 MW. On Sunday September 28, at 03:00 a.m., the Italian load was 27.7 GW (about 3500 MW of pumped storage plants) and the import level was 6650 MW (excluding Greece). Therefore, almost 26% of the load was supplied by the

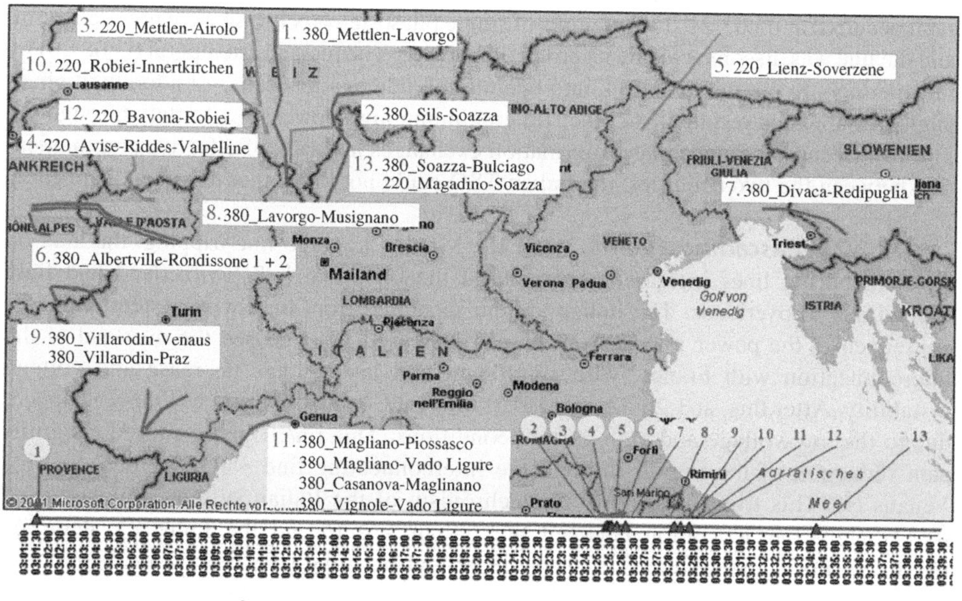

Figure 13.5. Overview of sequence of events [5].

import and some of the most important Italian power stations were off-line for economical reasons. The DC connection to Sardinia was out of service for maintenance, while the 500 MW DC cable to Greece was in operation, importing about 300 MW [4].

During the time before the blackout, the Swiss grid was also in a highly stressed condition, operated very close to the N-1 security limit. In its turn, the Italian system had a large amount of active power reserve, greater than the power import. Moreover, 1200 MW of "without notice" interruptible load was available within 1–2 min delay.

(iii) *Conclusions.* The condition of the Italian system is especially. About 26% of the load is supplied by the import. This level is beyond about 5400 MW (NTC computed according to N-1 security level). This condition is the same that 26% generators cannot be controlled by Italy power system. In other word, Italy power system should prepare 6500 MW of load shedding or generation reserve for emergency condition. In fact, the power reserve of Italian generators in operation was more than 5000 MW and some load shedding plan was designed. But due to its slow action, the generation reserve could not prevent angle instability. The load shedding plan was not either efficient in this case. Angle instability is often combined with fast voltage collapse and spreads widely through the power system. Thus, when the angle instability happened, the system could not stop the blackout.

13.2.2.2 Initiating Events. The causal event of the Italy blackout is particular as it happened in Switzerland. The Swiss 380 kV line Mettlen-Lavorgo was highly loaded at approximately 86% (100% limit at ambient temperature of 10°C being 2100 A) of its maximum capacity. The heating process of the conductors caused a progressively decrease of distance to the trees. A flashover occurred as the increased line sag with the consequent reduced distance from the trees and because of the environmental circumstances with high humidity and movement of the conductors by wind. The event of single phase to ground fault occurred at 03:01:21. The attempts of single phase auto-reclosing were not successful, and the line was disconnected by its protection device. The manual attempt (at 03:08:23) by the operators to put this line back into operation failed again because of a too high phase angle (42°), which resulted from the still continuing high power flow to Italy through the grid [3]. Because of inadequate information exchanges and lack of effective communication between the two countries, the Italian TSO could not know that this event happened.

13.2.2.3 Cascading Events. After the Mettlen-Lavorgo line tripping, the load on the neighboring lines increased. This caused the lines that link Switzerland and Italy tripped due to overloads. The Italian system separated from the Swiss system and, as a consequence, the power import from France increased. This caused the overload of the interconnection with France, with significant and fast voltage decrease and voltage instability. After this, at 3:25:32, the Albertville-La Coche 400 kV line (France) tripped, due to the low-voltage and high-current conditions. This line tripping caused a significant reduction of the import through the interconnecting line 380 kV Villarodin (F)-Venaus (I). This trigged the loss of synchronism of the Italian system, which caused immediately the separation between the Italian system and the French system. At 03:26:24, the Italian system was totally separated from the UCTE network. Separation from UCTE was equal to the loss of important generations. The Italian system launched an automatic plan to prevent blackout and 10,000 MW of load was disconnected.

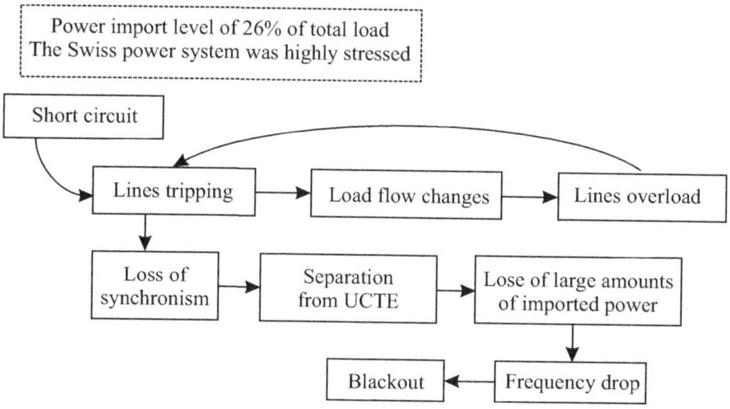

Figure 13.6. The mechanism of Italy blackout.

Unfortunately, it was not effective. The decay process still pursued and the blackout happened about 2 min and 30 s after islanding.

Figure 13.5 represents location and time of the sequence of events. We can see that the first and the second events happened in Switzerland.

The detail time line of the cascading events is shown in Table 13.3.

Figure 13.6 illustrates a simple flowchart of the Italy blackout mechanism.

There are some points that should be noticed:

1. Many events happened outside of the Italian power system. The Italian system lacked awareness of the interconnection lines. This condition made that the Italian system could not take the effective and fast actions for the emergency condition. If the Italian system has got phasor measurements and online load flow monitoring of the interaction lines, GRTN operators could find some problem in the load flow change and could prepare themselves for the emergency condition.
2. The Italian system lacks a real-time simulation tool for unpredictable events. For example, in the precondition at 03:10:47, ETRANS called GRTN and asked to decrease load to solve the line overload. If the Italian system has got a real-time simulation tool, GRTN operators could know that the system will enter into a danger condition. An effective load shedding could be made before the second line tripped.
3. There was no effective defense plan for the emergency condition, even if GRTN has monitored some critical sections. For example, if there are overloads in the interconnection lines, the load shedding plan takes place before the protection devices action. In fact, this load shedding plan was not launched because the overload did not occur on the interconnection lines.

The automatic load shedding scheme is not sufficient to stop the frequency decay. At 03:25:30, if the load shedding would have been about 60% of the total load (about 13,800 MW), the system would have been returned in the normal condition.

The main types of instability in this blackout concern are as follows:

- cascading line tripping by overloads;
- loss of synchronism due to angle instability;

TABLE 13.3. Time Line of the September 28, 2003 Italy Blackout

Period	No.	Time	Event	Equipment	Observation
Steady-State progression: about 20 min	1	03:01:21	Single phase to ground fault at 380 kV line Mettlen-Lavorgo (CH)	L	NO
	2		380 kV line Lavorgo: starting of zero sequence protection (CH)	L	NO
	3	03:01:42	380 kV line Mettlen-Lavorgo definitely switched off at Lavorgo; after two unsuccessful automatic attempts to reconnect the line, line at Mettlen still connected and under tension (CH)	L	NO
	4	03:05	Load alarm at Lavorgo 220 kV transformer 1(CH)		NO
	5	03:06:12	Heavy load on the 220 kV line Mettlen-Airolo was spotted (CH)		NO
	6	03:08:23	Attempt at Lavorgo to get 380 kV line Mettlen-Lavorgo into operation again (CH)		NO
	7	03:10:47	Communication (phone call) between ETRANS and GRTN (CH)		O
	8	03:18:00	EGL (Electricité de Laufenbourg) decided to switch off one 380/220 kV transformer in Soazza after the coordination with ETRANS (CH TSO) and ATEL (CH independent producer)		NO
	9	03:22:02	Tap change of 220 kV transformer 1 at Lavorgo (CH)	T	NO
High-speed cascade: about 2.5 min	10	03:25:21	Protection of 220 kV line Mettlen-Airolo triggered at Airolo (CH)	L	NO
	11		Trip of 380 kV line Sils-Soazza line (1783 MW) after single phase to ground fault (CH)	L	NO
	12	03:25:25	Trip of 220 kV line Mettlen-Airolo at Airolo (740 MW); at Mettlen the line stays connected and under tension (CH)	L	NO
	13	03:25:26	Start of automatic disconnecting device at Lienz (A)		NO
	14	03:25:28	Trip of bus bar coupler at Lienz (A)		NO
	15		Trip 220 kV line Cislago-Sondrio (I)	L	O
	16		Trip 220 kV line Riddes-Avise (281 MW) (CH-I)	L	O
	17		Trip 220 kV line Riddes-Vallpelline (299 MW) (CH-I)	L	O
	18	03:25:32	Trip of 400 kV line Albertville-La Coche (F)	L	NO
	19		Trip of storage pump at Malta (145 MW) by voltage drop, tripping of storage pump within 110 kV network (35 MW) by voltage drop, tripping of several small generators (within 20 kV networks) (A)	S	NO
	20	03:25:33	Trip of 220 kV line Lienz-Soverzene at Lienz (209 MW that is, >548 A) (A-I)	L	O
	21		Trip of 220 kV line Le Broc-Carros-Menton-Camporosso (248 MW) (F-I)	L	O
	22	03:25:34	Trip of 400 kV line Albertville-Rondissone 1 at both substations (841 MW) (F-I)	L	O
	23		Trip of 400 kV line Albertville-Rondissone 2; disconnected only at Rondissone (682 MW) (F-I)	L	O
	24	03:25:35	Trip of 380 kV line Divaca-Redipuglia (646 MW) (SI-I)	L	O
	25		Trip of 380 kV line from Redipuglia to Planais (I)	L	O

26		Trip of 220 kV line from Redipuglia to Safau (I)	L	O
27	03:25:42	Trip of 220 kV line Divaca-Klece including 220 kV synchronous compensator in Divaca without tension (SI)	L	NO
28	03:26	All circuits at 220 kV Fiesch disconnected (CH)		NO
29	03:28:08	All connections at 380 kV Lavorgo without tension, including Lavorgo-Musignano (503 MW) (CH-I)		O
30		220 kV line Gorduno-Mese (125 MW) without tension (CH-I)	L	O
31		220 kV line Airolo-Ponte (191 MW) without tension (CH-I)	L	O
32		Trip of 220 kV line Robbia-Sondrio (253 MW) (CH-I)	L	O
33		Trip of 220 kV line Pallanzeno-Serra (110 MW) (CH-I)	L	O
34		Trip of 220 kV line Padriciano-Divacia; disconnected at Padriciano (199 MW) (SI--I)	L	O
35	03:28:10	Trip of 400 kV lines Villarodin-Praz (F) and Villarodin-Venaus (712 MW) (F-I)	L	O
36	03:28:14	Trip of 220 kV line Innertkirchen-Robiei (CH)	L	NO
37	03:28:28	Trip of 380 kV lines Casanova-Magliano, Magliano-Vado Ligure, Vignole-Vado Ligure and Magliano-Piossasco (I)	L	O
38	03:28:29	Trip of 220 kV line Robiei-Bavona (CH)	L	NO
39	03:34:11	Trip of 380 kV line Soazza-Bulciago; disconnected at Soazza (1205 MW) (CH-I)	L	O

In this table, (CH) means Swiss, (I) is Italy, (F) is France, (SI) is Slovenia and (A) is Austria.

- oscillatory instability causing self-exciting interarea-oscillations;
- exceeding of the allowed frequency range (over and underfrequency);
- voltage collapse.

13.2.2.4 Final State. This blackout affected the whole country (excluding Sardinia, already disconnected that night) and the load was not supplied for a time interval ranging from 1.5 to 9–19 h. The total load shedding can be estimated in about 8000 MW, about 35% of the total load supplied [21].

TABLE 13.4. A Summary of Events of the Restoration Process of the Italian System

Time	Events
03:42	Start of successive reconnection of interconnectors to Italy
	Reconnection of the 380 kV line Soazza-Sils (CH)
03:47	Reconnection of the 220 kV line Airolo-Ponte (CH-I)
03:48	Reconnection of the 220 kV line Pallanzeno-Serra (CH-I)
03:47	Reconnection of the 220 kV line Airolo-Ponte (CH-I)
04:05	Reconnection of the 220 kV line Le Broc-Carros-Menton-Camporosso (F-I)
04:21	Reconnection of the 220 kV line Divaca-Padrciano-Divacia (SI-I)
04:37	Reconnection of the 380 kV line Soazza-Bulciago (CH-I)
04:52	Reconnection of the 380 kV line Divaca-Redipuglia (SI-I)
05:17	Reconnection of the 400 kV line Altbertville-Rondissone 1 (F-I)
05:30	Reconnection of the 380 kV line Lavorgo-Musignano (CH-I)
06:00	*Import from UCTE system to Italy: 2100 MW*
06:18	Reconnection of the 400 kV line Villarodin-Venaus (F-I)
06:27	Reconnection of the 220 kV line Gorduno-Mese (CH-I)
06:48	Reconnection of the 220 kV line Robbia-Sondrino (CH-I)
07:00	*Import from UCTE system to Italy: 3490 MW*
08:00	*Import from UCTE system to Italy: 3800 MW*
08:05	Reconnection of the 220 kV line Riddes-Vallpelline (CH-I)
08:23	Reconnection of the 220 kV line Lienz-Soverzene (A-I)
08:48	Reconnection of the 220 kV line Riddes-Avise (CH-I)
08:00	*Import from UCTE system to Italy: 5620 MW*
	Load shedding of interruptible customers with and without advance notice by means of remote control in the regional control centers of Milano, Torino, and Venezia
11:00	From 11:00 to 17:00 50 MWh/h reserve power delivery support from ELES (Elektro Slovenija TSO) to GRTN
12:45	Reconnection of the 400 kV line Albertville-Rondissone 2 (F-I)
16:00	*Import from UCTE system to Italy: 6545 MW*
16:40 ÷ 23:52	Load shedding of interruptible customers with and without advance notice by means of remote control in the Middle-South of Italy to cover the load diagram and due to high power flows on the network section "North-Florence". In total 60 MW
16:48	Energisation of the bus bar in Brindisi Cerano 380 kV substation
16:50	Agreement with HTSO (Greece TSO) on an import of 500 MW until 7:00 at September 29, 2003. After that scheduled programs will be followed with a possible opportunity to import 500 MW extra
17:10 ÷ 23:52	Load shedding of interruptible customers with and without advance notice by means of remote control in Tuscany to cover the load diagram and due to high power flows on the network section "North-Florence." In total 47 MW
17:30	Reconnection of Sicily due to switching on of the line Sorgente-Corriolo (it was inoperable before)
21:40	Request to supply customers in Sicily
23:00	All customers supplied

TABLE 13.5. Root Causes and the Actions

Identified Root Causes	Impact on Events	Root Causes	Action
1. Unsuccessful re-closing of the Mettlen-Lavorgo line because of a too high phase angle difference	Decisive	Large phase angle due to power flows and network topology	Study settings of concerned protection devices Reassess possible consequences for NTC to Italy Coordination of emergency procedures
2. Lacking a sense of urgency regarding the San Bernandino line overload and call for inadequate countermeasures in Italy	Decisive	Human factor	Operator training for emergency procedures Reassess acceptable overload margins Study real-time monitoring of transmission line capacities
Angle stability and voltage collapse in Italy	Not the cause to the origin of the events but was the cause that successful islanding operation of Italy after its disconnection did not succeed	General tendency toward grid use close to its limits	Further studies necessary on how to integrate stability issues in UCTE security and reliability policy
Right-of-way maintenance practices	Possible	Operational practices	Perform technical audit If necessary, improve tree-cutting practices

13.2.2.5 Restoration. Table 13.4 shows a summarized sequence of events happened during the restoration process of the Italian system.

13.2.2.6 Root Causes of the Blackout. First, let us have a look on Table 13.5. These root causes should be seen in the context that the interconnected network was developed with a view to ensure mutual assistance between national subsystems and, to some extent, to optimize the use of energy resources by allowing exchanges between these systems, but not in view of the present high level of cross-border exchanges. The development of the market has led to operators using parts of the network continuously to its limits, as far as is allowed by the security criteria. The blackout must be seen in this context [4].

13.2.2.7 Recommendations to Prevent Blackouts. With the opening of the energy market in Europe, the customers can buy cheaper energy from other countries and producers can sell energy to other countries. The load flows and exchanges between areas controlled by different TSOs become complex. This is a challenge for system operators. From the Italy blackout, we can get some experiences to prevent blackouts:

1. The critical element of the system should be efficiently monitored. An efficient measurement system should be installed in the Italian power system;

2. Real-time power flow calculations and security assessment tools should be used by the Italian operator;
3. Established efficient automatic emergency plan to prevent blackout;
4. Established wide-area fault tolerant control system (WAFTCS)

13.2.3 September 23, 2003 Eastern Denmark and Southern Sweden Blackout

13.2.3.1 Precondition. On September 29, the load level was low in Eastern Denmark and Southern Sweden. The system had big security margins. Unless some generators and line were out of service, it was for planned maintenance reasons and this had no effects on the course of events. In fact, there was no reason for a blackout.

The total power generation in Eastern Denmark was around 2250 MW, which consisted of 1800 MW of power plants production and 450 MW of wind turbine production. The load was about 1850 MW and 400 MW was exported to Southern Sweden. There was also 775 MW of reserves, which was largely enough to face a tripping problem.

From the Sweden point of view, the system operation was also normal.

13.2.3.2 Initiating Events. At 12:30, the unit 3 at Oskarshamn nuclear power station tripped, due to problems with a valve in the feed-water circuit, and so the system lost 1200 MW of production (53% of the total production). As a consequence, the frequency of the system dropped, but it was stabilized at 49.9 Hz by activating the reserves, not only in Eastern Denmark but also in Sweden, Norway, Finland, and rest of Denmark. There were no voltage problems. The system was bringing back in safety condition [19].

13.2.3.3 Cascading Events. When the initiating event occurred, the system lost 1200 MW of production but it was stabilized using large power reserves. This constitutes the steady-state progression, which had duration of 5 min because at 12:35, a second event occurred and this one was the critical event that trigged the high-speed cascade. The critical event was a double bus bar fault at the Horred 400 kV substation, in the South Swedish transmission grid. A consequence was the disconnection of two critical lines and two units at Ringals power station. Therefore, the system lost more than 1800 MW of production. At this time, the total generation loss was about 4000 MW and the system, in this configuration, did not have enough reserves anymore.

Some lines in Sweden became heavily loaded and the voltage began to fall, which led to others lines tripping by serious overload.

Because of the weakening of the transmission grid and the lack of reserves, it was impossible to restore the voltage that felled to zero at 12:37.

The duration of the high-speed cascade was about 2 min (Figure 13.7).

13.2.3.4 Final State. This blackout leaved the entire subarea without electricity. The two largest plants of Zealand were damaged by the voltage collapse.

13.2.4 January 12, 2003 Blackout in Croatia

13.2.4.1 Precondition. The power system in Croatia was seriously damaged due to war in the previous decade, and its operational performances were degraded. The system

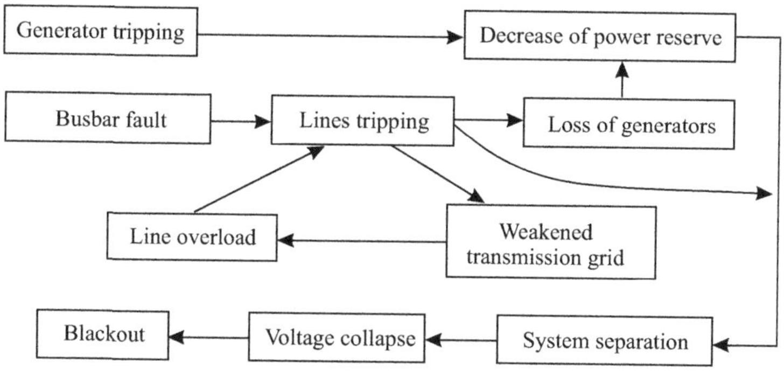

Figure 13.7. Mechanism of Eastern Denmark and Southern Sweden blackout.

was small and weak, as seven transmission lines were out of operation in Dalmatia before the blackout, due to extremely bad weather conditions (low temperatures, icing rain, snow, and high wind). The bad weather led also to a high load level for the whole system, and such bad conditions supported the appearance of the blackout.

13.2.4.2 Initiating Events. The event that initiated the blackout was a three phases short circuit on the Konjsko-Velebit 400 kV line at 16:43:58. Unfortunately, one pole of the circuit breaker did not work. This caused the system to operate in unsymmetrical conditions [16].

13.2.4.3 Cascading Events. As the protection system was not complete, compare to others transmission systems, and that it was moreover failing, the large power unbalance due to the breaker malfunctioning had increased power flows on some lines that tripped by overload. In the same time, the generators have been also disconnected due to the loss of excitation caused by asymmetry of thyristor valves. Because of the unbalance conditions, there was no steady-state progression and the initiating event trigged directly the high-speed cascade, which led to blackout in 33 s.

In Figure 13.8, a simple flowchart of the Croatia blackout mechanism is presented.

13.2.4.4 Final State. All generators in Dalmatia were out of order within 30 s from the initial disturbance.

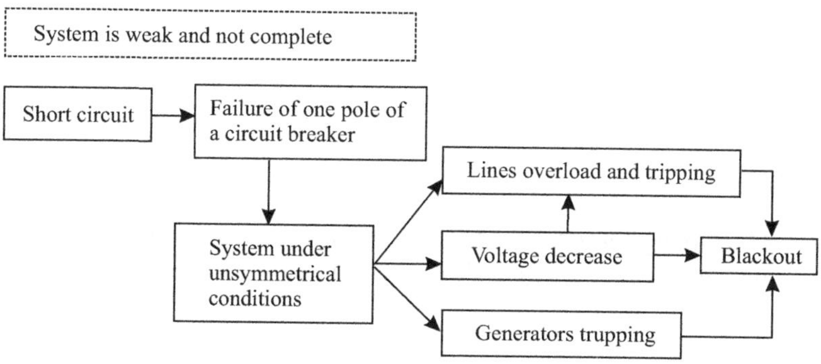

Figure 13.8. Mechanism of Croatia blackout.

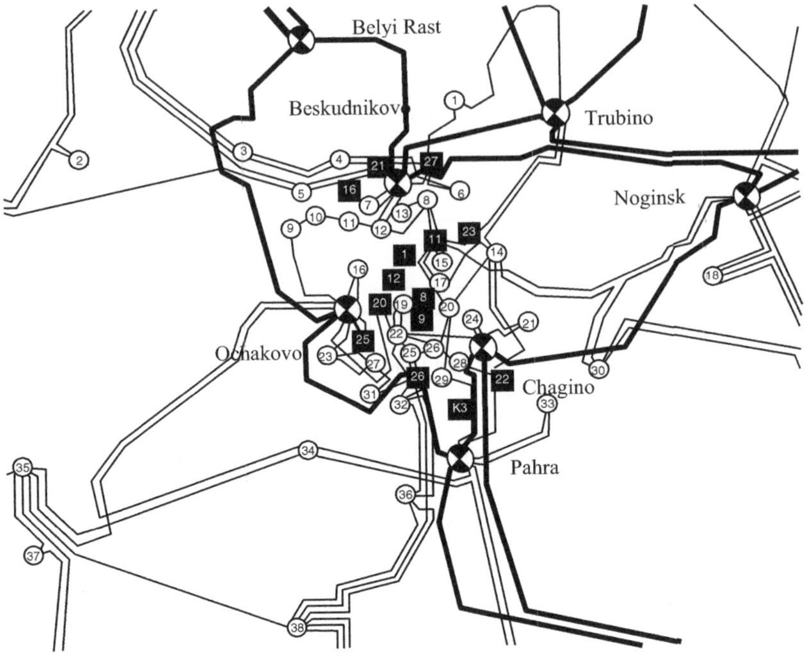

Figure 13.9. Diagram of Moscow power supply system [27].

13.2.5 May 25, 2005 Blackout in Moscow[1]

13.2.5.1 Precondition. Diagram of Moscow power supply system is shown in Figure 13.9.

(i) *The power system around Moscow was in a state of advanced decay.* More than 70% of the Moscow 220 kV electric power substations had exceeded their planned lifetime for a long time. Moreover, three more 500 kV substations were loaded to such extent that it would be very hard to repair them. In addition to that, some transformers exploded on May 25 and this led to heavily loading a 110 kV line, followed by the failure of a transformer on this line.

(ii) At the time of the blackout, *the weather was exceptionally hot* and there was a strong demand of power for air-conditioning.

With these conditions, the power system had many risks to break down.

The main precondition events during May 23, 24 and early morning 25 are shown in Table 13.6 [28].

13.2.5.2 Initiating Events. On May 25, 2005 at 10:12, five 220 kV lines tripped due to short circuits and another one was disconnected due to overload (see Table 13.6).

[1] The analysis of Moscow blackout mechanisms was made together with D.N. Efimov [28].

TABLE 13.6. Time Line of May 23–25, 2005 Moscow Blackout

Period	No.	Time	Event
Precondition	1	May 23 19:57	A failure and fire occurred on the current transducers installed at the 110 kV breaker of the autotransformer at the Chagino substation
	2		Disconnecting the autotransformer, the second 500 kV bus bar, and the first 110 kV bus bar
	3	20:26	The fire was extinguished
	4	23:40	The second 500 kV bus bar was reconnected
	5	May 24 0:30	The first 110 kV bus bar was reconnected through the 110 kV cable line at the Chagino substation
	6	20:57	One more current transducer on the breaker between 110 kV sections exploded, leading the damage and short circuit
	7		Disconnecting the second 110 kV bus bar at the Chagino substation
	8		The 110 kV breaker connecting another autotransformer with the second 110 kV bus bar, the first 500 kV bus bar and the autotransformer were isolated
	9	21:17	The 110 kV transformer explodes from overheating, damaging nearly equipment. Fires put out on four transformers at the Chagino substation
	10		220 kV power supply disrupted to Moscow oil refinery five adjoining Moscow districts and three factories
	11	21:30	Service to the oil refinery (which is normally supplied at 220 kV for its major electricity loads) is restored using just a 110 kV line from Chagino (and also fed by Heat Power Plants #22)
	12		Other loads adjacent to the refinery are served from this line as well, so the line is heavily loaded
	13	May 25 05:31	Third Chagino transformer on the 110 kV line fails
	14		Supply to the 110 kV line taken from another 500/110 kV transformer
	15	05:33	That transformer is disconnected
	16	09:00–10:00	Extra demand for power to run air conditioning. Chagino substation is operating at reduced capacity. Transmission lines in the system become heavily loaded
Initiating event	17	May 25 10:12	Short circuit at 220 kV line from Ochakovo high-voltage substation
Cascading events	18	May 25 10:12	220 kV line tripped after short circuits
	19		Five more lines tripped due to overloading
	20		The 220 kV lines coming from Tula and Kaluga overload
	21	10:47	Unsuccessful connection of 220 kV line from cogeneration power plant (CGPP) #23
	22		Successful connection of 220 kV line Baskakovo-Golianovo
	23	10:48	Unloading of CGPP #20 in Moscow
	24	10:53	Two 220 kV lines tripped due to overloading
	25	10:54	Unloading of CGPP #23 in Moscow
	26	10:55	Unsuccessful connection of breaker on CGPP #2 in Moscow
	27	10:56–11:04	Overloading and disconnection of four 110 kV and 220 kV lines
	28	11:04	Unsuccessful connection of 220 kV line from Ochakovo substation
	29	11:06–11:12	Overloading and disconnection of 11 110–220 kV lines. Voltage decreasing in 110–220 kV network. Disconnections of units on CGPP #8, #26 in Moscow
	30	11:12–11:16	Overloading and disconnection of 11 110–220 kV lines. Voltage decreasing in 110–220 kV network. Disconnections of units on CGPP #1, #4, #8, #17, #20, #22, #26
	31	11:16–11:32	Overloading and disconnection of 7 110–220 kV lines. Voltage collapse in 110–220 kV network. Disconnections of units on CGPP #2, #9, #11, #20, #22, #26

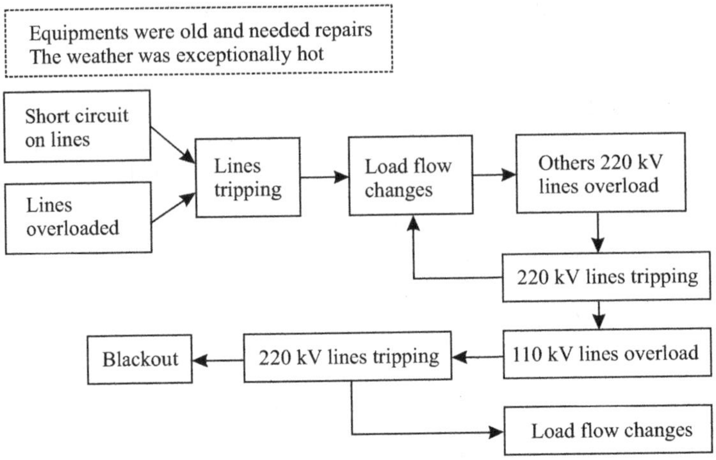

Figure 13.10. Mechanism of Moscow blackout.

13.2.5.3 Cascading Events. The cascade progression lasted 2 h and 13 min, but unfortunately, we are not able to determine if there was a high-speed cascade because we did not find enough information on it. Nevertheless, cascading outages were limited to Moscow and surrounding regions [23].

Figure 13.10 describes a flowchart of the Moscow blackout mechanism.

13.2.5.4 Final State. As we aforementioned, the blackout was limited around Moscow. At the worst moment of the incident, a quarter of Moscow's consumption was cut off, 22% in the Kaluga region and 90% in the Tula region. About 2500 MW of load was interrupted in the Moscow area.

13.2.6 July 12, 2004 Greece Blackout

13.2.6.1 Precondition.

(i) Many new equipments were planned to install in the period 2003–2004 in the Greece power system, but a significant number of these upgrades have been carried out only after the yearly peak of 2004, which occurred on July 12.

(ii) One 125 MW generating unit in the Peloponnese peninsula and another generating unit in Northern Greece were out of service the day before the incident.

Before the blackout, the failure of auxiliaries causes the disconnection of the unit 2 of the Lavrio power station in the Athens area. Thus, the system lost 300 MW of production. Unit 2 was repaired and was synchronized at 12:01.

(iii) At this time, the load peak reached 9160 MW and voltages in the Athens area were declined significantly. The voltage drop stopped as soon as Lavrio-2 synchronized and started generating.

13.2.6.2 Initiating Events. The Lavrio-2 unit was lost again at 12:12, due to high drum level.

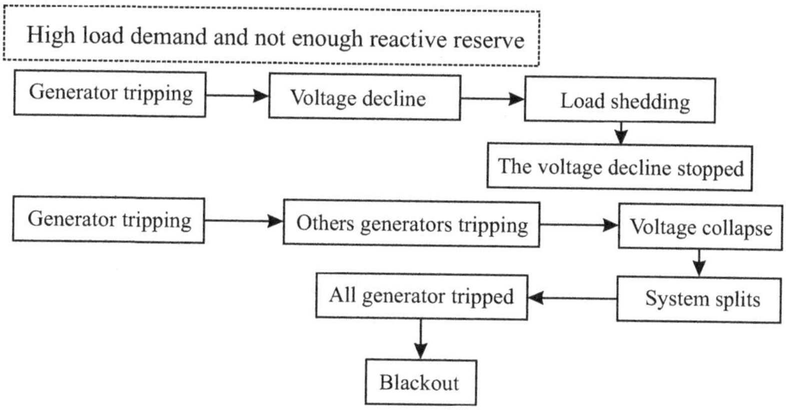

Figure 13.11. Mechanism of July 12, 2004 Greece blackout.

13.2.6.3 Cascading Events. After the loss of the Lavrio-2 unit, the system needed more reactive power for voltage support. A load shedding of 80 MW was used at 12:30 to stop the voltage decline. At 12:37, unit 3 of Aliveri power station in the weak area of Central Greece was automatically tripped. One minute later, the remaining unit in Aliveri was manually tripped at 12:39, the voltage collapse started. In the same time, the system was split by the undervoltage protection of the north–south 400 kV lines [21]. Figure 13.11 illustrates the flowchart of the Greece blackout mechanism.

13.2.6.4 Final State. After the split of the system, all the remaining generations in the areas of Athens and Peloponnese peninsula were disconnected, leading to the blackout.

13.2.7 July 2, 1996 Northwest U.S. Blackout

13.2.7.1 Precondition. On July 2, the Western Systems Coordinating Council (WSCC) system load was in condition of peak summer consumption in Idaho and Utah. There were some important power transfers through the grid and the system was stressed as:

- high north-to-south electricity transfers on the California-Oregon AC and DC interconnections;
- maximum power flow conditions in the Pacific Northwest;
- high volumes of electricity transfers from Canada to the Northwest;
- transfers from the Northwest to Idaho and Utah.

The voltage support capabilities were also insufficient in the Northwest and Idaho.

13.2.7.2 Initiating Events. The first significant event was a single phase-to-ground fault at 14:37:18 on the 345 kV Jim Bridger-Kinport line. The conductor sagged close to the trees and the result was a flashover. The fault was cleared by the system protection, which removed the line from service [13].

13.2.7.3 Cascading Events. This blackout was constituted only by a high-speed cascade, which led to the system separation in five islands only in 60 s. After the first

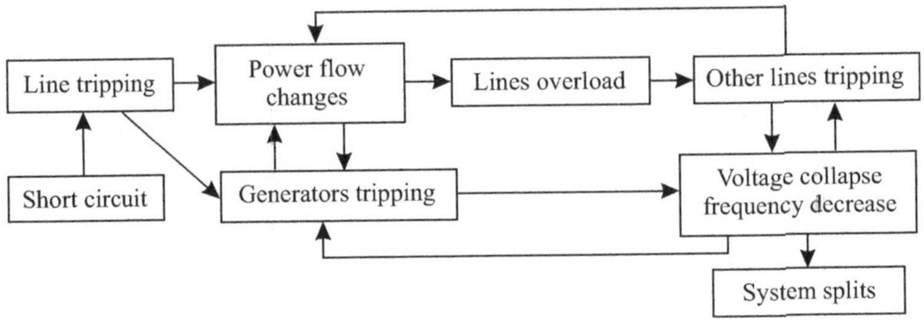

Figure 13.12. Mechanism of July 2, 1996 U.S. blackout.

significant event, another 345 kV line tripped due to a protection relay malfunctioning and this has initiated a remedial action scheme (RAS) that disconnected two-generation units. Consequently, the system lost 1040 MW of production. Another protection relay malfunctioning caused a 230 kV line tripping and during this time, the BPA (Bonneville Power Administration) system saw its voltage drop. The induced changes of power flows made that some 230 kV lines were heavily loaded and the 230 kV Mill Creek-Antelope line tripped at 14:25:01. During this time, another two-generation units had tripped due to a field excitation overcurrent. Then, the voltage began to collapse rapidly and more and more lines were overloaded and tripped. The frequency also began to decrease and under-frequency load shedding was operated in some areas. Figure 13.12 shows the flowchart of the U.S. blackout mechanism.

13.2.7.4 Final State. Finally, the WSCC system was separated into five islands.

13.2.8 August 10, 1996 Northwest U.S. Blackout

13.2.8.1 Precondition. In this case too, most of the region suffered from high summer temperatures. There were also heavy exports from the Pacific Northwest to California and from Canada to the Pacific Northwest. Moreover, the loss of some 500 kV lines in Oregon has weakened the system.

13.2.8.2 Initiating Events. The 500 kV Big Eddy-Ostrander line tripped at 14:06:39 due to a tree flashover.

13.2.8.3 Cascading Events. Fifty minutes after the initiating event, one other 500 kV line was removed following a tree flashover and the same process started again 46 min later with another line. At this point of the cascade, several hundred MVAr of reactive power support were removed and some lines were heavily loaded because they got the power flows of the tripped lines. Some generators were asked to produce their maximum reactive power. Following the steady-state progression, which lasted 1 h and 38 min, the high-speed cascade started at 15:47:36 with the disconnection of a 230 kV line due to a tree flashover. The system lost several generators and lines and growing power oscillations appeared. When the system oscillations had increased to a level of about 1000 MW and 60 kV, the voltage collapsed, which caused more and more lines tripping. The high-speed cascade had duration of 7 min [13]. The mechanism of the Northwest U.S. blackout is shown in Figure 13.13.

13.2.8.4 Final State. The system was separated into four electrical islands at 15:54.

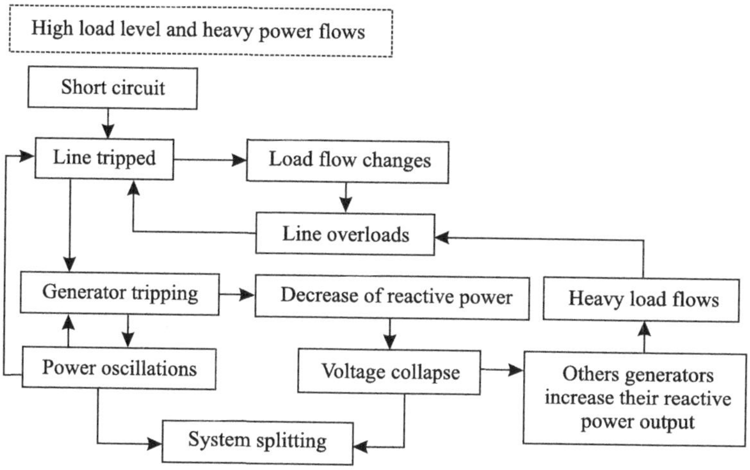

Figure 13.13. Mechanism of August 10, 1996 Northwest U.S. blackout.

13.2.9 December 19, 1978 National Blackout in France

This incident is the most serious that Electricité de France has experienced (if we except the 1999 blackout, due to a storm which has brought down to the ground most of the infrastructures)—both with regard to the duration and geographical extent—since the end of the postwar period of shortage.

13.2.9.1 Precondition. December 19 was cold and overcast, and the load escalation proved to be more rapid and higher than forecast (38,500 MW dispatched). All available generating facilities were used to the maximum of their possibilities (active and reactive) and over 3500 MW were imported from neighboring countries, mainly Germany.

13.2.9.2 Initiating Events. The rise of the load increased the already high transits, from the east to the Paris region and, as a result, voltages turned out to be extremely low in a large part of the network (Paris region, west). Starting at 08:00 a.m., overloads were observed on the grid and, at 08:06 a.m., an alarm "overload 20 min" appeared on the Bézaumont-Creney 400 kV line in the east of France. Despite various switching operations on the network topology, this overload was unable to be reduced and the line tripped due to the action of its protection at 08:26 a.m. [9].

13.2.9.3 Cascading Events. The load transfer on the remaining transmission lines resulted in three 225 kV lines tripped due to overload. Subsequently, the four-generation units at Revin were disconnected from the grid due to their current protection. A 400 kV interconnection with Belgium then tripped and the voltage dropped further. The tripping, which was hard to explain, of a new 400 kV interconnection with Belgium was accompanied by a further voltage drop and the loss of stability of a large part of the grid. The next phase witnessed the opening of lines and tripping of generation units (notably due to their voltage minimum and frequency minimum protection devices). Isolated subnetworks formed, where the generation-load balance could not be restored (insufficient load shedding, generation unit outages).

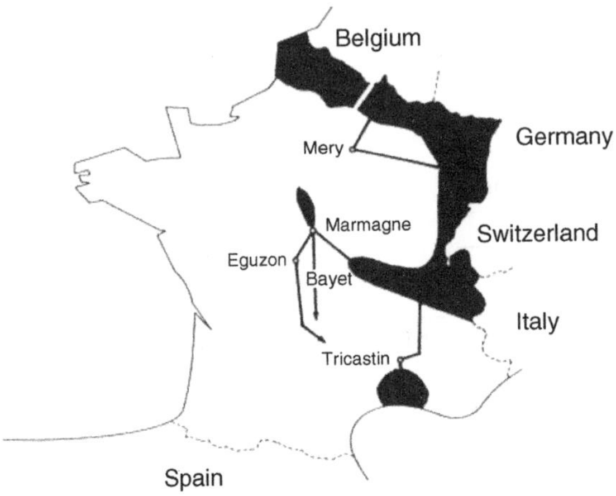

Figure 13.14. Part of the French grid still live at 8.26 a.m. on December 19, 1978 immediately after the first collapse [9].

13.2.9.4 Final State. Finally, 75% of consumers were disconnected, although the South-Eastern part of France and areas in the vicinity of the northern and eastern borders remained connected (Figure 13.14). Many generation units did not successfully trip to house load.

13.2.9.5 Restoration. An initial power restoration was too rapid and led to a further collapse of the network at 09:08 a.m. A more careful recovery relying on hydro generation units and imported power permitted almost complete network restoration at about midday. Customers sustained power cuts lasting between 30 min and 10 h.

13.2.9.6 Causes of Blackout. This incident, resulting from a tight management of the network (high transits and low voltages in some areas) and a cascade of overloads, clearly showed that the defense plan at that time was unable to cope: automatic load shedding operations were insufficient, the tripping of generation units occurred too soon on voltage minimum criteria, and the dividing up of the network was not successful. Service restoration was also not satisfactory. Many great actions were undertaken after this incident to improve all these deficient areas.

13.2.10 January 12, 1987 Western France Blackout

13.2.10.1 Precondition. Although January 12 was a particularly cold day (the "extreme cold" alert had been sent out since the previous Friday), all available generation units started up and managed to ensure a satisfactory generation margin (5900 MW) and normal voltage in the West (405 kV at Cordemais).

13.2.10.2 Initiating Events. In the space of a single hour, from 10:55 to 11:42 a.m., Cordemais generation units 1, 2, and 3 broke down due to independent causes (failure of a sensor, explosion of an electric coupler pole, shutdown due to a fire). The last unit available, which would have been sufficient to maintain the voltage in the zone, tripped as a result of inappropriate adjustment of the maximum rotor current protection and the disturbance created by the outage of unit number 3.

13.2.10.3 Cascading Events. The loss of Cordemais generation led to a sudden drop of voltage to 380 kV in the area. It stabilized for about 30 s, but the automatic on-

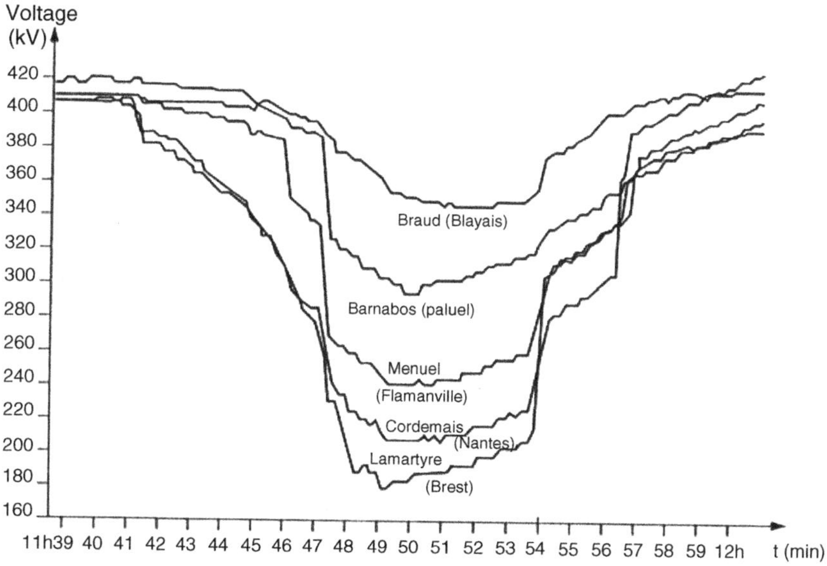

Figure 13.15. Voltage trend on the 400 kV grid in the West of France during the incident of January 12, 1987 [9].

load tap changers of the 225 kV/HV and HV/MV transformers, while attempting to restore a normal voltage, triggered a rise in load and the voltage again began to drop quickly. In a matter of minutes, nine thermal generation units near the area tripped successively, giving rise to a loss of power of 9000 MW and maintaining the voltage downward trend.

13.2.10.4 Emergency Actions. Load shedding was carried out and, at 11:50 a.m., the voltage steadied, but at a very low level in the west, less than 300 kV. In view of this extremely precarious situation, which jeopardized the national grid, the dispatching center decided to shed a load of 1500 MW in Brittany and the region of Angers, bringing the network voltage back to its normal level (Figure 13.15).

In the noon, the situation was well under control and the voltage of the network could be restored. The restoration was lengthy because of the difficulty in reconnecting a sufficient number of generation units close to Brittany and Normandy, which did not trip to house load during the incident. It took until nighttime to fully restore the power supply to these regions, when two and subsequently three-generation units were again in operation at Cordemais. At the most serious time, the outages reached a capacity of about 8000 MW.

13.2.10.5 Causes of Blackout. The main cause of this incident can be attributed to a lack of quality of adjustment of some system components, in particular of the voltage regulators and associated protection devices of generation units. The corresponding functions are now dealt with under quality assurance. The automation of some actions (blocking of on-load tap changers), the reduction of load shedding execution time (by means of remote load shedding) also appeared indispensable following this incident.

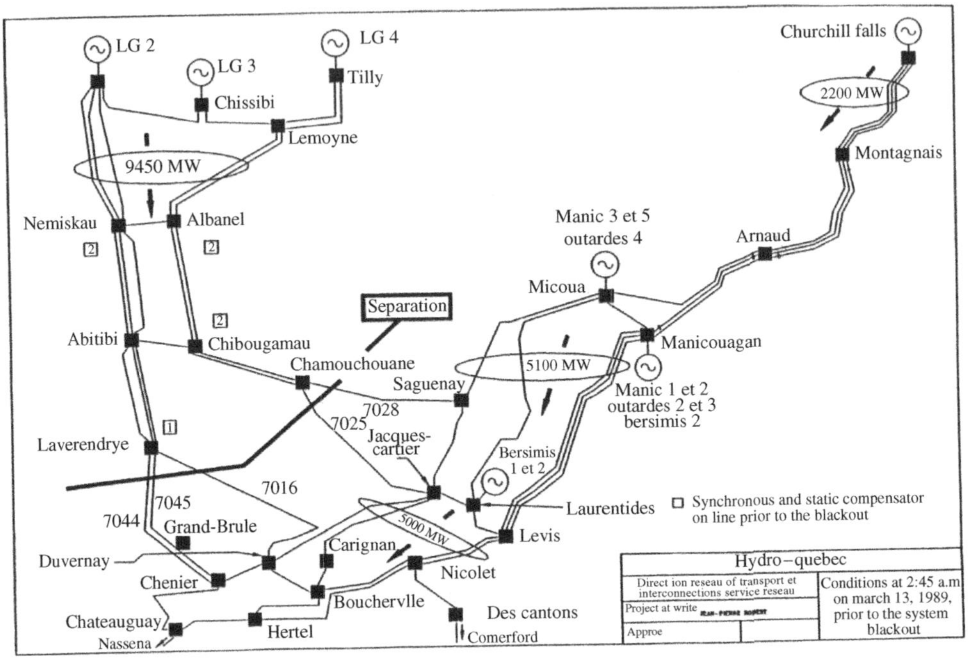

Figure 13.16. Diagram of the Hydro-Quebec system prior to the system blackout [10].

13.2.11 March 13, 1989 Hydro-Quebec System Blackout Response to Geomagnetic Disturbance

Hydro-Quebec's 735 kV transmission network mainly consists of two sets of transmission lines that carry power from major generating centers more than 1000 km away from the province's main load centers. There are 10 lines in all, each about 1000 km long; five run from the La Grande Complex in the James Bay area and five from Churchill Falls in Labrador and Manicouagan complex in the north shore of the St. Lawrence River (Figure 13.16). Given the length of the lines, La Grande system stability is achieved with static compensation, although the Churchill Falls system has no such compensation. Indeed, successful operation of such a vast network depends on reliable operation of static compensators and shunt reactors to guarantee stability and voltage control.

13.2.11.1 Precondition. When the first low-intensity magnetic disturbances[2] began on the evening of March 12, 1989, the La Grande network was loaded to 90% of its loading

[2] Magnetic storms generally induce earth surface potentials (ESP) that can give rise to circulation of quasi direct currents of as much as several hundred amperes between points on a power system that are far apart. The quasi direct current flows through transformer grounds, causing asymmetrical saturation of transformer cores and substantially increasing harmonic currents—mainly second-, third-, and fourth-order harmonic currents. The result is voltage distortion on the system, distortion that varies with time and with the length of the transmission lines.

A distorted harmonic voltage waveform can have various harmful effects. Among the major possibilities are the following: overloads on equipment such as capacitors and filters; disoperation of protection systems; damage to transformers; disturbance of the control systems of semiconductor-based equipment such as AC/DC converters, static VAr compensators, and so forth. A distorted harmonic voltage waveform can have various harmful effects. Among the major possibilities are the following: overloads on equipment such as capacitors and filters; disoperation of protection systems; damage to transformers; disturbance of the control systems of semiconductor-based equipment such as AC/DC converters, static VAr compensators, and so forth.

limit. Total system generation was 21,500 MW—with 9500 MW coming from the power stations of the La Grande Complex—and exports to neighboring systems totaled 1949 MW, of which 1352 MW was along DC interconnections (Figure 13.16) [10].

Operating staff did have some difficulty controlling voltage along the La Grande network when the first disturbances hit, but they were still able to perform the necessary switching of shunt reactors in time.

13.2.11.2 Initiating and Cascading Events. Although Hydro-Quebec staff were successful in maintaining voltage control with the first low-intensity disturbances, at 02:45 a.m. on the morning of March 13, 1989, an exceptionally intense magnetic storm generated harmonic currents that tripped or shutdown one after the other all seven static VAr compensators (SVC) on line within less than a minute, well before any preventive measures could be taken. The sequence of events was as follows:

02:44:17 a.m.	Tripping of SVC 12 at Chibougamau
02:44:19 a.m.	Tripping of SVC 11 at Chibougamau
02:44:33 a.m.to	Shutdown of the four SVCs at the Albanel
02:44:46 a.m.	Nemiscau substations
02:45:16 a.m.	Tripping of SVC 12 at La Verendrye

Nine seconds after the loss of the last SVC at La Verendrye, all five 735 kV transmission lines of the La Grande network tripped because of an out-of-step condition, completely separating the La Grande corridor from the Manicouagan-Churchill Falls network. Frequency dropped rapidly, as a result, triggering automatic load shedding systems to compensate for the loss of generation from the La Grande Complex. No amount of load shedding, however, could offset the 9500 MW generation loss; although local and remote automatic load shedding systems performed well under the circumstances. Within seconds of the onset of the event, the remainder of the Hydro-Quebec system had collapsed.

13.2.11.3 Causes of the SVC Tripping.

(i) Case of SVC installations at the La Veren drye and Chibougamau substations. Figure 13.17a shows a typical SVC installation one-line diagram. These SVCs were subjected to severely distorted voltage caused by geomagnetically induced DC currents (GIC, ground-induced currents).

Spectrum analysis of the waveforms recorded indicates predominance of second- and fourth-order harmonics resulting from DC saturation of transformer cores. Table 13.7 shows the harmonic distortion content of voltage and current at La Verendrye prior to system shutdown.

TABLE 13.7. Voltage and Current Harmonic Distortion Content at La Verendrye Before System Shutdown

Harmonic Order	AC Voltage at 735 kV	Secondary 16 kV Bus Voltage	Current (TSC)
1	100%	100%	100% (2371 A)
2	7.2%	16.7%	32%
3	2.1%	4.6%	1.8%
4	5.9%	0.9%	3.4%
5	1.8%	0.6%	3.4%

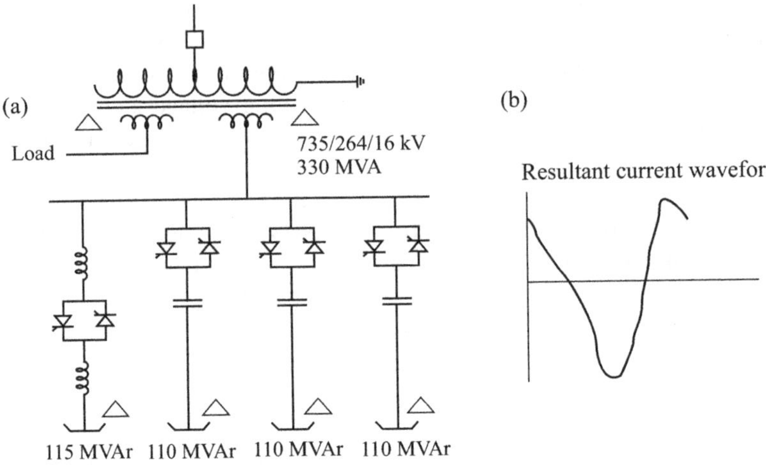

Figure 13.17. Simplified one line diagram La Verendrye/Chibougamau SVC's (a); current harmonics substation 16 kV side (b) [10].

The equipment protection scheme was originally designed for normal conditions; the possibility of intense geomagnetic storms was not considered. With the exceptional disturbance of March 13, 1989, overload protection systems of the capacitive branches initiated tripping of the SVCs at the Chibougamau site. Figure 13.17b shows the resultant current waveform measured in the thyristor-switched capacitor branch prior to protection system operation.

At the La Verendrye site, over voltage protection on the 16 kV bus side was responsible for tripping the only SVC in service. The components most sensitive to ground-induced currents are the capacitors, the thyristor-switched capacitor (TSC) reactors, and the power transformers. Because of the low impedance of the capacitors for higher order harmonics, exposure the harmonic current has a greater impact on the TSC branch than on the TCR branch. Given the abnormal conditions, the relays had to be readjusted since the protection systems were set to values that allowed only a fraction of inherent overload capacity to be used. Peak-value overload and over voltage protection are, in fact, provided for these installations, but when harmonics are present, the risk margin for improper protection system operation increases.

(ii) Case of SVC installations at the Nemiscau and Albanel sites. Figure 13.18 shows typical SVC installation one-line diagram. These SVCs were tripped by capacitor unbalance and resistor overload protection devices of the third harmonic filter branch.

As Table 13.8 shows, substantial 2nd and 4th harmonic distortions were recorded on the 735 kV side of the Albanel substation.

Unfortunately, lack of readings for the secondary 22 kV side of the static compensators at the time of the collapse makes exact assessment of the stress on the SVC components impossible. The impact of voltage and current distortions on the 22 kV side was, therefore, determined theoretically. The findings thus obtained indicate values in excess of the settings of the protection devices that operated.

TABLE 13.8. Harmonic Distortion Recorded at the Albanel Substation

Harmonic Order	AC Voltage on the 735 kV Side	AC Current on the 735 kV Side
1	100%	100%
2	5.1%	145%
3	3.4%	39%
4	0.5%	90%
5	0.9%	28%
6	0.4%	8%
7	0.2%	3%

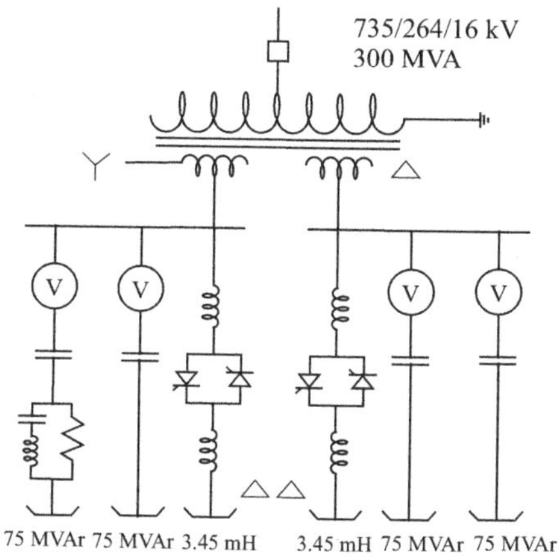

Figure 13.18. Simplified one line diagram Nemiskau/Albanel SVC's [10].

13.2.11.4 Equipment Damage. The system blackout caused loss of all static compensators on the La Grande network, damaged some strategic equipment and rendered other major pieces of equipment unavailable. As a result, it took over 9 h to restore 17,500 MW, that is, 83% of full power.

Among the major pieces of damaged equipment were two La Grande 4 generating station step-up transformers damaged by over voltage when the network separated and a shunt reactor at Nemiscau that requires factory repair. The SVCs at the Albanel and Nemiscau substations suffered only minor damage: thyristors burned at Nemiscau and capacitor bank units failed at Albanel. The SVC phase C transformer at the Chibougamau substation was also damaged by over voltage following system separation.

Hydro-Quebec's telecommunication network operated satisfactorily throughout the magnetic storm, as did all special protection systems.

13.2.11.5 Lessons Learned. The blackout of the Hydro-Quebec system on March 13, 1989 was caused by an exceptionally intense magnetic storm. The storm induced DC ground current that saturated transformers and generated even order harmonic currents that caused seven static compensators on the 735 kV network to trip or shutdown. Loss of the static compensators gave rise to system instability that culminated in separation of the La Grande network. Automatic load shedding was not able to offset the loss of the 9500 MW of generation from the La Grande generating stations, and the rest of the system collapsed within seconds.

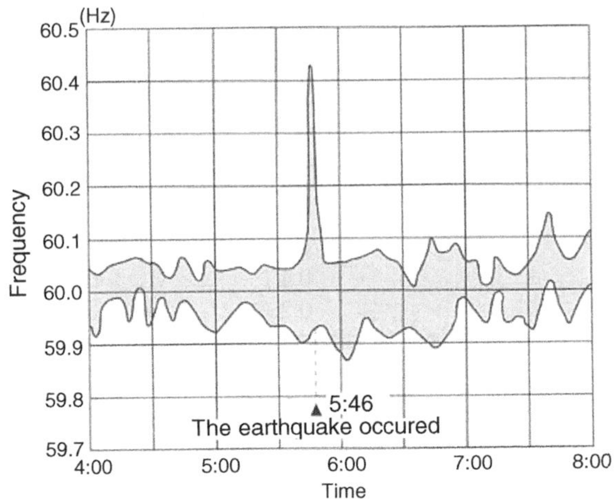

Figure 13.19. Frequency at January 17, 1995 [11].

The La Grande's vast transmission network relies on static VAr compensators to maintain system stability and voltage control. Since this type of equipment is particularly sensitive to magnetic storms, Hydro-Quebec has made great efforts to improve SVC performance under magnetic storm conditions. Remedial action was taken immediately to increase the reliability of the static compensators, and two task forces were set up to make recommendations for short term as well as for long term. Some of these recommendations have already been implemented, guidelines for solar magnetic disturbance operating procedures have been developed and an automatic alerting system has been devised [10].

13.2.12 January 17, 1995 Japan Blackout After Hanshin Earthquake

13.2.12.1 Precondition. Immediately prior to the Hanshin earthquake[3], the Kansai Electric Power Company's total power demand was on the order of 13,000 MW. The earthquake took roughly 20% of the total power system out of operation, but it did not damage any nuclear-power generation plants or hydroelectric-power generation plants.

The communication system sustained some damage typically from the fire following the earthquake. However, the system was still operable. The other facilities were damaged by the earthquake and necessitated repairing or replacement.

13.2.12.2 Supply and Demand.

(i) *Generation and Load Balance Immediately After the Earthquake*
The Kansai Electric Power Company lost 1760 MW of power generation immediately after the earthquake. The loss of load because of damage to the power system was, however, greater than the lost generation. As a result, the frequency increased by 0.45 Hz (Figure 13.19), and the voltage at the 500 kV

[3] The Hanshin-Awaji earthquake struck Southern Hyogo Prefecture at 5:46 a.m. on January 17th, 1995. The epicentre was located in the Northern tip of Awaji island, and the earthquake focus was at a depth of about 20 km. The quake registered a magnitude of 7.2 on the Richter scale. The epicentre was only 20 km away from the centre of Kobe city, which has a population of 1.5 million people. The damages of Hanshin-Awaji earthquake were the greatest in Japan's post-war history.

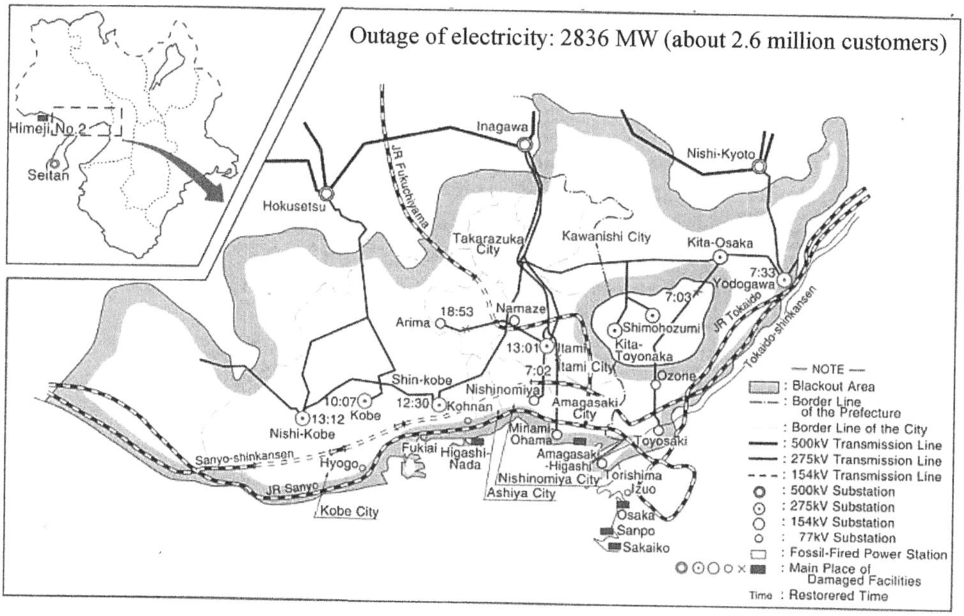

Figure 13.20. Blackout area just after the earthquake [11].

Hokusetu and Inagawa substations, which supply with electricity to the greater Kobe region, went up by 20 kV.

(ii) *Interruption of Electricity Supply*

Six 275 kV substations, the Yodogawa, Kita-Osaka, Itami, Shin-Kobe, Kobe, and Nishi-Kobe substations, and two 154 kV substations, the Torishima and Minami-Ohama substations, were damaged by the earthquake vibrations and completely out of service. As a result, power supply was cut off throughout a wide-area stretching from Akashi city in Hyogo prefecture to Southwestern region of Kyoto prefecture. The area affected by the outage of electric power represents a total demand of 2836 MW from approximately 2.6 million customers (Figure 13.20).

13.2.12.3 Damage to Electric Power Facilities.

(i) *Fossil-Fired Power Generation Plants*

The Kansai Electric Power system operates 21 fossil-fired thermal power plants, comprising 64 units among these, 20 units located at 10 different power stations suffered damage, typically to the boiler tubes.

The Higashi-Nada gas turbine power station having two 60 MW generators located near the epicenter suffered serious damage from the earthquake. Land subsidence occurred through the power station and caused the uneven settlement of the soil beneath the foundations supporting major equipment.

(ii) *Transmission Lines and Substations*

The earthquake did not cause any damage to the 500 kV facilities, which form the backbone of the power system. There was, however, substantial damage to the 275 kV and lower voltage facilities including minor incidents, damages occurred at 50 substations, and along 112 transmission lines:

- The overhead transmission lines suffered damages along 23 lines, mostly in the Kobe region. This included damage to structural components of the steel transmission-line towers and the damage to long-rod insulators used to fix jumpers:
  - The earthquake motion did not directly cause the damage to the steel towers. Rather, this damage was caused by fissures, slippage, and other land dislocation in adjacent area, which displaced the tower foundations, and thus caused damage to the structural components.
  - The earthquake motion caused the damage to long-rod insulators. These were broken on the 275 kV Hokushin line.
  - Damage to underground transmission lines has been found along 95 lines, and the damage surveys are still underway. The greatest damage was apparently located in those areas subject to the most intense earthquake movement, primarily in the flat areas in the Kobe and Amagasaki regions along the waterfront of Osaka bay. A large number of the cable suffered deformations and other abnormalities but were still capable of power transmission. Only three of the cables were completely damaged and could no longer carry electricity.
- The damage to the substations was also widespread, with a total of 181 incidents reported at 50 substations including the 275 kV Itami, Shin-Kobe, and Nishi-Kobe substations. These included damage to 17 transformers, which slipped out of place when the bolts, which anchor these transformers to their foundations, broke off; oil leakage from the bushings at eight circuit breakers because the bushings slipped out of position; the breakage of the support insulators of 22 disconnected switches.

(iii) *Distribution Lines*

Troubles were reported at 649 circuits because of earthquake damage to the distribution lines:

- Distribution poles fell over or were broken when buildings collapsed. The distribution poles tilted over and the distribution lines were broken off because of problems in the underlying soil structure including liquefaction and subsidence. Approximately 8000 distribution poles were damaged by the earthquake, and indeed the greatest damage occurred in those areas which were subject to the intensity of “7” earthquake vibrations.
- The underground distribution lines and their auxiliary equipment also suffered various types of damage:
  - The transformers and the other ground-mounted equipment for these lines were subject to tilting and deformations from shifts in structure of the underlying soil;
  - The underground ducts that the distribution cables run through suffered various damages, including collapse of the ducts’ side walls.

13.2.12.4 Restoration of Electricity Supply. Immediately after the earthquake, the restoration of the power system began by switching properly functioning equipment at disabled substations over to systems that were still running:

- At 07:30 a.m., 2 h after the earthquake hit, the number of customers left without electricity had been reduced to approximately 1.0 million, mostly in Kobe and Nishinomiya cities.

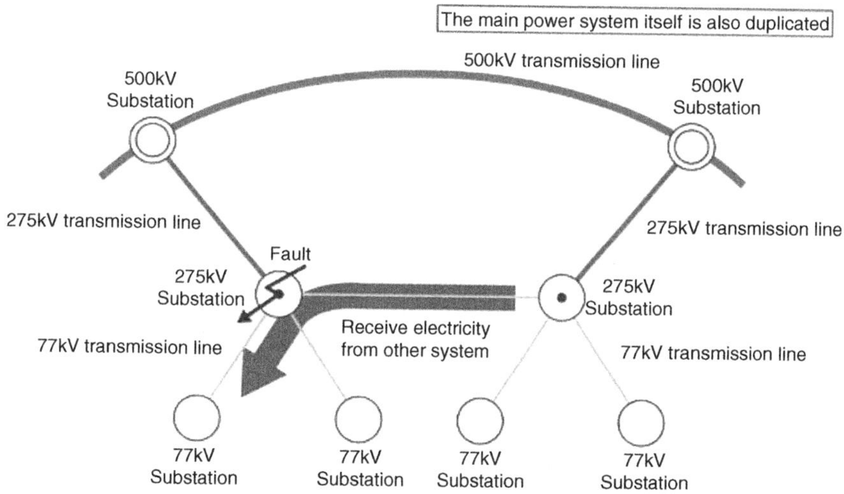

Figure 13.21. Switching over to 77 kV system when a 275 kV substation fails [11].

- At 08:00 a.m. on January 18, one day after the earthquake occurred, all of the substations were restored to a temporary operating condition, and the number of customers left without electricity was further reduced to about 0.4 million.
- Three days after the earthquake, at 06:00 a.m. on January 20, the distribution lines had been temporarily restored to all but about 110,000 customers in the Sannomiya, Hyogo, and the Nishinomiya districts.
- The temporary restoration works were completed at 03:00 p.m. on January 23, demand recovered to 70% of the pre-earthquake level.

The rapid restoration of electricity supply after the earthquake can be explained through the following key aspects:

(i) *A Robust and Flexible Power System*
It was possible to switch over the damaged 275 kV system to the properly functioning 77 kV system.
The 275 kV system suffered extensive damages leading to power outages over a wide area, but the 275 kV substations are linked by the 77 kV system (Figure 13.21). By switching over to this 77 kV system, it was possible to reduce the area suffering from power outages within a relatively short period of time.

(ii) *Restoration by Overhead Distribution Lines*
The overhead distribution lines were used for temporary restoration, and this is one of the main reasons why it was possible to restore electricity supply quickly. Temporary restoration was conducted utilizing high-voltage power generation vehicles, and all the undamaged equipment and materials remaining at the damage sites to the greatest possible extent.
In particular, because it takes long time to identify the damaged sections of the underground sections of underground cables, overhead lines were used to bypass the damaged cables for temporary restoration at several locations.

(iii) *A Private Communications System with Strong Aseismic Design*
An important factor contributing to the rapid restoration of power supply was the fact that the earthquake did not damage the communication system.

The Kansai Electric Power Company maintains a microwave communications system linking the central load-dispatching center with all of the main power stations and substations. Also, it exists an independent telephone system using microwaves and fiber-optic cables. Because these systems suffered no damages from the earthquake, it was possible to conduct all the communications necessary for the restoration works.

In the aftermath of the earthquake, the public telephone systems failed, as the huge number of calls jammed them. If the dispatching center had been forced to rely on the public phone system, it probably would not have been possible to carry out the restoration works in a timely manner.

(iv) *Nationwide System for Assistance from Other Electric Power Companies*
Invaluable assistance was received from companies based on all the way from Hokkaido in the north to Okinawa in the south. In addition to dispatching a total of 319 personnel, different corporations in the electric power industry provided diverse types of material assistance including 52 high-voltage power generation vehicles; 77 working vehicles; materials, food, and water for the restoration works; and vehicles equipped with satellite communications facilities.

(v) *Aseismic Design and Other Measures Against Earthquakes*
The aseismic designs adopted for different facilities based on past experiences with earthquakes did effectively minimize the earthquake damages.

The Kansai Electric Power Company has been systematically incorporating aseismic design for all their main facilities ever since the 1978 Off-Miyagi earthquake. Although the Hanshin-Awaji earthquake damaged their facilities, the application of this aseismic design made possible to avert any fatal damage from this massive quake.

13.2.13 European Incident of November 4, 2006

13.2.13.1 Precondition. On the evening of November 4, the European power system (Figure 13.22) was operated as a whole, in secure conditions with a system frequency near the reference value of 50 Hz. As usual during a weekend when the consumption is lower, some transmission lines are not in operation due to maintenance or other works. And some substations in Germany were operated with two bus bars, either for works or for limitation of short circuit current. The generation was estimated as 274,000 MW with around 15,000 MW of wind generation, mainly located in North Germany and in Spain [25] (Figure 13.23).

The exchange programs and the pattern of the physical flows between countries were not unusual, with significant differences between exchange values and physical flows at some borders. The only point to be underlined was the high flows from Germany to the Netherlands and to Poland, due to the high level of wind generation.

13.2.13.2 Initiating Events. In September, the local TSO was asked by a shipyard to disconnect a double circuit 380 kV line Conneforde-Diele in North Germany, for the transfer of a ship on the river Ems to the North Sea, on November 5 at 01:00, a type of operation already carried out several times in the past. The TSO informed its neighboring

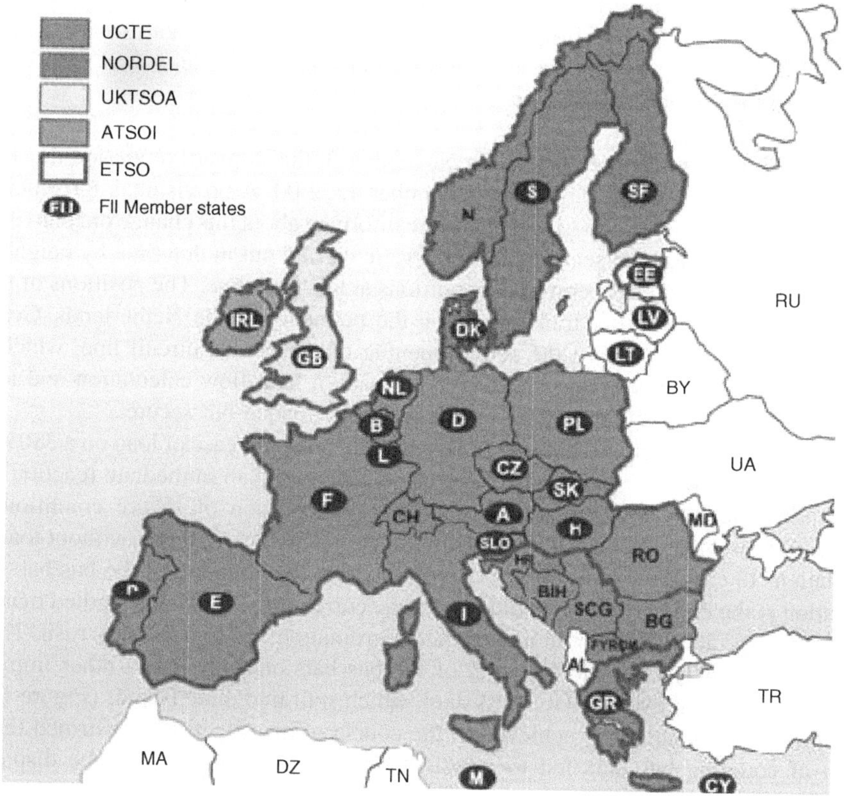

Figure 13.22. The UCTE (ENTSO-E since 2009) system.

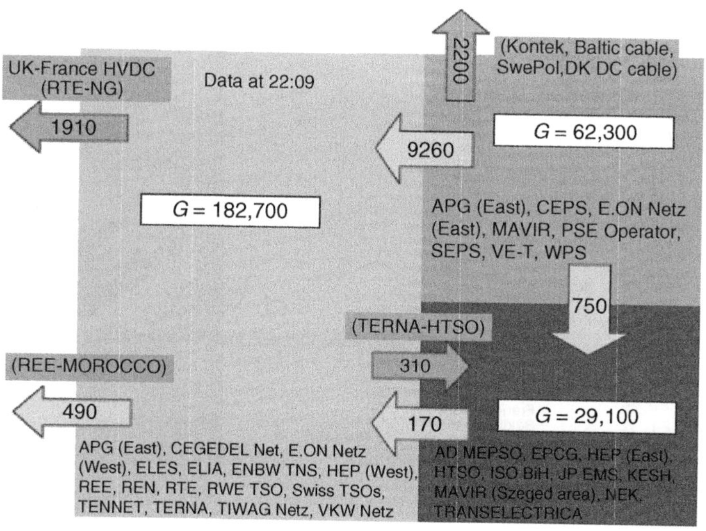

Figure 13.23. Generation and power flows between the three areas just before splitting November 4, 22:09 [25,29].

TSOs of its provisional agreement, hence they could carry out N-1 security analysis on their network. The results confirmed a high loading of the grid, but with secure conditions at this time of the night; still the cross-border transmission capacity was reduced from Germany to the Netherlands.

13.2.13.3 Cascading Events. On November 3, the shipyard requested the TSO to schedule the disconnection earlier, on November 4, 22:00. A provisional agreement was given by the TSO. The neighboring TSOs were informed about this change only at 19:00 on November 4, so no special security analyses were carried out in due time by neighboring TSOs to take into account operational conditions at this new time. The positions of the tap changer of the phase shifter transformer on the boundary of the Netherlands–Germany were modified; 10 min before the actual opening of the double circuit line, which took place at 21:39, the neighboring German TSO made a load flow calculation and an N-1 analysis and concluded that its grid would be highly loaded but secure.

According to the report, between 22:05 and 22:07, the increase of load on a 380 kV tie-line between the two German areas triggered an alarm with an immediate reaction of the neighboring German TSO, requesting an urgent restoration of secure conditions. An empirical assessment of corrective switching measures was carried out, without load flow calculations to check the N-1 criterion, expecting that the coupling of the bus bars in the substation at the end of the line would reduce the current on it. This mishandled maneuver was effectuated at 22:10 without any further coordination due to necessary rush. The line tripped immediately after the coupling of the bus bars and this led to other immediate cascade trippings all over the UCTE system, which split into three islands (Figure 13.24).

The ex-post simulations achieved in the course of investigations confirmed that this action of coupling bus bars led to a result quite the opposite to what the dispatchers

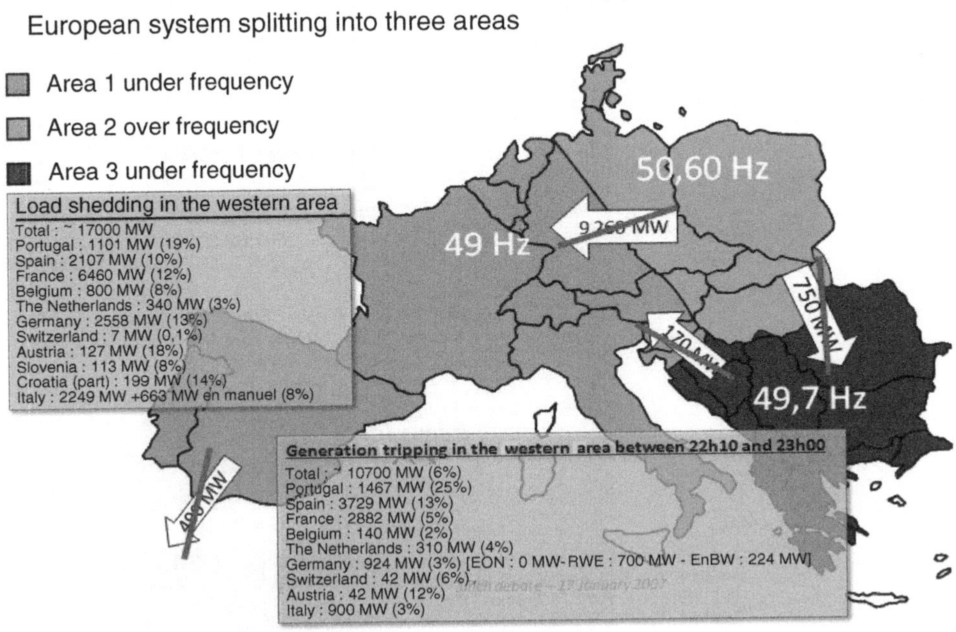

Figure 13.24. Schematic map of UCTE area split into three areas and load shedding in the western area [25,26,29].

expected; the current on the line increased and the line was automatically tripped by the distance protection relays, as a result of the overload.

13.2.13.4 Final State. The tripping of the tie-lines between the two German areas and the subsequent cascading effect resulted in a division of the European system along a north to south-east line, with an additional separation of the South Eastern countries. At 22:10:28, the European system was split into three separated islands: Western, North-Eastern, and South-Eastern.

Some national grids suffered also a split of their internal networks, in Austria, Hungary, and Croatia. The magnitude of the rapid frequency drop led to the tripping of the connection between Morocco and Spain.

The generation distribution just after the split amounted to 182,700 MW (6500 MW of wind generation) in the Western island, 62,300 MW (8600 MW of wind generation) in the North-Eastern island, and about 29,100 MW in the South-East island (no wind generation).

- *The Western area* was composed of Spain, Portugal, France, Italy, Belgium, Luxemburg, The Netherlands, Switzerland, and a part of Germany, Austria, Slovenia and Croatia. It faced a significant supply–demand imbalance, of 8940 MW, due to the no longer available imports from Eastern side.

  This severe imbalance caused a quick drop (within 8 s) of system frequency from the normal value of 50 Hz to about 49 Hz (Figure 13.25), which activated automatic load shedding (rejection of load). Eventually, the total shedding was some 17,000 MW of consumption and of 1600 MW of pumps (pumped storage plants).

  According to UCTE security rules, these automatic actions are designed to prevent any system collapse as a result of significant power imbalance. They helped to limit the frequency drop and the action of the load frequency control (LFC) started restoring its nominal value.

  Unfortunately immediately after the frequency drop, some generation units tripped, thus increasing the demand–supply imbalance in the area. About 40% of the units that

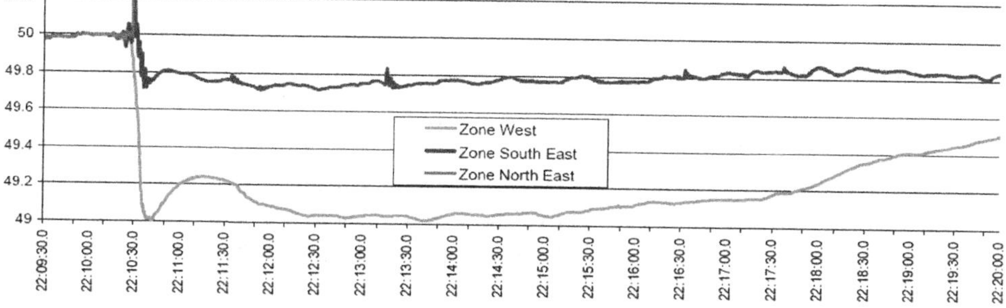

Figure 13.25. Frequency recordings after the split [29].

tripped during the incident were wind power units. Around 30% of CHP (combined heat and power: cogeneration production) in operation just before the event tripped during the frequency drop. Wind generation and CHP connected to the distribution grid are not directly controlled by TSOs; these small units were automatically reconnected to the grid when the accepted conditions of voltage and frequency resumed.

TSOs started generation units (mainly hydro units) in order to quickly restore the frequency to 50 Hz, as directed in the TSOs' restoration plans; no special coordination between TSOs was sought at that time and each TSO acted according to its own rules. A total of about 16,800 MW of generation was started in the Western area.

A few minutes after the incident, some TSOs stopped their load frequency control (LFC) in order to assess the situation. Around 22:30, some TSOs were requested to switch LFC into pure frequency mode.

The restoration of power supply for customers in most of the countries was achieved without coordination and without an accurate knowledge of the status of the split network.

- *North-Eastern area* also faced serious imbalance conditions, with in this case an over-generation of more than 10,000 MW. The excess was due to the fact that before splitting there was a huge transit of electricity from this area to West and South Europe, a typical load flow situation in this region, but higher than usual due to wind conditions in the north of Germany.

  It resulted in a rapid increase of frequency up to about 51.4 Hz, quickly reduced to about 50.3 Hz by the automatic actions of primary control—standard and emergency range, of the activation of speed control of certain generating units and by the automatic tripping of windmills sensitive to high frequency. This automatic tripping of some 6200 MW of wind generation contributed to limit the increase of frequency during the first seconds of the disturbance.

  The windmills that tripped at the time of collapse started automatically to be reconnected to the power systems (in Germany and Austria) and thus gradually increased generation in those control areas, behavior contrary to the required generation decrease in this area. These actions included some TSOs instructions to generation companies to decrease the output, stopping some of them and starting pumps in pumped storage plants. Therefore, the frequency slowly increased again from 50.3 Hz at 22:13 up to 50.45 Hz at 22:28 and then slowly came back to around 50.3 Hz.

  The reconnection of windmills with the gradual increase of generation in the north of Germany and the decrease of thermal generation, mainly in Poland and the Czech Republic, led to significant changes of power flows. The flows profile between Germany and, respectively, Poland, and Czech Republic exceeded the transfer capacities on these borders, to reach "levels unacceptable even for emergency conditions" with overloads of some internal lines. The N-1 rule was not fulfilled at that time in this region and tripping of any element would have caused further overloads and possible cascade tripping, a real danger of further collapse.

- *The South Eastern area*, including a small part of South Hungary, supported a lightly negative balance of around 770 MW. As the frequency value during the whole disturbance was significantly above the first threshold for load shedding (49 Hz) neither automatic actions nor load shedding took place during the event and the defense plans were not activated. The frequency dropped to 49.79 Hz and came back rapidly to an acceptable normal value of 49.98 Hz.

13.2.13.5 Resynchronization. The resynchronization process was performed in Germany, Austria, Croatia, Romania, and West Ukraine. The TSOs started quickly to reconnect the tripped lines with a minimal coordination. Some attempts at resynchronization failed, some resulted in actual interconnection but failed after a few seconds, and then a successful resynchronization process took place.

Successful resynchronization took place first on the 380 kV line between Western and North-Eastern areas, in Germany at 22:47, with a recorded difference of frequencies of about 180 mHz. The resynchronization process with the South-Eastern area took place 2 min later with the closing of the 400 kV line between Ukraine and Romania.

The full resynchronization of the three areas started about 40 min after the triggering event—mishandled coupling of a bus bar—and was completed in less than 2 h. It was carried out in a fully decentralized way with a minimum of coordination between directly affected countries.

13.2.14 Some Lessons Learned

Taking a look at the three latest major disturbances, U.S. and Canada blackout on August 14, 2003, the Italian blackout on September 28, 2003 and this European incident, common denominators could be highlighted as follows:

a. the inappropriate estimation of the situation leading to nonrespect of the security rules;
b. the weakness of the inter-TSO coordination;
c. conflicting interests with separated ISO and TSO (maintenance, costs, investments);
d. bad operator decisions in critical situations and lack of cross-border coordination;
e. lack of obligations in power system generation standards and for dispersed generation (performances for frequency control and voltage, real-time monitoring);
f. lack of transmission infrastructures (authorization procedures and delays).

13.3 ANALYSIS OF BLACKOUTS

Blackouts seem to progress with some regularity. A previous study showed that the progression of blackouts after the occurrence of initiating events could be divided into steady-state progression and transient progression [30]. In our study, the progression of eight blackouts (see Table 13.1, no. 18, 20, 28, 29, 32, 33, 36, 37), from which detailed information are available, was investigated. The results suggested that the progressions of blackouts can be divided into several phases. Figure 13.26 clearly describes these phases, which are precondition, initiating events, cascade events, final state, and restoration. Among these five phases, cascade events can be further divided into three phases in the process of some blackouts: steady-state progression, triggering events, and high-speed cascade.

A high-speed cascade usually follows the critical point, that is, the occurrence of triggering events. But, not all of the blackouts have all the phases listed above. For example, the steady-state progression was skipped in some previous blackouts. In these cases, the initiating events were also the triggering events that started the high-speed cascade. For example, in the case of Croatia blackout on January 12, 2003, the initiating event at 16:43:58 triggered the high-speed cascade, and the blackout happened within 30 s.

In the following sections, the blackouts are discussed according to these sequential phases.

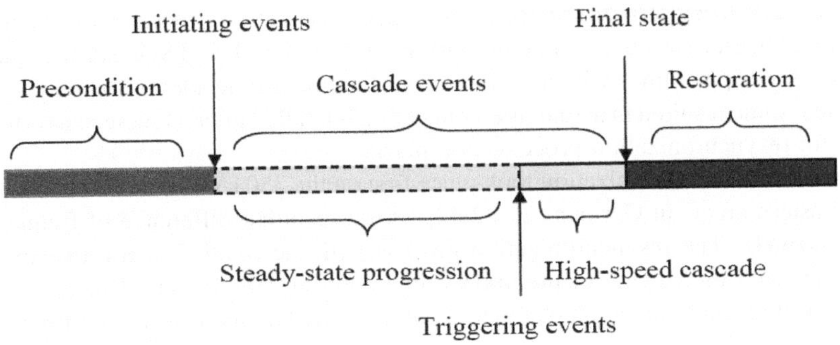

Figure 13.26. Phases of blackout [31].

13.3.1 Classification of Blackouts

13.3.1.1 Precondition. As stated previously, 39 blackouts, with comparative detailed information, were collected in this chapter. Different preconditions happened in these blackouts, but we can classify them according to their common characteristics. The classification is carried out as follows:

1. *System Condition is Stressful in Summer Peak and Winter Peak*
   Table 13.9 shows that 13 blackouts happened in summer peak and 11 blackouts in winter peak. Moreover, 61.5% of the blackouts (24/39) happened in summer peak and winter peak when there was a high electrical demand, and 38.5% of the blackouts (15/39) happened in normal system condition.
2. *Aging Equipments*
   The Russian power system was a system with a high security level. No blackout happened from 1975 to 2005. However, a blackout occurred in Moscow on May 25,

TABLE 13.9. Blackouts in Summer and Winter Peak

Blackouts in Summer Peak	Blackouts in Winter Peak
07/1977, New York City	09/11/1965, United States
23/07/1987, Japan	19/12/1978, France
24/08/1994, Italy	01/1981, Idaho, Utah and Wyoming, United States
08/06/1995, Israel power system	27/12/1983, Sweden
02/07/1996, 14 states in the United States	12/01/1987, France
07/08/1996, Big Rivers Electric Corporation, United States	12/03/1996, Florida, United States
10/08/1996, California Pacific Northwest	01/1998, Canada, New York and New England
26/08/1996, New York, United States	12/1998, San Francisco, California Bay Area
07/1999, New York City	01/12/2003, Southeastern Massachusetts from New Bedford to Provincetown to the islands, United States
14/08/2003, United States and Canada	12/01/2003, Southern part of Croatia and a part of Bosnia Herzegovina
28/08/2003, South London	14/12/1994, Arizona and Washington state, United States
05/09/2003, West Midlands	
12/07/2004, Athens and Southern Greece	

2005, because more than 70% of the Moscow 220 kV power substations were working over their lifetime, and such system became unstable during the emergency condition.

3. *Inadequate Reactive Power Reserve*
Inadequate reactive power reserve was the reason of the blackout in Northeast United States and Canada on August 14, 2003 and in France on December 19, 1978. Reactive power is related to the voltage. Lacking of reactive power decreased the flexibility of the voltage control, which may increase the risk of voltage collapse.
4. *Some Important Equipments Out of Service*
Before the blackout in Athens and Southern Greece on July 12, 2004, one 125 MW generating unit in Peloponnesus and one generating unit in Northern Greece were out of service. This led the system to a stressful condition.
5. *Natural Reasons Such As Wind, Thunderstorms, Earthquakes, Fog, Geomagnetic Disturbances, and Fire*
Before the blackout in the United States on April 16, 1996, the area had an unusually high amount of dust and soot from prairie fires due to drought conditions. This increased the possibility of flashovers.
The voltage collapse in Quebec in March 1989 was caused by the side effects of a geomagnetic storm during a period of heightened solar activity. The blackout in Japan in January 1995 was caused by an earthquake.

13.3.1.2 Initiating Events. Initiating events were various in different blackouts. These events can directly cause blackout or can worsen the system condition, which may indirectly lead to blackout. Short circuit, overload, and protection hidden failure are the usual initiating events, and other events such as loss of generator sometimes can also be initiating events. The initiating events of some blackouts, of which related information could be obtained, are described in Table 13.10.

13.3.1.3 Cascading Events. The cascade is a dynamic phenomenon. It can be triggered by the initiating events. These initiating events can cause power oscillations and voltage fluctuations that may result in high currents and low voltages. The high currents and low voltages can be detected by other lines and be treated as faults. The lines and the generators can trip to protect themselves from damage, which may lead more and more lines and generators to become out of order. Besides power oscillations and voltage fluctuations, line overloads also can cause cascade. When a line is tripped, due to an overload, the neighboring lines can become overloaded and be tripped.

By analyzing the eight blackouts cited in the introduction of Section 13.3, we can divide the period of cascade events into steady-state progression and high-speed cascade (see Figure 13.26). In the period of steady-state progression, the progression of the cascade events is slow, and the system can keep balance between the generation and the consumption. During this period, the major incident is the cascade overload. Because of the slow speed of the worsening of the situation in the period of steady-state progression, it may be a good opportunity for the system operator to take actions to stop the spread of the cascade overload and then, prevent the occurrence of blackout. While the triggering events have triggered the high-speed cascade, the balance between the generation and the consumption may be broken, series of system equipments may be tripped rapidly, and system collapse can happen in a very short time. In the period of

TABLE 13.10. Initiating Events of Blackouts

	Initiating Events			
Blackouts	1	2	3	4
09/11/1965 United States		√	√	
07/1977 New York			√	
27/12/1983 Sweden		√		
19/12/1978 France		√		
12/01/1987 Western France				√
08/06/1995 Israel	√			
12/03/1996 Florida		√		
16/04/1996 United States	√	√		
02/07/1996 United States	√			
10/08/1996 California Pacific Northwest		√		
26/08/1996 New York			√	
21/09/1996 Allegheny			√	
11/03/1999 Brazilian	√			
12/01/2003 Croatia	√		√	
14/08/2003 Northeast United States and Canada	√			
28/08/2003 London			√	
23/09/2003 Eastern Denmark and Southern Sweden				√
28/09/2003 Italy	√	√		
12/07/2004 Athens and Southern Greece				√
14/03/2005 South Australia	√			
Total: 18	**8**	**6**	**6**	**22**

1—Short circuit, 2—Overload, 3—Protection hidden failure, 4—Loss of power plants.

high-speed cascade, it is generally too late for the system operator to take actions to stop the rapid progression of blackout.

The period of cascade events was studied by analyzing the accumulated number of the tripped lines, transformers, and generators in this period. Data of the U.S. blackout on August 14, 2003, the Italy blackout on September 28, 2003, and the Croatia blackout on January 12, 2003 were cited to describe the progression of blackout during the cascade.

In the U. S. blackout on August 14, 2003, the triggering event was the tripping of the East Lima-Fostoria Central 345 kV line at 16:09:06, and this event caused large power oscillations through New York and Ontario into Michigan. The high-speed cascade was triggered and led to blackout (Figure 13.27).

In the Italy blackout on September 28, 2003, we can clearly see the period of the steady-state progression and the period of the high-speed cascade (Figure 13.28). The triggering events tripped the Mettlen-Airolo 220 kV line and Sils-Soazza line in Switzerland at 03:25:21. These triggering events directly led to 17 lines tripped within 21 s. In the Croatia blackout on January 12, 2003, the steady-state progression was skipped (Figure 13.29). The initiating event, which was also the triggering event, triggered the high-speed cascade, and the blackout happened within 30 s.

The duration of the periods of steady-state progression, high-speed cascade, and restoration of eight blackouts are listed in Table 13.11.

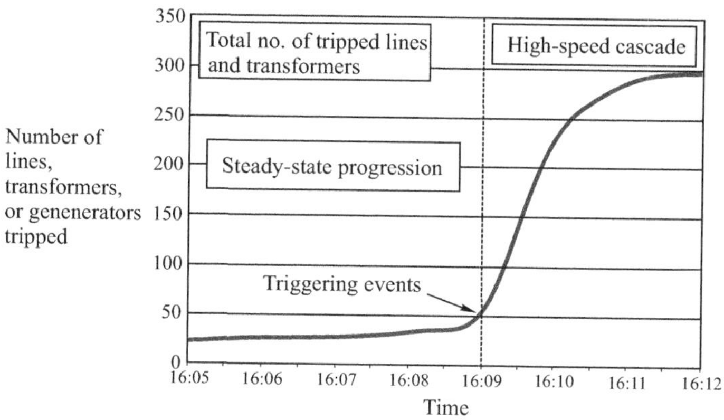

Figure 13.27. Accumulated lines and generators tripping during the cascade in the U.S. blackout on August 14, 2003 [1].

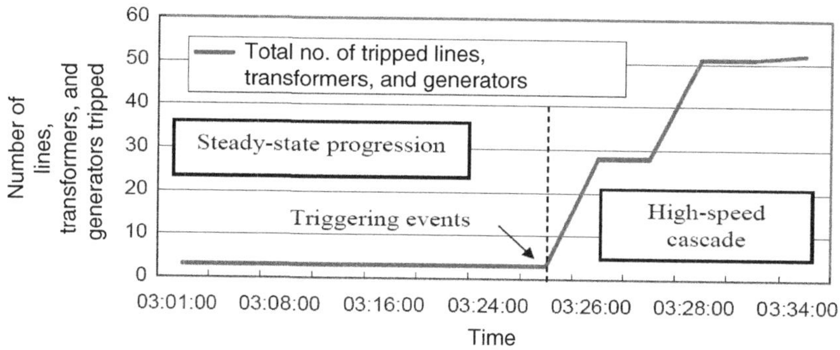

Figure 13.28. Accumulated lines and generators tripping during the cascade in the Italy blackout on September 28, 2003 [3].

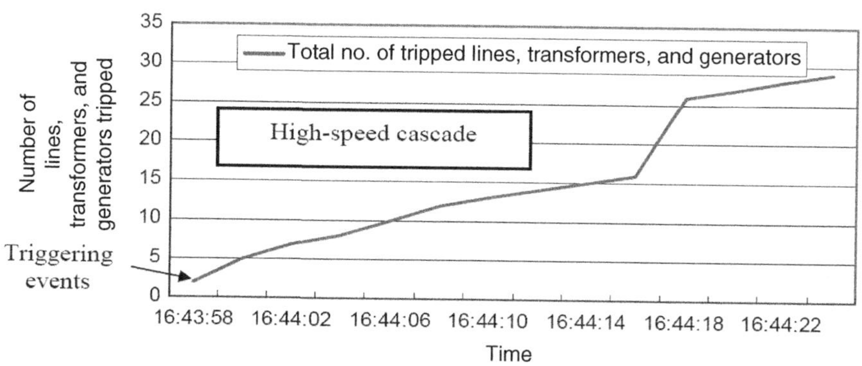

Figure 13.29. Accumulated lines, transformers and generators tripping during the cascade in Croatia blackout on January 12, 2003 [16].

TABLE 13.11. Periods of the Blackouts

Blackouts	Steady-State Progression	High-Speed Cascade	Restoration
14/08/2003 United States and Canada	1 h 5 min	3 min	~24 h
28/09/2003 Italy	24 min	9 min	20 h
12/01/2003 Croatia	Without	30 s	>3 h 15 min
14/03/2005 South Australia	Without	6 min	1.5 h
12/07/2004 Greece	13 min	2 min	3 h
02/07/1996 United States	Without	60 s	>6 h
10/08/1996 United States	1 h 38 min	7 min	~9 h
19/12/1978 France	47 min	6 min	10 h

From Table 13.11, we can find that:

(i) The progression of the high-speed cascade is very fast. Numerous of lines and generators can trip in several seconds or several minutes. Of course, it is a too short time for system operators to take effective actions to stop the cascade. If we want to stop the blackout, we need to take adequate actions before the period of high-speed cascade.

(ii) In some blackouts, the period of the steady-state progression was skipped. After the initiating event happened, the system quickly went into the period of high-speed cascade. This kind of blackout is more difficult to stop because of its rapid progress.

13.3.2 Blackouts: Types of Incidents

From available data, some serious incidents in 12 blackouts from 1965 to 2005 were analyzed in this section (Table 13.12).

TABLE 13.12. Blackout: Types of Incidents

	Types of Incidents				
Blackouts	1	2	3	4	5
9/11/1965 United States			✓		
19/12/1978 France	✓		✓		
12/01/1987 Western France	✓				
2/7/1996 United States	✓		✓		
07/08/1996 United States	✓		✓		
12/01/2003 Croatia				✓	
14/08/2003 Northeast United States and Canada	✓		✓		
23/09/2003 Eastern Denmark and Southern Sweden	✓		✓		
28/09/2003 Italy		✓	✓		✓
12/07/2004 Athens and Southern Greece	✓				
14/03/2005 South Australia					✓
04/11/2006 European power system			✓		
Total: 12	**7**	**1**	**8**	**1**	**2**

1—Voltage collapse, 2—Frequency collapse, 3—Cascade overload, 4—System unsymmetrical, 5—Loss of synchronism.

Referring to Table 13.12, voltage collapses (7/12) and cascade overloads (7/12) happened with a higher frequency in these blackouts. It suggests that voltage collapse and cascade overload are the major incidents in the progression of blackouts. Finding out effective methods to avoid these types of incidents in emergency condition might be a good way to stop most of progressions of blackouts.

The system separation is also a blackout incident, but it is a consequence of the incidents that are represented in Table 13.12.

We have identified the phases of blackout, analyzed the phenomena occurred in these phases and found some common characteristics of blackouts. But how did the blackout progress in each phase? In the next part, the mechanisms of blackouts are analyzed.

13.3.3 Mechanisms of Blackouts

The power system may enter into an emergency condition due to some critical events that may happen in the system. Usually, the system can be pulled back to normal condition by its protection and control system. But, sometimes, the system cannot return to the normal conditions in a good time and some new events can trigger the cascade incidents, which may interact and rapidly worsen the situation. Finally, blackout can happen.

In this study, we analyzed the mechanisms of the eight blackouts cited in the introduction of Section 13.3, from which we could get detailed information. Combined with a previous study, which suggested a common generic scenario of cascading processes for blackouts [23], we describe the mechanism of blackouts in Figure 13.30.

There are five types of faults that cause blackouts: voltage collapse, frequency collapse, cascading overload, system separation, and loss of synchronism. The major

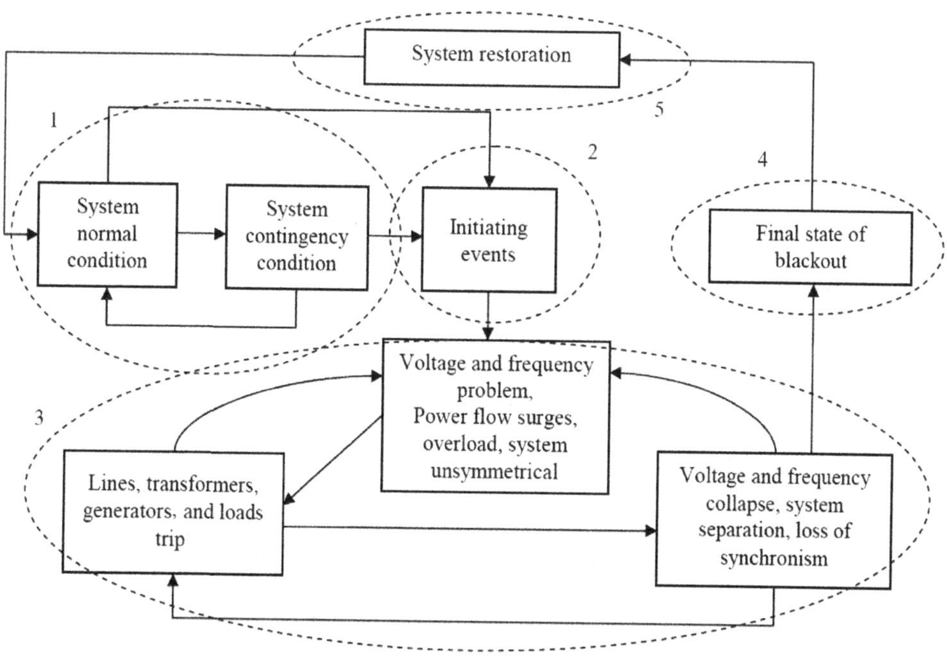

Figure 13.30. Mechanism of blackouts.

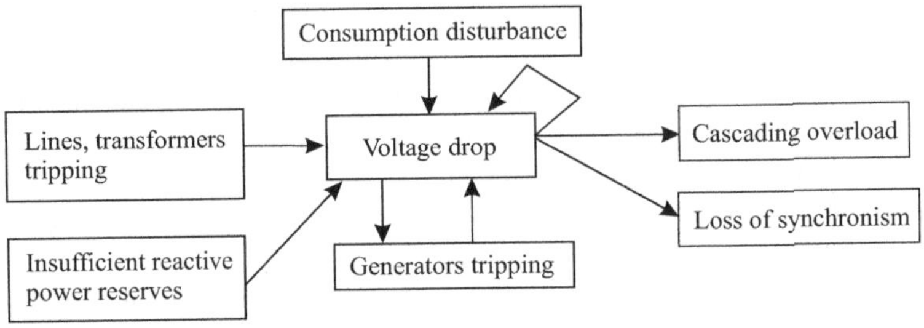

Figure 13.31. The mechanism of voltage collapse.

reasons are voltage collapse and cascading overload. The mechanisms of these faults are presented in the following sections.

13.3.3.1 Voltage Collapse. The voltage collapse is caused by lines, transformers and/or generators tripping, and consumption disturbance. If there are not enough reactive power reserves in the system, the voltage drops, which cause cascading overload of the lines and transformers and accentuates the voltage to decline. Indeed, the voltage drop in the transmission system leads to low voltages in the distribution systems. Therefore, the ULTCs (under load tap changer) try to increase the voltage profile on the lower voltage levels by modifying their turns ratio.

The apparent impedance of low voltage levels (including ULTC) then decrease, and more current is requested from transmission lines. This leads to line overloads.

When the voltage is below the thresholds of undervoltage protections of the generators, they trip and the lack of reactive reserves are worsened. At the same time, overcurrent protections of lines may also trip and the voltage drops more and more: the blackout happens. The duration of voltage collapse is about some minutes. FACTS, reactive resources and undervoltage load shedding, can help voltage return to normal condition (Figure 13.31).

13.3.3.2 Frequency Collapse. Unbalance between production and consumption appears, insufficient active power reserve in the system and generator tripping cause the frequency collapse. Facing these events, the power system uses the primary active reserve to keep the frequency in the limited operating range. If there is not enough primary reserve, the frequency can go out of the limits. The underfrequency protections of generators provoke their cascade tripping and, as a consequence, the acceleration of the frequency collapses. It happens in some seconds and can be stopped by underfrequency load shedding (Figure 13.32).

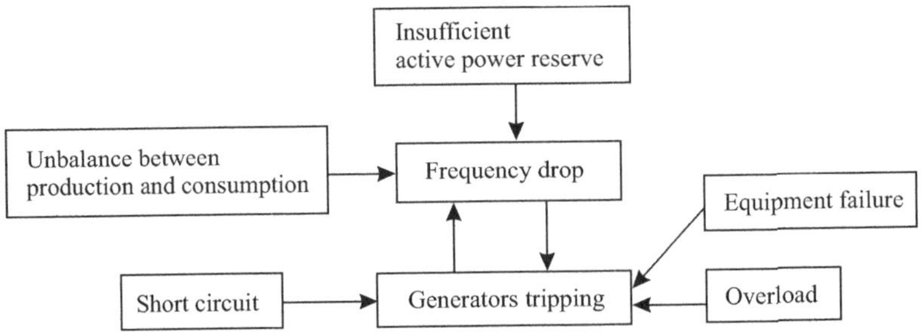

Figure 13.32. The mechanism of frequency collapse.

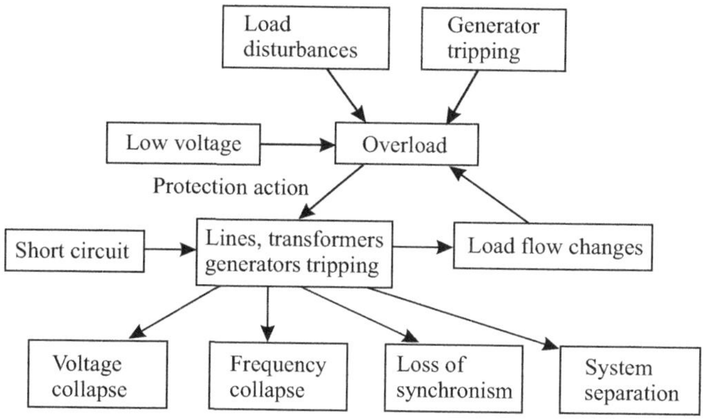

Figure 13.33. The mechanism of cascading overload.

13.3.3.3 Cascading Overload. When the power system is stressed, the power flows are close to the limit of the lines transmission capacity. A heavily loaded line can sag close to trees due to overheating which may, in the long run, provoke a flashover. So, the protection relays of the line disconnect it and its power flow is shifted onto others lines in the neighborhood.

A line can also be tripped by its overload protection. The power transfer on the other lines may cause them to be overloaded and also tripped by their protection devices and the cascade begins. The system enters into voltage collapse, frequency collapse then loss of synchronism or system separation and finally, the blackout happens. The duration of the cascading overload is from some minutes to some hours. A way to avoid the cascading overload is to use FACTSs, which can change the load flows and relieve the power flows in some critical lines. Load shedding can also help relieve the stress of the system and stop the cascade (Figure 13.33).

13.3.3.4 System Separation. When the grid loses some critical lines or transformers, the system is separated. In each isolated subsystem, an unbalance between production and consumption may appear. If the system operator cannot keep the system balance in these subsystems, a voltage or frequency collapse causes a blackout. Load shedding can keep the balance between load and generation in the isolated subsystems (Figure 13.34).

13.3.3.5 Loss of Synchronism. Let us consider two power systems linked by interconnections lines. If one of these lines trips, the others may be overloaded. Unbalance

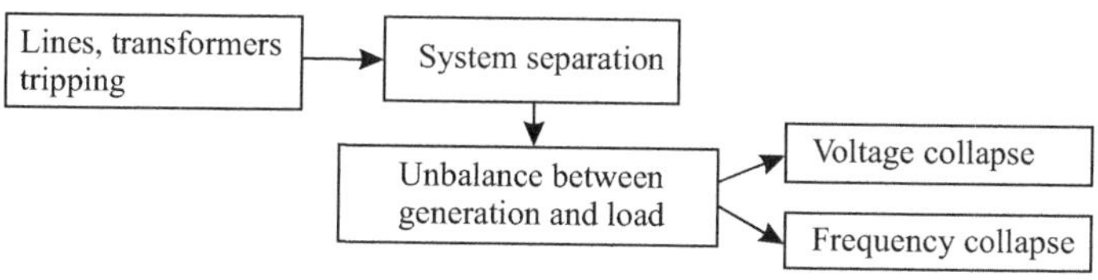

Figure 13.34. The mechanism of system separation.

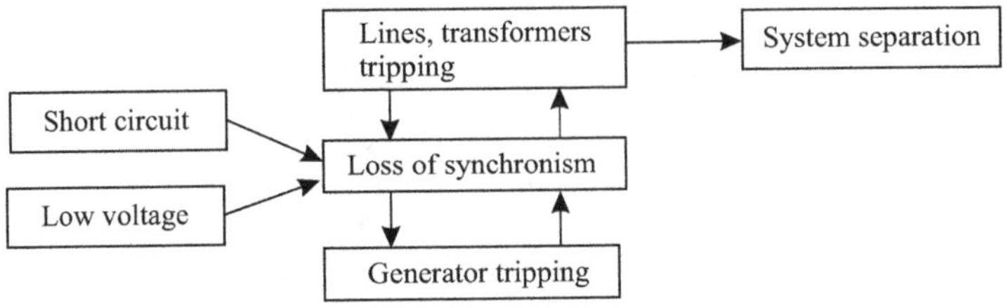

Figure 13.35. The mechanism of loss of synchronism.

between production and consummation appears and the frequency became different between the power systems. Power oscillations appear on the interconnection lines and the protection system reacts. Finally, the systems split. The loss of synchronism can also be caused by a short circuit that makes the generators angle exceed 90°. To avoid loss of synchronism, load shedding can keep the balance between load and generation and FACTS can be installed in some critical interconnection lines to stop power oscillation spreading (Figure 13.35).

***13.3.3.6 Generalization*[4].** During operation, the electrical power systems (EPS) are subjected to various events. These events differ by their origin and their qualitative and quantitative characteristics.

The common way of retrospective analysis of emergency contingencies in power systems is the detachment of a certain time-sequence (chain) from the set of occurred events, only that which caused triggering and development of emergency contingency in a characteristic way.

This section presents some definitions based on which all possible events can be subdivided into three groups according to their qualitative characterization. Description of those groups and generalization of their interrelation are given and an example of the groups separation from the events sequence is presented [28,32].

A. *Definitions*

Let us define "change of power system state" as a process of redistribution of active and reactive power flows, which has feed forward and feedback with the change of voltages in the network nodes and, possibly, with the change of the system frequency.

Let us introduce three types of changes of power system state coming from the reliability of its further operation (i.e., from the risk of blackout occurrence):

- Negative change (i.e., deterioration) of the state, which appears as decreasing the reserves of transmission capability of the main network and reserves of generating power;
- Positive change (i.e., improvement) of the state, which appears as increasing the reserves of transmission capability of the main network and reserves of generating power;
- Imperceptible change—that is,, insignificant (negligible) change of those reserves.

[4] This section was prepared together with D.N. Efimov

- Let us define an "event" as the cause of change of the system state or the obstacle in the state change.
- Furthermore, an event is identified by three characteristics:
- Probability of the event occurrence;
- Orientation of the event to deterioration or improvement of the system state (to negative or positive change of the system state);
- System effect of the event (the event influence on the system state)—that is, quantitative measure of change in the systems state under the event influence.

Coming from above definitions, all the events taking place in the power system can be subdivided into three groups as follows.

B. *Group I: Accidental Events*
Accidental events may be classified as:

- Disturbances—mainly short circuits on the transmission lines, but also conductors break, unplanned disconnection of network elements, and load/generation on/off. These disturbances represent accidental changes in the system state.
- Wrong actions—that is, misoperations of relay protection or automatic control devices or erroneous commutations accomplished by personnel.
- Failures—that is, relay protection or emergency control device failures or missing the necessary personnel actions for the proper change of the system state. Failure preventions from the power system state change are addressed by event of Group II or other events of Group I.

An accidental event is directly related with certain system element (generation unit, load, or transmission line) and results in change of the element capability/reliability as a local effect of the event. An accidental event is indirectly (through an event of Group III) related with the change of entire system—that is, characterized as manifestation of the system wide effect of event.

C. *Group II: Purposeful Events (Control Actions)*
Let us refer to purposeful events as correct and successful, those control actions for planned changing of the system state or as a response to events of Groups I and III. According to proposed classification, incorrect or unsuccessful control actions belong to Group I since they are considered as having accidental nature.

Purposeful events improve the system state. They are accomplished by relay protection, emergency control devices and/or personnel by commutating certain generation, load and/or transmission elements of the system—this is the local effect of event (i.e., change of the element capability/reliability). As with accidental event, the control action is indirectly (through the event of Group III) related with the change of entire power system—that is manifestation of the system wide effect of event.

D. *Group III: Regular (Natural) Events*
Let us refer to regular (natural) events as the actions of nature on the power system, which manifests as natural response of the system to aggregate of all previous events. Regular events can result both worsening and improvement of the system state and also imperceptible (negligible) change of the state. A regular event manifests locally (e.g., transmission line overloading) or system wide (e.g., change of voltages in the system nodes).

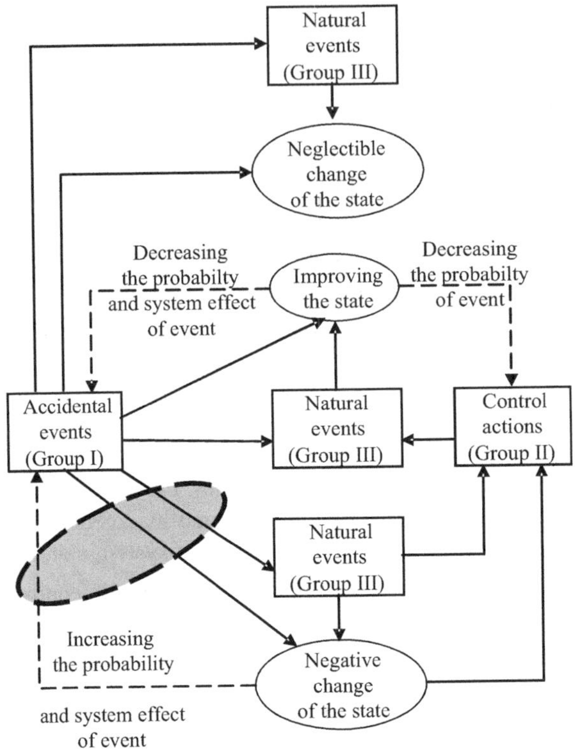

Figure 13.36. Cause–effect relationships between EPS events and its state changes.

E. *Generalization of Mechanisms*

Notice that while events of Groups I and II influence certain elements of the system, the events of Group III are responses of entire system to those influences.

Events of Groups I and II always lead to events classified in Group III, which in turn are direct cause of the system state change. Therefore, the system effects caused by any event directly results from an event of Group III.

Cause–effect relationships between foregoing three groups of events are sketched in Figure 13.36.

Probability and system effect of event are the values depending on the previous events, first of all on those events of Group III, which result in worsening the system state. In more detail, these events lead to

- First, probability of Group I events occurrence (e.g., the probability of short circuit of overloaded transmission lines is much higher than of normally loaded or low loaded lines).
- Second, negative effects of Group I events on the system (e.g., accidental redistribution of power flows in "heavy" operation conditions results in larger overloading of transmission lines than similar redistribution in "easier" conditions).

It can be seen from Figure 13.36 that the most dangerous (potentially most of the worsening of the system state and hence the risk of blackout occurrence) case is a

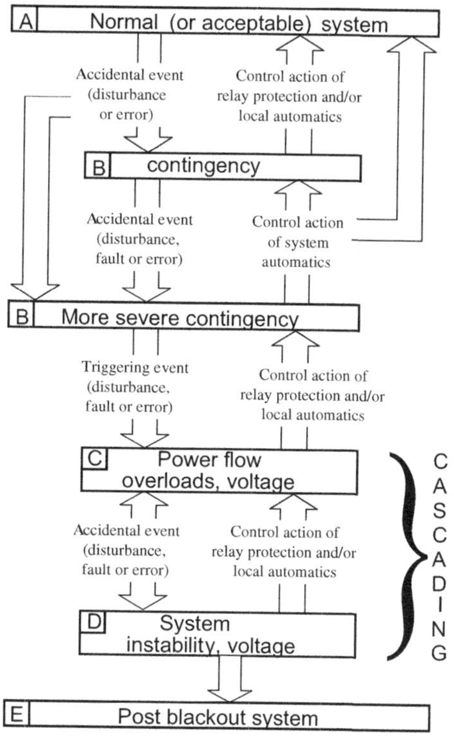

Figure 13.37. Generic scenario of a cascading failure.

combination of Group I events and of the worsening of the system state due to Group III events. Just under that combination, a cause–effect cycle formation is possible which, if considered in timescale, is a process of cascading deterioration of the power system state—that is, cascading development of emergency situation.

Appearance of such dangerous cycle means that the events sequence in some instants of operation can bring the power system to a marginal state, from which the next event occurrence becomes *triggering*—that is, an event that starts uncontrollable cascading process of further events (firstly tripping of elements) with disastrous consequences (system blackout).

A triggering event separates a period, in which multiple "undirected" factors finally contributing but not directly connected to a blackout are accumulated, from the "blackout-directed" sequence of events with clear cause–effect relationships between the subsequent phases [23].

By analyzing the sequences of events of recent blackouts, the following common scenario of the cascading process, containing cyclically repeating change of the system states, can be suggested (see Figure 13.37).

13.4 ECONOMICAL AND SOCIAL EFFECTS

Table 13.13 presents the economical and social impacts for some important blackouts.

This table emphasizes the important consequences of blackouts in terms of a number of impacted consumers and financial losses.

TABLE 13.13. Economical and Social Impacts of Blackouts

No.	Name of Blackout	Economy	Society
1	09/11/1965 10 states in Northeast of United States	Loss of 20,000 MW of load	30 million people
2	05/1977 Miami, United States	Not available	1 million people 15,000 squares miles
3	13/07/1977 New York City	6,000 MW of demand	10 million people
4	19/12/1978 France	29,000 MW of the 39,000 MW demand lost	Not available
5	03/1982 Oregon, United States	Not available	900,000 people
6	27/12/1983 Sweden	11,400 MW of 18,000 MW demand lost	4.5 million people
7	12/01/1987 France	Loss of 8,000 MW of load	Not available
8	13/03/1989 Québec, Canada	Loss of 21,500 MW of generation	6 million people
9	24/08/1994 Italy	Loss of 4,500 MW of load	Not available
10	14/12/1994 Arizona and Washington state, United States	Loss of 9,336 MW of load	2 million people
11	17/01/1995 Japan	Loss of 2,836 MW of load	2.6 million people
12	08/06/1995 Israel power system	Not available	70% of the consumers: 5 million people
13	02/07/1996 14 states in the United States	Loss of 11,743 MW of load	2 million people
14	10/08/1996 California Pacific Northwest, United States	$1 billion loss of 30,500 MW of load	7.5 million people
15	01/1998 In Canada, New York and New England	Not available	3 million people
16	07/1999 New York City	Not available	300,000 people
17	21/01/2002 Brazilian power system	Loss of 61.3 GWh of load with 50 USD$/MWh	Not available
18	12/01/2003 Southern part of Croatia and a part of Bosnia Herzegovina	2,375,000 dollars (1,270 MWh)	5 million people
19	14/08/2003 Northeast of United States and Canada	Between $7 and $10 billion lost	50 million people
20	28/08/2003 South London	Loss of 724 MW of load	476,000 customers
21	02/09/2003 Cancun, Mexico	Not available	3 million people
22	23/09/2003 Eastern Denmark and Southern Sweden	Loss of 8 GWh of load	2.4 million people
23	23/09/2003 Most of Chile		5 million people
24	28/09/2003 Italian power system	Loss of 180 GWh of load	57 million people
25	01/12/2003 Southeastern Massachusetts from New Bedford to Provincetown to the islands, Unites States	Not available	300,000 people
26	21/12/2003 San Francisco, United States	Not available	120,000 customers
27	12/07/2004 The Athens and Southern Greece	9,000 MW	250,000 homes 7 million people
28	23/08/2004 Bahrain	Not available	650,000 people 700 square kilometers
29	25/05/2005 Moscow, Russia	$1 billion lost	4 million people

13.5 RECOMMENDATIONS FOR PREVENTING BLACKOUTS

Following the description of the phases and the mechanisms of blackouts, the major incidences of each phase of blackouts are listed and the possible resolution methods are suggested in Table 13.14. As traditional means for blackouts prevention were widely discussed in the literature, our purpose is to discuss about the use of relatively recent technologies.

Indeed, some new technologies for monitoring and system control became more and more mature in recent years. To realize real-time control in the power systems, the time delay of the measurement system typically needs to be limited in a time around 100–200 ms [33]. SCADA/EMS, which was traditionally used in power systems, provides 1–5-s measurement intervals, but this is not sufficient for real-time control [33], because some incidents in the power system can lead to serious problems only in several hundreds of milliseconds. The wide-area measurement system (WAMS), which is based on the phasor measurement units (PMU) and the global positioning system (GPS), can provide more effective and quick real-time system information to system operators and realize real-time control [33–36]. More and more new control systems and control methods based on WAMS, such as wide-area stability and voltage control system (WACS) and wide-area monitoring and control systems (WAMC) [34,35], have been developed.

Besides real-time control system, flexible AC transmission system (FACTS), which can be used for voltage control and load flow control, can make the power system more stable and flexible [36]. These equipments, based on power electronics, can react quickly to a disturbance.

The application of these techniques in power systems would decrease the incidence of blackouts.

New techniques combined with the traditional ones could be applied in different phases of blackout.

1. *Precondition.* In the period of precondition, the security margin of the power system is often limited. The WAMS can be used to detect the system condition. When the system is approaching its security limit, the system operator could rapidly know this information through WAMS and quickly take effective actions to keep

TABLE 13.14. Major Incidences of Each Phase of Blackout and Possible Resolution Methods

Phase	Precondition	Initiating Event	Steady-State Progression	Triggering Events	High-Speed Cascade	Final State and Restoration
Time	Hours to minutes	Milliseconds	Hours to minutes	Milliseconds	Minutes to seconds	Minutes to days
Types of fault	Heavy load flows	Oscillation	Cascade overload, system separation	Oscillation	Voltage collapse, frequency collapse, loss of synchronism	Restoration
Resolution method	FACTS, power reserves	FACTS, PSS	FACTS, PSS, load shedding, generator rescheduling	FACTS, PSS, load shedding	Load shedding, system islanding	Restoration plan

the system in a safe condition. The system operator should adjust reactive power output of the generators, use Static VAr compensators for preventing voltage decrease, reschedule generation scheme, change load flows, shed load, and use FACTS in some critical lines to prevent overload.

2. *Initiating Events.* The initiating events of blackouts are usually short circuit, lines or generators tripping, and so on, and they can lead to power oscillations in the power system. Power system stabilizers (PSSs) and FACTS could be utilized to damp these oscillations and to prevent the inappropriate actions of the protection system.
3. *Steady-State Progression.* The period of the steady-state progression is about 10 min to 1.5 h. Cascade overload is the major event that occurs during this period. The system operator could change the topology of the power network, utilize generator rescheduling, load shedding and FACTS to avoid cascade overload in this period.
4. *Triggering Events.* The triggering events of blackouts are usually short circuits, important lines and generators tripping, and so on. These triggering events could result in large oscillations that could cause the action of the protection system, and more and more lines and generators would trip and become out of order. Power system stabilizers (PSS) and FACTS could be used to damp oscillations in a coordinated manner, and the protection system could split the network and isolate the transient oscillations to keep other parts of the network safe.

   The period before triggering events is a critical period for the system operator to take actions to pull the system back to safe condition. When the triggering events happen, it is difficult for the system operator to stop the rapid worsening of the system condition, and the blackout may be unavoidable.
5. *High-Speed Cascade.* In the period of high-speed cascade, numerous lines and generators trip in several seconds or several minutes. When the system enters this period, it is too late to take the effective actions to stop the blackout. In this period, massive load shedding may be a way to reduce the impact of a blackout.
6. *Final State and Restoration.* When the blackout has happened, the system operator must restart the system as soon as possible.

13.6 ON SOME DEFENSE AND RESTORATION ACTIONS

The purpose of this part is not to describe all the possible plans because each TSO has its own methods. The goal is to give general ideas with examples to fix minds. Nevertheless, we think that the reader should find other various approaches, means, and methods in the literature or in TSOs' operating manuals.

The objective of a power system is to ensure the continuity of load supplying, guarantying a certain level of voltage quality. This can be decline in four criteria that are respect of operation constraints, safe-keeping of equipments, safe-keeping of the network, and system restoration. Functions of these various criteria, the system tasks to be satisfied are shown in Figure 13.38.

The consequences of an electric breakdown are so heavy that each power system, function of its own characteristics, has to develop procedures including decisions and actions to take, automatically or not, to defend the system against incidents. These procedures must be able to restore the normal operation of the system as quick as possible

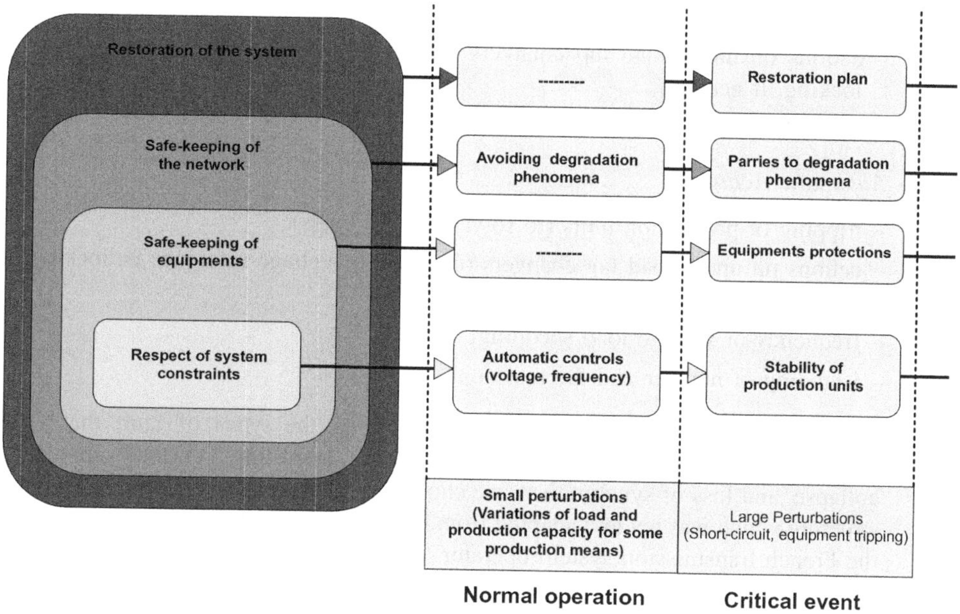

Figure 13.38. Objectives of a power system in terms of security.

if all the protection means that have been requested did not succeed to save it. The objectives are the followings:

- *Detection.* The system must have the necessary means to detect its deterioration.
- *Security.* We need adequate procedures to stop the incident spreading and its propagation on the whole network, sacrificing some customers if needed and/or preparing the splitting plan in several subnetworks to protect the safety areas. This takes part of the *defense plan*.
- *Rapidity.* for the areas which are in extreme state, we need automatic or manual procedures to do to allow coming back quickly at the normal situation. This is the *restoration plan*.

13.6.1 Defense Actions

In a first approach, the objectives of a defense plan are to detect the degraded state of the network, to launch ultimate parries to avoid the propagation of the incident (sacrificing non priority customer and/or splitting the system if needed), to allow a rapid come-back to a safety situation and to plan the restoration.

The defense plan includes preventive and curative actions that are carried out by operators in control centers before and during an incident, when it is allowed by the dynamic of the phenomena. We can decline curative action as follows:

- *Human Timescale Actions:*
  - fast redispatching of active/reactive productions;
  - modifications of the topology of the network;

- starting of production units;
- actions on under load tap changers (decrease of voltage reference or manual locking, if needed);
- manual load shedding.

- *Automatic Actions:*
  - tripping of production units (to solve line overloads);
  - actions on under load tap changers (decrease of voltage reference or locking, if needed);
  - frequency or voltage load shedding;
  - islanding of nuclear and thermal units.

To present possible defense actions regarding the types of fault that were previously described, that is, voltage collapse, cascading overload, frequency collapse, and loss of synchronism, we choose, as an example, a simplified defense structure which was get and adapted from RTE (Réseau de Transport d'Electricité), the French transmission system operator [9]:

- *Prevention/Planning:*
  - reliability, availability, and equipment performances: it is the purpose of the preventive maintenance;
  - redundancy of critical equipments;
  - N-k criterion;
  - reserves, reactive compensation.
- *Supervision/Action:*
  - detection and correction of deviations;
  - automatic controls;
  - normal actions of operators.
- *Ultimate Parries:*
  - avoiding the system collapse;
  - facilitating the restoration.

Defense Against Voltage Collapse.

- *Prevention/Planning:*
  - adequate means of reactive compensation;
  - dispose of waiting compensation units, capacity banks, reactance banks, and so on;
  - dispose of reactive reserves.
- *Supervision/Action:*
  - control the voltage profile: that is the purpose of the primary and secondary automatic voltage controls and tertiary manual voltage control.

- *Ultimate Parries:*
  - modify voltage references of HV/MV ULTC transformers;
  - lock ULTCs at the VHV/HV and HV/MV levels;
  - reactive overload of production units;
  - start fast production units such as gas turbines;
  - load shedding.

DEFENSE AGAINST CASCADING OVERLOAD.

- *Prevention/Planning:*
  - perfectly coordinated and selective protection plan. The protection plan must trip only the equipments that are needed to eliminate the fault;
  - robust operating schemes: N-k criterion.
- *Supervision/Action:*
  - supervise power flows on heavily loaded lines and ensure that there is no unauthorized constraints in case of N-k power flow transfers;
  - eliminate overloads on lines and/or transformers by switching operations or action on production units.
- *Ultimate Parries:*
  - load or production shedding.

DEFENSE AGAINST FREQUENCY COLLAPSE.

- *Prevention/Planning:*.
  - precise and reliable forecast of load and exchanges on interconnection lines;
  - the production plan represent the sum of load forecast, exchanges forecast, and margins. Active reserves must be well-sized;
  - ensure that reserves will be available in required time.
- *Supervision/Action:*.
  - control the frequency: that is the purpose of the primary, secondary automatic frequency controls, and tertiary manual frequency control;
  - verify, in real time, the availability of reserves.
- *Ultimate Parries:*.
  - change to maximum active power of operating production units;
  - fast load shedding (manual);
  - frequency load shedding (automatic).

DEFENSE AGAINST LOSS OF SYNCHRONISM.

- *Prevention/Planning:*
  - adequate speed and voltage control loops on production units

- efficient protection plan: the critical clearing times have to be as small as possible;
- avoid topologies that facilitate collapses (long antennas).

- *Supervision/Action:*
  - control the acceleration of production units (with threshold accelerometers).
- *Ultimate Parries:*
  - SPS (special protection schemes): they split the network and shed loads to return to the equilibrium;
  - islanding of nuclear and thermal units on their auxiliary services.

13.6.2 Restoration Actions

The restoration plan is launched when all the means related to the defense plan failed to stop the system collapse. It includes all actions of power system to be done whenever a major incident in order to restore as soon as possible the system to the equilibrium normal conditions. The priorities take place in ensuring and consolidating the re-energizing of large production centers for their participation in the reconstitution of networks; then restoring progressively the totality of customers. Two major restoration strategies in use today are well known: *buildup* and *build-down* strategies [37–39]. The first one is to re-energize the bulk power network before synchronizing most generators, whereas the second one is to restore system by islands that will then be mutually interconnected. A mixed solution called *build-together* is used in specific cases. They are all made for bulk transmission system.

On the lower voltage level, system restoration service can be always seen as network reconfiguration. Whenever a fault, the distribution system operator tries to transfer the maximum of loads from the faulted feeder to the healthy feeders by closing the normally opened switches (tie switches). In case of a larger incident, distribution system must stay in black until transmission bulk would be restored and become available.

A typical restoration plan comprises the following essential actions:

- the system operators identify the system status including the circuit breakers, connection possibilities, black-start unit capacity, location of critical loads, and so on;
- after starting at least one black-start unit, emergency energy from black-start units have to be sent to the large nuclear and/or thermal power plants within critical minimum interval and, then, to the noncritical minimum interval units. Critical loads are defined and used to stabilize the system;
- supply progressively the equipments of the transmission network avoiding overvoltage or undervoltage problems due to long lines. Other problems such as ferroresonance must be avoided to prevent another collapse;
- the more units are taken in operation, the more quickly the subsystem and its load will be energized. Loads at various level voltages are then reconnected.

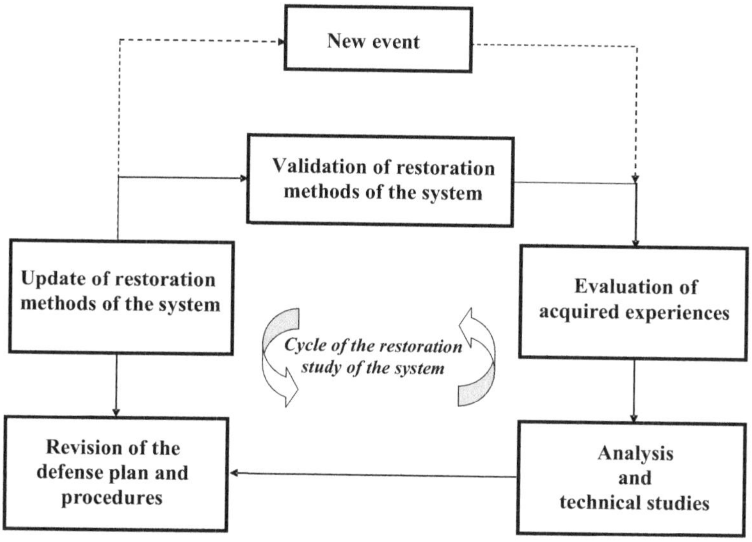

Figure 13.39. Cycle of the restoration study of the system.

The most part of the process duration concerns the integration of thermal units and their power generation growth. The system operator has to face some problems during the process:

- the choice of critical loads;
- stability problems at the beginning;
- reactive power limits of generators, especially concerning absorption;
- high and low voltages.

The restoration plan depends on the system structure and available technologies. The network evolutions need an adequate adaptation of the restoration plan. These evolutions can be the integration of new technologies in terms of generation, energy transmission, protection, or operating means of the system. On the contrary, they can also produce new critical events and/or worsening factors that have been not already experienced on the power system.

Each time, the corresponding phenomenon has to be rigorously analyzed. It will allow getting operating experiences that will provide necessary means to the operator to update the system defense plan. This analysis procedure forms a cycle showed in Figure 13.39. It is called cycle of the restoration study of the system.

A new restoration concept has been recently proposed [40–43] using dispersed generation (DG) located at the distribution level. The idea is to profit of the actual and forecasted increase of DG penetration rate in distribution networks. This restoration process works simultaneously in transmission (downward stream) and distribution (upward stream) networks with the aid of DG and it is called *deep build-together* strategy. This approach considers DG not only as additional sources in normal conditions but possible to be active and flexible support ones in case of major failure, that is, black-out, should one occur. So, DG is used to help the power system to reduce blackout consequences in terms of out-of service's load volume, its duration and the system restoration duration. Main objectives are as follows:

- to profit efficiently the available electrical energy of DG units in critical situations;
- to accelerate the network re-energizing;
- to restore clients as much and rapid as possible.

This strategy is to re-energize by islanding in transmission network while distribution network cells are being formed and expanded based on the availability of DG black-start and power unit's capacity without support by the main power system. The priority tasks in transmission level are to black-start successfully at least one unit; to restart up the nonblack-start capacity units in order to take more units in operation and then, to pick up the load in adequacy with the decrease of the load demand that is recovered by the DG support as soon as possible. At the distribution level, instead of staying in black till the transmission system would be restored and available (normally during for several hours), DG could facilitate the apparition of a lot of autonomous areas or cells using DG black-start capacity, providing the local service continuity and energizing the grid as large as possible by switching operation. This concept is called intentional islanding multilevel. As a result, the volume of load restored during the restoration process is more important at any time and the collapsed time of many customers is shorter.

On the other hand, reinforcements on distribution network infrastructures are required for this kind of restoration strategy. Depending on the DG capacity and its localization, the equipment of more sectionalizing switches is necessary because they will be useful both in reconfiguration service and restoration service. Since the main information and communication control system is now applied in the transmission and distribution levels (SCADA, EMS/DMS), in the future, the introduction of new information and communication technology (NICT) components in the specific points of distribution networks is also required.

13.7 SURVIVABILITY/VULNERABILITY OF ELECTRIC POWER SYSTEMS

13.7.1 Introduction

The survivability problem is typical for complex systems of any character including electric power systems. The power system survivability is its property to withstand disturbances, preventing from their cascading development with large-scale interruption of consumer supply, and to restore quickly the normal state or the state close to it. The survivability problem is of particular priority for large extended power systems and is related to the rise of heavy systems emergencies (of the cascading nature as well), which can lead to substantial unfavorable consequences for consumers and to violation of the whole economy operation [44].

A great number of cascading system emergencies is characteristic for bulk electric power grids. For instance, in the 1970s–1980s in such large interconnections as the unified power systems of the former USSR countries and interconnection of the United states and Canada, the statistics recorded tens of severe system emergencies yearly [45]. Particularly, severe cascading system emergencies with disturbed power supply on a large territory and for a long time happen relatively seldom (once per several years) but they become national disasters (see above).

The first attempt to formalize the power system survivability problem was made by Kitushin [46]. The methods for survivability analysis were devised by Avramenko [47], Koshcheev [48] considered the problem of selection of emergency control means to improve the system survivability. Voropai [49] introduced the notion of system limiting state from the survivability standpoint. It proved to be constructive for devising the methodology to study survivability improvement independently of the disturbance character. Fouad, Zhou, and Vittal [50] analyzed the power system vulnerability to large cascading emergencies, treating it as an important component of security [47].

This subchapter describes the methodology for studying and improving survivability of large power systems.

13.7.2 Conception

Let S_i, $i = 1, \ldots, m$ be different states of the power system after disturbance. It is possible to distinguish some critical "limiting" state S_{lim} which can be reached by the system in the course of emergency development and after which the probability of irreversible consequences for the power system and its ability to perform the set functions, that is, to supply consumers with electric power of the set quality and in the required volume, is high. Hence, development of the emergency disturbance with large-scale interruption of consumer supply is inadmissible. In Figure 13.40, S_o, S_j are pre-emergency and postemergency states; $S_{\text{lim}b}$, $S_{\text{lim}s}$ are limiting states with big and small margins; S_i is intermediates state.

The limiting state for a power system is determined by inadmissible decrease in frequency and voltage in the system, insufficient values of spinning reserve of the generating capacity and transfer capability margins of the tie-lines. The power system limiting state for consumers is determined by the minimum admissible (from the conditions of their technology) value of power supplied by the system and the limiting length of interruption of their supply with power.

The limiting state of the power system in terms of survivability can also be determined by the level of emergency load shedding that exceeds the value of load disconnected by automatic frequency load shedding devices or by the systems state when the auxiliary power supply of power plants is not violated by emergency disconnections and hence, the complete power supply can be restored sufficiently quickly [45,47]. On the whole, the quantitative characteristics of the limiting state are determined to a great extent by specific

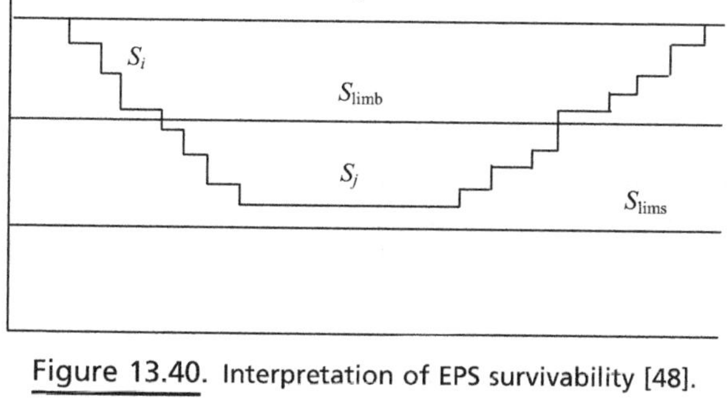

Figure 13.40. Interpretation of EPS survivability [48].

requirements to power systems from the survivability standpoint, which can differ depending on specific conditions.

It is necessary to make the sense of the functions fulfilled by the power system from more concrete survivability viewpoint. These functions must consist in power supply only to the most important consumers. Less important consumers may be disconnected using the emergency control, including automatic frequency load shedding, that opens greater possibilities for the power system control to improve survivability, preventing from emergency cascading, and facilitates the system restoration.

With such an approach the power system survivability level is characterized by the extent of remoteness of the system postemergency state S_j, determined by the operating parameters, the spinning reserve value, the transfer capability margins, and so on, from the limiting state $S_{\lim}$. The emergency character turns out to be important only as much as it brings the power system also depends on the character of the pre-emergency state S_o: less intensive disturbances can turn out to be dangerous for heavy conditions of the power system than for normal conditions with the sufficient transfer capability margins of tie-lines, the spinning reserve value, and so on.

The indicated aspects of the power system survivability problem are supported by the experience accumulated in the study of other complex systems of different nature, when the process of system sophistication is accompanied by the growing probability of large-scale fluctuations, decrease of its reliability and survivability which in turn can be improved using new means and ways for the structural arrangement of the system and its control. It reflects objective contradictions in development of complex systems. For a power system, they consist in the appearing discrepancy between the changing structural properties of the system in the process of its sophistication and development, which determine the variation in the conditions of its operation and dynamic properties, and the remaining principles of the power system formation and control of its operating conditions.

Accumulation of these contradictions leads to negative consequences, such as appearance of "weak places" in the system, deterioration of the EPS controllability, and decrease of its survivability. Resolution of the arising contradictions is the key goal of the power system development management that should comprehensively consider conceptual aspects in the main structure formation of the system, conditions of its operation, principles of dispatching and automatic control.

Hence, the power system survivability is characterized by endurance, vitality of its structure and reflects vulnerability to disturbances. Whereas survivability is an internal property of the power system, vulnerability indicates as if an external response of the system to disturbances. Consideration of survivability as an internal property of the power system provides constructive possibilities and ways for its improvement.

13.7.3 Technology of Study

The sense and contents of the power system survivability study are to reveal weak places in the system in terms of survivability and to identify measures on their elimination. The problem complexity consists in the fact that due to difficulties in adequate probabilistic description of disturbances determining survivability (as a result of their rare occurrence and uniqueness) the estimates of the power system survivability level and the impacts of insufficient survivability level can be only relative in most cases. At the same time, inaccurate economic estimates of the indicated impacts (specific damages of consumers) as well as absence of the survivability standards lead to its treatment as an independent criterion determining the power system development and operation and to orientation to comparative studies on substantiation of survivability improvement decisions.

From the standpoint of quantitative estimation of the power system survivability and substantiation of means for its improvement, the choice of the applied survivability indexes is important. Such indexes should characterize to some or other extent the closeness of power system to the limiting state. Most of the power system survivability indexes suggested by different authors are connected to the load that is tripped as a result of the specific emergency. Among other indexes worth to be mentioned is the extent of emergency "propagation" in the system that is measured, for example, by the number of subsystems isolated as a result of emergency development, the number of possible contingencies at the emergency cascading, and the value of critical disturbances in terms of the power system survivability.

The risk concept can validly underlie the methodological approaches to study the power system survivability. The quantitative risk estimation is determined by the possibility for the system to reach the limiting state S_{lim}, that is, it is necessary to estimate the risk of power supply interruption of the most important consumers.

The models for the risk analysis are usually intended for the solution of the following problems: construction of scenarios of emergency situations and determination of the associated probabilities of different contingencies; quantitative determination of risk by modeling the chains of possible contingencies in the system; and estimation of feasibility of the determined risk level and identification of measures on its decrease.

When constructing the scenarios of heavy emergencies that can lead to the EPS limiting state, account should be taken of different reasons for such situations: cascading emergencies, large single disturbances, coincidence of relatively light emergency situations with heavy pre-emergency conditions, and so on.

As was noted earlier, the quantitative estimation of risk is connected to determination of probabilities of power supply interruption of important consumers as a result of implementation of the constructed scenarios by modeling the states and processes in the system. To estimate feasibility of the determined risk level, the corresponding standards should be developed.

13.7.4 Concluding Remarks

Growth and complexity of interconnections increase the urgency of the power system survivability problem.

Introduction of the notion of the limiting state of EPS makes the understanding of the survivability property more certain, specifies the composition of survivability problems. The main goal of survivability study is to identify the weak points in power system and to select means for their elimination. The technology of studies is based on the risk notion (with respect to interruption of power supply to important consumers).

The multiaspect nature of the survivability improvement problem is explained by the complexity of this property. The appropriate set of means for its improvement should embrace the stages of system development planning and its operation. These means should guarantee the required "security margin" of the power system in terms of survivability, on the one hand and improve the system controllability, that is, "intensity" of its response to disturbances to prevent undesirable development of the emergency process and to accelerate the power system restoration, on the other hand.

13.8 CONCLUSIONS

Large-scale blackouts cause enormous economical and social damages. Above all, to prevent blackouts, we must well know the previous blackouts and find their characteristics.

In this chapter, by analyzing 39 previous blackouts, we divided the progression of blackout into five phases: precondition, initiating events, cascade events, final state, and restoration. During the period of cascade events, the triggering events follow the period of steady-state progression, and the triggering events, being a critical point, lead to start the high-speed cascade. In some blackouts, the period of steady-state progression was skipped and the initiating events became the triggering events.

It was found that 35.1% of the blackouts happened in system normal condition; the period of steady-state progression is long ($t > 10\,\text{min}$), whereas the period of high-speed cascade is very short (several seconds $< t < 10\,\text{min}$), and the blackout usually happened quickly after the high-speed cascade was triggered. So, we suggested that effective actions should be taken before the triggering events happened, because when the high-speed cascade has been triggered, the situation can become uncontrollable and the blackout can happen within a very short time.

According to the analysis of 10 blackouts (from which detailed information was available) from 1965 to 2005, voltage collapse and cascade overload, which happened with a higher frequency, can be considered as the major types of incidents.

Some new technologies could be utilized to prevent blackouts. FACTS, which could decrease not only the possibility of voltage collapse but also the possibility of cascade overload, would be a good method to strengthen the power system and decrease the risk of blackouts. Wide-area measurement system (WAMS) could be used for system supervision and control, and wide-area stability and voltage control system (WACS) and wide-area monitoring and control systems (WAMC) could improve the reliability and robustness of the power system. Based on these technologies, good strategies for the system security would largely decrease the frequency of blackout.

How to perfectly organize and utilize the new methods and the traditional ones in modern power systems would be an interesting goal in this field.

Acknowledgments

We thank Wei Lu, whose works [31,51] have greatly contributed to the elaboration of this chapter.

REFERENCES

[1] NERC Steering Group *Technical analysis of the August 14, 2003, blackout: what happened, why, and what did we learn*? Report to the NERC Board of Trustees, July 13, 2004.

[2] U.S. – Canada Power System Outage Task Force, *Final report on the August 14, 2003 blackout in the United States and Canada: causes and recommendations*, U.S. – Canada Power System Outage Task Force, April, 2004.

[3] Corsi, S., Sabelli, C. General blackout in Italy September 28, 2003, *IEEE PES General Meeting*, June 6–10, 2004.

[4] Berizzi, A. The Italian 2003 blackout, *IEEE PES General Meeting*, June 6–10, 2004.

[5] UCTE, *Interim report of the investigation committee on the 28 September 2003 blackout in Italy*, UCTE, October 27, 2003.

[6] Vassell, G.S. Northeast Blackout of 1965, *IEEE Power Eng. Rev.*, Vol. 11, No. 1, pp. 4, January 1991.

[7] Amin, M. North America's electricity infrastructure, *IEEE Secur. Privacy Mag.*, Vol. 1, No. 5, pp. 19–25, September–October 2003.

[8] Novosel, D., Begovic, M.M., Madani, V. Shedding light on blackouts, IEEE, *Power Energy Mag.*, Vol. 2, No. 1, pp. 32–43, January–February 2004.

[9] RTE *Mémento de la sureté du système électrique*, Edition 2004, http://www.rte-France.com.

[10] Czech, P., Chano, S., Huynh, H., Dutil, A. The Hydro-Quèbec system blackout of 13 March 1989: system response to geomagnetic disturbance, *Proceedings of the Geomagnetically Induced Currents Conference*, November 8–10, 1989.

[11] Morii, K. Electric power systems and natural disasters, *CIGRE Tokyo Symposium*, May 22–24, 1995.

[12] Hain, Y., Schweitzer, I. Analysis of the power blackout of June 8, 1995 in the Israel Electric Corporation, *IEEE Trans. Power Syst.*, Vol. 12, No. 4, pp. 1752–1758, November 1997.

[13] NERC, *System disturbances review of selected 1996 electric system disturbances in North America*, NERC, August 2004.

[14] Filho, V.X., Pilotto, L.A.S., Martins, N., Carvalho, A.R.C., Bianco, A. Brazilian defense plan against extreme contingencies, *IEEE PES Summer Meeting*, July 15–19, 2001.

[15] Dass, R. Grid disturbance in India on 2nd January 2001 *Electra*, No. 196. pp. 6–15, 2001.

[16] Dizdarevic, N., Majstrovic, M. Causes, analysis and countermeasures with respect to blackout in Croatia on January 12, 2003, *The CRIS International Workshop on Power System Blackout*, Lund, Sweden, May 3–5, 2004.

[17] OFGEM *Report on support investigations into recent blackouts in London and west midlands*, Vol. 1, Main Report, OFGEM, February, 2004.

[18] OFGEM *Report on support investigations into recent blackouts in London and west midlands*, Vol. 2, Supplement Report, protection commissioning & performance, OFGEM, February, 2004.

[19] Ekrft System *Power failure in Eastern Denmark and Southern Sweden on 23 September 2003.* Final report on the course of events, November 4, 2003.

[20] IREC, *Chile – Power blackout hits most of country*, Interstate Renewable Energy Council, http://www.solarstorms.org/Chile2002.html, November 7, 2003.

[21] Vournas, C. *Technical summary on the Athens and Southern Greece blackout of July 12, 2004* http://www.pserc.org/Greece_Outage_Summary.pdf.

[22] NEMMCO, *Power system incident 14 March 2005 final report*, National Electricity Market Management Company Limited ABN 94 072 010 327.

[23] Makarov, Y.V., Reshetov, V.I., Strojev, V.A., Voropai, N.I. Blackout prevention in the United States, Europe, and Russia, *Proc. IEEE*, Vol. 93, No. 11, November 2005.

[24] PSERC, *Resources for understanding Moscow blackout of 2005*, Power Systems Engineering Research Center, http://www.pserc.org/Mowcow Blackout.htm.

[25] Merlin, A., Desbrosses, J.P. European incident of 4th November 2006. The events and the first lessons drawn, *Electra*, No. 230, February 2007.

[26] ERGEG, *ERGEG Interim Report on the lessons to be learned from the large disturbance in European power supply on 4 November 2006*, European Regulators Group for Electricity and Gas, E06-BAG-01-05, 20 December 2006.

[27] Voropai, N.I., Efimov, D.N, Reshetov, V.I *The analysis of system emergencies development mechanisms in electric power systems, /Elektrichestvo/*, No 10, pp.12–14, 2008 (in Russian).

[28] Efimov, D.N., Voropai, N.I. Blackouts analysis and generalization, *The International Workshop on Liberalization and Modernization for Asset Management*, Irkutsk, Russia, August 14–18, 2006.

[29] UCTE, final report on system disturbance on 4 november 2006, UCTE, 2007.

[30] McCalley, J.D. Operational defence of power system cascading sequences: probability, prediction, and mitigation, *PSerc Seminar*, October 7, 2003.

[31] Lu, W. *Le délestage optimal pour la prévention des grandes pannes d'électricité - Optimal load shedding for blackouts prevention*, Ph.D. Dissertation, Grenoble Institute of Technology, 2009 (in French) http://tel.archives-ouvertes.fr/tel-00405654/en/.

[32] Voropai, N.I., Efimov, D.N. *Analysis of blackout development mechanisms in electric power systems*,IEEE PES General Meeting, Pittsburgh, USA, July 21–24, 2008.

[33] Cai, J.Y., Huang, Z., Hauer, J., Martin, K. Current status and experience of WAMS implementation in North America, *2005 IEEE PES Transmission and Distribution Conference & Exhibition: Asia and Pacific Dalian,* China, August 15–18, 2005.

[34] Taylor, C.W., Erickson, D.C., Martin, K.E., Wilson, R.E., Wenkatasubramanian, V. WACS – Wide-area stability and voltage control system: R&D and on-line demonstration, *Proc. IEEE*, Vol. 93, No. 5, May 2005.

[35] Zima, M., Larsson, M., Korba, P., Rehtanz, C., Anderson, G. Design aspects for wide-area monitoring and control systems, *Proc. IEEE*, Vol. 93, No. 5, May 2005.

[36] Hingorani, N.G. *Flexible AC transmission*, IEEE Spectrum, April 1993.

[37] Fink, L.H., Liou, K.L., Liu, C.C. From generic restoration actions to specific restoration strategies, *IEEE Trans. Power Syst.*, Vol. 11, No. 2, pp. 745–751, May 1995.

[38] Adibi, M.M., Fink, L.H. Power system restoration planning, *IEEE Trans. Power Syst.*, Vol. 9, No. 1, pp. 22–28, February 1994.

[39] Andrews, C.J., Arsanjani, F., Lanier, M.W., Miller, J.M., Volkmannn, T.A., Wrubel, J. Special consideration in power system restoration, *IEEE Trans. Power Syst.*, Vol. 7, No. 4, November 1992.

[40] Pham, T.T.H., Bésanger, Y., Hadjsaid, N., Ha, D.L. Optimizing the re-energizing of distribution systems using the full potential of dispersed generation, *IEEE-PES General Meeting*, Montreal, Canada, July 18–22, 2006.

[41] Pham, T.T.H., Bésanger, Y., Hadjsaid, N. Intelligent distribution grid solution to facilitate expanded use of dispersed generation potential in critical situations, *Power System Infrastructure: CRIS Conference on Interdependencies and Applications,* Springer Science/Business Media, October, 2008.

[42] Pham, T.T.H., Bésanger, Y., Andrieu, C., Hadjsaid, N., Fontela, M., Enacheanu, B. *A new restoration process in power systems with large scale of dispersed generation, IEEE-PES Transmission and Distribution Conference*, Dallas, USA, May 21–24, 2006.

[43] Pham, T.T.H., Bésanger, Y., Hadjsaid, N. New challenges in power system restoration with large scale of dispersed generation insertion, *IEEE Trans. Power Syst.*, Vol. 24, No. 1, pp. 398–406, February 2009.

[44] Voropai, N.I. Survivability and emergency control of electric power systems in a market environment, *World Engineers Convention*, Shanghai, China, November 2–6, 2004.

[45] Voropai, N.I., Ershevich, V.V., Luginsky, Ya.N., et al. *Control of powerful power grids*. Energoatomizdat, Moscow, 1984 (in Russian).

[46] Kitushin, V.G. Survivability of power systems, *Proceedings of the 3rd All-Union Scientific-Technical Conference on Stability and Reliability of Power Systems in the USSR*, Leningrad, Energiya, pp. 405–412, 1973 (in Russian).

[47] Avramenko, V.N. Analysis of electric power system survivability. *Collection: Reliability Problems in Electric Power System Operation and Development Management*, Energoatomizdat, Leningrad pp. 59–67, 1986 (in Russian).

[48] Koshcheev, L.A. Efficiency estimation and choice of emergency control measures with regard for the reliability and survivability indexes of an electric power system. *Collection: reliability*

Problems in Electric Power System Operation and Development Management, Energoatomizdat, Leningrad, pp. 24–33, 1986 (in Russian).

[49] Voropai, N.I. The problem of large electric power system survivability, *IEEE Stockholm PowerTech Conference,* Stockholm, Sweden, June 18–22,1995.

[50] Fouad, A.A., Zhou Q., Vittal, V. System vulnerability as a concept to assess power system dynamic security, *IEEE Trans. Power Syst.*, Vol. 9, No. 2, pp. 1009–1015, 1994.

[51] Lu, W., Bésanger, Y., Zamaï, E., Radu, D. Analysis of large scale blackouts and recommendations for prevention, *WSEAS Trans. Power Syst.*, Vol. 1, No. 7, pp. 1189–1195, July 2006.

14

RESTORATION PROCESSES AFTER BLACKOUTS

Alberto Borghetti, Carlo Alberto Nucci, and Mario Paolone

14.1 INTRODUCTION

This chapter aims at illustrating the power system dynamic behavior during restoration scenarios in transmission networks, with particular reference to the analysis of thermal power plant black-start capabilities. This chapter is structured into three parts. Section 14.2 briefly reviews the requirements and the structure of power system restoration processes. The unbundling and liberalization of electrical power systems justifies the reassessment of power system restoration strategies, also by increasing the number of black-start resources, traditionally only represented by hydro stations and gas turbines (GTs). Hence, Section 14.3 deals with the analysis of the black-start capabilities of thermal stations equipped with both steam turbines (STs) and GTs. Moreover, accurate modeling of the dynamic interaction between the power plant and the restoring electrical power system, throughout the use of modern computer simulation tools, appears to be of considerable help to study the early stage of the restoration plans and to design specific control systems to support the feasibility and reliability of black-start maneuvers. Therefore, Section 14.3 also includes the analysis of the typical transients occurring during load restoration maneuvers and islanding formation for two representative types of power plants, namely a once-through boiler steam group repowered by a gas turbine and a combined combined-cycle unit. The models of the two relevant simulators and their identification by means of experimental data are described in Section 14.4.

Handbook of Electrical Power System Dynamics: Modeling, Stability, and Control. Edited by Mircea Eremia and Mohammad Shahidehpour.

14.2 OVERVIEW OF THE RESTORATION PROCESS

Following the typical classification [1], the restorative state is one of the different power system operating conditions (the others being the normal, alert, and emergency-in extremis one). The restoration process after a blackout is therefore a typical problem of power system operation and control.

As in all systems there is a certain risk of more or less large blackouts, it is necessary to develop plans for a rapid and secure recovery of a normal operating condition able to coordinate the actions of all the operators involved in the process.

The objective is to achieve a restoration that is optimal with respect to maximum restored energy and minimum elapsed time. In this context, the problem of restoration can be viewed as a complex optimization problem [2,3].

Being the process divided into several consecutive stages, the general objective, namely the minimization of the time required for the restoration of all customer services, is divided for each stage into multiobjective functions that include the minimization of restoration time of key network components or plants and the maximization of customer load restoration at each stage.

The constraints include:

- power flow constraints (i.e., power balance between generation and load, line flows, and voltage values within the limits);
- frequency constraints mainly related to the power balance and to the dynamics of operating power plants;
- dynamic constraints related to electromagnetic, electromechanical and control transient responses of the system to switching maneuvers and perturbations;
- time limits for generation restarts related to the difference in time requirements between cold restarts and hot restarts typical of steam power plants (SPPs);
- generator load pick-up capability constraints;
- transmission and tie-line switching sequence constraints and priority ranking.

The main control and optimization variables are the schedule of available generation, with particular reference to the plants that are able to autonomously restore part of the system (i.e., are characterized by black-start capabilities) and switching sequence.

The main decision maker is the system operator, as far as computer/communication systems of the energy management systems (EMS) are able to provide the knowledge of the current status of the system, the availability of viable alternatives, and to allow the implementation of the selected restoration strategies.

The difficulties in the problem solution are due to its combinatorial characteristic, to the need of a large amount of data from various origin and of different type and to the variety of the several criteria to be satisfied.

Due to such characteristics of the power system restoration problem, the use of both optimization algorithms and expert systems has been investigated in order to assist the operators. A CIGRÉ brochure deals with this subject [4]. These approaches have been applied particularly for the restoration of distribution systems, characterized by a priority ranking that determines which feeders with priority loads should be energized first in accordance to their importance (e.g. hospitals, airports, police stations) and by a more definite network structure than the case of transmission networks that is the main objective of this chapter (see for example [5] which contains also an comprehensive literature review).

The restoration process highly depends on the characteristics of the power system, on the extension of the network affected by the blackout and also on the actual transient behavior of the power system to the specific blackout condition. In particular, the dynamic phenomena leading the system to collapse determine, in general, the set of power plants and network components that could be used for the restoration.

The following paragraphs briefly review the typical stages, duration, and main tasks of a restoration process.

14.2.1 System Restoration Stages, Duration, Tasks, and Typical Problems

Most power systems have certain characteristics in common, and therefore some general procedure and guidelines may be defined. Obviously, a detailed plan must be developed specifically to meet the particular requirements of an individual power system. As already mentioned, further improvement may be provided by online restoration guidance tools [4–10].

Figure 14.1 illustrates a typical two-stage restoration plan. The preparation actions are very important because they permit to define the starting point of the restoration process, that is the area affected by the blackout, the available resources, and the most appropriate type of restoration strategy that needs to be followed. During the first stage, the main objective is to restore the main power network functionalities, by also using the energization of load blocks as a control means to maintain the system stability. Only during the second stage, the complete and fast load restoration becomes the real objective of the maneuvers.

Two basic strategies for the restoration process are [11]

- "Build-Down" strategy: based on the re-energization of the bulk power network as a first step, followed by balanced, step-by-step restoration of loads and generation;
- "Build-Up" strategy: the system is first divided into subsystems, each with black-start capability. Then, each subsystem is stabilized, and eventually the subsystems are interconnected.

A more detailed picture of various restoration strategies can be found in [12], which also defines, as constituent elements of the process, the following concepts: "targeted system," namely the state of the system to be achieved at the end of the process or by some significant earlier stage; "restoration building blocks," namely the minimal configurations of an autonomous stable source of power together with any associated transmission line adequate to serve as elements of the restoration process, and "generic restoration actions," namely a limited set of predefined actions that operators can take during system restoration.

Actual wide area restoration processes result, in general, in a complex combination of various strategies, based on the decisions that system operators make step by step taking

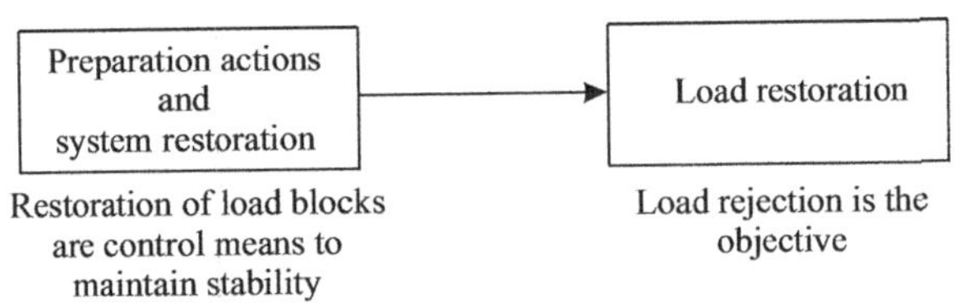

Figure 14.1. Typical two-stage restoration plan.

into account available resources and the online estimation of the system operating conditions. For example, the predefined Italian restoration plant is traditionally structured in the following build-up phases (e.g., [13,14]):

- blackout recognition;
- organization of restoration service via paths, reliant on electrical interconnections between black-start-up units (traditionally, hydro plants and open cycle combustion turbines) and large thermal power plants;
- synchronization of subsystems and load pick-up to stabilize all units;
- closing of ties for the interconnection and subsequent remeshing of the 420-kV network;
- reconnection of the remaining generating units and gradually supplying lower voltage grids at 245, 165, 145 kV, up to the progressive supplying of all the customers connected to the distribution grids.

However, in the aftermath of the September 28, 2003 system wide blackout, most part of the transmission grid was restored by using the links energized from the neighboring systems, namely from France and later from Switzerland, Slovenia, and Greece, as described in [15,16].

The restoration plan provides in general information on the following points:

- blackout recognition criteria, at each operating level and organization of the restoration service via restoration paths, connecting black-start-up units and large thermal power plant;
- selection of production units capable of carrying out black-start-up procedures in the absence of external supply. Such units have to exhibit stable islanded operation at reduced power as well, with precise voltage and frequency control and rapid power output capability;
- prioritization of some end user power supply, through pick-up load distribution paths at distribution substations along the restoration path;
- establishment of communication lines between various control levels, reliant on redundant communication systems;
- installation of automatic synchronization parallel devices in network substations.

The activities during restoration can be subdivided into the following general areas:

- load balancing and frequency regulation (see for example [17]), based on the amount of rapidly active reserve available to preserve the system frequency. Such a reserve consists of two components: (i) reserve on generators available from generators and (ii) system load already restored that is subject to tripping by underfrequency relays;
- transmission line energization and voltage regulation (see for example [18], based on the reactive reserve mainly provided by underexcitation capability limits of generators, shunt compensation, and early energization of selected loads (preloading);
- switching strategies (e.g., [19]), which take into account the occurrence of severe transient overvoltages particularly in the case of harmonic resonance due to the nonlinear magnetizing characteristic of the energized transformers [20];
- protective system issues and reclosing maneuvers between restored islands by means of synchroscope and check synchronizing relays (e.g., [21]).

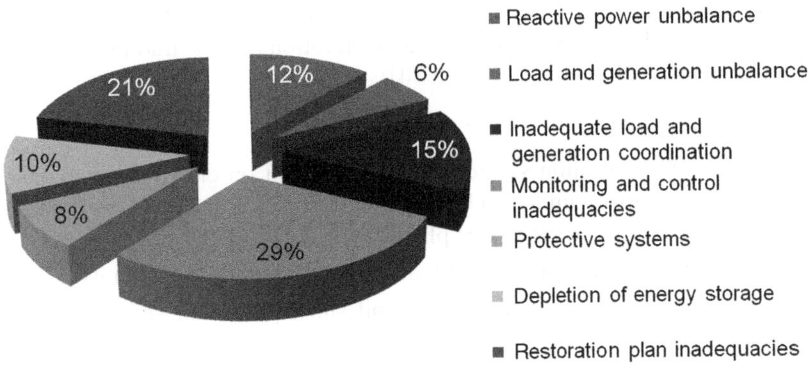

Figure 14.2. North American Electric Reliability Council (NERC)—10-year data (From the results provided in [24]).

A detailed description of each of the previous activities can also be found in two reports of a specific IEEE working group, namely [22,23].

Typical restoration issues and problems during the process are illustrated in Figure 14.2.

14.2.2 New Requirements

The liberalization of the electrical power market has also affected the design and organization of the restoration plan. This is mainly due to the following consequence of the new market framework:

- the operating conditions of the system are often different from those for which the system and its components have been designed;
- power system components operated near their capability limits more often than in the past;
- multiplicity of ownership may increase risks of emergency, due to conflicting interests among committed parties.

In this framework, more efficient defense measures against blackouts and restoration plans, which minimize recovery time and loss of load, appear to be required [25]. As an example of the new issues caused by the liberalized market in [26] is key aspects of the annual competitive process for the selection of black-start resources by an independent system operator to meet a set of system selection criteria consisting of system reliability and minimum cost parameters.

As earlier mentioned, black-start maneuvers are in general started from selected hydro power stations or open cycle gas turbines equipped with special speed governors, and are devoted to start-up thermal units through the restoration lines. The use of additional generation plants that have black-start capability becomes an important issue. This justifies the interest in assessing and enhancing the black-start capabilities of thermal power plants equipped with both steam units and gas turbines. The size of these power stations is in general significant and is therefore suitable to energize the local load and hence, for the buildup of so-called restoration nucleuses. The relevant restoration procedure may then avoid the difficulty in energization of long lines, with increased speed and reliability.

The following paragraph deals with the analysis of the black-start capability of thermal power plant and description of coordination procedures between gas turbine and steam unit productions that appear to improve such a capability.

14.3 BLACK-START-UP CAPABILITIES OF THERMAL POWER PLANT: MODELING AND COMPUTER SIMULATIONS

Several studies have been carried out on the steam plant start-up during system restoration [27] and also on the black-start-up capabilities of steam units [28,29] for assessing the possible contribution to the restoration plan from the conventional thermal units, which are also equipped with once-through boilers. The conclusion was that, in general, even when the thermal units have successfully completed the load rejection maneuver and been successfully islanded on their own auxiliaries, still the contribution of hydro plants to pick-up load first with their fast regulation appears to be necessary.

Modern thermal power plants are often equipped with both GTs and steam units, whose complementary characteristics may be exploited to improve the black-start capabilities of the complete power plant during the initial phase of the restoration plan. We present here two illustrative case studies relevant to two different power plants.[1] The first is composed of a large once-through boiler steam unit (320 MW) repowered by a smaller GT (120 MW), the exhausted gases of which are used for the high-pressure (HP) feedwater heating of the steam unit. The structure of the second power plant is based on a typical combined cycle with two 30-MW aeroderivative combustion turbines. The exhausted gases are used in two heat recovery steam generators feeding a 33-MW steam turbine.

Both studies are carried out through the development of computer simulators of the power plants and the nearby network and loads. A brief description of the two simulators and the implemented models is also reported in Section 14.4. As often mentioned in the literature on this subject [30–32]), the accurate modeling of the dynamic interaction between the power plant and the restoring electrical power system, throughout the use of modern computer simulation tools, is indeed of considerable help to study the early stage of the restoration plans and to design specific control systems to support the feasibility and reliability of black-start maneuvers.

14.3.1 Black-Start-Up of a Steam Group Repowered by a Gas Turbine

In the 1990s, some Italian thermal power stations, most of which consist of 320-MW steam turbine units equipped with once-through UP boilers, have been repowered in combined cycle. The repowering technique used by ENEL consists of topping the existing steam section SPP with a 120-MW gas turbine GT and recovering heat from the gas turbine exhaust flow by means of an exchanger, which replaces part of the high-pressure feedwater heaters [33].

An electro diesel generator is able to feed the auxiliaries of the GT to perform the start-up maneuver, and the gas turbine can, in turn, start the auxiliaries of the SPP. This is made easier by the presence of the bypass valve that diverts the gas turbine exhaust flow from the exchanger, making the two sections (GT and SPP) completely independent in this phase.

[1] The case studies are based on the work carried out as a part of collaborative projects involving the University of Bologna (A. Borghetti, M. Bosetti, C.A. Nucci, M. Paolone), CESI (G. Migliavacca, S. Spelta, F. Tarsia), and HERA (G. Ciappi, A. Solari).

The following two paragraphs deal with the black-start capability of a single SPP, without the help of other resources, and then with the black-start capability of the couple of units composed of the SPP and repowering GT. The third paragraph provides a description of a control system specifically designed to help the restoration maneuver and improves its reliability.

14.3.1.1 Black-Start-up Capability of a Single Steam Group. It is worth mentioning that for the considered steam unit equipped with once-through boilers, a specific load rejection procedure has been developed, tested, and used in several occasions in the past. The procedure consists of tripping all fuel to the boiler without tripping the turbine, switching the boiler to the start-up and bypass circuit, and then refiring it after a certain period of time (up to 30 min) during which the house load is maintained on the energy and mass stored in the boiler [34]. Only a limited number of attempts are allowed to perform new ignition, which in case of failure forces the unit to a cold restart because of inadmissible thermodynamic conditions.

If the complete load rejection process is successfully accomplished, the thermal unit can remain in operation feeding a restored island of the network, but its chance to autonomously collect load still remains questionable, since it strongly depends on prime mover-specific characteristics.

This is illustrated in Figures 14.3–14.5. These figures show some results obtained by means of an engineering simulator of a 320-MW SPP equipped with a once-through universal pressure (UP) boiler. A brief description of the simulator is given in Section 14.4.1, while additional details can be found in [35,36].

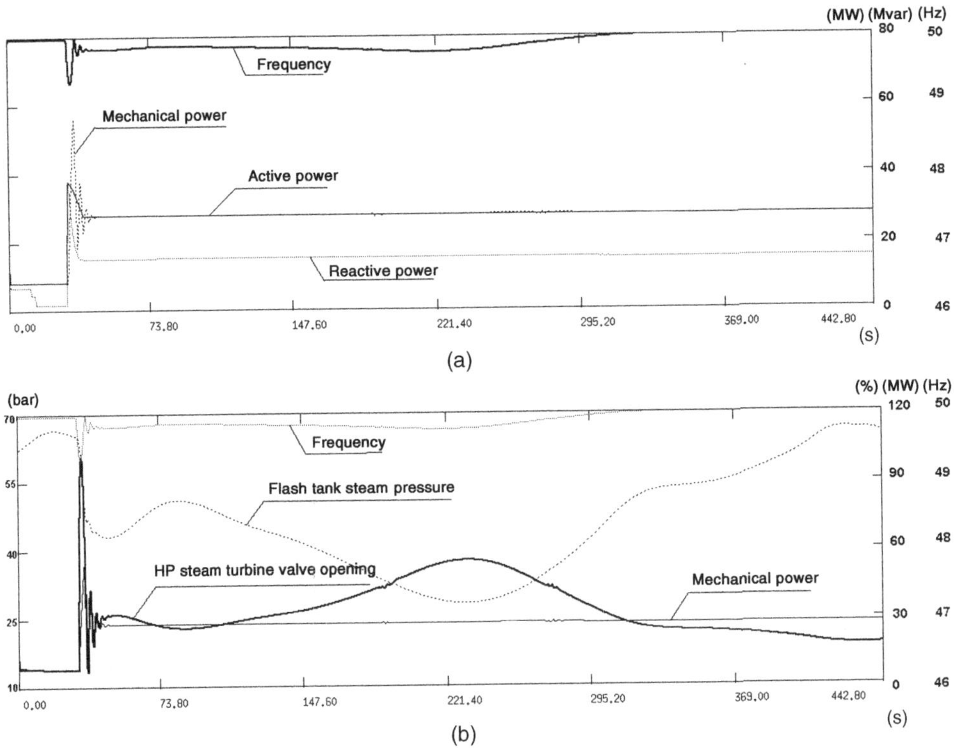

Figure 14.3. Simulation of the pick-up of a load of 18 MW by a 320-MW SPP.

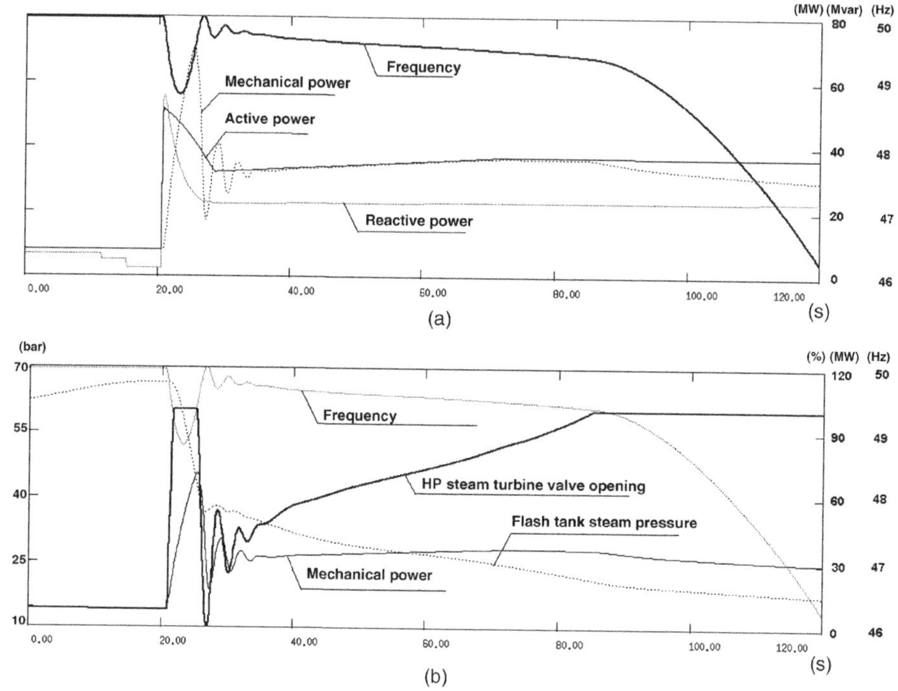

Figure 14.4. Simulation of the pick-up of a load of 30 MW by a 320-MW SPP.

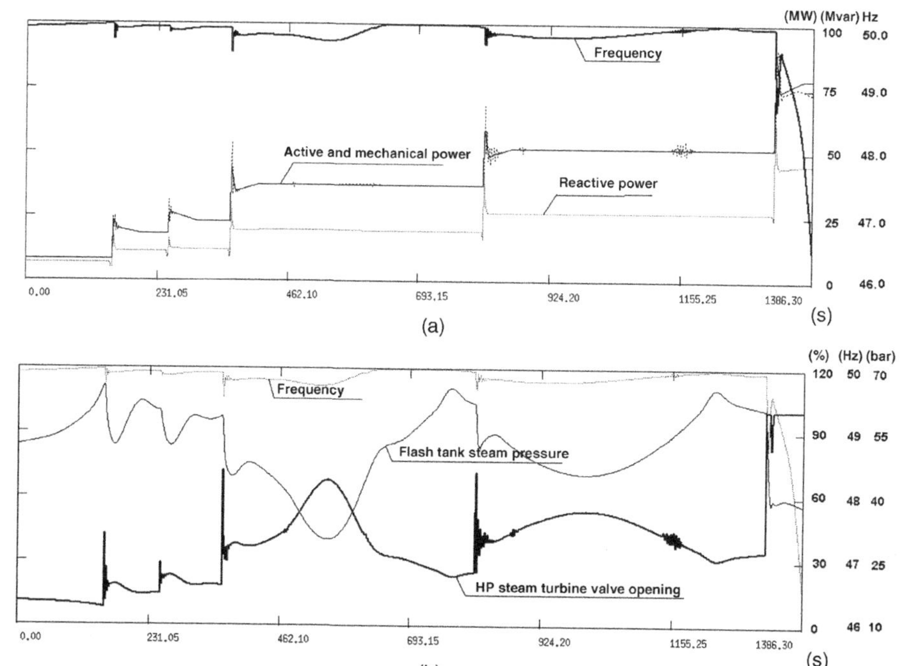

Figure 14.5. Simulation of the pick-up of four ballast load connections by a 320-MW SPP.

The results of Figure 14.3 show that the steam section, when feeding its auxiliary systems, can overcome the frequency transient due to ballast load connection, provided that the load power is lower than 18 MW. The main reason for the maneuver failure is the fall of the throttle pressure due to the speed regulation that opens the admission valve.

Figure 14.4 shows the results of the simulation of the failing of a 30-MW load pick-up. As shown in Figure 14.4, the critical condition for the SPP occurs when the admission valve is fully open and the steam pressure is reduced, so the mechanical power decreases. The mechanical torque balance being negative, the frequency drops rapidly causing turbine trip.

The simulation of Figure 14.5 shows that SPP is in a critical control speed condition throughout the whole start-up procedure, so from 0 up to 110 MW. The first ballast loads are smaller than 15 MW and the generator is able to overcome the frequency transient. In this way, the plant reaches a power of about 50 MW. The last ballast load connection is of 30 MW and the corresponding HP valve opening attains its full opening rapidly. The pressure drop is nearly 20 bar and the mechanical power cannot equilibrate the load demand, so the frequency decreases causing a turbine trip.

14.3.1.2 Black-Start-Up Capability of a Steam Group Repowered by a Gas Turbine. The simulator of the repowered power plant also includes the model of the gas turbine (GT) section.

The effect of the presence of the GT section in the repowered power plant is illustrated in Figures 14.6 and 14.7. The GT and SPP are both synchronized, performing the black-

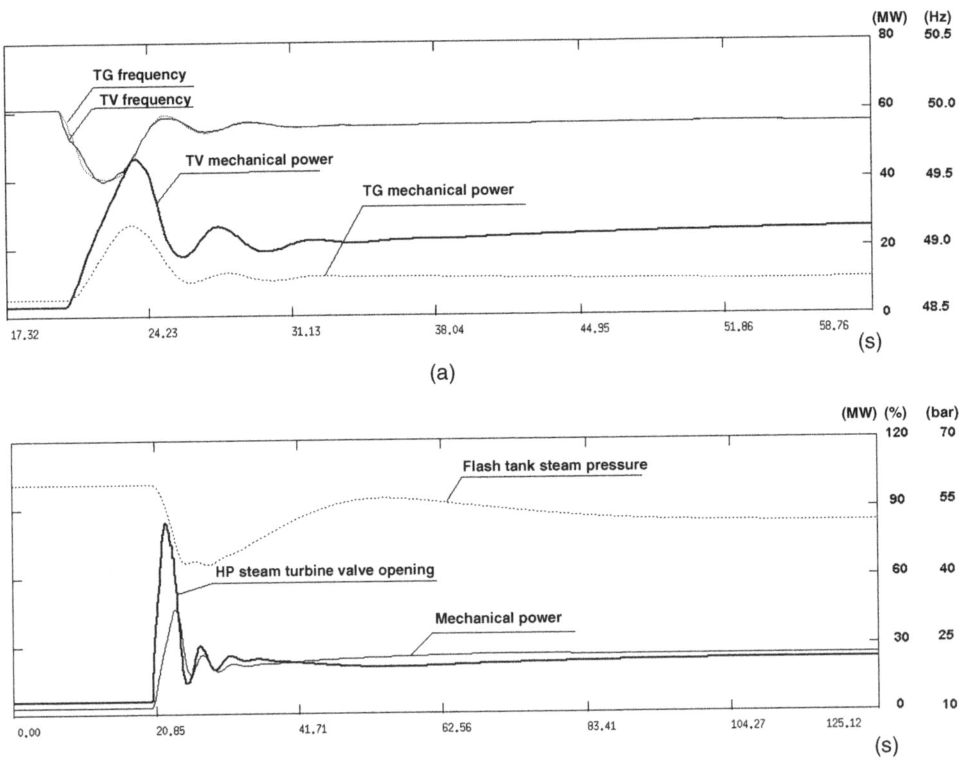

Figure 14.6. GT and SPP both synchronized and 30-MW ballast load connection: (a) GT and SPP comparison; (b) SPP results.

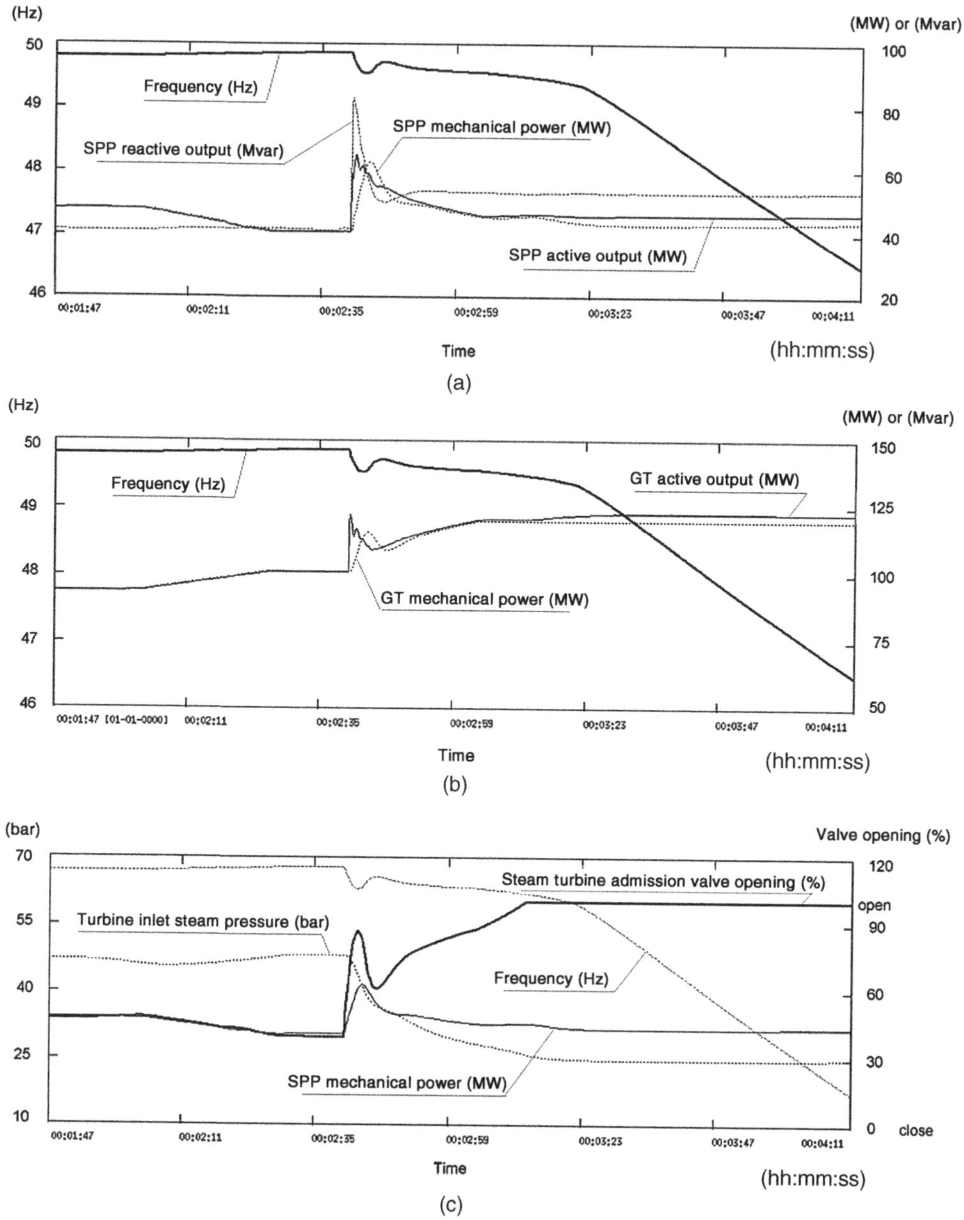

Figure 14.7. Unit trip on underfrequency after a 30-MW ballast load connection, with GT and SPP synchronized: (a) SPP outputs; (b) GT outputs; (c) turbine inlet steam pressure and HP valve opening.

start-up maneuver. The first transient depicted in Figure 14.6 is a ballast load pick-up of 30 MW, the same already presented in Figure 14.4 for the steam section alone. Figure 14.6a shows that the maximum speed variation is 0.55 Hz, while the GT bears 5 MW and the SPP bears the remaining 25 MW, reaching 27 MW after 2 min.

The comparison between the GT and SPP mechanical power shows that the SPP speed governor is faster than the GT one. As shown in Figure 14.6b, the steam turbine valve does not reach the full opening and the steam pressure drop is of about 17 bar. The conclusion is that the GT speed governor intervention aids the SPP to overcome the ballast load demand in the first seconds, thereby saving the boiler steam production.

However, it is important to note that the presence of the GT is not always sufficient to guarantee the black-start-up capability of the SPP. As an example, Figure 14.7 shows the failure of a black-start maneuver due to a 30-MW load connection when the GT power output is more than 80% of the rated value and the SPP steam pressure value is lower than 70% of the rated value. Figure 14.7b shows an important SPP pressure decrease and saturation of the HP steam turbine valve position. This condition demonstrates how critical is the black-start maneuver without the aid of a specifically designed automatic control system.

As shown also in Figure 14.7, one of the most critical problems during the restoration of a power system is the frequency control. It is necessary to avoid that load energization transients cause such significant frequency degradation involving generators protection intervention. In case the gas turbine and steam thermal unit perform contemporarily the restoration maneuver, the two relevant frequency regulators must be suitably coordinated. As a matter of fact, the GT is able to control the frequency error without significant delays, while the behavior of the SPP depends on the thermal inertia of the boiler, very high at low load. This can impair the black-start-up maneuver, and, therefore, as will be illustrated in the following paragraph, a load scheduler can certainly help.

14.3.1.3 Control System Modifications to Improve Black-Start-Up Capabilities. The basic function of the load scheduler is to maintain the gas turbine load as low as possible, to make it sustain the whole frequency transient caused by the energization of ballast loads. Only after the end of the frequency transients and if the boiler pressure is high enough, the load scheduler increases progressively the load request to the SPP section, contemporarily unloading the GT. Another important role of the load scheduler is to provide the appropriate moment for connecting additional loads, taking into account both the gas turbine power output level and the steam unit conditions.

The scheme of Figure 14.8 highlights the interactions between the load scheduler and the functional control blocks of both GT and SPP sections. The 320-MW steam unit (SPP) and the 120-MW gas turbine (GT) load programmers elaborate operator requests, transforming them in ramps with a given gradient (3 MW/min for SPP and 9 MW/min for GT). Being the primary frequency regulation a droop speed (proportional) control, a secondary frequency regulation is required to compensate the steady-state frequency error. This regulation consists of a frequency local integrator (FLI). The FLI device is associated with the GT section because, at very low loads, its dynamics provides a faster and more reliable response than the SPP one. When the local frequency error exceeds a given threshold, the FLI supersedes the GT load programmer and remains active, until the frequency error is adequately near to zero.

Moreover, as previously stated, to operate the GT in the part-load range during the whole restoration maneuver, so to allow an efficient action of the primary frequency governor and of the FLI, it is advisable to transfer progressively the load request from the GT section to the SPP one. The role of the simulated load scheduler is to modify the amount of power produced by the two machines, by increasing the SPP output while contemporarily decreasing the GT output, so to maintain the total required power constant. This indicates that the two units (GT and SPP) are now operated as a single unit. The load scheduler plays the same role as a tertiary frequency regulation: at each moment it decides the power to be produced by the SPP and GT units and translates it into a load request for

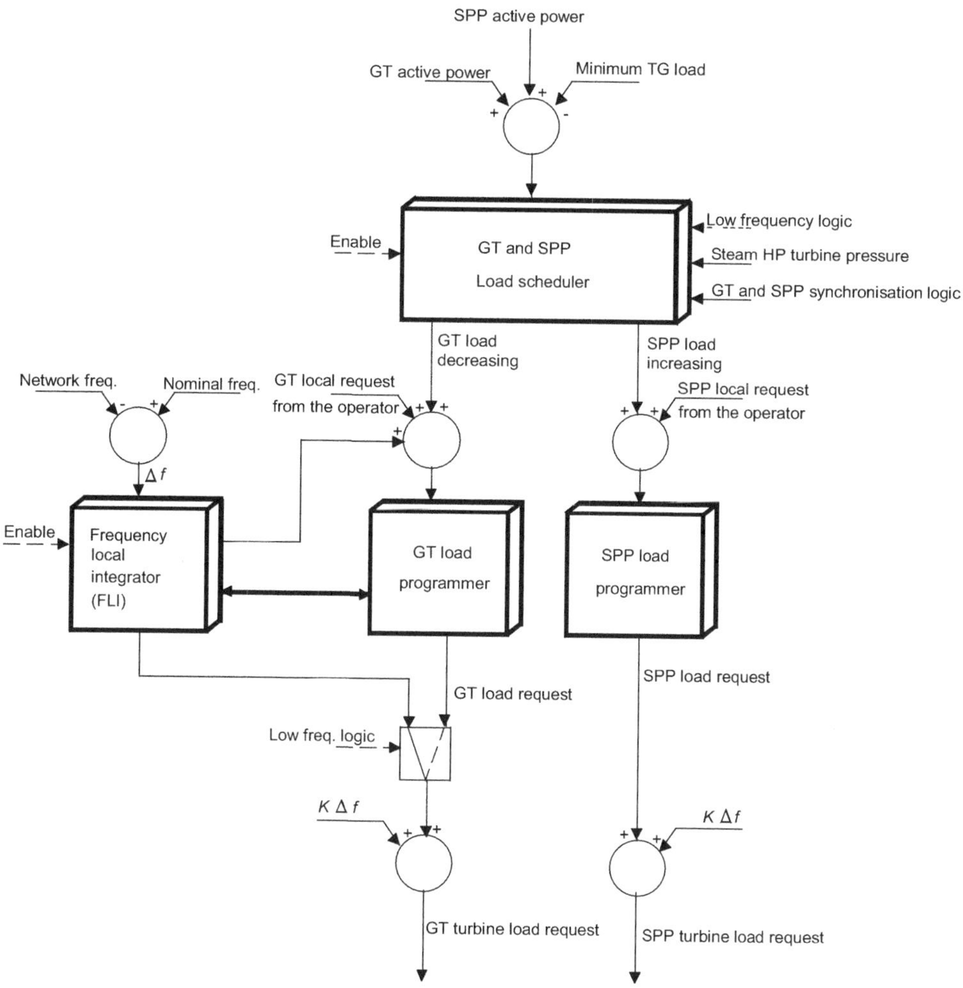

Figure 14.8. Scheme of the load scheduler for the black-start-up maneuver.

each unit. Moreover, the load scheduler halts the loading–unloading process when a low-frequency logical signal is active (that is, while FLI is operating) or when the steam HP turbine pressure is not high enough. As a consequence, another important role of the load scheduler is to advise the operator when the plant reaches the good conditions for additional load connection, that is when the GT power output is sufficiently low (less than 70% of the rated value), the FLI device is not active and the HP steam pressure of SPP is high enough (more than 80% of the rated value).

The behavior of this load scheduler is illustrated in Figure 14.9, in which the simulation of the pick-up of several loads by a 320-MW thermoelectric unit repowered with a 120-MW gas turbine is shown.

During every load pick-up, the steam section participates with the GT to the frequency regulation only during the first seconds. In the subsequent minutes, the FLI controller increases the fuel request of the gas turbine, thereby satisfying the total load demand. For this reason, the steam turbine does not increase its power, saving steam production in the boiler. When the

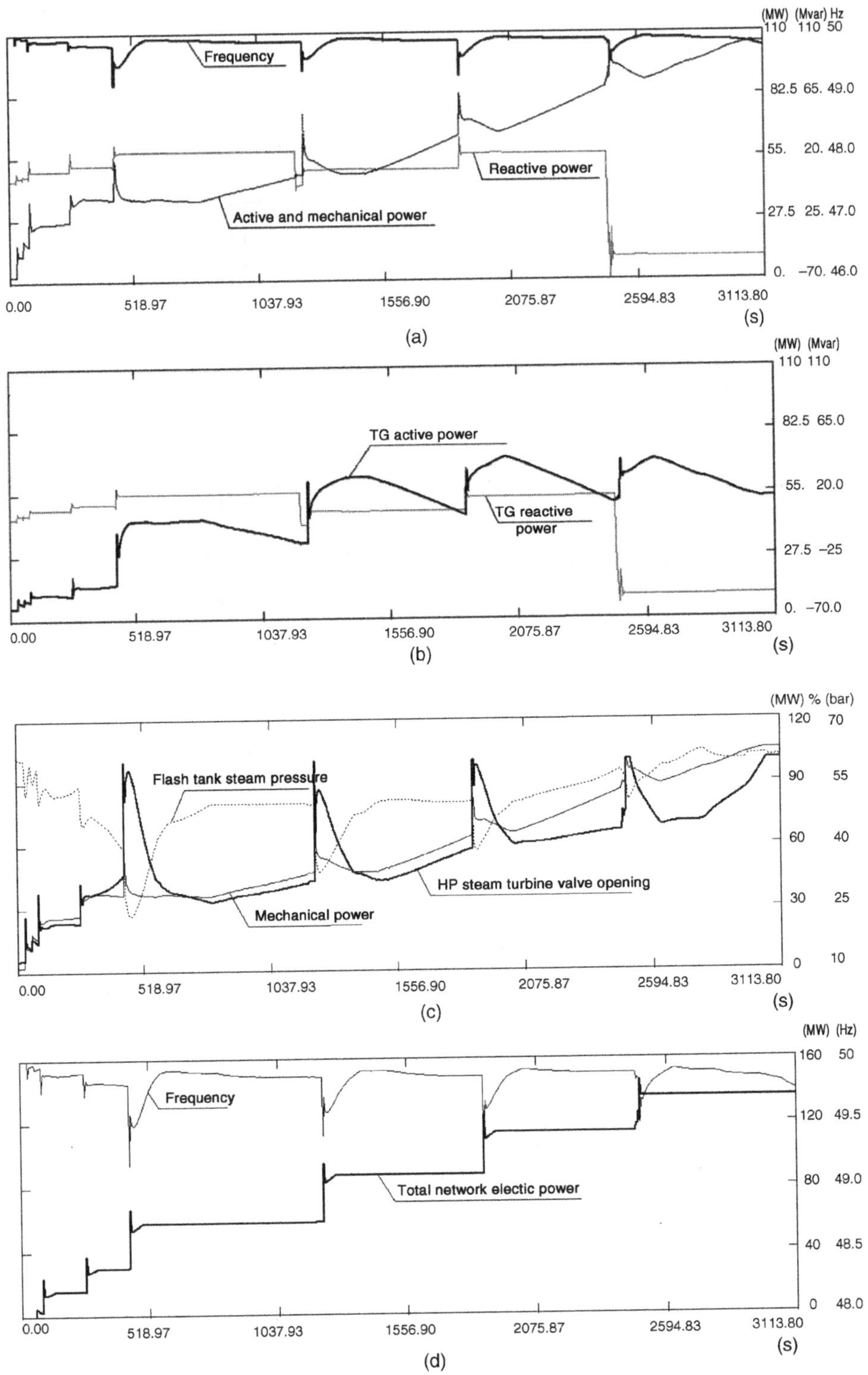

Figure 14.9. Simulation of the pick-up of several loads: (a) steam section outputs; (b) gas turbine outputs; (c) pressure in the flash tank of the SPP start-up steam flow circuit and HP valve opening; (d) network variables.

pressure attains 80% of its nominal value and when the frequency error is less than 0.1 Hz, the load scheduler progressively discharges the gas turbine and increases the load request of the steam turbine. This maneuver is performed at 2-MW/min slope. This simulation shows that the steam section reaches its minimum operating load of 110 MW in 15 min.

14.3.2 Black-Start-Up of a Combined-Cycle Power Plant

For the second example, we refer to the case of an 80-MW power plant composed of two gas turbines (GT1 and GT2) and a steam turbine unit (ST) in combined cycle, as illustrated in Figure 14.10.

The synchronous generators of the three units are connected to a 132-kV power plant substation through 15/132 kV step-up transformers. The power plant substation is linked, by means of a cable line, to the 132-kV substation that feeds 15 feeders of the local medium-voltage (15 kV) distribution network and also provides the connection with the external transmission network throughout circuit breaker BR1.

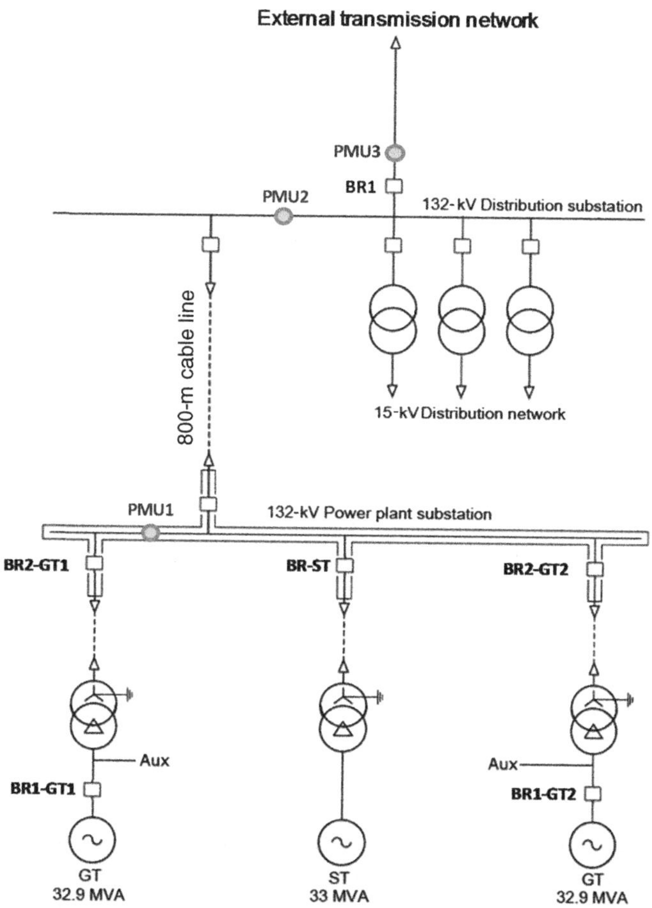

Figure 14.10. Scheme of the simulated power plant and the local network. (PMU1, PMU2, and PMU3 indicate the position of the three-phasor measurement units installed during the experimental tests described in Section 14.3.2.3.)

The two GTs are aeroderivative industrial RB211 packages and are equipped with a speed governor, characterized by a 5% droop and an acceleration limiter and a local frequency integrator (LFI). Each turbine delivers 32-MW mechanical power (at 4850-rpm rated speed and 15°C reference temperature) to a 50-Hz 32.9 MVA (at 40°C cooling) synchronous generator and 93 kg/s exhaust mass flow at 509°C to a heat recovery steam generator (HRSG) that generates 10.5 kg/s high pressure (HP) steam flow at 52 bar and 487°C, together with 2.5 kg/s low pressure (LP) steam flow at 6.5 bar and 231°C. The main control valve of the 22-MW ST can be regulated in three different operation modes, namely: (i) no-load speed control, (ii) HP pressure control to a constant value, and (iii) power and speed regulation. The no-load speed control mode is used at the start-up and synchronizing phases, while the other two modes used in normal operating conditions are the pressure or power–speed regulation.

The link between each GT and the ST to its generator is realized by means of a gearbox and each generator is equipped with a brushless exciter.

The analysis is carried out by means of a computer simulator aimed at reproducing the islanding and black-start-up energization transients of the combined plant and the local distribution network with the relevant loads, as described in [37,38]. The main characteristics of the implemented models are summarized in Section 14.4.2.

The following sections describe the analysis of the black-start capabilities and the comparison between different islanding strategies. Section 14.3.2.1 focuses on the assessment of the feasibility of the energization path from a GT, with autonomous black-start capabilities to the local distribution network loads. Section 14.3.2.2 shows that the contribution to the frequency regulation and load balance provided by the ST is effective for successful islanding maneuver performed by opening circuit breaker BR1 at large export power levels to the external transmission network, in addition to the frequency regulation provided by the GTs. Section 14.3.2.3 describes some experimental results obtained during experimental intentional islanding and reconnection tests by using specifically developed phasor measurement units [39].

14.3.2.1 Analysis of the Energization Maneuvers. The simulation results of Figure 14.11 show the transients during the subsequent energization of the components along the path from unit GT1, assumed with autonomous black-start capabilities to the distribution network loads after a blackout. In particular, Figure 14.11a shows the generator armature currents during the energization of step-up transformer TR-GT1.

Figure 14.11b shows the line to ground transient voltages at the 132-kV distribution substation during the energization of the 800-m cable line. Figure 14.11c shows the GT generator armature currents during the energization of distribution transformer TR-D1 feeding the aggregate load of one of the distribution feeders.

Figure 14.12 compares the GT1 generator active and reactive (*P*–*Q*) power trajectory during the TR-GT1 energization with the capability limits provided by the manufacturer. As power transformer residual fluxes decay very slowly and transformers can retain high levels of residual flux for long periods [35], the comparison is carried out for two values of residual flux, namely 0 and 0.8 pu.

To verify the minimum intervention current of BR-GT1 breaker relay during TR-GT1 energization, statistical studies can help. This can be performed by making reference to the random nature of the closing time of the generator breaker poles. Closing time T_c of each pole is assumed a random variable characterized by a Gaussian distribution with typical mean value $\mu_{Tc} = 36$ ms and standard deviation $\sigma_{Tc} = 0.75$ ms [40–42]. An additional time delay T_d, uniformly distributed along a 20-ms time window (50 Hz), is added to T_c to take into account the random instant within the period of the steady-state voltage waveform in which the breaker closes.

Figure 14.13 shows the cumulative distribution function of the maximum root mean square (RMS) values of the GT1 generator currents obtained by 200 simulations of the TR-GT1 energization with reference to both 0 and 0.8 pu residual flux [43].

14.3.2.2 Analysis of the Islanding Maneuvers. The analysis of the capability of system islanding is usual in power restoration studies [44]. The success of the islanding

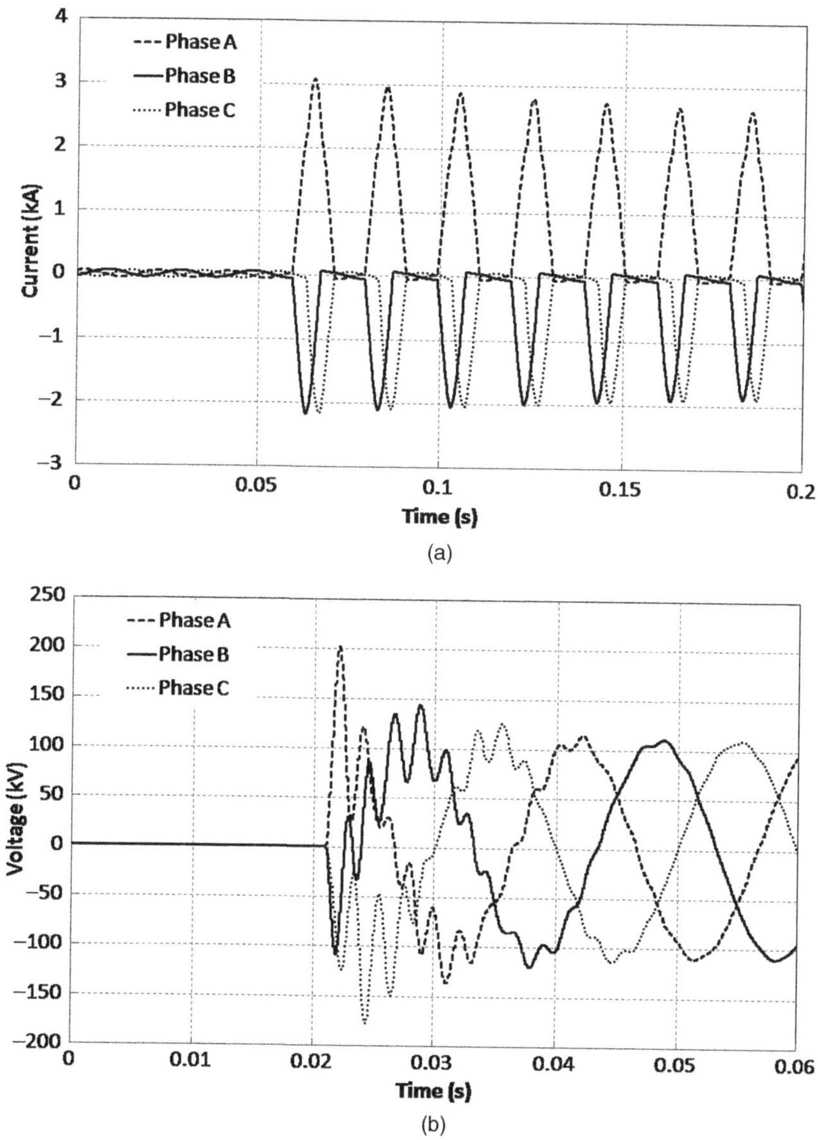

Figure 14.11. Transients during network component energization: (a) generator currents during TR-GT1 energization; (b) 132-kV distribution substation phase voltages during cable line energization; (c) generator currents during the energization of TR-D1 feeding a 0.58-MW and 0.28-MVAr load.

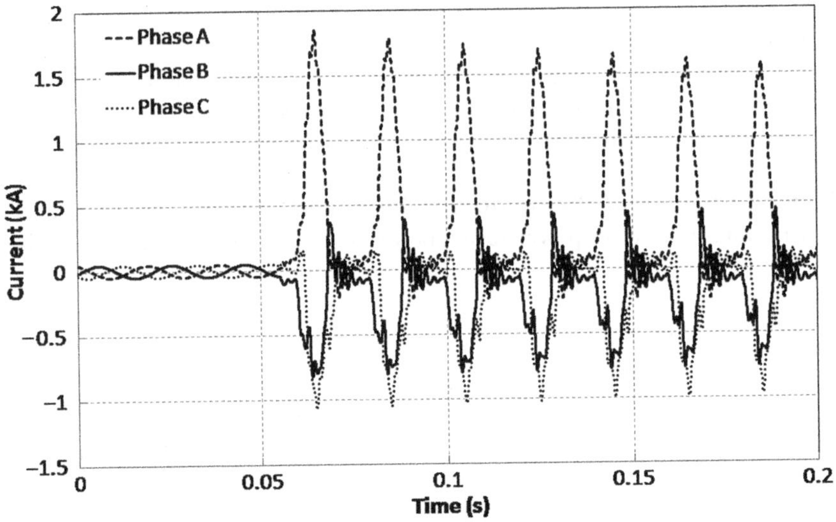

(c)

Figure 14.11. *(Continued)*

maneuver is related to the balance between the power production and local network load. Due to the structure and characteristics of the considered system, initial scenarios of both import and export power flows to the transmission network should be taken into consideration.

The islanding maneuver for the case of an initial import scenario requires the fast reduction of the local network load. On the other hand, an initial export scenario, which is

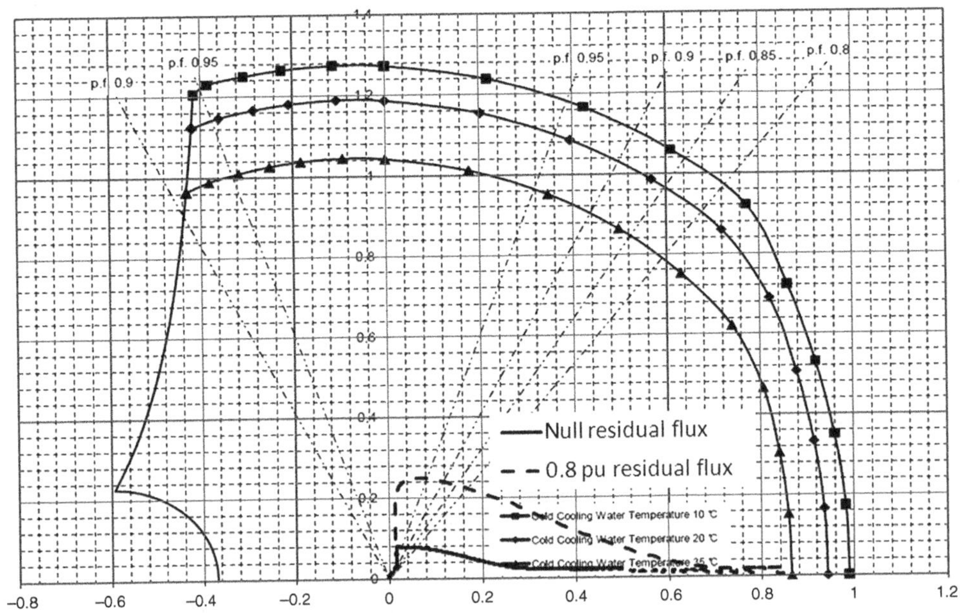

Figure 14.12. Comparison between GT1 *P–Q* trajectories during the TR-GT1 energization and the synchronous machine capability limits.

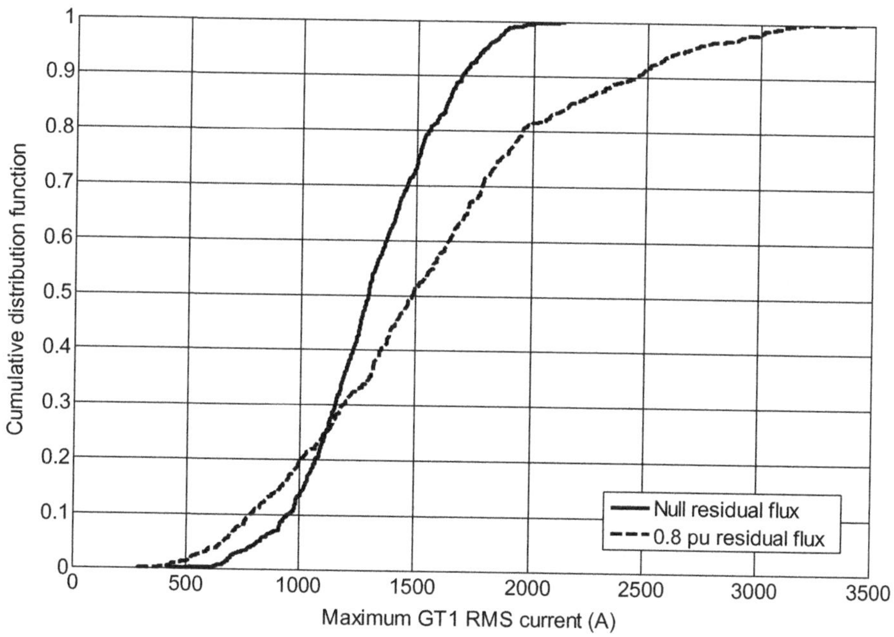

Figure 14.13. Cumulative distribution functions of the maximum GT1 generator current RMS values during the TR-GT1 energization, with and without the presence of the residual flux.

the case analyzed here, requires a proper control strategy of the three production units to preserve stable and secure operating conditions of the system.

For the case in which the system is exporting power to the transmission grid, the most critical constraints are

- overspeed limits of the GT and ST units (e.g. 110%),
- GT combustion constraint,
- HRSG overpressure limits.

The frequency regulation is manly performed by the GTs. At rated power output, the dry low emission (DLE) combustion system of the GTs is controlled to minimize NOx emission. The combustor control switches to the conventional combustion mode when the GT load is lower than around 55–60% (15°C reference temperature) [45]. To preserve the combustion stability, such a transition may be performed only at low GT output change rate and therefore it represents a critical constraint for the system frequency control when the islanding maneuver causes a significant power surplus. If the ST unit also contributes to load balancing and frequency regulation, the HRSG overpressures and HP collector pressure p_{HP} must be limited by means of the bypass valve opening.

To successfully perform the islanding maneuver in the case of a rather large power export, two basic techniques are compared here: shutdown of one of the two GTs or the fast reduction of the ST control valve opening. In particular, for the case of a power export equal to half the power plant generating capacity, namely 40 MW, we compare the results obtained for three different operation modes of GTs and ST units

- ST pressure controlled, GT1 frequency controlled with LFI and GT2 frequency controlled without LFI;

- ST pressure controlled, GT1 frequency controlled with LFI and GT2 shutdown;
- GT1 frequency controlled with LFI, GT2 frequency controlled without LFI and fast reduction of ST output power to the lower limit.

The GT2 shutdown is achieved by the load reference fast removal and by the subsequent action of the reverse active power relay with a 10-s delay. The ST output fast reduction is obtained by setting the reference of the power-controlled ST to the minimum value (3.4 MW). The closing rate limit of the ST control valve servomotor is assumed equal to −1 p.u./s. An even faster ST action could be achieved by the intervention of a load drop anticipator (LDA) relay.

OPERATING MODE **1.** The islanding maneuver transient when the system is exporting 40 MW is calculated by assuming that the initial GT outputs equal to 30.5 MW and ST outputs equal to 22.7 MW. The transient simulation is started by the opening of BR1 at 1 s from the reference time. (The automatic power management system may command the islanding maneuver within 0.2 s after a violation of the transmission network frequency decay limit).

Figure 14.14 shows the transients of the GTs and ST mechanical powers P_m and the corresponding active power outputs P_e.

ST is controlled to maintain p_{HP} at the rated value. GT1 and GT2 participate, instead, in the frequency regulation with a 5% droop. The islanding maneuver is assumed to activate the GT1 FLI device that integrates the frequency error to bring it back to the 50-Hz rated value.

As shown in Figure 14.14, the GT combustion constraint is violated. This is due to the fact that the islanding maneuver is performed with all the three units in operation with the mechanical output of the pressure-controlled ST that does not significantly change, as it follows the slow pressure dynamics shown in Figure 14.15.

As the HP collector pressure is lower than the rated value, the bypass control valves at the HRSG outputs stay closed. Therefore, the frequency is controlled only by the fast reduction of the GTs output, so as to compensate the significant initial positive power imbalance. The different behavior of the two GTs is due to the fact that only the LFI of GT1 is active, while GT2 is controlled only by its droop speed governor.

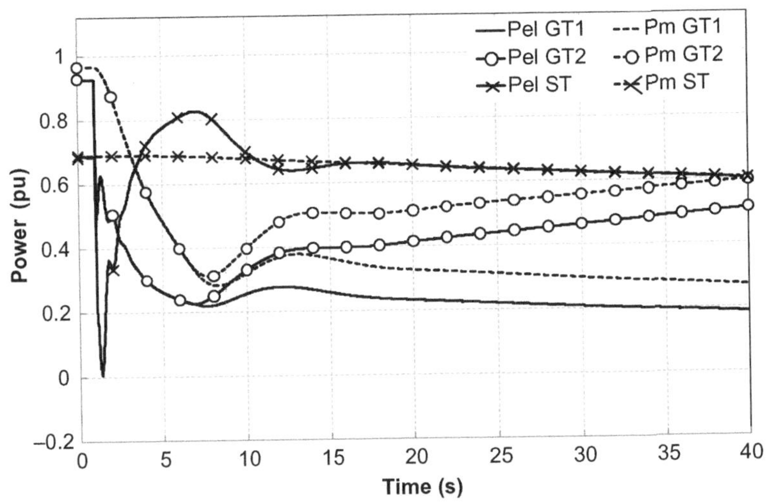

Figure 14.14. GTs and ST mechanical powers and active power outputs (in pu of the corresponding synchronous generator rated power) during islanding maneuver for the case of 40-MW power flow export and operating mode 1.

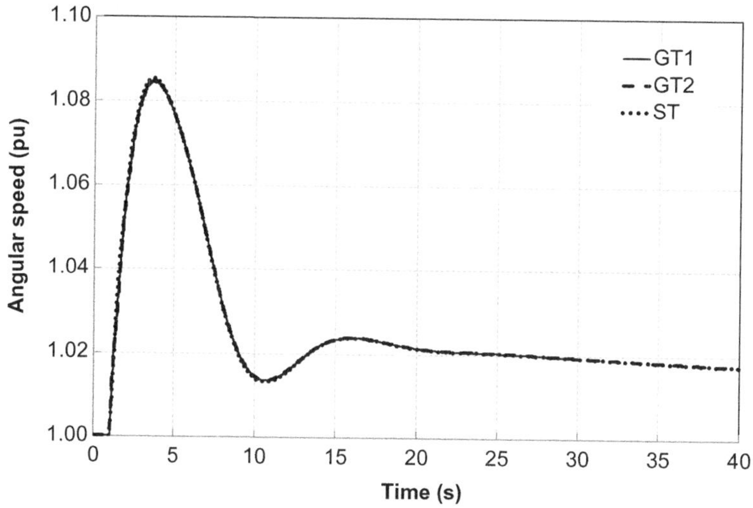

Figure 14.15. GTs and ST rotor angular speeds (in pu of the rated value) during islanding maneuver for the case of 40-MW power flow export and operating mode 1.

The transient does not reach a new steady state in the considered 40-s time interval after the islanding maneuver due to the action of the GT1 ILF and the time constants related to HRSGs and steam collector dynamics which cause the pressure slow droop of the steam collector pressure (Figure 14.16) and the corresponding ST output (Figure 14.14).

OPERATING MODES 2 AND 3. The islanding maneuver transient is calculated assuming the initial steady state of the simulation of 1. Both the commands of GT2 shutdown and ST output fast reduction are simultaneous to the BR1 opening.

Figures 14.17 and 14.18 show the transients of GT and ST mechanical powers P_m and active power outputs P_e obtained by applying operation modes 2 and 3, respectively.

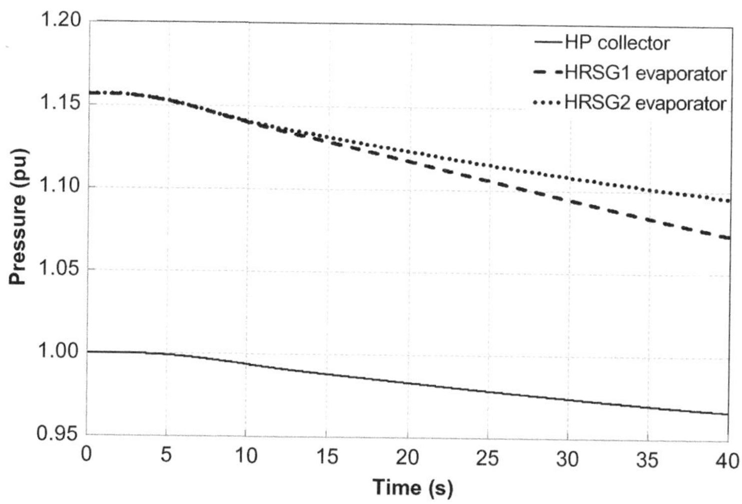

Figure 14.16. Pressures at the HP collector and at the HRSG evaporators (in pu of the HP main steam collector rated pressure, namely 48 bar) during islanding maneuver for the case of 40-MW power flow export and operating mode 1.

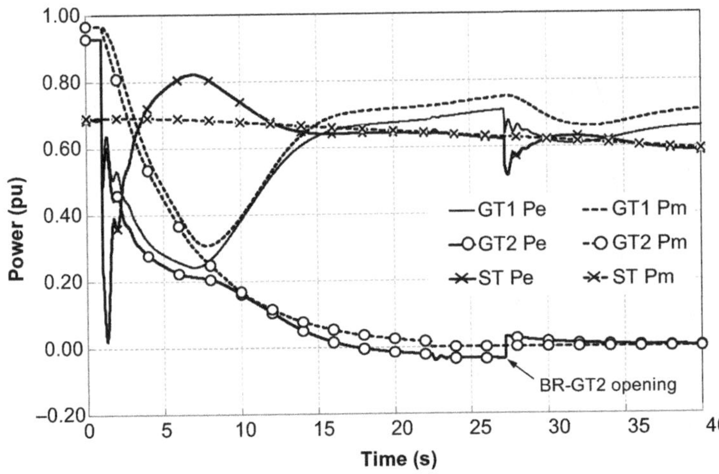

Figure 14.17. GTs and ST mechanical powers and active power outputs during islanding maneuver for the case of 40-MW power flow export and operating mode 2.

Figure 14.17 shows that with the GT2 shutdown—operating mode 2—the GT1 mechanical power transient violates the combustion constraint. The GT2 shutdown is realized by the fast closing of the fuel valve in correspondence with the BR1 opening, then the relevant mechanical power decreases and, at 17 s, the unit starts to absorb active power from the network. The reverse power relay, used to protect the turbine against motoring, operates by opening generator breaker BR-GT2 at 27 s. Within the first 7–8 s after the BR1 opening, the GT2 active output reduction is similar to the one that would be forced by the speed governor (Figure 14.14) with, therefore, limited additional benefits for the combustion constraints of GT1.

Figure 14.18 shows that by an ST output fast reduction to the minimum limit—operating mode 3—both the GT1 and GT2 mechanical power transients do not violate the

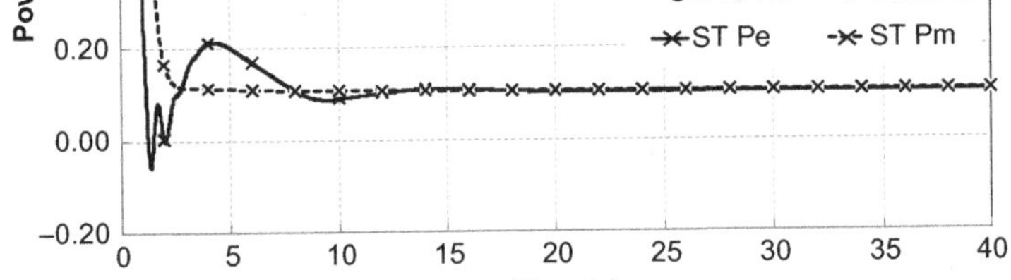

Figure 14.18. GTs and ST mechanical powers and active power outputs during islanding maneuver for the case of 40-MW power flow export and operating mode 3.

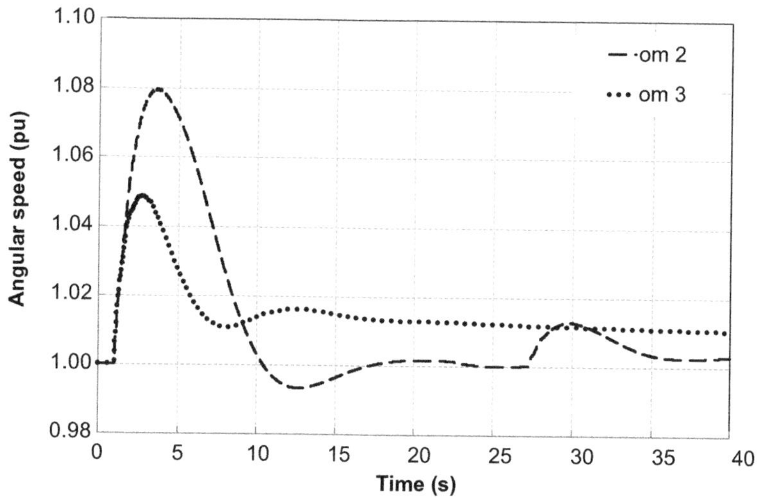

Figure 14.19. GT1 shaft angular speed transients, obtained by using operating modes 2 and 3.

combustion constraint. GT1 mechanical power crosses the combustion constraint only at low change rate by the action of the LFI device. This behavior may result in a successful islanding maneuver.

Figure 14.19 shows that by applying operating mode 3 the frequency regulation also improves. However, the fast reduction of the ST power requires that the pressure of the steam collector is regulated by the action of the bypass control valves located at the HRSG outputs, as shown in Figure 14.20.

The bypass valve opening rate limit is assumed equal to 0.5 p.u./s. By using operating modes 1 and 2, the bypass valves stay closed.

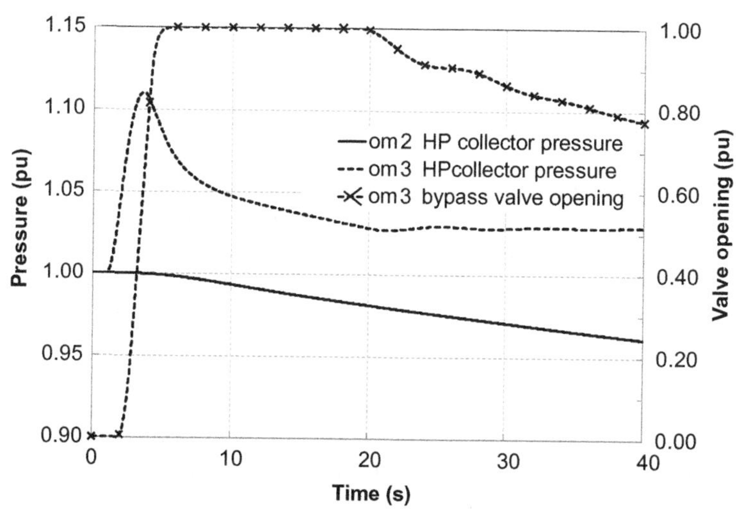

Figure 14.20. Steam collector pressure transients, obtained by using operating modes 2 and 3, and bypass valve opening for the case of mode 3.

14.3.2.3 Description of Some Islanding Tests and Obtained Experimental Results. This section shows the experimental results obtained during one of the islanding tests carried out at a power system with the characteristics described in the previous sections. Additional results have been presented in [39].

Three PMUs have been installed in the system as indicated in Figure 14.10: one (PMU1) at the power plant substation and two (PMU2 and PMU3) at two sides of circuit breaker BR1. The characteristics of the specific developed PMUs are described in [39,46]. As proved by the laboratory experimental characterization of the developed PMUs, very low values of total vector error (TVE) and both root mean square (RMS) and phase errors of the synchrophasor estimates are obtained, as required by their application in distribution networks. Moreover, a substantial independence of the PMU performances from the distortion level of the input signal has been obtained.

The islanding test started at 6:09 a.m., after other tests, when unit TG2 output level was around 28.46 MW, while TG1 output was 29.81 MW and the export to the external grid was 30.42 MW.

The power plant is equipped with a computer-based power management system (PMS) that: (a) operates circuit breaker BR1 for the disconnection of the network from the external grid, (b) communicates the so-called Load Droop Anticipator command to the ST control system in case of islanding manoeuvres accomplished at rather large power exported levels to the transmission network, (c) operates a disconnection of feeders following a predefined priority list to guarantee the load balance, (d) selects the operation control mode of the two gas turbines (master and slave) for the frequency regulation of the network in islanded conditions, and (e) controls the power plant units to allow a reliable reconnection manoeuvre of the network to the external grid.

During the considered tests, the power plant PMS was set so to consider TG2 as the master unit. After the islanding maneuver at 6:09, therefore, TG1 was automatically disconnected by the PMS together with two light-loaded distribution feeders. Unit TG2 remained in stable operation with an output at first of about 28.46 MW, decreased to 25.78 MW at 6:11.

Figure 14.21 shows the results provided by the installed PMUs during the islanding maneuver.

To provide a more concise presentation of the results, the figure makes reference to the positive-sequence components of bus voltages only. We have verified that negative-sequence and zero-sequence components are negligible.

Figure 14.21a shows both the angle deviation between the positive-sequence components measured by PMU1 and by PMU2 and the analogous angle deviation between the positive-sequence components measured by PMU2 and PMU3. Figure 14.21b shows the trend of the magnitudes of the positive-sequence components measured by PMU1, PMU2, and PMU3 and Figure 14.21c the corresponding frequency transients.

At first, the frequency of the islanded network decreases with respect to the frequency of the external grid, as shown in Figure 14.21c. The TG2 speed governor reacts by increasing the power output as shown in Figure 14.21a by increasing the angle deviation between PMU1 and PMU2 phasors. Then, the LFI action stabilizes the frequency of the islanded network at the predetermined value of about 50.1 Hz.

The phase angle difference between PMU2 and PMU3 shown in Figure 14.21a is related to the frequency transient shown in Figure 14.21c. Indeed, as the frequency difference between the islanded and external networks is negative (Figure 14.21c between 22 and 30 s), the relevant phase angle difference is decreasing. As this frequency difference becomes positive (Figure 14.21c between 30 and 60 s) the relevant phase angle difference starts increasing.

Figure 14.21c also shows that the frequency transients measured by the developed PMUs are in good agreement with those recorded by the power plant SCADA, demonstrating the PMU ability to monitor transients also characterized by non-negligible deviations from the nominal frequency. In this framework, the information provided by the PMUs appears to be of great help for the development of improved control and management systems aimed at making these maneuvers more straightforward and reliable.

It is worth noting that the islanding maneuver shown in Figure 14.21 makes reference to severe operation conditions of the power plant in which, the contemporary disconnection

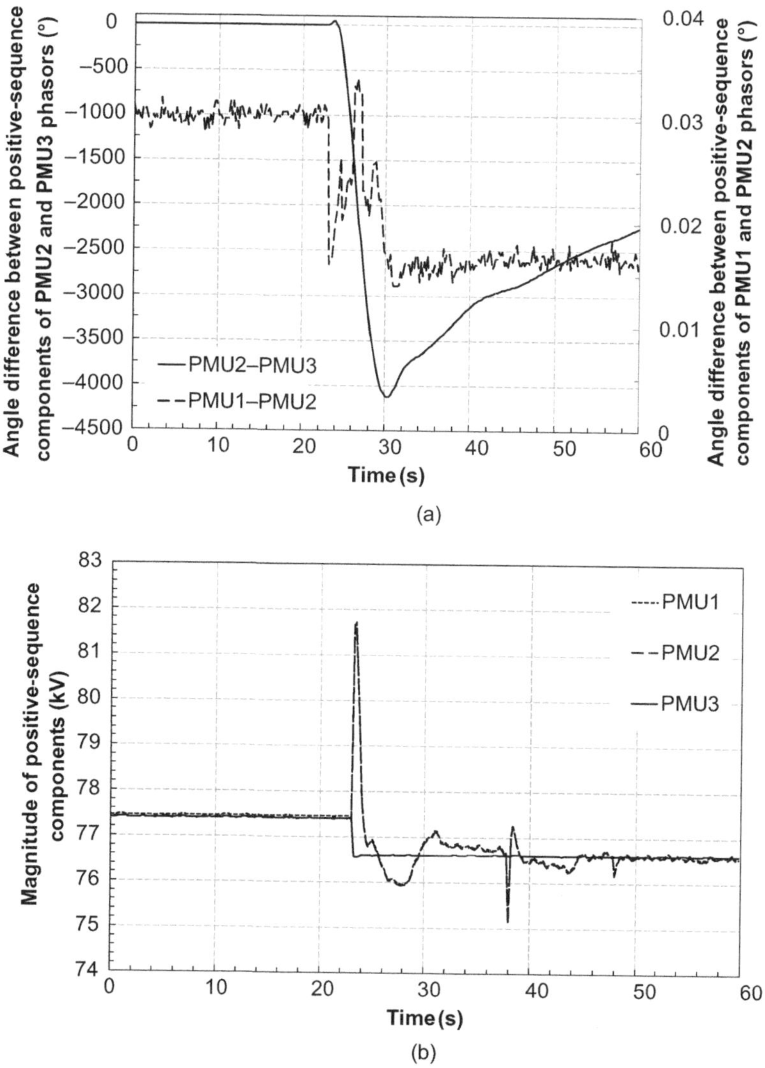

Figure 14.21. PMU data collected during the islanding maneuver: (a) angle deviations between positive-sequence components of voltage phasors; (b) transient of the positive-sequence component magnitudes; (c) frequency transients.

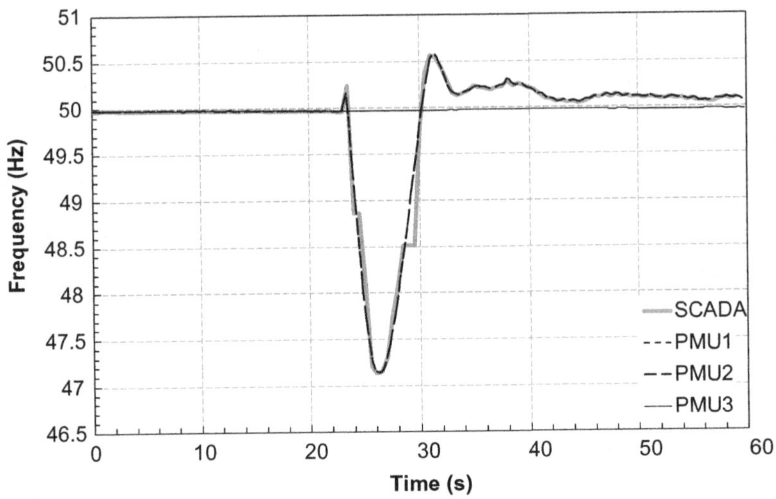

(c)

Figure 14.21. (*Continued*)

of the TG1 together with two distribution feeders has resulted into a frequency transient characterized by a minimum frequency of 47.138 Hz very close to the tripping threshold of the unit minimum frequency relay.

At 6:17, the reconnection maneuver of the islanded network to the external grid was performed by the closing of circuit breaker BR1 under the reliable PMS control, as shown also by the PMU results of Figure 14.22.

14.4 DESCRIPTION OF COMPUTER SIMULATORS

The computer results shown in Section 14.3 have been obtained by two simulators, whose models are described in this section: the first simulator refers to the steam power plant (SPP) unit repowered with a gas turbine (GT) dealt with in Section 14.3.1; the second one refers to the case of a combined-cycle thermal power plant equipped with two gas turbines (GT1 and GT2), each with a heat recovery steam generator (HRSG1 and HRSG2) that feed a steam turbine (ST), dealt with in Section 14.3.2. Additional details are reported in [35–38], respectively.

14.4.1 Simulator of a Steam Group Repowered with a Gas Turbine

The simulator has been built using the modular code Lego [47], developed at the Centro Ricerca Automatica of ENEL. The simulator refers to a group of the an Italian power station which consists of four 320-MW units with once-through universal-pressure boilers burning fuel oil, each topped with a 120-MW gas turbine (GT) unit. The following paragraphs illustrate the main features of the dynamic mathematical models developed for the simulator, namely the models of the gas turbine, the boiler, the steam turbine with their relevant regulations, the generators, the power station auxiliaries, and the model of part of the transmission network around the station.

14.4.1.1 Gas Turbine Model and Its Validation. The model of the 120-MW gas turbine reproduces essentially speed and load regulation, fuel feeding combustor, and air compression dynamics [48,49].

Figure 14.23 represents the simplified block diagram for the single-shaft gas turbine, together with its control and fuel systems. This figure also shows the parameters also obtained by the comparison between computer results and field measurements.

14.4.1.2 Steam Section Modeling and Its Validation. During the black-start-up maneuver, the most important phenomena involving boiler dynamics are those concerning frequency variations. In the start-up phase, frequency regulation mainly concerns high-pressure

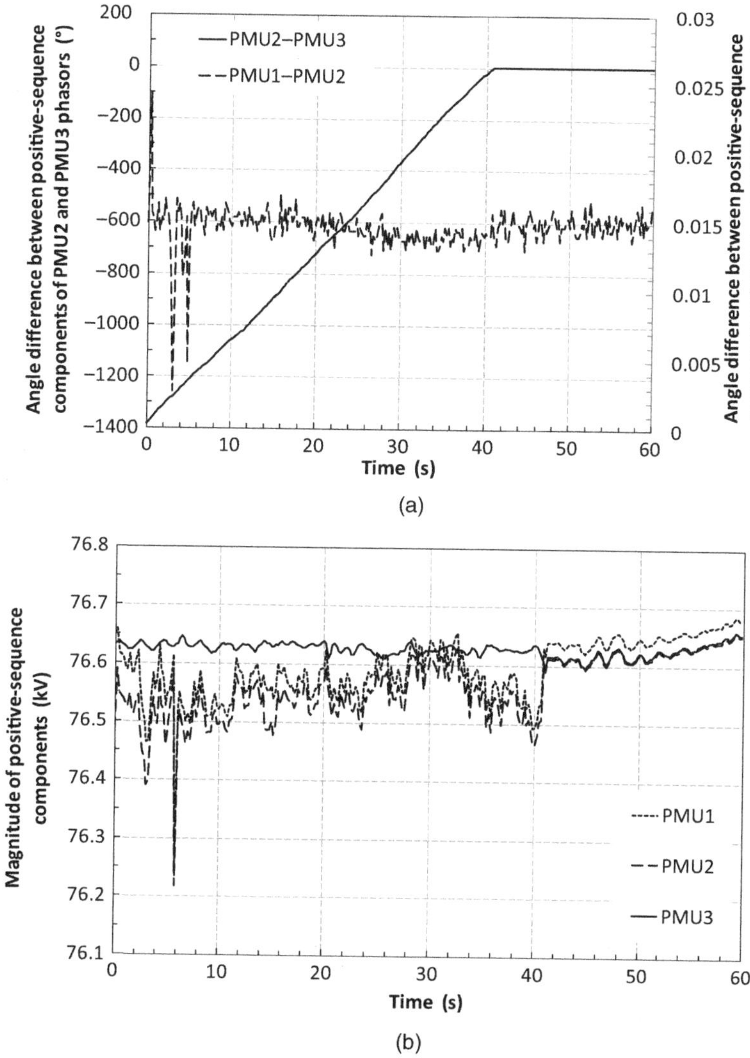

Figure 14.22. PMU data collected during the reconnection maneuver on August 14 at 6:17 a.m.: (a) angle deviations between positive-sequence components of voltage phasors; (b) transient of the positive-sequence component magnitudes; (c) frequency transients.

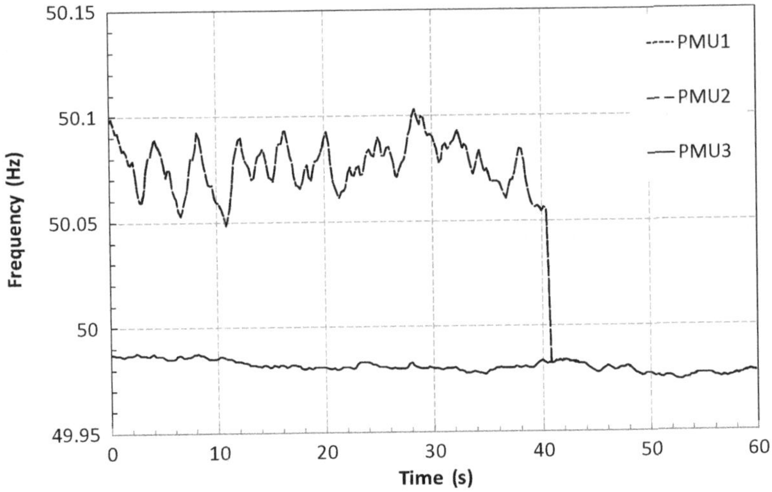

(c)

Figure 14.22. (*Continued*)

turbine steam admission valves (hereafter simply called HP valves), while the intercept valve is kept completely open.

The simplified models of the SPP and the values of their parameters have been inferred by comparison with more detailed models of a simulator developed by ENEL and some experimental measurements. The simplified models take into account the mass and momentum conservation equations as well as pressure regulation dynamics. Temperature regulation effects, instead, have been neglected, as they involve time constants much greater than those relevant to pressure dynamics, as shown by the results obtained by using the more detailed simulator.

The scheme of the normal and start-up steam flow circuit of the SPP is shown in Figure 14.24 and the simplified block diagram of the boiler and turbine models introduced in the simulator are represented in Figures 14.25 and 14.26, respectively. Figure 14.27 shows the block diagram of the start-up mode control system.

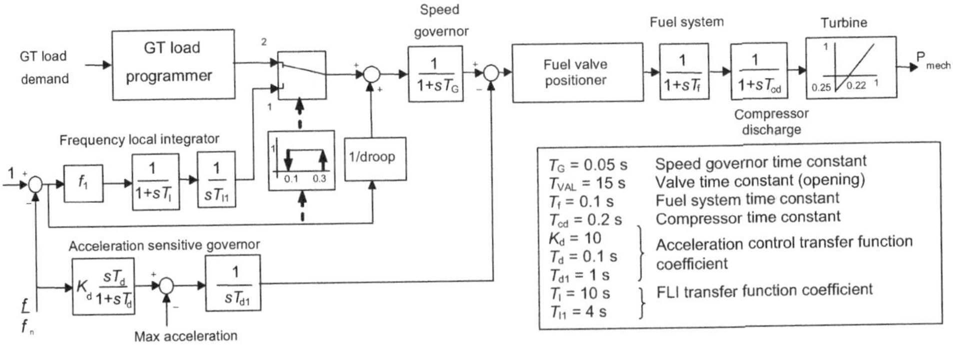

Figure 14.23. Simplified single-shaft gas turbine block diagram.

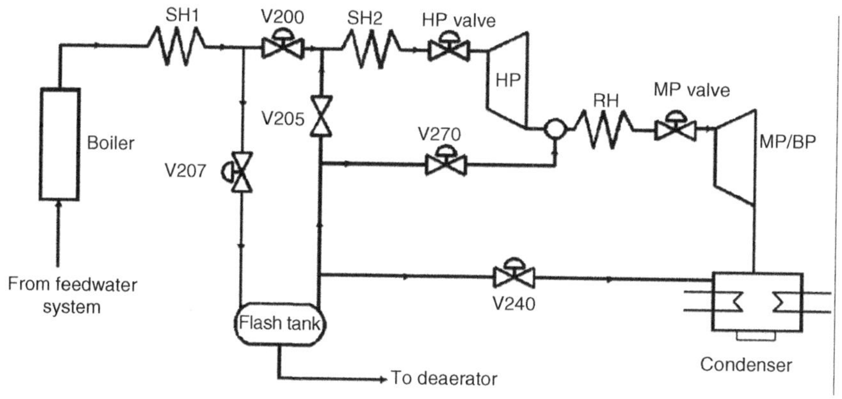

Figure 14.24. Scheme of the start-up steam flow circuit of a once-through boiler.

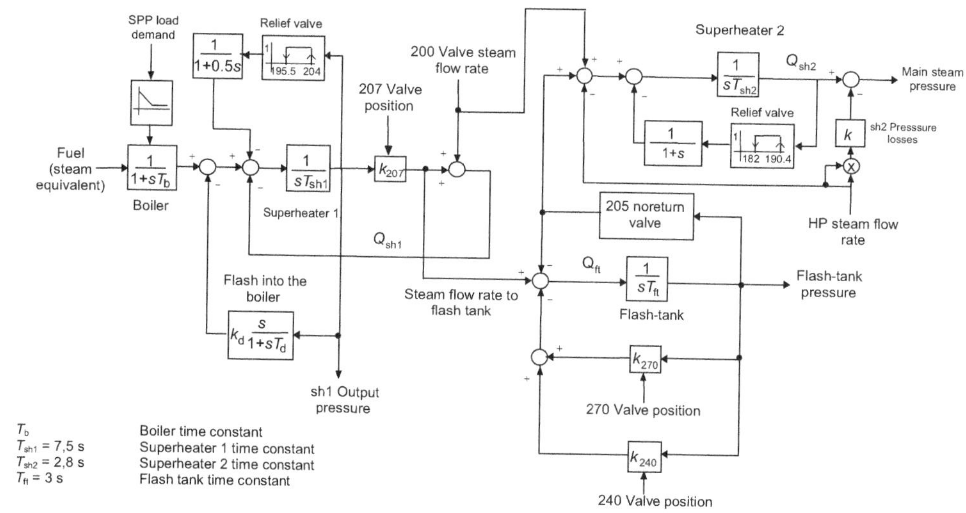

Figure 14.25. Simplified once-through boiler block diagram with start-up circuit.

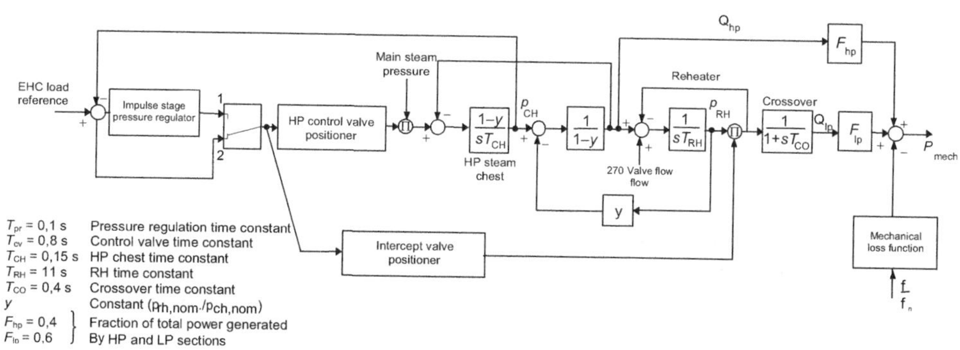

Figure 14.26. Simplified steam turbine block diagram.

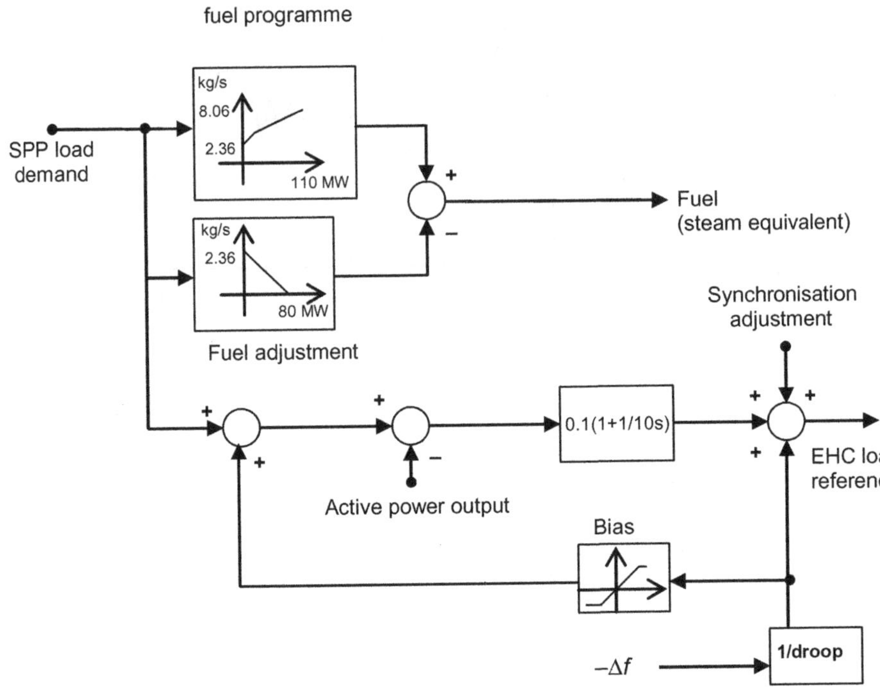

Figure 14.27. Block diagram of the start-up control mode.

14.4.2 Simulator of a Combined-Cycle Power Plant

The simulator used for the second case study dealt with in Section 14.3.2 refers to a combined heat and power (CHP) station connected to a town local network. The simulator has been developed by using the EMTP-RV simulation environment described in [50–52].

Figure 14.28 shows the scheme of the model that represents the dynamic behavior of the two GTs, corresponding two HRSGs feeding the high-pressure (HP) main steam collector, ST, and their control systems. The simulator also includes the synchronous generator models along with their exciters and automatic voltage regulators (AVRs). The model of the electrical network also included in the simulator represents the step-up unit transformers, the cable link between the power plant substation and the distribution substation, connection to the external transmission network, and local distribution network loads. Figure 14.29 also shows the main variables connecting the various parts of the power plant model.

The EMTP-RV synchronous machine modeling method is described in [53]. Each of the three synchronous generators has two inputs: excitation voltage E_f and mechanical power Pm. The mechanical power values are provided by the models of the GTs (Pm_{GT}) and the one of the STs (Pm_{ST}).

The model used for the two GTs is based on a transfer function that represents the dynamic link between fuel flow rate and output mechanical power and includes the fuel metering valve (FMV) dynamic and a speed governor. The used GT model is illustrated in Figure 14.29. The GT dynamics is represented by 4 poles-4 zeros transfer function that

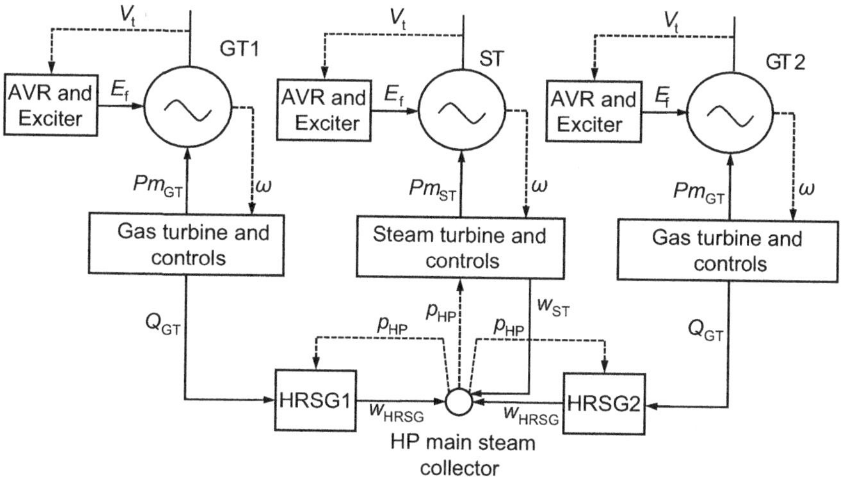

Figure 14.28. Block diagram of the simulated power plant and main variables connecting the various blocks: V_t terminal voltage of the synchronous generators; E_f field voltage; ω angular speed of the generators rotors; Pm_{GT} mechanical output power of a GT; Pm_{ST} mechanical output power of the ST; Q_{GT} thermal power provided by the GT exhaust gas; w_{HRSG} output steam mass flow rate of an HRSG; w_{ST} inlet steam mass flow to the ST; p_{HP} pressure in the HP main steam collector.

represent the dynamic link between fuel flow rate and output mechanical power

$$\mathrm{GT}(s) = K_{\mathrm{GT}} \frac{(\tau_{z1}s + 1)}{(\tau_{p1}s + 1)} \cdot \frac{(\tau_{z2}s + 1)}{(\tau_{p2}s + 1)} \cdot \frac{(\tau_{z31}s^2 + \tau_{z32}s + 1)}{(\tau_{p31}s^2 + \tau_{p32}s + 1)} \tag{14.1}$$

A feedback algebraic lookup table block determines the required correction at partial loads. The FMV dynamics is represented by a first-order transfer function with time constant T_{FMV} equal to 0.1 s. The GT speed governor is characterized by a 5% droop, an acceleration limiter. The droop of the speed governor is assumed equal to 5%. The model also includes an acceleration limiter and a local frequency integrator (LFI), not shown in this figure.

The GT is connected to the rotor of the synchronous machine by means of a mechanical gearbox with a transmission ratio equal to 1500/4850. The GT mechanical

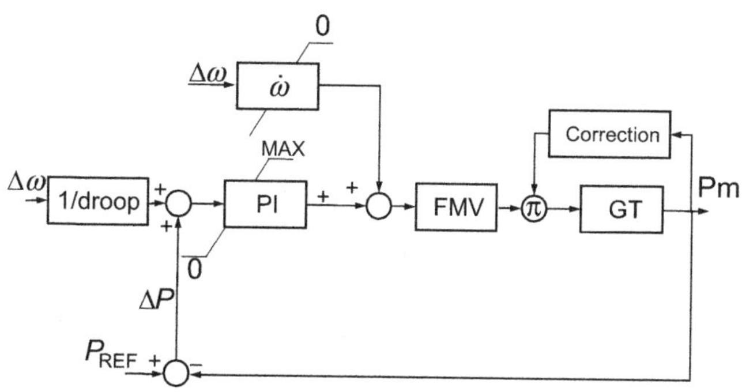

Figure 14.29. Model of the GT and its speed governor.

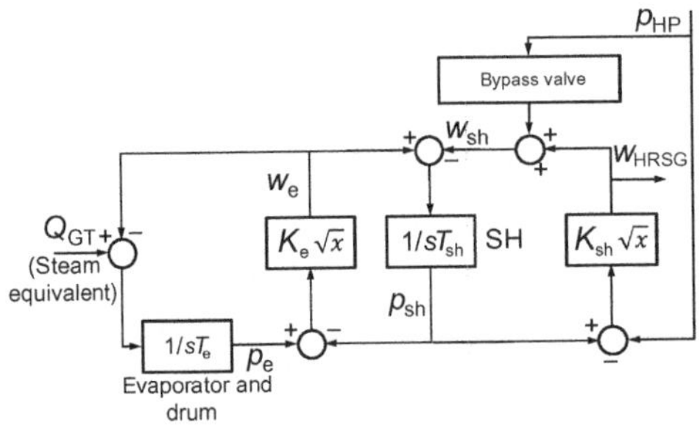

Figure 14.30. Model of the HRSG.

drive train is represented by a two-mass model: the first represents the inertia of the low-speed GT shaft and the gearbox, while the second represents the generator rotor. The two masses are linked by an elastic connection and an absolute speed selfdamping coefficient is used to take into account the power losses associated with the gearbox (1.2 MW at synchronous speed of 1500 rpm).

The thermal power Q_{GT} provided by each GT to the HRSG is assumed to be a linear function of GT mechanical power Pm_{GT} (a first approximation also suggested in [54]). As shown in Figure 14.30, Q_{GT} and pressure p_{HP} in the HP main steam collector are the inputs of the HRSG model.

Such a model, which represents the HP section equipped with a bypass control valve at the HRSG output, is adapted from the one proposed in [55], relevant to the HP section with a bypass control valve at the HRSG output. The model is based on the following main assumptions: fast feedwater adjustments, negligible effects of temperature control and water flows to the attemperators, and constant enthalpy value of the steam at the ST inlet.

Figure 14.30 shows the structure of the implemented HP section HRSG model, which takes into account the evaporator time constant T_e for the calculation of the pressure p_e in the drum according to the energy balance equation, a time constant T_{sh} relevant to the superheaters (SH) storage capacity. The steam flow rates between drum and SH and between SH and the collector are determined from the pressure drop relationship with flow rate being proportional to square root of pressure drop.

Concerning p_{HP}, it is calculated from the difference between the steam mass flow rate at the HRSG outputs (w_{HRSG}) and the steam turbine inlet mass flow rate (w_{ST}) through a transfer function that takes into account the time constant associated with the steam storage capacity into the collector volume.

Pressure p_{HP} is also an input of the ST model, which represents the inlet steam chest charging time delay and the main control valve operation modes, namely: (i) no-load speed control, (ii) p_{HP} pressure control to a constant value, and (iii) power and speed regulation. The no-load speed control mode is used at the start-up and synchronizing phases, while the two modes in normal operating conditions are the pressure or power–speed regulation.

The model of the steam turbine, shown in Figure 14.31, represents the time delay associated with the steam store in the inlet chest, the main valve dynamic implements three control operation modes: (i) no-load speed control, (ii) control to keep a constant value of upstream pressure p_{HP}, and (iii) power and speed regulation. The no-load speed control mode is used at the start-up and synchronizing phases, while in normal conditions the two modes are the pressure control or power–speed regulation.

The exciters of the GT units are of brushless type. This kind of exciters supplies the field circuit of the synchronous generator by a DC voltage produced using the combined

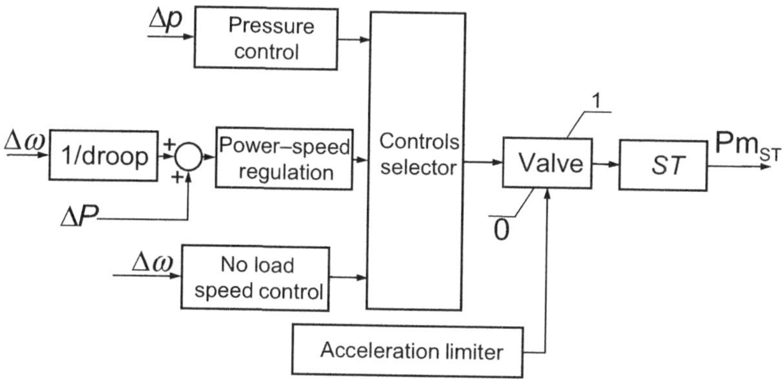

Figure 14.31. Model of the ST and its controls.

action of an AC generator and a noncontrolled diode rectifier. The model used for the exciter corresponds to Type AC8B of IEEE Std. 421 [56] as illustrated in Figure 14.32. The AVR is a PID control with independent values of the proportional (K_P), integral (K_I), and derivative (K_{DIFF}) gains and derivative time constant T_d. The behavior of brushless exciters depends on the generator loading condition and also requires the field current as input. This kind of exciters does not allow negative values for the field voltage and current. As illustrated in Figure 14.32 the exciter voltage V_E is corrected by a feedback ring that takes into account the sum of three contributions: (i) the product between V_E and the saturation function $S_E(V_E)$, (ii) the product between V_E and exciter constant K_E, and (iii) the product between demagnetizing factor K_D and field current I_f.

The output field voltage E_f is calculated taking into account also the reduction due to the commutation reactance of the rectifier, by means of the rectifier regulation block. The impedance of the AC source supplying the AC side of rectifiers is characterized by predominantly inductive impedance. The impedance causes a strongly nonlinear decrease of the rectifier voltage output. This effect depends on the value of the current supplied by the rectifier and is represented by means of a rectifier loading factor K_c proportional to commutating reactance (assumed equal to 0.29) and the rectifier regulation characteristic.

The ST AVR is based on a simple PI regulator with a lead-lag compensator.

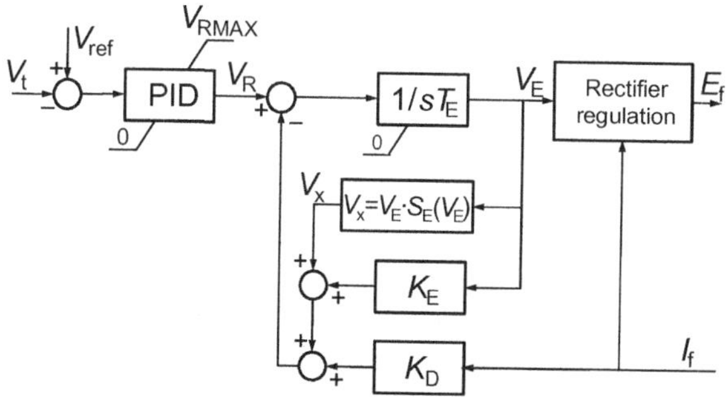

Figure 14.32. Model of GT AVR and excitation system (with parameter values $V_{RMAX} = 11.9$, $K_D = 1.03$, $K_E = 1.0$, $T_E = 0.31$, $S_E(E_1) = 0.58$ and $S_E(E_2) = 0.61$ for $E_1 = 5.2$ p.u. and $E_2 = 7$ p.u., respectively).

The step-up unit transformers model includes also a nonlinear magnetization branch whose parameters are inferred from experimental data. The loads of the distribution substation are represented by the constant RLC devices of the EMTP-RV library.

For the cable line, two different models have been used for the simulation of electromagnetic transients and electromechanical ones, respectively: the frequency-dependent model [57] and the exact Π-model, both obtained by making reference to the manufacturer data of the 132-kV three-phase cable.

The initial states of the GT models, the excitation systems, and the AVR models are defined by the results of the initial network power flow calculation. The states of the HRSGs and the states of the ST model are initialized from the initial values of Q_{GT1}, Q_{GT2}, and those of w_{HRSG1}, w_{HRSG2} respectively, assuming pressure p_{HP} be equal to the rated value and that the bypass valves be closed.

The main parameters of the models are identified based on the available experimental transients by means of a numerical procedure based on least-square optimization.

14.5 CONCLUDING REMARKS

The unbundling and liberalization of electrical power systems justifies the reassessment of power system restoration strategies in competitive electricity markets. In this respect, the investigation of potential black-start-up capabilities of various types of plants is certainly of interest. This chapter has mainly focused on two study cases: the first aimed at assessing the capability of steam groups repowered by means of gas turbines of playing the role of restoration plants, role currently performed, in some countries, only by hydro power stations; and the second aimed at assessing that the capability of a GT unit of a combined-cycle power station to perform a successful energization maneuver and at demonstrating that the contribution to the frequency regulation and load balance provided by the ST, in addition to the frequency regulation provided by the GTs, is effective for a successful islanding maneuver performed at large export power levels.

Additional control systems represent a powerful aid. They can coordinate the load requests to improve both the reliability and speed of the energization and islanding maneuvers. Computer simulators appear to be a useful tool for the design of the main features of these automatic systems.

REFERENCES

[1] Dy Liacco, T.E. Real-time computer control of power systems, *Proc. IEEE*, Vol. 62, pp. 884–891, July 1974.

[2] Wu, F.F., Monticelli, A. Analytical tools for power system restoration-conceptual design, *IEEE Trans. Power Sys.*, Vol. 3, No. 1, pp. 10–26, February 1988.

[3] Nadira, R., Dy Liacco, T.E., Loparo, K.A. A hierarchical interactive approach to electric power system restoration, *IEEE Trans. Power Syst.*, Vol. 7, No. 3, pp. 1123–1131, August 1992.

[4] Task Force 38.06.04 The use of expert systems for power system restoration, Cigré Brochure, No. 90, 1994.

[5] Pérez-Guerrero, R., Heydt, G.T., Jack, N.J., Keel, B.K., Castelhano, A.R. Optimal restoration of distribution systems using dynamic programming, *IEEE Trans. Power Syst.*, Vol. 23, No. 3, pp. 1589–1596, July 2008.

[6] Kostic, T., Cherkaoui, R., Pruvot, P., Germond, A.J. Decision aid function for restoration of transmission power systems: conceptual design and real time considerations, *IEEE Trans. Power Syst.*, Vol. 13, No. 3, pp. 923–929, August 1998.

[7] Matsumoto, K., Sakaguchi, T., Kafka, R.J., Adibi, M.M. Knowledge-based systems as operational aids in power system restoration, *Proc. IEEE*, Vol. 80, No. 5, pp. 689–697, May 1992.

[8] Kirschen, D.S., Volkmann, T.L. Guiding a power system restoration with an expert system, *IEEE Trans. Power Syst.*, Vol. 6, No. 2, pp. 558–566, May 1991.

[9] Liu, C.-C., Liou, K.-L., Chu, R.F., Holen, A.T. Generation capability dispatch for bulk power system restoration: a knowledge-based approach, *IEEE Trans. Power Syst.*, Vol. 8, No. 1, pp. 316–325, February 1993.

[10] Nagata, T., Sasaki, H., Yokoyama, R. Power system restoration by joint usage of expert system and mathematical programming approach, *IEEE Trans. Power Syst.*, Vol. 10, No. 3, pp. 1473–1479, August 1995.

[11] Adibi, M.M., Fink, L.H. Power system restoration planning, *IEEE Trans. Power Syst.*, Vol. 9, No. 1, February 1994.

[12] Fink, L.H., Liou, K.L., Liu, C.C. From generic restoration actions to specific restoration strategies, *IEEE Trans. Power Syst.*, Vol. 11, No. 2, pp. 745–752, May 1995.

[13] Salvati, R., Sforna, M., Pozzi, M. Restoration project. Italian power restoration plan, *IEEE Power Energy Mag.*, Vol. 2, No. 1, pp. 44–51, January–February 2004.

[14] Delfino, B., Denegri, G.B., Invernizzi, M., Morini, A., Cima Bonini, E., Marconato, R., Scarpellini, P. Black-start and restoration of a part of the Italian HV network: modelling and simulation of a field test, *IEEE Trans. Power Syst.*, Vol. 11, No. 3, pp. 1371–1379, August 1996.

[15] UCTE *Final Report of the Investigation Committee on the 28 September 2003 Blackout in Italy*, April 2004.

[16] CRE—AEEG *Report on the Events of September 28th, 2003 Culminating in the Separation of the Italian Power System from the Other UCTE Networks*, April 22, 2004.

[17] Adibi, M.M., Borkoski, J.N., Kafka, R.J., Volkmann, T.L. Frequency response of prime movers during restoration, *IEEE Trans. Power Syst.*, Vol. 14, No. 2, pp. 751–756, May 1999.

[18] Corsi, S., Pozzi, M. A multivariable new control solution for increased long lines voltage restoration stability during black startup, *IEEE Trans. Power Syst.*, Vol. 18, No. 3, pp. 1133–1141, August 2003.

[19] Liu, Y., Gu, X. Skeleton-network reconfiguration based on topological characteristics of scale-free networks and discrete particle swarm optimization, *IEEE Trans. Power Syst.*, Vol. 22, No. 3, pp. 1267–1274, August 2007.

[20] Morin, G. Service restoration following a major failure on the Hydro-Quebec power system, *IEEE Trans. Power Deliv.*, Vol. PWRD-2, No. 2, pp. 454–463, April 1987.

[21] Sidhu, T.S., Tziouvaras, D.A., Apostolov, A.P., Castro, C.H., Chano, S.R., Horowitz, S.H., Kennedy, W.O., Sungsoo, K., Martilla, R.J., McLaren, P.G., Michel, G.L., Mustaphi, K.K., Mysore, P., Nagpal, M., Nelson, B., Plumptre, F.P., Sachdev, M.S., Thorp, J.S., Uchiyama, J.T. Protection issues during system restoration, *IEEE Trans. Power Deliv.*, Vol. 20, No. 1, pp. 47–56, January 2005.

[22] IEEE, WG Special considerations in power system restoration, *IEEE Trans. Power Syst.*, Vol. 7, No. 4, pp. 1419–1427, November 1992.

[23] IEEE, WG Special consideration in power system restoration. The second working group report, *IEEE Trans. Power Syst.*, Vol. 9, No. 1, pp. 15–21, February 1994.

[24] Adibi, M.M., Fink, L.H. Overcoming restoration challenges associated with major power system disturbances—Restoration from cascading failures, *IEEE Power Energy Mag.*, Vol. 4, No. 5, pp. 68–77, September–October 2006.

[25] Barsali, S., Borghetti, A., Delfino, B., Denegri, G.B., Giglioli, R., Invernizzi, M., Nucci, C.A., Paolone, M. Guidelines for ISO operator aid and training for power system restoration in open electricity markets, *Proceedings of of the IREP 2001 Bulk Power System Dynamics and Control V*, August 26–31, Onomichi, Japan, 2001.

[26] Saraf, N., McIntyre, K., Dumas, J., Santoso, S. The annual black start service selection analysis of ERCOT grid, *IEEE Trans. Power Syst.*, Vol. 24, No. 4, pp. 1867–1874, November 2006.

[27] De Mello, F.P., Westcott, J.C. Steam plant startup and control in system restoration, *IEEE Trans. Power Syst.*, Vol. 9, No. 1, pp. 93–101, February 1994.

[28] Fusco, G., Venturini, D., Mazzoldi, F., Possenti, A. Thermal units contribution to the electric power system restoration after a blackout, *Proceedings of CIGRE*, Paris, France, Paper 32-21, 1982.

[29] Mariani, E., Mastroianni, F., Romano, V. Field experiences in reenergization of electrical networks from thermal and hydro units, *IEEE Trans. Power ApparatusSyst.*, Vol. PAS-103, No. 7, pp. 1707–1713, 1984.

[30] Fountas, N.A., Hatziargyriou, N.D., Orfanogiannis, C., Tasoulis, A. Interactive long-term simulation for power system restoration planning, *IEEE Trans. Power Syst.*, Vol. 12, No. 1, pp. 61–68, February 1997.

[31] Sancha, J. L., Llorens, M.L., Moreno, J.M., Meyer, B., Vernotte, J.F., Price, W.W., Sanchez-Gasca, J.J. Application of long term simulation programs for analysis of system islanding, *IEEE Trans. Power Syst.*, Vol. 12, No. 1, pp. 189–197, February 1997.

[32] CIGRE TF38.02.14 Analysis and modeling needs of power systems under major frequency disturbances, *CIGRE Brochure* No. 148, 1999.

[33] Anzano, L., Guagliardi, A., Pastorino, M., Pretolani, F., Ruscio, M. Repowering of Italian power plants: control systems design and overall dynamic verification by means of a mathematical model, *Proceedings of the International Symposium on Performance Improvement, Retrofitting, and Repowering of Fossil Fuel Power Plants*, Washington, USA, 1990.

[34] Gadda, E., Radice, A. Load rejection operation in conventional power plants, *IEEE Trans. Energy Conv.*, Vol. 4, No. 3, pp. 382–391, September 1989.

[35] Borghetti, A., Migliavacca, G., Nucci, C.A., Spelta, S. The black-startup simulation of a repowered thermoelectric unit, *Control Eng. Pract.*, Vol. 9/7, pp. 791–803, 2001.

[36] Borghetti, A., Migliavacca, G., Nucci, C.A., Spelta, S., Tarsia, F. Simulation of the load following capability of a repowered plant during the first phase of the system restoration, *Proceedings of 14th IFAC World Congress*, Beijing, P.R. China, pp. 115–124, 1999.

[37] Borghetti, A., Bosetti, M., Nucci, C.A., Paolone, M. Parameters identification of a power plant model, *Proceedings of Electrimacs*, Québec City, Canada, June 8–11, 2008.

[38] Borghetti, A., Bosetti, M., Nucci, C.A., Paolone, M., Ciappi, G., Solari, A. Analysis of black-startup and islanding capabilities of a combined cycle power plant, *Proceedings of 43rd International Universities Power Engineering Conference (UPEC)*, Padua, September 1–4, 2008.

[39] Borghetti, A., Nucci, C.A., Paolone, M., Ciappi, G., Solari, A. Synchronized phasors monitoring during the islanding maneuver of an active distribution network, *Proceedings of IEEE PES Conference on Innovative Smart Grid Technologies*, Washington, DC, USA, January 19–21, 2010.

[40] Nunes, P., Morched, A., Correia de Barros, M.T. Analysis of generator tripping incidents on energizing nearby transformers, *Proceedings of the 6th International Conference on Power Systems Transients, IPST 2003*, New Orleans, USA, September 28–October 2, 2003.

[41] CIGRE, Working Group, 13.05, The calculation of switching surges (i). a comparison of transient network analyzer results, *Electra*, No. 19, pp. 67–78, 1971.

[42] CIGRE Working Group 13-02 Switching Surges Phenomena in EHV SystemsSwitching Overvoltages in EHV and UHV Systems with Special Reference to Closing and Re-closing Transmission Lines, *Electra*, No. 30, pp. 70–122, 1973.

[43] CIGRE Working Group 33.02 *Guidelines for representation of network elements when calculating transients, brochure*, Paris, 1990.

[44] Archer, B.A., Davies, J.B. System islanding considerations for improving power system restoration at Manitoba Hydro, *IEEE CCECE*, Vol. 1, pp. 60–65, 12–15 May 2002.

[45] James, D., A solution for noise associated with a series staged DLE combustion system, *Proceedings of 4th International Pipeline Conference IPC'02*, Calgary, Alberta, Canada, September 29–October 3, 2002.

[46] Paolone, M., Borghetti, A., Nucci, C.A. Development of an RTU for synchrophasors estimation in active distribution networks, *Proceedings of the 2009 IEEE Bucharest PowerTech*, Bucharest, Romania, June 28–July 2, 2009.

[47] Marcocci, L., Spelta, S. Computer aided modelling of complex processes: a program package, *Proceedings of IMACS International Symposium on Simulation in Engineering Sciences*, Nantes, France, pp. 61–66, 1983.

[48] Rowen, W.I., Simplified mathematical representation of heavy-duty gas turbines, *Trans. ASME—J. of Eng. Power*, Vol. 105, pp. 865–869, 1983.

[49] Kiat Yee, S., Milanovic, J.V., Hughes, F.M. Overview and comparative analysis of gas turbine models for system stability studies, *IEEE Trans. Power Syst.*, Vol. 23, No. 1, pp. 108–118, February 2008.

[50] Mahseredjian, J., Dubé, L., Gérin-Lajoie, L. New advances in the simulation of transients with EMTP: computation and visualization techniques, *Proceedings of 7th International Conference on Modeling and Simulation of Electric Machines, Converters and Systems*, Montreal, August 2002.

[51] Mahseredjian, J., Dennetière, S., Dubé, L., Khodabakhchian, B. On a new approach for the simulation of transients in power systems, *Proceedings of the International Conference on Power Systems Transients IPST'2005*, Montreal, June 2005.

[52] Mahseredjian, J., Dube, L., Zou, M., Dennetiere, S., Joos, G. Simultaneous solution of control system equations in EMTP, *IEEE Trans. Power Syst.*, Vol. 21, No. 1, pp. 117–124, February 2006.

[53] Dommel, H., Brandwain *Three-phase Synchronous Machine, in the EMTP Theory Book*, Microtran Power System Analysis Corporation, April 1996.

[54] Cigre Task Force C4.02.25 *Modelling of Gas Turbines and Steam Turbines in Combined Cycle Power Plants*, Brochure, 2003.

[55] Kunitomi, K., Kurita, A., Tada, Y., Ihara, S., Price, W.W., Richardson, L.M., Smith, G. Modeling combined-cycle power plant for simulation of frequency excursions, *IEEE Trans. Power Syst.*, Vol. 18, No. 2, pp. 724–729, May 2003.

[56] IEEE Std 421.5^{TM}*IEEE Recommended Practice for Excitation System Models for Power System Stability Studies*, 2005.

[57] Marti, J. Accurate modelling of frequency dependent transmission lines in electromagnetic transient simulations, *IEEE Trans. Power Apparatus Syst.*, Vol. PAS-101, pp. 147–157, January 1982.

15

COMPUTER SIMULATION OF SCALE-BRIDGING TRANSIENTS IN POWER SYSTEMS

Kai Strunz and Feng Gao

Electromagnetic transients result from changes of the electric and magnetic energies that are stored in lumped as well as distributed capacitances and inductances of the electric power network. These transients cause temporary disturbances of voltage and current waveforms. The frequency range of electromagnetic transients typically extends from several hertz up to hundreds of kilohertz. Electromechanical transients also involve oscillatory interactions due to the mechanical energy stored in the rotating shafts of the generating units. These transients cause oscillations of the electric power over the lines of the electric power networks at frequencies that typically range from fractions of a hertz up to several hertz.

In a power system blackout, both types of transients do appear. In the 2003 North America blackout, three 345-kV transmission lines operated by First Energy (FE) were tripped due to the contact of trees, which led to electromagnetic transients. As the transmission lines failed, "power flows shifted to other transmission paths, voltages on the rest of FE's system degraded further" [1]. This involved the electromechanical transients. The following cascading events were similar and scale-bridging transients appeared throughout the blackout.

In the following sections, it is shown how diverse scale-bridging phenomena ranging from electromagnetic to electromechanical transients are modeled efficiently. In Section 15.1, the critical role of bridging instantaneous and phasor signals in this context is explained. In Section 15.2, the construction of the network model is discussed. In Section 15.3, the modeling of transformers, lines, and synchronous machinery is elaborated upon in detail. In Section 15.4, the efficient simulation of a power system blackout scenario, which covers diverse transients, is studied using a four-machine two-area power system.

Handbook of Electrical Power System Dynamics: Modeling, Stability, and Control. Edited by Mircea Eremia and Mohammad Shahidehpour.

15.1 BRIDGING OF INSTANTANEOUS AND PHASOR SIGNALS

When studying the effects of lightning strikes, switching or other phenomena that trigger electromagnetic transients, it is of central interest to simulate the instantaneous values of the distorted voltages and currents. The electromechanical transients are slower and effectively lead to angle modulations of the generator terminal voltages since the machine rotors do not turn at their steady-state angular frequencies. Here, it is most illustrative to simulate the envelopes of AC voltages and currents as well as the average power transfers. In Figure 15.1, the difference between tracking the instantaneous values of the natural waveform of an AC voltage and the tracking of its envelope are illustrated. On the left, the AC voltage is zoomed in for one cycle of duration $1/f_c$ with the carrier frequency f_c of the voltage being equal to 50 Hz in European and 60 Hz in North American electric power networks. On the right, five cycles and the envelope are shown. From inspection, it is obvious that a much smaller time step size is needed to accurately track the instantaneous values of the natural waveform than it is for tracking the envelope. In simulators of the type of the EMTP (electromagnetic transients program) [2–4], the natural waveforms are represented. In simulators for electromechanical transients, phasor techniques are used to represent the envelopes [5,6].

In the sequence of a power system blackout, both electromagnetic and electromechanical transients can appear. For a scale-bridging simulation that can track both electromagnetic and electromechanical transients, the application of analytic signals is appropriate as they enable the representation of both natural and envelope waveforms. An analytic signal is obtained by adding a quadrature component as an imaginary part to the original real signal [7,8]. This quadrature component is obtained through the Hilbert transform of the original real signal $s(t)$ denoted as:

$$\mathcal{H}[s(t)] = \frac{1}{\pi}\int_{-\infty}^{\infty} \frac{s(\tau)}{t-\tau}\mathrm{d}\tau \tag{15.1}$$

The analytic signal, marked by an underscore to indicate that it is complex, is then obtained as follows:

$$\underline{s}(t) = s(t) + j\,\mathcal{H}[s(t)] \tag{15.2}$$

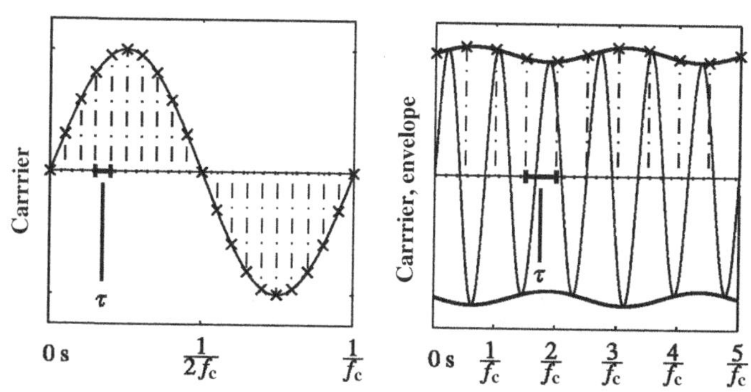

Figure 15.1. Tracked AC voltages; solid light: natural waveform; solid bold: envelope waveform; ×: sampling points.

The effect of the creation of the analytic signal is particularly illustrative when the Fourier spectrum of $s(t)$ shows bandpass character, that is, when it is narrowly concentrated about the carrier frequency f_c. This is typical for voltages and currents that are subject to electromechanical transients. While the Fourier spectrum of the real signal $s(t)$ extends to negative frequencies, it can be seen from Figure 15.2 that this is not the case for the Fourier spectrum $\mathcal{F}[\underline{s}(t)]$ of the corresponding analytic signal $\underline{s}(t)$.

The analytic signal can be shifted by the frequency f_s, which is hereafter referred to as shift frequency, as follows:

$$\mathcal{S}[\underline{s}(t)] = \underline{s}(t)e^{-j2\pi f_s t} \tag{15.3}$$

Inserting the angular frequency $\omega_s = 2\pi f_s$, (15.3) becomes

$$\mathcal{S}[\underline{s}(t)] = \underline{s}(t)e^{-j\omega_s t} \tag{15.4}$$

For the special case where the shift frequency is equal to the carrier frequency, $f_s = f_c$ or $\omega_s = \omega_c$, the complex envelope [7] is obtained:

$$\mathcal{E}[\underline{s}(t)] = \underline{s}(t)e^{-j2\pi f_c t} \tag{15.5}$$

Since $|e^{-j2\pi f_s t}| = 1$, the magnitude is not changed through the shift operation. It follows from (15.5) that the magnitude can readily be derived from the complex envelope [7]:

$$|\mathcal{E}[\underline{s}(t)]| = |\underline{s}(t)| \tag{15.6}$$

The impact of the shifting on the Fourier spectrum is given in Figure 15.2. It can be seen that the complex envelope is a low-pass signal whose maximum frequency is lower than the one of the original real bandpass signal. In accordance with Shannon's sampling theorem, a lower sampling rate can be chosen when tracking the complex envelope rather than the original bandpass signal.

The method frequency-adaptive simulation of transients (FAST) [9,10] processes analytic signals to enable a scale-bridging simulation. Compared to simulators of the EMTP-type that process instantaneous signals or simulators for electromechanical transients that process phasor signals, FAST comprises the shift frequency as a simulation parameter in addition to the time step size. If the shift frequency is set equal to the carrier frequency, then the envelope is obtained and electromechanical transients are emulated efficiently as it is for phasor signals. If the shift frequency is set to zero, then the instantaneous values of the natural waveforms are obtained as it is the case in simulators

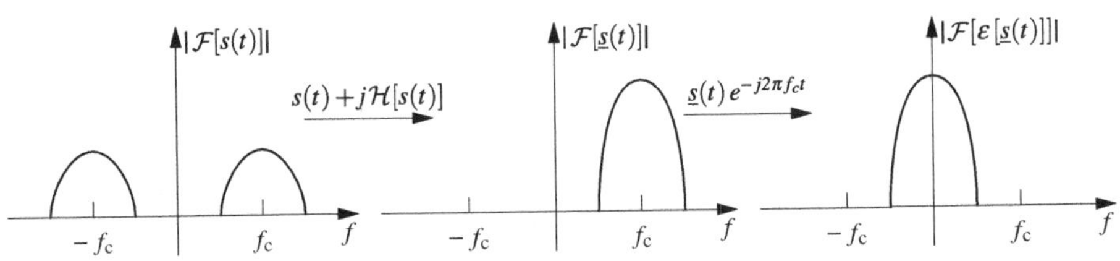

Figure 15.2. Creation of the analytic signal and frequency shifting.

of the EMTP-type. The shiftable analytic signals can so bridge instantaneous and phasor signals.

15.2 NETWORK MODELING

An electric network consists of branches, which are interconnected at nodes. Through the application of nodal analysis techniques, the derivation of a network model is based on straightforward rules. In a first step, the principle of modeling network branches is introduced in Section 15.2.1. The construction of the nodal admittance matrix is explained in Section 15.2.2. The key steps of the simulation procedure are summarized in Section 15.2.3.

15.2.1 Companion Model for Network Branches

While the equations relating voltages and currents of resistive network branches are governed by Ohm's law, the modeling of inductive and capacitive characteristics involves differential equations. For the inductance in Figure 15.3a, the differential equation when using analytic signals is

$$\frac{\mathrm{d}\,\underline{i}_L(t)}{\mathrm{d}t}=\frac{\underline{v}_L(t)}{L} \tag{15.7}$$

Expressing the current through (15.4), $\boldsymbol{S}[\,\underline{i}_L(t)]=\underline{i}_L(t)\mathrm{e}^{-j\omega_s t}$, and insertion into (15.7) yield:

$$\frac{\mathrm{d}(\boldsymbol{S}[\,\underline{i}_L(t)]\mathrm{e}^{j\omega_s t})}{\mathrm{d}t}=\frac{\underline{v}_L(t)}{L} \tag{15.8}$$

which can be expanded as follows using the chain rule of differentiation:

$$\frac{\mathrm{d}(\boldsymbol{S}[\,\underline{i}_L(t)])}{\mathrm{d}t}=\mathrm{e}^{-j\omega_s t}\left(-j\omega_s\,\underline{i}_L(t)+\frac{\underline{v}_L(t)}{L}\right) \tag{15.9}$$

Since simulations performed on digital computers require that all differential equations are transformed into difference equations, a numerical integration method needs to be applied. Very popular in the simulation of electric networks is the trapezoidal method as described in Section 10.3.3. On the left side of (15.9), the time differential dt is replaced by the time step size τ, which separates two instants $t=k\tau$ and $(k-1)\tau$ at which a network solution is established. The differential $\mathrm{d}(\boldsymbol{S}[\,\underline{i}_L(t)])$ is replaced by $\boldsymbol{S}[\,\underline{i}_{Lk}]-\boldsymbol{S}[\,\underline{i}_{L(k-1)}]$, where k appears as the time step counter. On the right of (15.9), the average obtained at

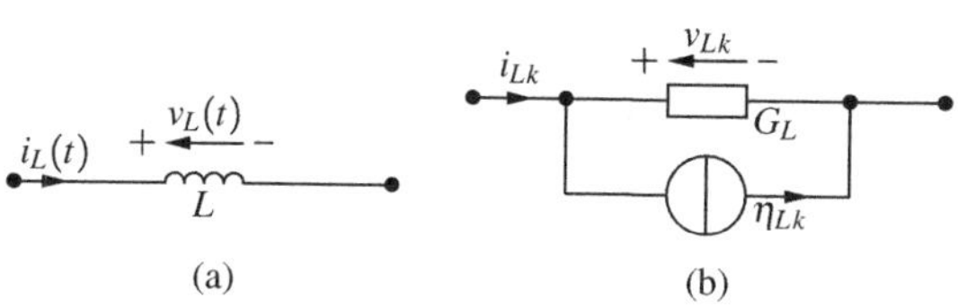

Figure 15.3. Single-phase inductance and associated companion model.

instants $k\tau$ and $(k-1)\tau$ is taken

$$\frac{\mathcal{S}[\underline{i}_{Lk}] - \mathcal{S}[\underline{i}_{L(k-1)}]}{\tau} = \frac{1}{2}\mathrm{e}^{-j\omega_s k\tau}\left(-j\omega_s\,\underline{i}_{Lk} + \frac{\underline{v}_{Lk}}{L}\right) + \frac{1}{2}\mathrm{e}^{-j\omega_s (k-1)\tau}\left(-j\omega_s\,\underline{i}_{L(k-1)} + \frac{\underline{v}_{L(k-1)}}{L}\right) \tag{15.10}$$

Back substitution of analytic signals for $\mathcal{S}[\underline{i}_{Lk}]$ and $\mathcal{S}[\underline{i}_{L(k-1)}]$ yields:

$$\frac{\underline{i}_{Lk}\mathrm{e}^{-j\omega_s k\tau} - \underline{i}_{L(k-1)}\mathrm{e}^{-j\omega_s (k-1)\tau}}{\tau} = \frac{1}{2}\mathrm{e}^{-j\omega_s k\tau}\left(-j\omega_s\,\underline{i}_{Lk} + \frac{\underline{v}_{Lk}}{L}\right) + \frac{1}{2}\mathrm{e}^{-j\omega_s (k-1)\tau}\left(-j\omega_s\,\underline{i}_{L(k-1)} + \frac{\underline{v}_{L(k-1)}}{L}\right) \tag{15.11}$$

Multiplying both sides of (15.11) by $\mathrm{e}^{j\omega_s k\tau}$ and gathering like terms lead to

$$\underline{i}_{Lk} = \frac{\tau}{L(2+j\omega_s\tau)}\,\underline{v}_{Lk} + \mathrm{e}^{j\omega_s\tau}\left(\frac{2-j\omega_s\tau}{2+j\omega_s\tau}\,\underline{i}_{L(k-1)} + \frac{\tau}{L(2+j\omega_s\tau)}\,\underline{v}_{L(k-1)}\right) \tag{15.12}$$

This can be more clearly summarized as [9]:

$$\underline{i}_{Lk} = \underline{Y}_L\,\underline{v}_{Lk} + \underline{\eta}_{Lk} \tag{15.13}$$

with

$$\underline{Y}_L = \frac{\tau}{L(2+j\omega_s\tau)} \tag{15.14}$$

$$\underline{\eta}_{Lk} = \mathrm{e}^{j\omega_s\tau}\left(\frac{2-j\omega_s\tau}{2+j\omega_s\tau}\,\underline{i}_{L(k-1)} + \frac{\tau}{L(2+j\omega_s\tau)}\,\underline{v}_{L(k-1)}\right) \tag{15.15}$$

Equations (15.13), (15.14), and (15.15) model a single-phase inductor through a companion model that consists of an admittance $\underline{Y}_L$ in parallel with a history current source $\underline{\eta}_L$ as shown in Figure 15.3b. The term history current source stems from the fact that it only contains information calculated at the preceding time step $k-1$.

For the special unshifted case of $f_s = 0\,\mathrm{Hz}$, both the admittance and the history current source are real: $\underline{Y}_L = \tau/(2L)$, $\underline{\eta}_{Lk} = \underline{i}_{L(k-1)} + \tau/(2L)\,\underline{v}_{L(k-1)}$. The companion model then has exactly the same format as its counterpart developed for the EMTP [3] and is suitable for simulating electromagnetic transients. For $f_s = f_c$, the frequency-adaptive companion model is suitable for simulating electromechanical transients. This can be understood by applying the Z-transform to (15.12):

$$\underline{I}_L(Z) = \frac{\tau}{L(2+j\omega_s\tau)}\,\underline{V}_L(Z) + \mathrm{e}^{j\omega_s\tau}\left(\frac{2-j\omega_s\tau}{2+j\omega_s\tau}z^{-1}\,\underline{I}_L(Z) + \frac{\tau}{L(2+j\omega_s\tau)}z^{-1}\,\underline{V}_L(Z)\right) \tag{15.16}$$

where $\underline{I}_L(Z)$ is the Z-transform of $\underline{i}_{Lk}$, $z^{-1}\underline{I}_L(Z)$ is the Z-transform of $\underline{i}_{L(k-1)}$, $\underline{V}_L(Z)$ is the Z-transform of $\underline{v}_{Lk}$, and $z^{-1}\underline{V}_L(Z)$ is the Z-transform of $\underline{v}_{L(k-1)}$. Equation (15.16) can be rearranged as:

$$\underline{I}_L(Z) = \frac{1}{j\omega_s L + \frac{2L}{\tau}\frac{z - e^{j\omega_s\tau}}{z + e^{j\omega_s\tau}}}\underline{V}_L(Z) \tag{15.17}$$

Substitution of the exponential form of $z = e^{j\omega_c\tau}$ in (15.17) and setting $\omega_s = \omega_c$ lead to:

$$\underline{I}_L(Z) = \frac{1}{j\omega_c L}\underline{V}_L(Z) \tag{15.18}$$

Thus, for $f_s = f_c$, the admittance is $\underline{Y}_L = 1/(j\omega_c L)$ and is suitable for processing phasors as it is desirable in the simulation of electromechanical transients. The frequency-adaptive companion model is so suitable for the scale-bridging simulation of transients as a function of the setting of the shift frequency.

Using analytic signals, the differential equation describing the behavior of a capacitance is as follows:

$$\underline{i}_C(t) = C\frac{\mathrm{d}\,\underline{v}_C(t)}{\mathrm{d}t} \tag{15.19}$$

Expressing the voltage through (15.14), $\boldsymbol{S}[\underline{v}_C(t)] = \underline{v}_C(t)e^{-j\omega_s t}$, and insertion into (15.17) yield:

$$\underline{i}_C(t) = C\frac{\mathrm{d}\boldsymbol{S}[\underline{v}_C(t)]e^{j\omega_s t}}{\mathrm{d}t} \tag{15.20}$$

which can be expanded to:

$$\frac{\mathrm{d}(\boldsymbol{S}[\underline{v}_C(t)])}{\mathrm{d}t} = e^{-j\omega_s t}\left(-j\omega_s\,\underline{v}_C(t) + \frac{\underline{i}_C(t)}{C}\right) \tag{15.21}$$

Similar to the inductance model, the differential $\mathrm{d}(\boldsymbol{S}[\underline{v}_C(t)])$ on the left of (15.21) is replaced by $\boldsymbol{S}[\underline{v}_{Ck}] - \boldsymbol{S}[\underline{v}_{C(k-1)}]$, On the right of (15.21) the average obtained at instants $k\tau$ and $(k-1)\tau$ is taken as follows:

$$\frac{\boldsymbol{S}[\underline{v}_{Ck}] - \boldsymbol{S}[\underline{v}_{C(k-1)}]}{\tau} = \frac{1}{2}e^{-j\omega_s k\tau}\left(-j\omega_s\,\underline{v}_{Ck} + \frac{\underline{i}_{Ck}}{C}\right) + \frac{1}{2}e^{-j\omega_s(k-1)\tau}\left(-j\omega_s\,\underline{v}_{C(k-1)} + \frac{\underline{i}_{C(k-1)}}{C}\right) \tag{15.22}$$

Back substitution of analytic signals for $\boldsymbol{S}[\underline{v}_{Ck}]$ and $\boldsymbol{S}[\underline{v}_{C(k-1)}]$ and rearranging lead to:

$$\underline{i}_{Ck} = \frac{C(2 + j\omega_s\tau)}{\tau}\underline{v}_{Ck} + e^{j\omega_s\tau}\left(-\underline{i}_{C(k-1)} - \frac{(2 - j\omega_s\tau)C}{\tau}\underline{v}_{C(k-1)}\right) \tag{15.23}$$

This can be written as:

$$\underline{i}_{Ck} = \underline{Y}_C\, \underline{v}_{Ck} + \underline{\eta}_{Ck} \tag{15.24}$$

with

$$\underline{Y}_C = \frac{C(2 + j\omega_s \tau)}{\tau} \tag{15.25}$$

$$\underline{\eta}_{Ck} = e^{j\omega_s \tau}\left(-\underline{i}_{C(k-1)} - \frac{(2 - j\omega_s \tau)C}{\tau}\, \underline{v}_{C(k-1)} \right) \tag{15.26}$$

Equations (15.24), (15.25), and (15.26) model a single-phase capacitance through a companion model that consists of an admittance $\underline{Y}_C$ in parallel with a history current source $\underline{\eta}_C$.

15.2.2 Direct Construction of Nodal Admittance Matrix

The nodal equation system for the network model is of the following form:

$$\boldsymbol{Y}_k \boldsymbol{v}_k = \boldsymbol{j}_k \tag{15.27}$$

where $\boldsymbol{Y}$ is the nodal admittance matrix, $\boldsymbol{v}$ is the nodal voltage vector, and $\boldsymbol{j}$ is the nodal current injection vector that includes the history current sources. If besides the ground node there are N nodes in the network under study, then $\boldsymbol{Y}$ is of size $N \times N$, and $\boldsymbol{v}$ and $\boldsymbol{j}$ are of size $N \times 1$. Using analytic signals, $\boldsymbol{Y}$, $\boldsymbol{v}$, and $\boldsymbol{j}$ become complex as indicated through underlining:

$$\underline{\boldsymbol{Y}}_k\, \underline{\boldsymbol{v}}_k = \underline{\boldsymbol{j}}_k \tag{15.28}$$

In the following, it is introduced how to directly construct the network nodal admittance matrix $\boldsymbol{Y}$ and nodal current injection vector $\boldsymbol{j}$ through the stamping method [11]. The time step counter k is dropped for the sake of clarity of explanation.

For a resistive branch with conductance G connected to nodes m and n, the admittance matrix stamp contributed by this branch is as follows:

$$\Delta Y = \begin{array}{c} \\ \\ m \\ \\ n \\ \\ \end{array} \begin{array}{c} \begin{array}{ccccc} \phantom{\mathbf{0}} & m & & n & \phantom{\mathbf{0}} \end{array} \\ \begin{pmatrix} \mathbf{0} & \cdots & \mathbf{0} & \cdots & \mathbf{0} \\ \vdots & G & \vdots & -G & \vdots \\ \mathbf{0} & \cdots & \mathbf{0} & \cdots & \mathbf{0} \\ \vdots & -G & \vdots & G & \vdots \\ \mathbf{0} & \cdots & \mathbf{0} & \cdots & \mathbf{0} \end{pmatrix} \end{array} \tag{15.29}$$

The admittance matrix stamp of (15.29) is an $N \times N$ matrix and contains only four nonzero entries at positions (m, m), (m, n), (n, m), and (n, n). The sign is positive at the diagonal positions and negative at the off-diagonal positions. If a branch is connected from node m to the ground, then the admittance stamp will only have one nonzero entry at (m, m):

$$\Delta Y \; = \; m \begin{pmatrix} \overset{}{\mathbf{0}} & \cdots & \overset{m}{\mathbf{0}} & \cdots \\ \vdots & G & \vdots & \vdots \\ \mathbf{0} & \cdots & \mathbf{0} & \cdots \\ \vdots & \cdots & \vdots & \ddots \end{pmatrix} \tag{15.30}$$

The nodal admittance matrix $\boldsymbol{Y}$ is obtained by first conceptually removing all branches from the network and then successively stamping the contributions of the companion models into the emerging nodal admittance matrix in accordance with the network topology:

$$\boldsymbol{Y} = \sum_{\mathrm{b}} \Delta \boldsymbol{Y}_{\mathrm{b}} \tag{15.31}$$

where b refers to all branches in the network. The construction of the network current injection vector $\boldsymbol{j}$ is based on the same concept. If a current source injects a current i into node n, then the corresponding current injection vector stamp is as follows:

$$\Delta \boldsymbol{j} = n \begin{pmatrix} \vdots \\ 0 \\ i \\ 0 \\ \vdots \end{pmatrix} \tag{15.32}$$

The stamping process for constructing the network admittance matrix and the current injection vector is illustrated for a simple circuit in Figure 15.4 . Since there are three nodes in the circuit, that is, $N = 3$, all the admittance matrix stamps are 3×3 matrix contributions.

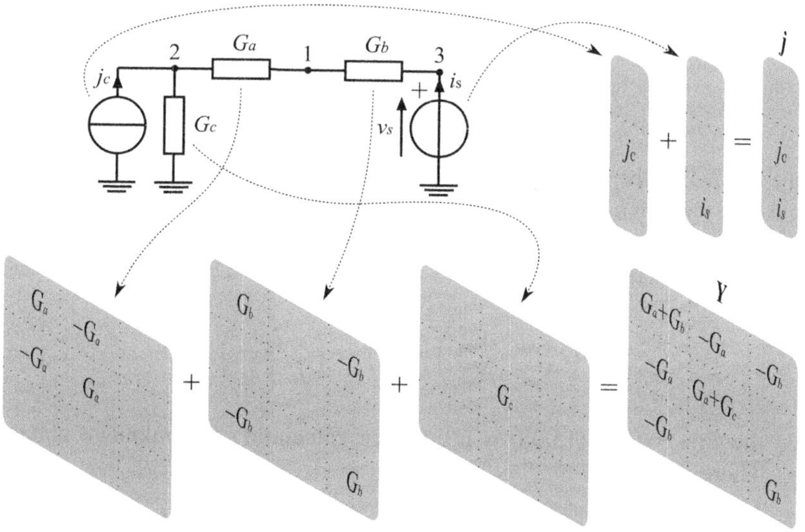

Figure 15.4. Example of stamping method.

The conductance G_a is connected to nodes 1 and 2 and therefore the corresponding admittance stamp in accordance with (15.29) is

$$\Delta \boldsymbol{Y}_{G_a} = \begin{pmatrix} G_a & -G_a & 0 \\ -G_a & G_a & 0 \\ 0 & 0 & 0 \end{pmatrix} \tag{15.33}$$

Similarly, the admittance matrix stamp of G_b is as follows:

$$\Delta \boldsymbol{Y}_{G_b} = \begin{pmatrix} G_b & 0 & -G_b \\ 0 & 0 & 0 \\ -G_b & 0 & G_b \end{pmatrix} \tag{15.34}$$

The conductance G_c is connected to node 2 and the ground, so with the application of (15.30) its admittance matrix stamp becomes:

$$\Delta \boldsymbol{Y}_{G_c} = \begin{pmatrix} 0 & 0 & 0 \\ 0 & G_c & 0 \\ 0 & 0 & 0 \end{pmatrix} \tag{15.35}$$

With all the admittance matrix stamps built, the network admittance matrix can then be obtained by adding the stamps according to (15.31):

$$\boldsymbol{Y} = \Delta \boldsymbol{Y}_{G_a} + \Delta \boldsymbol{Y}_{G_b} + \Delta \boldsymbol{Y}_{G_c} = \begin{pmatrix} G_a + G_b & -G_a & -G_b \\ -G_a & G_a + G_c & 0 \\ -G_b & 0 & G_b \end{pmatrix} \tag{15.36}$$

The current source connected to node 2 in Figure 15.4 provides excitation in that it injects current j_c into node 2. Therefore, its corresponding current injection vector stamp is $\Delta \boldsymbol{j}_c = (0 \quad j_c \quad 0)^T$. The current i_s through the ideal voltage source v_s cannot be calculated solely as a function of the source voltage element v_s but also depends on other nodal voltages. It contributes a stamp $\Delta \boldsymbol{i}_s = (0 \quad 0 \quad i_s)^T$. The current injection vector is therefore:

$$\boldsymbol{j} = \Delta \boldsymbol{j}_c + \Delta \boldsymbol{i}_s = (0 \quad j_c \quad i_s)^T \tag{15.37}$$

In the following, the nodal voltage vector is partitioned into a set of voltages $\boldsymbol{v}_d$ that are dependent and unknown and need to be calculated, and a set of voltages $\boldsymbol{v}_e$ that provide excitation and are given through source terms. The partitioning leads to the following modified version of equation (15.27):

$$\begin{pmatrix} \boldsymbol{Y}_{dd} & \boldsymbol{Y}_{de} \\ \boldsymbol{Y}_{ed} & \boldsymbol{Y}_{ee} \end{pmatrix} \begin{pmatrix} \boldsymbol{v}_d \\ \boldsymbol{v}_e \end{pmatrix} = \begin{pmatrix} \boldsymbol{j}_d \\ \boldsymbol{j}_e + \boldsymbol{i}_e \end{pmatrix} \tag{15.38}$$

Only the first row of (15.38) is needed for calculating the unknown nodal voltage vector $\boldsymbol{v}_d$. Therefore

$$\boldsymbol{Y}_{dd}\boldsymbol{v}_d + \boldsymbol{Y}_{de}\boldsymbol{v}_e = \boldsymbol{j}_d \tag{15.39}$$

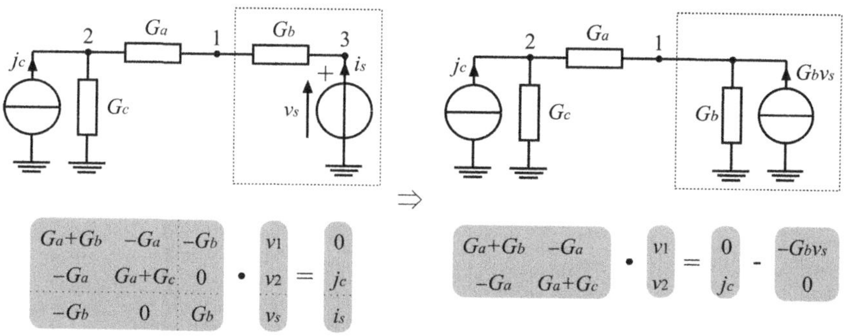

Figure 15.5. Example of the network equation reformulation.

or

$$\boldsymbol{Y}_{\mathrm{dd}}\boldsymbol{v}_{\mathrm{d}} = \boldsymbol{j}_{\mathrm{d}} - \boldsymbol{Y}_{\mathrm{de}}\boldsymbol{v}_{\mathrm{e}} \tag{15.40}$$

Equation (15.40) is solved for $\boldsymbol{v}_{\mathrm{d}}$. The second term $-\boldsymbol{Y}_{\mathrm{de}}\boldsymbol{v}_{\mathrm{e}}$ on the right-hand side of (15.40) is a multiterminal Norton equivalent source that makes the excitation voltage sources appear as current injections.

The vector $\boldsymbol{j}_{\mathrm{e}}$ in the second row of (15.38) contains the excitation current injections into the nodes at which the ideal voltage sources are incident, whereas $\boldsymbol{i}_{\mathrm{e}}$ contains the unknown currents through the ideal voltage sources.

The reformulation of (15.27) leading to the partitioned counterpart (15.40) can be illustrated by the example shown in Figure 15.5.

The voltages at nodes 1 and 2 are dependent while the voltage at node 3 is known. The Norton equivalent source is here given by $-G_{\mathrm{b}}v_{\mathrm{s}}$.

15.3 MODELING OF POWER SYSTEM COMPONENTS

In the following, the scale-bridging modeling of multiphase inductors, transformers, transmission lines, and synchronous machinery is discussed [10].

15.3.1 Multiphase Lumped Elements

In the continuous time domain, the magnetically coupled inductances shown in Figure 15.6 can be described in vector–matrix notation as follows:

$$\frac{\mathrm{d}\boldsymbol{i}_L(t)}{\mathrm{d}t} = \boldsymbol{L}^{-1}\boldsymbol{v}_L(t) \tag{15.41}$$

where

$$\frac{\mathrm{d}\boldsymbol{i}_L(t)}{\mathrm{d}t} = \begin{pmatrix} \frac{\mathrm{d}i_{L1}}{\mathrm{d}t} \\ \frac{\mathrm{d}i_{L2}}{\mathrm{d}t} \\ \vdots \\ \frac{\mathrm{d}i_{LM}}{\mathrm{d}t} \end{pmatrix}; \quad \boldsymbol{L}^{-1} = \begin{pmatrix} L_{11} & L_{12} & \cdots & L_{1M} \\ L_{21} & L_{22} & \cdots & L_{2M} \\ \vdots & \vdots & \ddots & \vdots \\ L_{M1} & \cdots & \cdots & L_{MM} \end{pmatrix}^{-1}; \quad \boldsymbol{v}_L(t) = \begin{pmatrix} v_{L1}(t) \\ v_{L2}(t) \\ \vdots \\ v_{LM}(t) \end{pmatrix}$$

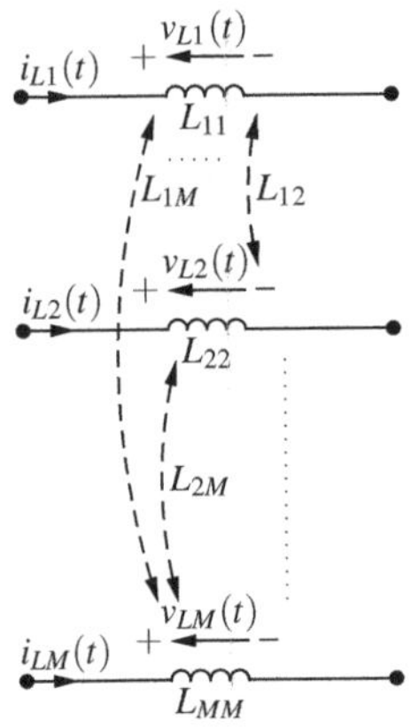

Figure 15.6. Current and voltage conventions for magnetically coupled inductances.

where M is the number of inductors, L_{mm} with $m = 1, 2, \ldots, M$ is the self inductance of the mth inductor, and L_{mn} with $m, n = 1, 2, \ldots, M$ and $m \neq n$ is the mutual inductance between mth inductor and nth inductor.

The frequency-adaptive companion model of the magnetically coupled inductances can be developed in an analogous way as for the single inductance already described in Section 15.1.2. Using analytic signals, (15.41) becomes:

$$\frac{\mathrm{d}\,\underline{\boldsymbol{i}}_L(t)}{\mathrm{d}t} = \boldsymbol{L}^{-1}\,\underline{\boldsymbol{v}}_L(t) \tag{15.42}$$

Substitution of the shifted analytic signals in analogy to (15.9) results in the vector–matrix format:

$$\frac{\mathrm{d}(\mathcal{S}[\,\underline{\boldsymbol{i}}_L(t)])}{\mathrm{d}t} = \mathrm{e}^{-j\omega_s t}\left(-j\omega_s\,\underline{\boldsymbol{i}}_L(t) + \boldsymbol{L}^{-1}\,\underline{\boldsymbol{v}}_L(t)\right) \tag{15.43}$$

Application of the trapezoidal rule to differentiate (15.43) yields the companion model of the magnetically coupled inductances [9]:

$$\underline{\boldsymbol{i}}_{Lk} = \underline{\boldsymbol{Y}}_L\,\underline{\boldsymbol{v}}_{Lk} + \underline{\boldsymbol{\eta}}_{Lk} \tag{15.44}$$

with

$$\underline{\boldsymbol{Y}}_L = \boldsymbol{L}^{-1}\frac{\tau}{2 + j\omega_s\tau} \tag{15.45}$$

$$\underline{\boldsymbol{\eta}}_{Lk} = \mathrm{e}^{j\omega_s\tau}\left(\frac{2 - j\omega_s\tau}{2 + j\omega_s\tau}\,\underline{\boldsymbol{i}}_{L(k-1)} + \boldsymbol{L}^{-1}\frac{\tau}{2 + j\omega_s\tau}\,\underline{\boldsymbol{v}}_{L(k-1)}\right) \tag{15.46}$$

While in (15.14) the admittance Y_L is a scalar, it is a matrix in (15.45) for the multiphase case. Similarly, while in (15.15) the history current source η_L is a scalar, it is a vector in (15.46).

15.3.2 Transformer

A transformer made of an ideal core with infinite permeability and a finite leakage inductance is modeled by the equivalent circuit consisting of an ideal transformer with turns ratio a of primary side to secondary side and a series inductance as depicted in Figure 15.7. The series inductance that represents the leakage effects is shown on the primary side with the higher voltage because the winding supporting the higher voltage is usually further away from the transformer core. Alternatively, the leakage inductance may be split up to model the respective leakage contributions on both sides.

In the discrete time domain, the leakage inductance can be represented by the companion model developed in Section 15.1.2. This leads to the circuit model illustrated by Figure 15.8.

For the currents shown in Figure 15.9, the following equations apply

$$\underline{i}_{T1k} = \underline{Y}_L(\underline{v}_{Pk} - a\,\underline{v}_{Sk}) + \underline{\eta}_{Lk} = \underline{Y}_L(\underline{v}_{T1k} - \underline{v}_{T3k} - a(\underline{v}_{T2k} - \underline{v}_{T4k})) + \underline{\eta}_{Lk} \tag{15.47}$$

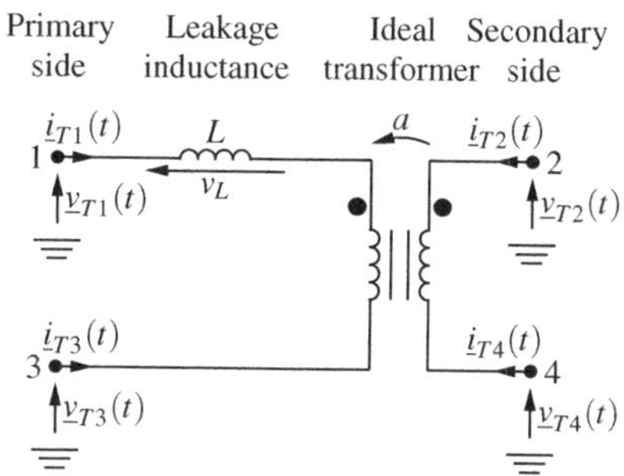

Figure 15.7. Conventions for single-phase transformer with leakage inductance.

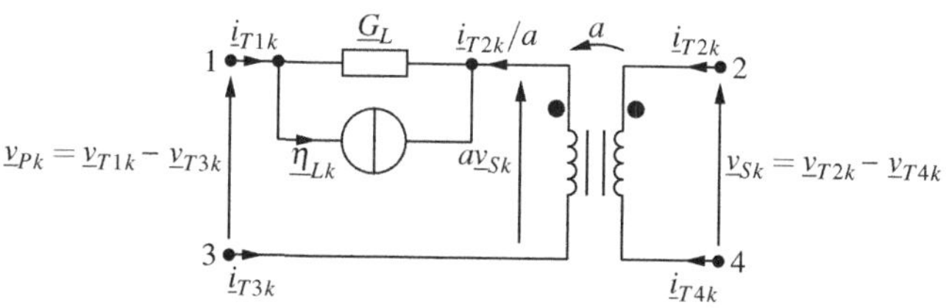

Figure 15.8. Circuit model of single-phase transformer with leakage inductance.

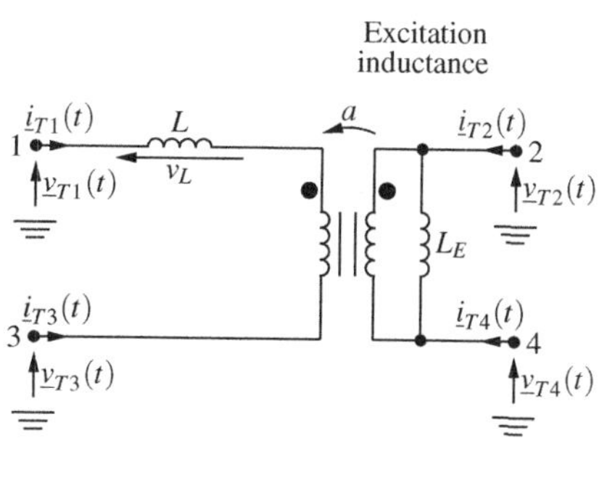

Figure 15.9. Conventions for single-phase transformer with leakage and excitation inductance.

$$\underline{i}_{T2k} = -a\,\underline{i}_{T1k} \tag{15.48}$$

$$\underline{i}_{T3k} = -\,\underline{i}_{T1k} \tag{15.49}$$

$$\underline{i}_{T4k} = a\,\underline{i}_{T1k} \tag{15.50}$$

The admittance $\underline{Y}_L$ and history source term $\underline{\eta}_L$ are provided by (15.14) and (15.15). Equations (15.47) to (15.50) can be reorganized in the vector–matrix format:

$$\underline{\boldsymbol{i}}_{Tk} = \underline{\boldsymbol{Y}}_T\,\underline{\boldsymbol{v}}_{Tk} + \underline{\boldsymbol{\eta}}_{Tk} \tag{15.51}$$

where $\underline{\boldsymbol{i}}_{Tk} = (\underline{i}_{T1k}, \underline{i}_{T2k}, \underline{i}_{T3k}, \underline{i}_{T4k})^{\mathrm{T}}$ is the vector of transformer terminal currents and $\underline{\boldsymbol{v}}_{Tk} = (\underline{v}_{T1k}, \underline{v}_{T2k}, \underline{v}_{T3k}, \underline{v}_{T4k})^{\mathrm{T}}$ is the vector of transformer terminal voltages. The admittance stamp of the transformer model is therefore:

$$\underline{\boldsymbol{Y}}_T = \begin{pmatrix} \underline{G}_L & -a\,\underline{G}_L & -\,\underline{G}_L & a\,\underline{G}_L \\ -a\,\underline{G}_L & a^2\,\underline{G}_L & a\,\underline{G}_L & -a^2\,\underline{G}_L \\ -\,\underline{G}_L & a\,\underline{G}_L & \underline{G}_L & -a\,\underline{G}_L \\ a\,\underline{G}_L & -a^2\,\underline{G}_L & -a\,\underline{G}_L & a^2\,\underline{G}_L \end{pmatrix} \tag{15.52}$$

The history source vector is $\underline{\boldsymbol{\eta}}_{Tk} = (\underline{\eta}_{Lk}, -a\,\underline{\eta}_{Lk}, -\,\underline{\eta}_{Lk}, a\,\underline{\eta}_{Lk})^{\mathrm{T}}$.

More detailed models of the transformer can be developed in accordance with the simulation needs. Excitation effects, for example, can be emulated by connecting an inductance in parallel with the above transformer model to account for the magnetization of the transformer core. The circuit model is illustrated in Figure 15.9. Multiphase transformers can be modeled by connecting primary and secondary sides of multiple single-phase transformer models depending on the case under study in star- or delta-configurations.

15.3.3 Transmission Line [10]

In order to enable the scale-bridging simulation of electromagnetic and electromechanical transients on transmission lines, the line model is to represent both traveling wave effects at time step sizes on the order of microseconds as well as slowly changing voltage and current envelopes due to electromechanical transients at much larger time step sizes as illustrated in Figure 15.1. In what follows, the scale-bridging line model is elaborated upon.

15.3.3.1 Single-Phase Line Model. The propagation of voltages and currents on a single-phase lossless line with length ℓ, distributed inductance L', and capacitance C' as shown in Figure 15.10 is given through the general solution by d'Alembert [12]:

$$i_{l1}(t) = v_{l1}(t)/Z_0 - i_{l2}(t - T_{\mathrm{wp}}) - v_{l2}(t - T_{\mathrm{wp}})/Z_0 \tag{15.53}$$

$$i_{l2}(t) = v_{l2}(t)/Z_0 - i_{l1}(t - T_{\mathrm{wp}}) - v_{l1}(t - T_{\mathrm{wp}})/Z_0 \tag{15.54}$$

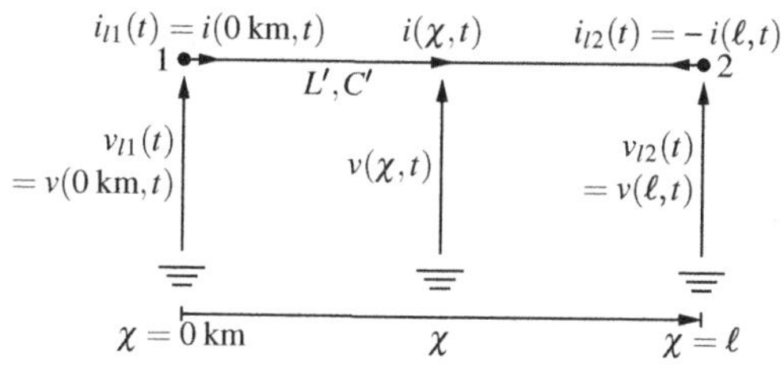

Figure 15.10. Current and voltage conventions for lossless single-phase line.

where $Z_0 = \sqrt{L'/C'}$ is the characteristic impedance and $T_{\text{wp}} = \ell\sqrt{L'C'}$ is the propagation time that the waves need to travel from one to the other end of the line. From (15.53), (15.54) it can be seen that there is no topological coupling through a lumped element between both ends of the line in the model because the voltages and currents at the opposite end date from instant $t - T_{\text{wp}}$ rather than t. In digital simulation, this can only be modeled if the wave propagation time exceeds the time step size, that is, $\tau < T_{\text{wp}}$. If $\tau \geq T_{\text{wp}}$, the line can be modeled with a π-circuit giving a topological coupling between both ends.

A scale-bridging transmission model needs to be valid for any τ, that is, for both $\tau < T_{\text{wp}}$ and $\tau \geq T_{\text{wp}}$. The switches are key to the model that is shown in Figure 15.11 [13]. For electromagnetic transients and $\tau < T_{\text{wp}}$, the switches are open. For electromechanical transients and $\tau \geq T_{\text{wp}}$, the switches are closed to provide the topological lumped coupling via a π-circuit cell, the parameters of which are developed in the following.

As all other models, the line model is formulated through a companion model with all time-varying quantities represented through analytic signals:

$$\underline{\boldsymbol{i}}_{lk} = \underline{\boldsymbol{Y}}_l\, \underline{\boldsymbol{v}}_{lk} + \underline{\boldsymbol{\eta}}_{lk} \tag{15.55}$$

where $\underline{\boldsymbol{i}}_{lk} = (\underline{i}_{l1k}, \underline{i}_{l2k})^{\text{T}}$ and $\underline{\boldsymbol{v}}_{lk} = (\underline{v}_{l1k}, \underline{v}_{l2k})^{\text{T}}$ are the terminal quantities identified in Figure 15.11. The solution by d'Alembert in (15.53) and (15.54) is the starting point for deriving $\underline{\boldsymbol{Y}}_l$ and $\underline{\boldsymbol{\eta}}_{lk}$. In order to calculate values of the time-varying quantities $\underline{i}_{l1}$, $\underline{i}_{l2}$, $\underline{v}_{l1}$, and $\underline{v}_{l2}$ at instant $(t - T_{\text{wp}})$, the usage of linear interpolation is appropriate whenever T_{wp} is not an integer multiple of τ. To enable such interpolation at any time step size, that is, for $\tau < T_{\text{wp}}$ when the switches in Figure 15.11 are open and for $\tau \geq T_{\text{wp}}$ when these switches are closed, a variable κ is defined as follows:

$$\kappa = \text{ceil}\left(\frac{T_{\text{wp}}}{\tau}\right), \quad \kappa \geq 1 \tag{15.56}$$

where ceil is a function that rounds a variable to the nearest integer greater than or equal to itself. Then $(\kappa - 1)\tau < T_{\text{wp}} \leq \kappa\tau$ where $\kappa > 1$ corresponds to the case where $\tau < T_{\text{wp}}$, and

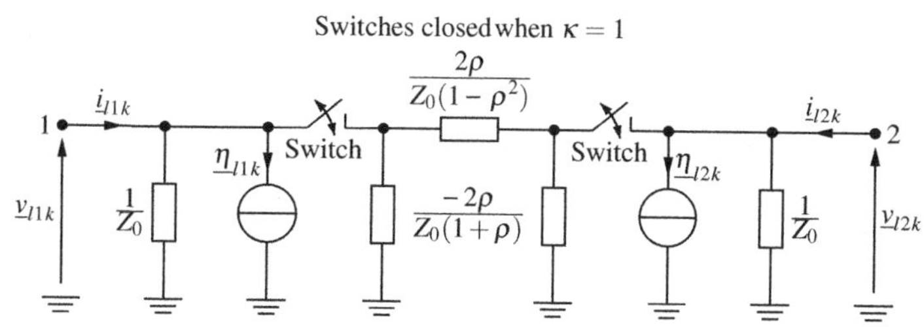

Figure 15.11. Frequency-adaptive model of single-phase line model.

$\kappa = 1$ corresponds to the case where $\tau \geq T_{\text{wp}}$. Application of linear interpolation to shifted analytic signals of $\mathcal{S}[\underline{i}_l(t - T_{\text{wp}})]$ gives:

$$\mathcal{S}[\underline{i}_l(t - T_{\text{wp}})] = \left(\kappa - \frac{T_{\text{wp}}}{\tau}\right)\mathcal{S}[\underline{i}_l(t - (\kappa - 1)\tau)] + \left(1 - \kappa + \frac{T_{\text{wp}}}{\tau}\right)\mathcal{S}[\underline{i}_l(t - \kappa\tau)] \tag{15.57}$$

Back substitution of the definition of shifted analytic signals in (15.4) and solving for $\underline{i}_l(t - T_{\text{wp}})$ lead to:

$$\underline{i}_l(t - T_{\text{wp}}) = \rho\ \underline{i}_l(t - (\kappa - 1)\tau) + \sigma\ \underline{i}_l(t - \kappa\tau) \tag{15.58}$$

with

$$\rho = \left(\kappa - \frac{T_{\text{wp}}}{\tau}\right)e^{j\omega_s((\kappa-1)\tau - T_{\text{wp}})} \tag{15.59}$$

$$\sigma = \left(1 - \kappa + \frac{T_{\text{wp}}}{\tau}\right)e^{j\omega_s(\kappa\tau - T_{\text{wp}})} \tag{15.60}$$

The interpolation for $\mathcal{S}[\underline{v}_l(t - T_{\text{wp}})]$ is performed in exactly the same way. Inserting the results of these interpolations in the analytic signal versions of (15.53) and (15.54) yields:

$$\underline{i}_{lk} = \frac{1}{Z_0}\begin{pmatrix} 1 & 0 \\ 0 & 1 \end{pmatrix}\underline{v}_{lk} + \begin{pmatrix} 0 & -1 \\ -1 & 0 \end{pmatrix}\left(\frac{\rho}{Z_0}\underline{v}_{l(k-\kappa+1)} + \frac{\sigma}{Z_0}\underline{v}_{l(k-\kappa)} + \rho\,\underline{i}_{l(k-\kappa+1)} + \sigma\,\underline{i}_{l(k-\kappa)}\right) \tag{15.61}$$

For $\kappa > 1$, the terms $\underline{v}_{l(k-\kappa+1)}$, $\underline{v}_{l(k-\kappa)}$, $\underline{i}_{l(k-\kappa+1)}$, $\underline{i}_{l(k-\kappa)}$ on the right-hand side of (15.61) refer to values calculated in past time steps so that no topological coupling through a lumped element is given between both ends of the line. Therefore, the history term $\underline{\eta}_{lk}$ as well as the admittance stamp $\underline{Y}_{lk}$ in (15.55) when $\kappa > 1$ can be described by:

$$\underline{\eta}_{lk} = \begin{pmatrix} 0 & -1 \\ -1 & 0 \end{pmatrix}\left(\frac{\rho}{Z_0}\underline{v}_{l(k-\kappa+1)} + \frac{\sigma}{Z_0}\underline{v}_{l(k-\kappa)} + \rho\,\underline{i}_{l(k-\kappa+1)} + \sigma\,\underline{i}_{l(k-\kappa)}\right) \quad \text{if } \kappa > 1 \tag{15.62}$$

$$\underline{Y}_l = \underline{Y}_{lD} = \frac{1}{Z_0}\begin{pmatrix} 1 & 0 \\ 0 & 1 \end{pmatrix} \quad \text{if } \kappa > 1 \tag{15.63}$$

For $\kappa = 1$, both $\underline{v}_{lk}$ and $\underline{i}_{lk}$ appear on the right-hand side of (15.61), which implies a topological coupling between both ends of the line. That is to say when $\kappa = 1$, (15.61) is

$$\underline{i}_{lk} = \frac{1}{Z_0}\begin{pmatrix} 1 & 0 \\ 0 & 1 \end{pmatrix}\underline{v}_{lk} + \begin{pmatrix} 0 & -1 \\ -1 & 0 \end{pmatrix}\left(\frac{\rho}{Z_0}\underline{v}_{lk} + \frac{\sigma}{Z_0}\underline{v}_{l(k-1)} + \rho\,\underline{i}_{lk} + \sigma\,\underline{i}_{l(k-1)}\right) \tag{15.64}$$

which can be recast to:

$$\begin{pmatrix} 1 & \rho \\ \rho & 1 \end{pmatrix} \underline{\boldsymbol{i}}_{lk} = \frac{1}{Z_0}\begin{pmatrix} 1 & -\rho \\ -\rho & 1 \end{pmatrix} \underline{\boldsymbol{v}}_{lk} + \begin{pmatrix} 0 & -1 \\ -1 & 0 \end{pmatrix}\left(\frac{\sigma}{Z_0} \underline{\boldsymbol{v}}_{l(k-1)} + \sigma\, \underline{\boldsymbol{i}}_{l(k-1)}\right) \quad (15.65)$$

Since

$$\begin{pmatrix} 1 & \rho \\ \rho & 1 \end{pmatrix}^{-1} = \frac{1}{1-\rho^2}\begin{pmatrix} 1 & -\rho \\ -\rho & 1 \end{pmatrix} \quad (15.66)$$

the left multiplication of (15.65) by (15.66) leads to:

$$\begin{aligned} \underline{\boldsymbol{i}}_{lk} &= \frac{1}{Z_0}\frac{1}{1-\rho^2}\begin{pmatrix} 1 & -\rho \\ -\rho & 1 \end{pmatrix}\begin{pmatrix} 1 & -\rho \\ -\rho & 1 \end{pmatrix} \underline{\boldsymbol{v}}_{lk} \\ &\quad + \frac{1}{1-\rho^2}\begin{pmatrix} 1 & -\rho \\ -\rho & 1 \end{pmatrix}\begin{pmatrix} 0 & -1 \\ -1 & 0 \end{pmatrix}\left(\frac{\sigma}{Z_0} \underline{\boldsymbol{v}}_{l(k-1)} + \sigma\, \underline{\boldsymbol{i}}_{l(k-1)}\right) \\ &= \frac{1}{Z_0}\begin{pmatrix} \dfrac{1+\rho^2}{1-\rho^2} & \dfrac{-2\rho}{1-\rho^2} \\ \dfrac{-2\rho}{1-\rho^2} & \dfrac{1+\rho^2}{1-\rho^2} \end{pmatrix} \underline{\boldsymbol{v}}_{lk} + \frac{\sigma}{1-\rho^2}\begin{pmatrix} \rho & -1 \\ -1 & \rho \end{pmatrix}\left(\frac{1}{Z_0} \underline{\boldsymbol{v}}_{l(k-1)} + \underline{\boldsymbol{i}}_{l(k-1)}\right) \end{aligned} \quad (15.67)$$

Therefore, the history term $\underline{\boldsymbol{\eta}}_{lk}$ as well as the admittance stamp $\underline{\boldsymbol{Y}}_{lk}$ in (15.55) when $\kappa = 1$ can be described by:

$$\underline{\boldsymbol{\eta}}_{lk} = \frac{\sigma}{1-\rho^2}\begin{pmatrix} \rho & -1 \\ -1 & \rho \end{pmatrix}\left(\frac{1}{Z_0} \underline{\boldsymbol{v}}_{l(k-1)} + \underline{\boldsymbol{i}}_{l(k-1)}\right) \quad \text{if } \kappa = 1 \quad (15.68)$$

$$\begin{aligned} \underline{\boldsymbol{Y}}_l &= \frac{1}{Z_0}\begin{pmatrix} \dfrac{1+\rho^2}{1-\rho^2} & \dfrac{-2\rho}{1-\rho^2} \\ \dfrac{-2\rho}{1-\rho^2} & \dfrac{1+\rho^2}{1-\rho^2} \end{pmatrix} = \frac{1}{Z_0}\begin{pmatrix} 1 & 0 \\ 0 & 1 \end{pmatrix} + \frac{1}{Z_0}\begin{pmatrix} \dfrac{1+\rho^2}{1-\rho^2} & \dfrac{-2\rho}{1-\rho^2} \\ \dfrac{-2\rho}{1-\rho^2} & \dfrac{1+\rho^2}{1-\rho^2} \end{pmatrix} - \frac{1}{Z_0}\begin{pmatrix} 1 & 0 \\ 0 & 1 \end{pmatrix} \\ &= \frac{1}{Z_0}\begin{pmatrix} 1 & 0 \\ 0 & 1 \end{pmatrix} + \frac{1}{Z_0}\begin{pmatrix} \dfrac{2\rho^2}{1-\rho^2} & \dfrac{-2\rho}{1-\rho^2} \\ \dfrac{-2\rho}{1-\rho^2} & \dfrac{2\rho^2}{1-\rho^2} \end{pmatrix} = \underline{\boldsymbol{Y}}_{lD} + \underline{\boldsymbol{Y}}_{lC} \quad \text{if } \kappa = 1 \end{aligned} \quad (15.69)$$

The matrix $\underline{\boldsymbol{Y}}_{lC}$ in (15.69) provides the topological coupling for $\kappa = 1$, that is, for $\tau \geq T_{\mathrm{wp}}$, as used when electromechanical transients are studied. It can also be described through the π-circuit cell as shown in Figure 15.11.

In summary, the two cases when $\kappa > 1$ and $\kappa = 1$ can be combined together:

$$\underline{\boldsymbol{\eta}}_{lk} = \begin{cases} \begin{pmatrix}0 & -1\\ -1 & 0\end{pmatrix}\left(\dfrac{\rho}{Z_0}\,\underline{\boldsymbol{v}}_{l(k-\kappa+1)} + \dfrac{\sigma}{Z_0}\,\underline{\boldsymbol{v}}_{l(k-\kappa)} + \rho\,\underline{\boldsymbol{i}}_{l(k-\kappa+1)} + \sigma\,\underline{\boldsymbol{i}}_{l(k-\kappa)}\right) & \text{if } \kappa > 1 \\ \dfrac{\sigma}{1-\rho^2}\begin{pmatrix}\rho & -1\\ -1 & \rho\end{pmatrix}\left(\dfrac{1}{Z_0}\,\underline{\boldsymbol{v}}_{l(k-\kappa)} + \underline{\boldsymbol{i}}_{l(k-\kappa)}\right) & \text{if } \kappa = 1\end{cases} \tag{15.70}$$

$$\underline{\boldsymbol{Y}}_l = \begin{cases} \underline{\boldsymbol{Y}}_{lD} & \text{if } \kappa > 1 \\ \underline{\boldsymbol{Y}}_{lD} + \underline{\boldsymbol{Y}}_{lC} & \text{if } \kappa = 1 \end{cases} \tag{15.71}$$

15.3.3.2 Multiphase Line Model. In order to model a fully transposed m-phase line, the m coupled phase quantities are transformed into m decoupled modal quantities using eigenvalue analysis as, for example, discussed in [2,12,14]. In the special case of the balanced three-phase line as shown in Figure 15.12, the decoupling can be achieved using $0\alpha\beta$ components [2,12,14] for both voltages and currents: $\underline{\boldsymbol{v}}_{l1abc} = \boldsymbol{T}_L\,\underline{\boldsymbol{v}}_{l10\alpha\beta}$, $\underline{\boldsymbol{v}}_{l2abc} = \boldsymbol{T}_L\,\underline{\boldsymbol{v}}_{l20\alpha\beta}$, $\underline{\boldsymbol{i}}_{l1abc} = \boldsymbol{T}_L\,\underline{\boldsymbol{i}}_{l10\alpha\beta}$, and $\underline{\boldsymbol{i}}_{l2abc} = \boldsymbol{T}_L\,\underline{\boldsymbol{i}}_{l20\alpha\beta}$ with the transform matrix [12]:

$$\boldsymbol{T}_L = \frac{1}{\sqrt{3}}\begin{pmatrix} 1 & \sqrt{2} & 0 \\ 1 & -\dfrac{1}{\sqrt{2}} & \sqrt{\dfrac{3}{2}} \\ 1 & -\dfrac{1}{\sqrt{2}} & -\sqrt{\dfrac{3}{2}} \end{pmatrix} \tag{15.72}$$

and $\underline{\boldsymbol{v}}_{l1abc} = (\underline{v}_{l1a}, \underline{v}_{l1b}, \underline{v}_{l1c})^{\mathrm{T}}$, $\underline{\boldsymbol{v}}_{l10\alpha\beta} = (\underline{v}_{l10}, \underline{v}_{l1\alpha}, \underline{v}_{l1\beta})^{\mathrm{T}}$, $\underline{\boldsymbol{v}}_{l2abc} = (\underline{v}_{l2a}, \underline{v}_{l2b}, \underline{v}_{l2c})^{\mathrm{T}}$, $\underline{\boldsymbol{v}}_{l20\alpha\beta} = (\underline{v}_{l20}, \underline{v}_{l2\alpha}, \underline{v}_{l2\beta})^{\mathrm{T}}$, $\underline{\boldsymbol{i}}_{l1abc} = (\underline{i}_{l1a}, \underline{i}_{l1b}, \underline{i}_{l1c})^{\mathrm{T}}$, $\underline{\boldsymbol{i}}_{l10\alpha\beta} = (\underline{i}_{l10}, \underline{i}_{l1\alpha}, \underline{i}_{l1\beta})^{\mathrm{T}}$, $\underline{\boldsymbol{i}}_{l2abc} = (\underline{i}_{l2a}, \underline{i}_{l2b}, \underline{i}_{l2c})^{\mathrm{T}}$, and $\underline{\boldsymbol{i}}_{l20\alpha\beta} = (\underline{i}_{l20}, \underline{i}_{l2\alpha}, \underline{i}_{l2\beta})^{\mathrm{T}}$. The decoupling scheme is shown in Figure 15.13.

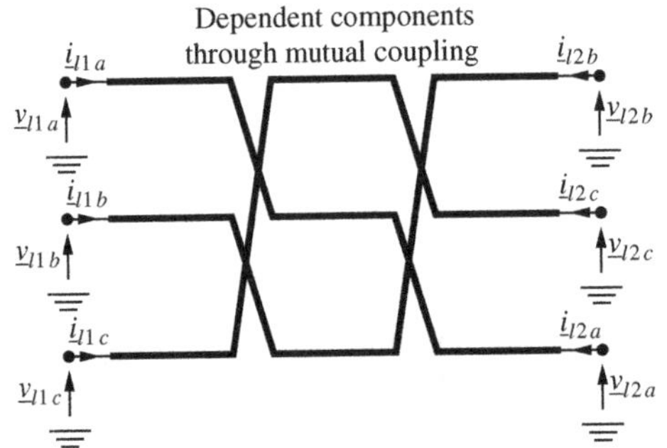

Figure 15.12. Current and voltage conventions for balanced three-phase line.

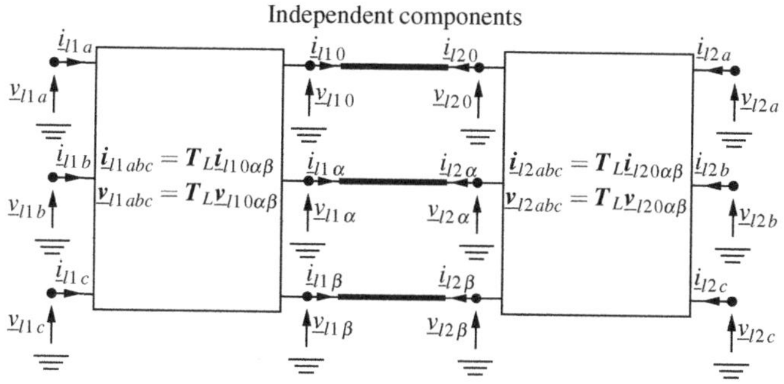

Figure 15.13. Decoupling of balanced three-phase line based on $0\alpha\beta$ components.

The three decoupled $0\alpha\beta$ modes are dealt with in the same way as a single phase in the single-phase model (15.55). In analogy to (15.55), a vector–matrix form is obtained as follows

$$\begin{pmatrix} \underline{i}_{l10\alpha\beta k} \\ \underline{i}_{l20\alpha\beta k} \end{pmatrix} = \underline{Y}_{l0\alpha\beta} \begin{pmatrix} \underline{v}_{l10\alpha\beta k} \\ \underline{v}_{l20\alpha\beta k} \end{pmatrix} + \begin{pmatrix} \underline{\eta}_{l10\alpha\beta k} \\ \underline{\eta}_{l20\alpha\beta k} \end{pmatrix} \tag{15.73}$$

where $\underline{Y}_{l\alpha\beta 0}$ is formed by stamping the elements of three 2×2 matrices $\underline{Y}_{l0}$, $\underline{Y}_{l\alpha}$, and $\underline{Y}_{l\beta}$ into the corresponding positions:

$$\underline{Y}_{l\alpha\beta 0} = \begin{pmatrix} \underline{Y}_{l0}(1,1) & & & \underline{Y}_{l0}(1,2) & & \\ & \underline{Y}_{l\alpha}(1,1) & & & \underline{Y}_{l\alpha}(1,2) & \\ & & \underline{Y}_{l\beta}(1,1) & & & \underline{Y}_{l\beta}(1,2) \\ \underline{Y}_{l0}(2,1) & & & \underline{Y}_{l0}(2,2) & & \\ & \underline{Y}_{l\alpha}(2,1) & & & \underline{Y}_{l\alpha}(2,2) & \\ & & \underline{Y}_{l\beta}(2,1) & & & \underline{Y}_{l\beta}(2,2) \end{pmatrix} \tag{15.74}$$

For example, the two ends of the α component correspond to the second and fifth rows in the current and voltage vectors; therefore, $\underline{Y}_{l\alpha}$ is stamped into $\underline{Y}_{l\alpha\beta 0}$ at the four intersections of the second and fifth rows and columns.

Transforming back to the original *abc* quantities with T_L^{-1} gives

$$\begin{pmatrix} T_L^{-1}\,\underline{i}_{l1abck} \\ T_L^{-1}\,\underline{i}_{l2abck} \end{pmatrix} = \underline{Y}_{l0\alpha\beta} \begin{pmatrix} T_L^{-1}\,\underline{v}_{l1abck} \\ T_L^{-1}\,\underline{v}_{l2abck} \end{pmatrix} + \begin{pmatrix} T_L^{-1}\,\underline{\eta}_{l1abck} \\ T_L^{-1}\,\underline{\eta}_{l2abck} \end{pmatrix} \tag{15.75}$$

which can be recast as:

$$\begin{pmatrix} \underline{i}_{l1abck} \\ \underline{i}_{l2abck} \end{pmatrix} = \underline{Y}_{labc} \begin{pmatrix} \underline{v}_{l1abck} \\ \underline{v}_{l2abck} \end{pmatrix} + \begin{pmatrix} \underline{\eta}_{l1abck} \\ \underline{\eta}_{l2abck} \end{pmatrix} \tag{15.76}$$

where the 6×6 matrix $\underline{Y}_{labc}$ for representing the three-phase balanced line in the phase domain is

$$\underline{Y}_{labc} = \begin{pmatrix} T_L & 0 \\ 0 & T_L \end{pmatrix} \underline{Y}_{l0\alpha\beta} \begin{pmatrix} T_L^{-1} & 0 \\ 0 & T_L^{-1} \end{pmatrix} \tag{15.77}$$

Similar procedures can be used for transmission lines with more than three phases. When losses are represented by lumped multiphase resistances at the ends and the middle of the lines, then the modal transform is also applied to the multiphase resistances.

15.3.4 Synchronous Machine in *dq0* Domain

The modeling of synchronous machinery is essential to the study of electromechanical transients. The Park transformation from *abc* to *dq0* variables is popular in synchronous machine modeling because it allows for the representation of inductance terms independent of the present rotor angle [15]. Adopting the Park transformation, the synchronous machine modeling can be organized as shown in Figure 15.14 [16]. The three-phase machine terminal voltage $\underline{\boldsymbol{v}}_{Mabc}$ of the *abc* domain is mapped into the voltage $\underline{\boldsymbol{v}}_{Mdq0}$ of the *dq0* domain by the Park transform. The machine model provides as output the three-phase terminal current $\underline{\boldsymbol{i}}_{Mabc}$ following a back transform from $\underline{\boldsymbol{i}}_{Mdq0}$ in the *dq0* domain to the *abc* domain to ensure compatibility with the network model, which is also represented in the *abc* domain.

The exciter model uses a reference voltage U_{ref} and the real part of terminal voltage $\underline{\boldsymbol{v}}_{Mabc}$ to provide the field voltage U_{f} as input to the machine equations. The model of the governor-turbine stage compares the rotor electrical angular speed ω_{r} with the speed reference ω_{ref} and provides the mechanical torque C_{m} as an input to the machine equations. The modeling of those is described in Chapters 2 and 3.

The real part of $\underline{\boldsymbol{i}}_{Mdq0}$ is obtained from the electromagnetic and mechanical machine equations which has as inputs the real part of $\underline{\boldsymbol{v}}_{Mdq0}$, U_{f}, and C_{m}, and the imaginary part of $\underline{\boldsymbol{i}}_{Mdq0}$ can be constructed as will be explained later in this section.

15.3.4.1 Electromagnetic and Mechanical Machine Equations. In Section 2.1.3.4, the electromagnetic and mechanical machine equations make use of the per unit system are described. The electromagnetic equations (2.39′), (2.40′), (2.32′) and (2.34′) can be written in matrix form and the time *t* is not per unit here:

$$\frac{1}{\omega_0}\frac{\mathrm{d}\boldsymbol{\psi}_{M*}}{\mathrm{d}t} = \boldsymbol{Z}_{M^*}\boldsymbol{i}_{M^*} + \boldsymbol{v}_{M^*} \tag{15.78}$$

$$\frac{1}{\omega_0}\frac{\mathrm{dRe}[\underline{\psi}_{0^*}]}{\mathrm{d}t} = R_*\mathrm{Re}[\underline{i}_{0^*}] + \mathrm{Re}[\underline{v}_{0^*}] \tag{15.79}$$

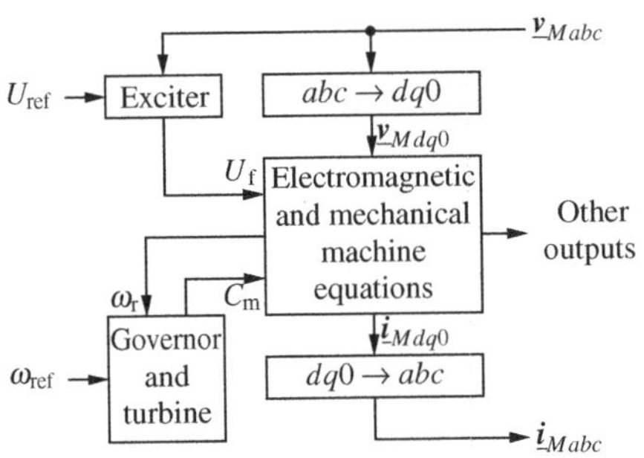

Figure 15.14. Block diagram organization of synchronous machine model in *dq0* domain.

where

$$\Psi_{M^*} = \left(\mathrm{Re}[\underline{\psi}_{d^*}] \quad \psi_{f^*} \quad \psi_{D^*} \quad \mathrm{Re}[\underline{\psi}_{q^*}] \quad \psi_{Q^*}\right)^{\mathrm{T}} \tag{15.80}$$

$$\boldsymbol{i}_{M^*} = \left(\mathrm{Re}[\underline{i}_{Md^*}] \quad i_{f^*} \quad i_{D^*} \quad \mathrm{Re}[\underline{i}_{Mq^*}] \quad i_{Q^*}\right)^{\mathrm{T}} \tag{15.81}$$

$$\boldsymbol{v}_{M^*} = \left(\mathrm{Re}[\underline{v}_{Md^*}] \quad U_{f^*} \quad 0 \quad \mathrm{Re}[\underline{v}_{Mq^*}] \quad 0\right)^{\mathrm{T}} \tag{15.82}$$

$$\boldsymbol{Z}_{M^*} = \begin{pmatrix} R_* & 0 & 0 & -\omega_{\mathrm{r}^*} L_{q^*} & \omega_{\mathrm{r}^*} L_{mQ^*} \\ 0 & -R_{f^*} & 0 & 0 & 0 \\ 0 & 0 & -R_{D^*} & 0 & 0 \\ \omega_{\mathrm{r}^*} L_{d^*} & -\omega_{\mathrm{r}^*} L_{md^*} & -\omega_{\mathrm{r}^*} L_{mD^*} & R_* & 0 \\ 0 & 0 & 0 & 0 & -R_{Q^*} \end{pmatrix} \tag{15.83}$$

The flux linkage equations (2.32′) and (2.34′) (see Section 2.1.2) can also be written in matrix form:

$$\psi_{M^*} = \boldsymbol{L}_{M^*} \boldsymbol{i}_{M^*} \tag{15.84}$$

where

$$\boldsymbol{L}_{M^*} = \begin{pmatrix} -L_{d^*} & L_{md^*} & L_{mD^*} & 0 & 0 \\ -L_{md^*} & L_{f^*} & L_{fD^*} & 0 & 0 \\ -L_{mD^*} & L_{fD^*} & L_{D^*} & 0 & 0 \\ 0 & 0 & 0 & -L_{q^*} & L_{mQ^*} \\ 0 & 0 & 0 & -L_{mQ^*} & L_{Q^*} \end{pmatrix} \tag{15.85}$$

and

$$\mathrm{Re}[\underline{\psi}_{0^*}] = -L_{0^*} \mathrm{Re}[\underline{i}_{0^*}] \tag{15.86}$$

The electromechanical equations given in (2.12) (see Section 2.1.2) are given here in per unit form:

$$\frac{\mathrm{d}\omega_{\mathrm{r}^*}}{\mathrm{d}t} = \frac{1}{2H}\left(C_{\mathrm{m}^*} - C_{\mathrm{e}^*} - D(\omega_{\mathrm{r}^*} - 1)\right) \tag{15.87}$$

$$\frac{\mathrm{d}\delta}{\mathrm{d}t} = \omega_0 \omega_{\mathrm{r}^*} \tag{15.88}$$

The angle δ (in electric radians) is the rotor position, at any instant t, with respect to a synchronously rotating reference frame (see Figure 2.4). The air-gap torque is calculated by:

$$C_{\mathrm{e}^*} = \mathrm{Re}[\underline{\psi}_{d^*}]\mathrm{Re}[\underline{i}_{Mq^*}] - \mathrm{Re}[\underline{\psi}_{q^*}]\mathrm{Re}[\underline{i}_{Md^*}] \tag{15.89}$$

15.3.4.2 Calculation of Real Part of Stator Current. By substitution of (15.84) and (15.86) into (15.78) and (15.79), respectively, and converting the differential equations into difference equations, the real part of $\underline{\boldsymbol{i}}_{Mdq0}$ at time step k can be obtained. To have a noniterative solution, the values of $\boldsymbol{v}_{M^*}$ and $\mathrm{Re}[\underline{v}_{0^*}]$ are taken at time step $k-1$. The value of Z_{M^*} is also taken at time step $k-1$ because Z_{M^*} is a function of the rotor electrical angular speed ω_{r^*}, which is a function of time. For $\boldsymbol{i}_{M^*}$ and $\mathrm{Re}[\underline{i}_{0^*}]$, it is possible to take their values at time step k because they can be moved to the left of the equations and combined to the corresponding values at time step k. Therefore, the difference equations for solving the real part of stator currents are as follows:

$$\frac{\boldsymbol{L}_{M^*}}{\omega_0}\frac{\boldsymbol{i}_{M^*k}-\boldsymbol{i}_{M^*(k-1)}}{\tau}=\boldsymbol{Z}_{M^*(k-1)}\boldsymbol{i}_{M^*k}+\boldsymbol{v}_{M^*(k-1)} \tag{15.90}$$

$$\frac{-L_{0^*}}{\omega_0}\frac{\mathrm{Re}[\underline{i}_{0^*k}]-\mathrm{Re}[\underline{i}_{0^*(k-1)}]}{\tau}=R_*\mathrm{Re}[\underline{i}_{0^*k}]+\mathrm{Re}\left[\underline{v}_{0^*(k-1)}\right] \tag{15.91}$$

They can be reorganized as:

$$\left(\boldsymbol{L}_{M^*}-\tau\omega_0\boldsymbol{Z}_{M^*(k-1)}\right)\boldsymbol{i}_{M^*k}=\boldsymbol{L}_{M^*}\boldsymbol{i}_{M^*(k-1)}+\tau\omega_0\boldsymbol{v}_{M^*(k-1)} \tag{15.92}$$

$$(-L_{0^*}-\tau\omega_0R_*)\mathrm{Re}[\underline{i}_{0^*k}]=-L_{0^*}\mathrm{Re}[\underline{i}_{0^*(k-1)}]+\tau\omega_0\mathrm{Re}[\underline{v}_{0^*(k-1)}] \tag{15.93}$$

Therefore

$$\boldsymbol{i}_{M^*k}=\left(\boldsymbol{L}_{M^*}-\tau\omega_0\boldsymbol{Z}_{M^*(k-1)}\right)^{-1}\left(\boldsymbol{L}_{M^*}\boldsymbol{i}_{M^*(k-1)}+\tau\omega_0\boldsymbol{v}_{M^*(k-1)}\right) \tag{15.94}$$

$$\mathrm{Re}[\underline{i}_{0^*k}]=\frac{L_{0^*}\mathrm{Re}[\underline{i}_{0^*(k-1)}]-\tau\omega_0\mathrm{Re}[\underline{v}_{0^*(k-1)}]}{L_{0^*}+\tau\omega_0R_*} \tag{15.95}$$

15.3.4.3 Calculation of Imaginary Part of Stator Current. With the models presented above, the real part of $\underline{\boldsymbol{i}}_{Mdq0}$ is obtained. In order to generate $\underline{\boldsymbol{i}}_{Mdq0}$ in Figure 15.14, it is necessary to also construct the imaginary part of $\underline{\boldsymbol{i}}_{Mdq0}$. For analytic signals, the imaginary part is obtained by the Hilbert transform of the real part. To find an equation for obtaining the imaginary part, it is helpful to consider the Park transformation:

$$\underline{\boldsymbol{i}}_{Mabc}=\boldsymbol{P}_M\,\underline{\boldsymbol{i}}_{Mdq0} \tag{15.96}$$

where

$$\boldsymbol{P}_M=\begin{pmatrix}\cos\theta & -\sin\theta & 1\\ \cos\left(\theta-\dfrac{2\pi}{3}\right) & -\sin\left(\theta-\dfrac{2\pi}{3}\right) & 1\\ \cos\left(\theta+\dfrac{2\pi}{3}\right) & -\sin\left(\theta+\dfrac{2\pi}{3}\right) & 1\end{pmatrix} \tag{15.97}$$

The angle θ is the angle by which the d axis leads the axis of phase a, as illustrated in Figure 2.6.

Phase a of the transformed signal becomes:

$$\underline{i}_{Ma} = \underline{i}_{Md} \cos\theta - \underline{i}_{Mq} \sin\theta + \underline{i}_{M0} \qquad (15.98)$$

or

$$\mathrm{Re}[\underline{i}_{Ma}] = \mathrm{Re}[\underline{i}_{Md}]\cos\theta - \mathrm{Re}[\underline{i}_{Mq}]\sin\theta + \mathrm{Re}[\underline{i}_{M0}] \qquad (15.99)$$

and

$$\mathrm{Im}[\underline{i}_{Ma}] = \mathrm{Im}[\underline{i}_{Md}]\cos\theta - \mathrm{Im}[\underline{i}_{Mq}]\sin\theta + \mathrm{Im}[\underline{i}_{M0}] \qquad (15.100)$$

From the definition (15.2), $\underline{\boldsymbol{i}}_{Mdq0}$ is an analytic signal when

$$\mathrm{Im}[\underline{i}_{Ma}] = \mathcal{H}[\mathrm{Re}[\underline{i}_{Ma}]] \qquad (15.101)$$

Substitution of (15.99) in (15.101) yields:

$$\mathrm{Im}[\underline{i}_{Ma}] = \mathcal{H}[\mathrm{Re}[\underline{i}_{Md}]\cos\theta] - \mathcal{H}[\mathrm{Re}[\underline{i}_{Mq}]\sin\theta] + \mathcal{H}[\mathrm{Re}[\underline{i}_{M0}]] \qquad (15.102)$$

In steady state, $\mathrm{Re}[\underline{i}_{Md}]$, $\mathrm{Re}[\underline{i}_{Mq}]$ and $\mathrm{Re}[\underline{i}_{M0}]$ are constant. In steady state (15.102) becomes:

$$\mathrm{Im}[\underline{i}_{Ma}] = \mathrm{Re}[\underline{i}_{Md}]\mathcal{H}[\cos\theta] - \mathrm{Re}[\underline{i}_{Mq}]\mathcal{H}[\sin\theta] = \mathrm{Re}[\underline{i}_{Mq}]\cos\theta + \mathrm{Re}[\underline{i}_{Md}]\sin\theta \qquad (15.103)$$

Subtracting (15.103) from (15.100) yields:

$$(\mathrm{Im}[\underline{i}_{Md}] - \mathrm{Re}[\underline{i}_{Mq}])\cos\theta - (\mathrm{Im}[\underline{i}_{Mq}] + \mathrm{Re}[\underline{i}_{Md}])\sin\theta + \mathrm{Im}[\underline{i}_{M0}] = 0 \qquad (15.104)$$

Equation (15.104) should hold for any θ, therefore:

$$\mathrm{Im}[\underline{i}_{Md}] = \mathrm{Re}[\underline{i}_{Mq}] \qquad (15.105)$$

$$\mathrm{Im}[\underline{i}_{Mq}] = -\mathrm{Re}[\underline{i}_{Md}] \qquad (15.106)$$

$$\mathrm{Im}[\underline{i}_{M0}] = 0 \qquad (15.107)$$

The same conditions also need to hold when phases b and c are considered.

Equations (15.105), (15.106), and (15.107) exactly hold for constructing the imaginary part of $\underline{\boldsymbol{i}}_{Mdq0}$ from the real part of $\underline{\boldsymbol{i}}_{Mdq0}$. In practical application, these equations are also applicable when the system is subject to slow transients or fast transients. For slow transients, $\mathrm{Re}[\underline{i}_{Md}]$, $\mathrm{Re}[\underline{i}_{Mq}]$, and $\mathrm{Re}[\underline{i}_{M0}]$ change slowly, and (15.103) provides an acceptable approximation of (15.102); for fast transients, natural waveforms are tracked so that the imaginary part of an analytic signal is of little importance.

15.3.4.4 Calculation of Rotor Speed and Angle. With the obtained stator currents, the flux linkage at time step k, ψ_{M^*k}, can be calculated by (15.84), and the air-gap torque at time step k, C_{e^*k}, can be calculated by (15.89). By converting differential equations (15.87) and (15.88) into difference equations, the rotor electrical angular speed ω_{r} and rotor angle δ at time step k can be obtained. Using the trapezoidal rule, (15.87) and (15.88) become:

$$\frac{\omega_{\mathrm{r}^*k} - \omega_{\mathrm{r}^*(k-1)}}{\tau} = \frac{(C_{\mathrm{m}^*k} - C_{\mathrm{e}^*k} - D(\omega_{\mathrm{r}^*k} - 1)) + (C_{\mathrm{m}^*(k-1)} - C_{\mathrm{e}^*(k-1)} - D(\omega_{\mathrm{r}^*(k-1)} - 1))}{4H} \tag{15.108}$$

$$\frac{\delta_k - \delta_{k-1}}{\tau} = \omega_0 \frac{\omega_{\mathrm{r}^*k} + \omega_{\mathrm{r}^*(k-1)}}{2} \tag{15.109}$$

Equations (15.108) and (15.109) can be reorganized as:

$$\omega_{\mathrm{r}^*k} = \frac{4H - \tau D}{4H + \tau D}\omega_{\mathrm{r}^*(k-1)} + \frac{\tau}{4H + \tau D}(C_{\mathrm{m}^*k} + C_{\mathrm{m}^*(k-1)} - C_{\mathrm{e}^*k} - C_{\mathrm{e}^*(k-1)} + 2D) \tag{15.110}$$

$$\delta_k = \delta_{k-1} + \frac{\tau\omega_0}{2}(\omega_{\mathrm{r}^*k} + \omega_{\mathrm{r}^*(k-1)}) \tag{15.111}$$

On the right of (15.110), C_{m^*k} is obtained from the output of the governor and turbine. After ω_{r^*k} is obtained by (15.110), δ_k is calculated by (15.111).

15.3.4.5 Integration with AC Network. It can be seen from Figure 15.14 that the model accepts the three-phase terminal voltage that it shares with the network as input, and injects the three-phase terminal current as output into the network. Therefore, from the network side the synchronous machine model described above is in fact a voltage-controlled current source. If a noniterative simulation process is adopted, then three-phase terminal current injection at time step k, $\underline{i}_{Mabck}$, is obtained based on terminal voltage information available in time step $k-1$. To attain a fast simulation process with good numerical stability and accuracy, an adjustment current source $\underline{i}_{Aabck}$ and the adjustment three-phase resistance $\boldsymbol{G}_{Aabc}$ are introduced at the machine terminal. This leads to the interface as illustrated in the circuit model of Figure 15.15.

The per-phase resistance G_A of $\boldsymbol{G}_{Aabc}$ is selected to reflect the characteristic impedance [17] when the machine operates near open-circuit conditions:

$$G_A = \frac{\tau}{L''_{\mathrm{d}} + L''_{\mathrm{q}}} \tag{15.112}$$

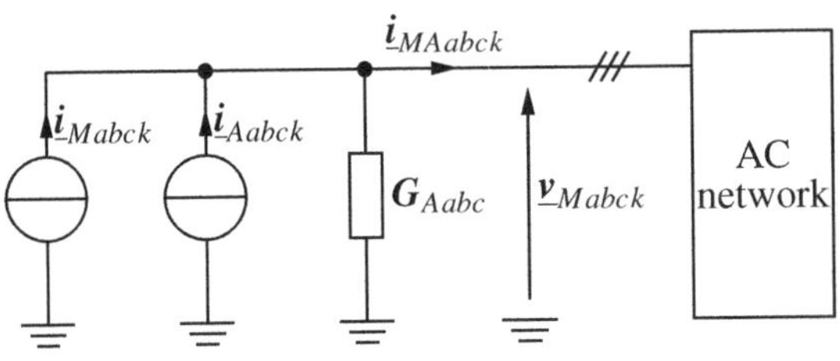

Figure 15.15. Network interfacing of synchronous machine model in *dq0* domain.

where L''_d and L''_q are the subtransient inductances of the stator circuit. The adjustment current source is as follows:

$$\underline{\boldsymbol{i}}_{Aabck} = G_A \, \underline{\boldsymbol{v}}_{Mabck-1} e^{j\omega_{rk}\tau} \tag{15.113}$$

where ω_{rk} is obtained at the same time as $\underline{\boldsymbol{i}}_{Aabck}$. The net current $\underline{\boldsymbol{i}}_{MAabck}$ flowing into the network considering $\underline{\boldsymbol{i}}_{Mabck}$, $\underline{\boldsymbol{i}}_{Aabck}$ and the current flowing through $\boldsymbol{G}_{Aabc}$ is given as the sum of $\underline{\boldsymbol{i}}_{Mabck}$ and an adjustment term:

$$\underline{\boldsymbol{i}}_{MAabck} = \underline{\boldsymbol{i}}_{Mabck} + G_A \left(\underline{\boldsymbol{v}}_{Mabck-1} e^{j\omega_{rk}\tau} - \underline{\boldsymbol{v}}_{Mabck} \right) \tag{15.114}$$

The interface allows the characteristic impedance G_A to be seen from the network for sudden voltage changes.

From (15.112), it can be seen that G_A increases as the time step size decreases. When electromagnetic transients are studied at very small time step sizes, the adjustment term on the right of (15.114) is therefore very small. When electromechanical transients are studied, the adjustment term is small too, even though the time step size is much larger. This is because the three-phase terminal voltage retains its sinusoidal shape for such low-frequency transients and therefore $\underline{\boldsymbol{v}}_{Mabck-1} e^{j\omega_{rk}\tau} \approx \underline{\boldsymbol{v}}_{Mabck}$. In steady state, the adjustment term is zero because $\underline{\boldsymbol{v}}_{Mabck-1} e^{j\omega_{rk}\tau} = \underline{\boldsymbol{v}}_{Mabck}$ for any time step size.

15.3.4.6 Initialization. A simulation can start with zero initial conditions and can then reach steady state fast. However, when the simulated network contains synchronous machines, the following initialization method needs to be used so that the simulated network can reach to steady state fast enough. The first step is to use the steady-state calculation from power flow results to obtain machine quantities as described in Section 2.1.3.6. Then the machines act as sinusoidal current sources until the entire network reaches steady state. After that, the electromagnetic equations in the machine model are taken into account while the electromechanical equations are not. This means the rotation speed of the machine is fixed. When the transients fade away, the electromechanical equations can also be included and now the machines are simulated by the full model described above.

15.4 APPLICATION SIMULATION OF BLACKOUT

The network depicted in Figure 15.16 is identical in structure to the two-area system described in [15]. Two areas, each with two synchronous machines and loads, are

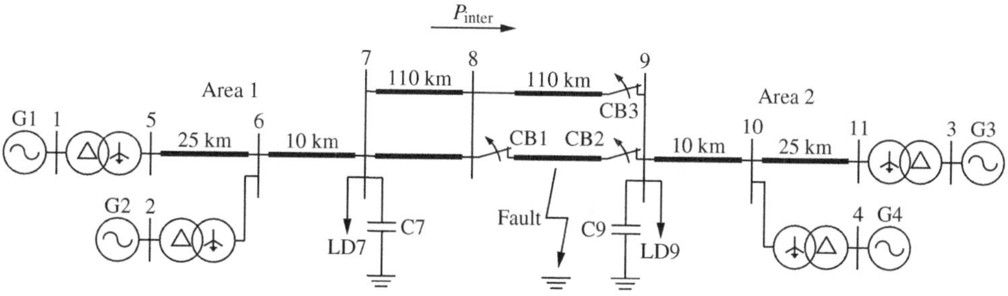

Figure 15.16. Four-machine two-area test system.

TABLE 15.1. Generator Parameters

$R_* = 0.0025$ p.u.	$X_{d^*} = 1.8$ p.u.	$X'_{d^*} = 0.3$ p.u.	$X''_{d^*} = 0.25$ p.u.	$T'_{d0^*} = 8.0$ s	$T''_{d0^*} = 0.03$ s
$X_{l^*} = 0.2$ p.u.	$X_{q^*} = 1.7$ p.u.	$X'_{q^*} = 0.55$ p.u.	$X''_{q^*} = 0.25$ p.u.	$T'_{q0^*} = 0.4$ s	$T''_{q0^*} = 0.05$ s
$D = 0$	$V_b = 20$ kV	$S_b = 900$ MVA	$H = 6.5$ (G1, G2),	$H = 6.175$ (G3, G4)	

TABLE 15.2. Transmission Line Parameters

$L'_* = 2.653\text{e-}6$ p.u./km	$C'_* = 4.642\text{e-}6$ p.u./km	$R'_* = 0.0001$ p.u./km	$V_b = 230$ kV	$S_b = 100$ MVA

TABLE 15.3. Transformer Parameters (per Phase)

$L_* = 0.15$ p.u.	$a = (230/\sqrt{3})/20 = 6.64$
$V_{b\ \text{primary}} = 230/\sqrt{3} = 132.8$ kV, $V_{b\ \text{secondary}} = 20$ kV	$S_b = 900$ MVA

connected through weak tie lines. In steady state, each generator provides a real power of the order of 700 MW, while constant-impedance loads LD7 and LD9, respectively, consume about 1000 and 1800 MW. About $P_{\text{inter}} = 400$ MW flows over the tie lines. Reactive power also flows from area 1 to area 2. The scenario considered hereafter shows a sequence of diverse phenomena that are common to blackouts [1]. It starts with a short circuit, which can be caused by a line touching a tree. Such a short circuit triggers electromagnetic transients. The missing line leads to an increase of the impedance between the two areas weakening the tie further. This gives rise to electromechanical transients involving inter-area power oscillations over the one remaining tie line. If no other actions to stop the overload condition of the remaining tie line are undertaken in time, then a tripping of this line to decompose the system into two islands may be necessary. If the formerly importing area experiences a lack of power generation capability, then load shedding can prevent further problems and stop the cascading events.

The inter-area tie lines were represented through the scale-bridging line model introduced in Section 15.3.3. Other lines were represented by the π-models with lumped parameters. The passive resistive and reactive loads were also modeled as lumped RLC components. Each transformer was modeled by connecting three single-phase transformer models as described in Section 15.3.2. The synchronous machines were represented by the models presented in Section 15.3.4. Network parameters are shown in Tables 15.1, 15.2, and 15.3 [15] where R'_* is the distributed resistance.

The results obtained are depicted in Figures 15.17, 15.18, and 15.19. From the voltage of load LD9 in Figure 15.17, it can be seen that the simulation alternately tracks the envelope or the natural waveform.

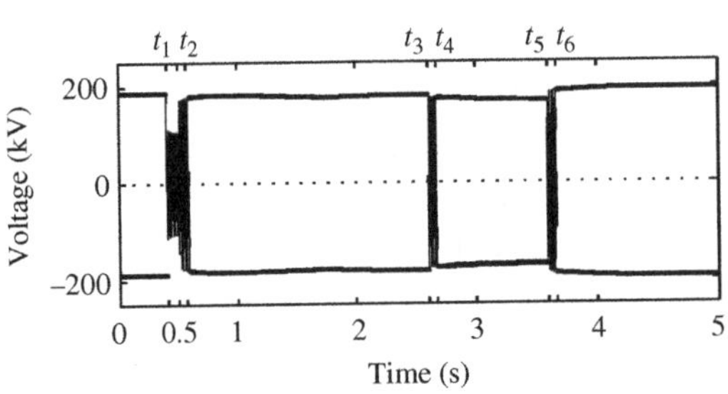

Figure 15.17. Phase a voltage of load LD9: solid light: natural waveform; solid bold: envelope.

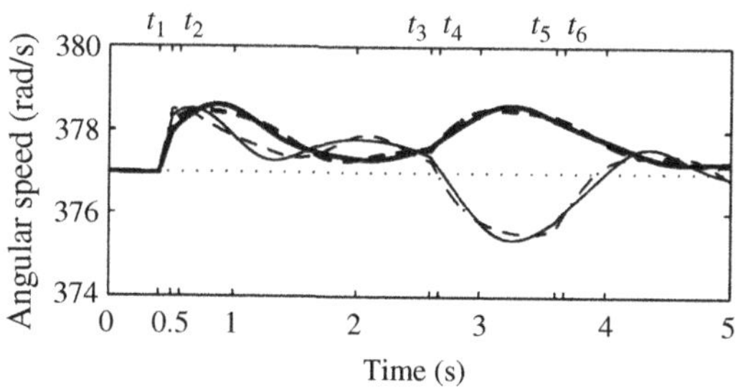

Figure 15.18. Machine angular speed; solid bold: G1; dash dot bold: G2; solid light: G3; dash dot light: G4.

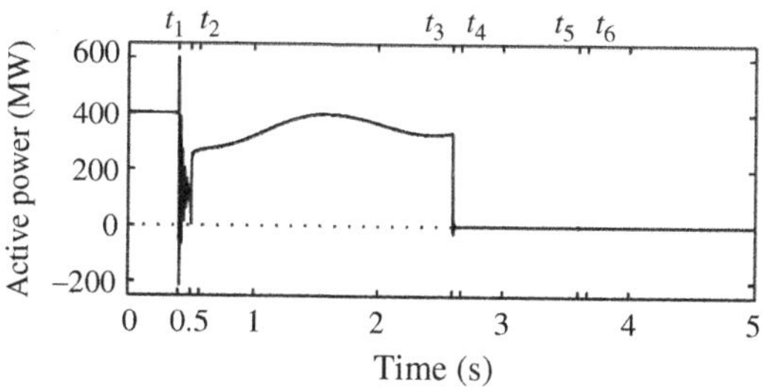

Figure 15.19. Interarea active power P_{inter}.

The two-dimensional parameter control of the FAST method for the test case at different times is illustrated in Figure 15.20, where each cross is an operating point representing the simulation setting for a given time period, and the arrows with time points show the transitions of the operating points. The x-axis stands for the time step size τ, and the y-axis stands for the shift frequency f_s.

The simulation starts out in steady state at $t = 0$ s. Therefore, the FAST method tracks envelope waveforms at the beginning of the simulation. As shown in Figure 15.20, initially the shift frequency f_s is equal to the carrier frequency of 60 Hz and the time step size τ equals 2 ms. As the permanent three-phase fault occurs at $t_1 = 0.4$ s at the center of one of the lines interconnecting Bus 8 and Bus 9, the simulation parameters are reset to $f_s = 0$ Hz and $\tau = 50\ \mu$s, and natural waveforms are now being tracked since electromagnetic transients are expected. The simulation results in Figure 15.17 show a drop of the voltage of load LD9, which therefore consumes less power. More energy is therefore absorbed by the generation units' rotating shafts, which show speed-ups in Figure 15.18. At $t = 0.5$ s, the

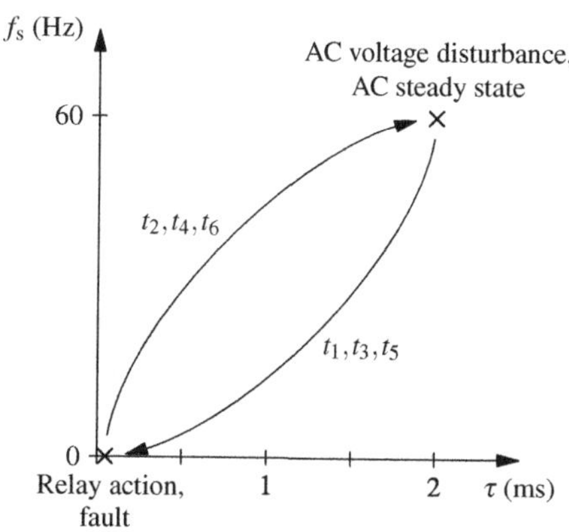

Figure 15.20. Obtained two-dimensional settings of shift frequency f_s and time step size τ in study of transients in two-area system.

breakers CB1 and CB2 open to take out the faulted line. The tracking of instant natural waveforms continues. By $t_2 = 0.57$ s the electromagnetic transients due to the actions of the breakers have sufficiently damped out to resume envelope tracking at $f_s = 60$ Hz and $\tau = 2$ ms. However, the interconnection between both areas is now weakened due to the loss of one of the tie lines. This triggers inter-area oscillations of the power P_{inter} as visible in Figure 15.19. The oscillation can also be noticed from Figure 15.18. The angular speeds of generators G1 and G2 within area 1 are in phase, and so are the angular speeds of generators G3 and G4 within Area 2. As in inter-area oscillations, the generators in area 1 swing in anti-phase to those in Area 2. In FAST, these electromechanical transients are emulated with envelope tracking at time step size $\tau = 2$ ms.

In the official report investigating the August 14, 2003 blackout in the USA and Canada [1], it is said that "commonly used protective relays that measure low voltage and high current cannot distinguish between the currents and voltages seen in a system cascade from those caused by a fault. This leads to more and more lines and generators being tripped". If for the same reason the remaining line in service between Bus 8 and Bus 9 is taken out because of its overload, then an opening of CB3 at $t_3 = 2.6$ s decouples area 1 and area 2 into two islands. The zero power flow P_{inter} confirms this. After a brief period of electromagnetic transient and tracking of natural waveforms following the opening of CB3, envelope tracking resumes at $t_4 = 2.67$ s at a larger time step as illustrated by Figure 15.20. The voltage envelope across LD9 in Figure 15.17 decreases steadily. This points to a lack of reactive power at the load LD9 following the islanding. At $t_5 = 3.60$ s, a quarter of LD9 is shed to avoid further voltage drop. This leads to the partial blackout in area 2. To simulate the shedding action accurately, the natural waveforms are captured until $t_6 = 3.67$ s. Then, envelope tracking resumes.

Over the entire simulated scenario, the simulation algorithm of FAST processes analytic signals and can so adapt to diverse types of waveforms for an efficient simulation of electromagnetic and electromechanical transients.

REFERENCES

[1] Final report on the August 14, 2003 blackout in the United States and Canada: causes and recommendations, U.S.-Canada Power System Outage Task Force, Technical report, U.S. Secretary of Energy and Minister of Natural Resources, Canada, 2004.

[2] Watson, N., Arrillaga. J. *Power systems electromagnetic transients simulation*, Institution of Electrical Engineers, London, 2003.

[3] Dommel, H.W. Digital computer solution of electromagnetic transients in single- and multi-phase networks, *IEEE Transactions on Power Apparatus and Systems*, Vol. 102, No. 6, pp. 388–399, April 1969.

[4] Woodford, D.A., Gole, A.M., Menzies, R.W. Digital simulation of DC links and AC machines, *IEEE Transactions on Power Apparatus and Systems*, Vol. 102, No. 6, pp. 1616–1623, June 1983.

[5] Ilic, M., Zaborszky, J. *Dynamics and control of large electric power systems*, John Wiley & Sons, New York, 2000.

[6] Stanković, A.M., Lesieutre, B.C., Aydin, T. Modeling and analysis of singlephase induction machines with dynamic phasors, *IEEE Transactions on Power Systems*, Vol. 14, No. 1, pp. 9–14, February 1999.

[7] Lüke, H.D. *Signalübertragung*, 4th edition, Springer-Verlag, Berlin, 1990.

[8] Mitra, S.K. *Digital signal processing: a computer-based approach*, 2nd edition, McGraw-Hill, New York, 2001.

[9] Strunz, K., Shintaku, R., Gao, F. Frequency-adaptive network modeling for integrative simulation of natural and envelope waveforms in power systems and circuits, *IEEE Trans. Circuits Syst. I Fundam. Theory Appl.*, Vol. 53, No. 12, pp. 2788–2803, December 2006.

[10] Gao, F., Strunz, K. Frequency-adaptive power system modeling for multi-scale simulation of transients, *IEEE Trans. Power Syst.*, Vol. 24, No. 2, pp. 561–571, 2009.

[11] Chua, L.O., Lin, P.M. *Computer aided analysis of electronic circuits: algorithms and computational techniques*, Prentice-Hall, Englewood Cliffs, NJ, 1975.

[12] Dommel, H.W. *EMTP Theory book. Microtran power system analysis corporation*, 2nd edition, Vancouver, British Columbia, 1992.

[13] Gao, F., Strunz, K. Modeling of constant distributed parameter transmission line for simulation of natural and envelope waveforms in power electric networks, *37th North American Power Symposium (NAPS)*, Ames, USA, October, 2005.

[14] Clarke, E. *Circuit analysis of AC power systems*, Vol. 1, John Wiley & Sons, New York, USA, 1943.

[15] Kundur, P. *Power system stability and control*. McGraw-Hill, New York, 1993.

[16] Gao, F., Strunz, K. Multi-scale simulation of multi-machine power systems. *Int. J. Elect. Power Energy Syst.*, Vol. 31, No. 9, pp. 538–545, 2009.

[17] Gole, A.M., Menzies, R.W., Turanli, H.M., Woodford, D.A. Improved interfacing of electrical machine models to electromagnetic transactions programs, *IEEE Trans. Power Apparatus Syst.*, Vol. 103, No. 9, pp. 2446–2451, 1984.

INDEX

IEEE Press Series on Power Engineering

Series Editor: M. E. El-Hawary, Dalhousie University, Halifax, Nova Scotia, Canada

The mission of IEEE Press Series on Power Engineering is to publish leading-edge books that cover the broad spectrum of current and forward-looking technologies in this fast-moving area. The series attracts highly acclaimed authors from industry/academia to provide accessible coverage of current and emerging topics in power engineering and allied fields. Our target audience includes the power engineering professional who is interested in enhancing their knowledge and perspective in their areas of interest.

1. *Principles of Electric Machines with Power Electronic Applications, Second Edition*
M. E. El-Hawary

2. *Pulse Width Modulation for Power Converters: Principles and Practice*
D. Grahame Holmes and Thomas Lipo

3. *Analysis of Electric Machinery and Drive Systems, Second Edition*
Paul C. Krause, Oleg Wasynczuk, and Scott D. Sudhoff

4. *Risk Assessment of Power Systems: Models, Methods, and Applications*
Wenyuan Li

5. *Optimization Principles: Practical Applications to the Operations of Markets of the Electric Power Industry*
Narayan S. Rau

6. *Electric Economics: Regulation and Deregulation*
Geoffrey Rothwell and Tomas Gomez

7. *Electric Power Systems: Analysis and Control*
Fabio Saccomanno

8. *Electrical Insulation for Rotating Machines: Design, Evaluation, Aging, Testing, and Repair*
Greg Stone, Edward A. Boulter, Ian Culbert, and Hussein Dhirani

9. *Signal Processing of Power Quality Disturbances*
Math H. J. Bollen and Irene Y. H. Gu

10. *Instantaneous Power Theory and Applications to Power Conditioning*
Hirofumi Akagi, Edson H. Watanabe, and Mauricio Aredes

11. *Maintaining Mission Critical Systems in a 24/7 Environment, Second Edition*
Peter M. Curtis

12. *Elements of Tidal-Electric Engineering*
Robert H. Clark

13. *Handbook of Large Turbo-Generator Operation Maintenance, Second Edition*
Geoff Klempner and Isidor Kerszenbaum

14. *Introduction to Electrical Power Systems*
Mohamed E. El-Hawary

15. *Modeling and Control of Fuel Cells: Disturbed Generation Applications*
M. Hashem Nehrir and Caisheng Wang

16. *Power Distribution System Reliability: Practical Methods and Applications*
Ali A. Chowdhury and Don O. Koval

17. *Introduction to FACTS Controllers: Theory, Modeling, and Applications*
Kalyan K. Sen and Mey Ling Sen

18. *Economic Market Design and Planning for Electric Power Systems*
James Momoh and Lamine Mili

19. *Operation and Control of Electric Energy Processing Systems*
James Momoh and Lamine Mili

20. *Restructured Electric Power Systems: Analysis of Electricity Markets with Equilibrium Models*
Xiao-Ping Zhang

21. *An Introduction to Wavelet Modulated Inverters*
S.A. Saleh and M. Azizur Rahman

22. *Probabilistic Transmission System Planning*
Wenyuan Li

23. *Control of Electric Machine Drive Systems*
Seung-Ki Sul

24. *High Voltage and Electrical Insulation Engineering*
Ravindra Arora and Wolfgang Mosch

25. *Practical Lighting Design with LEDs*
Ron Lenk and Carol Lenk

26. *Electricity Power Generation: The Changing Dimensions*
Digambar M. Tagare

27. *Electric Distribution Systems*
Abdelhay A. Sallam and Om P. Malik

28. *Maintaining Mission Critical Systems in a 24/7 Environment, Second Edition*
Peter M. Curtis

29. *Power Conversion and Control of Wind Energy Systems*
Bin Wu, Yongqiang Lang, Navid Zargan, and Samir Kouro

30. *Integration of Distributed Generation in the Power System*
Math Bollen and Fainan Hassan

31. *High Voltage Protection for Telecommunications*
Steven W. Blume

32. *Doubly Fed Induction Machine: Modeling and Control for Wind Energy Generation*
Gonzalo Abad, Jesús Lopéz, Miguel Rodríguez, Luis Marroyo, and Grzegorz Iwanski

33. *Smart Grid: Fundamentals of Design and Analysis*
James Momoh

34. *Electromechanical Motion Devices, Second Edition*
Paul Krause, Oleg Wasynczuk, and Steven Pekarek

35. *Arc Flash Hazard and Analysis and Mitigation*
J. C. Das

36. *Electrical Energy Conversion and Transport: An Interactive Computer-Based Approach, Second Edition*
George G. Karady and Keith E. Holbert

37. *Analysis of Electric Machinery and Drive Systems, Third Edition*
Paul Krause, Oleg Wasynczuk, Scott Sudhoff, and Steven Pekarek

38. *Direct Transmission: Advances in Research and Development*
Giovanni Mazzanti and Massimo Marzinotto

39. *Handbook of Electrical Power System Dynamics: Modeling, Stability, and Control*
Mircea Eremia and Mohammad Shahidehpour

Printed and bound by CPI Group (UK) Ltd, Croydon, CR0 4YY
21/09/2026
14965806-0002